Fourth Edition

Remote Sensing

Principles, Interpretation, and Applications

Floyd F. Sabins, Jr.

James M. Ellis

Long Grove, Illinois

For information about this book, contact:
Waveland Press, Inc.
4180 IL Route 83, Suite 101
Long Grove, IL 60047-9580
(847) 634-0081
info@waveland.com
www.waveland.com

Front Cover Image:

Perspective view of Salt Lake City, Utah, looking north (path/row 38/32).
Landsat 8 OLI bands 6-5-2 as R-G-B (reflected SWIR1-NIR-blue light).
Image acquired on November 11, 2018 and draped over SRTM 30 m DEM.
Vertical exaggeration 2.5X. Image and DEM courtesy USGS.

Legend

Blue to black: water
Bright green: irrigated vegetation (agriculture and golf courses)
Green: healthy vegetation in mountains
Brown: exposed soil, fallow agricultural fields
Purple to light violet: moist soil along lake edge
Gray: urban core and major roads

10-digit ISBN 1-4786-3710-2
13-digit ISBN 978-1-4786-3710-3

Printed in the United States of America

7 6 5 4 3 2

THE FOURTH EDITION IS DEDICATED TO
THE YOUNGER GENERATIONS:

Gen X, Millennials, and Gen Z

Floyd F. Sabins, Jr., PhD

(January 5, 1931–February 4, 2019)

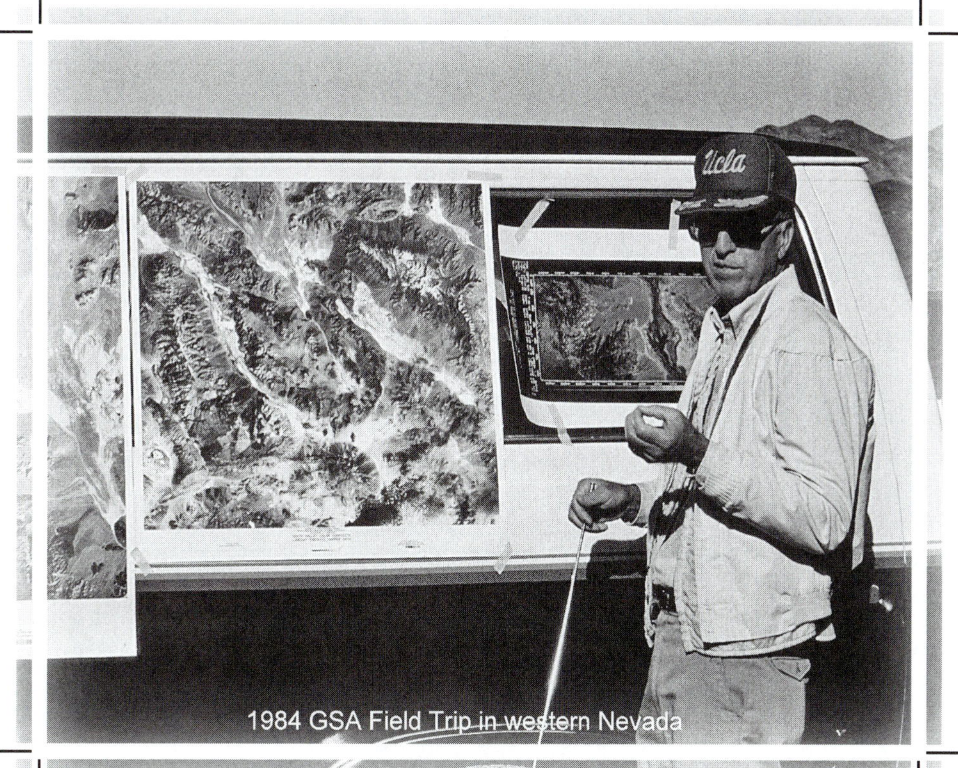

Photo courtesy G. Byran Bailey

Floyd F. Sabins, Jr. led an enlightening Geological Society of America (GSA) Annual Meeting field trip to western Nevada in 1984, explaining his new interpretations using satellite images and maps taped to the side of a vehicle. Floyd was a knowledgeable, insightful, and entertaining teacher and mentor to hundreds of professional and aspiring remote sensors and geologists during his long and distinguished career. His coauthor, James Ellis, traveled to the GSA meeting in Reno, Nevada, in order to meet Floyd (for the first time), plan how they could work together going forward, and learn how Floyd was creating leading edge geologic maps with the new Landsat TM satellite. Floyd and James worked together for the next 35 years. It was an excellent and memorable journey for both of them with many highlights along the way.

In 2019, to honor his legacy, his family created the Dr. Floyd F. Sabins, Jr. Endowment for Remote Sensing at his alma mater the Jackson School of Geosciences, University of Texas at Austin to fund faculty and students employing remote sensing to advance geoscience and environmental research and education.

contents

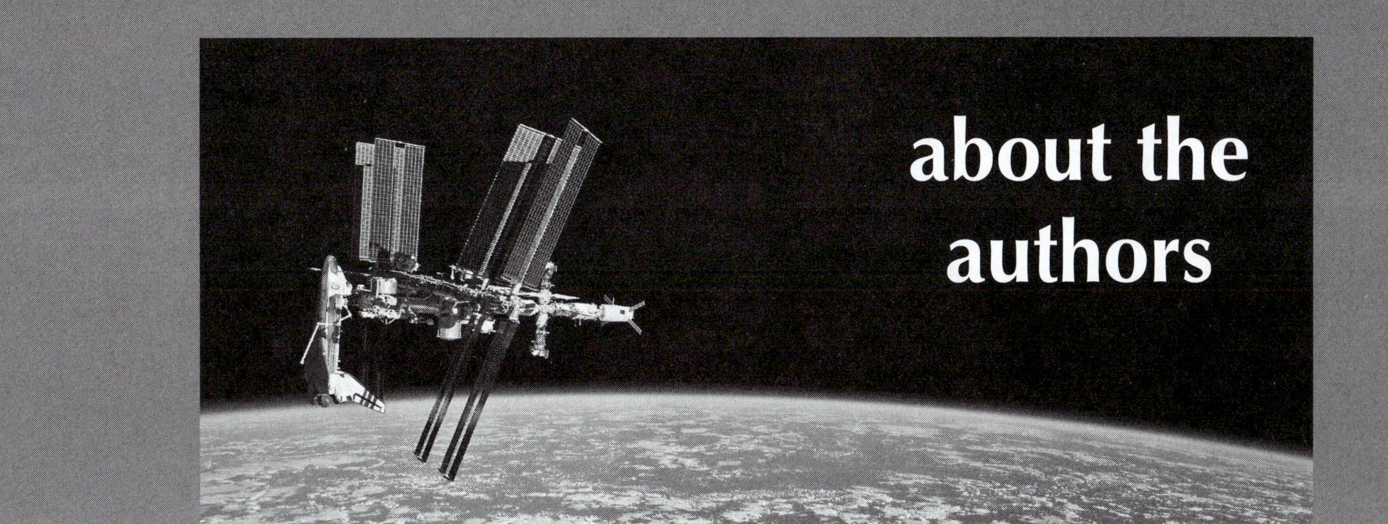

about the authors

FLOYD F. SABINS, JR., PhD

(January 5, 1931–February 4, 2019)

Floyd was an American petroleum geologist, educator, and author. He was a pioneer in the development, application, and advocacy of the field of geological remote sensing. He had a BS from the University of Texas and a PhD in Geology from Yale University. Floyd was a research scientist for Chevron for 37 years and employed remote sensing to support the discovery of new oil fields in Saudi Arabia and Papua New Guinea. As a mineral explorationist he used remote sensing to enable copper and gold discoveries in Chile, boron and lithium discoveries in Bolivia, and most recently (2010–2013) target identification across Afghanistan.

Floyd was a Regent's Professor with the Earth and Space Sciences Department at UCLA and led many field trips under the auspices of the Geological Society of America, Environmental Research Institute of Michigan, NASA, and JPL. Floyd received a number of honors and professional awards—notably the William T. Pecora Award from NASA and the US Department of Interior in 1983 for "His outstanding contributions in education, science, and policy formulation to the field of remote sensing." His landmark text *Remote Sensing: Principles and Interpretation* was published in 1978 with the Third Edition published in 1997. Floyd was totally engaged in the process of clarifying principles, improving exploration examples, and creating new interpretation maps for the Fourth Edition when he passed.

JAMES M. ELLIS, PhD

James has implemented remote sensing and GIS for environmental and geological applications around the globe with Gulf Oil and Chevron (15 years), The MapFactory (5 years), and through his consulting firm Ellis GeoSpatial (18 years). Much of his work establishes environmental baselines for proposed and ongoing development, generates land use maps, and monitors change with satellite and airborne images. Recent projects include image processing, interpretation, and GIS database development in Afghanistan, Tanzania, Nigeria, Angola, China, Haiti, and Turkey. A multiyear project that analyzed modern carbonate environments resulted in many publications and educational DVD sets that include extensive GIS databases and visualizations. He conducted airborne radar surveys in Papua New Guinea, Colombia, and Congo and helped coordinate and process data from the first airborne hyperspectral surveys of Mongolia and Edwards Air Force Base. He realized the potential of remote sensing as a US Navy officer in the 1970s while tracking submarine movements with classified satellite images and ship- and seafloor-based sonar.

James has a BA from the University of Rochester and a PhD in Geology from University at Buffalo. He was a National Science Foundation Postdoctoral Fellow and visiting Assistant Professor at Rice University. For 12 years he taught remote sensing and GIS cartography lecture/lab courses in Diablo Valley College's GIS/GPS certificate program. In 2017 he taught Environmental Remote Sensing at the Department of Geology, University at Buffalo. He is a State of California Professional Geologist (Certificate No. 7391). Links to publications are available at ellis-geospatial.com.

When the Third Edition of this book was published in 1997, there were only a few dozen earth observation (EO) satellites available to the civilian remote sensing community, bulky film cameras were flown in manned aircraft, and computer storage was limited and expensive. Today, hundreds of EO satellites collect remote sensing data used by governments, academia, nongovernment organizations, and industry; digital images have completely replaced film-based photographs; and digital data storage is inexpensive, growing exponentially, and migrating to the cloud. Since 1972, the US Landsat program has provided exceptional global imagery that spans the visible and infrared wavelength regions. The European Space Agency recently launched two radar satellites that enable precise ground deformation measurements. Thermal IR images enable accurate mapping of hot spots in urban areas to improve tree-planting programs that reduce unhealthy rising temperatures. Civilians can now access satellite imagery that detects objects 30 cm (1 ft) across.

Miniaturization has enabled hundreds of microsatellites to be built and launched during the past few years, expanding low-cost and highly repetitive global coverage. Small unmanned aerial systems (sUAS or "drones") have exploded onto the remote sensing landscape. Smartphone technology and miniaturization of sensors enable drones to carry the most sophisticated remote sensing technologies at a fraction of the cost when compared with manned aircraft. Using lasers and overlapping images acquired from different directions, advanced software and powerful computers enable the generation of accurate 3-D models of terrain, buildings, and trees. Users of remote sensing technology and data now have many software options on different types of platforms, including mobile, desktop, and cloud-based.

This is a very exciting time for the dynamic field of remote sensing. We are the stewards of our planet and remote sensing provides valuable tools and resources to manage our world responsibly.

OVERVIEW OF THE FOURTH EDITION

This book was designed for an introductory university course in remote sensing. No prior training in remote sensing is required. Courses in introductory physics, physical geography, and physical geology provide useful background, but are not essential for users of this book. There is enough depth in the text for both undergraduate and graduate students, and enough structure to support individuals outside of academics who are eager to learn how remote sensing can help them improve their scientific knowledge and impact in the workplace.

The approach of the text follows the format of the remote sensing courses we have taught in the Earth, Planetary, and Space Sciences, Geology, and Geography Departments at several institutions. It emphasizes a basic understanding of remote sensing concepts and systems followed by the interpretation of images and their application to a range of disciplines. The Fourth Edition includes six new chapters:

- Chapter 7: Digital Elevation Models and Lidar
- Chapter 8: Drones and Manned Aircraft Imaging
- Chapter 10: Geographic Information Systems
- Chapter 12: Renewable Resources
- Chapter 16: Climate Change
- Chapter 17: Other Applications

It also provides over 500 figures and color plates. More than half of the figures found in the chapters are new. Throughout the text this symbol 🌐 indicates that a color version of a figure or a digital image is available online for viewing and further study (waveland.com/Sabins-Ellis).

ORGANIZATION OF THE TEXT

The first eight chapters focus on major remote sensing concepts and systems, the next two chapters review digital image processing and GIS, and the last seven chapters provide case histories from many disciples. Each chapter has extensive references that support more in-depth learning.

The first chapter introduces the major remote sensing systems and the interactions between electromagnetic energy and materials that are the basis for remote sensing. The seven following chapters describe the major remote sensing systems: photographs, Landsat, multispectral satellites, thermal IR, radar, DEMs and lidar, and drones and manned aircraft. For each imaging and surface elevation measuring system the following topics are covered:

1. Physical properties of materials and their interactions with electromagnetic energy,
2. Design and operation,
3. Characteristics, and
4. Guidelines and examples for interpretation.

The chapter on image processing emphasizes enhancement techniques and information extraction while the following GIS chapter demonstrates how images and DEMs benefit from GIS integration and analysis. The remaining chapters describe environmental, renewable and nonrenewable resources, land use/land cover, natural hazards, climate change, and other applications for remote sensing, such as public health and archeology.

Each chapter includes a series of questions that will help promote understanding of key concepts presented in the chapter. An Answer Key is available to instructors (waveland.com/Sabins-Ellis). Students will need a pocket lens stereoscope and a pair of red/cyan anaglyph glasses to view stereo models in 3-D. A Glossary provides basic definitions of important remote sensing terms and acronyms. There are four appendices:

- Appendix A: Basic Geology for Remote Sensing describes earth science concepts that are employed in some image interpretations.
- Appendix B: Location of Images provides a world map and a conterminous United States map that display the location of images that are discussed and shown in figures, plates, and digital images.
- Appendix C: Remote Sensing Digital Database provides information on how the database is organized as well as the types of data that are supplied.
- Appendix D: Digital Image Processing Lab Manual outlines the goals of each of the 12 exercises and the processes that will be executed during each lab.

ADDITIONAL AND COMPLEMENTARY FEATURES

For this edition we have created two new features that coordinate with the topics and locations discussed throughout the text: the Remote Sensing Digital Database and the Digital Image Processing Lab Manual. These features were designed to give instructors the flexibility they require to integrate the materials into their courses or act as a springboard to create their own projects that are tailored to their needs. Introductory and instructional videos are provided to guide users on ways to access and utilize these features. Information contained in the videos is also outlined in PDF form (waveland.com/Sabins-Ellis).

The Remote Sensing Digital Database is a user-friendly resource that instructors and students can download to digitally explore 27 examples of satellite and airborne imagery. The database includes descriptions, georeferenced images, DEMs, maps, and metadata so users can display, process, and interpret images with open-source and commercial image processing and GIS software (waveland.com/Sabins-Ellis).

We also created the Digital Image Processing Lab Manual so that instructors and students have ready-to-use exercises to support computer analysis, visualization, integration, and interpretation of data acquired by a broad range of sensors. The lab manual provides 12 step-by-step exercises that employ ENVI image processing software by L3Harris.

COURSE SCHEDULE

A proposed sequence for integrating the textbook, lab manual, and digital database into a semester-long course that includes lectures, labs, and student projects is shown on the next page. The instructor may choose to cover only one or two application chapters when the course is constrained by a short semester or to assign application chapters to students to support class presentations. Chapter 9 (Digital Image Processing), Chapter 10 (Geographic Information Systems), and Chapter 14 (Land Use/Land Cover) can be integrated into a lab schedule. If special projects are included in a course, the digital database can be used to help students jump-start a project or support a topic.

ACKNOWLEDGEMENTS

Much of the data used in the figures, plates, digital images, and the Remote Sensing Digital Database came from government and commercial sources. We gratefully acknowledge their efforts and contributions to the discipline of remote sensing.

We are most appreciative of the remarkable open source and commercial remote sensing and GIS software currently available that enables users with a broad range of expertise and interests to develop informative images and maps from complex and sophisticated digital data. We do not endorse specific brands of geospatial software, but want to acknowledge that ENVI, ERDAS Imagine, and ArcGIS Desktop software was used to create our textbook examples, Remote Sensing Digital Database, and Digital Image Processing Lab Manual.

We thank Waveland Press for publishing this textbook, in particular Don Rosso for encouraging us to undertake the Fourth Edition, Diane Evans for exceptional copy editor work, Deborah Underwood for graphics preparation, and Peter Lilliebridge for typesetting the text. Timely and insightful reviews by Rebecca Dodge (Midwestern State University), Brian Hausbeck (California State University, Sacramento), and Dan Taranik (Exploration Mapping Group) materially improved the focus of our chapters. The interaction with

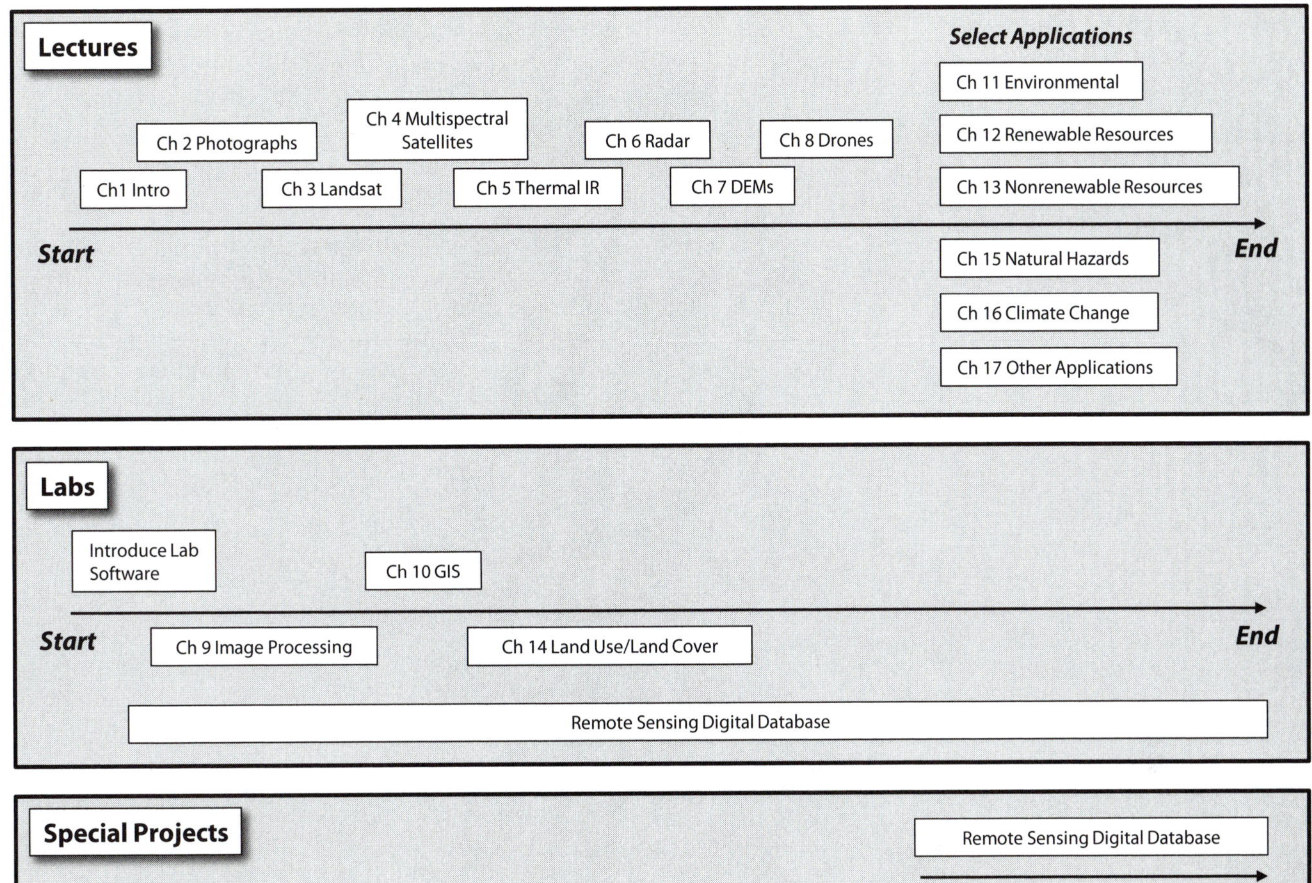

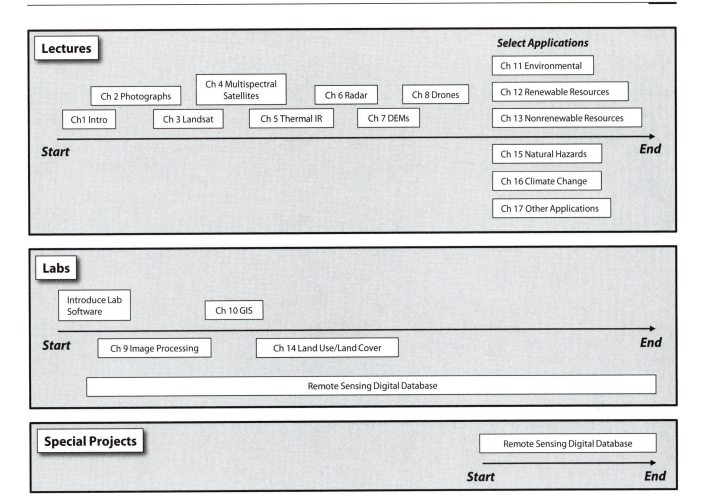

and feedback from students and faculty have improved the instructional aspects of the textbook.

The experience, research opportunities, and recognition that Chevron provided Floyd, and the many opportunities that a diverse group of clients provided to James to conduct remote sensing and GIS projects around the world, are gratefully acknowledged. We are also very thankful for our many associates, business and research partners, and clients who have helped us understand and use remote sensing to deliver timely and effective solutions to solve real-world problems. The remote sensing team at Chevron's research lab—Bill Kowalik, Linda Fry, Hans Beck, Scott Hills, Brad Dean, Todd Battey and Dan Harris—provided cutting-edge technology and software to support our early geological and environmental work. Joseph Zamudio, the late Ward Kilby, Matt Levey, Sam Purkis, Michael Quinn, Mark Choiniere, Pat Caldwell, Peter Goodwin, Dan Taranik, and Chris Baynard increased James's knowledge and application of geospatial technologies. Binita Sinha, Michael Quinn, John Karachewski, and Ted Wieden at Diablo Valley College and Bea Csatho at University at Buffalo supported James's academic efforts while Hattie Davis of Artistic Earth provided years of image processing expertise.

James thanks his wife Lynn for her infinite patience and encouragement. Floyd was very appreciative of his family's continual support. We welcome comments and suggestions on this edition. Send an email to James Ellis at jellis@ellis-geospatial.com.

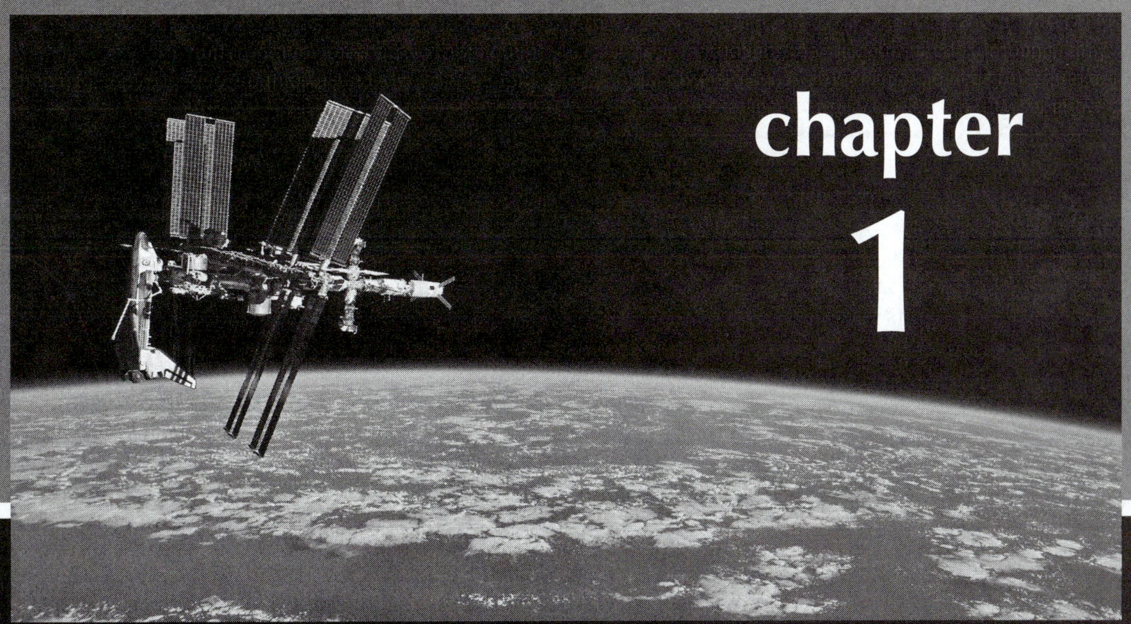

chapter 1

Introduction to Concepts and Systems

In this book we define *remote sensing* as the science of

- acquiring,
- processing, and
- interpreting

images, and related data, that are typically acquired from aircraft and satellites with sensor systems that digitally record the interaction between electromagnetic energy and matter. *Acquiring* images refers to the technology employed, such as electro-optical framing systems or scanning systems. *Processing* refers to the digital procedures that convert the raw data into images. *Interpreting* the images is, in our opinion, the most important step because an interpreter, such as the reader of this book, converts an image into information, such as land use/land cover maps or prospects for oil or minerals, that is meaningful and valuable for a wide range of applications. The *interaction between matter and electromagnetic energy*, such as light, heat, and microwaves, is determined by

- the physical properties of the matter and
- the wavelength of electromagnetic energy that is remotely sensed.

The science of remote sensing excludes geophysical methods such as seismic, electrical, magnetic, and gravity surveys that measure force fields rather than electromagnetic radiation. Underwater surveys that use pulses of sonic energy for imaging (*sonar*) are included as a remote sensing method. A later section in this chapter describes the various wavelength regions and the types of interactions. The geospatial technologies Geographic Information Systems (GIS) and Global Navigation Satellite System (GNSS) provide geographic projections, coordinates, and location of images (as described in Chapter 9 and Chapter 10). Online three-dimensional (3-D) virtual globes, such as those created by Google Earth, NASA World Wind, and ArcGIS Earth, are made possible by integrating remote sensing images with geospatial technologies.

Aerial film photography was the original form of remote sensing. Over the past two decades, however, film has been replaced by digital images (as described in Chapter 2). Archives of aerial film photographs are widely used in change-detection studies, which utilize such older media. Interpretations of aerial photographs are used to inventory and monitor forests, build land use/land cover maps, develop soil maps, and identify wetland species. Geologic interpretations of aerial photographs have led to the discovery of many oil and mineral deposits. These successes, using only the narrow visible and near infrared (VNIR) regions of the electromagnetic spectrum, suggest that additional discoveries could be made by using other wavelength regions. Beginning in the 1940s, technology was developed to acquire images in the infrared (IR) and microwave (radar) regions, which greatly expanded the scope and applications of remote sensing. The development and deployment of manned and unmanned satellites began in the 1960s and provides an orbital vantage point for acquiring repetitive images of the Earth.

This chapter introduces the basic physical concepts and the imaging systems that are employed in remote sensing. Subsequent chapters describe the major types of remote sensing and are followed by chapters that describe applications of the technology.

UNITS OF MEASURE

This book employs the metric system with the basic units and symbols shown in Table 1-1. Distance is expressed in multiples and fractions of meters, as shown in Table 1-2. Where appropriate for clarity, English units for distance will be used, with metric equivalents shown in parentheses.

Frequency (v) is the number of wave crests passing a given point in a specified period of time. Frequency was formerly expressed as cycles per second, but today we use *hertz* (Hz) as the unit for a frequency of 1 cycle per second. Table 1-3 lists the terms for designating frequencies.

Temperature is given in degrees Celsius (°C) or in degrees Kelvin (°K). The Kelvin scale is also known as the *absolute temperature* scale. A temperature of 273°K is equivalent to 0°C. The metric system omits the degree symbol for Kelvin temperatures; however, the letter K is also used to designate other constants, so °K is used in this text. A few temperatures commonly given in degrees Fahrenheit (°F) will remain in that scale where conversion to degrees Celsius is inconvenient.

TABLE 1-1 Basic metric units and symbols.

Unit	Symbol
meter	m
second	sec
gram	gm
radian	rad
hertz	Hz
watt	w

TABLE 1-2 Metric units for distance.

Unit	Symbol	Equivalent
kilometer	km	1,000 m = 10^3 m
meter[a]	m	1.0 m = 10^0 m
centimeter	cm	0.01 m = 10^{-2} m
millimeter	mm	0.001 m = 10^{-3} m
micrometer[b]	μm	0.000001 m = 10^{-6} m
nanometer	nm	0.000000001 m = 10^{-9} m

[a] Basic unit.
[b] Formerly called micron (μ).

TABLE 1-3 Metric units for frequencies.

Unit	Symbol	Frequency (cycles · sec^{-1})
hertz	Hz	1
kilohertz	kHz	10^3
megahertz	MHz	10^6
gigahertz	GHz	10^9

ELECTROMAGNETIC ENERGY

Electromagnetic energy refers to all energy that moves with the velocity of light in a harmonic wave pattern. A harmonic pattern consists of waves that occur at equal intervals in time. The wave concept explains how electromagnetic energy propagates (moves), but this energy can only be detected as it interacts with matter. In this interaction, electromagnetic energy behaves as though it consists of many individual bodies called *photons* that have such particle-like properties as energy and momentum. When light bends (refracts) as it propagates through media of different optical densities, it behaves like waves. When a light meter measures the intensity of light, however, the interaction of photons with the light-sensitive photodetector produces an electrical signal that varies in strength proportional to the number of photons. Suits (1983) describes the characteristics of electromagnetic energy that are significant for remote sensing.

PROPERTIES OF ELECTROMAGNETIC WAVES

Electromagnetic waves can be described in terms of their velocity, wavelength, and frequency. All electromagnetic waves travel at the same velocity (c). This velocity is commonly referred to as the *speed of light*, since light is one form of electromagnetic energy. For electromagnetic waves moving through a vacuum, $c = 299{,}793\ \text{km} \cdot \text{sec}^{-1}$ or, for practical purposes, $c = 3 \times 10^8\ \text{m} \cdot \text{sec}^{-1}$.

The *wavelength* (λ) of electromagnetic waves is the distance from any point on one cycle or wave to the same position on the next cycle or wave. The *micrometer* (μm) is a convenient unit for designating wavelength of both visible and IR radiation. In order to avoid decimal numbers, optical scientists commonly employ nanometers (nm) for measurements of very short wavelengths, such as visible light.

Unlike velocity and wavelength, which change as electromagnetic energy is propagated through media of different densities, frequency remains constant and is therefore a more fundamental property. Electronic engineers use frequency nomenclature for designating radio and radar energy regions. This book uses wavelength, rather than frequency, to simplify comparisons among all portions of the electromagnetic spectrum. Velocity (c), wavelength (λ), and frequency (v) are related by

$$c = \lambda v \qquad \textbf{(1-1)}$$

or $\lambda = c/v$, which means that the longer the wavelength, the lower the frequency, and vice versa.

Also important for remote sensing is the relationship of electromagnetic energy and wavelength proposed by Max Planck over a century ago. We can associate the wavelength associated with a quantum of energy as

$$\lambda = \frac{hc}{Q} \quad \textbf{(1-2)}$$

where h is Planck's constant and Q is the energy of the quantum (a photon or a particle of light). The longer the wavelength collected by the remote sensing system, the lower its energy content. For example, this means that sensors collecting IR energy must be more sensitive than sensors that collect visible light with its higher energy content.

INTERACTION PROCESSES

Electromagnetic energy that encounters matter, whether solid, liquid, or gas, is called *incident* radiation. Interactions with matter can change the following properties of the incident radiation: intensity, direction, wavelength, polarization, and phase. The science of remote sensing detects and records these changes. We then interpret the resulting images and data to determine the characteristics of the matter that interacted with the incident electromagnetic energy.

During interactions between electromagnetic radiation and matter, mass and energy are conserved according to basic physical principles. Figure 1-1 illustrates the five common results of these interactions. The incident radiation may be

1. *Transmitted*, that is, passed through the substance. Transmission of energy through media of different densities, such as from air into water, causes a change in the velocity of electromagnetic radiation. The ratio of the two velocities is called the *index of refraction* (n) and is expressed as

$$n = \frac{c_a}{c_s} \quad \textbf{(1-3)}$$

   where c_a is the velocity in a vacuum and c_s is the velocity in the substance.
2. *Absorbed*, giving up its energy largely to heating the matter.
3. *Emitted* by the substance, usually at longer wavelengths, as a function of its structure and temperature.
4. *Scattered*, that is, deflected in all directions. Surfaces with dimensions of *relief*, or roughness, comparable to the wavelength of the incident energy produce scattering. Light waves are scattered by molecules and particles in the atmosphere whose sizes are similar to the wavelengths of light.
5. *Reflected*, that is, returned from the surface of a material with the angle of reflection equal and opposite to the angle of incidence. Reflection is caused by surfaces that are smooth relative to the wavelength of incident energy. *Polarization*, or direction of vibration, of the reflected waves may differ from the polarization of the incident wave.

Emission, scattering, and reflection are called *surface phenomena* because these interactions are determined primarily by properties of the surface, such as color and roughness. Transmission and absorption are called *volume phenomena* because they are determined by the internal characteristics of matter, such as density and conductivity. The particular combination of surface and volume interactions with any particular material depends on both the wavelength of the electromagnetic radiation and the specific properties of that material.

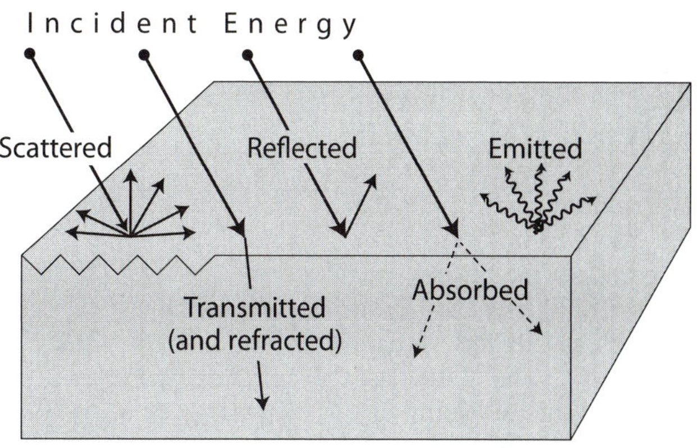

FIGURE 1-1 Interaction processes between electromagnetic energy and matter.

These interactions between matter and energy are recorded, processed, and displayed as *soft copy* on screens, or as *hard copy* on paper. The remote sensing images are interpreted to extract the desired information, from which one may interpret the characteristics of matter. Individual interaction mechanisms are described more completely in later chapters. For example, Chapter 2 describes the interaction of solar energy with the Earth and its atmosphere. Chapter 5 describes the interactions with thermal IR (TIR) energy. Chapter 6 describes the interactions with active microwave energy (radar).

ELECTROMAGNETIC SPECTRUM

The *electromagnetic spectrum* is the continuum of energy that ranges from nanometers to meters in wavelength, travels at the speed of light, and propagates through a vacuum such as outer space. All matter radiates a range of electromagnetic energy such that the peak intensity shifts toward progressively shorter wavelengths with increasing temperature of the matter.

WAVELENGTH REGIONS AND BANDS

Figure 1-2 shows the electromagnetic spectrum, which ranges from the very short wavelengths of the gamma ray region (measured in fractions of nanometers) to the long wavelengths of the radio region (measured in meters). The spectrum is divided on the basis of wavelength into *regions*, which may be further subdivided into *ranges* and *bands*. Table 1-4 lists and summarizes the spectral regions, ranges, and bands. The term band refers not only to a wavelength interval, but also to the image that is acquired of that interval. The horizontal scale in Figure 1-2 is logarithmic in order to portray adequately the shorter wavelengths. Note that the visible region (0.4 to 0.7 μm) occupies only a small portion of the spectrum. Visible colors with their wavelengths are shown in Plate 1. Energy received from the 6000°K sun and reflected from the Earth during the daytime may be recorded as a function of wavelength. The maximum amount of energy received and reflected is at the 0.5 μm wavelength, which corresponds to the green wavelengths of the visible region and is called the *reflected energy peak*. The Earth also radiates energy both day and night, with the maximum energy radiating at the 9.7 μm wavelength. This *radiant energy peak* occurs in the thermal portion of the IR region (Figure 1-2).

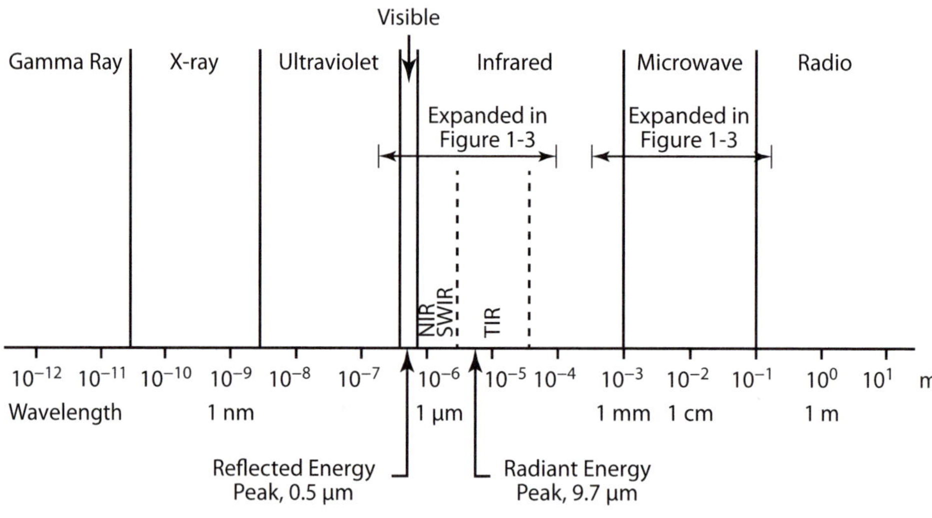

FIGURE 1-2 Electromagnetic spectrum. Expanded diagrams of the visible, infrared, and microwave regions are shown in Figure 1-3.

The Earth's atmosphere absorbs energy in the gamma ray, X-ray, and most of the ultraviolet (UV) regions; therefore, these wavelength regions are not used in remote sensing. Terrestrial remote sensing records energy in the visible, infrared, and microwave regions, as well as the long wavelength portion of the UV region. Figure 1-3 shows details of these regions. The horizontal axis shows wavelength on a logarithmic scale; the vertical axis shows the percentage of incoming electromagnetic energy that is transmitted through

TABLE 1-4 Spectral regions, ranges, and bands.

Spectral Regions, Ranges, and Bands	Wavelength	Comments
Gamma ray region	< 0.03 nm	Incoming radiation is absorbed by upper atmosphere. Not employed in remote sensing.
X-ray region	0.03 to 30 nm	Incoming radiation is absorbed by upper atmosphere. Not employed in remote sensing.
Ultraviolet region	0.03 to 0.4 μm	Incoming radiation < 0.3 μm is absorbed by ozone layer.
▪ Far UV band	0.3 to 0.4 μm	Transmitted through atmosphere, but scattering is severe.
Visible region	0.4 to 0.7 μm	Formerly imaged with film; now imaged with photodetectors.
▪ Blue band	0.4 to 0.5 μm	Scattered by atmosphere. Absorbed by chlorophyll in plants.
▪ Green band	0.5 to 0.6 μm	Includes peak reflected energy of Earth.
▪ Red band	0.6 to 0.7 μm	Absorbed by chlorophyll in plants.
▪ Lidar	0.4 to 0.7 μm	Also operates in UV and NIR wavelengths.
Infrared region	0.7 to 1,000 μm	Atmospheric transmission windows are separated by absorption bands.
▪ NIR range	0.7 to 0.9 μm	Solar radiation that is strongly reflected by vegetation.
▪ SWIR range	0.9 to 3.0 μm	SWIR spectra are used to identify minerals.
▪ TIR range	3.0 to 5.0 μm	Useful for "hot" targets (e.g., fires and volcanoes) (medium wavelength IR [MWIR]).
	8.0 to 14.0 μm	Useful for "warm" targets (e.g., land and oceans) (long wavelength IR [LWIR]).
Microwave region	0.1 to 100 cm	Largely used in active radar bands.
▪ Passive methods	0.1 to 100 cm	Used for soil moisture and other investigations.
▪ Active methods (radar)	0.8 to 100 cm	Different wavelength bands are used for a wide range of applications.
Radio region	> 100 cm	Primarily used for communication.

the Earth's atmosphere. Wavelength intervals with high transmission of energy are called *atmospheric windows* and are used to acquire remote sensing images. Some atmospheric gases such as ozone, water vapor, and carbon dioxide absorb energy at certain wavelength intervals, which are called *atmospheric absorption bands*. These bands are not useable for remote sensing of the Earth. The major remote sensing regions (visible, infrared, and microwave) are further subdivided into ranges and *bands*, such as the blue, green, and red bands of the visible region (Figure 1-3). Horizontal lines in the center of the diagram show wavelength bands in the visible through TIR regions that are recorded by major imaging systems. For example, the Landsat Thematic Mapper (TM) records bands 1, 2, and 3 in the blue, green, and red visible spectral region; band 4 in the near IR (NIR) range; bands 5 and 7 in the shortwave IR (SWIR) range; and band 6 in the TIR range.

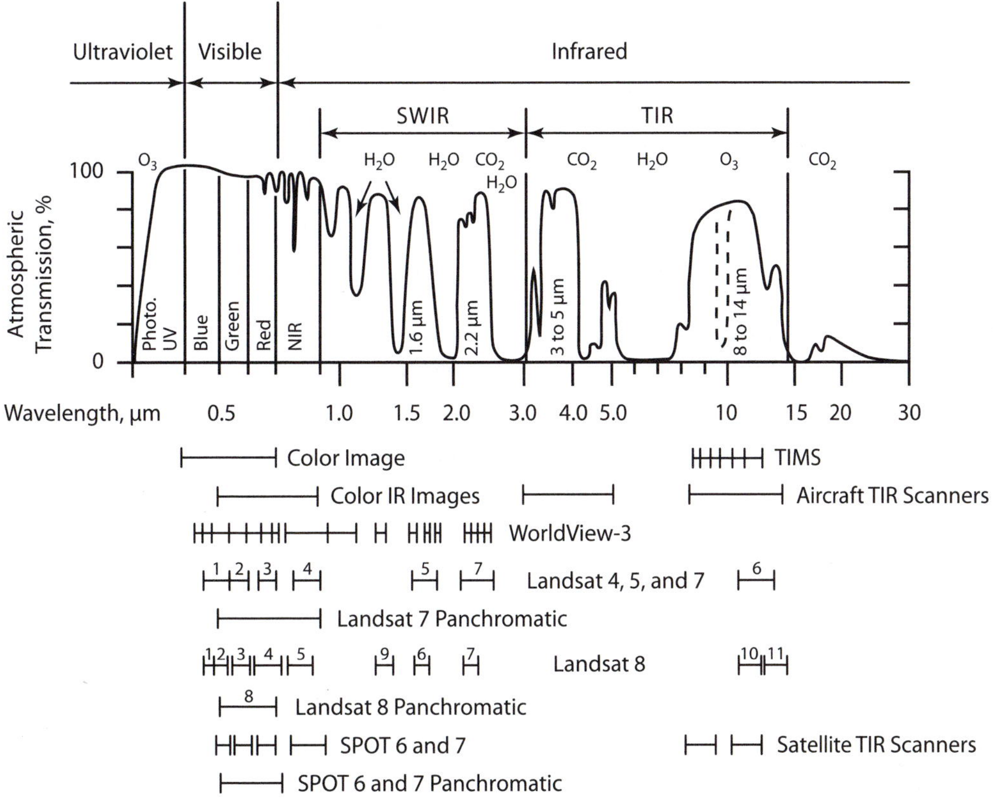

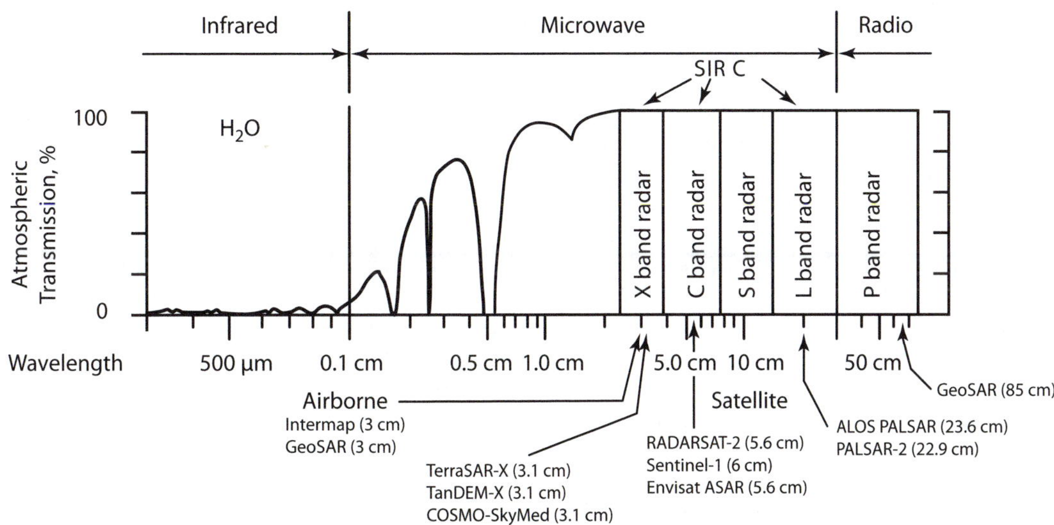

FIGURE 1-3 Expanded diagrams of the visible, infrared, and microwave spectral regions showing transmission of electromagnetic energy through the atmosphere. Gases responsible for atmospheric absorption bands are indicated. Wavelength bands recorded by commonly used remote sensing systems are shown.

Passive remote sensing systems record the energy that naturally radiates or reflects from an object. *Active* systems provide their own source of energy, directing it at the object and recording the returned energy. Other forms of active remote sensing are lidar (VNIR) and radar (microwave). Sonar systems transmit pulses of sonic energy.

ATMOSPHERIC EFFECTS

Our eyes inform us that the atmosphere is essentially transparent to light, and we tend to assume that this condition exists for all electromagnetic energy. In fact, the gases of the atmosphere absorb electromagnetic energy at specific wavelength intervals called absorption bands. Figure 1-3 shows the wavelengths of the absorption bands and names of the atmospheric gases that cause the absorption.

Wavelengths shorter than 0.3 μm are completely absorbed by the ozone (O_3) layer in the upper atmosphere. This absorption is essential to life on Earth, because prolonged exposure to the intense energy of UV wavelengths destroys living tissue. For example, sunburn occurs more readily at high mountain elevations than at sea level. Sunburn is caused by UV energy, which is largely absorbed by the atmosphere at sea level. At higher elevations, however, there is less atmosphere to absorb the UV energy.

Clouds consist of aerosol-sized particles of liquid water that absorb and scatter electromagnetic radiation at wavelengths less than about 0.1 cm. Only radiation of microwave and longer wavelengths is capable of penetrating clouds without being scattered, reflected, or absorbed. Active microwave systems (radar) at longer wavelengths penetrate clouds and rain cells. "Weather radars" operate at shorter wavelength radars and record active rain cells.

IMAGE CHARACTERISTICS

We use *image* as a noun that only means a portrayal of a scene or subject that is acquired by a digital system. The *digital* requirement is significant as it distinguishes an image from a *photograph*, which is a portrayal of a scene or subject acquired by an analog system that employs a lens and a photo-sensitive film medium. The term *imagery* is occasionally used as a synonym for image or for a group of images.

Originally the term photograph was defined as an image that records wavelengths from 0.3 to 0.9 μm that were acquired on photographic film. Over the past few decades, digital images have replaced film, but the term photograph is commonly applied to digital images at 0.3 to 0.9 μm wavelengths (VNIR wavelengths). Archives of aerial film photographs are valuable resources. Images can be described in terms of certain fundamental characteristics regardless of the wavelength at which the image is recorded. These characteristics are: scale, brightness, contrast ratio, and resolution. The tone and texture of images are functions of the fundamental characteristics.

SCALE

Scale is the ratio of the distance between two points on an image to the corresponding distance on the ground. A common scale on US Geological Survey (USGS) topographic maps is 1:24,000, which means that one unit on the map equals 24,000 units on the ground. Thus 1 cm on the map represents 24,000 cm (240 m) on the ground, or 1 in represents 24,000 in (2,000 ft). The maps and images of this book show scales graphically as bars.

The deployment of imaging systems on satellites has changed the concepts of image scale. In this book, scales of images are designated as follows:

Small scale (greater than 1:500,000)	1 cm = 5 km or more (1 in = 8 mi or more)
Intermediate scale (1:50,000 to 1:500,000)	1 cm = 0.5 to 5 km (1 in = 0.8 to 8 mi)
Large scale (less than 1:50,000)	1 cm = 0.5 km or less (1 in = 0.8 mi or less)
Very large scale (less than 1:5,000)	1 cm = 50 m or less (1 in = 420 ft or less)

Today sensing systems on satellites, aircraft, and drones can acquire digital images of excellent quality and with relatively small pixels (1 cm to 2 m). These high spatial resolution images can be enlarged to very large scales and still maintain clarity and detail of features in the scene.

Our experience indicates that geologic and environmental maps of a certain scale can be interpreted from clear images acquired with the following specific pixel dimensions:

Mapping Scale	Pixel Size
1:100,000	30 m
1:50,000	15 m
1:5,000	2 m
1:2,000	0.5 m (50 cm)

Mapping man-made infrastructure and features in urban landscapes typically requires pixels smaller than 5 m.

BRIGHTNESS

Remote sensing systems record the intensity of electromagnetic radiation that an object reflects, emits, or scatters at particular wavelength bands. Variations in intensity of electromagnetic radiation from the terrain are displayed as variations in brightness on images. On positive images, such as those in this book, the brightness of objects is directly proportional to the intensity of electromagnetic radiation that is recorded from that object.

Brightness is the magnitude of the response produced in the eye by light; it is a subjective sensation that differs between observers and can be determined only approximately. *Luminance* is a quantitative measure of the intensity of light from a source and is measured with a device called a photometer, or light meter. Variations in brightness may be calibrated with a grayscale such as those shown later in this chapter.

On images in the visible region the brightness of an object is primarily determined by the ability of the object to reflect incident sunlight, although atmospheric effects are also factors. On images acquired at other wavelength regions, brightness is determined by other physical properties of objects. On TIR images the brightness of an object is proportional to the heat radiating from the object. On active images (lidar or radar) the brightness of an object is determined by the intensity at which the transmitted beam of lidar or radar energy is returned to the receiving antenna.

Contrast Ratio

Contrast ratio (CR) is the ratio between the brightest and darkest parts of the image and is defined as

$$CR = \frac{B_{max}}{B_{min}} \quad (1\text{-}4)$$

where B_{max} is the maximum brightness of the scene and B_{min} is the minimum brightness. Figure 1-4 shows images of high, medium, and low contrast, together with profiles of brightness variation across each image. On a brightness scale of 0 to 10, these images have the following contrast ratios:

- High contrast: CR = 9/2 = 4.5
- Medium contrast: CR = 5/2 = 2.5
- Low contrast: CR = 3/2 = 1.5

Note that when $B_{min} = 0$, CR is infinity; when $B_{min} = B_{max}$, CR is unity. This discussion is summarized from the extensive review by Slater (1983), which describes other terms for contrast. In addition to describing an entire scene, contrast ratio is also used to describe the ratio between the brightness of an object on an image and the brightness of the adjacent background. Contrast ratio is a vital factor in determining the ability to resolve and detect objects.

Images with a low contrast ratio are commonly referred to as washed out, with monotonous, nearly uniform tones of gray. Low contrast may result from the following causes:

1. The objects and background of the scene may have a nearly uniform electromagnetic response at the particular wavelength band that the remote sensing system recorded. In other words, the scene has an inherently low contrast ratio.
2. Scattering of electromagnetic energy by the atmosphere can reduce the contrast of a scene. This effect is most pronounced in the shorter wavelength portions of the photographic remote sensing region, as described in Chapter 2.
3. The remote sensing system may lack sufficient sensitivity to detect and record the contrast of the terrain. Incorrect recording techniques can also result in low contrast images even though the scene has a high contrast ratio.

A low contrast ratio, regardless of the cause, can be improved by digital enhancement methods, as described in Chapter 9.

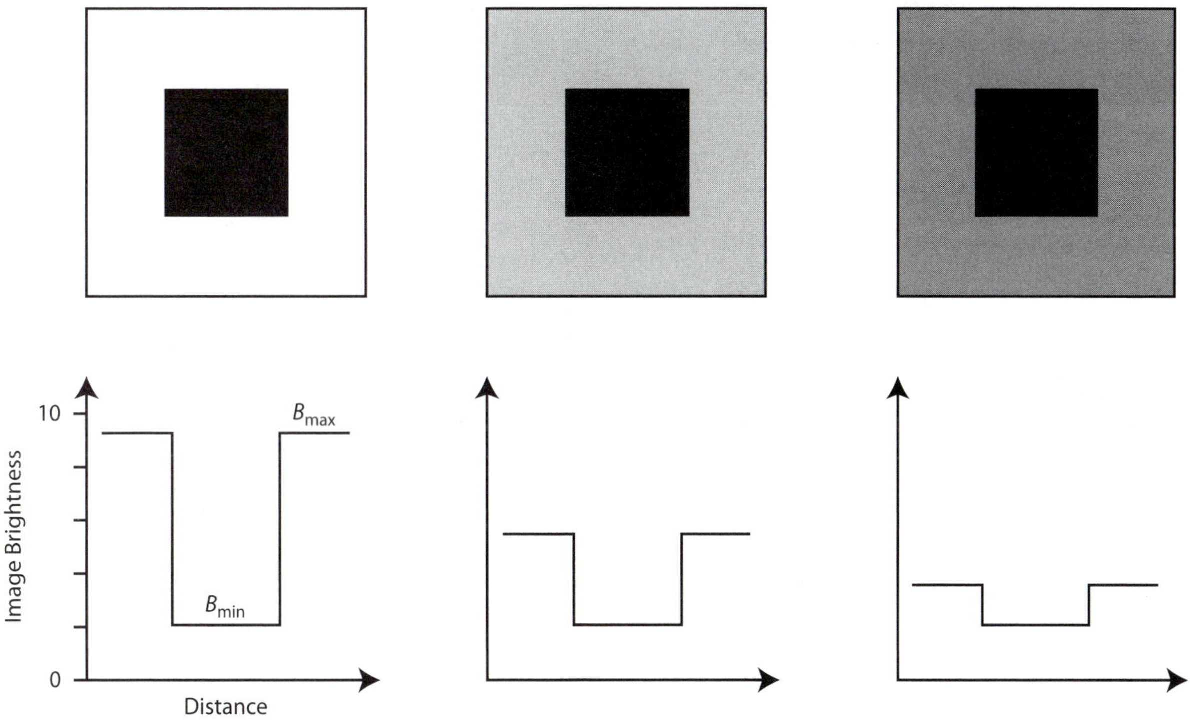

Figure 1-4 Images with different contrast ratios and corresponding brightness profiles.

RESOLUTION

For remote sensing, resolution is the ability to distinguish fine details in images. Four types of resolution are associated with remote sensing: (1) spatial, (2) spectral, (3) radiometric, and (4) temporal. Each type impacts the interpretability of an image, as explained below.

A. 0.6 m pixels.

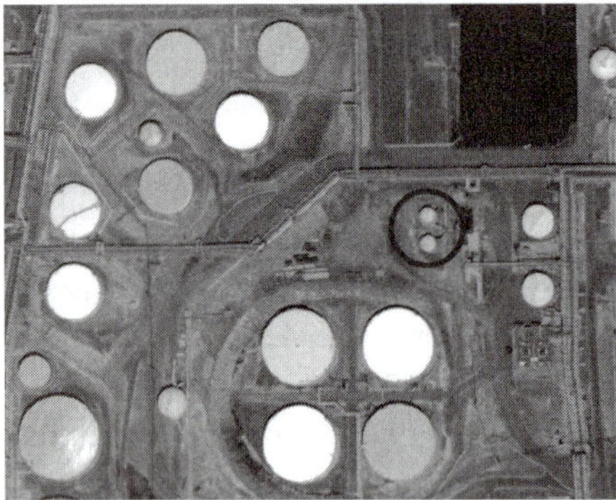

B. 5.0 m pixels.

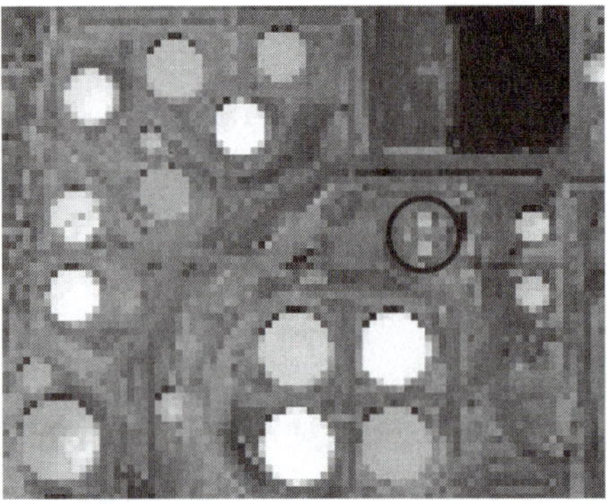

C. 10.0 m pixels.

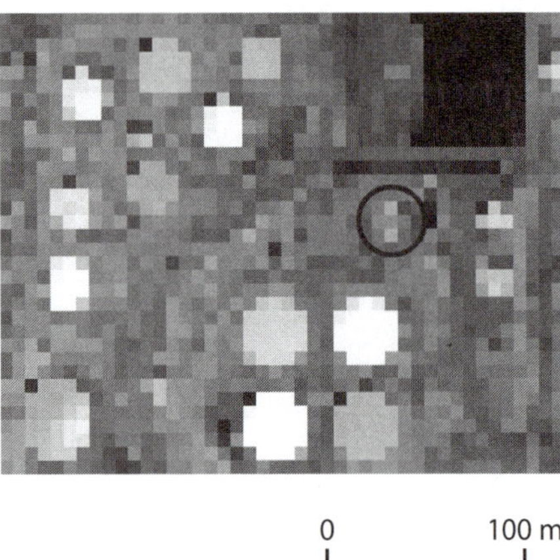

FIGURE 1-5 Images with different levels of spatial resolution. The scene is a tank farm at an oil refinery. The scene is 0.4 km wide. Courtesy D. Ruiz, Quantum Spatial, Inc., Novato, California.

Spatial Resolution

Spatial resolution is the ability to distinguish between two closely spaced objects on an image. More specifically, it is the minimum distance between two objects at which the objects are distinct and separate on the image. Objects spaced together more closely than the resolution limit will appear as a single object on the image. For remote sensing images spatial resolution is determined by the size of the ground resolution cell, which is determined by (1) the size of the detector and (2) the elevation of the imaging system. Figure 1-5A is a scanner image of oil storage tanks that was acquired with 0.6 m ground resolution cells. The tanks range from 10 to 50 m in diameter. The black circle encloses two tanks that are aligned in a north–south direction. The tanks are ~10 m in diameter and separated by ~10 m. In Figure 1-5B the image has been digitally resampled to show 5.0 m pixels. Each small tank is shown as an array of four (2 by 2) pixels that form a 10 m square on the image. The tanks are separated by ~10 m. The larger tanks retain their circular outlines, which are now separated by the 5 m pixels. In Figure 1-5C the original image has been resampled to show 10 m pixels. Each of the smaller tanks is now shown as a single pixel separated by a single pixel. The larger tanks are not recognizable as circular shapes.

Spectral Resolution

Spectral resolution refers to the wavelength interval that a detector records. In Figure 1-6 the vertical scale shows the response, or signal strength, of a detector as a function of wavelength, which is shown in the horizontal scale. As the wavelength increases, the detector response increases to a maximum, or peak, and then decreases. The peak for this detector is 0.55 μm. Spectral resolution, or bandwidth, is defined as the wavelength interval that is recorded at 50% of the peak response of a detector. For the detector in Figure 1-6 the 50% limits occur at 0.50 and 0.60 μm, which corresponds to a bandwidth of 0.10 μm.

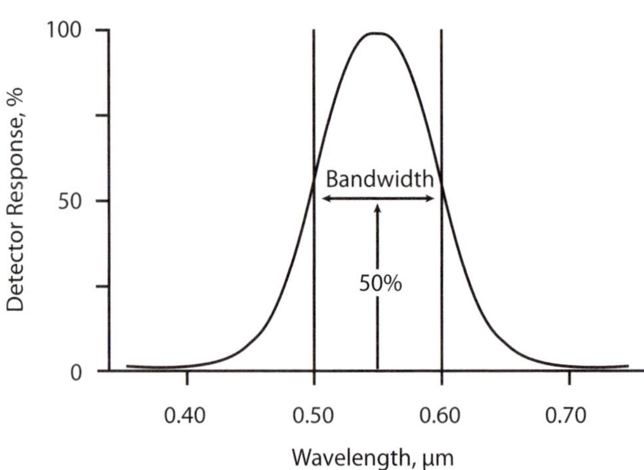

FIGURE 1-6 Spectral resolution, or bandwidth, of a detector. Bandwidth of this detector is 0.10 μm.

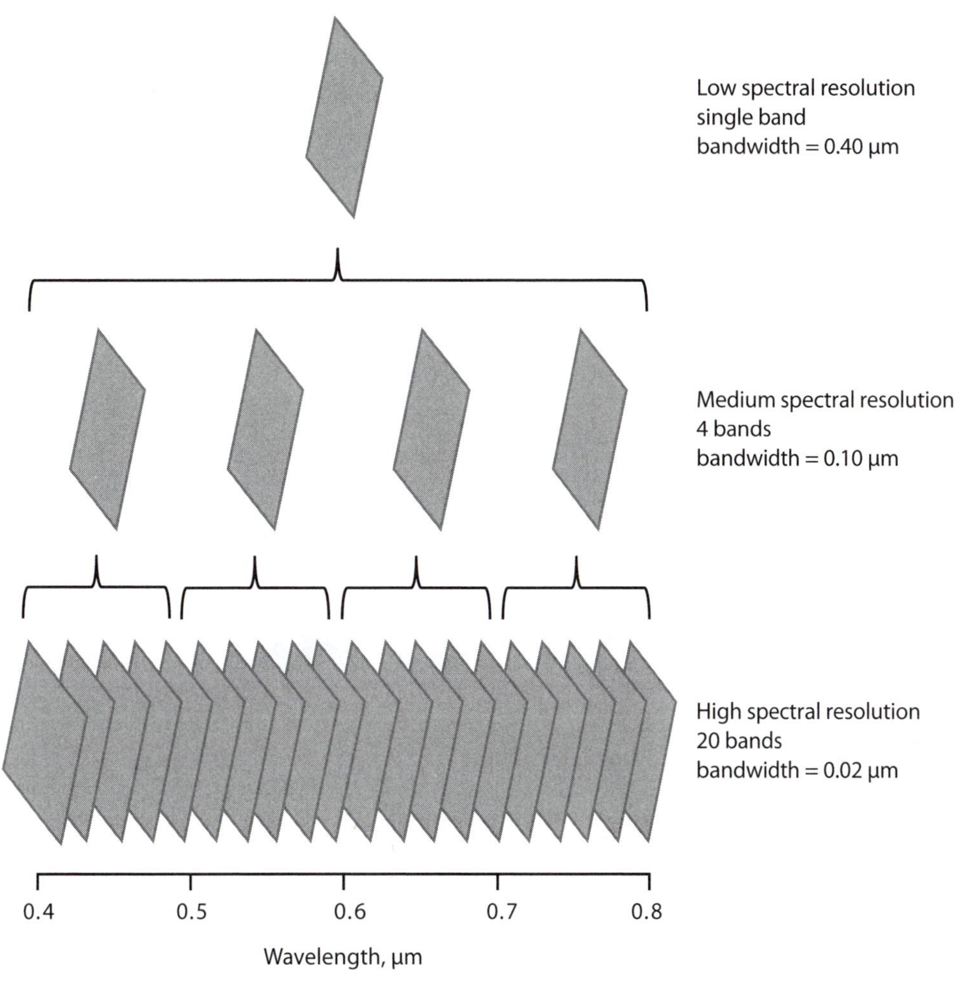

Figure 1-7 Spectral resolution. The gray shapes represent spectral bands (images) displayed at high, medium, and low levels of spectral resolution for a multispectral system that records 20 spectral bands ranging from 0.4 to 0.8 μm in wavelength.

Figure 1-7 shows the number of "images" acquired by three systems with different levels of spectral resolution that cover the same blue to NIR (0.40 to 0.80 μm) wavelengths, which is a range of 0.40 μm. The system with high spectral resolution records 20 images, each with a bandwidth of 0.02 μm. The system with medium spectral resolution records four images, each with a bandwidth of 0.10 μm. The system with low spectral resolution records one image with a bandwidth of 0.40 μm.

Figure 1-8 is an alternate way to describe spectral resolution as spectral reflectance curves recorded by three systems with different levels of spectral resolution. Information in the three curves is summarized in Table 1-5.

As the wavelength of electromagnetic energy being collected by a sensor becomes longer, the energy associated with the longer wavelengths decreases (Equation 1-2). To deal with the decreased energy, instrument designers can:

1. Improve the sensitivity of the sensor,
2. Increase the bandwidth to gather more energy,
3. Increase area of the ground resolution cell or pixel, or
4. Increase the dwell time.

For example, in Landsat 7, band 3 (red) has a bandwidth of 0.061 μm and a ground resolution cell of 30 by 30 m (900 m^2). Band 6 (TIR), however, has a bandwidth of 2.05 μm and a ground resolution cell of 60 by 60 m (3,600 m^2). In band 6 the bandwidth is expanded by a factor of 33X that of band 3; the ground resolution cell of band 6 is enlarged by a factor of 4X that of band 3.

Radiometric Resolution

Radiometric resolution is the number of subdivisions, or *bits*, that an imaging system records for a given range of values. A bit (abbreviation for *binary digit*) is the smallest unit of data in a computer. One bit stores just two values: 0 or 1. The radiometric resolution (the number of brightness levels recorded by a sensor) doubles as a bit is added to the sensor. Mathematically, n bits yield 2^n radiometric levels. As the number of bits increases, the radiometric resolution increases. Figure 1-9 shows images and charts for three levels of radiometric resolution.

The human eye has a radiometric resolution of approximately five bits (Figure 1-9C), which enables us to resolve 25 or 32 gray levels ranging from black to white. Most remote sensing systems record at least eight bits for 256 or 2^8 radiometric levels (Figure 1-9E). Eight bits is equivalent to a byte. Figure 1-10 shows three grayscales that correspond to the radiometric levels shown in Figure 1-9.

Temporal Resolution

For images recorded repetitively of the same area, temporal resolution is the time interval between successive images. Four major satellite systems with repetitive coverage are listed below with their time interval.

System	Interval (Days)
AVHRR	Daily
Landsat	16
IRS	22
SPOT (nadir)	26

RESOLVING POWER

Resolving power applies to an imaginary system or a component of the system, whereas spatial resolution applies to the image produced by the system. For example, the lens and film or digital array of a camera system each have a characteristic resolving power that, together with other factors, determines the spatial resolution of the photographs.

Angular resolving power is defined as the angle subtended by imaginary lines passing from the imaging system and two targets spaced at the minimum resolvable distance. Angular resolving power is commonly measured in radians. As shown in Figure 1-11, a *radian* (rad) is the angle subtended by an arc (BC) of a circle having a length equal to the radius (AB) of the circle. Because the circumference of a circle has a length equal to 2π times the radius, there are 2π, or 6.28, rad in a circle. A radian corresponds to 57.3° or 3,438 min, and 1 milliradian (mrad) is 10^{-3} rad. In the radian system of angular measurement,

$$\text{angle} = \frac{L}{r}\text{rad} \quad \textbf{(1-5)}$$

where L is the length of the subtended arc and r is the radius of the circle. A convenient relationship is that at a distance r of 1,000 units, 1 mrad subtends an arc L of 1 unit.

The *instantaneous field of view* (IFOV) of a detector is the solid angle through which a detector is sensitive to radiation. Equation 1-5 can be rearranged to calculate the spatial resolution of a sensor (ground sampling distance or GSD) given the sensor's IFOV and the flying altitude above the ground (AGL). L (length of the subtended arc in

TABLE 1-5 Spectral resolution, bands, and bandwidth from Figure 1-8.

Spectral Resolution	Number of Bands	Bandwidth (μm)
Low	5	0.120
Medium	13	0.050
High	128	0.005

A. Low spectral resolution with 0.120 μm bandwidth.

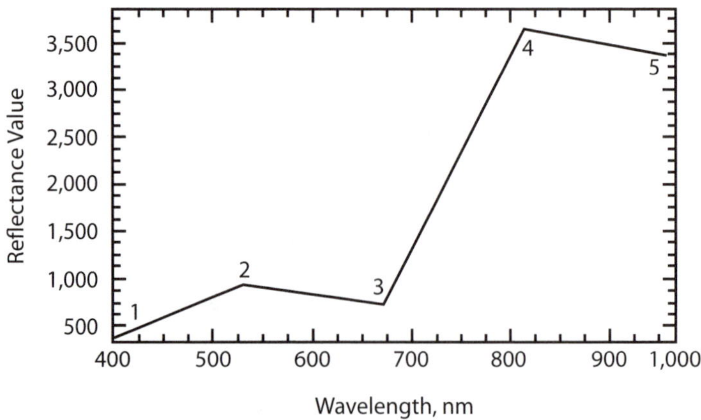

B. Medium spectral resolution with 0.050 μm bandwidth.

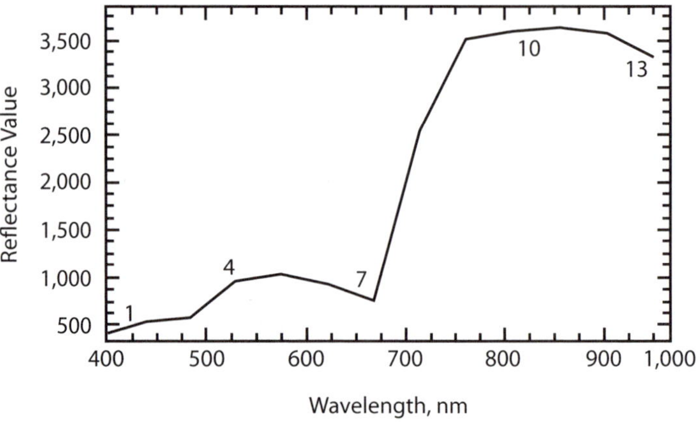

C. High spectral resolution with 0.005 μm bandwidth.

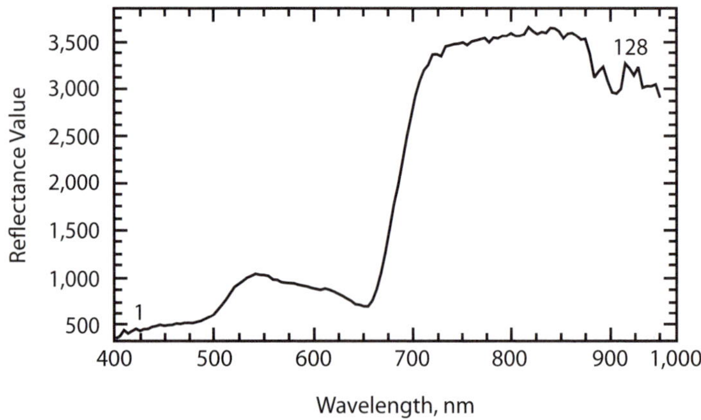

FIGURE 1-8 Spectral resolution shown as reflectance spectra for vegetation.

A. 3-bit radiometric resolution (8 gray levels).

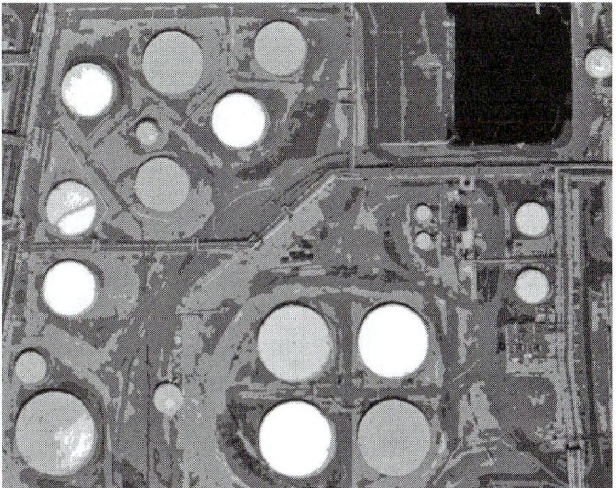

B. Chart of 3-bit radiometric resolution.

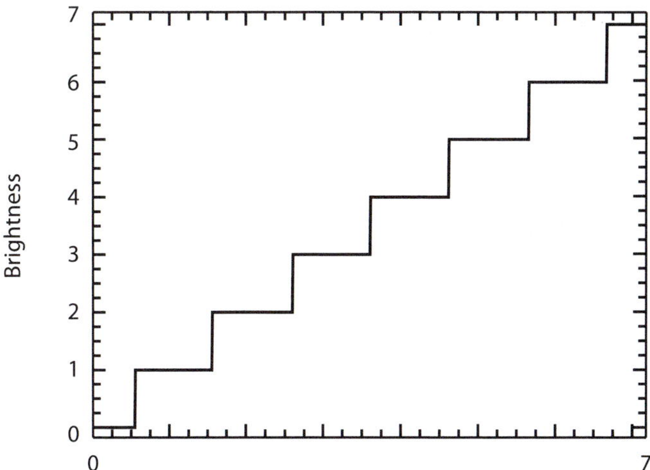

C. 5-bit radiometric resolution (32 gray levels).

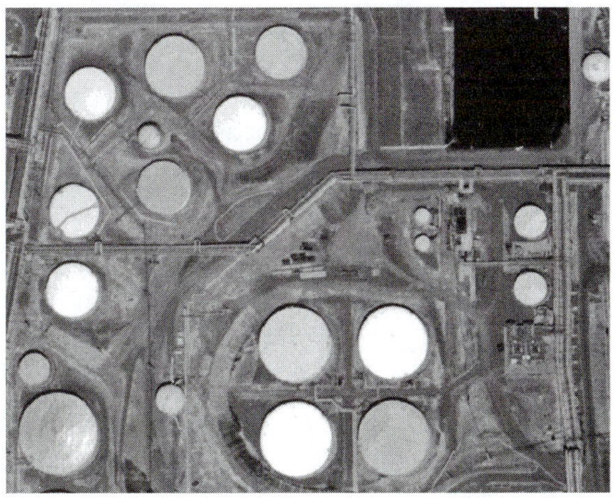

D. Chart of 5-bit radiometric resolution.

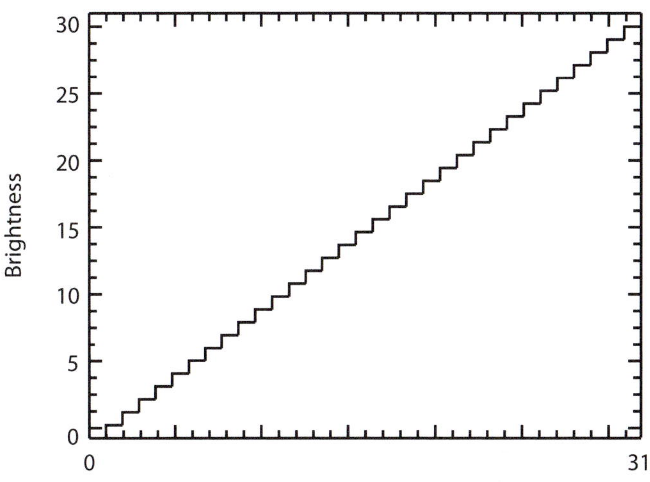

E. 8-bit radiometric resolution (256 gray levels).

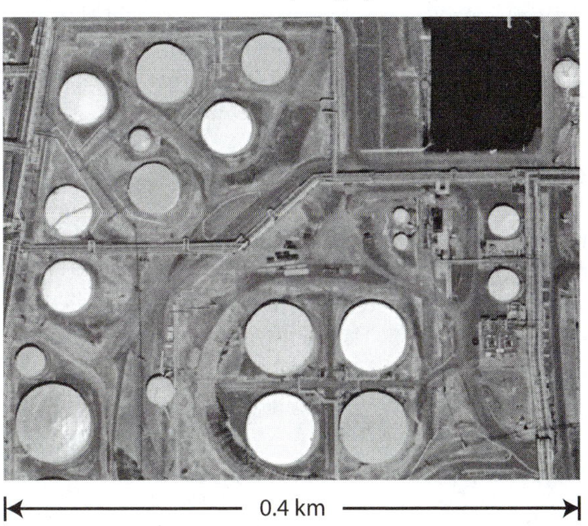

F. Chart of 8-bit radiometric resolution.

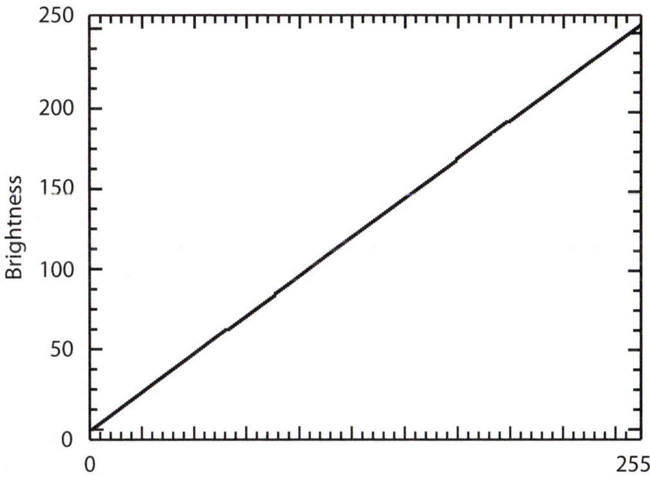

FIGURE 1-9 Radiometric resolution. Images and charts that illustrate 3, 5, and 8 bits of radiometric resolution. Locality is a refinery tank farm in northern California. Image covers a width of 0.4 km. Courtesy D. Ruiz, Quantum Spatial, Inc., Novato, California.

A. 3-bit grayscale (8 levels of gray).

B. 5-bit grayscale (32 levels of gray).

C. 8-bit grayscale (256 levels of gray).

FIGURE 1-10 Radiometric resolution. The grayscales show the 3, 5, and 8 bits of resolution that correspond to the images and charts in Figure 1-9.

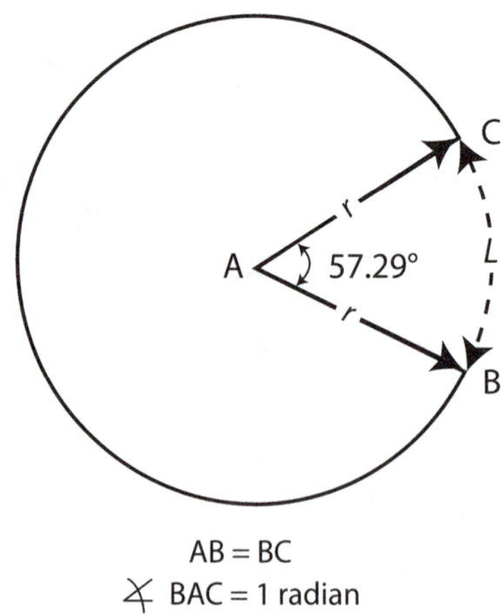

FIGURE 1-11 Resolving power—radian system of angular measurement.

Figure 1-11) represents the GSD of the sensor and is given by the r (radius) or sensor's flying height (AGL) times the sensor's IFOV. For instance, a sensor with an IFOV of 1 mrad that is flown at 1,000 m will have a GSD of 1 m.

$$\begin{aligned} L &= \text{IFOV} \times r \\ &= 1 \text{ mrad} \times 1{,}000 \text{ m} \\ &= 0.001 \text{ rad} \times 1{,}000 \text{ m} \\ &= 1 \text{ m} \end{aligned}$$

OTHER CHARACTERISTICS OF IMAGES

Detectability is the ability of an imaging system to record the presence or absence of an object, although the identity of the object may be unknown. An object may be detected even though it is smaller than the theoretical resolving power of the imaging system.

Recognizability is the ability to identify an object on an image or reflectance chart. Objects may be detected and resolved and yet not be recognizable. For example, roads on an image appear as narrow lines that could also be railroads or canals. Unlike resolution, there are no quantitative measures for recognizability and detectability. It is important for the interpreter to understand the significance and correct use of these terms. Rosenberg (1971) summarizes the distinctions between them.

A *signature* is the expression of an object on an image or reflectance chart that enables the object to be recognized. Characteristics of an object that control its interaction with electromagnetic energy determine its signature.

Texture is the frequency of change and arrangement of tones on an image. *Fine*, *medium*, and *coarse* are qualitative terms used to describe texture.

An *interpretation key* is a characteristic or combination of characteristics that enable an object to be identified on an image. Typical keys are size, shape, tone, and color. The associations of different characteristics are valuable keys. On images of cities, one may recognize single-family residential areas by the association of a dense street network, lawns, and small buildings. The associations of certain landforms and vegetation species are keys for identifying different types of rocks.

VISION

Of our five senses, two (touch and vision) detect electromagnetic radiation. Our sense of touch includes the ability of nerves in our skin to detect noncontact TIR radiation as heat. Vision is the most important sense and accounts for most of the information input to our brain. Vision is not only an important remote sensing system in its own right, but it is also the means by which we interpret the images produced by other remote sensing systems. The following section describes the human eye as a remote sensing system (Gregory, 1966).

STRUCTURE OF THE EYE

For such a complex structure, the human eye (Figure 1-12) appears deceptively simple. Light enters through the clear *cornea*, which is separated from the lens by fluid called the *aqueous humor*. The *iris* is the pigmented part of the eye that controls the variable aperture called the *pupil*. It is commonly thought that variations in pupil size allow the eye to function over a wide range of light intensities. However, the pupil varies in area over a ratio of only 16:1 (i.e., the maximum area is 16 times the minimum area), whereas the eye functions over a brightness range of about 100,000:1. The pupil contracts to limit the light rays to the central and optically best part of the lens, except when the full opening is needed in dim light. The pupil also contracts for near vision, increasing the depth of field for near objects.

A common misconception is that the lens *refracts* (bends) the incoming rays of light to form the image. The amount that light bends when passing through two adjacent media is determined by the difference in the refractive indices (n) of the two media; the greater the difference, the greater the bending. For the eye, the maximum difference is between air (n = 1.0) and the cornea (n = 1.3); therefore this interface is where the maximum light refraction occurs and the image is formed. Although the lens is relatively unimportant for forming the image, it is important in *accommodating*, or focusing, for near and far vision. In the human eye, the shape of the lens is changed by muscles that vary the tension on the lens. For near-vision tension, the muscles release, allowing the lens to become thicker in the center and assume a more convex cross section. With age, the cells of the lens harden and the lens becomes too rigid to accommodate for different distances; this is the time in life when bifocal glasses may become necessary to provide for near and far vision. Eventually the lens may become cloudy and be surgically replaced by an artificial lens.

An inverted image is focused on the *retina*, a thin sheet of interconnected nerve cells that line the interior of the eye ball. The retina includes the light receptor cells called rods and cones, that convert light into electrical impulses. The rods and cones receive their names from their longitudinal shapes when viewed microscopically. The cones function in daylight conditions to provide color vision, called *photopic vision*. The rods function under low illumination and give vision only in tones of gray, called *scotopic vision*. Rods and cones are not uniformly distributed throughout the retinal surface. The maximum concentration and organization of rods and cones is in the *fovea* (also called the *macula*), a small region at the center of the retina that provides maximum visual acuity (Figure 1-12). You can demonstrate the existence and importance of the fovea by concentrating on a single letter on this page. The rest of the page and even the nearby words and letters will appear indistinct because they are outside the field of view of the fovea. The eye is in continual motion to bring the fovea to bear on all parts of the page or scene. Near the fovea is the blind spot, where the optic nerve joins the eye and there are no receptor cells. The optic nerve transmits electrical impulses from the receptor cells to the brain, which interprets them as visual perception.

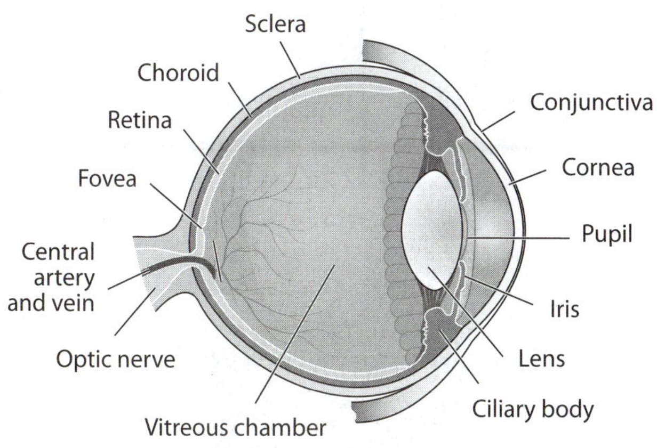

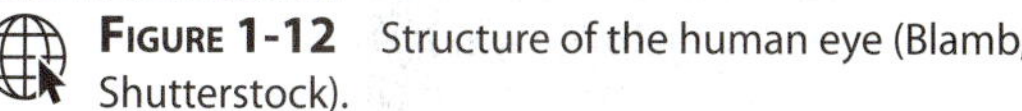
FIGURE 1-12 Structure of the human eye (Blamb, Shutterstock).

RESOLVING POWER OF THE EYE

The diameter of the largest receptor cells (3 μm) in the fovea determines the resolving power of the eye. Multiplying this diameter maximum by the refractive index of the vitreous humor (n = 1.3), which fills the eye ball, determines an effective diameter (4 μm) for the receptor cells. The *image distance*, or distance from the retina to the lens, is about 20 mm, or 20,000 μm. The effective width of the receptors is 4/20,000 (1/5,000) of the image distance. Image distance is proportional to *object distance*, which is the distance from the eye to the object. An object forms an image that fills the width of a receptor if the object width is 1/5,000 the object distance. Therefore, adjacent objects must be separated by 1/5,000 the object distance for their images to fall on alternate receptors and be resolved by the eye.

You can estimate the resolving power of your eyes in the following manner. View the resolution bar charts of Figure 1-13A at a distance of 5 m (16.4 ft), and determine the most closely spaced set of line-pairs that you can resolve. Also determine the narrowest of the bars in Figure 1-13B that you can detect. Make these determinations now, before reading further, because the following text may influence your perception of the targets.

For the high-contrast resolution targets of Figure 1-13A at a distance of 5 m, the normal eye should be able to resolve the middle set that has 5 line-pairs · cm^{-1}. The black and white bars are 1 mm wide. The IFOV of any detector is the solid angle through which a detector is sensitive to radiation. Equation 1-5 is used to calculate the IFOV of the eye, where the radius (r) is 5,000 mm and the length of the subtended arc (L) is 1 mm:

$$\begin{aligned} \text{IFOV} &= \frac{L}{r}\text{rad} \\ &= \frac{1\text{ mm}}{5{,}000\text{ mm}}\text{rad} \\ &= 0.2\times10^{-3}\text{rad} \\ &= 0.2\text{ mrad} \end{aligned}$$

The 0.2 mrad IFOV of the eye means that at a distance of 1,000 units, the eye can resolve high-contrast targets that are spaced no closer than 0.2 units. As an example, the eye can resolve high-contrast targets that are spaced no closer than 2 cm viewed from a distance of 100 m (based on a 0.2 mrad IFOV).

DETECTION CAPABILITY OF THE EYE

When the detection targets of Figure 1-13B are viewed from a distance of 5 m, most readers can detect the narrowest bar, which is 0.2 mm wide. Recall, however, that at this distance the minimum separation at which bar targets can

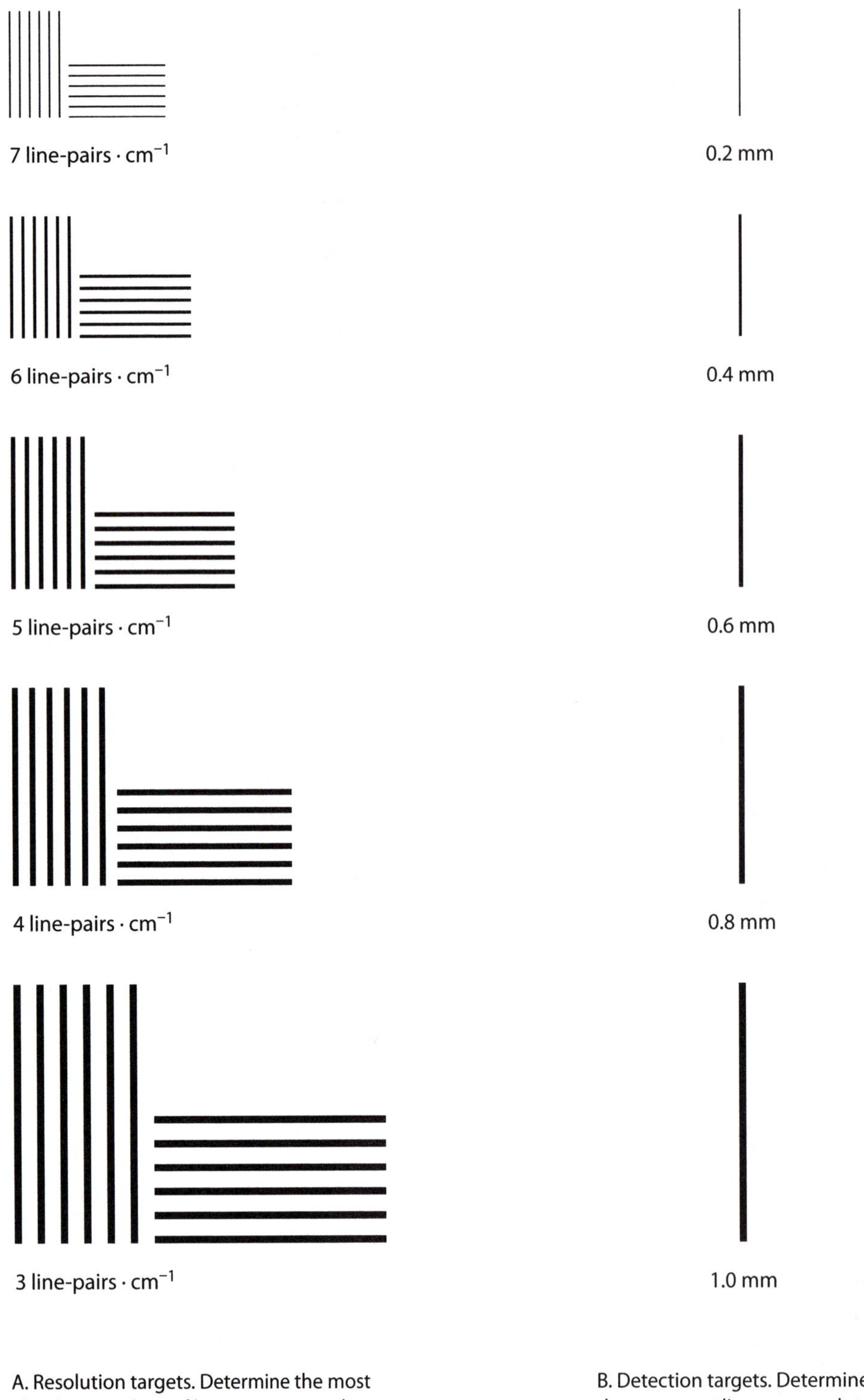

FIGURE 1-13 Resolution and detection targets with high contrast ratio. View this chart from a distance of 5 m (16.4 ft).

be resolved is 1.0 mm. This test illustrates the difference between resolution and detection. Detection is influenced not only by the size of objects but also by their shape and orientation. For example, if dots are used in place of lines in Figure 1-13B, the diameter of the smallest detectable dot would be considerably larger than 0.2 mm.

Effect of Contrast Ratio on Resolution and Detection

The resolution and detection targets in Figure 1-14 have the same spacing as those in Figure 1-13, but the contrast ratio has been reduced by the addition of a gray background. To evaluate the effect of the lower contrast ratio, view Figure 1-14 from a distance of 5 m (16.4 ft), and determine which targets can be resolved and detected. Using this figure, most readers can resolve only 3 line-pairs · cm^{-1}, and the smallest detectable target is the 0.6-mm wide line. These dimensions are larger than the 5 line-pairs · cm^{-1} and the 0.2 mm line of the high-contrast target and demonstrate the effect of a lower contrast ratio on resolution and detection.

LOCATION AND NAVIGATION SYSTEMS

Images acquired from aircraft or satellites must be located relative to a geographic coordinate system on the terrain, which is typically defined by latitude and longitude. The coordinates may be displayed with a variety of projections that are described in Chapter 9. The attitude (pitch, roll, and heading) of the imaging system is also required in order to georeference or orthorectify the image.

Global Navigation Satellite System and Global Positioning System

A global navigation satellite system (GNSS) collects accurate horizontal and vertical (*x*, *y*, *z*) coordinates for moving or fixed platforms. The United States deployed the first GNSS named NAVSTAR Global Positioning System (GPS). GPS is the term most often used in the United States when using or describing GNSS. However, Russia (GLONASS), a European group (Galileo), and China (BeiDou) have developed and launched their own GNSS that provide global coverage. For remote sensing there are three components to a GPS:

1. Receivers on the remote sensing system;
2. A constellation of satellites transmitting positioning signals; and
3. Ground control stations that track, communicate, and maintain the accuracy of the system.

A GPS satellite transmits a coded signal that is recorded by a receiver on the Earth-orbiting remote sensing system. The travel time for the signal, measured at the speed of light, is the distance between the GPS satellite and the remote sensing system. A minimum of three GPS satellites is required for an accurate location. Bolstad (2016) provides details of GPS and related topics. The same system is employed by our handheld GPS systems.

Inertial Navigation System

Airborne remote sensing platforms use GNSS with differential positioning to provide the horizontal and vertical location (*x*, *y*, *z*) of the platform while flying. To determine where the sensor is pointed, an *inertial navigation system* (INS) with an *inertial measurement unit* (IMU) is connected to the sensor and records the sensor's pitch, roll, and heading. Pixel locations are rapidly and accurately calculated with the data collected using GNSS and INS.

REMOTE SENSING SYSTEMS

Remote sensing systems belong to the two major categories: framing systems and scanning systems.

Framing Systems

Framing systems instantaneously acquire an image of an area, or *frame*, of the terrain. The imaging systems on smartphones are common examples of such systems. A framing system (Figure 1-15) employs a lens to form an image of the scene at the *focal plane*, which is the plane at which the image is sharply defined. A shutter opens at selected intervals to allow light to reach the focal plane.

Successive frames of images may be acquired with *forward overlap* (Figure 1-15). The overlapping portions of the two frames may be viewed with a stereoscope to produce a three-dimensional view, as described in Chapter 2. A framing system can instantaneously acquire an image of a large area because the system has a dense array of detectors located at the focal plane.

Scanning Systems

A simple *scanning system* employs a single detector with a narrow field of view that is swept across the terrain to produce an image. When photons of electromagnetic energy, that are reflected or radiated from the terrain, encounter the detector, an electrical signal is generated that varies in proportion to the number of photons. The electrical signal is amplified and recorded for later playback to produce an image. All scanning systems sweep the detector's field of view across the terrain in a series of parallel scan lines. Figure 1-16 shows the four common scanning modes: cross-track scanning, circular scanning, along-track scanning, and side scanning.

Cross-Track Scanners

The *cross-track scanners* employ a faceted mirror that is rotated by an electric motor, with a horizontal axis of rotation aligned parallel with the flight direction (Figure 1-16A). The mirror sweeps across the terrain in a pattern of parallel scan lines oriented normal (perpendicular) to the flight direction. Energy reflected or radiated from the terrain is focused

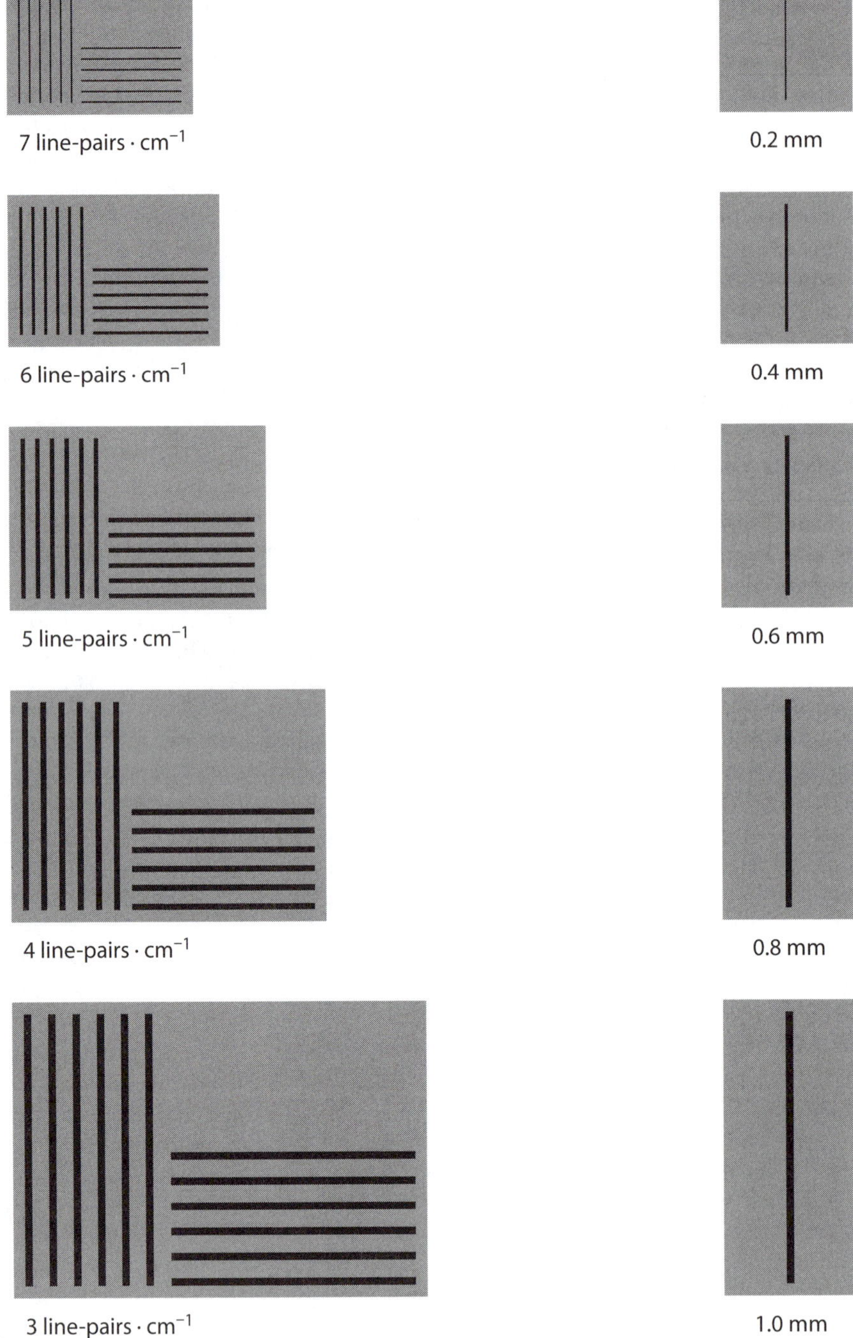

FIGURE 1-14 Resolution and detection targets with medium contrast ratio. View this chart from a distance of 5 m (16.4 ft). Compare your responses with those you determined for the high-contrast targets of Figure 1-13, which will demonstrate the effect of contrast ratio on vision.

onto the detector by secondary mirrors (not shown). Images recorded by framing and scanning systems are described by their characteristics: *spatial resolution* (Figure 1-5) and *spectral resolution* (Figure 1-6). The physical dimensions of a detector and the altitude of the system determine the size of the *ground resolution cell*, as shown by the small dark square in Figure 1-16A. Each ground resolution cell is recorded as a picture element or pixel. A cross-track scanner flying at an altitude of 10 km with an IFOV of 1 mrad provides ground resolution cells of 10 by 10 m (Equation 1-5). Cross-track scanners are also called whisk broom scanners because the movement of the detector across the terrain is comparable to the movement of a bristle in a whisk broom.

The *angular field of view* (Figure 1-16A) is that portion of the mirror sweep, measured in degrees, that is recorded as a scan line. The angular field of view and the altitude of the system determine the *ground swath*, which is the width of the terrain strip recorded on the image. Ground swath is calculated as

$$\tan\left(\frac{\text{angular field of view}}{2}\right)\times \text{altitude}\times 2 \qquad \textbf{(1-6)}$$

The distance between the scanner and terrain is greater at the margins of the ground swath than at the center of the swath. As a result, ground resolution cells are larger toward the margins than at the center of the image, which results in a geometric distortion. Modern scanner systems include software that automatically corrects this distortion. At the high altitude of satellites, a narrow angular field of view is sufficient to cover a broad swath of terrain. For this reason the rotating mirror is replaced by a flat mirror that oscillates back and forth through a narrow angle. An example is the scanner of the Thematic Mapper on Landsat described in Chapter 3. The strength of the signal generated by a detector is a function of the following factors:

- **Energy or radiant flux:** The amount of energy reflected or radiated from terrain is the *energy flux*. For detectors of visible light, this flux is lower on a dark day than on a sunny day.
- **Altitude:** For a given ground resolution cell, the amount of energy reaching the detector is inversely proportional to the square of the distance. At higher altitudes the signal strength is weaker.
- **Spectral bandwidth of the detector:** The signal is stronger for detectors that respond to a broader bandwidth of energy. For example, a detector that is sensitive to the entire visible range will receive more energy than a detector that is sensitive to a narrow band, such as visible red.
- **IFOV:** The physical size of the sensitive element of the detector determines the IFOV. A small IFOV is required for high spatial resolution but also restricts the signal strength, which is the amount of energy received by the detector.
- **Dwell time:** The time required for the detector IFOV to sweep across a ground resolution cell is the *dwell time*. A longer dwell time allows more energy to impinge on the detector, which creates a stronger signal.

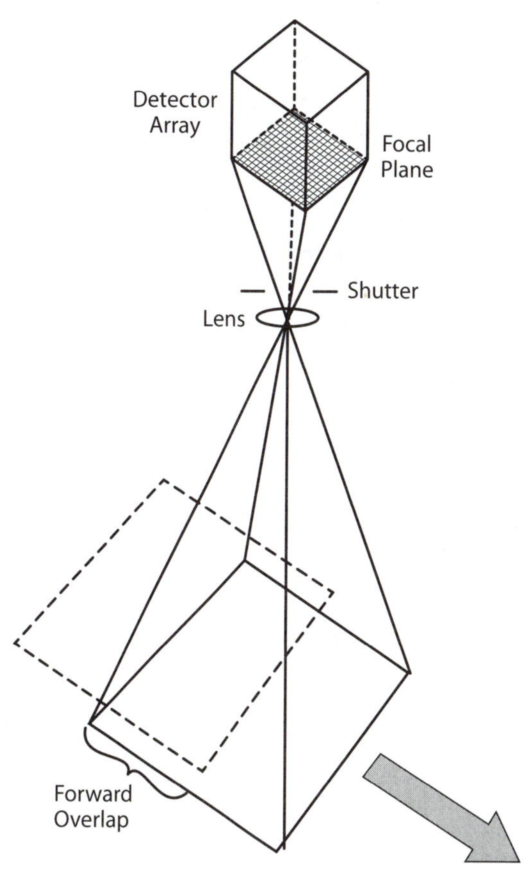

Figure 1-15 Digital framing system for acquiring stereo images. The focal plane is a dense array of detectors that acquires an image each time the shutter is cycled. The overlapping portions of successive images may be viewed stereoscopically. Digital framing systems have replaced film cameras.

For a cross-track scanner, the dwell time is determined by the detector IFOV and by the velocity at which the scan mirror sweeps the IFOV across the terrain. Figure 1-17A shows a typical cross-track scanner with a detector IFOV of 1 mrad, a 90° angular field of view, and operating at 2×10^{-2} sec per scan line at an altitude of 10 km. This scanner has a dwell time of 1×10^{-5} sec per ground resolution cell. It is instructive to compare the dwell time with the ground speed of the aircraft. At a typical ground speed of 720 km · hr^{-1}, or 200 m · sec^{-1}, the aircraft crosses the 10 m of a ground resolution cell in 5×10^{-2} sec. The cross-track scanner time of 1×10^{-5} is 5×10^{3} times faster than the ground velocity of the aircraft. The high scanner speed relative to ground speed prevents gaps between successive scan lines. The short dwell time of cross-track scanners imposes constraints on the other factors that determine signal strength. For example, the IFOV and spectral bandwidth must be large enough to record a signal of sufficient strength to overcome the inherent electronic noise of the system. The *signal-to-noise ratio* must be sufficiently high for the signal to be recognizable.

A. Cross-track scanner.

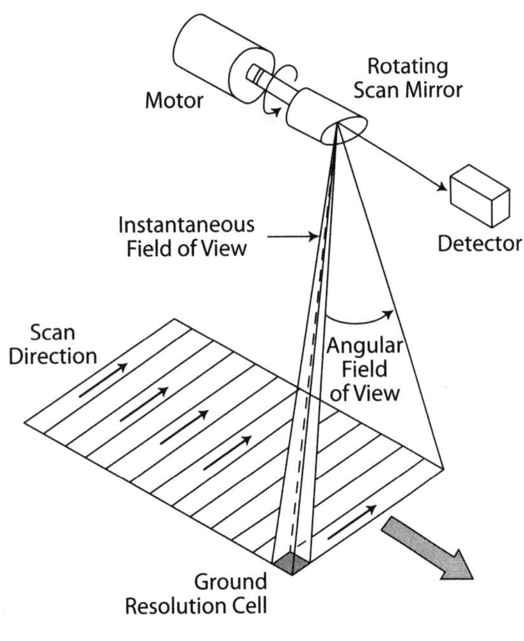

B. Circular scanner.

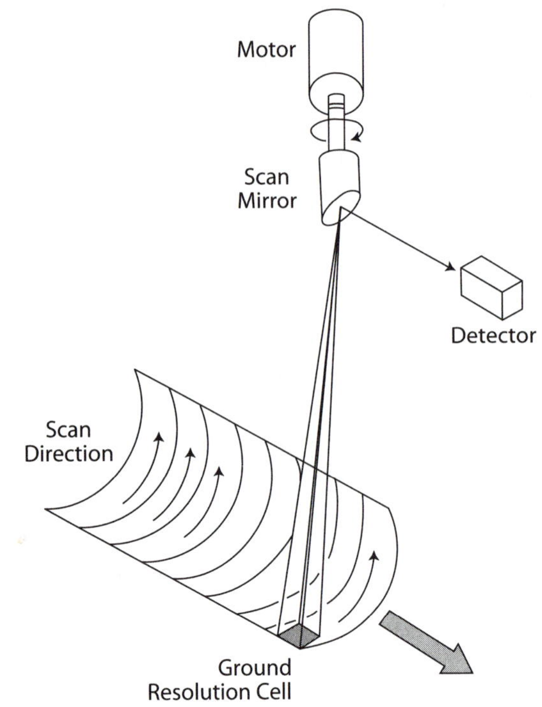

C. Along-track scanner.

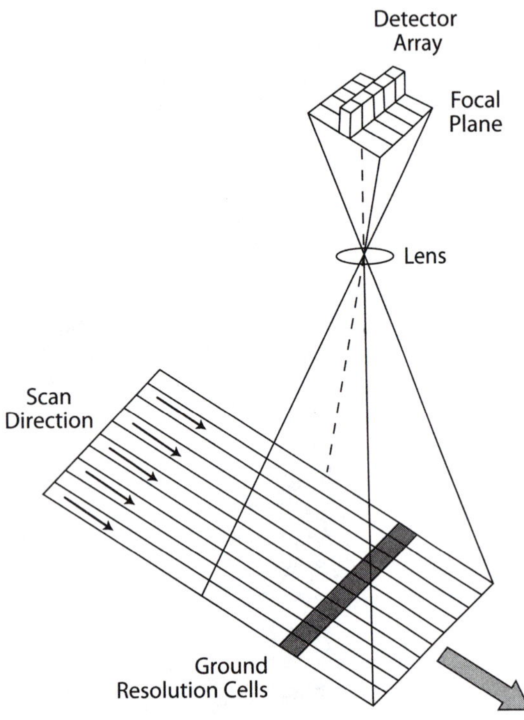

D. Side-scanning system.

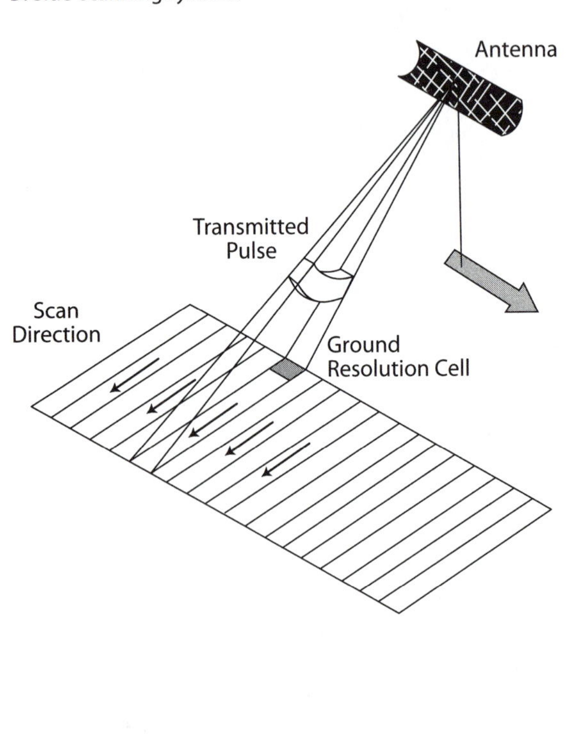

FIGURE 1-16 Scanning systems for acquiring remote sensing images. Passive systems (such as cross-track, circular, and along-track scanners) record energy that is reflected, scattered, or radiated from the target. Active systems (such as radar or side-scanning) provide their own source of energy and record the energy returned, or backscattered, from terrain.

Circular Scanners

In a *circular scanner* the scan motor and mirror are mounted with a vertical axis of rotation that sweeps a circular path on the terrain (Figure 1-16B). Only the forward portion of the sweep is recorded to produce images. An advantage of this system is that the distance between the scanner and terrain is constant and all the ground resolution cells are recorded with the same dimensions. Circular scanners are used for reconnaissance purposes in aircraft. The axis of rotation is tilted forward to acquire images of the terrain far ahead of the aircraft's position. The images are displayed in real time on a screen in the cockpit to guide the pilot. Airborne circular scanners with IR detectors are called FLIR (forward looking IR) systems.

Along-Track Scanners

For scanner systems to achieve finer spatial and spectral resolution, the dwell time for each ground resolution cell must be increased. One method is to eliminate the scanning mirror and provide an individual detector for each ground resolution cell across the ground swath (Figure 1-16C). The detectors are placed in a linear array in the focal plane of the image formed by a lens system. The long axis of the linear array is oriented normal to the flight path, and the IFOV of each detector sweeps a ground resolution cell along the terrain parallel with the flight track direction (Figure 1-16C). *Along-track scanning* refers to this movement of the ground resolution cells. These systems are also called push broom scanners because the movement of the detectors along the terrain is analogous to the movement of bristles of a push broom as it sweeps along the floor.

For along-track scanners, the dwell time of a ground resolution cell is determined solely by the ground velocity of the system, as Figure 1-17B illustrates. For a jet aircraft flying at 720 km $\cdot$ hr^{-1}, or 200 m $\cdot$ sec^{-1}, the along-track dwell time for a 10-m cell is 5×10^{-2} sec, which is 5×10^{3} times greater than the dwell time for a comparable cross-track scanner. The increased dwell time allows two improvements: (1) detectors can have a smaller IFOV, which provides finer spatial resolution, and (2) detectors can have a narrower spectral bandwidth, which provides higher spectral resolution. Airborne along-track scanners can operate with a spectral bandwidth of 0.01 μm. Typical cross-track scanners have bandwidths of 0.10 μm, which is a spectral resolution coarser by one order of magnitude. Landsat 8 employs an along-track system.

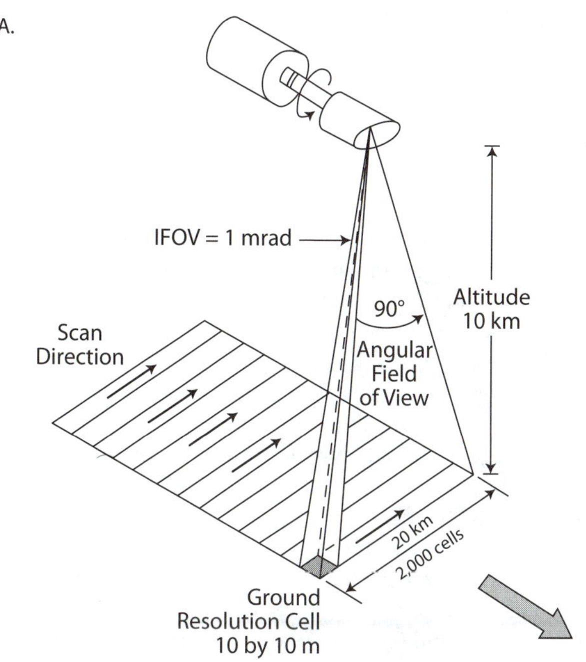

scan rate = 2×10^{-2} sec per scan line

$$\text{dwell time} = \frac{\text{scan rate per line}}{\text{number of cells per line}} = \frac{2\times10^{-2}\ \text{sec}}{2{,}000\ \text{cells}} = 1\times10^{-5}\ \text{sec}\cdot\text{cell}^{-1}$$

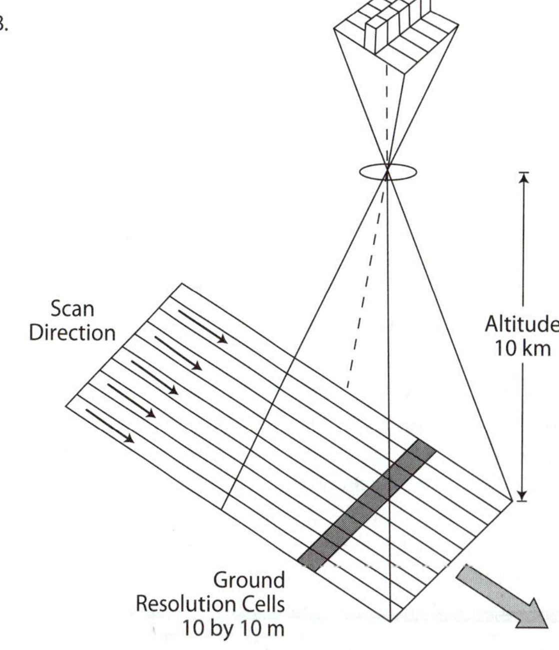

IFOV for each detector = 1 mrad

$$\text{dwell time} = \frac{\text{cell dimension}}{\text{velocity}} = \frac{10\ \text{m}\cdot\text{cell}^{-1}}{200\ \text{m}\cdot\text{sec}^{-1}} = 5\times10^{-2}\ \text{sec}\cdot\text{cell}^{-1}$$

Figure 1-17 Dwell time calculated for (A) cross-track and (B) along-track scanners.

Side-Scanning Systems

The cross-track, circular, and along-track scanners just described are *passive systems* that detect and record energy naturally reflected or radiated from the terrain. *Active systems*, such as radar, provide their own energy source. Figure 1-16D is a side-scanning radar system that transmits pulses of microwave energy to one side of the flight path and records the energy scattered from the terrain back to the antenna, as described in Chapter 6. Sonar and lidar are additional active systems. Side-scanning sonar transmits pulses of sonic energy in the ocean to map bathymetric features while lidar transmits pulses of laser energy and records the returns to produce detailed topographic maps (Chapter 7).

Scanner Systems Compared

Cross-track and along-track scanners have different characteristics that are summarized in Table 1-6. The selection of a scanner system involves a number of choices, or trade-offs. Cross-track scanners are generally preferred for reconnaissance surveys because the wider angular field of view records images that cover more terrain. Circular scanning systems, such as FLIR, acquire images of terrain and hazards located ahead of the aircraft. Along-track scanners acquire detailed spectral and spatial information because the dwell time is longer. The longer dwell time acquires sufficient energy to accommodate detectors with a narrow bandwidth or a small IFOV. Active side-scanning systems such as radar can acquire images at night and during inclement weather.

REFLECTANCE AND RADIANCE

Reflectance and radiance are terms that are commonly misused as synonyms. Figure 1-18 is a diagram of a passive remote sensing system with a detector that records solar energy reflected from a target on the terrain. This energy is called target reflectance. The diagram also shows that solar energy interacts with gases in the Earth's atmosphere in a process called *atmospheric scattering*. Some of the scattered energy, called path radiance, follows the path of the target reflectance toward the detector. The combined target reflectance and atmospheric path radiance that reaches the detector is called *total radiance*. Chapter 9 shows how to compensate for the scattered wavelengths that contaminate an image.

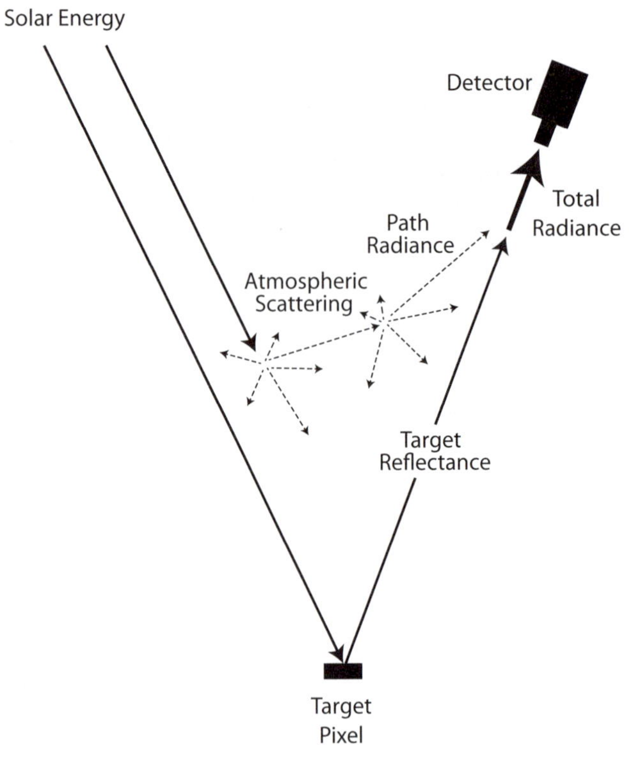

FIGURE 1-18 Simplified diagram showing how atmospheric scattering directs unwanted light (path radiance) into the sensor. The path radiance combines with the target reflectance to form an unwanted total radiance measurement from the target of interest.

The spectral (and polarization) characteristics of the total radiance recorded by the detector are a function of the (1) location of the illumination source, (2) orientation of the target pixel, and (3) location of the detector (Jensen, 2007). The *bidirectional reflectance distribution function* (BRDF) measures the impact caused by different angles of incoming and reflected energy on the interpretation of features in imagery. Understanding BRDF is important when comparing remote sensing images acquired on multiple dates with different illumination and viewing angles. Jensen (2007) provides examples of the BRDF effect on images of the same feature taken from different viewing angles.

Reflectance spectra are curves that record the intensity of energy reflected from a target as a function of wavelength. Spectrometers and spectroradiometers record spectra of

TABLE 1-6 Characteristics of cross-track and along-track scanners.

Characteristic	Cross-Track Scanner	Along-Track Scanner
Angular field of view	Wider	Narrower
Mechanical system	Complex	Simple
Optical system	Simple	Complex
Spectral resolution of detectors	Wider range	Narrower range
Dwell time	Shorter	Longer

rocks, minerals, vegetation, and other materials in the field (Figure 1-19) and laboratory (such as the curves shown in Figure 1-20). The curves are offset vertically to prevent confusing overlaps. The horizontal scale shows wavelengths ranging from blue through NIR and SWIR (0.40 to 2.50 μm or the VNIR-SWIR range). The vertical scale shows reflectance. The shale, limestone, and sandstone curves have different absorption features and reflectance brightness values that are used to identify these rocks with multispectral and hyperspectral data. The curve for healthy vegetation shows that blue and red wavelengths are absorbed during photosynthesis and green light is reflected by chlorophyll. The strong reflectance in the NIR region is caused by the cell structure of leaves, as shown in Chapter 2. The "red edge" describes the very sharp increase in reflectance from red to NIR wavelengths that is a unique spectral characteristic of healthy vegetation. The two striped bands in the SWIR region are absorption features caused by water vapor in the atmosphere. The horizontal lines shown above the spectra are the bands recorded by multispectral and hyperspectral systems.

Hunt (1980) published a number of mineral spectra and explained the interactions between energy and matter that cause the spectral features at different wavelengths. The USGS Spectral Library (Kokaly and others, 2017) has laboratory reflectance spectra for thousands of materials, including minerals, organic and volatile compounds, vegetation, and man-made substances (speclab.cr.usgs.gov/index.html). This library can be downloaded and used with image processing software to interpret multispectral and hyperspectral data. Clark (1999) described spectroscopy and the causes of absorption features. Published spectral libraries, including those compiled by the USGS, Jet Propulsion Laboratory (JPL), the Council of the International Geoscience Programme (IGCP), Johns Hopkins, or other labs, should be included in more advanced image processing software.

A spectral characteristic viewer is available online as part of the USGS Landsat Missions Program (landsat.usgs.gov/spectral-characteristics-viewer). This interactive viewer displays spectra for minerals, vegetation types, water classes, and rock coatings. The user can choose between eight different satellite systems to select the system with optimum bands to map the materials of interest.

MULTISPECTRAL IMAGING SYSTEMS

A *multispectral image* is an array of simultaneously acquired images that record separate wavelength intervals, or bands. Much of this book deals with multispectral images that are acquired in all the remote sensing spectral regions.

Multispectral systems differ in the following characteristics:

- Imaging technology—framing or scanning method,
- Total spectral range recorded,
- Number of spectral bands recorded, and
- Range of wavelengths recorded by each spectral band (bandwidth).

The upper portion of Figure 1-20 shows the wavelength bands recorded by representative aircraft and satellite multispectral systems, which are described in the following sections. Multispectral images are acquired by two methods: framing systems or scanning systems.

Multispectral Framing Systems

A multispectral framing system consists of several lenses and matching detector arrays mounted together and aligned to acquire simultaneous multiple images of an area. Filters are attached to the lenses that allow only specific wavelength

Figure 1-19 SVC HR-1024i VNIR-SWIR spectroradiometer on tripod for calibration panel measurement, Antarctic. Photo courtesy of Andrew Gray, NERC Field Spectroscopy Facility, University of Edinburgh.

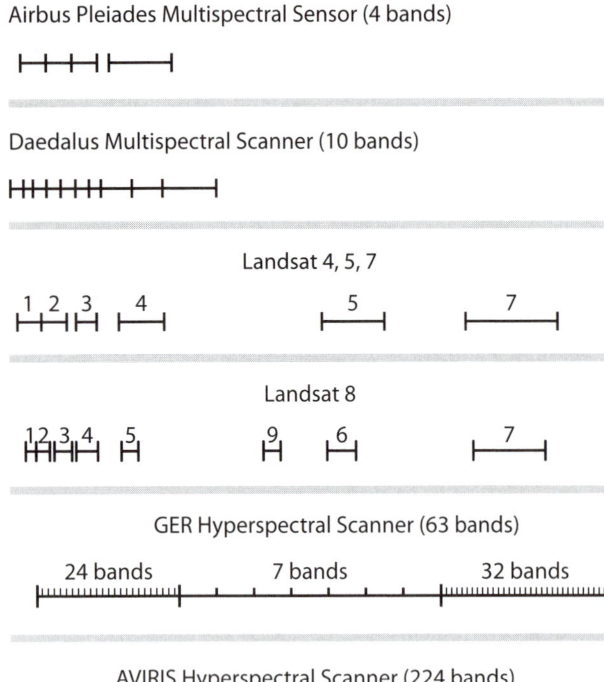

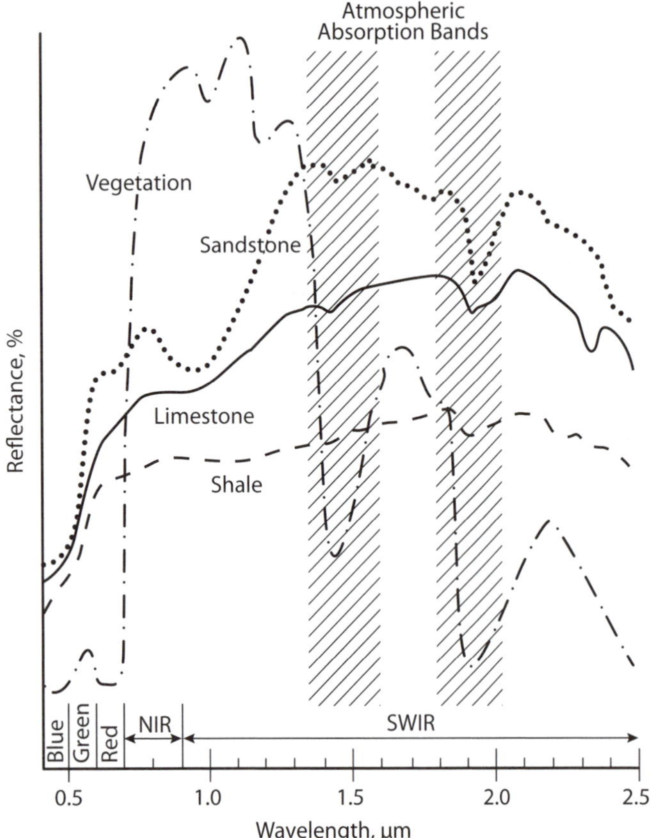

FIGURE 1-20 Reflectance spectra of rocks and vegetation. Spectral bands of typical multispectral and hyperspectral systems are shown.

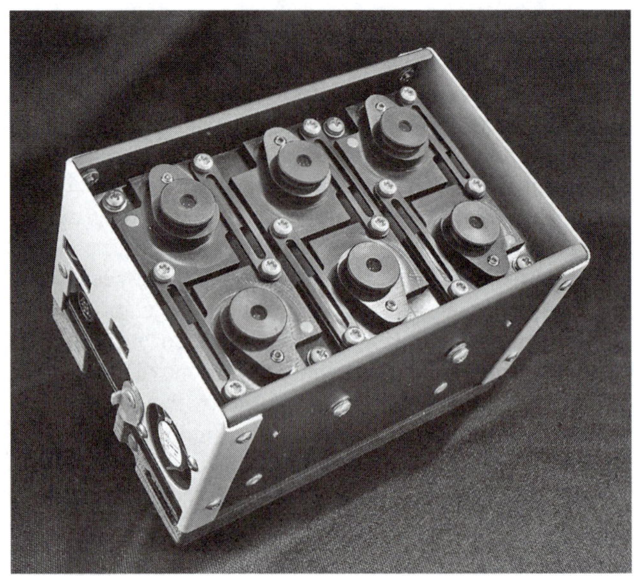

FIGURE 1-21 Airborne multispectral framing system designed for drones. The six lenses and filters record VNIR images. Multiple Camera Array Wireless System (MCAW) courtesy Tetracam (tetracam.com).

bands to reach the arrays. Electronic shutters are linked together and triggered simultaneously. The system with an array of six lenses in Figure 1-21 simultaneously acquires six VNIR images. This multispectral camera can be flown on drones (small unmanned aerial vehicles) as it only weighs 550 gm (19 ounces).

MULTISPECTRAL SCANNING SYSTEMS

Multispectral scanner systems are widely used to acquire images from aircraft and satellites. Both cross-track and along-track systems are used.

Cross-Track Multispectral Scanner Images

Cross-track multispectral scanners employ a spectrometer to disperse the incoming energy into a spectrum (Figure 1-22A). Detectors are positioned to record specific wavelength bands of energy (denoted λ_1, λ_2, λ_3, λ_4 in the figure). Figure 1-23 shows 10 images of San Pablo Bay, California, acquired by a cross-track multispectral scanner. There are seven visible bands, two NIR bands, and one SWIR band. The map in Figure 1-24 shows the categories of land use and land cover in the San Pablo Bay area, which is the northern extension of San Francisco Bay. **Digital Image 1-1** is a natural color image that was prepared by projecting bands 2 (blue), 4 (green), and 7 (red). Many other color combinations can easily be created.

The images of the individual spectral bands (Figure 1-23) illustrate the relationships among wavelength, atmospheric scattering, contrast ratio, and spatial resolution. Band 1 in the UV and blue spectral region records the shortest wavelengths of all the bands and has the maximum atmospheric scattering, which results in a low contrast ratio and poor spatial resolution. The network of streets in the city of Vallejo is a useful resolution target; as the wavelength of the images increases, the ability to resolve the streets improves and reaches a maximum in the NIR wavelengths (bands 8 and 9) and SWIR wavelengths (band 10).

Vegetation, water, and urban areas are the major types of land cover and land use in the San Pablo Bay area (Figure 1-24). In the spectral diagram (Figure 1-20) the wavelength of each band can be compared to the spectral reflectance curve for vegetation. Vegetation has a somewhat higher reflectance in the green bands (Figure 1-23D–E) than in the blue (Figure 1-23A–C) and red (Figure 1-23F–G) bands where chlorophyll absorbs energy. The reflectance curve of vegetation increases abruptly in the NIR and SWIR bands. Images of these bands (Figure 1-23H–J) show the bright signatures of vegetated hills in the northeast and southeast quadrants of the scene. The signature of water is also different in the various spectral bands. Along the shoreline in San Pablo Bay, bright signatures of suspended silt are obvious in the visible bands; in the IR bands, however, water has a uniform dark signature because these wavelengths are completely absorbed. Some of the salt evaporating ponds, along the Napa River in Figure 1-24, have red and pink signatures in **Digital Image 1-1** because of red microorganisms in the brine.

Along-Track Multispectral Scanner Images

Along-track (or push broom) multispectral scanners employ multiple linear arrays of detectors with each array recording a separate band of energy (Figure 1-17B). Because of the extended dwell time, the detector bandwidth may be narrow and still collect an adequate signal. The Operational Land Imager (OLI) of Landsat 8 (Chapter 3) is an along-track system as are the many SPOT systems (Chapter 4).

The Leica ADS series of airborne sensors uses an along-track system to acquire forward-looking, nadir (downward-looking), and rear-looking images of reflected VNIR light and a high resolution, panchromatic image. The system collects multispectral images with ground resolution cells of 10 cm when the aircraft flies at a maximum ground speed of 240 knots.

Side-Scanning Multispectral Images

Aircraft and satellite radar systems (Chapter 6) can record two or more wavelengths of microwave energy.

HYPERSPECTRAL SCANNING SYSTEMS

From the beginning of remote sensing, imaging technology has advanced in three major ways:

1. Improving the spatial resolution of images by decreasing the IFOV of detectors.
2. Improving the spectral resolution of images by increasing the number of spectral bands and decreasing the bandwidth of each band.
3. Improving the radiometric resolution by increasing the grayscale levels (current 12-bit sensors can record 4,096 levels of gray for each pixel).

A. Cross-track.

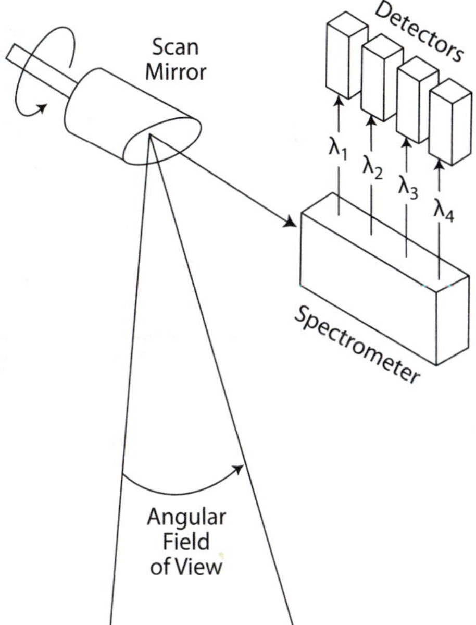

B. Along-track.

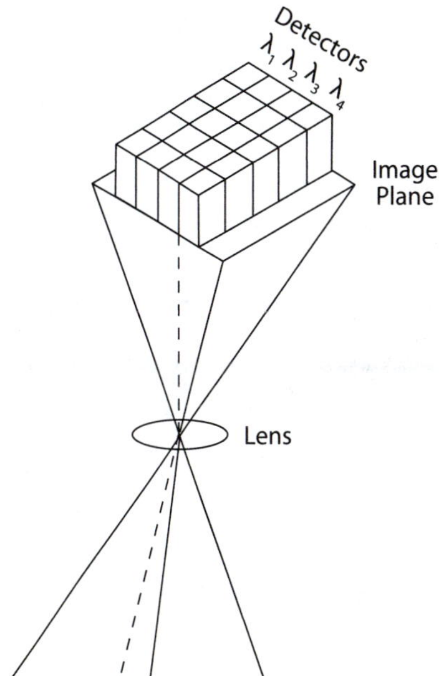

FIGURE 1-22 Multispectral scanning systems.

FIGURE 1-23 Aircraft multispectral scanner images of San Pablo Bay, California. Each image covers an area of 15 by 30 km. **Digital Image 1-1** is a natural color image composited from the blue, green, and red bands. Courtesy of NASA Ames Research Center.

Hyperspectral Technology

Hyperspectral systems collect a continuous spectrum of reflectance at many narrow, contiguous, and closely spaced wavelength bands (Shippert, 2004) in contrast to multispectral systems, which collect bands of different widths that are contiguous, separate, or overlap (Figure 1-3). The term *data cube* describes hyperspectral imagery because so many bands are recorded for each pixel. Plate 2 is a 128-band hyperspectral data cube of an industrial area in Martinez, California. The color IR image on the front panel displays healthy vegetation with shades of red. The edges of 128 bands (seen along the top and right side of the cube) are stacked from reflected blue light in the front to SWIR wavelengths toward the back of the cube.

On the sides of the data cube in Plate 2, the amount of reflection as a function of wavelength is shown with a rainbow color scheme. Wavelengths that are absorbed by a feature on the Earth's surface are depicted with black and purple colors while wavelengths that are reflected are seen with green, yellow, and red colors. For example, vegetated pixels along the edge of the image are highly reflective in the NIR bands and are shown with red, yellow, and green patterns in that wavelength range along the sides of the cube. The two sharp breaks in color continuity in the back half of the cube are the water absorption features (~1.5 and 1.9 µm, see Figure 1-20) where little airborne hyperspectral data of features on the ground can be collected.

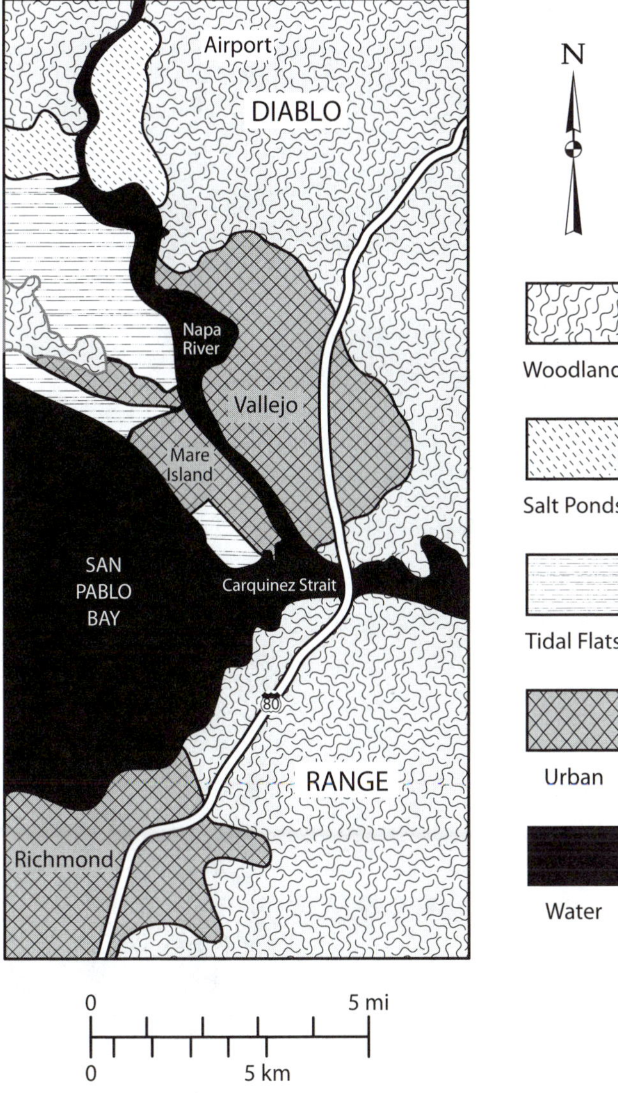

Figure 1-24 Land use and land cover types of San Pablo Bay, California, area interpreted from aircraft multispectral scanner images in Figure 1-23.

GER Hyperspectral Scanner

Figure 1-25 shows 63 hyperspectral bands of the Cuprite mining district in west-central Nevada that were acquired by the legacy GER airborne hyperspectral system. The band number and wavelength are shown for every fifth image. The bars for the GER scanner in the upper part of Figure 1-20 show the position of the image bands relative to reflectance spectra of vegetation and rocks. The blue band (0.40 to 0.50 µm) was not recorded because of strong atmospheric scattering at those wavelengths. Twenty-four images are recorded in the green through NIR wavelengths. Seven bands span the interval of 1.00 to 2.00 µm, which is dominated by atmospheric absorption caused by water vapor. This absorption effect causes the bands centered around band 25 to be wholly or partially obscured. Thirty-two bands record the longer SWIR wavelengths of 2.00 to 2.50 µm, which are clear of absorption effects.

AVIRIS Hyperspectral Scanner

JPL developed an airborne hyperspectral scanner system called the *airborne visible / infrared imaging spectrometer* (AVIRIS) that acquires 224 images, each with a spectral bandwidth of 10 nm in the spectral region that spans 0.4 to 2.5 µm (Figure 1-20). AVIRIS images are typically acquired from NASA's ER-2 jet aircraft at an altitude of 20 km with a swath width of 11 km and a spatial resolution of 17 m (aviris.jpl.nasa.gov). **Digital Image 1-2** 🌐 is a color image of the Cuprite mining district that was prepared from the following SWIR bands of AVIRIS: The band at 2.469 µm is shown in blue, 2.231 µm in green, and 2.01 µm in red. In the upper part of Figure 1-20 the spectral positions of these bands are indicated by the letters R, G, and B along the AVIRIS range. Figure 1-26 is a geologic sketch map of the image. In **Digital Image 1-2** 🌐 the blue-gray tones are unaltered younger volcanic rocks of the Thirsty Canyon Tuff. The orange and red tones are older volcanic rocks of the underlying Siebert Tuff that have been altered to opal. The light-toned rocks in the center of the image have been replaced by silica. The ability to recognize replacement minerals on hyperspectral images is important for mineral exploration, as described in Chapter 13.

Reflectance Spectra from Hyperspectral Data

Hyperspectral scanners are also called *imaging spectrometers* because the narrow spectral bands may be converted into reflectance spectra (Clark, 1999; van der Meer, 1994). In Figure 1-27 the three solid curves are spectra for three minerals that were calculated from the AVIRIS data of Cuprite, Nevada. Each curve represents a different array of 5 by 5 ground resolution cells that cover 100 by 100 m (10,000 m^2) where a particular mineral predominates. The three minerals are kaolinite, alunite, and buddingtonite. Kaolinite is a clay mineral, alunite is an aluminum sulfate mineral, and buddingtonite is an ammonium feldspar mineral. For each ground array the percentage of reflectance for each AVIRIS band is plotted as a function of wavelength. The values are connected to produce the solid curves in Figure 1-27. The dotted curves are laboratory spectra for the three minerals, which are similar to the AVIRIS spectra. The differences between the AVIRIS spectra and the spectrometer spectra are explained as follows: The 10,000 m^2 AVIRIS ground array includes a variety of materials in addition to the predominant mineral. These additional minerals contaminate an AVIRIS spectrum, whereas a laboratory spectrum represents a pure sample of each mineral.

SOURCES OF REMOTE SENSING INFORMATION

Tasking online search engines with key words such as multispectral, hyperspectral, spatial resolution, spectral resolution, radiometric resolution, remote sensing, remote sensing tutorials, and other terms used in this chapter will generate links to informative sources to improve your understanding of remote sensing concepts and systems. NASA, USGS, NOAA, JPL, and the European Space Agency (ESA) websites provide timely online information about remote sensing systems and applications.

The American Society of Photogrammetry and Remote Sensing (ASPRS), Geoscience and Remote Sensing Society

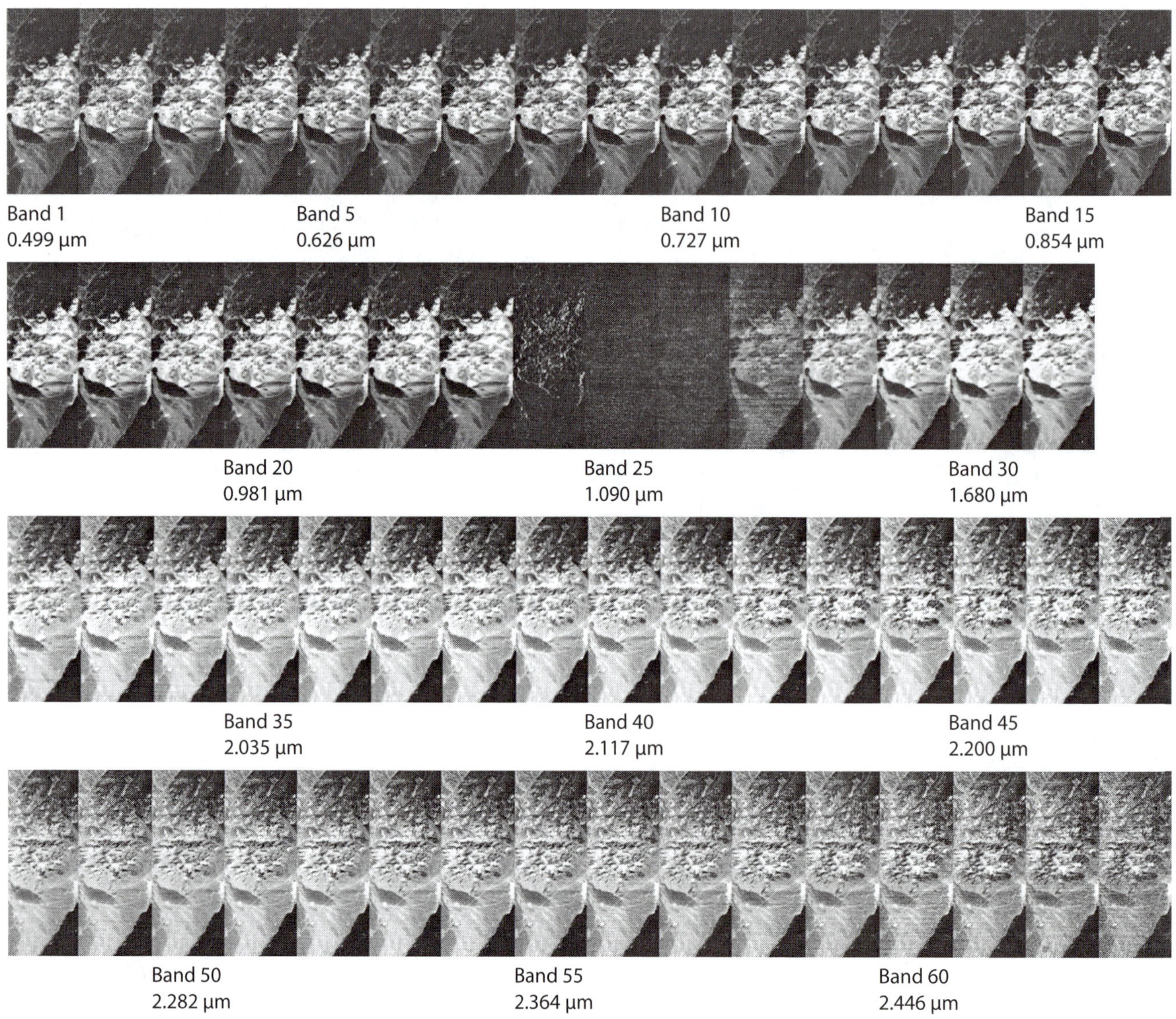

FIGURE 1-25 Hyperspectral scanner images of Cuprite, Nevada, acquired with a GER 63-band aircraft system.

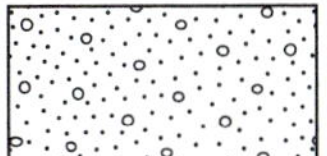
Sand and Gravel

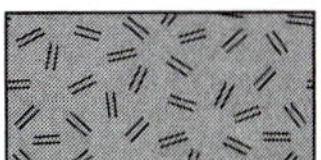
Younger Volcanic Rocks, unaltered

Older Volcanic Rocks, altered to clays

Older Rocks, altered to silica

Figure 1-26 Geologic interpretation map of Cuprite, Nevada (see **Digital Image 1-2**).

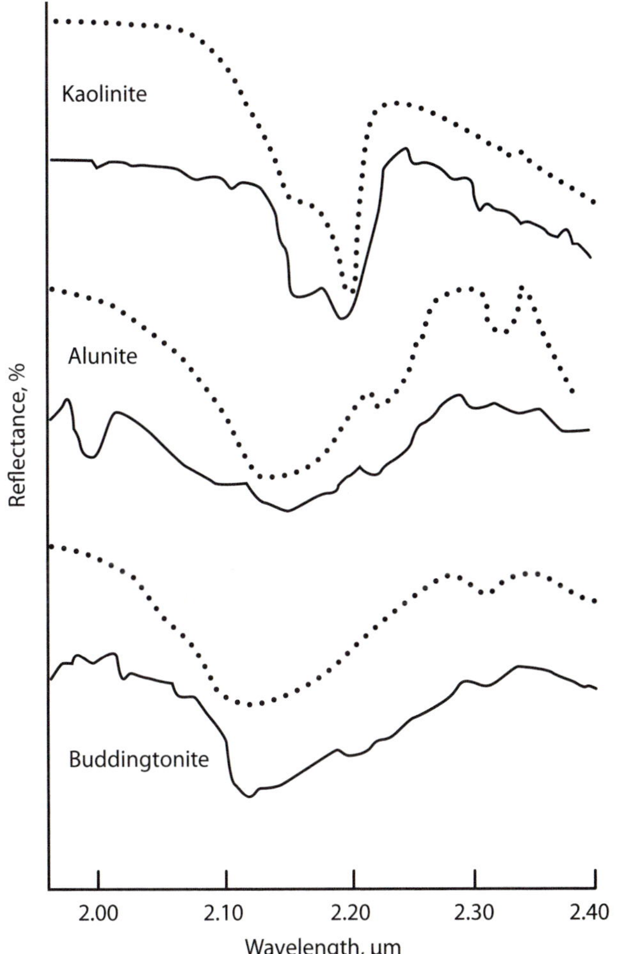

Figure 1-27 Spectra of minerals derived from AVIRIS hyperspectral data (solid lines) and spectra measured by a laboratory spectrometer (dotted lines). The close agreement for each mineral demonstrates that spectra can be derived from hyperspectral scanning systems. From F. van der Meer. 1994. Extraction of mineral absorption features from high-spectral resolution data using non-parametric geostatistical techniques. *International Journal of Remote Sensing*, 15(11), 2,193–2,214. Reprinted by permission of Taylor & Francis Ltd.

(GRSS), International Society for Photogrammetry and Remote Sensing (ISPRS), and Remote Sensing & Photogrammetry Society (RSPSoc) are a few of the professional societies that provide publications, webinars, and conferences to promote the use of remote sensing for education, research, and industry. Geological Remote Sensing Group (GRSG) is a special interest group of RSPSoc focused on the geological aspects of remote sensing.

QUESTIONS

1. Use Equation 1-1 to calculate the wavelength (in cm) of radar energy at a frequency of 10 GHz. What is the frequency in GHz of radar energy at a wavelength of 25 cm?
2. What is the temperature of boiling water at sea level in °K?
3. Distinguish between the Earth's radiant energy peak and the reflected energy peak.
4. The atmosphere is essential for life on Earth, but it causes problems for remote sensing. Describe these problems.
5. Use Equation 1-4 to calculate the contrast ratio between a target with a brightness of 17 and a background with a brightness of 8.
6. Why is $\lambda = hc/Q$ important for the design of remote sensing instruments?
7. What does scale = 1:4,000 mean?
8. What is a band in remote sensing?
9. Remote sensing has four resolutions associated with it. What are they and what does each one mean?
10. How many levels of gray (radiometric resolution) are collected by each sensor listed below:
    a. 3-bit sensor.
    b. 8-bit sensor.
    c. 11-bit sensor.
11. On images acquired from a satellite at an altitude of 910 km, targets on the ground separated by 80 m can be resolved. Use Equation 1-5 to calculate the angular resolving power (in mrad) of the scanning system.
12. Assume that your eyes have the normal resolving power (0.2 mrad) and that you are an airline passenger at an altitude of 9 km. For targets on the ground with high contrast ratio, what is the minimum separation (in m) at which you can resolve these targets?
13. An airborne cross-track scanner has the following characteristics: IFOV = 1.5 mrad; angular field of view = 45°; scan mirror rotates at 4,000 rpm. The aircraft altitude is 10 km. Calculate the following:
    a. Size of ground resolution cell (m).
    b. Width of ground swath (km).
    c. Dwell time for a ground resolution cell (sec).
14. An along-track scanner has detectors with a 2 mrad IFOV. The scanner is carried in an aircraft at an altitude of 15 km and a ground speed of 600 km · hr^{-1}. Calculate the following:
    a. Ground resolution cell (m).
    b. Dwell time for a ground resolution cell (sec).
15. Refer to the aircraft multispectral scanner images (Figure 1-23) and map of San Pablo Bay (Figure 1-24). Select the three bands that show maximum brightness (reflectance) for the following terrain categories:
    a. Vegetation in northeast corner of the scene.
    b. Silty water in San Pablo Bay adjacent to Mare Island.
    c. Urban areas of Vallejo.
    d. Salt ponds in northwest portion of the image.

REFERENCES

Bolstad, P. 2016. *GIS Fundamentals: A First Text on Geographic Information Systems* (5th ed.). Ann Arbor, MI: XanEdu Publishing.

Clark, R. N. 1999. Spectroscopy of rocks and minerals, and principles of spectroscopy. In A. N. Rencz (Ed.), *Manual of Remote Sensing*, Vol. 3, Remote Sensing for the Earth Sciences (pp. 3–52). New York: John Wiley.

Gregory, R. L. 1966. *Eye and Brain, the Psychology of Seeing*. New York: McGraw-Hill.

Hunt, G. L. 1980. Electromagnetic radiation—the communication link in remote sensing. In B. S. Siegal and A. R. Gillespie (Eds.), *Remote Sensing in Geology*. New York: John Wiley.

Jensen, J. R. 2007. *Remote Sensing of the Environment: An Earth Resource Perspective* (2nd ed.). Upper Saddle River, NJ: Pearson/Prentice Hall.

Kokaly, R. F., R. N. Clark, G. A. Swayze, K. E. Livo, T. M. Hoefen, N. C. Pearson, R. A. Wise, W. M. Benzel, H. A. Lowers, R. L. Driscoll, and A. J. Klein. 2017. USGS Spectral Library Version 7. US Geological Survey Data Series 1035. http://doi.org/10.3133/ds1035

Rosenberg, P. 1971. Resolution, detectability, and recognizability. *Photogrammetric Engineering*, 37, 1,244–1,258.

Shippert, P. 2004. Introduction to Remote Sensing. Research Systems, Inc. http://spacejournal.ohio.edu/pdf/shippert.pdf (accessed January 2018).

Slater, P. N. 1983. Photographic systems for remote sensing. In R. N. Colwell (Ed.), *Manual of Remote Sensing* (2nd ed.) (chapter 6, pp. 231–291). Falls Church, VA: American Society of Photogrammetry.

Suits, G. H. 1983. The nature of electromagnetic radiation. In R. N. Colwell (Ed.), *Manual of Remote Sensing* (2nd ed.) (chapter 2, pp. 37–60). Falls Church, VA: American Society of Photogrammetry.

van der Meer, F. 1994. Extraction of mineral absorption features from high-spectral resolution data using non-parametric geostatistical techniques. *International Journal of Remote Sensing*, 15(11), 2,193–2,214.

chapter 2

Aerial and Satellite Photographs

As explained in Chapter 1, framed image systems employ a lens and shutter system to focus an image on the focal plane (Figure 1-15). Formerly the image was recorded on film as a photograph. Today, however, the image is recorded as a two-dimensional digital array of lines and pixels. Photographs acquired from aircraft (aerial photographs) were the first form of remote sensing. In the early 1970s, interpretation of aerial photographs led to the discovery of several valuable oil fields in Indonesia. Because of the digital technology revolution, digital images have replaced film photographs. Most of our methods for interpreting digital images were originally developed from aerial photographs; these methods are now being used to interpret digital images.

DIGITAL TECHNOLOGY REVOLUTION

Younger readers accustomed to smartphones and digital cameras may be unaware of the digital revolution, and how it completely changed photographic technology over the past two decades. Consider the former film-based technology for photographs, such as the common 35 mm camera format. The photographer visited a vendor (camera shop or drug store) to select from the available film types (panchromatic, positive color, or negative color). When using the camera the photographer composed the scene, focused the lens, and set the exposure time. The exposed film was then returned to the vendor for developing into positive or negative transparencies together with proof prints. The photographer then selected the transparencies for printing and returned them to the vendor with instructions for enlargement, cropping, and color balancing. Some dedicated photographers/hobbyists had home darkrooms for processing and printing the photographs.

During the twentieth century, aerial photographs were widely used for a range of applications. Film-based aerial mapping cameras employed a single lens and acquired images on a 9-in (23-cm) wide roll of film that recorded either panchromatic, natural color, or color IR images. During World War II, aerial cameras were utilized to gather intelligence on military targets and munitions installations. In 1962, photographs taken from U-2 aircraft showed that the Soviets had secretly installed missiles in Cuba that threatened the United States. The photographs were displayed at the United Nations and forced the Soviets to withdraw their missiles. From 1964 to 1989, the U-2 was replaced by the manned SR-71 with electronic sensors. The SR-71 was replaced by the RB-57, which has been followed by a succession of classified systems, including unmanned aerial vehicles (UAVs), or drones.

In the twenty-first century, conventional film-based, hand-held cameras and aerial cameras are essentially obsolete. Film and film processing are increasingly difficult to obtain. Most camera factories and vendors have converted to the digital mode. Film as an analog recording medium has been replaced by two-dimensional arrays (framing devices) of photosensitive detectors (Figure 2-1) that capture an image in digital format.

Modern multispectral digital aerial cameras (Figure 2-2) can simultaneously acquire panchromatic, natural color, and color IR imagery. One pair of lenses is filtered to acquire blue and green images. The second pair acquires red and near IR (NIR) images. Each of the four lenses in a row acquires one-quarter of the scene at high spatial resolution

FIGURE 2-1 Example of a charge-coupled device (CCD) digital array. Courtesy Canon.

in panchromatic (black and white) format. The four panchromatic images may be digitally merged (mosaiced) to show the entire scene. Examples of these images are shown later in this chapter. Today, UAVs acquire a range of classified digital images for intelligence applications. Small, hand-launched UAVs are widely used to acquire multispectral images.

The geometric and spectral properties of both photographic and digital images are essentially the same. In this text we use both digital and film images, as there are extensive repositories of film-based photographs that are useful resources.

FIGURE 2-2 Multispectral camera with four lenses for color bands and four lenses for high resolution panchromatic images. Courtesy Vexcel Imaging.

INTERACTIONS BETWEEN LIGHT AND MATTER

As with other forms of electromagnetic energy, light may be reflected, absorbed, or transmitted by matter. Aerial photographs record the light reflected by a surface, which is determined by the property called albedo. *Albedo* is the ratio of the energy reflected from a surface to the energy incident on the surface. Dark surfaces have a low albedo, and bright surfaces have a high albedo. Light that is not reflected is transmitted or absorbed by the material. During its transmission through the atmosphere, light interacts with the gases and particulate matter in a process called scattering, which has a strong effect on aerial photographs.

ATMOSPHERIC SCATTERING

Atmospheric scattering results from multiple interactions between light rays and the gases and particles of the atmosphere, as shown in Figure 2-3. The two major processes, selective scattering and nonselective scattering, are related to the size of particles in the atmosphere. In *selective scattering* the shorter wavelengths of UV energy and blue light are scattered more severely than the longer wavelengths of red light and NIR energy. Selective scattering is caused by fumes and by gases such as nitrogen, oxygen, and carbon dioxide. The selective scattering of blue light causes the blue color of the sky. The red skies at sunrise and sunset are due to sunlight passing horizontally through the atmosphere, which scatters blue and green wavelengths so only red light reaches the viewer.

In *nonselective scattering* all wavelengths of light are equally scattered. Nonselective scattering is caused by dust, clouds, and fog in which the particles are much larger than the wavelengths of light. Clouds and fog are aerosols of very fine water droplets; they are white because the droplets scatter all wavelengths equally. The curves in Figure 2-4 show relative scattering as a function of wavelength. Nonselective scattering is shown by the horizontal line at the top of Figure 2-4.

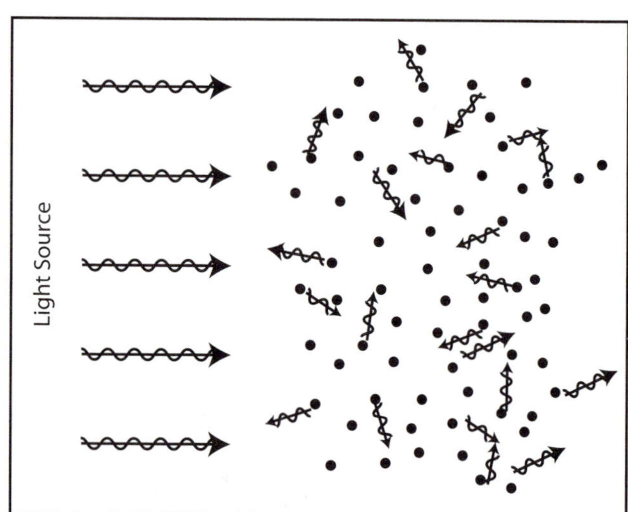

FIGURE 2-3 Scattering of light waves by particles in the atmosphere.

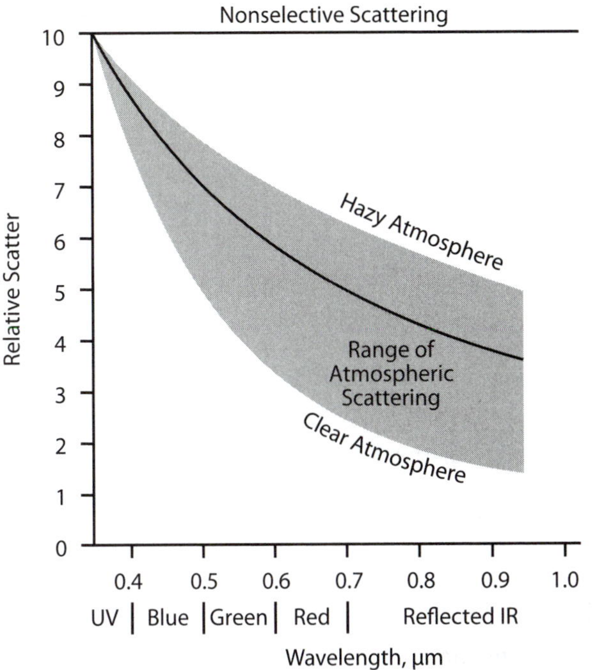

FIGURE 2-4 Atmospheric scattering as a function of wavelength. The shaded region shows the range of scattering caused by a typical atmosphere. From Slater (1983, Figure 6-15).

Scattering in the atmosphere results from a combination of selective and nonselective processes. The range of atmospheric scattering is shown by the shaded area in Figure 2-4. The lower curve of the shaded area represents a clear atmosphere and the upper curve a hazy atmosphere. Typical atmospheres have scattering characteristics that are intermediate between these extremes. The important point for aerial imaging is that the Earth's atmosphere scatters UV and blue wavelengths at least twice as strongly as red light.

Light scattered by the atmosphere illuminates shadows, which are never completely dark but are bluish in color. This scattered illumination is referred to as *skylight* to distinguish it from direct sunlight. A striking characteristic of photographs taken by Apollo astronauts on the surface of the moon is the black appearance of the shadows. The lack of atmosphere on the moon precludes any scattering of light into the shadowed areas.

EFFECTS OF SCATTERING ON AERIAL IMAGES

Scattered light that enters an imaging system is a source of illumination, but contains no information about the terrain. This extra illumination reduces the contrast ratio (CR) of the scene, thereby reducing the spatial resolution and detectability of the image. Figure 2-5 diagrams the effect of scattered light on the contrast ratio of a scene in which a dark area (brightness = 2) is surrounded by a brighter background (brightness = 5). For the original scene with no scattered light (Figure 2-5A and B), the contrast ratio is determined from Equation 1-4 as follows:

$$CR = \frac{B_{max}}{B_{min}}$$
$$= \frac{5}{2} = 2.5$$

Figure 2-5C shows the appearance of the original scene in conditions of heavy haze, where the atmosphere contributes 5 brightness units of scattered light. As the brightness profile of Figure 2-5D shows, scattered light adds uniformly to all parts of the scene and results in a contrast ratio of CR = 10/7 = 1.4. Thus atmospheric scattering has reduced the contrast ratio of the scene from 2.5 to 1.4, which lowers the spatial resolution on a photograph of that scene. Chapter 1 demonstrated this relationship between contrast ratio and resolving power. The effect of atmospheric scattering on aerial images is illustrated later in this chapter.

Filtering out the selectively scattered shorter wavelengths reduces the effects of atmospheric scattering. For color images a haze filter removes longer UV wavelengths. For many black-and-white images a "minus blue" filter is used to eliminate those strongly scattered wavelengths. There is a trade-off with filters: although they reduce haze, they also remove the spectral information contained in the wavelengths that are absorbed by the filter.

A. Original scene.

B. Brightness profile of image with no scattered light.

10

CR = 5/2 = 2.5

Image Brightness

0

Distance

C. Profile of image with five brightness units added by scattered light.

D. Brightness profile and contrast ratio of image with scattered light.

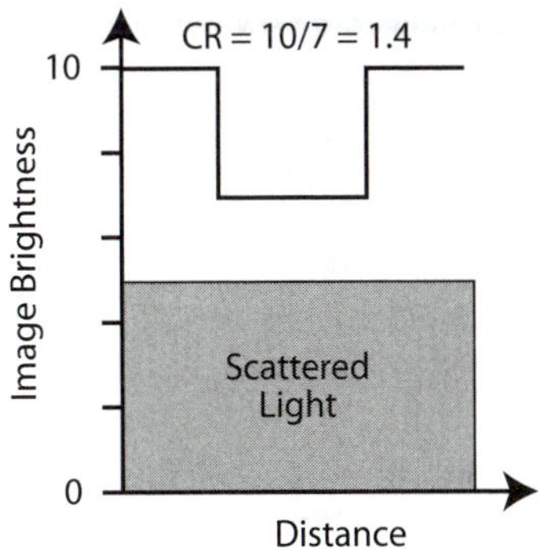

FIGURE 2-5 Effect of scattered light on the contrast ratio of an image.

HANDHELD DIGITAL CAMERAS

Most of us carry some version of a cell phone that includes a digital natural color camera. Because of their compact size, these cameras cannot employ the multiple lenses and detector arrays of aerial cameras. In 1976, Bryce Bayer of Eastman Kodak provided a solution. The Bayer filter is a sensing array for color imaging that includes individual *luminance-sensitive elements* and *chrominance-sensitive elements.* On a two-dimensional array of detectors, Bayer used twice as many green elements as red or blue to mimic the physiology of the human eye. During daylight vision, the luminance perception of the human retina uses M and L cone cells combined, which are most sensitive to green light. These elements are referred to as *sensor elements*, *sensels*, *pixel sensors*, or simply *pixels*; sample values sensed by them, after interpolation, become image pixels. Plate 3 shows the array of the Bayer filter and the resulting pattern.

The raw output of Bayer filter cameras is referred to as a *Bayer pattern* image. Since each pixel is filtered to record only one of three colors, the data from each pixel cannot fully specify each of the red, green, and blue values on its own. To obtain a full-color image, various de-mosaicing algorithms can be used to interpolate a set of complete red, green, and blue values for each pixel. These algorithms make use of the surrounding pixels of the corresponding colors to estimate the values for a particular pixel. Different algorithms requiring various amounts of computing power result in final images that vary in quality. This processing can be done in-camera, to produce a JPEG or TIFF image, or outside the camera using the raw data directly from the sensor. In addition to the Bayer filter, other digital imaging technologies are available for cameras and smartphones to produce panchromatic and color images (Paine and Kiser, 2012).

CHARACTERISTICS OF AERIAL IMAGES

Characteristics such as resolution, scale, and relief displacement are common to both digital and film aerial photographs. The following sections discuss these characteristics in some detail.

SPATIAL RESOLUTION OF IMAGES

Spatial resolution, or *resolving power*, of aerial images is influenced by several factors:

1. Atmospheric scattering;
2. Vibration and motion of the aircraft, which are minimized by vibration-free camera mounts and motion compensation devices;
3. Resolving power of lenses; and
4. Resolving power of detector arrays.

All of these factors combine to determine the spatial resolution of an image.

RESOLVING POWER OF LENSES

The resolving power of a lens is determined by its optical quality and size. When a lens images a resolution target, such as those shown in Chapter 1, there is an upper limit to the number of line-pairs within the space of a millimeter that can be resolved on the resulting image. This maximum number of resolvable line-pairs per mm is a measure of the resolving power of the lens.

GROUND RESOLUTION

Ground resolution expresses the ability to resolve ground features on aerial images. System resolution is converted into ground resolution by the formula

$$R_g = \frac{R_s f}{H} \tag{2-1}$$

where:

R_g = ground resolution in line-pairs · mm^{-1}
H = camera height above the ground in m (do not confuse this with aircraft altitude above mean sea level)
R_s = system resolution in line-pairs · mm^{-1}
f = lens focal length in mm

Figure 2-6 shows the geometric basis for this relationship. For a camera lens with a focal length of 152 mm producing images with a system resolution of 20 line-pairs · mm^{-1} acquired at a camera height of 6,100 m, the ground resolution, using Equation 2-1, is

$$\begin{aligned} R_g &= \frac{R_s f}{H} \\ &= \frac{20 \text{ line-pairs} \cdot \text{mm}^{-1} \times 152 \text{ mm}}{6{,}100 \text{ m}} \\ &= 0.5 \text{ line-pairs} \cdot \text{m}^{-1} \end{aligned}$$

Based on visual inspection of the aerial photograph, the most closely spaced resolution target on the ground that can be resolved on the image consists of 2.0 line-pairs · m^{-1}. The width of an individual line-pair in m is determined by the reciprocal

$$\frac{1.0 \text{ line-pair}}{R_g}$$

The width is 0.5 m in this example.

Minimum ground separation is the minimum distance between two objects on the ground at which they can be resolved on the image. As Figure 2-6 shows, it is the separation between lines or bars in the resolution target and is determined by

Table 2-1 Minimum ground separation on typical aerial images acquired at different heights (focal length of camera lens is 152 mm).

		Minimum Ground Separation	
Camera Height (H) (m)	Scale of Photographs	Medium System Resolution (m)	High System Resolution (m)
1,525	1:10,000	0.12	0.05
3,050	1:20,000	0.25	0.10
4,575	1:30,000	0.37	0.15
6,100	1:40,000	0.50	0.20

$$\begin{aligned}\text{min ground separation} &= \frac{1.0\ \text{line-pair} / R_g}{2} \\ &= \frac{1.0\ \text{line-pair} / 2.0\ \text{line-pairs} \cdot \text{m}^{-1}}{2} \\ &= 0.25\ \text{m}\end{aligned} \tag{2-2}$$

Table 2-1 lists minimum ground separation values in meters for typical aerial images acquired with imaging systems of medium (40 line-pairs · mm^{-1}) and high (100 line-pairs · mm^{-1}) resolutions. The aircraft heights and lens focal lengths of Table 2-1 correspond to the medium-resolution aerial photographs in Figure 2-7. Inspection of the photographs with a magnifier indicates that these values for minimum ground separation are appropriate, although the photographs lack a ground resolution target, which is necessary for precise measurement.

Table 2-2 lists features that may be identified on images with different ground separation values. These are only guidelines to illustrate the general relationship between ground resolution and recognition.

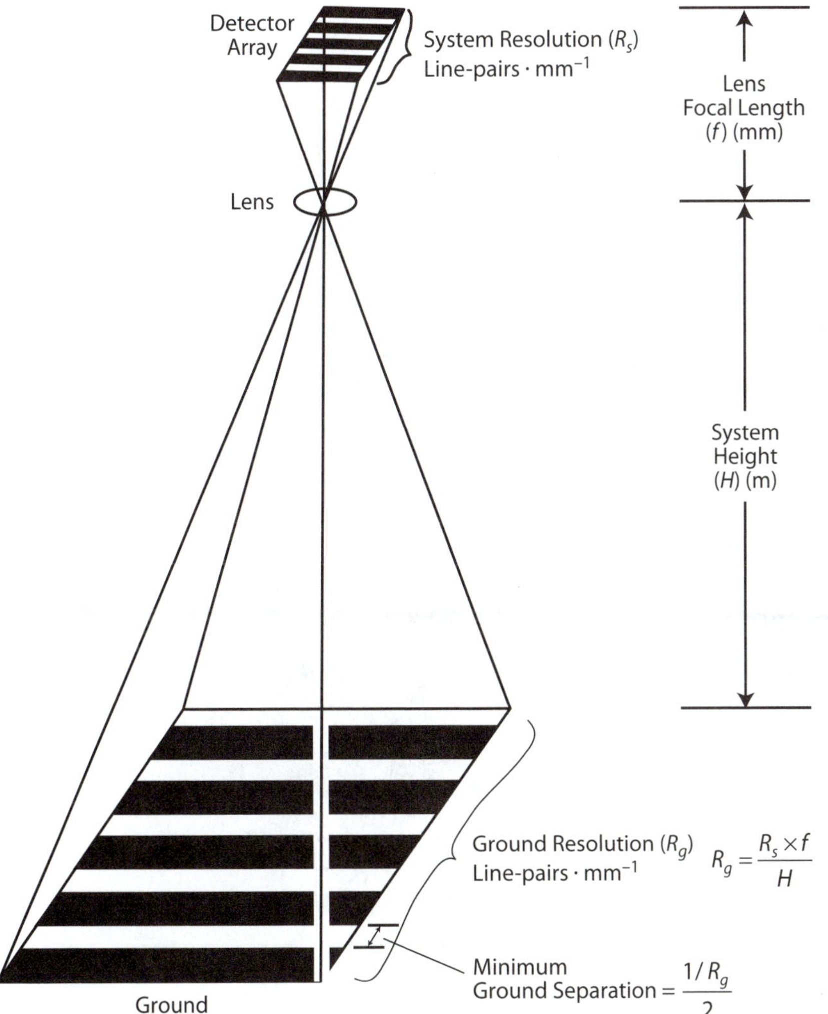

Figure 2-6 Ground resolution and minimum ground separation on aerial photographs.

A. Height 1,525 m.

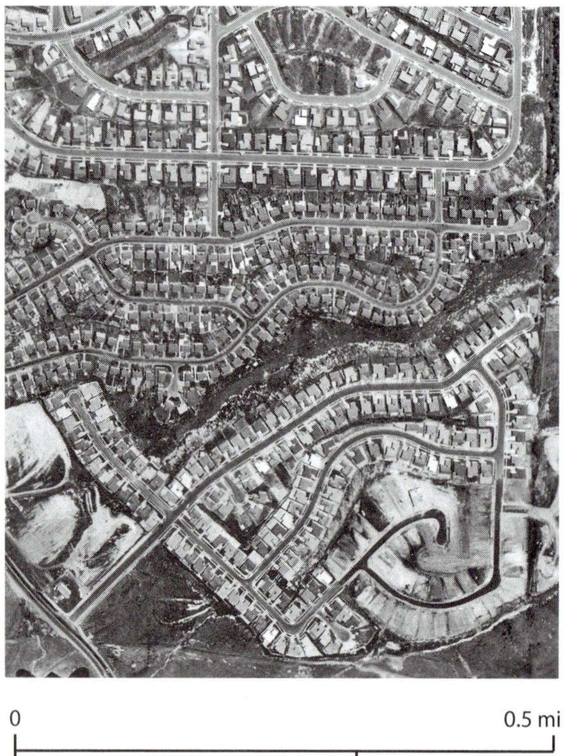

0 0.5 mi

0 0.5 km

B. Height 3,050 m.

0 0.5 mi

0 0.5 km

C. Height 4,575 m.

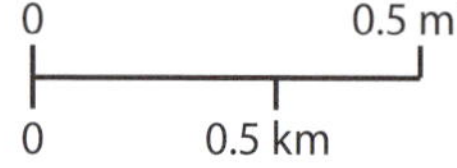

D. Height 6,100 m.

0 0.5 mi

0 0.5 km

Figure 2-7 Aerial photographs of Palos Verde Peninsula, California, acquired at different cameral heights with a 152 mm focal length lens. Table 2-1 lists minimum ground separation values for this medium resolution system. The southeastern corner is common to all photographs.

Table 2-2 Features recognizable on aerial images at different minimum ground separation values.

Minimum Ground Separation (m)	Recognizable Features
15.0	Identify geographic features such as shorelines, rivers, mountains, and water.
4.50	Differentiate settled areas from undeveloped land.
1.50	Identify roadways.
0.15	Distinguish front from the rear of automobiles.
0.05	Count people, particularly if there are shadows and if the individuals are not in crowds.

Source: After Rosenblum (1968, Table 2).

Photographic Scale

The scale of aerial images is determined by the relationship

$$\text{scale} = \frac{1}{H/f} \tag{2-3}$$

Both H and f must be given in the same units, typically meters. For example, the scale of an image acquired at an aircraft height of 3,050 m with a 152-mm lens (Figure 2-7B) is

$$\begin{aligned}\text{scale} &= \frac{1}{H/f} \\ &= \frac{1}{3{,}050\text{ m}/0.152\text{ m}} \\ &= \frac{1}{20{,}000} \text{ or } 1{:}20{,}000\end{aligned}$$

A scale of 1:20,000 means that 1 cm on the image represents 20,000 cm (200 m) on the ground (1 in = 20,000 in = 1,667 ft). Figure 2-7 illustrates the different scales that result from photographing the same area at different altitudes with the same camera.

Relief Displacement

Figure 2-8 illustrates the geometric distortion called *relief displacement*, which is present on all vertical aerial photographs and images that are acquired with a camera that is aimed directly downward. The tops of objects such as buildings appear to "lean" away from the *principal point*, or optical center, of the photograph. The amount of displacement increases at greater radial distances from the center and reaches a maximum at the corners of the photograph. Figure 2-9A shows the geometry of image displacement, where light rays are traced from the terrain through the camera lens and onto the film. Prints made from the film appear as though they were in the position shown by the plane of the photographic print in Figure 2-9A. The vertical arrows on the terrain represent objects of various heights located at various distances from the principal point. The light ray reflected from the base of object A intersects the plane of the photographic print at position A, and the ray from the top intersects the print at A′. The distance A–A′ is the relief displacement (d) shown in the plan view (Figure 2-9B).

The amount of relief displacement (d) on an aerial photograph is

1. Directly proportional to the height of the object (h). For objects A and C in Figure 2-9A, although they are at equal distances from the principal point, d is greater for A, which is the taller object.
2. Directly proportional to the radial distance (r), which is measured from the principal point to the top point on the displaced image corresponding to the top of the object (Figure 2-9B). For objects A and B, which are of equal height, d is greater for A because it is located farther from the principal point.
3. Inversely proportional to the height of the camera (H) above the terrain.

These relationships are expressed mathematically as

$$d = \frac{h \times r}{H}$$

which may be transposed to

$$h = \frac{H \times d}{r} \tag{2-4}$$

This equation may be used to determine the height of an object from its relief displacement on an aerial photograph. For the building in the lower right corner of Figure 2-8, d (40 m) and r (260 m) are measured using the scale of the photograph. The height of the camera above the terrain (H) is 500 m. The height of the building (h) is calculated from Equation 2-4 as

$$\begin{aligned}h &= \frac{H \times d}{r} \\ &= \frac{500\text{ m} \times 40\text{ m}}{260\text{ m}} = 77\text{ m}\end{aligned}$$

Orthophotographs (or *orthoimages*) are aerial photographs that have been computer processed to remove the radial distortion. These photographs have a consistent scale throughout the image and may be used as maps. Individual orthoimages can be color-balanced and digitally mosaiced to create almost seamless imagery that covers large areas.

PHOTOMOSAICS

Aerial photographs are typically acquired at scales of 1:80,000 or larger and therefore cover relatively small areas. Taking photographs on a series of parallel flight lines provides broader coverage. Along a flight line, successive photographs are often acquired with 60% forward overlap (Figure 2-10). Flight lines are commonly spaced to provide 30% *sidelap*, which is the overlap between adjacent strips of photographs. A *photomosaic* is a composite of these individual photographs that cover an extended area. Figure 2-11A is a photomosaic of aerial photographs acquired in 1967 of the northern Coachella Valley in southern California. Flight lines are oriented north–south. This is an amateur "homemade" mosaic that shows the borders of individual photographs and retains the radial distortion of each aerial photograph. This photomosaic is not an accurate base map, but has significant value for documenting the desert environ-

FIGURE 2-8 Vertical aerial photograph of Long Beach, California, showing relief displacement. Courtesy J. Van Eden.

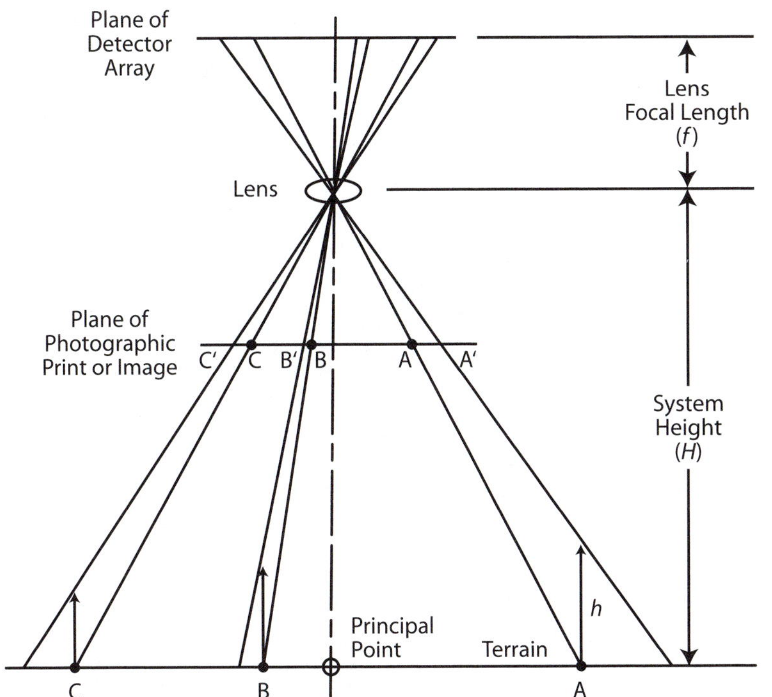

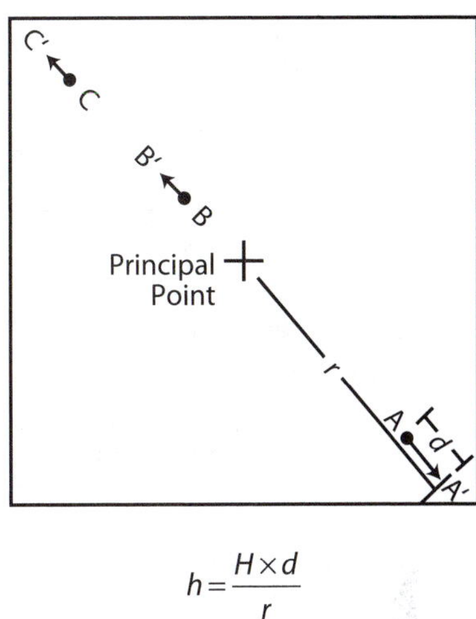

FIGURE 2-9 Geometry of relief displacement on a vertical aerial photograph.

ment before urban sprawl covered much of the southern and northern area.

Figure 2-11 shows a variety of natural and man-made features. Windblown sand, which covers much of the area, has a bright tone. Outcrops of sedimentary bedrock at Garnet Hill and Edom Hill are eroded into ridges and canyons. Vegetation has a dark signature as seen along the White Water River and the golf course on the northern edge of Palm Springs. The San Andreas fault strikes northwest through the area and is marked by a pronounced linear feature in the vicinity of Palm Drive, where the fault forms the boundary between windblown sand on the south and vegetated terrain on the north. The fault is a barrier to the southward movement of groundwater in the subsurface. The water table is shallower on the north side of the fault and thus supports the growth of native vegetation. This expression of geologic features by vegetation patterns is called a *vegetation anomaly*. The transportation network, urban areas, and other cultural features are clearly visible.

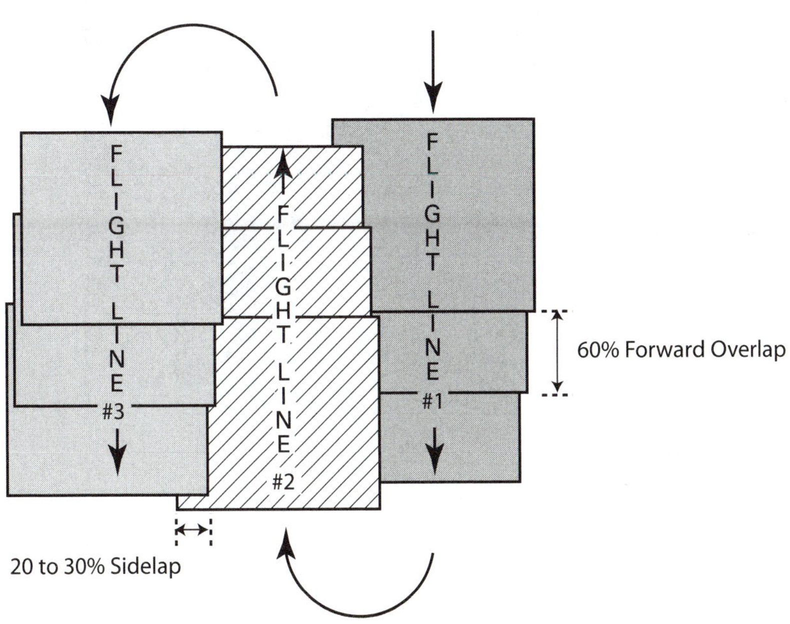

FIGURE 2-10 North–south flight lines showing forward overlap and sidelap.

A. Photomosaic.

B. Physical map.

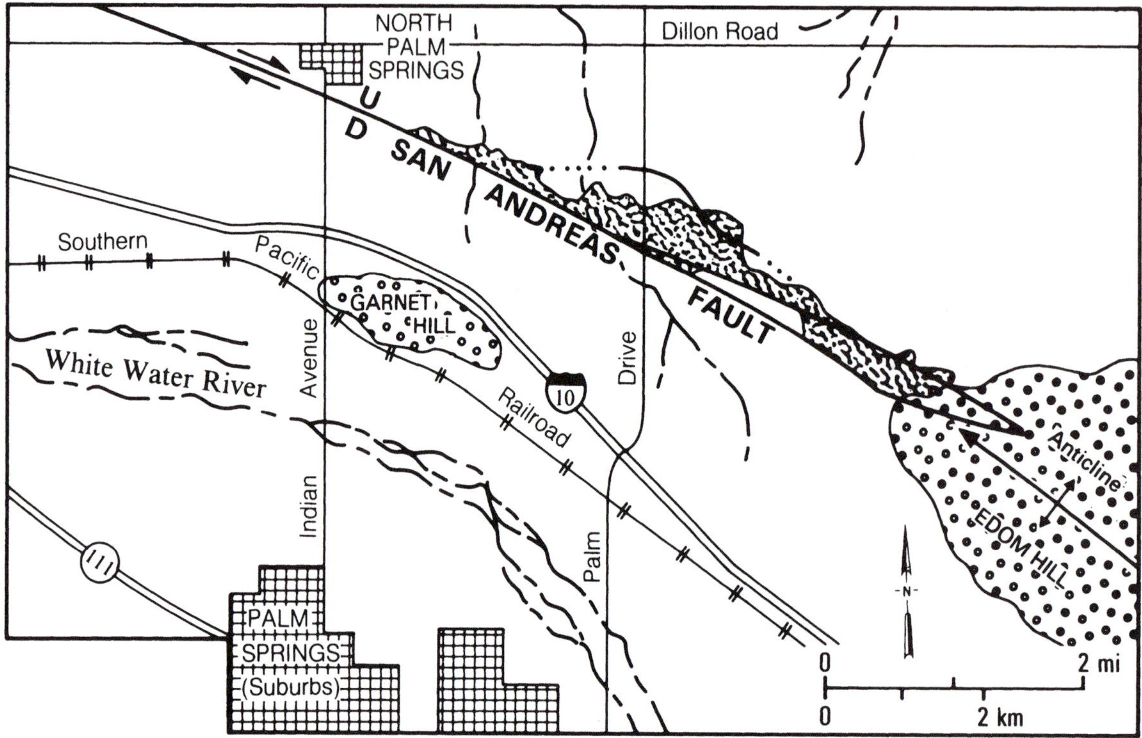

FIGURE 2-11 Coachella Valley, California.

STEREO-PAIRS OF AERIAL PHOTOGRAPHS

Figure 2-12 illustrates a flight line where two successive photographs are acquired with 60% forward overlap. The overlapping photographs are then used to form a *stereo-pair*. The stereo-pair may be viewed with a pocket lens stereoscope (Figure 2-13) to produce a three-dimensional image called a *stereo model*. Stereoscopes are also built with mirrors and lenses that provide a stereo model of the full stereo-pair.

Before viewing photographs in stereo, it is advisable to test one's ability to see in stereo.

Test of Stereo Vision

Figure 2-14 is a test of stereo vision. The two large circles (Figure 2-14A) are not exact duplicates. Within each circle the symbols are radially offset to produce different amounts of relief displacement. As a result, when viewed with a stereoscope, the symbols will appear to float in space at different levels above and below the plane of the page. Adjust the interpupillary distance on the stereoscope to suit your eyes; for most people approximately 6.4 cm is adequate. In normal vision the left and right eyes converge at a point and the scene appears as a planar surface. The stereoscope, however, causes the eyes to diverge, with the left eye viewing the left circle and the right eye viewing the right circle. The brain then merges the two images to produce the stereo model in which the individual symbols "float" at different heights. Complete the test by ranking the symbols on the basis of relative height, starting with 1 for the highest. Compare your results with the rankings shown in Figure 2-14B.

The next step is to view the stereo-pair of photographs in Figure 2-15 with the stereoscope. Figure 2-15 is a stereo-pair of aerial photographs of the Alkali anticline in the eastern part of the Bighorn Basin, Wyoming. When the stereoscope is properly adjusted, the image will appear as a series of alternating ridges and valleys. Figure 2-16 provides a topographic and geologic map of the area. The viewer will be impressed by the apparent extreme vertical relief, which is caused by a characteristic of stereo models called vertical exaggeration.

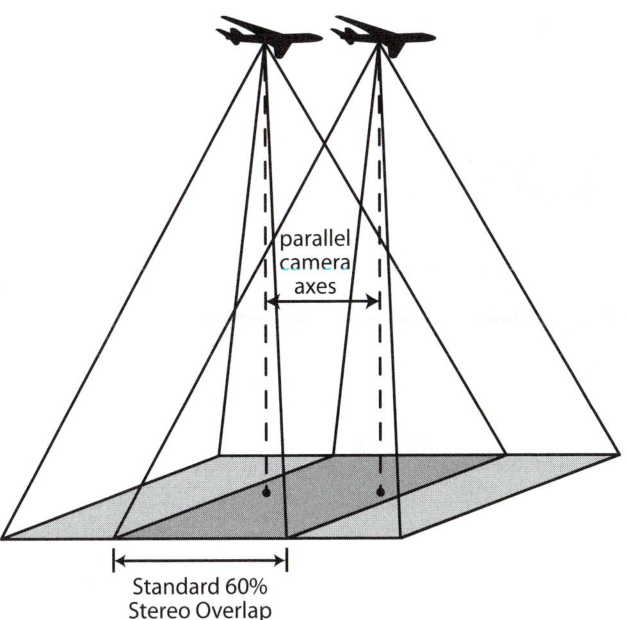

Figure 2-12 Perspective view of a flight line where two overlapping aerial photographs are taken to develop a stereo model.

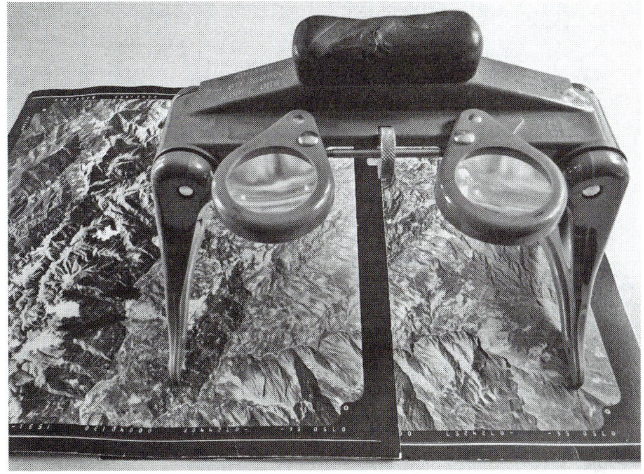

Figure 2-13 Pocket lens stereoscope setup to view a stereo-pair of overlapping aerial photographs.

Vertical Exaggeration

Vertical exaggeration (VE) results because the perceived vertical scale is larger than the horizontal scale in a stereo model. The amount of exaggeration may be approximately calculated by

$$\mathrm{VE} = \left(\frac{\mathrm{AB}}{H}\right)\left(\frac{\mathrm{AVD}}{\mathrm{EB}}\right) \tag{2-5}$$

where:

AB = air base, or ground distance, which is the distance on the ground between the centers of successive overlapping photographs.

H = height of the camera above terrain.

AVD = apparent stereoscopic viewing distance. This is the distance not from the stereoscope lenses to the photographs but from the lenses to the plane in space where the image appears to be. The stereo model always appears to be somewhere below the tabletop on which the stereoscope sits. Several interpreters using a variety of stereoscopes estimated an average value for AVD of 45 cm (Wolf, 1974).

EB = eye base, which is the distance between the interpreter's eyes. For the average adult, EB is 6.4 cm.

For the stereo-pair of Figure 2-15, with an air base of 1,700 m and height of 3,000 m, vertical exaggeration may be estimated from Equation 2-5 as

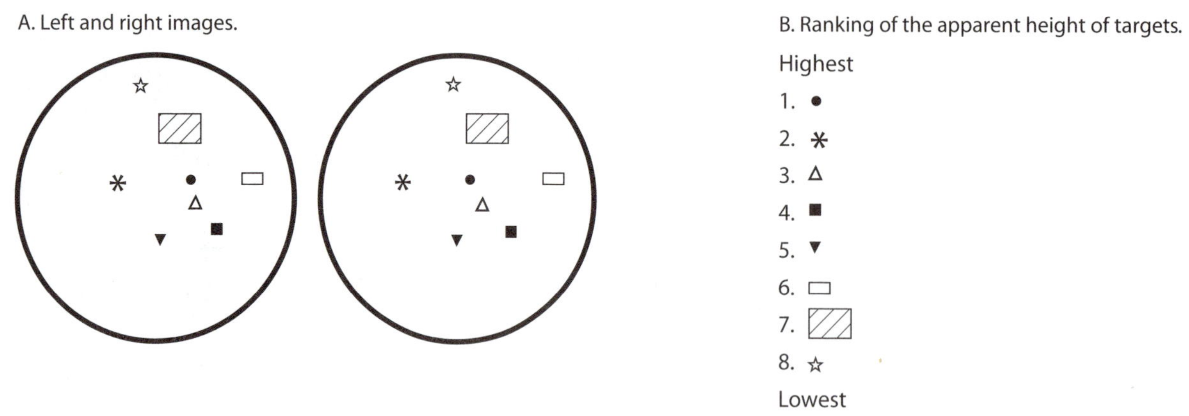

FIGURE 2-14 Test of stereo vision.

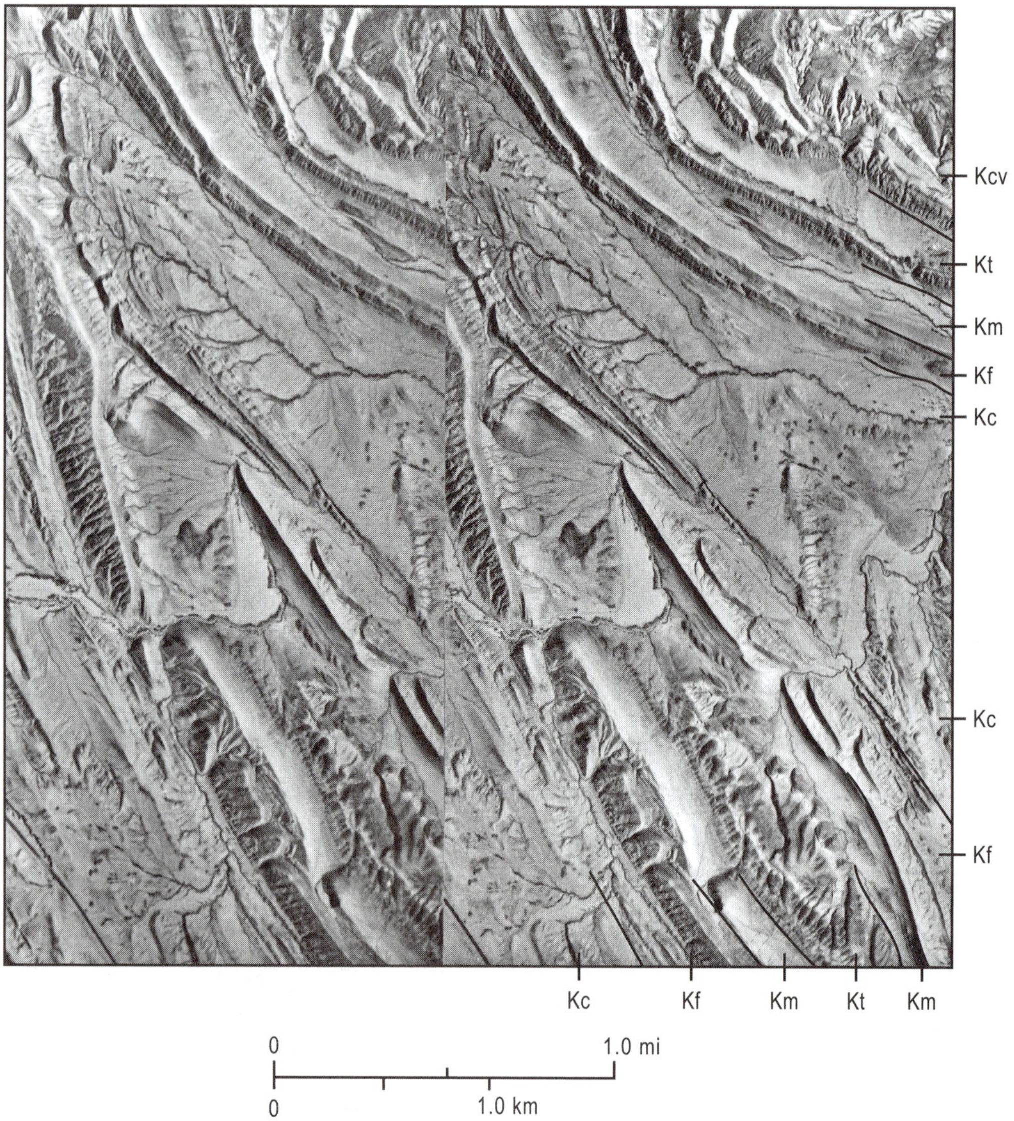

FIGURE 2-15 Stereo-pair of aerial photographs of the Alkali anticline, Bighorn Basin, Wyoming, with formation contacts indicated. Formation symbols are explained in Figure 2-16.

A. Topographic map. The contour interval is 20 ft (6 m).

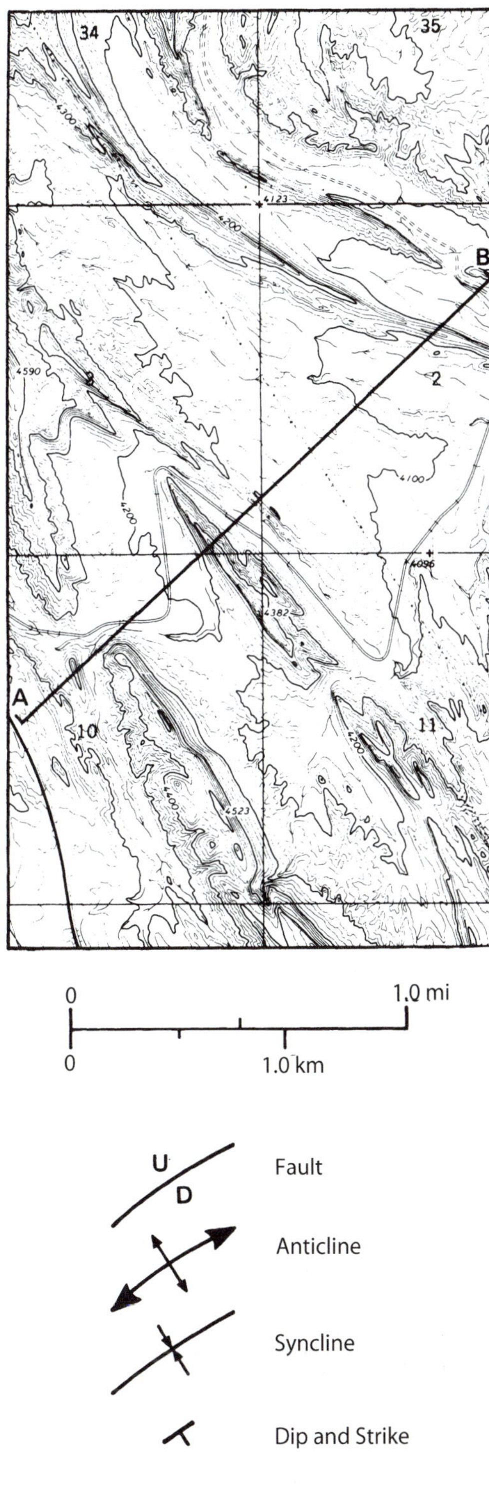

B. Geologic map interpreted with pocket lens stereoscope setup in Figure 2-15.

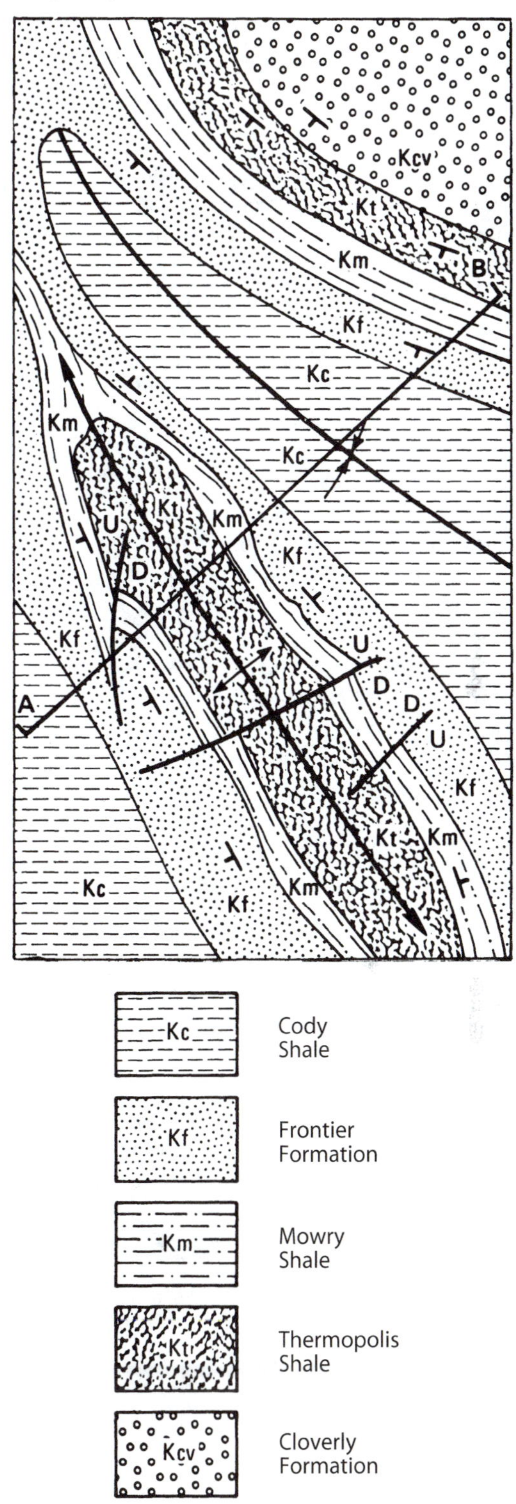

Figure 2-16 Alkali anticline, Bighorn Basin, Wyoming.

$$\begin{aligned}\text{VE} &= \left(\frac{\text{AB}}{H}\right)\left(\frac{\text{AVD}}{\text{EB}}\right) \\ &= \left(\frac{1{,}700\ \text{m}}{3{,}000\ \text{m}}\right)\left(\frac{45\ \text{cm}}{6.4\ \text{cm}}\right) \\ &= 4.0\text{X}\end{aligned}$$

On viewing Figure 2-15 with a stereoscope, the average interpreter should perceive vertical distances to be exaggerated four times the equivalent horizontal distances. In Equation 2-5 the terms AVD and EB are constant; therefore, the amount of vertical exaggeration is determined by the ratio AB/H, which is called the *base-height ratio*. The graph in Figure 2-17 shows the vertical exaggeration for stereo-pairs with various base-height ratios.

The effect of vertical exaggeration is illustrated in Figure 2-18, which shows two topographic profiles constructed along line A–B in the contour map of Figure 2-16A. The profile in Figure 2-18A has the same vertical and horizontal scales (1X), which is equivalent to no vertical exaggeration. The profile in Figure 2-18B was constructed with the vertical scale four times that of the horizontal scale (4X), which is the same vertical exaggeration as that of the stereo model. Comparing the topographic profiles demonstrates the effect of vertical exaggeration.

Although vertical relief is exaggerated by a factor of 4 on a stereo model, the angles of topographic slope and structural dip do not increase four times. Figure 2-19 illustrates the slope exaggeration of a hill with a height (BC) of 270 m and a width (AB) of 1,000 m. The tangent of the true slope angle (CAB) is BC/AB, which is 0.27, or 15°. On the stereo model, the exaggerated height DB is 1,080 m, and the tangent of the exaggerated slope angle (DAB) is BD/AB, which is 1.08, or 47°. The effect of the 4X vertical exaggeration on topographic slopes is illustrated by the cross section in Figure 2-18B. La Prade (1972, 1973) describes the geometry of stereoscopic photographs and reviews various hypotheses for explaining vertical exaggeration.

A. No vertical exaggeration.

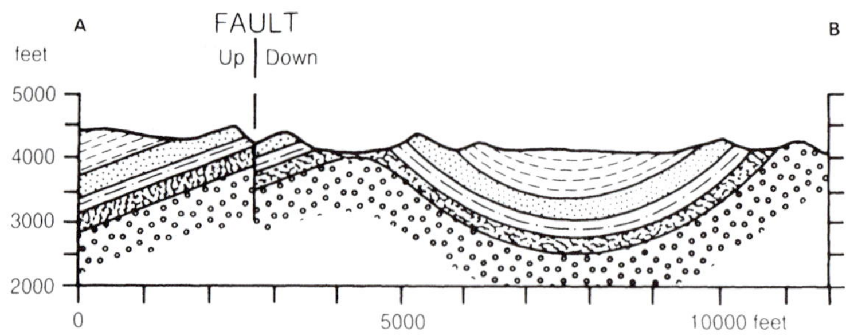

B. 4X vertical exaggeration.

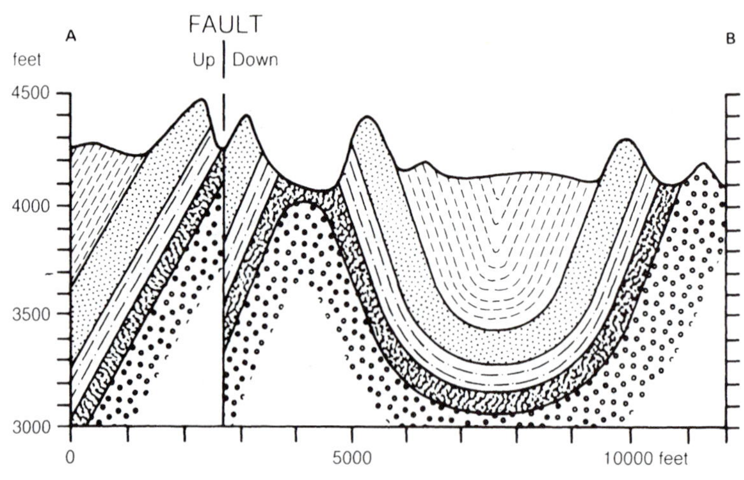

FIGURE 2-18 Geologic cross sections of the Alkali anticline along topographic profile A–B, showing effects of vertical exaggeration.

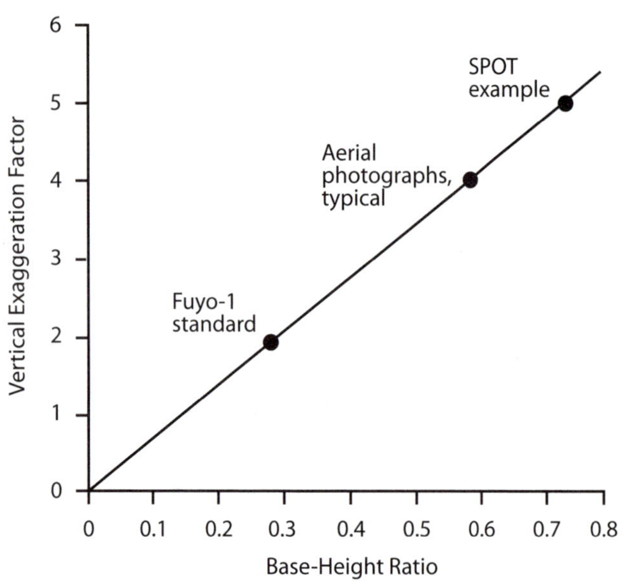

FIGURE 2-17 Relationship between the base-height ratio and the vertical exaggeration factor for stereo models. Modified from Thurrell (1953, Figure 5).

INTERPRETATION OF STEREO-PAIRS

Geologic maps may be prepared by interpreting stereo-pairs. For readers who are unfamiliar with geology, Appendix A provides a condensed description of geologic concepts employed in interpreting remote sensing images. The first step in interpreting an area underlain by sedimentary rocks, such as the Alkali anticline, is to define the geologic units, or formations. We will use Figure 2-16B as an example. The formations in this geologic map are described below.

Cody Shale (Kc): A thick unit of shale with a medium gray tone that is readily eroded and forms broad valleys cut by numerous streams. Cody Shale is the youngest formation exposed at the Alkali anticline.

Frontier Formation (Kf): Medium gray sandstone with shale interbeds. The sandstone is resistant to erosion and forms ridges. The nonresistant shale forms narrow valleys. The prominent dipslopes of the sandstone are incised by numerous closely spaced minor stream channels, producing a distinctive serrated appearance.

Mowry Shale (Km): Siliceous shale that weathers to a slope with a very light gray tone.

Thermopolis Shale (Kt): Very dark gray shale with thin, light-toned interbeds that cause a banded appearance.

Cloverly Formation (Kcv): Alternating beds of resistant sandstone and nonresistant shale that weather to ridges and valleys. The Cloverly is the oldest formation exposed in the area. These characteristics are used to map the contacts between formations.

At several localities, the contacts are interrupted and offset by normal faults (Appendix A), which are mapped with heavy lines and symbols to show the displacement along the faults (Figure 2-16B). Attitudes of the beds are readily seen in the stereo model and are recorded with symbols showing strike and dip. The amount of dip is vertically exaggerated in the same fashion as topographic slopes (shown in Figure 2-19). Techniques for converting exaggerated dips to true dips are described by Miller (1961). Outcrop patterns and the attitude of beds are used to locate the axial trace of the Alkali anticline and the adjacent syncline.

As mentioned earlier, the topographic map in Figure 2-16A was used to draw the topographic profile for the cross section with no vertical exaggeration (Figure 2-18A). The geologic cross section was filled in by adding the faults, formation contacts, and strike and dip information from the geologic interpretation (Figure 2-16B). Simple geometric techniques were used to produce the cross section with 4X vertical exaggeration (Figure 2-18B). This photogeologic interpretation should not be considered complete until the map has been checked in the field.

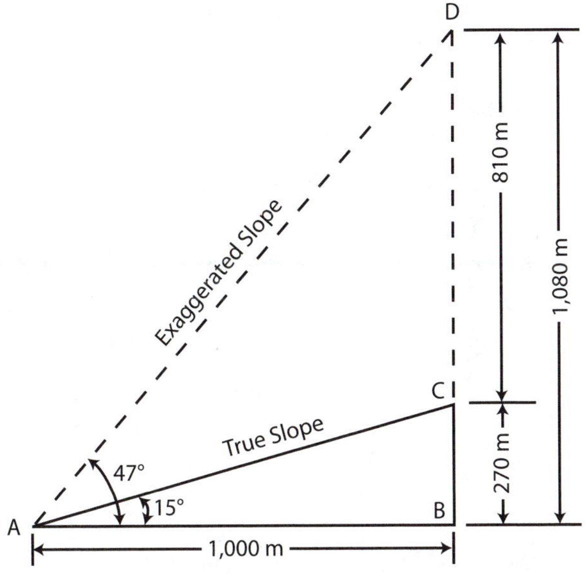

FIGURE 2-19 True slope and slope exaggeration of a stereo model with 4X vertical exaggeration.

INTERPRETATION OF OIL TANK STEREO-PAIRS

Overlapping aerial photographs were acquired in 1959 over the Shell Refinery in Martinez, California. A stereo-pair (Figure 2-20) was produced to support interpretation of the infrastructure, topography, and environment around the oil tanks. This stereo-pair can be viewed with a lens stereoscope (Figure 2-13); when the stereo model is displayed as a color anaglyph (Plate 4) it may be viewed with red/cyan glasses.

Anaglyphs are an alternate means of displaying and viewing stereo images. In Plate 4 the right image of a stereo-pair is printed in red and the left image is printed in cyan. Viewing the image through eyeglasses with corresponding color filters causes the viewer's left eye to see only the left image and the right eye to see only the right image. The brain merges the images to produce a stereo model. Inexpensive anaglyph glasses are mounted on cardboard frames and are suitable for classroom use (JPL, 2017). Not all of the stereo model can be seen with the lens stereoscope due to its limited field of view. Most of the stereo model can be viewed on the anaglyph (Plate 4).

Figure 2-21 shows a portion of the 1951 Port Chicago Quadrangle topographic map produced by the USGS at a scale of 1:24,000. The area of the stereo model seen in Figure 2-20 is outlined with a thick black rectangle. The topographic map's contour interval is 25 ft. The black circles are footprints of oil tanks built on a hill that rises 100 to 150 ft above the surrounding terrain. Referring to Figure 2-20, oil tank elevation above the ground, berms around the tanks to contain oil leaks, and orchard trees (black dots) of varying size and density on the southwest flank of the hill can be interpreted from the stereo model. In addition, sprawling suburbia on the southern and eastern margins of the model and new infrastructure north of the tank farm can be interpreted from the model, however, these features are not on the topographic map. A detailed environmental base map for 1959 can be built from the stereoscopic interpretation.

LOW SUN-ANGLE PHOTOGRAPHS

Aerial photographs are normally acquired between 10:00 AM and 3:00 PM, when the sun is at a high angle above the horizon and shadows have a minimum extent. These photographs are desirable for topographic mapping, which requires unobscured terrain, but for geologic interpretation, photographs acquired with lower sun angles are often valuable. The photograph in Figure 2-22A was acquired shortly after sunrise, when the sun was only 15° above the horizon. A photograph of the same area acquired at midday (Figure 2-22B) is shown for comparison. The area is a gravel-covered slope on the east flank of Carson Range, south of Reno, Nevada. The low sun-angle photograph shows several north-trending linear features with prominent bright or dark signatures. As shown on the map and cross section (Figure 2-23), these features are low topographic *scarps*, which are caused by active normal faults that cut the surface. In Figure 2-22A, the east-facing scarps are strongly illuminated by the early morning sun and have bright signatures, or highlights. The west-facing scarps are shadowed and have dark signatures. Orientation of the highlights and shadows provides information on the sense of displacement along the faults, which is shown in cross section (Figure 2-23B). Note, however, that if the scarps trended east–west, essentially parallel with the sun azimuth, the shadows and highlights would not be present in the photograph.

In the midday photograph (Figure 22-B), highlights and shadows are minimal and the fault scarps are inconspicuous. There are some subtle linear gray traces along the scarps that are not due to illumination, but are local concentrations of sagebrush. Runoff from rainfall is concentrated at the foot of the scarps and supports a higher density of vegetation. This pattern is one of several types of vegetation anomalies that aid in recognizing geologic features.

Acquiring good low sun-angle photographs in the summer is complicated by the limited number of hours in the morning and evening when the desired illumination occurs. Illumination values are low and change rapidly during these times; therefore, proper camera exposures may be difficult to achieve. At middle and high latitudes, the sun is at relatively low elevations throughout the day in the winter, which may be an optimum season for acquiring these photographs. Low sun-angle photographs are widely used to recognize subtle topographic features associated with active faults.

BLACK-AND-WHITE PHOTOGRAPHS

During the film era several types of black-and-white films were available for acquiring aerial photographs at wavelengths ranging from the UV, visible, and NIR spectral regions. These film-based photographs are still valuable resources.

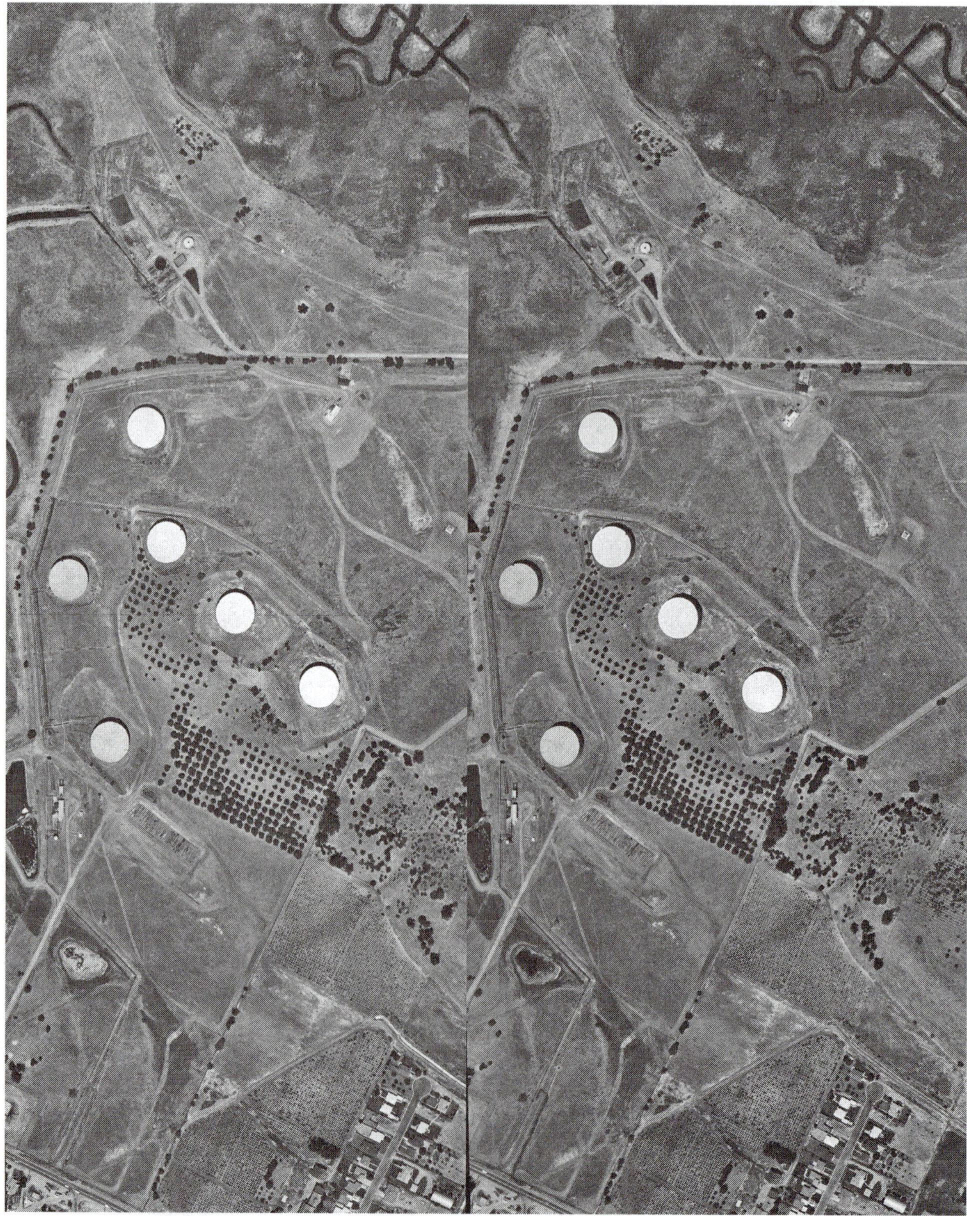

FIGURE 2-20 Stereo-pair of 1959 aerial photographs of the Shell Refinery oil tanks in Martinez, California. Courtesy D. Ruiz, Quantum Spatial, Inc., Novato, California.

PANCHROMATIC BLACK-AND-WHITE PHOTOGRAPHS

Black-and-white photographs that record visible light are called *panchromatic photographs*. Panchromatic aerial photographs are normally acquired with a minus blue filter over the lens, which eliminates the UV and blue wavelengths that are selectively scattered by the atmosphere. Examples of these minus blue photographs are shown in Figures 2-7, 2-8, and 2-11.

These photographs were a widely used and readily available remote sensing product. Stereo coverage of most of the United States is available from the agencies and companies listed later in this chapter. These photographs were used to compile topographic maps, geologic surveys, engineering studies, and crop inventories. Color photographs are superior for many applications, but panchromatic photographs are still a major source of remote sensing information.

NIR BLACK-AND-WHITE PHOTOGRAPHS

Prior to the digital revolution, panchromatic photographs in the NIR spectral band were acquired with IR-sensitive film and a filter that transmits only NIR energy (0.7 to 0.9 μm). NIR energy is reflected solar radiation and should not be confused with thermal IR (TIR) energy, which occurs at wavelengths of 3 to 14 μm. Chapter 5 deals with TIR remote sensing. The spectra in Figure 2-24 show transmittance of energy through water with different turbidity. Overall transmittance decreases with increasing turbidity from distilled water through bay water, which is turbid. The wavelength of peak transmittance shifts from blue to green with increasing turbidity. Irrespective of turbidity, there is no transmittance in the NIR region at wavelengths greater than 0.7 μm.

Figure 2-25 illustrates simultaneously acquired panchromatic and NIR black-and-white photographs of the

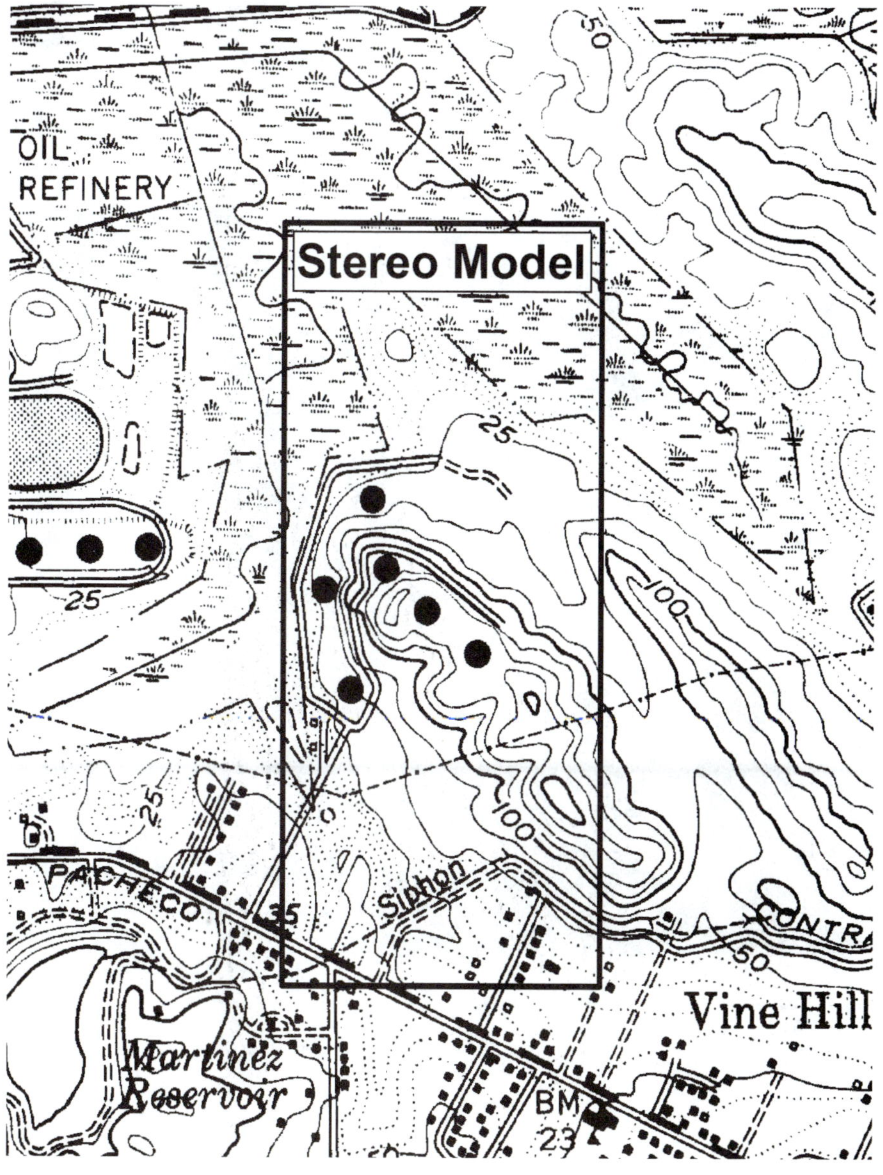

FIGURE 2-21 Portion of the topographic map of the Port Chicago Quadrangle showing the Shell Oil Refinery in Martinez, California (Figure 2-20). Courtesy USGS.

Massachusetts coast. This example demonstrates the following advantages of NIR photographs:

1. Haze penetration improves because the NIR filter eliminates the atmospheric scattering that occurs in the visible and UV regions (Figure 2-26A). Eliminating most scattered light results in a higher contrast ratio and therefore higher spatial resolution on the IR photograph, as discussed earlier.
2. Maximum reflectance from vegetation occurs in the NIR region, as shown by the bright tones in the NIR photograph. In addition, maximum spectral differences between vegetation types, such as birch and fir, occur in the NIR region (Figure 2-26B), which is advantageous for mapping plant communities.
3. NIR energy is almost totally absorbed by water, as shown on Figure 2-24, which causes water to have a dark tone on IR photographs. For this reason, boundaries between land and water show up more clearly on IR photographs than on panchromatic photographs. Note the bay in the north-central part of the scene.
4. The pan photograph of the bay includes two large medium-gray areas that appear to be islands. On the NIR photograph, however, the "islands" have the same dark gray signature as the water, which indicates that they are actually submerged banks from the land.
5. For the same reason, the shoreline of the bay is indistinct on the pan photograph, but is clearly shown on the NIR photograph.

Today panchromatic and NIR images are acquired by digital cameras (Figure 2-2). Color IR photographs, described in a later section, combine these properties of NIR black-and-white photographs with the advantages of color.

A. Low sun-angle photograph.

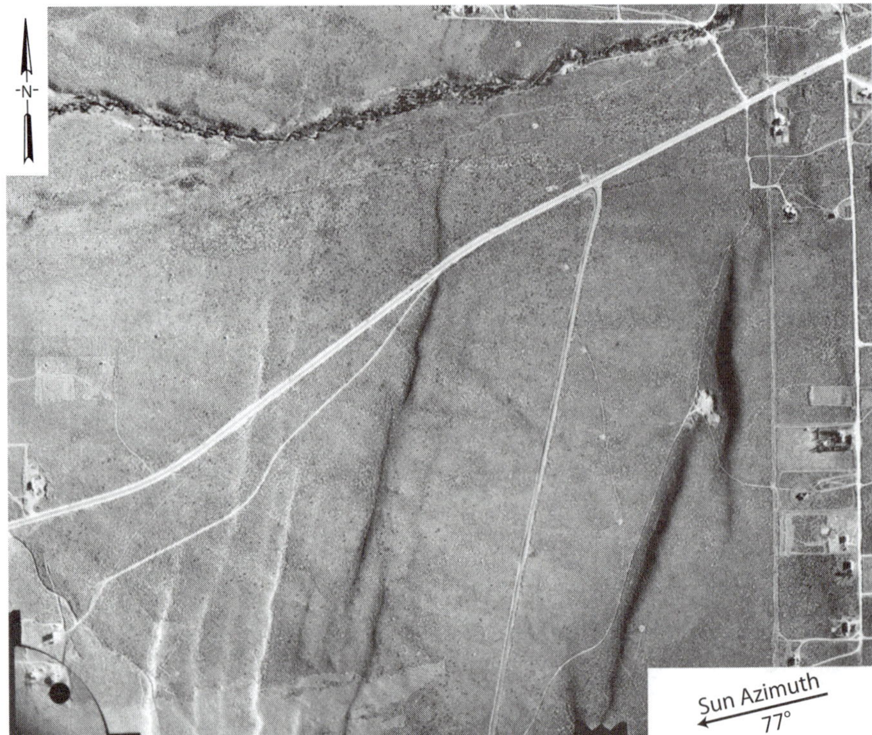

B. High sun-angle photograph.

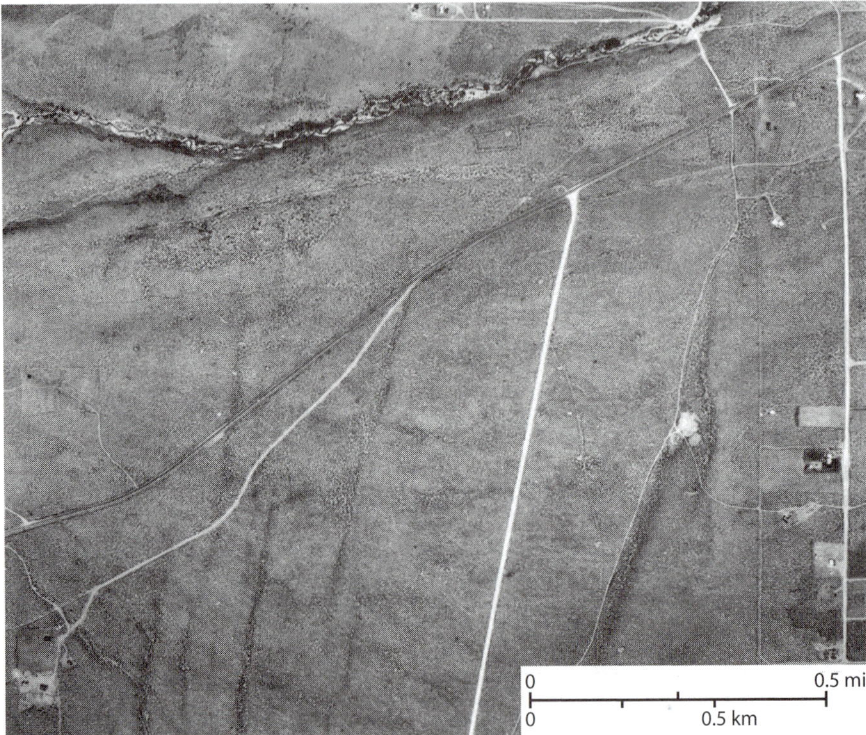

FIGURE 2-22 East flank of Carson Range, Nevada.

UV Photographs

The UV spectral region extends from 3 nm to 0.4 μm; however, the atmosphere only transmits UV wavelengths from 0.3 to 0.4 μm, which is known as the *photographic UV region*. Most camera lenses absorb UV energy of wavelengths less than about 0.35 μm, but special quartz lenses transmit shorter wavelengths. UV photographs have low contrast ratios and poor spatial resolution because of severe atmospheric scattering (Figure 2-4). As a result, the UV spectral region is rarely employed in remote sensing, except for special applications, such as monitoring oil films on water.

A. Interpretation map.

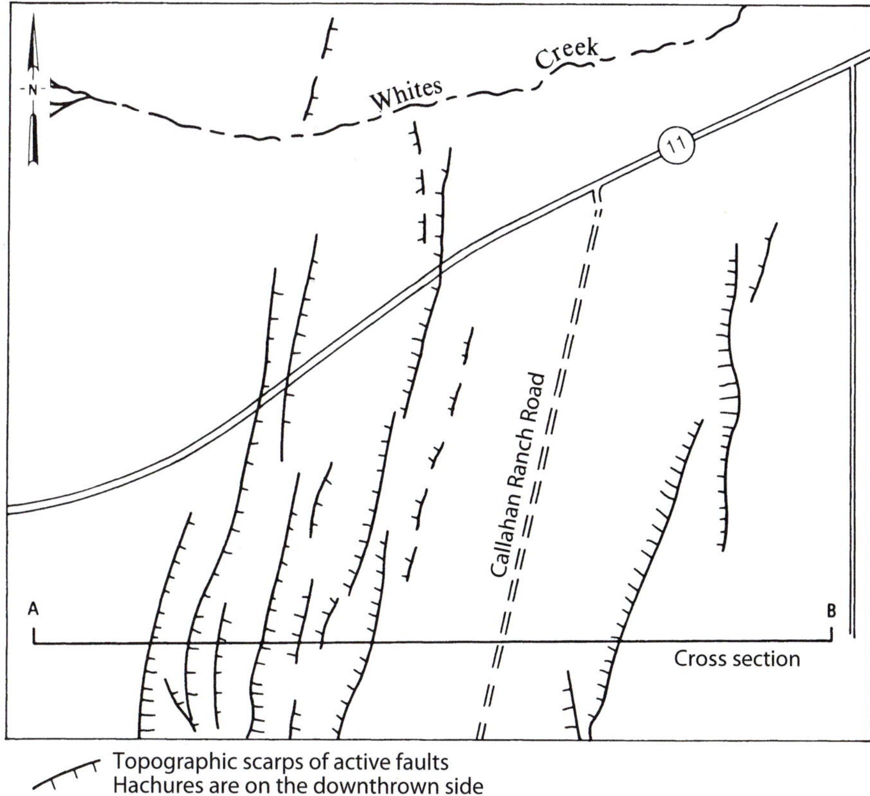

B. Vertically exaggerated cross section along line A–B.

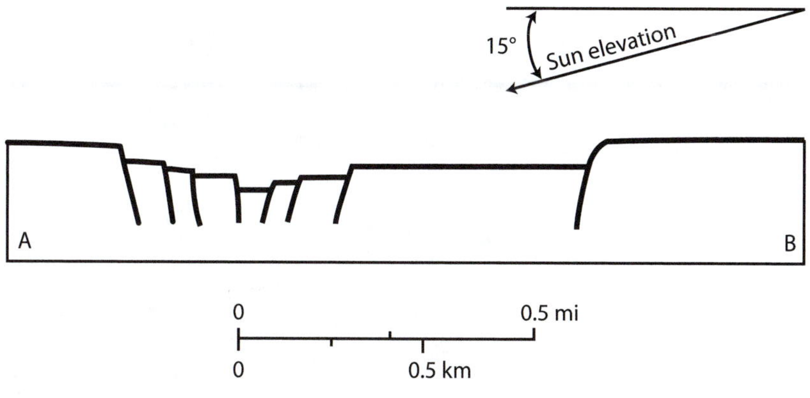

Figure 2-23 Topographic scarps as seen in the low sun-angle photograph of Figure 2-22A.

COLOR SCIENCE

The average human eye discriminates many more shades of color than tones of gray. This greatly increased information content is a major factor favoring the use of color photographs and images.

The visible region is divided into the blue, green, and red bands (or colors). These are called additive primary colors because white light is formed when equal amounts of blue, green, and red light are combined. Colors of the visible spectrum may be formed by combining various amounts of the additive colors. Most modern aerial cameras (Figure 2-2) employ four multispectral lens/filters to acquire digital images of the blue, green, and red spectral bands, plus the NIR spectral band

The images in Figure 2-27 were acquired on June 6, 2014 over the Concord Naval Weapons Station east of Oakland and San Francisco as part of the US Department of Agriculture's National Agricultural Imagery Program (NAIP). A location map is provided in Figure 2-28. The images and map are oriented with north up. The scene is characterized by wetlands to the north along the Sacramento River, hills to the east that are covered with dry grasses, a reservoir in the southern portion with a golf course to the southeast, a refinery with oil tanks to the southwest, and commercial buildings and suburbs adjacent to the golf course. The weapons station is now closed, but the railroad tracks, docks along the river, and concrete bunkers can be seen in the northeast portion of the scene.

Comparing the contrast ratio of the four images illustrates the effect of atmospheric scattering, which is strongest in the blue spectral region. Recall that contrast ratio is the ratio between the brightest and darkest elements of an image. The blue image (Figure 2-27A) is dominated by medium gray to black tones that form a

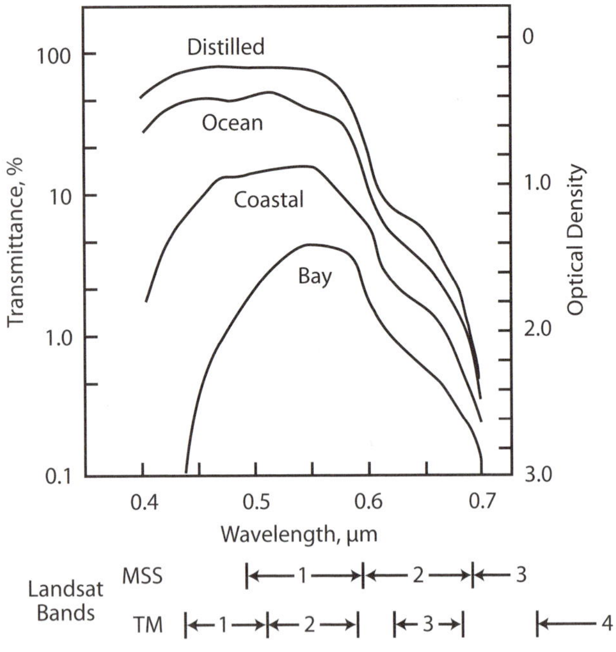

FIGURE 2-24 Spectral transmittance through 10 m of various types of water. Modified from Specht and others (1973, Figure 1).

A. Atmospheric scattering.

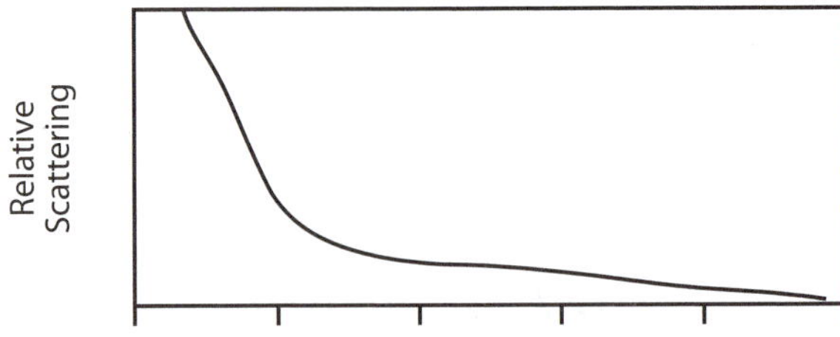

B. Vegetation reflectance spectra.

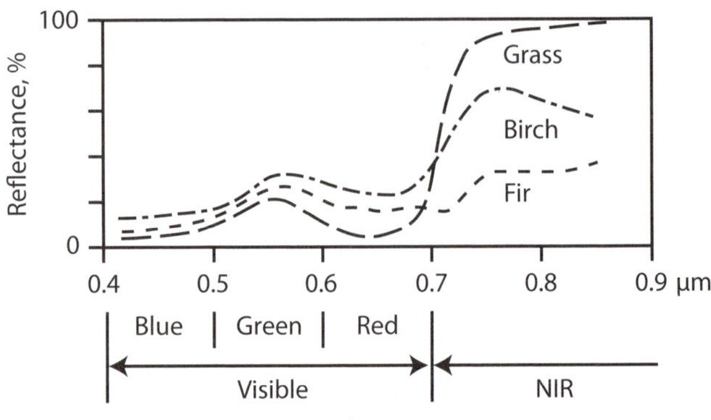

FIGURE 2-26 Atmospheric scattering and spectral reflectance curves of vegetation.

low-contrast image. Contrast improves greatly in the green band (Figure 2-27B) and reaches a maximum in the NIR band (Figure 2-27D).

When viewing grayscale images on a computer screen, pixels that are bright will transmit light when while those that are dark will block the light. Pixels in a band that are relatively dark indicate that the material in those pixels (water, vegetation, asphalt, etc.) is absorbing incoming light at the wavelength band of that image. Water bodies in the NIR image (Figure 2-27D) are very dark, indicating those longer wavelengths are strongly absorbed.

Pixels in an image that are bright indicate that the material in those pixels strongly reflects the incoming light with that image's wavelength. In the three visible light images (Figure 2-27A–C), commercial building rooftops in the south and many oil tank tops in the southwest are light gray to white. The irrigated grass on the golf course to the southeast displays white pixels only in the NIR image (Figure 2-27D), but the golf course pixels in the green image have a lighter shade of gray compared to those in blue and red bands (compare Figure 2-27B to Figure 2-27A and Figure 2-27C).

Any three wavelength bands can be combined and illuminated with blue, green, and red light (the three primary colors) to generate a color image. For instance, white or shades of light gray are displayed in the color image where the three bands have pixels with very high brightness or reflectance. Black or shades of dark gray are displayed where the three

A. Panchromatic.

FIGURE 2-25 Aerial photographs of the Massachusetts coast.

TABLE 2-3 Terrain signatures on natural color and color IR images.

Subject	Natural Color Image	Color IR Image
Healthy Vegetation		
Irrigated grass	Green	Red
Broadleaf type	Green	Red to magenta
Needle-leaf type	Green	Reddish brown to purple
Stressed Vegetation		
Previsual stage	Green	Pink to blue
Visual stage	Yellowish green	Cyan
Autumn leaves	Red to yellow	Yellow to white
Other Categories		
Clear water	Blue-green	Dark blue to black
Silty water	Light green	Light blue
Damp ground	Slightly darker than dry soil	Distinctly darker than dry soil
Shadows	Blue with details visible	Black with few details visible
Water penetration	Good	Moderate to poor
Contacts between land and water	Poor to fair discrimination	Excellent discrimination
Red bed outcrops	Red	Yellow

bands have pixels with very low brightness (low reflectance and high absorption). Green is displayed when the green band has relatively high reflectance compared to the blue and red bands. The reflectance (brightness) differences between any three bands control the additive colors that are seen on the digital color image.

B. NIR black and white.

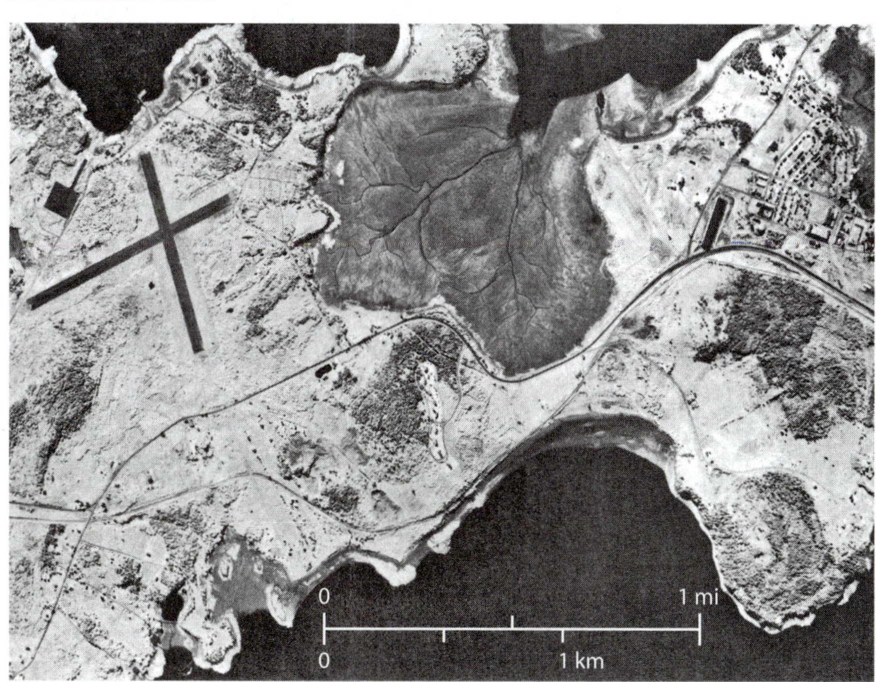

NATURAL COLOR AND COLOR IR PHOTOGRAPHS COMPARED

Natural color photographs record a scene in its true colors. The blue, green, and red bands in Figure 2-27A–C are combined in those colors to produce the natural color image shown in Plate 5A. The green, red, and NIR bands in Figure 2-27 are combined in blue, green, and red to produce the color IR image shown in Plate 5B. Table 2-3 compares the color signatures of common subjects on the two types of photographs.

SIGNATURES OF VEGETATION

The most striking difference between the two color images is the signature of healthy vegetation, which is green in the natural color image (Plate 5A), and red in the NIR image (Plate 5B). This difference is explained by the spectral reflectance curves of vegetation shown in Figure 2-26B. Spectral reflectance curves show the percentage of incident energy reflected by a material as a function of wavelength. Blue and red light are absorbed by foliage in the process of photosynthesis. Up to 20% of the incident green light is reflected, causing the familiar green color of leaves on natural

A. Reflected blue.

B. Reflected green.

C. Reflected red.

D. Reflected NIR light.

FIGURE 2-27 Images of the Concord Naval Weapons Station, California. Courtesy USDA Farm Service Agency.

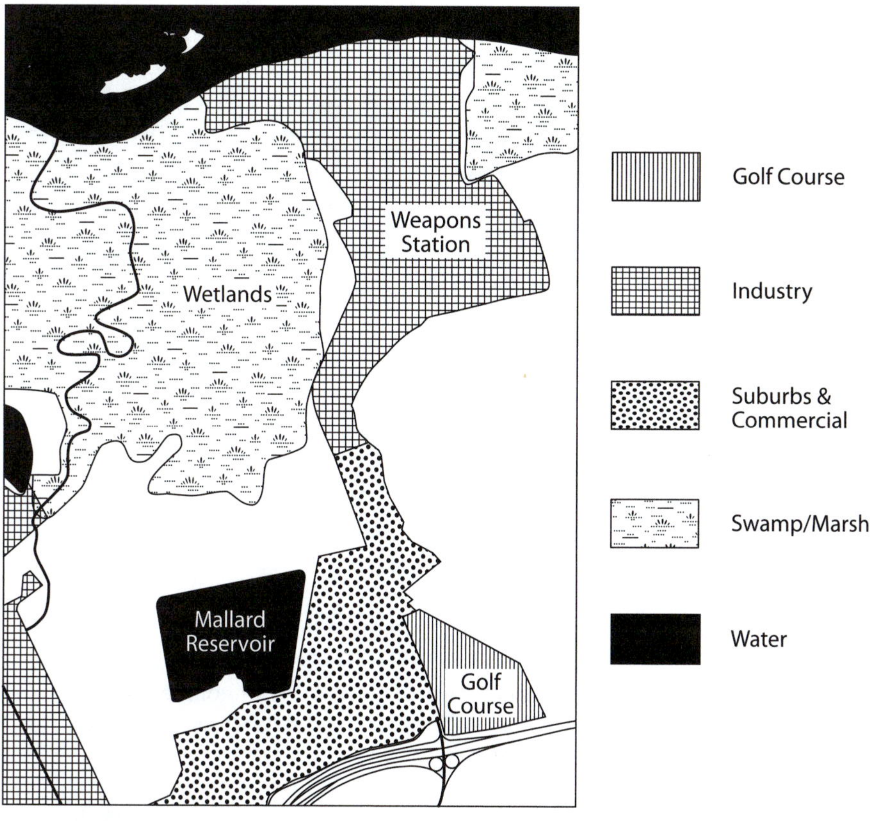

Figure 2-28 Location map of Concord Naval Weapons Station, California (Figure 2-27).

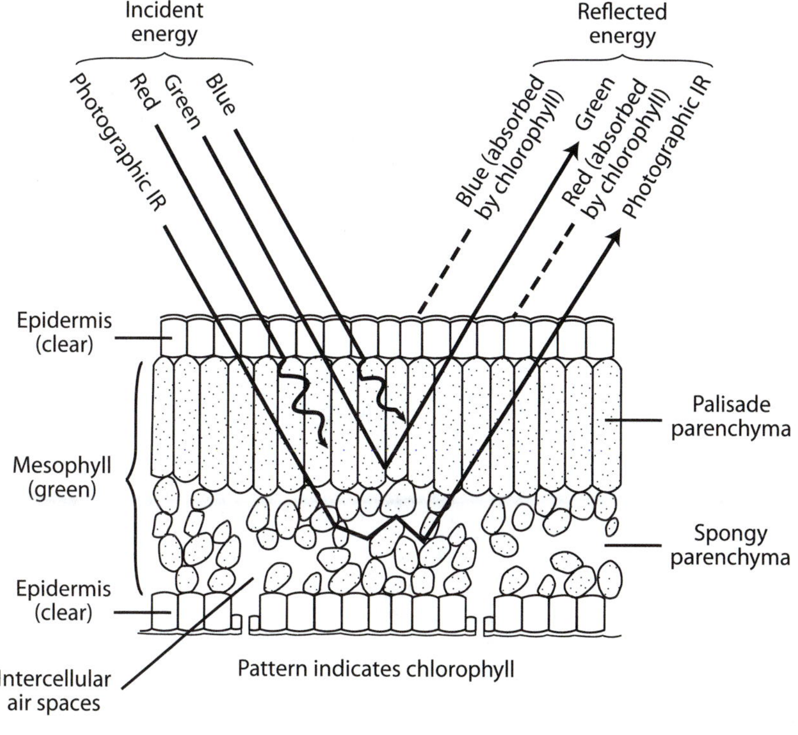

Figure 2-29 Diagrammatic cross section of a leaf, showing interaction with incident energy. Incident blue and red wavelengths are absorbed by chlorophyll in the process of photosynthesis. Incident green wavelengths are partially reflected by the chlorophyll. Incident NIR energy is strongly scattered and reflected by the cell walls in the mesophyll. Modified from Buschmann and Nagel (1993, Figure 9).

color photographs. The spectral reflectance of vegetation increases abruptly in the NIR region, which includes the wavelengths that are shown in red on color IR images.

Figure 2-29 is a diagrammatic cross section of a leaf that explains the spectral signatures of vegetation. The transparent epidermis allows incident sunlight to penetrate into the mesophyll, which consists of two layers: (1) the palisade parenchyma of closely spaced cylindrical cells, and (2) the spongy parenchyma of irregular cells with abundant interstices filled with air. Both types of mesophyll cells contain chlorophyll, which reflects part of the incident green wavelengths and absorbs all of the blue and red energy for photosynthesis. Chlorophyll causes the green signature of vegetation. The longer wavelengths of NIR energy penetrate into the spongy parenchyma, where the energy is strongly scattered and reflected by the boundaries between cell walls and air spaces. The strong, NIR reflectance of leaves is caused not by chlorophyll but by the internal cell structure. Gausman (1985) gives details of optical properties of plant leaves in the visible and reflected IR regions. Buschmann and Nagel (1993) describe the roles of chlorophyll and cell structure in spectral reflectance of leaves.

Detection of Stressed Vegetation

Vegetation may be stressed because of drought, disease, insect infestation, or other factors that deprive the leaves of water, which collapses the mesophyll. The photomicrographs in Figure 2-30 compare the internal structure of nonstressed and stressed leaves. The nonstressed leaf (Figure 2-30A) has a cell structure and reflectance characteristics comparable to those shown in the leaf diagram (Figure 2-29). In the stressed leaf (Figure 2-30B), the shortage of water causes the mesophyll cells to collapse, which strongly reduces NIR reflectance from the spongy parenchyma. This decreased reflectance diminishes

FIGURE 2-30 Photomicrographs of cross sections of nonstressed and stressed leaves. Collapse of cells in the mesophyll layer strongly reduces reflectance of incident NIR energy. From Everitt and Nixon (1986). Courtesy J. H. Everitt, US Department of Agriculture.

A. Nonstressed leaves.

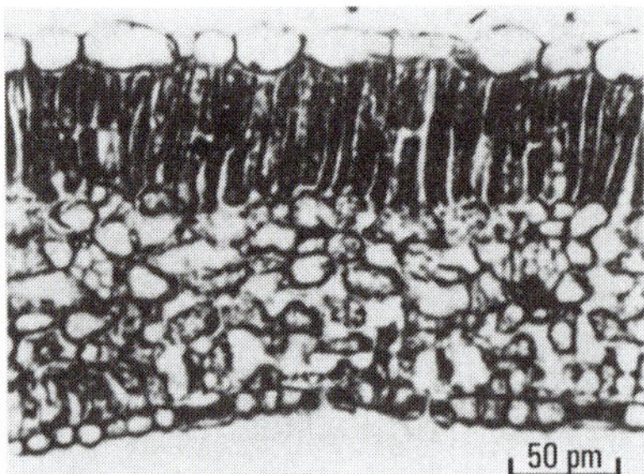

B. Stressed leaves.

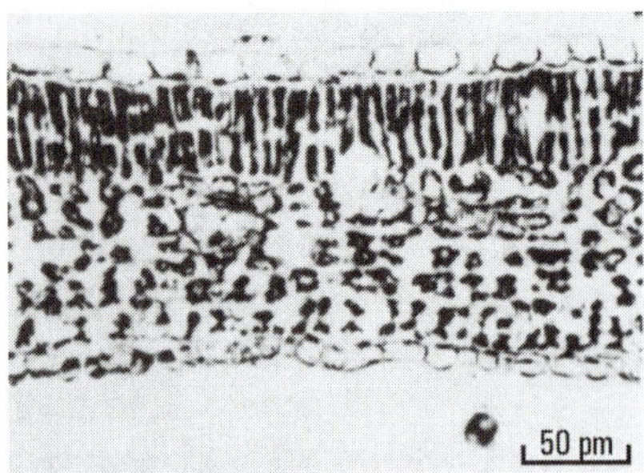

the red signature in color IR images. Chlorophyll is still present in the stressed foliage, which may have a green signature in natural color images for some time after the onset of stress. In color IR images, however, stressed foliage has a distinctive blue signature. The loss of NIR reflectance is a *previsual symptom* of plant stress because it often occurs days or even weeks before the visible green color begins to change. The previsual effect may be used for early detection of disease and insect damage in crops and forests. Artificial turf lacks the cell structure of vegetation and has a blue signature on color IR images.

AUTUMN SENESCENCE OF VEGETATION

In the autumn, leaves of deciduous trees undergo senescence and turn red, yellow, and brown. Figure 2-31 compares spectra of green and senescent foliage. As the green chlorophyll decays, red wavelengths are no longer absorbed. The organic compounds anthocyanin and tannin are formed, causing the familiar autumn colors (Boyer and others, 1988). The spectrum for senescent foliage (Figure 2-31) shows nearly equal reflectance values in the green, red, and photographic IR bands, which results in a white signature in color IR photographs. Boyer and others (1988) also describe the changes in leaf physiology and spectral reflectance during senescence.

The aerial imagery in Plate 5 was acquired in the dry season of northern California. Grasses that are not irrigated undergo senescence due to a lack of water. The dry grass on the eastern and southeastern slopes is illuminated by the morning sun in the east-central portion of Plate 5. On these sunny slopes the dry grass is light yellow to tan in the natural color image and whitish in the color IR image. The color of dry grass contrasts strongly with the color of healthy irrigated grass in the golf course (bright green in the natural color image and bright red in the color IR image). The spectra of dry and healthy grass are very similar to that of Boyer and others (1988) spectra for senescent and green foliage (Figure 2-31).

SIGNATURES OF OTHER TERRAIN FEATURES

Comparing the natural color image (Plate 5A) and the color IR image (Plate 5B) shows several advantages for the color IR image.

1. The color IR image has a better contrast ratio than the natural color image because strongly scattered blue wavelengths are eliminated.
2. Within the vegetated terrain there is a wider range of red signatures on the color IR image than among green signatures on the natural color image. Reflectance spectra for different vegetation types have a wider range in the NIR band than in the visible green band.
3. On the natural color image portions of the shore are indistinct due to the similar signatures of blue-green water and green vegetation. On the color IR image, however, the blue water is readily distinguished from the red vegetation. The small islands are more distinct on the color IR image.
4. In the east-central area, topography of the eroded bedrock is expressed by shadows and highlights. On the color IR image shadows are darker and more distinct because the strongly scattered blue light is eliminated.

SOURCES FOR PHOTOGRAPHS/IMAGES OF THE UNITED STATES

There are a number of libraries/repositories of aerial photographs/images that are operated by federal and state gov-

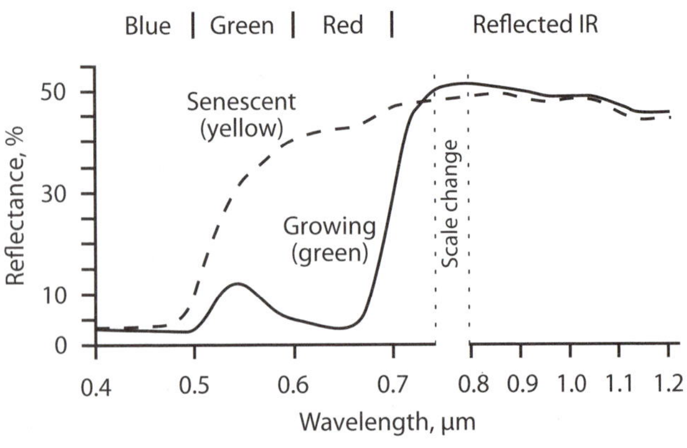

FIGURE 2-31 Reflectance spectra of green and senescent foliage. In the autumn, chlorophyll deteriorates, which reduces the absorption of incident red energy. The development of anthocyanin and tannin causes the yellow to red fall colors. From Schwaller and Tkach (1985, Figure 2).

ernment agencies. For example, the US Bureau of Land Management (BLM) has over 450,000 aerial photograph frames covering portions of the western United States. Many state agencies and universities maintain image libraries. Commercial aerial photography companies typically have libraries of local areas. The companies can also acquire new images on a contract basis. All of these sources may be located and contacted via the Internet.

This section summarizes several federal government programs that acquire photographs/images of the conterminous United States.

National Agriculture Imagery Program

The National Agricultural Imagery Program (NAIP) acquires 1-m aerial images of the conterminous United States during the peak growing season. Beginning in 2002 the program acquired images with film or digital cameras. Since 2013 acquisition is entirely with digital systems that record blue, green, red, and NIR bands. Each image covers a 3.75 by 3.75 minute quarter of a USGS 7.5-minute topographic quadrangle map plus a 300 m buffer on all four sides. NAIP acquires new imagery every three years. Figure 2-27 shows the four bands acquired over the southeast quadrant of the USGS Vine Hill, California Quadrangle topographic map. Plates 5A and 5B show the natural color and color IR images generated from these bands.

National Aerial Photography Program

The National Aerial Photography Program (NAPP) was an interagency project coordinated by the USGS that was operational from 1987 to 2007. The program acquired two types of photographs on 9-in film at a scale of 1:40,000:

1. Black-and-white panchromatic film of the visible wavelengths.
2. Color IR film of the green, red, and NIR wavelengths. Some of the photographs have an overall blue cast due to degradation prior to processing.

The program has more than 1.3 million photographs that cover the conterminous United States. Each pair of photographs is centered over quarters of USGS 7.5-minute topographic quadrangle maps. NAPP flight lines were oriented in a north to south direction through the east and west halves of 7.5-minute quadrangles with a 60% forward overlap, which allows for stereographic viewing.

National High Altitude Photography Program

The National High Altitude Photography (NHAP) Program, coordinated by USGS, began in 1978 to acquire coverage of the United States with a uniform scale and format. The program was operational until 1989, collecting approximately 500,000 photographs. From aircraft at an altitude of 12 km, two cameras (23 by 23 cm format) acquired black-and-white photographs and color IR photographs. The black-and-white photographs were acquired using a camera with a 152-mm focal length to produce photographs at a scale of 1:80,000, which covers 338 km^2. The color IR photographs were acquired using a camera with a 210-cm focal length to produce photographs at a scale of 1:58,000, which covers 178 km^2.

NASA High-Altitude Photographs

For a number of years, NASA has been acquiring photographs of the United States from U-2 and RB-57 reconnaissance aircraft at altitudes of 18 km above terrain with standard aerial cameras (152-mm focal length) on film with a 23 by 23 cm format. The resulting photographs cover 839 km^2 at a scale of 1:120,000. Black-and-white, normal color, or color IR film is used; many missions employ two cameras to acquire photographs with two different film types. Coverage of NASA photographs is concentrated over numerous large regional test sites for which repeated coverage over several years may be available. Many areas lack this coverage.

Online Sources for Aerial Photography

- American Society for Photogrammetry and Remote Sensing, Aerial Data Catalog (dpac.asprs.org)
- Bureau of Land Management (blm.gov/services/geospatial)
- L3Harris (harrisgeospatial.com/Data-Imagery)
- National Archives (archives.gov/research/order/maps.html)
- USDA Aerial Photography Field Office (fsa.usda.gov/programs-and-services/aerial-photography)
- USGS EarthExplorer (earthexplorer.usgs.gov)

PHOTOGRAPHS FROM SATELLITES

Handheld Photographs from Satellites

Over the past decades NASA has deployed a number of manned Earth-orbiting satellites from which astronauts have acquired more than 1 million handheld photographs and digital images; these images are accessible at earthobservatory.nasa.gov. The images are listed by year and month and are grouped into the following categories: atmosphere, heat, human, land, life, natural event, remote sensing, snow and ice, and water.

International Space Station

Digital Image 2-1 🌐 is an image of the Namib Desert that was acquired March 27, 2016 by an astronaut on the International Space Station (ISS) using a Nikon D4 digital camera with a 500 mm lens. ISS was positioned over the south Atlantic Ocean. The image was acquired near sunset looking inland (eastward). The low sun angle causes strong highlights and shadows that enhance subtle topographic features, as described earlier in this chapter. The image covers a geologically interesting portion of the hyperarid Namib Desert.

In the eastern third of the image, an older set of north-trending dunes is overlain by younger west–northwest-trending dunes. This superposed pattern is attributed to drier climates and shifts in wind direction (NASA Earth Observatory, 2016). In the northeast portion of the image the west-trending dunes become progressively narrower and fade out gradually on the underlying terrain. In the northwest portion of the image, however, the north–northwest-trending linear dunes terminate abruptly along an east-trending linear scarp of sand that faces north. The valleys between the linear dunes are filled with sand to form the scarp. We have noted similar scarps in satellite images of dune fields in the Mojave Desert of California that we interpret as second-order effects of active faults. The faults do not offset the very young sand. Instead, the faults form barriers to subsurface flow of groundwater, which causes the water table to be shallower on one side of the fault. Phreatophyte vegetation grows on the side of the fault with the shallow water table, which is the south side in **Digital Image 2-1**. The vegetation interferes with the northerly wind, which deposits its load of sand along the vegetated upwind (south) side of the fault and fills the interdune depressions. The terrain north of the fault lacks vegetation and consequently lacks windblown sand. A well drilled on the south side of the fault might produce water.

Handheld images from satellites are useful supplements to images from unmanned satellites such as Landsat (Chapter 3), which are acquired with a vertical aspect angle and a fixed footprint. Astronauts can acquire images in any direction with a range of aspect angles and a range of footprints. This flexibility enables astronauts to acquire images of transient events such as storms, forest fires, and volcanic eruptions. Landsat, however, captures such events only if they occur within the 185-km wide image swath directly beneath the satellite. Landsat images are acquired within a fixed time window centered around 10:00 AM in the morning. Astronauts can acquire images whenever there is sufficient sunlight.

For additional information on ISS and the astronaut image projects see Gebelein and Eppler (2006), Green and Jackson (2009), and Hornyak (2013). NASA also distributes a weekly collection of astronaut photographs at earthobservatory.nasa.gov/IOTD.

Space Shuttle

The Space Shuttle program, or *Space Transportation System* (STS), began in 1981 and ended with STS-135 in 2011. The vehicle was similar in size to a medium commercial jet airliner and accommodated a crew of up to seven on missions that lasted up to nine days. Many photographs have been acquired from the Shuttle by handheld cameras and by the large-format camera.

The Shuttle had several windows for acquiring photographs with handheld cameras. Most photographs were recorded on normal color film at formats of 35, 70, and 140 mm, using a variety of cameras and lenses. After each mission NASA prepared a catalog listing the location, features, and time for each photograph. For example, the catalog for STS Mission 37 in April 1991 lists 4,283 photographs of the Earth. More information is available from NASA's Gateway to Astronaut Photography of Earth website (eol.jsc.nasa.gov) and the Johnson Space Center's communications team in Houston, Texas.

DECLASSIFIED PHOTOGRAPHS FROM INTELLIGENCE SATELLITES

For many years the United States has employed satellites to acquire photographs of strategic areas, mainly the Sino–Soviet bloc, for intelligence purposes. These photographs have been highly classified and unavailable. In 1995 the United States announced the declassification of intelligence photographs acquired from 1960 to 1972 by the Corona camera, which is now obsolete. The Corona missions acquired over 800,000 photographs. Each photograph covers approximately 16 by 195 km at spatial resolutions ranging from 2 to 8 m. The transparencies were duplicated, indexed, and transferred to the USGS Earth Resources Observation and Science Center for unrestricted sale to the public by late 1996. McDonald (1995a, 1995b) has reviewed the history and specifications of the classified US satellite photography programs and has published a number of photographs.

Additional declassified satellite imagery was made available in 2002 and 2013. The Keyhole (KH) satellite systems were capable of recording high resolution photographs with spatial resolution ranging from 0.6 to 1.3 m. The film can be scanned at either 1,800 or 3,600 dpi at a cost of $30 per frame. They are available for download from the USGS's EarthExplorer website (earthexplorer.usgs.gov).

QUESTIONS

1. Normal color photographs taken of subjects in shaded areas have a bluish cast. Explain why.
2. Calculate the contrast ratio for a scene in which the brightest and darkest areas have brightness values of 6 and 2, respectively.
3. Suppose the scene in question 2 is covered by an atmosphere that contributes 4 brightness values of scattered light. What is the resulting contrast ratio? Panchromatic aerial photographs will be acquired of this scene. How can their contrast ratio be improved?
4. What is the ground resolution for aerial photographs acquired at a height of 5,000 m with a camera having a system resolution of 30 line-pairs · mm^{-1} and a focal length of 304 mm?
5. What is the minimum ground separation in the photographs of question 4?
6. What is the scale of the photographs of question 4?
7. For Figure 2-8, calculate the height of the highest portion of the building in the extreme lower left corner.
8. The airbase for two overlapping photographs is 1,500 m. The photographs were acquired from a height of 3,000 m. What is the base-height ratio of this stereo-pair? What is the vertical exaggeration of the stereo model?
9. Panchromatic, color, and color IR photographs are acquired from satellites and aerial platforms. Describe the advantages and disadvantages of satellite photographs relative to aerial photographs.

References

Boyer, M., J. Miller, M. Berlanger, and E. Hare. 1988. Senescence and spectral reflectance in leaves of northern pin oak (*Quercus palustris* Muenchh.). *Remote Sensing of Environment*, 25, 71–87.

Buschmann, C., and E. Nagel. 1993. *In vivo* spectroscopy and internal optics of leaves as basis for remote sensing of vegetation. *International Journal of Remote Sensing*, 14, 711–722.

Everitt, J. H., and P. R. Nixon. 1986. Canopy reflectance of two drought-stressed shrubs. *Photogrammetric Engineering and Remote Sensing*, 52, 1,189–1,192.

Gausman, H. W. 1985. Plant Leaf Optical Properties in Visible and Near-Infrared Light. Texas Tech University Graduate Studies, no. 29, Lubbock, TX.

Gebelein, J., and D. Eppler. 2006. How earth remote sensing from the International Space Station complements current satellite-based sensors. *International Journal of Remote Sensing*, 27(13), 2,613–2,629.

Green, K., and M. W. Jackson. 2009. Timeline of key developments in platforms and sensors for earth observations. In M. W. Jackson (Ed.), *Earth Observing Platforms & Sensors, Manual of Remote Sensing*, Vol. 1.1 (3rd ed., pp. 1–48). Bethesda, MD: American Society for Photogrammetry and Remote Sensing.

Hornyak, D. M. 2013. A Researcher's Guide to: International Space Station. Technology Demonstration (NP-2013-06-008-JSC). NASA, Johnson Space Center.

JPL. 2017. Mars 3D. http://mars.nasa.gov/mars3d (accessed December 2017).

La Prade, G. L. 1972. Stereoscopy—a more general theory. Photogrammetric Engineering and Remote Sensing, 38, 1,177–1,187.

La Prade, G. L. 1973. Stereoscopy—will data or dogma prevail? *Photogrammetric Engineering and Remote Sensing*, 39, 1,271–1,275.

McDonald, R. A. 1995a. CORONA. *Photogrammetric Engineering and Remote Sensing*, 61(6), 689–720.

McDonald, R. A. 1995b. Opening the cold war sky to the public: Declassifying satellite reconnaissance imagery. *Photogrammetric Engineering and Remote Sensing*, 61, 385–390.

Miller, C. V. 1961. *Photogeology*. New York: McGraw-Hill.

NASA Earth Observatory. 2016, March 17. Linear Dunes, Namib Sand Sea. http://earthobservatory.nasa.gov/IOTD/view.php?id=89136 (accessed December 2017).

Paine, D. P., and J. D. Kiser. 2012. *Aerial Photography and Image Interpretation* (3rd ed.). Hoboken, NJ: John Wiley.

Rosenblum, L. 1968. Image quality in aerial photography. *Optical Spectra*, 2, 71–73.

Schwaller, M. R., and S. J. Tkach. 1985. Premature leaf senescence remote sensing detection and utility for geobotanical prospecting. *Economic Geology*, 80, 250–255.

Slater, P. N. 1983. Photographic systems for remote sensing. In R. N. Colwell (Ed.), *Manual of Remote Sensing* (2nd ed.) (chapter 6, pp. 231–291). Falls Church, VA: American Society for Photogrammetry and Remote Sensing.

Specht, M. R., D. Needler, and N. L. Fritz. 1973. New color film for water penetration photography. *Photogrammetric Engineering and Remote Sensing*, 40, 359–369.

Thurrell, R. F. 1953. Vertical exaggeration in stereoscopic models. *Photogrammetric Engineering and Remote Sensing*, 19, 579–588.

Wolf, P. R. 1974. *Elements of Photogrammetry*. New York: McGraw-Hill.

chapter 3

Landsat Images

Since 1972, the Landsat program has operated a series of unmanned satellites that have acquired an uninterrupted multispectral record of the Earth's land surface on a global basis. The program is a joint venture of NASA, which designs and operates the satellites, and the US Geological Survey (USGS), which distributes the free digital image data to a broad, international user community. Landsat operates in the international public domain, which means that

1. Under an "open skies" policy, images are acquired of the entire Earth without obtaining permission from any government and
2. Users anywhere in the world may download all images at no cost. As of June 2019, over 94 million scenes have been downloaded by users worldwide.

Figure 3-1 is a time line for the eight Landsat satellites that have been placed in orbit using Delta rockets launched from Vandenberg Air Force Base on the California coast between Los Angeles and San Francisco. The eight Landsat satellites belong to three generations of the Landsat program; each generation can be identified with different imaging systems and different satellites (Table 3-1):

First Generation

- MultiSpectral Scanner (MSS) on Landsat 1, 2, and 3.

Second Generation

- Thematic Mapper (TM) on Landsat 4 and 5.
- Landsat 6 was lost shortly after launch. It was carrying the Enhanced Thematic Mapper (ETM).
- Enhanced Thematic Mapper Plus (ETM+) on Landsat 7.

Third Generation

- Operational Land Imager (OLI) and Thermal Infrared Sensor (TIRS) on Landsat 8.

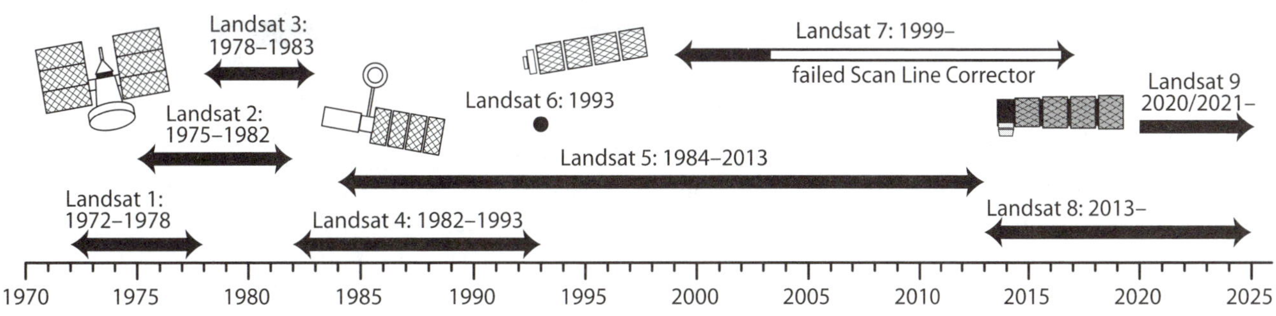

FIGURE 3-1 Time line showing three generations of the Landsat program. Courtesy USGS.

TABLE 3-1 Characteristics of imaging systems for the three generations of Landsat.

	Generation 1 MSS	Generation 2 TM and ETM+	Generation 3 OLI and TIRS
Spectral Region			
VNIR and SWIR	0.50 to 1.10 μm	0.45 to 2.35 μm	0.43 to 1.65 μm
TIR		10.5 to 12.5 μm	10.60 to 11.19 μm 11.50 to 12.51 μm
Spectral bands	4	8	11
Terrain Coverage			
Swath width	185 km	183 km	183 km
North–south direction	185 km	170 km	170 km
Instantaneous Field of View			
VNIR and SWIR	0.087 mrad	0.043 mrad	0.043 mrad
TIR		0.086 mrad	1.43 mrad
Ground Resolution Cell			
VNIR and SWIR	68 by 83 m	30 by 30 m	30 by 30 m
TIR		120 by 120 m 60 by 60 m[a]	100 by 100 m
Panchromatic		15 by 15 m[a]	15 by 15 m

[a] ETM+ only.

The Landsat program has been a major contributor to the growth and acceptance of remote sensing as a scientific discipline. Landsat provides the first repetitive worldwide database with adequate spatial and spectral resolution for many applications. The USGS and NASA's Goddard Space Flight Center (GSFC) provide much online information about the Landsat program. Present and future generations of remote sensing users are indebted to the late William T. Pecora and the late William Fischer of the USGS, who did so much to make Landsat a reality. We trust and expect that the Landsat program will continue to provide new and valuable images in the future.

FIRST GENERATION MSS: LANDSAT 1, 2, AND 3

The three satellites of the first Landsat generation were launched in 1972, 1975, and 1978; all have ceased operation, but they acquired hundreds of thousands of valuable images. The MSS was the primary imaging system in this first generation of Landsat. A return-beam vidicon (RBV) system was also carried, but failed to function. The MSS sensor was a cross-track scanning system that recorded four spectral bands of visible and near IR (VNIR) imagery with a ground resolution cell of 68 by 83 m (commonly resampled to 57 by 57 m or 60 by 60 m). Reflected green, red, and two NIR bands were collected by the MSS sensor. Figure 3-2 shows spectral ranges of the MSS bands together with reflectance spectra of vegetation and sedimentary rocks. MSS bands were originally designated 4, 5, 6, and 7, but were renumbered as 1, 2, 3, and 4 when carried on Landsat 4 and 5 platforms. We use both designations. When we describe how an image was acquired, we will provide the following information: the system used, the spectral bands that were used, and the colors (blue, green, red). This information will be provided in the following format: Landsat MSS bands 1-2-4 in B-G-R.

The archive of first generation Landsat MSS images provides a worldwide multispectral baseline from the early 1970s to the early 1980s. Images from the second and third generation extend the base line to the present and beyond. Temporal changes are accurately documented by comparing first-generation Landsat MSS images with newer images acquired by the second and third generation. In Saudi Arabia, for example, Plate 6 compares a 1979 Landsat 3, color IR image to a 2016 Landsat 8, color IR image of an area 80 km southwest of Riyadh. The area is now crossed by a northeast-to-southwest highway and a network of secondary roads. Water from wells drilled into an underground aquifer supports irrigated crops that are shown as shades of red in the color IR image (Plate 6B). The red circles are pivot-point fields with an artesian water well in the center connected to a radial pipe, mounted on wheels, that extends to the edge of the circle. Water pressure drives the wheels and the radial arm circles the field, typically on a 24-hour cycle. Large sprinklers mounted on the pipe irrigate the field. The 37-year time span clearly shows the increase in roads, agriculture, and infrastructure. The readily accessible archive of three generations of Landsat images provide a worldwide, geospatial base for monitoring changes dating back to 1972.

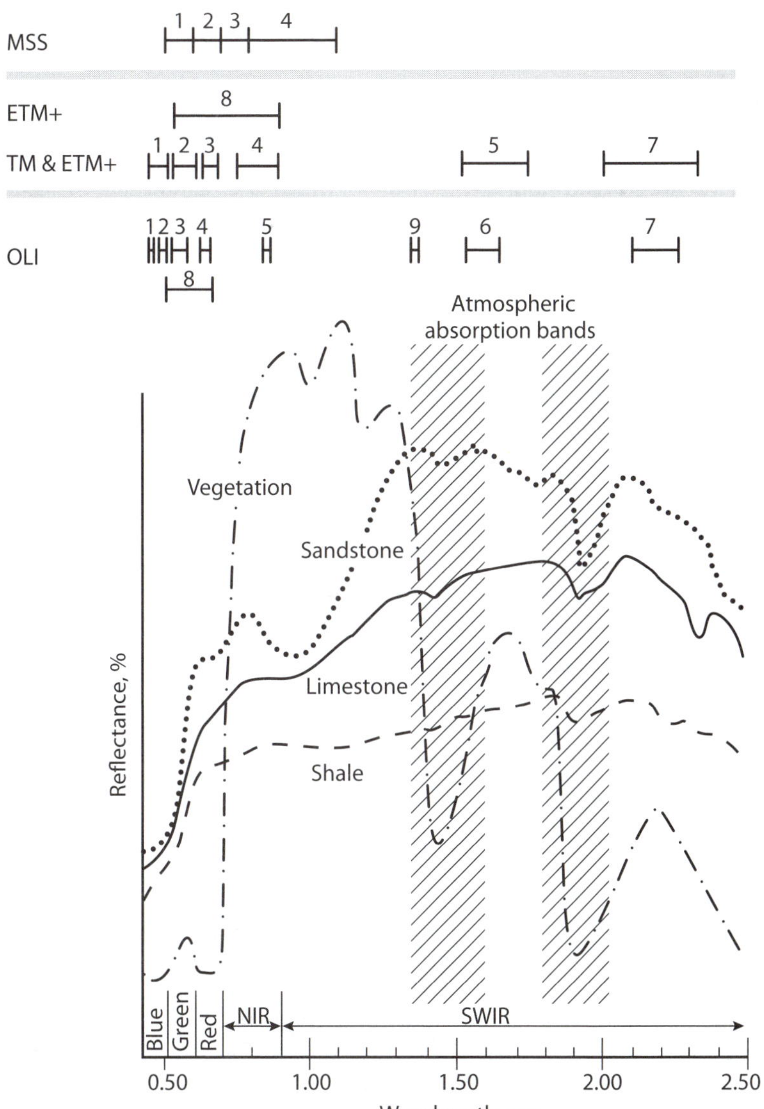

Figure 3-2 Reflectance spectra of vegetation and sedimentary rocks, showing spectral ranges of Landsat MSS, ETM+, and OLI bands.

SECOND GENERATION TM AND ETM+: LANDSAT 4, 5, AND 7

The second generation of Landsat consists of three satellites. Landsat 4 was launched in 1982 and ceased functioning in December 1993. Landsat 5 functioned for almost 30 years with the last image collected in January 2013. Landsat 6 was launched in September 1993, but failed to reach orbit, which is why it is not included in the second generation. Landsat 7 was launched in 1999, lost some capability in 2003 (described below), and, as of this writing, continues to function (Figure 3-1). Together, the second generation has acquired over 3,000,000 scenes around the globe.

Satellites

Figure 3-3 shows the second generation of Landsat satellites, which carried an improved imaging system called the Thematic Mapper (TM) and the MSS. The solar array generated electrical power to operate the satellite. The microwave antenna received instructions and transmitted image data to the ground receiving stations. When a satellite was within the receiving range of a station, TM images were scanned and transmitted simultaneously. Images of areas beyond receiving ranges were transmitted to *tracking and data relay satellites* (TDRS), which were placed in geostationary orbits. The TDRS system relayed the image data to a receiving station at Norman, Oklahoma.

Thematic Mapper

TM is a cross-track scanner with an oscillating scan mirror and arrays of 16 detectors for each of the VNIR bands (Figure 3-4). Data are recorded on both eastbound and westbound sweeps of the mirror, which allows a slower scan rate, longer dwell time, and higher signal-to-noise ratio than with MSS images of the first Landsat generation. At the satellite altitude of 705 km the 14.9° angular field of view covers a swath 185 km wide. Spectral ranges of the six visible and reflected IR bands are shown in Figure 3-2. Reflected, shortwave IR (SWIR) light is collected in bands 5 and 7 (SWIR1 and SWIR2, respectively). Band 6 (10.4 to 12.5µm) records emitted thermal IR (TIR) energy, which

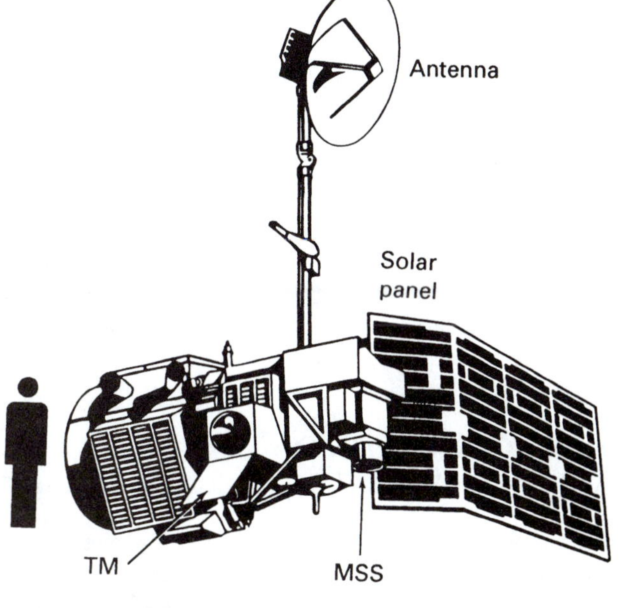

Figure 3-3 Landsat TM satellite. The human figure (2 m high) is added for scale.

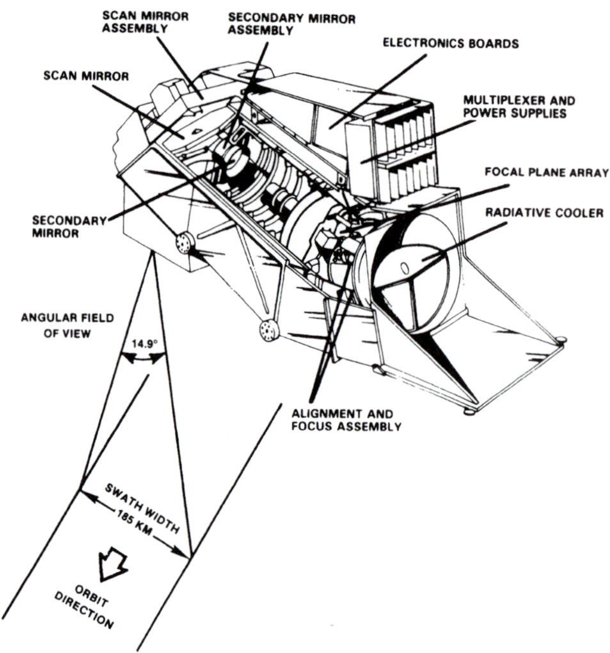

FIGURE 3-4 Thematic Mapper (TM) imaging system and cross-track image swath.

is beyond the spectral range covered by Figure 3-2. The TM was originally designed to include bands 1 through 5 (VNIR and SWIR1) and band 6 (TIR). Users pointed out that information in the spectral band from 2.1 to 2.4 µm had great value for geologic mapping and mineral exploration (Chapter 13). Band 7 (SWIR2) was added to acquire these data. The original system for numbering TM bands remained the same, however, which explains why band 7 is out of sequence on a spectral basis. Table 3-2 lists characteristics of the bands for the second generation of Landsat satellites.

ENHANCED THEMATIC MAPPER

Landsat 7 was launched on April 15, 1999 with the Enhanced Thematic Mapper Plus (ETM+) sensor that added panchromatic band 8. This panchromatic image records a broad spectral band in the VNIR wavelengths (0.52 to 0.90 µm) with 15 m spatial resolution (Table 3-2). It is used to sharpen the multispectral bands with 30 m spatial resolution to 15 m resolution, as described in Chapter 9. Landsat 7 suffered a partial instrument failure in 2003, resulting in images with systematic scan line dropouts. The ETM+ scene footprints use the same WRS-2 path/row system as Landsat TM 4 and 5.

THIRD GENERATION OLI AND TIRS: LANDSAT 8

Landsat 8 was launched on February 11, 2013 with a new OLI system for visible, NIR, and SWIR images and the new TIRS that acquires two TIR bands. The TIRS system is discussed in Chapter 5. Table 3-1 lists the characteristics of the Landsat 8 satellite. Figure 3-5 shows the Landsat 8 satellite and the OLI system, which uses an along-track (push broom) scanner instead of the cross-track (whisk broom) scanner of the first two generations of Landsat. The along-track scanner provides a longer dwell time for each ground resolution cell, which is a significant advantage (Chapter 1). Figure 3-2 graphically compares the bandwidths of OLI detectors with the corresponding ETM+ detectors. Because of the increased dwell time, all of the OLI detectors can have narrower bandwidths, which means higher spectral resolution. The 12-bit radiometric scale of Landsat 8 eliminates saturation of bright targets, which improves the detection and mapping of land degradation and the characterization of snow and ice (Loveland and Irons, 2016). Landsat 8 acquires up to 740 images per day (usgs.gov/land-resources/nli/landsat/landsat-acquisitions).

Table 3-2 lists the bandwidths and a brief description of the nine OLI bands. Bands 1 and 9 are new and record spectral bands that were not recorded by earlier Landsat satellites. Band 8 is a narrow bandwidth version of band 8 on ETM+. These new OLI bands are described below.

OLI BAND 1

This new OLI coastal aerosol band records the shorter wavelength portion of the blue band (0.43 to 0.45 µm) that is not recorded by band 1 of TM or ETM+. Chlorophyll, colored dissolved organic material, and suspended materials

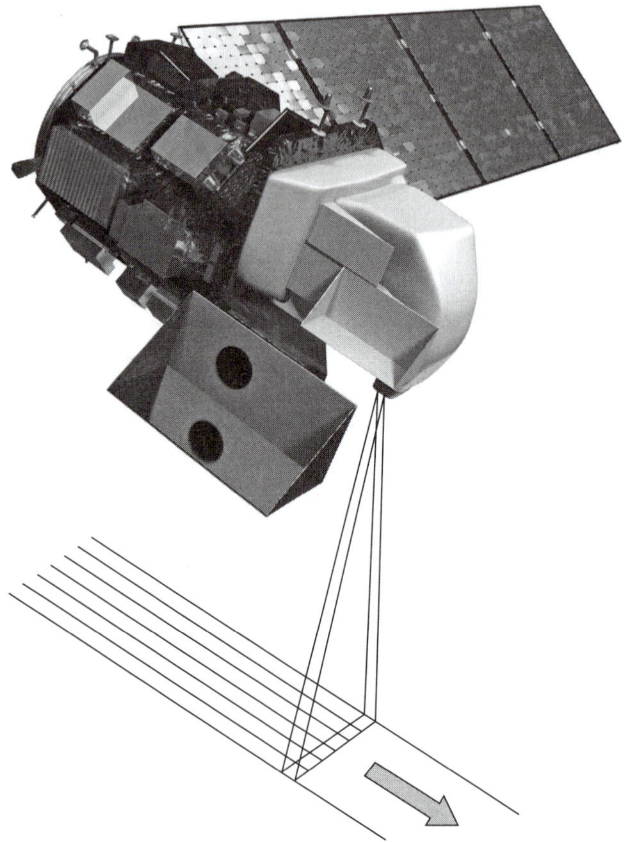

FIGURE 3-5 Landsat 8 showing OLI and its along-track image swath. Courtesy NASA.

Table 3-2 Spectral bands and characteristics for three generations of Landsat.

Landsat 1, 2, and 3 (MSS)	Landsat 4, 5, and 7 (TM and ETM+)	Landsat 8 (OLI and TIRS)	
Band (wavelength)	**Band (wavelength)**	**Band (wavelength)**	**Characteristics of Bands**
		1 (0.43 to 0.45 μm)	Light blue. Coastal/aerosol band for imaging shallow water and tracking fine particles like dust and smoke. Distinguishes colors in water.
	1 (0.44 to 0.51 μm)	2 (0.45 to 0.51 μm)	Blue-green. Maximum penetration of water. Distinguishes soil from vegetation and deciduous from coniferous plants. Absorbed by chlorophyll during photosynthesis.
4 (0.50 to 0.60 μm)	2 (0.52 to 0.60 μm)	3 (0.53 to 0.59 μm)	Green. Matches green reflectance peak of vegetation, which is useful for assessing plant vigor.
5 (0.60 to 0.70 μm)	3 (0.63 to 0.69 μm)	4 (0.64 to 0.67 μm)	Red. Absorbed by chlorophyll during photosynthesis. Discriminates vegetation types. TM band ratio 3/1 distinguishes secondary iron minerals.
6 (0.70 to 0.80 μm)	4 (0.77 to 0.90 μm)	5 (0.85 to 0.88 μm)	NIR1. Determines biomass content. Used to map shorelines. Used with red band to generate many vegetation indices to map vigor and stress.
7 (0.80 to 1.10 μm)			NIR2. Determines biomass content. Used to map shorelines. Used with red band to generate many vegetation indices to map vigor and stress.
	5 (1.55 to 1.75 μm)	6 (1.57 to 1.65 μm)	SWIR1. Indicates moisture content of soil and vegetation. Penetrates thin clouds. Provides good contrast between vegetation types. Discriminates between clouds, snow, and ice.
	6 (10.31 to 12.36 μm)	10 (10.60 to 11.19 μm) 11 (11.50 to 12.51 μm)	TIR. Records radiant temperature.
	7 (2.06 to 2.35 μm)	7 (2.11 to 2.29 μm)	SWIR2. Matches an absorption band caused by hydroxyl ions in minerals. Ratio of bands 5/7 is used to map hydrothermally altered rocks associated with mineral deposits.
		9 (1.36 to 1.38 μm)	Cirrus band. Detects high-altitude clouds that may "contaminate" other bands.
	8 (0.52 to 0.90 μm)	8 (0.50 to 0.68 μm)	Panchromatic. In the second generation, band 8 was only carried on Landsat 7. The 15 m pixels can be convolved with 30 m pixels of VNIR-SWIR bands to improve apparent spatial resolution.

in coastal and inland waters can the measured more accurately using band 1 of Landsat 8 (Concha and Schott, 2015; Loveland and Irons, 2016). In addition, the band is used to estimate the concentration of aerosols in the atmosphere and to track fine particles such as dust and smoke.

OLI Band 8

This panchromatic band records 15 m pixels whereas the other OLI multispectral bands record 30 m pixels as shown in Figure 3-6. The panchromatic band can be used to sharpen the coarser multispectral bands (Chapter 9).

OLI Band 9

New OLI band 9 is a very narrow spectral band positioned at the edge of an atmospheric absorption band shown in Figure 3-2. It records thin clouds that may not be detectable on image bands in the visible region, which is not a concern for most image interpreters. However, reflectance from undetected thin clouds degrades computerized statistical analyses and comparison of images acquired on different dates (*multi-date images*). Band 9 improves the accuracy of interpreting and mapping multi-date images.

A. OLI multispectral band 6 (SWIR1) with 30 m pixels.

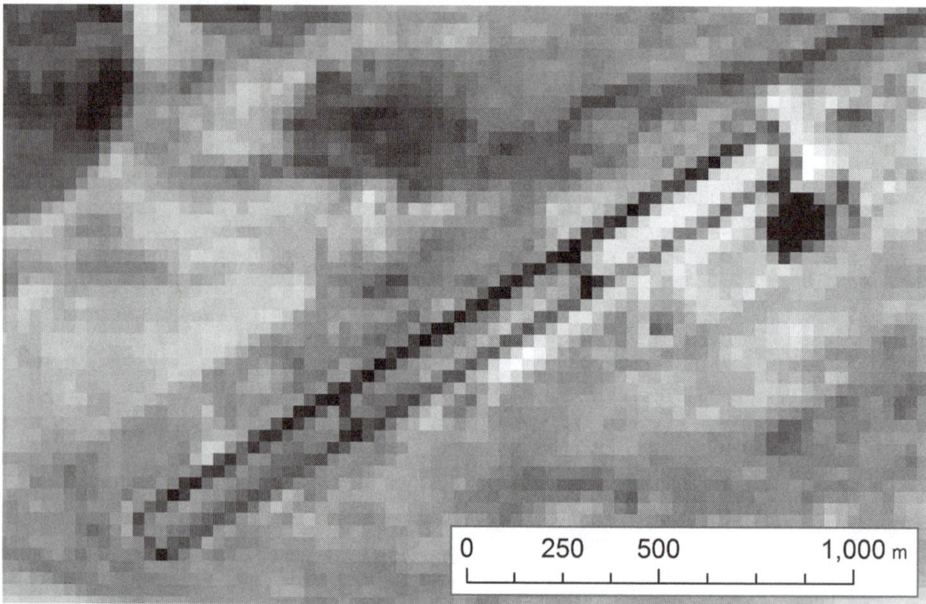

B. OLI panchromatic band 8 with 15 m pixels.

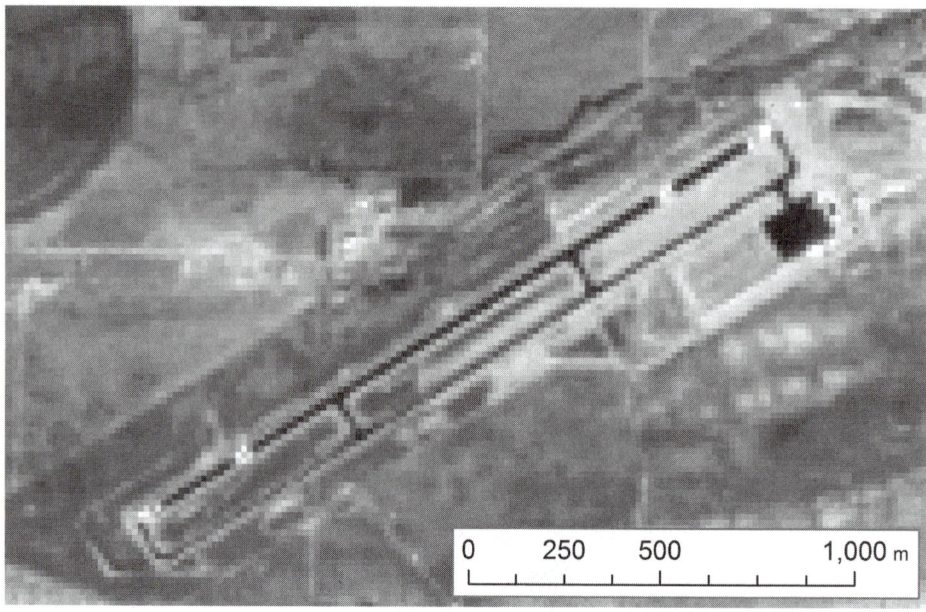

FIGURE 3-6 Comparison of the spatial resolution of Landsat OLI 30 m and 15 m pixels. Landsat courtesy USGS.

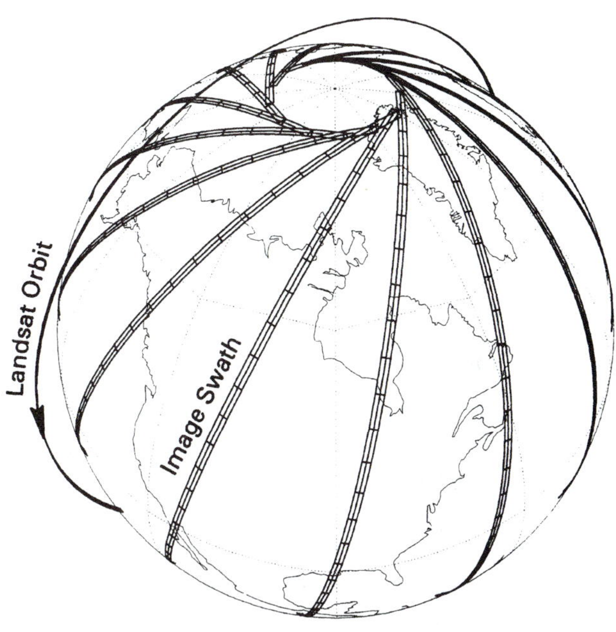

Figure 3-7 Landsat TM sun-synchronous orbit (solid circle) and image swaths (patterned lines) generated in one day. Courtesy D. N. Boosalis, Analytical Graphics, Inc., King of Prussia, Pennsylvania. Diagram plotted with Satellite Tool Kit software.

ORBIT PATTERNS

In order to obtain images of the entire Earth, the Landsat satellites are placed in *sun-synchronous orbits*. Figure 3-7 shows the circular orbit of the second- and third-generation Landsat satellites (solid line) at an altitude of 705 km. The orbit is southbound over the daylight hemisphere of the Earth when images are acquired. The circular orbit is fixed and the Earth rotates beneath it. Every 24 hours, 14.5 image swaths are acquired. The map in Figure 3-8 shows the southbound, daylight portion of the 14.5 image swaths (185 km wide) acquired during a 24-hour period. The northbound segment of each orbit covers the dark hemisphere. Polar areas at latitudes greater than 81° are the only regions not covered. Every 24 hours the Earth's rotation shifts the image swaths westward. After 16 days the Earth has been covered by 233 adjacent, sidelapping image swaths and the cycle begins again. This 16-day interval is called the *repeat cycle*. The sun-synchronous orbit pattern causes the corresponding orbits in each repeat cycle to occur at the same time. For example, every 16 days a southbound Landsat crosses Los Angeles at approximately 10:00 AM local sun time. The midmorning schedule results in intermediate to low sun elevations, which cause highlights and shadows that enhance subtle topographic features. Sun azimuth is from the southeast for images in the Northern Hemisphere and from the northeast in the Southern Hemisphere. Table 3-3 lists the orbital characteristics of the three generations of Landsat satellites.

The interactive USGS Landsat Acquisition Tool (landsat.usgs.gov/landsat_acq) allows users to see the daytime and/or nighttime paths of both the Landsat 7 and Landsat 8 satellites on any given day (a simplified version is shown in Figure 3-8). The tool also allows users to input a Landsat path/row or longitude/latitude data set in order to illustrate the next planned acquisition date for the satellite for that location.

Since the launch of Landsat 1 in 1972, the repeat cycles have imaged much of the Earth many times, which provides several advantages:

1. Areas with persistent clouds, such as rain forest terrain, may be imaged on the rare cloud-free days.
2. Images may be acquired at the optimum season for interpretation, as described later.
3. The repeated images record changes that have occurred since 1972, such as urbanization, deforestation, and desertification.

A network of ground stations in 14 countries download and distribute Landsat data. Ground stations in South Dakota and Alaska, as well as international ground stations in

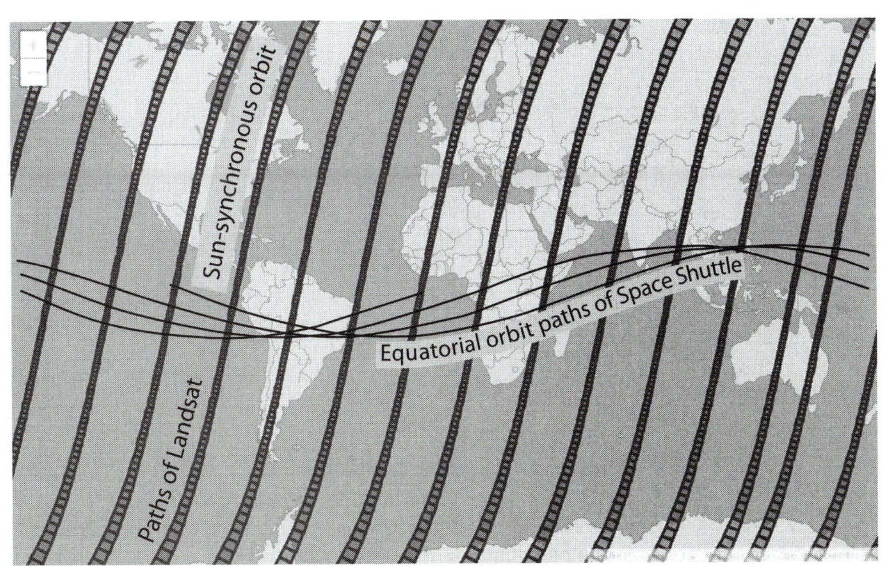

Figure 3-8 Orbit map for second and third Landsat generations. Map shows the 14.5 southbound, daytime image swaths (patterned lines) acquired during a single day. Each day the Earth's rotation shifts the pattern westward; after 16 days the Earth is covered and the cycle is repeated. For comparison the solid curves are equatorial orbits of a typical Space Shuttle mission and the International Space Station. Courtesy USGS. Modified from the Landsat Acquisition Tool (landsat.usgs.gov/landsat_acq).

TABLE 3-3 Orbit patterns and imaging systems of three Landsat generations.

	Generation 1 Landsat 1, 2, and 3	Generation 2 Landsat 4, 5, and 7	Generation 3 Landsat 8
Altitude	918 km	705 km	705 km
Orbits per day	14.0	14.5	14.5
Number of orbits (paths) per cycle	251	233	233
Repeat cycle	18 days	16 days	16 days
Image sidelap at equator	14.0%	7.6%	7.6%
Crosses 40°N latitude at approximate local sun time	9:30 AM	10:30 AM	10:11 AM
Operational from	1972 to 1983	1982 to present	2013 to present
Imaging system	MSS	TM or ETM+	OLI and TIRS
Radiometric resolution	6 bit	8 bit	12 bit
Multispectral spatial resolution	57 m	30 m	30 m

Norway, Germany, and Australia, serve as the primary data capture and telemetry, tracking, and control (TT&C) facilities for the USGS Landsat satellite missions.

PATH-AND-ROW INDEX MAPS

Landsat images are referenced to worldwide path-and-row index maps called the Worldwide Reference System (WRS). The first generation of Landsat used WRS-1. The second and third generations use WRS-2. Figure 3-9 shows a portion of the WRS-2 index map. The *paths* are the southbound segments of orbit paths. The paths are intersected by east–west, parallel *rows* spaced at intervals of 165 km. In Figure 3-9 the intersections are marked with circles, which are the centerpoints of the images. A TM image of Los Angeles, shown by the outlined A on the index map, has a centerpoint at path 041/row 036. All images of Los Angeles acquired on repeat cycles have the same centerpoint. Metadata for each scene includes cloud cover percent, image quality, level of processing, corner coordinates, sun elevation, and sun azimuth.

LANDSAT ARCHIVES

The long-term archive at the USGS Earth Resources Observation and Science (EROS) Center in Sioux Falls, South Dakota, is one of the largest civilian remote sensing data archives. It contains images with a comprehensive record of the past four decades of the Earth's changing land surface. Scientists from around the world depend on this archive to conduct research on changes that affect our environment, resources, health, and safety. Time-series images are a valuable resource for scientists, disaster managers, engineers, educators, and the general public. The EROS Center is a leader in preserving land remote sensing imagery and has archived, managed, and preserved land remote sensing data for more than 35 years. Landsat data (along with aerial photographs, other satellite images, DEMs, and others) also can be searched for and downloaded from the USGS EarthExplorer (earthexplorer.usgs.gov) and the Global Visualization Viewer (GloVis) (glovis.usgs.gov) websites.

During summer 2016, the USGS reorganized the extensive Landsat archive into a formal tiered data collection structure to ensure it will provide a stable environmental record (usgs.gov/land-resources/nli/landsat/landsat-collections).

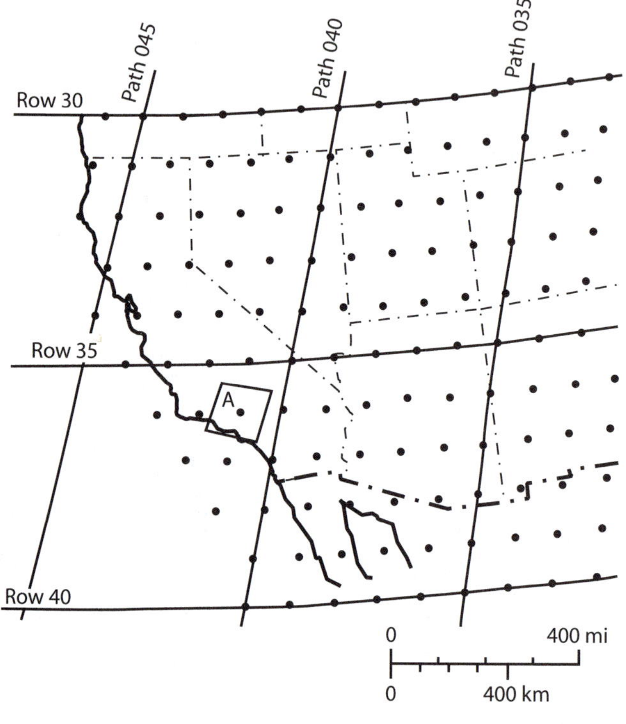

FIGURE 3-9 Path-and-row WRS-2 index map for Landsat TM and OLI images of the southwestern United States. The Landsat footprint for path 041/row 036 is shown at A.

Table 3-4 Evaluation of common color combinations for Landsat TM and OLI images.

Type of Image	TM Display Colors (B-G-R)	OLI Display Colors (B-G-R)	Advantages	Disadvantages
Natural color image	1-2-3	2-3-4	Optimum for mapping shallow bathymetric features. Easy to understand—viewers are familiar with colors.	Lower spatial resolution due to band 1. Limited spectral diversity because no reflected IR bands are used.
Color IR image	2-3-4	3-4-5	Moderate spatial resolution. Emphasizes vegetation (red signature) vigor and patterns.	Limited spectral diversity.
Total reflected IR image	4-5-7	5-6-7	Optimum for humid regions. Maximum spatial resolution. Displays spectral information not visible to the eye.	Limited spectral diversity because no visible bands are used. Can be difficult to understand.
Enhanced color image	2-4-7	3-5-7	Optimum for temperate to arid regions. Maximum spectral diversity. Healthy vegetation has shades of green.	Unfamiliar color display, but interpreters quickly adapt.

Landsat Level-1 data products are available for immediate download and have pixel brightness (the digital number or DN) in radiance values. Landsat Level-2 science products are processed within 2 to 5 days and have pixel brightness atmospherically corrected to reflectance values, enabling more sophisticated image processing and direct correlation of images acquired at different times. Landsat Level-3 science products represent biophysical properties of the Earth's surface, such as extent of water, snow cover, and burned area.

LANDSAT COLOR IMAGES

The first Landsat generation acquired only four bands; three of which were combined into color IR images with relatively coarse spatial resolution. The TM and ETM+ systems of the second generation acquired six bands in the VNIR and SWIR regions with 30 m resolution. The OLI system of the third generation acquires the six VNIR-SWIR bands of TM, plus four additional bands that were described earlier. The following discussion is based on the six bands common between the OLI and TM/ETM+ systems.

Any three of the six visible and reflected IR bands may be combined and illuminated with blue, green, and red light to generate a color image. There are 120 possible color combinations, which is an excessive number for practical use. Theory and experience, however, show that a small number of color combinations are suitable for most applications. Table 3-4 lists the four color combinations that our experience has found to be most useful. The optimum color combination is determined by the terrain, climate, and nature of the interpretation project. Users are encouraged to experiment with other combinations.

A new Spectral Characteristics Viewer (Figure 3-10) is available online as part of the USGS Landsat missions program (landsat.usgs.gov/spectral-characteristics-viewer). This interactive tool helps users determine which bands

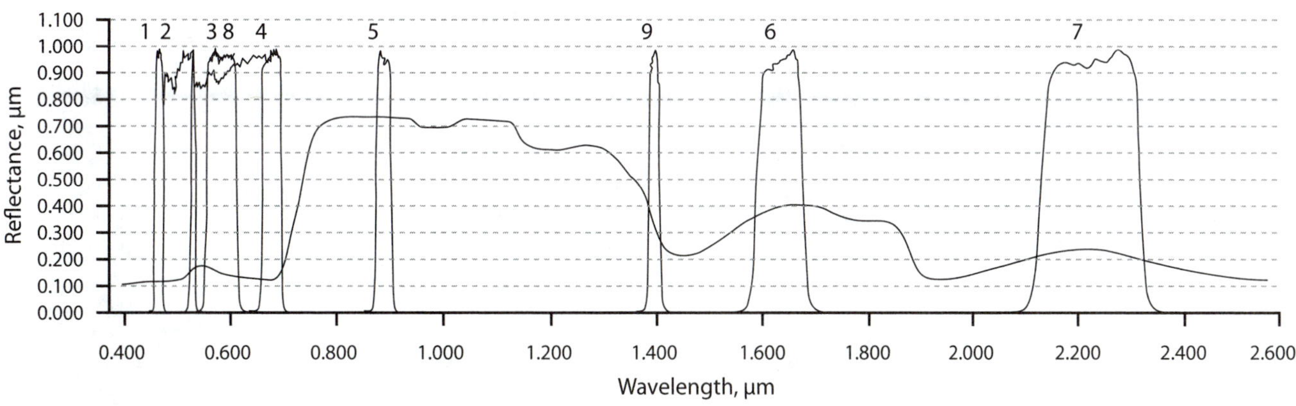

Figure 3-10 Interactive Spectral Characteristics Viewer displaying Landsat 8 OLI bands and the spectral signature for lawn grass. Courtesy USGS.

provide the most spectral information for the feature being mapped. The viewer has two components:

1. A library of spectral curves for materials, including nine minerals, nine vegetation types, and four water classes.
2. Band passes of Landsat MSS, TM, ETM+, and OLI (along with other multispectral sensors discussed in Chapter 4) that are superposed onto the spectral curve for a material to indicate the optimum color combination for that material.

A. Landsat OLI band 2 blue (0.45 to 0.51 μm).

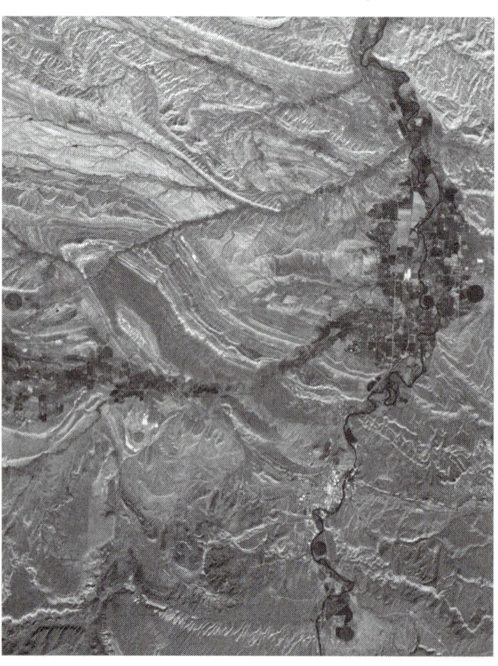

B. Landsat OLI band 3 green (0.53 to 0.59 μm).

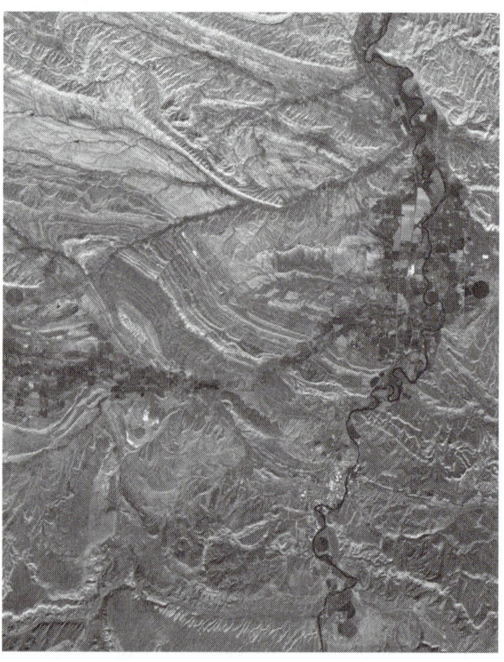

C. Landsat OLI band 4 red (0.64 to 0.67 μm).

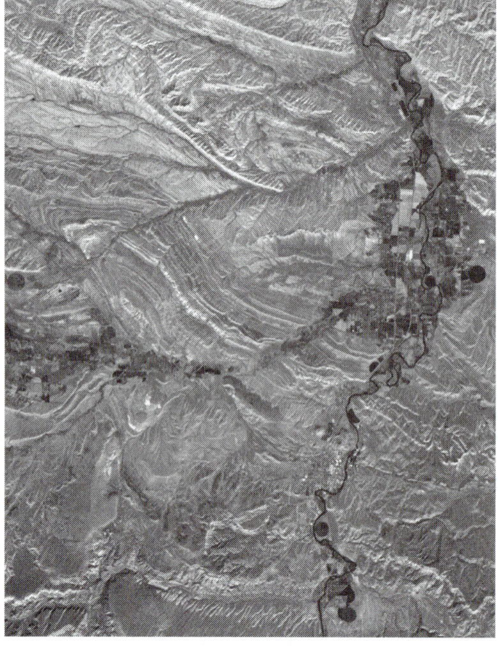

D. Landsat OLI band 5 NIR (0.85 to 0.88 μm).

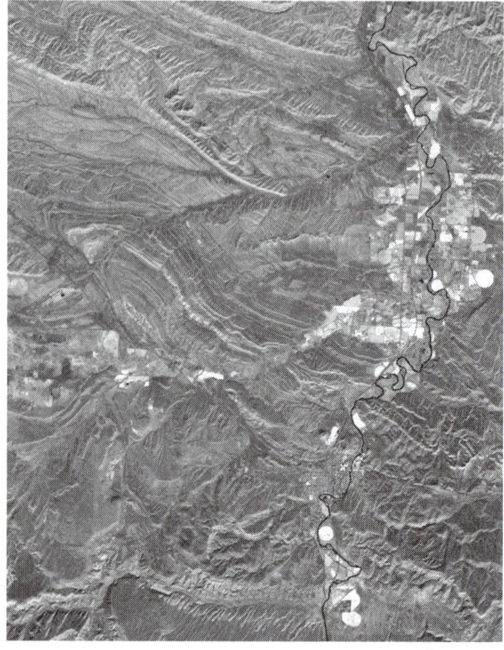

FIGURE 3-11 Landsat OLI bands for the Thermopolis, Wyoming, subscene. Landsat courtesy USGS.

Images of Semiarid Terrain: Thermopolis, Wyoming

Figure 3-11A–F shows six VNIR-SWIR bands and one TIR band acquired on September 21, 2015 by the Landsat 8 system for the Thermopolis subscene in central Wyoming. The subscene is located in the south flank of the Bighorn Basin and includes a stretch of the Wind River and the town of Thermopolis (Figure 3-11H). Most of the area is used for ranching. Some irrigated crops are grown in the stream valleys. The Gebo anticline and Little Sand Draw oil fields occur in the northern part of the area. Figure 3-11G is TIR band 10 (100 m spatial resolution) with bright signatures that show warmer radiant temperatures.

E. Landsat OLI band 6 SWIR1 (1.57 to 1.65 μm).

F. Landsat OLI band 7 SWIR2 (2.11 to 2.29 μm).

G. Landsat OLI band 10 TIR (10.6 to 11.19 μm).

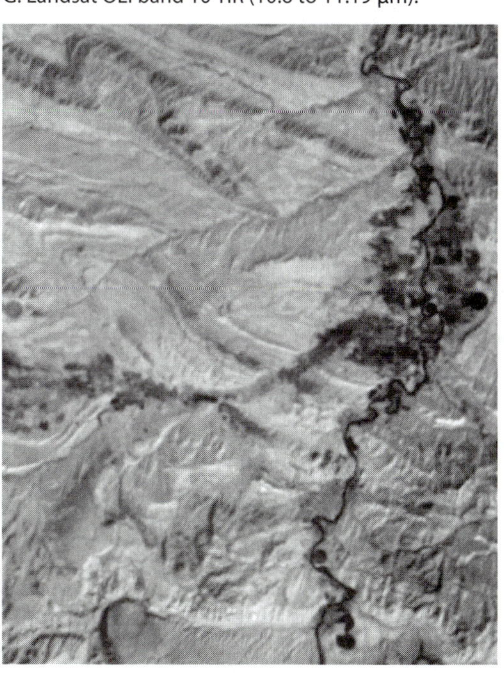

H. Interpretation map. Stippled areas are outcrops of Chugwater red beds.

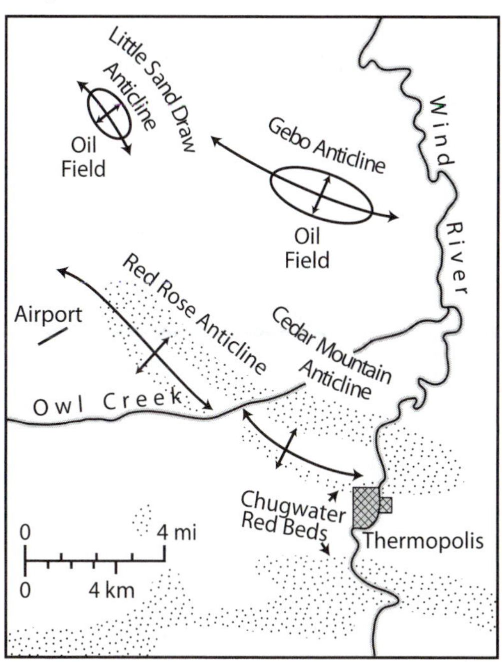

OLI Color Combinations

Plate 7 shows four OLI color combinations for the Thermopolis subscene that are listed and compared in Table 3-4. Vegetation has a red signature in the color IR image because band 5, which records the strong vegetation reflectance in the NIR band, is shown in red. Red beds of the Chugwater Formation (shown by the stippled pattern in the map of Figure 3-11H) have an orange signature in the natural color image (Plate 7A). In the color IR image (Plate 7B), Chugwater outcrops have a distinctive yellow signature that is typical for red rocks throughout the world when viewed with the color IR combination. The color IR image has a better contrast ratio and spatial resolution than the natural color image because the blue band was not used in Plate 7B. Plate 7C displays only the reflected IR (NIR and SWIR) bands 5-6-7 as B-G-R and has the best spatial resolution of all the combinations. There is little color contrast, however, between the different rock outcrops, which are monotonous shades of pale blue. Even the red beds of the Chugwater Formation are light blue and are indistinguishable from the other outcrops.

In Plate 7D the visible green band 3 is shown in blue, the NIR band 5 is shown in green, and the SWIR2 band 7 is shown in red. This enhanced color combination provides the maximum range of color signatures for the rock outcrops and is optimum for interpreting geology in this semiarid area. Our experience in other arid and semiarid regions throughout the world confirms that the OLI 3-5-7 enhanced color combination is optimum. Vegetation is green in Plate 7D because band 5 is shown in green. Some investigators prefer a 2-5-7 version of this combination, but we find 3-5-7 to be optimum because band 3 has less atmospheric scattering than band 2.

Image Interpretation

Figure 3-11H is a generalized geologic sketch of the enhanced color image (Plate 7D). (For terminology, nongeologists may refer to Appendix A, Basic Geology for Remote Sensing.) Regional dip of the beds is northward. Local reversals of dip toward the south form four anticlines. In the southern portion of the subscene the Red Rose and Cedar Mountain anticlines are deeply eroded and are outlined by outcrops of the Chugwater Formation, shown by the stippled pattern. In the north the Gebo and Little Sand Draw anticlines are less deeply eroded and are oil fields that produce from the Phosphoria Formation. The cross section (Figure 3-12) shows these structural relationships. The oil fields were discovered as a result of surface mapping long before the launch of Landsat. By studying images of such known oil fields we learn to recognize similar, but undrilled, structures on images of less well-explored regions of the world. Chapter 13 illustrates oil discoveries that were based on several types of remote sensing images.

IMAGES OF TROPICAL TERRAIN: WEST PAPUA, INDONESIA

Figures 3-13A and 3-13B show TM bands 1 (blue) and 5 (SWIR) of the Mapia subscene in the south-central portion of West Papua in Indonesia. Cloud-free images are notoriously rare in rain forest terrain such as this. Sabins has done field work in the Indonesian islands of Java and Borneo; he can attest to the persistent clouds and rainfall. The poor spatial resolution of the blue image (Figure 3-13A) is caused by the strong atmospheric scattering of visible wavelengths. Resolution is improved at the SWIR wavelengths of Figure 3-13B.

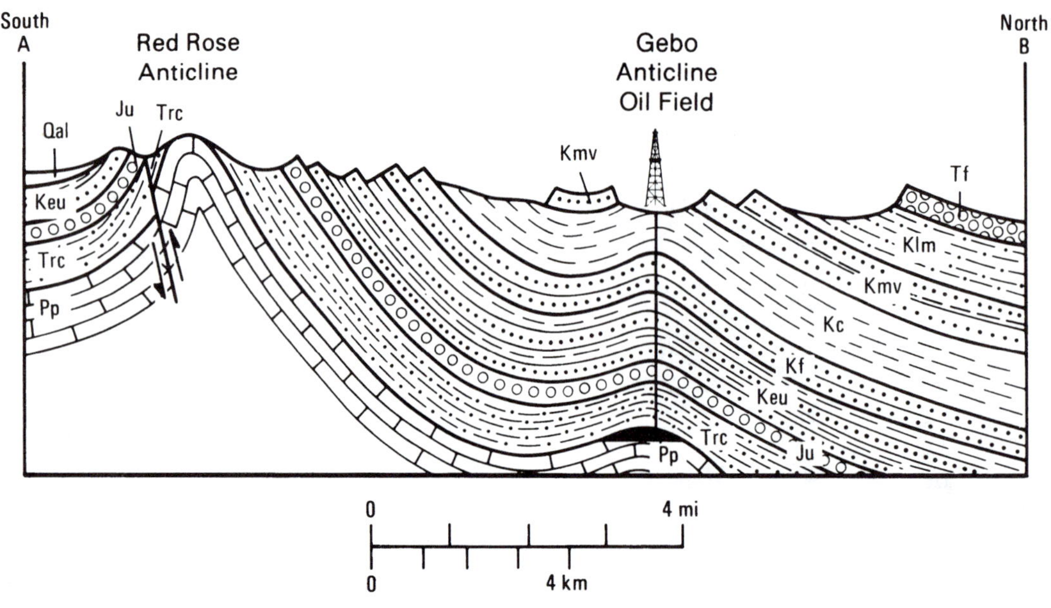

FIGURE 3-12 Cross section for Thermopolis map in Figure 3-11H.

The two south-trending white streaks in the visible blue band 1 (Figure 3-13A) are high-altitude thin clouds that cast no shadows. The cloud streaks are penetrated by the longer wavelengths of reflected IR energy and are absent in the SWIR band 5 (Figure 3-13B). The dark gray signatures on the geologic map (Figure 3-13C) are low-elevation rain clouds and shadows that are present at all wavelengths. Cross section A–B in Figure 3-14 shows the Makamo and Mapia anticlines with outcrops of the New Guinea Limestone, and karst topography, at their crests.

A. Band 1 blue (0.44 to 0.51 µm).

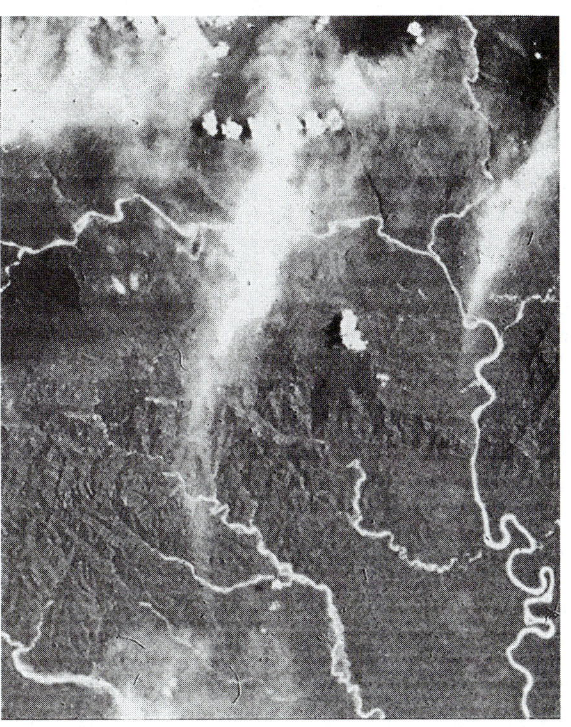

B. Band 5 SWIR (1.55 to 1.75 µm).

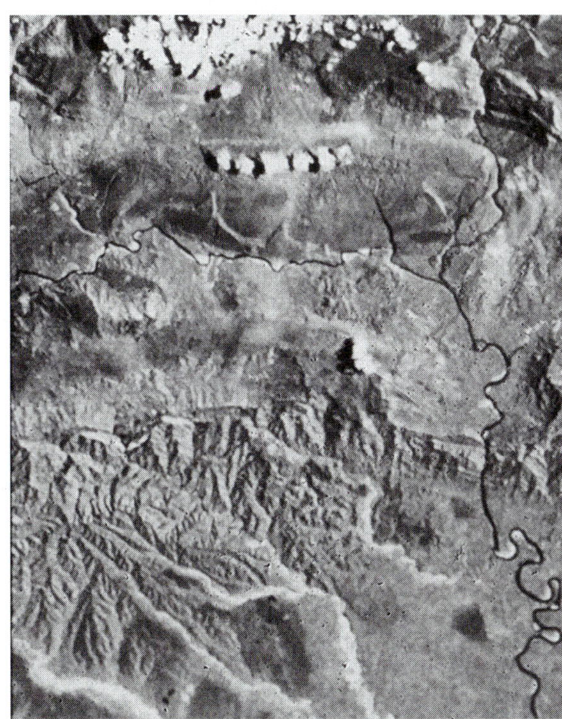

C. Geologic map.

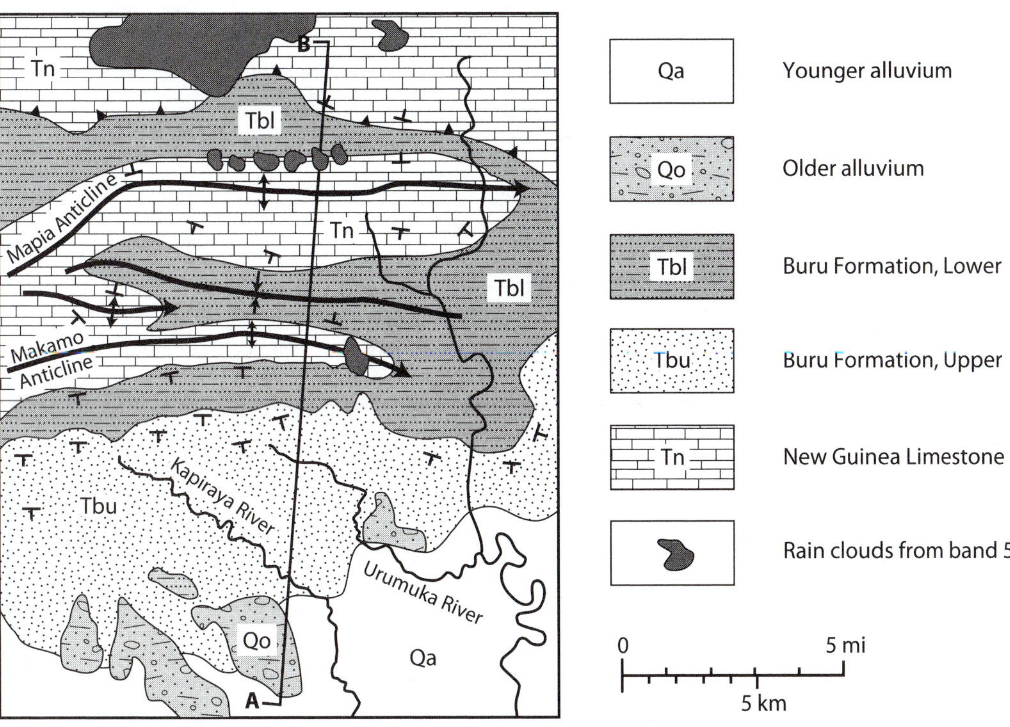

Figure 3-13 Landsat TM bands and map for the Mapia subscene, West Papua, Indonesia. Landsat courtesy USGS.

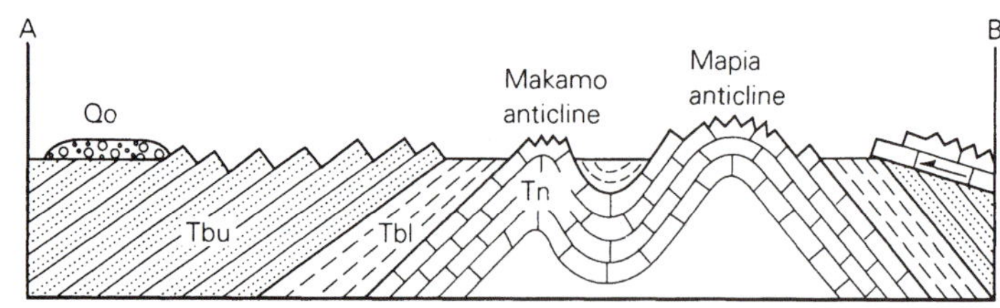

FIGURE 3-14 Cross section for the Mapia subscene. See Figure 3-13C.

TM Color Combinations

Plate 8 shows color images of Mapia in the same four band combinations as the Thermopolis color images in Plate 7. One objective is to demonstrate that different environments require different color combinations. In rain forest environments, images in the visible spectral region have lower contrast and lower spatial resolution because the high moisture content of the atmosphere strongly scatters these short wavelengths. The NIR band 4 and SWIR bands 5 and 7 have better contrast and resolution because these longer wavelengths are less susceptible to atmospheric scattering. Thanks to our computer image enhancement (Chapter 9), the natural color image (Plate 8A) has better color contrast than we anticipated, based on the appearance of the three original visible bands. Clouds obscure portions of the image, and terrain features are difficult to discern. The color IR image of bands 2-3-4 in B-G-R (Plate 8B) has improved contrast and detail because band 4 (NIR) replaces the low-contrast band 1 image (visible blue). The clouds are greatly diminished. The red signature indicates the extensive forest cover of the region. Plate 8C is compiled from NIR band 4 and SWIR bands 5 and 7. For tropical regions this combination provides optimum spatial resolution, color contrast, and reduction of clouds. In Plate 8C vegetation has a range of color signatures, which aids interpretation, whereas in Plate 8A vegetation is a saturated dark green. Vegetation patterns are more distinct in Plate 8C because bands 4, 5, and 7 coincide with major peaks in the reflectance spectrum of vegetation (Figure 3-2). Plate 8D consists of bands 2, 4, and 7 and is the second best of the four images. This TM 2-4-7 in B-G-R combination is optimum for the Thermopolis area, but in West Papua the vegetation cover and humid conditions reduce its effectiveness.

Image Interpretation

Figure 3-13C shows the geologic interpretation for the 4-5-7 TM color image of the Mapia subscene. Table 3-5 lists the formations together with their ages, lithology, and signatures on the TM 4-5-7 image. Figure 3-14 is a geologic cross section along line A–B in Figure 3-13C. Despite the vegetation cover, the different formations are mappable on the image because each unit erodes to a distinctive topographic pattern. The resistant sandstones of the upper member of the Buru Formation erode to form the rugged ledge-and-slope topography in the south portion of the image. The nonresistant shale of the lower member of the Buru Formation forms broad featureless strike valleys. The New Guinea Limestone crops out in the crest of the Mapia and Makamo anticlines where it forms broad arches. In this humid environment, solution and collapse of the limestone produce a distinctive terrain of closely spaced pits and pinnacles called *karst* topography. Karst topography is well developed on the New Guinea Limestone on the crest of Mapia and Makamo anticlines. At the small scale of the images, however, the karst pattern is somewhat difficult to recognize.

Geologic structure is interpreted for the Mapia subscene in the same manner as the Thermopolis image; however, there are significant differences in solar illumination for the two images. The differences are:

	Mapia Subscene	Thermopolis Subscene
Sun azimuth	From ENE	From SE
Sun elevation	45°	25°

In the Mapia area, shadows and highlights are subdued because the sun has a high elevation and an azimuth nearly parallel with the east–west regional strike. Despite these disadvantages, geologic structures can be interpreted. This tropical rain forest terrain provides a contrast with the semiarid rangeland of the Thermopolis image.

TABLE 3-5 Formations in the Mapia subscene.

Formation	Age	Lithology	Image Signature
Younger alluvium (Qa)	Recent	Soil and gravel	Valleys along major drainages
Older alluvium (Qo)	Recent	Gravel deposits	Eroded terraces and alluvial fans
Buru Formation, Lower (Tbl)	Early Tertiary	Shale	Strike valleys
Buru Formation, Upper (Tbu)	Early Tertiary	Sandstone and minor shale beds	Ridges and dipslopes
New Guinea Limestone (Tn)	Early Tertiary	Limestone	Karst topography

Landsat Mosaics

The broad regional coverage of individual Landsat images can be extended by combining adjacent images into a mosaic. The east–west sidelap of adjacent orbit swaths and the north–south forward overlap of consecutive images greatly facilitate compilation of mosaics. The uniform scale and minimal distortion of Landsat images also make mosaic compilation easier. In the early days of Landsat, mosaics were compiled from prints of adjacent images that were manually trimmed and assembled. Today mosaics are compiled digitally.

APPLICATIONS OF LANDSAT

Landsat images have a wide range of applications, a few of which are described below.

Snow and Ice Mapping: Northeast Afghanistan

Landsat images are used to map the geographic extent of snow and ice. Time sequences of images are used to calculate changes in glacier size. Time sequences are also used to determine the velocity of glaciers by measuring the downslope movement of features on the ice surface, such as crevasses, large boulders, and ice bands (Winsvold and Kaab, 2016). The VNIR and SWIR bands of Landsat (Table 3-2) record spectral characteristics that distinguish snow, ice, and debris-covered ice from clouds and surrounding unglaciated terrain (Dozier, 1989; Paul, 2000; Smith and others, 2015).

Landsat 7 acquired images of cirque glaciers in northeastern Afghanistan on July 30, 2002 (Figure 3-15A). The terrain is above the tree line and ranges between 3,000 to 5,000 m above sea level. Landsat ETM+ band 2 displays glacier ice and snow in white and rock deposits on the downslope tongues of ice with medium gray tones. The surrounding unglaciated terrain has darker gray tones.

Glacier snow and ice, and downslope rock deposits, can be mapped with spectral classification of the Landsat bands (Figure 3-15B). The classification map of the two landforms was converted to polygons so that it could be loaded into a GIS and symbolized with white features representing glacier ice and patterned features showing glacier deposits (these techniques are covered in Chapters 9 and 10). The area covered with glacial ice and snow, and downslope deposits, can be calculated in the GIS.

A. Landsat ETM+ band 2 (reflected green light).

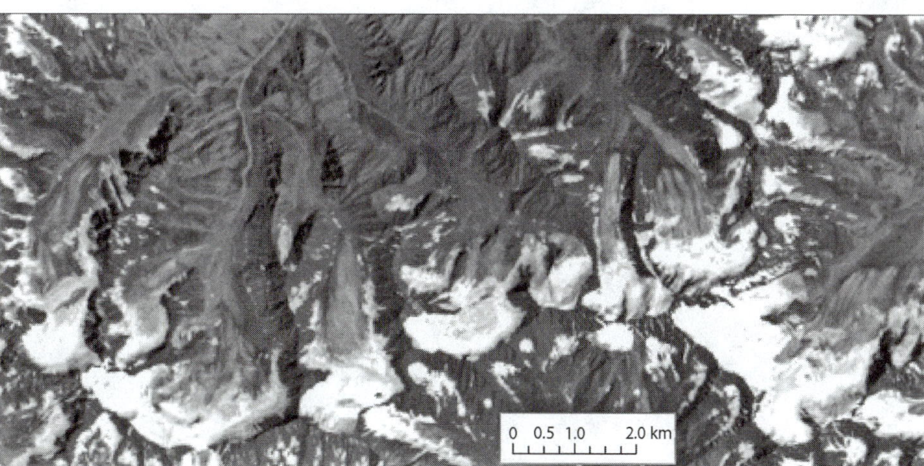

B. Computer-classification map of glacier ice and snow and downslope deposits.

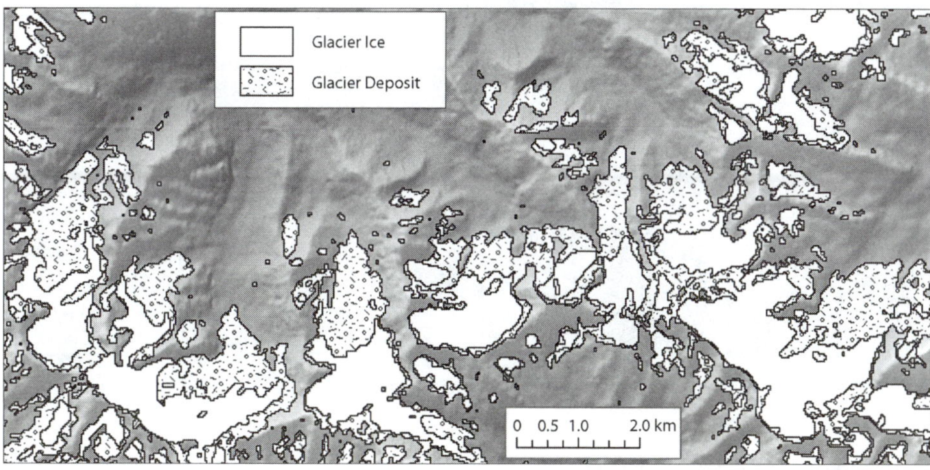

Figure 3-15 Landsat image and interpretation map of cirque glaciers, northeast Afghanistan. Landsat courtesy USGS.

Bathymetry: Exumas Islands, The Bahamas

Bathymetric mapping uses the three visible light bands of Landsat with the blue band, providing maximum penetration of the water column (Figure 2-24). Landsat TM bands 1, 2, and 3 are processed to map and monitor shallow water environments that support coral reefs, seagrass, other submerged aquatic vegetation, fisheries, and development while the infrared bands accurately delineate the land-water boundary (Figure 2-25).

Light-colored to white carbonate sands characterize the shallow waters throughout the

Exumas Islands of The Bahamas (Harris and Ellis, 2009). Color Landsat imagery (Plate 9A) displays the clear water with shades of blue. Beaches and crests of shallow sand bars are light blue to white. Deeper water is dark blue. The clear water and relatively consistent, light seafloor color means that the digital numbers of each offshore pixel in the Landsat TM band 1 (reflected blue light) is an approximation of water depth (Harris and Kowalik, 1994). The brightness value of 8-bit Landsat pixels varies between 0 (pure black) and 255 (pure white); the deeper the water, the less bright the pixel.

In Figure 3-16, the black shapes are islands. The bright arcuate features are shallow sand banks and the gray areas are deeper water. Navigation charts were superimposed on the satellite image in a GIS to determine water depth for select offshore pixels (Harris and Ellis, 2009). Based on this calibration, the 8-bit brightness values in the Landsat TM band 1 image were converted to water depth values in meters and a color-coded bathymetric map was generated (Plate 9B). Changes in the shape, size, and water depth of shallow water reefs, sand bars, and beaches can be monitored using images acquired at different times. More sophisticated processing and calibration techniques are available for bathymetric and habitat mapping using Landsat and other imagery (Purkis and Klemas, 2011; Purkis and Pasterkamp, 2004).

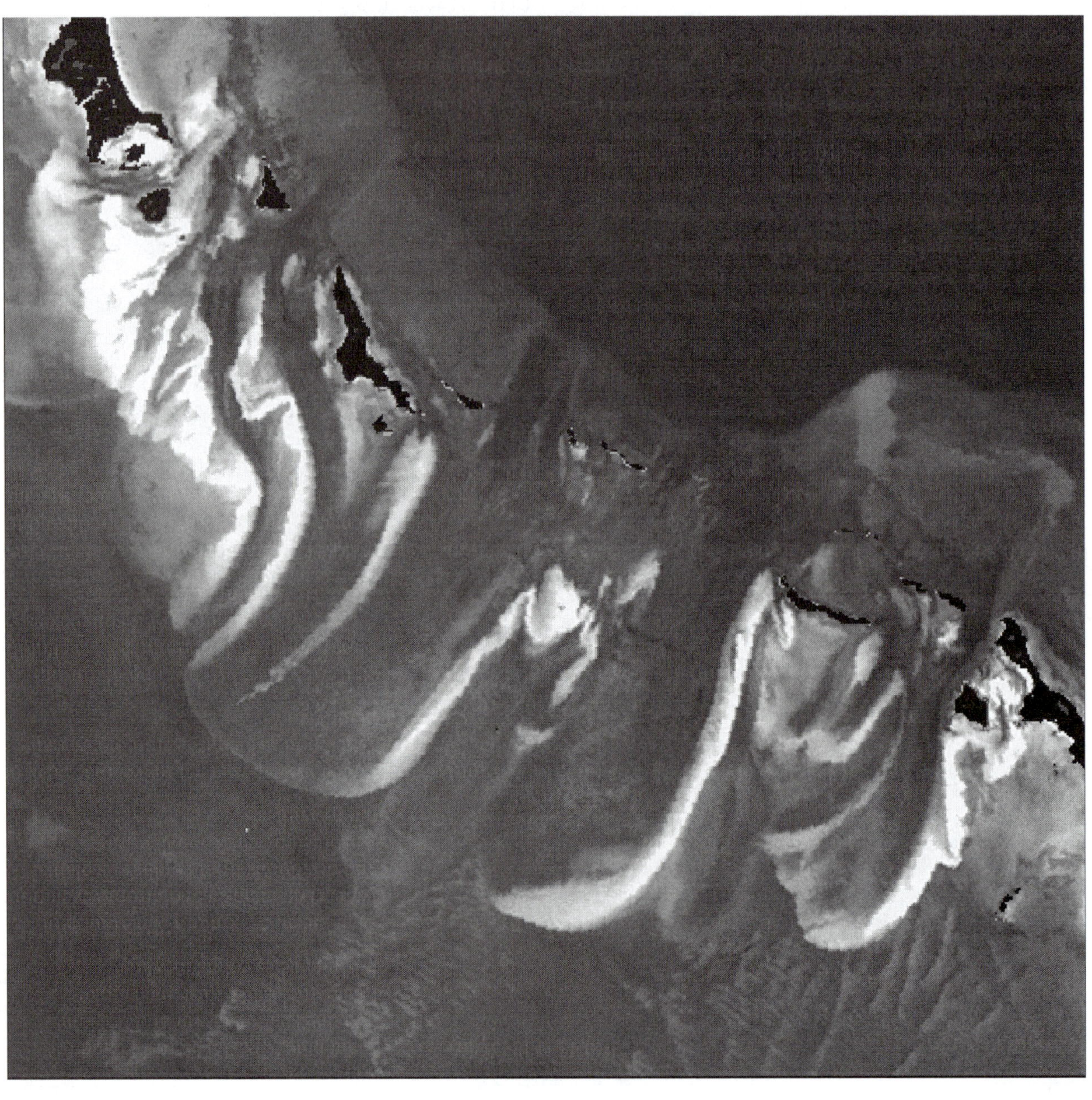

0 1 2 4 km

FIGURE 3-16 Landsat TM band 1 (reflected blue light) of carbonate sand bars submerged in shallow clear water, Exumas Islands, The Bahamas (Plate 9). Landsat courtesy USGS.

Deforestation: Cambodia

According to NASA Earth Observatory (2017a), Cambodia has one of the world's fastest rates of forest loss. Broad swaths of densely forested landscape have been clear-cut over the past decade—even in protected areas. Between 2001 and 2014 the annual rate of forest loss increased by 14.4%. In other words, the country lost a total of 1.44 million ha (5,560 mi^2).

Digital Image 3-1A 🌐 is a Landsat ETM+ natural color image, acquired on December 31, 2000, of the area near the border of Kampong Thom and Kampong Cham provinces. Except for a small developed area in the southwest corner, the dense forest is intact. **Digital Image 3-1B** 🌐 is a Landsat OLI natural color image of the area acquired on October 30, 2015. Over a 15-year period, much of the forest has been replaced by a rectangular pattern of large-scale rubber plantations. Large, formerly forested areas in the east are now a maze of small landholdings. Interestingly, in 2015, the southwestern area that appeared developed in 2000 is mostly covered by second-growth vegetation. It appears that the population has migrated to the now deforested areas to the north (NASA Earth Observatory, 2017a).

Water Quality: Lake Erie, United States

Early in July 2015, NOAA and University of Michigan scientists predicted that the 2015 season for harmful algal blooms would be severe in western Lake Erie (Lynch, 2015). On July 28, 2015, the OLI on Landsat 8 captured an image of extensive algal blooms in western Lake Erie (NASA Earth Observatory, 2015). The natural color image displays the bloom is visible as swirls of green across much of the water body (**Digital Image 3-2A** 🌐). The shoreline is difficult to discern in the natural color image as algae in the water and vegetation onshore are both displayed with shades of green. A Landsat color IR image (**Digital Image 3-2B** 🌐) sharply delineates the shoreline. The color IR image displays clear water as black and water with the algae bloom as dark blue to violet.

Algae in this basin thrive when there is an abundance of nutrients (mainly from agricultural runoff) and sunlight, as well as warm water temperatures. Harmful algal blooms can lead to fish kills. They can also render the water unsafe for recreation and for consumption (as was the case in Toledo, Ohio, and southeast Michigan during a 2014 bloom). The toxin microcystin is a component of the algae that has become an annual occurrence in the lake's western basin during the past decade. On July 28 the City of Toledo's Water Quality website reported "Microcystin has been detected in the intake crib 3 miles out on Lake Erie, but not in drinking water. . . . Our water is safe to drink" (Lynch, 2015).

Research confirmed that in 2011, phosphorus from farm runoff combined with favorable weather and lake conditions to produce a bloom three times larger than previously observed. The researchers noted that if land management practices and climate change trends continue, the lake is likely to see more blooms similar to the 2011 event (NASA Earth Observatory, 2015).

Mapping Wildfires: Chile

Wildfires continued to ravage Chile's countryside in early February 2017, weeks after they flared up in mid-January. The blazes thwarted firefighters' efforts to control them, with new hot spots emerging daily. Satellite data and scientific analysis suggest the fires were among the worst the country has seen in decades (NASA Earth Observatory, 2017b).

A Landsat 8 image was acquired on December 23, 2016 of the agricultural and forested terrain near Empedrado, Chile an inland rural community 15 km east of the Pacific Ocean (Plate 10A). One month later, on January 24, 2017, Landsat 8 acquired an image of a massive burn scar with active fires and smoke along the margins of the burned land (Plate 10B). Plate 10A and 10B combine reflected green, NIR, and SWIR2 (OLI bands 3-5-7 in B-G-R) to distinguish burned land (brown) from unburned and vegetated land (green). Active fires at the margins of burned land are red. Temperatures in Chile reached 40°C (104°F) and strong winds created optimum conditions for fires to spread (Watts, 2017).

Up-to-date imagery of active fires and burned terrain enables firefighters and emergency response teams to more effectively and safely deploy their resources.

Geology: Saharan Atlas Mountains, Algeria

Plate 11 is a Landsat TM enhanced color image in the Saharan Atlas Mountains, Algeria, where lithologic units and geologic structure are well exposed and suitable for demonstrating interpretation techniques. The two major steps in geologic interpretation are:

1. Define and map lithologic units, and
2. Map geologic structure.

These steps are described in the following sections.

Map Lithologic Units

The first step for geologic interpretation of the image is to identify the mappable lithologic units; layered units are called formations. The next task is to identify the vertical sequence of formations (or stratigraphic column in "geospeak"). Figure 3-17 is a stratigraphic column with the oldest rocks at the base and the youngest at the top. We identify each unit with a number, e.g., with 1 (oldest) at the base and 6 (youngest) at the top. Traditionally, the units are assigned geographic names, such as Dakota Sandstone or Monterey Shale, but in this example we will omit this tradition. The profile on the left margin of the column shows the topographic expression (ridges or slopes) of the units. Each unit is described on the right margin of the column, together with its color signature on the image. In the column each unit has a distinctive pattern that corresponds to patterns in the geologic map (Figure 3-18). Unit 6 consists of salt from a deeper layer that has penetrated upward through the strata as plugs, or diapirs. In the northwest portion of the image two salt diapirs have dark signatures in the Landsat image (Plate 11), which is anomalous because salt is white.

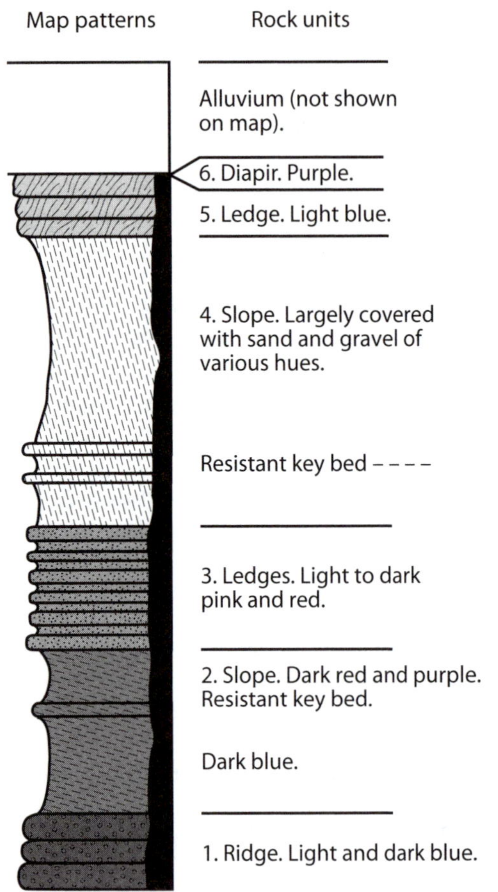

FIGURE 3-17 Stratigraphic column for interpreting the TM image (Plate 11) of the Saharan Atlas Mountains, Algeria.

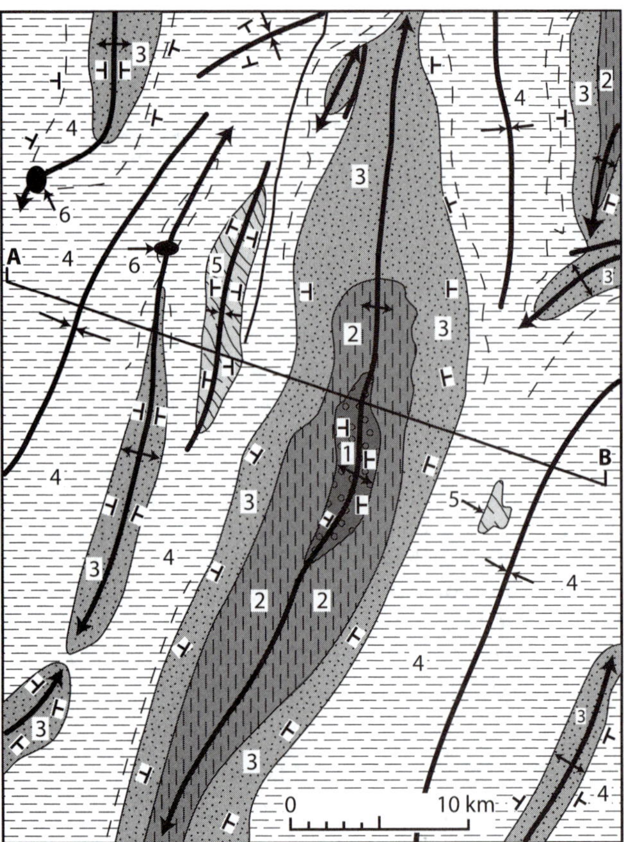

FIGURE 3-18 Geologic map of the TM image (Plate 11) of the Saharan Atlas Mountains, Algeria. Numbers and patterns are keyed to Figure 3-17.

The anomalous dark signature of the diapirs was explained by our Chevron colleague, Peter Verall (personal communication), who did field geology in Iran several decades ago. Similar salt diapirs with dark signatures crop out in Iran. As the diapirs migrate upward they carry blocks of the adjacent bedrock, called xenoliths. At the surface, the infrequent rains dissolve the salt and leave a mantle, or residue of xenoliths. With time the xenoliths become coated with dark desert varnish, which explains their dark signature on the image.

After mapping the lithologic units, the next step is to map geologic structure.

Determine Attitude of Beds (Dip and Strike)

The first step in mapping geologic structure is to determine attitude of the beds (dip and strike). Traditionally this task employs stereo images, such as overlapping aerial photographs that provide vertical exaggeration of the scene. An example is the stereo-pair of the Alkali anticline, with 4X vertical exaggeration in Chapter 2. Landsat images lack forward overlap that provides stereo capability. At high and low latitudes there is sidelap of images that may be viewed stereoscopically. The low base-height ratio, however, precludes any vertical exaggeration on the stereo models. Interpretation of attitudes is also facilitated by highlights and shadows caused by the low to moderate sun elevation of many TM images.

TM images are acquired at midmorning times. The Saharan Atlas Mountains of Algeria are located in the Northern Hemisphere; therefore, the midmorning sun shines toward the northwest. We can employ the resulting highlights and shadows on the image to interpret dip and strike. The method is explained in Appendix A. It is difficult to estimate the amount of dip (vertical angle) from stereo images. Experienced interpreters, however, can estimate categories of dip amount, such as low, gentle, moderate, and steep. The highlight and shadow method for interpreting attitude is applicable to any image acquired with low to intermediate illumination angles, such as Landsat, SPOT, ASTER, WorldView, and radar images.

The attitudes of beds are plotted on the map with dip-and-strike symbols.

Map Folds and Faults

The next step is to interpret fold axes based on outcrop patterns and attitudes of beds. Anticlines have older beds in the center surrounded by successively younger beds that dip away from the center; the pattern is reversed for synclines. Faults are recognized by offsets, truncations, or repetitions of units. Two small faults are shown by lines in the northeast portion of the map. These relationships are explained and illustrated in Appendix A.

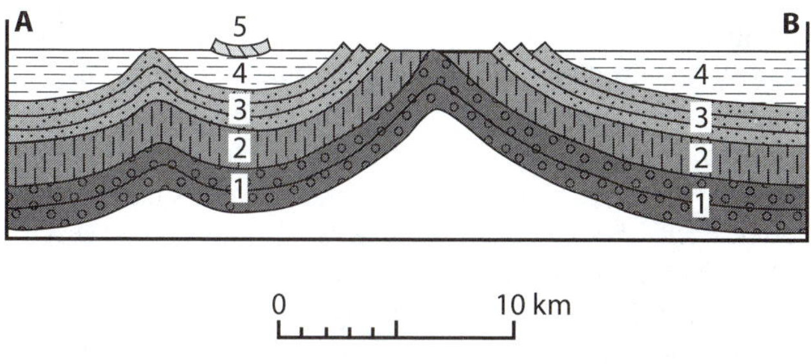

FIGURE 3-19 Cross section (A–B) of the TM image (Plate 11) of the Saharan Atlas Mountains, Algeria.

The final step is to construct a cross section (Figure 3-19) that shows the subsurface view of the structure.

Summary

The Saharan Atlas Mountains project demonstrates the following sequence of interpretation steps:

1. Establish a sequence of mappable rock units. If published information is lacking, the sequence may be established directly from the image.
2. Determine attitudes of beds. Highlights and shadows associated with dipslopes and antidip scarps are vital clues.
3. Interpret folds and faults. Outcrop patterns and attitudes of beds are keys.
4. Prepare a cross section to accompany the interpretation map.
5. Check the interpretation in the field.

This sequence of interpretation steps is applicable to images of all types in addition to Landsat. Inexperienced interpreters often try to identify folds and faults before compiling the basic data, which leads to problems.

SEASONAL INFLUENCE ON IMAGES

The repetitive coverage by Landsat can provide images that were acquired at the optimum season for interpretation. The optimum season depends on the terrain and climatic conditions and is different for various regions. In regions with seasonal rainfall patterns, images acquired in the wet season are markedly different from those acquired in the dry season. At high latitudes, there are major differences between summer and winter images.

DRY SEASON AND WET SEASON IMAGES: TRANSVAAL BASIN, SOUTH AFRICA

Digital Image 3-3 shows two seasonal Landsat MSS images of the south flank of the Transvaal Basin; **Digital Image 3-3A** was acquired during the winter dry season and **Digital Image 3-3B** during the summer wet season. The area is a grass-covered plateau with ridges of resistant rock units.

Geologic Setting

Figure 3-20 is a generalized geologic sketch map for the images in **Digital Image 3-3**. In the southeast a northeast-trending anticline separates the Transvaal Basin in the north from the Potchefstroom Basin in the south. The oldest rocks are the Ventersdorp system, Witwatersrand system, and the Archaean basement that crop out in the anticline (labeled "Older Rocks" in Figure 3-20). These massive rocks are overlain by the layered rocks of the Transvaal system that dip gently northeastward into the Transvaal Basin. The Black Reef series at the base of the dolomite series is a thin clastic unit, shown in black in Figure 3-20, with a distinctive dark brown signature on the Landsat images. The overlying dolomite series consists of massive dolomitic limestone with alternating white and light orange signatures. The Pretoria series consists of alternating quartzites and shales with some volcanic layers. The Bushveld complex consists of layered mafic rocks

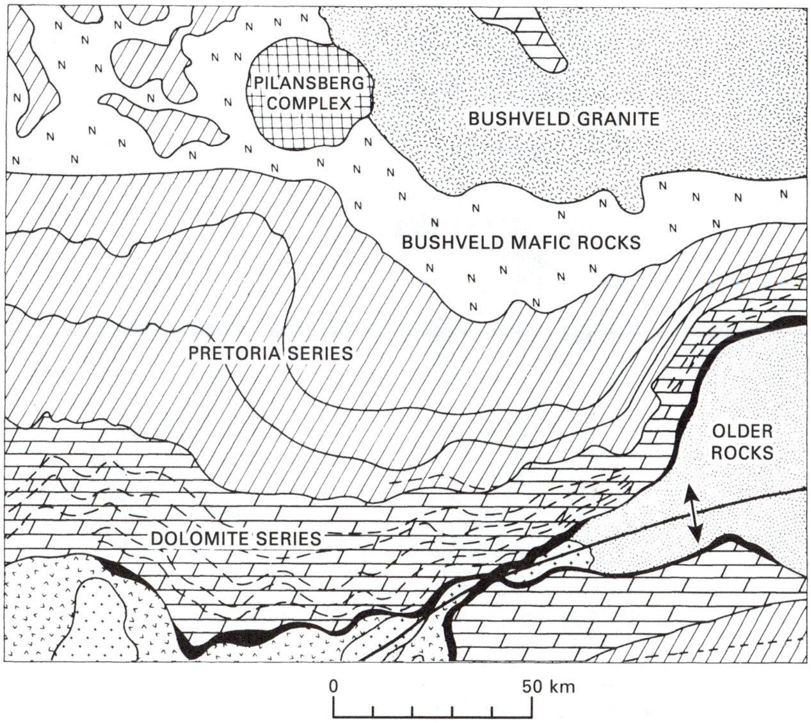

FIGURE 3-20 Geologic interpretation map of an MSS image (**Digital Image 3-3B**) acquired during the rainy season showing the south flank of the Transvaal Basin, South Africa. From Grootenboer (1973, Figure 1).

at the base that have a distinctive blue signature on the wet season image. The Bushveld Granite occupies the northeast portion of the Transvaal Basin. In the northwest portion of Figure 3-20 the Bushveld mafic complex is intruded by the Pilansberg complex, a circular pluton of silicic igneous rocks. Most of the area is covered by residual soil that conceals much of the bedrock. The only significant outcrops are the Pilansberg complex, quartzites of the Pretoria series, and scattered exposures of the Bushveld mafic complex.

Comparison of Seasonal Images

When the dry season color IR image (**Digital Image 3-3A** 🌐) was acquired, the area was covered with dry, brown grass, the indigenous vegetation was leafless, and the cornfields were fallow. Black patches on the image mark areas of recent burning. Slight tonal variations enable recognition of the major stratigraphic units to a degree comparable to that on 1:1,000,000 scale geologic maps published prior to Landsat (Grootenboer, 1973). The wet season color IR image (**Digital Image 3-3B** 🌐) was acquired at the height of the summer rainy season, when the perennial vegetation was in full leaf. The strong color variations are directly related to bedrock lithology, particularly in the area underlain by the Dolomite series, Pretoria series, and the Bushveld mafic complex (Figure 3-20). Of particular interest are the seven zones of tonal variations within the outcrop of the Dolomite series. Field checks by Grootenboer and others (1973) established that the four darker zones correspond to dark-toned, chert-free dolomite and the three lighter zones to light-toned dolomite with abundant chert. During the previous 90 years of geologic investigation in the area, no such stratigraphic subdivisions had been recognized in the Dolomite series.

Several factors contribute to the superiority of the wet season image:

1. Windblown dust causes nonselective (Mie) atmospheric scattering, which severely scatters light in the dry season. Rainfall in the wet season removes dust from the air, producing a clearer atmosphere and good image contrast. The rain also washes away the surface dust layer from the outcrops.
2. Greater soil moisture enhances tonal and color differences between rock types.

A. Summer image acquired June 18, 1973, with a 45° sun elevation.

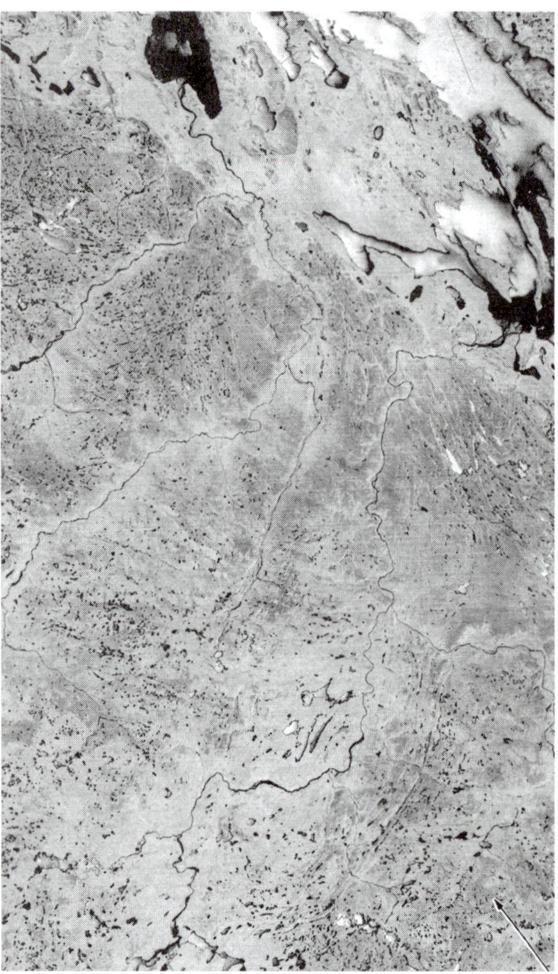

B. Winter image acquired April 2, 1974, with a 27° sun elevation.

FIGURE 3-21 Landsat MSS band 4 (NIR) seasonal images of Bathurst Inlet, Canada. Landsat courtesy USGS.

3. Vegetation grows preferentially on belts of soil with higher moisture. The red stripes of vegetation in the wet season image help delineate geologic trends.

These advantages have also been observed in other arid areas.

Winter and Summer Images: Bathurst Inlet, Canada

At high and moderate latitudes, winter and summer images differ in sun elevation and in snow cover. The advantages of low sun elevation for aerial photographs are demonstrated in Chapter 2 and also apply to Landsat and other optical satellite images. Figure 3-21 shows MSS band 4 (NIR) images of Bathurst Inlet in Arctic Canada that were acquired in the summer and late winter. In the summer image (Figure 3-21A), there is no snow; vegetation growth is vigorous, as shown by the bright signatures, and many of the small lakes have thawed, as shown by their dark signatures. Some of the large lakes in the northeast are still frozen. Few geologic features are recognizable. Most of the lakes are only a few hundred meters in size and tend to obscure the geologic features that are thousands of meters in size. All of the lakes are frozen and concealed by the snow in the winter image (Figure 3-21B), which enhances the geologic structures.

The relatively high sun angle (45°) of the summer image causes only minimal highlights and shadows. The low sun angle (27°) of the winter image, however, causes highlights and shadows that emphasize subtle topographic features that show strike ridges of sedimentary rocks, folds, faults, fractures, and igneous dikes. The geologic map (Figure 3-22) shows these features, which were interpreted from the winter image.

In the southern half of the winter image a major north-trending fold is outlined by strike ridges of stratified rocks (Figure 3-21B). The fold is surrounded and underlain by highly fractured, unstratified crystalline basement rocks. The highlight and shadow method shows that the fold is a syncline with younger strata preserved in the center. A linear feature trending northwest across the northern part of the image is a major fault that juxtaposes different terrain on either side. North of the fault is a small anticline. In the northeast corner of the scene a north-striking prominent narrow ridge is an igneous dike. In the northwest corner of the

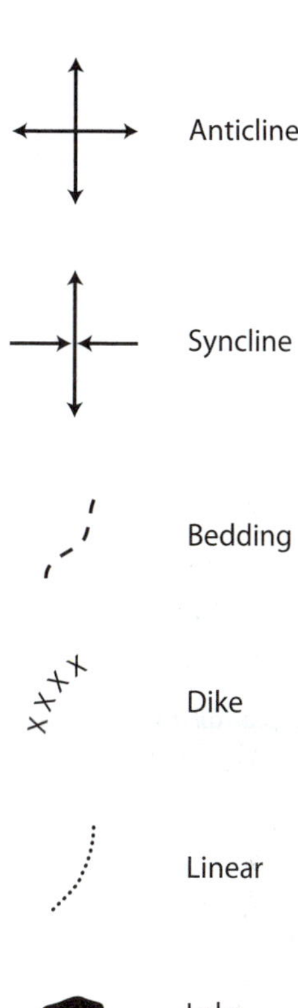

Figure 3-22 Geologic map interpreted from MSS image acquired in winter of Bathurst Inlet, Canada. Dashed lines are resistant strata. Reproduced with permission from A. F. Gregory and H. D. Moore, Recent advances in geologic applications of remote sensing from space. Proceedings of 24th International Astronautical Congress (pp. 153–170). Copyright 1976, Pergamon Press, Ltd.

scene, two distinct circular features (8 km in diameter) may be impact features. These geologic features are mappable only from the winter image. Linears can be interpreted from straight segments of drainage, aligned water bodies, and aligned topographic valleys, crests, and ridges. These linears may be related to fractures, faults, and outcrop patterns.

This Arctic example demonstrates the advantages of snow cover and low sun angle for geologic mapping. These conditions also occur in winter images of areas at intermediate latitudes.

QUESTIONS

1. You plan to use TM images to map a strip 150 km wide spanning the border between the United States and Mexico from the Pacific Ocean through Arizona. The northern edge of the map will be the border between Utah and Arizona. List the path and row numbers for the Landsat scenes that will cover this area. Refer to Figure 3-9 to select the minimum number of images for the project.
2. Assume that your eyes have normal spatial resolution (Chapter 1) and that you are an astronaut traveling on the Landsat 1 satellite (orbit altitude of 918 km). What is the dimension of the ground resolution cell you observe? How does the resolving power of your eye compare with that of MSS images?
3. Calculate the ground resolution cell of your eyes if you were an observer on the Landsat 5 satellite. How does this resolving power compare with that of TM images?
4. Compare the spectral range of the eye with the spectral ranges of the Landsat MSS, TM, and OLI systems.
5. Based on the analyses for questions 3 and 4, discuss the relative merits of Earth observations made by an astronaut versus the Landsat TM.
6. What Landsat 8 OLI bands would be most useful for mapping in cloudy tropical areas?
7. Why is Landsat 8 OLI band 9 included in the instrument?
8. Which TM and OLI band(s) would you use for bathymetric mapping in clear water?
9. Critique and correct this statement by an image interpreter: "In the Landsat image, straight and aligned stream segments over 10 km in length form linears that are undoubtedly faults, although no geologic maps are available and I have not field-checked the area."

REFERENCES

Concha, J. A., and J. R. Schott. 2015. Atmospheric Correction for Landsat 8 over Case 2 Waters. Proceedings of the SPIE, Optics and Photonics, Earth Observing Systems XX (Volume 9607), San Diego, CA. http://doi: 10.1117/12.2188345

Dozier, J. 1989. Spectral signature of alpine snow cover from the Landsat Thematic Mapper. *Remote Sensing of Environment*, 28, 9–22.

Grootenboer, J. 1973. The Influence of Seasonal Factors on the Recognition of Surface Lithologies from ERTS Imagery of the Western Transvaal. Third ERTS Symposium (SP-351, Vol. 1, pp. 643–655), NASA.

Grootenboer, J., K. Eriksson, and J. Truswell. 1973. Stratigraphic subdivision of the Transvaal Dolomite from ERTS imagery. Third ERTS Symposium (SP-351, Vol. 1, pp. 657–664), NASA.

Harris, P. M., and J. M. Ellis. 2009. Satellite Imagery, Visualization and Geologic Interpretation of the Exumas, Great Bahamas Bank: An Analog for Carbonate Sand Reservoirs. In SEPM Short Course Notes 53 (pp. 1–49). Society for Sedimentary Geology.

Harris, P. M., and W. S. Kowalik (Eds.). 1994. *Satellite Images of Carbonate Depositional Settings*. AAPG Methods in Exploration Series, Number 11. Tulsa, OK: American Association of Petroleum Geologists.

Loveland, T. R., and J. R. Irons. 2016. Landsat 8: The plans, the reality, and the legacy. *Remote Sensing of Environment*, 185, 1–6. http://doi.org/10.1016/j.rse.2016.07.033

Lynch, J. 2015. Toxic Lake Erie algae spotted but drinking water safe. *The Detroit News*, July 28 (accessed February 2017).

NASA Earth Observatory. 2015, August 4. Algae Bloom in Lake Erie. http://earthobservatory.nasa.gov/IOTD/view.php?id=86327 (accessed December 2017).

NASA Earth Observatory. 2017a, January 10. Cambodia Forests are Disappearing. NASA Earth Observatory with USGS and Global Forest Watch. http://earthobservatory.nasa.gov/IOTD/view.php?id=89413 (accessed December 2017).

NASA Earth Observatory. 2017b, February 3. Satellites Capture Different Views of Devastating Fires in Chile. NASA Earth Observatory with US Geological Survey. http://earthobservatory.nasa.gov/IOTD/view.php?id=89570 (accessed December 2017)

Paul, F. 2000, June 16–17. Evaluation of Different Methods for Glacier Mapping Using Landsat-TM Data. Proceedings of EARSeL-SIG-Workshop on Remote Sensing of Land Ice and Snow (pp. 239–245), Dresden/FRG.

Purkis, S. J., and V. V. Klemas. 2011. *Remote Sensing and Global Environmental Change*. West Sussex, UK: Wiley Blackwell.

Purkis, S. J., and R. Pasterkamp. 2004. Integrating in situ reef-top reflectance spectra with Landsat TM imagery to aid shallow-tropical benthic habitat mapping. *Coral Reefs*, 23(1), 5–20.

Smith, T., B. Bookhagen, and F. Cannon. 2015. Improving semi-automated glacier mapping with a multi-method approach: Applications in central Asia. *The Cryosphere*, 9, 1,747–1,759.

Watts, J. 2017. Chile battles devastating wildfires: "We have never seen anything on this scale." *The Guardian*, January 25. http://www.theguardian.com/world/2017/jan/25/chile-fire-firefighting-international-help (accessed December 2017).

Winsvold, S. H., and A. Kaab. 2016, August. Regional glacier mapping using optical satellite data time series. *IEEE Journal of Selected Topics in Applied Earth Observations and Remote Sensing*, 9(8), 1,939–1,404.

Multispectral Satellites

OVERVIEW

Chapter 3 covered Landsat satellites, which have the longest and most continuous history of a repetitive sequence of Earth images. This chapter describes multispectral Earth observation satellites that were deployed post-Landsat (post-1972). Similar to Landsat, they operate in the passive mode and acquire imagery in the reflected VNIR to SWIR spectral regions. Many of these satellites are operated by government agencies, but some are operated by commercial companies. These satellites acquire images from one of two alternate orbit patterns (Figure 4-1):

1. *Polar-orbiting satellites* are in sun-synchronous orbit patterns that were described in the Landsat chapter. They acquire images that range from high to low spatial resolution with swath widths that range from a few tens of kilometers to a few hundred kilometers.
2. *Geostationary satellites* are "parked" above the Earth's equator at an altitude of approximately 36,000 km. They periodically acquire images of the same scene that typically covers a hemisphere with spatial resolutions coarser than several hundred meters.

A few systems (SPOT, DMSP, GOES) were first deployed in the early to mid-1970s and remain active today due to periodic replacements of older satellites. Beginning in 2000, numerous new multispectral systems were launched with different levels of spatial and spectral resolution. In 2019, there were 767 satellites orbiting the Earth with the main purpose of Earth observation (Union of Concerned Scientists, 2019).

For more information, Chapter 5 covers satellites that acquire images in the TIR spectral region. Chapter 6 covers

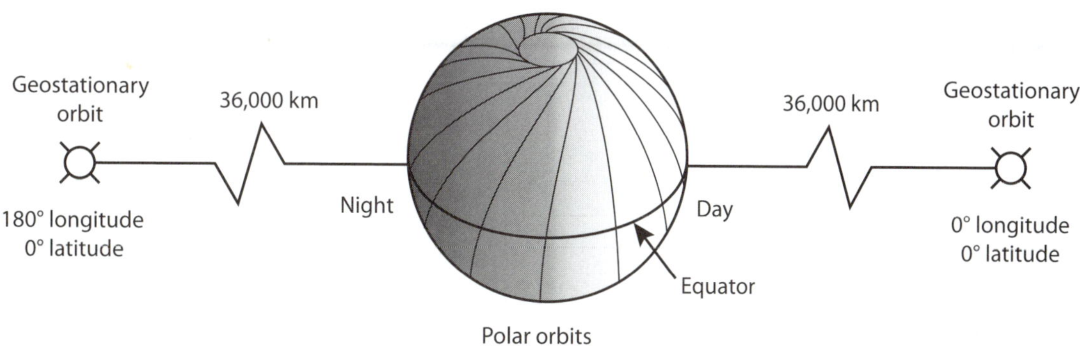

FIGURE 4-1 Orbit patterns for polar and geostationary satellites.

microwave (radar) satellites. Prost (2014) summarizes over 100 launched satellites and over 20 future satellites. The Satellite Imaging Corporation provides overviews of commercial and government systems with high to medium spatial resolution (satimagingcorp.com). SpaceNews is a current information source that reports on a broad range of satellites and remote sensing activities (spacenews.com).

RESOLUTION OF SATELLITE IMAGES

When we conversationally speak of "image resolution" we typically intend "spatial resolution." In remote sensing, however, we recognize four categories of "resolution":

1. Spatial resolution,
2. Spectral resolution,
3. Temporal resolution, and
4. Radiometric resolution.

We summarize each resolution category in the following sections.

Spatial Resolution

Spatial resolution was defined in Chapter 1 as the ability to distinguish between two closely spaced objects. Specifically, it is the minimum distance between two objects at which the objects are distinct and separate on the image. For remote sensing images, spatial resolution is determined by the dimensions of the ground resolution cell, which are determined by the size of the detector and the altitude of the satellite. Figure 4-2A shows the dimensions of ground resolution cells for multispectral satellite systems that range from WorldView-2 to Landsat TM/OLI.

Spectral Resolution and Reflectance Spectra

Spectral resolution, or bandwidth, is defined as the wavelength interval that is recorded at 50% of the peak response of a detector (Figure 1-6). Figure 4-3 shows the spectral resolution for selected satellite systems. The VNIR-SWIR wavelength range from 0.4 µm to 2.50 µm is shown on the horizontal scale. For each satellite system the horizontal lines and vertical bars show the number of detectors and the wavelength band recorded by each detector. For example, band 12 of Sentinel-2 ranges from 2.07 to 2.32 µm for a bandwidth of 0.25 µm. ASTER acquires five bands in the same wavelength interval with an average bandwidth of 0.05 µm, which is a significantly higher spectral resolution than Sentinel-2.

Figure 4-3 shows reflectance spectra for vegetation and for three categories of sedimentary rocks. (As geologists we are hard-wired to cite rocks.) In Figure 4-3 the rock spectra show pronounced diagnostic spectral features in the atmospheric window 2.00 to 2.50 µm. Sentinel-2 band 12 acquires only a single wide, nondiagnostic reflectance value in the

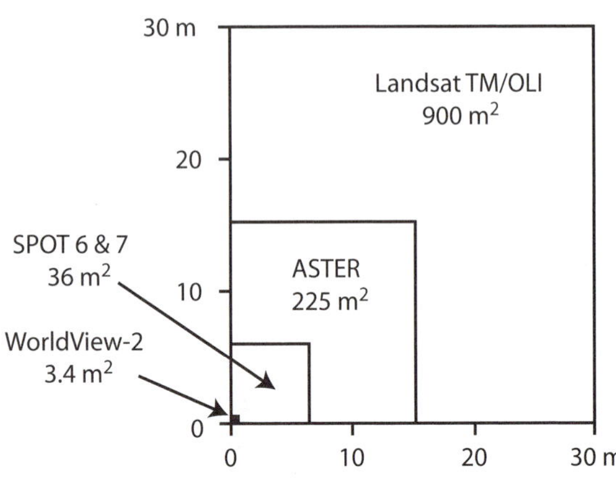

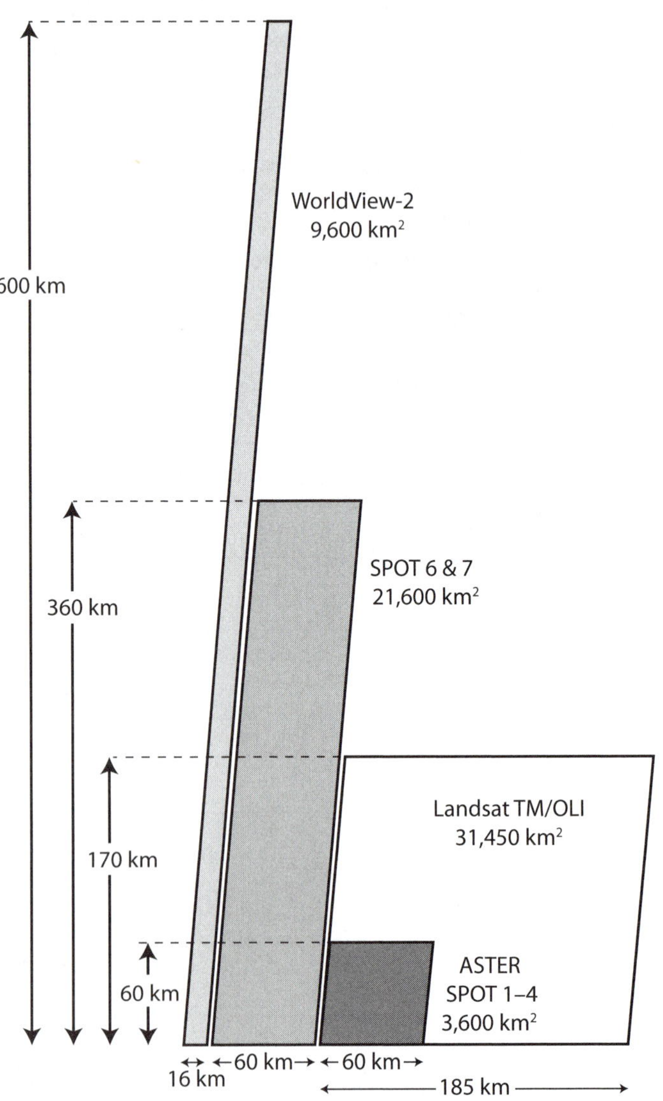

FIGURE 4-2 Spatial resolution and terrain coverage for selected multispectral satellites.

window and does recognize the rock types. ASTER, however, acquires five narrow bands that can be digitally processed to classify each rock category.

Temporal Resolution

Temporal resolution is the time interval between successive images of the same area. Rotation of the Earth causes polar-orbiting satellites to eventually re-occupy the image swaths that are viewed in the nadir mode from the preceding orbital cycle. The elapsed time (in days) for each cycle is determined by the swath width and sidelap at the equator. For Landsat OLI the orbital cycle is 16 days.

Radiometric Resolution

Features of interest may have a low contrast ratio with their background, which makes the features difficult to detect. Examples are black bears within shadows or polar bears on sea ice. Such features ae difficult to detect with imagery that has only 8-bit radiometric resolution (256 levels of gray) (Figure 1-10). These challenges may be resolved by systems with high radiometric resolution that acquire 11-bit to 14-bit data (2,048 to 16,384 levels of gray).

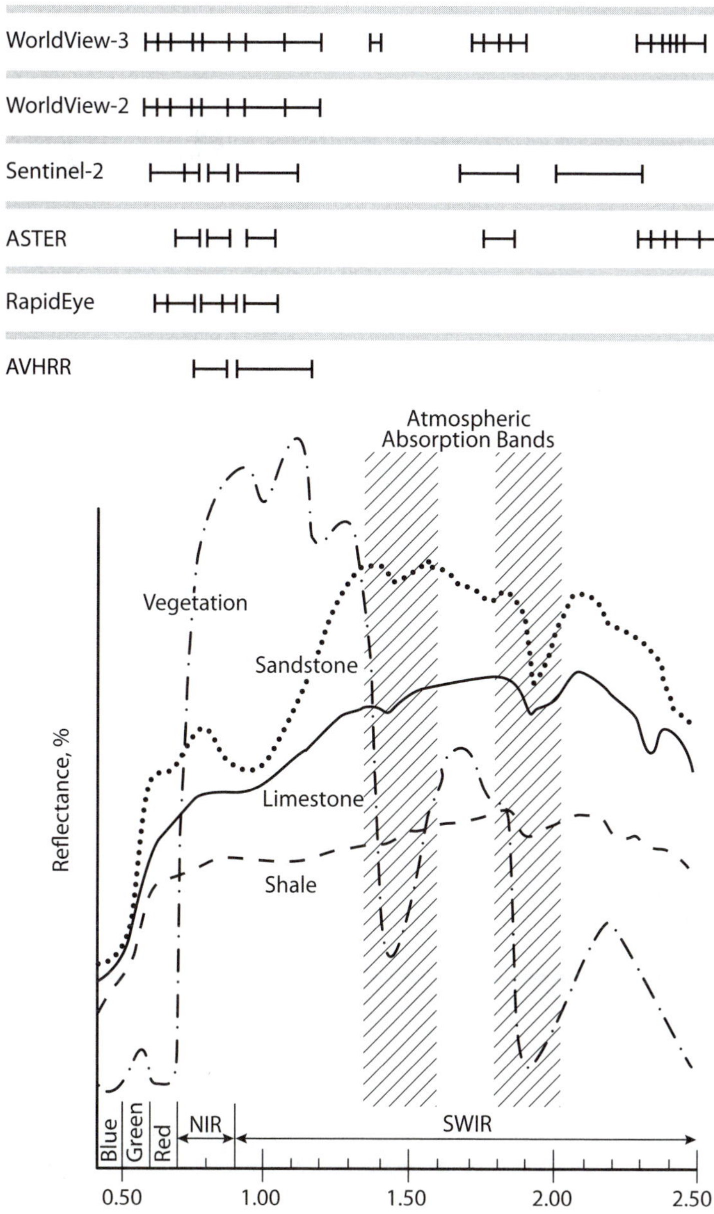

FIGURE 4-3 Spectral resolution. The horizontal lines and vertical bars show spectral bands recorded by selected satellites. The curves show spectral reflectance of vegetation and rocks.

TERRAIN COVERAGE OF SATELLITE IMAGES

We show terrain coverage as the dimensions of a single image expressed in kilometers. Figure 4-2B shows terrain coverage for selected satellite systems. The systems comprise of two categories based on swath width:

1. Narrow swaths (< 50 km wide), such as Pleiades and WorldView.
2. Wide swaths (> 50 km wide), such as Sentinel-2, Landsat, ASTER, and SPOT.

Terrain coverage is determined by:

- Swath width (west to east), which is determined by the angular field of view, which ranges from 16 km wide (WorldView-2) to 290 km wide (Sentinel-2). Swath width for cross-track scanners may also be determined by the number of ground resolution cells per scan line multiplied by the width of a cell.
- Ground distance (north to south), which typically matches the swath width dimension.

Having reviewed the basics for satellite image systems we can now deal with specific systems that fall into one of two categories: polar-orbiting satellites and geostationary satellites.

POLAR-ORBITING SATELLITES

Figure 4-4 lists three categories of polar-orbiting satellites based on the spatial resolution of the images. The categories are:

1. High spatial resolution (≤ 1.5 m)
2. Medium spatial resolution (> 1.5 to ≤ 30.0 m)
3. Low spatial resolution (> 30.0 m)

To categorize multispectral satellites we use spatial resolution as the dimension of the ground resolution cell recorded by the panchromatic band; or, if the satellite lacks a panchromatic image, the highest resolution VNIR band. Figure 4-4 shows the launch year, as well as functioning (lines with arrows) and nonfunctioning (lines with vertical bars) satellites. Lines terminating in arrows indicate satellites that were functioning in 2018. The satellites have a range of spatial resolution and spectral resolution that are summarized in the following sections.

HIGH SPATIAL RESOLUTION SATELLITES

Commercial satellite images and data with high spatial resolution (≤ 1.5 m) were introduced in 1999 and have expanded the application of remote sensing for a range of users. For example, Facebook uses the Maxar Technologies (Maxar) mosaic of the globe at 50 cm spatial resolution to identify man-made structures as a proxy for population density. Facebook is using this mosaic to plan the deployment of terrestrial networks, satellites, and drones to reach the 4.2 billion people that remain offline (DigitalGlobe, 2016a). High resolution satellite imagery is streamed around the world with online mapping services such as Google Maps and ESRI's base maps, and with virtual 3-D globes such as Google Earth and ESRI's Earth. In 2014 Google purchased a microsatellite company, perhaps to increase the amount and quality of global information that the Google search engine is able to accumulate (Mims, 2014). In February 2017, Google sold this microsatellite enterprise to Planet (formerly Planet Labs), which operates the largest constellation of microsatellites. Table 4-1 lists characteristics of representative, high spatial resolution satellites (Figure 1-20).

FIGURE 4-4 Time line for selected satellites. The satellites are assigned to groups according to the highest spatial resolution recorded by their sensor.

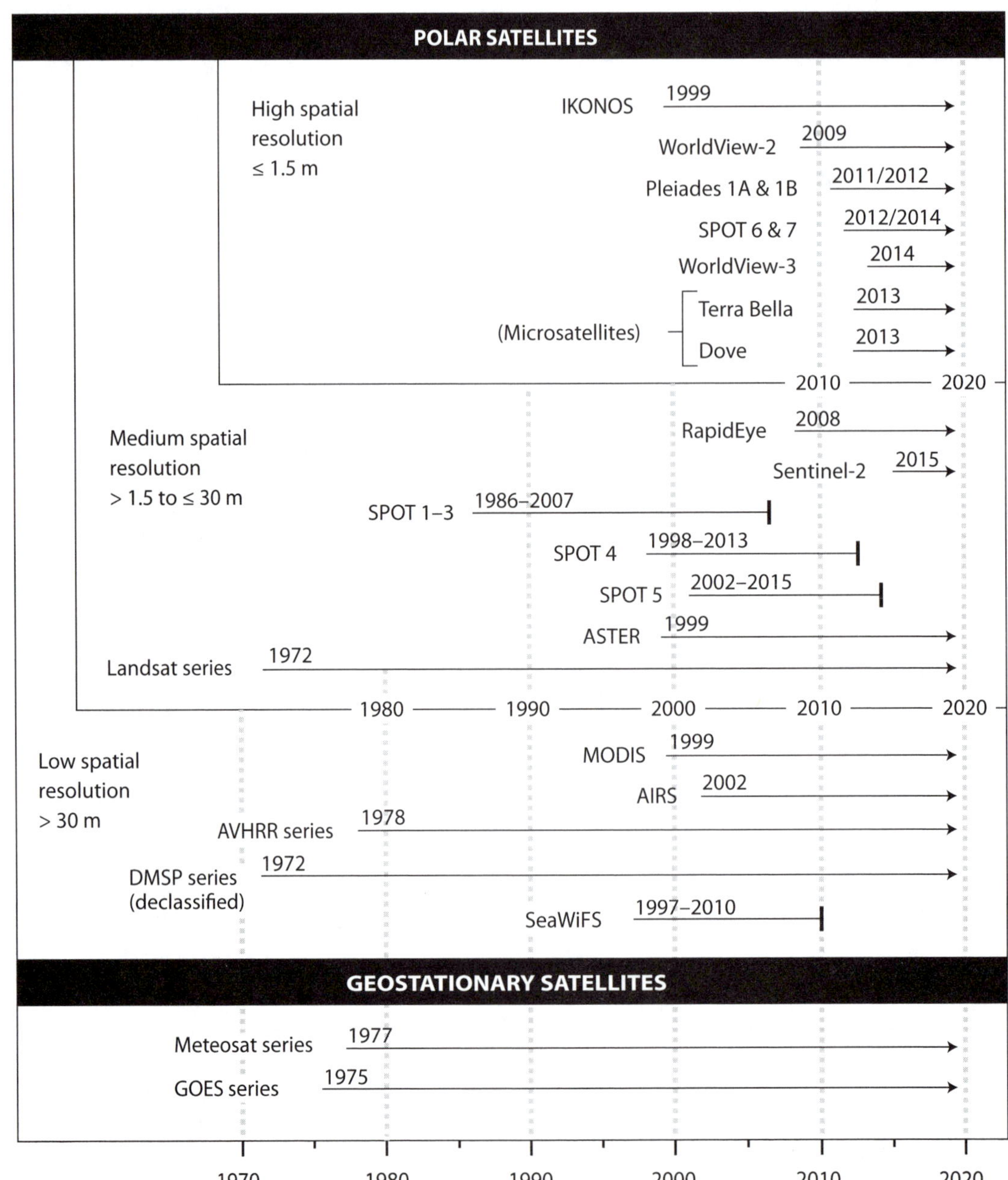

Agile Satellites

The commercial high resolution satellites are agile. They can direct their image systems to acquire images of localities distant from the nadir orbit position of the satellite (Figure 4-5A). The image systems may be tilted to the right or left to acquire overlapping stereo-pairs or aimed forward or backward while the satellite is moving along its orbital path to sequentially acquire multiple targets of varying sizes and at different locations (as shown in Figure 4-5B). The systems can also acquire parallel, side-lapping image strips for generating a mosaic of larger areas. Agile satellites have significantly greater temporal resolution compared with systems that can only acquire imagery in a nadir (downward-looking) mode.

Sources of High Spatial Resolution Imagery

Maxar and *Airbus Defence and Space* (DS) are the two dominant, commercial companies that provide global, high resolution imagery and multispectral data. Maxar's satellites include *IKONOS*, *Quickbird*, *GeoEYE*, and the *WorldView* series. The WorldView and GeoEYE-1 systems can collect strips of imagery with swath widths of ~16 km and maximum lengths exceeding 300 km. GeoEYE-1 is capable of stereo coverage over a 28 by 228 km area using its agile pointing capability. Airbus DS satellites include the SPOT (Satellite pour l'Observation de la Terre) series and the Pleiades 1A and 1B systems. Other sources of high resolution imagery are:

Table 4-1 High spatial resolution satellites (≤ 1.5 m ground resolution cells).

Operator	Maxar	Maxar	Maxar	Airbus DS	Airbus DS
Name	**WorldView-3**	**WorldView-2**	**GeoEye-1**	**Pleiades 1A, 1B**	**SPOT 6/7**
Spatial Resolution (m)					
Pan	0.30	0.46	0.41	0.50	1.5
VNIR	1.24	1.85	1.65	2.80	6.0
SWIR	7.50				
Spectral Resolution (μm)					
Pan	0.45 to 0.8	0.45 to 0.8	0.45 to 0.8	0.47 to 0.83	0.45 to 0.74
VNIR	0.40 to 0.45	0.40 to 0.45	0.45 to 0.51	0.43 to 0.55	0.45 to 0.52
	0.45 to 0.51	0.45 to 0.51	0.51 to 0.58	0.50 to 0.62	0.53 to 0.59
	0.51 to 0.58	0.51 to 0.58	0.65 to 0.69	0.59 to 0.71	0.62 to 0.69
	0.58 to 0.62	0.58 to 0.62	0.78 to 0.92	0.74 to 0.94	0.76 to 0.89
	0.63 to 0.69	0.63 to 0.69			
	0.70 to 0.74	0.70 to 0.74			
	0.77 to 0.89	0.77 to 0.89			
	0.86 to 1.04	0.86 to 1.04			
SWIR	1.19 to 1.22				
	1.55 to 1.59				
	1.64 to 1.68				
	1.71 to 1.75				
	2.145 to 2.185				
	2.185 to 2.225				
	2.235 to 2.285				
	2.295 to 2.365				
Radiometric resolution	VNIR 11-bit SWIR 14-bit	11-bit	11-bit	12-bit	12-bit
Stereo	Yes	Yes	Yes	Yes	Yes
Swath (km)	13.1	16.4	15.3	20	60
Launched	2014	2009	2008	2011	2014

- ImageSat International operates the *EROS* series of satellites that acquires 70 cm panchromatic imagery on an exclusive basis.
- Indian Space Research Organisation's *Cartosat-3* satellite that acquires 30 cm panchromatic and 70 cm multispectral green, red, and NIR data.
- Korea Aerospace Research Institute's *KOMPSAT-3* satellite that acquires 1 m panchromatic and 4 m multispectral blue, green, red, and NIR data.
- China's SpaceWill Info. Co., Ltd.'s *GaoJing* series of satellites (also known as the SuperView series) collects 0.5 m panchromatic and 2 m multispectral data.

SPOT 6 and 7, Pleiades 1A and 1B, WorldView-2, and WorldView-3 are reviewed here as representative examples of high spatial resolution satellite imagery.

SPOT 6 and 7

SPOT 6 was launched on September 9, 2012 and SPOT 7 was launched on June 30, 2014 from the Satish Dhawan Space Center in India. Airbus DS is the commercial operator of SPOT 6. In December 2014, SPOT 7 was sold to Azerbaijan's space agency (Azercosmos), who renamed it Azersky (De Selding, 2014). SPOT 6 and 7 collect a panchromatic band with 1.5 m spatial resolution and four VNIR multispectral bands with 6.0 m spatial resolution (Table 4-1) that support vegetation, geologic, and environmental mapping.

SPOT 6 and 7 can acquire stereoscopic images with vertical exaggeration ranging between 1.5X and 5X by varying the orbital baseline distance between the fore and aft viewing images (which varies the base-height ratio in Figure 2-17). The system also can integrate nadir views to achieve tri-stereo coverage to support more accurate 3-D models. SPOT 6 and 7 can image a strip up to 360 km in length (Figure 4-2B). The 60-km wide image swath is an advantage over systems that acquire narrow swaths. SPOT 6 and 7 are useful systems for large areas where submeter spatial resolution is not required.

A. Nadir and off-nadir viewing angles.

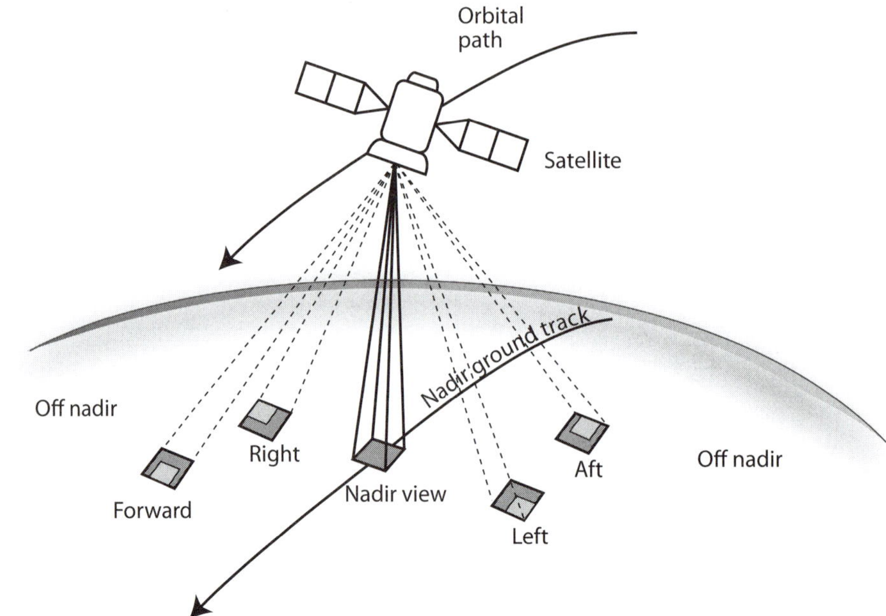

B. Examples of image footprints.

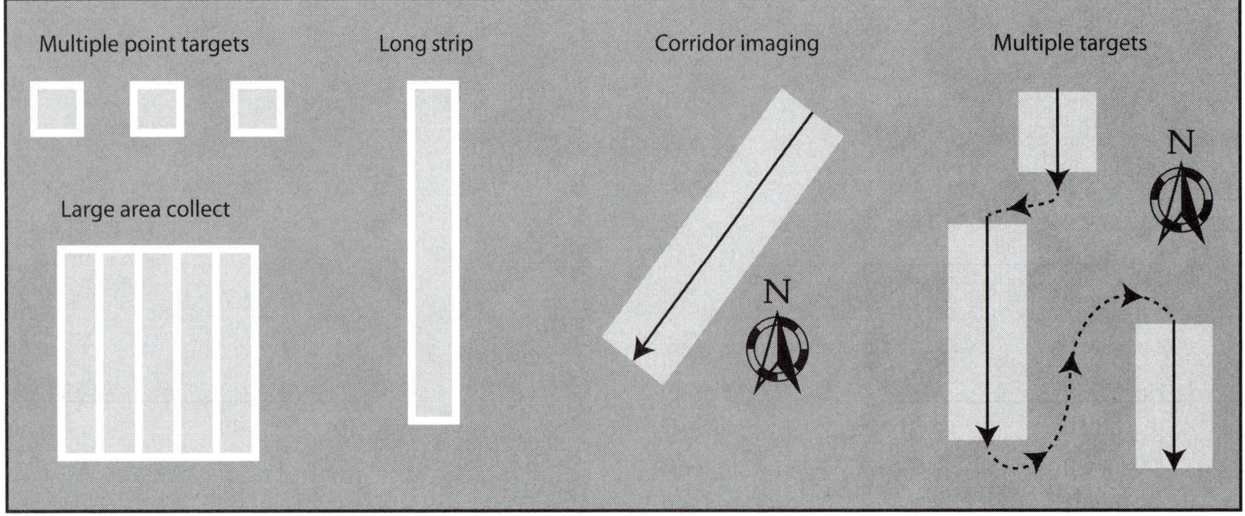

FIGURE 4-5 Collection modes for agile satellites.

Pleiades 1A and 1B

The Pleiades 1A spacecraft was launched on December 16, 2011 on a Russian Soyuz ST rocket from Europe's spaceport in Kourou, French Guiana (Astrium, 2012). Pleiades 1B was launched on December 2, 2012. The satellites have a 26-day repeat cycle for nadir viewing of the same site on the Earth's surface, compared with Landsat 8's 16-day temporal resolution. In addition to nadir views, Pleiades systems can tilt up to an oblique viewing angle of 45°, to acquire images far from their orbital path. This ability can decrease the time between revisits to localities with dynamic changes, such as forest fires. Image swaths are 20 km wide at nadir, but because of their pointing capability a 100 by 100 km image can be acquired in the large-area collect mode. Pleiades 1A and 1B orbits are 180° apart in polar, sun-synchronous orbits at an altitude of 694 km.

Table 4-1 shows that Pleiades 1A and 1B satellites have high spatial resolution and record bands comparable to those of SPOT 6 and 7. Figure 4-6 is a vertical view of the Northern Hemisphere that shows the positions of these four satellites relative to the equator. SPOT 6 and 7 are separated by 180° of longitude. Pleiades 1A and 1B are separated by 90° of longitude from SPOT 6 and 7. The four satellites are in polar, sun-synchronous orbits, which enables twice daily revisits (12 hour temporal resolution) to any location on the planet.

Pleiades acquires panchromatic images with 70 cm spatial resolution. The data are re-sampled to 50 cm pixels (Astrium, 2012). Figure 4-7A is a panchromatic Pleiades image of Melbourne, Australia. Figure 4-7B is an enlargement that shows individual boats and ships tied up along the docks. Multispectral images are collected in the blue, green, red, and NIR wavelengths with 2.0 m spatial resolution. Both pan and multispectral bands are acquired with 12-bit radiometric resolution, which provides 4,096 levels of gray that can detect dark objects concealed shadows and can detect bright objects bright environments such as sand, ice, and snow (Astrium, 2012). Pleiades also acquires stereo images.

Pleiades 1A

SPOT 6

SPOT 7

Pleiades 1B

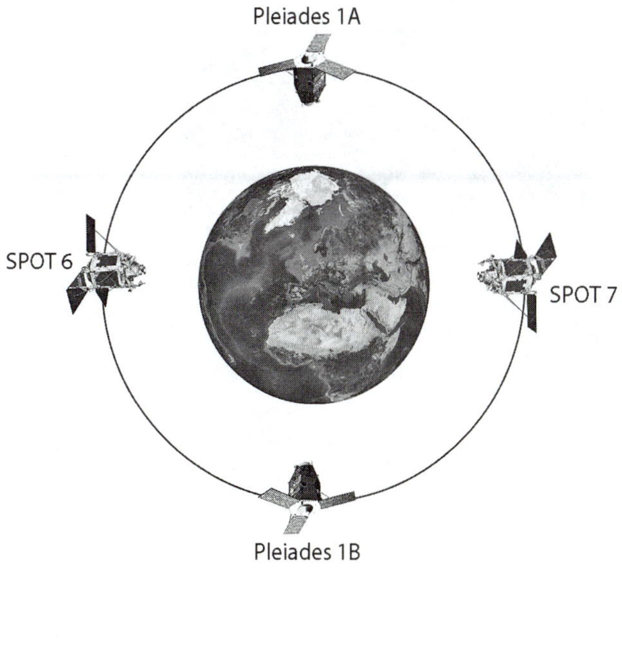

Figure 4-6 Earth orbits of SPOT and Pleiades satellite constellations. Orbits of alternate Pleiades and SPOT satellites are separated by 90° of longitude, which provides a twice daily revisit over any site and provides high temporal resolution. Courtesy Airbus DS.

WorldView-2

WorldView-2 was launched October 8, 2009 from Vandenberg Air Force Base. It was the first satellite with high spatial resolution that acquired eight bands of multispectral data in the visible to near IR (VNIR) region (DigitalGlobe, 2013). In addition to the standard four VNIR bands, WorldView-2 acquires the following bands:

1. Shortwave blue (0.40 to 0.45 μm), or coastal band, provides deeper penetration in clear water to improve bathymetric mapping.
2. Yellow band (0.58 to 0.62 μm) improves mapping of yellow signatures such as polluted water bodies, disposal sites, and rocks containing the iron minerals limonite and jarosite.
3. Red edge band (0.77 to 0.89 μm) provides additional spectral information on vegetation type and vigor or stress.
4. NIR2 band (0.86 to 1.04 μm) provides additional spectral information for vegetation and biomass studies.

The WorldView-2 satellite can acquire four side-lapping strips of stereo coverage that measure 112 km by 63 km using its agile pointing capability. It can collect nearly 1 million km^2 of imagery per day and has a temporal resolution of 1.1 days (DigitalGlobe, 2016b). The 8-band, VNIR multispectral images were used to interpret the presence, cover, and biomass of the Sahara mustard plant, an invasive species that is common in the Mojave and Sonoran Deserts of the southwestern United States (Sankey and others, 2014).

The high spatial resolution images provide detail that is useful for mapping settlements, infrastructure, small agricultural plots, and local environments in remote areas. Figure 4-8A is a WorldView-2 panchromatic image of a small settlement and its surroundings in the upper Rio Yanatile Valley in the Andes Mountains of Peru. Figure 4-8B is a land cover interpretation map that shows buildings, fences, roads and trails, vegetation, and a river channel. Fences, probably made of stone, are recognized by their dark, narrow linear

A. City-wide image.

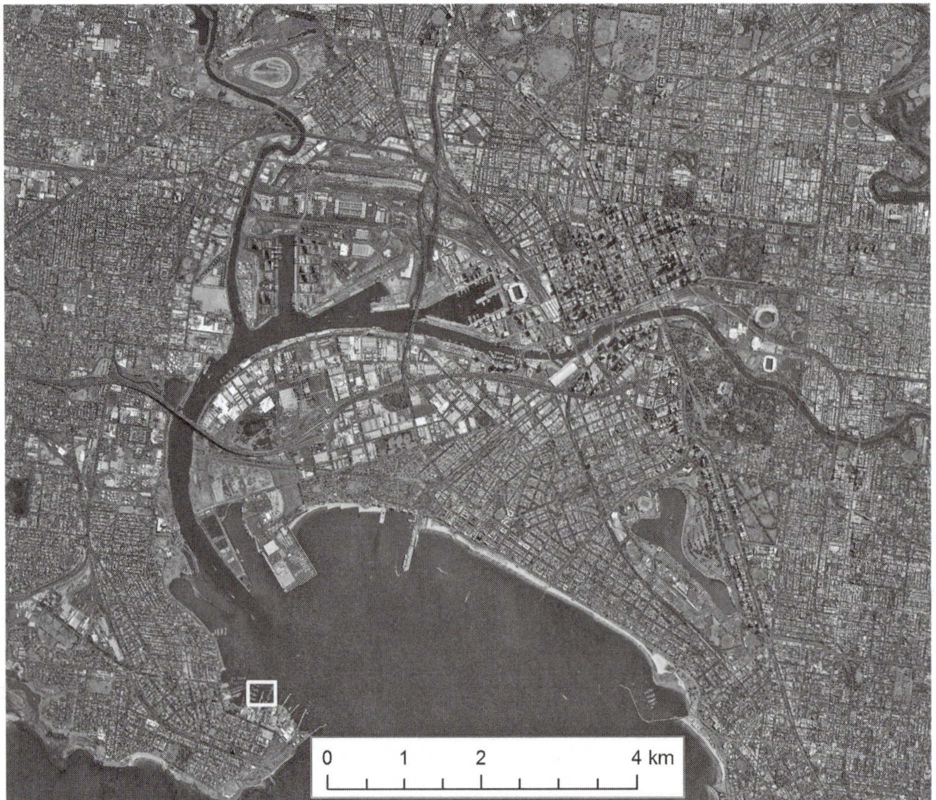

B. Detail image of area outlined in (A).

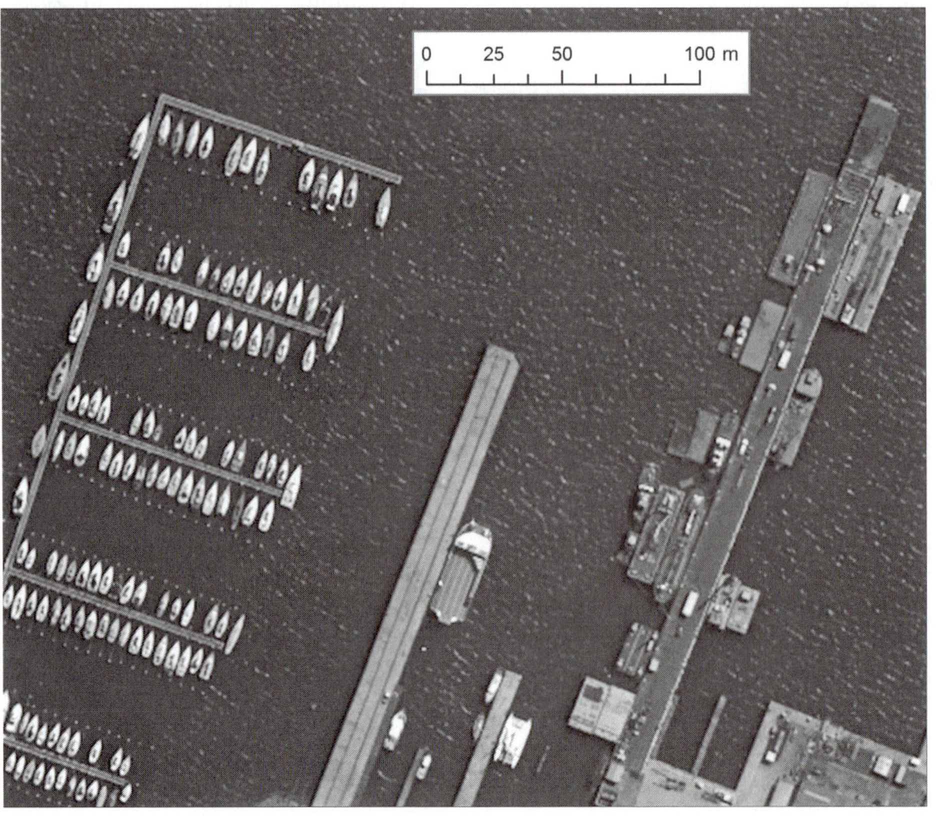

Figure 4-7 Pleiades images of Melbourne, Australia. © Airbus DS 2019.

A. WorldView-2 satellite panchromatic image with 50 cm spatial resolution. Satellite image © 2019 Maxar Technologies.

B. Interpretation map.

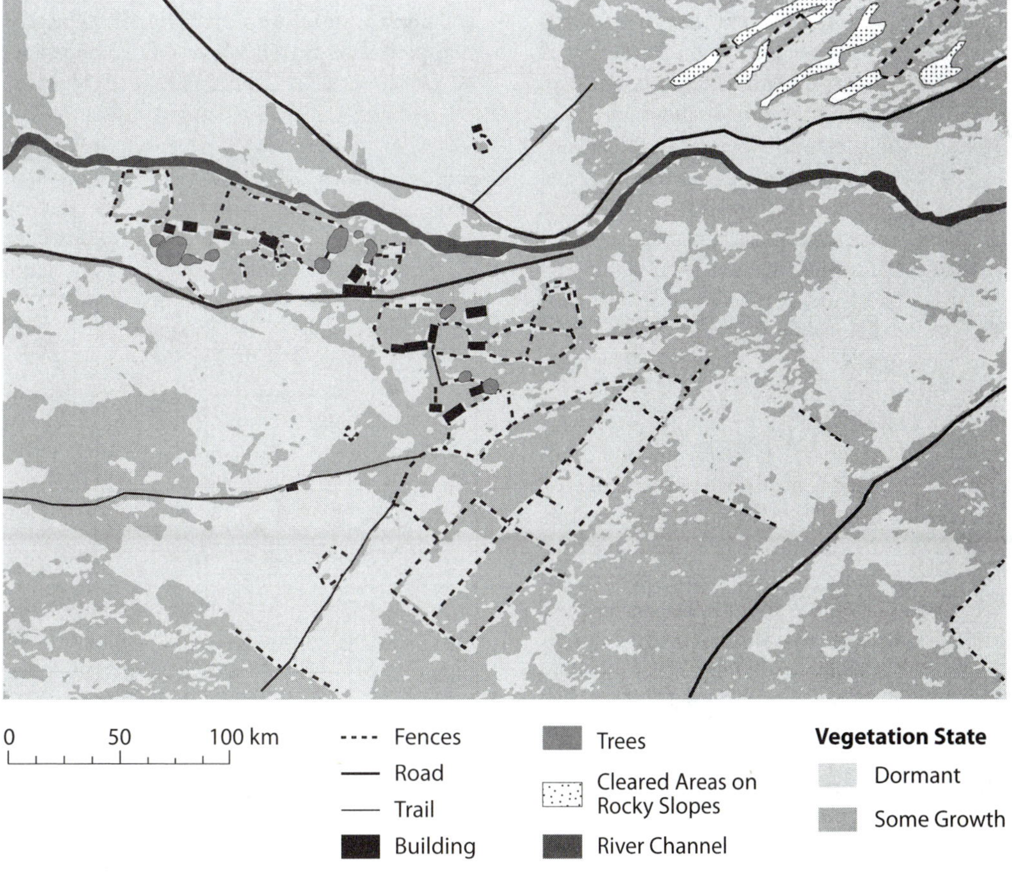

Figure 4-8 An agricultural settlement in Peru.

signatures. Buildings have small rectangular shapes with bright signatures. Roads and trails have extended narrow, bright signatures. Rocky slopes in the northeast corner of the image have irregular, bright signatures up to 10 m wide.

Most of the area is covered with cultivated and natural vegetation with a range of panchromatic tones from light to dark gray (Figure 4-8A). Areas of cultivated vegetation are distinguished by largely linear boundaries and enclosing fences. The multispectral data that accompanied the panchromatic image was computer processed to distinguish dormant vegetation from growing vegetation. The two conditions of vegetation are plotted as the background on the land cover map (Figure 4-8B). Images and maps such as these can be used by local agencies and landowners for planning and monitoring, and to develop a detailed land use/ land cover map.

WorldView-3

WorldView-3 was launched on August 13, 2014 as the first high resolution satellite with eight SWIR bands. WorldView-3 also acquires the same eight VNIR bands as WorldView-2. WorldView-3 panchromatic and pan-sharpened VNIR bands provide 30 cm pixels; the highest spatial resolution available from commercial satellites. Figure 4-9A is a 30 cm pan-sharpened grayscale band (reflected green light) acquired around 10:00 AM local time on November 29, 2017 of two tennis courts in Tokyo, Japan. At this time of day the sun is rising in the southeast. The image has been rotated 180° so shadows from the sun are cast toward the bottom of the page to minimize viewer issues with topographic inversion (discussed in Chapter 6). Figure 4-9B is our diagram of the courts that we label as left court and right court. In the image the eight doubles players are clearly shown as tiny bright pixels. Their relatively long shadows are due to image acquisition during mid-morning in late Autumn. Each player is given a label on Figure 4-9B.

The action on each court is frozen in an instant in time. On the left court player 1L is preparing to serve to the forecourt that is received by 4L. Her partner 3L is conservatively positioned behind her forecourt. Action is underway on the right court where 3R and 4R are poorly aligned on the upper side of the court. On the bottom side 2R has aggressively charged the net, but 1R is poorly positioned behind her. This brief interval suggests that the players in the left court are more experienced than those in the right court.

The SWIR bands are used to identify rooftop material as part of a fire-risk mapping program and to map post-fire burn severity (Warner and others, 2017). The longer wavelength SWIR bands penetrate smoke better than VNIR bands and pinpoint sites of active burning. For example, in Figure 4-10A, the original true color image shows smoke, haze, and clouds, which obscure the terrain and hinder detection of active fires. However, in Figure 4-10B, SWIR band 6 (2.215 μm wavelength) penetrates the smoke and haze to reveal the active fires and differentiates white smoke and haze from a moisture-rich cloud.

Moisture in crops and natural vegetation can be assessed with SWIR bands (Jensen, 2007) to improve monitoring of crop health. SWIR bands are also used to map and quantify how much crop residue is left behind, which improves the soil quality for future crops (DigitalGlobe, 2015). Land cover mapping is improved when SWIR bands are processed with VNIR bands. Clays, carbonates, and oil seeps and stains may

A. Pan-sharpened band (green). Satellite image © 2019 Maxar Technologies.

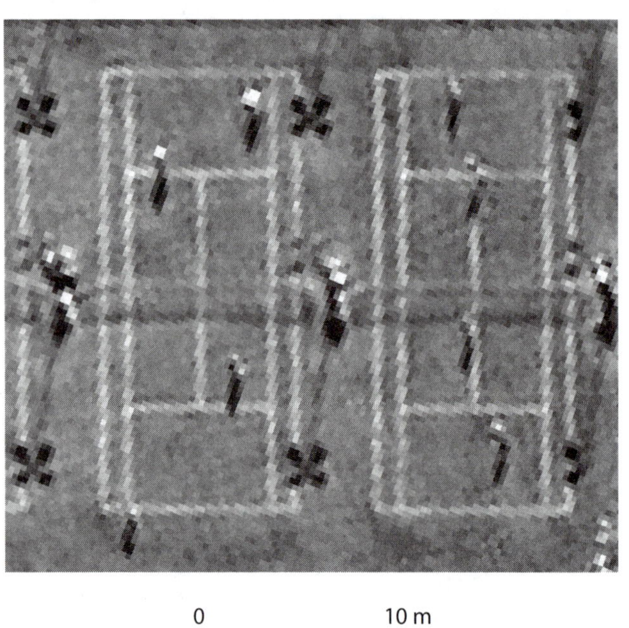

B. Diagram of two tennis courts. Squares and stipples are players and shadows.

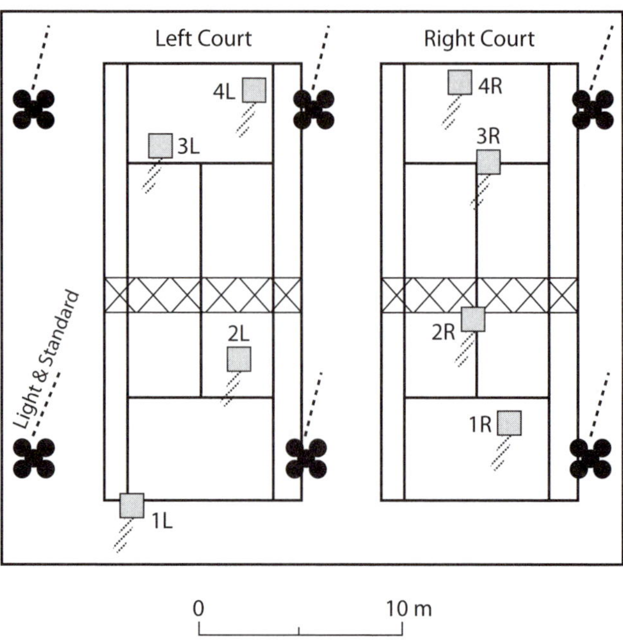

FIGURE 4-9 Pan-sharpened WorldView-3 image with 30 cm spatial resolution and diagram of two tennis courts, Tokyo, Japan.

A. True color image.

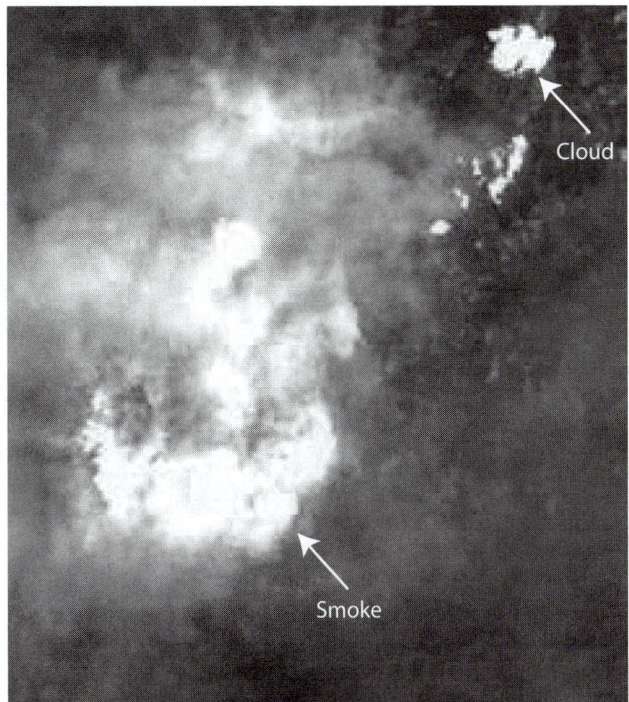

B. SWIR (2.215 μm).

FIGURE 4-10 WorldView-3 uses SWIR bands to penetrate smoke to aid in active fire detection. From Navulur and others (2014). Satellite images © 2019 Maxar Technologies.

be recognized and mapped with WorldView-3 SWIR bands, as these materials have unique spectral absorption features in the SWIR wavelengths (Figure 4-3). Kruse and others (2015) compared geologic mineral maps from hyperspectral AVIRIS and WorldView-3 images of hydrothermally altered terrain in Cuprite, Nevada. The eight SWIR bands of WorldView-3 provide new remote mineral mapping capabilities.

Microsatellites

Miniaturization, the manufacture of smaller mechanical, optical, and electronic instruments, has enabled a new class of remote sensing satellites and sensors, called *microsatellites*, to be developed and launched by the hundreds. Standards such as *CubeSat* are established and maintained for microsatellite design that lower cost, accelerate production time, and ensure alignment with the many requirements for launching payloads into space. The CubeSat standard was created by California Polytechnic State University, San Luis Obispo and Stanford University's Space Systems Development Lab in 1999 to facilitate access to space for university students. The CubeSat standard facilitates frequent and affordable access to space with most launch vehicles (cubesat.org). The standard has been adopted by hundreds of organizations worldwide.

Constellations of microsatellites are in orbit and more are being planned (Table 4-2). *Axelspace* is a Japanese company that plans a 50-satellite constellation. *UrtheCast* is a Vancouver, Canada company that plans a 16-satellite constellation. Currently UrtheCast has two high definition video cameras (1 m and 6 m spatial resolution) on the International Space Station (ISS). UrtheCast acquires pan (0.75 m spatial resolution) and 4-band multispectral (4 m spatial resolution) images with the 310 kg Deimos-2 satellite that was launched by Spain in 2014.

Planet Planet has a large operational constellation of satellites that consists of PlanetScope, RapidEye, and SkySat (planet.com). The PlanetScope constellation of over 120 satellites can image the Earth's entire landmass at 3 m spatial resolution every day with a capacity of 200 million km^2 per day. Planet also maintains the RapidEye constellation of five satellites with medium spatial resolution. In 2017, Planet acquired Terra Bella's 13 SkySat satellites that can image 400,000 km^2 per day. Planet has a global image archive for over 10 billion km^2 dating back to the 2009 launch of RapidEye.

Planet developed the *Dove* class of microsatellites that only weigh ~5 kg and measure 10 by 10 by 30 cm (Figure 4-11). The satellites follow the CubeSat 3U design standard (Planet, 2018). This San Francisco based company has deployed Doves from the ISS at an altitude of 400 km and an equatorial orbital path that is inclined 51.6° from the Earth's north–south axis (Figure 3-8) where most human populations live. The temporal resolution is better than for satellites that are in near-polar orbits. Doves are also placed into sun-synchronous, near-polar orbits to cover more northerly and southerly latitudes. On February 14, 2017, 84 Doves were simultaneously launched. Each Dove acquires multispectral images with 12-bit radiometric resolution and 3 to 5 m spatial resolution. Both natural color and color IR images are acquired.

Terra Bella, also owned by Planet, launched their first satellite in November 2013. Their *SkySat* satellite is a compact package (60 by 60 by 80 cm) that weighs less than 100 kg. The sensor acquires panchromatic and 4-band multispectral data (blue, green, red, NIR) with sub-90 cm spatial resolution (Table 4-2). Terra Bella launched three SkySat satellites between November, 2013 and June 2016. As of September 2016, the three SkySats had collected over 100 million images. In September 2016, four SkySats were simultaneously launched on an Arianespace Vega rocket from French Guiana and now orbit as a line of four satellites. The new generation SkySat-3 system includes a propulsion module to maintain orbit position and to improve spatial resolution (Planet, 2018).

FIGURE 4-11 Planet's Dove microsatellites on a rack, ready for launch. Image by Planet Labs, Inc. (planet.com).

MEDIUM SPATIAL RESOLUTION SATELLITES

Landsat (see Chapter 3), several other government systems, and some commercial systems acquire panchromatic and multispectral imagery with medium spatial resolutions between > 1.5 to ≤ 30 m (Table 4-3). The European Space Agency (ESA), USGS, and NASA provide their data at no cost. *Hyperion* collects 30 m hyperspectral data that is available from NASA via the USGS EarthExplorer website (earthexplorer.usgs.gov). Planet's RapidEye has a spatial resolution of 6.5 m (Planet, 2018).

Several countries operate their own medium resolution satellites. Nigeria's *NigerSAT-2* has been operating since 2011. Japan's *ALOS PRISM* (panchromatic remote-sensing instrument for stereo mapping) acquired imagery from 2006 to 2011.

TABLE 4-2 **Microsatellites.**

Operator	Planet	Planet	Axelspace	UrtheCast
Name	**Planetscope**	**Terra Bella**	**GRUS**	**Theia Iris**
Spatial Resolution (m)				
Pan	3.0 to 3.7	0.72 to 0.86	2.5	
VNIR		1.0	5.0	5
Video		1.1 (pan)		1 m and 6 m
Spectral Resolution (µm)				
Pan		0.40 to 0.90	0.450 to 0.900	
VNIR	0.455 to 0.515	0.45 to 0.515	0.450 to 0.505	0.47 to 0.57
	0.5 to 0.59	0.515 to 0.595	0.515 to 0.585	0.5 to 0.6
	0.59 to 0.67	0.605 to 0.695	0.620 to 0.685	0.6 to 0.7
	0.78 to 0.86	0.74 to 0.90	0.705 to 0.745	0.78 to 0.88
			0.770 to 0.900	
Radiometric resolution	12-bit	16-bit	12-bit	
Stereo	No	No	No	No
Swath (km)	24.6	8 (imagery)	50	50
		2 (video)		
First launch	2013	2013	2018	Theia on ISS
				Video on ISS
Proposed constellation	Over 120	24	50	16

Chile's *Sistema Satelital para Observación de la Tierra* (SSOT) was launched in 2011 and acquires 1.45 m panchromatic and 4-band multispectral VNIR images with 5.8 m spatial resolution. Spain launched *Deimos-1* in 2009. It carries a multispectral system that acquires green, red, and NIR images with spatial resolution of 22 m, and a 600 km swath. Deimos-1 is owned and operated by Deimos Imaging, an UrtheCast company.

Many medium resolution satellites are built by *Surrey Satellite Technology Ltd* (SSTL) of the United Kingdom. A subsidiary of SSTL, DMC International Imaging (DMCii) provides a commercial data service that processes large volumes of imagery downloaded from the multinational *Disaster Monitoring Constellation* (DMC). The DMC consists of low-cost small satellites that provide free satellite imagery for humanitarian use for major international disasters (dmcii.com). The United Kingdom, China, Nigeria, Turkey, and Algeria have satellites in the constellation.

Table 4-3 Medium spatial resolution satellites (> 1.5 to ≤ 30 m).

Operator	ESA	USGS	NASA	NASA	Planet
Name	**Sentinel-2**	**ASTER**	**ALI**	**Hyperion**	**RapidEye**
Spatial Resolution (m)					
Pan			10		
VNIR	10 and 20	15	30	30	6.5
SWIR	20	30	30		
Spectral Resolution (μm)					
Pan			0.48 to 0.69		
VNIR	0.445 to 0.550	0.52 to 0.60	0.433 to 0.453	0.4 to 1.2	0.44 to 0.51
	0.537 to 0.582	0.63 to 0.69	0.450 to 0.515	(35 visible)	0.52 to 0.59
▪ 10 m	0.646 to 0.684	0.76 to 0.86	0.525 to 0.605	(35 NIR)	0.63 to 0.685
	0.762 to 0.904				0.69 to 0.73
					0.76 to 0.85
▪ 20 m	0.693 to 0.713		0.630 to 0.690		
	0.731 to 0.749		0.775 to 0.805		
	0.766 to 0.794		0.845 to 0.890		
	0.848 to 0.880				
SWIR	1.54 to 1.68	1.600 to 1.700	1.2 to 1.3	1.2 to 2.5	
	2.07 to 2.32	2.145 to 2.185	1.55 to 1.75	(172 SWIR)	
		2.185 to 2.225	2.08 to 2.35		
		2.235 to 2.285			
		2.295 to 2.365			
		2.360 to 2.430			
TIR		8.125 to 8.475			
		8.475 to 8.825			
		8.925 to 9.275			
		10.250 to 10.950			
		10.950 to 11.650			
Radiometric resolution	12-bit	8-bit VNIR-SWIR	16-bit	16-bit	16-bit
		12-bit TIR			
Stereo	No	Yes	No	No	No
Swath (km)	290	60	37	7.5	77
Launched	2015	1999	2000	2000	2008

SPOT

Since 1986, the SPOT 1 to 7 series of satellites has acquired panchromatic and multispectral bands with a swath of 60 km (Figure 4-2). The early SPOT satellites (1, 2, 3, 4, and 5) had medium spatial resolution (Table 4-4). SPOT 6 and 7 have high spatial resolution (Table 4-1). For continuity, we are treating SPOT 1 to 5 satellites as a unit. All SPOT satellites acquire stereoscopic images. The SPOT satellites are in sun-synchronous orbits with a nadir-viewing temporal resolution of 26 days. Although SPOT 1 to 5 are no longer in operation, their images are a useful data source.

SPOT 1 to 4 SPOT 1 was launched in 1986 on an Ariane rocket from French Guiana. The identical SPOT 2 and SPOT 3 satellites were launched in 1988 and 1993. They collected a panchromatic image with 10 m spatial resolution and three "XS" color bands (green, red, and NIR) with 20 m spatial resolution. The images cover an area of 60 by 60 km. These three satellites ceased operation between 1990 and 2007. SPOT 4 was operational from 1998 to 2013 with spatial resolution of the color bands improved to 10 m (Table 4-4).

SPOT 5 Between 2002 and 2015 SPOT 5 acquired: panchromatic images with 2.5 m spatial resolution; three VNIR images (10 m resolution); and one SWIR image (20 m resolution). These images with 8-bit radiometric resolution are cost-effective for regional environmental monitoring and

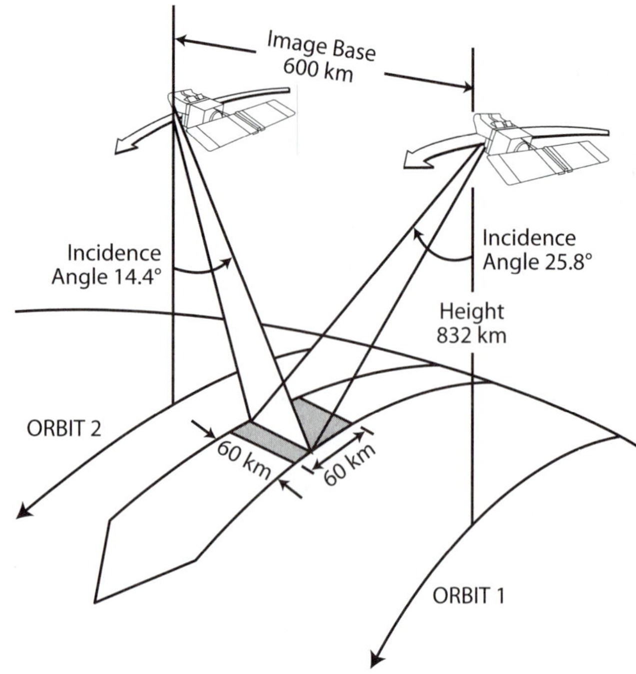

$$\frac{\text{Base}}{\text{Height}} = \frac{600\text{ km}}{832\text{ km}} = 0.72 \sim 5\text{X Vertical Exaggeration}$$

FIGURE 4-12 SPOT system for acquiring stereo images.

TABLE 4-4 SPOT satellites.

Operator	Airbus DS	Airbus DS	Airbus DS
Name	**SPOT 1, 2, and 3**	**SPOT 4**	**SPOT 5**
Spatial Resolution (m)			
Pan	10	10	2.5
VNIR	20	20	10
SWIR		20	20
Spectral Resolution (μm)			
Pan	0.51 to 0.73	0.61 to 0.68	0.48 to 0.71
VNIR		0.50 to 0.59 0.61 to 0.68	0.50 to 0.59 0.61 to 0.68
▪ 10 m		0.78 to 0.89	0.78 to 0.89
▪ 20 m	0.50 to 0.59 0.61 to 0.68 0.79 to 0.89		
SWIR		1.58 to 1.75	1.58 to 1.75
Radiometric resolution	8-bit	8-bit	8-bit
Stereo	Yes	Yes	Yes
Swath (km)	60	60	60
Launched–Ended	1986–2007	1998–2013	2002–2015

geologic mapping projects. SPOT 5 carried a high resolution stereoscopic imaging system that simultaneously recorded stereopairs with a swath width of 120 km and a swath length of up to 600 km. The panchromatic stereo-pairs were acquired with a spatial resolution of 10 m. These stereo-pairs are used to generate DEMs.

Stereo Images The SPOT 1 to 4 imaging systems included mirrors that could be tilted to acquire images as far as 475 km to the east or west of the orbit path. A typical stereo-pair was collected from two orbit paths over southern California with an incidence angle from the east of 25.8° and from the west at 14.4° (Figure 4-12). The incidence angle is the angle between the vertical and a line from the satellite to the center of the image. For this stereo-pair, the image base (the distance between the two orbit paths) is 600 km. At the SPOT height of 832 km, the base-height ratio is 0.72. The chart in Figure 2-17 shows that this ratio results in a 5X vertical exaggeration in the stereo model.

Figure 4-13 is a stereo-pair of SPOT pan images that were acquired with the stereo viewing geometry shown in Figure 4-12. We encourage the reader to view this stereo-pair with a stereoscope, which will display the 5X vertical exaggeration. Figure 4-14 is an interpretation map of the stereo-pair that covers a syncline called the Devil's Punchbowl located in southern California on the north flank of the San Gabriel Mountains, which are separated by the San Andreas fault from the Antelope Valley to the north. The San Andreas is an active fault with right-lateral, strike-slip displacement. The stereo model shows scarps, shutter ridges, and offset drainages that indicate active strike-slip faults. The last movement, and a magnitude 8 earthquake, on this stretch of the fault occurred in 1857. The Punchbowl syncline is located between the San Andreas fault and the Punchbowl fault to the south. The attitudes (dip and strike) of the steeply dipping beds in the flanks of the syncline are clearly seen in the stereo model. South of the Devil's Punchbowl, the Punchbowl and San Jacinto faults are recognizable on the stereo-pair by north-facing linear scarps and aligned northwest-trending valleys.

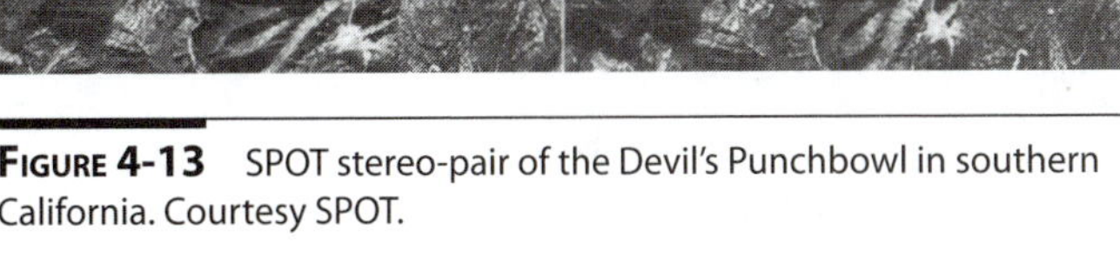

FIGURE 4-13 SPOT stereo-pair of the Devil's Punchbowl in southern California. Courtesy SPOT.

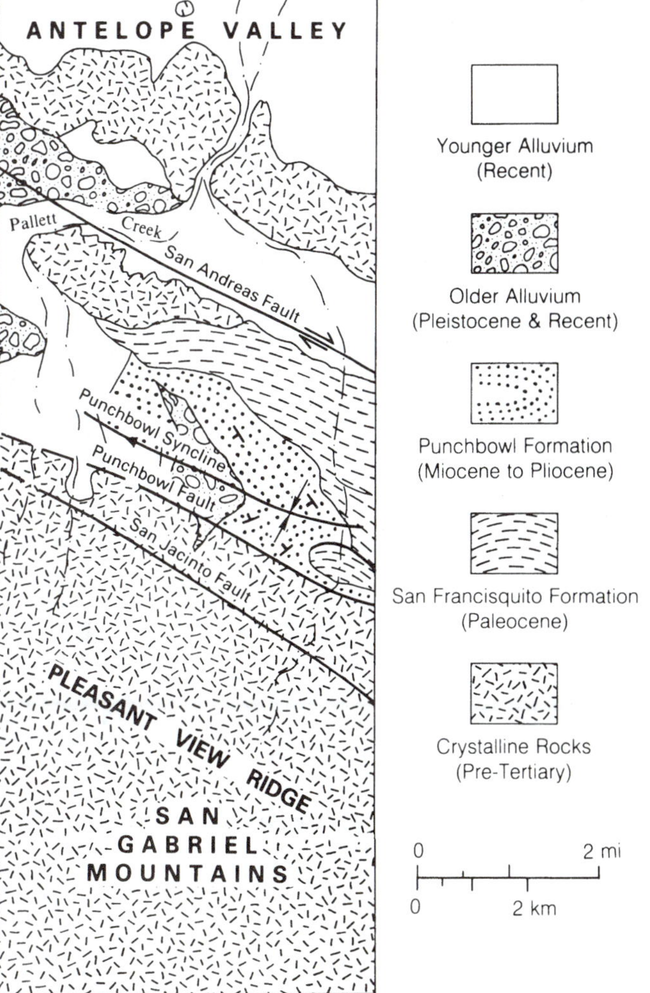

FIGURE 4-14 Interpretation map of the Devil's Punchbowl syncline.

SPOT XS Images Plate 12 is a SPOT XS color IR image of the Djebel Amour area in the Saharan Atlas Mountains of Algeria. In this arid region, the rare vegetation occurs in the *wadis*, or dry streambeds, and has red signatures on the color IR image. Figure 4-15 is a geologic map that Sabins interpreted from the image at a scale of 1:62,500. The map and image are greatly reduced for this book and some of the detailed geology may not be apparent. No geologic information was available when the units were used to portray them on the geologic map. The first task, therefore, was to define rock units for mapping.

Rock Units Figure 4-16 shows the vertical sequence of mappable rock units that was derived from the color IR image. The map symbols identify these rock units on the geologic map. Characteristics of the rock units are summarized below.

Recent Sand and Gravel: Unconsolidated deposits of recent sand and gravel with yellow and white signatures that partially obscure bedrock outcrops in the southeast portion of the image.

Sandstone Unit: As shown in the rock column, the youngest and uppermost bedrock formation consists of sandstone with minor shale interbeds. The Sandstone Unit crops out in the troughs of synclines and has a distinctive orange and grayish pink signature. In the southeast portion of the image it is mantled by a thin cover of sand that imparts a white and light yellow signature, but we recognize the unit by its topography and stratigraphic position. The Sandstone Unit erodes to broad dipslopes with steep antidip scarps. The base of the unit is mapped at the contact between the basal sandstone ledge and the slope formed on the underlying unit.

Sandstone and Shale Unit: This thin unit consists of an upper nonresistant shale and an underlying resistant sandstone. The shale weathers to slopes and the sandstone weathers to ridges. In the northwest the unit has a dark blue signature that is distinct from the orange of the overlying Sandstone Unit. The Sandstone and Shale Unit is relatively thin, but mapping it as a separate formation helps to define several structural features. The base of the unit is mapped at the contact between the ledge-forming sandstone and the slope of the underlying Shale Unit.

Shale Unit: This nonresistant unit weathers to broad valleys and lowlands with a distinctive light to medium blue signature. Three thin resistant beds, probably sandstone, are interbedded with the Shale Unit and weather to form distinctive narrow strike ridges. In the rock column these sandstones are designated as key beds A, B, and C. On the geologic map (Figure 4-15) each key bed is shown in a dashed line pattern, which helps define folds within the Shale Unit. The base of the Shale Unit is mapped at the contact between the slope-forming shale and the ledges and ridges of the underlying Carbonate Unit.

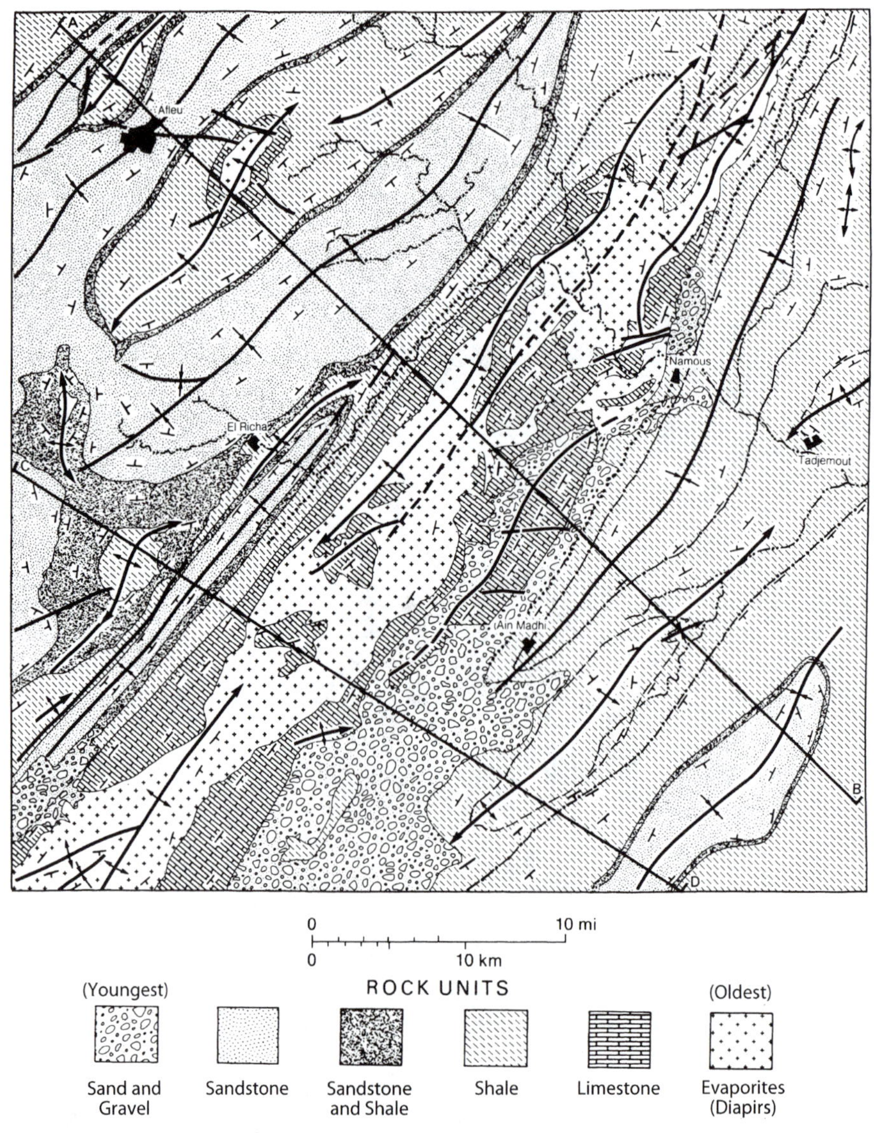

FIGURE 4-15 Geologic interpretation map of SPOT XS color IR image of the Djebel Amour area, Saharan Atlas Mountains, Algeria, shown in Plate 12.

Carbonate Unit: The Carbonate Unit weathers to rugged ridges with a dark brown signature. The carbonate lithology is inferred from the resistant nature and dark colors of the unit. The Carbonate Unit that crops out on the flanks of a major northeast-trending anticline where it is intruded shown diagrammatically in the rock column (Figure 4-16). The depositional, or stratigraphic, base of the Carbonate Unit is not exposed in the area, because all contacts with the older Evaporite Unit appear to be intrusive in nature.

Evaporite Unit: The oldest rocks in the area are the Evaporite Unit, which consists of massive, nonstratified, irregular outcrops with distinctive purple and very dark blue signature colors. The Evaporite Unit crops out in the core of a major northeast-trending, complex anticlinal belt. Between the towns of El Richa and Ain Madhi (Figure 4-15), large blocks of Carbonate Unit rocks are surrounded by evaporite and appear to be roof pendants within the diapiric mass. In the northeast corner of the map, at the northeast plunge of two major anticlines, the Evaporite Unit diapirs have completely penetrated the Carbonate Unit and intrude the overlying Shale Unit. These relationships are shown diagrammatically in the rock column (Figure 4-16). The base of the Evaporite Unit is not exposed in this area.

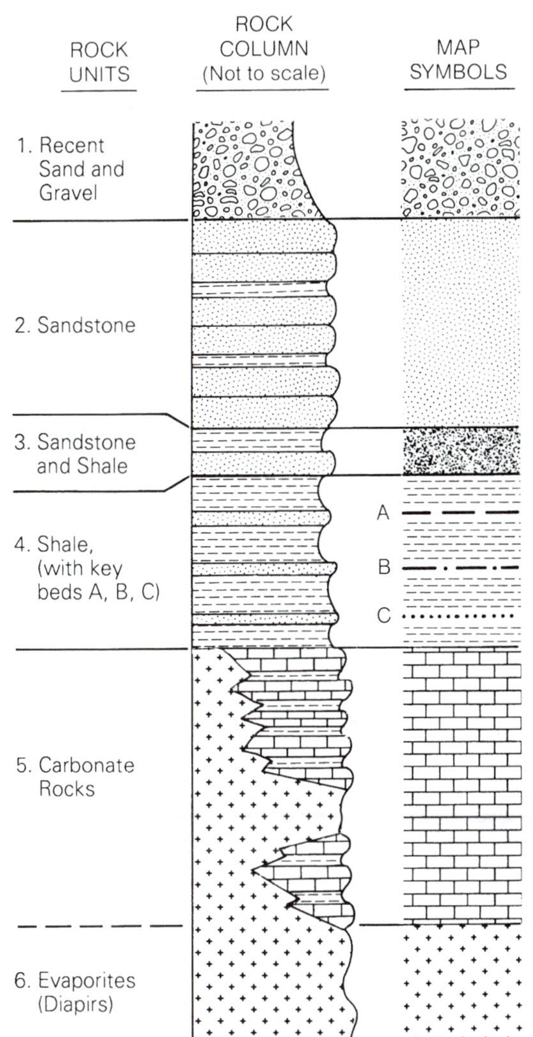

FIGURE 4-16 Stratigraphic column derived from SPOT XS color IR image of the Djebel Amour area, Saharan Atlas Mountains, Algeria, shown in Plate 12.

Structural Features Structural features (dip and strike, folds, and faults) were interpreted from the SPOT XS color IR image (Plate 12) using the same techniques described in Chapter 2. Dips and strikes are readily interpreted using the highlight and shadow method. The cross sections in Figure 4-17 have similar vertical and horizontal scales. The geologic structure is dominated by northeast-trending folds. In the northwest and southeast portions of the image the simple anticlines and synclines are broad and open, with gentle dips and few faults. These simple folds are separated by a complex northeast-trending belt of narrow, steeply dipping anticlines separated by strike faults that cut out the synclines. Crests of the central anticlines are extensively intruded by diapirs of the Evaporite Unit, as shown in the cross sections (Figure 4-17). These features resemble salt-cored anticlines in regions such as the Punjab Salt Range of Pakistan, Sverdrup Basin of the Canadian Arctic Islands, and the Salt Valley anticlines of Utah and Colorado.

Sentinel-2

ESA's Sentinel-2 is an along-track (push broom) system that acquires 13 bands in the VNIR and SWIR wavelengths with a swath width of 290 km. Four of the VNIR bands have spatial resolutions of 10 m while two of the SWIR bands have spatial resolutions of 20 m (Figure 4-3). The multispectral data for each Sentinel-2 scene is distributed in reflectance and as 100 by 100 km^2 orthoimage tiles that average 500 MB each. Each scene contains 12 tiles. Sentinel-2 is designed to support land monitoring, emergency management, security, and climate change activities (European Space Agency, 2015). The mission has twin polar-orbiting satellites (Sentinel-2A and 2B) in the same orbit, 180° apart. The temporal resolution (time interval between images of the same location) with two satellites is five days. The mission objectives are to provide:

- Systematic global acquisitions of high resolution, multispectral images allied to a high revisit frequency;
- Continuity of multispectral imagery provided by SPOT and Landsat instruments; and
- Observation data for the next generation of image-based products, such as land cover maps, land change detection maps, and geophysical variables.

Figure 4-18A includes Sentinel-2 satellite imagery acquired on February 10, 2016 with 10 m spatial resolution of an agricultural area surrounding Benue, Nigeria. The image illustrates the relative age and the geographic extent of burn scars on the landscape. The tone of burn scars changes from black to medium gray to light gray in the Sentinel-2, reflected red light (band 4) grayscale image as burn scars age and vegetation grows back on the soil (Figure 4-18A). The aging of burn scars on color IR images (not shown) transitions from black to medium gray to increasing shades of brown and red as vegetation grows back on the soil. The Remote Sensing Digital Database (waveland.com/Sabins-Ellis) contains the Sentinel-2 digital data used to generate the burn scar map.

The margins of burn scars are often defined by sharp linear boundaries many kilometers in length (Figure 4-18B). These boundaries represent roads, fences, and edges of property that are used by the local farmers to limit the extent of the burn. These controlled burns return nutrients to the soil as crop residue is recycled on site and rids the land of excess straw and stubble. Burn scars are interpreted and classified into five age-dependent categories (oldest, old, recent, most recent, and burning) with areas determined using GIS. Based on the amounts of area per category calculated below, approximately 402 km^2 of land has been subjected to controlled burns with the oldest burn being the largest.

Type	Area (km^2)
Burning	0.2
Most recent	16.1
Recent	23.9
Old	70.5
Oldest	291.0

The four more recent categories (old, recent, most recent, and burning) have more irregular boundaries compared to the oldest burn. In the northern portion of the image (Figure 4-18A), smoke can be seen drifting toward the west–southwest from an active fire (mapped as the small black polygon on Figure 4-18B). Involving the local farming community with satellite-borne observations and maps provides timely information for managing agriculture.

ASTER

The *Advanced Spaceborne Thermal Emission and Reflection Radiometer* (ASTER) is a multispectral system onboard NASA's Earth Observing System (EOS) Terra platform. The

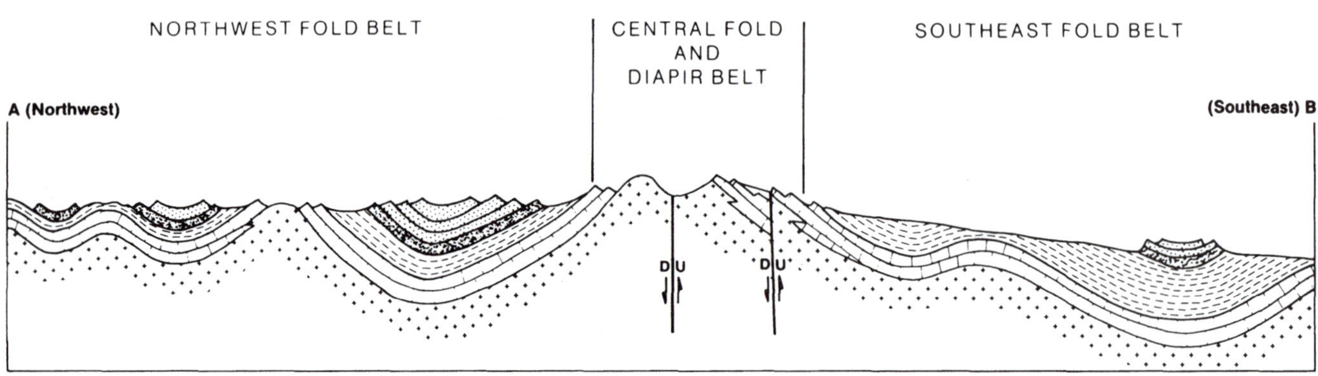

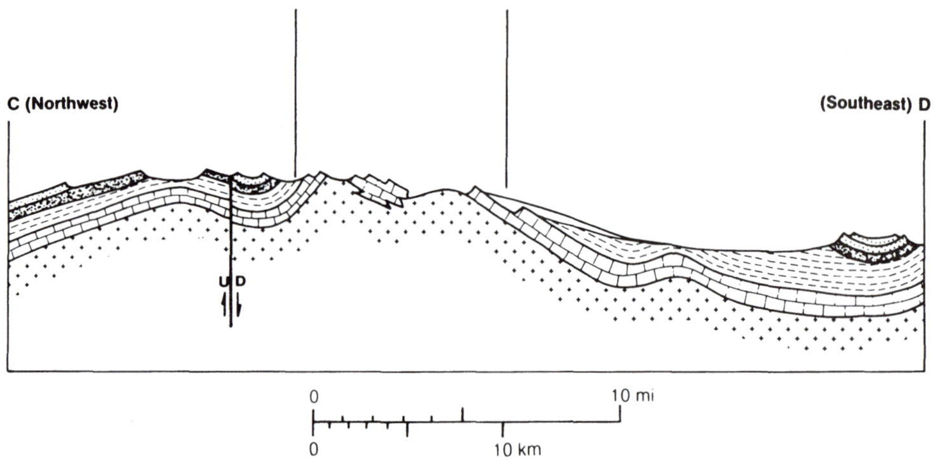

FIGURE 4-17 Cross sections derived from SPOT geologic map (Figure 4-15).

A. Sentinel-2 band 4 (red) image of burn scars. Courtesy European Space Agency.

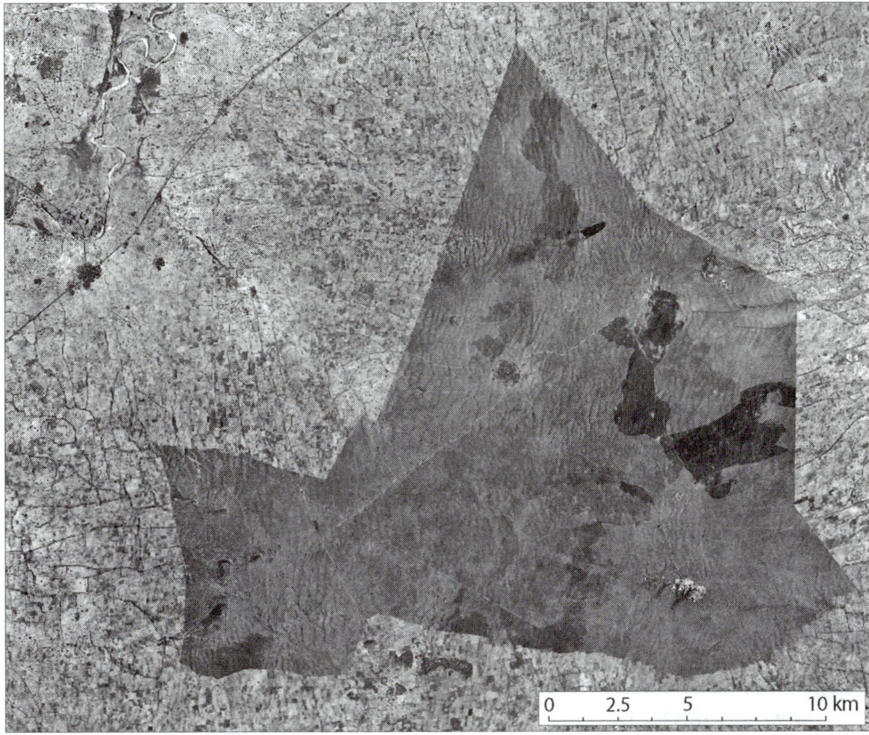

B. Interpretation map for the relative age of agricultural burn scars.

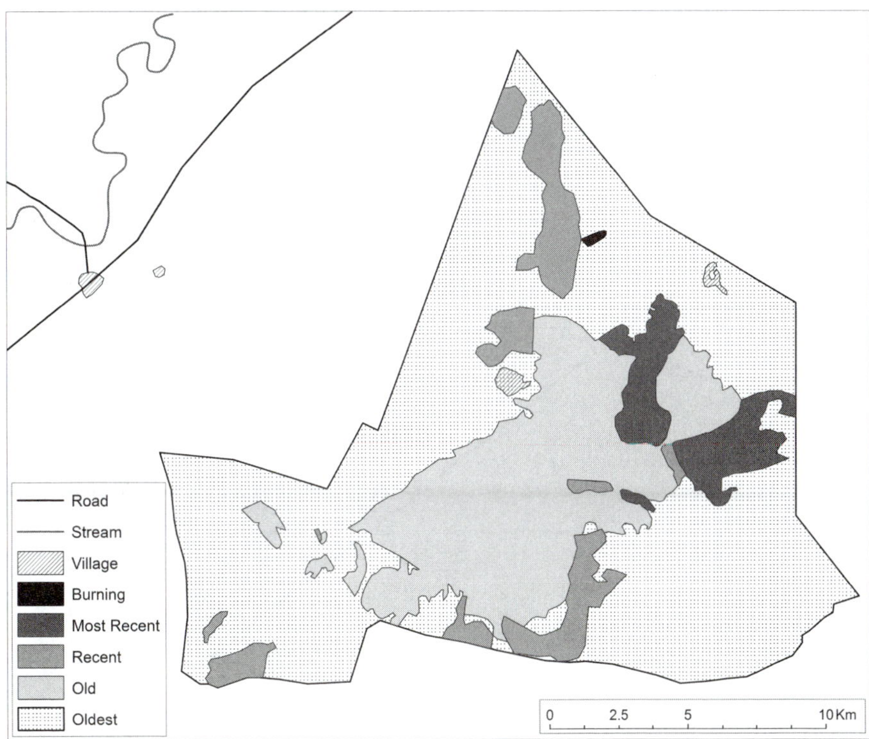

Figure 4-18 Sentinel-2 image of Benue, Nigeria illustrating the relative age and the geographic extent of burn scars on the landscape.

ASTER instrument was provided by the Japanese Ministry of Economy, Trade, and Industry to a collaborative science program supported by NASA, JPL, and the GSFC. ASTER acquires three VNIR bands and six SWIR bands (Figure 4-19). Spatial resolution is 15 m for the VNIR bands and 30 m for the SWIR bands. ASTER also acquires five TIR bands that are described in Chapter 5.

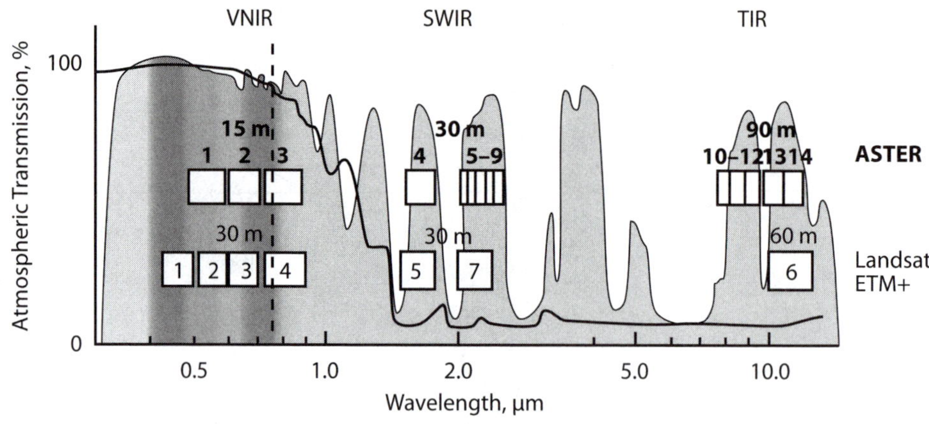

FIGURE 4-19 Spectral bands acquired by ASTER. The curve shows the atmospheric transmission spectrum. Courtesy NASA.

The system uses two imaging systems to acquire stereo-pairs along the orbital track (Figure 4-20). One system points directly downward and acquires a nadir image. The second system points to the rear of the satellite with a depression angle of 27.6° to acquire a second image of the scene. The resulting stereo-pair has a fixed base-height ratio of 0.6 for a 4X vertical exaggeration. Hundreds of thousands of these stereo pairs were used to build a global DEM with 30 m spatial resolution that covers 99% of the Earth's landmass.

The following six ASTER global products were developed between 2000 and 2015 (Abrams and others, 2015):

1. Global Digital Elevation Model (GDEM)—the most complete, highest resolution DEM available (reviewed in Chapter 7).
2. Global Emissivity Dataset (ASTER GED)—a global 5-band emissivity map of the land surface.
3. Global Urban Area Map (AGURAM)—a 15 m spatial resolution database of over 3,500 cities.
4. Volcano Archive (AVA)—an archive of over 1,500 active volcanoes.
5. ASTER Geoscience Map of Australia.
6. Global Land Ice Measurements from Space (GLIMS) Project.

Hyperion and ALI NASA's Earth Observing-1 (EO-1) satellite was launched in November 2000 from Vandenberg Air Force Base, California, carrying the *Hyperion* hyperspectral sensor and the *Advanced Land Imager* (ALI) multispectral sensor into a sun-synchronous orbit that follows one minute behind Landsat 7 ETM+. This close orbital proximity enabled

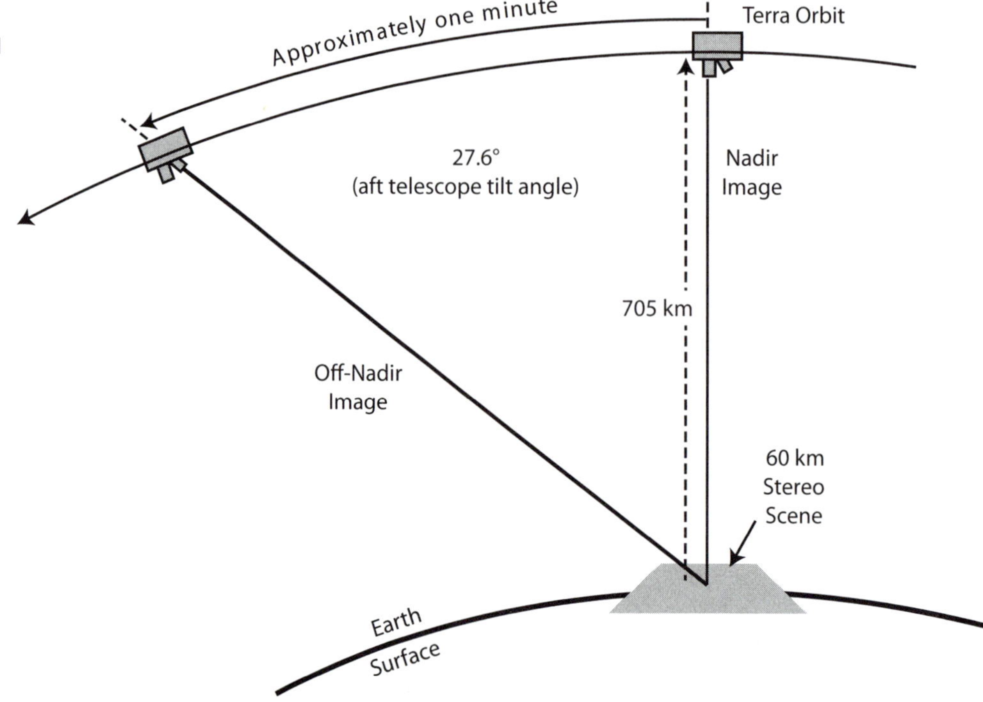

FIGURE 4-20 ASTER system for acquiring stereo images. Modified from Hirano and others (2003, Figure 1).

A. HMS *Beagle* in the Straits of Magellan, 1834. Frontispiece from Charles Darwin (1890), *Journal of Researches into the Natural History and Geology of the Various Countries Visited by HMS* Beagle *etc.* London: John Murray.

B. Military Sealift Command's USNS *Maury* (108 m [354 ft]) in the Gulf of Mexico, 2017. Photo by Bill Mesta, US Navy.

Figure 4-21 Oceanographic survey ships.

cross comparisons of the proven ETM+ instrument with the research-oriented Hyperion and ALI sensors (Beck, 2003). Hyperion and ALI data are available for download from the USGS's EarthExplorer website (earthexplorer.usgs.gov).

Hyperion collected 35 visible, 25 NIR, and 172 SWIR bands with 30 m spatial resolution (Table 4-3) (TRW, 2001). Because of the narrow swath width and limited off-nadir pointing capability, the temporal resolution is 200 days for Hyperion to revisit the same area on the Earth. Hyperion acquired improved data for Earth observation.

The ALI system was focused on technologies to reduce costs for future Landsat missions. ALI was an along-track (push broom) instrument with compact multispectral detector arrays, a new detector array for SWIR bands, and other new technologies (Lencioni and others, 2005). Landsat 8 implemented many of ALI technologies, including designing the OLI as an along-track sensor.

RapidEye Planet's *RapidEye* system uses five identical sensors in the same orbital plane that collect the standard four VNIR bands (blue, green, red, NIR), along with the red edge band that refines vegetation vigor/and stress measurements (RapidEye, 2008). RapidEye's five satellites were the first commercial Earth observation constellation and together can capture 4 million km^2 of imagery per day. Planet operates the RapidEye constellation and supports interactive data searches extending back to 2008.

Satellite Images, Coral Reefs, Darwin, and the *Beagle*

Darwin developed theories on the evolution of coral reefs almost 200 years ago. He also developed a classification of reefs that is still in use. Today, his insights can be explored and better understood with satellite images that have high to medium spatial resolution.

Coral reefs are organic structures that consist of calcium carbonate (calcite) deposited by polyps that grow in the euphotic zone of tropical and subtropical oceans. The euphotic zone is the sunlit upper ~30 m where food is produced for the polyps. With time and a favorable environment the reefs grow to wave-breaking features that can extend laterally for tens of kilometers.

When Britain ruled the seas in the nineteenth century the admiralty sent ships on scientific missions to collect information on their far-flung possessions and to "fly the flag." In 1831, Charles Darwin sailed as the newly graduated young naturalist/geologist on the epic five-year voyage of the 10-gun, 27 m (90 ft) brig HMS *Beagle* (Figure 4-21A), which circled the globe with numerous shore expeditions (Darwin, 1842). Among his possessions, Darwin carried the first volume of Charles Lyell's newly printed book, *Principles of Geology*.

We now fast-forward to the early twenty-first century with the satellite images found in Plate 13. These images of reefs were acquired by four imaging systems with medium (30 m) to high (1.5 m) spatial resolution. Tables 4-1 and 4-3 list the characteristics of each system. The following discussion will focus on interpreting the images to illustrate Darwin's three classes of coral reefs. Cross sections of the reef classes are also provided.

Plate 13A is an ALI image of the coral reef atoll Cocos (Keeling) Islands, which the *Beagle* visited for 10 days on its homeward-bound route in April of 1836. Atolls are narrow, irregular, circular rims of growing coral that enclose relatively shallow lagoons. The columnar atolls occur in depths down to several thousands of meters. British geologists were aware of atolls from the reports of whalers and other seamen. In his chapter on atolls Lyell used the conventional explanation that atolls grew on the rims of submerged volcanic calderas. Darwin used the *Beagle*'s small launch and a sounding line to acquire depth readings in and around the atoll. The weight at the end of the line was cup-shaped on the bottom and filled with tallow, which collected a sample of bottom sediment. The samples confirmed that live coral was confined to the euphotic zone and was underlain by dead coral, rather than by a volcanic caldera, which disproved the conventional theory. Darwin then needed an alternative for Lyell's explanation for the origin of atolls. Previously the *Beagle* had made several port calls along the east and west coasts of southern South America, which gave him opportunities to record local geology using his trusty compass and geology hammer. The field excursions were also reprieves from his endemic seasickness. On high mountain peaks he collected marine fossils that proved the enclosing sedimentary rocks had emerged from the ocean. Darwin then reasoned that if the Earth could rise it could also sink to make space for sediment. The concept of subsiding islands is the basis for his classification system for reefs.

Darwin's Three Classes of Reef Development

In his 1842 report, Darwin recognized three classes of reef: fringing reefs, barrier reefs, and atolls (Figure 4-22). Reefs in the Pacific Ocean are typically associated with volcanic islands that gradually sink and disappear beneath the ocean. Darwin's three reef classes are identified by the dwindling footprint and eventual disappearance of the subsiding central islands.

A. Fringing reef.

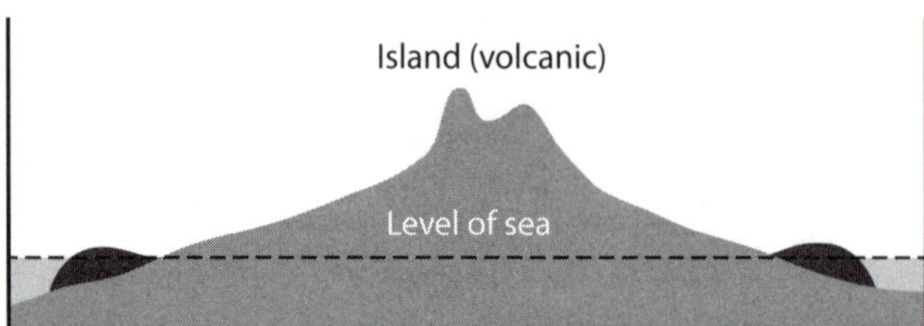

B. Barrier reef.

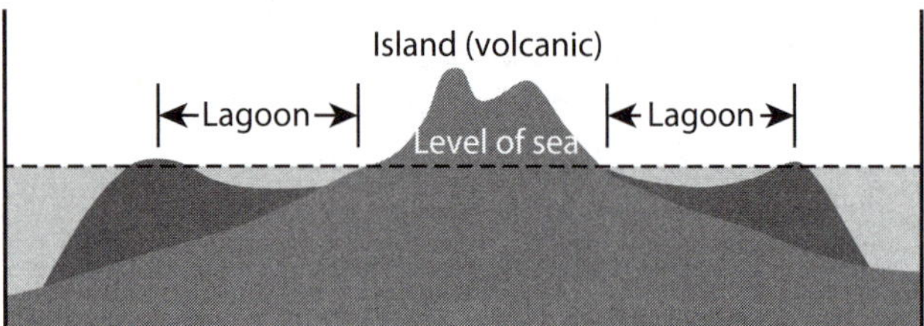

C. Atoll.

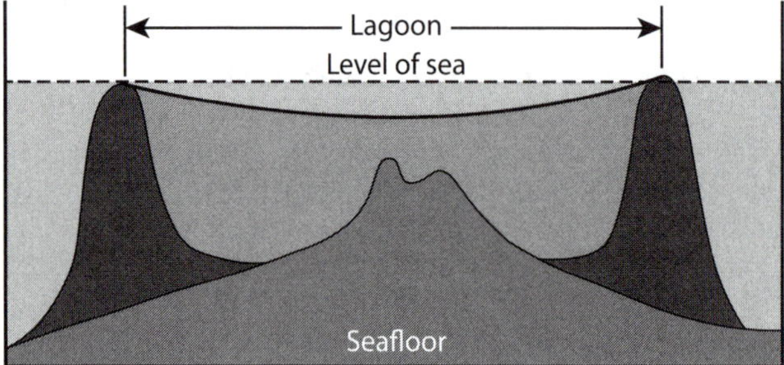

FIGURE 4-22 Darwin's (1842) three reef classes, which are used today.

Stage 1: Fringing Reefs Plate 13B shows the island of Mo'orea, which represents the fringing reef class. Plate 13B is an ASTER image with 15 m spatial resolution that shows the abundant vegetation cover. The older, established reefs in the region periodically reproduced and ejected swarms of tiny polyps into the ocean. Currents carried the polyps until some polyps reached islands such as Mo'orea, where they took hold and expanded to form a fringing reef. A fringing reef is the narrow, medium blue band in Plate 13B that is attached to the north, southwest, and south margins of island. The thin, outermost white line shows ocean waves breaking on the front of the reef. On the east margin of the island a lagoon (dark blue) separates the reef (medium blue) from the island (green). The lagoon is the initial phase of subsidence that leads to the barrier reef class. The interpretation map (Figure 4-23B) and the profile (Figure 4-23C) show fringing reefs.

Stage 2: Barrier Reefs Plate 13C shows the island of Bora Bora, which represents the barrier reef class. Plate 13C is a SPOT 6 image pan-sharpened to 1.5 m spatial resolution. Originally the island was larger with a fringing reef like Mo'orea. As Bora Bora subsided, the former fringing reef grew vertically to remain in the euphotic zone. The reef also maintained its geographic position. As subsidence continued, the footprint of the island became smaller, which caused a lagoon to appear between the reef and the island and expand to a width of several kilometers.

Figure 4-24B is our interpretation map of the Bora Bora image in Plate 13C. The island is surrounded on the northwest, northeast, and east by the barrier reef that is up to 0.8 km wide and covered with vegetation. The southern and western barrier reef is narrow and lacks vegetation. The narrow white line on the seaward border is the surf zone on the reef front. The reef is separated from the island by a lagoon that is floored by two concentric color zones. The zone with light to medium blue signatures are back-reef deposits that largely consist of detritus from the reef produced by storm waves crashing onto and over the reef. The belt of back-reef sediment is narrow on the north and east where vegetation helps protect the reef. On the west and south the broad belt of back-reef deposits extends almost to the island because the narrow unvegetated reef is unprotected from storm waves. These relationships are shown in our cross section (Figure 4-24C), which matches the features shown on Darwin's profile of his barrier reef class (Figure 4-22B).

Stage 3: Atolls Plate 13D shows the Maldives atoll, which represents the atoll class. Plate 13D was acquired by Sentinel-2 with 10 m spatial resolution. The original volcanic island and most back reef deposits have subsided below the limit of visibility within the lagoon. The reef, however, has continued to grow upward in the euphotic zone and ranges from 1 to 3 km wide. A narrow crescent of back reef deposits with a medium blue signature does occur inside the southwest bend of the atoll. There is a 5 km gap in the southwest portion of the atoll. A small group of patch reefs occur just inside the gap. The profile of the atoll (Figure 4-25C) shows the third and final class of Darwin's three reef classes.

A. ASTER satellite image with 15 m spatial resolution (Plate 13B). Courtesy USGS.

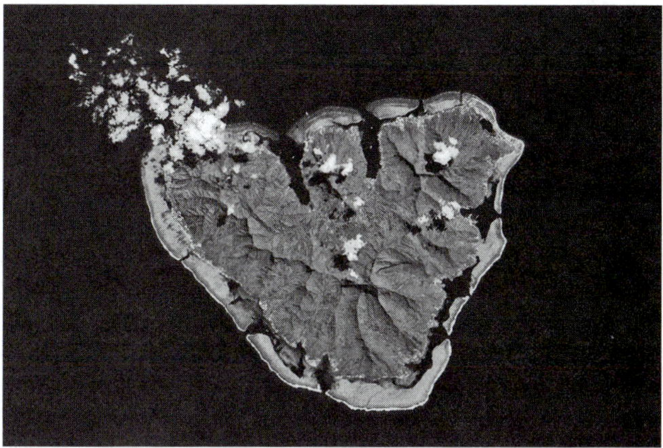

B. Interpretation map.

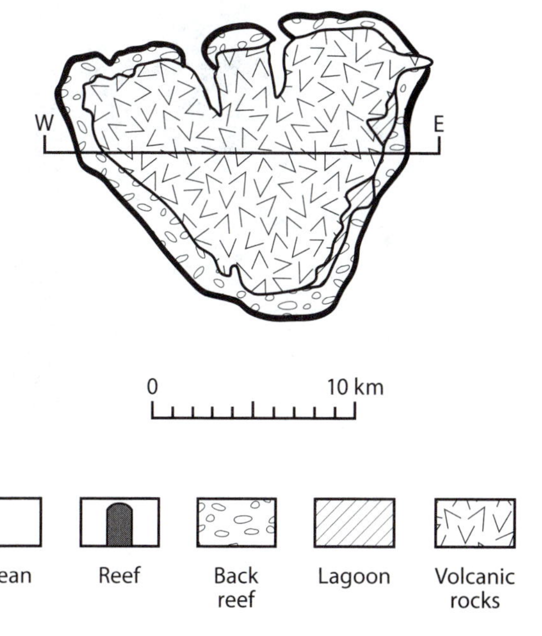

C. Cross section and legend.

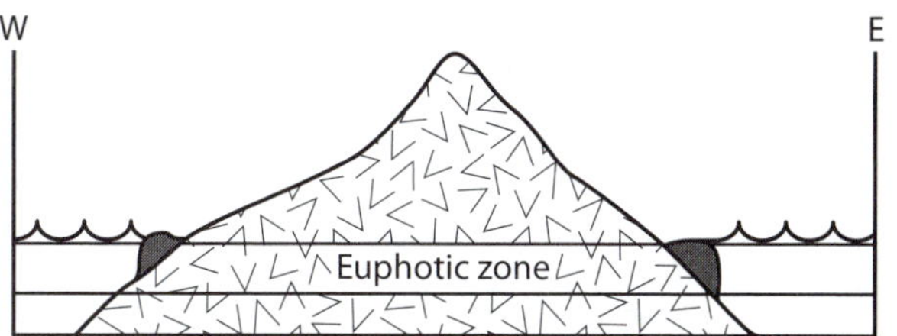

Ocean | Reef | Back reef | Lagoon | Volcanic rocks

Figure 4-23 Fringing reef, Mo'orea Island.

A. SPOT 6 satellite image with 1.5 m spatial resolution (Plate 13C). © Airbus DS 2019.

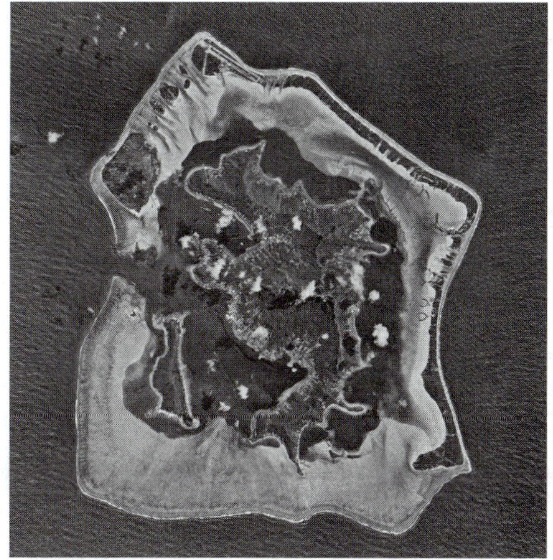

B. Interpretation map.

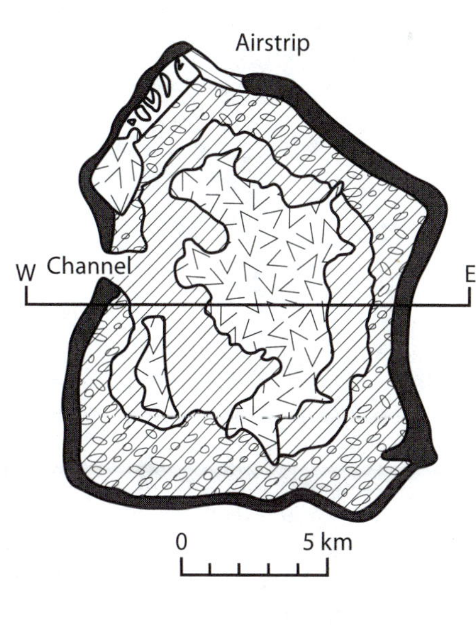

C. Cross section and legend.

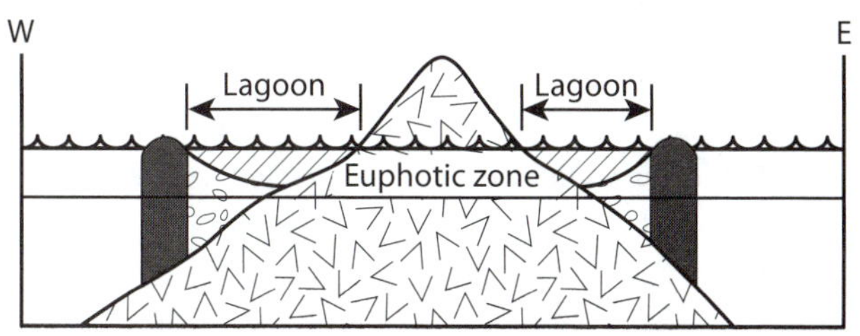

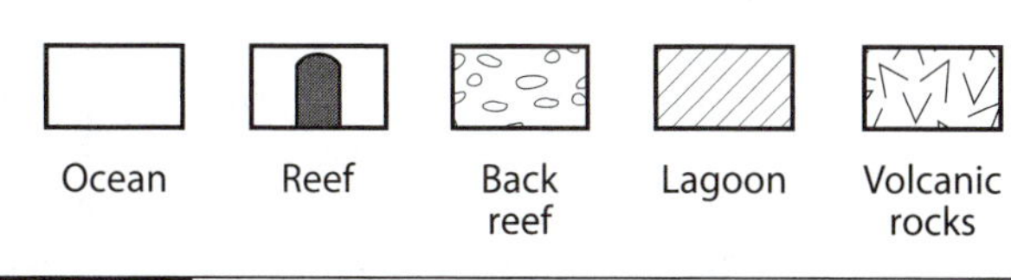

Figure 4-24 Barrier reef. The island of Bora Bora is partially submerged.

A. Sentinel-2 satellite image with 10 m spatial resolution (Plate 13D). Courtesy European Space Agency.

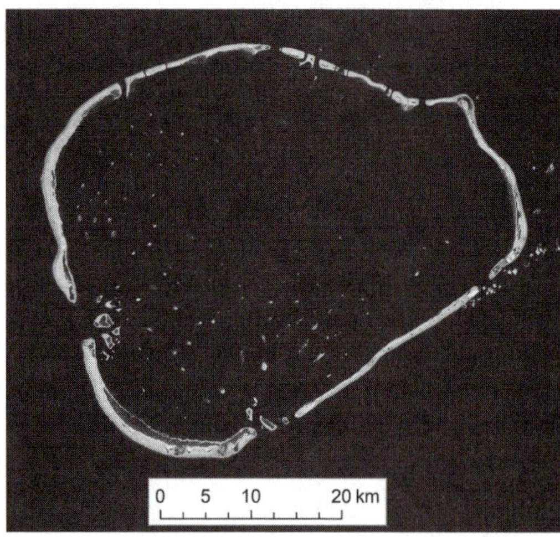

B. Interpretation map.

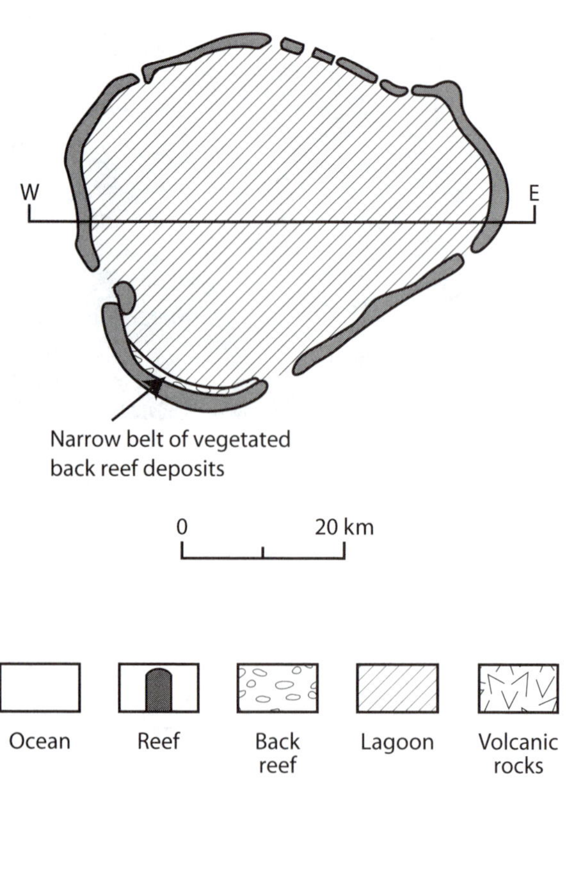

C. Cross section and legend.

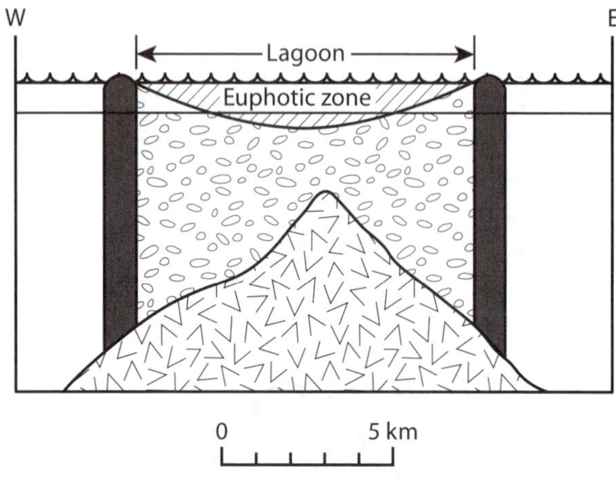

FIGURE 4-25 Atoll. A Maldives island is completely submerged.

LOW SPATIAL RESOLUTION SATELLITES

NASA, NOAA, the US Department of Defense (DOD), the former Soviet Union, India, and Japan all launched low spatial resolution (> 30 m ground resolution cells), multispectral satellites into polar orbit starting in the 1970s. NASA's EOS program was conceived in the early 1980s and took shape in the early 1990s (eospso.nasa.gov).

EOS is a coordinated series of remote sensing satellites for long-term global observations of the land surface, biosphere, solid Earth, atmosphere, and oceans. As of early 2019, there are 26 EOS missions in orbit. NOAA's National Environmental Satellite, Data, and Information Service provides reliable and timely access to global meteorologic and oceanographic data collected by an international array of polar and geostationary satellites. The onboard instruments collect specific wavelengths to support the scientific objectives of the satellites. The polar-orbiting satellites are summarized in Table 4-5.

Terra

Terra was launched in 1999 into a sun-synchronous orbit that travels in formation with Landsat 7 and 8 and NASA's EOS *Aqua* satellite. Aqua is described in a later section. Terra and Aqua function together, with Terra crossing the equator in the morning and Aqua crossing the equator in the afternoon. Terra is a joint mission between the United States, Japan, and Canada that carries the following five instruments: (1) ASTER, (2) CERES, (3) MISR, (4) MOPITT, and (5) MODIS. The instruments are described as follows:

1. *ASTER* (*Advanced Spaceborne Thermal Emission and Reflection Radiometer*): ASTER is a medium spatial resolution system. The VNIR and SWIR systems were discussed earlier in this chapter. The TIR system is described in Chapter 5.
2. *CERES* (*Clouds and the Earth's Radiant Energy System*): CERES measures energy from the Earth's surface to the top of the atmosphere to study clouds and their role in the Earth's heat budget. The first CERES instrument

was launched aboard the Tropical Rainfall Measuring Mission (TRMM) satellite in November 1997. Several CERES systems are in orbit on different platforms to provide temporal sampling of clouds and radiative fluxes that vary throughout the day (CERES is described further in Chapter 11).

3. *MISR* (*Multi-angle Imaging SpectroRadiometer*): MISR has nine sensors that collect multispectral data (blue, green, red, and NIR bands). One sensor is nadir-looking, four are forward-looking, and four are backward-looking, which acquires data for stereo models. The nine sensors measure heights of clouds, smoke, and volcano plumes. MISR also measures the scattering of sunlight by forests, deserts, clouds, smoke, and pollution.
4. *MOPITT* (*Measurements of Pollution in the Troposphere*): MOPITT is a Canadian instrument that divides the globe into approximately 1,000,000 cells with an area of ~22 km^2 each. MOPITT measures carbon monoxide and methane in the lower atmosphere of each cell every four days.
5. *MODIS* (*Moderate Resolution Imaging Spectroradiometer*): MODIS is carried on the Terra and Aqua satellites. The most comprehensive EOS sensor is MODIS, which:
   - Detects a wide spectral range of electromagnetic energy.
   - Supports a broad range of atmosphere, land, and water data products.
   - Takes measurements at three spatial resolutions:

	Spatial Resolution (m)
Bands 1 and 2	250
Bands 3 to 7	500
Bands 8 to 36	1,000

   - Takes measurements 24/7, with a 2,330 km wide swath (Figure 4-26).

MODIS records 36 bands that range from visible blue (band 8, 0.405 to 0.420 μm) to TIR (band 36, 14.085 to 14.385 μm). The bands are assigned to 11 primary use categories (Table 4-6).

MODIS acquires a comprehensive collection of global images every two days that complements other imaging systems, such as Landsat (NASA, 2002). Monitoring and assessing conditions on the Earth's surface is critical to understanding the environmental impacts of weather, climate change, and human activities. MODIS provides global maps of several land surface characteristics, including surface reflectance, albedo (the percent of total solar energy that is reflected back from the surface), land surface temperature, and vegetation indices. Vegetation index maps show density and vigor of vegetation. The maps provide the basis for MODIS's real-time global monitoring of subtle changes in vegetation that may signal biosphere stress, such as pollu-

Table 4-5 Low spatial resolution satellites (> 30 m).

Operator	NASA	NASA	NOAA	NASA	DOD
Name	MODIS[a]	AIRS[b]	AVHRR	SeaWiFS	DMSP F17 and F18
Spatial resolution (m)	250		1,090	1,100	560
	500			4,500	2,700
	1,000				
Spectral resolution (number of bands)	36	4 (VNIR)	5	8	2
		2,378 (TIR)			
Wavelength regions	VNIR	VNIR	Green	VNIR	VNIR
	SWIR	TIR	NIR		TIR
	TIR		SWIR		
			TIR		
Swath (km)	2,330	1,650	2,700	2,800	3,000 (VNIR)
Launched or started series	1999 and 2002	2002	1978	1997	1972
Number of satellites in series	2	1	15	1	18
Temporal resolution (days)	1	1	2× per day	1	4× per day

[a] Instrument onboard Terra and Aqua satellites.
[b] Instrument onboard Aqua satellite.

tion, drought, or temperature extremes, which in turn could be used to predict and prevent wildfire danger or crop failure (NASA, 2002).

Aqua and the AIRS Instrument

Aqua was launched in 2002 as a joint EOS mission between the United States, Brazil, and Japan. Aqua collects information about the Earth's water cycle, including evaporation from the oceans, water vapor in the atmosphere, clouds, precipitation, soil moisture, sea ice, land ice, and snow cover on the land and ice. Additional variables also being measured by Aqua include radiative energy fluxes, aerosols, vegetation cover on the land, phytoplankton and dissolved organic matter in the oceans, and air, land, and water temperatures. Aqua transmits high-quality data from four of its six instruments, *Atmospheric Infrared Sounder* (AIRS), *Advanced Microwave Sounding Unit* (AMSU), CERES, and MODIS, and reduced quality data from a fifth instrument, the *Advanced Microwave Scanning Radiometer on EOS* (AMSR-E). The sixth Aqua instrument, the *Humidity Sounder for Brazil* (HSB), collected approximately nine months of high-quality data but failed in February 2003 (aqua.nasa.gov).

AIRS works with the AMSU as the primary observing system to study the global water and energy cycles, climate variations and trends, and the response of the climate system to increased greenhouse gases (NASA, 2001). AIRS uses cutting-edge infrared technology to create three-dimensional maps of air and surface temperature, water vapor, and cloud properties. AIRS measures upwelling radiance in 2,378 spectral channels (bands) in the infrared from 3.74 to 15.4 μm. A set of four channels in the VNIR observes wavelengths from 0.4 to 1.0 μm to provide cloud cover and spatial variability information. AIRS can also measure trace greenhouse gases such as ozone, carbon monoxide, carbon dioxide, and methane.

TABLE 4-6 MODIS primary use categories.

Primary Use	Band
Land/cloud/aerosols boundaries	1 and 2
Land/cloud/aerosols properties	3 to 7
Ocean color/phytoplankton/ biogeochemistry	8 to 16
Atmospheric water vapor	17 to 19
Surface/cloud temperature	20 to 23
Atmospheric temperature	24 and 25
Cirrus clouds/water vapor	26 to 28
Cloud properties	29
Ozone	30
Surface/cloud temperature	31 and 32
Cloud top attitude	33 to 36

Source: NASA Level-1 and Atmosphere Archive Distribution System, Distributed Active Archive Center (ladsweb.modaps.eosdis.nasa.gov/missions-and-measurements/modis).

Advanced Very High Resolution Radiometer

The *Advanced Very High Resolution Radiometer* (AVHRR) is a cross-track multispectral scanner that acquires images with a swath width of 2,700 km (Figure 4-26). This wide

A. Terrain coverage for AVHRR and MODIS compared with Landsat.

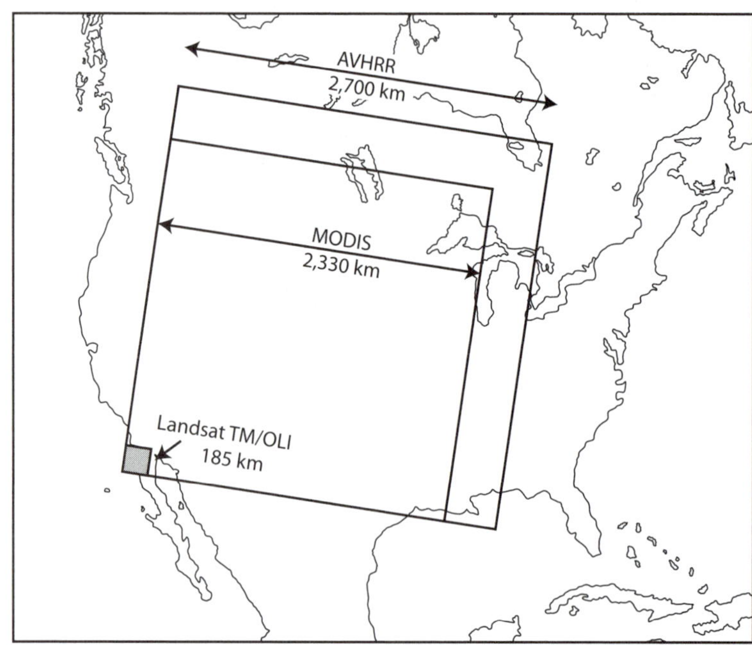

B. Ground resolution cells for AVHRR, MODIS, and Landsat.

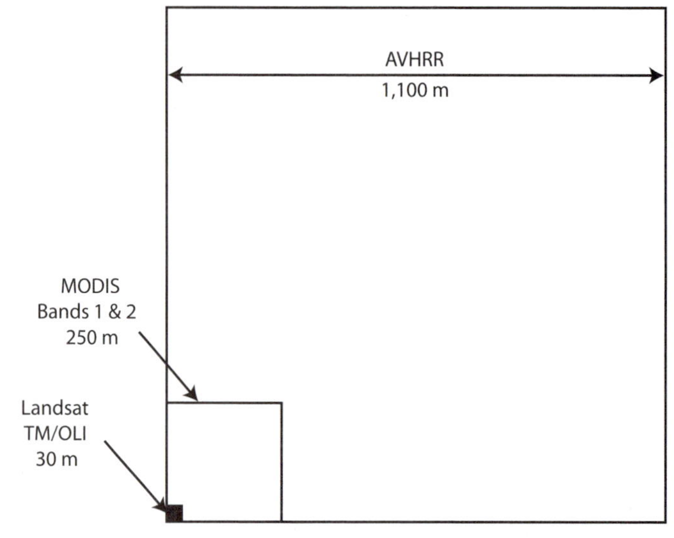

FIGURE 4-26 Terrain coverage patterns acquired by Landsat, MODIS, and AVHRR.

swath combined with the 14 daily sun-synchronous orbits of a NOAA satellite provides complete coverage of the Earth each day. The first AVHRR instrument was launched in 1978 with the most recent system (AVHRR/3) carried on NOAA-15 and launched in 1998. NASA has at least two AVHRR satellites in orbit at all times, with one satellite crossing the equator in the early morning and early evening and the other crossing the equator in the afternoon and late evening. Together the satellites provide twice daily global coverage for any region on the Earth. The European Organisation for the Exploitation of Meteorological Satellites (EUMETSAT) added the AVHRR/3 instrument to their MetOp satellites that were launched in 2006, 2012, and 2018.

AVHRR acquires one visible band, one reflected NIR band, one reflected SWIR band, and three TIR bands (Table 4-7) (Figure 4-3), all with a spatial resolution of 1.09 km. All six spectral channels of the AVHRR/3 are registered so that they all measure energy from the same spot on the Earth at the same time. All six channels are also calibrated so that the signal amplitude in each channel is a measure of the scene radiance.

AVHRR data provide opportunities for studying and monitoring vegetation conditions in ecosystems including forests, tundra, and grasslands. Applications include agricultural assessment, land cover mapping, producing image maps of large areas such as countries or continents, and tracking regional and continental snow cover. AVHRR data are also used to retrieve various geophysical parameters such as sea surface temperatures and energy budget data.

Figure 4-27 is a subscene of southern Louisiana that shows the three commonly used bands of AVHRR images. Band 1 (Figure 4-27A) records visible light. In the land areas, variation in gray tones relates to differences in vegetation. The cities of Baton Rouge and New Orleans have distinct dark signatures. In the water areas, brighter tones correlate with higher concentrations of suspended silt and mud. The Mississippi River and its discharge into the Gulf of Mexico are notably bright. Plumes of sediment-laden water discharged into Atchafalaya Bay and Lake Pontchartrain form conspicuously bright patches. Clouds are bright.

Band 2 (Figure 4-27B) records reflected IR energy. Water bodies are clearly distinguished from land by their very dark signatures, which are caused by absorption of these wavelengths. The thin clouds present in the band 1 image are readily penetrated by reflected IR energy and are not visible in band 2.

Band 4 (Figure 4-27C) records TIR energy. Bright signatures represent relatively cool radiant temperatures, and dark signatures are relatively warm. In this early spring season, water in the Gulf of Mexico is warm (dark tones) relative to the freshwater discharged at the Mississippi Delta and Atchafalaya Bay (bright tones). Lake Pontchartrain and adjacent lakes are also cool. The stringers of clouds are cool.

SeaWiFS

Sea-viewing Wide Field-of-view Sensor (SeaWiFS) was launched in 1997 and ceased operating in 2010. The SeaWiFS instrument collected eight bands ranging from blue to NIR (Table 4-8) with a spatial resolution of 1.1 km and 4.5 km. The swath width was 2,800 km with 1.1 km pixels.

The purpose of the SeaWiFS Project was to provide quantitative data on global ocean bio-optical properties to the oceanographic community. The concentration of microscopic marine plants, called phytoplankton, can be derived from satellite observation through ocean color and correlated with ocean color. Subtle changes in ocean color signify various types and quantities of marine phytoplankton, the knowledge of which has both scientific and practical applications (NASA SeaWiFS Project, 2017). Ocean color is discussed in Chapter 11.

Defense Meteorologic Satellite Program

Satellites of the *Defense Meteorological Satellite Program* (DMSP) provide global visible, infrared, and passive microwave data as well as other specialized meteorological, oceanographic, and solar-geophysical data in support of the DOD, Department of Commerce, and NASA operations (Hall, 2001; Strom and Iwanaga, 2005). In December 1972, DMSP data were declassified and made available to the civil and scientific communities.

A pair of polar-orbiting, sun-synchronous DMSP satellites cover the Earth in similar fashion to NOAA's AVHRR system. The existing generation of DMSP satellites (F17

Table 4-7 AVHRR/3 characteristics.

Channel Number	Resolution at Nadir (km)	Wavelength (μm)	Typical Use
1	1.09	0.58 to 0.68	Daytime cloud, surface, and vegetation mapping.
2	1.09	0.725 to 1.00	Land-water boundaries, vegetation mapping.
3A	1.09	1.58 to 1.64	Snow and ice detection.
3B	1.09	3.55 to 3.93	Night cloud mapping, sea surface temperature, hot targets such as fires and volcanoes.
4	1.09	10.30 to 11.30	Night cloud mapping, sea surface temperature.
5	1.09	11.50 to 12.50	Sea surface temperature.

Source: NOAA Satellite Information System (noaasis.noaa.gov/NOAASIS/ml/avhrr.html).

and F18) began operation in 2006 with four instruments that have the following swath widths:

- Visible and infrared images: ~3,000 km.
- Microwave images: ~1,700 km.
- Temperature sounder: ~1,500 km.
- Water vapor profiler: ~1,500 km.

The *Operational Linescan System* (OLS) is the primary sensor on DMSP satellites. OLS measures visible (0.4 to 1.1 µm) and TIR (10.25 to 12.6 µm) wavelengths to provide day and night cloud cover images. The satellite collects data at a 0.56 km spatial resolution, which is averaged on board, to produce global coverage at 2.7 km spatial resolution. All of

TABLE 4-8 SeaWiFS bands.

Band	Wavelength (nm)
1	402 to 422
2	433 to 453
3	480 to 500
4	500 to 520
5	545 to 565
6	660 to 680
7	745 to 785
8	845 to 885

A. Band 1, red (0.58 to 0.68 µm).

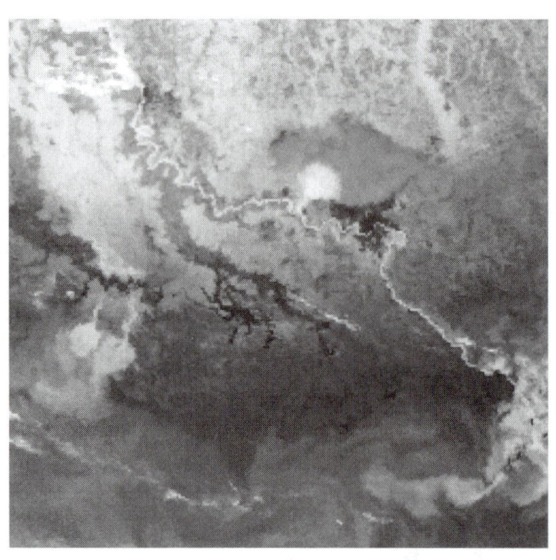

B. Band 2, NIR and SWIR (0.72 to 1.10 µm).

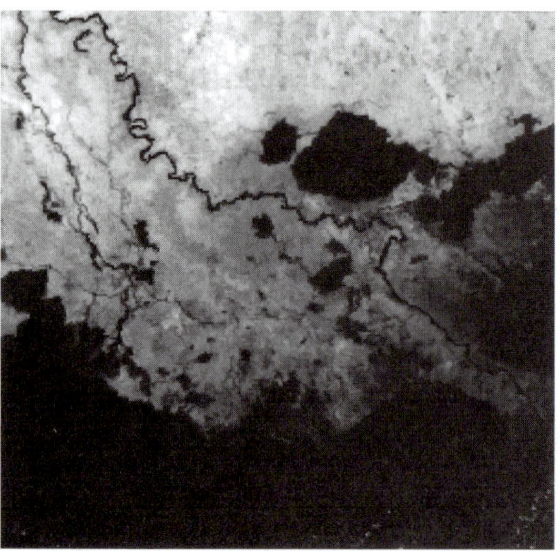

C. Band 4, TIR (10.5 to 11.5 µm).

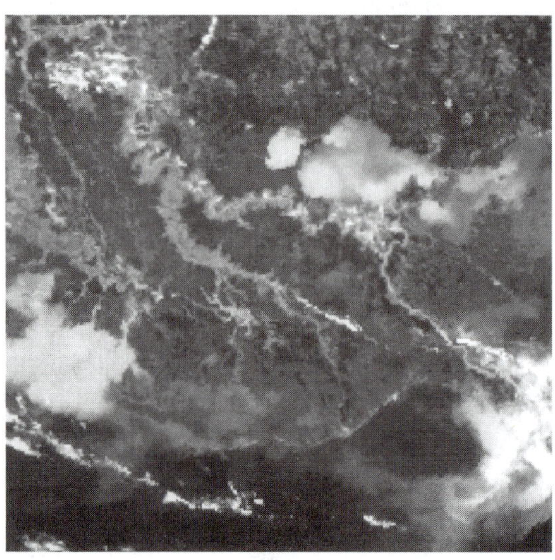

D. Location map.

FIGURE 4-27 AVHRR bands of a southern Louisiana subscene. Courtesy EROS Data Center.

the 2.7 km spatial resolution (resampled) data is downlinked to the ground sites while a small amount of the 0.56 km spatial resolution (original) data is stored and downlinked. DMSP data are available for a fee from NOAA's Earth Observation Group (ngdc.noaa.gov/eog/services.html).

GEOSTATIONARY SATELLITES

Geostationary satellites travel at the same angular velocity at which the Earth rotates; as a result, they remain above the same point on Earth at all times (Figure 4-1). A typical orbit has an altitude of almost 36,000 km and a velocity of 11,000 km $\cdot$ hr^{-1}. Each day numerous images are acquired that cover the full Earth disk and continent-sized areas. Table 4-9 lists the characteristics of the following geostationary satellites that are described in the following sections: GOES, Meteosat, and Himawari.

Geostationary Operational Environmental Satellites

NOAA operates the *Geostationary Operational Environmental Satellites* (GOES), which are essentially "parked" above the equator. Two GOES satellites provide continuous coverage of North America, South America, and adjacent oceans. GOES-East is positioned above the equator at 75° west longitude and covers the East Sector (including the eastern United States), and GOES-West is positioned at 135° west longitude and covers the West Sector (including the western United States).

GOES-8 to -12 have been decommissioned. GOES-13 is in orbit and provides storage. GOES-14 and -15 are in orbit and standing by as backups. GOES-12 through -15 collect five bands of imagery (Table 4-10) (NOAA–NASA, 2010).

The new GOES-R series, NOAA's next-generation geostationary weather satellites, provides continuous imagery and

Table 4-9 Geostationary satellite characteristics.

Operator	NOAA	NOAA	EUMETSAT	Japan
Name	GOES-12 to -15	GOES-16 GOES-17	Meteosat (2nd generation)	Himawari-8 Himawari-9
Spatial resolution (m)	1,000	500	1,000	500
	4,000	1,000	3,000	2,000
		2,000		
Spectral resolution (number of bands)	5	16	16	16
Wavelength regions	Red	VNIR	VNIR	VNIR
	TIR	SWIR	TIR	SWIR
		TIR		TIR
Swath (km)	full disk	full disk	full disk	full disk
Launched (series started)	1975	2016	2004	2015 and 2016
Number of satellites in series	15	1	4	2
Temporal resolution (min)	20 min full disk	15 min full disk; 5 min CONUS	15 min full disk	10 min full disk

Table 4-10 GOES-12 to -15 imagery bands and applications.

GOES Imager Band	Name	Wavelength Range (μm)	Objective
1	Visible	0.52 to 0.71	Cloud cover and surface features during the day, smoke, etc.
2	Shortwave window	3.73 to 4.07	Low cloud/fog, fire detection, winds, nighttime clouds, etc.
3	Water vapor	5.80 to 7.30	Upper-level water vapor, winds, etc.
4	Longwave window	10.7	Sea surface temperature and water vapor, surface or cloud-top temperature, precipitation, etc.
5	N/A	N/A	N/A
6	Carbon dioxide band	13.3	Carbon dioxide, cloud cover and height, etc.

atmospheric measurements of Earth's Western Hemisphere, total lightning data, and space weather monitoring. The first of this new generation, GOES-16, was launched on November 19, 2016. The GOES-R series collects 34 bands that provide atmospheric, land, ocean, solar, and space weather products for the forecasting and warning community (goes-r.gov). A GOES-R satellite carries six instruments, including the Advanced Baseline Imager (ABI), which collects 16 spectral bands, including two visible bands, four reflected IR bands, and 10 TIR bands compared to the five bands on GOES-12 to -15 satellites (Table 4-11). GOES-R provides three times more spectral information, four times the spatial resolution, and more than five times faster temporal coverage than the previous system (NOAA–NASA, 2016). On March 12, 2018 the second satellite in the GOES-R series reached geostationary orbit and was renamed GOES-17. As of this writing, GOES-16 covers the East Sector and GOES-17 covers the West Sector with GOES-14 as the on-orbit spare.

TABLE 4-11 GOES-R series ABI bands and characteristics.

ABI Band Number	Approximate Central Wavelength (μm)	Band Name
1	0.47	Blue
2	0.64	Red
3	0.86	NIR
4	1.37	Cirrus
5	1.6	Snow/ice
6	2.2	Cloud particle size
7	3.9	Short TIR
8	6.2	Upper-level tropospheric water vapor
9	6.9	Mid-level tropospheric water vapor
10	7.3	Lower-level water vapor
11	8.4	Cloud-top phase
12	9.6	Ozone
13	10.3	"Clean" TIR window
14	11.2	TIR window
15	12.3	"Dirty" TIR window
16	13.3	"CO_2" TIR

Source: Modified from NASA–NOAA GOES-R, ABI Bands Quick Information Guides (goes-r.gov/education/ABI-bands-quick-info.html).

The visible band image in Figure 4-28A was acquired by the GOES-13 satellite over the GOES East Sector simultaneously with the TIR image in Figure 4-28B. In GOES TIR images, the temperature signatures are reversed from those of other TIR images in this book (see Chapter 5). Bright tones record cool radiant temperatures and dark tones record warm temperatures. These reversed signatures cause clouds, which are cooler than water and land, to have familiar bright signatures on GOES TIR images.

The visible light image (Figure 4-28A) shows the late afternoon sun illuminating the western United States while the sun has already set on the east coast. No patterns can be

A. Visible light image.

B. TIR image. In this image, dark signatures are warm radiant temperatures and bright signatures are cool temperatures.

FIGURE 4-28 GOES-East full disk image of the Western Hemisphere acquired on December 4, 2016. Courtesy NOAA Geostationary Satellite Server.

seen in the nighttime, east portion of the visible light disk. In Figure 4-28B, however, thermal energy emitted from the clouds, land, and ocean can be clearly seen, even on the nighttime, east side of the disk.

GOES full disk images are updated every three hours. GOES images, along with Meteosat and Himawari-8 images, are available at NOAA's Geostationary Satellite Server website (goes.noaa.gov/index.html).

Meteosat

The Meteosat program is operated by the EUMETSAT. The first generation Meteosat (1 to 7) collected images on the half-hour in three bands (visible, TIR, and water vapor). Meteosat-7 transmitted the final first generation image on March 31, 2017.

The initial second generation Meteosat-8 was launched in 2004, followed by Meteosat-9 (2005), Meteosat-10 (2012), and Meteosat-11 (2015). Meteosat-8 is stationed at longitude 41.5°E over the Indian Ocean. Both Meteosat-9 and -10 are positioned over Africa. Meteosat-11 was placed into a storage orbit as a reserve. Table 4-9 shows the characteristics of second generation Meteosat satellites.

Himawari (Sunflower)

Himawari-8 is the eighth of the geostationary weather satellites operated by the Japan Meteorological Agency. Himawari-8 entered service in 2015 and succeeded Himawari-7, which was launched in 2006. Himawari-8 is positioned at longitude 140.7°E, over the Pacific Ocean north of the island of New Guinea. Himawari-9 was launched in 2016 and is in stand-by orbit until it replaces Himawari-8 in 2022. The main system on Himawari-8 is the Advanced Himawari Imager (AHI), which is a 16-band multispectral system that is summarized in Table 4-9.

Questions

1. Refer to Figure 4-26 and calculate how many Landsat images are required to cover one MODIS image.
2. Refer to Figure 4-26 and calculate how many Landsat pixels are required to cover
   a. One MODIS 250 m pixel.
   b. One AVHRR 1,100 m pixel.
3. If a WorldView-2 panchromatic (grayscale) pixel is 0.5 m and a Landsat-8 OLI panchromatic pixel is 15 m, how much more area is covered by one Landsat pixel compared to one WorldView-2 pixel?
4. SPOT 1 to 3 collected only nadir, left-looking or right-looking images while orbiting. What are the modes of acquisition for today's agile, high spatial resolution satellites?
5. Figure 4-7 is a Pleiades pan image of an urban scene. Plate 12 is a SPOT color image of a geologic scene. Describe the advantages and disadvantages of both pan and color images for interpreting urban and geologic scenes.
6. What is the spatial resolution of ASTER's VNIR bands?
7. What is the spatial resolution of ASTER's SWIR bands?
8. How does ASTER collect stereo images?
9. What's the difference between polar-orbiting and geostationary satellites?
10. How many bands does MODIS collect?
11. In what years were the two MODIS instruments launched aboard the Terra and Aqua satellites?
12. What is the temporal resolution of MODIS?
13. GOES means what?
14. GOES-R transitioned to GOES-16. What are the spatial resolutions of GOES-16?
15. GOES-16 has how many bands?
16. What is the GOES-16 swath (or geographic extent of the field of view)?

References

Abrams, M., H. Tsu, G. Hulley, K. Iwao, D. Pieri, T. Cudahy, and J. Kargel. 2015. The Advanced Spaceborne Thermal Emission and Reflection Radiometer (ASTER) after fifteen years: Review of global products. *International Journal of Applied Earth Observation and Geoinformation*, 38, 292–301. doi: http://dx.doi.org/10.1016/j.jag.2015.01.013

Astrium. 2012. Pleiades Imagery: Users' Guide (USRPHR-DT-125-SPOT-2.0). Astrium Services. http://www.intelligence-airbusds.com/en/65-satellite-imagery (accessed December 2017).

Beck, R. 2003. EO-1 User Guide (v. 2.3). Sioux Falls, SD: USGS Earth Resources Observation Systems Data Center, Satellite Systems Branch.

Darwin, C. R. 1842. *The Structure and Distribution of Coral Reefs*. London: Steward and Murray.

De Selding, P. B. 2014, December 4. Airbus sells in-orbit SPOT 7 imaging satellite to Azerbaijan. SpaceNews. http://spacenews.com/42840airbus-sells-in-orbit-spot-7-imaging-satellite-to-azerbaijan (accessed December 2017).

DigitalGlobe. 2013. The Benefits of the Eight Spectral Bands of WorldView-2 (WP-8SPEC Rev 01/13). http://www.digitalglobe.com (accessed December 2017).

DigitalGlobe. 2015. Exploring the Benefits of SWIR Satellite Imagery (WP-RAD 03/15). http://www.digitalglobe.com (accessed December 2017).

DigitalGlobe. 2016a, February 22. Helping Facebook Connect the World with Deep Learning. Blog. http://blog.digitalglobe.com/news/helping-facebook-connect-the-world-with-deep-learning (accessed December 2017).

DigitalGlobe. 2016b. WorldView-2 Data Sheet (DS-WV2-rev2 02/16). http://www.digitalglobe.com (accessed December 2017).

European Space Agency. 2015. Sentinel-2 User Handbook. ESA Standard Document, Issue 1, Rev. 2. Paris: European Space Agency.

Hall, R. C. 2001. A History of the Military Polar Orbiting Meteorological Satellite Program. National Reconnaissance Office, Center for the Study of National Reconnaissance.

Hirano, A., R. Welch, and H. Lang. 2003. Mapping from ASTER stereo image data: DEM validation and accuracy assessment. *Journal of Photogrammetry and Remote Sensing*, 57(5–6), 356–370.

Jensen, J. R. 2007. *Remote Sensing of the Environment: An Earth Resource Perspective* (2nd ed.). Upper Saddle River, NJ: Pearson.

Kruse, F. A., W. M. Baugh, and S. L. Perry. 2015. Validation of DigitalGlobe WorldView-3 Earth imaging satellite shortwave infrared bands for mineral mapping. *Journal of Applied Remote Sensing*, 9, 096044. http://remotesensing.spiedigitallibrary.org/journal.aspx (accessed December 2017).

Lencioni, D. E., D. R. Hearn, C. J. Digenis, J. A. Mendenhall, and W. E. Bicknell. 2005. The EO-1 Advanced Land Imager: An overview. *Lincoln Laboratory Journal*, 15(2).

Mims, C. 2014. Amid stratospheric valuations, Google unearths a deal with Skybox. *The Wall Street Journal*, June 15.

NASA. 2001. AIRS/AMSU/HSB: The Atmospheric Infrared Sounder, with Its Companion Advanced Microwave Sounding Unit and Humidity Sounder for Brazil (NP-2001-5-248-GSFC). Greenbelt, MD: NASA, Goddard Space Flight Center.

NASA. 2002. MODIS: Moderate Resolution Imaging Spectroradiometer (NP-2002-1-423-GSFC). Greenbelt, MD: NASA, Goddard Space Flight Center.

NASA SeaWiFS Project. 2017. Background of the SeaWiFS Project. http://oceancolor.gsfc.nasa.gov/SeaWiFS/BACKGROUND/SEAWIFS_BACKGROUND.html (accessed December 2017).

Navulur, K., W. Baugh, G. Hammann, and V. Leonard. 2014. Moving from Pixels to Products . . .and Data to Insight (WP-P2P 10/14). DigitalGlobe.

NOAA–NASA. 2010. GOES N Series Data Book. http://goes.gsfc.nasa.gov/text/GOES-P_Databook.pdf (accessed December 2017).

NOAA–NASA. 2016. GOES-R Advanced Baseline Imager (ABI) Factsheet. http://www.goes-r.gov/education/ABI-bands-quick-info.html (accessed December 2017).

Planet. 2018, December. Planet Imagery Product Specifications. http://assets.planet.com/docs/Combined-Imagery-Product-Spec-Dec-2018.pdf.

Prost, G. 2014. *Remote Sensing for Geoscientists: Image Analysis and Integration* (3rd ed.). Boca Raton, FL: CRC Press.

RapidEye. 2008, September. RapidEye Standard Image Product Specifications v. 2.0. RapidEye AG, Germany.

Sankey, T., B. Dickson, S. Sesnie, O. Want, A. Olsson, and L. Aachmann. 2014. WorldView-2 high spatial resolution improves desert invasive plant detection. *Photogrammetric Engineering & Remote Sensing*, 80(9), 885–893.

Strom, S. R., and G. Iwanaga. 2005. Overview and history of the Defense Meteorological Satellite Program. *Crosslink*, 6(1).

TRW. 2001. EO-1/Hyperion Science Data User's Guide, Level 1-B (HYP.TO.01.077). Redondo Beach, CA: TRW Space, Defense & Information Systems.

Union of Concerned Scientists. 2019, January 9. UCS Satellite Database. http://www.ucsusa.org/nuclear-weapons/space-weapons/satellite-database (accessed May 2019).

Warner, T. A., N. Skowronski, and M. R. Gallagher. 2017. High spatial resolution burn severity mapping of the New Jersey Pine Barrens with WorldView-3 near-infrared and shortwave infrared imagery. *International Journal of Remote Sensing*, 38(2), 598–616.

Thermal Infrared Images

The thermal IR spectral domain is that portion of the electromagnetic spectrum ranging in wavelength from 0.7 to 1,000 μm. Remote sensing deals with the wavelengths from 0.7 to 14 μm that are subdivided as follows:

- Near IR (NIR) domain: 0.7 to 1.0 μm (reflected solar radiation).
- Shortwave IR (SWIR) domain: 1.0 to 3.0 μm (reflected solar radiation).
- Thermal IR (TIR) domain: 3.0 to 14 μm (radiated energy).

This chapter deals with the TIR domain, which is important because it records information about the internal composition of materials. Special detectors and optical-mechanical scanners detect and record images in the TIR spectral region. The ability to detect and record thermal radiation as images at night takes away the cover of darkness and has obvious reconnaissance applications. For these reasons, beginning in the 1950s government agencies funded the early development of TIR imaging technology. The developments were classified for security purposes. Military interpreters recognized that geologic and terrain features greatly influenced the signatures of the background against which strategic targets were displayed. Word of these benefits and their potential nonmilitary applications created interest in the civilian geologic community. In the mid-1960s some manufacturers received approval to acquire images for civilian clients using the classified systems. In 1968, the US government declassified systems that did not exceed certain standards for spatial resolution and temperature sensitivity. In 1978, the NASA Heat Capacity Mapping Mission (HCMM) acquired daytime and nighttime TIR images with a 600 m spatial resolution for geologic applications. Today, TIR scanner systems are available for unrestricted use.

TIR images record data within two wavelength ranges: 3 to 5 μm (*medium wavelength IR* [MWIR]) and 8.0 to 14.0 μm (*long wavelength IR* [LWIR]). Today many satellites, including Landsat 4 to 8, ASTER, AVHRR, MODIS, and GOES, acquire TIR imagery. TIR sensors are available for smartphones, drones, and handheld cameras due to miniaturization and sophisticated built-in, image processing algorithms. Drones currently acquire increasing amounts of the imagery that were formerly acquired by manned aircraft.

THERMAL PROCESSES

To interpret TIR images, one must understand the basic physical processes that control the interactions between thermal energy and matter, as well as the thermal properties of matter that determine the rate and intensity of the interactions.

Kinetic Heat, Temperature, and Radiant Flux

Kinetic heat is the energy of particles of matter moving in a random motion. The random motion causes particles to collide, resulting in changes of energy state and the emission of electromagnetic radiation from the surface of materials. The internal, or kinetic, heat energy of matter is thus converted into *radiant energy*. The amount of heat is measured in calories. A *calorie* is the amount of heat required to raise the temperature of 1 g of water by 1°C. *Temperature* is a measure of the concentration of heat. On the Celsius scale, 0°C and

100°C are the temperatures of melting ice and boiling water, respectively. On the Kelvin, or absolute, temperature scale, 0°K is *absolute zero*, the point at which all molecular motion ceases. The Kelvin and Celsius scales correlate as follows: 0°C = 273°K, and 100°C = 373°K. The electromagnetic energy radiated from a source is called *radiant flux* (F) and is measured in watts per square centimeter ($W \cdot cm^{-2}$).

The concentration of kinetic heat of a material is called the *kinetic temperature* (T_{kin}) and is measured with a thermometer placed in direct contact with the material. The concentration of the radiant flux of a body is the *radiant temperature* (T_{rad}). Radiant temperature may be measured remotely by nonimaging devices called *radiometers*. The radiant temperature of materials is always less than the kinetic temperature because of a thermal property called emissivity, which is defined later.

HEAT TRANSFER

Heat energy is transferred from one place to another by three means:

1. *Conduction* transfers heat through a material by molecular contact. The transfer of heat through a frying pan to cook food is one example.
2. *Convection* transfers heat through the physical movement of heated matter. The circulation of heated water and air are examples of convection.
3. *Radiation* transfers heat in the form of electromagnetic waves. Heat from the sun reaches the Earth by radiation. In contrast to conduction and convection, which can only transfer heat through matter, radiation can transfer heat through a vacuum.

Materials at the surface of the Earth receive thermal energy primarily in the form of radiation from the sun. To a much lesser extent, heat is also conducted from the interior of the Earth. There are daily and annual cyclic variations in the duration and intensity of solar energy. Energy from the interior of the Earth reaches the surface primarily by conduction and is relatively constant at any locality, although there are regional variations in this heat flow. Hot springs and volcanoes are local sources of convective heat. Regional heat-flow patterns may be altered by geologic features such as salt domes and faults.

ATMOSPHERIC TRANSMISSION

The atmosphere does not transmit all wavelengths of TIR radiation uniformly. The *atmospheric transmittance curve* in Figure 5-1 shows a series of wavelength intervals or peaks, called *atmospheric windows*, where most of the radiant energy is transmitted through the Earth's atmosphere. The windows are separated by intervals where limited or no energy is transmitted. These intervals are called *atmospheric absorption bands* that are caused by gases (carbon dioxide, ozone, and water vapor) in the atmosphere. The narrow absorption band from 9 to 10 μm (shown as a dashed curve in Figure 5-1) is caused by the ozone layer at the top of the Earth's atmosphere. To avoid the effects of this absorption band, satellite TIR systems typically record wavelengths from approximately 8.0 to 9.0 μm and from 10.5 to 12.5 μm. Systems on aircraft and drones, which fly beneath the ozone layer, are not affected and may record the full window from 8 to 14 μm.

RADIANT ENERGY PEAKS AND WIEN'S DISPLACEMENT LAW

For an object at a constant kinetic temperature, the radiant energy, or flux, varies as a function of wavelength. The *radiant energy peak* (λ_{max}) is the wavelength at which the maximum amount of energy is radiated. Figure 5-2A shows radiant energy curves for objects ranging in temperature from

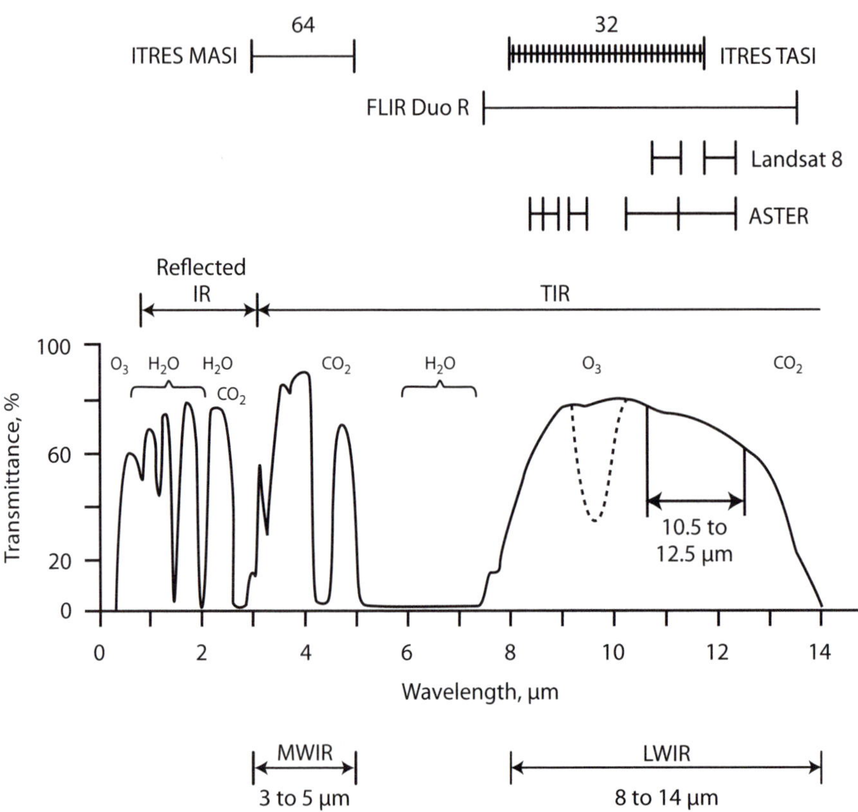

FIGURE 5-1 Electromagnetic spectrum showing spectral bands used in the TIR region with select sensors. Gases that cause atmospheric absorption are indicated.

300 to 700°K while Figure 5-2B shows radiant energy curves for objects ranging in temperature from 3000 to 6000°K. The 6000°K sun has a radiant energy peak with a dominant wavelength of 0.48 μm (green light), which is the reflected energy peak for Earth.

As temperature increases, the total amount of radiant energy (the area under each curve) increases and the radiant energy peak shifts to shorter wavelengths. The dotted lines in Figures 5-2A and 5-2B pass through the radiant energy peaks and indicate the shift. This shift, or displacement, to shorter wavelengths with increasing temperature is described by *Wien's displacement law*, which states that

$$\lambda_{max} = \frac{2{,}897\ \mu m \cdot {}^{\circ}K}{T_{rad}} \tag{5-1}$$

where T_{rad} is radiant temperature in Kelvin and 2,897 μm · °K is a physical constant. The wavelength of the radiant energy peak of an object may be determined by substituting the value of T_{rad} into Equation 5-1. For example, the average radiant temperature of the Earth is approximately 300°K (27°C or 80°F). Substituting this temperature into Equation 5-1 results in the following:

$$\lambda_{max} = \frac{2{,}897\ \mu m \cdot {}^{\circ}K}{300^{\circ}K}$$
$$= 9.7\ \mu m$$

which is the peak for the 300°K radiant energy curve in Figure 5-2A and is the radiant energy peak of the Earth. Figure 5-2A is used to evaluate optimum wavelength bands for detecting targets at various temperatures.

Wien's displacement law also applies to hot objects that glow at visible wavelengths, such as an iron poker in a fire. As the poker heats up, the color progresses from dark red to bright red to orange to yellow at successively shorter wavelengths. The changing colors represent shifts in λ_{max} with increasing temperature.

IMAGING SYSTEMS AND PLATFORMS

TIR imaging systems are deployed on platforms that range from handheld cameras to satellites. The platforms are listed in Table 5-1 together with representative imaging systems deployed on the platforms.

A. Radiant energy peak for TIR. Note atmospheric transmission bands at 3 to 5 μm and 8 to 14 μm. Modified from Colwell and others (1963, Figure 2).

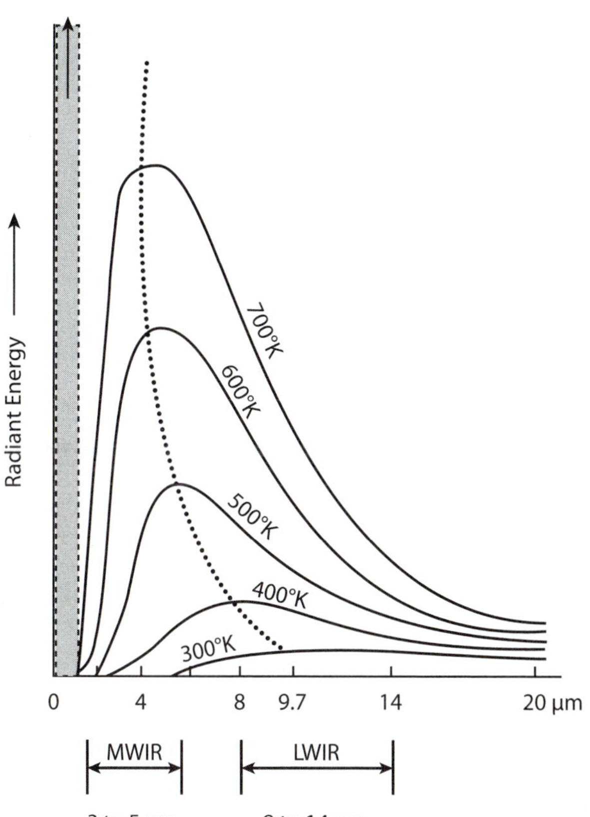

B. Radiant energy peak for visible light. Modified from Douma (2008).

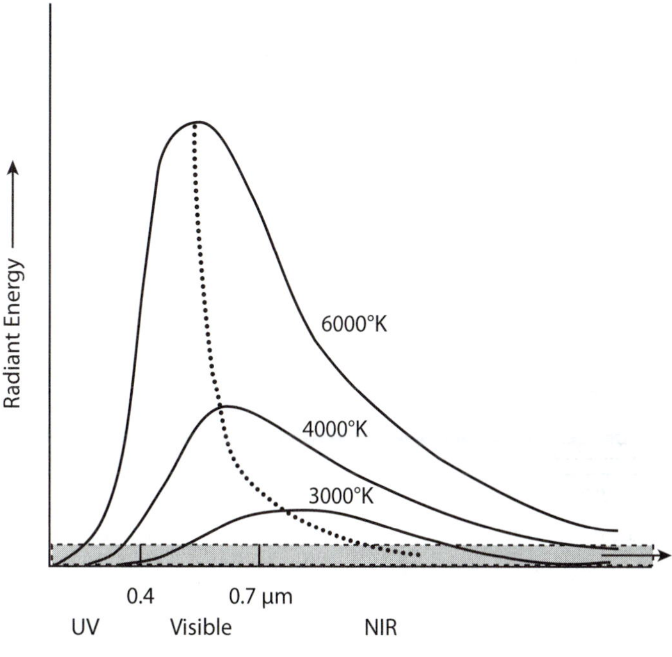

Figure 5-2 Spectral distribution curves of energy radiated from objects at different temperatures. The dotted line represents the radiant energy peak at different temperatures. Overlap of (A) and (B) is shown by the gray area along one axis on each plot.

HANDHELD CAMERAS

TIR sensors are available for smartphones and handheld cameras. The FLIR ONE TIR attachment is adapted for smartphones and blends thermal and visible imaging using a patented algorithm named MSX (FLIR Systems, 2015). As discussed in Chapter 1, the longer the wavelength, the less energy received at the detector. Therefore, TIR images require larger pixels and arrays with fewer pixels compared to visible light images of the same object from the same distance. The FLIR ONE thermal array is 80 by 60 pixels and the visible camera array is 640 by 480 pixels. The MSX algorithm extracts the high-contrast highlights from the visible camera image and adds the highlights into the FLIR infrared image in real time.

Plate 14A shows the TIR attachment on a smartphone. Plate 14B is the color thermal image with red being warmest and blue being coolest. Plate 14C is the high spatial resolution visible light image showing a pipe with fittings alongside walls and a ceiling. Plate 14D is the detail extracted from the high spatial resolution image (Plate 14C). The extracted detail is merged with the thermal infrared image, which increases the sharpness and information content of the TIR image (Plate 14E). The smartphone attachment enables users to employ TIR imaging for day and night applications (FLIR Systems, 2015).

FLIR's handheld GF320 detects methane, hydrocarbons, and volatile organic compounds using the 3.2 to 3.4 μm wavelength range (FLIR Systems, 2016). The array is 320 by 240 pixels. The camera measures temperatures from –20 to –350°C (–4 to –662°F). It detects and images escaping gases venting from pressure relief valves and leaking from valves and pipe joints that can collect and ignite. Hydrologists use handheld TIR cameras in the field to locate groundwater discharge, characterize local hydrogeologic conditions, and to define sampling and monitoring locations (USGS, 2018).

TABLE 5-1 TIR systems.

Sensor	Recording Mode	Detector	3 to 5 μm	8 to 14 μm	Type	Spatial Resolution
Handheld—Stationary [a]						
FLIR ONE (for smartphones)		Frame array		X	Single band	Submeter
FLIR T-Series		Frame array		X	Single band	Submeter
FLIR GF320		Frame array	X		Single band	Submeter
Small Unmanned Aircraft System (sUAS)/Drone						
FLIR Duo R	Along-track	Frame array		X	Single band	Submeter
FLIR Vue Pro R	Along-track	Frame array		X	Single band	Submeter
Manned Aircraft						
FLIR Star SAFIRE 380-HD	Along-track	Linear array	X		Single band	
ITRES TABI-1800	Along-track	Linear array		X	Single band	0.1 to 1.25 m
ADSI Spectra-View	Along-track	Frame array	X	X	Single band	0.15 m +
NASA TIMS	Cross-track	Scanner		X	Multispectral	50 m
ITRES MASI-600	Along-track	Linear array	X		Hyperspectral	1 to 3.5 m
ITRES TASI-600	Along-track	Linear array		X	Hyperspectral	1 to 3.5 m
Satellites						
Polar Orbiting						
Landsat TM 4, 5, and 7	Cross-track	Scanner		X	Single band	60 m
Landsat TIRS 8	Along-track	Linear array		X	Two bands	100 m
ASTER	Cross-track	Scanner		X	Multispectral	90 m
AVHRR	Cross-track	Scanner		X	Two bands	1 km
MODIS	Cross-track	Scanner	X	X	Multispectral	1 km
Geostationary						
GOES-16 GOES-17	Along-track	Linear array	X	X	Multispectral	2 km

[a] Can be adapted for aerial platforms.

Small Unmanned Aircraft System (sUAS)/ Drone Cameras

TIR imagery may be acquired by systems mounted on sUAS/drones (Table 5-1). Figure 5-3 is a quadcopter drone equipped with the FLIR Duo R camera that acquires TIR imagery within the 7.5 to 13.5 μm wavelength range onto a 160 by 120 array. The FLIR Duo R camera weighs 84 gm (3 oz) with dimensions of 41 by 59 by 29.6 mm (1.6 by 2.3 by 1.2 in). The system includes a visible light camera with a 1,920 by 1,080 array. As with the FLIR ONE camera, the MSX algorithm extracts the high-contrast highlights from the visible camera's image and then virtually etches the skeletonized details onto the TIR image in real time. The operator can switch between visible, TIR, and blended imagery while flying the drone (FLIR Systems, 2017).

Manned Aircraft Systems

Large TIR systems are carried by manned aircraft. Table 5-1 lists some of these sensors. The ITRES MASI-600 is a 64 band, hyperspectral system that collects TIR data in the 3 to 5 μm wavelength range (ITRES, 2011a). The sensor has dimensions of 48.3 by 17.8 by 52.3 cm and weighs 16 kg (35 lb). The MASI-600 acquires images with an along-track linear array 600 pixels wide. The spectral resolution, or bandwidth, is 32 nm. Biochemical gas detection, invasive species mapping, surface landmine detection, and pipeline leakage detection are some of the applications.

Satellite Systems

Several satellites collect TIR data (Table 5-1). ASTER is part of the payload carried on the Terra satellite shown in Figure 5-4. ASTER uses a cross-track scanner to acquire multispectral VNIR, SWIR, and TIR images. The capabilities of the ASTER system are shown in Table 5-2 (Abrams and others, 2002).

ASTER is important because it was the first satellite to acquire multispectral TIR images. ASTER TIR images are acquired during the day (~10:00 AM local time) and night (~10:00 PM local time). The USGS and NASA JPL provide the ASTER data as Level 1 orthorectified images with the digital numbers in radiance values. The Level 1 product is named "AST_L1T" and is available at no cost from the USGS EarthExplorer (earthexplorer.usgs.gov), GloVis (glovis.usgs.gov), and NASA's Earthdata Search (search.earthdata.nasa.gov/search) websites. Relative surface temperature is indicated with the AST_L1T data.

Figure 5-3 FLIR Duo R TIR camera mounted on a quadcopter drone. Courtesy FLIR Systems.

Table 5-2 Capabilities of the ASTER system.

ASTER Data Set	Number of Bands	Spectral Range (μm)	Ground Resolution (m)
VNIR	3	0.52 to 0.86	15
SWIR	6	1.60 to 2.36	30
TIR	5	8.13 to 10.65	90

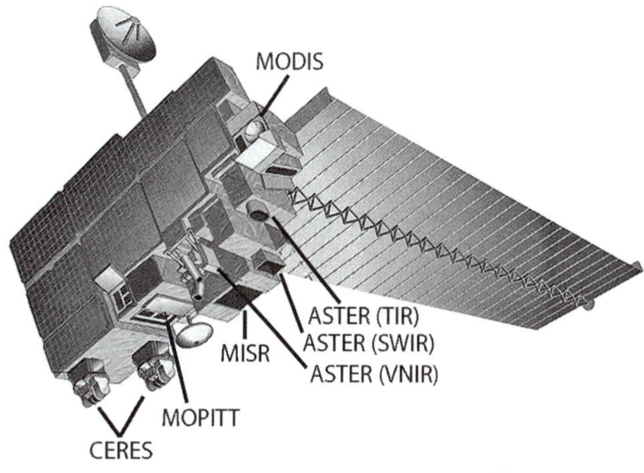

Figure 5-4 ASTER VNIR, SWIR, and TIR sensors onboard the Terra platform. Courtesy NASA JPL.

THERMAL PROPERTIES OF MATERIALS

Radiant energy striking the surface of a material is partly reflected, partly absorbed, and partly transmitted through the material. Therefore,

$$\text{Reflectivity} + \text{absorptivity} + \text{transmissivity} = 1 \quad \textbf{(5-2)}$$

Reflectivity, absorptivity, and transmissivity are determined by properties of matter and vary with the wavelength of the incident radiant energy and with the temperature of the surface. As discussed in Chapter 2, reflectivity is expressed as *albedo* (A), which is the ratio of reflected energy to incident energy. For materials in which transmissivity is negligible, Equation 5-2 reduces to

$$\text{Reflectivity} + \text{absorptivity} = 1 \quad \textbf{(5-3)}$$

The absorbed energy causes an increase in the kinetic temperature of the material.

BLACKBODY CONCEPT

The concept of a blackbody is fundamental to understanding heat radiation. A blackbody is a theoretical material that absorbs all the radiant energy that strikes it, which means

$$\text{Absorptivity} = 1 \tag{5-4}$$

A blackbody also radiates all of its energy in a wavelength distribution pattern that is dependent only on the kinetic temperature. According to the *Stefan–Boltzmann law*, the radiant flux of a blackbody (F_b) at a kinetic temperature of T_{kin} is

$$F_b = \sigma \times T_{\text{kin}}^4 \tag{5-5}$$

where σ is the *Stefan–Boltzmann constant*, which is 5.67×10^{-12} W · cm^{-2} · °K^{-4}.

For a blackbody with a T_{kin} of 10°C (283°K), the radiant flux may be calculated from Equation 5-5 as

$$\begin{aligned} F_b &= \sigma \times T_{\text{kin}}^4 \\ &= \left(5.67\times10^{-12}\ \text{W}\cdot\text{cm}^{-2}\cdot{}^{\circ}\text{K}^{-4}\right)\left(283^{\circ}\text{K}\right)^4 \\ &= \left(5.67\times10^{-12}\ \text{W}\cdot\text{cm}^{-2}\cdot{}^{\circ}\text{K}^{-4}\right)\left(6.41\times10^{9}\ {}^{\circ}\text{K}\right)^4 \\ &= 3.6\times10^{-2}\ \text{W}\cdot\text{cm}^{-2} \end{aligned}$$

A blackbody is a physical abstraction, because no material has an absorptivity of 1 and no material radiates the full amount of energy given in Equation 5-5. The following section describes a property called *emissivity*, which expresses the relationship between radiant flux and blackbody temperature.

EMISSIVITY OF MATERIALS

For real materials a property called emissivity (ε) is defined as

$$\varepsilon = F_r / F_b \tag{5-6}$$

where F_r is radiant flux from a real material. The emissivity for a blackbody is 1, but for all real materials it is less than 1. Emissivity is wavelength dependent, which means that the emissivity of a real material (not a blackbody) will be different when measured at different wavelengths of radiant energy.

Tables 5-3 and 5-4 list the emissivities of various materials in the 8 to 12 μm (LWIR) wavelength region. Water has a high emissivity, but a thin film of petroleum lowers the emissivity, which is a significant relationship for remote sensing of oil slicks (Chapter 11). Emissivity for man-made land cover types (artificial turf, metal roof, roof tile) is higher than emissivity for natural land cover types (lawn, trees, and green/vegetated roofs).

Emissivity Experiment

Combining Equations 5-5 and 5-6 produces the following equation for the radiant flux of a real material:

$$F_r = \varepsilon \times \sigma \times T_{\text{kin}}^4 \tag{5-7}$$

where ε is the emissivity for that material. Emissivity is a measure of the ability of a material both to radiate and to absorb energy. Materials with high emissivities absorb large amounts of incident energy and radiate large quantities of kinetic energy. Materials with low emissivities absorb and radiate lower amounts of energy.

TABLE 5-3 Emissivity of natural materials measured at wavelengths of 8 to 12 μm (LWIR).

Natural Material	Emissivity (ε)
Granite, typical	0.815
Dunite	0.856
Obsidian	0.862
Feldspar	0.870
Granite, rough	0.898
Silicon sandstone, polished	0.909
Sand, quartz, large-grain	0.914
Dolomite, polished	0.929
Basalt, rough	0.934
Dolomite, rough	0.958
Water, with a thin film of petroleum	0.972
Water, pure	0.993

Source: From K. J. K. Buettner and C. D. Kern. 1965. Determination of infrared emissivities of terrestrial surfaces. *Journal of Geophysical Research*, 70(6), 1,329–1,337. Copyrighted by American Geophysical Union.

TABLE 5-4 Emissivity of land cover materials measured at wavelengths of 8 to 12 μm (LWIR).

Land Cover Material	Emissivity (ε)
Lawn	0.930
Green (vegetated) roof	0.930
Wooden deck	0.944
Trees	0.946
Tile	0.951
Gravel	0.955
Sidewalk brick	0.962
Asphalt	0.972
Artificial turf	0.979
Concrete roof (gray)	0.980
Metal roof	0.980
Urethane	0.987
Roof tile (green)	0.990
Roof tile (black)	1.000

Source: Modified from Song and Park (2015, Table 2).

Figure 5-5 is an experiment on the effect different emissivities have on radiant flux. The aluminum block has a uniform kinetic temperature of 10°C (283°K). The portion of the block that is painted dull black has a high emissivity of 0.97. The radiant flux for this material is calculated from Equation 5-7 as

$$\begin{aligned} F_r &= \varepsilon \times \sigma \times T_{\text{kin}}^4 \\ &= 0.97\left(5.67 \times 10^{-12}\ \text{W} \cdot \text{cm}^{-2} \cdot {}^\circ\text{K}^{-4}\right)(283^\circ\text{K})^4 \\ &= 3.5 \times 10^{-2}\ \text{W} \cdot \text{cm}^{-2} \end{aligned}$$

For the shiny portion of the aluminum block with a low emissivity of 0.06, the radiant flux may be calculated from Equation 5-7 as

$$\begin{aligned} F_r &= \varepsilon \times \sigma \times T_{\text{kin}}^4 \\ &= 0.06\left(5.67 \times 10^{-12}\ \text{W} \cdot \text{cm}^{-2} \cdot {}^\circ\text{K}^{-4}\right)(283^\circ\text{K})^4 \\ &= 2.2 \times 10^{-3}\ \text{W} \cdot \text{cm}^{-2} \end{aligned}$$

Although the aluminum block has a uniform kinetic temperature of 283°K, the radiant flux from the black surface with high emissivity is more than 10 times greater than from the shiny surface with low emissivity.

Most TIR remote sensing systems record the radiant temperature (T_{rad}) of terrain rather than radiant flux. In order to determine T_{rad}, consider a blackbody and a real material that have different kinetic temperatures but the same radiant flux, so that $F_b = F_r$. For a blackbody $T_{\text{rad}} = T_{\text{kin}}$, therefore, Equation 5-5 may be written as

$$F_b = \sigma \times T_{\text{rad}}^4$$

This equation and Equation 5-7 may then be combined as follows:

$$\begin{aligned} F_r &= \varepsilon \times \sigma \times T_{\text{kin}}^4 \\ F_b &= \sigma \times T_{\text{rad}}^4 \\ F_b &= F_r \\ \sigma T_{\text{rad}}^4 &= \varepsilon \times \sigma \times T_{\text{kin}}^4 \\ T_{\text{rad}} &= \varepsilon^{1/4} T_{\text{kin}}^4 \end{aligned} \tag{5-8}$$

For a real material of known emissivity and kinetic temperature, Equation 5-8 may be used to calculate the radiant temperature. Radiant temperature is measured with radiometers. For the portion of the aluminum block in Figure 5-5 with an emissivity of 0.97, radiant temperature is calculated from Equation 5-8 as

$$\begin{aligned} T_{\text{rad}} &= \varepsilon^{1/4} T_{\text{kin}}^4 \\ &= 0.97^{1/4} \times 283^\circ\text{K} \\ &= 281^\circ\text{K (or } 8^\circ\text{C)} \end{aligned}$$

For the portion of the block with an emissivity of 0.06,

$$\begin{aligned} T_{\text{rad}} &= \varepsilon^{1/4} T_{\text{kin}}^4 \\ &= 0.06^{1/4} \times 283^\circ\text{K} \\ &= 140^\circ\text{K (or } -133^\circ\text{C)} \end{aligned}$$

which is 141°K lower than the radiant temperature for the portion of the aluminum block with high emissivity. An alternate way to understand the low radiant temperature of the shiny surface involves Equation 5-3. Because the emissivity is low, the absorptivity is also low, and the reflectivity is therefore high. In the out-of-doors with no clouds, the very low temperature of outer space is reflected by the shiny aluminum surface. For this reason, metallic objects such as airplanes and metal-roofed buildings have cold radiant temperatures.

Having established the theoretical and experimental basis for emissivity, the next step is to evaluate the impact of emissivity in the real world of images.

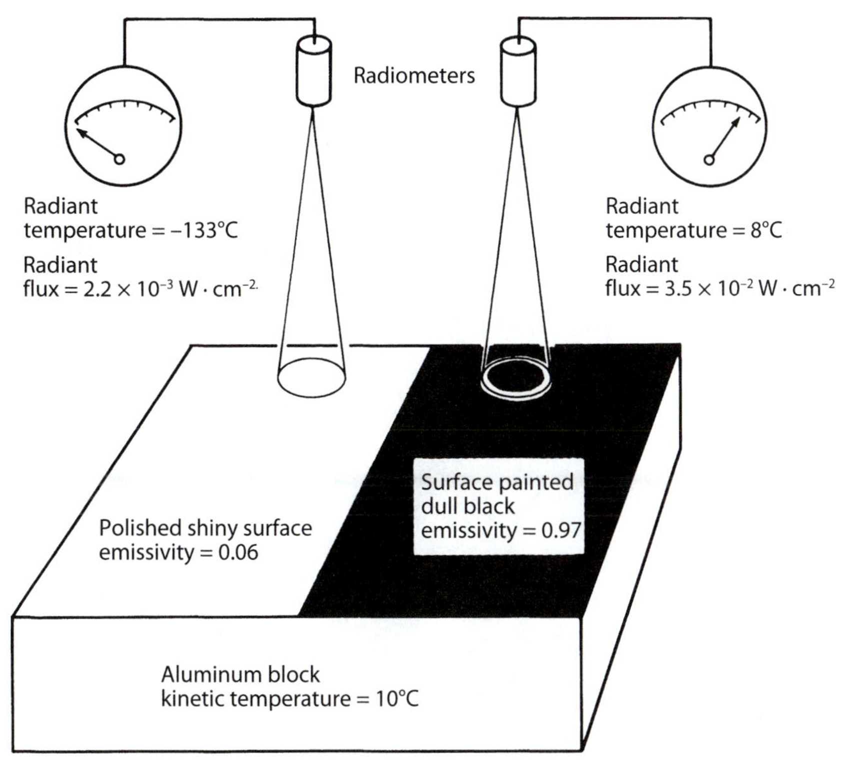

FIGURE 5-5 Effect of emissivities on radiant temperature. The kinetic (internal) temperature of the aluminum block is uniformly 10°C. Different emissivities cause different radiant temperatures, which are measured by radiometers.

Emissivity and Radiance Components of ASTER Images

ASTER is a unique spaceborne TIR sensor because it collects five bands that enable *direct* surface emissivity estimates after correction for atmospheric conditions (Gillespie and others, 1998). Surface emissivity is required to derive *accurate* land surface temperature data.

For an example, we selected the Ghazni area in east-central Afghanistan to illustrate radiance, emissivity, and surface temperature because we had earlier conducted a

A. Hillshade DEM illuminated from southeast.

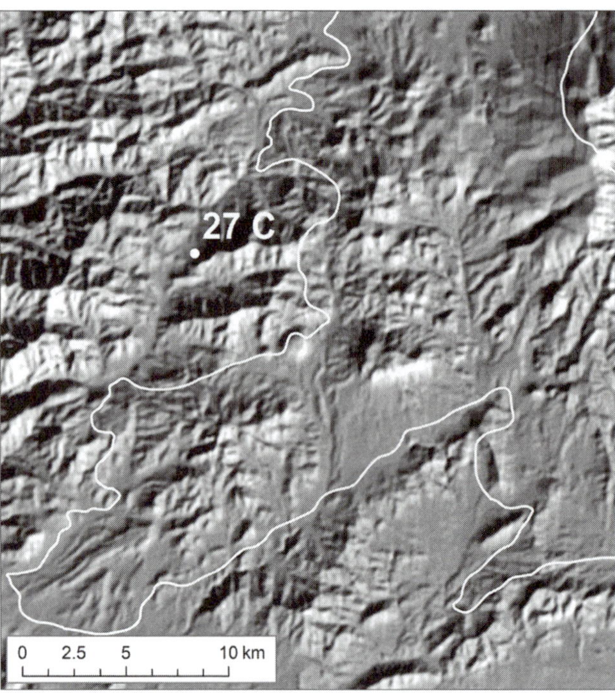

B. Radiance image.

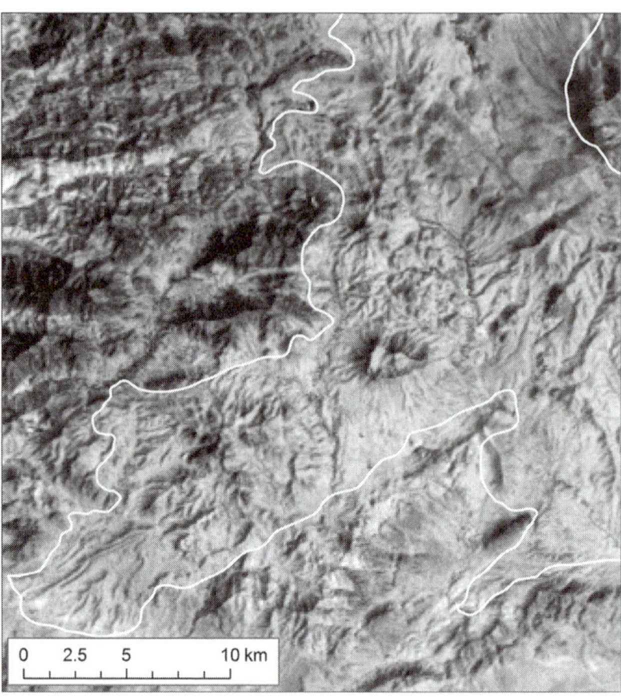

C. Emissivity image.

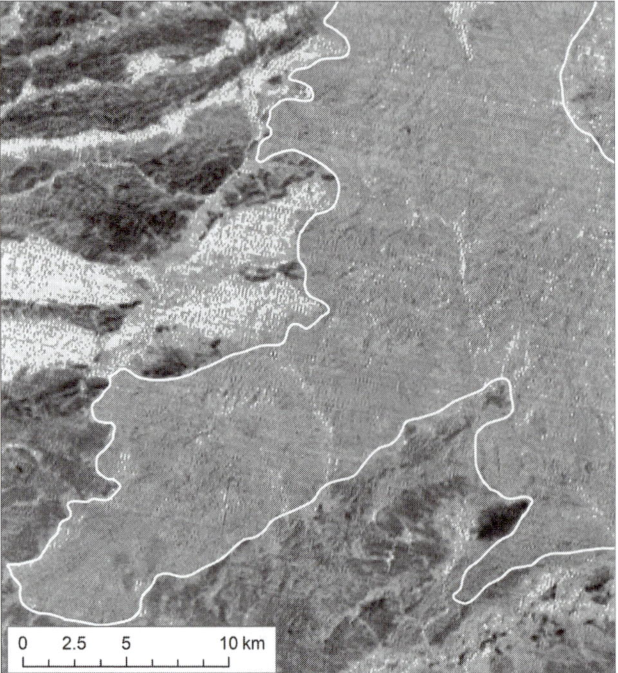

D. Surface temperature image.

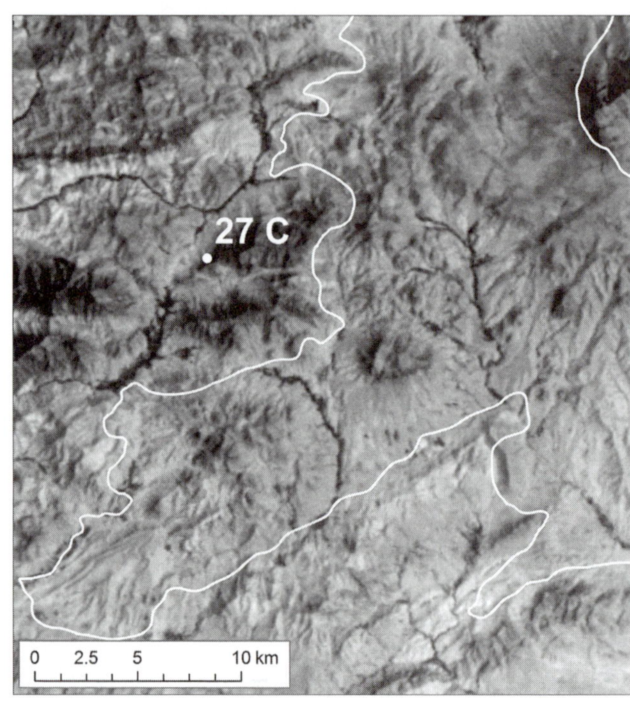

FIGURE 5-6 Radiance and emissivity, components of the ASTER band 11 (8.475 to 8.825 μm) daytime image, along with surface temperature image, of the Ghazni area, Afghanistan. Emissivity image was processed using the method of Gillespie and Rokugawa (2001) and provided courtesy B. Hubbard, USGS. ASTER and SRTM DEM courtesy USGS.

mineral exploration project in the area. The hillshade model in Figure 5-6A is compiled from satellite topographic data and shows the rugged topography of the Ghazni area. The hillshade model is illuminated from the southeast to simulate the mid-morning solar illumination of the ASTER radiance image in Figure 5-6B. The radiance image shows ASTER band 11 (8.475 to 8.825 μm) in the longwave TIR domain (Figure 5-1). The image was acquired around 10:00 AM local sun time. Bright signatures are relatively warm and dark signatures are relativity cool. Shadows and highlights are less evident on the image because it lacks the vertical exaggeration of the model. The white line is the contact between bedrock and alluvium, which consists of gravel, sand, and silt. Figure 5-7 is a geologic sketch of the area.

B. Hubbard of the USGS used the method of Gillespie and Rokugawa (2001) to extract the emissivity component (Figure 5-6C) and the surface temperature image (Figure 5-6D) from the five radiance bands. The radiance image closely resembles the temperature image because the radiant flux relationship (Equation 5-7) elevates temperature to the fourth power while emissivity is only unity. The surface temperature near a mountain ridge approximately 3,900 m above sea level is 27°C (white dot on Figure 5-6A and Figure 5-6D). The emissivity image (Figure 5-6C) clearly distinguishes bedrock with strong topographic highlights and shadows from alluvium with a uniform medium gray signature. The difference is attributed to the relatively uniform topography and composition of alluvium and the varied topography and composition of the bedrock.

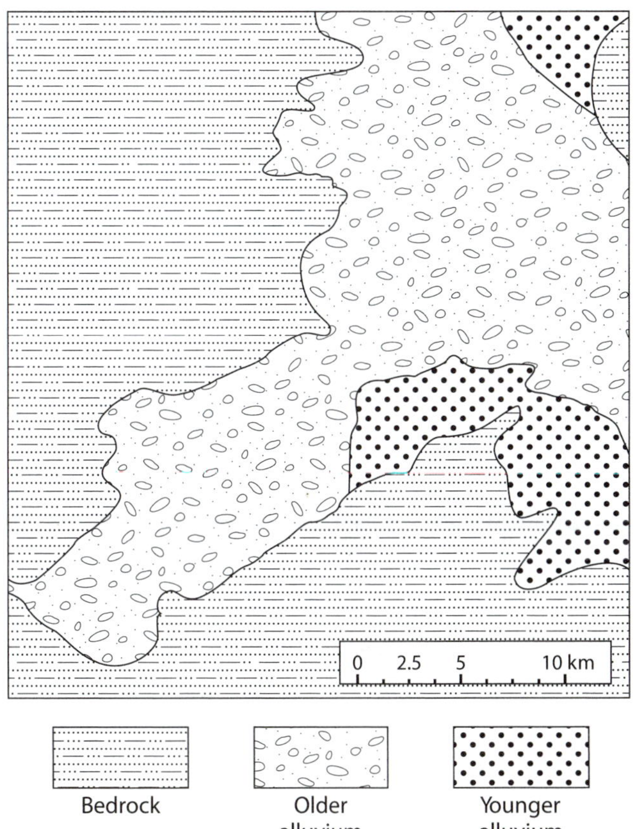

FIGURE 5-7 Geologic sketch map of Ghazni area shown in Figure 5-6.

The ASTER program of Gillespie and Rokugawa (2001) generates five thermal emissivity bands (AST-05 product) and land surface temperature data (AST-08 product) as Level 2 products from the Level 1 AST-L1T data. Level 2 ASTER TIR emissivity and surface temperature data can be downloaded at no cost and are used to monitor volcano eruptions, surface water temperature, gas (SO_2) emissions, fires, and urban heat islands (discussed below) (Abrams, 2005).

THERMAL CONDUCTIVITY

Thermal conductivity (K) is the rate at which heat passes through a material. It is expressed as calories per centimeter per second per degree Celsius (cal · cm^{-1} · sec^{-1} · $°C^{-1}$), which is the number of calories that will pass through a 1-cm cube of material in 1 sec when two opposite faces are maintained at a 1°C difference in temperature. Table 5-5 gives thermal conductivities for geologic materials. For any rock type the thermal conductivity may vary by up to 20% from the value given. Thermal conductivities of porous materials may vary by up to 200% depending on the nature of the substance that fills the pores. Rocks and soils are relatively poor conductors of heat. The average thermal conductivity of the materials in Table 5-5 is 0.006 cal · cm^{-1} · sec^{-1} · $°C^{-1}$, which is two orders of magnitude lower than the thermal conductivity of such metals as aluminum, copper, and silver.

THERMAL CAPACITY

Thermal capacity (c) is the ability of a material to store heat. Thermal capacity is the number of calories required to raise the temperature of 1 g of a material by 1°C and is expressed in calories per gram per degree Celsius (cal · g^{-1} · $°C^{-1}$). In Table 5-5 note that water has the highest thermal capacity of any substance. Figure 5-8 shows the difference between thermal capacity and kinetic temperature. Spheres of the same volume made from rhyolite, limestone, and sandstone are heated to a temperature of 100°C. The values for thermal capacity and density, taken from Table 5-5, are multiplied to determine the number of cal · cm^{-1} · $°C^{-1}$ that each rock stores. The rocks are assumed to have a uniform density of 2.5 g · cm^{-3}; therefore, the different values are determined solely by differences in thermal capacity. As Figure 5-8A shows, sandstone stores the greatest amount of heat and rhyolite stores the least. The heated spheres are simultaneously placed on a sheet of paraffin. Melting ceases when the spheres and paraffin have reached a uniform temperature. As Figure 5-8B shows, the amount of melting is related to the thermal capacity of the rocks and not to their temperature.

THERMAL INERTIA

Thermal inertia (P) is a measure of the thermal response of a material to temperature changes and is given in calories per square centimeter per second square root per degree Celsius (cal · cm^{-2} · $sec^{-1/2}$ · $°C^{-1}$).

A. Spheres of rock heated to 100°C and placed on a sheet of paraffin. The value for each rock is the product of its thermal capacity (c) and density (ρ) in cal · cm^{-3} · $°C^{-1}$.

B. After the rocks and paraffin have reached the same temperature.

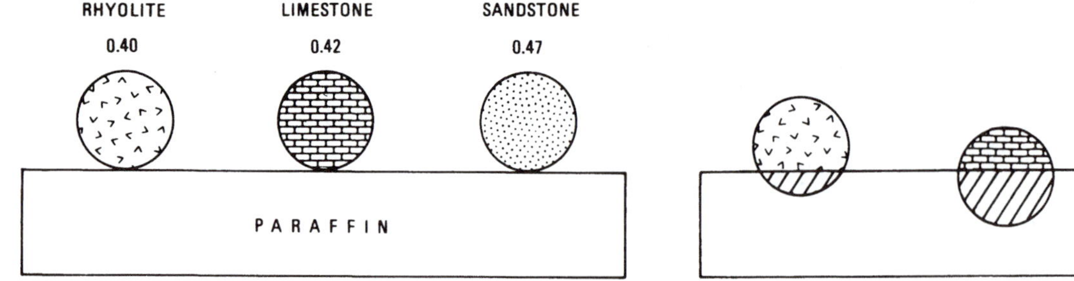

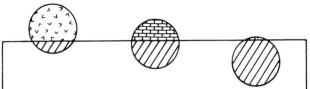

FIGURE 5·8 Effect of differences in the thermal capacity of various rock types.

TABLE 5-5 Thermal properties of geologic materials and water at 20°C.

Material	Thermal Conductivity (K) (cal · cm^{-1} · sec^{-1} · $°C^{-1}$)	Density (ρ) (g · cm^{-3})	Thermal Capacity (c) (cal · g^{-1} · $°C^{-1}$)	Thermal Diffusivity (k) (cm^2 · sec^{-1})	Thermal Inertia (P) (cal · cm^{-2} · $sec^{-1/2}$ · $°C^{-1}$)
1. Basalt	0.0050	2.8	0.20	0.009	0.053
2. Clay soil, moist	0.0030	1.7	0.35	0.005	0.042
3. Dolomite	0.012	2.6	0.18	0.026	0.075
4. Gabbro	0.0060	3.0	0.17	0.012	0.055
5. Granite	0.0075	2.6	0.16	0.016	0.056
6. Gravel	0.0030	2.0	0.18	0.008	0.033
7. Limestone	0.0048	2.5	0.17	0.011	0.045
8. Marble	0.0055	2.7	0.21	0.010	0.056
9. Obsidian	0.0030	2.4	0.17	0.007	0.035
10. Peridotite	0.011	3.2	0.20	0.017	0.084
11. Pumice, loose, dry	0.0006	1.0	0.16	0.004	0.009
12. Quartzite	0.012	2.7	0.17	0.026	0.074
13. Rhyolite	0.0055	2.5	0.16	0.014	0.047
14. Sandy gravel	0.0060	2.1	0.20	0.014	0.050
15. Sandy soil	0.0014	1.8	0.24	0.003	0.024
16. Sandstone, quartz	0.0120	2.5	0.19	0.013	0.075
17. Serpentine	0.0063	2.4	0.23	0.013	0.059
18. Shale	0.0042	2.3	0.17	0.008	0.041
19. Slate	0.0050	2.8	0.17	0.011	0.049
20. Syenite	0.0077	2.2	0.23	0.009	0.062
21. Tuff, welded	0.0028	1.8	0.20	0.008	0.032
22. Water	0.0013	1.0	1.01	0.001	0.036

Source: Janza and others (1975, Table 4-1).

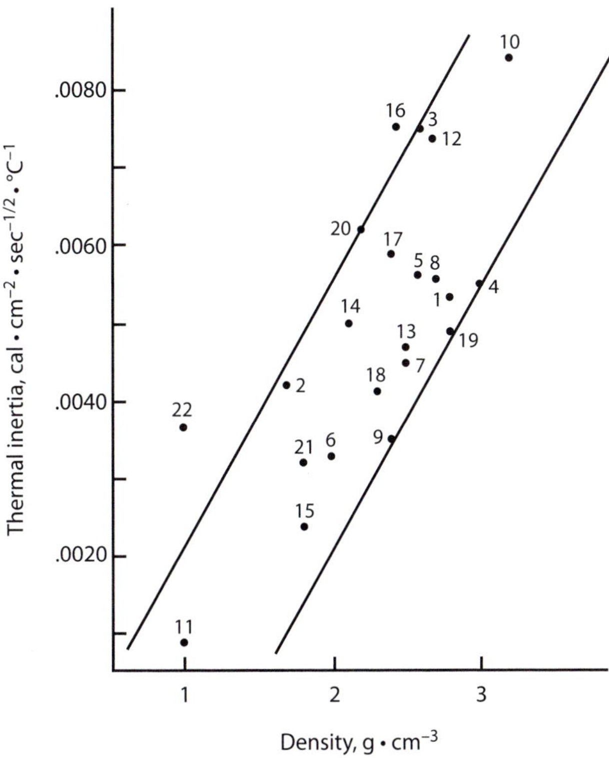

FIGURE 5-9 Relationship between thermal inertia and density for rocks and water. Numbers refer to the materials in Table 5-5.

Thermal inertia is expressed as

$$P = (K \times \rho \times c)^{1/2} \tag{5-9}$$

where K is thermal conductivity, ρ is density, and c is thermal capacity. Of the three properties that determine thermal inertia, density is the most important. For the most part, thermal inertia increases linearly with increasing density, as shown in Figure 5-9, which plots the values for the materials listed in Table 5-5.

Figure 5-10 illustrates the effect of differences in thermal inertia on surface temperatures. The difference between maximum and minimum temperature occurring during a diurnal solar cycle is called ΔT. Materials with low thermal inertia, such as shale, siltstone, and volcanic cinders, have low resistance to temperature change and have a relatively high ΔT. These materials reach a high maximum surface temperature in the daytime and a low minimum temperature at night. Materials with high thermal inertia, such as sandstone, granite, and basalt, strongly resist temperature changes and have a relatively low ΔT. These materials are relatively cool in the daytime and warm at night.

APPARENT THERMAL INERTIA

Thermal inertia cannot be measured by remote sensing methods because conductivity, density, and thermal capacity must be measured by contact methods. Maximum and minimum radiant temperature, however, can be measured from digitally recorded daytime and nighttime images. For corresponding ground resolution cells, ΔT is determined by subtracting the nighttime temperature from the daytime temperature. ΔT is low for materials with high thermal inertia and high for those with low thermal inertia (Figure 5-10). This relationship is used to determine the *apparent thermal inertia* (ATI) by the formula

$$\text{ATI} = \frac{1-A}{\Delta T} \tag{5-10}$$

where A is the albedo in the visible band. Albedo is employed to compensate for the effects that differences in absorptivity have on radiant temperature. During the day, dark materials (with low albedo) absorb more energy from sunlight than light materials (with high albedo). The absorbed solar energy increases kinetic temperature, which increases the radiant thermal energy. Therefore, a dark material typically has a

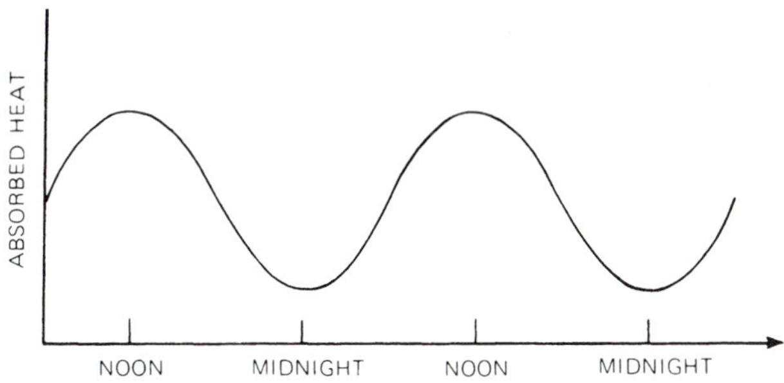

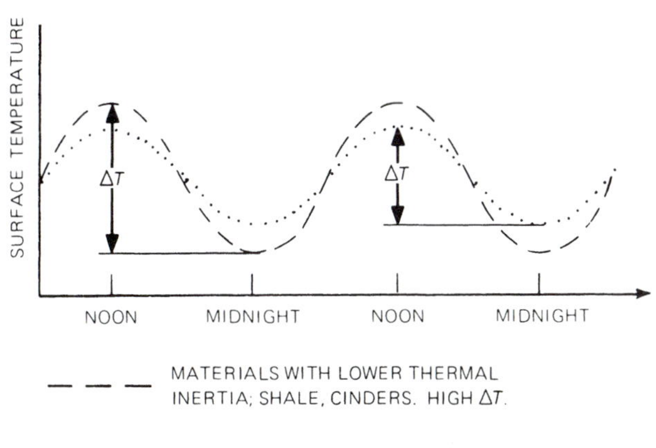

FIGURE 5·10 Effect of differences in thermal inertia on surface temperatures during diurnal solar cycles. Note differences in ΔT for materials with high and low thermal inertia.

higher ΔT than an otherwise identical material that has a light color. The term $1 - A$ corrects for some of these effects. A typical ATI image produced from a visible (albedo) image and daytime and nighttime thermal images is illustrated and interpreted in **Digital Images 5-1** and **5-2**, respectively.

ATI images must be interpreted with caution because ΔT may be influenced by factors other than thermal inertia. Consider an area of uniform material, such as granite, that has high topographic relief. In a daytime TIR image, the shadowed areas have a lower radiant temperature than sunlit areas. In a nighttime image the sunlit and shadowed areas have similar temperatures. As a result the shadowed granite has a lower ΔT and a higher ATI than the same granite that is sunlit. Some ATI computer programs compensate for shadows by using topographic data together with information on solar elevation and azimuth. Water and vegetation typically have ATI values that are determined by factors other than their thermal inertias, as discussed later in this chapter.

ATI may be measured with a thermal inertia meter, which employs a radiometer and standard rock samples of known emissivity and thermal inertia (Kahle and others, 1981). Electric lamps heat the target material and the standards to a uniform level, the radiometer measures radiant temperatures, and the meter then calculates the ATI for the target material. Table 5-6 lists ATI values determined in the field using this meter. These relative values are useful for discriminating among different materials and are not intended as absolute measures of thermal inertia.

THERMAL DIFFUSIVITY

The same values used to determine thermal inertia may be used to determine *thermal diffusivity* (k), which is given as

$$k = \frac{K}{c \times \rho} \quad \textbf{(5-11)}$$

Thermal diffusivity, given in centimeters squared per second ($cm^2 \cdot sec^{-1}$), governs the rate at which temperature changes within a substance. More specifically, it states the ability of a substance during a daytime period of solar heating to transfer heat from the surface to the interior and during a nighttime period of cooling to transfer stored heat to the surface.

DIURNAL TEMPERATURE VARIATION

Figure 5-11 shows typical diurnal variations in radiant temperature. The most rapid temperature changes, shown by steep curves, occur near dawn and sunset. At the intersection of two curves (called *thermal crossover*), radiant temperatures are identical for both materials.

TABLE 5-6 Apparent thermal inertia values measured in the field with a thermal inertia meter.

Material	ATI[a]
Sandy alluvium	0.014
Sand, windblown	0.015
Rhyolite tuff	0.022
Clay-silt playa	0.024
Basalt, pahoehoe	0.039
Basalt, olivine	0.042
Basalt, aa	0.042
Barite	0.043
Andesite	0.044
Rhyodacite, silicified	0.048
Chert	0.053

[a] ATI values are relative to a dolomite standard with a thermal inertia of 0.984 cal $\cdot$ $cm^{-2} \cdot sec^{-1/2} \cdot {}^{\circ}C^{-1}$. Error limits are approximately ±10%.
Source: Kahle and others (1981, Table 4).

CHARACTERISTICS OF TIR IMAGES

On most TIR images, the brightest tones represent the warmest radiant temperatures, and the darkest tones represent the coolest ones.

EFFECTS OF WEATHER ON IMAGES

Clouds typically show the patchy warm-and-cool pattern illustrated in Figure 5-12A, where the dark signatures are relatively cool and the bright signatures are relatively warm. Scattered rain showers produce a pattern of streaks parallel with the scan lines on the image. A heavy overcast layer reduces thermal contrasts between terrain objects because of reradiation of energy between the terrain and cloud layer.

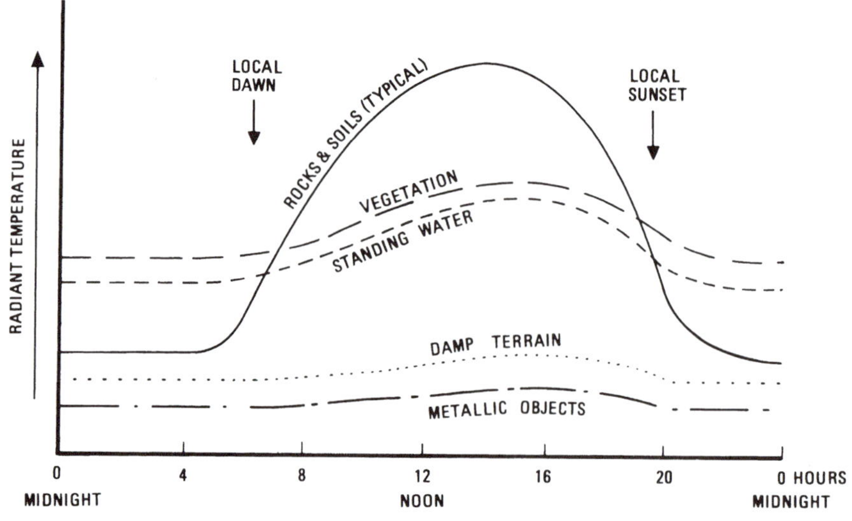

FIGURE 5-11 Diurnal radiant temperature curves (diagrammatic) for typical materials.

Images may be acquired by flying below the cloud layer, but the resulting thermal contrast is relatively low.

Wind produces characteristic patterns of smears and streaks on images. *Wind smears* (Figure 5-12B) are parallel curved lines of alternating lighter and darker signatures that may extend over wide expanses of the image. *Wind streaks* occur downwind from obstructions on flat terrain and typically appear as the warm (bright) patterns shown in Figure 5-12C. In this example the wind is blowing from right to left across obstructions, which are clumps of trees with warm signatures. Wind velocity is lower downwind from obstructions, which reduces the cooling effect; thus terrain in the sheltered areas is warmer than terrain exposed to the wind. Wind smears and streaks may be avoided by acquiring images only on calm nights, but, in many regions, surface winds persist for much of the year and their effects must be endured. Interpreters must be alert to avoid confusing wind-caused signatures with terrain features.

A. Clouds.

B. Surface wind smears.

C. Surface wind streaks.

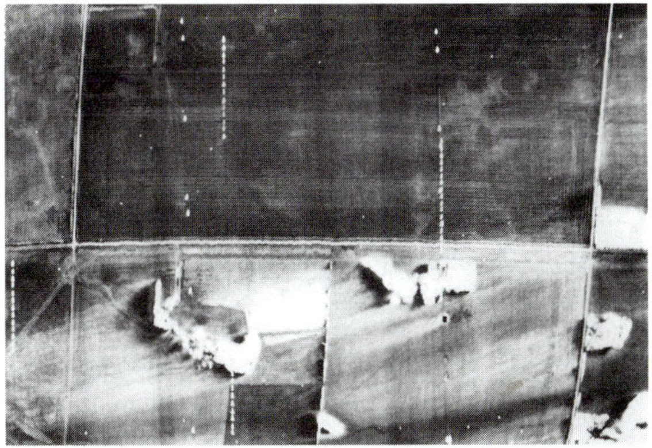

Figure 5-12 Effects of weather on TIR images.

Penetration of Smoke Plumes

Clouds consist of tiny particles of ice or water that have the same temperature as the surrounding air. As shown in Figure 5-12A, images acquired from aircraft or satellites above cloud banks record the radiant temperature of the clouds. Energy from the Earth's surface does not penetrate the clouds but is absorbed and reradiated. Smoke plumes, however, consist of ash particles and other combustion products so fine that they are readily penetrated by the relatively long wavelengths of TIR radiation.

Figure 5-13 shows a visible image and a TIR image that were acquired simultaneously during a daytime flight over a forest fire. The smoke plume completely conceals the ground in the visible image (Figure 5-13A), but terrain features are clearly visible in the TIR image (Figure 5-13B), where the burning front of the fire has a bright signature. The US Forest Service uses aircraft equipped with TIR scanners that process images in flight, which are transmitted to firefighters on the ground. These images provide information about the fire location that cannot be obtained by visual observation through the smoke plumes. TIR images are also acquired after fires are extinguished in order to detect hot spots that could reignite.

In northern Alberta, Canada the airborne TABI-1800 sensor (Table 5-1) acquired rapid response, high altitude, TIR images of wildfires (ITRES, 2015). The system was flown at 160 knots to collect images with 1 m spatial resolution (ITRES, 2011b). Once deployed, TIR images were collected during the night and analyzed, then delivered by 6:00 AM to assist fire suppression crews. The active fire perimeter and residual hot spots are shown as bright white pixels in the grayscale image of **Digital Image 5-3A** 🌐. A color density slice was applied to the grayscale image to show the hottest pixels in red and the next warmest pixels in yellow (**Digital Image 5-3B** 🌐). As discussed in Chapter 10, density slices can be converted to a vector polygon file and superimposed on high spatial resolution imagery in Google Earth to provide a map.

Influence of Water and Vegetation

The thermal inertia of water is similar to that of soils and rocks (Table 5-5), but in the daytime, water bodies have a cooler surface temperature than soils and rocks. At night the relative surface temperatures are reversed, so that water is warmer than soils and rocks (Figure 5-11). Convection currents cause the relatively uniform temperature at the surface of a water body. Convection does not operate to transfer heat in soils and rocks; therefore, heat from solar flux is concentrated near the surface of these solids in the daytime, causing a higher surface temperature. At night this heat radiates into the atmosphere and is not replenished by convection currents in these solid materials, causing surface temperatures to be lower than in adjacent water bodies (K. Watson, personal communication). Some images may not be annotated for the time of day at which they were acquired. The thermal signatures of water bodies are a reliable index to the time of image acquisition. If water bodies have warm signatures relative to the adjacent terrain, the image was acquired at night; relatively cool water bodies indicate daytime imagery.

A. Visible image.

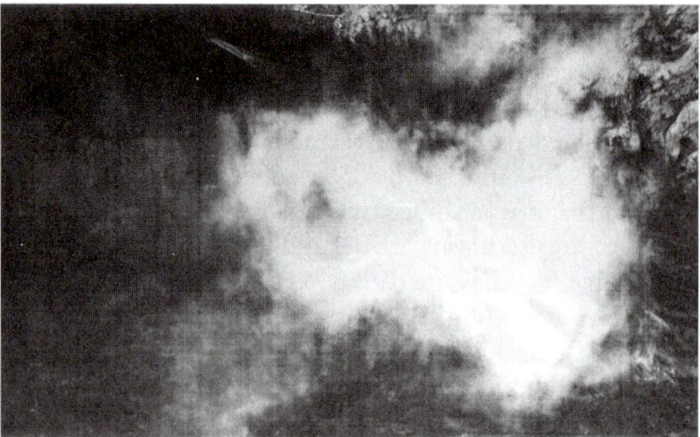

B. TIR image.

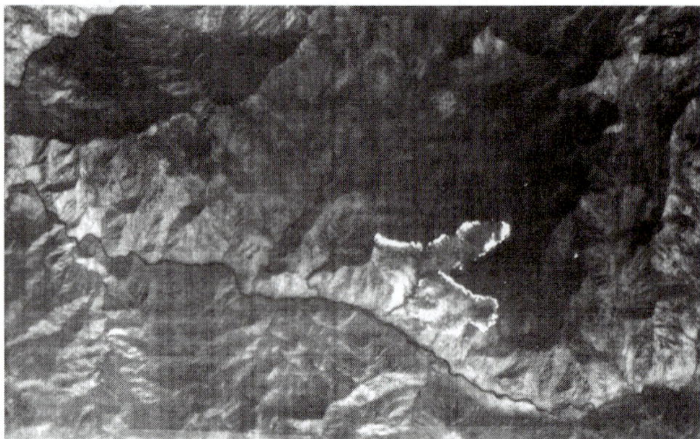

C. Location map.

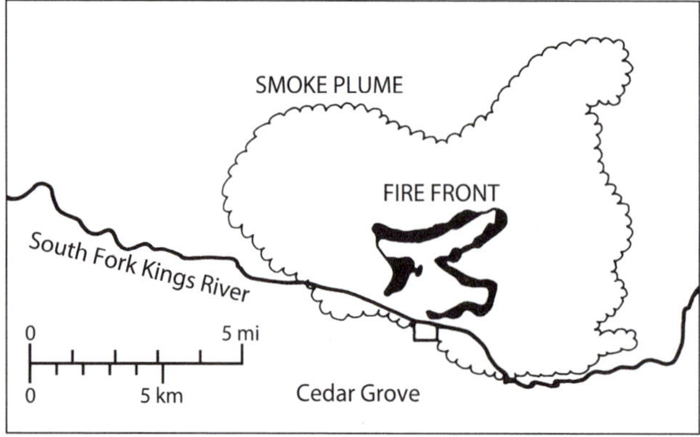

FIGURE 5-13 Penetration of smoke by a TIR image. Forest fire in King's Canyon, Sequoia National Forest, California. Courtesy NASA, Ames Research Center.

As shown in the diurnal temperature curves (Figure 5-11), damp soil is cooler than dry soil, both day and night. As absorbed water evaporates, it cools the soil. Figure 5-14 clearly shows this *evaporative cooling* effect. The aerial photograph and the TIR image were simultaneously acquired over an orchard of immature trees where adjacent rows were irrigated on successive days. In the photograph only the wettest, most recently irrigated soil has a discernibly darker tone than the surroundings. In the TIR image, however, the moisture pattern is clearly visible as variations in radiant temperature. The grayscale of the image is calibrated to show radiant temperature, which in turn correlates with moisture content. Some researchers have noted that adding water to dry soil increases the thermal inertia of the soil to values comparable to those of rocks. The effect of evaporative cooling, however, dominates the radiant temperature signature of damp ground. Many geologic faults and fractures are recognizable in TIR images because of evaporative cooling. Examples from California and South Africa are described later in this chapter.

Green deciduous vegetation has a cool signature on daytime images and a warm signature on nighttime images compared to rocks and soils (Figure 5-11). During the day, transpiration of water vapor lowers leaf temperature, causing vegetation to have a cool signature relative to the surrounding soil. At night the insulating effect of leafy foliage and the high water content retain heat, which results in warm nighttime temperatures. The relatively high nighttime and low daytime radiant temperature of conifers, however, does not appear to be related to their water content. The composite emissivity of the needle clusters making up a whole tree approaches that of a blackbody. Dry vegetation, such as crop stubble in agricultural areas, appears warm on nighttime imagery in contrast to bare soil, which is cool. The dry vegetation insulates the ground to retain heat and causes the warm nighttime signature.

For reasons discussed earlier, water, moist soil, and vegetation have relatively low values of ΔT when their signatures are compared on daytime and nighttime IR images. When these ΔT values are used in Equation 5-10, the resulting high ATI values differ significantly from the actual thermal inertias for these materials, which have low to intermediate values. Therefore, one must be cautious when interpreting areas of water, moist soil, or vegetation in ATI images.

CONDUCTING AIRBORNE TIR SURVEYS

sUAS/drones can be equipped with miniature TIR systems. Small GIS systems provide location data. Individual flight lines may be repeated to evaluate seasonal changes and day/night changes. Manned aircraft fly longer, higher, and farther and carry larger sensors capable of imaging larger areas than sUAS. Multispectral and hyperspectral TIR systems are carried on manned aircraft (Table 5-1); these sensors also are being miniaturized to deploy on sUAS platforms. For any aerial TIR survey the following factors must be considered: time of day, wavelength band, spatial resolution, orientation and altitude of flight lines, and ground measurement.

A. Aerial photograph.

B. Daytime TIR image (8 to 14 μm).

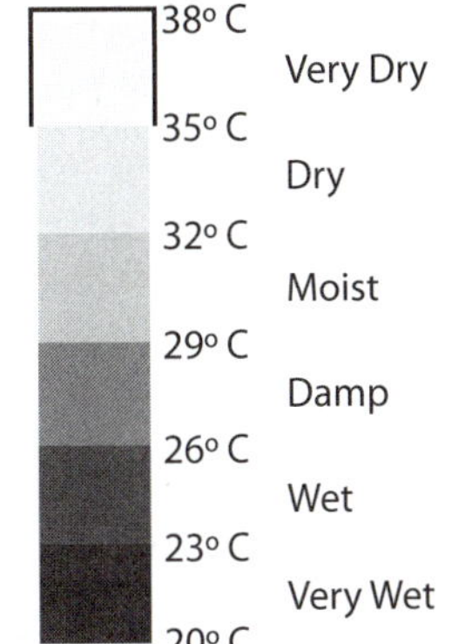

FIGURE 5-14 Relationship between soil moisture and radiant temperature in an irrigated orchard. Courtesy Daedalus Enterprises, Inc.

TIME OF DAY

On daytime images, topography is the dominant feature because of solar heating and shadowing. On nighttime images, however, the effects of solar heating and shadowing have been radiated to the sky and the images show thermal properties of materials. The daytime and nighttime images of the Caliente and Temblor Ranges, California, illustrate these differences (Figure 5-15). On the daytime image (Figure 5-15A), the ridges and slopes that face south and east are heated by the morning sun and have bright (warm) signatures; those facing north and west are shadowed and have dark (cool) signatures. The daytime image is dominated by these topographic effects. On the nighttime image (Figure 5-15B), solar heating and cooling effects have radiated to the sky and topographic highlights and shadows are largely eliminated. Geologic features are emphasized at night, as shown by comparing Figure 5-15B with the map in Figure 5-15C. The narrow warm signatures in the Caliente Range are basalt outcrops. In the Temblor Range the bands with cool signatures are outcrops of shale and siltstone. These differences are attributed to the relatively high thermal inertia of basalt (0.053) and low thermal inertia of shale (0.041), as listed in Table 5-5. The broad belts of warm signature are sandstone (thermal inertia 0.075) and conglomerate outcrops. These geologic features are obscure or invisible in the daytime image that is dominated by topographic signatures.

WAVELENGTH BANDS

TIR images may be acquired at wavelength bands of 3 to 5 μm (MWIR) and 8 to 14 μm (LWIR), which are atmospheric windows (Figure 5-1). Wien's displacement law shows the temperatures at which the maximum energy will radiate for each of these bands. Figure 5-2A shows that the MWIR band corresponds to the radiant energy peak for temperatures of 600°K and greater, which are associated with fires, lava flows, and other hot features. The LWIR band spans the radiant energy peak for a temperature of 300°K; this is the ambient temperature of the Earth, which has a radiant energy peak at 9.7 μm. Figure 5-16 shows nighttime images in central Michigan acquired by a multispectral scanner that recorded images in both the MWIR and LWIR bands. The area consists of pastures, fields, and woodlands cut by a divided highways and a few streams. In the 8 to 14 μm image (Figure 5-16B), terrain features are well expressed and have the following signatures: trees are relatively warm, fields have intermediate temperatures, and marshy areas along the left margin of the image are cool. The overall radiant temperature level of the 3 to 5 μm image (Figure 5-16A) is lower and the thermal contrasts among terrain features are much lower than in the 8 to 14 μm image. Of special interest are localities A through D in the location map (Figure 5-16C), which represent the following features:

A Three small fires of glowing charcoal briquettes are located within a grove of trees. On the 8 to 14 μm image the warm signature of the trees effectively masks the fires, but on the 3 to 5 μm image the fires are clearly visible and the signature of the trees is subdued.

B A large campfire in an open field is visible on both images.

C In an open field, a pit containing a small charcoal fire is concealed beneath a pile of brush. On the 8 to 14 μm image this target could be mistaken for vegetation, but on the 3 to 5 μm image it is clearly recognizable as a hot target.

D A large campfire and three vehicles with warm engines are located in an open field. The four targets are resolvable on the 3 to 5 μm image, but only the campfire is shown on the 8 to 14 μm image.

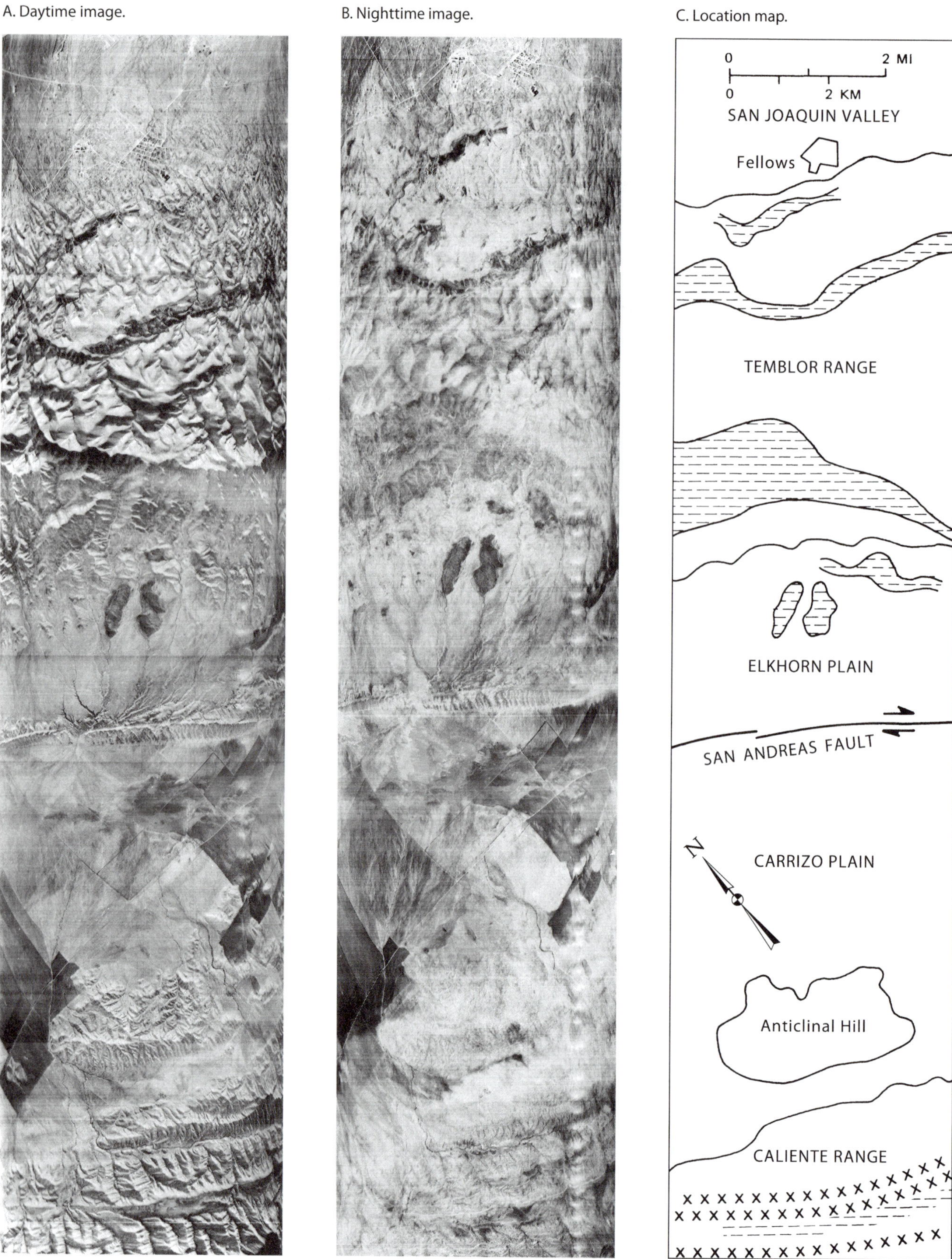

FIGURE 5-15 Comparison of daytime and nighttime TIR images (8 to 14 μm), Caliente and Temblor Ranges, California. Crosses (X) are basalt outcrops; dashes (–) are shale outcrops. From Wolfe (1971, Figures 3 and 4).

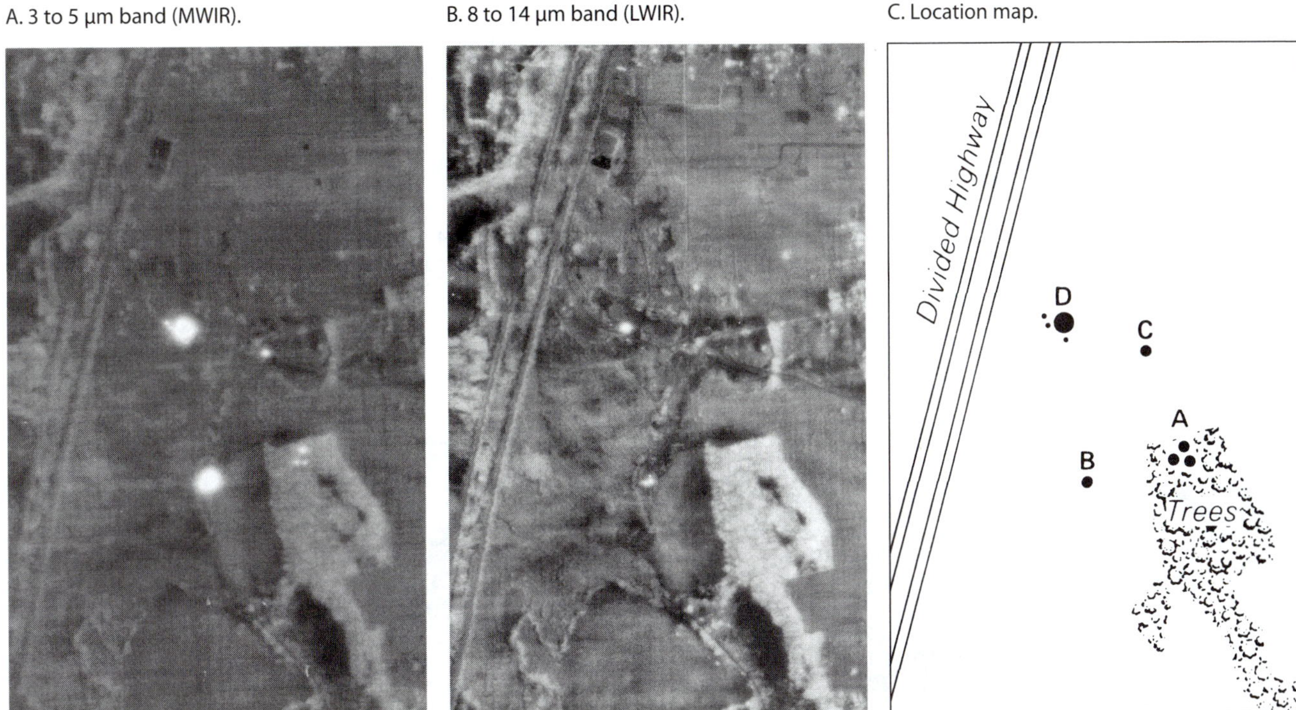

FIGURE 5-16 Different spectral bands of nighttime TIR images, central Michigan. Letters found on the map are explained in the text. Courtesy Daedalus Enterprises, Inc.

This example illustrates that LWIR images are optimum for terrain mapping, whereas MWIR images are optimum for mapping hot targets, such as fires and volcanic eruptions.

Findlay and Cutten (1989) used a mathematical modeling approach to compare systems operating at 3 to 5 μm and 8 to 12 μm in a tropical maritime environment. They also concluded that the 8 to 12 μm band is superior for all but the hottest targets.

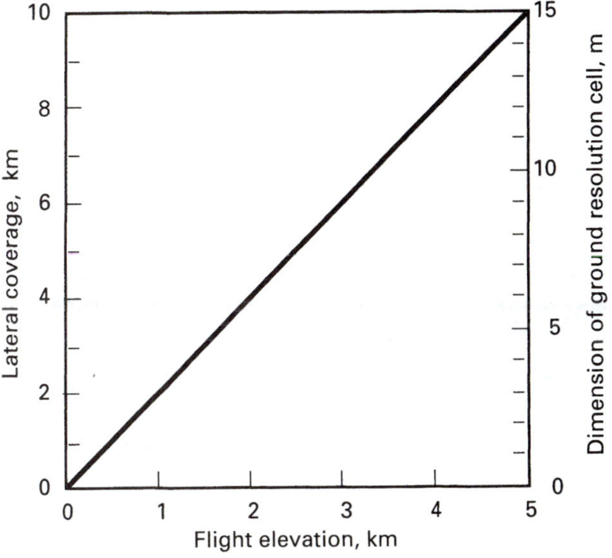

FIGURE 5-17 Lateral coverage and spatial resolution of scanner images as a function of flight elevation. For this scanner the angular field of view is 90° and the instantaneous field of view is 3.0 mrad.

SPATIAL RESOLUTION

Angular resolving power is defined as the angle subtended by imaginary lines passing from the imaging system and two targets spaced at the minimum resolvable distance. Angular resolving power is commonly measured in radians (Figure 1-11). Equation 1-5 enables the ground sampling distance to be calculated from IFOV and platform altitude (Figure 1-17A). A sensor with an IFOV of 1 mrad that is flown at 500 m elevation will have a ground sampling distance of 50 cm.

ELEVATION OF FLIGHT LINES

Flight elevation, which is height above average terrain, influences swath width and the spatial resolution of images. Figure 5-17 shows image swath width as a function of flight elevation, For example, a cross-track scanner system with a 90° angular field of view at an elevation of 2 km acquires images with a swath width of 4 km.

GROUND MEASUREMENTS

As described earlier, weather and surface conditions play a large role in determining terrain expression in TIR images. It may be useful to collect ground information on weather conditions, soil moisture, and vegetation at the time of an airborne TIR survey. Ground measurements are most practical and useful for surveys of relatively small areas that can be covered with a single flight line. If flight lines are repeated, ground data on changing weather conditions and solar flux may be valuable in comparing and interpreting the images. For larger areas covered by image mosaics, ground measurements can be made at only a limited number of localities during the 3 to 4 hours required to acquire the images. Ground measurements are most valuable if they are made at localities that can be identified on the images. The ground measurements may also help to explain anomalous features. In practice, however, it is virtually impossible to anticipate where anomalous features will occur; therefore, most regional surveys omit contemporaneous ground measurements. The availability of calibrated TIR images also reduces the need for ground measurements.

Ground measurements to help explain image signatures may be made some time after the image flight, as in the Indio Hills example shown later in the chapter. Ground information is useful for understanding other forms of remote sensing imagery in addition to TIR images.

A. Aerial photograph with overlay of heating lines.

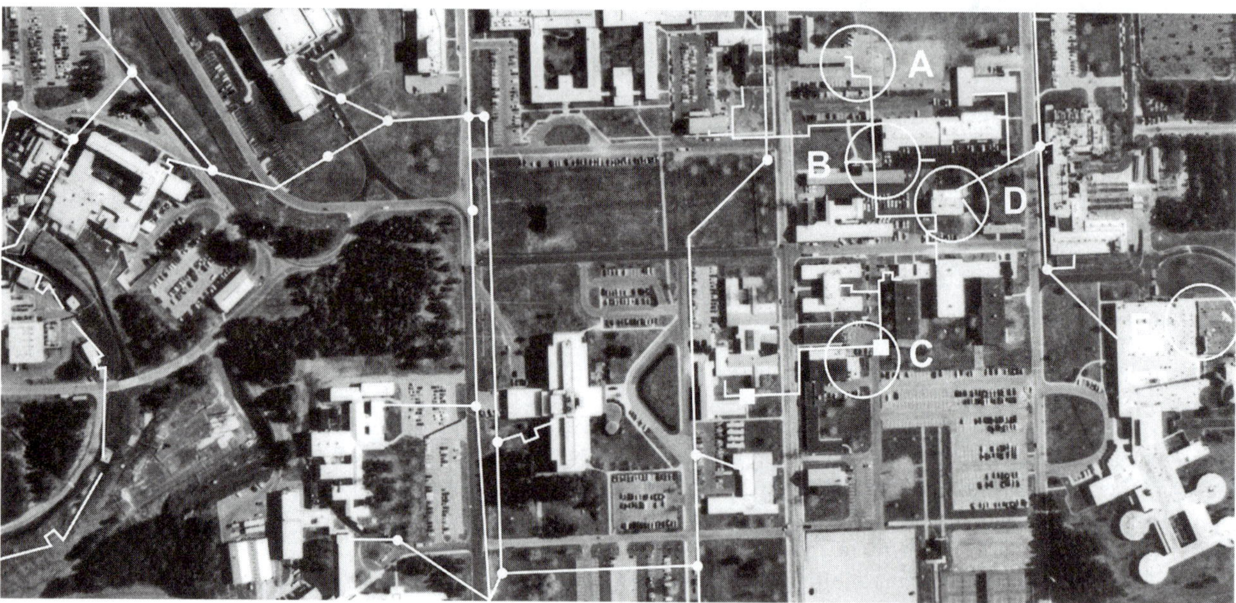

B. Nighttime TIR image (8 to 14 μm).

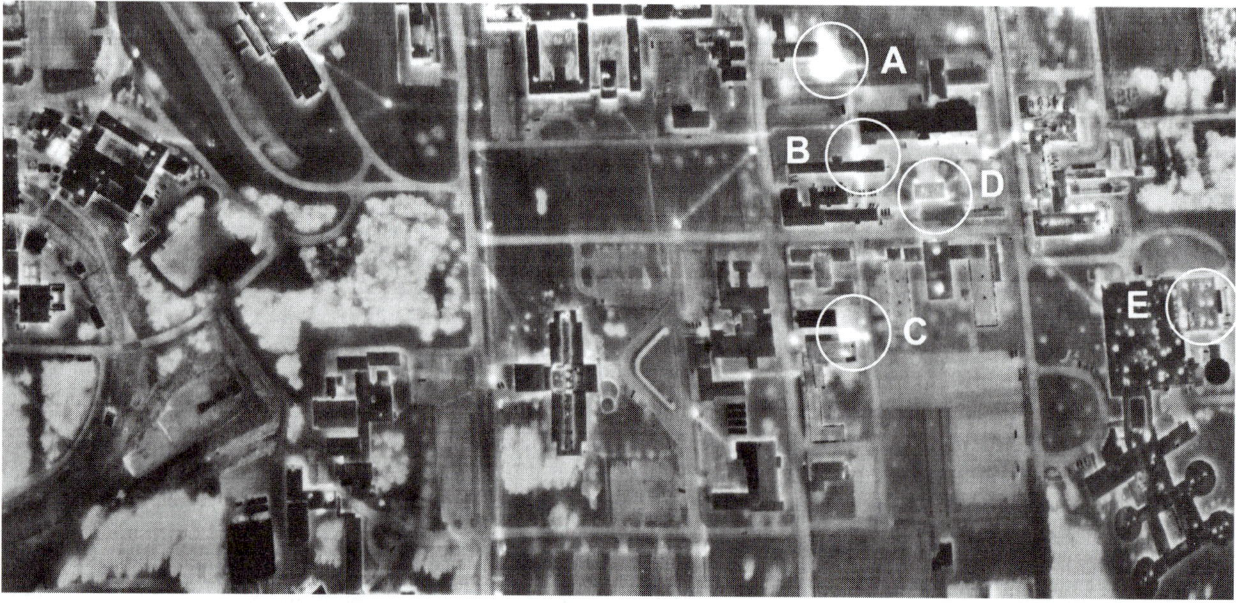

FIGURE 5-18 Heat-loss survey of Brookhaven National Laboratory, Long Island, New York. Localities are explained in the text. Courtesy Daedalus Enterprises, Inc.

HEAT-LOSS SURVEYS

An obvious application of TIR technology is to survey heated buildings, factories, and buried steam lines for anomalous hot spots that may indicate poorly insulated roofs and steam leakage. These small target areas are well-suited for drone surveys. By locating and correcting heat losses, the heating fuel saved in a few months can repay the cost of a survey drone. In the northern and central United States, many building complexes, such as university campuses and industrial complexes, are heated by steam distributed through buried pipelines from a central generating station. As the steam heats the buildings, it condenses into hot water, which returns via condensate lines to the steam plant, where it is used to generate more steam. Many of the pipeline systems are several decades old and have developed leaks that are difficult to detect because the lines are buried and many stretches are covered by sidewalks, streets, and parking lots.

Brookhaven National Laboratory on Long Island, New York, contracted for a heat-loss survey with a manned aircraft. Daytime aerial photographs and nighttime TIR images were acquired in November. The aerial photograph in Figure 5-18A includes an overlay of buried steam and condensate lines and manholes that was provided by the Brookhaven maintenance staff. On the nighttime TIR image (Figure 5-18B), the trees, standing water, and pavement are relatively warm. Roofs of well-insulated buildings are cool with warm spots formed by exhaust ventilators. Sides of buildings are relatively warm because of heat radiated from windows. The buried heating lines and manholes form bright lines and spots. Some TIR images revealed locations of steam lines for which the engineering records had been lost. Localities A, B, and C in the image and photograph are anomalous hot spots that proved to be major leaks or areas where pipe insulation had deteriorated. Localities D and E are building roofs that are significantly warmer than other roofs in the laboratory complex. Significant energy costs can be saved by improving the ceiling insulation in these buildings. These aircraft surveys were conducted in the pre-drone era. Today the same types of images are readily acquired from drones.

CONSERVATION—AFRICA

Since 2016 the World Wildlife Fund (WWF) has partnered with FLIR Systems to use TIR imaging technology for conservation of wildlife in Africa. The technology is used at night to:

1. Help park rangers detect and apprehend poachers, who operate primarily at night;
2. Mitigate potential conflict between elephants and neighboring villagers;
3. Enable ranger patrols to avoid large and dangerous animals; and
4. Monitor, count, and identify, with artificial intelligence, different large animals (lions, elephants, giraffes, buffalos, and hippopotamus) (WWF, 2017).

In 2016, WWF installed several long-range surveillance cameras (each with a 640 by 480 array and 4X zoom capability) on steel poles for perimeter security at Lake Nakuru National Park in Kenya to identify illegal intrusions and monitor animal behavior. The unattended, fixed-mount TIR cameras stream real-time video images to control rooms where operators can recognize poachers and direct park rangers to apprehend them. Within days of installing the system, poachers were identified and apprehended.

Also in 2016, WWF provided park rangers at the Maasai Mara National Reserve, located adjacent to Nairobi, with TIR systems mounted on vehicles to serve as mobile observation posts (WWF, 2016). Figure 5-19A shows the truck-mounted system. The ranger in the truck cab uses a keyboard with a joystick to operate the mobile camera. The camera has a range of approximately 1 mi (~1.6 km). Prior to implementing TIR technology, the rangers relied on flashlights to find poachers who avoided arrest by hiding in the shadows and the background vegetation. The mobile TIR systems enable the rangers to detect the warm bodies of concealed poachers.

A. Upper arrow indicates long-range TIR camera. Lower arrow indicates viewing display in cab of truck.

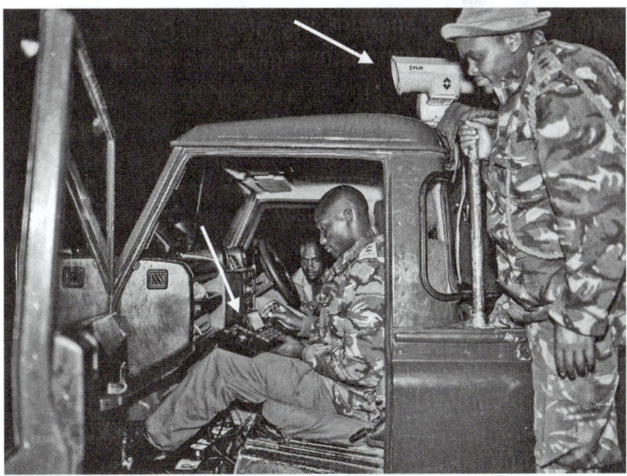

B. Nighttime TIR image of elephant herd. The warm (white) signature of the elephants contrasts with the relatively cool (dark) background. Any poachers would also have warm signatures.

Figure 5-19 Park rangers of the Mara Conservancy use nighttime TIR images to monitor and protect wildlife in Masai Mara National Reserve, Kenya. © James Morgan/WWF. WWF and FLIR Systems.

WWF has also deployed a TIR still-image and digital video camera (the FLIR Vue Pro) (Table 5-1) on a fixed wing sUAS at a national park in Malawi. The sensor has a 336 by 256 array, is 5.8 by 4.5 cm (including lens) in size, and weighs 113 gm (4 oz). The drone camera successfully patrolled a large, illegally fished river at night. The poachers soon learned they were no longer able to fish in the park without being detected, and almost all of the poaching has stopped. Aerial observation with drones also enables park rangers to direct elephants safely back to the park when they are heading toward a village and may damage the villagers and their crops (Figure 5-19B) (WWF, 2017).

Burke and others (2019) address some of the technical challenges and strategies in drone-mounted, TIR surveys of animals. Challenges include background "noise" from the atmosphere and the terrain, and camouflage of targets

A. Aerial photograph acquired May 5, 1953. White dashed outline is the location of Figure 5-20B.

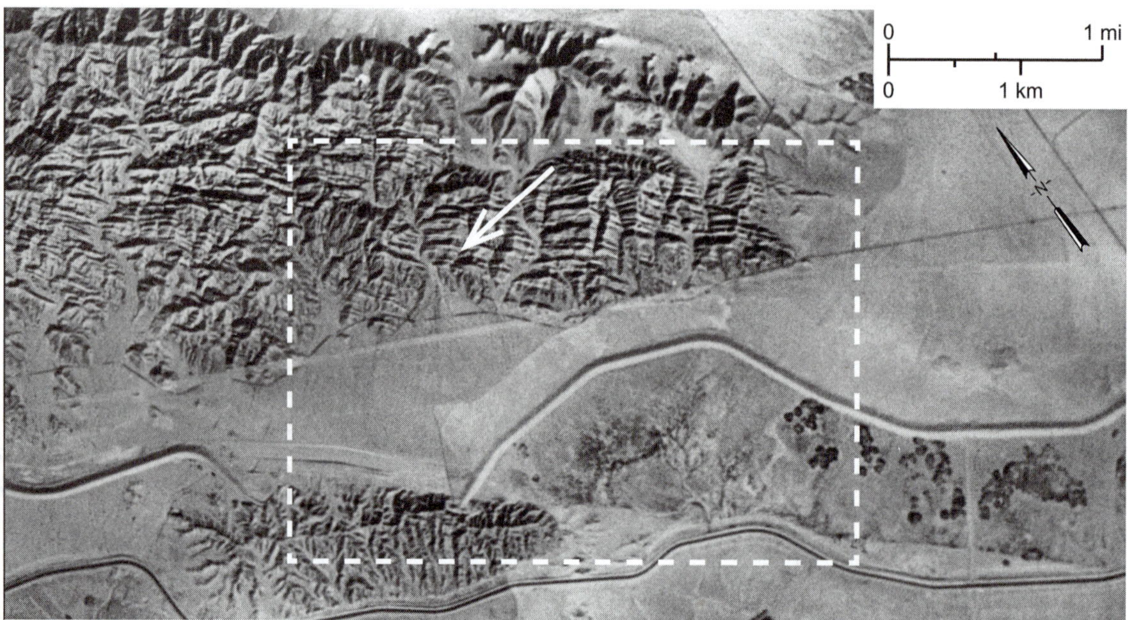

B. Nighttime TIR image (8 to 14 μm) acquired October 1963.

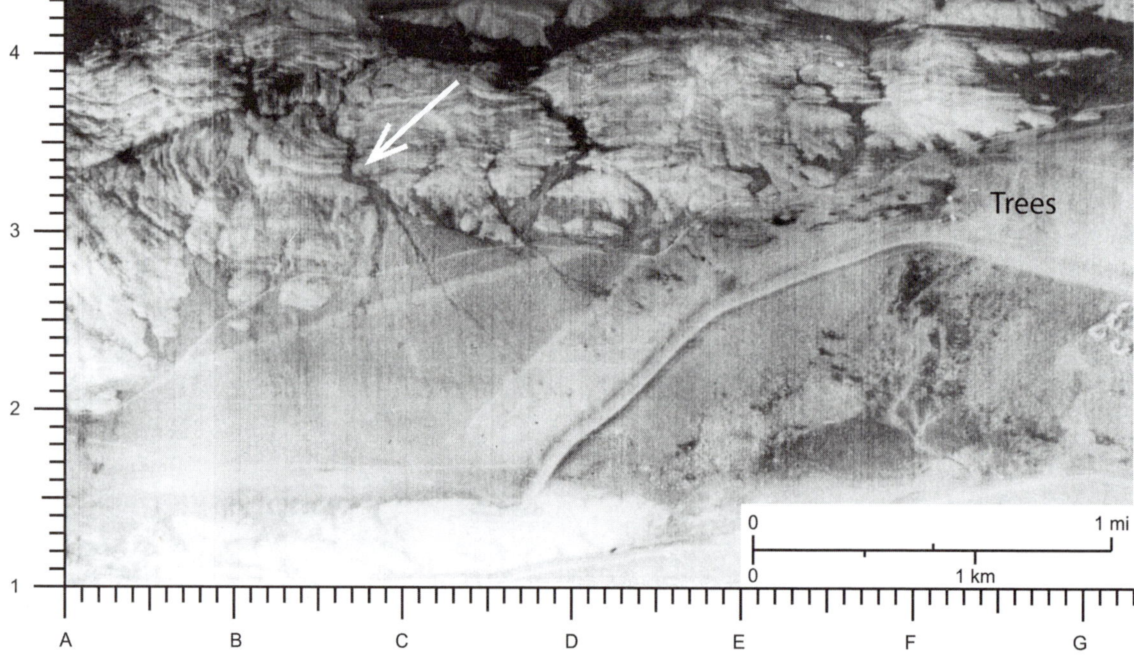

FIGURE 5-20 Images of the southern portion of Indio Hills, California. The white arrows point to Radiometer Gulch. From Sabins (1967, Figures 3 and 4).

(poachers) by vegetation. They provide an online calculator to determine the optimum flying height for covering a large area with adequate spatial resolution for recognizing the animals of interest (astro.ljmu.ac.uk/~aricburk/uav_calc).

In January 2019, WWF and FLIR Systems began the Kifaru Rising Project. This multiyear collaboration will deploy handheld, fixed, and mobile long-range imaging technology along with drone TIR imaging technology across 10 parks and game reserves in Kenya. The goal is to improve wildlife ranger safety and stop poaching of endangered species such as rhinos (FLIR Systems, 2019).

THERMAL INERTIA FIELD EXPERIMENT

Rocks with different thermal inertia values also have different relative radiant signatures on daytime and nighttime images. Denser rocks with higher thermal inertias, such as sandstone, have cooler daytime signatures and warmer nighttime signatures than less-dense rocks, such as siltstone. Less-dense rocks have reversed diurnal temperature patterns, with warmer daytime signatures and cooler nighttime temperatures. In the early 1970s, these anecdotal relationships were known, but quantitative data were lacking. In 1971, Sabins and colleagues conducted a field experiment to collect diurnal variations in radiant temperature from outcrops of two rock types with different thermal properties. The objective was to provide a database to analyze TIR images. The southern portion of the Indio Hills, California, which includes outcrops of the Palm Spring Formation, was selected for the experiment.

Figure 5-20A is an aerial photograph of the Indio Hills in the Coachella Valley of southern California acquired on May 5, 1953. Figure 5-20B is an 8 to 14 μm nighttime image that was acquired in October 1963 with an early experimental airborne system developed by a southern California aerospace company. The nighttime image covers the area shown by the dashed outline in the aerial photograph. In this uncalibrated analog TIR image, bright signatures are relatively warm radiant temperatures and dark signatures are cool. Both margins of the image show the data compression that is inherent in unrectified cross-track scanners. Despite these deficiencies we were pleased to have the image because it showed geologic features that were obscure or absent on aerial photographs (Figure 5-20A). Because the TIR image was acquired far below the ozone layer at the top of Earth's atmosphere, the full 8 to 14 μm spectral band was recorded.

Figure 5-21 is a geologic map interpreted from the aerial photograph of the Indio Hills. Outcrops of the Palm Spring Formation occupy much of the TIR image. The ground photograph (Figure 5-22) shows typical outcrops of the formation, which consist of alternating strata of sandstone and siltstone that dip gently to the northwest. The sandstone resists erosion and forms ledges. The siltstone is readily eroded and forms slopes. The aerial photograph (Figure 5-20A) shows these topographic and lithologic differences as alternating bright highlights (sandstone ridges) and parallel dark signatures (siltstone slopes). The natural dark albedo of the siltstone is enhanced by shadows cast by the sandstone ledges. By late nighttime, however, the effects of solar illumination and shadowing have radiated to the sky and the

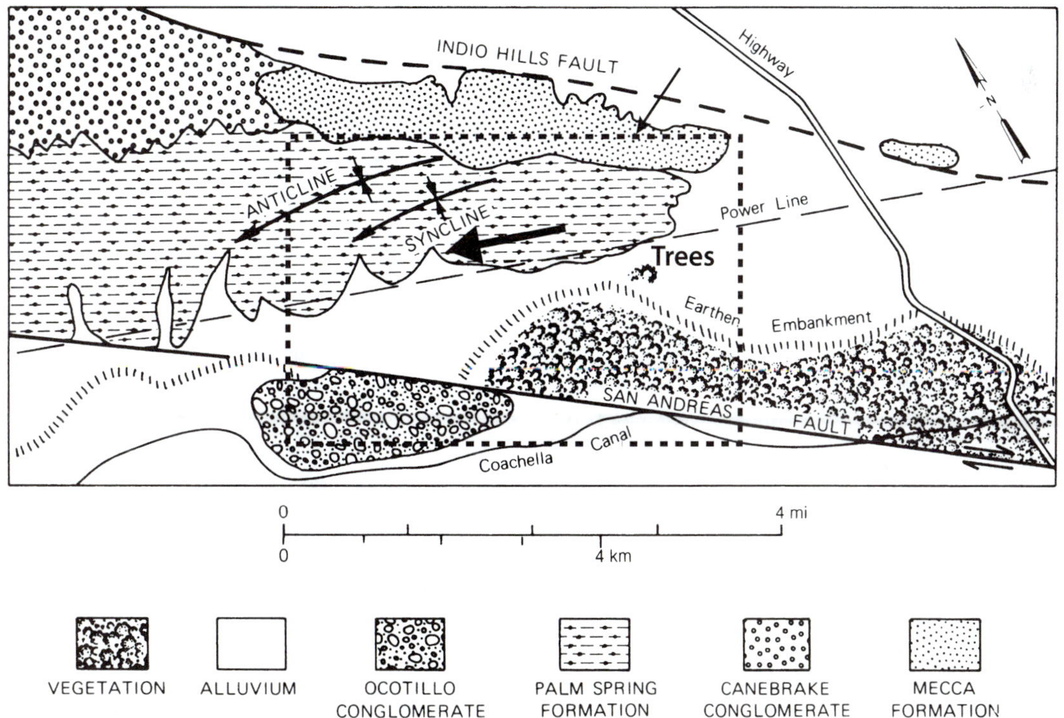

Figure 5-21 Geologic map of the southern portion of the Indio Hills, Riverside County, California. The black arrow points to Radiometer Gulch. The black dashed outline is the location of Figure 5-20B.

Figure 5-22 Sandstones (ledges) and siltstones (slopes) of the Palm Spring Formation in Radiometer Gulch. Radiant temperatures of these outcrops are included in Figure 5-23.

TIR image shows radiant thermal temperatures. For the project we selected a dry wash that we informally named Radiometer Gulch, which is indicated in Figure 5-20 with a white arrow and in Figure 5-21 with a black arrow. The ground photograph shows the alternating sandstone ledges and siltstone slopes. Within the Gulch we marked a spot on each of eight sandstone outcrops and on eight siltstone outcrops that we could recognize and re-occupy day or night.

Diurnal Radiant Temperature Measurements

IR radiometers are instruments that measure radiant temperature. We used a portable radiometer operating at 8 to 14 µm, which matched the band-pass of the TIR image. In Radiometer Gulch we selected eight pairs of sandstone and siltstone outcrops that are shown at the base of Figure 5-23. Throughout a 24-hour diurnal cycle we made measurements at approximately two-hour intervals. For analysis we separated the data into a daytime set and a nighttime set. The average of the daytime measurements is plotted for each sandstone outcrop as a black circle and for each siltstone outcrop as a black triangle in Figure 5-23A. For the nighttime set we repeated the process and show the results in Figure 5-23B. We combined the measurements to calculate the average temperature for sandstones and siltstones during the daytime and nighttime.

A. Daytime temperatures. The siltstones (25°C) are on average warmer than the sandstones (21°C).

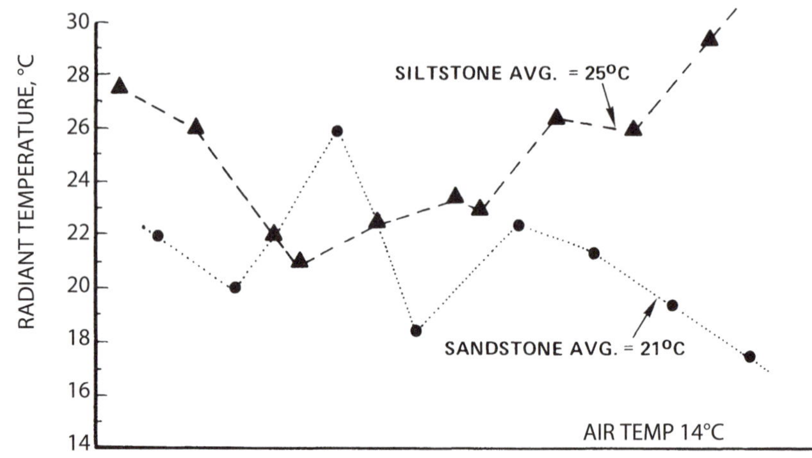

B. Nighttime temperatures. The sandstones (10°C) are on average warmer than the siltstones (8°C).

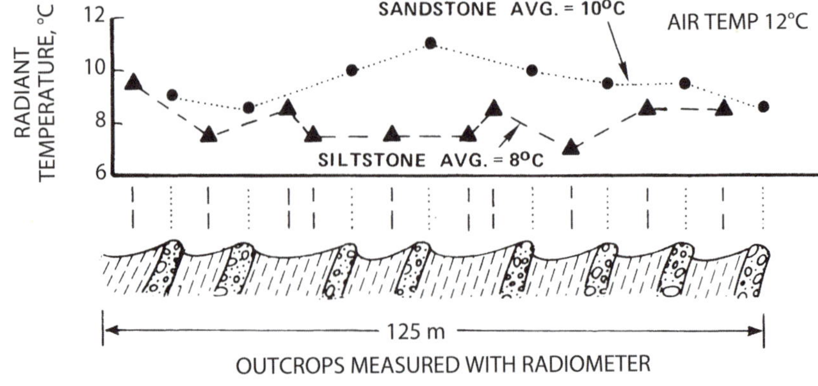

Figure 5-23 Daytime and nighttime radiant temperatures of the sandstones and siltstones of Palm Spring Formation at Radiometer Gulch (locality 3.4, B.7 of Figure 5-20B).

Data Analysis

In the plot of daytime temperatures (Figure 5-23A) the siltstones (with one exception) are consistently warmer (with an overall average of 25°C) than the sandstones (21°C). In the nighttime set (Figure 5-23B) the temperatures are inverted with the sandstones being warmer (average 10°C) than siltstones (average 8°C). Table 5-7 shows the average temperature values, together with ΔT.

Vegetation Signatures and Diurnal Cycle

During the temperature experiment we also measured diurnal radiant temperatures of vegetation and soil. In the nighttime TIR image (Figure 5-20B), the sparse vegetation has a distinct warm signature and soil and alluvium are cool. In daytime TIR images of similar areas (not illustrated) the signatures are reversed. Vegetation is cool on a daytime TIR image because of a process known as transpiration. At small pores on the underside of leaves water is vaporized and released to the atmosphere, which cools the plant. Transpiration ceases at night and vegetation has a warm signature due to the moisture content. In order to evaluate these relationships quantitatively, we made diurnal radiometer measurements at the locality labeled "Trees" on the nighttime TIR image. A small clump of salt cedar trees is shown as two white dots to the left of the "Trees" label on Figure 5-20B. The trees have distinctly warmer signatures than the surrounding bare soil. Radiant temperatures were measured periodically during a diurnal cycle for the salt cedars plus smaller creosote bushes. The values for all of the vegetation were averaged and plotted on Figure 5-24. For each observation period, radiant temperature measurements of six soil exposures were also averaged. Figure 5-24 shows the average radiant temperature values for vegetation and for soil at each observation time, together with air temperature. All temperatures drop abruptly at sunset, on the left side of Figure 5-24, with several crossovers. Soil, with its low density and low apparent thermal inertia, goes from the warmest to the coolest target within an hour. Figure 5-24 shows that vegetation at night, in the center of the plot, is consistently warmer than soil, with a maximum temperature difference of 5°C. The temperature relationships are reversed during the day, when soil is much warmer than vegetation. Note that the thermal crossovers of the various curves occur within less than one hour both in the evening and morning. These diurnal temperature relationships of soil and vegetation have since been confirmed on daytime and nighttime TIR images at many localities.

Table 5-7 Temperatures of sandstone and siltstone.

	Average Temperature (°C)		
Rock Type	**Day**	**Night**	ΔT
Sandstone	21	10	11
Siltstone	25	8	17

THE IMLER ROAD AREA, CALIFORNIA

The Imler Road area (Figures 5-25 and 5-26) is located in the west margin of the Imperial Valley in the southern California desert. The area is a featureless plain where bedrock is partially covered by gravel and windblown sand that largely conceal geologic structures. The featureless nature of the area is shown in the aerial photograph of Figure 5-25B. The nighttime TIR image in Figure 5-25A, however, displays a wealth of geologic and terrain information that is explained in the following sections.

Terrain Expression

Radiant temperatures of the irrigated fields in the south part of the area indicate different soil-moisture content. The cool fields with dark signatures were probably damp from recent irrigation, which resulted in evaporative cooling of the soil. The small bright area (1.7, C.1) probably has standing crops.

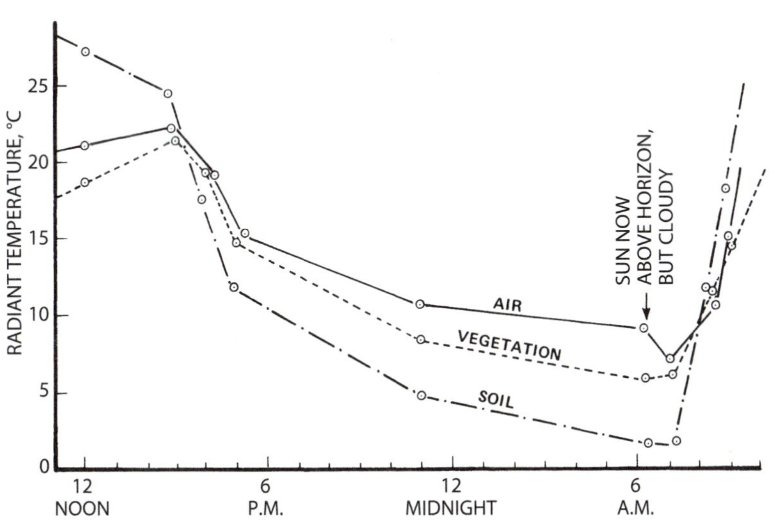

Figure 5-24 Diurnal radiant temperature curves of vegetation, soil, and air at Indio Hills. Locality 3.1, F.3 of Figure 5-20B.

A. Nighttime TIR image (8 to 14 μm) acquired August 1961.

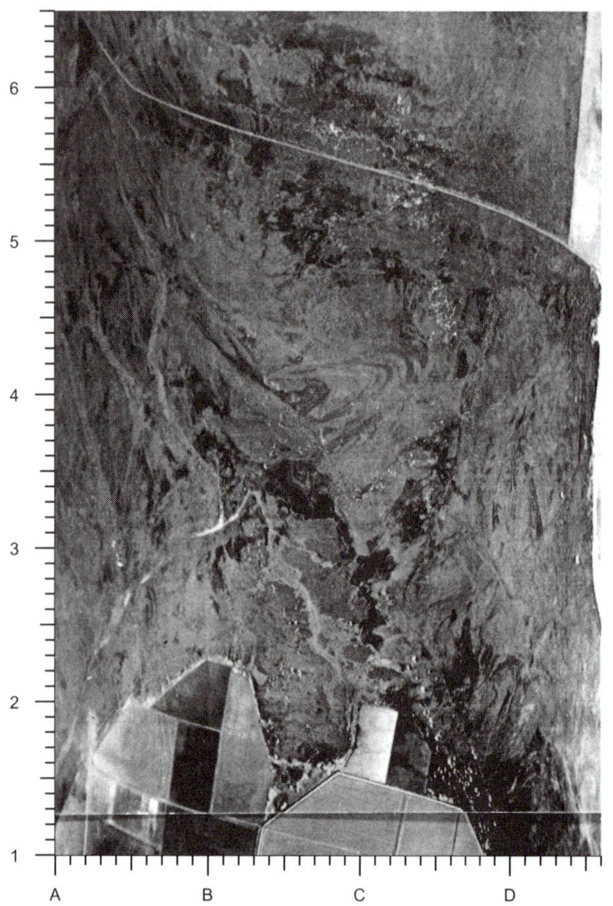

B. Aerial photograph acquired May 5, 1953. From Sabins (1969, Plate 1).

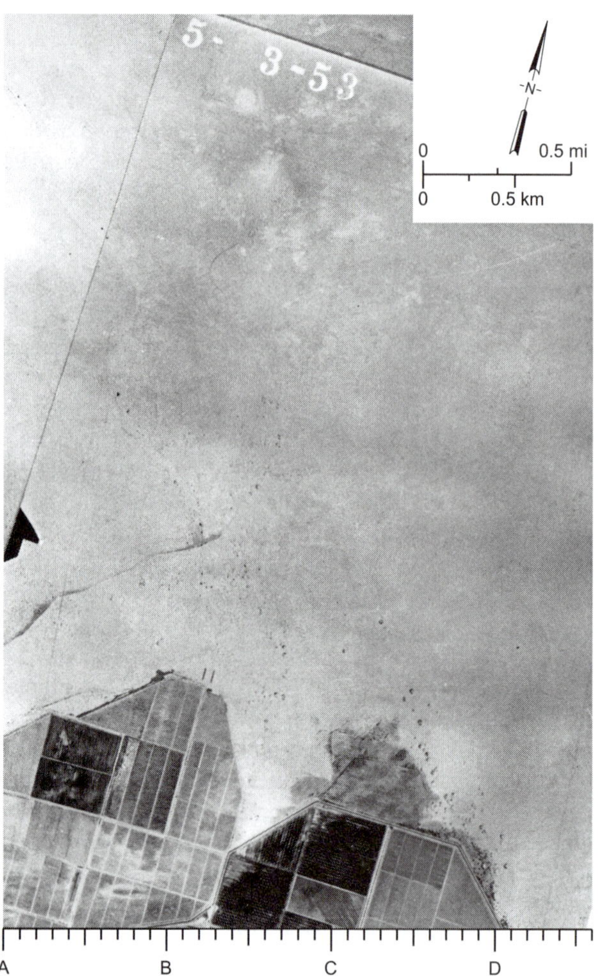

FIGURE 5-25 Images of the Imler Road area, Imperial Valley, California.

Most of the area is flat desert terrain with sparse clumps of native vegetation and patches of windblown sand. Figure 5-26 is a geologic interpretation map of the TIR image that is supplemented with field observations. A number of sand dunes, stabilized by mesquite trees, appear warm on the image and dark on the aerial photograph. The very warm, Y-shaped feature at locality 3.2, B.1 in Figure 5-25A is a thick accumulation of windblown sand lodged against an earthen embankment. Imler Road, with the warm signature in the northern part of the area is actually straight, but it appears curved because of distortion caused by the TIR scanner. The road was surfaced with hard-packed sand when the image was acquired but today is paved with asphalt.

Bedrock in the area is the Borrego Formation (Pleistocene age), which consists of brownish gray lacustrine siltstone with thin interbeds of well-cemented brown sandstone. Slabs and concretions of sandstone litter the surface where the Borrego Formation crops out. Light-colored, nodular, thin layers in the siltstone help to define bedding trends within this monotonous sequence. On the nighttime image the siltstone is relatively cool and the sandstone is warm, which corresponds to thermal signatures of similar rock types at Indio Hills.

ANTICLINE

An east-plunging anticline forms a conspicuous arcuate feature in the center of the TIR image (4.1, C.0 in Figure 5-25A). Without the TIR image, one could walk across the anticline without recognizing it, for there are no notable lithologic or topographic patterns. Once the anticline was located in the field by referring to the TIR image, the limbs and plunge were defined by walking along the subtle outcrops of individual beds. Structural attitudes are obscure in the siltstone, but dips up to 45° were measured in subtle outcrops of the sandstone beds. Careful tracing of key beds in the field established the dip-reversal that defined the fold axis and the northeast plunge of the anticline.

In the geologic map (Figure 5-26) the anticline is mapped as solid bedrock, but there are numerous thin patches of windblown sand that cause the local gray tones on the

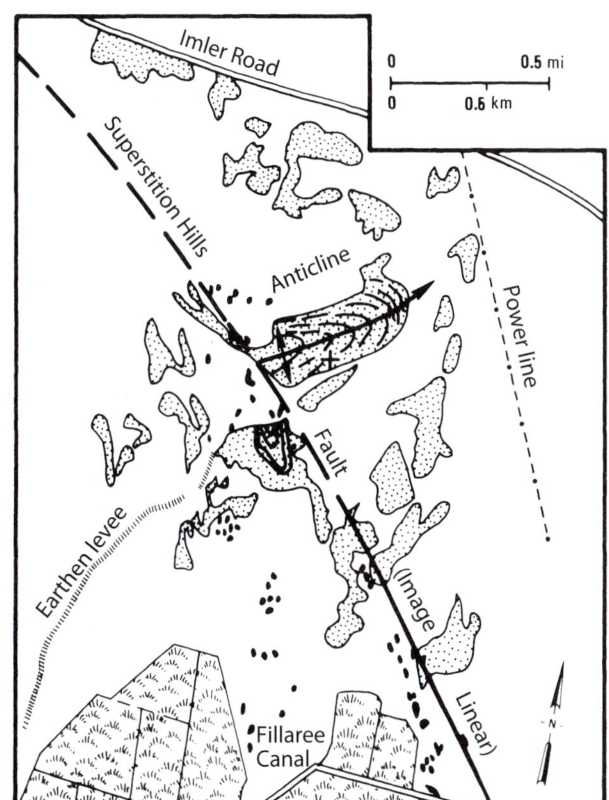

FIGURE 5·26 Geologic interpretation map of the TIR image of the Imler Road area. From Sabins (1969, Figure 5).

Outcrops of deformed tertiary lake deposits, showing trends of bedding.

Recent windblown sand cover.

Cultivated areas.

Stabilized sand dunes and tufa-coated boulders.

+ Brass cap marking SE corner Sec. 25, T.14S, R12E.

nighttime TIR image. The core of the anticline consists of contorted siltstone with a very cool signature. The narrow alternating warm and cool bands that outline the anticline correlate with outcrops of sandstone and siltstone, respectively. The west end of the anticline is truncated by the southeastward projection of the Superstition Hills fault. The inferred trace of the fault is obscured by windblown sand, but siltstone outcrops in the immediate vicinity of the fault are strongly deformed, as a result of fault movement.

SUPERSTITION HILLS FAULT

This right-lateral, strike-slip fault was named for exposures in the Superstition Hills, 14 km to the northwest and projected into this area on the San Diego–El Centro Sheet of the *Geologic Map of California*. The fault alignment shown in Figure 5-26 differs from that on the San Diego–El Centro Sheet. In addition to truncating the anticline, the fault is marked in the southeast part of the image by a southeast-trending linear feature that is cooler on the east and warmer on the west (2.2, C.5 to 1.2, D.0 in Figure 5-25A). The trend of the linear feature is parallel with, and about 0.2 km to the east of, the row of prominent sand dunes. On April 9, 1968, the Borrego Mountain earthquake caused surface breaks along the trace of the Superstition Hills fault that were mapped by A. A. Grantz and M. Wyss (Allen and others, 1972, Plate 2). In the area covered by Figure 5-25A, their map shows a series of breaks with less than 2.5 cm of right-lateral displacement. The trend of the breaks closely coincides with the linear feature on the TIR image (Figure 5-25A). The image, which was acquired seven years prior to the earthquake, shows an important structural feature that was obscure both on aerial photographs and in the field.

TIR IMAGE AND AERIAL PHOTOGRAPH COMPARED

The striking difference in tonal contrast and geologic information content between the TIR image and the aerial photograph is not caused by the eight-year difference in their ages. This desert area is a stable environment in which natural changes occur slowly. During annual field trips of Sabins' UCLA remote sensing class over a period of 25 years, no significant changes were noted in the area. The contrast and resolution of the aerial photograph are good and accurately record the low contrast of this area in the visible spectral region. Color and color IR aerial photographs of the anticline (not shown) are not significantly better than the black-and-white aerial photographs.

In the visible wavelengths there is little reflectance difference between the various rocks and surface materials. In the TIR band, however, there are marked differences in the thermal inertia of the materials, which explains the higher information content of the TIR image.

WESTERN TRANSVAAL, SOUTH AFRICA

South Africa is well suited for TIR surveys because of the dry climate and sparse vegetation, as shown by the Stilfontein area in western Transvaal. The aerial photograph

(Figure 5-27A) shows a featureless surface of low relief with a thin soil cover of 0.5 m or less and scattered trees with dark signatures. A tailings pond for a gold mine occupies the southeast part of the image. No significant geologic information can be interpreted from the photograph. The nighttime TIR image (Figure 5-27B), however, contains a wealth of information on geologic structure and lithology that is also shown in the geologic interpretation map (Figure 5-27C). The area is underlain by dolomite that includes a number of beds rich in chert, which is a silica-rich sedimentary rock. In the TIR image the dolomite has a bright (warm) signature, which is attributed to its relatively high density and high thermal inertia. The chert-rich beds have distinctly darker (cooler) signatures caused by the lower density and lower thermal inertia of these rocks. A belt of alternating dolomite beds and chert-rich beds trends northeastward across the image and is bounded on the northwest and southeast by broad areas of dolomite with uniform warm radiant temperature.

The numerous linear features with very dark (cool) signatures are the expression of faults and joints that have been enlarged by erosion and filled with sand and soil. The low thermal inertia of sand and soil contrasts with the high thermal inertia and warm signature of the dolomite bedrock. The two major sets of fractures trend approximately north to south and east to west. In much of the area, the bedrock is cut by closely spaced joints that produce a fine network of cool lines in the image.

The dolomite and chert have little contrast in the visible domain but are easily distinguished in the TIR image because of their different thermal properties. The geology of the Stilfontein area differs from that of the Indio Hills and Imler Road areas, but in all of these areas the TIR image is superior to the aerial photograph for geologic interpretation.

A. Aerial photograph.

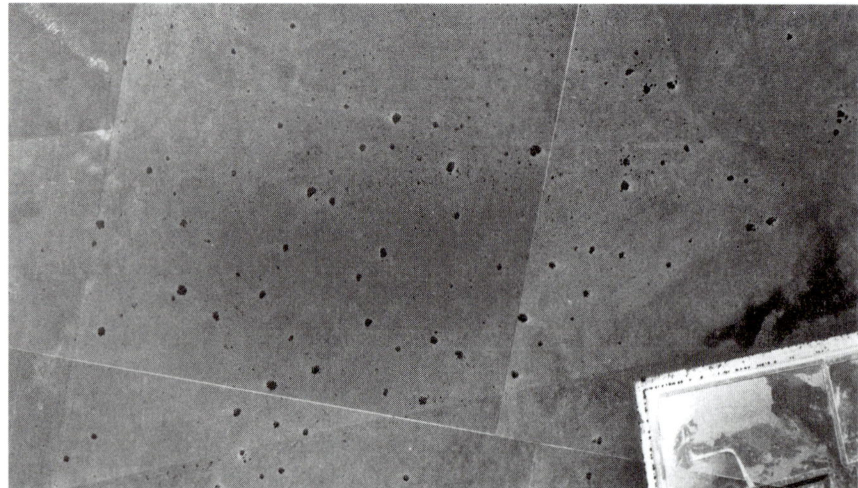

B. Nighttime TIR image (8 to 14 μm).

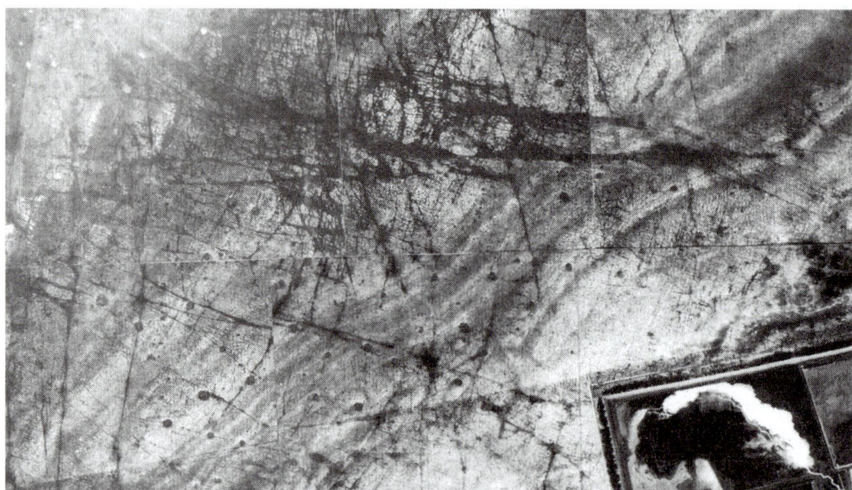

C. Geologic interpretation map of TIR image.

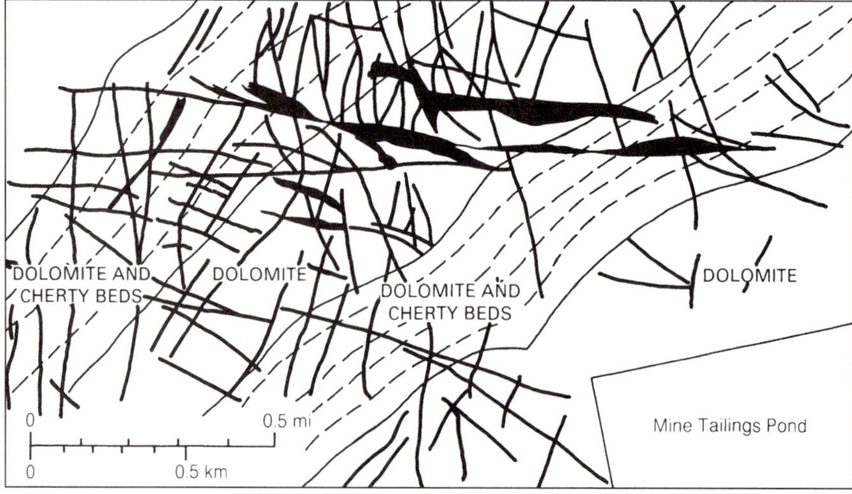

FIGURE 5-27 Stilfontein area, western Transvaal, South Africa. Republished with permission of *Quarterly Journal of Engineering Geology*, from Application of the thermal infrared line scanning technique to engineering geological mapping in South Africa, D. Warwick, P. G. Hartopp, and R. P. Viljoen, 12, 159–179, 1979; permission conveyed through Copyright Clearance, Inc.

SATELLITE TIR SYSTEMS

TIR images are acquired by a number of satellites, including Landsat 4 to 8, ASTER, MODIS, AVHRR, and GOES (Table 5-1). Table 5-8 lists the characteristics of some of these systems.

LANDSAT DAYTIME AND NIGHTTIME TIR IMAGES

Figure 5-28 compares Landsat TM band 6 (TIR) images recorded on daytime and nighttime orbits over West Virginia. The images cover portions of the Appalachian Mountains in the east and the Cumberland Plateau in the west. Figures 5-28A and 5-28B show simultaneously recorded daytime images for band 6 (TIR) and band 3 (visible red). Thin clouds in the northeast and snow cover in the northwest portions of these images produce dark (cool) signatures in the TIR image and bright signatures in the visible image. In both images topography is accentuated by solar illumination from the southeast. In the visible image (Figure 5-28B), the southeast-facing slopes are highlighted and northwest-facing slopes are shadowed. In the daytime TIR image, southeast-facing slopes are warm (bright) and those facing northwest are cool (dark). In the nighttime TIR band 6 image (Figure 5-28C), the effects of differential heating and shadowing have dissipated. In the Appalachian Mountains the ridges are warm and the adjacent valleys are cool. The ridges are outcrops of resistant dense rocks, such as quartzite and conglomerate, which have high thermal inertia values. Valleys are eroded in less-resistant limestone and shale and are covered with soil, which has a low thermal inertia. These differences in materials and thermal inertia explain the diurnal temperature patterns. We are reminded of the similar diurnal temperature patterns in Indio Hills. The elevation difference between Appalachian ridges and valleys is only a few hundred meters; therefore, adiabatic cooling is not a significant factor. Water in the rivers and lakes has the typical cool signature on the daytime TIR image and warm signature on the nighttime image.

The West Virginia Landsat images and images from aircraft show that TIR images in the 8 to 14 μm band should be acquired at night to be useful for most interpretations. However, only daytime Landsat TIR images are available from the USGS archives. Fortunately, daytime and nighttime ASTER TIR images are available for download at no cost from USGS and NASA archives.

AVHRR

The NOAA-16/AVHRR (N-16) satellite detects and maps active wildfires across North American forest ecosystems using two fire detection algorithms developed for daytime and nighttime daily imagery (Abuelgasim and Fraser, 2002). The satellite images the same area on the Earth twice a day, once during the day and once during the night, with 1 by 1 km pixels. The algorithms exploit both the multispectral and thermal information from the AVHRR daily images. During the daytime, N-16 records red, NIR, SWIR bands, and two TIR bands centered at 10.8 μm and 12.0 μm. During nighttime data collection the SWIR band is switched to the TIR band centered at 3.7 μm. A serious limitation with the SWIR band is that small fires are not readily detected on daytime imagery. Night detection is not limited and has significantly higher detection rates.

The day and night algorithms generate daily active fire maps across North America from the AVHRR data. This combined approach for fire detection improves the detection rate. Satellite-based remote sensing of wildfires provides global coverage and continuous observations of wildfires (Abuelgasim and Fraser, 2002).

ASTER URBAN HEAT ISLAND—MANILA, PHILIPPINES

The urban heat island (UHI) effect is caused by the higher temperatures of cities and urban areas relative to the surrounding rural areas. Urban areas are relatively warm because of the high radiance and emissivity of asphalt, cement, and brick (Table 5-4) plus heat generated by human

TABLE 5-8 Satellite TIR imaging systems.

System (launch date)	Band Numbers	Spectral Range (μm)	Spatial Resolution
Landsat 8 TIRS (2013)	10	10.60 to 11.19	100 m
	11	11.50 to 12.51	100 m
ASTER (1999)	10	8.125 to 8.475	90 m
	11	8.475 to 8.825	90 m
	12	8.925 to 9.275	90 m
	13	10.250 to 10.950	90 m
	14	10.950 to 11.650	90 m
AVHRR (1978)	3	3.55 to 3.93	1.1 km
	4	10.50 to 11.50	1.1 km
	5	11.50 to 12.50	1.1 km
MODIS (2000)	17 TIR bands	3.66 to 14.38	1.0 km

activity. The building materials absorb and store solar heat during the day, which is then released at night.

Tiangco and others (2008) used nighttime ASTER TIR data to map the surface temperature of the metro area of Manila and the adjacent rural areas. Manila is the capital of the Philippines with a population of 10,000,000. Plate 15A is the nighttime surface temperature image for May 4, 2002. Plate 15B is a perspective view of Plate 15A that displays the nighttime surface temperature profile along the north–south B–B′ transect. Manila Bay and the lake called Bay de Bays are masked and colored white in the surface temperature image. Rural areas adjacent to the city have blue and green cool temperatures (25 to 27°C) that grade laterally into the yellow intermediate temperatures (27 to ~29°C) of the suburbs. The commercial and business district of central Manila is a roughly circular red urban heat island (> 29°C).

A. Daytime TIR image (band 6) acquired November 16, 1982.

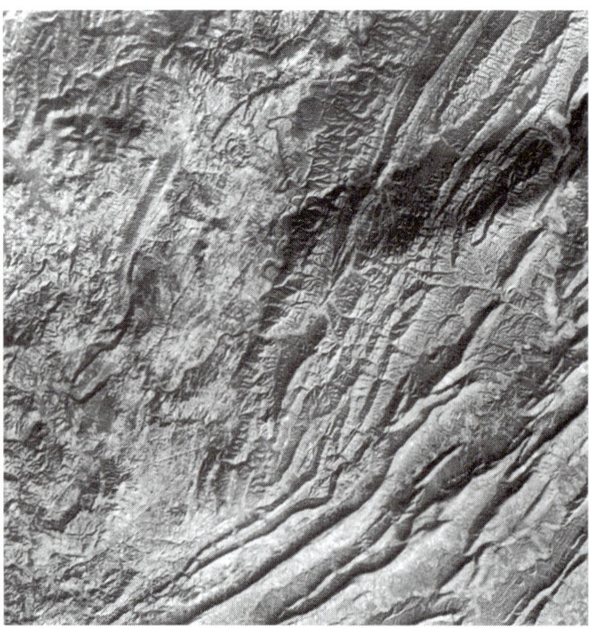

B. Daytime visible image (band 3) acquired November 16, 1982.

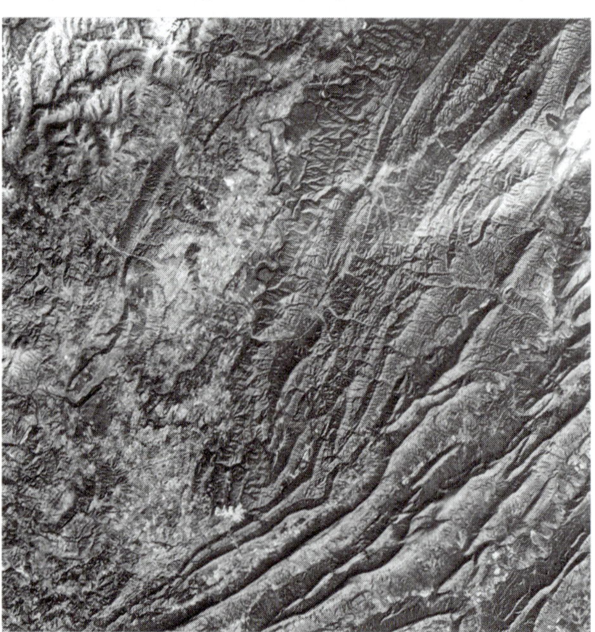

C. Nighttime TIR image (band 6) acquired November 8, 1982.

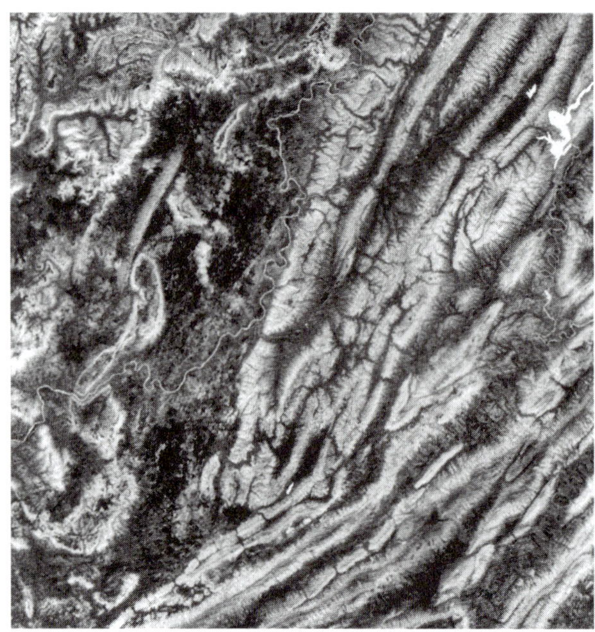

D. Location map.

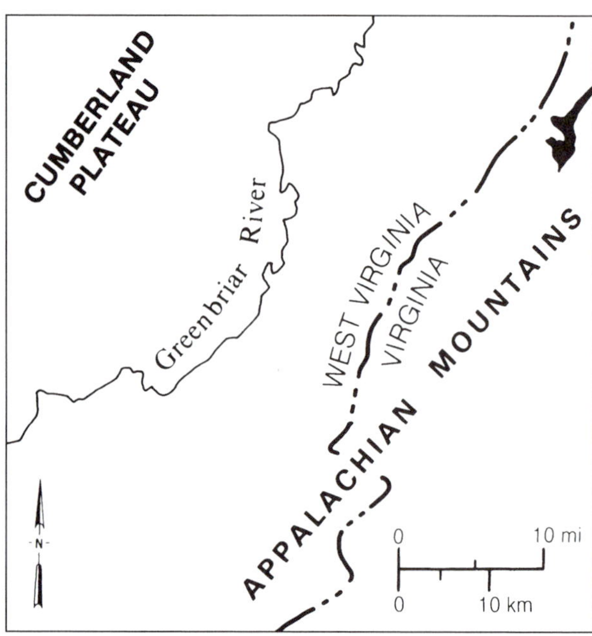

FIGURE 5-28 Images acquired by Landsat TM. Greenbriar River, West Virginia. Landsat courtesy USGS.

The spikes and valleys in the temperature profile are caused by distinct land use features such as major highways, runways, parks, golf courses, and lakes. The terms Cliff, Plateau, and Peak on Plate 15B relate to characteristics of rural-urban temperature profiles and are explained by Tiangco and others (2008).

MODIS Sea Surface Temperature—North Atlantic

NASA and NOAA satellites have been monitoring the sea surface temperature (SST) of the oceans since 1981. SST data are used for predicting climate change, meteorological forecasts, ocean currents maps, and monitoring the potent El Niño and La Niña cycles. Heat is one of the main drivers of global climate, and the ocean is a huge reservoir of heat. The top 1.9 m of ocean water stores the amount of heat equal to that contained in the atmosphere (NASA Science Beta, 2017).

The MODIS instrument on the Aqua satellite measures TIR radiance in the 3 to 5 μm and 8 to 14 μm wavelength regions across a 2,300 km swath with 1 km pixels (Figure 4-26). MODIS acquires an SST image using bands 31 and 32 that record emitted energy at 11 and 12 μm wavelengths. Plate 16 is the MODIS SST of the Gulf Stream that flows northeast through the northwest Atlantic Ocean at an average rate of 6.4 km · hr^{-1}. The United States is masked and shown in dark gray. Cooler temperatures in the southeast are masked and shown in light gray. Warmest water temperatures (25 to 30°C) are shown in yellow. Intermediate temperatures (15 to 25°C) are shown in orange and red. Coldest temperatures (1 to 15°C) are shown in blue and purple.

The Gulf Stream originates as a clockwise circulating current in the Gulf of Mexico, with input from the Caribbean Sea. Warm water (yellow and orange) of the Gulf Stream moves around the tip of Florida and north to northeast along the southeast coast of the United States. At Cape Hatteras the Gulf Stream enters the Atlantic Ocean with a sharp contact between the cooler (blue) ocean water and the warmer (yellow and orange) Gulf Stream water. Two prominent southeast deflections in the boundary are caused by two current gyres in the north Atlantic Ocean. Similar deflections are present on many MODIS images of the Gulf Stream that have been acquired on alternate dates.

TIR SPECTRA

TIR spectrometers typically record spectra of energy emitted from materials in the 8 to 14 μm band. Figure 5-29 shows laboratory spectra for igneous rocks with silica contents ranging from 50 to 71%. The spectra show a broad emission minimum called the *reststrahlen band*. In silicate rocks this minimum is due to stretching vibrations between silicon and oxygen atoms bonded in the silicate crystal lattice. The position and depth of the emission minimum are related to the crystal structure of the constituent minerals and are especially sensitive to the quartz content of the rocks. The spectra in Figure 5-29 are arranged in the order of decreasing silica (and quartz) content, which is shown for each rock. An arrow marks the geometric center of each emission minimum, which is not necessarily the position of least emission. For the leucogranite, which has the maximum percentage of silica, the arrow is located at a wavelength of slightly less than 9.0 μm. For the anorthosite, which has the minimum percentage of silica, the arrow is located at a wavelength of greater than 10.0 μm. Figure 5-30 is a plot of silica content as a function of the minimum emission wavelength for igneous rocks similar to those in Figure 5-29. The plot shows a linear relationship between decreasing silica and increasing wavelength for the emission minimum.

Figure 5-31 shows spectra of additional silicate rocks, together with clay minerals and carbonate rocks. The silicate rocks show the typical emissivity minimum at wavelengths of 8 to 12 μm. The silicate spectra are arranged in order of decreasing silica content from quartzite through basalt, and show the shift toward longer wavelengths of the emissivity minimum. In the clay minerals (kaolinite and montmorillonite), spectral features in the 8 to 14 μm region are attrib-

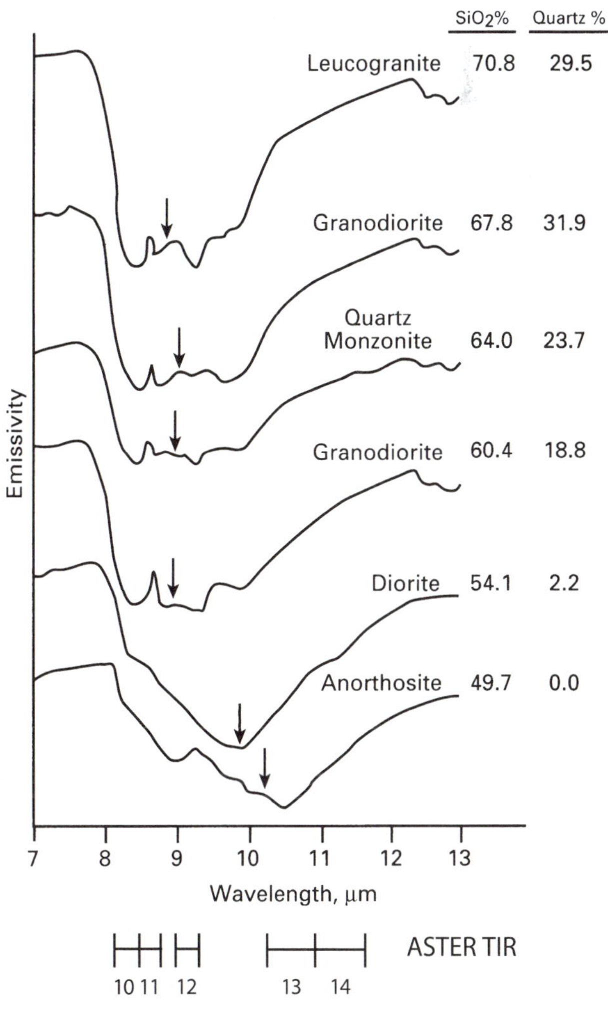

Figure 5-29 Emissivity spectra of igneous rocks with different silica and quartz contents. ASTER TIR bands are shown. Arrows show centers of absorption bands. Spectra are offset vertically. Modified from Sabine and others (1994, Figure 3).

uted to various Si-O-Si and Si-O stretching vibrations and to an Al-O-H bending mode. The spectral features in montmorillonite are less distinct than in kaolinite because the numerous exchangeable cations and water molecules in the montmorillonite structure allow many different vibrations.

Spectra of the carbonate rocks (limestone and dolomite) in Figure 5-31 have a major emissivity minimum at 6 to 8 μm due to internal vibrations in the carbonate anion. This wavelength region, however, coincides with an atmospheric absorption band and is not usable in remote sensing of the Earth. The spectra of pure carbonate rocks are featureless in the region of 8 to 11 μm, but the spectrum of limestone with clay and quartz has an emissivity minimum near 9.0 μm caused by clay.

ASTER TIR BANDS

Figures 5-29 to 5-31 show the position of the ASTER TIR bands and the spectra of rocks. ASTER TIR band 12 is designed to detect rocks with silica content of 60 to 75%. TIR band 13 detects the ~10.6 μm absorption feature associated with igneous rocks (mafic and ultramafic) that have little silica content (Mars and Rowan, 2011). Calcite ($CaCO_3$) has an absorption feature at 11.2 μm that is detected by band 14 (Mars and Rowan, 2011).

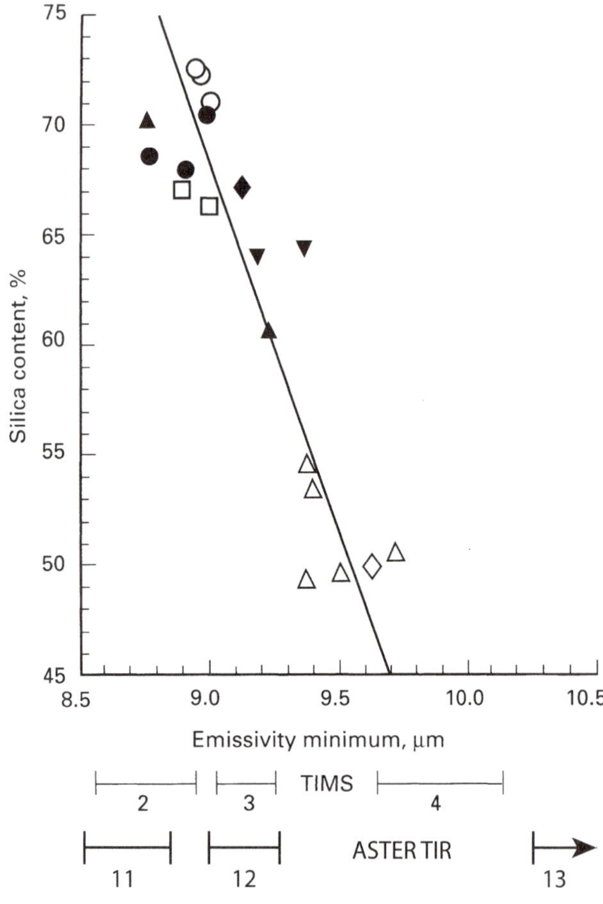

FIGURE 5-30 Plot showing the relationship between emissivity minimum and silica content with ASTER and TIMS bands. Horizontal axis shows the wavelength of emissivity minima (indicated by arrows in Figure 5-29). Vertical axis shows the percentage of silica. Symbols represent various types of igneous rocks. Modified from Sabine and others (1994, Figure 4).

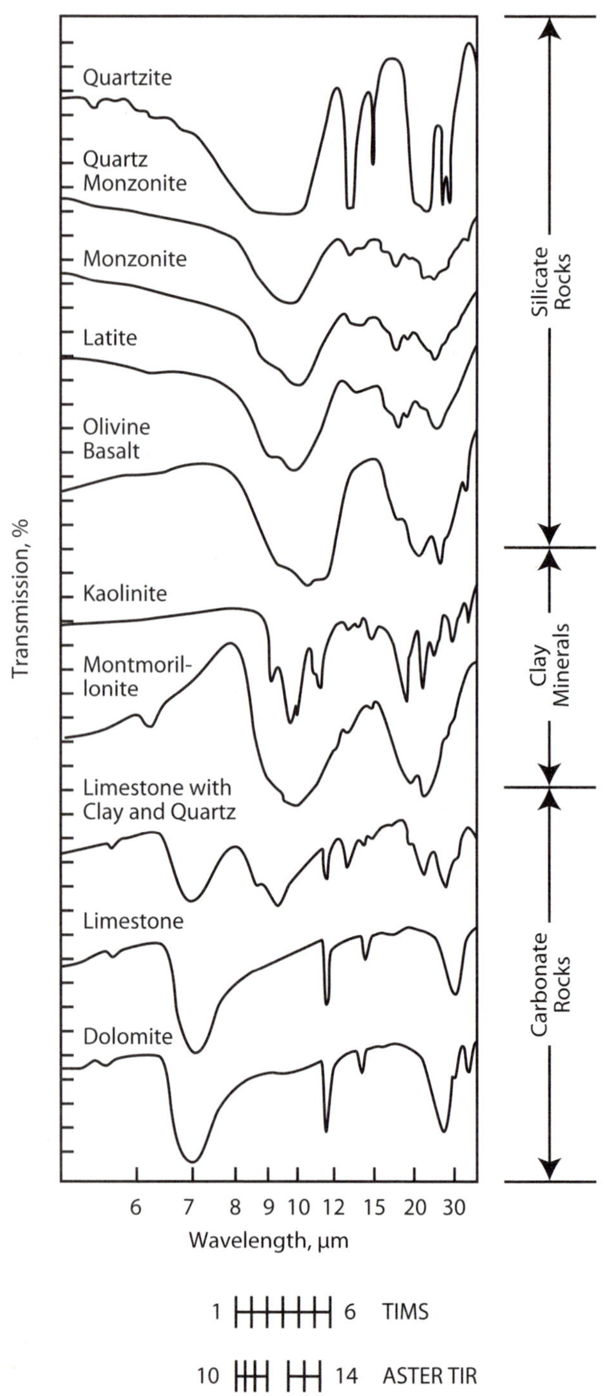

FIGURE 5-31 Laboratory TIR spectra of rocks and minerals. ASTER and TIMS bands are shown. Spectra are offset vertically. Modified from Kahle (1984, Figure 4).

TIR Multispectral Scanner

NASA's Stennis Space Center, JPL, and Daedalus Corporation developed the Thermal Infrared Multispectral Scanner (TIMS) for exploiting mineral signature information. The TIMS is a Daedalus multispectral scanning system using a six-element, mercury cadmium telluride detector array to acquire six discrete bands in the 8.2 to 12.2 μm region (8.19 to 8.55, 8.58 to 8.96, 9.01 to 9.26, 9.65 to 10.15, 10.34 to 11.14, and 11.29 to 11.56 μm). Figures 5-30 and 5-31 show the position of these bands and the spectra of rocks. TIMS acquires mineral signature data that permits the discrimination of silicate, carbonate, and hydrothermally altered rocks (Kahle and Goetz, 1983). TIMS data have been used extensively in volcanology research in the western United States, Hawaiian Islands, and Europe.

Hyperspectral TIR for Agriculture

The ITRES TASI-600 is an airborne hyperspectral TIR sensor that covers the 8 to 11.5 μm wavelength range with 32 bands (ITRES, 2010). The system was flown over bare soils in three agricultural fields in southern Italy to assess its ability to predict topsoil properties (Pascucci and others, 2014) (Plate 17). Six flight lines were flown at an altitude of 850 m above the land for a spatial resolution of 1 m. The flight lines were orthorectified and digitally mosaicked; the surface temperature was estimated from the TASI-600 data that range from 14 to 62°C.

The three agricultural fields were mapped with the TIR system. Each field covered about 50 by 250 m. Two fields were within 50 m of each other while the third field was approximately 1.5 km east–northeast. The fields showed different temperature patterns, with one relatively cool, one relatively warm, and the third warm along the southern margin but cooler in other areas.

The TIR data were processed to provide emissivity spectra that were used to calibrate prediction models for clay, sand, and soil organic carbon (SOC). For example, as the amount of sand (silica) decreases in the topsoil the wavelength of the emissivity minimum increases. The mean emissivity for the 32 bands over the three agricultural fields was plotted versus wavelength (8.0 to 11.5 μm). Distinct absorption features (emissivity minima) were documented at 9.5, 11, and 11.4 μm. Absorption features in the mean emissivity spectra are related to clay (9.6 to 10.0 μm), iron oxide (11.0 to 11.5 μm), and SOC content in the topsoil. An additional correlation was confirmed between SOC content and soil temperature.

The study by Pascucci and others (2014) showed that the warmest field (field 1) had the highest SOC content. They concluded that the SOC content in the topsoil can be estimated from hyperspectral TIR emissivity data. Additional research is needed to improve atmospheric correction algorithms and to understand the effects of soil roughness, moisture, and mineralogy on soil property modeling.

Questions

1. Is TIR energy reflected off or emitted from the Earth's surface features?
2. Why are TIR sensors more expensive than visible light sensors?
3. What are the two major atmospheric windows between 2.5 μm and 15 μm that allow TIR remote sensing from aircraft and satellites?
4. Why are satellite TIR sensors unable to collect data between approximately 9 to 10 μm?
5. The tungsten filament in a light bulb is heated to a radiant temperature of 3000°K. Use Wein's displacement law (Equation 5-1) to calculate the wavelength (μm) at which maximum energy radiates from the filament.
6. Calculate the radiant flux (F_b) for a blackbody with a kinetic temperature of 21°C using the Stefan–Boltzmann law (Equation 5-5).
7. Calculate radiant flux (F_r) from a block of rough dolomite (Table 5-3) with a kinetic temperature (T_{kin}) of 15°C.
8. Calculate radiant temperature (T_{rad}) of a block of rough dolomite (Table 5-3) with a kinetic temperature (T_{kin}) of 13°C.
9. Calculate the thermal inertia of a rock with the following properties:
   - a thermal conductivity (K) of 0.005 $cal \cdot cm^{-1} \cdot sec^{-1} \cdot °C^{-1}$
   - a density (ρ) of 2.7 $g \cdot cm^{-3}$
   - a thermal capacity (c) of 0.19 $cal \cdot g^{-1} \cdot °C^{-1}$
10. You are interpreting daytime and nighttime ASTER TIR images and a visible image of an area. A rock outcrop has a daytime radiant temperature of 20°C and a nighttime temperature of 10°C. The albedo from the visible image is 0.50. Calculate the ATI for this rock.
11. Describe the appearance of clouds and of smoke plumes on TIR images. Explain the differences in appearance.
12. List the type of signature (warm or cool) of the following targets on daytime and nighttime TIR images acquired in the southwest United States during summer: dry soil, damp soil, and standing water. Explain these signatures.
13. You are trying to differentiate and map soils, vegetation, and water with a thermal sensor for a particular area. What time(s) of the day would you avoid and why?

References

Abrams, M. 2005. ASTER Talk. Jet Propulsion Laboratory/California Institute of Technology ASTER website, Presentations & Documents, Microsoft PowerPoint presentation (129 MB), 83 slides. http://asterweb.jpl.nasa.gov/bibliography.asp (accessed December 2017).

Abrams, M., S. Hook, and B. Ramachandran. 2002. ASTER User Handbook Version 2. Jet Propulsion Laboratory/California Institute of Technology.

Abuelgasim, A., and R. Fraser. 2002, June 24–28. Day and Night-Time Active Fire Detection Over North America Using NOAA-16 AVHRR Data. IEEE International Geoscience and Remote Sensing Symposium, Toronto, Ontario, Canada.

Allen, C. R., M. Wyss, J. N. Brune, A. Grantz, and R. E. Wallace. 1972. *Displacements on the Imperial, Superstition Hills and San Andreas Faults Triggered by the Borrego Mountain Earthquake of April 9, 1968* (Professional Paper 787, pp. 87–104). Washington, DC: US Geological Survey.

Burke, C., M. Rashman, S. Wiche, A. Symons, C. Theron, and S. Longmore. 2019. Optimizing observing strategies for monitoring animals using drone-mounted thermal infrared cameras. *International Journal of Remote Sensing*, 40(2), 439–467. doi:10.1080/01431161.2018.1558372

Colwell, R. N., W. Brewer, G. Landis, P. Langley, J. Morgan, J. Rinker, J. M. Robinson, and A. L. Sorem. 1963. Basic matter and energy relationships involved in remote reconnaissance. *Photogrammetric Engineering*, 29(5), 761–799.

Douma, M. 2008. Seeing Heat. Causes of Color: Institute for Dynamic Educational Development. http://www.webexhibits.org/causesofcolor/3.html (accessed December 2017).

Findlay, G. A., and D. R. Cutten. 1989. Comparison of performance of 3 to 5 and 8 to 12 μm infrared systems. *Applied Optics*, 28, 5,029–5,037.

FLIR Systems. 2015. User Guide. http://www.flir.com (accessed December 2017).

FLIR Systems. 2016. Intrinsically Safe FLIR GFx320. http://www.flir.com (accessed December 2017).

FLIR Systems. 2017, March 3. How the World Wildlife Fund Uses Thermal-Equipped Drones in Africa. http://www.flir.com/home/news/details/?ID=83173 (accessed January 2019).

FLIR Systems. 2019. The Kifaru Rising Project: Ending Rhino Poaching in Kenya. http://www.flir.com/corporate/wwf (accessed January 2019).

Gillespie, A. R., T. Matsunaga, S. Rokugawa, and S. J. Hook. 1998. Temperature and emissivity from Advanced Spaceborne Thermal Emission and Reflection Radiometer (ASTER) images. *IEEE Transactions on Geoscience and Remote Sensing*, 36, 1,113–1,126. doi:10.1109/36.700995

Gillespie, A. R., and S. Rokugawa. 2001. Surface Emissivity ASTER Product AST-05. http://asterweb.jpl.nasa.gov/content/03_data/01_Data_Products/Emissivity.pdf (accessed December 2017).

ITRES. 2010. TASI-600 Specification Sheet. ITRES Research Limited.

ITRES. 2011a. MASI-600 Specification Sheet. ITRES Research Limited.

ITRES. 2011b. TABI-1800 Specification Sheet. ITRES Research Limited.

ITRES. 2015, July 8. TABI-1800 Deployed for 2015 Wildfire and Hotspot Mapping Season. http://www.itres.com/tabi-1800-deployed-for-2015-wildfire-and-hotspot-mapping-season (accessed October 2, 2018).

Janza, F. J., and others. 1975. Interaction mechanisms. In R. G. Reeves (Ed.), *Manual of Remote Sensing* (chapter 4, pp. 75–179). Falls Church, VA: American Society for Photogrammetry and Remote Sensing.

Kahle, A. B. 1984. Measuring spectra of arid lands. In F. El-Baz (Ed.), *Deserts and Arid Lands* (chapter 11, pp. 195–217). The Hague, Netherlands: Martinus Nijhoff.

Kahle, A. B., and A. F. H. Goetz. 1983. Mineralogic information from a new airborne thermal infrared multispectral scanner. *Science*, 222, 24–27.

Kahle, A. B., J. P. Schieldge, M. J. Abrams, R. E. Alley, and C. J. LeVine. 1981. *Geologic Applications of Thermal Inertia Imaging Using HCMM Data* (Publication 81-55). Pasadena, CA: Jet Propulsion Laboratory.

Mars, J. C., and L. C. Rowan. 2011. ASTER spectral analysis and lithologic mapping of the Khanneshin carbonatite volcano, Afghanistan. *Geosphere*, 7(1), 276–289.

NASA Science Beta. 2017. Temperature. NASA Programs Research and Analysis. http://science.nasa.gov/earth-science/oceanography/physical-ocean/temperature (accessed December 2017).

Pascucci, S., R. Casa, C. Belviso, A. Palombo, S. Pignatti, and F. Castaldi. 2014. Estimation of soil organic carbon from airborne hyperspectral thermal infrared data: A case study. *European Journal of Soil Science*, 65, 865–875.

Sabine, C., V. J. Realmuto, and J. V. Taranik. 1994. Quantitative estimation of granitoid composition from thermal infrared multispectral scanner (TIMS) data, Desolation Wilderness, northern Sierra Nevada, California. *Journal of Geophysical Research*, 99, 4,261–4,271.

Sabins, F. F. 1967. Infrared imagery and geologic aspects. *Photogrammetric Engineering and Remote Sensing*, 29, 83–87.

Sabins, F. F. 1969. Thermal infrared imagery and its application to structural mapping in southern California. *Geological Society of America Bulletin*, 80, 397–404.

Song, B. G., and K. H. Park. 2015. Analysis of Surface Temperature Accuracy in ASTER Images According to Land-Use Type. Proceedings of the Joint 9th International Conference on Urban Climate and the 12th Symposium on the Urban Environment.

Tiangco, M., A. M. F. Lagmay, and J. Argete. 2008. ASTER-based study of night-time urban head island effect in Metro Manila. *International Journal of Remote Sensing*, iFirst Article, pp. 1–20.

USGS. 2018. Handheld Thermal Imaging Cameras for Groundwater/Surface-Water Interaction Studies. USGS Water Resources. http://water.usgs.gov/ogw/bgas/thermal-cam (accessed April 2018).

Wolfe, E. W. 1971. Thermal IR for geology. *Photogrammetric Engineering and Remote Sensing*, 37, 43–52.

WWF. 2016, November 21. Anti-Poaching Technology in Africa Leads to Dozens of Arrests. World Wildlife Fund News and Press. http://www.worldwildlife.org/press-releases/anti-poaching-technology-in-africa-leads-to-dozens-of-arrests (accessed January 2019).

WWF. 2017, November 17. Wildlife Crime Technology Project. World Wildlife Fund News & Press. http://www.worldwildlife.org/projects/wildlife-crime-technology-project (accessed January 2019).

chapter 6

Radar Images

Radar is the acronym for "radio detection and ranging"; it operates in the microwave band of the electromagnetic spectrum, ranging from millimeters to meters in wavelength. Radar is an *active* remote sensing system because it provides its own source of energy. The system "illuminates" the terrain with electromagnetic energy, detects the energy returning from the terrain (called *radar return* or *backscatter*), and records the return energy as an image. *Passive* remote sensing systems, such as photography and thermal IR (TIR), detect the available energy reflected or radiated from the terrain. Radar systems operate independently of lighting conditions and are largely independent of weather. Images can be acquired with the optimum viewing geometry to enhance features of interest.

The reflection, or return, of radio waves from objects was noted in the late 1800s and early 1900s. Definitive investigations of radar began in the 1920s in the United States and Great Britain for the detection of ships and aircraft. Radar was employed during World War II for navigation and target location using rotating antennas and circular *cathode-ray tube* (CRT) displays. The continuous-strip mapping capability of *side-looking airborne radar* (SLAR) was developed in the 1950s to acquire reconnaissance images without the necessity of flying over politically unfriendly regions. Fischer (1975) prepared a comprehensive history of radar development. Henderson and Lewis (1998), Moore (1983), and Moore and others (1983) give details of radar theory and practice.

This chapter describes the technology and terrain interactions that produce radar images from a range of environments that are acquired from both aircraft and satellites. This chapter also introduces radar-based digital elevation models (DEM) and ground deformation measurements, which are expanded upon in Chapter 7.

Ground penetrating radar (GPR) is not covered in this textbook. GPR is a geophysical method that uses radar pulses to image the subsurface. The radar energy is emitted with wavelengths ranging from 10 to 300 cm. GPR systems operating at longer radar wavelengths penetrate the ground more deeply and can collect radar returns reflecting from features 20 m or more below the surface under optimum conditions. The GPR antenna is generally in contact with the ground, which removes GPR from the remote sensing category.

RADAR TECHNOLOGY

Imaging radars are complex systems that are continually evolving. For example, phased array radars and synthetic aperture radars (SAR) have replaced earlier conventional antenna radars. The radar antennas now used on satellites are steered to expand geographic area and repeat terrain coverage, with a range of spatial resolutions, swath widths, depression angles, and transmit/receive modes. This chapter begins with basic components and principles to explain how radar images are acquired. The following sections demonstrate a range of image types and their interpretation for different applications.

Components

Figure 6-1 shows the basic components of a radar imaging system. The timing pulses from the pulse-generating device serve two purposes: (1) they control the bursts of energy from the transmitter and (2) they synchronize the recording of successive energy returns to the antenna. The pulses of electromagnetic energy from the transmitter are of specific wavelength and duration, or pulse length, which is defined later in the chapter.

The same antenna transmits the radar pulse and receives the return from the terrain. An electronic switch, or *duplexer*, prevents interference between transmitted and received pulses by blocking the receiver circuit during transmission and blocking the transmitter circuit during reception. The *antenna* focuses the pulse of energy into the desired form for transmission and also collects the energy returned from the terrain. The duplexer sorts out each return pulse and sends it to the *receiver*, which amplifies the weak return pulse while preserving the variations in intensity. The receiver also records the timing of the return pulse, which determines the position of terrain features on the image. The return pulse is stored on a *digital recorder* and made available for computer processing and display as images.

FIGURE 6-2 NASA radar survey Gulfstream III. The L-band synthetic aperture, side-looking radar antenna is housed in the pod beneath the fuselage. Courtesy NASA.

IMAGING SYSTEMS

Figure 6-2 is a radar survey aircraft with the antenna housed in a pod mounted with its long axis parallel with the aircraft fuselage. Pulses of energy transmitted from the antenna illuminate strips of terrain in the *look direction* (also called the *range direction*). The look direction is oriented normal to the *azimuth direction* (aircraft flight direction). Figure 6-3 illustrates such a strip of terrain and the shape and timing of the energy pulse that is returned to the antenna. The return pulse is displayed as a function of two-way travel time on the horizontal axis. The shortest times are shown to the right, at the *near range*, which is the closest distance to the aircraft flight path. The longest travel times are shown at the far range. Travel time is converted to distance by multiplying travel time by the speed of electromagnetic radiation (3×10^8 m $\cdot$ sec^{-1}). The amplitude of the returned pulse is a complex function of the interaction between the terrain and the transmitted pulse.

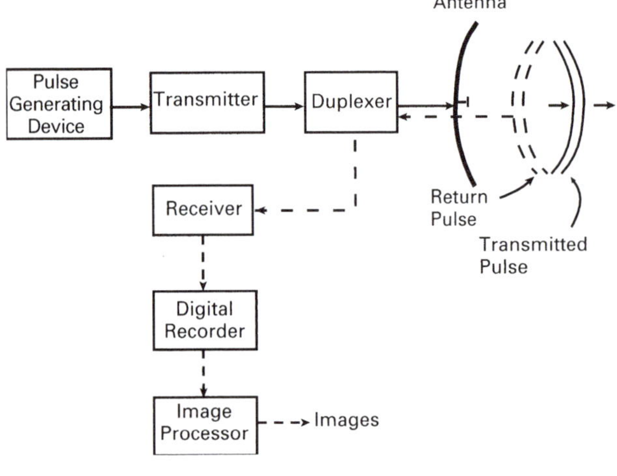

FIGURE 6-1 Diagram of a side-looking radar system. The duplexer both transmits the outgoing radar pulse and receives the incoming return pulse.

The long dimension of a radar image typically represents the flight direction or orbit direction. Returns are processed and plotted as shown in Figure 6-3, assigning the darkest tones of a grayscale to returns of the lowest intensity and the brightest tones to returns of the highest intensity.

In addition to being an active system radar differs from other remote sensing systems, such as framing systems and scanners, by recording electromagnetic energy as a function of time rather than angular distance. Time is much more precisely measured and recorded than angular distance; hence radar images can be acquired at longer ranges with higher resolution than images from other remote sensing systems. Atmospheric absorption and scattering are negligible except at the very shortest radar wavelengths.

Table 6-1 lists the terrain features illustrated in Figure 6-3 together with their signatures and tones in a radar image. The table also summarizes the causes of the signatures and tones. The following section illustrates and describes the appearance of these terrain features on a typical radar image.

TYPICAL IMAGES

Figure 6-4 is an aircraft radar image and map of Weiss Lake and vicinity in northeastern Alabama that includes examples of the terrain features shown diagrammatically in Figure 6-3 and listed in Table 6-1. The image was acquired with the look direction toward the west, which is toward the left margin of the image (this is the same orientation as the look direction in Figure 6-3). The east-facing slopes of the mountains and ridges in Figure 6-4A face the radar look direction and form bright signatures, or *highlights*, caused by the strong radar returns. The west-facing slopes are oriented away from the radar look direction and form dark *shadows* because no energy reached these areas.

The flat terrain adjacent to Weiss Lake and the Coosa River is covered with various types of crops and native vegetation that have *diffuse* signatures with intermediate gray

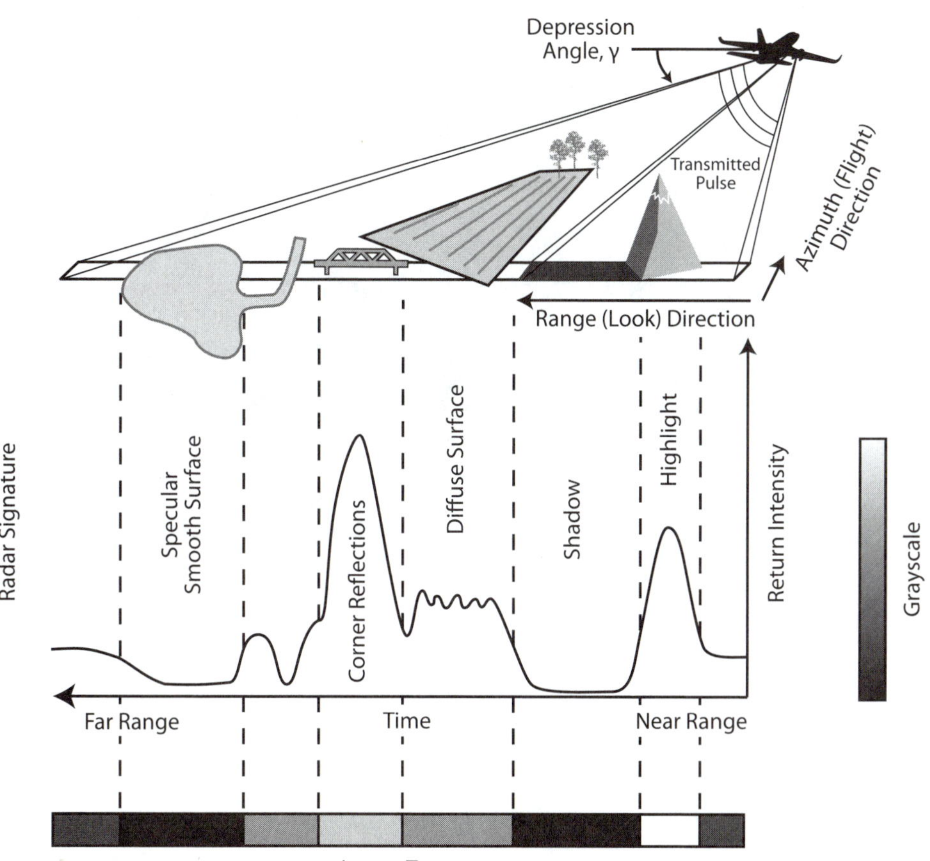

Figure 6-3 Terrain returns and image signatures for a pulse of radar energy.

tones. The bridge southwest of Cedar Bluff (see the location map in Figure 6-4B) has a bright signature caused by *corner reflectors*, which in this case are formed by intersecting structural supports. Energy encountering corner reflectors is strongly reflected toward the antenna. The towns of Cedar Bluff and Centre likewise have very bright signatures because of abundant corner reflectors. The calm waters of Weiss Lake and the Coosa River are *specular*, or smooth, surfaces that reflect all incident energy such that the angle of reflection is equal but opposite to the angle of incidence. Hence no energy returns to the antenna, and a very dark signature results. Rough water scatters the returns and has bright signatures.

The preceding descriptions show that tone alone (bright, intermediate, or dark) is insufficient to identify terrain features on radar images. Topographic scarps facing away from the antenna and specular surfaces both have dark tones but are completely different features. Radar signatures, however,

Table 6-1 Typical terrain features and signatures on radar images.

Image Signature	Image Tone	Terrain Feature	Cause of Signature
Highlights	Bright	Steep slopes and scarps facing toward antenna	Much energy is reflected back to antenna.
Shadows	Very dark	Steep slopes and scarps facing away from antenna	No energy reaches terrain; hence there is no return to antenna.
Diffuse surfaces	Intermediate	Vegetation	Vegetation scatters energy in many directions, including returns to antenna.
Corner reflectors	Very bright	Bridges and cities	Intersecting planar surfaces strongly reflect energy toward antenna.
Specular surfaces	Very dark	Calm water, pavement, dry lake beds	Smooth, horizontal surfaces totally reflect energy, with angle of reflectance opposite to angle of incidence.

FIGURE 6-4 Weiss Lake and vicinity, northeastern Alabama.

A. Radar image.

B. Location map.

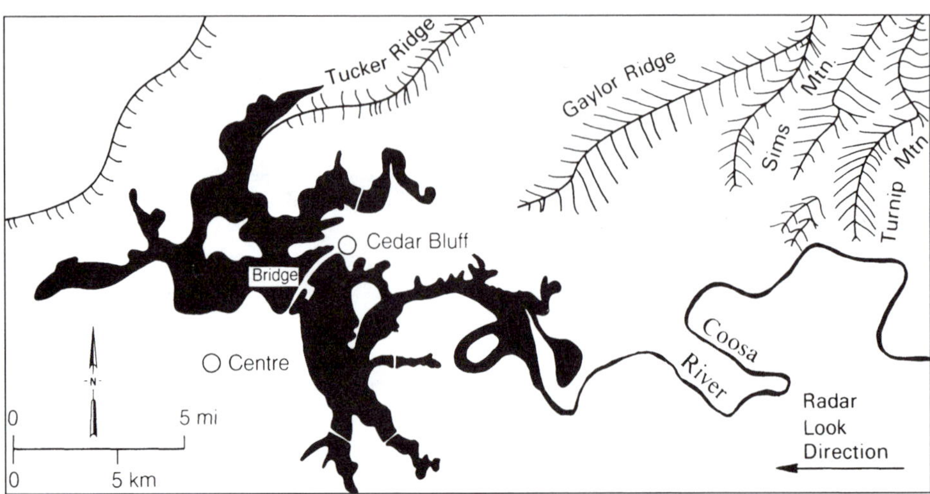

are determined not only by tone but also by size, shape, texture, and associations of the image feature. The size and shape of specular features are different from radar shadows. Radar shadows are generally associated with highlights, which are absent from specular features.

RADAR WAVELENGTHS

Table 6-2 lists the various radar wavelengths, or bands, and their corresponding frequencies. The bands are designated by letters that were assigned randomly rather than in order of increasing wavelength. These random designations are a relic of World War II security policies to avoid mentioning any wavelengths. Frequency is a more fundamental property of electromagnetic radiation than is wavelength. As radiation passes through media of different densities, frequency remains constant, whereas velocity and wavelength change. Most interpreters, however, can visualize wavelengths more readily than frequencies; also, wavelengths are used to describe the visible and infrared spectral regions. For continuity and consistency, therefore, we use wavelengths to designate various radar systems. Equation 1-1 (Chapter 1) converts frequency (ν) into wavelength (λ) in the following manner:

$$c = \lambda \nu$$

$$\lambda = \frac{3 \times 10^8 \text{ m} \cdot \text{sec}^{-1}}{\nu} \qquad \textbf{(1-1)}$$

where c is the speed of electromagnetic radiation. A convenient version of Equation 1-1 for converting radar frequencies into wavelength equivalents is

$$\lambda \text{ (cm)} = \frac{30}{\nu \text{ (GHz)}} \qquad \textbf{(6-1)}$$

In the 1960s and early 1970s, the first unclassified airborne radar was deployed. Modern aircraft and satellite radar systems operate at X-band and longer wavelengths.

TABLE 6-2 Radar wavelengths and frequencies used in remote sensing.

Band Designation[a]	Wavelength (λ) (cm)	Frequency (ν) (GHz)
K (1.5 cm)	0.8 to 2.4	40.0 to 12.5
X (3.0 cm)	2.4 to 3.8	12.5 to 8.0
C (6.0 cm)	3.8 to 7.5	8.0 to 4.0
S (8.0 cm, 12.6 cm)	7.5 to 15.0	4.0 to 2.0
L (23.5 cm, 25.0 cm)	15.0 to 30.0	2.0 to 1.0
P (68.0 cm)	30.0 to 100.0	1.0 to 0.3

[a] Wavelengths commonly used in imaging radars are shown in parentheses.

DEPRESSION ANGLE

Spatial resolution in both the range (look) direction and azimuth (flight) direction is determined by the engineering characteristics of the radar system. In addition to wavelength, a key characteristic is the *depression angle* (γ), defined as the angle between a horizontal plane and a beam from the antenna to a target on the ground (Figure 6-5A). The depression angle is steeper at the near-range side of an image strip and shallower at the far-range side. The average depression angle of an image is measured for a beam to the midline of an image strip. An alternate geometric term is *incidence angle* (θ), defined as the angle between a radar beam and a line perpendicular to the surface. For a horizontal surface, θ is the complement of γ (Figure 6-5A), but for an inclined surface there is no correlation between the two angles (Figure 6-5B). The incidence angle more correctly describes the relationship between a radar beam and a surface than does depression angle; however, in actual practice, the surface is usually assumed to be horizontal and the incidence angle is taken as the complement of the depression angle. Several other terms have been applied to this complementary angle, including *look angle* and *off-nadir angle*. We use *depression angle* to describe radar viewing geometry.

SLANT-RANGE AND GROUND-RANGE IMAGES

Side-scanning radar systems acquire images as *slant-range displays* that are geometrically distorted as a function of the depression angle. The cross section in Figure 6-6A shows that locations on the ground (ground range) are projected into the slant-range display, which is the plane from the antenna to the far range of the image swath. As a result, features in the near-range portion of the swath are compressed relative to features in the far range. On Figure 6-6A, features A and B are of equal size, but on the slant-range display image A′ is much smaller than image B′. Figure 6-6B is a slant-range image that illustrates this distortion. Figure 6-6C is the same image as 6-6B, now transformed into a ground-range display. This transformation assumes a horizontal ground surface and does not correct for the topographic variations discussed in the following section.

A. Horizontal surface.

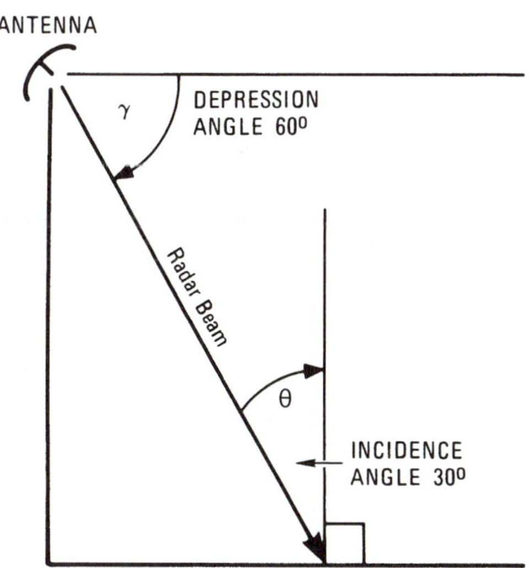

B. Inclined surface.

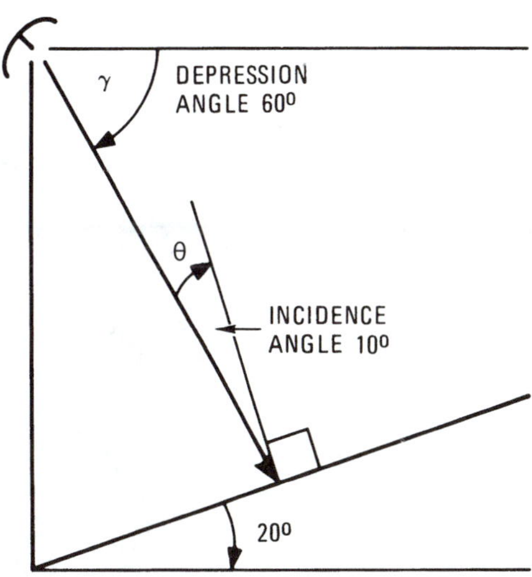

FIGURE 6-5 Depression angle and incidence angle.

A. Cross section.

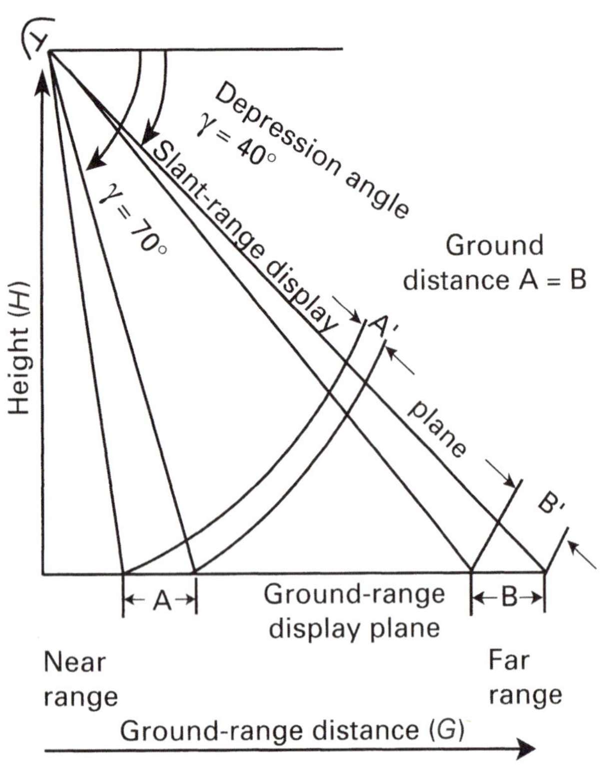

B. Slant-range display.

C. Ground-range display. Courtesy J. P. Ford, JPL (Retired).

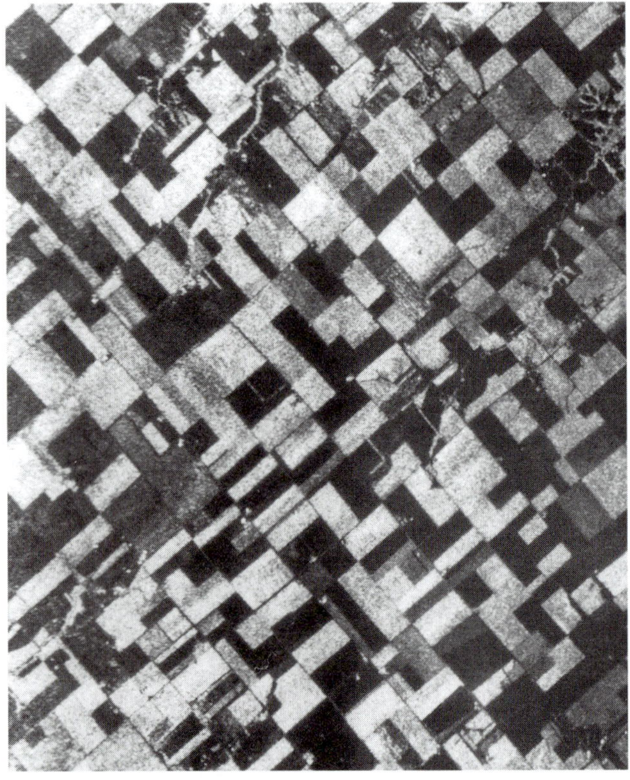

FIGURE 6-6 Slant-range image corrected into ground-range image. Aircraft L-band image in central Illinois. Image covers an area of 10 by 10 km.

All modern radar systems automatically project images into ground-range displays based on transformation calculations. Equation 6-2 is an example of a basic formula. For highly accurate results, more complex algorithms can be applied.

$$G = H\left(\frac{1}{\sin^2 \gamma} - 1\right)^{1/2} \qquad \textbf{(6-2)}$$

where:

G = ground-range distance
H = height of the antenna

SPATIAL RESOLUTION

The spatial resolution of a radar image is determined by the dimensions of the ground resolution cell, which are controlled by the combination of range resolution and azimuth resolution.

Range Resolution

Range resolution (R_r), or resolution in the radar look direction, is determined by the depression angle and by the pulse length. *Pulse length* (τ) is the duration of the transmitted pulse and is measured in microseconds (μsec, or 10^{-6} sec); it is converted from time into distance by multiplying by the speed of electromagnetic radiation ($c = 3 \times 10^8$ m · sec^{-1}). Range resolution is determined by the relationship

$$R_r = \frac{\tau \times c}{2 \cos \gamma} \qquad (6\text{-}3)$$

Figure 6-7 shows this relationship graphically. Targets A and B are spaced 20 m apart and are imaged with a depression angle of 50° and a pulse length of 0.1 μsec. For these targets range resolution is calculated from Equation (6-3) as

$$\begin{aligned} R_r &= \frac{\tau \times c}{2 \cos \gamma} \\ &= \frac{(0.1 \times 10^{-6}\ \text{sec})(3 \times 10^8\ \text{m} \cdot \text{sec}^{-1})}{2 \cos 50^\circ} \\ &= \frac{30\ \text{m}}{2 \times 0.64} \\ &= 23.4\ \text{m} \end{aligned}$$

Targets A and B are not resolved because they are closer together (20 m) than the range resolution distance (23.4 m). In other words, they are within a single ground resolution cell and cannot be separated on the image.

In Figure 6-7 targets C and D are also spaced 20 m apart but are imaged with a depression angle of 35°. For this depression angle, range resolution is calculated as 18 m. Therefore targets C and D are resolved because they are more widely spaced (20 m) than the ground resolution cell (18 m).

One method of improving range resolution is to shorten the pulse length, but this reduces the total amount of energy in each transmitted pulse. The energy and pulse length cannot be reduced below the level required to produce a sufficiently strong return from the terrain. Electronic techniques have been developed for shortening the apparent pulse length while providing adequate signal strength.

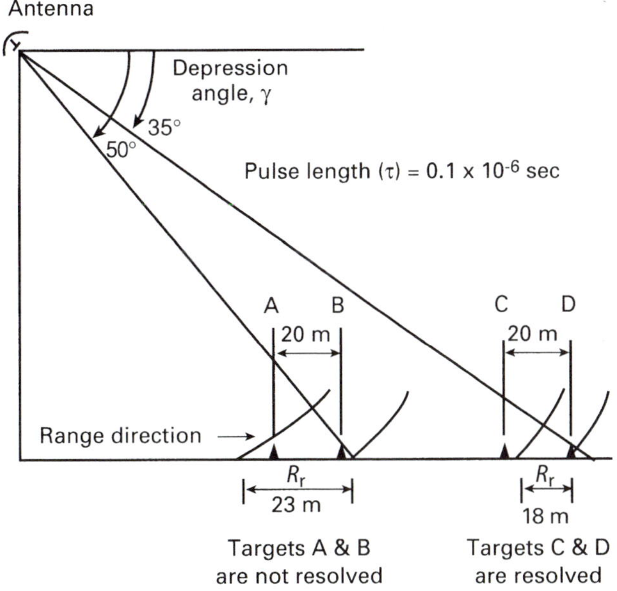

FIGURE 6-7 Range resolution for different depression angles.

Azimuth Resolution

Azimuth resolution (R_a), or resolution in the azimuth direction, is determined by the width of the terrain strip illuminated by the radar beam. To be resolved, targets must be separated in the azimuth direction by a distance greater than the beam width as measured on the ground. As shown in Figure 6-8, the fan-shaped beam is narrower in the near range than in the far range, causing azimuth resolution to be smaller in the near-range portion of the image. *Angular beam width* is directly proportional to the wavelength of the transmitted energy; therefore, azimuth resolution is higher for shorter wavelengths, but the short wavelengths lack the desirable weather penetration capability. Angular beam width is inversely proportional to *antenna length*; therefore, resolution improves with longer antennas, but there are practical limitations to the maximum antenna length.

The equation for azimuth resolution (R_a) is

$$R_a = \frac{0.7 \times S \times \lambda}{D} \qquad (6\text{-}4)$$

where S is the slant-range distance and D is the antenna length. For a typical X-band system ($\lambda = 3.0$ cm; $D = 500$ cm) with a slant-range distance (S_{near} in Figure 6-8) of 8 km, R_a is calculated from Equation 6-4 as

$$\begin{aligned} R_a &= \frac{0.7 \times S \times \lambda}{D} \\ &= \frac{0.7(8.0\ \text{km} \times 3.0\ \text{cm})}{500\ \text{cm}} \\ &= 33.6\ \text{m} \end{aligned}$$

Therefore, targets in the near range, such as A and B in Figure 6-8, must be separated by approximately 35 m to be

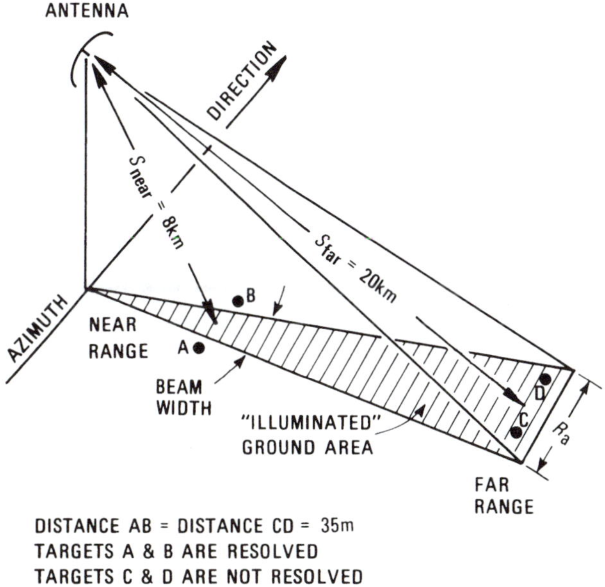

FIGURE 6-8 Azimuth resolution and beam width for a real-aperture system. From Barr (1969).

resolved. At the far-range position the slant-range distance (S_{far} in Figure 6-8) is 20 km, and R_a is calculated as 84 m; thus targets C and D, also separated by 35 m, are not resolved. Synthetic-aperture radar systems achieve much finer range resolutions than given in Equation 6-3, as described in the following section.

REAL-APERTURE AND SYNTHETIC-APERTURE SYSTEMS

The two basic systems are real-aperture radar and synthetic-aperture radar, which differ primarily in the method used to achieve resolution in the azimuth direction. *Real-aperture radar* (the "brute force" system) uses an antenna of the maximum practical length to produce a narrow angular beam width in the azimuth direction, as illustrated in Figure 6-8.

The *synthetic-aperture radar* (SAR) employs a small antenna that transmits a relatively broad beam (Figure 6-9A). The Doppler principle and data-processing techniques are employed to synthesize the azimuth resolution of a very narrow beam. Using the familiar example of sound, the Doppler principle states that the frequency (pitch) of the sound heard differs from the frequency of the vibrating source whenever the listener or the source are in motion relative to one another. The rise and drop in pitch of the siren as an ambulance approaches and recedes is a familiar example. This principle is applicable to all harmonic wave motion, including the microwaves employed in radar systems.

Figure 6-9A illustrates the apparent motion of a target through the successive radar beams from points A to C as a consequence of the forward motion of the aircraft. The curve shows changes in Doppler frequency during the elapsed time between targets A, B, and C. The frequency of the energy pulse returning from the targets increases from a minimum at A to a maximum at B normal to the aircraft. As the target recedes from B to C, the frequency decreases.

A digital system records the amplitude and phase history of the returns from each target as the repeated radar beams pass across the target from A to C. The digital record is computer processed to produce an image. The record of Doppler frequency changes enables each target to be resolved on the image as though it had been observed with an antenna of length L, as shown in Figure 6-9B. This synthetically lengthened antenna produces the effect of a very narrow beam with constant width in the azimuth direction, shown by the shaded area in Figure 6-9B. Comparing this narrow, constant beam with the fan-shaped beam of a real-aperture system (Figure 6-8) demonstrates the advantage of synthetic-aperture radar, especially for satellite systems. At range distances of hundreds of kilometers for satellites, a real-aperture beam is so wide that azimuth resolution is in hundreds of meters. Synthetic-aperture systems carried on satellites have an azimuth resolution ranging between sub-meter to tens of meters. For both real-aperture and synthetic-aperture systems, resolution in the range direction is determined by pulse length and depression angle (Figure 6-7). All modern aircraft and satellite radars operate in the SAR mode.

POLARIZATION

The electric field vector of the transmitted energy pulse may be polarized (or vibrating) in either the vertical or horizontal plane (Figure 6-10). On striking the terrain, most of the energy returning to the antenna usually has the same polarization as the transmitted pulse. This energy is recorded as *parallel-polarized* (or like-polarized) imagery and is designated HH (*horizontal transmit, horizontal return*) or VV (*vertical transmit, vertical return*). *Single polarization* radar systems operate with the same polarization for transmitting and receiving the signal (Parker, 2012). Many mapping radar systems operate in the HH mode because this mode produces the strongest return signals.

A. Shift in Doppler frequency caused by relative motion of target through repeated radar beams.

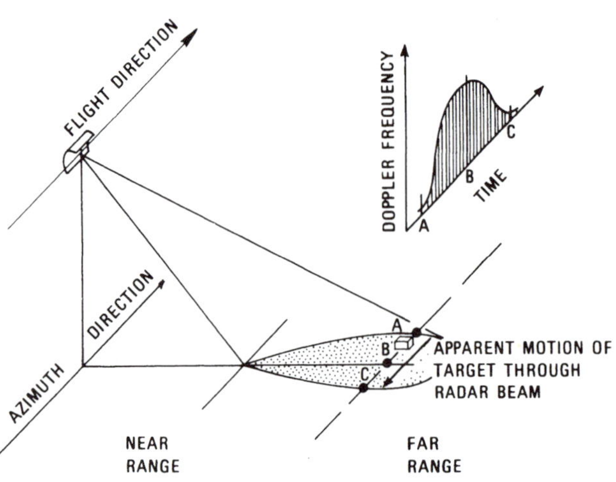

B. Azimuth resolution of synthetic-aperture radar. The physical antenna length D is synthetically lengthened to L.

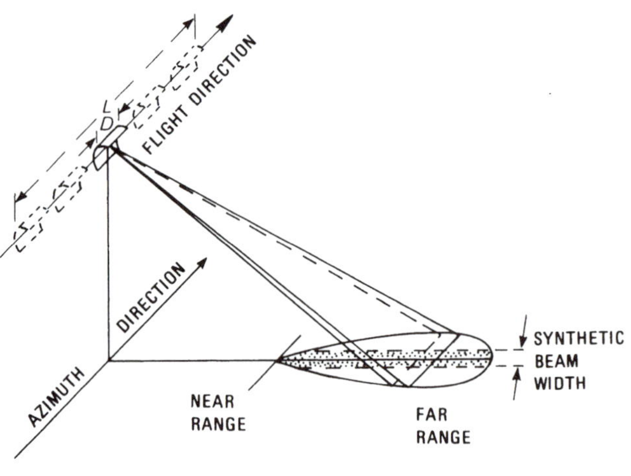

FIGURE 6-9 Synthetic-aperture radar system. From Craib (1972, Figures 3 and 5).

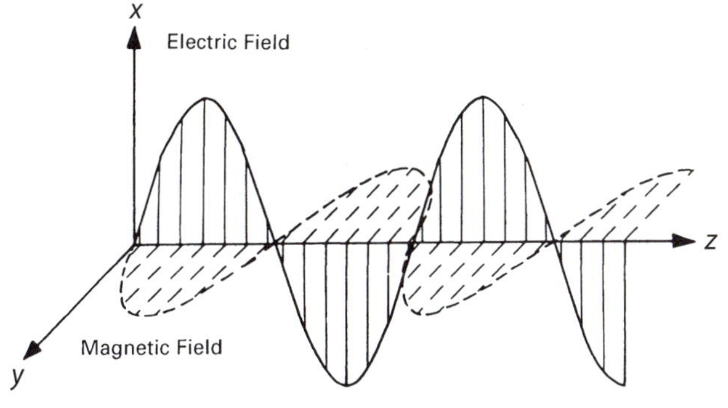

Figure 6-10 Polarization of radar energy (and other electromagnetic energy).

A portion of the returning energy is *depolarized* by the terrain features and vibrates in various directions. The mechanisms responsible for depolarization are not definitely known, but the most widely accepted theory attributes it to multiple reflections at the surface. This theory is supported by the fact that depolarization effects are much stronger from vegetation than from bare ground. Leaves, twigs, and branches are believed to cause the multiple reflections responsible for depolarization. Parker (2012) notes that vertically polarized waves interact with the vertical stalks of plant canopy, while horizontally polarized waves penetrate through plant canopy. *Cross polarization* radar systems receive the depolarized energy vibrating at right angles to the plane of the transmitted pulse. The resulting imagery may be HV (*horizontal transmit, vertical return*) or VH (*vertical transmit, horizontal return*).

Dual polarization radar systems transmit the signal with one polarization and simultaneously receive both polarizations of the signal. For example, a dual polarization system can transmit with H and receive both H and V, which generates two radar images—an HH and an HV image. Some dual systems transmit V and generate VV and VH radar images. *Quad polarization* radar systems transmit alternate pulses with H and V polarizations and simultaneously receive the parallel-polarized and cross-polarized HH, VV, HV, and VH returns. Dual and quad polarization images show the polarization properties of terrain and not simply the backscatter at a single polarization, which can improve classification and mapping of features (MDA, 2016). Images of Death Valley are provided here as examples.

Plate 18 and Figure 6-11 display PALSAR-2 radar images and a Landsat OLI panchromatic image of Death Valley, California. The Furnace Creek alluvial fan is located at the north end of Death Valley. Figures 6-11A and 6-11B are parallel-polarized (HH) and cross-polarized (HV) L-band images of the Furnace Creek alluvial fan, which is located along the west flank of the Funeral Mountains. Figure 6-11C is the Landsat panchromatic image and Figure 6-11D is a geologic map. Furnace Creek Ranch in the north portion of the fan is the Visitor Center of the US National Park Service. The fan consists of sand and gravel eroded from the mountains and deposited by a radial pattern of ephemeral streams to form a broad cone that slopes gently into the valley floor.

HISTORIC RADAR SYSTEMS

SAR systems acquire images from satellites at distances of hundreds of kilometers with good spatial resolution. Many radar satellites with X-, C-, and L-band sensors (Table 6-2) were launched over the past three decades, beginning with NASA's Seasat in 1978 (Figure 6-12). Several L-band radar systems were launched between 1980 and 1995, including SIR-A, SIR-B, and SIR-C, that were carried into orbit by the NASA Space Shuttle. By mid-2017, most active radar satellites are X-band, followed by two C-band systems and one L-band system.

Table 6-3 lists characteristics of decommissioned satellite radar systems. Images acquired by several of these decommissioned systems are interpreted later in this chapter. The images collected by these legacy systems are useful for environmental baseline and change-detection studies. The following section summarizes decommissioned satellite radars that led the way for more recent radar systems.

Seasat

Seasat was designed primarily to investigate oceanic phenomena (such as roughness, current patterns, and sea ice conditions), the images were also valuable for terrain observations. The unmanned satellite was launched in June 1978 and prematurely ceased operation in October 1978 because of a major electrical failure.

NASA Shuttle Imaging Radar Missions

Over the next 20 years, NASA expanded the assets of Seasat by conducting three *Shuttle Imaging Radar* (SIR) missions. The Space Shuttle was utilized for the launch and deployment of each mission. Some details of each mission are provided in Table 6-3. Because of communication and recovery requirements, all Shuttle missions are placed in near-equatorial orbits and do not provide the global coverage of polar-orbiting satellites. Figure 3-8 compares the orbit pattern of a typical Shuttle mission with the pattern of Landsat TM.

The first radar mission (SIR-A) was conducted in November 1981. At launch, the SIR-A antenna was folded and carried in the cargo bay. After reaching orbit, the Shuttle was inverted to orient the cargo bay toward the Earth, and the antenna was deployed to acquire images. Figure 6-13 (on p. 154) shows the configuration of the SIR-A system, which acquired 50-km wide image swaths with an average depression angle of 40°. Cimino and Elachi (1982) published a collection of SIR-A images. Ford and others (1983) published additional examples and an index map of the 10 million km^2 of imagery that is confined to the mid-latitudes covered by the SIR-A orbits.

A second SIR experiment (SIR-B) was conducted in October 1984 with the antenna modified to change the depression angle during the mission within a range from 30 to 75°. Multiple images of certain areas were acquired at different depression angles for two purposes: (1) to evaluate changes in radar return as a function of depression angle for different

A. PALSAR-2 HH polarization. Courtesy JAXA.

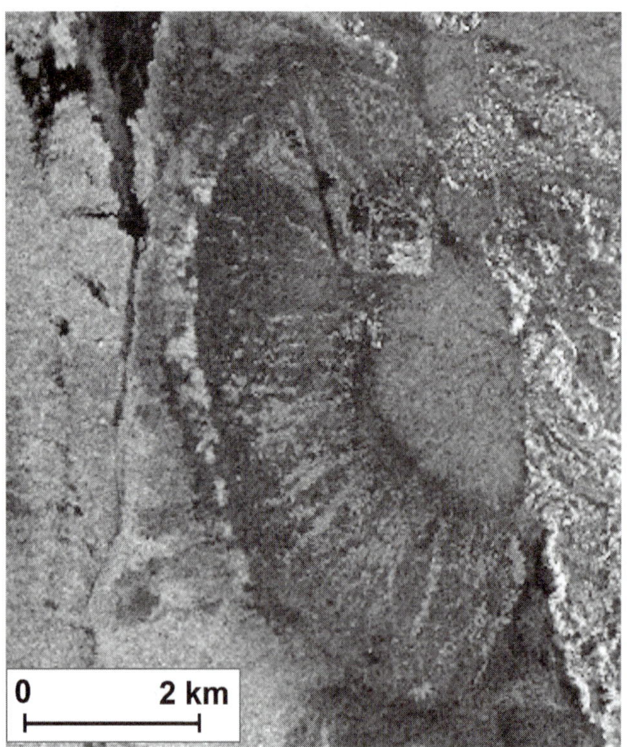

B. PALSAR-2 HV polarization. Courtesy JAXA.

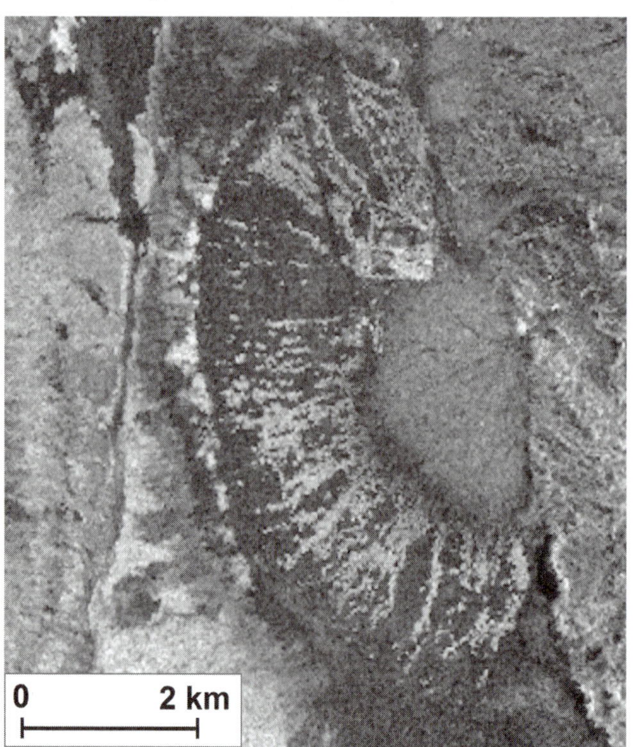

C. Landsat 8 panchromatic image. Landsat courtesy USGS.

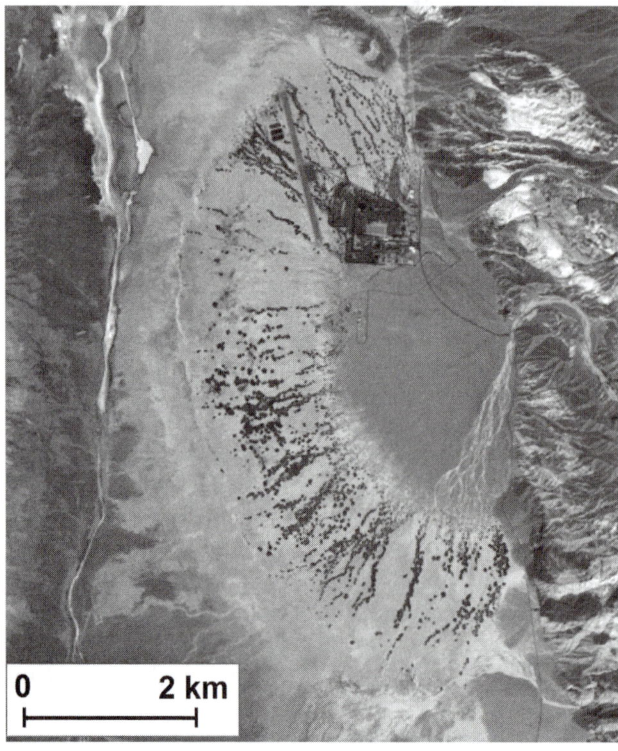

D. Geologic map. Symbols are explained in Figure 6-31. From Hunt and Mabey (1966, Plate 1).

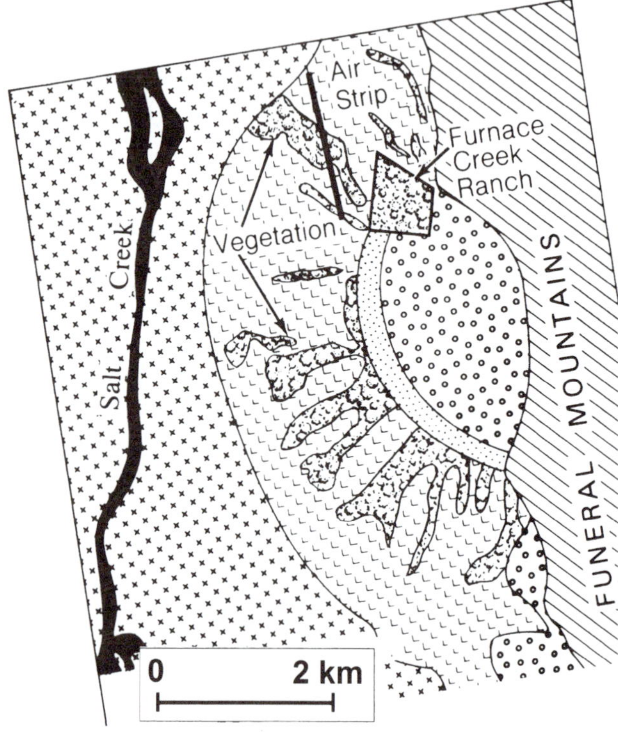

FIGURE 6-11 HH and HV L-band SAR images of Furnace Creek alluvial fan and vicinity, Death Valley, California.

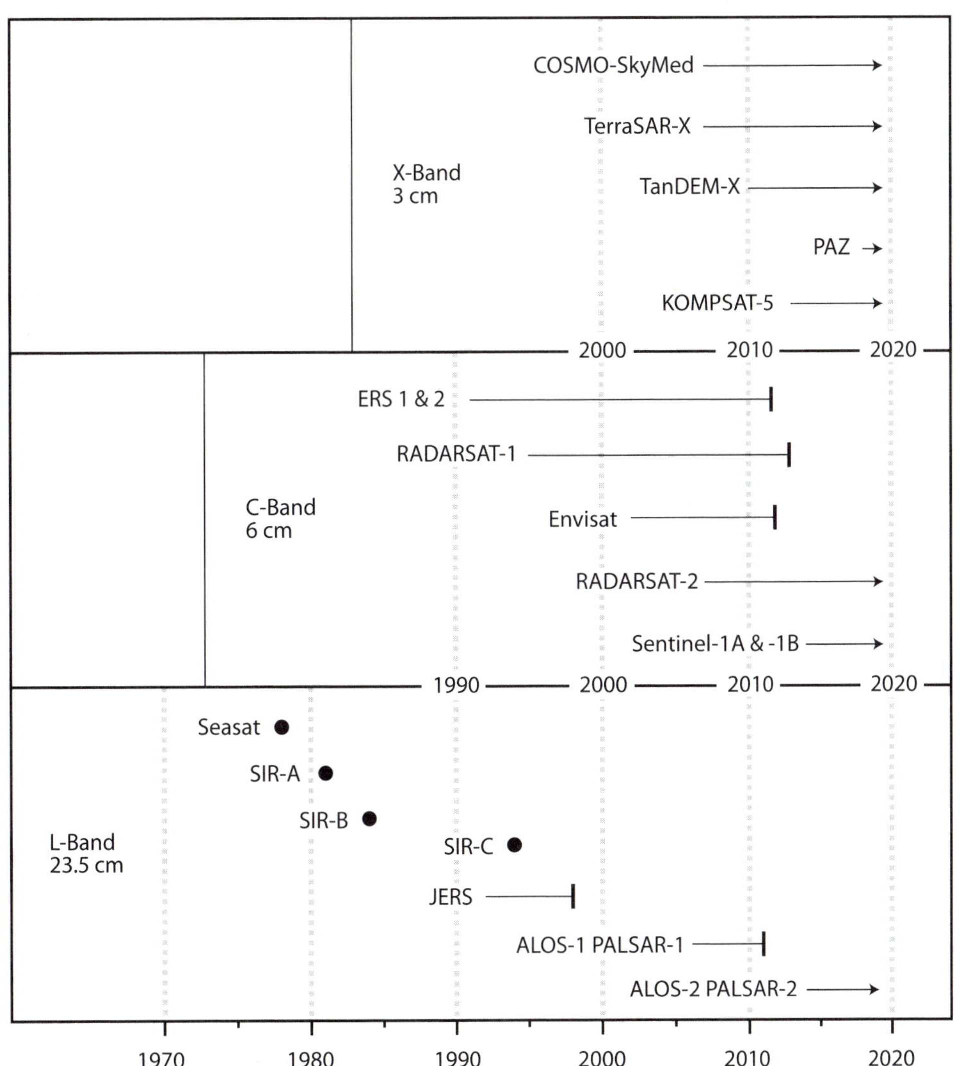

Figure 6-12 Time lines of radar satellites from 1978 to 2018. Functioning (lines with arrows) and nonfunctioning (lines with vertical bars) satellites are shown.

Table 6-3 Decommissioned satellite radar systems.

	Seasat	SIR-A	SIR-B	SIR-C	ERS-1	JERS-1	Almaz-1	Magellan	RADARSAT
Launch date	1978	1981	1984	1994	1991	1992	1991	1989	1995
Wavelength, cm (band)	23.5 (L)	23.5 (L)	23.5 (L)	3.0 (X)	5.7 (C)	23.5 (L)	9.6 (S)	12.6 (S)	5.7 (C)
				6.0 (C)					
				24.0 (L)					
Depression angle	70°	40°	30 to 75°	Variable	67°	55°	40 to 58°	65°	40 to 70°
Spatial resolution (m)	25	38	25	Variable	28	18	15	100	Variable
Polarization	HH	HH	HH	Multiple	VV	HH	HH	HH	HH
Swath width	100	50	40	30 to 60	100	75	50 to 100	20	50 to 100
Altitude (km)	790	250	225	225	785	568	350	290 to 2,000	800
Latitude covered	72°N to 72°S	50°N to 50°S	58°N to 58°S	57°N to 57°S	Polar	Polar	Polar	Polar	Polar
Operator	United States	United States	United States	United States	Europe	Japan	Russia	United States	Canada

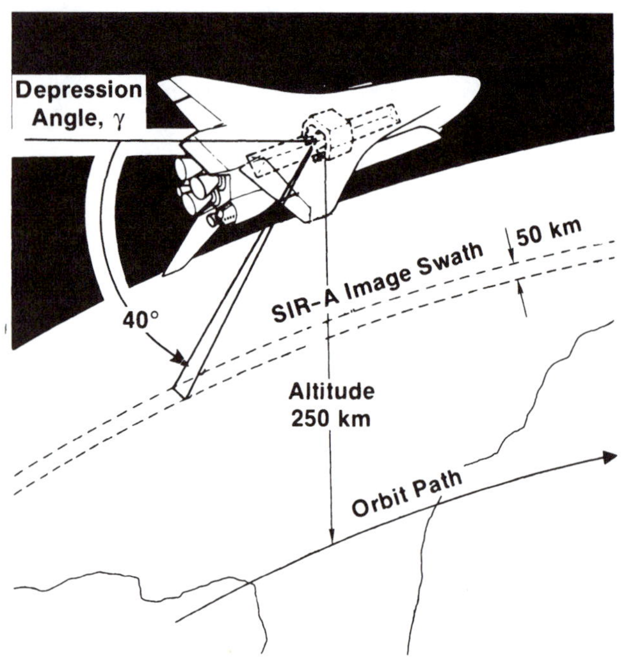

FIGURE 6-13 SIR-A system.

terrain types and (2) to produce stereo images based on the parallax provided by differences in viewing geometry.

Cimino and others (1988) provide details of the mission and an index to the images that were acquired. Ford and others (1986) published an annotated collection of representative SIR-B images. The *International Journal of Remote Sensing* (May 1988) published a collection of 14 papers on different aspects of the mission. Ford (1988) edited a series of SIR-B research papers. Ford and Sabins (1986) interpreted SIR-B image strips across the island of Borneo and recognized categories of terrain and structure similar to those on earlier SIR-A images of West Papua.

The SIR-C mission was carried on two Shuttle flights in 1994 and was a major technologic advance over earlier missions. Three antennas recorded X- (3.0 cm), C- (6.0 cm), and L- (24.0 cm) bands of imagery at multiple polarizations and variable depression angles (Table 6-3). Evans and others (1994) published a collection of multicolor SIR-C images.

OTHER EARTH ORBITING RADAR SATELLITES

The Japanese Earth Resources Satellite (JERS-1) carries a radar imaging system that is summarized in Table 6-3. The 55° depression angle of JERS-1 is intermediate between those of SIR-A (40°) and the European Remote Sensing Satellite (ERS-1) (67°). Therefore, the foreshortening on JERS-1 images is intermediate between that for SIR-A and ERS-1 images. ERS-1 acquires images with VV polarization and a steep depression angle (67°) to emphasize oceanographic features. The Russian Almaz-1 satellite (launched March 31, 1991) recorded images on holographic film that was converted into image film. Individual images cover 40 by 40 km. RADARSAT, Canada's first Earth observation satellite, was launched November 4, 1995 and decommissioned on May 10, 2013. The satellite completed 14 orbits daily in a sun-synchronous pattern that covered the Earth (north of latitude 80°S) every six days. One objective was to monitor sea ice conditions in the Arctic Ocean during periods of winter darkness and seasonal cloud cover.

PLANETARY EXPLORATION

The surface of planets covered by dense atmosphere are imaged and mapped with radar. The Cassini spacecraft was launched in 1997 to study Saturn, its rings, and moons. The onboard radar penetrated the thick, nitrogen-rich atmosphere of Saturn's largest moon, Titan, revealing lakes of methane and ethane, icy sand dunes composed of solid water coated with hydrocarbons, and ridges with jagged peaks, likely created by methane rainfall erosion. Cassini performed more than 100 flybys of Titan. Each flyby produced one strip of imagery over different areas of Titan. A global mosaic of Titan's surface was generated from the strips. In April 2017 the Cassini spacecraft was placed on an impact course with Saturn. On September 15, 2017 Cassini plunged into Saturn's atmosphere (NASA Science, 2019).

The Magellan satellite was carried on Space Shuttle *Atlantis*, which was launched May 4, 1989 from Cape Canaveral. Once *Atlantis* was in Earth orbit it launched Magellan on its mission to Venus. Magellan began transmitting Venus images on September 15, 1990. Magellan has a minimum altitude of 290 km and a maximum of 2,000 km at polar latitudes. Magellan also carried a radar altimeter for topographic mapping and a radiometer for measuring emissivity at radar wavelengths. Ford and others (1993) provide details of the imaging system, orbit pattern, and typical images.

The Magellan satellite acquired radar images of more than 98% of Venus where the dense carbon dioxide atmosphere prevents visible observation. All of the terrestrial planets have been bombarded by bolides (meteorites), which have cratered their surfaces. The older impact craters on Earth have been obliterated or obscured by erosion, deposition, and tectonism. Venus and the other planets have more complete cratering records that tell us about their histories and that of our solar system.

Schaber and others (1992) and Strom and others (1994) interpreted Magellan images of Venus impact features. Friction in the dense Venus atmosphere destroys the smaller incoming bolides capable of forming the missing craters. Venus impact craters are largely unchanged by post-impact volcanism, erosion, and tectonism. Magellan radar images of the craters show both morphology and materials.

RECENT AIRBORNE AND SATELLITE RADAR SYSTEMS

Table 6-4 lists the characteristics of two airborne and five satellite systems that were operational in mid-2017; they acquire X-, C-, L-, and P-band imagery for numerous applications. Typical acquisition parameters are listed in Table 6-4 for the airborne systems, as they can acquire imagery at different altitudes to vary the spatial resolution and swath width. The radar beam on satellite systems can be electronically tilted to change the depression angle to enable collection of images of different spatial resolution and swath width. Modern radar systems measure the target's polarization properties along with the backscatter intensity to improve classification information (Parker, 2012).

Figure 6-14 RADARSAT-2 satellite with antenna located beneath solar panels. Courtesy Canadian Space Agency.

Airborne Radar Systems

Intermap and GeoSAR are airborne systems that provide orthorectified grayscale images and digital elevation models (Table 6-4) (see discussion in Chapter 7). Intermap acquires radar images with 0.5 to 1.25 m spatial resolution (Intermap, 2016) while GeoSAR acquires images with 1.25 to 5.0 m spatial resolution. GeoSAR simultaneously acquires both X-band and P-band images (Sharp and Morton, 2007).

Satellite Radar Systems

Table 6-4 summarizes five satellite systems launched between 2007 and 2014. Electric power is generated by large solar panels (Figure 6-14).

RADARSAT-2

RADARSAT-2 orbits the Earth 14.3 times daily to acquire image swaths left or right along the orbital path with a range of depression angles, swath widths, and spatial resolutions

Table 6-4 Recent airborne and satellite radar systems.

	Airborne		Satellite				
Sensor	**Intermap**	**GeoSAR**	**RADARSAT-2**	**TerraSAR-X**	**ALOS-2 PALSAR-2**	**Sentinel-1**	**COSMO-SkyMed**
Launch date	~1996	~2002	2007	2007	2014	2014	2007
Wavelength, cm (band)	3.0 (X)	3.0 (X) 85 (P)	5.4 (C)	3.0 (X)	23.5 (L)	5.6 (C)	3.0 (X)
Depression angle	55 to 35°		Variable	Variable	Variable	Variable	Variable
Spatial resolution (m)	0.5 to 1.25	X-band: 1.25 to 3.0 P-band: 1.25 to 5.0	1.5, 3, 5, 8, 25, 50, 100	0.25, 1, 3, 18.5, 40	1, 3, 6, 10, 60, 100	5, 20, 40	1,100
Polarization[a]	S	X-band: S P-band: C	S, D, Q	S, D, Q	S, D, Q	S, D	S, D
Swath width (km)	11.5	12 to 14	18, 50, 150, 300, 500	4, 10, 30, 100, 270	25, 50, 350, 490	20, 80, 250, 400	11, 30, 40, 200
Altitude (km)	10.4	10 to 12.5	798	514	628	693	620
Number of satellites			1	2	1	2	4
Operator	Canada	United States	Canada	Europe	Japan	Europe	Italy

[a] Polarization: S = single, C = cross, D = dual, Q = quad.

(Livingstone and others, 2005; MDA, 2016). The SAR system operates in one of three modes: Spotlight, Single Beam, or ScanSAR. Spatial resolution and swath width of each mode are shown in Table 6-5.

TABLE 6-5 Characteristics of RADARSAT-2.

Operating Mode	Spatial Resolution (m)	Swath Width (km)
Spotlight	1.5	18 to 25
Single Beam	2.5 to 25	18 to 200
ScanSAR	30 to 150	250 to 500

TerraSAR-X

TerraSAR-X operates in Spotlight, StripMap, and ScanSAR image modes (Figure 6-15). These modes with their acquisition characteristics are shown in Table 6-6 (generalized from TerraSAR-X, 2014). The radar system acquires single and dual polarized imagery.

The TerraSAR-X satellite orbits in close formation with the identical TanDEM-X satellite. Together they generate a new global DEM (WorldDEM) with a 12 m grid (discussed in Chapter 7). Figure 6-16 is a TerraSAR Spotlight image with 0.25 m pixels of the typically cloudy Panama Canal. The radar beam penetrates any cloud cover to show water with its characteristic dark signature caused by specular reflection from the smooth surface. Two large ships are in the locks in the northwest portion of the image. In the southeast portion a single ship is escorted by two tugboats. Bright signatures of the ships and tugs are caused by strong corner reflectance from their superstructures. Bright lines are berms along drainage channels. Bright lines in the southwest portion are outlines of buildings and artificial terraces. Forested terrain has an uneven grayscale texture compared to cleared industrial areas with their smooth texture.

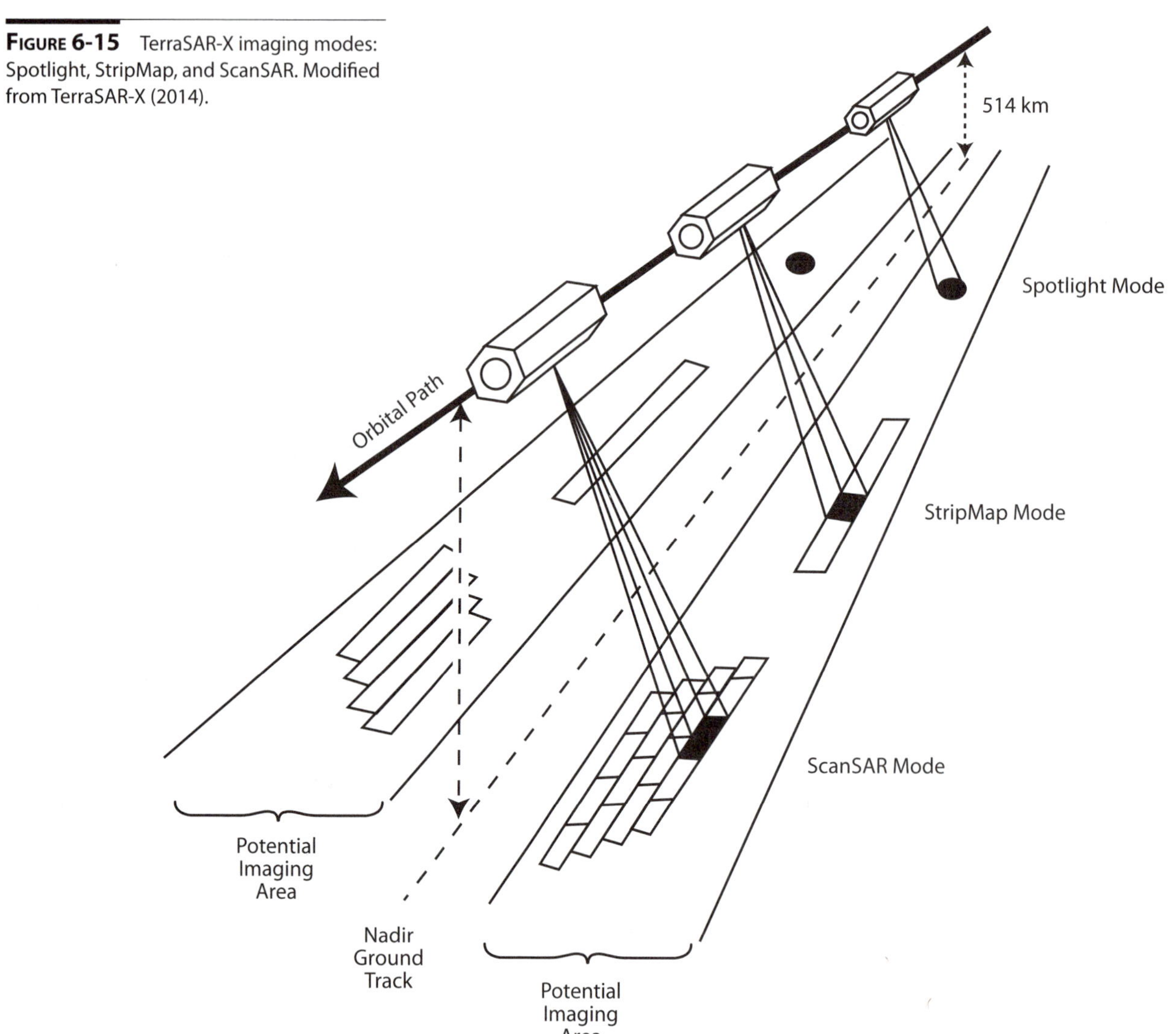

FIGURE 6-15 TerraSAR-X imaging modes: Spotlight, StripMap, and ScanSAR. Modified from TerraSAR-X (2014).

Table 6-6 Characteristics of TerraSAR-X.

Imaging Mode	Acquisition Characteristics		
	Azimuth Spatial Resolution with Single Polarization (m)	**Azimuth Spatial Resolution with Dual Polarization (m)**	**Swath Width (km)**
Staring Spotlight	0.24		3.7
High Resolution Spotlight (HS)	1.1	2.2	5
Spotlight (SL)	1.7	3.4	10
StripMap (SM)	3.3	6.6	30
ScanSAR (SC)	18.5		100
Wide ScanSAR (WS)	40		200

PALSAR-2

Japan's Advanced Land Observing Satellite-2 (ALOS-2) carries the Phased Array L-band Synthetic Aperture Radar-2 (PALSAR-2) system that acquires L-band (23.5 cm wavelength) images. The L-band energy penetrates into partially open forest canopy (small leaves are not resolved) and interacts with the larger components such as the trunks and larger branches of trees, which provides some indication of aboveground biomass (Committee on Earth Observation Satellites, 2016).

Sentinel-1

The European Space Agency's (ESA) Sentinel-1 mission consists of two polar-orbiting satellites that acquire C-band radar swaths along the right side of the orbit path (Sentinel-1 Team, 2013). Both satellites share the same orbit plane but are 180° apart, which provides a six-day repeat cycle. Sentinel-1 has four acquisition modes, which are shown in Figure 6-17 (Sentinel-1 Team, 2013).

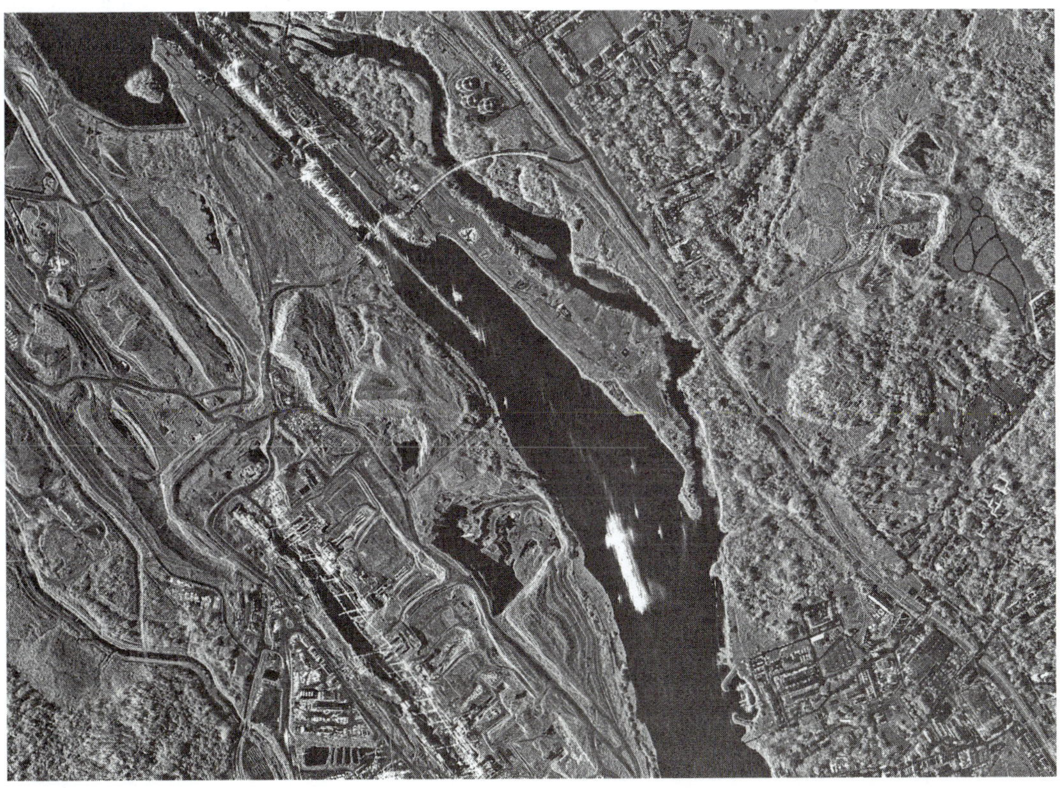

Figure 6-16 TerraSAR-X Spotlight image of Panama Canal locks and ships. © Airbus DS 2019.

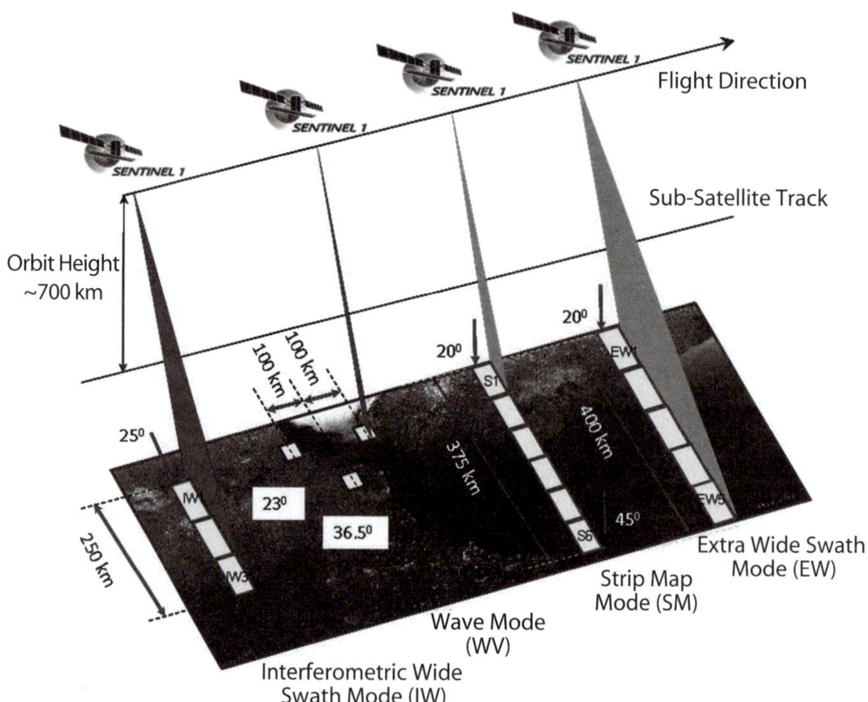

FIGURE 6-17 ESA Sentinel-1 and four image acquisition modes. From Sentinel-1 Team (2013).

1. **Strip Map**: Acquires images with 5 m spatial resolution and 80 km swath width, which provides image continuity with decommissioned radar satellites.
2. **Interferometric Wide Swath**: Used for generating DEMs and monitoring ground deformation.
3. **Extra-Wide Swath**: Used for maritime, ice, and polar zone operations that require wide coverage and short revisit times.
4. **Wave**: Supports ocean wave modeling

COSMO-SkyMed

The Italian COSMO-SkyMed constellation consists of four radar satellites with X-band radar systems. Three of the satellites are in the same orbit plane, separated by 120° to support rapid revisits of target sites on Earth by using different depression angles. The fourth satellite orbits with a slightly different path and separation from one of the other satellites. This separation results in a 151 km, along-track separation between the fourth satellite and one of the satellites in the same orbit plane. The separation enables interferometric measurements of topography and ground deformation (COSMO-SkyMed Team, 2010).

CHARACTERISTICS OF RADAR IMAGES

The nature of radar illumination causes specific geometric characteristics in the images that include shadows and highlights, slant-range distortion, and image displacement. The geometric characteristics provide stereo models and the capability for topographic mapping.

HIGHLIGHTS AND SHADOWS

Figure 6-3 illustrated how topographic features interact with a radar beam to produce highlights and shadows. Figure 6-18A is an image of the White Mountains in eastern California that was acquired with a look direction toward the west (bottom margin of image) and a depression angle of 17°. This gentle depression angle causes the pronounced highlights on the east flank of the range and shadows on the west flank. The shadow outline suggests the profile of the range. Even in the low-relief terrain west of the mountains (lower portion of Figure 6-18A) subtle topographic features are clearly expressed by highlights and shadows. For comparison, Figure 6-18B is a Landsat TM image acquired with a sun elevation of 40°, which results in minimal highlights and shadows. Instead of topography, the bright and dark signatures are largely the expression of differences in albedo of the surface. Figure 6-18C is a hillshade Shuttle Radar Topography Mission (SRTM) DEM (illuminated at 30° above the horizon from the top of the page) with 500 m topographic contours to show scale and variations in relief of the area.

The cross section in Figure 6-19A shows that the radar depression angle is steeper in the near range and shallower in the far range. Therefore, terrain features of constant elevation have proportionally longer shadows in the far-range portion of an image, as illustrated in the map view of Figure 6-19B. For terrain with low relief, radar images with a shallow depression angle can accentuate subtle topographic features. In terrain with high relief, an intermediate depression angle is desirable, because the extensive shadows caused by a low depression angle may obscure much of the image. Images such as aerial photographs and those acquired by

A. Aircraft radar image acquired with a depression angle of 17°.

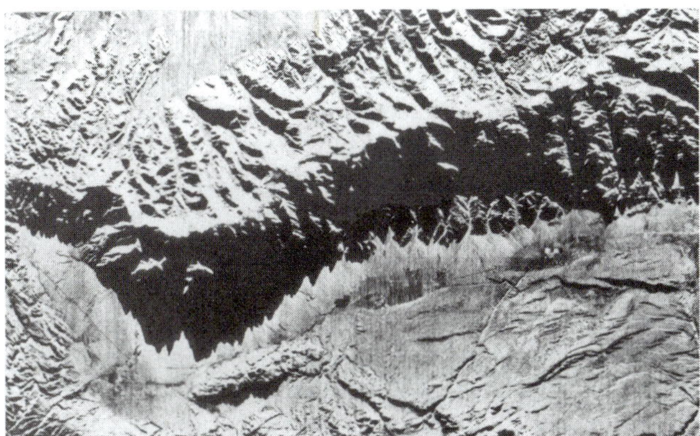

B. Landsat TM band 4 image acquired with a solar elevation of 40°.

C. Hillshade SRTM DEM with 500 m topographic contours. DEM courtesy USGS.

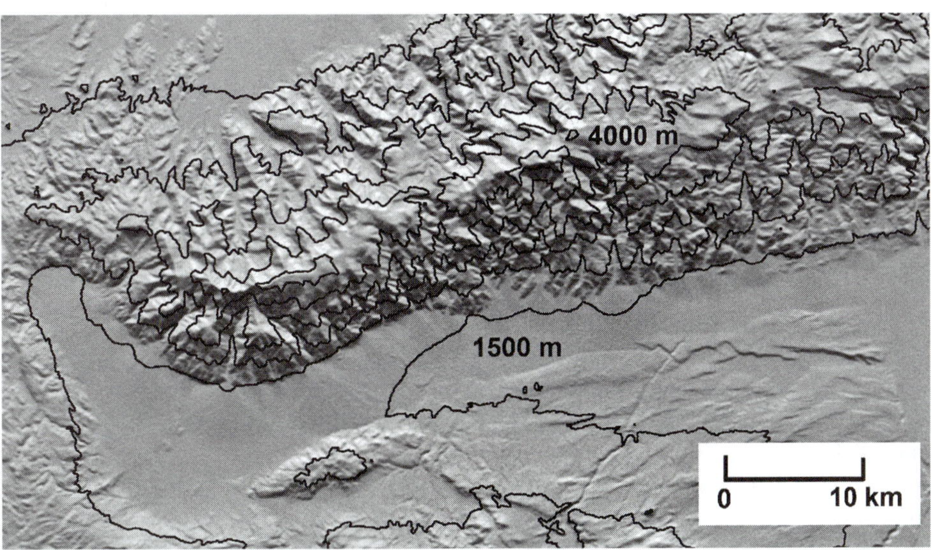

Figure 6-18 Radar shadows and highlights, White Mountains, California.

A. Cross-section view of depression angles.

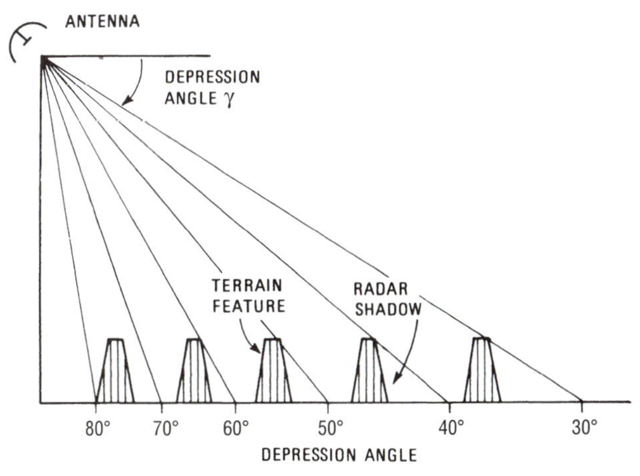

B. Map view of shadows.

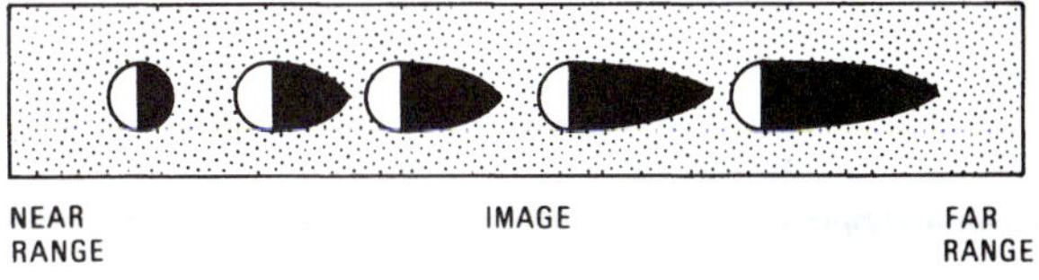

Figure 6-19 Radar illumination and shadows at different depression angles.

Landsat are acquired with solar illumination, which has a constant depression angle; thus, shadow lengths are directly proportional to topographic relief throughout these images. When interpreting a radar image one must recall that depression angles and shadow lengths are not constant throughout the image.

LOOK DIRECTION AND TERRAIN FEATURES

Many natural and cultural linear features of the terrain, such as fractures or roads, have a strong *preferred orientation* and are expressed as lines on images. The geometric relationship between the preferred orientation and the radar look direction strongly influences the signatures of linear features.

Geologic Features, Venezuela

Faults, joints, and outcrops of layered rocks are natural linear features that may be enhanced or subdued in radar images depending on their orientation relative to the look direction. Features trending normal or at an acute angle to the look direction are enhanced by highlights and shadows. Features trending parallel with the look direction produce no highlights or shadows and are suppressed in an image. These relationships are illustrated by an area in Venezuela where separate X-band images (3 cm wavelength) were acquired with look directions toward the south (Figure 6-20A) and toward the west (Figure 6-20B). The geologic interpretation map (Figure 6-20C) shows a high mesa capped by relatively horizontal beds of resistant quartzite. The mesa is underlain by deeply eroded lowlands of folded metamorphosed sedimentary rock that clearly express the bedding trends.

In the large fold in the eastern part of the area, one limb trends north and the second limb trends northwest. The north-trending limb is enhanced by the west look direction (Figure 6-20B). The northwest-striking limb trends at an acute angle to either look direction and is equally enhanced in both images. In the western part of the area, a major linear feature (probably a fault) strikes south through the metamorphic terrain and cuts the west portion of the mesa where it forms a linear valley. The linear feature is strongly enhanced by the west look direction (Figure 6-20B) and subdued by the south look direction (Figure 6-20A). Other linear features in the mesa that trend generally east or west of north are enhanced or subdued in the same fashion on the two images. In the southeast part of the area, a number of parallel linears trend slightly north of east in the metamorphic terrain; these are clearly visible with the south look direction but are not recognizable with the west look direction. The west-trending, parallel light and dark bands in Figure 6-20B are minor defects in the image. If this area were to be imaged with only a single look direction, an orientation toward the southwest or northeast would enhance most of the linear features.

In the southeast portion of Figure 6-20B a series of horizontal black bars obscure the terrain in the second radar strip from the east margin. From his experience on a Chevron radar project in Nicaragua, Sabins recognized these bars as shadows formed in the down-scan direction by cloud cells with active precipitation of rain or hail. The raindrops or hail pellets are large enough to scatter the X-band radar so that no energy reaches the terrain. X-band energy does penetrate clouds, which are water vapor. In the early 1980s, Sabins was the Chevron client-observer on a contractor four-engine piston DC-6B that was acquiring a radar mosaic of eastern Nicaragua. The plane was struck by lightning with a loud bang, a blue flash, and a strong ozone smell. The flight engineer came charging back to the radar compartment where we assured him that we were still with the program. After landing we found an irregular blackened hole about 6 in in diameter at the trailing edge of the starboard wing elevator. Upon processing the radar images, there was no evidence of the lightning strike, which indicated that the lightning cloud cell was not located within the narrow field of view of the radar system. It also indicated that the radar system was well isolated from the aircraft mainframe that absorbed the strike.

Significance of the Look Direction

The X-band images of Venezuela demonstrate that there is an important relationship between radar look direction, image signatures, and orientation of linear features in the terrain. Linear geologic features, such as faults, fractures, joints, and bedding planes, that are oriented at a normal or oblique angle to the radar look direction are enhanced by highlights and shadows in the image. Features oriented parallel with the look direction lack highlights and shadows; they are suppressed and are difficult to recognize in the image. The importance of radar look direction is comparable to the importance of sun azimuth in Landsat images acquired at low sun angles. A radar survey should be oriented with the look direction at a normal or acute angle relative to the known geologic trends of the area.

In urban areas the orientation of streets and buildings relative to the look direction also influences the signatures. In addition, linear metal features, such as fences and railroads, have bright signatures where they are oriented normal to look directions but are subtle where they are oriented parallel with the look direction. Interpreters of radar images of geologic or urban scenes must be aware of these geometric relationships.

TOPOGRAPHIC INVERSION

Ramachandran (1988) noted that the human visual experience of the world is based on two-dimensional images: namely, the flat patterns of varying light intensity and color that fall on the single layer of cells in the retina of the eye. Yet our brains come to perceive solidity and depth. A number of cues about depth are available in the retinal image: shading, perspective, occlusion of one object by another, and stereoscopic vision. The brain utilizes these cues to perceive the three-dimensional shapes of objects. Experiments show that our visual system assumes that an image is illuminated by only one light source and that the light comes from above, that is, from the upper margin toward the lower margin of the image. This assumption gives us a strong bias when we view images with pronounced shadows, such as shallow depression angle radar images and low sun angle Landsat images.

This illumination bias is demonstrated by the radar image in Figure 6-21A. Do you perceive these circular features as uplifts (domes) or as depressions (basins)? Most viewers see them as basins. For example, 24 students in a Sabins' UCLA class viewed the image; 19 students (79%) perceived the features as basins, while 5 students (21%) correctly perceived domes. The image (Figure 6-21A) is shown with the illumination and radar shadows oriented toward the upper margin, which is the reverse of our assumption about illumination. The circular features are actually uplifts (domes), as shown in the map and cross section of Figure 6-21B and 6-21C. Rotate the image (Figure 6-21A) 180° so that look direction and shadows are oriented toward the lower margin. The features should now appear in their correct perspective as

A. Look direction toward the south.

B. Look direction toward the west.

C. Interpretation map.

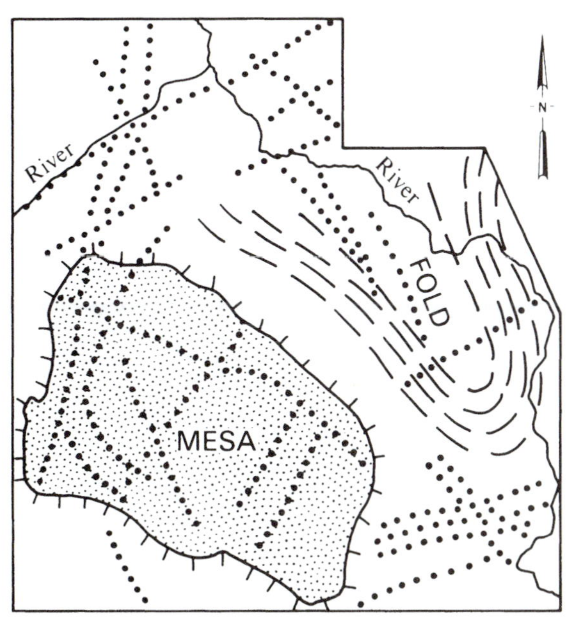

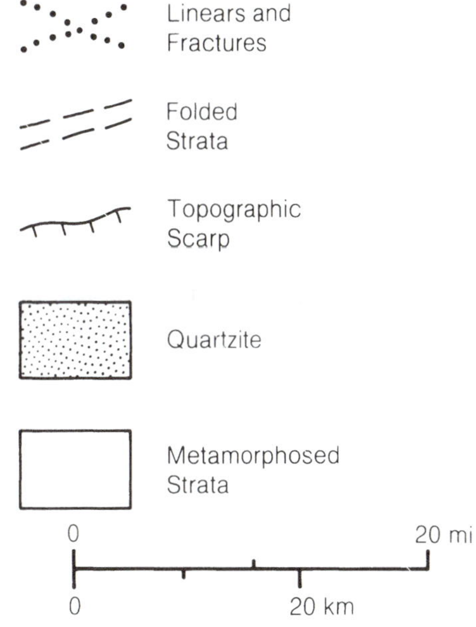

Figure 6-20 Expression of linear geologic features on images with different look directions. Aircraft X-band images in southeast Venezuela.

domes. This phenomenon is called *topographic inversion* and is common to all images with pronounced highlights and shadows. Interpreters should determine the illumination direction and orient the image with illumination and shadows toward the lower margin.

RADAR AND LANDSAT IMAGES COMPARED: INDONESIA

Figure 6-22 covers the Mapia anticline in the southern portion of West Papua, which is an Indonesian province in the western part of the island of New Guinea. Figure 6-22A is a TM subscene acquired with a 45° solar elevation, which results in minimal shadows. Figure 6-22B is a radar image acquired with a depression angle of 17°. Figure 6-22C is a geologic interpretation map of the area shown in Figure 6-22B. On the images and geologic map, north is toward the lower margin, in order to orient the radar look direction and shadows toward the lower margin. In the upper portion of the images resistant sandstones of the upper member of the Buru Formation dip south and form prominent dipslopes and antidip scarps. Nonresistant shale of the lower member forms a broad valley. The resistant New Guinea Limestone forms the crest of the Mapia anticline. In the northern (lower) portion of the image a thrust plate of New Guinea Limestone overrides the upper member of the Buru Formation and rests on the lower member. These geologic features are relatively obscure on the TM image, which has only minor shadows oriented westward, parallel with the geologic strike. The features are pronounced in the radar image, which has prominent shadows that are favorably oriented normal to the geologic strike.

The cross sections in Figure 6-23 show the relationship between vertical illumination angle and geologic structure for both images in Figure 6-22. The constant 45° solar elevation of the TM image produces minimal highlights and shadows (Figure 6-22A). The 17° radar depression angle produces strong highlights and shadows (Figure 6-22B). Beds dipping toward the antenna have broad highlights on the dipslopes and narrow shadows on the antidip scarps. These radar highlights and shadows are used to interpret dip and strike orientation in the same manner that solar highlights and shadows were used in the TM image of the Saharan Atlas Mountains in Chapter 3. The north look direction of the radar image is ideally oriented to enhance the west-trending Mapia anticline. An east or west look direction would have obscured the structural pattern.

This example demonstrates that in tropical regions radar is a valuable mapping system, not only because it penetrates cloud cover, but also because it can provide optimal illumination geometry.

A. Magellan radar image. The look direction is toward the upper margin at a depression angle of 51°. From Ford and others (1993, Figure 9-18).

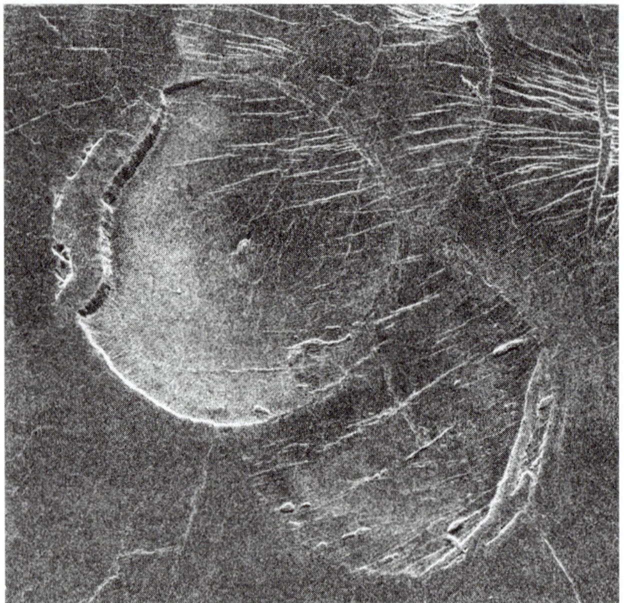

B. Interpretation map.

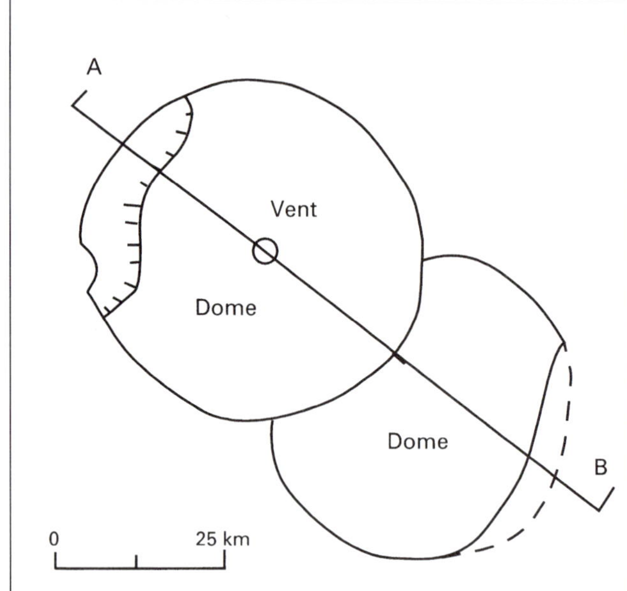

C. Cross section.

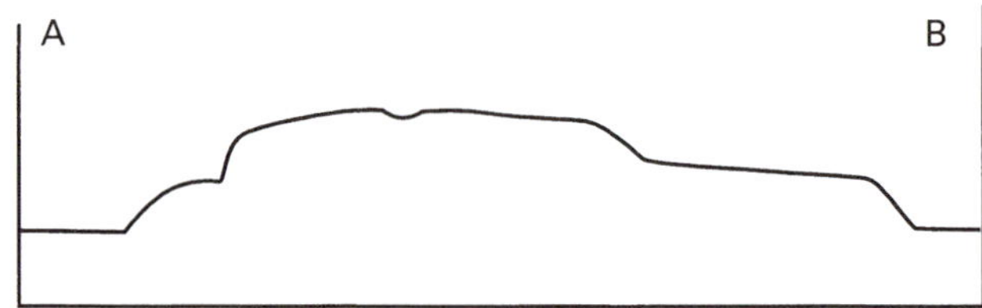

FIGURE 6-21 Topographic inversion caused by orientation of the radar look direction, Guinevere Planitia lowland, Venus.

A. Landsat TM band 4 image with a solar elevation angle of 45°.

B. X-band radar image with a depression angle of 17°.

C. Geologic map of Mapia anticline.

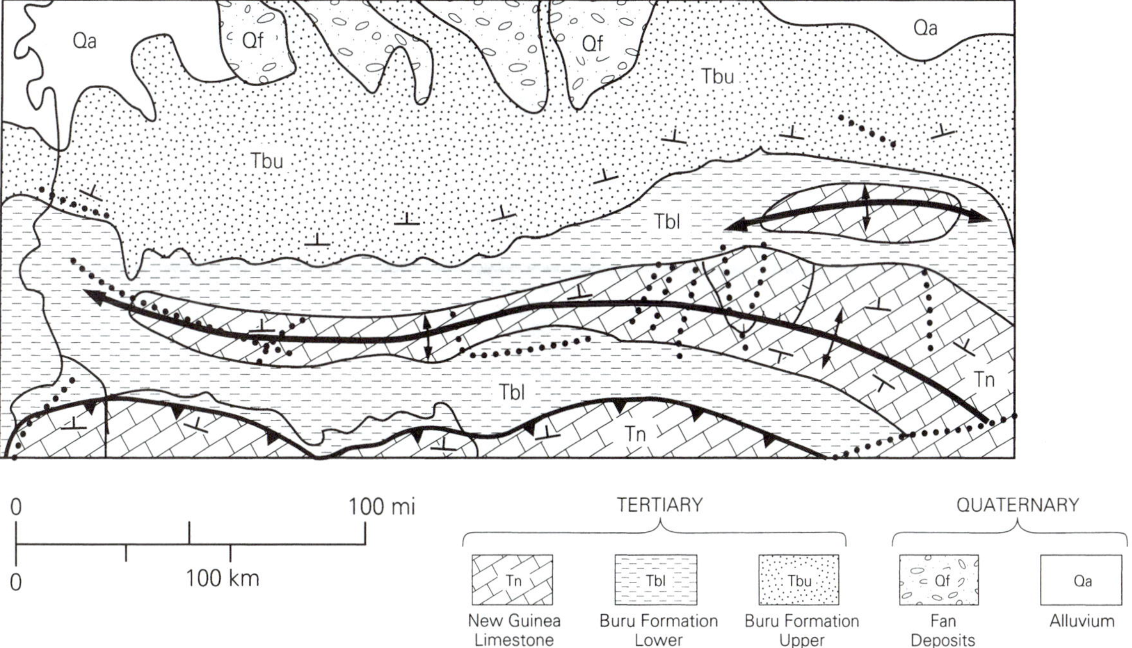

FIGURE 6-22 Comparison of terrain illumination by Landsat and a radar image, Mapia anticline, West Papua, Indonesia.

A. Illumination by the sun with a steep and constant 45° angle.

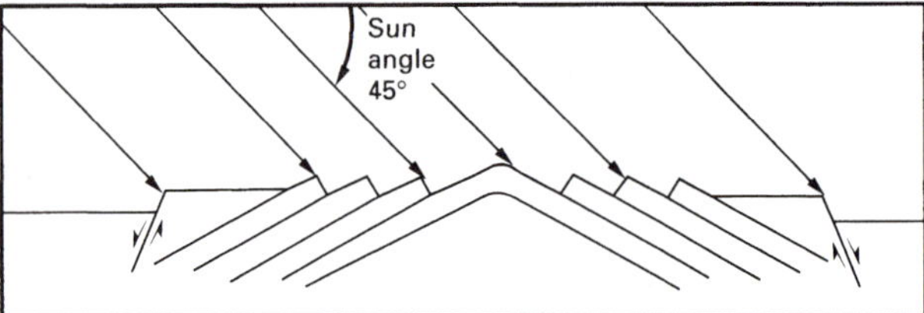

B. Illumination by radar with a low central depression angle of 17°.

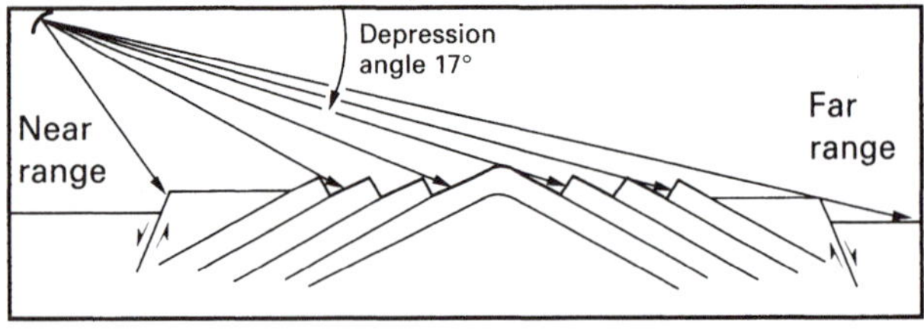

FIGURE 6-23 Generalized profiles comparing illumination geometry of Mapia anticline (Figure 6-22) for Landsat and radar images.

IMAGE FORESHORTENING

The curvature of a transmitted radar pulse causes the top of a tall feature to reflect energy in advance of its base, which results in displacement of the top toward the near range on the image. This distortion is called *foreshortening* or *image layover*. Figure 6-24C shows an image of Mount Saint Helens, acquired prior to the eruption, that is an extreme example of foreshortening. The profile (Figure 6-24A) shows that the wavefront encounters the mountaintop at B in advance of the base A. Therefore, the return from the top is received in advance of the return from the base, which foreshortens the mountain (Figure 6-24B). On the image (Figure 6-24C) the foreshortened mountain appears to lean toward the near-range direction. The extended backslope is obscured by shadows. Comparing the image to the topographic map (Figure 6-24D) illustrates the extent of foreshortening on the image. Foreshortening cannot be readily corrected because Equation 6-3 assumes a horizontal ground surface. Complex algorithms are required and have been developed to correct foreshortening.

The amount of foreshortening is influenced by the following factors:

1. **Height of targets**: Higher features are foreshortened more than lower features.
2. **Radar depression angle**: Features imaged with steep (large) depression angles are foreshortened more than those acquired with shallow (small) depression angles.
3. **Location of targets**: Features located in the near-range portion of the image swath are foreshortened more than comparable features located in the far range because the depression angle is steeper in the near range.

A. Profile of terrain and radar illumination.

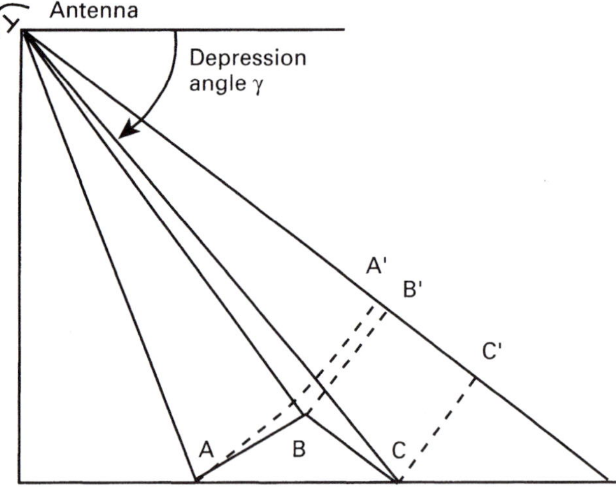

B. Profile of terrain with radar foreshortening.

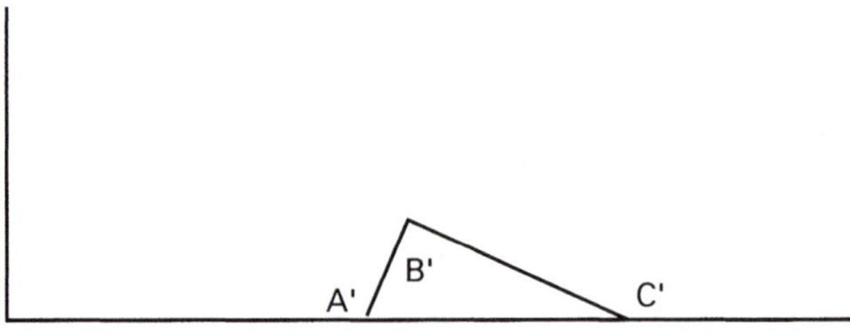

C. Radar image of Mount Saint Helens, Washington, showing foreshortening.

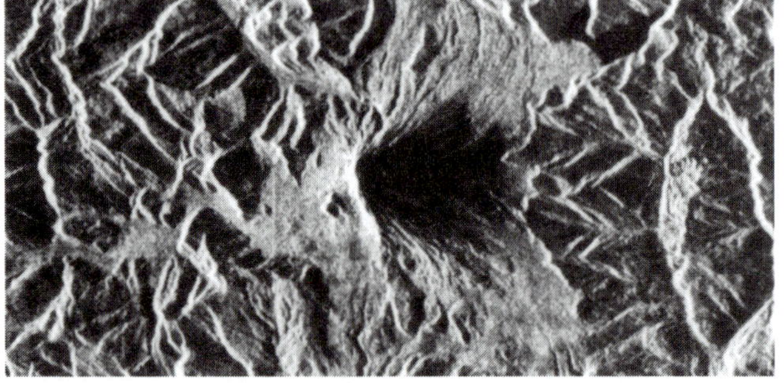

D. Topographic map showing symmetry of Mount Saint Helens.

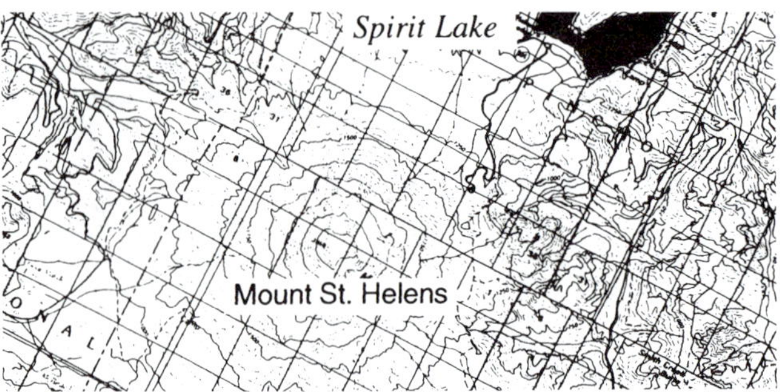

FIGURE 6-24 Foreshortening of topographic features on radar images.

Foreshortening can be minimized by acquiring images at shallow depression angles, but the resulting radar shadows may be excessive for terrain with high topographic relief.

RADAR RETURN AND IMAGE SIGNATURES

Stronger radar returns produce brighter signatures on an image than do weaker returns, as shown diagrammatically in Figure 6-3 and in the images of this chapter. The intensity of the radar return, for both aircraft and satellite systems, is determined by:

1. Radar system properties
   - Wavelength
   - Depression angle
   - Polarization
2. Terrain properties
   - Dielectric properties (including water content)
   - Surface roughness
   - Feature orientation

The following sections discuss these properties in more detail.

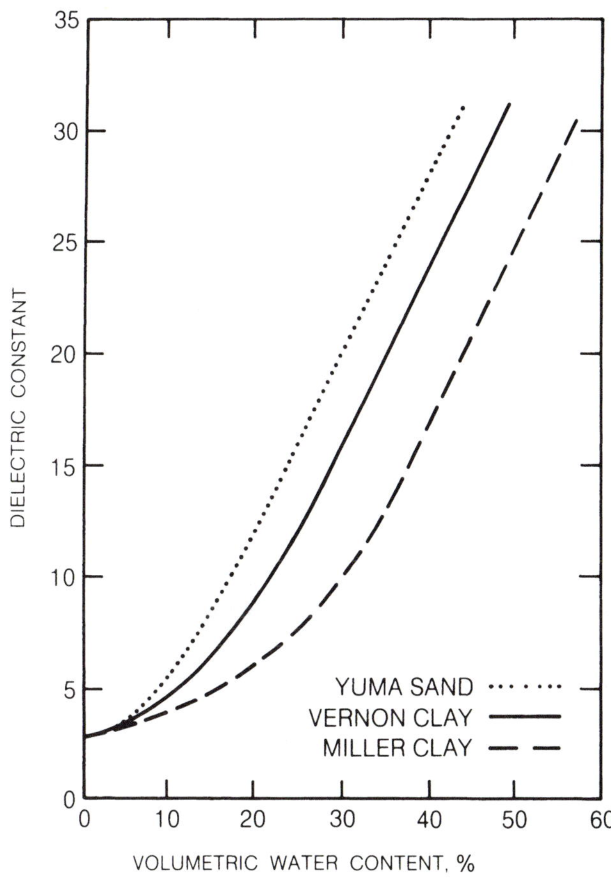

Figure 6-25 Variation of dielectric constant (at 27 cm wavelength) as a function of moisture content in sand and clay. Modified from Wang and Schmugge (1980, Figure 3).

Dielectric Properties and Water Content

The *dielectric constant* is an electrical property that influences the interaction between matter and electromagnetic energy, especially at radar wavelengths. At radar wavelengths the dielectric constant of dry rocks and soils ranges from about 3 to 8, while water has a value of 80. Therefore, as moisture content of a material increases, the dielectric constant can reach values of 25 or more, as shown for sand and clay in Figure 6-25.

Backscatter coefficient is a quantitative measure of the intensity of energy returned to the antenna. The chart in Figure 6-26 plots the relationship between backscatter coefficient and moisture content for fields of corn, bare soil, and milo, a grain resembling millet. As moisture increases, the backscatter coefficient also increases, which in turn produces brighter image signatures.

The experimental relationships between moisture content and image tone are illustrated by a radar image in Iowa (Figure 6-27A) where a rainstorm moved northeastward across an agricultural region 12 hours before the image was acquired. Figure 6-27B is a map of rain gauge measurements for the storm. The map shows that the bright signature of the western portion of the image correlates with areas that received up to 0.35 in (0.89 cm) of rain, whereas the dark eastern area was dry. The bright streaks in the otherwise dark eastern portion of the image are attributed to small rainstorms located northeast of the main weather front. Similar signatures of damp and dry agricultural areas occur in images acquired at other dates and localities (Ulaby and others, 1983).

At a test area in Oklahoma, soil moisture was measured at a number of sites at the same time a radar image was acquired (Blanchard and others, 1981). For each site the scattering coefficient was plotted as a function of soil moisture. The

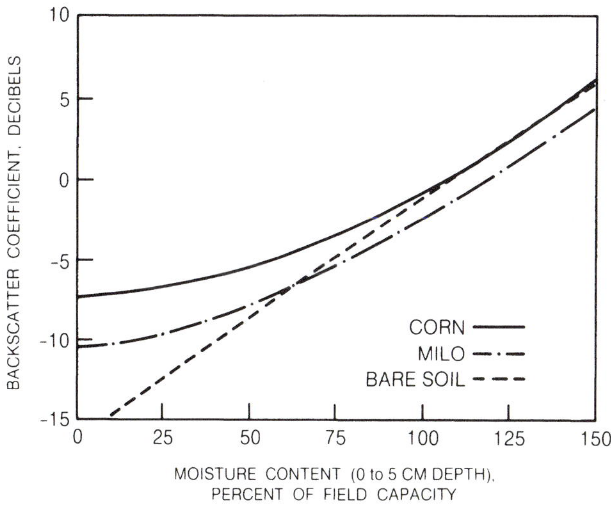

Figure 6-26 Graph showing variation of backscatter (at 6.7 cm wavelength) as a function of moisture content for individual fields in a test site at Guymon, Oklahoma. Milo is a grain resembling millet. From Blanchard and others (1981).

A. Radar image. Courtesy J. P. Ford, JPL (Retired).

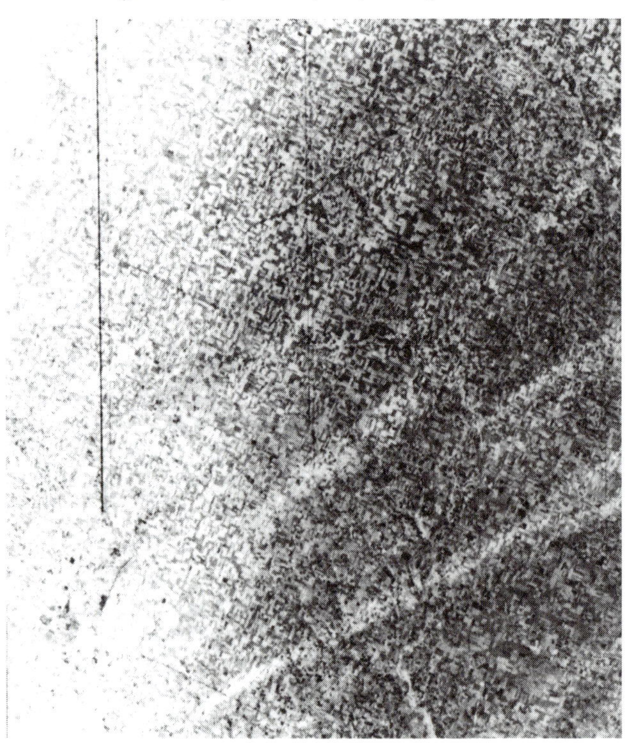

B. Map showing rain gauge data.

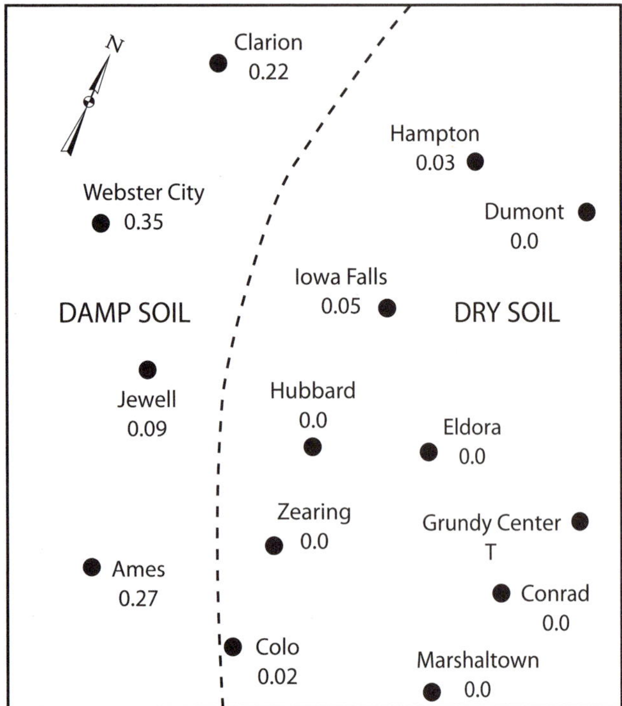

FIGURE 6-27 L-band image and map of rain gauge data for Ames, Iowa, and vicinity acquired 12 hours after a rainfall. From Ulaby and others (1983, Figure 2).

resulting graph (Figure 6-28) shows that radar backscatter, or image brightness, increases linearly with increasing soil moisture. The test sites plotted in Figure 6-28 include bare soil, alfalfa, and milo. Backscatter measurements were also made for cut and standing cornfields, which showed no correlation with soil moisture. Apparently both the cut and standing corn effectively masked the underlying soil from any interaction with the radar beam. Increasing soil moisture reduces the penetration of radar energy beneath the surface.

These image signatures and experimental data suggest that radar images may be used to estimate soil moisture, which would be valuable information for hydrology and agronomy. Radar backscatter, however, is also strongly influenced by other characteristics of the scene, such as surface roughness, which is discussed next. A number of researchers are working at sorting out the effects of these various characteristics of a scene.

The preceding comments refer to absorbed water. Standing water (fresh or salt) is highly reflective of radar energy and has dark or bright signatures depending upon whether the surface is calm or agitated.

SURFACE ROUGHNESS

Surface roughness is measured in centimeters and is determined by textural features comparable in size to the radar wavelength, such as leaves and twigs of vegetation and sand, gravel, and cobble particles. Surface roughness strongly influences the strength of radar returns and is distinct from *topographic relief*, which is measured in meters and hundreds of meters. Topographic relief features include hills, mountains, valleys, and canyons that are expressed on images by highlights and shadows. The average surface roughness within a ground resolution cell determines the intensity of the return for that cell. Surface roughness is a composite measure of the vertical and horizontal dimensions and spacing of the small-scale features, together with the geometry

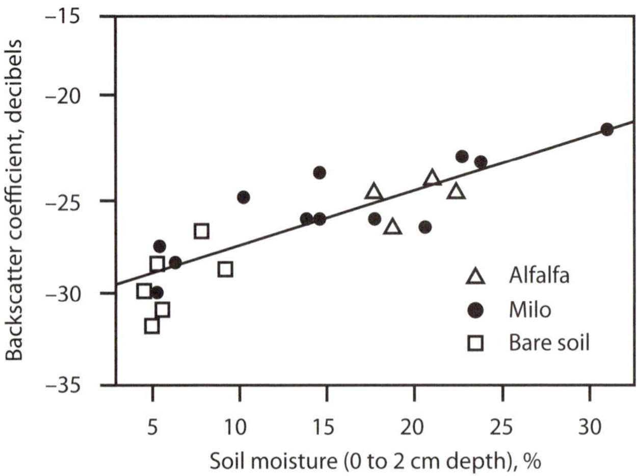

FIGURE 6-28 Graph showing increase in backscatter coefficient (L-band) with increasing soil moisture for individual fields in a test site at Guymon, Oklahoma. From Blanchard and others (1981).

of the individual features. Because of the complex geometry of most natural surfaces, it is difficult to characterize them mathematically, particularly for the large area of a radar ground resolution cell. For most surfaces the *vertical relief*, or average height of surface irregularities, is an adequate approximation of surface relief. Surfaces may be grouped into the following three roughness categories:

1. A *smooth surface* reflects all the incident radar energy with the angle of reflection equal and opposite to the angle of incidence (Snell's law).
2. A *rough surface* diffusely scatters the incident energy at all angles. The rays of scattered energy may be thought of as enclosed within a hemisphere, the center of which is located at the point where the incident wave encounters the surface.
3. A surface of *intermediate roughness* reflects a portion of the incident energy and diffusely scatters a portion.

The roughness of a surface return is determined by the relationship of surface relief, at the scale of centimeters, to radar wavelength and to the depression angle of the antenna.

Roughness Criteria

The Rayleigh criterion considers a surface to be smooth if

$$h < \frac{\lambda}{8 \sin \gamma} \tag{6-5}$$

where:

h = vertical relief

λ = radar wavelength

γ = depression angle

Both h and λ are given in the same units, usually centimeters. For a system with a wavelength of 23.5 cm and a depression angle of 70°, the surface relief below which the surface will appear smooth is determined by substituting into Equation 6-5:

$$\begin{aligned} h &< \frac{23.5 \text{ cm}}{8 \sin 70^\circ} \\ &< \frac{23.5 \text{ cm}}{8 \times 0.94} \\ &< 3.1 \text{ cm} \end{aligned}$$

Therefore, a vertical relief of 3.1 cm is the theoretical boundary between smooth and rough surfaces for the given wavelength and depression angle. Peake and Oliver (1971) modified the Rayleigh criterion by defining upper and lower values of h for surfaces of intermediate roughness. By their *smooth criterion*, a surface is smooth if

$$h < \frac{\lambda}{25 \sin \gamma} \tag{6-6}$$

Substituting for a system in which λ = 23.5 cm and γ = 70° results in

$$\begin{aligned} h &< \frac{23.5 \text{ cm}}{25 \sin 70^\circ} \\ &< \frac{23.5 \text{ cm}}{25 \times 0.94} \\ &< 1.0 \text{ cm} \end{aligned}$$

Therefore, a vertical relief of 1.0 cm is the boundary between smooth surfaces and surfaces of intermediate roughness for the given wavelength and depression angle.

Peake and Oliver (1971) also derived a *rough criterion* that considers a surface to be rough if

$$h > \frac{\lambda}{4.4 \sin \gamma} \tag{6-7}$$

Substituting for a system in which λ= 23.5 cm and γ = 70° results in

$$\begin{aligned} h &> \frac{23.5 \text{ cm}}{4.4 \sin 70^\circ} \\ &> \frac{23.5 \text{ cm}}{4.4 \times 0.94} \\ &> 5.7 \text{ cm} \end{aligned}$$

Therefore, a vertical relief of 5.7 cm is the boundary between intermediate surfaces and rough surfaces for the given radar wavelength and depression angle. Note that the value determined earlier from the Rayleigh criterion (h < 3.1 cm) is intermediate between those derived for the smooth criterion and the rough criterion.

The model cross sections in Figure 6-29 illustrate the interaction between a transmitted pulse of energy and surfaces that are smooth, intermediate, and rough relative to the radar wavelength. The smooth surface (Figure 6-29A) reflects all the energy, and no energy is returned to the antenna, which results in a dark signature. The surface with intermediate roughness (Figure 6-29B) reflects part of the energy and scatters the remainder; the backscattered component results in an intermediate signature. The rough surface (Figure 6-29C) diffusely scatters all the energy, causing a relatively strong backscattered component, which produces a bright signature.

Equations 6-5, 6-6, and 6-7 were used to calculate the values of vertical relief (h) that define smooth, intermediate, and rough surfaces for three radar systems listed in Table 6-7. This information is shown in an alternate fashion in Table 6-8, where the *roughness response* for different values of h is given for different radar wavelengths. For example, a surface with a vertical relief (h) of 1.40 cm, typical of fine gravel, appears rough on X-band images, intermediate on C-band images, and smooth on L-band images. The image signature of this surface is bright, medium gray, and dark at these wavelengths. Table 6-8 also illustrates the advantage of acquiring radar images at more than one wavelength for terrain analysis. By comparing the image signatures at two or more wavelengths,

A. Smooth surface, no return.

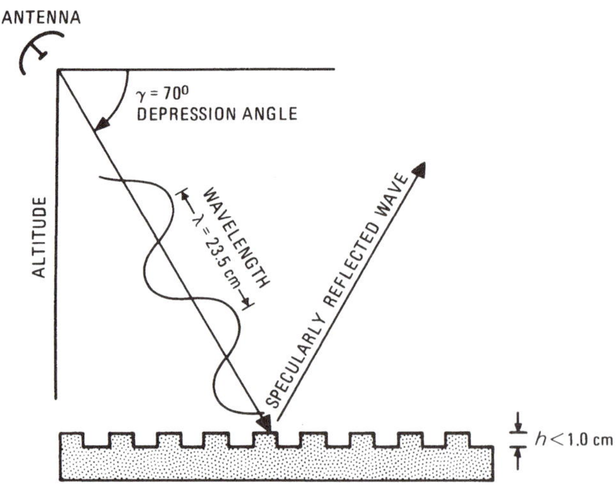

B. Intermediate relief, intermediate return.

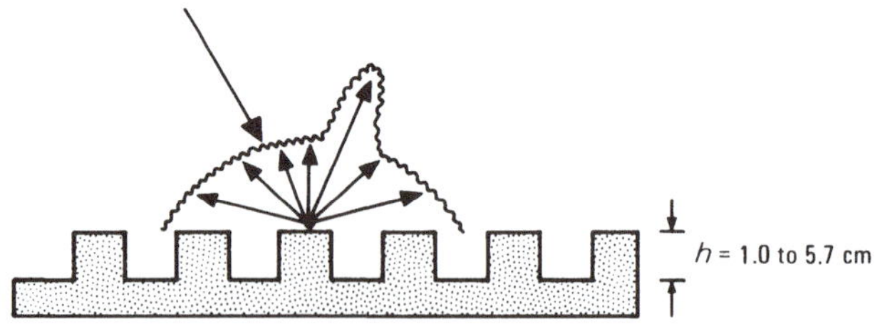

C. Rough surface, strong return.

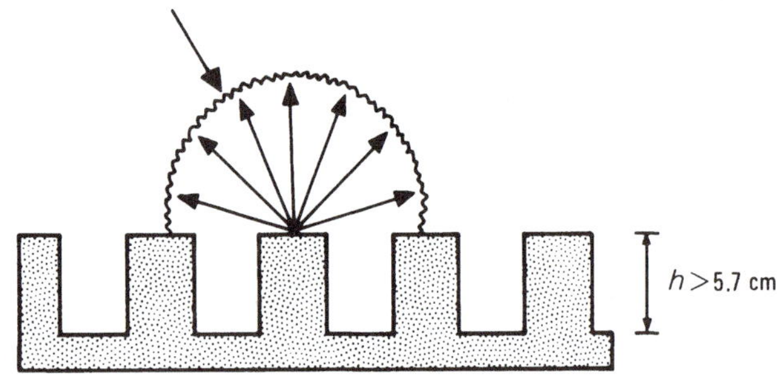

FIGURE 6-29 Models of surface roughness criteria and return intensity for radar images at 23.5 cm wavelength.

one can estimate the surface roughness more accurately than by looking at the image signature of a single wavelength. Images acquired at different wavelengths of the Copper Canyon alluvial fan are compared later in this chapter.

Depression Angle and Surface Roughness

As discussed earlier, the depression angle (γ) affects the smooth and rough criteria. Figure 6-30A shows that at low to intermediate depression angles the specular reflection from a smooth surface returns little or no energy to the antenna. At very high depression angles (80 to 90°), however, the specularly reflected wave may be received by the antenna and produce a strong return. A rough surface causes diffuse scattering of relatively uniform strong intensity for a wide range of depression angles (Figure 6-30B), which results in strong returns at all depression angles. Figure 6-30C compares the relative return intensity for smooth and rough surfaces at different depression angles. The relatively uniform return from the rough surface decreases somewhat at low depression angles because of the greater two-way travel distance. Smooth surfaces produce strong returns at depression angles near vertical, but little or no return at lower angles.

SURFACE ROUGHNESS ON LANDSAT AND RADAR IMAGES

The relationship between radar backscatter and surface roughness was initially based on theoretical analyses supplemented by laboratory studies of artificial surfaces. Subsequently, field

TABLE 6-8 Roughness response for different values of vertical relief (*h*) at different radar wavelengths. Depression angle (γ) is 40° for all systems.

Vertical Relief (*h*) (cm)	X-band Image (λ = 3 cm)	C-band Image (λ = 6 cm)	L-band Image (λ = 23.5 cm)
0.10	Smooth	Smooth	Smooth
0.40	Intermediate	Smooth	Smooth
1.40	Rough	Intermediate	Smooth
5.00	Rough	Rough	Intermediate
10.00	Rough	Rough	Rough

TABLE 6-7 Surface roughness categories for representative radar systems. Depression angle (γ) is 40° for all systems.

Roughness Category	X-band (λ = 3 cm)	C-band (λ = 6 cm)	L-band (λ= 23.5 cm)
Smooth	$h < 0.19$ cm	$h < 0.37$ cm	$h < 1.46$ cm
Intermediate	$h = 0.19$ to 1.06 cm	$h = 0.37$ to 2.12 cm	$h = 1.46$ to 8.35 cm
Rough	$h > 1.06$ cm	$h > 2.12$ cm	$h > 8.35$ cm

studies in Death Valley established the relationship between radar roughness criteria and natural surfaces with different degrees of roughness. Key papers are by Schaber, Berlin, and Brown (1976) and Schaber, Berlin, and Pitrone (1975). The following discussion of Cottonball Basin compares roughness signatures on Landsat and L-band radar images. The following discussion of Copper Canyon compares roughness signatures on X-band and L-band radar images.

A. Smooth surface with specular reflection.

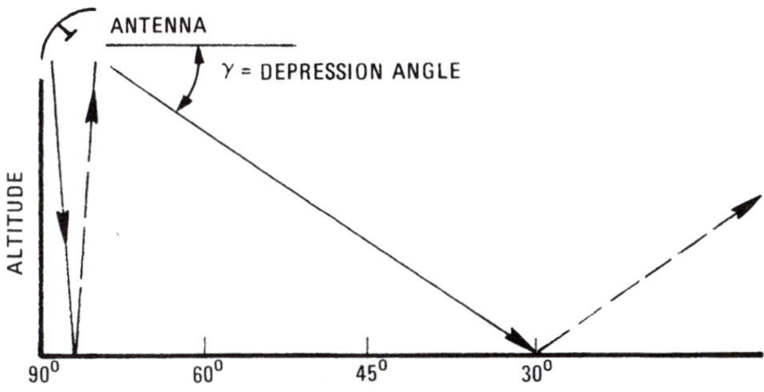

B. Rough surface with diffuse scattering.

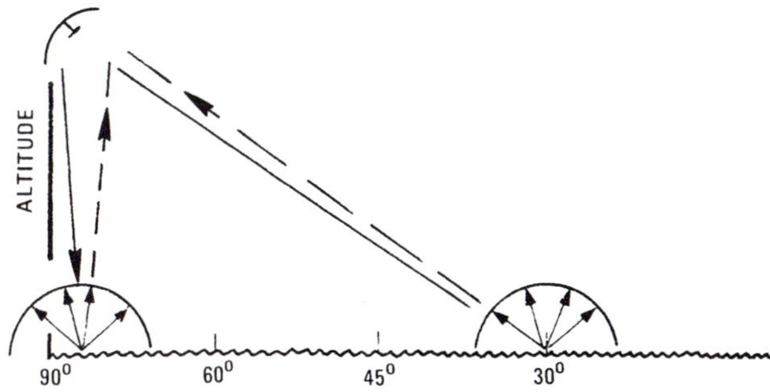

C. Return intensity as a function of depression angle.

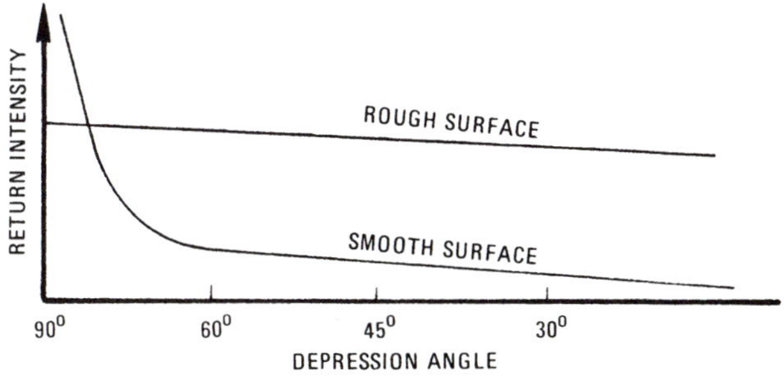

FIGURE 6-30 Radar return from smooth and rough surfaces as a function of depression angle.

COTTONBALL BASIN, DEATH VALLEY, CALIFORNIA

Plate 18 shows the location of Cottonball Basin, at the north end of Death Valley. The basin is a desiccated salt lake with a variety of saline deposits on the floor of the old lake. Figure 6-31A shows a Landsat TM band 3 (red) image and Figure 6-31B shows a Seasat L-band radar image of Cottonball Basin. Figure 6-31C is a map showing that the deposits have a wide range of surface roughness values. The gravel surfaces of the alluvial fans bordering the basin provide additional degrees of surface roughness. The diversity of surface materials makes this an excellent site for correlating radar signatures with materials of different degrees of surface roughness. Table 6-9 lists the roughness categories calculated for the image. Table 6-9 also lists the materials with surface relief that correspond to roughness criteria. Figure 6-32 shows field photographs of the materials, which are described in the following sections.

Coarse Gravel (h = 12.0 cm)

Cobbles and boulders eroded from the mountains surrounding Death Valley form alluvial fans on the east and west margins of Cottonball Basin (Figure 6-32A). Tucki Wash Fan in the southwest part of the map (Figure 6-31) is a good example. The gravel is deposited by intermittent streams and slope wash during the infrequent rainstorms in the mountains. The gravel consists of a wide range of rock types that are determined by the lithology of their bedrock source. In the radar image (Figure 6-31B) the rough gravel has a light to medium gray signature that contrasts with the dark signatures of desert pavement and sand that also occur on the fan.

Rough Halite (h = 29.0 cm)

Rough halite (Figure 6-32B) is the roughest of all materials in the Cottonball Basin. At the end of the glacial period the salt lake that filled Death Valley evaporated, depositing a laycr of halitc (rock salt). The salt crystals have partially dissolved and recrystallized, causing stresses that break the salt into jumbled slabs 1 m or more in diameter. The slabs

TABLE 6-9 Signatures of materials on a Seasat L-band image (γ = 70°) of Cottonball Basin, Death Valley, California.

Image Signature	Roughness Category	Roughness Criteria (cm)	Materials
Dark	Smooth	$h < 1.0$	Desert pavement, sand and fine gravel, floodplain deposits
Intermediate	Intermediate	$h = 1.0$ to 5.7	Intermediate halite, carbonate, and sulfate deposits
Bright	Rough	$h > 5.7$	Coarse gravel, rough halite

A. Landsat TM panchromatic image. Landsat courtesy USGS.

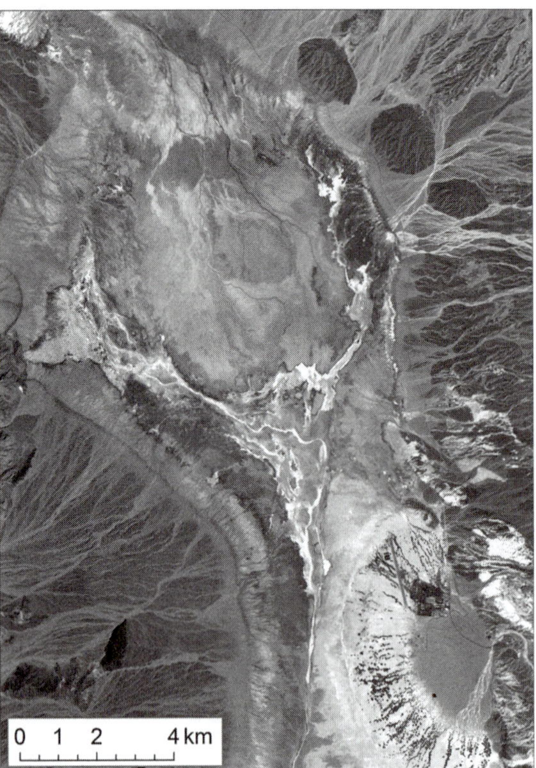

B. Seasat L-band image with look direction toward the east. Courtesy NASA.

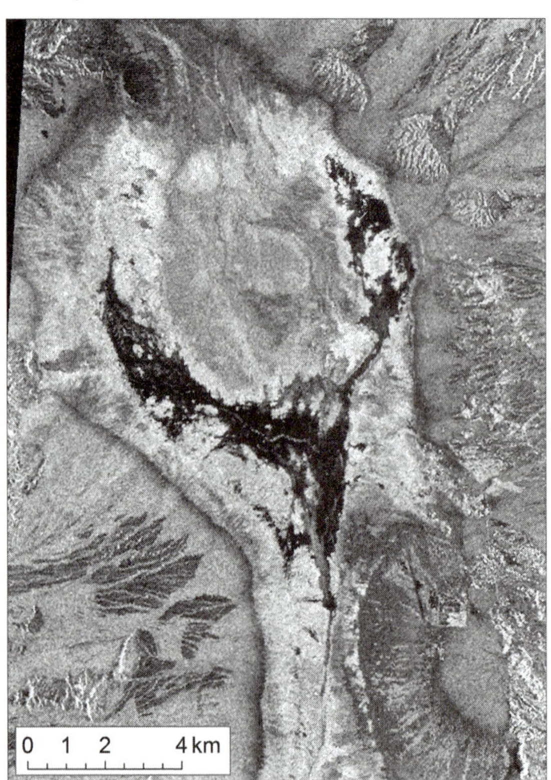

C. Map showing surface materials at Cottonball Basin. Based on geologic map by Hunt and Mabey (1966, Plate 1).

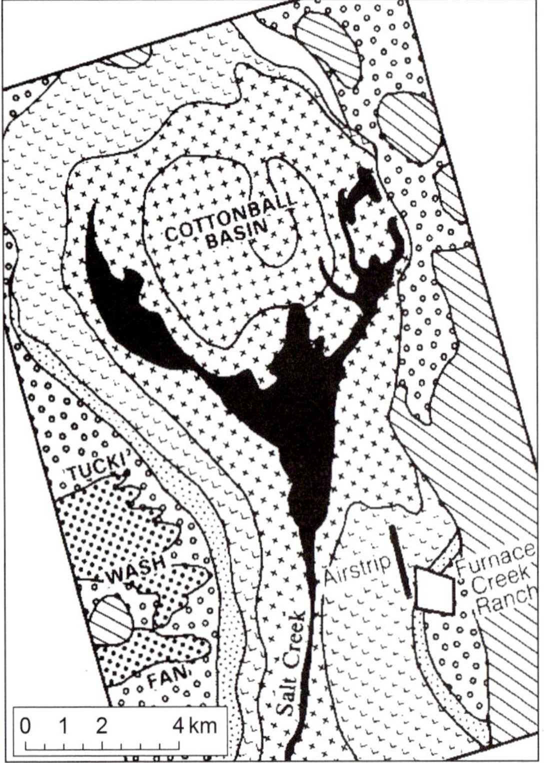

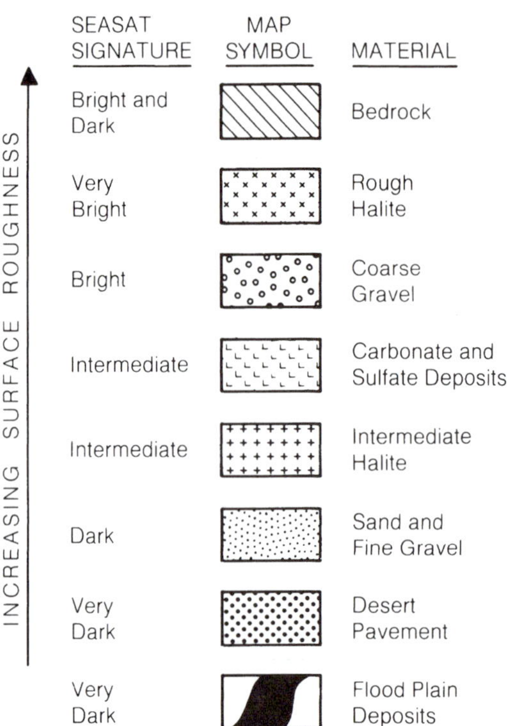

SEASAT SIGNATURE
MAP SYMBOL
MATERIAL
INCREASING SURFACE ROUGHNESS
Bright and Dark
Bedrock
Very Bright
Rough Halite
Bright
Coarse Gravel
Intermediate
Carbonate and Sulfate Deposits
Intermediate
Intermediate Halite
Dark
Sand and Fine Gravel
Very Dark
Desert Pavement
Very Dark
Flood Plain Deposits

FIGURE 6-31 Landsat and radar images of Cottonball Basin, Death Valley, California. Location of the basin at the north end of Death Valley is shown in Plate 18.

A. Coarse gravel, $h = 12.0$ cm.

B. Rough halite, $h = 29.0$ cm.

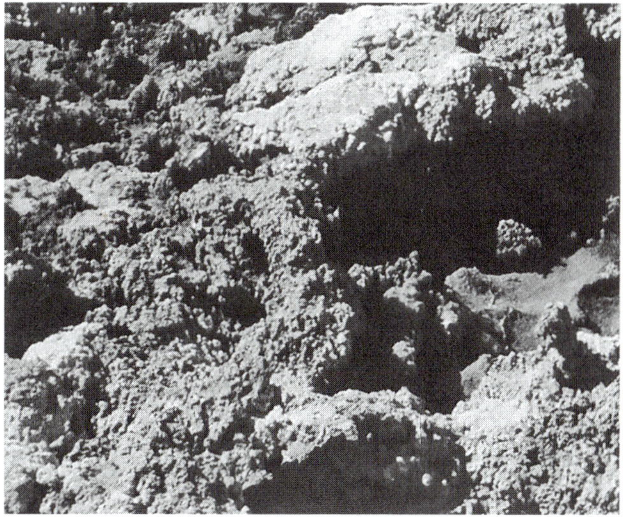

C. Intermediate halite, $h = 6.0$ cm.

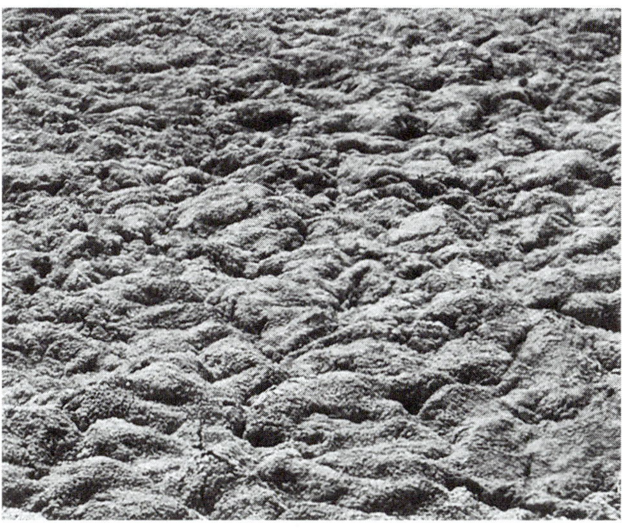

D. Carbonate and sulfate deposits, $h = 6.0$ cm.

E. Desert pavement, $h = 1.0$ cm.

F. Floodplain deposits, $h = 0.2$ cm.

Figure 6-32 Oblique ground photographs of materials at Cottonball Basin, with values for typical vertical relief. Distribution of these materials is shown in Figure 6-31. Areas are approximately 50 cm wide.

are covered with sharp pinnacles formed by solution during the infrequent rains. This material has the brightest signature in the radar image. In the Landsat image (Figure 6-31A) the signature of rough halite ranges from bright to dark, depending on the amount of dark silt included in the salt.

Intermediate Halite (h = 6.0 cm)

Halite deposits in the lower elevations of the basin are periodically flooded, a process that reduces surface relief to a hummocky surface of intermediate roughness (Figure 6-32C). This surface has an intermediate gray signature in the radar image and a similar gray signature in the Landsat image.

Carbonate and Sulfate Deposits (h = 6.0 cm)

Carbonate and sulfate deposits (Figure 6-32D) form a belt around the margins of the basin, where they are mixed with sand. Periodic wetting and drying produces a puffy surface with intermediate relief that has a gray signature in the radar image. In the Landsat image the carbonate and sulfate deposits have a light gray tone.

Desert Pavement (h = 1.0 cm)

Desert pavement (Figure 6-32E) forms on alluvial fans where older gravel deposits have been subjected to prolonged weathering, which disintegrates the surface layer of cobbles and boulders into slabs and chips. These fragments form a smooth surface that resembles a tile mosaic, or pavement. Areas of desert pavement are common at the Tucki Wash Fan and other fans on the west side of Death Valley. Much of the desert pavement is coated with desert varnish, which causes a dark signature in the Landsat image that contrasts with the slightly brighter signature of the younger deposits of coarse gravel (Figure 6-31A). In the radar image (Figure 6-31B) the smooth desert pavement has a distinctive dark signature.

Sand and Fine Gravel (h = 1.0 cm)

A narrow belt of sand and fine gravel (not shown in Figure 6-32) occurs at the margin of the Tucki Wash Fan and forms a narrow dark band in the radar image. The signature is consistent with the relatively fine grain size and low relief of this material. The belt of sand and fine gravel has a distinctive gray signature in the TM image.

Floodplain Deposits (h = 0.2 cm)

Rare floods from the ephemeral streams in Cottonball Basin have formed floodplains of silt (Figure 6-32F) that are saturated with brine and are extensively coated with thin salt crusts. These floodplains are the smoothest surfaces in the basin and have distinctive dark signatures in the radar image. In the Landsat image the tone ranges from bright to medium gray depending on the thickness of the white salt crust and the degree of water saturation.

Bedrock

Bedrock of the Black Mountains is exposed in the eastern portion of the images and the geologic map (Figure 6-31C). Faulting and weathering have resulted in rugged terrain where relief is measured in tens of meters. Signatures on both Landsat and radar images are dominated by topographic highlights and shadows, rather than by degrees of surface roughness.

Conclusions

Images and field measurements at Cottonball Basin demonstrate that calculated roughness criteria correlate with the surface relief of materials in the field and with signatures on a radar image. This close correlation between radar signatures and surface relief points out another important fact—namely, that radar signatures alone cannot be used to identify the composition of materials. For example, coarse gravel and rough halite have completely different compositions and origins, but both have similar bright radar signatures. Floodplain deposits, sand, desert pavement, and standing water are very different materials, but all have dark radar signatures because of their smooth surfaces. These materials can be distinguished in radar images because of their distribution and associations, but not solely on the basis of their radar signature.

The correlations between roughness and radar signature demonstrated at Cottonball Basin have been confirmed on radar images covering all of Death Valley and other regions.

Copper Canyon X-Band and L-Band Images Compared

Plate 18 shows the location of the Copper Canyon alluvial fan at the base of the Black Mountains on the east side and in the south end of Death Valley. The fan is an instructive site for comparing the signatures of two radar images acquired at different wavelengths. The comparison is somewhat subjective because the two images were acquired and processed separately and were not calibrated to a known standard. Figure 6-33A is an X-band image (3.0 cm wavelength) with the look direction toward the west, which causes strong shadows from the west-facing scarp of the Black Mountains. Figure 6-33B is an L-band radar image (23.5 cm wavelength) with the look direction toward the east, which causes strong highlights and shadows from the mountains. Figure 6-33C is a map that shows the distribution of materials at the Copper Canyon Fan with a legend that compares the radar signatures on the X-band and L-band images with the material on the ground. Our onsite inspections of the fan confirm that the materials are similar to those at Cottonball Basin. The X-band image (Figure 6-33A) has a smooth criterion of 0.3 cm and a rough criterion of 1.6 cm. The Seasat L-band image (Figure 6-33B) has a smooth criterion of 1.0 cm and a rough criterion of 5.7 cm. The X-band image (5 m pixels) has finer spatial detail than the L-band image (20 by 25 m pixels).

A. X-band image (3.0 cm wavelength).

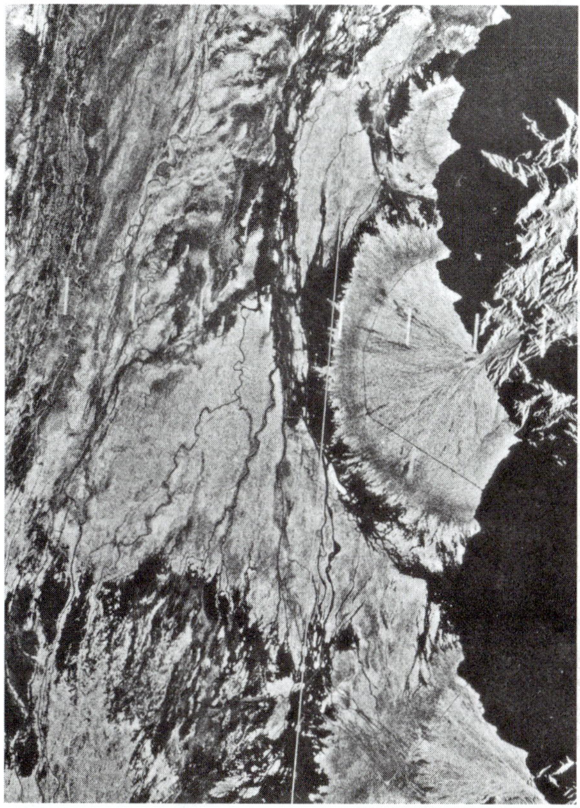

B. Seasat L-band image (23.5 cm wavelength). Courtesy NASA.

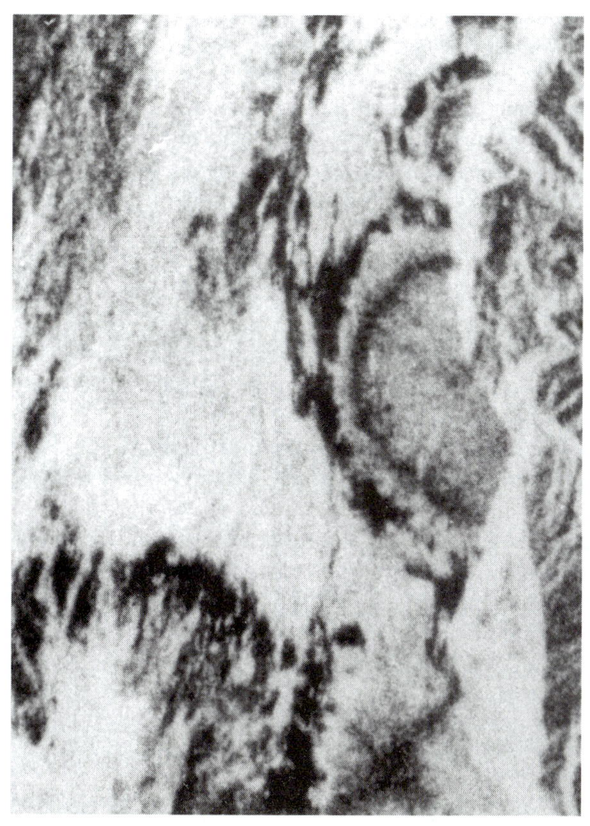

C. Geologic interpretation map.

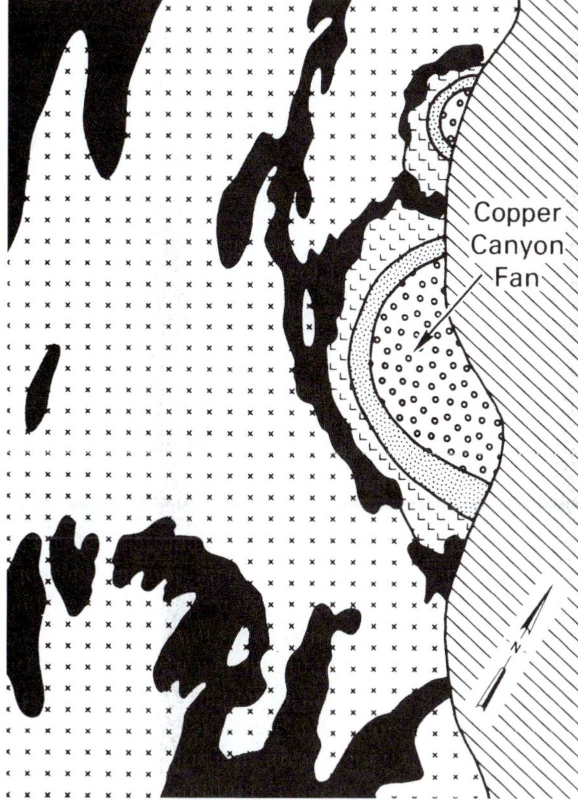

MATERIAL	SYMBOL	RADAR SIGNATURES X-BAND	L-BAND
Bedrock and Shadows		Bright and Dark	Bright and Dark
Rough Halite		Very Bright	Very Bright
Coarse Gravel		Bright	Intermediate
Carbonate and Sulfate		Bright	Intermediate
Sand and Fine Gravel		Intermediate	Dark
Flood Plain Deposits		Dark	Dark

0 4 mi

0 4 km

Figure 6-33 X-band and L-band radar images of the Copper Canyon alluvial fan and vicinity, Death Valley. Location of the basin is shown in Plate 18.

Rough halite exceeds the rough criterion of both images and has bright signatures in both images. Floodplain deposits have less relief than either of the smooth criteria and are dark in both images. The gravel of the Copper Canyon Fan is finer grained than that at the Tucki Wash Fan; its relief of 5 to 6 cm is rough for the X-band image (bright signature) but intermediate for the L-band image (gray signature). The belt of sand and fine gravel near the margin of the fan (Figure 6-33C) has a vertical relief of 1 cm, which is intermediate for the X-band image (gray signature) and smooth for the L-band image (dark signature). A belt of carbonate and sulfate deposits, with a vertical relief of 6 cm, occurs between the sand and the floodplain deposits. These deposits are bright in the X-band image and intermediate in the L-band image.

This example demonstrates the relationship of radar wavelength to roughness of materials and to their radar signatures. Images acquired at different wavelengths provide better definitions of surface relief. Consider the gravel at the Copper Canyon Fan, which has a relief greater than 1.6 cm in the X-band image and 1 to 5.7 cm in the L-band image. The combination of images provides an estimated roughness range of 1.6 to 5.7 cm for this gravel.

RADAR SIGNATURES IN SAND-COVERED DESERT TERRAINS

Over the years, many aircraft and Seasat images have been acquired of sand-covered desert areas, largely in North America. Sheets of sand and fine gravel have dark signatures because of their smooth surfaces. Sand dunes have highlights and shadows caused by their topography. The radar signatures are clearly determined by the surface roughness and landforms. Because of this experience, investigators were startled by SIR-A images of the Selima sand sheet of the eastern Sahara in Egypt and the Sudan, for reasons described below.

SELIMA SAND SHEET

Figure 6-34A is a Landsat MSS image of a portion of the Selima sand sheet in the Sudan that shows a featureless surface of windblown sand. Based on this signature in a visible spectral band, one would predict the sand sheet to have a uniform dark signature on SIR-A images. Instead, the SIR-A image (Figure 6-34B) shows details of ancient drainage patterns eroded into the bedrock that underlies the sand sheet. The SIR-A radar energy has clearly penetrated the sand

FIGURE 6-34 SIR-A and Landsat images of the Selima sand sheet, northwest Sudan. From McCauley and others (1982). Courtesy G. G. Schaber, USGS.

A. Landsat MSS band 2 image showing surface sheet of sand.

B. SIR-A image that penetrates the sand and shows old drainage patterns on bedrock that underlie the sand.

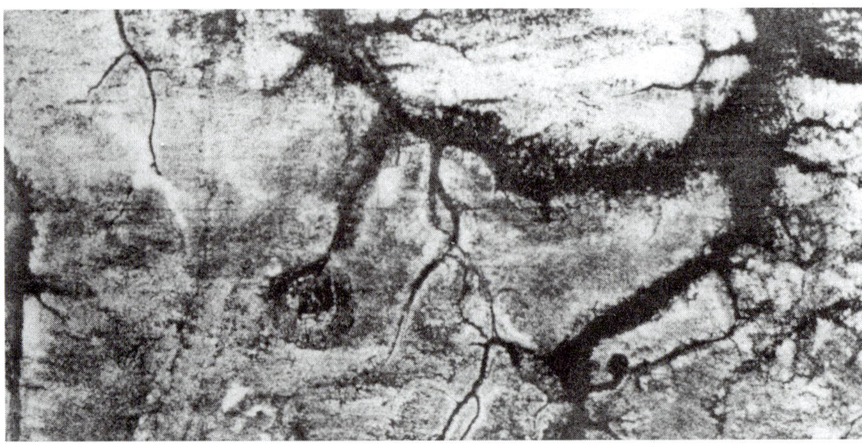

sheet and returned from the underlying bedrock terrain. McCauley and others (1982) investigated the images of the Selima sand sheet in the field. The sand sheet is so devoid of features that satellite navigation systems were used in field vehicles to locate specific sites on the SIR-A images. A number of pits were excavated in the sand and established the following:

1. The bright areas in the SIR-A images are the rough, eroded surface of the flat-lying Nubian Sandstone that is bedrock in the region.
2. The dark signatures are from smooth surfaces of alluvium that fills extinct drainage channels.

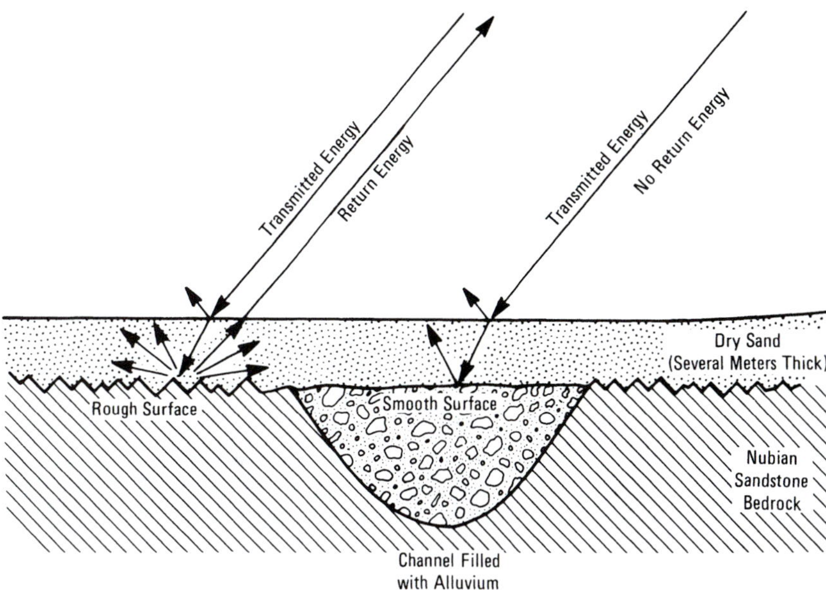

FIGURE 6-35 Cross section showing SIR-A energy that penetrates the thin Selima sand sheet of hyperarid sand in northwest Sudan. The underlying rough bedrock strongly backscatters the energy, which produces bright signatures on the image. The smooth surface of the channel fill deposits reflects the energy away from the antenna, which produces a dark signature. From information provided by G. G. Schaber, USGS.

These relationships are shown in the geologic interpretation map (Figure 6-34C). The diagrammatic cross section (Figure 6-35) shows that some of the transmitted radar energy is reflected by the smooth sand surface. However, much of the energy penetrates the sand and is refracted to a steeper depression angle as it enters this denser medium. The energy that encounters the rough bedrock surface is strongly backscattered, and a relatively large proportion returns to the antenna to produce a bright signature. The transmitted energy that encounters the smooth surface of a channel is specularly reflected (Figure 6-35), and no energy is returned to the antenna, which produces the dark signature of the channels.

The radar penetration of the Selima sand sheet is due to the hyperarid environment, where rainfall occurs at intervals of 30 to 50 years and the sun evaporates any absorbed moisture. Field excavations in the eastern Sahara (Schaber and others, 1986) showed that 1.5 m was the maximum sand thickness through which images of the substrate could be recorded. Most of the modern sand sheet in the eastern Sahara Desert is "transparent" to L-band radar.

OTHER SAND-COVERED DESERTS

Radar images of other desert regions have been examined for evidence of sand penetration, but only a few additional examples have been reported. Blom and others (1984) detected a sand-covered igneous dike on a Seasat image of an arid valley in the Mojave Desert, California. Berlin and others (1986) interpreted SIR-A and SIR-B images of the Nafud sand sea in northern Saudi Arabia. In this hyperarid area they reported that bedrock was imaged through sand up to 1.24 m thick.

C. Geologic interpretation map.

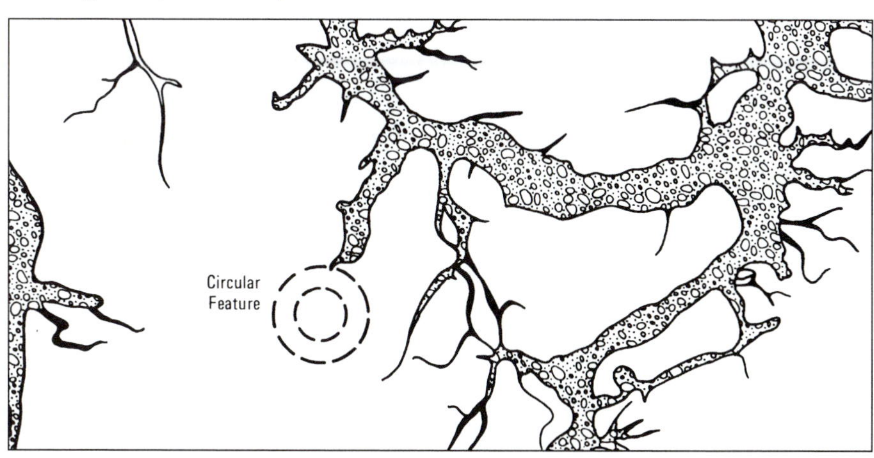

The general lack of radar penetration of sand in other deserts is due to:

1. Excessive thickness of sand and
2. The presence of subsurface moisture.

Most deserts have rainfall every few years. The water soaks into the sand and later evaporates from a thin surface layer that insulates and retains moisture in the deeper sand. In many deserts the sand has appreciable moisture at depths of less than 1 m.

RADAR INTERACTION WITH TROPICAL TERRAIN, INDONESIA

The SIR-A mission acquired images of portions of Indonesia. Persistent cloud cover in this tropical region has hampered acquisition of aerial photographs and Landsat images, but the SIR-A images are of excellent quality. Sabins (1983) interpreted the SIR-A images in five steps:

1. Produced base maps by tracing shorelines, drainage patterns, and the sparse cultural features (roads, cities, and airports) from the images. The contrasting signatures of water and vegetation made this a straightforward task.
2. Defined terrain categories that are recognizable throughout the region and mapped their distribution.
3. Mapped geologic structures including dips and strikes, faults, folds, and linears.
4. Constructed cross sections.
5. Evaluated the completed maps by comparing them with existing geologic maps.

The remainder of this section describes steps 2 and 3.

TERRAIN CATEGORIES

Much of Indonesia is densely forested, but the forest canopy conforms to the underlying topography because the trees have a relatively uniform height. Topography is controlled by erosional characteristics of the surface rocks and by the underlying geologic structure. The six terrain categories shown in Figure 6-36 were recognized on the basis of their expression on the SIR-A images. These categories are not restricted to images of Indonesia but are recognizable on radar images of forested regions throughout the world.

Carbonate Terrain

In humid environments, solution and collapse of carbonate rocks produce karst topography, which is readily recognized by the distinctive pitted surface (Figure 6-36A). In Indonesia, carbonate terrain generally occurs as uplands surrounded by lowlands eroded from less resistant rocks. Because of the relatively coarse spatial resolution of SIR-A images, small patches of karst terrain may not be recognizable. Also, faulting and stream erosion may obscure the expression of karst topography.

Clastic Terrain

Terrain formed on clastic sedimentary rocks, primarily sandstone and shale, is recognized by the stratification that forms asymmetric ridges, called cuestas and hogbacks, where the rocks are dipping, as seen in Figure 6-36B. Flat-lying clastic rocks form mesas, terraces, and associated erosional scarps. The lack of karst topography generally distinguishes clastic terrain from carbonate terrain in humid regions.

Volcanic Terrain

Young volcanic rocks form irregular flows associated with cinder cones or eroded volcanic necks (Figure 6-36C). Because of erosion and deformation, older volcanic terrains lack these distinctive features and therefore cannot be recognized on the images without additional information.

Alluvial and Coastal Terrain

Alluvial and coastal terrain is characterized by low relief, a uniform bright signature of heavily vegetated floodplains, and numerous estuaries and meandering streams (Figure 6-36D). Despite the generally featureless nature of this terrain, careful interpretation of the images often reveals subtle linear features and drainage anomalies that may be expressions of geologic structure.

Mélange Terrain

Mélange refers to rocks formed in subduction zones as a mixture of clastic sediments and oceanic crustal and mantle rocks. Lenticular rock fragments with a wide range of sizes, up to kilometers in length, are enclosed in a matrix of clay. Erosion of these rocks produces an irregular, rounded terrain with unsystematic drainage patterns (Figure 6-36E).

Metamorphic Terrain

Older sedimentary rocks that have been metamorphosed into slate, quartzite, and schist occur in portions of West Papua (Figure 6-36F). The original stratification is no longer recognizable and the rocks are resistant to erosion. The dissected metamorphic terrain has high relief and angular ridges that distinguish it from the lower relief and rounded appearance of mélange terrain. Foliation trends are not discernible in the scale of SIR-A images.

Undifferentiated Bedrock

There are areas where several terrain types are juxtaposed in such a complex manner that individual categories are not recognizable at the scale and resolution of SIR-A images. The category "undifferentiated bedrock" is used for these areas.

STRUCTURAL INTERPRETATION

Figure 6-37 shows six representative structural features at a uniform scale. Figure 6-38 shows the geologic interpretation for each feature. Appendix A, Basic Geology for Remote Sensing, explains the structural features.

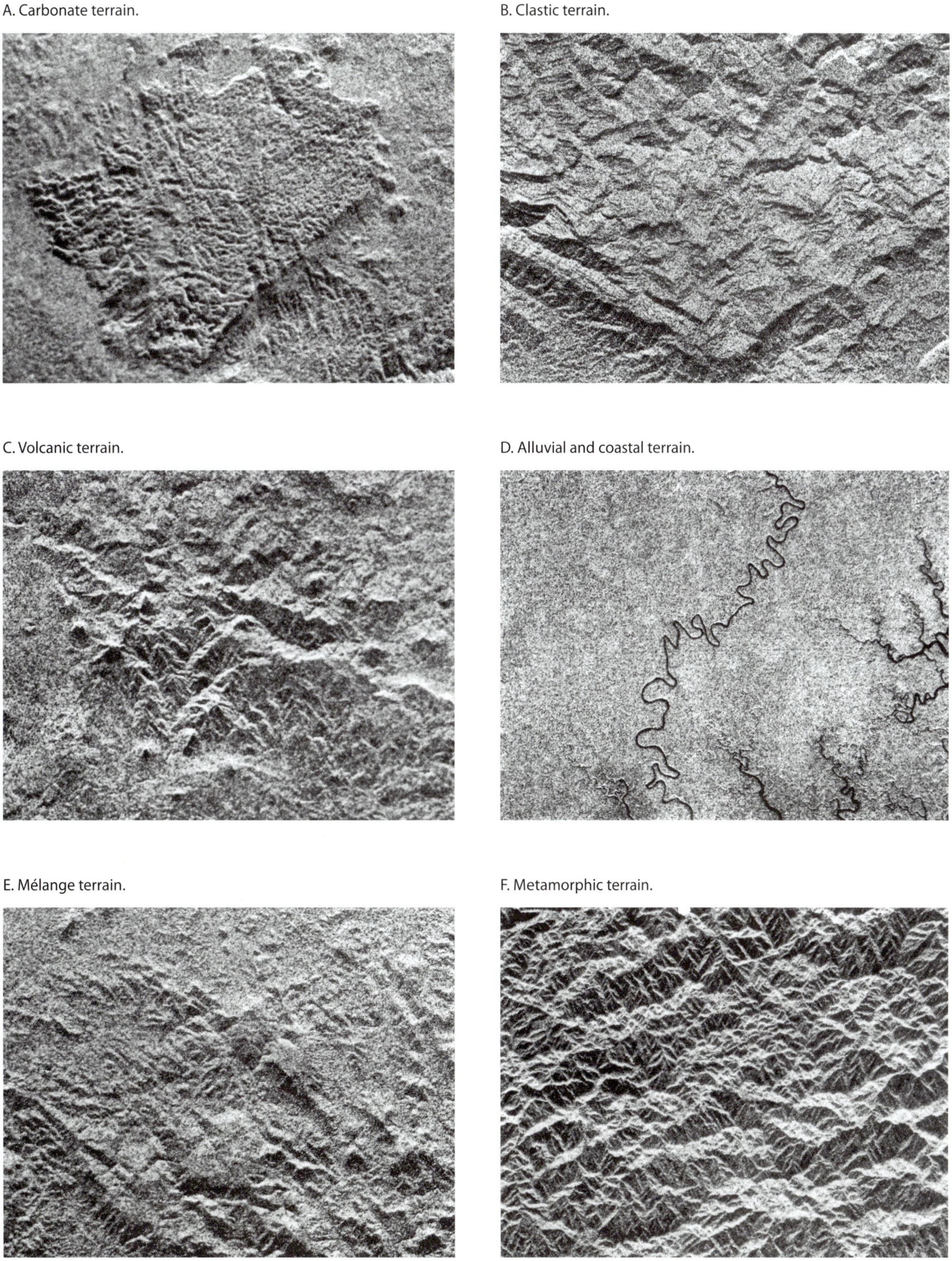

A. Carbonate terrain.

B. Clastic terrain.

C. Volcanic terrain.

D. Alluvial and coastal terrain.

E. Mélange terrain.

F. Metamorphic terrain.

Figure 6-36 SIR-A images of typical Indonesian terrain categories. Areas covered are 28 km wide (left to right). Look direction is toward the lower margin of each image. From Sabins (1983, Figure 4).

A. Strike and dip.

B. Thrust faults.

C. Folds, moderately eroded.

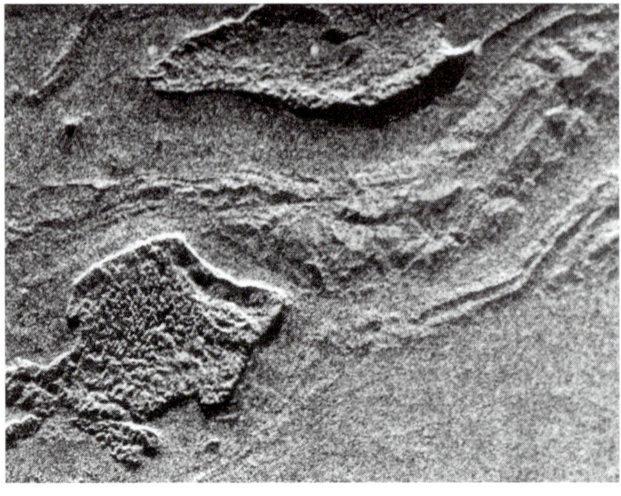

D. Folds, deeply eroded.

E. Linear features.

F. Strike-slip fault.

FIGURE 6-37 SIR-A images of typical structural features in Indonesia. Areas covered are 28 km wide (left to right). Look direction is toward the lower margin of each image. From Sabins (1983, Figure 5).

A. Strike and dip.

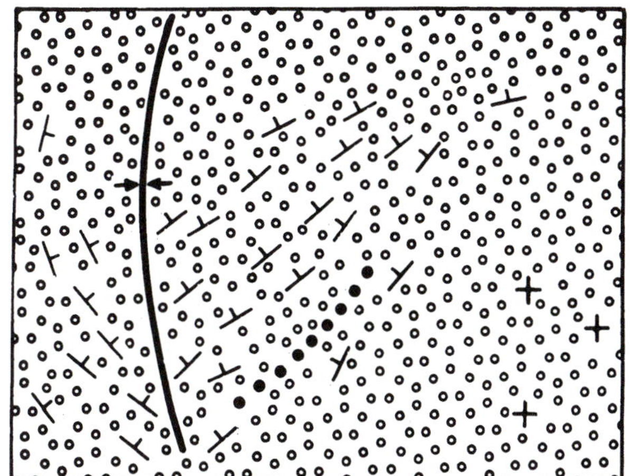

B. Thrust faults.

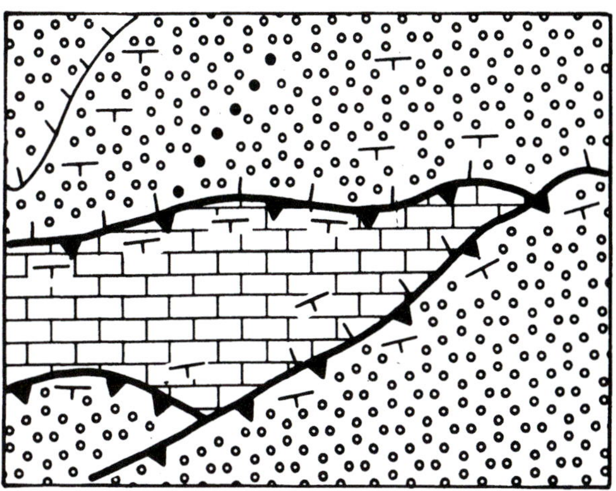

C. Folds, moderately eroded.

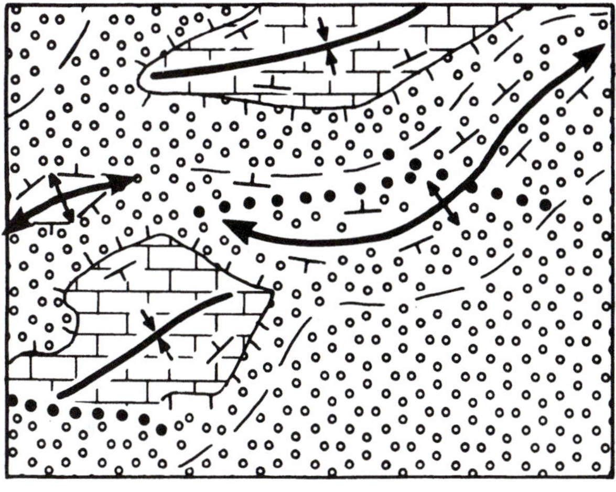

D. Folds, deeply eroded.

E. Linear features.

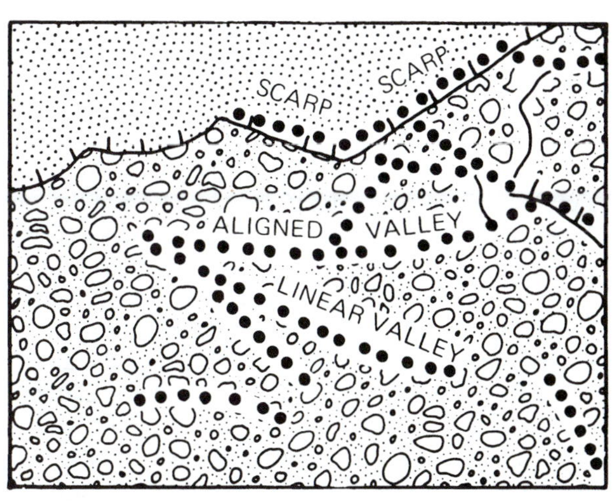

F. Strike-slip fault.

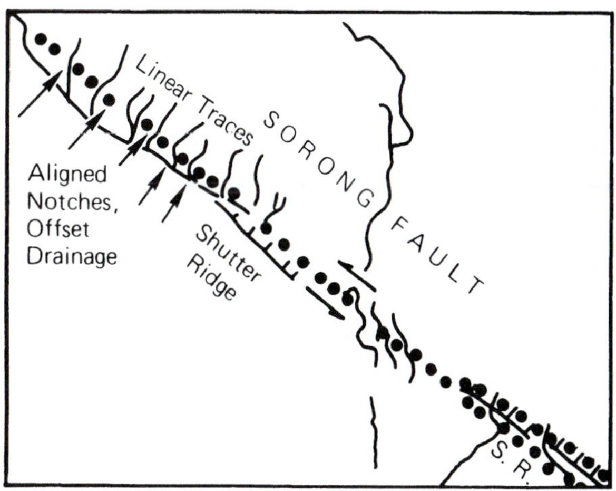

Figure 6-38 Interpretation maps of SIR-A images of Indonesian structural features shown in Figure 6-37; areas covered are 28 km wide (left to right). From Sabins (1983, Figure 6).

Strike and Dip (Attitude)

Strike and dip, or attitude, information can be interpreted in terrains of stratified rocks and in some volcanic terrains where flow surfaces are well expressed. The radar signature of dipping layers depends upon the relationship between the dip direction and the radar look direction. In the image (Figure 6-37A) and the map (Figure 6-38A), the strata are dipping generally toward the upper margin of the image, which is toward the radar antenna. The dipslopes therefore have bright signatures, and the antidip scarps, which face away from the look direction, have dark signatures. Where beds dip away from the radar antenna, the antidip scarps are bright and the dipslopes are dark, as seen in the flanks of the folds in Figure 6-37C.

Thrust Faults

Thrust faults are a challenge to interpret from radar and other remote sensing images for the following reasons:

1. The planes of thrust faults are commonly parallel or nearly parallel with bedding planes of associated strata. Therefore, most thrust faults do not cause the discordant geometric relationships that are associated with many normal and strike-slip faults.
2. Thrust faults are commonly recognized by anomalous rock relationships such as older beds over younger beds, repetition of beds, and omission of beds. Actual field mapping is the sure way to resolve these issues. Both authors are field geologists by training and experience: we embrace this approach.

Despite these challenges, thrust faults were interpreted from the SIR-A image of the Paniai Lakes region, mainland West Papua, where several imbricate (multiple) thrust plates occur. Figures 6-37B and 6-38B show thrust plates that dip gently toward the lower margin of the image and are terminated updip at eroded antidip scarps. The trend of these scarps is locally discordant with the trend of the underlying strata. At places such as the right margin of Figure 6-37B, an upper thrust plate overrides and truncates underlying thrust plates. Elsewhere in the Paniai region, thrust faults are recognized by the repetition of carbonate formations with distinctive karst topography.

Folds, Moderately Eroded

Folds may be recognized by attitudes of beds, outcrop patterns, and topographic expression. Figures 6-37C and 6-38C show moderately eroded folds where the strike and dip attitudes are readily interpreted using the criteria described earlier. Outcrop patterns are also diagnostic in Figure 6-37C, where the youngest beds (carbonate rocks with characteristic karst topography) are preserved as mesas in the troughs of synclines, but the carbonate strata are eroded from the crest of the anticline. These erosion patterns commonly result in topographic reversal, in which topographic elevations (mesas) correspond to structural depressions (synclines). In the upper right portion of Figures 6-37C and 6-38C carbonate strata have been eroded from the crest of the anticline to expose underlying, older strata.

Folds, Deeply Eroded

Deep erosion may produce a nearly planar surface, where plunging folds are marked by subdued parallel ridges with arcuate patterns that are formed by outcrops of resistant strata (Figure 6-37D). The noses and axes of these folds can be interpreted (Figure 6-38D) from subtle highlights and shadows. Anticlines cannot be distinguished from synclines, however, because dip attitudes cannot be interpreted from the subdued ridges.

Linear Features

Linear features are expressed as scarps, linear valleys, and aligned valleys. In the upper part of the image (Figure 6-37E) and map (Figure 6-38E), two intersecting linear scarps form the boundary between mélange terrain and alluvial terrain and may be the expression of faults. Linear valleys may be eroded into zones of nonresistant crushed rocks along faults. Two or more separate valleys may be aligned end to end to form a linear feature.

Strike-Slip Faults

The image (Figure 6-37F) and map (Figure 6-38F) cover a segment of the Sorong fault, which forms a linear valley with the following characteristics of a strike-slip fault: (1) aligned notches; (2) shutter ridges; (3) linear terraces; and (4) offset stream channels, which indicate left-lateral fault displacement. The preservation of these tectonic features in this region of heavy rainfall and rapid erosion indicates that the Sorong fault is active, that is, movement has occurred in the past 10,000 years. These characteristics remind us of the San Andreas fault in southern California, although the terrain is semi-arid and the displacement is right-lateral.

RADAR INTERACTION WITH VEGETATED TERRAIN

Despite persistent cloud cover and rain the long wavelength of SIR-A acquired excellent images of Indonesia. The humid, tropical environment supports a dense rain forest that conceals the underlying terrain and hampers/precludes geologic mapping. Despite the forest cover, however, Sabins was able to interpret the geology as shown by Figures 6-37 and 6-38. These examples may lead to the conclusion that radar penetrates the forest cover and acquires images of the underlying terrain. This incorrect conclusion assumes that radar energy penetrates the vegetation twice; first on the transmitted pulse and again when the weaker energy backscattered from the terrain must penetrate vegetation on its route back to the antenna. In fact, radar signatures of vegetated terrain are dominated by interactions at the top of the canopy and within the vegetation cover, rather than with the ground surface.

Seasat and SIR-A radars operate at the relatively long L-band wavelength of 23.5 cm. The rough criterion for SIR-A is 8.4 cm, which is exceeded by most vegetation. Growing vegetation has a high water content, which increases the dielectric constant and the radar reflectivity. Thus, radar theory predicts that because of its rough surfaces, high moisture content, and complex branch structure, dense vegetation will scatter and reflect incident radar energy and little or no energy will be returned from the underlying terrain.

We can test radar theory with SIR-A and Seasat L-band images (both 23.5 cm wavelength) of agricultural areas in the Great Plains of Texas and Kansas (Figure 6-39), where centerpoint irrigation is practiced. Clusters of bright circles are fields, predominantly wheat, with small dark interstices caused by smooth, bare soil. Differences in brightness of the fields are caused by differences in the growth, degree of canopy closure, and harvest status of the crops. The relevant point is that if the L-band radar energy had penetrated the vegetation the two images would show only the dark signature of the underlying soil. The bright patterns within the circular wheat fields seen on the SIR-A and Seasat images demonstrate that radar did not penetrate dense stands of wheat; therefore, it is improbable that radar of these wavelengths could penetrate a closed canopy rain forest, such as in Indonesia.

Figure 6-40 explains how features of the ground surface are recognizable in images of closed canopy, forested terrain. Figure 6-40A shows two faults and their associated topographic scarps in bedrock covered by forest. The trees are of relatively uniform height and the top surface of the canopy is controlled by the bedrock topography. The radar system produces an image of the canopy (Figure 6-40B) in which the underlying geology is enhanced by three mechanisms:

1. The inclined illumination produces highlights and shadows that enhances linear topographic features associated with the faults. Other structural features, such as folds and dipping strata, are similarly enhanced.
2. Because of the relatively large ground resolution cells (10 to 15 m for older aircraft radar and 38 m for SIR-A), individual trees are not resolved, which improves the topographic expression of geologic features with dimensions of hundreds and thousands of meters. In other words, the large resolution cell of radar acts as a filter to remove the high frequency spatial detail of vegetation "noise," thereby enhancing the lower frequency geologic "signals."
3. Radar penetrates the cloud cover that is associated with most forested regions, especially in the tropics.

For these reasons, radar is an essential tool for mapping and resource exploration in vegetated terrains with closed canopies. Chapter 13 shows how Chevron used radar images to help support the first oil discoveries in Papua New Guinea.

A. SIR-A image in northwest Texas.

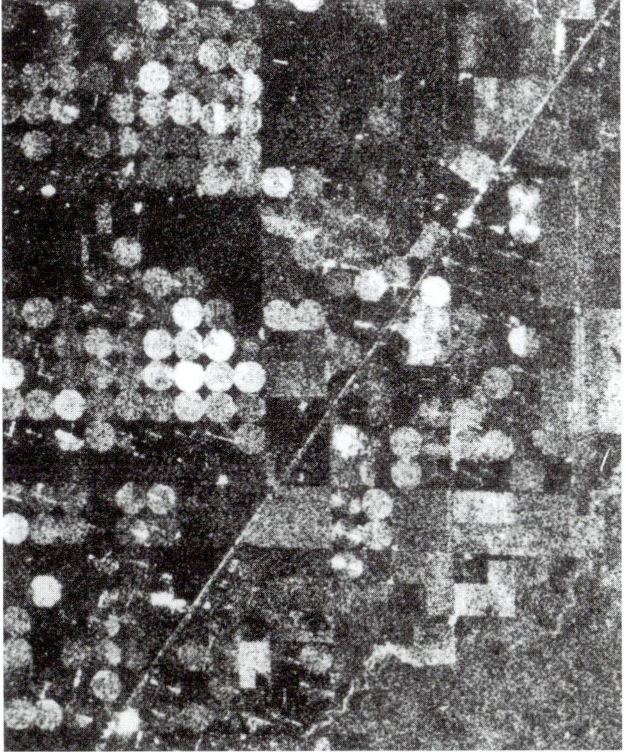

B. Seasat image in southwest Kansas.

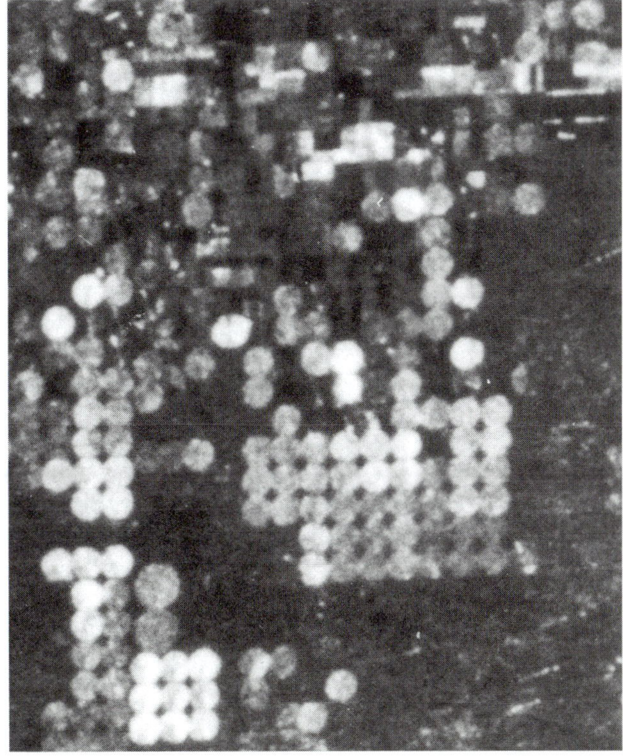

Figure 6-39 Satellite radar images of agricultural fields. Bright fields are growing crops that strongly backscatter the incident energy. Gray signatures are probably stubble and uncultivated areas with moderate backscattering. Dark areas are relatively smooth bare soil with specular reflectance and no radar returns. Areas covered are 8 km wide (left to right).

A. Block diagram.

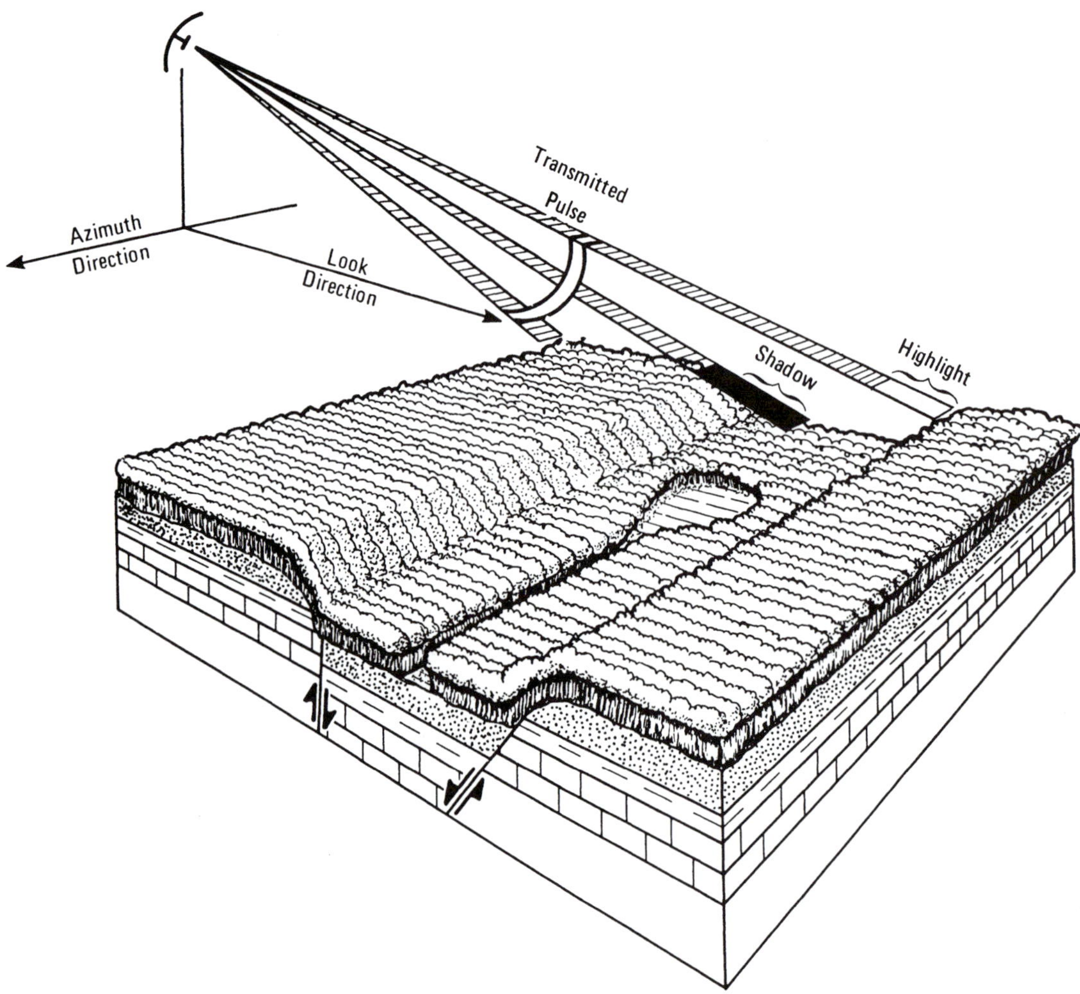

B. Radar image.

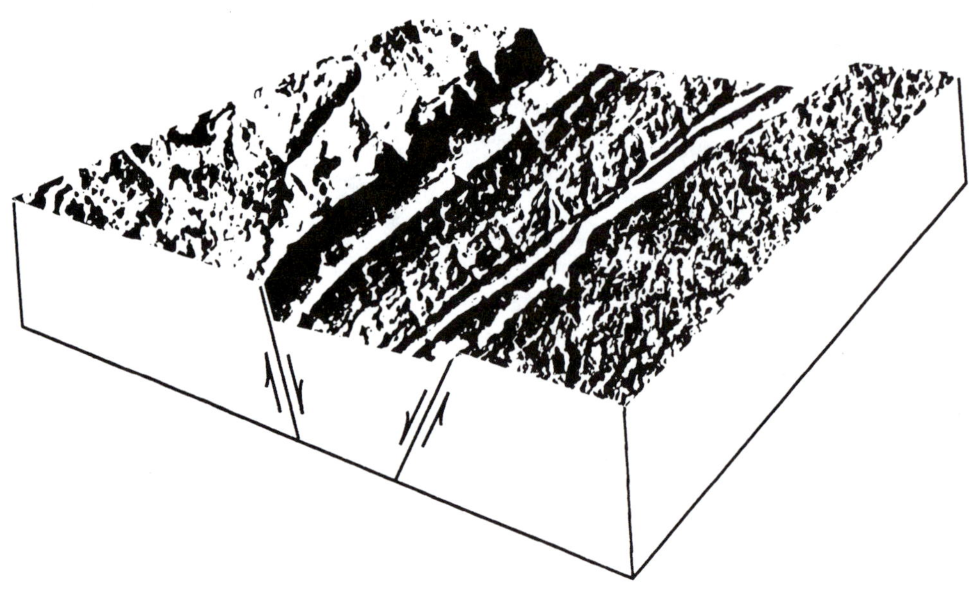

FIGURE 6-40 Enhanced expression of terrain and faults on radar images of forested terrain.

INTERFEROMETRY

Interferometry is the field of physics that deals with the interaction between superimposed wave trains; it is the principle that enables radar to acquire accurate images of targets at great distances. Interferometry enables radar systems to collect more than just images for interpretation of surface features; it also enables radar systems to measure ground deformation and the topography of the Earth's land areas. Rosen (2014) and Bamler and Hart (1998) provide detailed principles and theory of radar interferometry. The following offers simplified concepts of radar interferometry.

We know the velocity of electromagnetic radiation is 3×10^8 m · sec^{-1}; therefore, elapsed time is proportional to distance traveled. Figure 6-41A shows the positions of two receiving antennas (1 and 2) located at different distances (elapsed times) from the origin of the radar echo (returned signal) on the surface of the Earth. Antenna 1 is nearest the source and is separated from Antenna 2 by distance *B*. The distance d_1 equals d_2. Δd is the difference between the distance from Antenna 1 and Antenna 2 to the ground reflector. The radar echo travels a distance of d_1 to the closer Antenna 1 and a longer distance ($d_2 + \Delta d$) to Antenna 2. This difference in distance traveled results in a *phase difference* (θ) of the radar wave trains received at the two antennas. The phase difference is shown on Figure 6-41B by two thin curves (d_1 and $d_2 + \Delta d$) that show the history of a single wave as it progresses from left to right and is received consecutively by the two antennas.

One cycle of phase difference (also termed *interference fringe*) represents half the radar wavelength because this corresponds to a whole wavelength in two-way travel distance (Figure 6-41B). Each cycle of an interference fringe represents a height. For instance, one interference fringe cycle developed from the ERS-1 (Table 6-3) and ERS-2 radar satellites represent a height of 22 m (Duchossois and Martin, 1995). Wave phase is cyclic so phase differences that exceed one cycle repeat the interference fringe pattern on the interferometric image. Five cycles of fringes generated from the ERS-1 and ERS-2 radar systems would represent a height of 110 m.

A. The radar echo received by two antennas (1 and 2) is separated by distance *B*. The radar return originates from the same area on the Earth's surface.

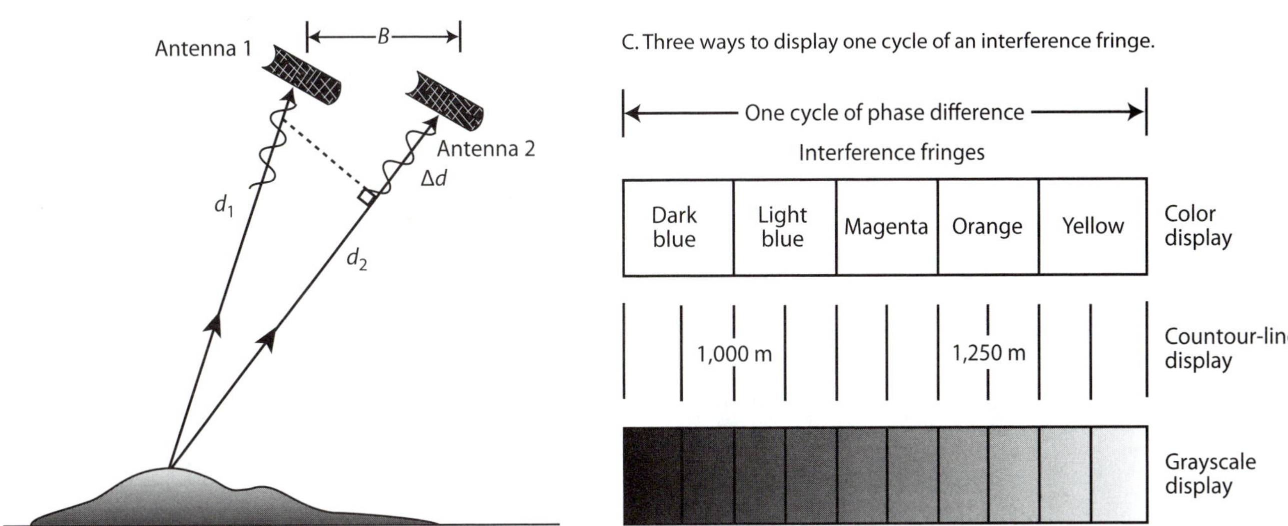

C. Three ways to display one cycle of an interference fringe.

B. Diagram showing the origin of a phase difference (θ) from a radar wave received by Antennas 1 and 2 (shown in Figure 6-41A).

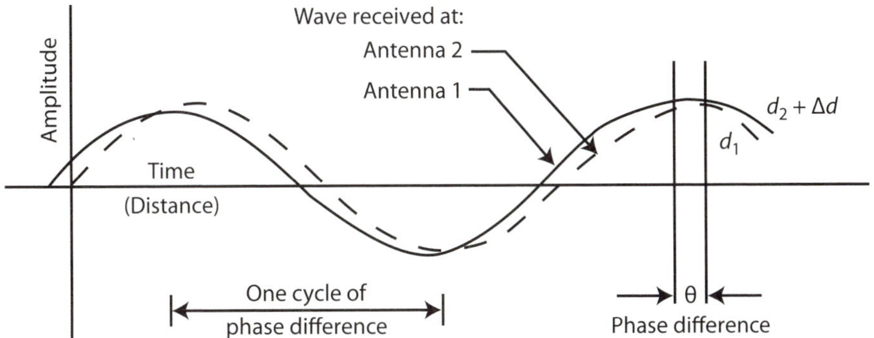

FIGURE 6-41 Origin and display of interference fringes.

There are three methods for displaying interference fringes: grayscale, contour line, and color. Figure 6-41C illustrates these methods.

1. **Grayscale display**: Each cycle of phase difference can be displayed as a continuous grayscale. The white tones typically show high values and dark tones show low values.
2. **Contour-line display**: The grayscale can be digitized into 10 equal divisions that are shown in the grayscale display. The interface between each division is shown as a contour line. In this example the elapsed time of a cycle represents a distance of 500 m; the lowest contour on the left is 900 m, the highest contour on the right is 1,400 m. Each of the 10 contours represents a vertical distance of 50 m.
3. **Color display**: The grayscale display can be displayed in colors that range from dark blue for the lowest elevation to yellow for the highest. Each sequence/cycle from yellow to blue is commonly called a color fringe. Alternate color schemes are also employed.

GRAYSCALE DISPLAY OF FRINGES

Figure 6-42A is an ERS-1 satellite SAR image of the Chilcotin Mountains, British Columbia, Canada (Xu and Cumming, 1997). The scene is illuminated from the left. Cliffs that face the radar beam are bright (white arrows) while surfaces that slope away from the beam are shades of gray. South-flowing rivers are located at the base of the cliffs. The relatively smooth medium gray feature in the lower right (black arrow) is a hillside that slopes toward the northwest. Figure 6-42B is a radar interferogram that measures topography (Xu and Cumming, 1997). The double-ended black arrow designates a single fringe that represents a range of elevation values. The dipslope in the southeast portion is covered by five complete interference fringes. If each fringe represents a vertical range of 50 m, the area has topographic relief of 250 m. Figure 6-42C is a west to east topographic profile across the interferometric fringes.

A. ERS-1 radar image illuminated from the west. White arrows are strong returns from west-facing scarps. The black arrow indicates a steep west-facing slope. From Xu and Cumming (1997, Figure 1).

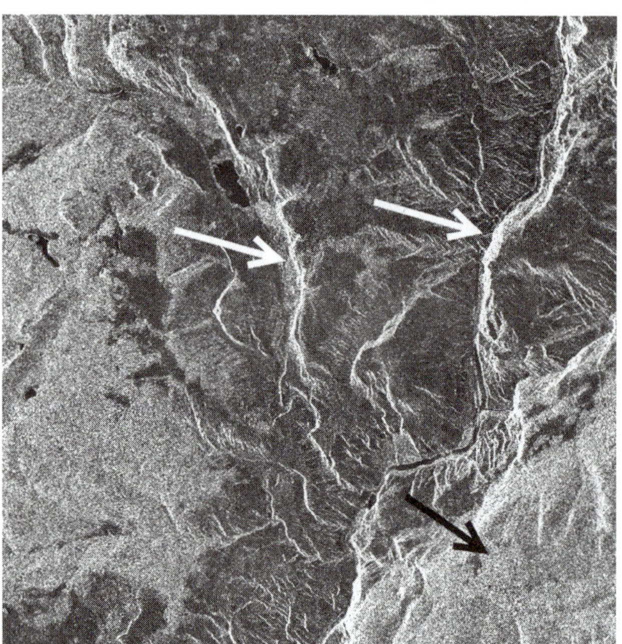

B. Interference fringes derived from ERS-1 radar data with white topographic profile line. From Xu and Cumming (1997, Figure 4).

C. Topographic profile interpreted from fringes in (B). Scale is arbitrary.

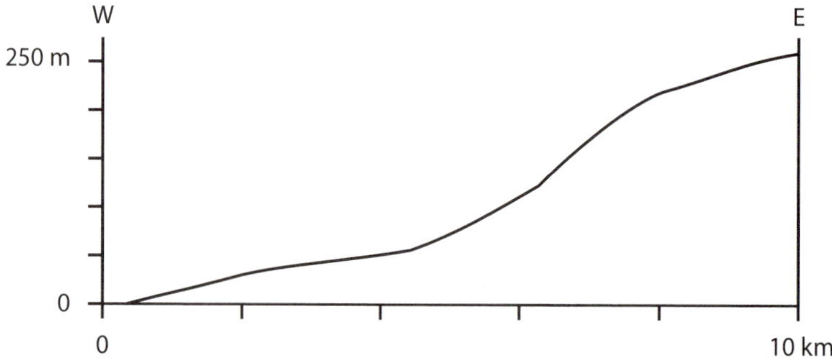

FIGURE 6-42 ERS-1 radar image, interference fringes, and terrain profile. Chilcotin Mountains, Canada.

Color Display of Fringes

Plate 19A is a color image of the Big Island of Hawaii. Plate 19B is an interference fringe image of the Kilauea volcanic caldera that is located on the island. The data were acquired by an X-band radar system onboard NASA's Space Shuttle *Endeavor* on April 13, 1994 and October 4, 1994 (JPL, 1999). The interference fringes from the two radar observations of Kilauea are shown in a basic three-color cycle that demonstrates the method. Each fringe ranges from:

- Green—highest elevation
- Red—intermediate elevation
- Blue—lowest elevation

Five cycles show the range of elevations in the scene. Figure 6-43A is a location sketch map and Figure 6-43B is a profile of the features shown in the image.

In Plate 19B, fringe 1 shows the topographic "bulge" in the south portion of the image. Fringes 2 and 3 include the terrain that slopes downward from the bulge toward the caldera. Narrow curving highlights and shadows on the background image mark the outline of the Kilauea crater. Fringe 4 is the floor of the caldera. Halemaumau pit crater in the south portion of Kilauea is outlined by highlights and shadows; the floor is defined by fringe 5 at a lower elevation than the floor of Kilauea. These topographic features are shown on the cross section in Figure 6-43B. As of 2018, a new volcanic cycle is underway; Halemaumau is filled to the brim with lava and a small "new" pit crater is present near the east rim of Halemaumau (Figure 6-43A). A series of fissures to the south (not shown) have erupted; lava has flowed downhill to the sea, where it generates a toxic gas called "laze" (lava haze). In route to the sea the flows have wiped out highways, villages, and a power plant. The characteristic slow progress of the flows has enabled an orderly evacuation of residents, but homes and infrastructure are gone.

RADAR APPLICATIONS

All weather radar penetrates clouds and darkness to provide up-to-date images of floods, forests, polar sea ice and ice bergs, and ground deformation. Radar is used to generate DEMs and topographic maps. In particular, the SRTM collected radar data to enable a DEM that covered much of the globe.

Shuttle Radar Topography Mission

In February 2000, the 11-day SRTM for the first time acquired topographic data for 99.97% of the Earth's landmass between 57°N and 60°S (JPL, 2017). It employed interferometry to acquire topography for Earth's terrain from satellite altitudes. The images were collected at both C (5.6 cm) and X (3.0 cm) wavelengths. The radar system that previously orbited on SIR-C in 1994 was equipped with an additional receive-only antenna mounted on a 60-m long boom (Figure 6-44). The boom provides the necessary spatial offset of antennas to produce the phase difference in the received waves that is required for interferometry. The C-band images were acquired with a swath width of 225 km that provided continuous geographic coverage at the equator. The SRTM orbited the Earth in a near-equatorial, tail-first orbit pattern to acquire the side-lapping image swaths. The swaths were then digitally mosaiced to produce a DEM of the world's land areas.

Plate 20A is the SRTM DEM of the Big Island of Hawaii illuminated from the northwest and color-coded for elevation. Volcanic deposits younger than approximately 200 years are colored red on the DEM (Sherrod and others, 2007). The north–northeast to south–southwest line is the location of the topographic profile in Figure 6-45 that crosses the majestic volcanic peaks of Mauna Kea and Mauna Loa.

A. Interpretation map.

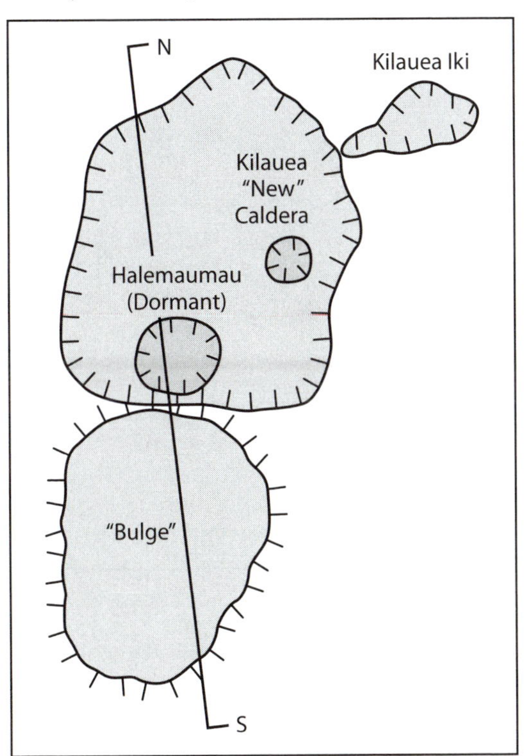

B. Cross section.

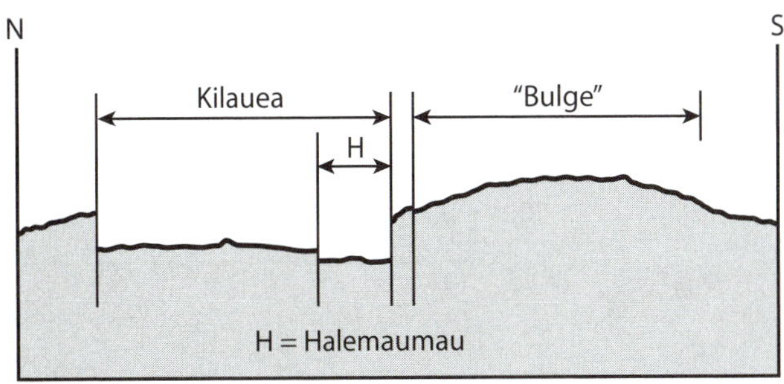

Figure 6-43 Interpretation of 1994 interferometer image in Plate 19B, Kilauea, Hawaii.

Both volcanic peaks are over 4,000 m high, but Mauna Kea is dormant and Mauna Loa, located ~40 km south, is active. The black square on the southeast flank of the Big Island is the location of the active Kilauea volcanic caldera that is described in the previous section.

Plate 20B shows the geologic map of volcanic rocks deposited from 5,700,000 years ago to the present (Sherrod and others, 2007) faded and draped over the SRTM DEM. The map graphically records the northwestward drift of the Pacific tectonic plate over the "Hawaii" hotspot. The oldest volcanic rocks in the northwest tip of the island (blue pattern) are followed, both laterally and chronologically, by intermediate age rocks (green) to the southeast and the youngest rocks (yellow, orange, and red) to the southeast and south. Today's volcanism occurs on the southeast flank of the island. Remote sensing did not discover these age and movement relationships, but provides a useful graphic medium for expressing the relationships.

RADARSAT-2: IMAGES OF FLOODS

For 40 days between April and June 2011, heavy rain and melting of a thick layer of snow caused flooding in the area surrounding

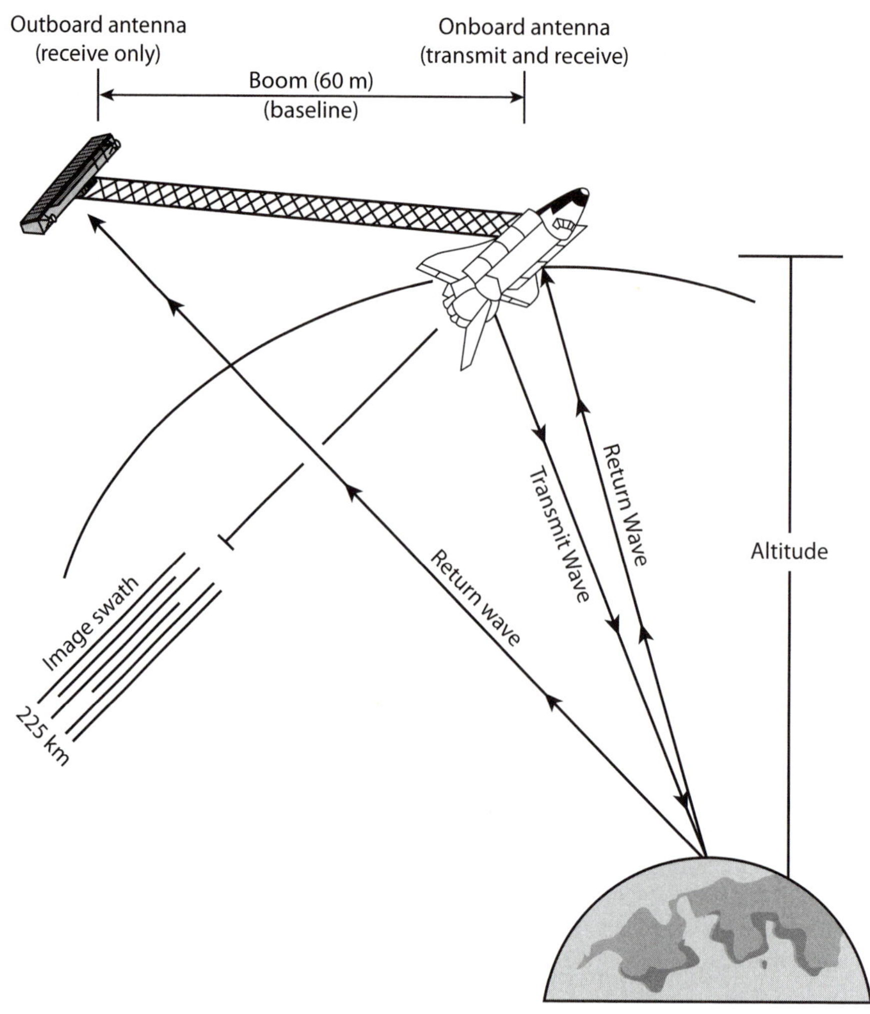

FIGURE 6-44 Schematic diagram of radar interferometry on SRTM. The onboard antenna transmits a radar pulse and receives a return pulse from nearly beneath the satellite. The same return pulse is also received by the outboard antenna at the end of the boom. The 60 m separation of the two antennas causes a phase difference (θ) and an interference fringe when the two received pulses (that are out of phase) are processed.

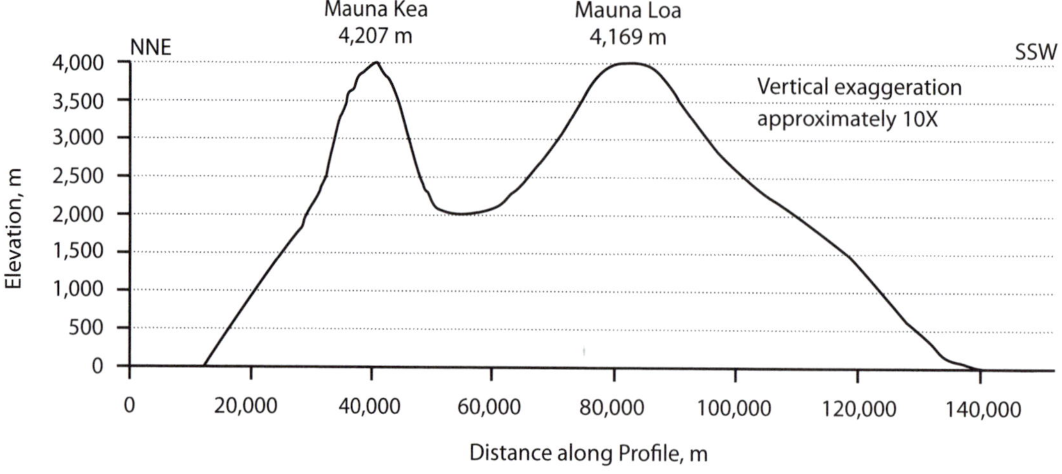

FIGURE 6-45 North–northeast to south–southwest topographic profile from the SRTM DEM of Hawaii shown in Plate 20A. SRTM courtesy USGS.

the Lake Champlain Basin and the Richelieu River, Quebec Province, Canada (Canadian Space Agency, 2011). The flood affected approximately 3,000 residences. **Digital Image 6-1A** is a natural color image acquired before the flooding that shows the normal stage of the rivers. Agricultural land, housing, dock facilities (upper left portion), and a historic fort on the island can be interpreted from the preflood image. A RADARSAT-2 image acquired during the flood event on May 1, 2011 shows the extent of submerged land and flood waters (colored blue in **Digital Image 6-1B**). RADARSAT-2 images covering April 11 to June 29, 2011 were used to develop maps showing the flood extent through time. These maps were used by Public Safety Canada to support their field operations (Canadian Space Agency, 2011). Radar enables reliable and repetitive monitoring of disaster sites because of its all-weather and day-night capability to image the Earth.

Airborne Radar Images: Forestry

Radar accurately monitors forest clear-cutting operations because of the difference in roughness, moisture, and other backscatter characteristics between forest and exposed soil, grasses, and shrubs in clear-cut areas. Airborne imagery was acquired in 1991 and 1992 over forested terrain in Whitecourt, Alberta, Canada using the C-SAR system (C-band with single HH polarization) in nadir mode, with a pixel spacing of 3.89 m (azimuth) by 4.0 m (range) (Canadian Center for Remote Sensing, 1993). The May 18, 1991 radar image (**Digital Image 6-2A**) shows forested terrain with characteristic uneven texture and medium gray tones, some access roads (linear dark features), and a small clearing with a probable man-made feature with very bright pixels in the middle of the clearing (white pixels surrounded by dark pixels). The February 8, 1992 radar image (**Digital Image 6-2B**) reveals an extensive smooth and dark gray area with linear margins matching the access roads in Digital Image 6-2A along the west and south margins. This area represents forest clear-cut between 1991 and 1992. The bright, reflective man-made feature in the east-central portion of the clear-cut area remained in place between 1991 and 1992, as it is visible in both images. To highlight the clear-cut area and the edge of the forest in color, a color composite image was generated with the 1991 image illuminated with red light and the 1992 image illuminated with blue light (**Digital Image 6-2C**).

Sentinel-1A Radar: Ice Bergs and Sea Ice

Radar satellites in near-polar orbits acquire images of the polar regions through cloud cover and many months of winter darkness. On April 13, 2014 the ESA Sentinel-1A radar satellite acquired an image of the Thwaites Glacier, Antarctica. The Sentinel-1A satellite occupies a near-polar orbit and records a 250-km wide swath of C-band (5.5 cm wavelength) radar imagery. The radar image was analyzed by NASA Earth Observatory (2014). Figure 6-46A is an enlarged segment of an image that covers a portion of the seaward margin of Thwaites Glacier and Pine Island Bay, which is the southeast extension of the Amundsen Sea, Antarctica. Thwaites Glacier is one of the fastest-moving glaciers in Antarctica.

Differences in gray tones in the radar image are an indication of how well each surface reflects microwave energy. "Radar signals are sensitive to surface roughness, salinity, and air bubbles in the ice," says NASA scientist Walt Meier. Sea ice contains bits of salty water (brine), which scatters the microwave energy enough that little gets back to the satellite. As a result, sea ice is medium gray to dark. Glacial ice—and the icebergs that come from it—is made of compacted snow. Air bubbles trapped in the ice reflect microwave energy extremely well, so more of the signal makes it back to the satellite and the ice appears bright (NASA Earth Observatory, 2014).We classified the image into the five categories (Figure 6-46B). Each category is summarized below.

1. **Sea ice**: The northern one-third of the image consists of sea ice with a uniform medium to light gray signature. Sea ice consists of a range of ice and water categories that are described below. The main body of sea ice is "permanent" year-round, but the southern margin, shown in Figure 6-46A, expands and retreats with Antarctic seasons. The Sentinel-1 image was acquired near the end of Antarctic summer (April 13). During the ensuing winter the sea ice will expand southward to enclose the bergs (bright signatures) that are lodged along the south margin of the ice.
2. **Old bergs**: In the original radar image the sea ice in the north portion of the image had relatively subtle variations of medium gray tones. We applied contrast enhancement, described in Chapter 9, to accentuate a number of subrounded shapes with a range of sizes and darker gray tones enclosed within light gray sea ice. We interpret these shapes as old bergs from past seasons that are enclosed by young sea ice. Summer thawing and winter freezing have reduced their original roughness. The northeast and northwest portions of sea ice show recently enclosed bright bergs that still retain their original roughness. The small black feature at the center of the north margin of the image is a polynya, which is an opening in the ice where smooth sea water is exposed.
3. **New bergs**: These bright (rough) angular blocks were calved into Pine Island Bay from the glacier during the summer. North-flowing currents and winds have lodged the bergs against the south margin of the sea ice, which is expanding to enclose them.
4. **Glacier, floating**: The very bright signature of Thwaites Glacier records a very rough surface at radar wavelengths that results from a myriad of fractures at a range of dimensions. Surges of the underlying sea expand the fractures into crevasses that are outlines of future bergs.
5. **Open water**: Near the end of the Antarctic summer a stretch of open sea water remains in the south-central and southwest portions of the image (Figure 6-46A). As pointed out earlier, calm water is a smooth surface that reflects all incident radar energy in specular fashion; there are no returns, which results in a dark signature.

SOURCES OF RADAR DATA

- Airbus Defence and Space (intelligence-airbusds.com)
- Alaska Satellite Facility (asf.alaska.edu)
- European Space Agency (sentinel.esa.int/web/sentinel/home)
- Japan Aerospace Exploration Agency (eorc.jaxa.jp/en)
- MDA Geospatial Services (mdacorporation.com/geospatial/international)

A. Enhanced Sentinel-1A image acquired on April 13, 2014. Original image courtesy of European Space Agency.

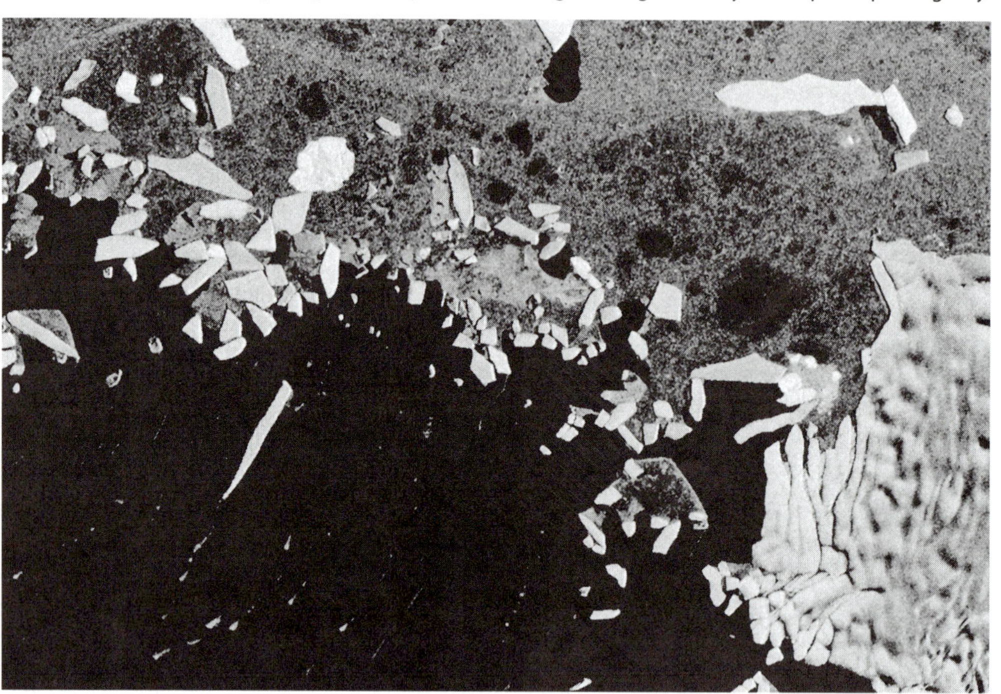

B. Interpretation map.

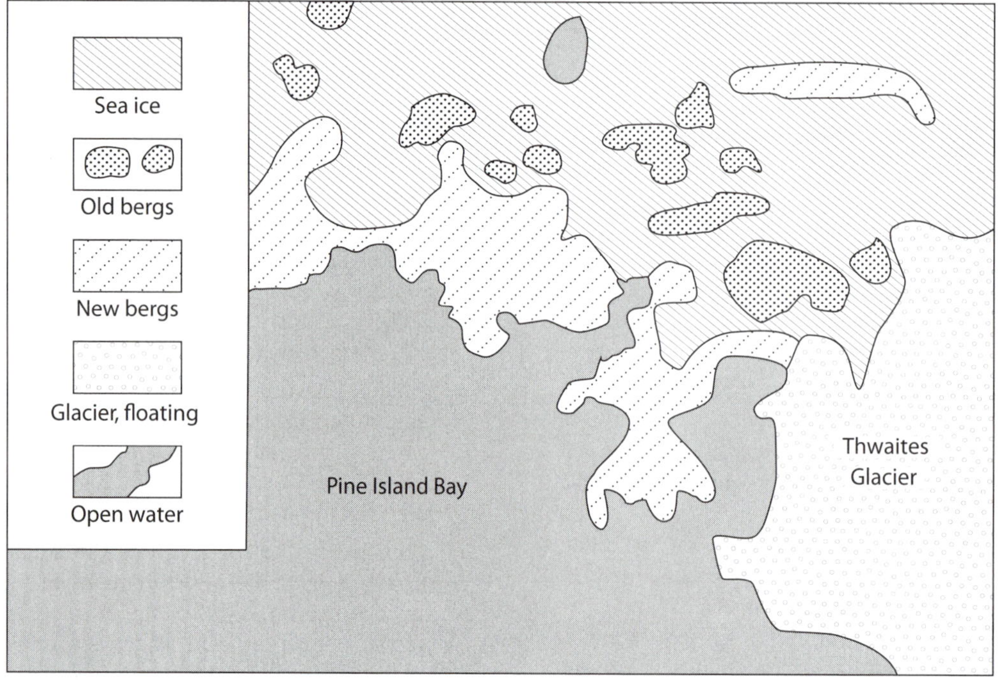

FIGURE 6-46 Thwaites Glacier and Pine Island Bay, Antarctica (area approximately 20 by 30 km).

QUESTIONS

1. Describe the advantages of radar remote sensing (for instance, compared to Landsat or aerial photography).
2. The military has led the way on radar remote sensing. Today, many of the common wavelengths used by civilians are nicknamed after their military designation. What are the wavelengths (in cm) associated with radar bands X, C, L, and P?
3. A typical radar system operates at a frequency of 5 GHz. What is the wavelength of this radar?
4. Calculate the range resolution for a radar system with a pulse length of 0.2 μsec and a depression angle of 30°.
5. What is the principal difference between real-aperture and synthetic-aperture radar systems?
6. Define and discuss the optimum depression angle (relatively steep or relatively shallow) for imaging terrain
   a. with low relief, such as coastal plains.
   b. with high relief, such as mountain chains.
7. Refer to Figure 6-25. What is the dielectric constant for Yuma sand, Vernon clay, and Miller clay at 20% and at 40% volumetric water content?
8. Refer to Table 6-7. For the X-band and L-band, calculate the vertical relief for smooth and rough surfaces (using Equations 6-6 and 6-7) and note the range of vertical relief for intermediate surfaces; the depression angle is 30°.
9. What are the four polarization configurations of radar (send/receive)?
10. Learn more about Sentinel-1. Visit the ESA's website (esa.int/Our_Activities/Observing_the_Earth/Copernicus/Sentinel-1/Instrument) and answer the following questions:
    a. How long is the Sentinel-1 SAR antenna?
    b. What radar wavelength band is this sensor using?
    c. What imaging and mapping applications are Sentinel-1 designed to accomplish?
11. The SRTM collected interferometric data that was used to generate a DEM of most of the Earth in 2000. This data collection took
    a. 2 years.
    b. 1 year.
    c. 11 days.
    d. 4 hours.

REFERENCES

Balmer, R., and P. Hart. 1998. Synthetic aperture radar interferometry. *Inverse Problems*, 14, R1–R54.

Barr, D. J. 1969. *Use of Side-Looking Airborne Radar (SLAR) Imagery for Engineering Studies* (Technical Report 46-TR). Fort Belvoir, VA: US Army Engineer Topographic Laboratories.

Berlin, G. L., M. A. Tarabzouni, A. H. Al-Naser, K. M. Sheikho, and R. W. Larson. 1986. SIR-B subsurface imaging of a sand-buried landscape: Al Labbah Plateau, Saudi Arabia. *IEEE Transactions on Geoscience and Remote Sensing*, GE-24(4), 595–602.

Blanchard, B. J., A. J. Blanchard, S. Theis, W. D. Rosenthal, and C. L. Jones. 1981. *Seasat SAR Response from Water Resources Parameters* (Final Report 3891). Washington, DC: US Department of Commerce/NOAA Contract 78-4332.

Blom, R. G., R. E. Crippen, and C. Elachi. 1984. Detection of subsurface features in Seasat radar images of Means Valley, Mojave Desert, California. *Geology*, 12, 346–349.

Canadian Center for Remote Sensing. 1993. Airborne C-SAR Mapping of Forest Clear-Cutting. Canadian Center for Remote Sensing.

Canadian Space Agency. 2011. RADARSAT-2 Featured Images Archives in North America. http://www.asc-csa.gc.ca/eng/satellites/radarsat2/featured-image/featured-north-america.asp (accessed December 2017).

Cimino, J. B., and C. Elachi (Eds.). 1982. *Shuttle Imaging Radar-A (SIR-A) Experiment Report* (82-77). Pasadena, CA: Jet Propulsion Laboratory.

Cimino, J. B., B. Holt, and A. H. Richardson. 1988. *The Shuttle Imaging Radar-B (SIR-B) Experiment Report* (88-10). Pasadena, CA: Jet Propulsion Laboratory.

Committee on Earth Observation Satellites. 2016, October 1. Interpretation Guide for ALOS PALSAR/ALOS-2 PALSAR-2 Global 25 m Mosaic Data (Version 1.1). CEOS Systems Engineering Office and Global Forest Observations Initiative.

COSMO-SkyMed Team. 2010. COSMO-SkyMed System Description & User Guide. Italian Space Agency.

Craib, K. B. 1972. Synthetic-aperture SLAR systems and their application for regional resources analysis. In E. Sahrokhi (Ed.), *Remote Sensing of Earth Resources* (Vol. 1, pp. 152–178). Tullahoma: University of Tennessee Space Institute.

Duchossois, G., and P. Martin. 1995. ERS-1 and ERS-2 Tandem Operations. European Space Agency Bulletin Nr. 83. http://www.esa.int/esapub/bulletin/bullet83/duc83.htm (accessed August 2019).

Evans, D. L., E. R. Stofan, T. D. Jones, and L. M. Godwin. 1994. Earth from sky. *Scientific American*, 271(6), 70–75.

Fischer, W. A. 1975. History of remote sensing. In R. G. Reeves (Ed.), *Manual of Remote Sensing* (chapter 2, pp. 27–50). Falls Church, VA: American Society of Photogrammetry.

Ford, J. P. (Ed.). 1988. *Advances in Shuttle Imaging Radar-B Research*. Bristol, PA: Taylor and Francis.

Ford, J. P., J. B. Cimino, and C. Elachi. 1983. *Space Shuttle* Columbia *Views the World with Imaging Radar—*the SIR-A Experiment (82–95). Pasadena, CA: Jet Propulsion Laboratory.

Ford, J. P., J. B. Cimino, B. Holt, and M. R. Ruzek. 1986. *Shuttle Imaging Radar Views the Earth from* Challenger*—the SIR-B Experiment* (86–10). Pasadena, CA: Jet Propulsion Laboratory.

Ford, J. P., and F. F. Sabins. 1986. Satellite Radars for Geologic Mapping in Tropical Regions. In Proceedings, Environmental Research Institute of Michigan Fifth Thematic Conference, Remote Sensing for Geology, Ann Arbor, MI.

Ford, J. P., C. M. Weitz, T. G. Farr, D. A. Senske, E. R. Stofan, G. Michaels, and T. J. Parker. 1993. *Guide to Magellan Image Interpretation* (93–24). Pasadena, CA: Jet Propulsion Laboratory.

Henderson, F. M., and A. J. Lewis (Eds.). 1998. *Manual of Remote Sensing* (3rd ed.). Principles and Applications of Imaging Radar (Vol. 2). New York: John Wiley.

Hunt, C. B., and D. R. Mabey. 1966. *Stratigraphy and Structure of Death Valley, California* (Professional Paper 494-A). Washington, DC: US Geological Survey.

Intermap. 2016. Intermap Product Handbook & Quick Start Guide (Edit Rules Edition, v4.5). http://www.intermap.com (accessed December 2017).

JPL. 1999. PIA01762: Space Radar Image of Kilauea, Hawaii. Jet Propulsion Laboratory http://photojournal.jpl.nasa.gov/catalog/PIA01762 (accessed December 2017).

JPL. 2017. Interferometry Explained—More Detail. Jet Propulsion Laboratory http://www2.jpl.nasa.gov/srtm/instrumentinterfmore.html (accessed December 2017).

Livingstone, C. E., I. Sikaneta, C. Gierull, X. Chiu, and P. Beaulne. 2005. RADARSAT-2 System and Mode Description. In Proceedings, Integration of Space-Based Assets within Full Spectrum Operations Meeting (RTO-MP-SCI-150, Paper 15). Neuilly-sur-Seine, France.

McCauley, J. F., G. G. Schaber, C. S. Breed, M. J. Grolier, C. V. Haynes, B. Issawi, C. Elachi, and R. Blom. 1982. Subsurface valleys and geoarcheology of the eastern Sahara revealed by Shuttle radar. *Science*, 318(4576), 1,004–1,020.

MDA. 2016, March 21. RADARSAT-2 Product Description (RN-SP-52-1238 Issue 1/13). MacDonald, Dettwiler and Associates, Ltd.

Moore, R. K. 1983. Radar fundamentals and scatterometers. In R. N. Colwell (Ed.), *Manual of Remote Sensing* (2nd ed.) (chapter 9, pp. 369–427). Falls Church, VA: American Society of Photogrammetry.

Moore, R. K., L. J. Chastant, L. Porcello, and J. Stevenson. 1983. Imaging radar systems. In R. N. Colwell (Ed.), *Manual of Remote Sensing* (2nd ed.) (chapter 10, pp. 429–474). Falls Church, VA: American Society of Photogrammetry.

NASA Earth Observatory. 2014. Thwaites Glacier from Sentinel-1A. http://visibleearth.nasa.gov/view.php?id=83538 (accessed December 2017).

NASA Science. 2019. Titan. Solar System Exploration: Cassini. http://solarsystem.nasa.gov/missions/cassini/science/titan (accessed May 2019).

Parker, W. V. 2012, October 5. Discover the benefits of radar imaging. *Earth Imaging Journal*. http://eijournal.com/print/articles/discover-the-benefits-of-radar-imaging (accessed December 2017).

Peake, W. H., and T. L. Oliver. 1971. *The Response of Terrestrial Surfaces at Microwave Frequencies* (Technical Report AFAL-TR70-301). Columbus: Ohio State University ElectroScience Laboratory.

Ramachandran, V. S. 1988. Perception of shape from shading. *Nature*, 331, 163–266.

Rosen, P. A. 2014, August 4. Principles and Theory of Radar Interferometry. Jet Propulsion Lab UNAVCO Short Course. http://www.unavco.org/education/professional-development/short-courses/course-materials/insar/2014-insar-isce-course-materials/InSARPrinciplesTheory_UNAVCO_14.pdf.

Sabins, F. F. 1983. Geologic interpretation of Space Shuttle radar images of Indonesia. *American Association of Petroleum Geologists Bulletin*, 67, 2,076–2,099.

Schaber, G. G., G. L. Berlin, and W. E. Brown. 1976. Variations in surface roughness within Death Valley, California: Geologic evaluation of 25-cm wavelength radar images. *Geological Society of America Bulletin*, 87, 29–41.

Schaber, G. G., G. L. Berlin, and D. J. Pitrone. 1975. Selection of Remote Sensing Techniques: Surface Roughness Information from 3-cm Wavelength SLAR Images. In Proceedings, American Society of Photogrammetry 42nd Annual Meeting (pp. 103–117), Washington, DC.

Schaber, G. G., J. F. McCauley, C. S. Breed, and G. R. Olhoeft. 1986. Shuttle imaging radar: Physical controls on signal penetration and subsurface scattering in the eastern Sahara. *IEEE Transactions on Geoscience and Remote Sensing*, GE-24, 603–623.

Schaber, G. G., R. G. Strom, H. J. Moore, L. A. Soderblom, R. L. Kirk, D. J. Chadwick, D. D. Dawson, L. R. Gaddis, J. M. Boyce, and J. Russell. 1992. Geology and distribution of impact craters on Venus: What are they telling us? *Journal of Geophysical Research*, 97(E8), 13,257–13,301.

Sentinel-1 Team. 2013. Sentinel-1 User Handbook (GMES-S1OP-EOPG-TN-13-0001). European Space Agency.

Sharp, B., and B. Morton. 2007. Introduction to the New GeoSAR Interferometric Radar Sensor. Fugro EarthData, Inc. http://web.nps.edu/Academics/Centers/RSC/documents/Fugro_GeoSAR.pdf (Accessed December 2017).

Sherrod, D. R., J. M. Sinton, S. E. Watkins, and K. M. Brunt. 2007. Geologic Map of the State of Hawaii (Open-File Report 2007-1089). Washington, DC: US Geological Survey.

Strom, R. G., G. G. Schaber, and D. D. Dawson. 1994. The global resurfacing of Venus. *Journal of Geophysical Research*, 99, 10,899–10,926.

TerraSAR-X. 2014. TerraSAR-X Image Product Guide: Basic and Enhanced Radar Satellite Imagery (Issue 2.0). Airbus Defence and Space.

Ulaby, F. T., B. Brisco, and M. C. Dobson. 1983. Improved spatial mapping of rainfall events with spaceborne SAR imagery. *IEEE Transactions on Geoscience and Remote Sensing*, GE-21, 118–121.

Wang, J. R., and T. J. Schmugge. 1980. An empirical model for the complex dielectric permittivity of soils as a function of water content. *IEEE Transactions on Geoscience and Remote Sensing*, GE-18, pp. 288–295.

Xu, W., and I. Cumming. 1997, January. InSAR Simulation for the Chilcotin Area. MDA Quarterly Review. http://sar.ece.ubc.ca/people/weix/paper/mda97_1 (accessed June 2018).

Digital Elevation Models and Lidar

A *digital elevation model* (DEM) describes the Earth's topography with *x*-coordinates (longitude or easting), *y*-coordinates (latitude or northing), and *z*-values that specify elevation at each *x*, *y* point. DEMs can be generated from (1) overlapping aerial images (film or digital) using rigorous photogrammetric techniques, (2) Structure from Motion (SfM), which has found broad use in the drone community, (3) lidar, (4) sonar for bathymetry, and (5) radar (Chapter 6). DEM is a general term for 3-D models of the Earth that are more accurately categorized as digital terrain models (DTMs) or digital surface models (DSMs). DTMs, or "bare earth" models, do not depict vegetation or man-made structures. DSMs depict the elevations of buildings, vegetation canopies, and exposed ground (Jensen, 2007).

Lidar is the acronym for "light detection and ranging"; it operates in the visible and near-infrared portions of the electromagnetic spectrum. Lidar is an active sensor that uses a laser that produces a very narrow, highly concentrated beam of light where all the wavelengths are in phase (coherent) and of one wavelength. Lidar is very similar to radar—both are active sensors that are used to measure distances at specific points using a complex and powerful collection system. Both lidar and radar can operate at night.

Lidar has revolutionized the way features on the Earth's surface are rendered as 3-D models, and has greatly expanded the use of 3-D models in engineering and in geological and environmental applications. Close-up photogrammetry, mobile and terrestrial lidar, and SfM are versatile 3-D modeling technologies that can be applied to any object (e.g., geologic outcrops, caves, trees, buildings, statues, bridges, etc.). In addition, these technologies are not constrained to only DEM generation.

PHOTOGRAMMETRY

Photogrammetry is the art and science of obtaining reliable information about physical objects and the environment through the process of recording, measuring, and interpreting photographic images and other remote sensing data. Perspective viewing of overlapping images to determine distance from image to object is fundamental to photogrammetry. Alspaugh (2004) summarizes the history of photogrammetry, which reaches back centuries. In 1525, Albrecht Durer constructed mechanical devices to make true perspective drawings and produce stereoscopic drawings. In the late 1800s, balloons and kites were carrying cameras aloft to capture photographs in an attempt to make topographic maps. World War I and World War II greatly accelerated aerial camera, film, lens, and plotter technology. Photogrammetry was transformed from analog to digital after the 1950s with the advent of computer programming, electronic computing, digital imaging, and the mapping requirements associated with space exploration technology and reconnaissance.

Today photogrammetry is a highly developed measuring technology that utilizes GNSS (GPS) to locate the airborne platform, an Inertial Navigation System (INS) with an inertial motion system (IMU) to determine where the sensor is pointed (Chapter 1), rigorous calibration procedures, advanced sensors and software, and sophisticated computer programming. Accurately measuring elevation of topography and objects on the Earth's surface from a stereo model of overlapping aerial images is a fundamental work process. Until the advent of lidar technology, almost all topographic maps were generated with overlapping images by photogrammetrists.

A. The aerial camera is looking straight down (nadir) at the tanks.

B. Sequential aerial photograph with 60% forward lap.

Flight Line

FIGURE 7-1 Sequential aerial photographs over the Shell Refinery oil tanks in Martinez, California. Courtesy D. Ruiz, Quantum Spatial, Inc., Novato, California.

Stereoscopic interpretation of overlapping images and estimate of object height from *one* aerial photograph was covered in Chapter 2. Overlapping sets of two images are used by photogrammetrists to more accurately measure elevation of the ground and objects on the ground. A brief summary of how photogrammetrists determine elevation from a stereo model of overlapping images follows (for more detail see Jensen, 2007; McGlone, 2013; Paine and Kiser, 2012).

Photogrammetric stereo models are acquired with a passive optical sensor with a flight plan that has a specific amount of *forward lap* (overlap between sequential images along a flight line) and *sidelap* (between adjacent flight lines) designed into the acquisition program (Figure 2-10). Typical overlap and sidelap is 60% and 20%, respectively, ensuring no data gaps and good stereo models with 4X vertical exaggeration if a base-height ratio of 0.6 is used in the survey (Figure 2-17). In urban areas with tall buildings, the overlap and sidelap would be increased, perhaps to 85% and 50%, respectively, to ensure features on the ground are not obscured by tall buildings. The more overlap between images, the less tall objects lean, and the more visible are features at the base of these tall objects (on the side away from the image center, or principal point). A large percentage of overlap decreases the base-height ratio, decreasing the vertical exaggeration (Figure 2-17).

Figure 7-1 provides an example. Two aerial photographs were sequentially imaged on a flight line with a 60% forward lap (Figures 7-1A and 7-1B). They were acquired in 1959 over the Shell Refinery oil tanks in Martinez, California. The aerial camera was looking straight down (nadir) at the tanks when Figure 7-1A was acquired. In Figure 7-1B, the cluster of six tanks that are in the center of Figure 7-1A are now on the left margin. These two photographs image the six tanks and other objects on the surface from two different angles where the photographs overlap. Figure 7-1C shows these two aerial photographs overlapped. A stereo model is created in the area of overlap. The overlapping images in the stereo model can be seen in 3-D in Figure 2-20 and Plate 4.

Parallax (Ø) is the angle between the same object in a stereo model and the camera position when each photograph was acquired. The closer the object is to the two camera positions, the greater the parallax angle (Figure 7-2). The two camera positions are along a flight line that is designed to be horizontal (maintains the same elevation above sea level). When we stereoscopically view a stereo-pair with our two eyes, we are replicating the aircraft camera positions when the stereo-pair was acquired. We perceive objects in the stereo model that have more parallax as closer, and those that have less parallax as farther away Figure 7-2 shows the geometric relationship in a stereo model of a tower between parallax ($Ø_{top}$ and $Ø_{bottom}$) and elevation above the ground (D_{top} and D_{bottom}). Objects with more parallax (larger Ø angle) have higher elevations and those with less parallax have lower elevations in the stereo model and in our brain.

C. Stereo model of the area.

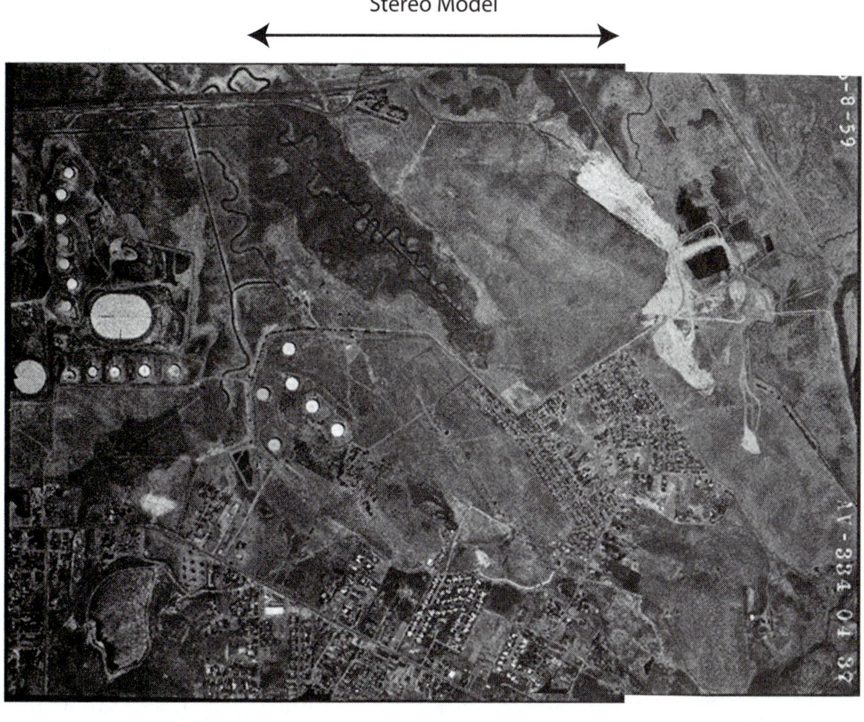

Inherent radial distortion causes tall buildings (and trees and other objects that project above the ground) to lean away from the center of the image (Figure 2-8). The correlation between elevation and parallax enables people to perceive and measure the elevation of the ground and objects on the Earth's surface (buildings, towers, trees, etc.) in a stereo model of overlapping aerial photographs. In Figure 7-3, D_{top} is the distance between the top of the object and D_{bottom} is the distance between the bottom of the object, measured along the flight line direction. The height of the tower is determined when the stereoscopic viewer measures the distance between the bottom of the tower (D_{bottom}) and the top of the tower (D_{top}) to calculate the amount of differential parallax (dp).

$$dp = D_{bottom} - D_{top} \tag{7-1}$$

Differential parallax is directly proportional to the height of objects in the stereo model and to ground elevation.

Geologists can measure heights of outcrops and foresters can measure heights of trees by viewing stereo models of imagery. Photogrammetrists stereoscopically view overlapping images to accurately measure the differential parallax of a feature in a stereo model to build a topographic map.

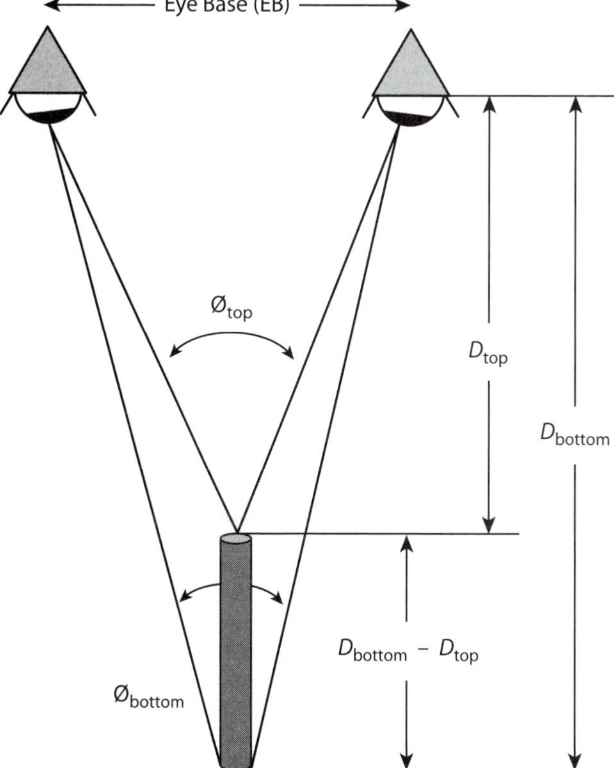

FIGURE 7-2 Parallax (Ø) is created by viewing an object from two eye or camera positions. Stereoscopic viewing of a stereo model with a different parallax angle at the top and bottom of the tower enables a perception of depth for the viewer.

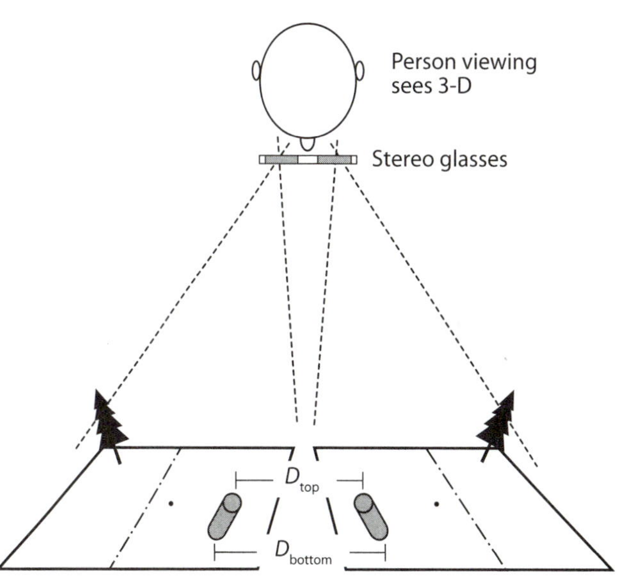

FIGURE 7-3 Stereoscopic viewing of overlapping images with objects on the ground showing radial distortion. The distance between the top and bottom of objects along the flight line direction enables the measurement of object height and ground elevation.

Before computers and digital imagery, topographic contours were based on photogrammetric measurements of terrain elevation from overlapping aerial photographs. In addition to the contours, photogrammetrists interpret and manually *compile* the road networks, utility corridors, wetlands, water bodies, and many other features into the map as points, lines, and polygons.

Photogrammetrists make differential parallax measurements and construct three types of elevation-control features that are used to build a DEM: mass points, breaklines, and

A. Mass points.

B. Breaklines.

C. Contours.

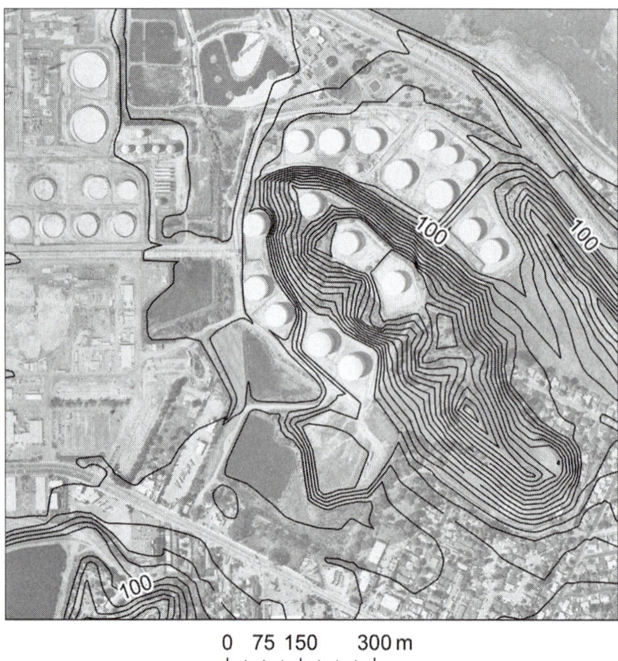

D. Mass points, breaklines, and contours combined.

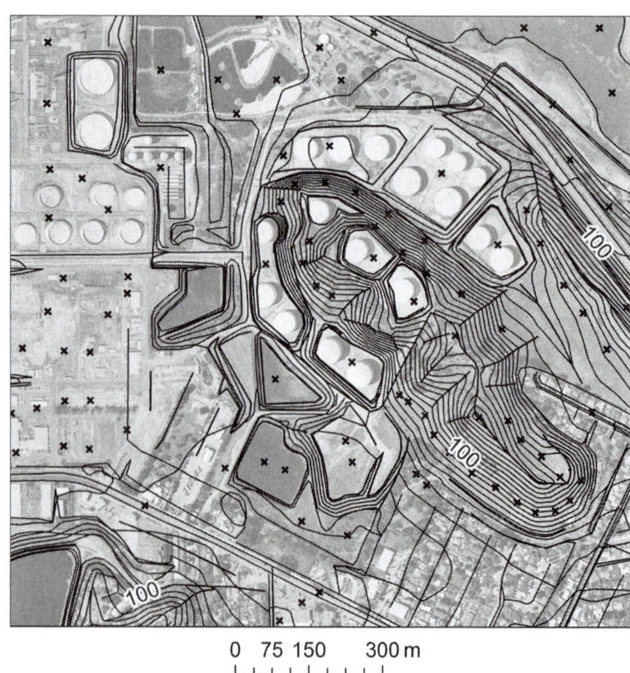

FIGURE 7-4 Photogrammetric stereoscopic mapping of aerial images to build a DEM for the Shell Refinery in Martinez, California. Courtesy D. Ruiz, Quantum Spatial, Inc., Novato, California.

topographic contours. *Mass points* are ground elevations at specific points with *x*, *y*, *z* values. The small black "x" symbols in Figure 7-4A are mass points. Each mass point has a horizontal coordinate (*x* and *y*) and an elevation (*z*) value. *Breaklines* are lines that represent sharp topographic edges such as ridge crests, valley channels, and edges of berms and levees (Figure 7-4B). *Topographic contours* are lines of equal elevation. Topographic contours are constructed at 10 ft (3.3 m) intervals. The 100 ft contour is labeled with "100" in three places on Figures 7-4C and 7-4D.

E. TIN model of terrain with triangles connecting mass points, breaklines, and contours.

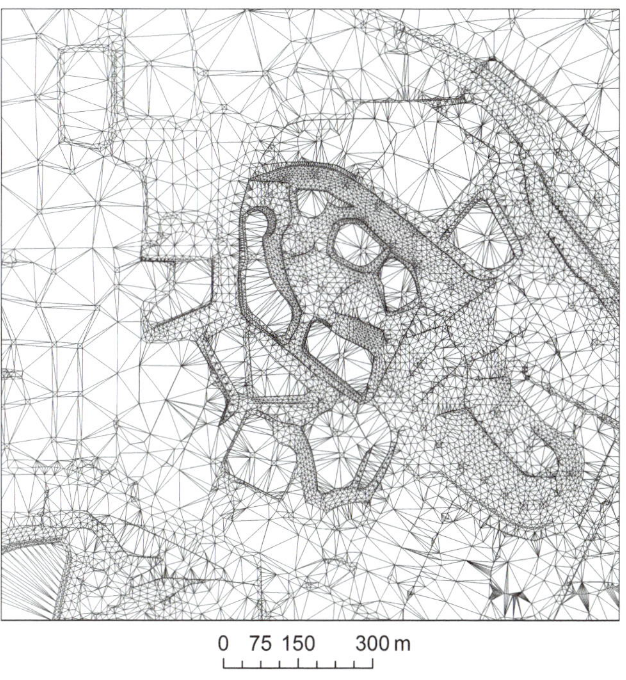

F. DEM with rectangular grid interpolated from TIN triangles.

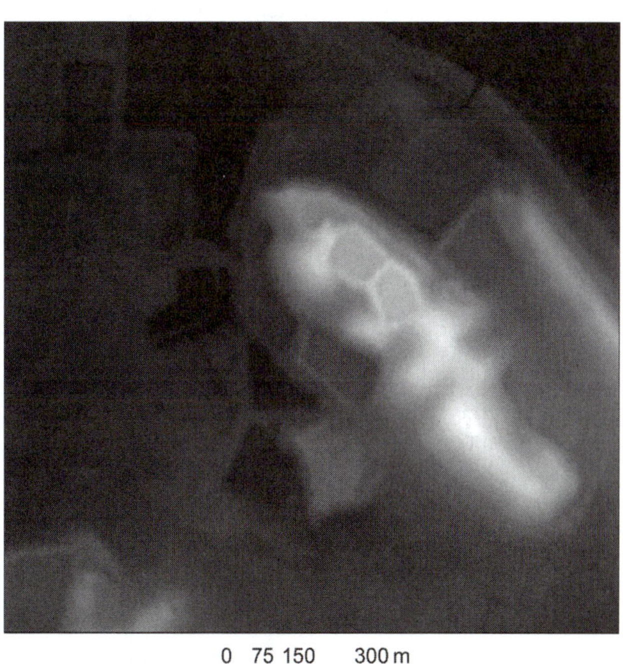

G. Hillshade DEM illuminated from the northeast.

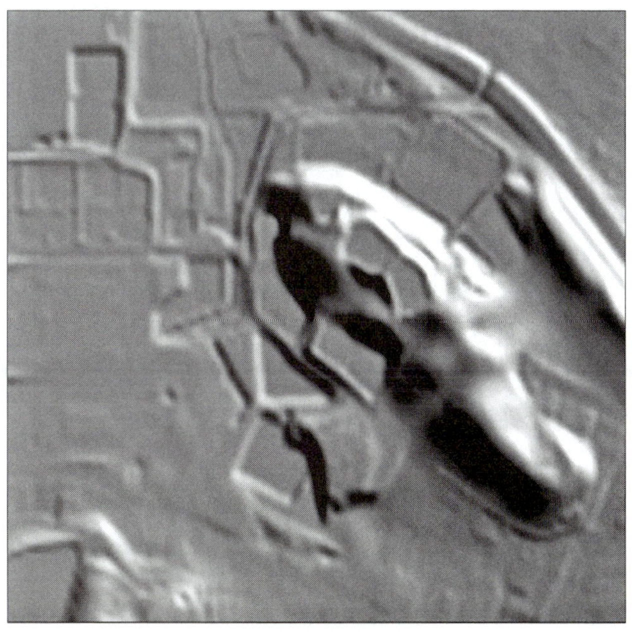

H. Photogrammetric compilation of roads, streams, and water bodies.

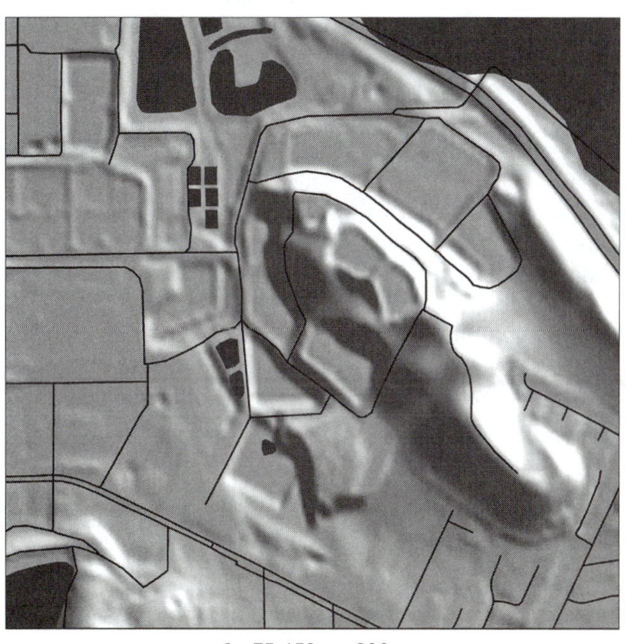

The contours, breaklines, and mass points are used to generate a triangular irregular network (TIN) model of the terrain (Figure 7-4E). A TIN is a vector-based representation of the terrain, composed of a network of triangles (see close-up in Figure 7-5). Three nodes define each triangle. Each node has an *x*, *y* coordinate and a *z* elevation value that is derived from the photogrammetrically measured contours, breaklines, and mass points. The TIN model was first developed in the early 1970s as a way to build a surface from a set of irregularly spaced points (Poiker, 1990).

TIN models are more effective at representing an irregular surface with sharp breaks as more points are in areas with rough terrain and fewer in areas with smooth terrain. TINs are typically used for high-precision modeling of smaller areas, such as in engineering applications, because they allow calculations of planimetric area, surface area, and volume. TIN models are vector-based, in contrast to raster-based DEMs. TINs are converted to raster grid DEMs with interpolation algorithms (Figure 7-4F). There can be some loss of detail in this conversion, but raster DEMs are more efficient for several applications due to their regular grid structure. In Figure 7-4G, the raster DEM is illuminated from the northeast to generate a hillshade DEM. Hillshade DEMs offer an excellent visualization of the topography and also are used for DEM quality assessment as data errors are displayed as spikes and holes across the smooth terrain surface (for more discussion see Chapter 10). The photogrammetrist compiles infrastructure and natural features as digital vector points, lines, and polygons (Figure 7-4H). These new map features have elevation values attached from the DEM, which enables them to be used for 3-D modeling.

Photogrammetric DEMs are used to remove radial distortion in aerial photographs, thereby creating orthorectified photographs or orthoimages with consistent scale. These planimetrically correct orthoimages can be digitally mosaicked to form an accurate base image for a very large geographic area, as is done with NAIP imagery (Plate 5).

Stereo models from passive optical sensors are not restricted to airborne platforms. The ASTER sensor onboard the Terra platform collects stereo models with 15 m ground resolution and a fixed 4X vertical exaggeration (Figure 4-20). Other satellites have optical sensors that can be aimed fore and aft and side-to-side to view the Earth's surface from different angles while orbiting the Earth. Sophisticated acquisition programming enables stereo-pairs and stereo models to be collected by these orbiting satellites to support stereoscopic interpretation, orthorectification of the imagery, and DEM generation.

STRUCTURE FROM MOTION

Structure from motion (SfM) generates high resolution DEMs and coregistered orthoimages based on principles that build on traditional stereo photogrammetry (Johnson and others, 2014). SfM produces the DEM using the *multi-ray photogrammetry* technique that images the same feature from multiple angles and distances. Figure 7-6 shows a camera moved in a circular path at a constant elevation around a feature of interest. Figure 7-7 illustrates four camera positions and the lines of sight from each camera position to two targets (one a triangle and the other a circle). The location of each target can be seen on the four images. The collection of overlapping images and changing perspective are used by SfM software to build a 3-D model of the feature of interest.

SfM can create DEMs because of increased computer power and sophisticated algorithms. As Johnson and others (2014) explain, unlike traditional photogrammetry, SfM algorithms support large changes in camera perspective and photographic scale through use of a feature recognition algorithm (scale invariant feature transform) (Lowe, 2004; Snavely and others, 2008), which eliminates the need for grid-like image acquisition (Figure 2-10). The SfM process is most suited to sets of images with a high degree of overlap that capture full three-dimensional structure of the scene viewed from a wide array of positions (Westoby and others, 2012). Microsoft's Photosynth started the SfM movement and allowed users to create 3-D models from photos taken with consumer cameras and smartphones (Miller, 2016; Snavely and others, 2008). The motion for the camera can be provided by drones, manned aircraft, kites, and balloons. The motion of the camera provides the depth information to features on the ground, which is recorded in a sequence of photographs. The photographs should be GNSS/GPS geotagged with the *x*, *y*, *z* location of the camera as each photograph is taken. Hayakawa and Obanawa (2015) had their camera acquiring photographs at one second intervals, indicating the GNSS/GPS geotag for each photograph had an accuracy of several meters.

Ground control points (GCPs) are established and their locations surveyed so that the point cloud can be georeferenced to a coordinate system. The lower the altitude of the camera, the denser the SfM point cloud density and the higher the spatial resolution of the DEM, but more time is spent in the field to image the area of interest as the photo-

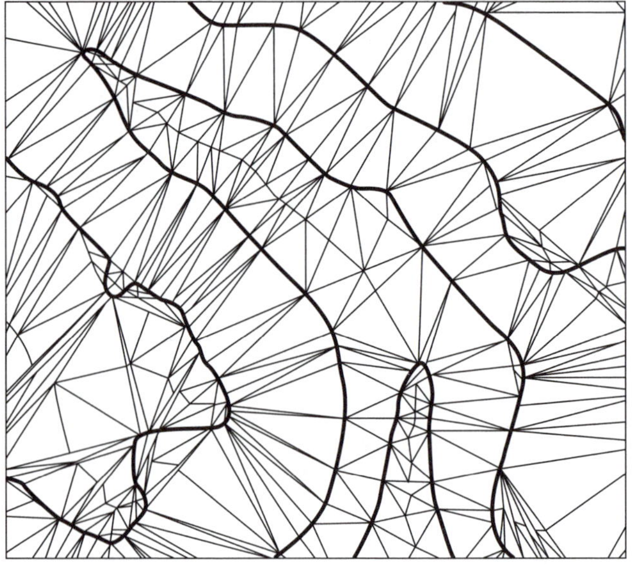

Figure 7-5 Close-up of a TIN model created from mass points, breaklines, and contours.

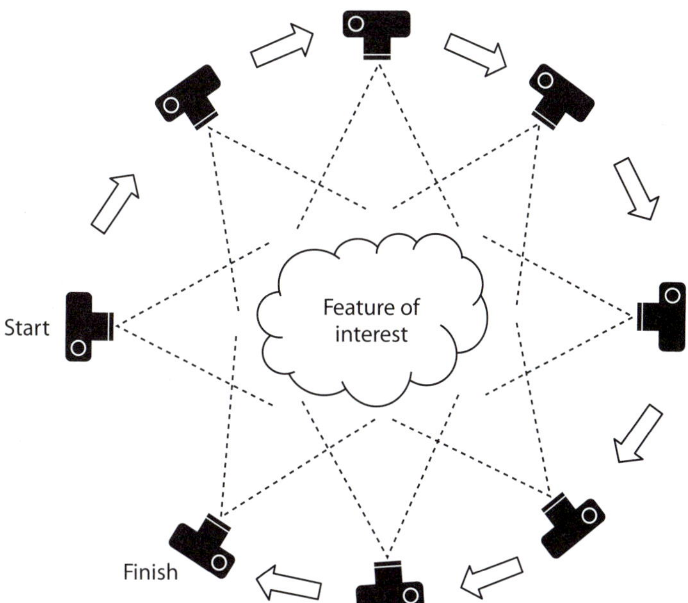

Figure 7-6 Instead of a single stereo-pair, the SfM technique requires multiple, overlapping images. Here the camera is moved along a circular path. Reprinted from Westoby and others. 2012. "Structure from Motion" photogrammetry: A low cost, effective tool for geoscience applications. *Geomorphology*, 179, 333–314, with permission from Elsevier.

graphs have a smaller footprint and the overlap covers a smaller area on the ground (Johnson and others, 2014).

Anderson (2016) reports that as part of the USGS monitoring efforts of the Oso landslide in Washington, regular overflights in a small plane with a wing-mounted camera acquired images of the area every few seconds. The resulting hundreds of aerial photos were then fed into a SfM software package that extracted an extremely detailed DSM of the imaged area and then projected the photos onto the DSM to produce a seamless, aerial mosaic with 7 cm spatial resolu-

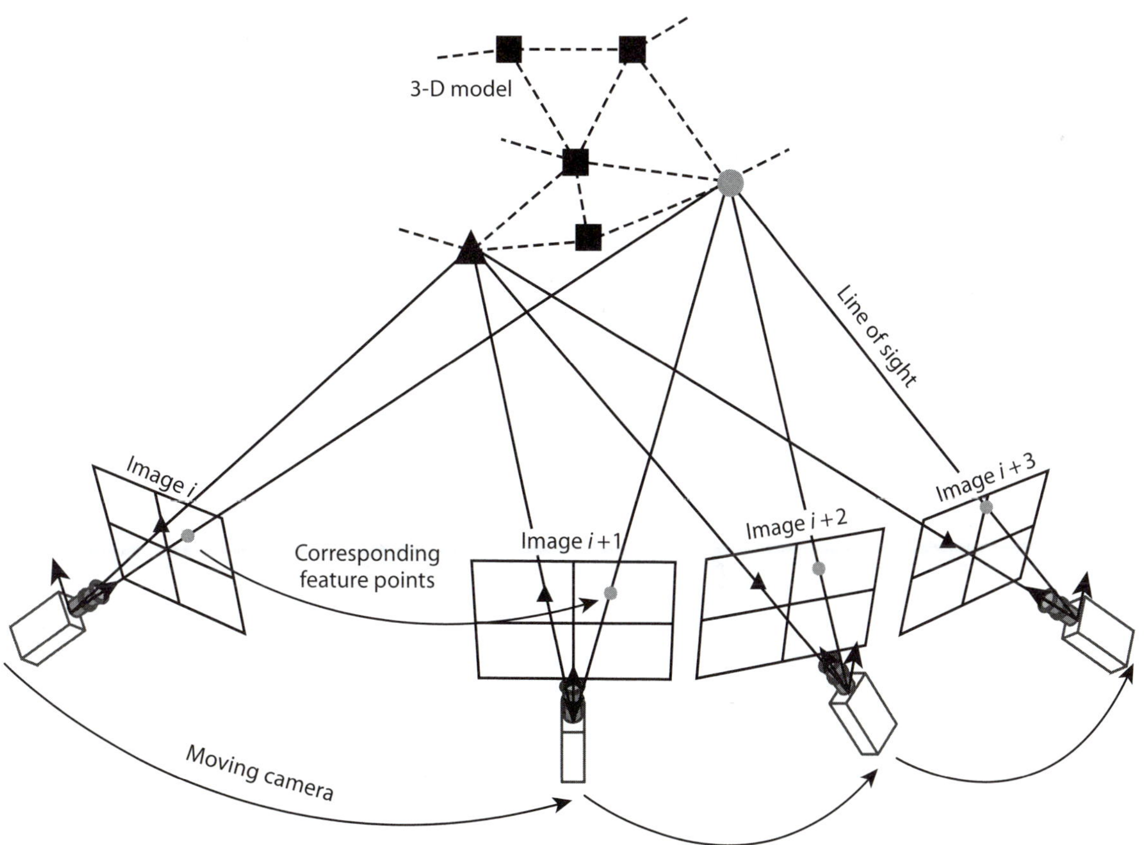

Figure 7-7 A camera is moved to four locations with lines of sight to two targets shown on each image. © Chris Sweeney, Theia Vision Library (theia-SfM.org).

tion (Figure 7-8A). The plane flew 500 to 1,000 ft above the land surface to enable a DSM with 25 cm resolution and vertical uncertainties on the order of ~10 cm to be generated with the SfM method (Figure 7-8B). Fallen tree trunks, tree limbs, small boulders, and channels can be seen on the high resolution DSM. Volume of sediment eroded from the landslide can be calculated from the DEM, along with parameters for hydraulic and sediment transport modeling (Anderson, 2016).

SfM provides an inexpensive and relatively accurate solution for building high resolution DEMs, and has been embraced by drone operators. Abdullah (2019) provides guidance on how photogrammetric techniques and mapping standards (ASPRS, 2014) can enhance the accuracy of DEMs and maps generated with drones and the SfM method. Higher accuracy can be attained with terrestrial and mobile lidar, and larger areas can be acquired with airborne lidar, but the costs can be much greater for these lidar solutions. A comparison of three of these methods is shown in Figure 7-9. Hayakawa and Obanawa (2015) note that an advantage of SfM is that the photographs can give a much better representation of RGB colors for the 3-D model as textures compared to terrestrial lidar.

eRock is a collaborative educational tool that provides virtual 3-D models of geologic outcrops to enhance teaching and research (Cawood and Bond, 2019). The models have coordinates to enable accurate measurements and quantitative analysis as well as metadata and links to key references to place the virtual outcrop in context. They are built with SfM processing of handheld and drone imagery or lidar (Cawood and others, 2017). All eRock models are open access, free to download, and 3-D viewable through a standard web browser, with no need for specialist software packages (e-rock.co.uk).

Open source software and commercial software is available to implement SfM. Westoby and others (2012) used SfMToolkit3 (Astre, 2010). Commercial packages include PhotoScan Professional by Agisoft, Bundler Photogrammetry Package, PhotoModeler, and Pix4D. Cloud-based services that import user data and create DEMs and other geospatial products using SfM and other software include Maps Made Easy, DroneDeploy, PrecisionHawk, and Skycatch (see Chapter 10 for more discussion)

SURVEYING

Field survey data are used as control points to improve the *x*, *y*, *z* accuracy of DEMs generated with other technologies. In addition, points other than those used for control can be used to verify the accuracy of DEMs. Projects that involve property legal maps, boundary line adjustments, fixed engineering works, etc. often require the participation of licensed professional surveyors. The laws vary state by state in the

A. Aerial image of river channel with fallen trees.

B. Shaded DSM generated by the SfM method revealing both land topography and 3-D shapes and heights of fallen tree trunks above the ground surface.

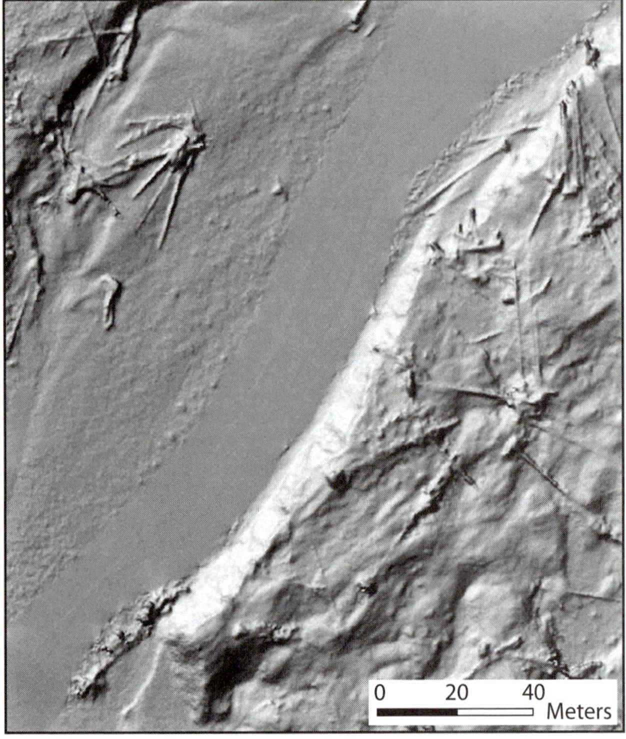

FIGURE 7-8 Comparison of aerial image and shaded DSM, Oso landslide, Washington. From Anderson (2016, Figure 2).

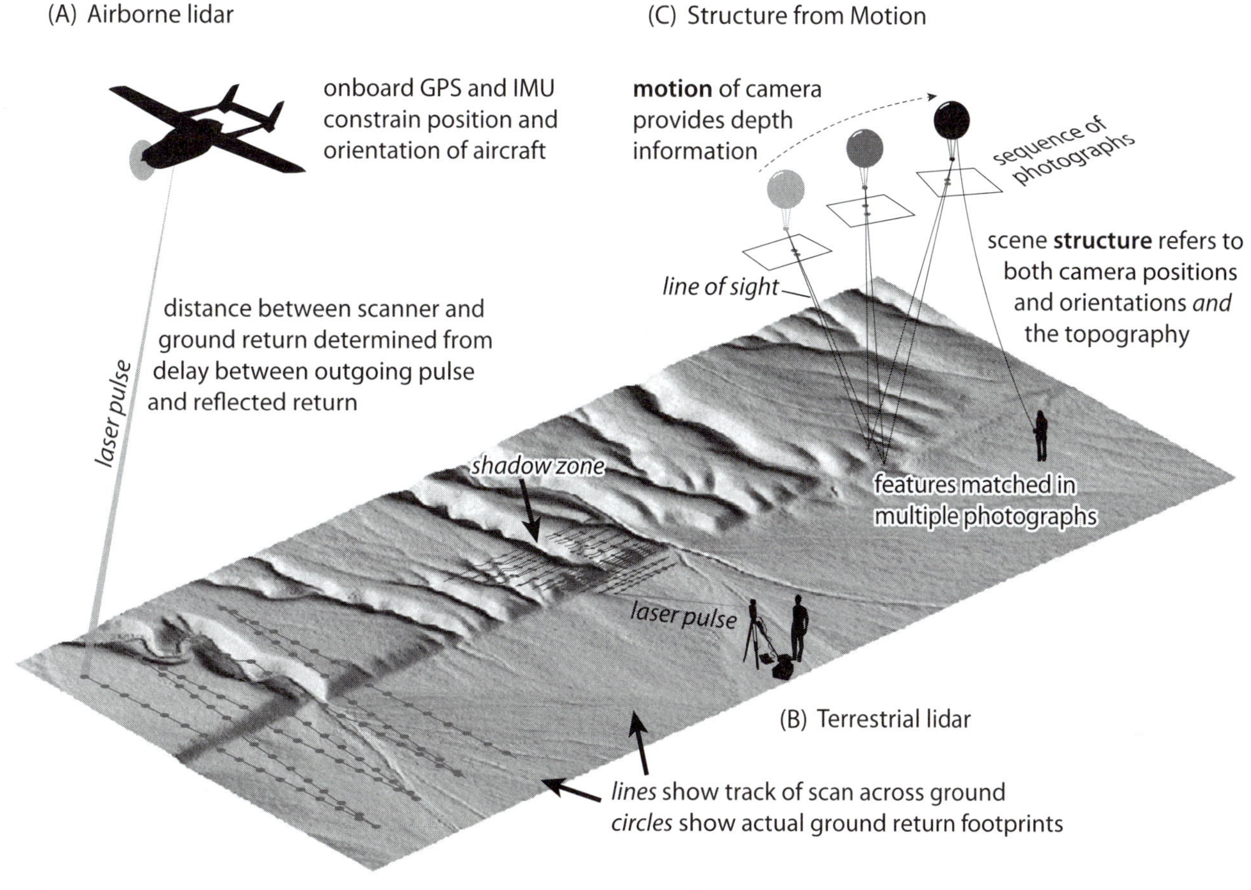
(A) Airborne lidar
onboard GPS and IMU constrain position and orientation of aircraft
laser pulse
distance between scanner and ground return determined from delay between outgoing pulse and reflected return
(C) Structure from Motion
motion of camera provides depth information
sequence of photographs
scene structure refers to both camera positions and orientations and the topography
line of sight
shadow zone
features matched in multiple photographs
laser pulse
(B) Terrestrial lidar
lines show track of scan across ground
circles show actual ground return footprints

FIGURE 7-9 Three methods of producing high resolution digital topography. From Johnson and others (2014, Figure 1).

United States. *Point of Beginning* is an informative website and magazine that is focused on surveying technologies, business, and education (pobonline.com).

LIDAR

Lidar has been used in surveying and mapping since the mid-1990s (Maune and Nayegandhi, 2018). In the last two decades, development of key enabling technologies, such as satellite navigation systems (GNSS), airborne INS and IMU technologies, lasers, and optical detectors, has made lidar the most accurate and highest resolution geodetic imaging and mapping method available (Diaz, 2011).

The lidar sensor transmits laser light to illuminate a target and then records the reflections of that light with a receiver. Lidar operates at the speed of light, measuring the traveling time from the transmitter to the target and back to the receiver. The range measurement process results in a collection of elevation data points (mass points or a cloud) with an x, y, and z location for each point. Lidar sensors can collect 3-D models of features from different platforms, including airborne (aircraft and drones), mobile (vehicles), and terrestrial (tripods). Lidar is acquired by different technologies, including cross-track line scanners, multiwavelength, Geiger-mode, and Flash (Romano, 2015). IceSAT-2 is a satellite-based, laser altimetry system launched in September 2018 to measure the heights of the Earth's ice, vegetation, land surface, water, and clouds (reviewed in Chapters 11 and 16).

LIDAR PRINCIPLES

A lidar instrument emits a pulse of laser light energy with a typical duration of a few nanoseconds and a *pulse repetition frequency* (PRF) that ranges from 10,000 to > 150,000 pulses per second (10 to > 150 kHz). *Scan frequency* is another term used to specify the number of pulses emitted in 1 second. At the same time, the lidar instrument is receiving reflections from these pulses (Figure 7-10). These laser pulses travel at the speed of light (c) (3×10^8 m $\cdot$ s^{-1}).

The lidar instrument calculates the distance (or *range*) to a target by measuring the travel time (t) of the laser pulse from the transmitter to the target and back to the receiver.

$$\text{range} = \tfrac{1}{2}\, tc \qquad \textbf{(7-2)}$$

Lidar instruments emit near-infrared laser light for topographic mapping and blue-green light for bathymetric applications. The cross-track, line scanner pulses of lidar light are moved along the flight line by the forward motion of the aircraft or drone and from side to side (perpendicular to the flight line) by a scanning mechanism in the lidar instrument

(Figure 7-10), typically a scanning or oscillating mirror (Jensen, 2007)). The *scan angle* is the distance the scanner moves from side to side and is measured in degrees. The scan angle is adjusted depending on the application and the accuracy of the desired product (Young, 2011). The combination of forward aircraft motion and side-to-side scanning generates a scanning pattern, typically sawtooth, of laser beam encounters with the ground (Figure 7-10). The *swath width* is the area covered from side to side of the ground track and is determined by the flying height and the selected scan angle.

The beam of laser light diverges as it leaves the transmitter, forming a narrow cone as it descends to the ground (Figure 7-11). The higher the aircraft flies, the larger the *beam divergence* cone becomes. Typical beam divergence settings range from 0.1 to 1.0 millirads (Gatziolis and Andersen, 2008). A beam directed vertically to the ground would form a circular *footprint* if the ground surface was horizontal. Footprints toward the margins of the swath would have slightly elliptical shapes (Figure 7-11), as would footprints that encountered ground that was not horizontal. "Ground" is used as a general term—a footprint could encounter man-made structures or a dense vegetation canopy that has varying slope and aspect, distorting the footprint shape and the amount of energy returned to the receiver. The amount of energy is not uniform over the extent of the footprint—it decreases radially from the center (Gatziolis and Andersen, 2008).

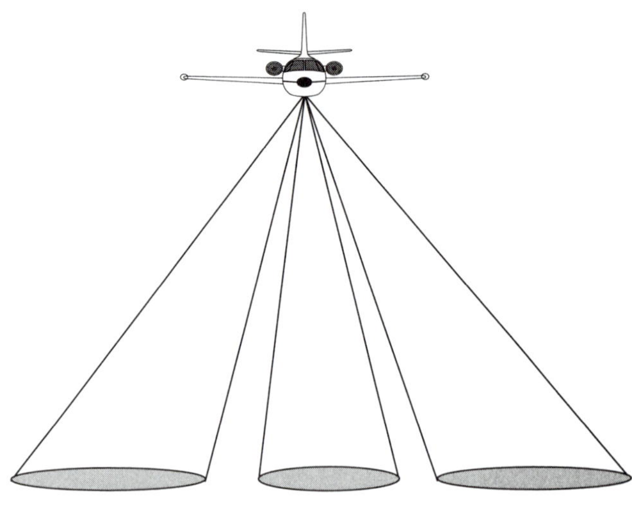

FIGURE 7-11 Nadir and off-nadir beam divergence of a pulse of laser light changes the shape of the footprint.

The size of the footprint is determined by the flying height and the laser beam divergence. At a flying altitude of 1,000 m above ground level, a laser beam divergence of 1 millirad results in a 1 m diameter footprint. At a flying altitude of 1,000 m above the ground, a beam divergence setting of 0.3 millirads results in a 30 cm footprint. A larger beam divergence leads to a lower signal-to-noise ratio

FIGURE 7-10 Airborne lidar acquisition with cross-track scanner. From Gatziolis and Andersen (2008, Figure 3).

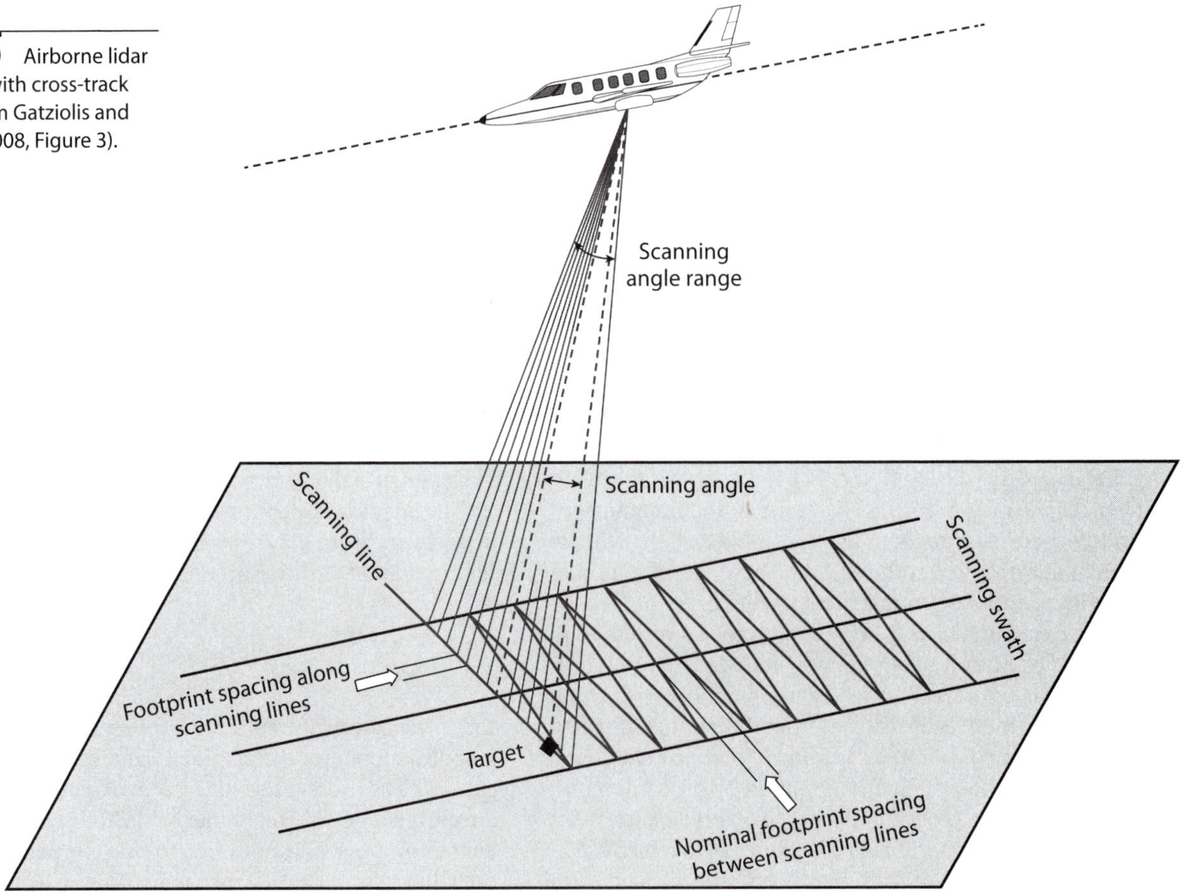

(Gatziolis and Andersen, 2008) and less detailed data reaching the receiver.

Footprint spacing (also *nominal point spacing* [NPS]) is the nominal distance between the centers of consecutive beam footprints along the scan lines and across the scan lines (Figure 7-10). Footprint spacing (*across-track resolution* and *along-track resolution* [Diaz, 2011]) is a function of scanning frequency, flying height, and aircraft velocity (Gatziolis and Andersen, 2008). Decreasing the footprint spacing by flying lower and slower will increase the number of times your target areas could be hit with laser beams, increasing the amount of information received and improving the definition of your targets.

Lidar Returns

The lidar beam interacts with features on the surface of the Earth within each footprint—tree canopies, trunks and branches, understory shrubs, grass cover, bare ground, and man-made objects (buildings, bridges, electric transmission lines and towers, etc.). Portions of the energy within the beam are returned to the receiver as the laser light encounters surfaces at different distances from the transmitter.

Two common types of lidar systems are *discrete return* and *full waveform* (Gatziolis and Andersen, 2008) (Figure 7-12). Discrete return systems can record one to five returns for each pulse during a flight while full waveform records the returned energy in a series of equal time intervals (Lim and others, 2003). Figure 7-12 shows a lidar beam intersecting the canopy along the left portion of a tree. The locations of five discrete returns within the canopy are displayed as dots on the right side of the figure. The full waveform return is displayed on the left side of Figure 7-12. The waveform shape reflects the forest structure from the top of canopy, through the crown volume and understory layers, to the ground surface (Lim and others, 2003). The number of recording intervals determines the amount of detail in a full waveform laser footprint. Waveform lidar yields a vertical summation of the returns from the pulse but usually does not achieve the fine horizontal resolution of point discrete lidar. It is considered more useful for some biological measurements.

If the beam only encounters one surface, and is reflected back to the receiver with no other earlier or later returns, this return is termed *first and last return*. First and last returns are most often associated with bare earth and tops of buildings. When lidar interacts with vegetation, multiple returns are common (Figure 7-12). For example, some of the light within the beam could reflect off of the top of a forest canopy (*first return*), then another reflection could occur as the beam backscatters off of a branch below the canopy (*intermediate return*), closer to the ground some of the remaining light could be backscattered by an understory of shrubs (another *intermediate return*), and finally light within the beam that penetrated the tree canopy and branches and the understory could reflect off the ground (*last return*). Complex post-processing algorithms and skilled analysts are required to accurately extract these multiple returns from lidar surveys and to convert them into useful information and models.

The returns are collected by the lidar system as a *cloud* of data points, referred to as mass points, each with an *x*, *y*, *z* location. The horizontal location of the mass points is aligned with the acquisition scan lines. First returns from the top of tree canopies are seen on Figure 7-13. Jensen (2007, p. 338) lists and reviews the most important variables

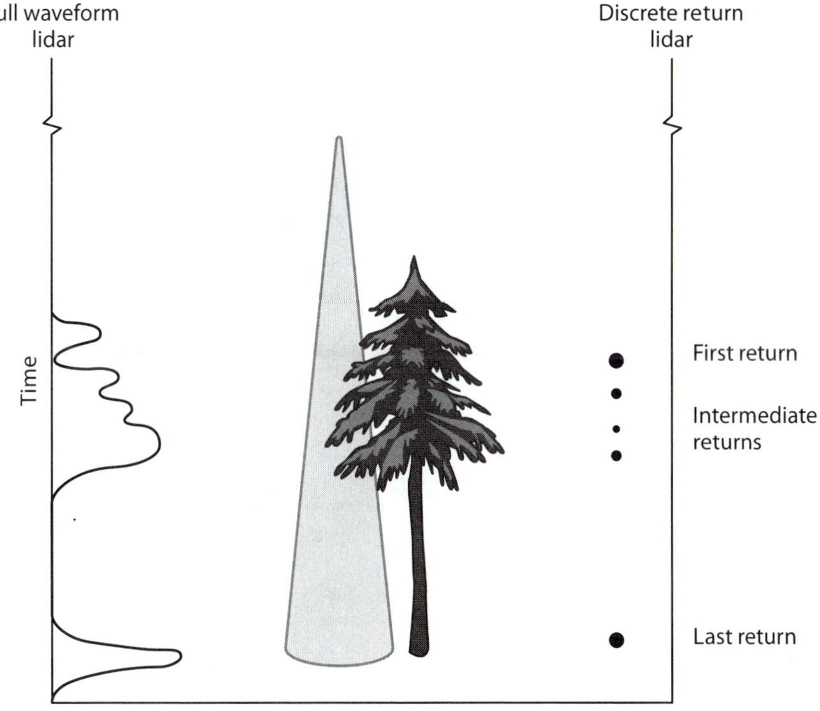

Figure 7-12 Comparison of discrete return and full waveform vertical sampling. From Lim and others. 2003. LiDAR remote sensing of forest structure. *Progress in Physical Geography*, 27(1), 88–106. Reprinted by permission of Sage Publications, Ltd.

controlling the accurate conversion of laser range information into georeferenced mass points. These include:

- The *x*, *y*, *z* location of the laser in 3-D space at the time of the laser pulse,
- The attitude (roll, pitch, and heading) of the laser at the time of the laser pulse,
- The scan angle of the lidar at the time of the laser pulse,
- The effect of atmospheric refraction on the speed of light, and
- The laser pulse travel time from the lidar instrument to the target and back.

LIDAR INTENSITY

In addition to being capable of recording multiple returns, lidar systems collect an *intensity file* (Jensen, 2007). The intensity file is most often the maximum returned echo from all the pulse returns (Baltsavias, 1999). In topographic mapping, the intensity image is generated with the NIR laser light. Interpreting an intensity image can be complicated as correlation between pixel brightness values in an intensity image that is generated by an active NIR laser beam and in a visible light or NIR grayscale image that is created with reflected sunlight is not 1 to 1 (Jensen, 2007, pp. 343–346). The intensity image of the Haiti National Palace (Figure 7-14A) displays rooftops and hardscape road surfaces with tones ranging from dark gray to white while the visible light image (Figure 7-14B) displays rooftops and hardscape with medium to light gray tones. Both images have 50 cm spatial resolution and were acquired to aid in relief efforts within weeks of a 7.0 earthquake that devastated Haiti's capital city of Port-au-Prince on January 12, 2010 (IPLER/RIT LIAS, 2010; Rochester Institute of Technology, 2010).

AIRBORNE LIDAR SYSTEMS

Airborne lidar can be flown on fixed-wing aircraft, helicopters, and drones. There are two basic types of airborne lidar: topographic and bathymetric. In addition, multiwavelength lidar is being implemented and different lidar technologies are being declassified and entering the commercial marketplace. Horizontal and vertical accuracy of lidar data depends on many factors. The accuracy is lower the farther the platform is from the target, the larger the footprint, and the larger the footprint spacing.

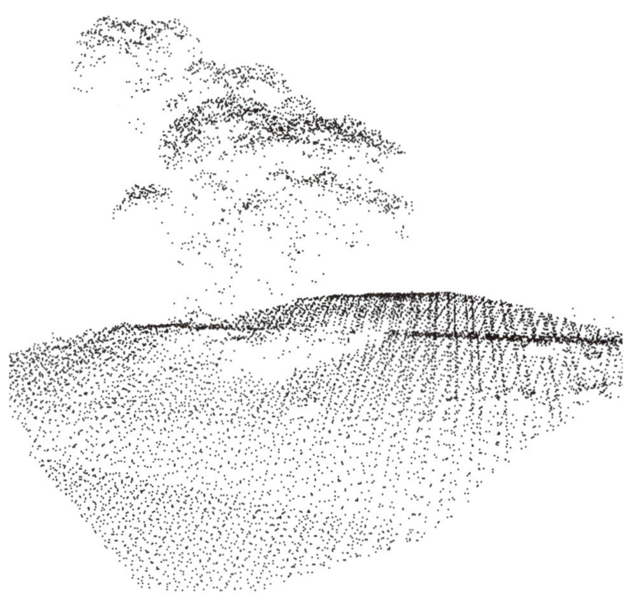

FIGURE 7-13 Point cloud generated with lidar. Note alignment of mass points with scan lines. Data courtesy Pacific Gas & Electric Company, Diablo Canyon Power Project, San Simeon, California.

A. Lidar intensity image.

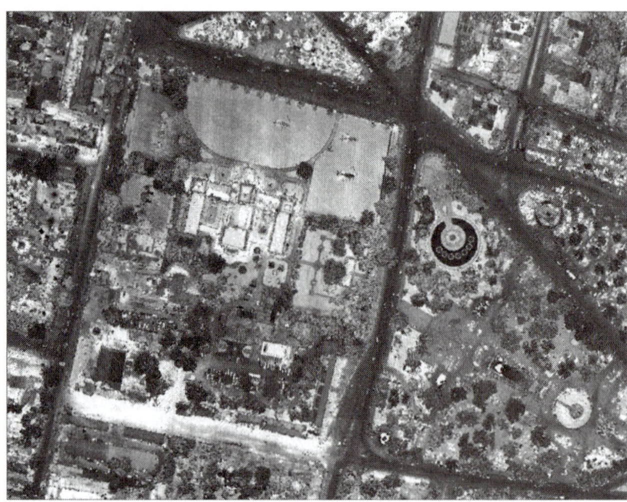

B. Visible light image.

FIGURE 7-14 Comparison of intensity and visible light images, Haiti National Palace, Port-au-Prince, Haiti. Courtesy IPLER/RIT LIAS (2010).

Topographic Lidar

Topographic lidar uses eye-safe, NIR laser light, typically with a wavelength of 1,064 nm. Small footprints range from 10 to 50 cm and are useful for detailed local mapping, edge detection, and vegetation canopy studies. Large footprints range from 10 to 30 m, cover a large swath, and are useful for regional mapping.

In the following discussion lidar data collected by the 2008 Contra Costa County Orthoimagery Project in Pleasant Hill, California, are used to explain and visualize topographic lidar products. Aerial imagery with 7.6 cm (3 in) pixels was acquired in color and color IR. In Figure 7-15, educational campuses (Diablo Valley College and College Park High School) cover the east-central terrain, a golf course in hilly terrain is in the northwest, and suburbs dominate the western, southern, and eastern margins of the area. Lidar and imagery data from four tiles are digitally mosaicked in Figures 7-15 to 7-23.

Digital Surface Model Typical flying altitude over the area is 720 m with lidar spot spacing of 1.37 m. Mass points associated with first return are visualized using an interpolation algorithm to create a raster DEM with 1.34 m (4.4 ft) grid cells. The interpolation process of first return lidar generates a DSM. The elevation range (above sea level) in the DSM is 7.6 to 95 m (25 to 310 ft) (Figure 7-16) with higher elevations displayed with lighter shades of gray. To better interpret topographic relief, buildings, and trees, the DSM is hillshaded with an illumination source from the northwest at 30° above the horizon (Figure 7-17). In Figure 7-17, hilly topography can be seen in the northwest, trees and houses in the suburbs, sport fields as flat terrain, and larger buildings in the educational campuses as large rectangles. The DSM can also be color-coded for elevation and the hillshade layer made partially transparent and draped on the DSM to enhance viewer understanding of topography, buildings, and trees (Plate 21).

Features of interest in the DEM can be analyzed with an interactive display that enables perspective viewing of the mass points in the point cloud as well as the raster surface that is created from the point cloud. In Figure 7-18, a subarea in the hillshade DSM is enlarged to reveal the surface that has been created from the point cloud. The model captures the canopy of large trees to the west and a building with a flat roof and four rectangular openings that are recessed below

FIGURE 7-15 Digital orthophotograph acquired with 7.6 cm (3 in) pixels in color and color IR, Pleasant Hill, California. Courtesy Contra Costa County Orthoimagery Project.

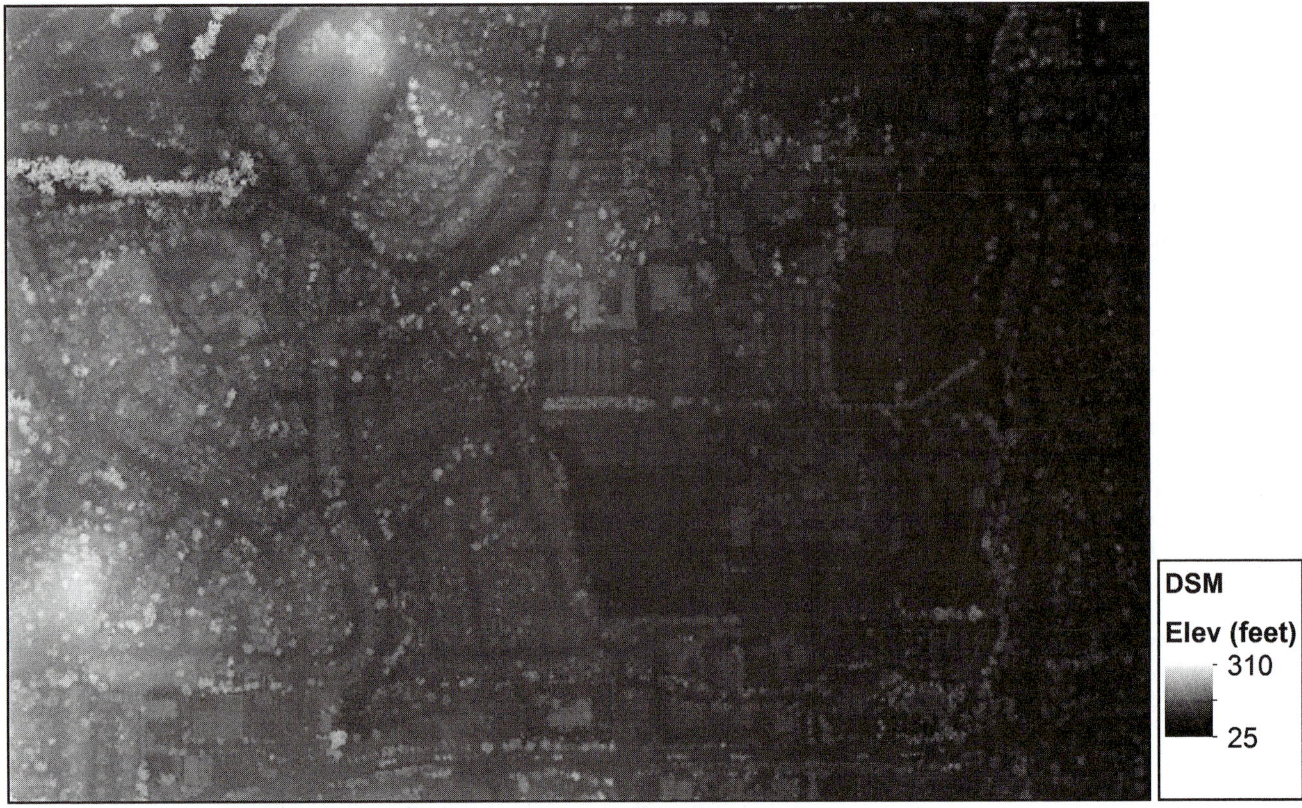

Figure 7-16 DSM generated by lidar first returns, Pleasant Hill, California. Courtesy Contra Costa County Orthoimagery Project.

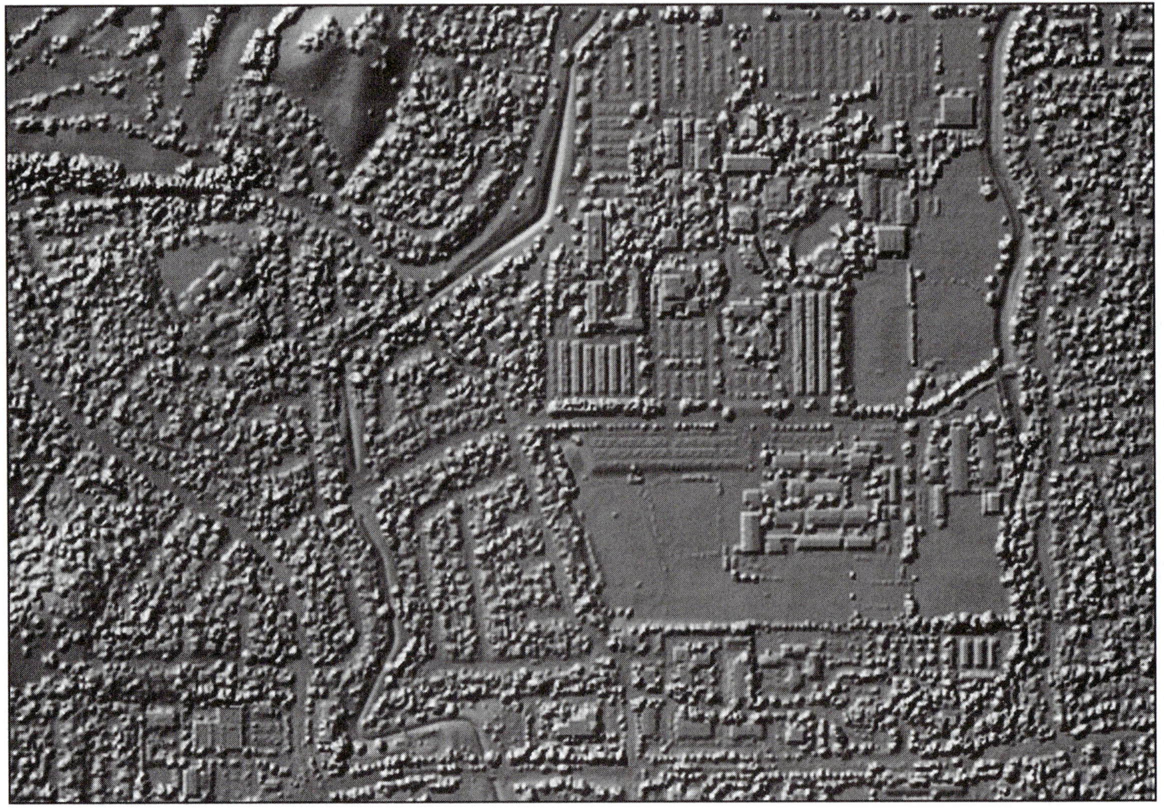

Figure 7-17 Hillshade of DSM (Figure 7-16) with illumination source from northwest and 30° above the horizon, Pleasant Hill, California. Courtesy Contra Costa County Orthoimagery Project.

the roof line. Elevations range from 10 to 32 m (33 to 104 ft) in the model. Elevations of points or surfaces are color-coded in the lidar-processing software to better interpret the top of the building, the surrounding ground, and the grove of trees to the west. Visualizing DSMs with hillshade models, point clouds, and mass point surfaces assists in assessing the quality of the elevation data. DSMs can be modified to minimize the presence of vegetation (trees) while retaining man-made structures to support drainage network modeling.

Digital Terrain Model DTMs, or bare earth models, are required for generating orthophotographs (McGlone, 2013). Both vegetation and man-made structures are eliminated from the DSM so aerial imagery can be rectified to the surface of the Earth. Lidar analysts and algorithms remove first and intermediate returns associated with vegetation and man-made structures. This process results in a collection of last return/bare earth mass points that have gaps where vegetation and man-made structures were located (described below). The last return/bare earth mass points are used to create the DTM. The DTM can be displayed with scaled elevation (Figure 7-19), hillshaded to improve visualization of the Earth's topography (Figure 7-20), and color-coded (Plate 22).

The hillshade DTM (Figure 7-20) clearly shows the terrain under the suburban houses in hilly terrain (western portion of the area) to be a flat surface. Cut and fill of the hilly terrain with earth-moving equipment prior to housing construction created the flat housing pads. In an enlargement of a suburban area (Figure 7-21), gaps are clearly visible as white features where vegetation and man-made structures (buildings and roads) are located in the last return/bare earth mass points. The flat DTM surfaces under buildings are interpolated from the elevations of the mass points surrounding each building footprint. The enlargement reveals the very high number of lidar mass points, each containing a set of horizontal coordinates and vertical elevation (x, y, z) that are used in the DTM model (Figure 7-20 and Plate 22).

Sharp breaks in the landscape (cliff edges, bottoms of valleys, narrow ridgelines, etc.) or in man-made structures (edges of buildings, curbs, canals, pools, bridges, etc.) are not captured by the lidar footprints, especially if the footprint spacing or footprint size is large. To include these sharp breaks in DSMs and DTMs, overlapping aerial imagery is

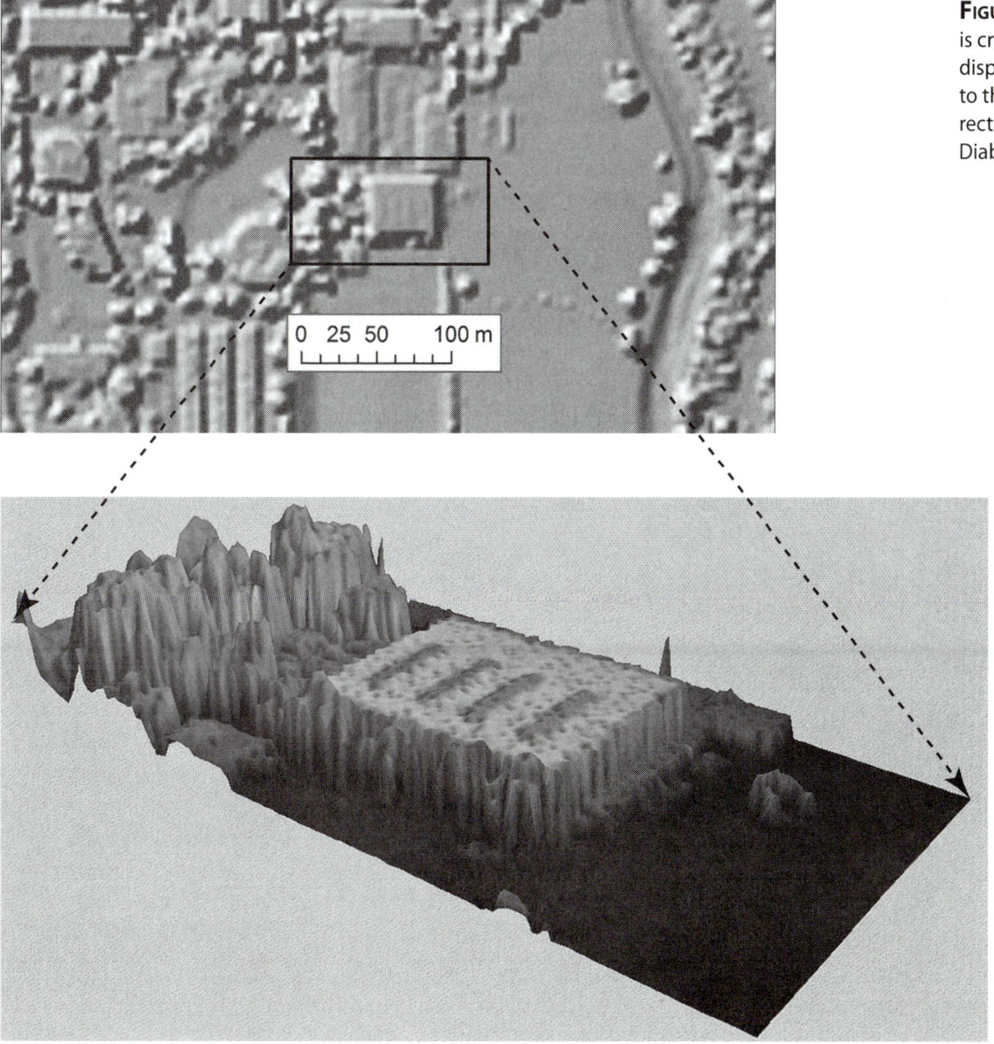

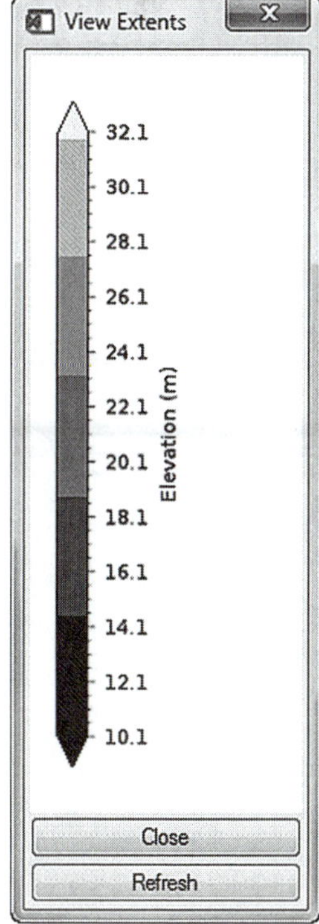

FIGURE 7-18 Raster surface that is created from the DSM point cloud displaying the canopy of large trees to the west and the flat roof with rectangular openings of a building at Diablo Valley College.

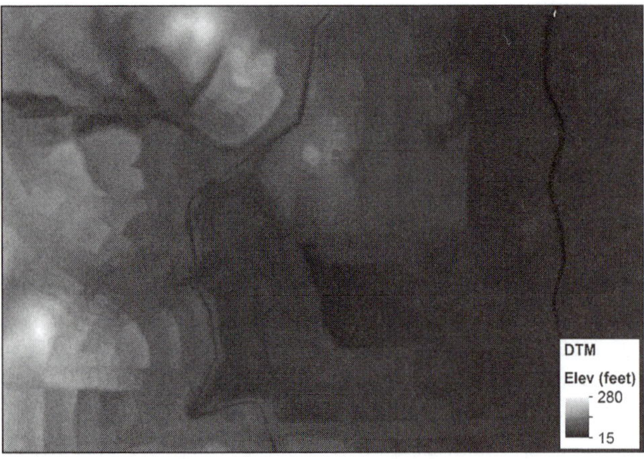

FIGURE 7-19 DTM of lidar last returns with buildings and trees in DSM removed. Courtesy Contra Costa County Orthoimagery Project.

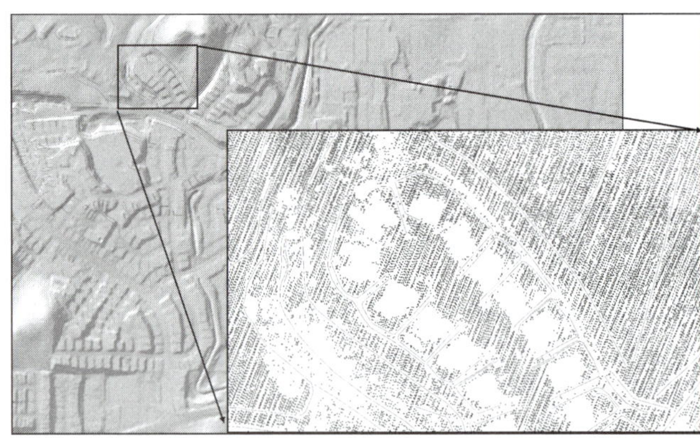

FIGURE 7-21 The enlarged area shows the point cloud of the DTM with lidar first return buildings and trees in subdivision removed (white areas).

acquired along with the lidar survey so that photogrammetric compilation can be accomplished using the imagery. The compiler visually maps out sharp breaks (breaklines) on the landscape and on man-made structures (Figure 7-22). The breaklines are integrated with the lidar mass points to generate a more accurate DTM. This DTM is then used to remove the effects of lens (radial) and relief displacement on overlapping aerial images so that a photogrammetric *orthoimage* (or *orthorectified* image) with uniformly scaled pixels is generated (Figure 7-15).

Topographic Contours The DTM is contoured and a new topographic map is generated with contours using geospatial software (Figure 7-23). Contour intervals should not exceed the stated vertical accuracy of the lidar survey. The 2008 Contra Costa County Orthoimagery Project adhered to published horizontal and vertical accuracy standards (discussed below). The lidar data met or exceeded the US Federal Emergency Management Agency guidelines for 0.6 m (2 ft) contour interval mapping.

Bathymetric Lidar

Bathymetric lidar uses green laser light, typically 532 nm, to penetrate water. Green light has a short enough wavelength to minimize absorption by water (which increases with longer wavelengths) and is long enough to minimize absorption by chlorophyll (which increases in the shorter blue wavelengths). Bathymetric lidar is of high value for filling the 0 to 10 m depth gap in coastal mapping (Nayegandhi, 2006). Onshore terrain is mapped with topographic lidar (and other DEM-generating technologies) while deeper seabeds are mapped with multibeam echo sounders. The theory behind using lidar for mapping bathymetry is provided by Guenther (2001). Water clarity is the most significant limitation for airborne lidar bathymetry as it limits the maximum depths for laser penetration (Guenther, 2001).

Multiwavelength Lidar

Multiwavelength lidar is being developed to improve 3-D land cover classification, the separation between vegetation

FIGURE 7-20 Hillshade of DTM (Figure 7-19) with illumination source from northwest and 30° above the horizon. Courtesy Contra Costa County Orthoimagery Project.

FIGURE 7-22 Breaklines manually digitized from stereo models of aerial images to capture sharp breaks in topography. Courtesy Contra Costa County Orthoimagery Project.

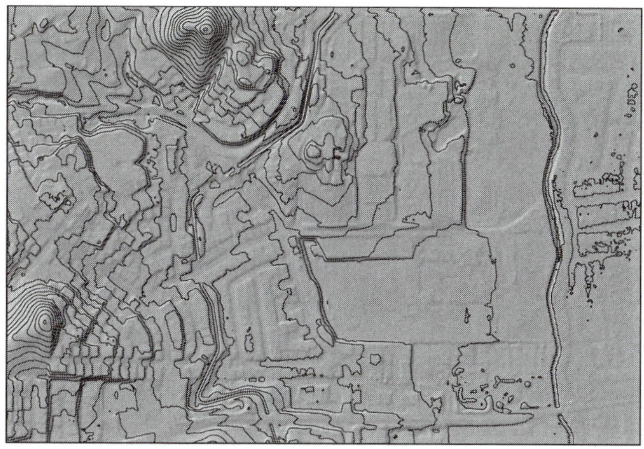

FIGURE 7-23 Topographic map generated with 10 ft contour intervals derived from the DTM.

and nonvegetation, utilization of vegetation indices, shallow water bathymetry, and dense topography mapping (Thomas, 2015). A commercial system, the Teledyne Optech Titan, has a single sensor with three active lasers at 532 nm, 1,064 nm, and 1,550 nm. The system also has passive digital cameras to gather coincident imagery. Multispectral image processing techniques use the three intensity bands (532 nm, 1,064 nm, and 1,550 nm), along with other geospatially aligned raster layers, to improve land cover classification (Thomas, 2015).

Geiger-Mode Lidar

A recent entry into the commercial lidar marketplace is the Geiger-mode lidar system by L3Harris (Romano, 2015). This lidar system is designed to fly higher and collect data faster using a technology that is different when compared with the lidar systems discussed previously that transmit laser light as single pulses systematically arranged along scan lines and within beam footprints. The Geiger-mode lidar collects returns from multiple angles, sampling the same spot on the ground multiple times. The sampling rate is very high, resulting in a higher density of returns that improves the detail of infrastructure, characterization of biomass, and penetration of vegetation canopy. In addition, multiple looks at features from different angles minimizes shadows (data gaps) that can occur with conventional single-look lidar systems, especially if the scan angle is excessive along thc margins of thc data collection swath. A huge amount of data is collected with this system—ranging from terabytes to petabytes—which requires specialized hardware, software, and data management to process and deliver the high-end lidar products. The Geiger-mode is very sensitive to ambient light and does not detect the intensity of the return (Higgins, 2014).

Flash Lidar

Small, solid state lidar systems with no moving parts are being developed. The "flash" or focal plan array (FPA) lidar uses a solid state array of pixels that detect one photon each. The technology significantly reduces the mass and power requirements of airborne lidar as well as promises to improve geometric accuracy and enable parallel processing (Higgins, 2014). As an example, Advanced Scientific Concepts' Peregrine system is a lightweight, low power video camera that illuminates an area of interest by the field of view of the lens with a single, short (5 nanosecond), eye-safe laser pulse per frame and captures the reflected laser light in the form of a 3-D range point cloud and coregistered intensity data. The array is 128 by 21 and the camera weighs 680 g (24 oz). Flash lidar is used for real-time vehicle crash avoidance systems, autonomous navigation, and object tracking. Lightweight flash lidar has not yet reached the quality and accuracy of conventional scanning technology.

Miniaturization

Miniaturization of hardware and more sophisticated software and data management systems is enabling small lidar systems to be developed and deployed in operational settings. As an example, the Routescene LidarPod® provides GNSS(GPS)/INS location information and generates a georeferenced point cloud using 32 laser sensors, operating at 905 nm. The instrument weights 2.5 kg (including GPS antennas and cables), is 32 cm long with a diameter of 10 cm (Figure 7-24). These miniaturized lidar systems are ideal for deployment on drones.

Mobile Lidar

Mobile lidar is a mapping system that integrates lidar instruments, cameras, and location/navigation technology on a moving vehicle (Figure 7-25). The vehicles can include cars, trains, and boats. Very high data density levels can be achieved (e.g., 150 points/ft^2) and 100,000s of points can be collected per second. The system is driven through urban landscapes and collects a cloud of mass points related to all visible, aboveground street features, including pavement, signs, utility poles, transmission wires, building facades, tree trunks and canopies, and floor elevations for all building entrances (Roy, 2012).

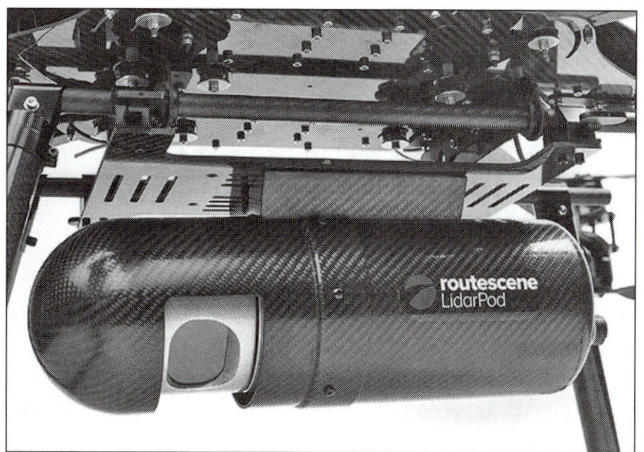

FIGURE 7-24 The Routescene LidarPod®—an example of the miniaturization of lidar systems. Courtesy Routescene (Mapix Technologies).

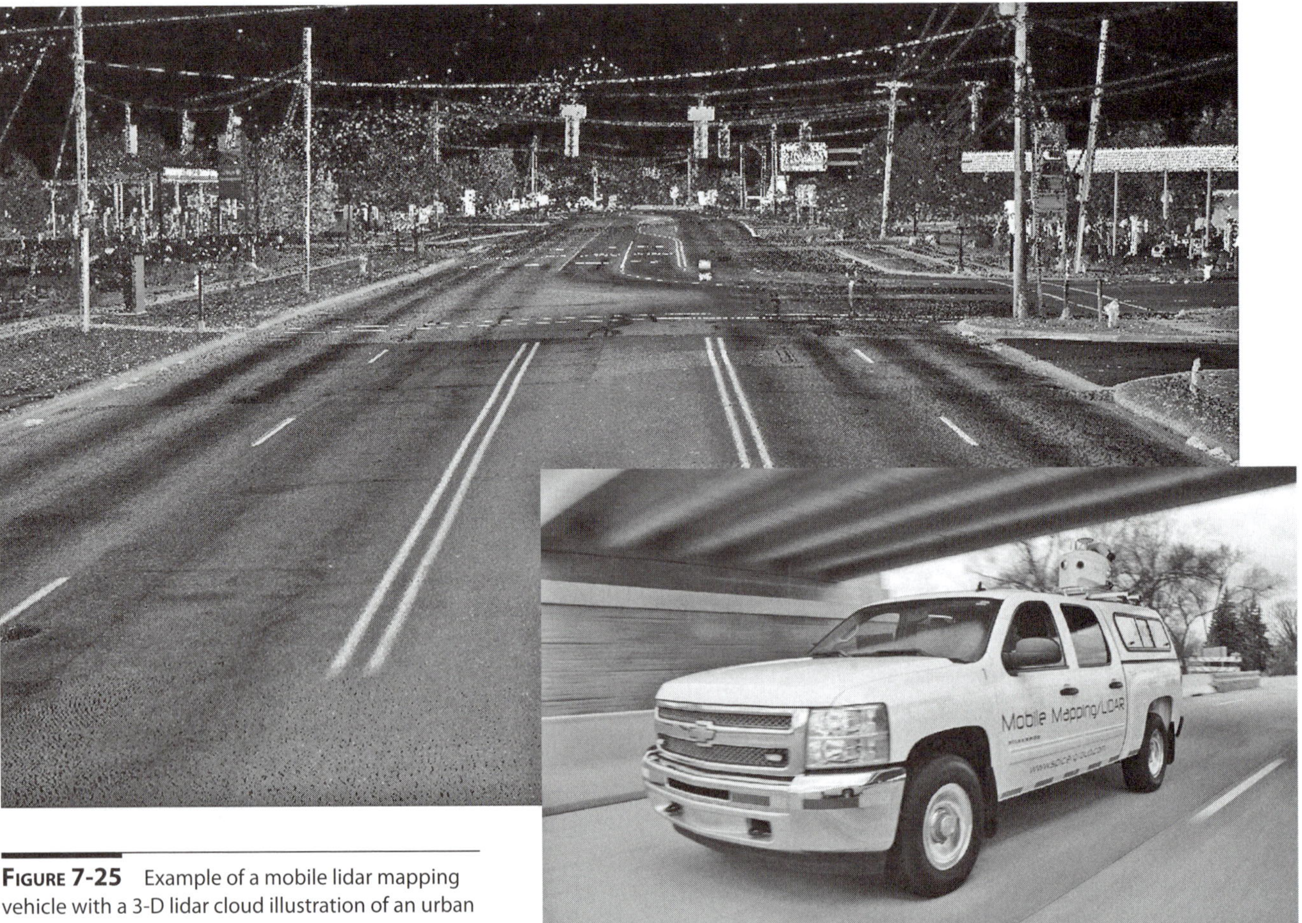

FIGURE 7-25 Example of a mobile lidar mapping vehicle with a 3-D lidar cloud illustration of an urban street intersection. Courtesy Spicer Group, Inc.

Terrestrial Lidar

A terrestrial lidar instrument is typically placed on a tripod where the lidar beam can be directed over a wide horizontal and vertical range. The scanner is on a rotary table that enables horizontal motion with a mirror that tilts vertically to transmit and receive returns above or below the horizontal plane. The 3-D point clouds can be matched with digital images of the scanned area to create realistic, 3-D models of outcrops, landslides, statues, or historic buildings. Reestablishing the survey site through time enables change detection, e.g., retreat of shoreline cliffs, movement of sand dunes, subsidence and uplift, mass wasting, and changes in vegetation cover. Terrestrial lidar can collect survey-quality point data quickly and accurately.

Bellian (2003) used terrestrial lidar to create high resolution, virtual, geologic outcrop models (Figure 7-26). A lidar point cloud was generated with a 1 cm x, y, z spacing that approximated the 3-D shape of the outcrop (Figure 7-26A). The laser intensity image was blended with a digital photo and applied as a texture to the point cloud model (Figure 7-26B). For scale, a human figure 6 ft tall is at the bottom of the two model images. Image processing of the intensity image with other parameters distinguished between sands and shales as well as differences in sand grain mineralogy in the virtual outcrop model.

RADAR

Radar is the acronym for "radio detection and ranging"; it is an all-weather, active remote sensing system that operates in the microwave band of the electromagnetic spectrum, ranging from millimeters to meters in wavelength. Radar technology, including the use of *synthetic aperture radar* (SAR) to generate DEMs with *interferometry* (InSAR or IfSAR), is discussed in Chapter 6. Zhou and others (2009) provide a review of InSAR applications for Earth and environmental science research.

As demonstrated in Chapter 6, radar does not penetrate closed canopy vegetation (Figures 6-39 and 6-40). However, vegetation canopies with openings are penetrated by longer wavelength radar beams, which interact with limbs and trunks, and potentially can reach the ground surface and reflect back to the radar antenna (Figure 7-27). L-band (23.5 cm) and P-band (85 cm) radar provide additional information on vegetation biomass and topographic relief beneath vegetation cover through openings that are not penetrated with the shorter X- and C-bands.

Figure 7-28 shows a partially open forest canopy in the tropics imaged with X-band and P-band wavelengths recorded by the airborne GeoSAR system (Sharp and Morton, 2007). Both the X-band and P-band images (Figure 7-28A and B) show cleared areas with straight margins,

A. Laser point cloud at 1 cm *x*, *y*, *z* point spacing of the outcrop.

B. Laser intensity image blended with a digital photo and applied as a texture to the *x*, *y*, *z* model.

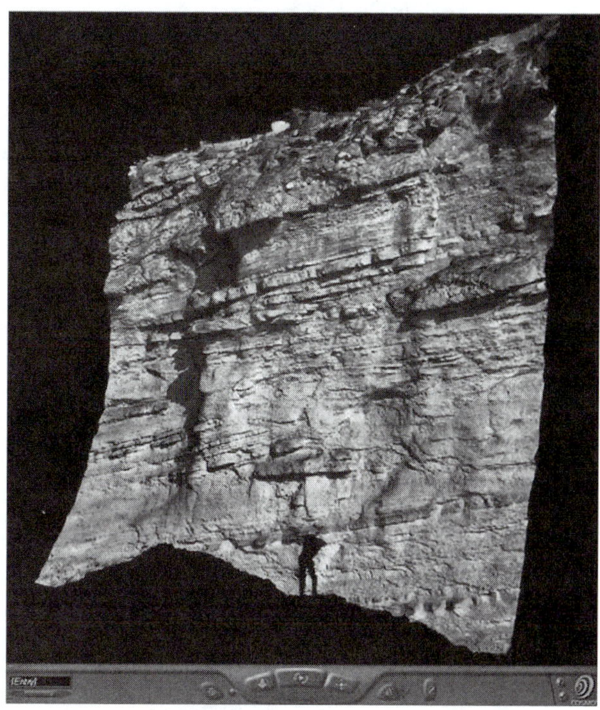

FIGURE 7-26 Terrestrial lidar is used to create high resolution, virtual, geologic outcrop models. Person in the foreground is approximately 6 ft tall. From Bellian (2003, Figures 1 and 3).

angular corners, and smooth dark gray to black tones while the forest is displayed with a mottled, uneven texture and medium to dark gray tones. Discontinuous, curvilinear features with a dark tone are difficult to interpret in the south-central portion of the X-band image (Figure 7-28A). In contrast, the P-band image (Figure 7-28B) clearly shows these to be continuous, meandering stream channels beneath the partially open forest canopy because the P-band wavelength penetrates the partially open forest canopy over the drainage channels. Figure 7-28C is a sketch map showing the drainage channels and clearings interpreted from the P-band image. The GeoSAR capability to image the top of the canopy with the X-band and the underlying bare earth with the P-band is used to generate DSMs and DTMs in forested Alaska, as discussed below.

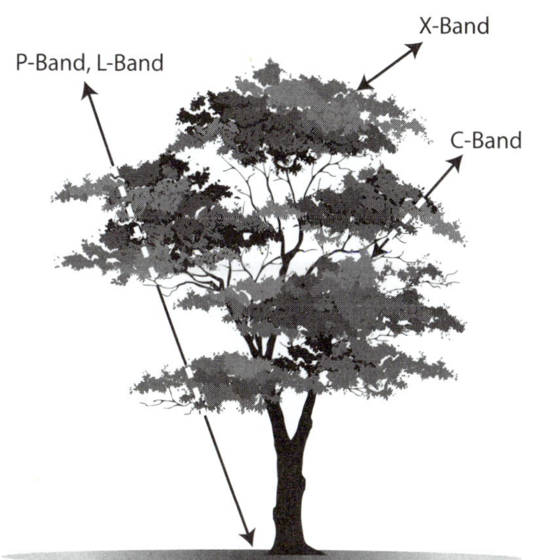

FIGURE 7-27 Canopy penetration varies with different wavelength radar.

ONE PLATFORM WITH TWO ANTENNAS (SINGLE PASS INSAR)

Aircraft and the 2000 NASA Shuttle Topography Mission (SRTM) (Chapter 6) are platforms that carry two radar antennas for generating DEMs in a single pass. NASA's airborne EcoSAR (Martin, 2014), Intermap's Star-3i, and Fugro's GeoSAR are airborne interferometric systems with dual antennas. The airborne systems provide more detailed DEMs compared to satellite systems and are especially useful in cloud-covered tropical terrains where clouds degrade images acquired with passive remote sensing systems. Aircraft can be configured for interferometric radar collection by installing the P-band antennas near the ends of the wings and the X-band antennas on each side of the fuselage. Figure 7-29 shows a profile of the GeoSAR aircraft flying 10 km above the terrain with the P-band antennas on the end of the wings and the X-band antennas on each side of the fuselage. Each antenna is simultaneously imaging a swath approximately 10 km wide.

The State of Alaska and USGS employed the GeoSAR system to generate X-band orthorectified radar imagery, DSM, and DTM with 5 m spatial resolution (postings) of central Alaska that included Fairbanks, Anchorage, and Mt.

A. GeoSAR X-band image does not display continuous drainage channels located beneath forest canopy. From Sharp and Morton (2007).

B. GeoSAR P-band image reveals drainage channels and networks located beneath forest canopy. From Sharp and Morton (2007).

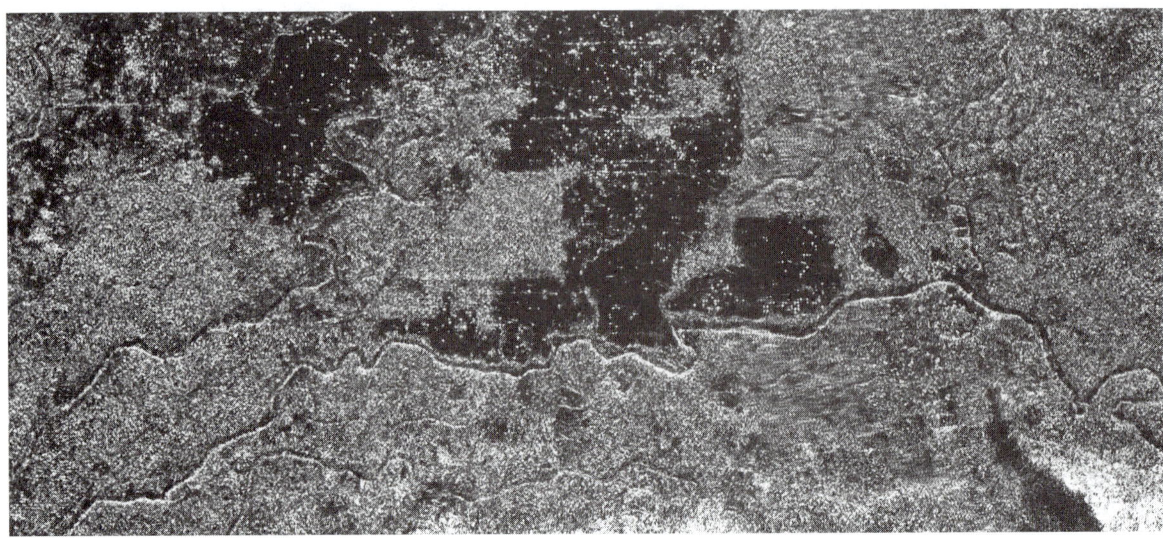

C. Sketch map of drainage channels and clearings interpreted from the P-band image.

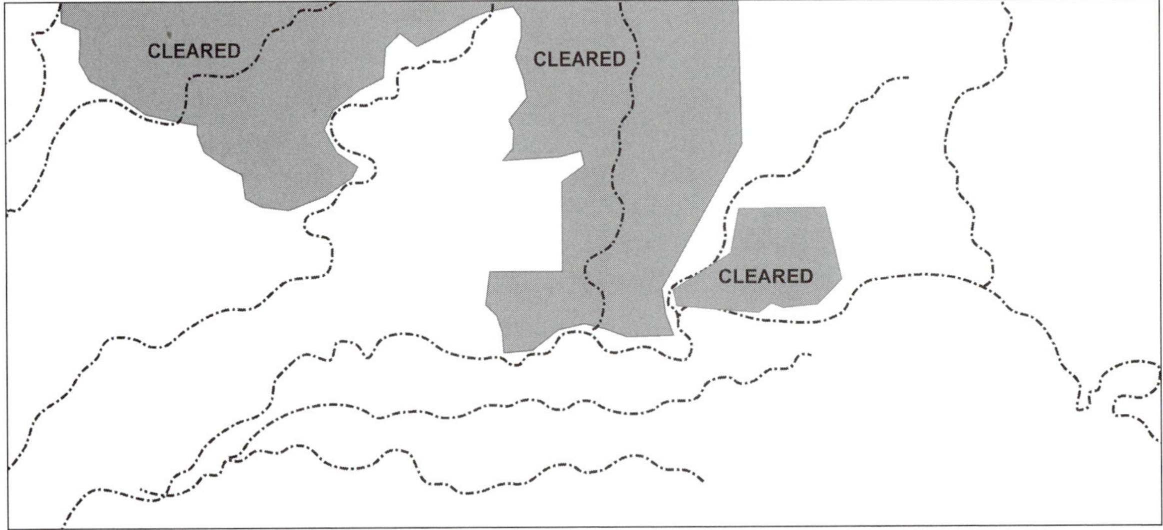

FIGURE 7-28 Penetration of partially open vegetation canopy with longer wavelength radar.

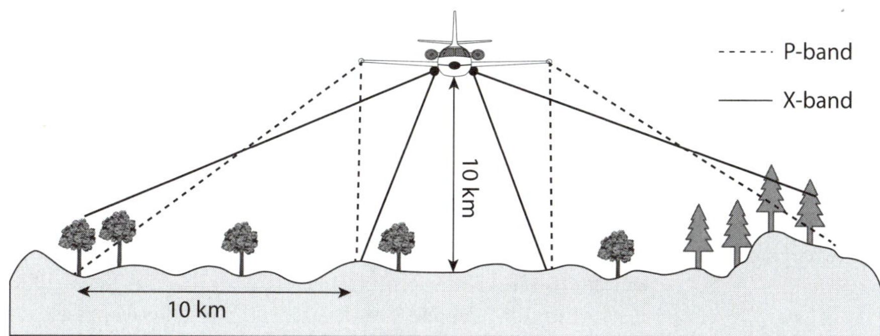

FIGURE 7-29 Example of Fugro's GeoSAR system with radar antennas near the end of the wings and along the fuselage for interferometric measurements. From Petrie (2007).

Denali. The landscape is highly variable with mixed land cover that ranges from flat and forested, to hilly, to mountainous with snow and ice cover (Kampes and others, 2011). The mapping program was designed to improve Alaska maps on a broad, statewide scale. An example of a GeoSAR orthorectified image, DSM, and DTM from northwestern Fairbanks is provided with Plate 23. The area is hilly and forest-covered with suburban development that has cleared the forest cover for roads, utility corridors, and buildings.

The GeoSAR X-band radar image displays a network of interconnected suburban roads and planned subdivisions cut through the forest. The forest canopy as imaged by the X-band has a smooth texture with a medium gray tone (Plate 23A). Radar shadows are cast from trees across roads that are cut through the forest. These shadows accentuate the narrow man-made cuts as dark linear features. Smooth man-made clearings are dark (approaching specular reflection). Many of the small bright objects are buildings (acting as corner reflectors).

Plates 23B and 23C are color-coded DEMs artificially illuminated from the northwest at an elevation of 30° above the horizon to highlight subtle surface (DSM) and bare earth (DTM) relief. The X-band DSM forest and shrub cover shows linear "grooves" caused by the rectangular grid of major roads, large sharp-edged depressions in the western portion of the area due to man-made clearings, and smaller hollows due to clearings for houses (Plate 23B). Features seen on the X-band image (Plate 23A) help interpret surface relief patterns evident on the DSM. The P-band DTM indicates the bedrock and unconsolidated sediments under the vegetation cover have relatively smooth and gently sloping surfaces (Plate 23C).

The X-band DSM and P-band DTM are contoured at 50 ft intervals. Two elevation points at the same location in the DSM and DTM are marked by yellow dots to measure the difference in elevation (above sea level) between the DSM and DTM. The higher elevation point in the north is 366.9 ft high in the DSM and 355.3 ft high in the DTM for a difference of 11.6 ft. The lower elevation point in the southwest is 215.2 ft in the DSM and 201.2 ft in the DTM for a difference of 14 ft. Both points are in vegetated areas (examine the radar image in Plate 23A); therefore, the difference in elevation between the DSM and DTM is due to vegetation approximately 11 and 14 ft in height.

An elevation profile on the image, DSM, and DTM is shown as a northwest–southeast line. The northwest portion of the profile includes two "holes" in the DSM and the southeast portion crosses a topographic ridge with roads. Plate 23D is a plot of the DSM and DTM elevation profile. The tallest vegetation approaches 17 ft (difference between the DSM and DTM elevations). Clearings that extend to the ground are revealed along the northwestern portion of the profile (at ~175 and 300 m distance from the northwest starting point). The vegetation around these clearings is approximately 15 ft tall. The vegetation thickness decreases as topographic elevation increases (Plate 23D).

The derived P-band DTM provides the State of Alaska and USGS with an accurate topographic base for planning, emergency response, and regional engineering. In addition, the height difference between the X-band DSM and P-band DTM reveals unique information about vegetation, land cover type, and the extent of development.

SATELLITE PLATFORMS WITH ONE ANTENNA EACH (MULTIPLE PASS INSAR)

Radar satellites orbiting the Earth illuminate the same site from slightly different viewing angles to generate a DEM. Multipass InSAR is possible because satellite platforms provide well-defined and stable orbits for interferometric measurements of the Earth's surface. Interferometric data has been collected by numerous SAR satellites since the early 1990s with the launch of ERS-1 (1991), JERS-1 (1992), and RADARSAT-1 and ERS-2 (1995). Two or more SAR images that view the same ground site from slightly different angles are coregistered and correlated to determine the phase difference and topographic relief.

Vegetation cover and characteristics need to be consistent at the time of image acquisition. For example, InSAR-derived elevations calculated for a field covered with wheat will be different for the same field after being plowed. Leaf-on and leaf-off differences between SAR images acquired at different times of the same area will lead to errors calculating elevation and changes in surface elevation. Changes with surface infrastructure over time will also lead to elevation and change detection errors. To minimize these errors, researchers have developed the *persistent scatterer InSAR* technique that focuses on pixels that remain coherent (provide a consistent and stable radar reflection) over a sequence of multiple pass InSAR images. The technique works best in areas with many permanent structures (dams, bridges, well pads, buildings).

The German TerraSAR-X and TanDEM SAR satellites were launched in 2007 and 2010, respectively, and orbit the Earth in close formation (separated by a few hundred meters) (Figure 7-30. The interferometric technique is applied to the SAR images collected by these two satellites to build the global WorldDEM with 2 m (relative) and 4 m (absolute) vertical accuracy in a 12 by 12 m raster grid. Post-processing removes vegetation and man-made objects to reveal the bare terrain of the Earth's surface as a DTM.

Ground Deformation

Analyzing many coregistered, satellite SAR images acquired at different times (ranging from days to years apart) enables *ground deformation* to be monitored and measured. Small vertical and horizontal differences in interferometric images acquired of the same area at different times indicate the ground surface is deforming by moving up, down, or sideways (Figure 7-31) (Helz, 2005). Horizontal movement of the ground can be caused by earthquakes, landslides, and slow surface creep downslope due to gravity. Subsidence can be due to many factors, including groundwater withdrawal, oil and gas extraction, and pressure reduction in a volcanic magma chamber. Uplift of the terrain can include the upward movement (intrusion) of magma underneath a volcano and swelling of terrain due to injection of water or steam.

Images from satellite InSAR reveal uplift of a broad ~10 by 20 km area in the Three Sisters volcanic center of the central Oregon Cascade Range, ~130 km south of Mt. St. Helens (Plate 24) (Wicks and others, 2002). The color bar shows a range increase from 0 to 28.3 mm that corresponds to one interference cycle or continuous color change from violet to red. The top image of Plate 24 has the 1996–2000 interferogram draped over a 30-m DEM. The overall elevation change during the five years is shown on the bottom map where the maximum decrease in range to the satellites approaches 15 cm (the land surface rose almost 15 cm in five years). The last eruption in the volcanic center occurred ~1,500 years ago (Wicks and others, 2002).

ESA's Sentinel-1A and 1B are a two-satellite SAR constellation launched in 2014 and 2016, respectively, that detects ground movement in millimeters and across large areas. A 7.8-magnitude earthquake struck New Zealand's South Island near the town of Kaikoura on November 14, 2016. The IfSAR map revealed that the ground rose by 8 to 10 m and offset roads and other features that crossed the fault by up to 12 m (European Space Agency, 2017; Hamling and others, 2017). Ground deformation associated

FIGURE 7-30 TerraSAR-X and TanDEM SAR satellites in close formation while orbiting the Earth. Rendering by EADS Astrium. German Aerospace Center (DLR) News Archive (dlr.de/en).

Pass 1
Pass 2
• Radio waves are bounced off the surface on two separate passes.
• Phase differences resolve elevation changes in fractions of the wavelength of the radar.
Earth's surface

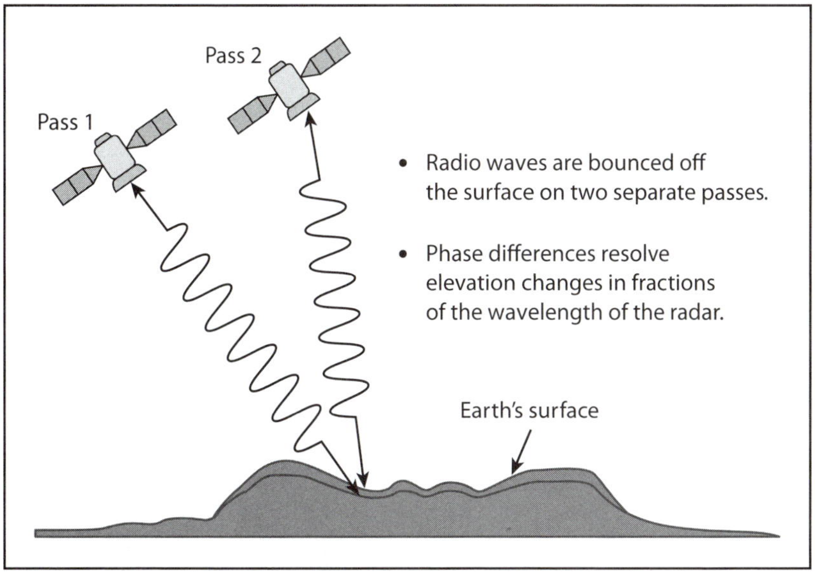

FIGURE 7-31 Diagram showing how radar interferometry detects uplift and subsidence of the Earth's surface. Both the distance between the satellite positions and the amount of deformation are exaggerated for clarity. From Helz (2005, Figure 1).

with the 2018 Kilauea eruptions and earthquakes in Hawaii are monitored with Sentinel-1 satellites as they orbit in both ascending and descending paths (Smith-Konter and others, 2018). The Pacific GPS Facility in Honolulu, Hawaii, reports ground displacement data toward and away from the satellite in a near real-time basis (see pgf.soest.hawaii.edu/Kilauea_insar). The section on earthquakes in Chapter 15 includes an interferogram that records topographic displacements caused by a recent earthquake in Napa, California.

GROUND-BASED INSAR

Landslides, rockfalls, avalanches, debris flows, and other movements on slopes can be monitored in all weather conditions, night and day, and in near real time with ground-based InSAR technology (Borron and Derby, 2019). The objectives include detection of the onset of slope displacement (failure), measurement of surface velocity and acceleration, and development of site-specific levels of risk. The technology enables timely safety alerts for communities and workers near monitored slopes. Ground-based InSAR is used to monitor industrial sites such as quarries, open mine pits, tailing dams, tunnels, and cut-slopes.

SONAR

Sound waves are readily transmitted through water and have long been used for detection of submarines and for *fathometers* (also referred to as *single beam echo sounders* or *single beam sonar*) that measure depth. The general term for this form of active remote sensing is *sound navigation ranging* (sonar). Sonar is an active remote sensing technology using *transducers*, which are devices that convert electrical energy into sound energy; they also convert received sound into electrical energy. The time interval between transmission and return of an acoustic pulse is used to determine the distance to the seafloor. The speed of sound in water can vary with temperature, pressure, and salinity. Sonar beams are also reflected by fish, debris, aquatic vegetation, and suspended sediment in the water column.

Single beam sonar transmits the acoustic pulses from a transducer on the bottom of the platform toward the seafloor. They are widely used for surveying. *Side-scan sonar* transmits a narrow pulse of sound ("single beam") at right angles (look direction) to the ship track (azimuth direction). The acoustic pulse pattern has the shape of a fan (Figure 7-32). The system can be located within the ship or, in order to get closer to the seafloor in deep water, it can be contained in a torpedo-shaped housing, called a *sonar fish*, that is towed near the seafloor by a cable. The cable also provides power and transmits data from the sonar fish to a recording system on the ship. The fish contains transducers on each side. The acoustic pulse encounters the seafloor and is reflected back to the transducer, where the received sound generates electrical signals that vary in amplitude and are proportional to the strength of the received sound. As the ship moves forward, the process continues, generating two strips of imagery separated by a narrow blank strip directly beneath the sonar fish.

As in radar terminology, the direction of travel is the azimuth direction and the direction of transmitted pulses is the look direction. The data are recorded digitally. The analogy between sonar and radar systems extends to the geometry of the images; sonar images are also subject to slant-range distortion, which is corrected during processing. In side-scan sonar images, strong returns are recorded as dark signatures and weak returns as bright signatures. Topographic scarps facing the sonic look direction produce strong returns, while surfaces facing or sloping away from the look direction produce weak returns. Smooth surfaces of mud or sand reflect the sonic pulse specularly; hence these surfaces have weak returns and are recorded with bright signatures. Rough surfaces such as boulder fields and lava scatter much of the incident energy back to the transducer and are recorded with dark signatures.

Multibeam sonar uses a specially designed transducer that transmits many beams of sound energy in a continuous fan-shaped pattern that records bathymetry along a

FIGURE 7-32 Side-scan sonar system.

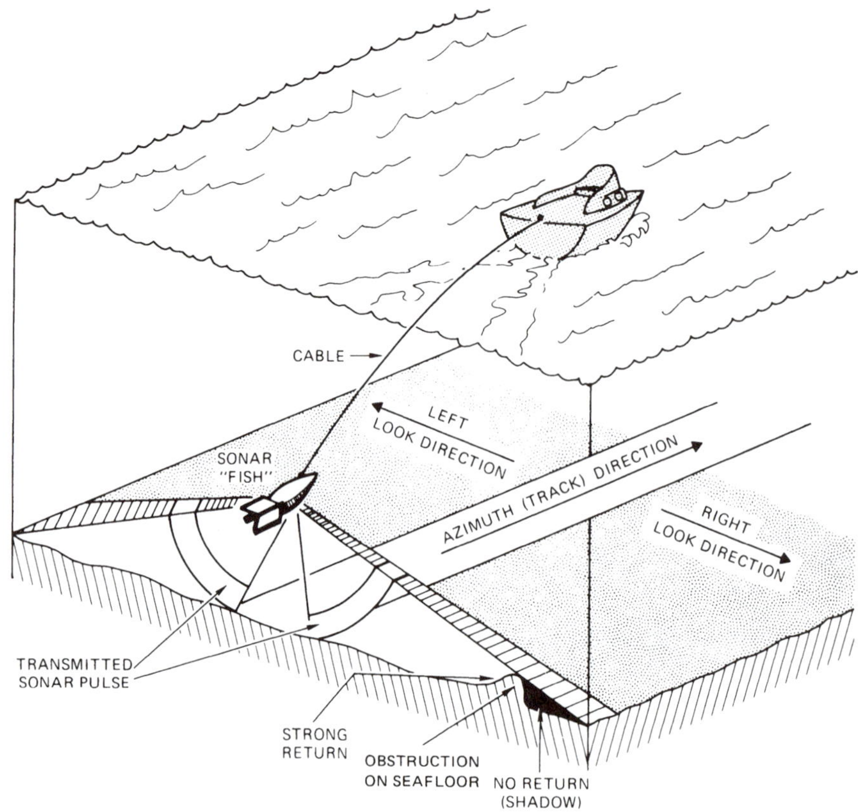

swath on both sides of the ship (Figure 7-33). The system can plot the bottom depths for dozens to hundreds of points in a line perpendicular to the heading of the ship from a single acoustic ping (AML Oceanographic, 2017). The coverage area on the seafloor depends on the depth of the water, typically two to four times the water depth (NOAA Office of Coast Survey, 2017). Excellent animations demonstrating the use of and differences between multibeam and side-scan sonar have been created by NOAA's Office of Coast Survey (nauticalcharts.noaa.gov).

NOAA's National Centers for Environmental Information (NCEI) provides downloadable grids of multibeam bathmetry for the coastal United States and select basins and sites in the oceans. NOAA's Office for Coastal Management's Marine Cadastre National Viewer (marinecadastre.gov/nationalviewer) provides interactive mapping of bathymetery and other offshore data. Figure 7-34 shows a grayscale version of the Marine Cadastre National Viewer that displays deepwater bathymetry with hillshade in the Gulf of Mexico offshore of New Orleans, Louisiana. The Sigsbee escarpment is evident as the dark cliff facing south along the southern margin of the hummocky seafloor in the Gulf of Mexico. The Sigsbee Escarpment is the leading edge of a mobile, subsurface salt layer that moves south. North

FIGURE 7-33 Multibeam sonar energy is transmitted into the water column below the ship in a fan-shaped pattern to map bathmetry in an along-track swath perpendicular to the ship's course direction. Modified from DeWitt and others (2010, Figure 7) (see nauticalcharts.noaa.gov/learn/hydrographic-survey-equipment.html).

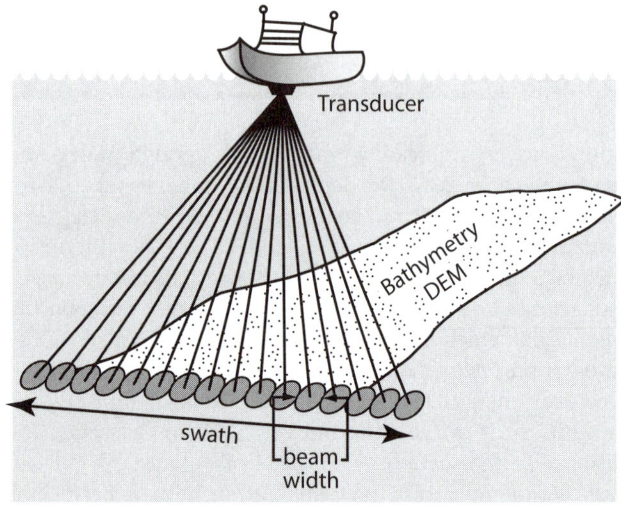

of the escarpment, salt movement from deep in the subsurface pierces thousands of meters of sediment, distorting the seabed and creating the hummocky surface (Garrison and Martin, 1973).

SATELLITE RADAR ALTIMETRY

Radar altimeters are nonimaging systems carried on aircraft and satellites to measure altitude with great precision. A pulse of microwave energy is transmitted vertically downward. The two-way travel time of the return pulse is recorded and converted into distance above the surface, or height. In 1978 the Seasat satellite (Chapter 6) carried a radar altimeter that recorded altitude profiles along orbits over the oceans. Dixon and others (1983) demonstrated that variations in sea surface elevation on these Seasat profiles correlate with bathymetric features on the seafloor. The right side of Figure 7-35 shows that bathymetric relief on the order of a few kilometers causes relief on the sea surface on the order of a few meters. This strong correlation between bathymetry and the shape of the sea surface is caused by differences in the gravity field that are associated with bathymetric features (Dixon and others, 1983).

SEAFLOOR BATHYMETRY

Most bathymetric maps were compiled only from sonar depth-sounding profiles recorded by oceanographic survey vessels until satellite altimetry data became available. Radar altimeters on the Geosat and ERS satellites have provided a wealth of data that are corrected and calibrated to produce bathymetric maps of the world's oceans. In the Southern Ocean, Geosat satellite orbits are 2 to 4 km apart, and very accurate bathymetric maps are generated. The fine sampling grid enables detection and mapping of seamounts and troughs across the ocean floor. In the Pacific Ocean, the Foundation Seamounts are a chain of undersea volcanoes that were discovered from satellite altimetry data. These seamounts were unknown prior to satellite altimetry data.

Recent radar altimeter data has been collected by NASA's TOPEX-Poseidon and Jason-1 satellites and many satellites operated by the ESA. Jason-2 was launched in 2008 and is still operating. Jason-3 was launched in January 2016 and orbits over different ocean surfaces compared to Jason-2, improving sampling and spatial coverage.

Smith and Sandwell (1997) describe the digital processing of the data and illustrate the resulting bathymetric maps. The accuracy of the bathymetric maps is 5 to 10 km in

FIGURE 7-34 NOAA's Marine Cadastre National Viewer of bathymetry and other data for the Gulf of Mexico. Courtesy NOAA, Office for Coastal Management.

location and 100 to 250 m in depth. The left side of Figure 7-35 shows how the maps are generated. The height above the sea surface (hs) is measured by the altimeter, which is carried on the Geosat satellite in this example. The height of Geosat above a datum (he), called the reference ellipsoid, is determined from precise laser measurements of the satellite orbit. Subtracting hs from he provides elevations for mapping the sea surface.

Marks and Smith (2006) evaluated six global bathymetric grids created by various organizations. Here, we will provide a brief overview of two grids: (1) the General Bathymetric Charts of the Oceans (GEBCO) by the British Oceanographic Data Centre and (2) the Smith and Sandwell grid, which was derived from satellite altimetry and ship data combined. In Figure 7-36, a portion of the Southern Ocean is described. Figure 7-36A is the GEBCO bathymetric model derived from only ship sonar depth-sounding profiles. The model is illuminated from the north. The ship tracks that collected the sounding profiles are shown in Figure 7-36B. Only seamounts and troughs traversed by ship tracks show up in the GEBCO bathymetry model. Most seamounts in the GEBCO grid were missed because they lie between ship tracks. In addition, the seamounts were aligned with the ship tracks and not in their true location on the seafloor.

Figure 7-36C is a bathymetric model by Smith and Sandwell (1997) that was generated from Geosat radar altimeter data combined with ship data. The model is illuminated from the north. Figure 7-36C is the same area that is shown in Figure 7-36A. More bathymetric detail is available and accurately located on the seafloor when satellite altimetry data is combined with ship data compared to bathymetry derived from only ship data (compare Figures 7-36A and 7-36C). Figure 7-36D is an interpretation map of the Smith and Sandwell bathymetry model that shows bathymetric scarps that are the expression of faults.

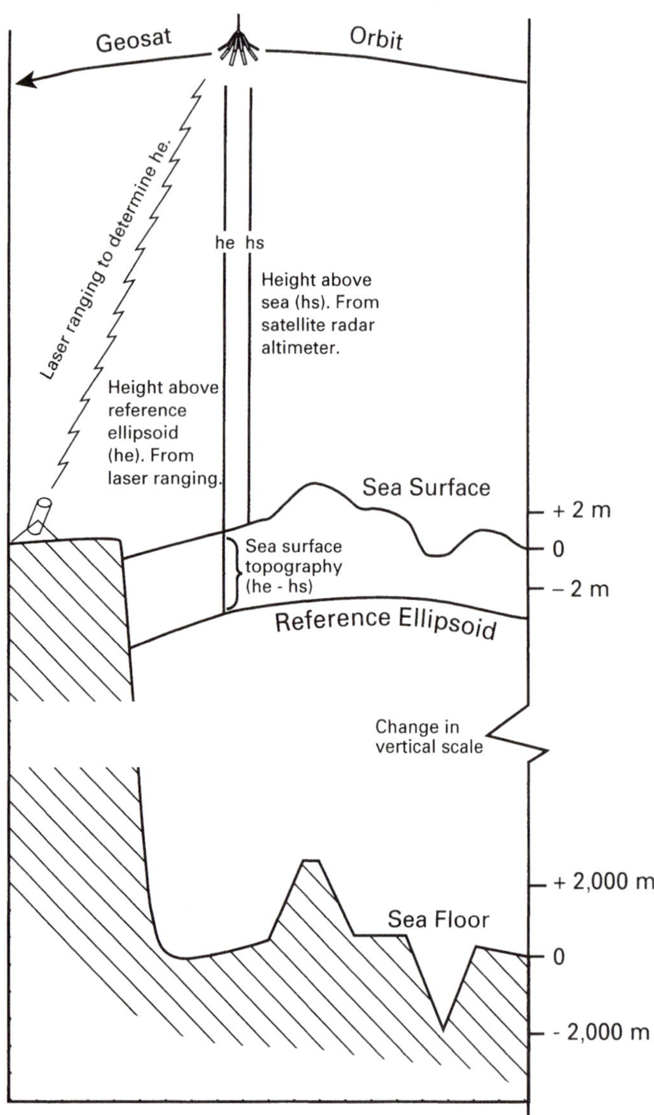

FIGURE 7-35 Mapping bathymetry from satellite radar altimeter data. Based on Gahagan and others (1988, Figure 2).

POLAR ICE

Surface-based, airborne, and satellite radar systems are used to measure and monitor the Greenland and Antarctica Ice Sheets and polar sea ice. Chapter 16 discusses satellite laser altimetry systems (ICESat and ICESat-2) that also measure the elevation of ice sheets, glaciers, and sea ice.

Cyrosat-2

ESA's CryoSat-2 was launched on April 8, 2010 to measure the thickness of polar sea ice and monitor elevation changes in the ice sheets that cover Antarctica and Greenland. The satellite carries the Synthetic Aperture Interferometric Radar Altimeter (SIRAL) instrument that emits pulses from one antenna and receives reflections with two antennas to record the phase difference between the returning radar waves. Onboard radio positioning and laser instruments determine the orbital position of the altimetry antennas within a few centimeters to enable very accurate measurements.

The radar altimeter is designed to measure ice sheet elevation and sea ice freeboard, which is the height of ice protruding from the water. Sea ice floats on the water with the height of freeboard determined by the water and ice density, along with the depth and density of snow on top of the sea ice. A freeboard of 0.5 m indicates an approximate sea ice thickness of 4 m.

Over sea ice the first returning energy in the radar echo comes from a point directly below the satellite, so there is no phase difference. On sloping surfaces, such as those found around the edges of the Greenland and Antarctica Ice Sheets, the first return comes from a point not directly beneath the CryoSat-2 satellite. The interferometric phase difference provides the elevation of, and compass direction to, the reflection point on the slope.

Parrinello and Drinkwater (2016) review the achievements of CryoSat-2. The system delivers sea ice thickness maps every 2, 14, and 30 days to polar operational agencies. Increasing mass loss in Antarctica and Greenland has been documented over recent years. CryoSat-2 provides the first

A. GEBCO bathymetry derived from sonar depth-sounding profiles.

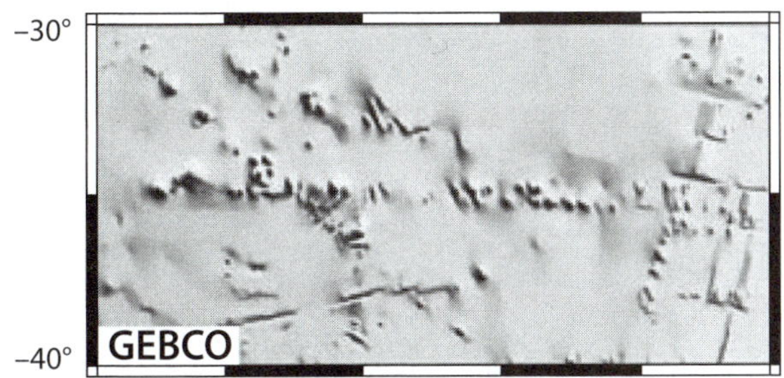

B. Ship tracks (black lines) illustrating the paths of the sounding profiles.

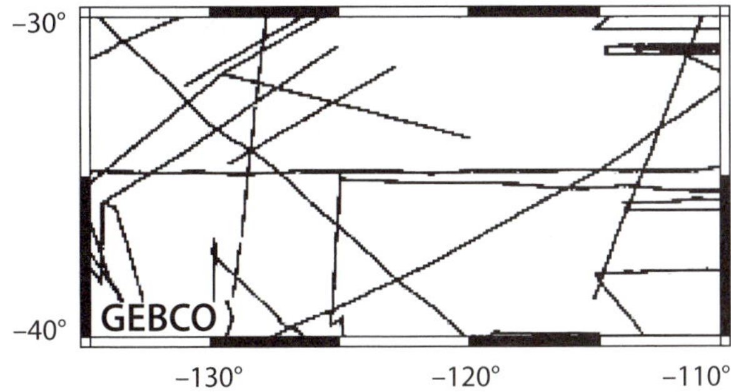

C. Smith and Sandwell bathymetry from satellite altimeter data combined with ship track soundings.

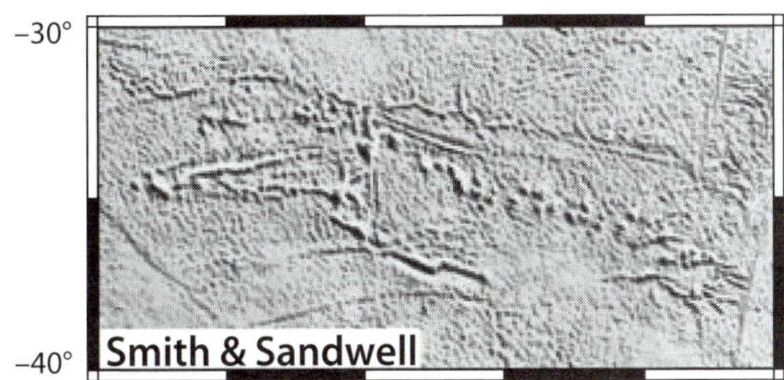

D. Interpretation map of the Smith and Sandwell bathymetry.

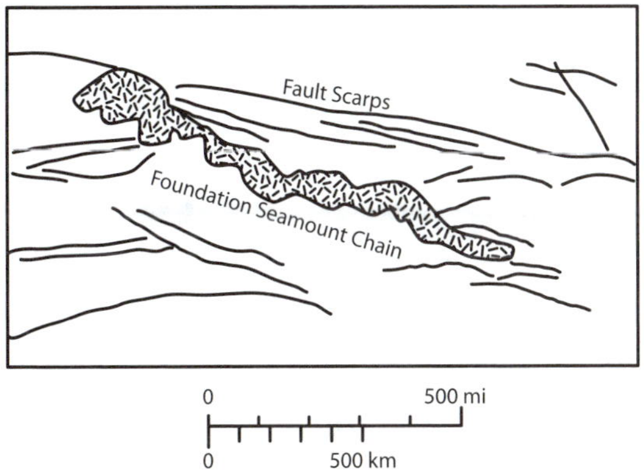

FIGURE 7-36 Seafloor bathymetry models of the Southern Ocean illuminated from the north. (A–C) K. M. Marks and W. H. F. Smith. 2006. An evaluation of publicly available global bathymetry grids. *Marine Geophysical Researches*, 27(1), 19–34. Reprinted by permission from Springer Nature.

assessment of mass balance of ice caps and mountain glaciers. A new algorithm has turned the SAR interferometric altimeter into an imaging sensor that displays topographic details along glacier and ice sheet margins.

Airborne SAR

Paden and others (2011) report on the use of radar to map bedrock beneath the Greenland Ice Sheet as follows. The interior of the Greenland and Antarctica Ice Sheets is characterized by a smooth surface underlain by cold dry ice that is transparent to downward-looking radar. Echo-sounding, surface-based, and airborne radar systems have allowed scientists to map the bedrock hidden far below the thick ice sheets. Radar also reveals internal layering within the ice, which can arise from the presence of air bubbles, density changes, liquid water, and even ancient dustings of the former ice surface with volcanic ash.

The margins of the ice sheets are difficult to survey because they have relatively warm ice near the bottom and lots of deep cracks on top. Also, some of the surface ice melts during the summer. The warm ice and liquid water attenuate radar waves, severely weakening the radar returns from the ice bed. The subtle radar echoes from the bed are often masked by strong waves bouncing back at odd angles from the rough spots on the surface. More complete and detailed information is needed about the bedrock surface buried beneath ice sheet margins to support computer models that predict ice loss and ice sheet stability. The new maps of ice margins need to show where subglacial terrain is flat or sloping and where there is liquid water lubricating the ice-bedrock contact.

Paden and others (2011) developed a special signal-processing algorithm for a new downward-looking SAR system that is flown in low altitude NASA research aircraft along ice sheet margins. The new system discriminates the incoming direction of the echoes and enables the mapping of bedrock beneath ice margins. The bedrock beneath the roughly 3-km thick Jakobshavn Glacier was mapped by the SAR system with one swath a few kilometers wide in a single pass, providing new and accurate information for modeling. The Jakobshavn Glacier is examined with other remote sensing tools in Chapter 16.

GLOBAL DEMS

Public domain global DEMs have been constructed from the technologies discussed, and include the 30-m optical ASTER GDEM3, the multisource and variable resolution GMTED2010, the 1 arc-minute (~1 km) ETOPO1 land and ocean bathymetry DEM, and the 30-m radar SRTM DEM. These global DEMs are downloadable and excellent for regional studies. The commercial WorldDEM discussed earlier in this chapter is also available with a 12 by 12 m raster grid. The Japan Aerospace Exploration Agency (JAXA) processed ALOS satellite radar data (Chapter 6) to generate a no-cost global DEM with a 30-m grid and a commercial global DEM with a 5-m grid.

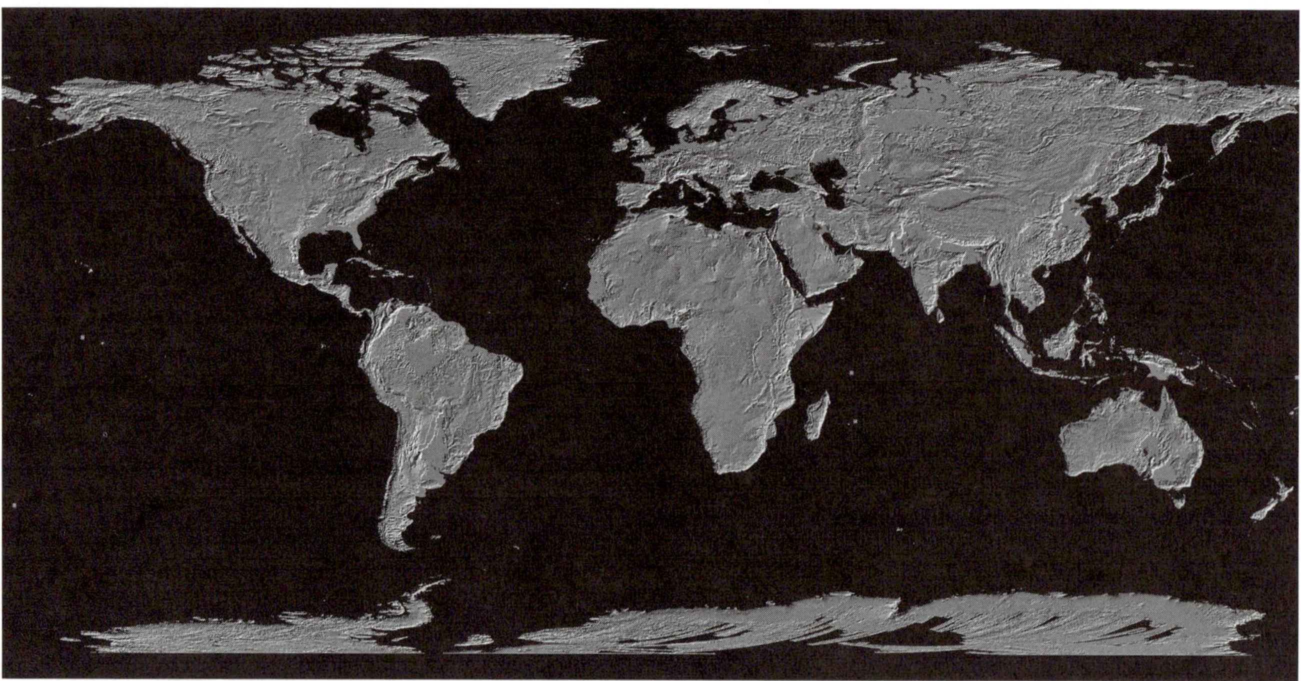

FIGURE 7-37 ASTER GDEM3 global land DEM. Courtesy METI and NASA.

ASTER GDEM3

The ASTER Global Digital Elevation Model (GDEM3) is provided by the Ministry of Economy, Trade, and Industry (METI) of Japan and NASA. It was constructed from hundreds of thousands of stereo-pair images collected by the ASTER instrument onboard Terra. The ASTER GDEM3 coverage spans from latitude 83°N to 83°S south, encompassing 99% of Earth's landmass (Figure 7-37). New with GDEM3 is the ASTER Water Body Dataset. It provides the only water mask covering nearly the entire surface of the Earth and identifies water bodies > 0.2 km^2 as either ocean, river, or lake. This DEM is available in GeoTIFF format with 30-m postings as 1 by 1° tiles (asterweb.jpl.nasa.gov/gdem.asp).

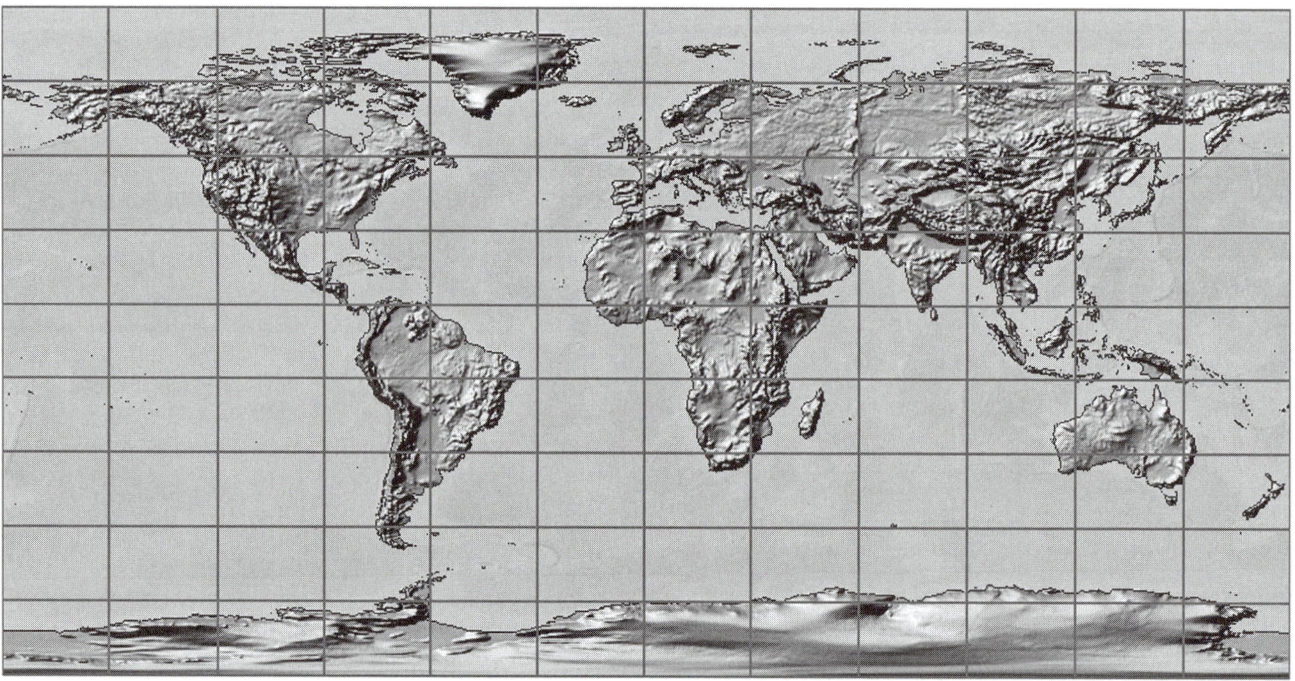

FIGURE 7-38 GMTED2010 global land DEM. Courtesy USGS and NGA.

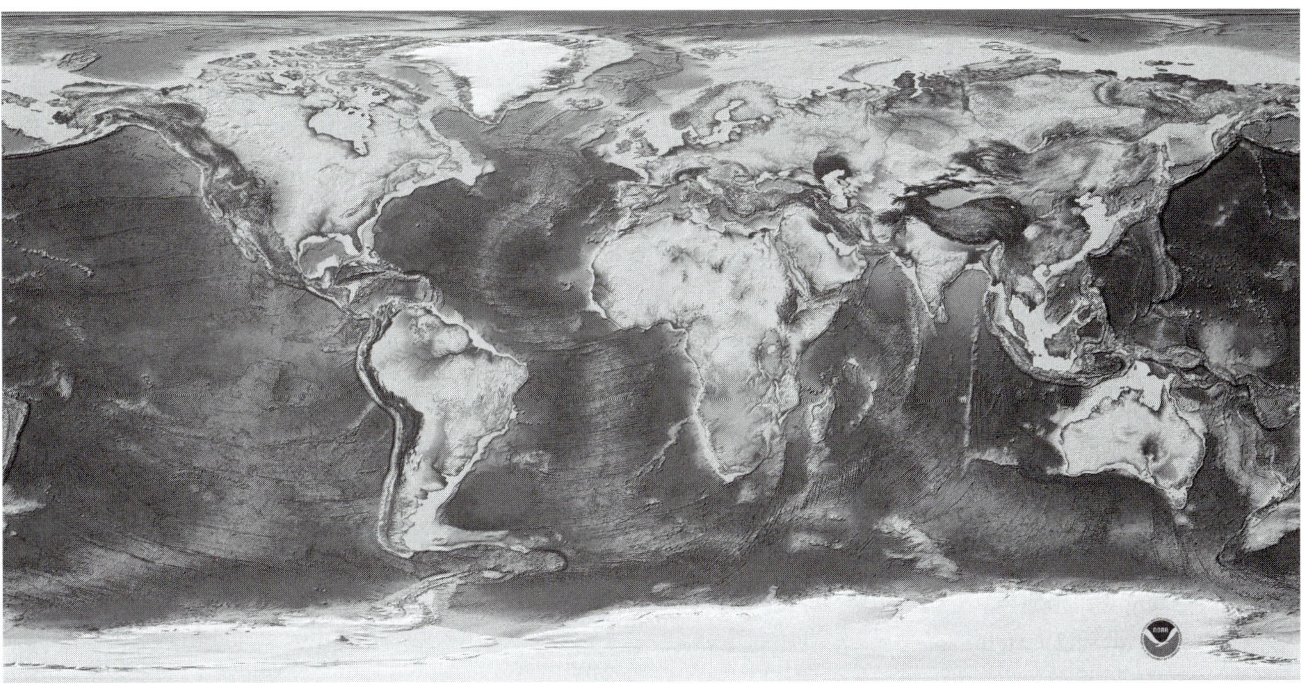

Figure 7-39 ETOPO1 global DEM of land and oceans. Courtesy NGDC.

GMTED2010

The USGS and the National Geospatial-Intelligence Agency (NGA) collaborated to develop Global Multi-resolution Terrain Elevation Data 2010 (GMTED2010). It provides a new level of detail in global topographic data (Figure 7-38). The GMTED2010 product suite contains seven new raster elevation products for each of the 30, 15, and 7.5 arc-second spatial resolutions and incorporates the current best available global elevation data. This product suite provides global coverage of all land areas from latitude 84°N to 56°S for most products, and coverage from 84°N to 90°S for several products. Many of these products will be suitable for various regional continental-scale land cover mapping, extraction of drainage features for hydrologic modeling, and geometric and radiometric correction of medium and coarse resolution satellite image data (lta.cr.usgs.gov/GMTED2010).

A. 90 m pixels. B. 30 m pixels.

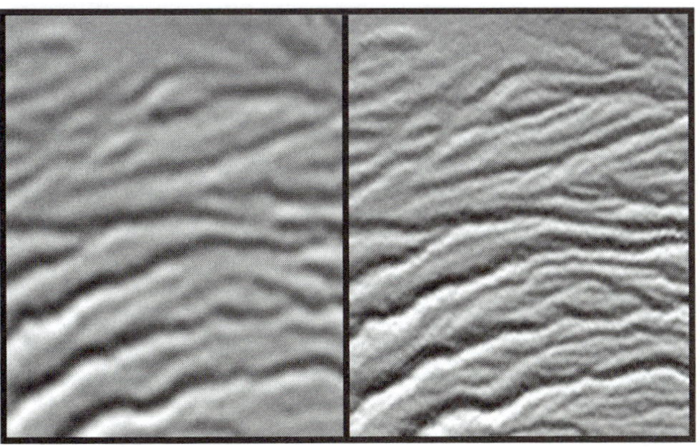

Figure 7-40 Comparison of SRTM shaded relief images at 90 and 30 m pixels. These DEMs show stream erosion patterns, Crater Highlands, Tanzania. The area covered is ~5.2 by 4.3 km. Courtesy NASA/JPL-Caltech/NGA.

ETOPO1

ETOPO1 is a 1 arc-minute (~1 km cells) global relief model of the Earth's surface that integrates land topography and ocean bathymetry (Figure 7-39). It was built from numerous global and regional data sets and is available in ice surface (top of Antarctic and Greenland Ice Sheets) and bedrock (base of the ice sheets) grid versions. The National Geophysical Data Center (NGDC) developed the ETOPO1 global relief model (Amante and Eakins, 2009) (ngdc.noaa.gov/mgg/global/global.html).

SRTM Global DEM

In 2000, NASA's SRTM collected interferometric data of the Earth. Originally, SRTM data for regions outside the United States were resampled for public release at 3 arc-seconds, which is equivalent to 1/1200th of a degree of latitude and longitude, or about 90 m (295 ft). In 2014, the topographic data for areas outside the United States was released to the public at 1 arc-second, or about 30 m (98 ft), sampling that reveals the full resolution of the original measurements. Figure 7-40 compares the 90-m grid with the 30-m grid. The SRTM global DEM covers terrain between latitude 60°N and 56°S.

SYNTHETIC STEREO DEMS

DEMs can be used to create a stereo model with variable vertical exaggeration by digitally introducing artificial parallax into a second, synthetic version of the DEM (Figure 7-41). Each pixel in the second DEM (a raster grid) is offset a distance in the x direction that is proportional to the pixel's elevation. The greater the value of the x-direction offset, the greater the parallax, and the greater the vertical exaggeration. Artificial parallax can also be introduced into both the left and right DEMs. Pavlopoulos and others (2009) provide a formula that expresses the value of the *artificial* parallax (DP) for a single pixel in the synthetic DEM:

$$\mathrm{DP} = \mathrm{Dh} \times \mathrm{K} \tag{7-3}$$

where:

Dh = elevation of the pixel above the minimum ground elevation

K = constant value determining the vertical exaggeration

Any raster image or map, and vector points, lines, or polygons, can be draped onto the DEM and visualized in the synthetic stereo model. Synthetic stereo DEMs have a left and right image that retain the grayscale or color values of the original image, so the two images can be plotted for viewing with a stereoscope or viewed on an interlaced computer display with synchronized stereoscopic glasses.

Interlaced displays show the left and right images on alternating lines (for instance, the left image is displayed only on odd numbered lines and the right image on even numbered lines of the screen). The display flickers between the alternating lines to show only the left or right image at one time. Glasses are worn that are synchronized with the flickering screen so the viewer perceives the scene in 3-D color. The flickering rate should be greater than ~60 times per second (hertz) or the viewer may suffer eye fatigue and headaches.

In addition, left and right grayscale images of the synthetic stereo DEM can be digitally superimposed, illuminated with red and blue light, and viewed as red/blue anaglyphs (Plate 25) on a computer screen or as a print. The viewer needs to wear red/cyan glasses to perceive the scene in 3-D. Anaglyphs are a low cost solution for presenting remote sensing images and maps integrated with topography and bathymetry. They can be displayed in the classroom on a projector, or printed on large sheets to hang on the side of a bus when out on field trips.

The advantages of synthetic stereo DEMs include:

1. The same DEM can have several vertical exaggerations applied so that the interpreter can increase vertical exaggeration in areas with low topographic relief and decrease vertical exaggeration in mountainous terrain, improving the interpretation.
2. A draped raster image can be made partially transparent so that the underlying, color-coded for elevation DEM is visible in color when viewed stereoscopically.

The disadvantages of synthetic stereo DEMs include:

1. Objects on the Earth's surface are not seen in 3-D if a DTM is used. For example, the synthetic stereo model in Plate 25 (which was based on a USGS DTM) does not elevate man-made structures (such as oil tanks) above the bare earth surface when viewed with red/cyan anaglyph

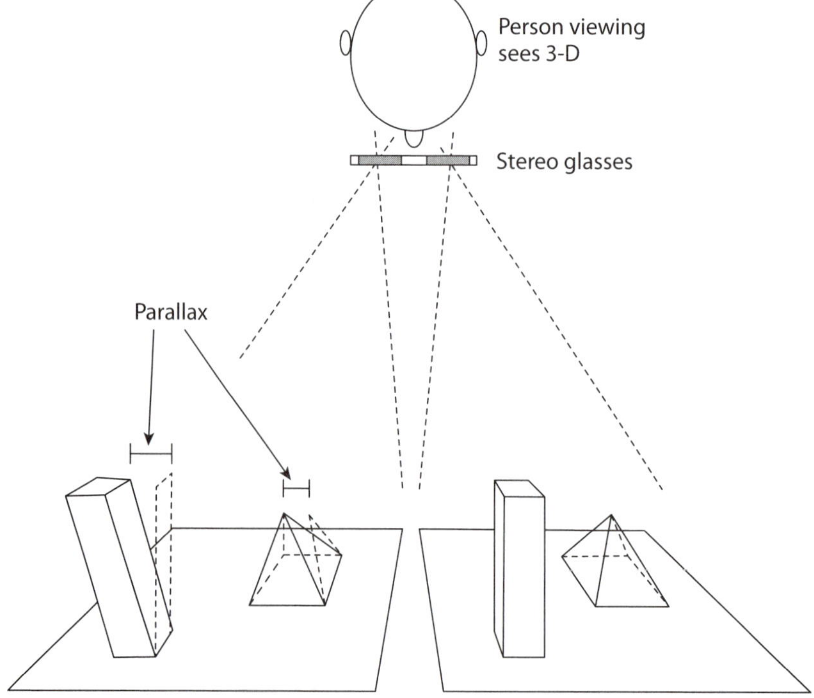

FIGURE 7-41 A synthetic stereo DEM has artificial parallax introduced into the model to create parallax in the x (left-right) direction so three dimensions are perceived when viewing the DEM stereoscopically.

glasses. In contrast, a stereo model of the same area that is generated with overlapping aerial photographs elevates the tank tops to their correct elevation above the terrain when viewed stereoscopically (Plate 4).

2. The draped image is often resampled to the grid size of the DEM. If the DEM has a 10 m grid and the original image had 1 m pixels, the synthetic DEM is displayed with 10 m pixels, degrading the level of detail on the draped image.
3. If the DEM topography does not match the topography when the image was acquired, or the DEM has errors of some significance, offsets, smears, and distorted man-made features will be visible on the draped image in the synthetic stereo DEM.
4. In areas with high topographic relief, pixels on the draped image can be stretched and smeared along cliff faces, especially if a high vertical exaggeration is applied to the DEM.

STANDARDS

The American Society of Photogrammetry and Remote Sensing, The Imaging & Geospatial Information Society (ASPRS), monitors, develops guidelines and best practices, and establishes standards for digital geospatial data, including lidar. ASPRS committees establish the standard digital format (.las) for lidar data. In 2014, ASPRS released the *Positional Accuracy Standards for Digital Geospatial Data.* These new standards provide more quantitative measures of accuracy to the National Map Accuracy Standards of 1947. The ASPRS Standards provide horizontal and vertical standards for digital elevation data (ASPRS, 2014; Maune and Nayegandhi, 2018, Chapter 3). For example, digital elevation data with a stated vertical accuracy of 5 cm would have an absolute accuracy of 5.0 cm in nonvegetated terrain and a relative vertical accuracy of 4 cm swath-to-swath. The *National Standard for Spatial Data Accuracy* (Federal Geographic Data Committee, 1998) also establishes accuracy guidelines.

Standards are important for developers, producers, and users of digital geospatial data. Standards enable the geospatial community to evaluate map accuracy versus cost and also to describe the accuracy of the product in a uniform manner that is reliable, defensible, and repeatable (ASPRS, 2014).

RESOURCES

Sources of DEM Data

- ASTER GDEM3 (asterweb.jpl.nasa.gov/gdem.asp)
- Global Multi-resolution Terrain Elevation Data 2010 (lta.cr.usgs.gov/GMTED2010)
- NOAA Multibeam Bathymetric Surveys (maps.ngdc.noaa.gov/viewers/bathymetry/)
- Shuttle Radar Topography Mission 30 m DEM (jpl.nasa.gov/srtm)
- State of Alaska, Division of Geological and Geophysical Surveys Elevation Portal (elevation.alaska.gov)
- USGS Elevation Products (eros.usgs.gov/products-data-available/elevation-products)
- USGS National Map Viewer (viewer.nationalmap.gov/viewer)
- USGS 3D Elevation Program (3DEP)—dynamic map service (elevation.nationalmap.gov/arcgis/rest/services/3DEPElevation/ImageServer)

Sources of Lidar Data

- Datta, A. 2018, May 31. Did You Know Which Are the Sources for Free LiDAR Data? Geospatial World, Blogs (geospatialworld.net/blogs/did-you-know-the-sources-for-free-lidar-data)
- Lidar Online (Lidar-online.com)
- National Ecological Observatory Network (neonscience.org)
- NOAA, Office of Coastal Management, Digital Coast (coast.noaa.gov/dataregistry/search/collection/info/coastalLidar)
- Open Topography (opentopography.org)
- United States Interagency Elevation Inventory (coast.noaa.gov/inventory)
- USGS Earth Explorer (earthexplorer.usgs.gov)
- Washington State Department of Natural Resources (dnr.wa.gov/lidar)

Information Sources

- American Society for Photogrammetry and Remote Sensing (asprs.org)
- *LIDAR* Magazine (lidarmag.com)

Questions

1. List three of the five technologies used to generate a new DEM.
2. How is a DEM generated from a grid different from a DEM generated with a TIN?
3. What do the following abbreviations mean?
   a. GCP, *b.* DEM, *c.* DSM, *d.* DTM,
   e. SfM, *f.* lidar, *g.* InSAR, and *h.* TIN.
4. Photogrammetry uses overlapping images. Where is the stereo model located?
5. What are photogrammetric breaklines and why are they collected?
6. Should images used for SfM models have a high degree of overlap? Why?
7. Why should photos of a feature on the ground that are used for SfM be acquired in a convergent and not a divergent mode?
8. Two wavelengths are routinely used by lidar systems. What are these wavelengths and for what application is each used?
9. Why is a lidar beam not a point on the Earth's surface, but a circular footprint?

10. Explain a lidar first return, intermediate return, and last return.
11. What are the three types of platforms used to collect lidar data?
12. Is X-band or P-band radar better able to penetrate a partially open forest canopy to generate a DTM? Why?
13. What is required on the radar platform (satellite or aircraft) to enable single-pass InSAR?
14. What is multipass InSAR? What is required of features on the ground that are measured through time with the multipass InSAR method?
15. What are the three types of sonar? Give a short description of each.
16. If the ocean surface is over a large and deep seafloor trench (a subduction zone), what does the radar altimeter record for the height of the ocean surface (no change, higher, or lower) compared with the area away from the trench? Why?
17. What is the CryoSat-2 satellite designed to measure?
18. What is ETOPO1?
19. What are two advantages and two disadvantages of synthetic stereo DEMs compared with DEMs created from overlapping images?

REFERENCES

Abdullah, Q. A. 2019. Harnessing drones the photogrammetric way. *Photogrammetric Engineering & Remote Sensing*, 85(5), 329–337.

Alspaugh, D. 2004. A brief history of photogrammetry. In J. C. McGlone (Ed.), *Manual of Photogrammetry* (5th ed.). Bethesda, MD: American Society for Photogrammetry and Remote Sensing.

Amante, C., and B. W. Eakins. 2009. *ETOPO1 1 Arc-Minute Global Relief Model: Procedures, Data Sources and Analysis* (NOAA Technical Memorandum NESDIS NGDC-24). Boulder, CO: National Geophysical Data Center.

AML Oceanographic. 2017. What is a Multi-Beam System? http://www.amloceanographic.com/CTD-Sound-Velocity-Environmental-Instrumentation-Home/Multibeam-Overview (accessed December 2017).

Anderson, S. 2016. Structure from Motion Photogrammetry (SR 530 Slide, Remote Sensing). USGS Washington Water Science Center. http://wa.water.usgs.gov/projects/sr530/remotesensing.htm (accessed December 2017).

ASPRS. 2014. ASPRS Positional Accuracy Standards for Digital Geospatial Data. *Photogrammetric Engineering & Remote Sensing*, 81(3). http://www.asprs.org/Standards-Activities.html (accessed December 2017).

Astre, H. 2010. SfMToolkit3. http://www.visual-experiments.com/demos/SfMtoolkit (accessed January 2019).

Baltsavias, E. P. 1999. Airborne laser scanning: Basic relations and formulas. *ISPRS Journal of Photogrammetry and Remote Sensing*, 54, 199–214.

Bellian, J. 2003, January. Laser intensity mapping of outcrop geology. *Oil IT Journal*. http://www.oilit.com/2journal/2article/0010/0002.htm (accessed December 2017).

Borron, S. E., and M. P. Derby. 2019, June 25–27. Ground-Based Interferometric Synthetic Aperture Radar Combined with a Critical Slope Monitoring Program Will Provide Early Detection of Slope Movement along Pipeline Corridors. In Proceedings of the International Pipeline Conference (IPG2019-5333), Buenos Aires, Argentina.

Cawood, A. J., and C. E. Bond. 2019. eRock: An open-access repository of virtual outcrops for geoscience education. *GSA Today*, 29, 2. http://doi.org/10.1130/GSATG373GW.1

Cawood, A. J., C. E. Bond, J. A. Howell, R. W. Butler, and Y. Totake. 2017. LiDAR, UAV or compass-clinometer? Accuracy, coverage and the effects on structural models. *Journal of Structural Geology*, 98, 67–82. http://doi.org/10.1016/j.jsg.2017.04.004

Dewitt, N. T., J. G. Flocks, W. R. Pfeiffer, J. N. Gibson, and D. S. Wiese. 2010, April. *Archive of Side Scan Sonar and Multibeam Bathymetry Data Collected during USGS Cruise 10CCT03 Offshore from the Gulf Islands National Seashore, Mississippi, from East Ship Island, Mississippi, to Dauphin Island, Alabama* (USGS Data Series 671). Washington, DC: US Geological Survey.

Diaz, J. C. F. 2011, Spring. Lifting the canopy veil—airborne lidar for archeology of forested areas. *Imaging Notes*, 26(2).

Dixon, W. S., M. Naraghi, M. K. McNutt, and S. M. Smith. 1983. Bathymetric prediction from Seasat altimeter data. *Journal of Geophysical Research*, 88, 1,563–1,571.

European Space Agency. 2017, March 24. Satellites Shed New Light on Earthquakes. http://www.esa.int/Our_Activities/Observing_the_Earth/Copernicus/Sentinel-1/Satellites_shed_new_light_on_earthquakes (accessed December 2017).

Federal Geographic Data Committee. 1998. *Geospatial Positioning Accuracy Standards Part 3: National Standard for Spatial Data Accuracy* (FGDC-STD-007.3-1998). http://www.fgdc.gov/standards/projects/FGDC-standards-projects/accuracy/part3/chapter3 (accessed December 2017).

Gahagan, L. M., J. Y. Royer, C. R. Scotese, D. T. Sandwell, J. K. Winn, R. L. Tomlins, M. I. Ross, J. S. Newman, R. D. Müller, C. L. Mayes, L. A. Lawver, and C. E. Heubeck. 1988. Tectonic fabric of the ocean basins from satellite altimetry data. *Tectonophysics*, 155, 1–26.

Garrison, L. E., and R. G. Martin. 1973. *Geologic Structures in the Gulf of Mexico Basin* (Professional Paper 773). Washington, DC: US Geological Survey.

Gatziolis, D., and H-E. Andersen. 2008. *A Guide to LIDAR Data Acquisition and Processing for the Forests of the Pacific Northwest* (PNW-GTR-768). Portland, OR: USDA Forest Service, Pacific Northwest Research Station.

Guenther, G. C. 2001. Airborne lidar bathymetry. In D. F. Maune (Ed.), *Digital Elevation Model Technologies and Applications: The DEM Users Manual* (pp. 237–306). Bethesda, MD: American Society for Photogrammetry and Remote Sensing.

Hamling, I. J., S. Hreinsdóttir, K. Clark, J. Elliott, C. Liang, E. Fielding, N. Litchfield, P. Villamor, L. Wallace, T. J. Wright, E. D'Anastasio, S. Bannister, D. Burbidge, P. Denys, P. Gentle, J. Howarth, C. Mueller, N. Palmer, C. Pearson, W. Power, P. Barnes, D. J. A. Barrell, R. Van Dissen, R. Langridge, T. Little, A. Nicol, J. Pettinga, J. Rowland, and M. Stirling. 2017, March. Complex multifault rupture during the 2016 M_w 7.8 Kaikōura earthquake, New Zealand. *Science*, 23 (epub). doi:10.1126/science.aam7194

Hayakawa, Y. C., and H. Obanawa. 2015, March 17–19. Mapping Cliff Face Changes around a Waterfall Using Terrestrial Laser Scanning and UAS-base SfM-MVS Photogrammetry. The International Symposium on Cartography in Internet and Ubiquitous Environments, Tokyo, Japan.

Helz, R. L. 2005, July. *Monitoring Ground Deformation from Space* (Fact Sheet 2005-3025). Washington, DC: US Geological Survey.

Higgins, S. 2014, November 19. Faster, More Powerful Lidar for Small UAVs, Airborne Mapping. Interview with Lewis Graham. SPAR 3D Blog, The Other Dimension. http://www.spar3d.com/blogs/the-other-dimension/vol12no47-faster-lighter-more-precise-Lidar-for-small-uavs-aerial-mapping (accessed December 2017).

IPLER/RIT LIAS. 2010. Haiti Earthquake. Information Products Lab for Emergency Response, Laboratory for Imaging Algorithms and Systems, Rochester Institute of Technology. http://ipler.cis.rit.edu/projects/haiti (accessed December 2017).

Jensen, J. R. 2007. *Remote Sensing of the Environment: An Earth Resource Perspective* (2nd ed.). Upper Saddle River, NJ: Pearson.

Johnson, K., E. Nissen, S. Saripalli, J. R. Arrowsmith, P. McGarey, K. Scharer, P. Williams, and K. Blisniuk. 2014. Rapid mapping of ultrafine fault zone topography with structure from motion. *Geosphere*, 10(5), 1–18.

Kampes, B., M. Blaskovich, J. J. Reis, M. Sanford, and K. Morgan. 2011, May 1–5. Fugro GeoSAR Airborne Dual-Band IfSAR DTM Processing. ASPRS 2011 Annual Conference, Milwaukee, Wisconsin. http://www.asprs.org/wp-content/uploads/2010/12/Kampes.pdf.

Lim, K., P. Treitz, M. Wulder, B. St-Onge, and M. Flood. 2003. LiDAR remote sensing of forest structure. *Progress in Physical Geography*, 27(1), 88–106.

Lowe, D. G. 2004. Distinctive image features from scale-invariant keypoints. *International Journal of Computer Vision*, 60, 91–110.

Marks, K. M., and W. H. F. Smith. 2006. An evaluation of publicly available global bathymetry grids. *Marine Geophysical Researches*, 27(1), 19–34. doi: 10.1007/s11001-005-2095-4

Martin, A. 2014. First Flights for EcoSAR, a New Instrument to Study Carbon-Rich Ecosystems. NASA Earth Science Technology Office. http://esto.nasa.gov/news/news_EcoSAR_7_2014.html (accessed December 2017).

Maune, D. F., and A. Nayegandhi (Eds.). 2018. *Digital Elevation Model Technologies and Applications: The DEM Users Manual* (3rd ed.). Bethesda, MD: American Society for Photogrammetry and Remote Sensing.

McGlone, J. C. (Ed.). 2013. *Manual of Photogrammetry* (6th ed.). Bethesda, MD: American Society for Photogrammetry and Remote Sensing.

Miller, J. C. 2016. The acquisition of image data by means of an iPhone mounted to a drone and an assessment of the photogrammetric products. MA Thesis, California State University at East Bay. http://iphonedroneimagery.com/project-methodology (accessed December 2017).

Nayegandhi, A. 2006. Green, Waveform Lidar in Topo-Bathy Mapping—Principles and Applications. US Geological Survey. http://www.ngs.noaa.gov/corbin/class_description/Nayegandhi_green_Lidar.pdf (accessed December 2017).

NOAA Office of Coast Survey. 2017. Multibeam Echo Sounders. Learn about Hydrography. https://nauticalcharts.noaa.gov/learn/hydrographic-survey-equipment.html (accessed December 2017).

Paden, J., D. Braaten, and P. Gogineni. 2011, August 24. A Next-Generation Ice Radar. IEEE Spectrum. http://spectrum.ieee.org/energy/environment/a-nextgeneration-ice-radar (accessed January 2018).

Paine, D. P., and J. D. Kiser. 2012. *Aerial Photography and Image Interpretation* (3rd ed.). Hoboken, NJ: John Wiley.

Parrinello, T., and M. Drinkwater. 2016, September 13–15. CryoSat: ESA's Ice Mission: 6 Years in Operations: Status and Achievements. Polar Space Task Group Meeting 6, ESTEC. http://www.wmo.int/pages/prog/sat/meetings/documents/PSTG-6_Doc_08_ESA-Drinkwater-CryoSat.pdf (accessed December 2017).

Pavlopoulos, K., N. Evelpidou, and A. Vassilopoulos. 2009. *Mapping Geomorphological Environments*. Berlin, Germany: Springer-Verlag.

Petrie, G. 2007, September 17–20. Airborne Digital Data Capture Systems. VII International Scientific & Technical Conference, From Imagery to Map: Digital Photogrammetric Technologies, Nessebar, Bulgaria (slide 47 of 63). http://www.slideshare.net/gpetrie/airborne-digital-data-capture-systems (accessed June 2018).

Poiker, T. K. 1990. The TIN model. NCGIA Core Curriculum in Geographic Information Science, Unit 39. http://www.ncgia.ucsb.edu/giscc/units/u056/u056.html (accessed December 2017).

Rochester Institute of Technology. 2010. RIT Haiti Response Mapping Mission, January 21–27, 2010. http://www.internet2.edu/presentations/spring10/20100427-haiti-casterline.pdf (accessed December 2017).

Romano, M. E. 2015. Commercial Geiger Mode Lidar. American Society for Photogrammetry and Remote Sensing IGTF 2015 Conference Online Proceedings. http://www.asprs.org/conference-proceedings/igtf-2015-conference-proceedings.html (accessed December 2017).

Roy, R. J. 2012, April. Mobile Lidar for urban streetscapes. *GIM International*, pp. 24–27. http://www.sam.biz/sites/default/files/GIM2012-Feature-Roy_0.pdf (accessed December 2017).

Sharp, B., and B. Morton. 2007. Introduction to the New GeoSAR Interferometric Radar Sensor. Fugro EarthData, Inc. http://web.nps.edu/Academics/Centers/RSC/documents/Fugro_GeoSAR.pdf (accessed December 2017).

Smith, W. H. F., and D. T. Sandwell. 1997. Bathymetric prediction from dense satellite altimetry and sparse shipboard bathymetry. *Journal of Geophysical Research*, 99, 21,803–21,824.

Smith-Konter, B., L. Ward, L. Burkhard, X. Xu, and D. Sandwell. 2018. 2018 Kilauea Eruption and M_w 6.9 Leilani Estates Earthquake: Line of Sight Displacement Revealed by Sentinel-1 Interferometry. University of Hawaii at Manoa and University of California. http://pgf.soest.hawaii.edu/Kilauea_insar (accessed June 2018).

Snavely, N., S. N. Seitz, and R. Szeliski. 2008. Modeling the world from internet photo collections. *International Journal of Computer Vision*, 80, 189–210.

Sweeney, C. 2016. Structure from Motion (SfM). Theia Vision Library. API Reference. http://www.theia-SfM.org/SfM.html (accessed June 2018).

Thomas, J. J. C. 2015. Terrain Classification Using Multi-Wavelength Lidar Data. Naval Postgraduate School, Monterey, California, Master's Thesis. http://www.nps.edu/Academics/Centers/RSC/documents/15Sep_Thomas_Judson.pdf (accessed December 2017).

Westoby, M. J., J. Brasington, N. F. Glasser, M. J. Hambrey, and J. M. Reynolds. 2012. "Structure from Motion" photogrammetry: A low cost, effective tool for geoscience applications. *Geomorphology*, 179, 333–314.

Wicks, C. W., Jr., D. Dzurisin, S. Ingebritsen, W. Thatcher, Z. Lu, and J. Iverson. 2002. Magmatic activity beneath the quiescent Three Sisters volcanic center, central Oregon Cascade Range, USA. *Geophysical Research Letters*, 29(7), 26-1 to 26-4.

Young, J. 2011. *Lidar for Dummies, Autodesk and DLT Solutions Special Edition*. Hoboken, NJ: Wiley.

Zhou, X., N. Chang, and S. Li. 2009. Applications of SAR interferometry in Earth and environmental science research. *Sensors*, 9, 1,876–1,912. doi: 10.3390/s90301876

chapter 8

Drones and Manned Aircraft Imaging

Manned and unmanned aircraft equipped with passive sensors provide remote sensing imagery with ground resolutions ranging from approximately 1 cm to 20 m and spectral ranges spanning from visible to TIR. The sensors have evolved from collecting single wavelength bands of reflected light to recording multispectral and hyperspectral data. Active lidar sensors are also flown on manned and unmanned aircraft. These versatile platforms can be flown at different altitudes, select times of the day, and with different camera lenses to acquire imagery that fulfills the resolution and timing requirements of a particular application and complies with airspace regulations.

Manned aircraft are capable of flying long duration and long distance surveys that can cover very large geographic areas. They provide very stable platforms, can carry large and complex sensors that require considerable electrical power (such as radar), and can be configured to simultaneously record remote sensing data from a variety of integrated, onboard instruments. Civilian small unmanned aerial systems (sUAS) (we interchangeably use sUAS and the more common term "drone") collect very high spatial resolution imagery of smaller areas compared with manned aircraft. They are typically flown at low altitudes and close to features of interest. Drones are less expensive and less complex to fly compared with manned aircraft. We focus on sUAS technology in this textbook. The US Federal Aviation Administration (FAA) states a sUAS must weigh less than 25 kg (55 lb) (faa.gov/uas).

Manned aircraft are used with sophisticated multispectral sensors for the USDA Farm Service Agency's National Agriculture Imagery Program (Plate 5). The program's maintenance plan designates that very three years the lower 48 states are imaged during the growing season with 1-m ground resolution and with four multispectral bands (blue, green, red, and NIR). Extensive technical information on aerial mapping (photogrammetry) with manned aircraft is available from the American Society for Photogrammetry and Remote Sensing (ASPRS) website (asprs.org). The ASPRS also publishes the *Manual of Photogrammetry* (McGlone, 2013).

New technology advances for unmanned aerial systems are accelerating the implementation of this remarkable remote sensing tool. Drone images and video footage are seen daily on news channels as these portable platforms are quickly carried into disaster sites to provide timely, low altitude, and very detailed imagery of damaged buildings, infrastructure, and the landscape. The availability of ready-to-use kits containing a drone, thermal sensor, flight controller, and software for first responders, building inspections, and utility applications has expanded. Drones are also used to quickly fly medicine into remote villages that can only be reached by foot (Anna, 2016).

More applications for data collected by drones and manned aircraft will develop as newer and better platforms, sensors, and software are developed. Colomina and Molina (2014), Johnson and others (2014), and Shervais (2016) provide excellent reviews of unmanned aerial systems.

HISTORY

Aerial photography was implemented during World War I as a new and accurate source of military information (Jensen, 2007). After the war, technological advances with cameras,

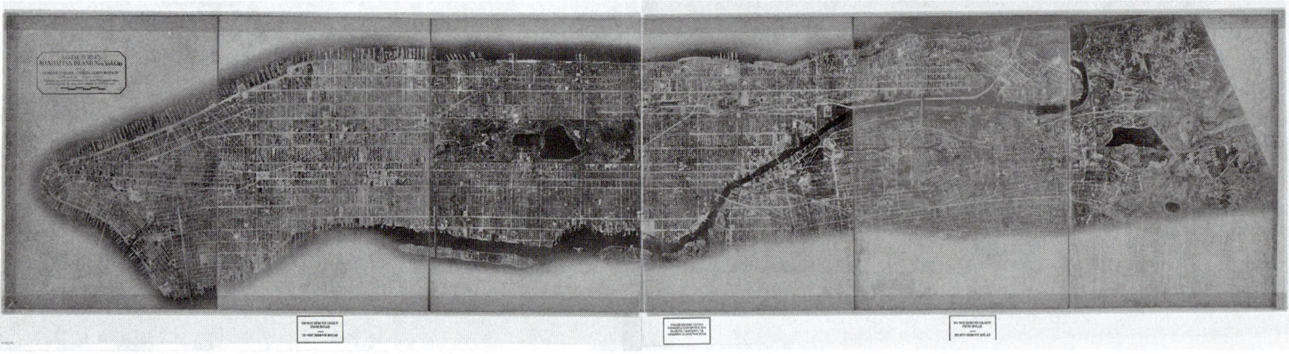

FIGURE 8-1 1921 mosaic of 100 aerial photographs, Manhattan Island, New York. Courtesy Fairchild Aerial Camera Corporation (1921). *Aerial survey, Manhattan Island, New York City.* [New York: The Corporation] [Map]. Retrieved from the Library of Congress, http://www.loc.gov/item/90680339/.

film, and photogrammetric equipment were centered in Europe, with US agencies applying the technology for aerial surveying and mapping (Alspaugh, 2004). On the commercial front, the Fairchild Aerial Camera Corporation constructed the first mosaic of Manhattan Island in 1921 (Figure 8-1) from 100 overlapping aerial photographs that proved aerial surveys to be faster and much less expensive than a ground survey (PAPA International, 2015). Photoreconnaissance and aerial navigation technology greatly accelerated during World War II as film photography and photo interpretation rose to the status of a tactical weapon (Alspaugh, 2004). Post World War II advances in electronic computing and space exploration moved the aerial imaging discipline to all digital technology and methods. In the twenty-first century, intelligence and defense research, development and commercialization, miniaturization, computer power, the Internet, mobile navigation, and consumer mass-market opportunities are exponentially expanding the remote sensing user community and embedding remote sensing into the fabric of modern society.

MANNED AIRCRAFT

Aircraft manned by licensed pilots have transitioned from film cameras to sophisticated digital instruments that collect large swaths of multispectral, hyperspectral, thermal, radar, and lidar data. The sensors can acquire imagery with pixels ranging from a few centimeters to many meters in size. The aircraft have become smaller and less expensive to operate as sensors and peripheral onboard equipment have become more compact and lighter. In addition, the smaller size of sensors facilitates onboard installation of multiple sensors to collect spatially integrated information at the same time, which is especially important in vegetation mapping. The National Ecological Observatory Network's (NEON) (neonscience.org) Airborne Observation Platform includes a VNIR-SWIR hyper-

A. The De Havilland DHC-6-300 Twin Otter aircraft that carries the NEON instrumentation.

B. NEON integrated airborne instrumentation, which consists of hyperspectral, lidar, RGB camera, and GPS/IMU. For scale, note the size of the cabin windows behind the instrument package.

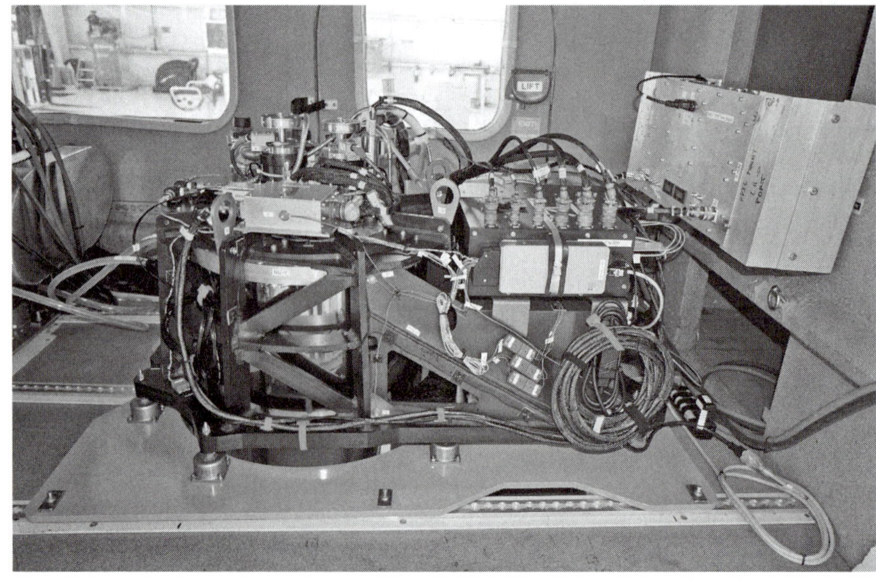

FIGURE 8-2 Instrumentation payload on the NEON Airborne Observation Platform. Courtesy Battelle Ecology, Inc., NASA JPL, and Teledyne Optech.

spectral instrument with 428 bands, a lidar system, and a high spatial resolution color camera for three-dimensional studies of ecosystems (Figure 8-2) (Asner and others, 2007; Krause, 2014). The NEON observatory is designed to collect high-quality, standardized data from 81 field sites (47 terrestrial and 34 aquatic) across the United States (including Alaska, Hawaii, and Puerto Rico).

UNMANNED AIRCRAFT

The use of aircraft without pilots ("drones") goes back to the early twentieth century, when radio-controlled vehicles for carrying bombs and artillery target training were developed and deployed (Shaw, 2016). During the 1960s, the drones transitioned from being targets to remote sensing platforms (Shaw, 2016) with a more common name of *unmanned aircraft vehicle* (UAV) or *unmanned aircraft system* (UAS). "Aerial" can replace "aircraft" in UAV and UAS definitions.

The US Federal Aviation Administration (FAA) prefers UAS, which is defined as "An unmanned aircraft system is an unmanned aircraft and the equipment necessary for the safe and efficient operation of that aircraft. An unmanned aircraft is a component of a UAS. It is defined by statute as an aircraft that is operated without the possibility of direct human intervention from within or on the aircraft" [Public Law 112-95, Section 331(8)]. The following are some of the components that make up the system: aircraft, data links, control station, and other components such as GNSS/GPS, accelerometer, barometer, magnetometer, gyroscope (IMU), sensor, and actuator for the sensor (SkyOp, 2016) (Figure 8-3).

Military and intelligence UASs combine cutting-edge vehicle design with imaging, navigation, and communication systems. They can carry payloads in access of 2,000 pounds at altitudes beyond 50,000 ft (Jensen, 2007). However, miniaturization has facilitated a trend toward smaller UASs that are embraced by civilians and aerial imaging and mapping companies.

PLATFORMS

MANNED AIRCRAFT

Manned aircraft include fixed-wing, propeller-driven airplanes, helicopters, and some jets with openings in the bottom of the fuselage for sensors. They are very efficient collectors of remote sensing data as the manned aircraft can fly fast, travel long distances, land at any airport then refuel and keep flying, and carry well stabilized and heavy sensors (Tulley, 2016). Rigorous accuracy standards (horizontal and vertical) have been established for image and map products generated with manned aircraft (ASPRS, 2014).

But fixed-wing aircraft are not ideal for many remote sensing applications, including flying low to increase ground resolution, flying small areas, and detailed mapping of infrastructure (Tulley, 2016). Over congested areas, fixed-wing aircraft have to fly 1,000 ft above the highest obstacle within a horizontal radius of 2,000 ft of the aircraft (FAA Aviation Regulations Part 91.119). These limitations open up many remote sensing opportunities for sUAS.

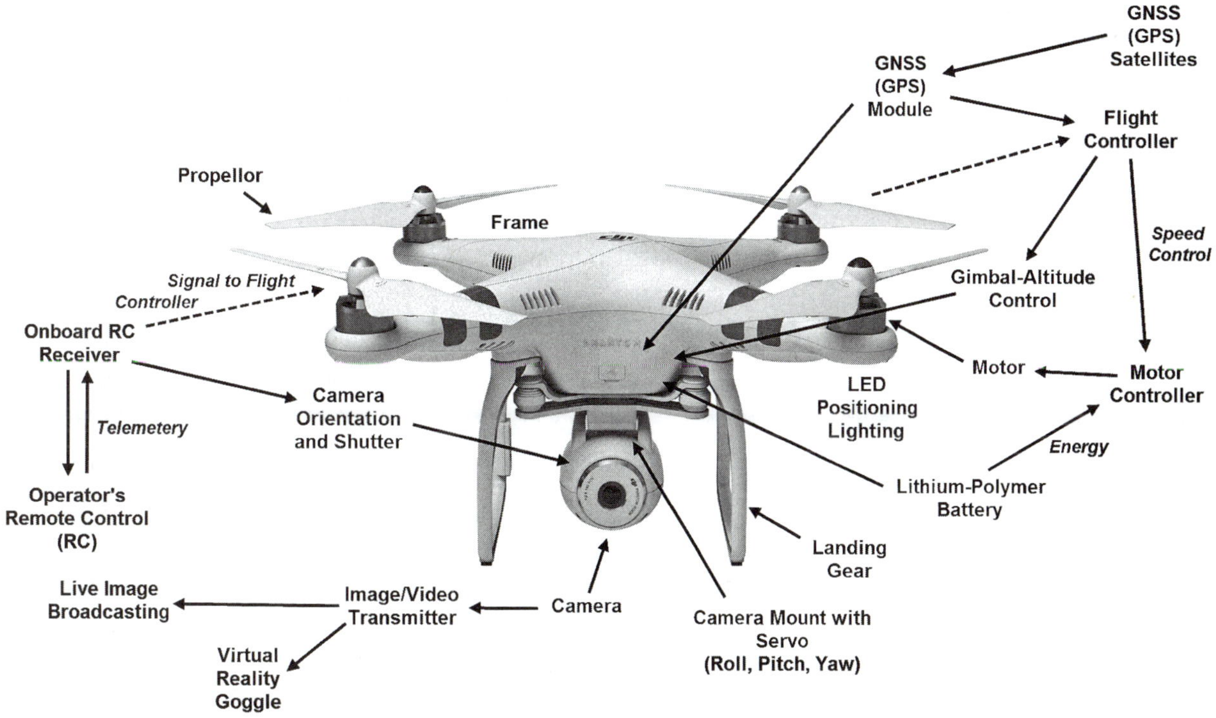

FIGURE 8-3 Main components of a sUAS system. Drone image courtesy of DJI (dji.com).

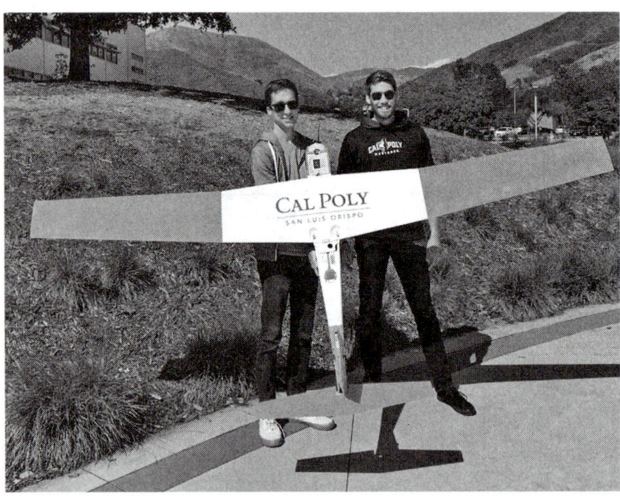

FIGURE 8-4 Students (Eric Belfield and Mike Nothem) at California Polytechnic State University, San Luis Obispo holding an AeroVironment RQ-20 Puma AE fixed-wing sUAS.

SMALL UNMANNED AIRCRAFT SYSTEMS

sUAVs are either rotating-wing (multi-rotor) copters or fixed-wing aircraft. Quadcopters are the most common multi-rotor copter (Figure 8-3). Few sUAS helicopters are deployed, as they are difficult to remotely pilot. Figure 8-4 is a larger fixed-wing system modeled after a conventional high-wing aircraft. This battery-powered drone is operated by the Aerospace Engineering Department of California Polytechnic State University at San Luis Obispo. The drone is hand launched and has an endurance of 2 to 3 hours on a single battery. The system acquires imagery to support agricultural experiments, including detection of feral pigs for population control. The image data is telemetered to a portable recording station on the ground, which eliminates the additional weight of an onboard recorder. This chapter does not review balloons and kites for collecting aerial imagery, but these are very effective platforms that are inexpensive and easily deployed (Shervais, 2016).

The choice of drone platform depends on the survey area and project requirements. Drones are chosen based on project area size and location, optimal flight time, and payload (sensors to be deployed). Multispectral VNIR and lidar sensors can be mounted on the smallest UAVs while hyperspectral systems typically require a more robust and powerful platform. Table 8-1 compares drone payload with endurance (flight time) to help determine the optimum UAS platform for a project.

Fixed-wing drones are more suitable for large scale aerial mapping and long distance inspections of pipelines, utility lines, and roads because they can fly larger areas in less time and with less vibration compared with multi-rotor drones (Table 8-2). Multi-rotor systems are more effective for close inspection from different viewpoints of infrastructure, buildings, and vegetation and for aerial photography of small sites. For a given payload, fixed wing are safer because they can glide to a landing if the motor fails. Each rotor on a multi-rotor sUAS has an electric motor; together they significantly shorten battery life compared to a fixed-wing sUAS with one propeller and motor.

TABLE 8-1 Payload and endurance of selected UAVs.

	UAV Weight	Payload	Endurance
Fixed Camera Payload			
Parrot Anafi Quadcopter	0.7 lb (315 g)	N/A	25 min
DJI Mavic 2 Quadcopter	2 lb (907 g)	N/A	31 min
DJI Phantom 4 Pro v2.0	3 lb (1,375 g)	N/A	30 min
Changeable Payload			
PrecisionHawk Lancaster 5 Fixed Wing	5.3 lb (2.4 kg)	2.2 lb (1 kg)	45 min
DJI Inspire 2 Quadcopter	7.58 lb (3.4 kg)	1.79 lb (812 g)	23 to 27 min
Wingtra VTOL Fixed Wing	8.1 lb (3.7 kg)	1.8 lb (800 g)	60 min
BirdsEyeView FireFLY6 Pro VTOL Fixed Wing	8.4 lb (3.8 kg)	1.5 lb (700 g)	50 to 59 min
Applied Aeronautics Albatross Fixed Wing	8.8 lb (4 kg)	22 lb (10 kg)	4 hr
Versadrones Heavy Lift Octocopter	8.8 lb (4 kg)	26 lb (12 kg)	8 to 30 min
Freefly Alta 6 Hexacopter	10 lb (4.5 kg)	15 lb (6.8 kg)	10 min
Hov Pod Transranger VTOL Fixed Wing	12.3 lb (5.6 kg)	2.2 lb (1 kg)	8 hr
Freefly Alta 8 Octacopter	13.6 lb (6.2 kg)	20 lb (9 kg)	20 min
DJI Matrice 600 Pro Hexacopter	22 lb (10 kg)	12 lb (5.4 kg)	16 min
FLōT Systems Prophex 50 VTOL Hexarotor	115 lb (52 kg)	50 lb (23 kg)	10 hr
Griff Guardian Octacopter	165 lb (78 kg)	331 lb (150 kg)	30 min

Vertical take-off and landing (VTOL) drones are becoming more common because autopilot, gyroscope, and accelerometer technology can be integrated into the sUAS to facilitate remote piloting of these complex systems. VTOLs combine many of the best attributes of fixed-wing and multi-rotor drones. The Arcturus-UAV Jump 15 system (Figure 8-5) has the vertical lift motors and rotors of a quadcopter attached to the wing and a large propeller attached to the front of the fuselage so the system has the motor-driven, winged capability of an airplane. This system has an endurance of 6 hours (Arcturus UAV, 2017).

SENSORS

Manned aircraft and sUAS sensors include 8-bit color cameras, 24-bit color cameras, multispectral and hyperspectral systems that span the VNIR-SWIR-TIR wavelength regions, and laser range finders. Some sensors are pointable with continuous optical zoom capability to support aerial reconnaissance and surveillance. Images are collected with frame sensors and cross-track and along-track scanners.

TABLE 8-2 Comparison of fixed-wing and multi-rotor sUAS factors.

Factor	Fixed Wing	Rotating Wing
Flight distance	High	Low
Time aloft	High	Low
Multi-image quality	Higher	Lower
Wind tolerance	Higher	Lower
Flight safety for same payload	Higher	Lower
Ability to hover	None	High
Room for take-off/landing	High	Low
Battery life	Higher	Lower
Vibration	Lower	Higher
Airborne mobility	Low	High
Cost	Higher	Lower

MANNED AIRCRAFT

Metric cameras are deployed in manned aircraft (Table 8-3). These cameras were developed for precise photogrammetric surveys and are characterized by stable and measured interior orientation (Kraus, 2008).

Single Band

Single-band, high resolution digital cameras are flown to capture color and color IR images that can be rapidly color-balanced, orthorectified, and digitally mosaicked. The NEON Airborne Observation Platform carries the Applanix DSS camera to provide high spatial resolution images that can help identify features imaged on the coarser hyperspectral and lidar sensors (Figure 8-2). The Applanix DSS is a metric camera that captures a framed image within 5,412 by 7,216 pixels (Table 8-3). Each image will cover 541 by 721 m when 10 cm pixels are collected.

Multispectral

Images that are collected as *multispectral* (MSS) data inherently provide more information about the features in the image that can be discovered and mapped with computer processing. Metric cameras with *panchromatic* (pan) + MSS capability have replaced metric film cameras for photogrammetric surveys. The large frame array and the wide linear array of these metric cameras enable large swaths of terrain to be imaged with precision and accuracy (Figure 8-6).

The color bands are sharpened with the pan band to generate high spatial resolution

FIGURE 8-5 Vertical take-off and landing sUAS. Image by Arcturus UAV (arcturus-uav.com).

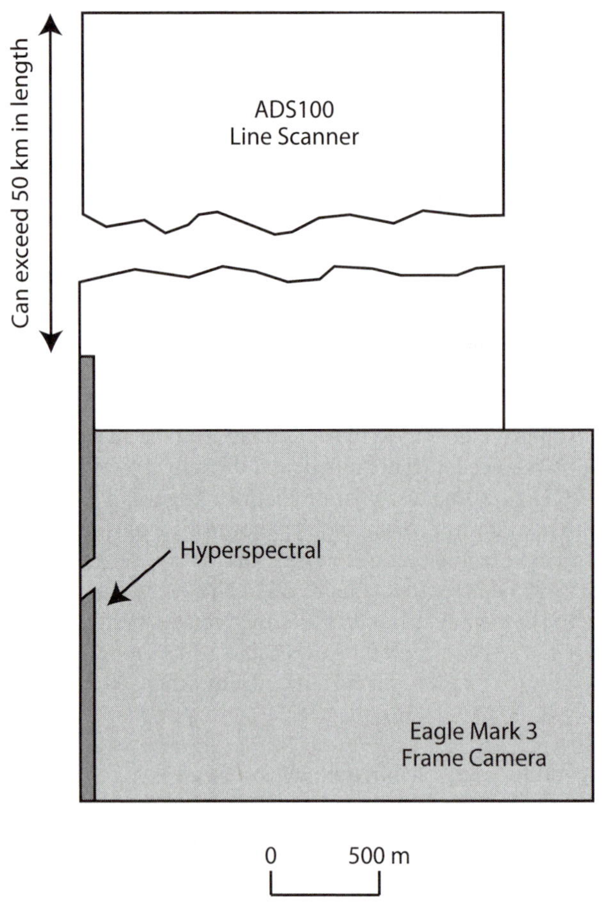

FIGURE 8-6 Comparison of footprint for multispectral frame camera (Vexcel Eagle Mark 3), multispectral line-scanning camera (Leica ADS100), and hyperspectral strip.

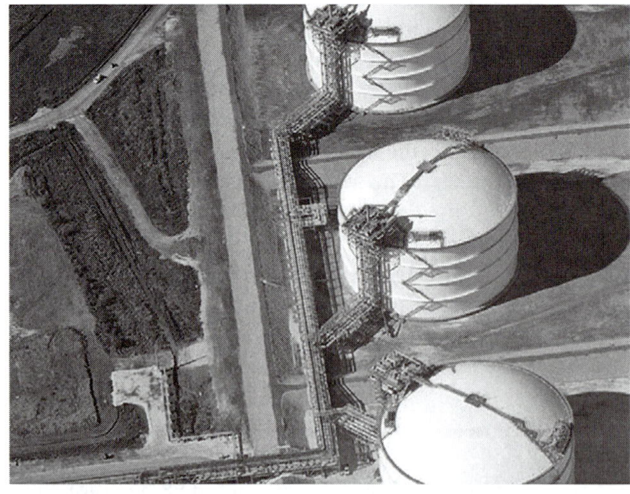

FIGURE 8-7 Oblique aerial photograph of oil tanks at a tank frame. Courtesy EagleView Technologies, Inc. (http://www.eagleview.com/casestudy/pictometry-imagery-creole-trail-pipeline).

color and color IR orthoimages. The Vexcel UltraCAM Eagle Mark 3 is a frame camera with a pan array of 26,460 by 17,004 cells. One pan image with 10 cm pixels will cover 2.6 by 1.7 km, as would the pan-sharpened color or color IR image (Figure 8-6). The Leica ADS100 line scanner collects 20,000 pixels across-track with along-track data collection only limited by onboard data storage capacity. An ADS100 image strip with 10 cm pixels is 2 km across and can exceed 50 km in length (Figure 8-6).

The MSS color bands are processed to measure vegetation vigor using the red and NIR bands, delineate the land/water boundary with the NIR band, and penetrate the water column with the blue band for bathymetric mapping. Some multispectral systems also collect MWIR and LWIR emitted TIR data (Table 8-3; Figure 1-3).

Pictometry®

While most aircraft imaging records overhead (nadir) views of features on the Earth's surface, a patented process named *Pictometry®* involves aircraft flight patterns and specialized cameras to acquire numerous oblique perspectives and an overhead view of every location flown. Oblique aerial

TABLE 8-3 Representative manned aircraft imaging systems.

System	Technology	Spectral Range (μm)	Bands	Array Size (pixels)
Applanix DSS	Frame	Color or color IR	1	5,412 by 7,216
ADSI Spectra-View	Frame	0.48 to 14.0 VNIR + TIR	6 MSS	4,872 by 3,248
Z/I DMC	Frame	0.4 to 0.85 VNIR + pan	4 MSS 1 pan	7,680 by 13,824 (pan) 3,000 by 2,000 (color)
Vexcel UltraCAM Eagle Mark 3	Frame	~0.45 to 0.85 VNIR + pan	4 MSS 1 pan	26,460 by 17,004 (pan) 8,220 by 5,668 (color)
Leica ADS100	Scanning (along-track)	.42 to .9 VNIR + pan	4 MSS 1 pan	20,000 in a line scanner
Daedalus AA3600DS	Scanning (cross-track)	0.43 to 12.5 8-VNIR 2-TIR	10 MSS	750 in a line scanner

photographs pointed in different directions show buildings, infrastructure, and land from all sides. Vertical pipes and sides of tanks not seen in overhead aerial photographs are seen in oblique aerial photos (Figure 8-7).

With specialized Pictometry® software, the typical perspective view with objects farther away (Figure 8-8A) is digitally warped so that scale is consistent across the oblique photograph (Figure 8-8B). Note that the same buildings are in both photographs. Users can accurately measure lengths, heights, and elevation and map shapes of features in the consistent scale view. For instance, length and pitch of roofs, heights of light posts and walls, sizes of gas tanks sheltered under an overhead canopy, and dimensions of elevated pipes in an industrial site can be measured by users with the aerial 3-D photography.

A. Typical aerial perspective view.

B. Digitally warped Pictometry® image where scale is constant across the photograph.

FIGURE 8-8 Comparison of oblique photography. Courtesy Pima County GIS (http://gis.pima.gov/pictometry/about.cfm) and EagleView Technologies, Inc.

Hyperspectral

Hyperspectral imaging is employed when detailed mapping of vegetation, soils, water, rock outcrops, and manmade materials (roofing, pavement, etc.) is required. The very large amount of data that is collected for each pixel (**Digital Image 1-2**), when combined with the forward motion of a manned aircraft and need for adequate dwell time, limits the number of pixels that can be acquired perpendicular to the flight line by both cross-track and along-track scanners. Hyperspectral sensors record long strips of data with typical widths of 300 to 600 pixels. A survey recording 1 m pixels generates flight strips 300 to 600 m in width. The length of the flight strip is limited by onboard data storage capacity and image processing system constraints (Figure 8-6).

Companies flying commercial VNIR-SWIR hyperspectral sensors include HyMap, Galileo Group, and SpecTIR, while sensor builders include Spectral Imaging Ltd (Specim), ITRES Research Ltd, and HyVista. In general, the commercial instruments collect 126 to 400 continuous bands with bandwidths between 10 to 20 nm and ground resolution between 0.5 to 20 m, depending on flight altitude (Table 8-4). The newest, NASA/JPL-engineered, hyperspectral sensor NEON has a very narrow bandwidth of 5 nm, enabling very fine detection of spectral absorption features and improved characterization and identification of materials (Figure 8-2). ITRES builds thermal infrared hyperspectral sensors that collect 64 bands in the 3 to 5 μm region and 32 bands in the 8 to 11.5 μm region (ITRES, 2017).

SMALL UNMANNED AIRCRAFT SYSTEMS

Miniaturization has accelerated the development of lightweight and small sensors that acquire single band, multispectral, hyperspectral, and thermal imagery. The sUAS

TABLE 8-4 Representative manned aircraft VNIR-SWIR hyperspectral imaging systems.

System	Technology	Spectral Range (μm)	Bands	Bandwidth (μm)	Typical Spatial Resolution (m)	Image Swath (pixels)
Commercial sensors	Scanning (cross-track and along-track)	0.4 to 2.5	126 to 400	10 to 20	1 to 20	~300 to 600
AVIRIS	Scanning (cross-track)	0.4 to 2.5	224	10	5 to 20	677
NEON	Scanning (along-track)	0.4 to 2.5	428	5	1 to 1.5	640

sensors are ideal for reconnaissance, disaster response, detailed imaging of infrastructure, and aerial mapping if appropriate planning, navigation, and processing are integrated with the image acquisition program. In general, the arrays are smaller on sUAS sensors compared with those on manned aircraft (Tables 8-3 and 8-4). The FAA flying height restriction of 400 ft and a smaller array results in a smaller footprint on the ground. Miller (2016) calculated that an image acquired with an iPhone 5S camera from a flying height of 29 m had a footprint of 32.6 by 24.5 m.

Smartphones mounted on drones are capable of capturing color aerial imagery with ground resolution smaller than 1 cm. Multispectral systems with a red and NIR sensor are used extensively for agricultural and vegetation mapping (Table 8-5). The MSS systems can use multi-camera mounts that hold separate cameras or use one camera body with several lenses and internal arrays (Figure 8-9). Each camera lens is covered with a blue, green, red, NIR, or red edge filter. Hyperspectral sensors are available for detailed spectral imaging. A major challenge for multi- and hyperspectral sensors is alignment of the pixels in the individual bands to ensure accurate spectral signatures over features of interest.

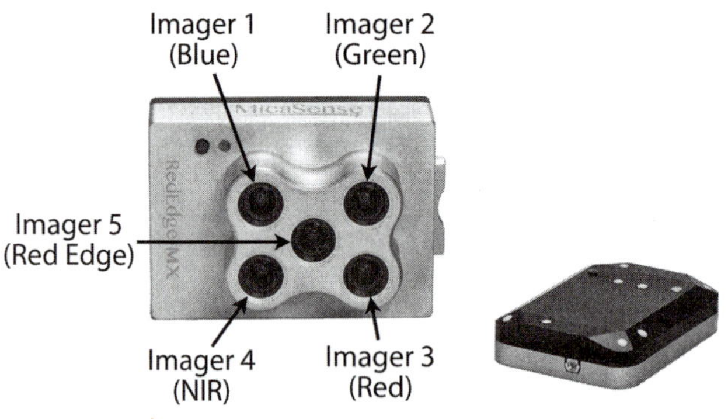

FIGURE 8-9 Multispectral camera (RedEdge-MX) with 5 bands for sUAS platform. Courtesy MicaSense, Inc.

AERIAL MAPPING

Transforming aerial images into a mapping coordinate system with an accurate location and elevation for each pixel is critical for integrating remote sensing with other geospatial data, including parcels, infrastructure (roads, pipelines, utility corridors), and surveyed field data. While aerial images are collected by manned aircraft and drones for many purposes (advertising, art, recreation, etc.), only those images that have been acquired to support aerial mapping (and integration with GIS) are considered below.

TABLE 8-5 Representative sUAS imaging systems.

System	Technology	Spectral Range (μm)	Bands	Array Size	Comments
Smartphone					
Samsung Galaxy S7	Frame	Visible	1	1,440 by 2,560	1.4 μm pixel size, geotags
Apple iPhone 5S	Frame	Visible	1	3,264 by 2,448	1.5 μm pixel size, geotags
Single Band					
MAPIR Survey 2	Frame	RGB or NIR	1	4,608 by 3,456	1 to 4 cameras can be mounted on sUAS platform; 4 cm pixels at 120 m flying height
Multispectral					
SlantRange 3p	Frame	RGB + NIR 0.41 to 0.95	4	1,200 by 1,220	4.8 cm pixel at 120 m flying height
PrecisionHawk MicaSense RedEdge-MX	Frame	0.46 to 0.86	5	1,280 by 960	8 cm pixel at 120 m flying height
Parrot Sequoia	Frame	0.55 to 0.79	5	1,280 by 960	13 cm pixel at 120 m flying height
Hyperspectral					
Corning MicroHSI 410-Shark	Along-track line scanner (push broom)	0.4 to 1.0	154	704	2 nm bandwidth, 12-bit; 3.7 kg including payload and battery
Specim AisaKESTREL 10	Along-track line scanner (push broom)	0.4 to 1.0	380	1,312 or 2,048	5 to 10 cm pixels, 4.5 kg

Figure 8-10 Manned aircraft capable of acquiring very accurate and high spatial resolution lidar and aerial imagery. Courtesy Quantum Spatial, Inc.

Manned Aircraft

Orthoimages and DEMs with highly accurate location and elevation values are generated with manned aircraft's proven flight plans, calibrated sensors and navigation instruments, surveyed ground control points (GCPs), and conventional photogrammetric software. The nadir aerial images are acquired along numerous parallel flight lines, typically with 60% forward lap and 20% side lap (Figure 2-10). Pixels are usually larger than ~5 cm due to minimum flying height restrictions for manned aircraft.

Manned aircraft imaging and mapping projects are complex operations employing sophisticated hardware and software to generate orthorectified imagery and DEMs of very large areas. Hundreds to thousands of overlapping aerial images are oriented to their proper geographic location and tied together with photogrammetric software. Aircraft, calibrated instruments and software, licensed pilots, and skilled staff (along with licensed surveyors for projects involving public safety, health, and engineering) cause aerial imaging and mapping with manned aircraft to be a substantial investment in money, equipment, and staff.

Quantum Spatial, Inc. (2015) acquired 4-band multispectral imagery over 5,800 mi^2 (9,340 km^2) along with lidar data over 1,400 mi^2 (2,250 km^2) of San Diego County, California, and a portion of Tijuana, Mexico during 2014. Multiple aircraft that provide a stable aerial base for lidar and orthoimagery acquisition were flown (Figure 8-10). The Vexcel UltraCam Eagle multispectral sensor was flown with a 100 mm lens focal distance. Each full resolution image contains 20,010 by 13,080 pixels. The Teledyne Optech lidar sensor used in the survey is capable of collecting up to 167,000 points per second and up to five returns per outgoing pulse from the laser (Figure 8-11).

The spatial resolution for the imagery was 30 cm for unpopulated areas and 10 cm for urban areas. In unpopulated areas the forward lap was 60% and sidelap was 30% while in urban areas the overlap forward lap was 80% and sidelap was 60% to ensure unobstructed views of the sides of tall buildings. For the 30 cm imagery the aircraft flew

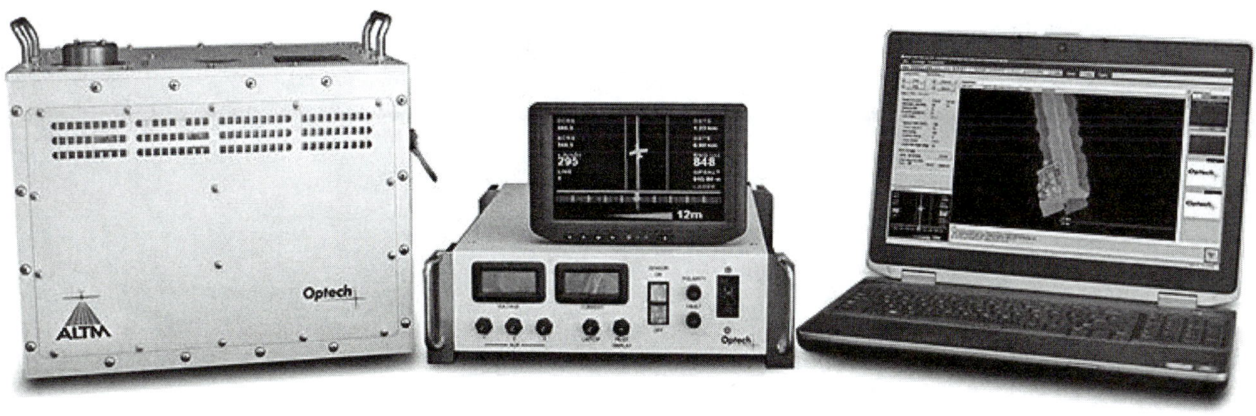

Figure 8-11 Lidar components capable of collecting up to 167,000 points per second from a manned aircraft. Courtesy Quantum Spatial, Inc.

at 18,932 ft (above ground level) and for the 10 cm imagery the aircraft flew at 6,311 ft. Forty-four flight lines with 1,719 exposures were planned for the 30 cm imagery and 85 flight lines and 5,821 exposures planned for the 10 cm imagery. The 2014 orthorectified aerial imagery is in the public domain and available for download from the San Diego County GIS website (sanGIS.org). Figure 8-12 is an orthorectified grayscale image of band 2 (red) with 1 ft pixels from the survey. The imagery is projected into the California State Plane Zone 6 coordinate system with the lines on the image representing this map projection grid at 200 ft intervals. The very accurate orthorectified aerial imagery supports detailed urban planning and map-making.

The average point density of the lidar was 2.43 pts/m^2 with 11.12 pts/m^2 in priority areas. The aircraft flew at 975 m for the coarser lidar density and 617 m for the denser data. For the lidar project area, 338 flight lines and almost 7,000 km of flight line were planned. Quantum Spatial collected over 100 GCPs in the field for calibration and verification of lidar data accuracy.

SMALL UNMANNED AIRCRAFT SYSTEMS

In contrast to manned aircraft, flying drones and collecting images can be a very low cost and low skill venture. However, to build accurate aerial maps from drone images requires comprehensive flight planning and knowledge about complex software that builds DEMs and orthoimage mosaics.

Images collected by sUAS require many of the same elements as needed by conventional photogrammetry, such as forward lap/sidelap and GCPs, to convert the individual scenes to orthoimage mosaics and DEMs that are in a mapping coordinate system (Miller, 2016). A new photogrammetric technique, Structure from Motion (SfM), has been embraced by the sUAS community because SfM is highly automated, improves with easily achieved, increased over-

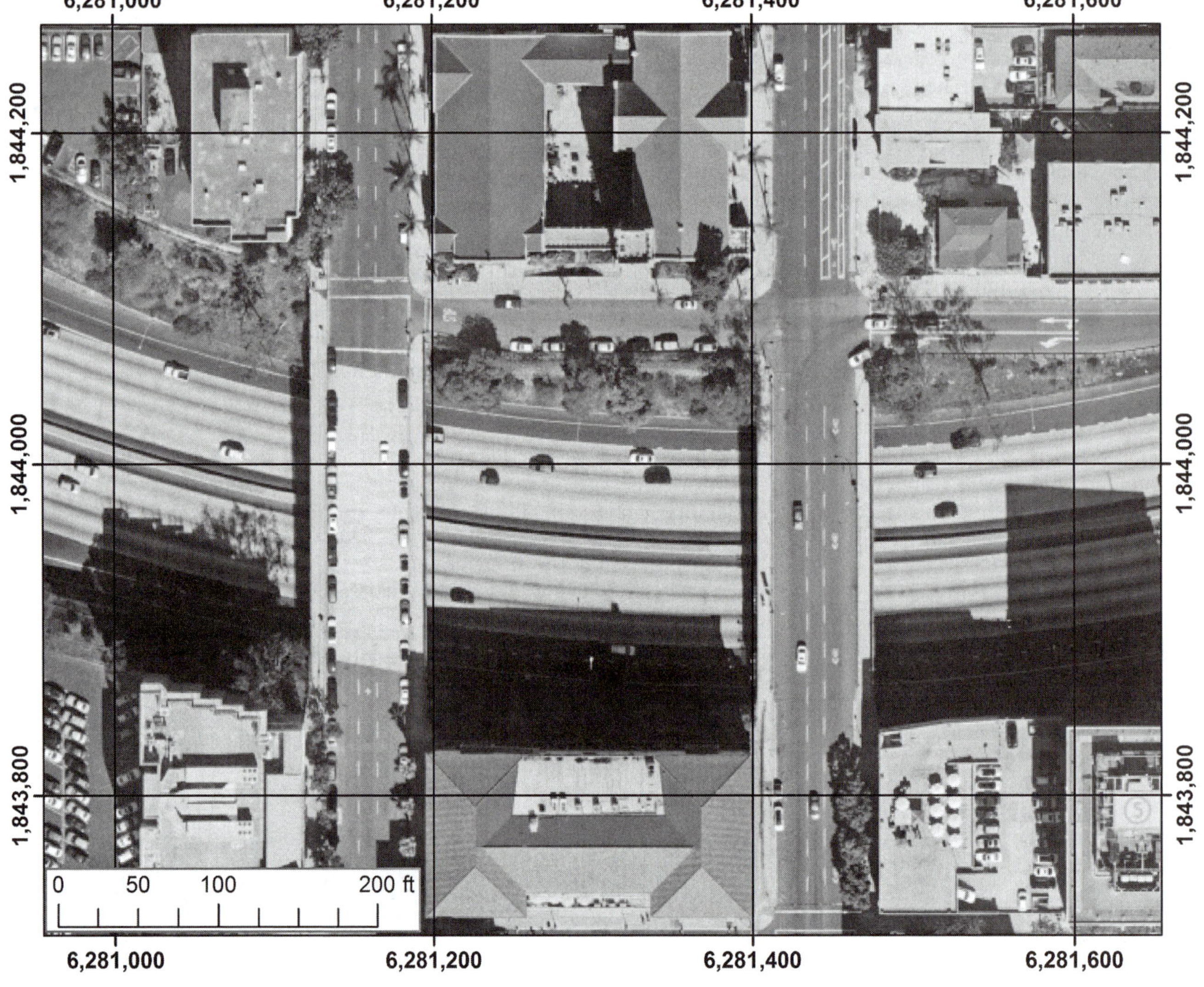

FIGURE 8-12 Orthorectified aerial image for detailed engineering, urban planning, and GIS applications acquired from manned aircraft. Courtesy SanGIS.

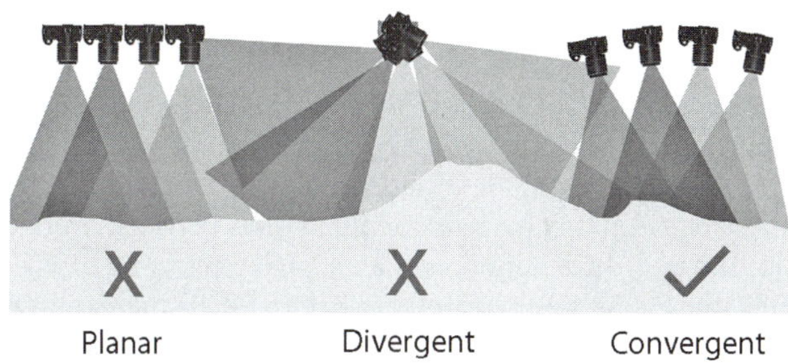

Figure 8-13 Do not acquire photos in a planar or divergent fashion. Acquire photos in a convergent manner. From Shervais (2016, Figure 11).

Figure 8-14 Acquire photos converging on a feature, at multiple distances and angles. From Shervais (2016, Figure 12).

lap between images during airborne acquisition, works with consumer-level cameras, and is inexpensive and relatively accurate (see discussion in Chapter 7).

Camera Orientation

James and Robson (2014) have shown photos from sUAS should be pointed at the terrain or object of interest in a convergent manner as divergent or planar orientations distort the SfM model (see also Raugust and Olsen, 2013; Shervais, 2016). For example, Figures 8-13 and 8-14 show a topographic profile of terrain with flat land to the left and elevated land in the center-right. The camera is shown in many positions and pointing in different directions as it is flown across the terrain, illustrating three methods for photo acquisition: planar, divergent, or convergent.

In the left portion of Figure 8-13, the camera is looking straight down (planar orientation) at the terrain. In the center of the profile, the camera is in a stationary position and pointed in diverging directions. In Figure 8-14, the camera is pointed off nadir but in converging directions, increasing overlap between images, and the camera is flown at two altitudes. Also, the camera-viewing direction converges on the topographic feature with higher elevation in the center-right, minimizing distortion in the model. Views of the same feature from multiple angles and different elevations ensure a more reliable SfM model.

Image Overlap

The amount of overlap/sidelap between individual images is another key factor for building a good SfM model, an accurate orthoimage mosaic, and a DEM. Shervais (2016) notes less than 70% overlap will affect the interpreted scene while more than 90% may significantly increase processing time. Miller (2016) imaged 2.5 acres of an abandoned quarry with 80% forward lap and 40% sidelap while Mapir (2017) covered 18.7 acres of a vineyard with 70% forward lap and 70% sidelap.

Mission Planning

Collecting sUAS imagery in the field that can support accurate aerial mapping requires an unmanned aerial vehicle, a ground control station with a communications data link (operator's remote control), flight planning interface, battery recharger, and onboard GPS/IMU (Figure 8-15). A GPS base station will increase the *x*, *y*, *z* accuracy of the

Figure 8-15 Components for operating a sUAS. Courtesy Black Swift Technologies.

imagery and the SfM 3-D models. In addition, Colomina and Molina (2014) emphasize the importance of mission planning and management, including design of the aircraft flight program (waypoints, flight lines, speed, altitude, direction, etc.) and sensor configuration. Surveying and marking GCPs that are large enough and distinct from surrounding material to be seen on the aerial imagery improves the location accuracy of orthoimage mosaics. Images that lack distinct features (tidal flats, monoculture crops, water, wetlands, sand dunes, etc.) are more difficult to digitally mosaic as the SfM software needs to find and match individual features in multiple images to build the 3-D geometry of the scene. The sUAS must have excellent GNSS/GPS capability that can assign an accurate location to each image during the acquisition to build a DEM and orthoimage mosaic of a relatively homogeneous landscape.

Mapping with a Smartphone

Miller (2016) describes building a DTM and orthoimage mosaic with 1 cm pixels across 2.5 acres (10,000 m^2) of abandoned quarry in Concord, California, with a quadcopter and iPhone 5S. The survey area was approximately 80 by 100 m (2.5 acres) with the base of the abandoned quarry ~72 m above sea level. The topographic elevation range between the floor of the quarry and the surrounding undisturbed land was approximately 18 m.

The drone was flown no higher than 29 m to achieve 1 cm pixels. Each iPhone photo had a 32.6 by 24.5 m footprint on the ground (Figure 8-16). Ten flight lines with 236 images were needed to achieve 80% forward lap and 40% sidelap (Plate 26A). The camera shutter was set to acquire a photo every 1 second using an onboard app. SfM processing technology was used to generate an orthoimage mosaic (Plate 26B), the DTM (Plate 26C), and the contours (Plate 26D).

Miller (2016) performed many accuracy and performance assessments based on nine surveyed, ground control targets (with absolute accuracy of 2 cm horizontal and vertical) that had markers visible on the imagery and 132 surveyed checkpoints. He concludes that a consumer grade drone with an iPhone 5S camera is capable of producing high-accuracy 3-D models useful in numerous applications (Miller, 2016).

Mapping in a National Park

The National Park Service (NPS) collaborated with Wohnrade Civil Engineers, Black Swift Technologies, and UAS Colorado to deploy a fixed-wing sUAS to build a detailed DEM and orthoimage mosaic of a 1 mi^2 (2.6 km^2) area of the Great Sand Dunes National Park and Preserve. The project documents several challenges with implementing sUAS technology, including an imaging site that cannot be accessed by vehicle, a launch site several miles from the imaging area, gusting winds, strongly contrasting shadows, and a relatively homogeneous landscape. The lessons learned, and the solutions offered in this NPS survey, are valuable for others considering sUAS for aerial mapping (Wohnrade Civil Engineers and Black Swift Technologies, 2016).

The area was flown with a hand-launched, Black Swift Technologies SwiftTrainer (Figure 8-17) in 2-1/2 hours with two flights that collected 1,755 images. The homogeneous ground surface texture of sand dunes, along with intense elevation changes, made it difficult for the photogrammetric software to identify common match points between overlapping images. In addition, the area of interest was inaccessible by vehicle, limiting the number of GCPs that could be surveyed and marked in the allotted time frame to provide x, y, z control for the project. Horizontal (17.5 cm) and vertical (33.3 to 66.7 cm) accuracy for this initial sUAS project was limited because of the match point and GCP issues. Nevertheless, an orthoimage mosaic with 1.2 in (3 cm) pixels, a 1 ft (30 cm) contour map, and a point cloud of the DEM with 145,000,000 points (each point representing ~1.5 in^2 [~10 cm^2] on the ground) were generated from the sUAS data.

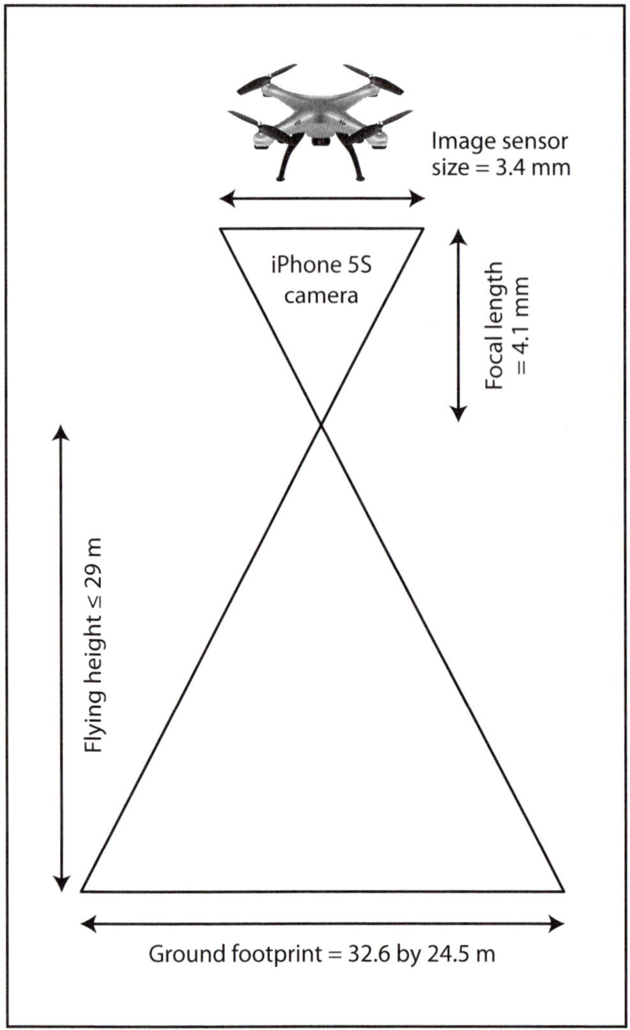

FIGURE 8-16 Profile of sUAS carrying an iPhone 5S camera with a focal length of 4.1 mm flying 29 m above the terrain for an image ground footprint of 32.6 by 24.5 m. From Miller (2016, Figure 18).

Figure 8-17 Hand launch of the fixed-wing SwiftTrainer sUAS. Courtesy Black Swift Technologies.

sUAS REGULATIONS

A brief review of US regulations follows to provide some critical guidance to readers who are flying or intend on flying a drone for fun or recreation or for nonrecreational (commercial or business) use. Key concerns are the safety of the National Airspace System and the safety of people and property on the ground. Part 107 of the Title 14 Code of Federal Regulations defines small UAS as UAS that use unmanned aircraft weighing less than 55 lb (25 kg). FAA guidelines for small UAS will evolve as the industry matures. Check with the FAA (faa.gov/uas) for the latest rules.

On October 5, 2018, the FAA Reauthorization Act of 2018 (Pub. L. 115-254) was passed. Recreational sUAS users must register their drone with the FAA if it weighs between 250 g (0.55 lb) and 25 kg (55 lb). The Act also requires recreational flyers to pass an aeronautical knowledge and safety test. However, as of August 2019, recreational users who fly as a hobby do not need a certification or drone pilot license to fly. Starting in September 2019, the FAA will develop the training and test content with input from private and public sectors. Recreational flyers should know that if they intentionally violate any of the safety requirements, and/or operate in a careless and reckless manner, they could be liable for criminal and/or civil penalties.

Operators flying sUAS for commercial applications must obtain a remote pilot certificate with a sUAS rating or be under the direct supervision of someone who holds such a certificate. To qualify for a remote pilot certificate you must pass an aeronautical knowledge test at an FAA approved testing center. If you already have a Part 61 pilot certificate other rules apply. If you are acting as pilot in command, you must make your drone available for FAA inspection and report to the FAA any operation that results in serious injury or property damage of at least $500.

sUAS operators must maintain visual line-of-sight (VLOS) between the remote pilot and the sUAS, maximum flight altitude of 400 ft (122 m) above ground level, maximum groundspeed of 100 mph (160 km $\cdot$ hr^{-1}), and operate only during the daytime. There are many drone no-fly zones in the United States, including airports, restricted or special use airspace, stadiums and sporting events, wildfires, and NOAA marine protected areas. B4UFLY is a free, easy-to-use smartphone app from the FAA that helps recreational drone operators learn where they can and can't fly (faa.gov/uas/recreational_fliers/where_can_i_fly/b4ufly).

NEW DEVELOPMENTS

Cloud-Based Services

Significant technical knowledge and software are required to convert drone images into products that can support aerial mapping and be integrated with other geospatial layers in a GIS. The steep learning curve and need for specialized software has encouraged many drone operators to subscribe to cloud-based, image processing services. These services give operators the option to fly their own drone when they need to acquire new imagery, but after the survey they upload their images with camera and navigation data to a cloud-based service for calibration, processing, storage, and sharing of their orthoimages and DSMs. In addition, these services derive new maps from multispectral data such as the Normalized Difference Vegetation Index (NDVI), other vegetation indices, row-based plant counting, ponded water, etc. (see PrecisionHawk, 2017). The orthoimages, DSMs, and maps are shared via the cloud or mobile devices.

Collision Avoidance

The "threat of accidents" slows the growth of consumer drones (Grand View Research, 2016). Collision avoidance systems are being integrated into drone platforms to improve the safety record of drones and to help remote control pilots avoid crashing into objects. The drone manufacturer DJI builds a front-facing collision avoidance system into their Mavic Pro drone that can spot obstacles up to 15 m (49 ft) away when travelling below 36 kmph (22 mph). Additional sensors detect the terrain below, allowing the Mavic Pro drone to avoid crashing into a topographic slope that rises up toward, and in the direction of, the flight path (BBC News, 2016). A primary goal for the industry is to deploy technology that will enable drones to autonomously avoid obstacles, making the skies much safer. Also implementing reliable, collision avoidance systems on drones will help drone operators obtain agency approval for *beyond visual-line-of-sight* (BVLOS) operations.

SWARMS

Swarms of drones can be deployed on the battlefield to overwhelm an enemy's defenses or to accomplish wide-ranging surveillance. Civilian applications include using swarms to monitor large agricultural fields and aid in search and rescue operations. At the Naval Postgraduate School, Timothy Chung defines swarming as a large collection of aerial robots working together to do something meaningful or interesting (Almanzan, 2015). One drone is programmed to be the leader and the others follow in a pattern, communicating via WiFi. Chung has deployed 30 fixed-wing drones as a swarm.

GEOREFERENCED VIDEO

Digital cameras can be configured to combine a sequence of images to form a moving picture or video. Aerial videos are often recorded by drones. They are very useful for early surveillance and monitoring of disaster areas. Damage caused by the massive, April 2015 earthquake in Kathmandu, Nepal was rapidly communicated to the world and emergency response teams using drones with video cameras. However, most of the airborne videos recorded on manned and unmanned aircraft are not georeferenced, so they are difficult to use as a base for mapping and interpretation.

The National GeoSpatial-Intelligence Agency developed metadata standards for capturing and recording aircraft and camera parameters essential for georeferencing video frames (Kalinski, 2015). Commercial geospatial software now uses this standard to display the video's georeferenced footprint on the ground as the aircraft moves along. This facilitates more accurate mapping and interpretation. Mapping applications using video remote sensing will significantly increase as visible light, reflected IR, and TIR images can now be captured.

Crutsinger and others (2019) evaluated the potential of low-cost consumer drones to capture georeferenced video of property along major roads in Paradise, California, to assess infrastructure damage after the devastating Camp Fire. To generate a georeferenced drone video, the timestamp of the video recording is synchronized with the timestamp of the drone telemetry, or where the drone was flown. The end result is a video overlain on a base map that is easy for a broad audience to understand. The georeferenced video was collected along 32 km of road in 75 minutes by three drone teams. The trial demonstrates that georeferenced drone video can provide situational awareness and damage assessment within the first few hours of an emergency.

STOPPING DRONES

Consumer drones that enter restricted airspace (airports, stadiums, military bases, parades, etc.) cause great safety and security concerns. Airspace regulators such as the FAA have to maintain strict regulations and establish large no-fly zones to minimize the hazard of errant drones flying into restricted areas because there has been a lack of tools to deter or stop a flying drone. Technology to detect, target, and deter (or destroy) drones in midair is rapidly being developed for civilian safety and for military and law enforcement applications. Boeing invented a laser weapon that melts drones midair, while three tech companies in the United Kingdom developed a system that aims radio waves at a flying drone to disrupt the drone's communication system and stop the drone's movement (Palermo, 2015). Having the capability to stop drones near restricted areas will encourage regulators to expand the available airspace for commercial and recreation drone operation.

QUESTIONS

1. What is the FAA definition of a UAS?
2. List five components found in a UAS.
3. What are three advantages of acquiring remote sensing data with manned aircraft?
4. What are three disadvantages of acquiring remote sensing data with manned aircraft?
5. What are three advantages of acquiring remote sensing data with sUAS (drones)?
6. What are three disadvantages of acquiring remote sensing data with sUAS?
7. List four advantages of fixed-wing sUAS.
8. List four advantages of multi-rotor sUAS.
9. A manned aircraft flying a Vexcel UltraCAM Eagle Mark 3 collects 5 cm pixels with the color arrays. How large of an area is acquired in one frame?
10. One of the technical challenges of collecting multispectral bands is coregistering the stack of images. Why is this important?
11. What is the range of recommended overlap and sidelap for building a good SfM model?
12. List five components and/or tasks included in mission planning and management that enable accurate DEMs and image mosaics to be built with sUAS technology.
13. What are four FAA requirements that must be followed by a commercial and recreational sUAS operator?
14. What are the advantages of cloud-based image processing and mapping services for sUAS operators?

REFERENCES

Almanzan, K. 2015, July 23. Follow the Leader: Drones Learn to Behave in Swarms. National Public Radio. All Tech Considered (4:21 AM EST). http://www.npr.org/sections/alltechconsidered/2015/07/23/424685529/follow-the-leader drones-learn-to-behave-in-swarms (accessed June 2018).

Alspaugh, D. 2004. A brief history of photogrammetry. In J. C. McGlone (Ed.), *Manual of Photogrammetry* (5th ed., pp. 1–14). Bethesda, MD: American Society for Photogrammetry and Remote Sensing.

Anna, C. 2016, October 9. Drones carrying medicines, blood face challenge: Africa. *U.S. News & World Report*. http://www.usnews.com/news/world/articles/2016-10-09/drones-carrying-medicines-blood-face-top-challenge-africa (accessed January 2018).

Arcturus UAV. 2017. Jump 15 VTOL. http://arcturus-uav.com/product/jump-15.

Asner, G. P., D. E. Knapp, T. Kennedy-Bowdoin, M. O. Jones, R. E. Martin, J. Boardman, and C. B. Field. 2007. Carnegie Airborne Observatory: In-flight fusion of hyperspectral imaging and waveform light detection and ranging (with lidar) for three-dimensional studies of ecosystems. *Journal of Applied Remote Sensing*, 1(1), 013536. doi: 10.1117/1.2794018

ASPRS. 2014. ASPRS Positional Accuracy Standards for Digital Geospatial Data. *Photogrammetric Engineering & Remote Sensing*, 81(3). http://www.asprs.org/Standards-Activities.html (accessed January 2018).

BBC News. 2016, September 27. DJI's Mavic Pro fold-up drone detects obstacles. http://www.bbc.com/news/technology-37475568 (accessed January 2018).

Colomina, I., and P. Molina. 2014. Unmanned aerial systems for photogrammetry and remote sensing: A review. *Journal of Photogrammetry and Remote Sensing*, 92, 79–97.

Crutsinger, G., R. Hasna, C. Tholborn, J. Ladner, R. Borkert, S. Reed, T. Wrangham, A. Maximow, and J. Cherbini. 2019. Georeferenced Drone Video for Rapid Damage Assessment Following the Camp Fire. http://www.scholarfarms.com/freewhitepaper.

Grand View Research. 2016. Consumer Drone Market Analysis by Product (Multi-Rotor, Nano), by Application (Prosumer, Toy/Hobbyist, Photogrammetry), and Segment Forecasts to 2024. http://www.grandviewresearch.com/industry-analysis/consumer-drone-market (accessed January 2018).

ITRES. 2017. Airborne Hyperspectral and Thermal Remote Sensing. http://www.itres.com/ (accessed January 2018).

James, M. R., and S. Robson. 2014. Mitigating systematic error in topographic models derived from UAV and ground-based image networks. *Earth Surface Processes and Landforms*, 39, 1,413–1,420. doi: 10.1002/esp.3609

Jensen, J. R. 2007. *Remote Sensing of the Environment: An Earth Resource Perspective* (2nd ed.). Upper Saddle River, NJ: Pearson.

Johnson, K., E. Nissen, S. Saripalli, J. R. Arrowsmith, P. McGarey, K. Scharer, P. Williams, and K. Blisniuk. 2014. Rapid mapping of ultrafine fault zone topography with structure from motion. *Geosphere*, 10(5), 1–18.

Kalinski, A. 2015, April 2. Georeferenced full motion video: Mitigating a difficult "big data" problem. *GeoSpatial Solutions*. http://geospatial-solutions.com/georeferenced-full-motion-video-mitigating-a-difficult-big-data-problem (accessed January 2018).

Kraus, K. 2008. Cameras (Vol. 1, Sections 3.1–3.4 and 3.6–3.7). Yıldız Teknik Üniversitesi. İstanbul, Türkiye, 39 slides. http://yildiz.edu.tr/~bayram/fotogrametri/6-Cam.pdf (accessed January 2018).

Krause, K. 2014. NEON Airborne Observation Platform (AOP) Overview. National Ecological Observatory Network, 18 slides. http://www.slideshare.net/jjparnell/neon-airborne-operations (accessed January 2018).

Mapir. 2017. Flower Field 1 (200 ft)—April 1, 2016. http://www.mapir.camera/pages/flower-field-1-april-1-2016 (accessed January 2018).

McGlone, J. C. (Ed.). 2013. *Manual of Photogrammetry* (6th ed.). Bethesda, MD: American Society for Photogrammetry and Remote Sensing.

Miller, J. C. 2016. The acquisition of image data by means of an iPhone mounted to a drone and an assessment of the photogrammetric products. MA Thesis, California State University at East Bay. http://iphonedroneimagery.com/project-methodology (accessed January 2018).

Palermo, E. 2015, October 10. Signal-scrambling tech "freezes" drones in midair. *LiveScience*. http://www.livescience.com/52448-new-tech-freezes-drones.html (accessed January 2018).

PAPA International. 2015. History of Aerial Photography. http://professionalaerialphotographers.com (accessed January 2018).

PrecisionHawk. 2017. DataMapper Algorithm Marketplace. http://www.datamapper.com/algorithms (accessed January 2018).

Quantum Spatial, Inc. 2015. Airborne Topographic LiDAR and Ortho Report, San Diego, CA, 2014. Post Flight Aerial Acquisition and Calibrated Report for USGS (Contract No. G10PC00026). Lexington, KY: Quantum Spatial, Inc.

Raugust, J. D., and M. J. Olsen. 2013. Emerging technology: Structure from Motion. *LiDAR Magazine*, 3(6).

Shaw, I. G. R. 2016. *Predator Empire: Drone Warfare and Full Spectrum Dominance*. Minneapolis: University of Minnesota Press.

Shervais, K. 2016. *Structure from Motion Guide for Instructors and Investigators*. Boulder, CO: UNAVCO.

SkyOp. 2016. Introduction to sUAS—Student Workbook. Canandaigua, NY: SkyOp.

Tulley, M. 2016, April 6. "Commercial" drones, but just barely. *Lidar News*. http://lidarnews.com/ (accessed January 2018).

Wohnrade Civil Engineers and Black Swift Technologies. 2016. *The Great Sand Dunes National Park and Preserve: Unmanned Aircraft System Aerial Survey*. Wohnrade Civil Engineers, Broomfield, Colorado, and Black Swift Technologies, Boulder, Colorado.

Digital Image Processing

Digital processing did not originate with remote sensing and is not restricted to these types of data. Many image processing techniques were developed in the medical field to process X-ray images and images from sophisticated body-scanning devices. For remote sensing, the initial impetus was the program of unmanned planetary satellites in the 1960s that *telemetered* (transmitted) images to ground receiving stations. The low quality of the images required the development of processing techniques to make the images useful. Another impetus was the Landsat program, which began in 1972 and provided the first sets of worldwide imagery in digital format. A third impetus was, and remains, the continued development of faster, more powerful, and less expensive computer systems for image processing.

The quality of data, and the diversity of platforms and sensors acquiring remote sensing images, has increased exponentially over the past few decades. The image processing community has also greatly expanded as more applications are found for remote sensing data. Many of the fundamental principles and algorithms used for processing images were established decades ago (Russ, 2011; Schowengerdt, 1983). Jensen (2016), Richards (2013), and Schowengerdt (2007) describe in detail the methods and the mathematical transformations. When compared to older images, the imagery generated by these established principles and algorithms are significantly improved because today's multispectral, hyperspectral, thermal, and radar sensors collect almost noise-free data.

This chapter describes and illustrates the major categories of image processing. Readers can use the many images and DEMs provided in the Remote Sensing Digital Database (available for download at waveland.com/Sabins-Ellis) to practice and compare their image processing results with the examples discussed in Chapters 2 through 10.

STRUCTURE OF DIGITAL IMAGES

One can think of any image as consisting of small, equal areas, or *picture elements*, arranged in regular rows and columns, called a *raster array*. The position of any picture element, or *pixel*, is determined on an *x, y* coordinate system. Each pixel also has a numerical value, called a *digital number* (DN), that records the intensity of the electromagnetic energy measured for the ground resolution cell represented by that pixel. Digital numbers range from zero to some higher number on a grayscale. This system records an image in strictly numerical terms in a three-coordinate system in which the *x*- and *y*-values locate each pixel and *z* gives the DN, which is displayed as a grayscale intensity value.

There are several ways of obtaining the data needed to compose an image. Scanner systems record images directly in a digital format where each ground resolution cell is represented by a pixel. Analog images, such as photographs and maps, are converted into digital format by a process known as *digitization*. In this chapter digital image processing is largely illustrated using Landsat examples because the data are readily available and the images are familiar. Digital processes, however, are equally applicable to all forms of digital image data.

SATELLITE SCENES

Satellites image the Earth in different modes, ranging from a fixed, downward-looking nadir view to adjustable fore and aft, side-to-side views.

Landsat

Landsat satellites view the Earth in a downward-looking, fixed nadir mode. The sensor does not point side-to-side or forward and aft. Landsat MSS and TM data are recorded by a cross-track (whisk broom) scanner while Landsat OLI/TIR data are recorded by an along-track (push broom) scanner (Figure 1-16). The Earth rotates from west to east as the Landsat satellites orbit from north to south during the day, resulting in a systematic shift of scan lines toward the west with satellite movement toward the south. Figure 9-1 shows the effect of the Earth's rotation during the acquisition of a Landsat TM scene. As it is acquired, the east and west margins of the Landsat image are not aligned north–south, but are offset ~9° from north to south. The east and west margins of Landsat scenes require digital image processing to minimize noise at the ends of the scan lines.

The Landsat TM cross-track scanners have a ground resolution cell of 30 by 30 m (Figure 9-1). An oscillating scan mirror sweeps the cell alternately east and west across the terrain to produce scan lines oriented at right angles to the satellite orbit path. Scanning is continuous along the orbit; the data are subdivided into scenes consisting of 5,667 scan lines that are 185 km long in the scan direction and 30 m wide in the orbit direction. The analog signal is sampled at intervals of 30 m to produce 6,167 pixels (measuring 30 by 30 m) per scan line. Each TM image band consists of 34.9×10^6 pixels. The seven bands have a total of 244.3×10^6 pixels. Each of the six TM visible and reflected IR bands employs an array of 16 detectors; each sweep of the scan mirror records 16 lines of data. The TIR band 6 employs four detectors with 120 by 120 m ground resolution cells; these are resampled to 30 by 30 m pixels. Landsat scenes have a consistent number of pixels, rows, and columns for each generation of sensor (MSS, TM, and OLI). The uniformly sized scenes are extracted from data collected continuously along an orbital path following the WRS-1 (Landsat MSS) or WRS-2 (Landsat TM and OLI) structure (Figure 3-9). Sentinel-2 and ASTER also have a standard size for each scene.

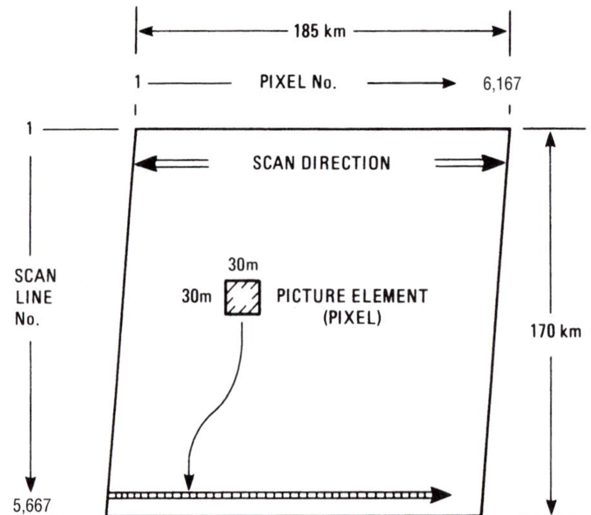

FIGURE 9-1 Arrangement of scan lines and pixels in a Landsat TM image.

Satellites with agile sensors can point side-to-side and forward and aft, collecting scenes that maintain a north–south orientation as the satellites move along their near polar orbit and the Earth rotates toward the east (Figure 4-5). These images do not have the east–west wedge of no data that characterizes Landsat scenes. The number of pixels in an agile satellite's image is fixed along east–west scan lines, but will vary along the north–south direction depending on the acquisition program. For example, a SPOT 6 scene will have 10,000 6-m pixels along each east–west scan line, but the number of pixels in the north–south direction can vary from less than 10,000 to 100,000 (Figure 4-2B).

DIGITAL IMAGE ARRAY

The structure of a digital image is illustrated in Figure 9-2, which is a greatly enlarged portion of Landsat 8 band 5 (NIR) from the Thermopolis, Wyoming, subscene that was described in Chapter 3. Figure 9-2A shows an image that contains 66 rows and 66 columns of pixels. Bright pixels have high DNs and dark pixels have lower DNs. Figure 9-2B is a map that identifies features in the satellite image and shows the location for the array of DNs in Figure 9-2D. A profile of DNs obtained along the A–A′ dotted line on Figure 9-2A is displayed in Figure 9-2C.

Referring again to the DN array (Figure 9-2D), note the low values associated with the Wind River and the high values of the adjacent vegetation, especially the agricultural irrigated field to the northeast. These values are consistent with the strong absorption by water and strong reflection by vegetation of the NIR energy recorded by OLI band 5. Along the river, note that the lowest digital numbers (DN = 12 to 42) correspond to the center of the stream; pixels along the stream margin have higher values (DN = 38 to 196). The 30 by 30 m ground resolution cells in the center of the river are largely occupied by water, whereas marginal cells are occupied partly by water and partly by vegetation, resulting in an intermediate DN value between water and vegetation. The mixed pixels form narrow gray bands along the river in Figure 9-2A. The TM format is typical of all multispectral digital data sets.

DIGITIZATION

Digitization is the process of converting an analog image into a digital raster array of pixels. Hardcopy maps and other information may also be digitized. Photogrammetric

A. Grayscale image of OLI band 5 (NIR).

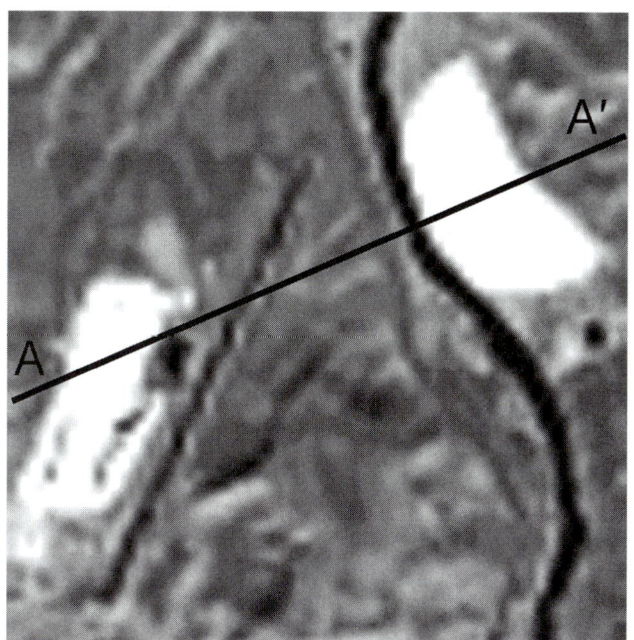

B. Location map.

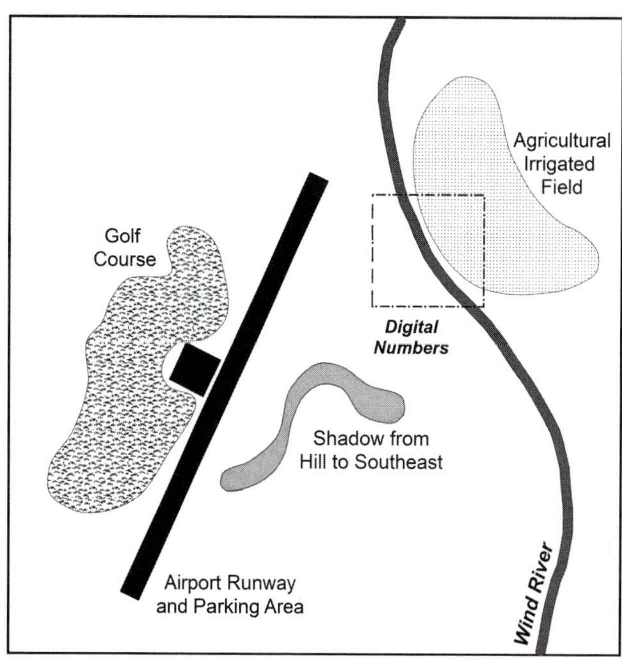

C. DN values along profile A–A' shown in Figure 9-2A.

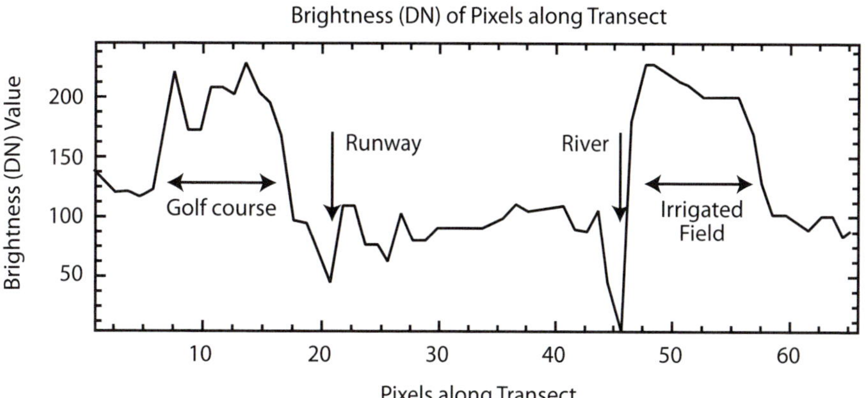

D. Array of DNs for the area shown in the location map. Gray cells show the relatively low digital numbers associated with water in the Wind River.

Array Columns

Array Rows	1				5					10		
1	94	96	90	30	38	202	246	243	235	233	222	215
	89	96	105	57	18	133	251	249	241	227	221	214
	93	88	95	85	15	86	231	246	234	217	214	215
	94	88	94	106	50	42	196	242	234	222	219	222
5	99	93	98	111	83	20	130	236	242	236	230	228
	102	98	99	102	103	23	63	188	238	241	236	234
	102	97	97	103	109	49	12	115	230	246	240	238
	106	102	94	96	115	85	21	26	141	233	243	242
	120	117	98	90	111	123	54	16	35	137	229	245
10	116	113	105	91	110	132	102	51	16	48	155	149
	95	99	101	98	98	126	132	112	63	19	59	149
	100	102	101	97	95	127	127	113	124	74	25	54

Figure 9-2 Digital structure of a Landsat OLI band 5 (NIR) image. The area is a portion of the Thermopolis, Wyoming, subscene shown in Chapter 3.

film scanners are precision instruments that can record over 4,000 dots per inch (dpi). A typical black-and-white aerial photograph measuring 23 by 23 cm (9 by 9 in) that is digitized at 3,000 dpi is converted into a 729 MB grayscale raster image (729,000,000 pixels). A digitized color film at 3,000 dpi becomes a 2.2 GB raster file. Inexpensive desktop scanners and digital cameras generate distorted raster versions of the original analog image or map due to inherent instrument limitations. These geometric distortions should be corrected with image processing software prior to using the converted image or map for mapping.

DATA FORMATS

Satellite and aerial remote sensing images have at least two components included with each scene: the raster array and a header (metadata) that contains essential information about the digital data in the scene. The header can include pixel size, shape, and orientation; coordinate system; data type (i.e., integer, floating point, byte, etc.); number of samples, lines, and bands; and how the bands are arranged in the digital file. Band sequential (BSQ), band interleaved by line (BIL), or band interleaved by pixel (BIP) are the formats used to organize multispectral and hyperspectral bands in a file.

Commercial and widely distributed government digital images are often available in the common and easily read GeoTIFF format or compressed to JPEG2000, MrSID, or ECW formats. Remote sensing data collected for research and military, by older systems (especially if they are no longer operating), and by radar platforms often have formats that are not supported by, and therefore cannot be read by, basic image processing software and consumer photo enhancement software. In contrast, advanced image processing software are expected to read many input formats and continually update their capabilities as new systems are launched with new data formats.

IMAGE PROCESSING OVERVIEW

Image processing methods are grouped into three functional categories: preprocessing, image enhancement, and information extraction. We will define these categories here, followed by a thorough discussion of typical processing routines within each category. The routines are illustrated with Landsat examples, but the techniques are equally applicable to other digital image data sets.

1. *Preprocessing* involves steps to understand the data quality (level of noise in different bands, artifacts, distortions, etc.) and to prepare the data to meet project requirements.
    a. Minimize raster data errors and noise
    b. Review data statistics
    c. Calibrate as needed (including atmospheric correction)
    d. Correct geometric errors
    e. Map projection
    f. Georeferencing or orthorectification
2. *Image enhancement* alters the visual impact that the image has on the interpreter in a manner that improves the information content.
    a. Contrast enhancement
    b. Density slicing
    c. Edge enhancement
    d. Intensity, hue, and saturation transformations
    e. Pan-sharpening
    f. Digital mosaics
    g. Masking
3. *Information extraction* utilizes the decision-making capability of the computer to recognize and classify pixels on the basis of their digital signatures.
    a. Principal component images
    b. Ratio images
    c. Multispectral classification
    d. Change-detection images
    e. Hyperspectral processing

PREPROCESSING

Preprocessing involves steps to understand the data quality and to prepare the data to meet project requirements.

MINIMIZE RASTER DATA ERRORS AND NOISE

In the early days of Landsat (mid-1970s) and airborne sensors, scan line, banding, and random noise defects were not uncommon. Today, however, defects are relatively rare as data providers have rigorous quality control programs and excellent instruments. Advanced image processing software has algorithms to restore defects in multispectral and hyperspectral data and to fill gaps and minimize spurious values in DEMs.

EVALUATION OF IMAGE STATISTICS

In this section the six Landsat 8 VNIR-SWIR bands of the Thermopolis, Wyoming, area (Figure 3-11) are evaluated for their statistical characteristics through the use of histograms and tables. Statistics developed from single bands are univariate while those that involve more than one band are multivariate. Univariate band statistics are important for evaluating individual band quality and potential for generating informative maps. Multispectral statistics enable remote sensing analysts to determine which bands are highly correlated (similar spectral pattern) and which are most different. Bands that are poorly correlated to others in a multispectral scene have the most potential for revealing new spectral information about features in the scene. The greater the range between minimum and maximum DN values for a band, the more variability in brightness values, and the more potential for new information being gleaned from the band during processing. Bands with little range (little DN dispersion about the mean) indicate a uniform spectral pattern across the image, minimizing the potential for detecting

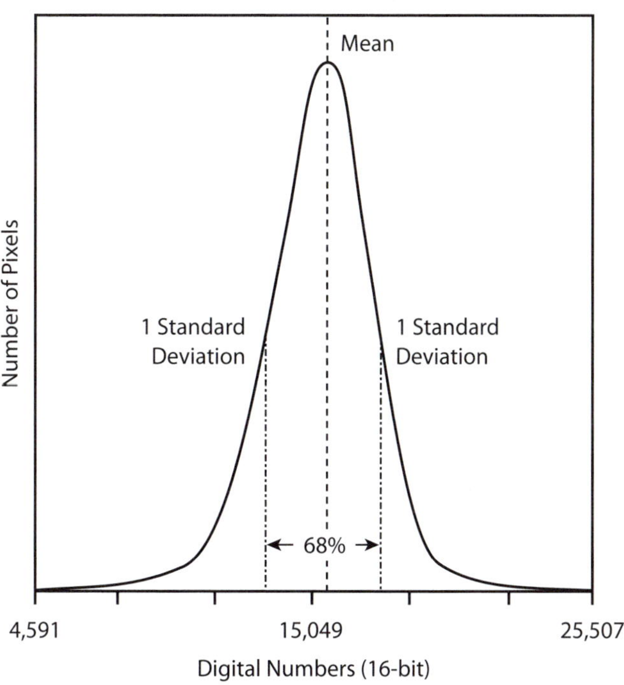

FIGURE 9-3 Landsat OLI band 6 (SWIR1) input histogram with a mean DN of 15,049.

anomalies and patterns of interest. Jensen (2016) provides an in-depth discussion of image statistics.

Univariate Statistics

The histogram for Landsat OLI band 6 (SWIR1) is shown in Figure 9-3. The horizontal axis represents the DNs recorded by the OLI instrument as it orbited across the Thermopolis area. Pixel brightness increases from left to right on the horizontal axis. The minimum brightness is 4,591 and the maximum brightness is 25,507 (the OLI instrument collects 16-bit data or 0 to 65,536 levels of gray). The vertical axis is the number of pixels with a DN value between 4,591 and 25,507. Table 9-1 compiles the minimum, maximum, and mean values for the six Landsat 8 bands in Figure 3-11. The SWIR1 data shown as a histogram in Figure 9-3 is highlighted in gray in Table 9-1. OLI band 2 (blue) has the lowest mean brightness while OLI band 6 (SWIR1) has the highest mean brightness. Contrast stretching will be different for bands with different mean brightness and brightness ranges.

The shape of the histogram is important. Normally distributed data has a bell shaped curve (Figure 9-3). The shape of this curve is expected by many image processing algorithms, however, data that has many dark or bright pixel outliers (far from the mean) is skewed and can degrade processing output. Standard deviation and variance are statistical measures of the dispersion of values about the mean (Jensen, 2016). Variance involves subtracting the DN value for each pixel (or a sample of pixels) from the band's mean DN value, and then squaring the difference. Variance is the sum of the squared differences divided by the number of samples. The SWIR1 band has 6.5 times the variance compared with the blue band (Table 9-1), indicating more spectral richness for the SWIR1 band. Standard deviation (σ) is the common statistical measure of dispersion around the mean and is derived from the positive square root of variance.

$$\text{standard deviation} = \sqrt{\text{variance}} \qquad \textbf{(9-1)}$$

One standard deviation on either side of the mean contains 68% of the population or sample. For our SWIR1 band, 68% (544,000 pixels) of the 800,000 pixels in the image are within one standard deviation of the mean and have DNs between 12,891 and 17,208.

Multivariate Statistics

Multispectral data can have a high band-to-band correlation as the images are collected at the same time over the same area, from the same altitude, and with the same illumination and atmospheric conditions. Data from bands that are highly correlated have less potential when used with image processing algorithms that are designed to bring out subtle spectral patterns, discover anomalies, or highlight spectral targets that cover a small portion of the imaged area.

One band can be compared to another band on a 2-D plot where the horizontal axis represents the range of DNs from one band and the vertical axis represents the range of DNs from the other band (Figure 9-4). This 2-D plot is termed a two-band *feature space plot* (Jensen, 2016). The feature space plot extracts the brightness value from two bands for every pixel in the scene and plots the occurrence in the Landsat OLI 16-bit feature space. The plot can display the frequency of occurrence of unique pairs of values with varying brightness

TABLE 9-1 **Band DN statistics associated with Landsat OLI (16-bit) data.**

	Band	Minimum	Maximum	Mean	Standard Deviation	Variance
Blue	2	7,442	22,807	9,619.6	845.7	716,018.5
Green	3	6,474	23,233	9,774.9	1,189.9	1,417,120.7
Red	4	5,778	25,165	10,470.3	1,540.4	2,374,469.4
NIR	5	4,420	28,517	13,397.6	1,912.0	3,657,625.1
SWIR1	6	4,591	25,507	15,049.1	2,158.5	4,661,414.4
SWIR2	7	4,777	23,767	12,991.3	2,077.2	4,316,661.9

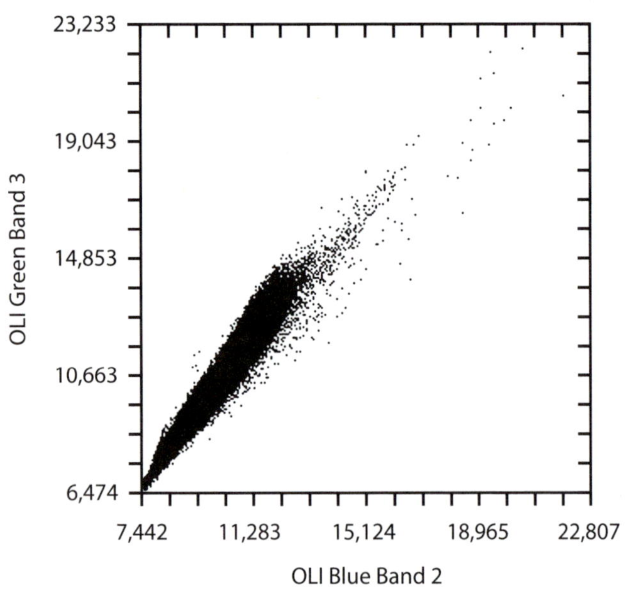

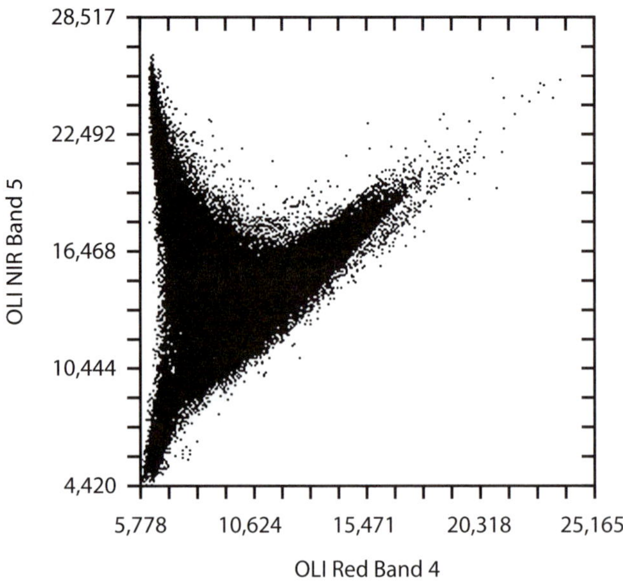

FIGURE 9-4 Feature space plots of Landsat OLI bands.

or color. However, Figure 9-4 is simplified and displays all pairs (whether one deep or hundreds deep) as only one black dot.

Landsat OLI blue band 2 is compared to the green band 3 in the feature space plot of Figure 9-4A. The DNs from the blue band are on the horizontal axis and the DNs from the green band are on the vertical axis. These two visible light bands are highly correlated—where one is bright the other is bright, and vice versa. The bands are highly redundant and that minimizes their potential for new spectral information. In Figure 9-4B, Landsat OLI NIR band 5 (vertical axis) is compared to the red band 4 (horizontal axis). Many pixels in the two bands are not well correlated. Many bright NIR pixels range from dark to bright in the red band. The low correlation indicates there is excellent potential for new spectral information in these two bands. For more information, Figure 9-4A can be evaluated alongside Figures 3-11A and 3-11B, while Figure 9-4B can be evaluated alongside Figures 3-11C and 3-11D.

A correlation matrix based on covariance and standard deviation measurements provides a unitless correlation ratio between bands that varies from –1 to +1. A correlation ratio of 1 indicates a positive, systematic relationship between the brightness values of two bands—as one band increases in brightness the other band's values also increase (Jensen, 2016). A correlation matrix provides a repeatable and defensible number to what is visually interpreted from feature space plots. Table 9-2 is a correlation matrix of the six Landsat bands of the Thermopolis area. The bands used in the feature space plots of Figure 9-4 are highlighted in gray in Table 9-2.

The high correlation seen in Figure 9-4A between the blue and green bands is confirmed by a correlation value of 0.978 in Table 9-2. In contrast, the low correlation seen in Figure 9-4B between the NIR and red bands has a correlation value of 0.423 in Table 9-2. The NIR band contains the most unique spectral information compared to the visible and SWIR bands in the Landsat scene of Thermopolis. The SWIR bands have only moderate correlation with the visible bands (ranging from 0.79 to 0.92 in Table 9-2). The correlation matrix and feature space plots confirm the NIR and SWIR bands will provide the most spectral information for image enhancement techniques.

TABLE 9-2 Correlation matrix of Landsat OLI data.

	Band	Band 2 (Blue)	Band 3 (Green)	Band 4 (Red)	Band 5 (NIR)	Band 6 (SWIR1)	Band 7 (SWIR2)
Blue	2	1					
Green	3	0.978681	1				
Red	4	0.927854	0.962711	1			
NIR	5	0.371338	0.469063	0.423231	1		
SWIR1	6	0.793064	0.837701	0.893752	0.485006	1	
SWIR2	7	0.832312	0.862618	0.926953	0.328594	0.951521	1

CORRECTING FOR ATMOSPHERIC SCATTERING

In the preprocessing stage it is important to confirm that your multispectral dataset is in radiance or reflectance. Imagery acquired by aerial and satellite sensors is affected by atmospheric scattering and absorption of light, especially in the shorter wavelengths (Figure 2-26A). Light reflected from the target interacts with scattered and absorbed light from the atmosphere and both are collected by the sensor (Figure 1-18). The combined target reflectance and atmospheric scattering that reaches the detector is called *total radiance*. If the user is processing single-date imagery, comparing maps generated from multi-date imagery, and spectral libraries are not being used, the imagery can remain in total radiance or original DN values provided by the sensor operator (Jensen, 2016). If the user is generating band ratios, comparing spectral measurements from imagery acquired on different dates, building spectral training sites from imagery acquired on different dates, and using spectral libraries, then the imagery needs to be atmospherically corrected and converted into reflectance (Jensen, 2016).

Visual inspection of spectra profiles from multispectral data can help decipher if the data has been atmospherically corrected. In Figure 9-5, the pixels are vegetated and were collected by the Landsat 8 OLI sensor over Haiti. The horizontal axis illustrates the six VNIR-SWIR Landsat bands with blue farthest to the left, while the vertical axis is the DN/brightness value of the pixels. Atmospheric scattering of blue light causes the brightness to be greater in radiance data compared with reflectance data. Green light is reflected by chlorophyll in healthy vegetation while blue and red light are absorbed (Figures 2-26B and 2-31). Reflectance imagery will confirm the absorption of blue and red light with darker DN values compared to the brighter DN values of reflected green light (Figure 9-5).

Absolute Atmospheric Correction

Many complex algorithms and atmospheric models have been developed to convert the digital brightness values recorded by an aerial or satellite multispectral and hyperspectral sensor into scaled surface reflectance values that can be compared to scaled surface reflectance values of any image collected elsewhere on the globe at a different time (Jensen, 2016). A major goal of absolute atmospheric correction is accurate modeling of the atmosphere's properties at the time of image acquisition. NASA's MODIS system collects data daily that can be used for this correction. Jensen (2016) provides details on the processes and software involved in atmospheric correction. Landsat Level-2 science products have pixel brightness atmospherically corrected to reflectance values.

Relative Atmospheric Correction

Chapter 2 described how the atmosphere selectively scatters the shorter wavelengths of light, which causes haze and reduces the contrast ratio of images. For Landsat TM images, band 1 (blue) has the highest component of scattered light and band 7 (reflected SWIR2) has the least. Figure 9-6 shows two techniques for determining the relative correction factor for different TM bands. Both techniques are based on the fact that band 7 is essentially free of atmospheric scattering, which can be verified by examining the DNs for shadows; these typically have very low DN values on band 7. The first technique (Figure 9-6A) employs data from an area within the image that has shadows caused by irregular topography. For each pixel the DN in band 7 is plotted against the DN in band 1, and a straight line is fitted through the plot using a least squares technique. If there was no haze in band 1, the line would pass through the origin. Because there is haze, the intercept is offset along the band 1

A. Landsat OLI radiance values for one vegetated pixel.

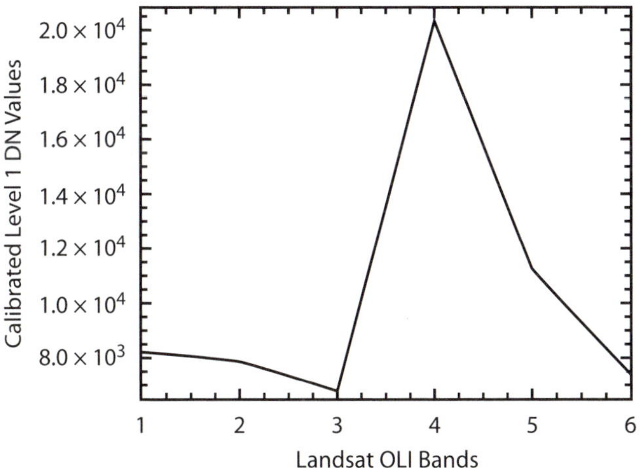

B. Atmospherically corrected Landsat OLI reflectance values for one vegetated pixel.

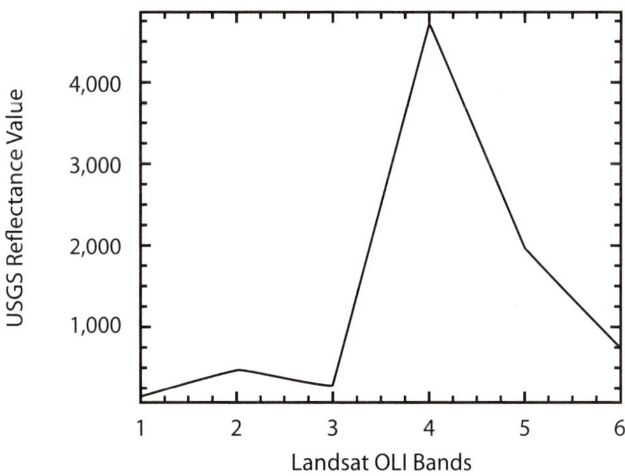

FIGURE 9-5 Calibrated image DN values compared to atmospherically corrected image reflectance values.

A. Plot of TM band 7 versus band 1 data for an area with shadows. The offset of the straight line is caused by atmospheric scattering in band 1.

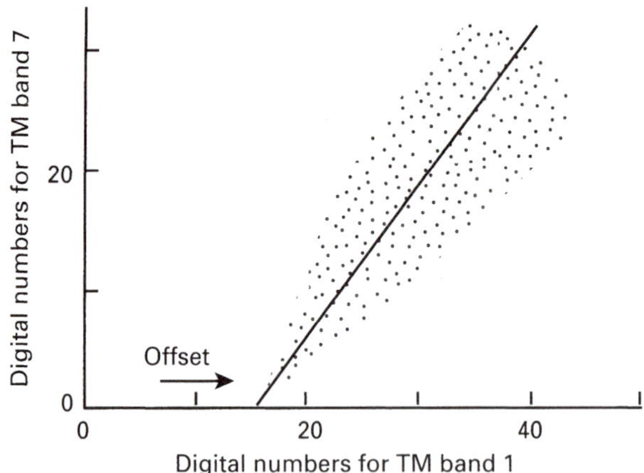

B. Histograms for TM bands 7 and 1. The lack of pixels with low DNs in band 1 is caused by illumination from light selectively scattered by the atmosphere.

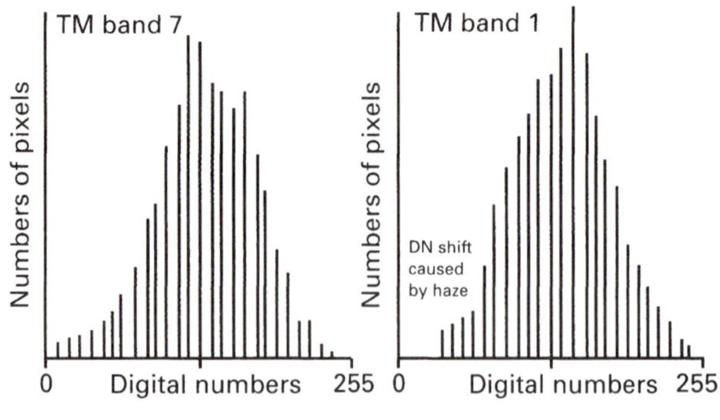

FIGURE 9-6 Methods for determining atmospheric corrections on individual TM bands. From Chavez (1975, Figures 2 and 3).

axis, as shown in Figure 9-6A. Haze has an additive effect on scene brightness. To correct the haze effect on band 1, the value of the intercept offset (in this case, 15) is subtracted from the DN of each band 1 pixel for the entire image. The procedure is repeated for the other TM bands.

Figure 9-6B shows a second correction technique, which also requires that the image have some shadows or bodies of clear water with very low DNs on band 7. The histogram of band 7 has pixels with DNs of 0. The histogram of band 1 lacks pixels in the range from 0 to approximately 20 because of light scattered into the detector by the atmosphere. The band 1 histogram also shows the characteristic abrupt increase in pixels at a DN of approximately 20. This value of 20 is subtracted from all the DNs in band 1 to minimize the effects of atmospheric scattering. The histograms of bands 2 and 3 are also restored in this manner. Band 2 (green) normally requires a subtraction of less than 10 DNs. Only a few DNs are typically subtracted from band 3 (red).

The amount of atmospheric correction depends upon the wavelength of the bands and the atmospheric conditions. As mentioned earlier, scattering is more severe at shorter wavelengths. Humid, smoggy, and dusty atmospheres cause more scattering than clear, dry atmospheres. These relative atmospheric correction methods are applicable to all multispectral data.

GEOMETRIC CORRECTION

Digital images can have a number of geometric errors that are classified as internal and external (Jensen, 2016). Most of these errors are now corrected by the sensor operator prior to data delivery because the operator has up-to-date and significant knowledge of the instrument characteristics, orbital or airborne flight path, and onboard/ground navigation information. In the preprocessing stage, the user should check the imagery for geometric errors that could be related to internal and external problems. Older imagery can have significant geometric errors (Sabins, 1997). Minimizing geometric errors is essential for integrating the images (and also scanned maps) with other geospatial layers in a GIS.

Internal

Internal geometric errors are generally introduced by the remote sensing system itself or in combination with the Earth's rotation or curvature characteristics (Jensen, 2016). Cross-track sensor (Figure 1-16) distortions result from sampling pixels along a scan line at constant time intervals. The width of a pixel (in the scan direction) is proportional to the tangent of the scan angle and therefore is wider at either margin of the scan line. The data are recorded and displayed at a constant rate, however, which causes the pixels at the margins of the image to be compressed relative to those at the center. Cross-track distortion occurs in all unrestored images acquired by cross-track scanners, whether from aircraft or satellites. Older data would require correction of cross-track distortion.

Sun-synchronous satellites in near polar orbits collect a swath of imagery along their north–south orbital path as the Earth rotates from west to east. The interaction between the fixed orbital path of the satellite and the Earth's rotation skews the geometry of the imagery collected (Jensen, 2016). Today the distortion is corrected by the sensor operator.

External

Variations in aircraft and spacecraft attitude (roll, pitch, and yaw), velocity, and altitude are corrected with tracking data and onboard navigation data. For older data collected without onboard navigation equipment, ground control points (GCPs) are needed to correct the distortions and digitally warp the image into a coordinate system. The GCPs can be surveyed features or features on a map that are identifiable on the image.

Map Projections

Images and maps contain inherent geometric distortions because they record the curved surface of the Earth on a flat display. Areas, distances, and angular relationships are distorted to varying degrees. A *map projection* is the systematic representation of a curved surface on a plane. Figure 9-7 shows the three basic projections, which are planar, conic, and cylindrical. Many different versions of the basic projections have been devised for different purposes. The history, characteristics, and mathematics of map projections are given by Bolstad (2016), Snyder (1987), and the US Geological Survey (2013), which form the basis for the following discussion.

The conic and cylindrical projections are "unrolled" onto flat surfaces. The cylindrical projection is developed by unrolling a cylinder wrapped around the globe, touching at a great circle, with meridians projected from the center of the globe.

Earth's Geographic Coordinate System

Geodesy measures and monitors the size and shape of the Earth. Geodesists have developed spatial reference systems that enable satellite and aerial images to be located accurately on the surface of the Earth. Horizontal and vertical datums are the foundation of spatial reference systems. Horizontal datums are a collection of specific points on the

A. Polar azimuth (planar).

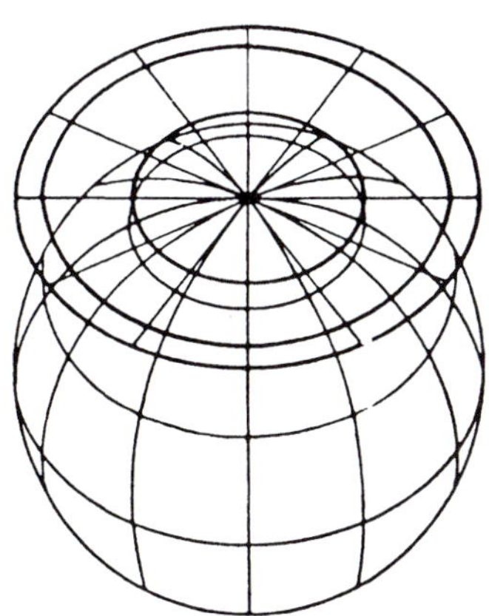

B. Regular conic.

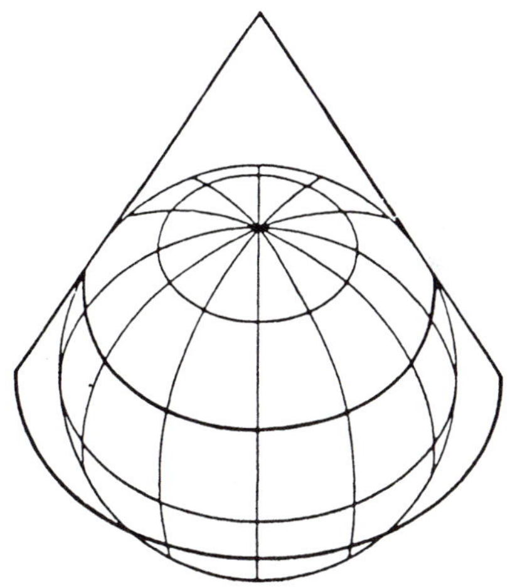

C. Regular cylindrical (Mercator).

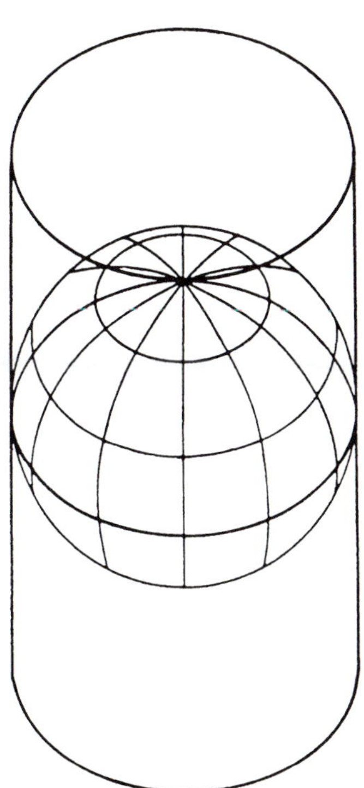

D. Transverse cylindrical (transverse Mercator).

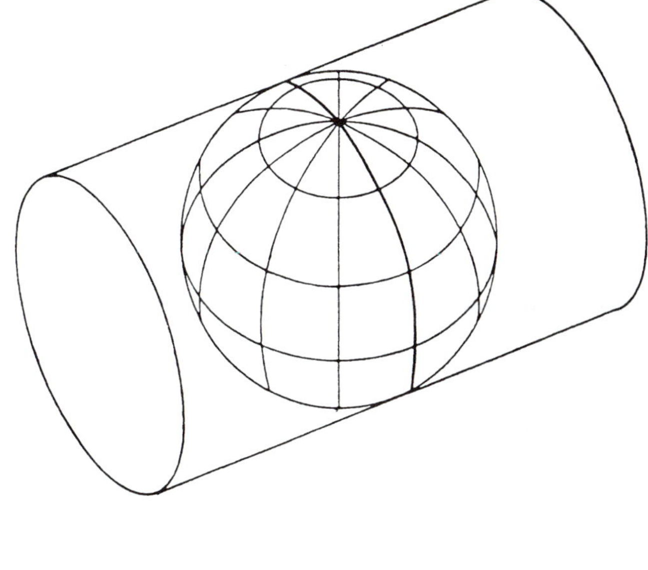

Figure 9-7 Basic map projections. From Snyder (1987, Figure 1).

Earth that have been identified according to their precise latitude and longitude location. North American Datum of 1927 (NAD 27), North American Datum of 1983 (NAD 83), and World Geodetic System 1984 (WGS 84) are common horizontal datums used for remote sensing images. These datums are associated with a "spheroid" that defines the shape and size of the Earth. Vertical datums are a collection of specific points on the Earth with known heights either above or below mean sea level. The North American Vertical Datum of 1988 (NAVD 88) is the most commonly used vertical datum in the United States. Ground deformation (subsidence and uplift) studies using interferometric radar, lidar, and photogrammetry require reference to an established vertical datum.

Images, maps, and features are located on the Earth's spherical surface with a *geographic coordinate system* of latitude (varying from north to south) and longitude (varying from west to east). Two common notation formats for latitude are *decimal degree* (DD) and *degree-minute-second* (DMS). The geographic center of the contiguous United States is located near Lebanon, Kansas, and has a DD location of latitude 39.83333°N and longitude 98.583333°W and a DMS location of latitude 39°50′N and longitude 98°35′W. This location has UTM Zone 14 North coordinates of Easting 535652.9478 and Northing 4409341.7866 (m). Geographic coordinate systems require a horizontal datum and a spheroid for an image or map to be correctly located in GIS.

Projected coordinate systems offer various systematic methods and mathematical transformations to transfer or "project" the maps from the Earth's spherical surface onto a flat surface. The Mercator projection and State Plane Coordinate System are two projected systems.

Mercator Projection

A well-known type of cylindrical projection is the *Mercator projection*, which is also important because it has similar characteristics to images acquired by Landsat. A Mercator projection has the following characteristics:

1. The great circle of contact with the cylinder is the equator.
2. Maps are *conformal*; that is, the relative local angles about every point are shown correctly. Although the shape of a large area is distorted, its small features are shaped essentially correctly.
3. Meridians are equally spaced parallel lines; parallels are unequally spaced parallel lines.
4. There is little error close to the equator. (The scale 10° north or south is only 1.5% larger than at the equator.)

Because of these useful attributes, other versions of the Mercator projection have been developed. In the *transverse Mercator projection* the cylinder is oriented at a right angle to the equator. The great circle of contact is a longitude line called the central meridian. Map scale is essentially true within a zone 10° east or west of the central meridian. A widely used map projection for satellite remote sensing data is based on the *Universal Transverse Mercator* (UTM) projection in which the Earth, between latitudes 84°N and 80°S, is divided into 60 zones, each generally 6° wide in latitude (Figure 9-8). Each zone is divided into 20 quadrangles generally 8° high in latitude. Each quadrangle is divided into grid squares 10,000 m on a side. Locations are readily identified by referring to the grid coordinates.

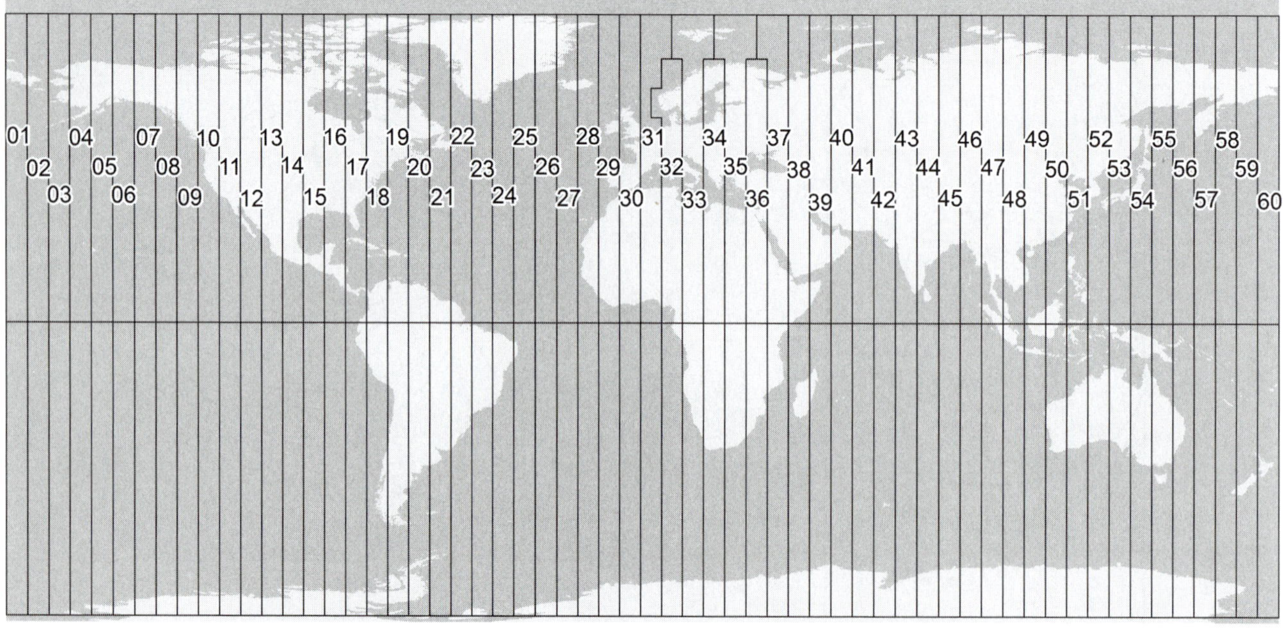

FIGURE 9-8 UTM map with 60 zones. Modified from National Geodetic Survey.

State Plane Coordinate System

UTM is not accurate enough for local surveying and engineering applications at the city and county level. Each state in the United States has its own State Plane Coordinate System (SPCS) that uses a map projection and varying number of SPCS zones within a state to minimize distortions and increase the accuracy of measurements. SPCS is designed to be four times more accurate than UTM (SPCS commonly uses feet as the primary unit of measure). Because California is long in the north–south direction, six SPCS zones were established. Aerial and satellite imaging, mapping, and surveying for local government in Shasta County of northern California would use SPCS Zone 1 (Figure 9-9).

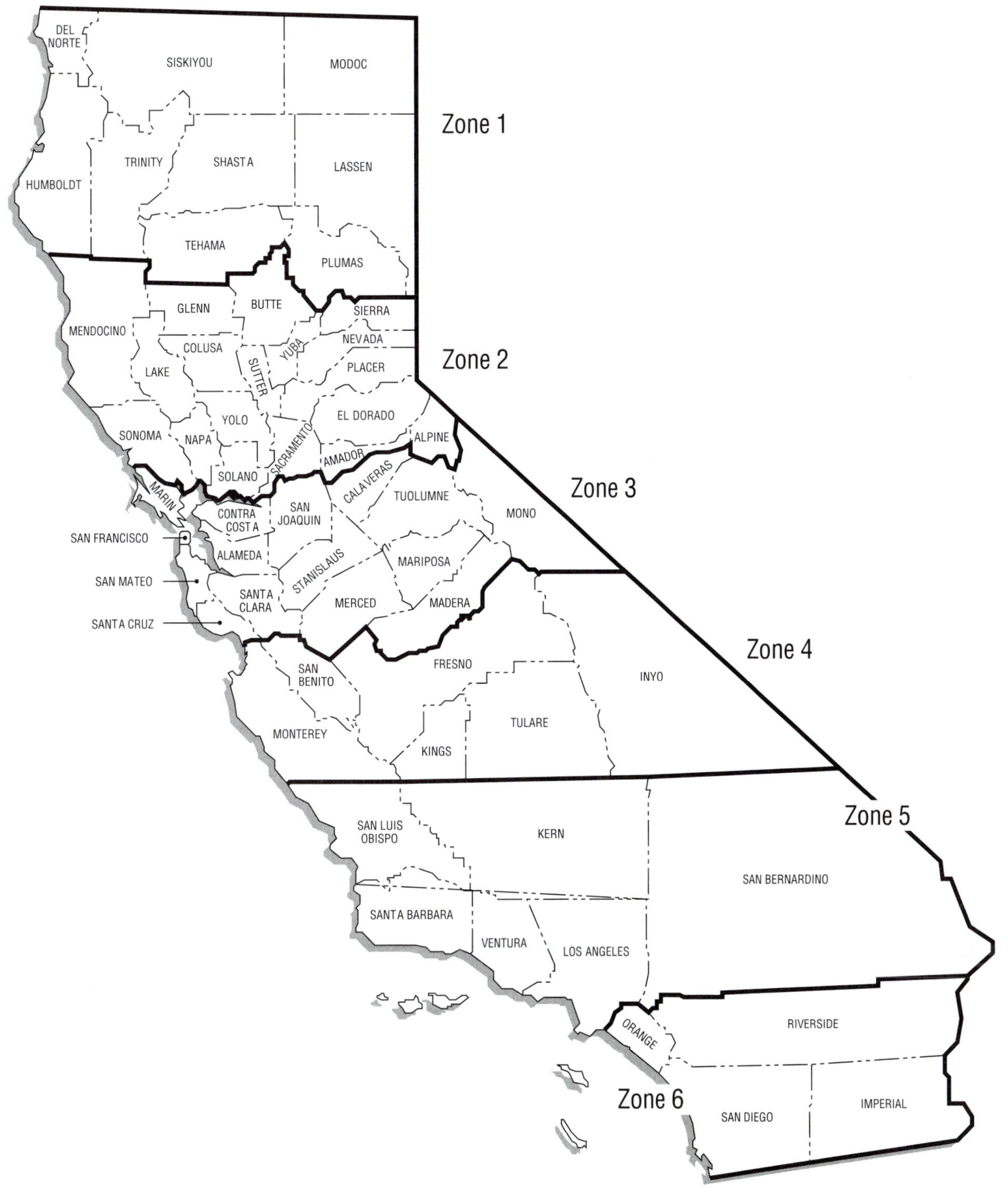

FIGURE 9-9 State Plane Coordinate System for California. Courtesy California Geologic Survey.

GEOREFERENCING

Remote sensing images and scanned maps are georeferenced (or *rectified*) to a standard map projection to enable their use with other geospatial layers in a GIS. Georeferencing adds *x*, *y* or latitude, longitude coordinates to each pixel in the image or scanned map.

Today's satellite imagery is acquired using onboard GNSS/GPS, ground tracking stations, star trackers, and inertial navigation systems (with accelerometers and gyroscopes) to monitor the platform location with respect to the Earth and the platform (and sensor) orientation. Imagery downloaded from these accurately positioned systems has excellent planimetric (*x*, *y*) characteristics, are already georeferenced or orthorectified into a coordinate system, and are ready for loading into a GIS and integrating with other geospatial layers for mapping projects. For example, Landsat 8 OLI imagery is delivered with each 30 m pixel having a 90% confidence that it is within 12 m of its true location on the surface of the Earth (i.e., 12 m circular error with 90% confidence level).

Satellite and aerial images that achieve appropriate horizontal accuracy levels (ASPRS, 2015; Federal Geographic Data Committee, 1998) can be used as a reliable base for georeferencing older satellite and aerial images and scanned maps. The image (or scanned map) is georeferenced to another image that has coordinates or a map with coordinates. The image with coordinates is named the *reference* or *base* image and the image with no coordinates is the *input*, *source*, or *warp* image.

Image-to-Image

Image-to-image georeferencing involves finding the same feature in the reference and input images and then adding a GCP point to each feature. This pair of GCPs generates one link between the reference and input image. Pairs of GCPs should be selected that provide a broad distribution of control across the input image. If GCPs are not well distributed across the input image, or the points in a pair do not link to the same feature, then the fit between the input image and the reference image will be poor. A table is generated by georeferencing software that lists the GCPs and the *x*, *y* error calculated for each GCP pair. GCPs with excessive error can be replaced with GCPs that reduce the error.

Image-to-Map

Input images can be georeferenced to a map that has coordinates using GCPs that can be seen on both the image and the map. The photoidentifiable GCP pairs are treated the same as those generated with the image-to-image process. After the GCPs are selected, the next step in the process is to digitally warp or rectify the input image to the reference so that coordinates and a map projection are assigned to the pixels in the input image.

Rectification

Only the selected GCPs in the input image and the corresponding GCPs in the reference image are used in the rectification process. All the other pixels in the input image are transformed into the reference's coordinate system based on the selected GCPs (Figure 9-10A). The pixels in the input image are resampled using one of three methods of interpolation: nearest neighbor, bilinear, and cubic convolution

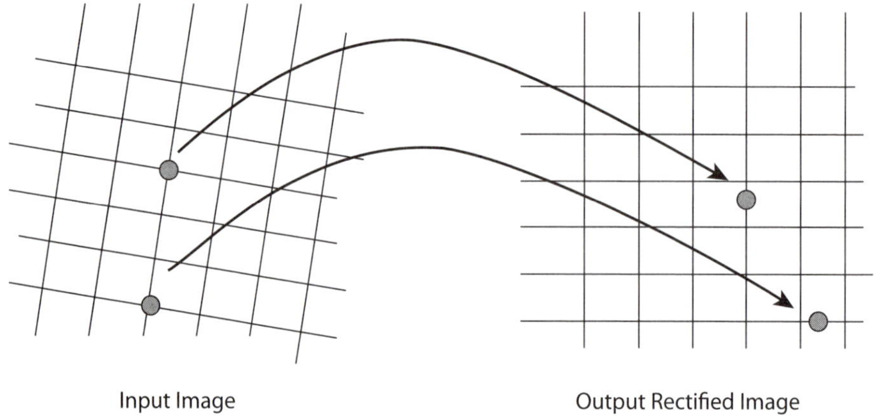

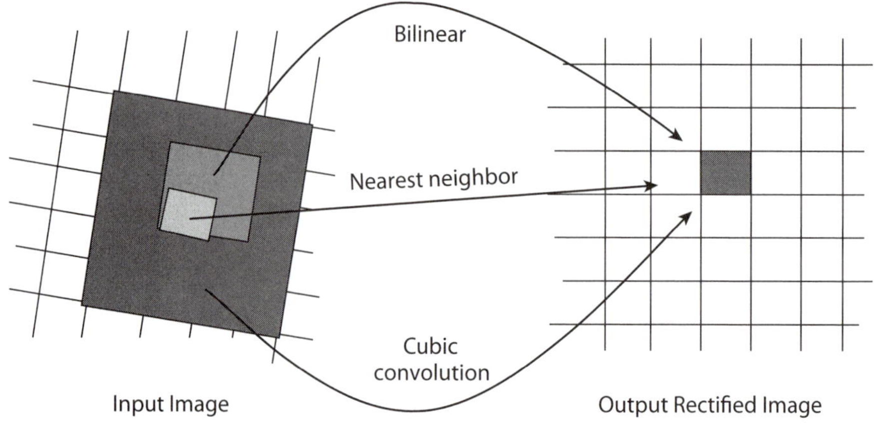

FIGURE 9-10 Rectification (or digital warping, georeferencing, transformation) of an image into a map coordinate space with resampling options.

(Figure 9-10B). The size of the input image's pixels will change during resampling to match the reference image's pixel size.

The left side of Figure 9-10A shows the input image in a raster grid coordinate space (based on the rows and columns of the raster image). The right side of Figure 9-10A shows the base image that is projected into a coordinate system with 10 m pixels. The dots represent the two GCPs as they are seen on the input image (left side) and *after* rectification as they are seen on the base image (right side). The GCPs are assigned *x*, *y* coordinates after rectification by the software.

Figure 9-10B shows the options the remote sensing analyst has for resampling the input image into the coordinate space of the base image. The resampling algorithm determines the DN value for each resampled pixel in the rectified image.

To simplify the discussion of resampling, we'll assume the input image has the same pixel size as the output georeferenced image (Figure 9-10B). Nearest neighbor resampling assigns a brightness value to each output pixel that is closest to the input image value. This method is computationally fast, doesn't change the DN values (which is preferred for scientific research), but results in a blocky looking output image. Bilinear resampling computes a brightness value for one output pixel based on the brightness of four pixels in the input image. Cubic convolution resampling assigns a brightness value to one output pixel based on 16 input pixels. Cubic convolution images have a smoother appearance compared with nearest neighbor and bilinear, but are computationally intense to generate and the input DN values are changed. Resampling raster arrays and digital grids with these three methods is very common in image processing. Pan-sharpening (discussed later in the chapter), image pyramid layers, image display on monitors, georeferencing, and orthorectification all involve resampling. Jensen (2016) provides more detail about resampling and rectification.

Digitally warping the input image to the reference image's coordinate system and map projection is commonly achieved using a transformation matrix computed from the GCPs and polynomial equations (Hexagon, 2019). The polynomial transformations change the shape of the output georeferenced image and can be visualized as lines (Figure 9-11A) or trend surfaces (Figure 9-11B). The three most basic transformations are 1st order (linear), 2nd order (parabola or quadratic), and 3rd order (cubic or "sine wave"). The polynomial equations for these three transformations are:

$$y = b + ax$$
$$y = c + bx + ax^2$$
$$y = d + cx + bx^2 + ax^3 \quad \textbf{(9-2)}$$

Figure 9-11A shows the three transformations as lines. It is helpful to visualize these lines as profiles of the georeferenced image so that the 1st order transformation is a sloping line, the 2nd order generates a simple parabola, and the 3rd order generates a sine wave. Figure 9-11B shows these three orders as trend surfaces with the 1st order as a sloping, tilted flat surface, the 2nd order as a sloping surface with a curvature down the slope, and the 3rd order as an undulating surface. These trend surfaces represent what can happen to

A. Linear, 2nd order polynomial, and 3rd order polynomial curves representing a profile of a trend surface. From FEATool Multiphysics (2016, Figure 2). Precise Simulation Ltd.

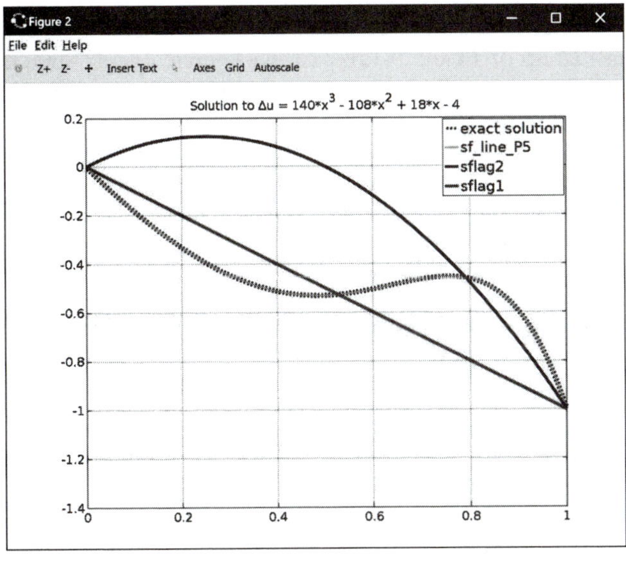

B. Trend surfaces of rectified image after a different order of transformation is applied. From Dempsey (2013).

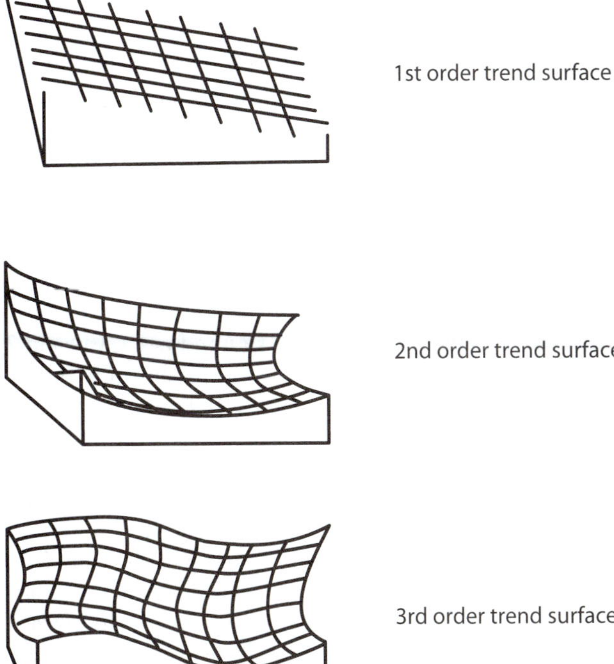

FIGURE 9-11 Effects of three transformations on the shape of a rectified image.

your input image when you rectify it with a different order transformation. A 1st order transformation can only change the input image's:

- location in x and/or y
- scale in x and/or y
- skew in x and/or y
- rotation

1st order warping is recommended for most georeferencing. The 2nd and 3rd order transformations are nonlinear and have the appearance of rubber sheeting. Aerial photographs with radial distortion caused by the curvature of the lens can be at least partially corrected with 2nd and 3rd order warping. Input images can be severely distorted with 2nd and 3rd order transformations if the GCPs are not well distributed.

ORTHORECTIFICATION

Georeferencing does not account for image distortions due to topographic relief. A DEM needs to be used in the orthorectification process to minimize input image distortions due to topographic relief. Some satellite operators provide metadata that includes a model of geolocation information for individual pixels in an image to enable the user to improve the horizontal accuracy of an image. These rational polynomial coefficients (RPCs) are downloaded with the imagery and provide a mean elevation value for the entire image. Users can import a DEM when processing an RPC to significantly increase the location accuracy of the image.

IMAGE ENHANCEMENT

Enhancement is the modification of an image to alter its impact on the viewer. Generally, enhancement changes the original digital values; therefore, enhancement is not done until preprocessing is completed.

CONTRAST ENHANCEMENT

Contrast enhancement modifies the grayscale to produce a more interpretable image. Virtually all bands of all images acquired by Landsat (and similar satellites) require contrast enhancement. Figure 9-12A is a Landsat 8 OLI band 2 (blue) image of the Thermopolis subscene shown in Figure 3-11 and Plate 7. The original unenhanced images are sometimes called *raw* images. The image is very dark and lacks contrast. The histogram (Figure 9-12B) shows the statistical distribution of the input data. Landsat 8 records reflected light with 4,096 levels of gray (12-bit), but the USGS provides 16-bit data (65,536 levels of gray) to users. The 16-bit DNs recorded by the OLI sensor are displayed on the horizontal axis of Figure 9-12B. The DNs in the scene cover a range from 7,495 (darkest pixels) to 20,949 (brightest pixels); however, the vast majority of DNs are in the lower range and have values less than 12,000. Most of the DNs occupy the lower 30% of the total range of DNs in the image. This explains the low contrast of the image.

The image in Figure 9-12A is representative of virtually all raw TM images because the sensor had a radiometric depth of only 8-bit (256 levels of gray). Contrary to appearances, the characteristics of raw images do not indicate a defect in the TM system. The sensitivity range of TM detectors was designed to record a wide range of terrain brightness, from black basalt plateaus to white sea ice, under a wide range of lighting conditions. No individual scene has a brightness range that covers the full sensitivity range of the TM detectors. Therefore, each original TM band requires contrast enhancement to produce useful images. The greater radiometric depth of the Landsat 8 OLI sensor reduces oversaturation issues, so most of the bands are not as dark as those in TM images.

Contrast enhancement is a subjective operation that can be strongly influenced by the personal preferences of the operator and user. Histograms of the enhanced data give an objective view of the process. Many routines have been developed for enhancing contrast; three useful methods are described in the following sections.

Linear Contrast Stretch

Although remote sensing image bands can be acquired with a 10-, 11-, or 12-bit radiometric range, most standard computer monitors can only display 8-bit (256 shades of gray). Histograms of enhanced bands show a grayscale range from 0 to 255, regardless of the input band's radiometric range (Figure 9-12D).

Contrast stretches are done with *lookup tables* (LUTs), which consist of an array of original (*input*) pixel values and a corresponding array of enhanced (*output*) values that are used to produce the stretched image. Figure 9-13 (on p. 257) is a graphic display of the LUTs that were used to display the original unsaturated linear stretch data shown in the histogram of Figure 9-12B and the saturated linear stretch with 2% saturation shown in Figure 9-12D. The vertical axis represents the DNs of the original pixels, and the horizontal axis represents the DNs of the enhanced pixels. The dashed lines show the enhancement transforms. In a *simple (unsaturated) linear contrast stretch*, the lowest original DN is assigned a new value of 0, the highest original DN is assigned a new value of 255, and the remaining original DNs are linearly reassigned new values ranging from 1 to 254. A disadvantage of this stretch is that the few percent of original pixels at the head and tail of the original histogram occupy an excessive portion of the new dynamic range.

This disadvantage is eliminated with the *linear saturated contrast stretch* shown by the solid line in Figure 9-13. The darkest 2% (DN = 7,495 to 8,072) and brightest 2% (DN = 11,550 to 20,949) of the original input pixels are assigned output values of 0 and 255; in other words, they are *saturated* to pure black and white. The remaining 96% of input pixels (DN = 8,073 to 11,549) are linearly reassigned to the output range from 1 to 254. The resulting output image and histogram are shown in Figures 9-12C and 9-12D. The dramatic impact of contrast enhancement is emphasized by comparing the output image (Figure 9-12C) with the input image (Figure 9-12A). For this example, a saturation of 2% was selected as optimum after experimenting with other saturation levels. Other images may require higher or lower saturation levels.

Gaussian Contrast Stretch

Variations in nature are commonly distributed in a normal (Gaussian) pattern, which is the familiar bell-shaped curve. This distribution is emulated for images by the *Gaussian contrast stretch*, in which the original pixels are reassigned to fit a Gaussian distribution curve. The image and histogram

A. Image from original data.

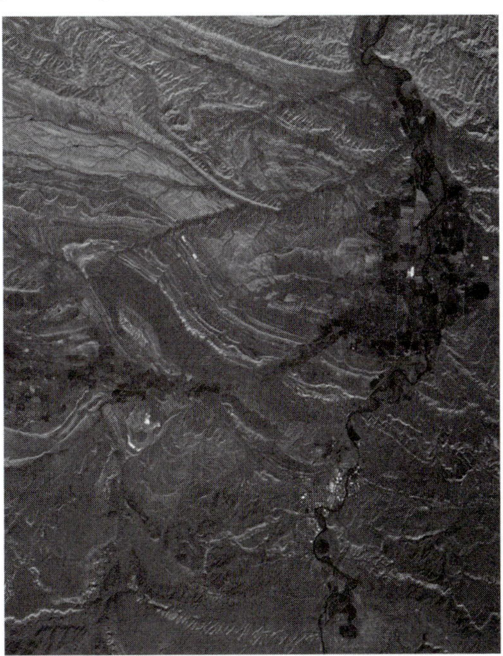

B. Histogram of original data.

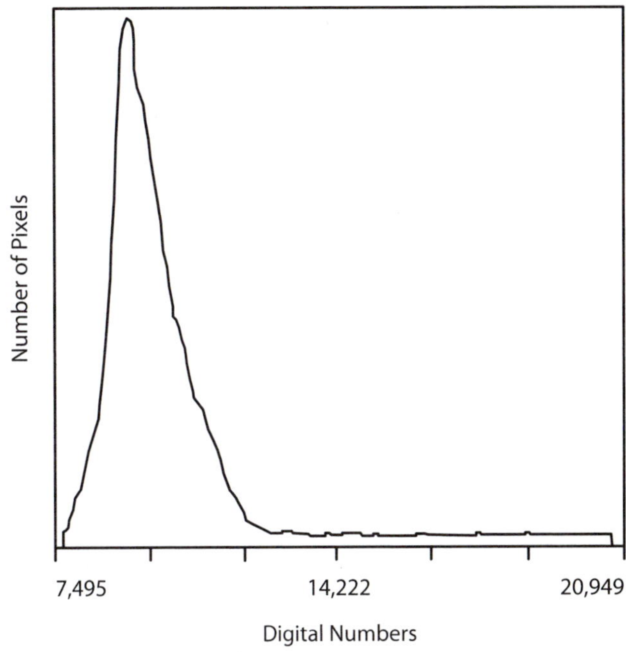

C. Image with linear contrast stretch with 2% saturation.

D. Histogram of linear contrast stretch and 2% saturation.

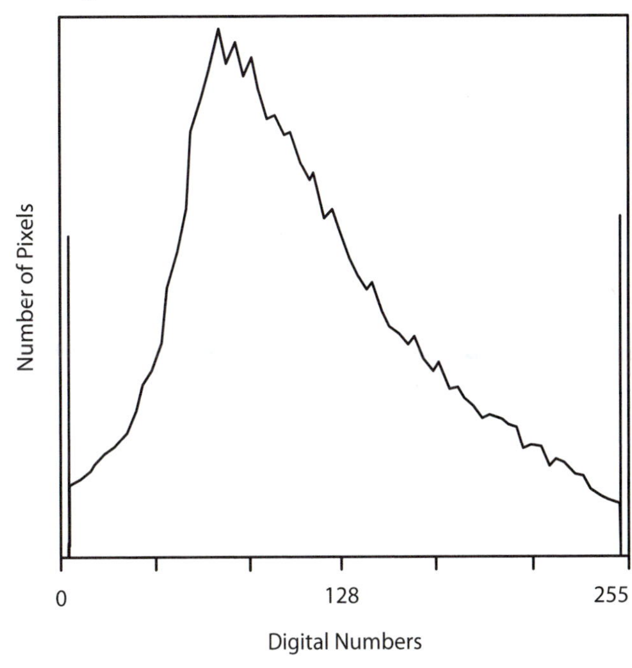

FIGURE 9-12 Contrast-enhancement methods. Landsat 8 OLI band 2 (blue) of the Thermopolis, Wyoming, subscene. Landsat courtesy USGS. (*cont.*)

in Figures 9-12E and 9-12F show the effects of applying a Gaussian stretch to the original data.

Uniform Distribution Contrast Stretch

A *bin* of pixels refers to all pixels having the same DN. The linear and Gaussian stretches assign each bin of pixels to a uniform new DN range regardless of the number of pixels in the bin. This disparity is compensated for in the *uniform distribution stretch* (or *histogram equalization stretch*), in which the input pixels are redistributed to produce a uniform population density of pixels along the output axis. The resulting output histogram (Figure 9-12H) has a wide spacing of bins in the center of the distribution curve and a close

E. Image with Gaussian contrast stretch.

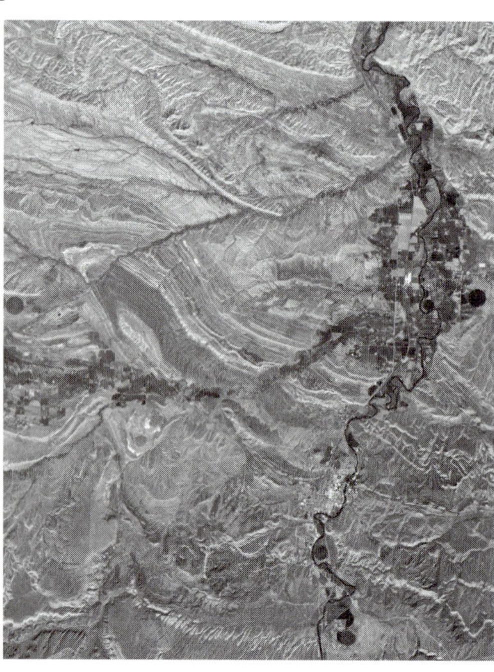

F. Histogram of Gaussian contrast stretch.

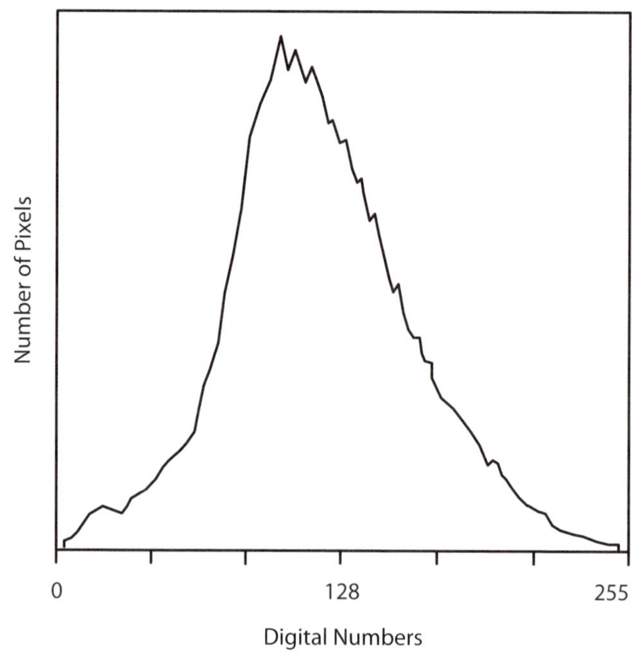

G. Image with uniform distribution stretch.

H. Histogram of uniform distribution stretch.

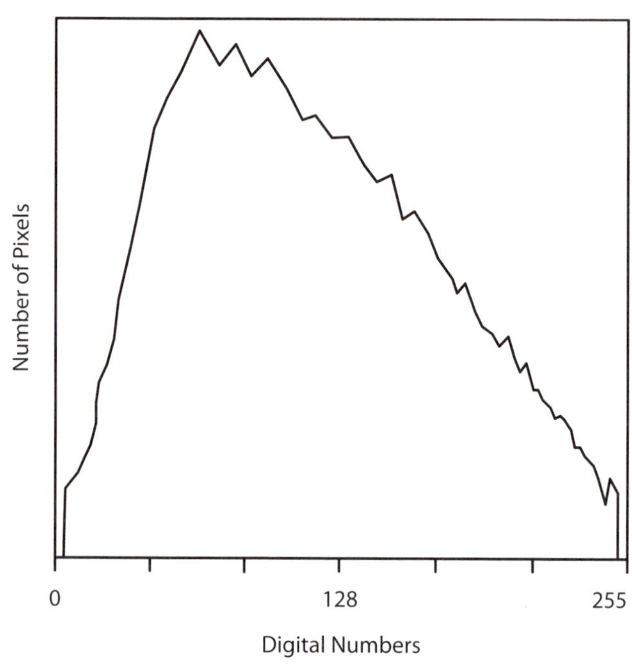

FIGURE 9-12 *Continued*

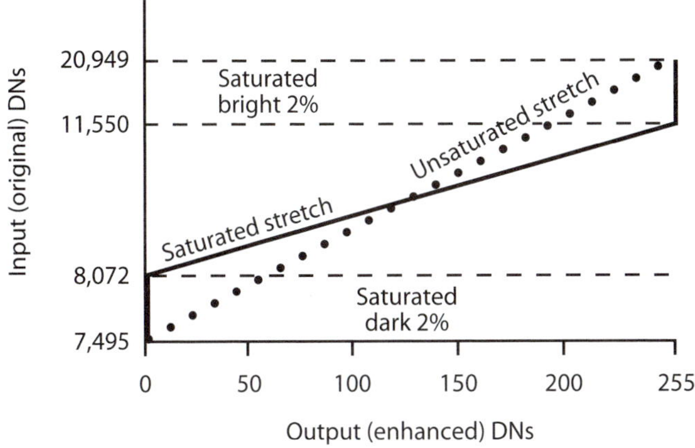

FIGURE 9-13 Graphic plot of the lookup tables for a linear unsaturated stretch and a linear contrast stretch with 2% saturation.

spacing of the less-populated bins at the head and tail of the histogram. In the resulting image (Figure 9-12G), the greatest contrast enhancement is applied to the most populated central range of DNs in the original image. Gray-level variations at the bright and dark extremes are compressed and become less distinct.

Other Contrast Stretches

In addition to the standard methods described above, most software packages for image processing provide additional methods, such as exponential and logarithmic stretches. The user can also define custom contrast stretches. An important step in the process of contrast enhancement is for the user to inspect the original histogram and determine the elements of the scene that are of greatest interest. The user then chooses the optimum stretch for his or her needs. Experienced operators of image processing systems bypass the histogram examination stage and interactively adjust the brightness and contrast of images that are displayed on a computer screen. For some scenes a variety of stretched images are required to fully display the original data. It also bears repeating that contrast enhancement should not be done until other processing is completed, because the stretching modifies the original values of the pixels.

DENSITY SLICING

Density slicing converts the continuous gray tone of an image into a series of density intervals, or slices, each corresponding to a specified range of DNs. Each digital slice is displayed as a separate color or outlined by contour lines. Qualitative analog displays are thus converted into quantitative digital displays, assuming that calibration data are available. This technique also emphasizes subtle grayscale differences that may be imperceptible to the viewer.

EDGE ENHANCEMENT

Many interpreters are concerned with recognizing linear features in images. Geologists map faults, joints, and linear features. Geographers map man-made linear features such as highways and canals. Some linear features occur as narrow lines against a background of contrasting brightness; others are the linear contact between adjacent areas of different brightness. In all cases, linear features are formed by edges. Some edges are marked by pronounced differences in brightness and are readily recognized. Typically, however, edges are marked by subtle brightness differences that may be difficult to recognize. *Edge enhancement* is the process of emphasizing the signatures of edges on images. Edges are enhanced in two ways:

1. Expanding the width of the linear feature and
2. Increasing the DN difference across the feature.

Two categories of digital filters are used for edge enhancement: nondirectional filters and directional filters.

Nondirectional Filters

Nondirectional filters (also called *Laplacian filters*) are named because they have no directional bias in enhancing linear features; almost all directions are enhanced. The only exception applies to linear features oriented parallel with the direction of filter movement; these features are not enhanced. Many image processing facilities routinely apply a nondirectional filter to most images, which imparts a "crisper" appearance to the final product.

Figure 9-14A shows a Laplacian filter that is a kernel of three lines and three columns operating on an 8-bit grayscale image. The kernel is a template with 4 as the central value, 0 at each corner, and –1 at the center of each edge. The Laplacian kernel is placed over a 3 by 3 array of original pixels, and each pixel is multiplied by the corresponding value in the kernel. The nine resulting values (four of which are 0 and four are negative numbers) are summed (0 + –40 + 0 + –40 + 160 + –35 + 0 + –40 + 0 = 5). The resulting value for the filter kernel (5) is combined with the original central pixel of the 3 by 3 data array (40 + 5 = 45), and this new number replaces the original DN of the central pixel. In Figure 9-14B, the original central pixel in the array (DN = 40) is now replaced with the new value of 45. The kernel then moves one column of pixels to the right, and the process repeats until the kernel reaches the right margin of the pixel array. The kernel then drops down one line of pixels, and continues the same process across the line. The outermost column and line of pixels are blank because they cannot form central pixels in an array.

The effect of the edge-enhancement operation can be evaluated by comparing profiles along one row (between A and B) of the original and the filtered data. Figure 9-14C is a profile of the original data. The regional background (DN = 40) is intersected by a darker linear (DN = 35) that is three pixels wide and has DN values of 35. The contrast ratio between the linear and background, as calculated from Equation 1-4, is 40/35, or 1.14. In the profile of the enhanced data (Figure 9-14D), the contrast ratio is 45/30, or 1.50, which is an enhancement of 32%. The original linear, which was three pixels wide, is five pixels wide in the filtered version.

The right-hand portion of the original profile (Figure 9-14C) has a second linear marked by a change in values from 40 to 45 along an edge that has no width. The original contrast ratio (45/40 = 1.13) is increased by 27% in the enhanced image (50/35 = 1.43). The original edge is expanded to a width of two pixels.

Figure 9-15A is a computer-generated synthetic 8-bit image with a uniform background (DN = 127). The left portion of the image is crossed by a dark band (DN = 107). The central portion is crossed by a light band (D = 147). In the right portion of the image, an edge is formed at the contact of the background with a brighter surface (DN = 137). These three linear features have subtle expressions in the original image. The linear features of the synthetic image are similar to those in the digital arrays of Figure 9-14A. The Laplacian filter of Figure 9-14A was applied to the synthetic image. After the value of the filter kernel has been calculated, and prior to combining it with the original central data pixel, the calculated value may be multiplied by a weighting factor. The factor weight may be less than 1 or greater than 1 in order to diminish or to accentuate the effect of the filter. The weighted filter value is then combined with the central original pixel to produce the enhanced image.

A. Original data and Laplacian filter kernel.

0	−1	0
−1	4	−1
0	−1	0

	40	40	40	40	35	35	35	40	40	40	40	40	40	40	45	45	45	45	
	40	40	40	40	35	35	35	40	40	40	40	40	40	40	45	45	45	45	
A	40	40	40	40	35	35	35	40	40	40	40	40	40	40	45	45	45	45	B
	40	40	40	40	35	35	35	40	40	40	40	40	40	40	45	45	45	45	

B. Enhanced data.

	40	40	45	30	35	30	45	40	40	40	40	40	35	50	45	45	45	
A	40	40	(45)	30	35	30	45	40	40	40	40	40	35	50	45	45	45	B
	40	40	45	30	35	30	45	40	40	40	40	40	35	50	45	45	45	

C. Profile of the original data.

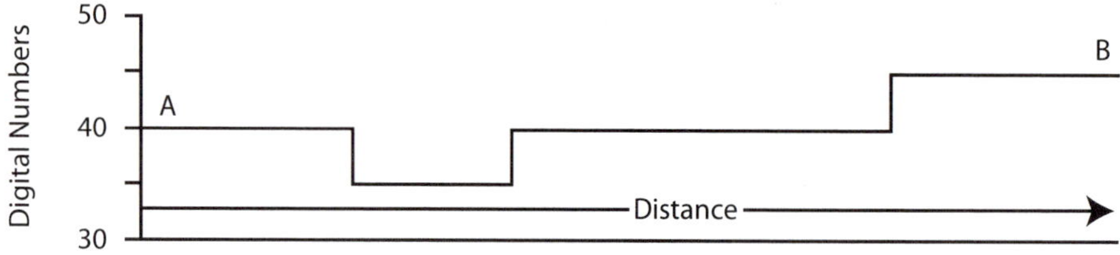

D. Profile of the enhanced data.

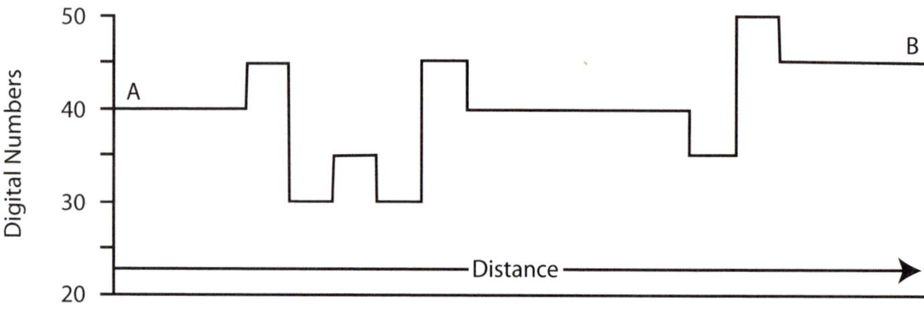

FIGURE 9-14 Nondirectional edge enhancement using a Laplacian filter.

In Figure 9-15B the calculated kernel value was weighted by a factor of 2.0 to produce the enhanced image. In Figure 9-15C a factor of 5.0 was used. In Figure 9-15D a factor of 10.0 was used, which saturates the brightest values to a DN of 255 and the darkest to a DN of 0. The filtering has significantly enhanced the expression of the original linears in Figure 9-15A.

Figure 9-16 shows nondirectional edge enhancement applied to a TM band 4 subscene in northwest Saudi Arabia. The area is a plateau of horizontal strata that is cut by fractures that trend northwest and northeast. Figure 9-17 is a map that shows the major fractures. Figure 9-16B shows the image after processing with the Laplacian filter kernel. The image and fractures are sharper in the enhanced image.

Directional Filters

Directional edge enhancement is a valuable technique for selectively improving both the brightness difference and the geometric width of linear features. *Directional filters* are used to enhance linear features that trend in a specific direction, such as N45°W. Figure 9-18A (on p. 262) shows four directional filters that are designed to enhance the four cardinal directions. Figure 9-18B shows an array of original pixels for terrain with a background DN of 25 that is cut by two linears that trend northeast (northeast–southwest) and northwest (northwest–southeast). The linears have DNs of 30 and are brighter than the background. Figure 9-18C is a profile along row A–B that crosses the linears in an east–west direction.

A. Original image with DNs.

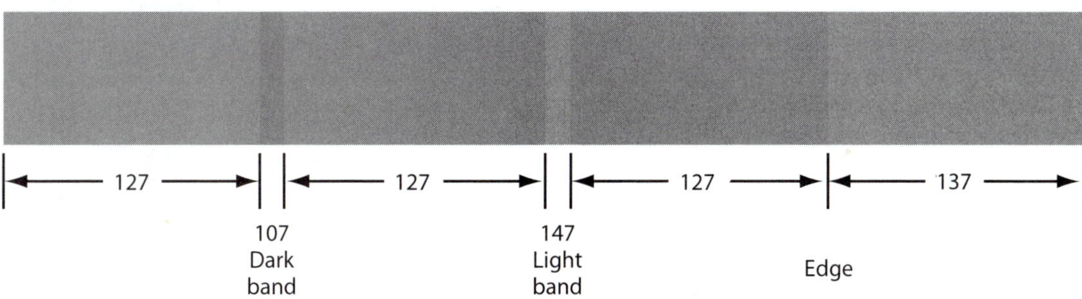

B. Enhanced by the factor 2.0.

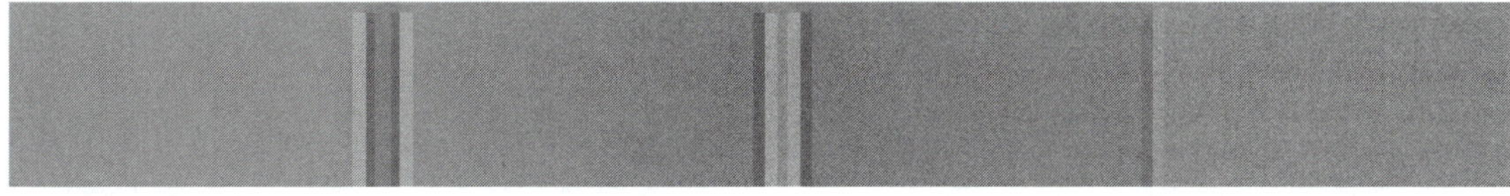

C. Enhanced by the factor 5.0.

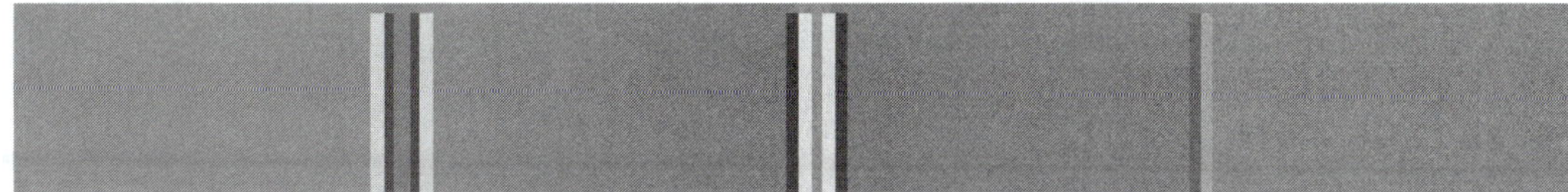

D. Enhanced by the factor 10.0.

Figure 9-15 Computer-generated images illustrating nondirectional edge enhancement with a Laplacian filter and different weighting factors.

A. Original image.

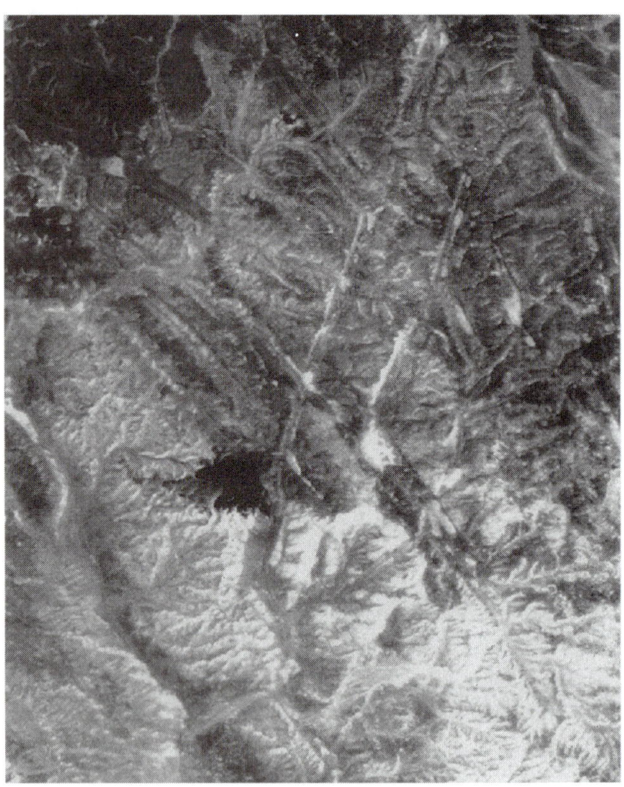

B. Nondirectional enhancement.

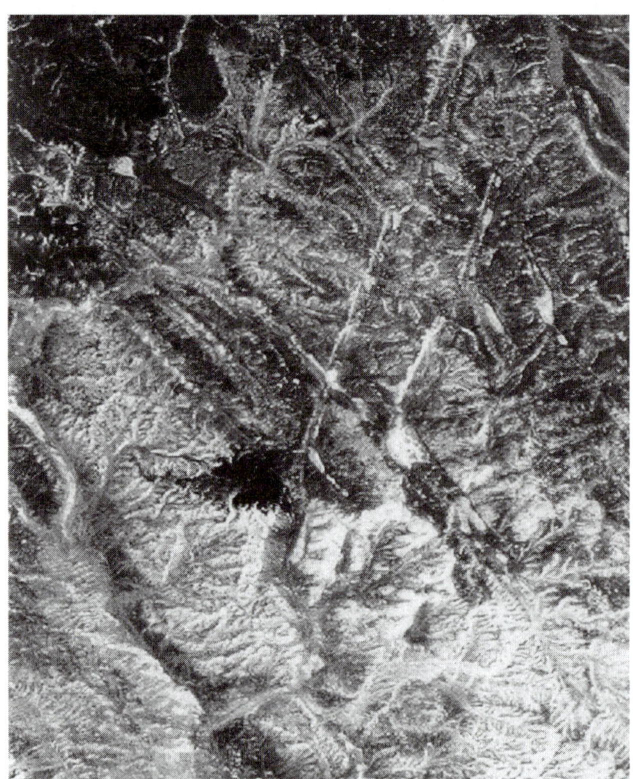

C. Directional enhancement of northeast-trending linear features.

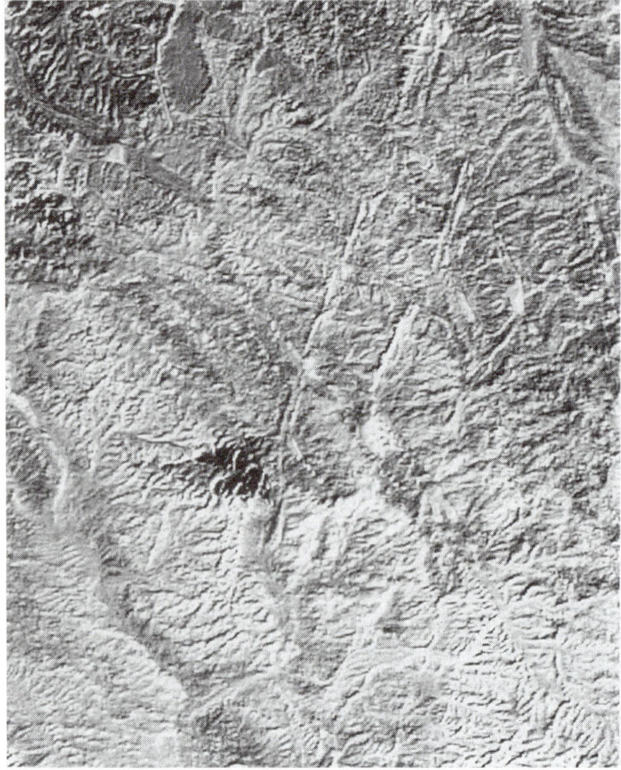

D. Directional enhancement of northwest-trending linear features.

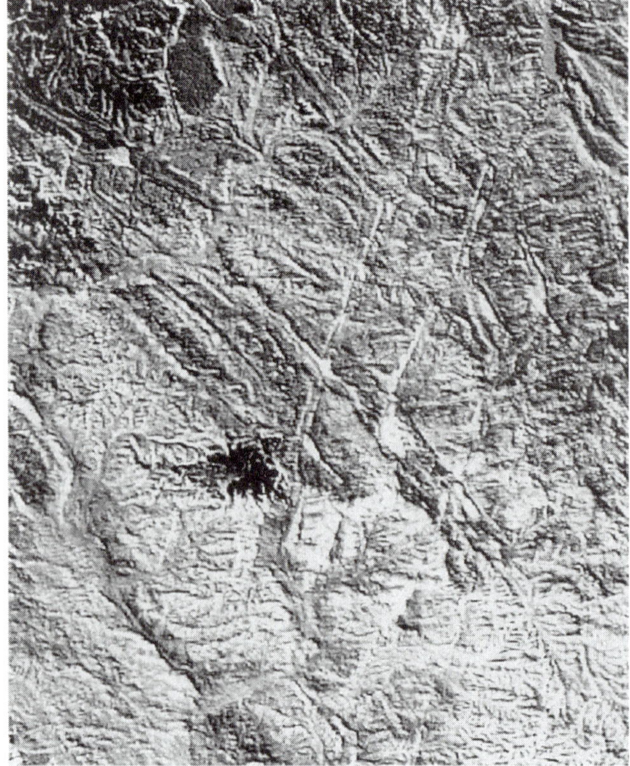

FIGURE 9-16 Edge enhancements of TM images of the Jabal an Naslah area, northwest Saudi Arabia.

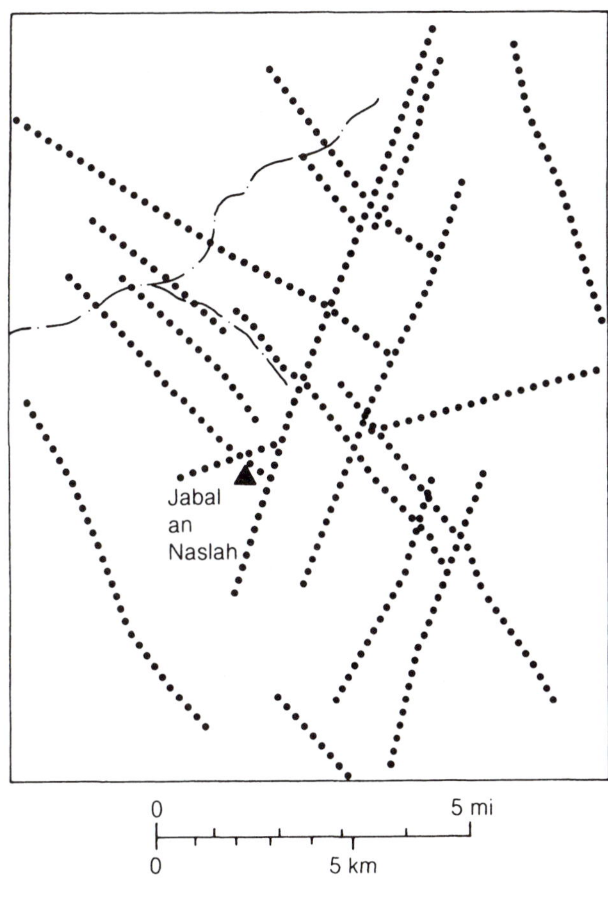

FIGURE 9-17 Linear features interpreted from edge-enhanced images of the Jabal an Naslah area, northwest Saudi Arabia (Figure 9-16).

The profile shows that each linear is one pixel wide with a brightness difference from the background of 5 DN.

The filter to enhance northeast–southwest edges is selected from Figure 9-18A and applied to the original data set. The filter kernel is placed over an array of nine original pixels (three rows by three columns). Each original pixel is multiplied by the corresponding kernel cell and the results are summed. The resulting kernel value is then combined with the value of the original pixel in the center of the array. Figure 9-18D shows the enhanced pixels that have been processed in this manner. The DNs of the northeast-trending linear are increased from 30 to 50. In addition, the filter has generated two bands of dark pixels (DN = 15), with a width of one pixel, that trend parallel with the original linear. Enhanced profile A–B (Figure 9-18E) graphically shows the effects of the directional edge enhancement. Compare this profile with the profile of the original data (Figure 9-18C). The DNs of the original linear differ from the background by 5 DN for a brightness difference of 20% (5/25 = .20). The three parallel bands of the enhanced linear differ from the background by a total of 35 DN (25 + 10 = 35) for a brightness difference of 140% (35/25 = 1.40). The brightness difference is enhanced by 700% (1.40/.20 = 7.00). The directional filter also enhances the geometric expression of the linear. The original linear is 1 pixel wide. The enhanced linear is 3 pixels wide. Figure 9-18D shows the enhanced linear (DN = 50) and the two parallel bands (DN = 15). Cross section A–B (Figure 9-18E) crosses the linear obliquely and therefore gives a misleading impression of the geometric relationships. A cross section (not shown) drawn normal to the linear (Figure 9-18D) shows that the enhanced linear is three pixels wide. The geometric expression is enhanced by 300% (3 pixels width/1 pixel width = 3). The key point is that the northeast-trending linear is strongly enhanced, but the northwest-trending linear is completely unchanged by this process.

In Figures 9-18F and 9-18G, the filter kernel to enhance northwest-trending edges is applied to the original array of pixels. The brightness difference and the geometric expression of the northwest-trending linear are enhanced in the same manner described above. The northeast-trending linear is completely unchanged. In summary, the two directional filters have selectively enhanced the northeast- and northwest-trending linears by 300% in width and 700% in brightness difference.

Filters can be modified to enhance linear features trending at directions other than the four cardinal directions, as described by Haralick (1984). Directional edge-enhancement filters were applied to the original Landsat subscene in Saudi Arabia (Figure 9-16). In Figure 9-16C, the northeast-trending linears are preferentially enhanced. In Figure 9-16D, the northwest-trending linears are preferentially enhanced. Comparing Figures 9-16C and 9-16D with the original image (Figure 9-16A) shows that both the brightness difference and the geometric width of the linears are selectively enhanced.

INTENSITY, HUE, AND SATURATION TRANSFORMATIONS

The additive and subtractive systems of primary colors were described in Chapter 2. An alternate approach to color is the *intensity, hue, and saturation* (IHS) *system*, which is useful because it presents colors more nearly as they are perceived by humans. IHS is also termed hue, saturation, and value (HSV), where intensity is referred to as value (V). The IHS system is based on the color sphere (Figure 9-19) in which the vertical axis represents intensity, the radius represents saturation, and the circumference represents hue. The intensity (I) axis represents brightness variations and ranges from black (0) to white (255); no color is associated with this axis. Hue (H) represents the dominant wavelength of color. Hue values commence with 0 at the midpoint of red tones and increase counterclockwise around the circumference of the sphere to conclude with 255 adjacent to 0. Saturation (S) represents the purity of color and is represented by the radius that ranges from 0 at the center of the color sphere to 255 at the circumference. A saturation of 0 represents a completely impure color in which all wavelengths are equally represented as a shade of gray. Intermediate values of saturation are pastel shades, whereas high values are purer and more intense colors.

Figure 9-20 shows graphically the relationship between the BGR and IHS systems. Numerical values may be

FIGURE 9-18 Edge enhancement using a directional filter.

A. Filters for directional edge enhancement.

0	0	0
−2	4	−2
0	0	0

Enhance N–S edge

0	−2	0
0	4	0
0	−2	0

Enhance E–W edge

−2	0	0
0	4	0
0	0	−2

Enhance NE–SW edge

0	0	−2
0	4	0
−2	0	0

Enhance NW–SE edge

B. Original data set.

25	25	30	25	25	25	25	25	25	25	30	25	25
25	25	25	30	25	25	25	25	25	30	25	25	25
25	25	25	25	30	25	25	25	30	25	25	25	25
25	25	25	25	25	30	25	30	25	25	25	25	25
25	25	25	25	25	25	30	25	25	25	25	25	25
25	25	25	25	25	30	25	30	25	25	25	25	25
25	25	25	25	30	25	25	25	30	25	25	25	25
25	25	25	30	25	25	25	25	25	30	25	25	25
25	25	30	25	25	25	25	25	25	25	30	25	25

A (left of row 2) ● — ● B (right of row 2)

C. Profile A–B of original data.

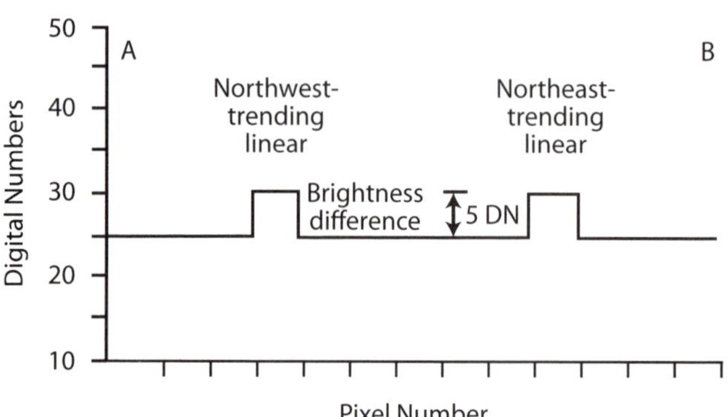

D. Northeast-trending edges enhanced.

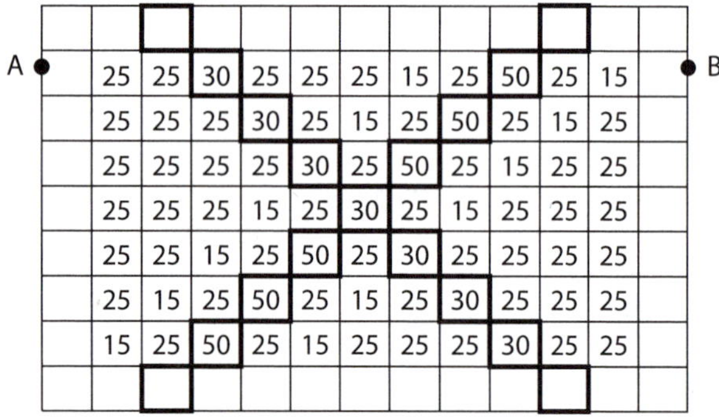

	25	25	30	25	25	25	15	25	50	25	15	
	25	25	25	30	25	15	25	50	25	15	25	
	25	25	25	25	30	25	50	25	15	25	25	
	25	25	25	15	25	30	25	15	25	25	25	
	25	25	15	25	50	25	30	25	25	25	25	
	25	15	25	50	25	15	25	30	25	25	25	
	15	25	50	25	15	25	25	25	30	25	25	

A ● — ● B

E. Profile A–B of enhanced northeast-trending edges.

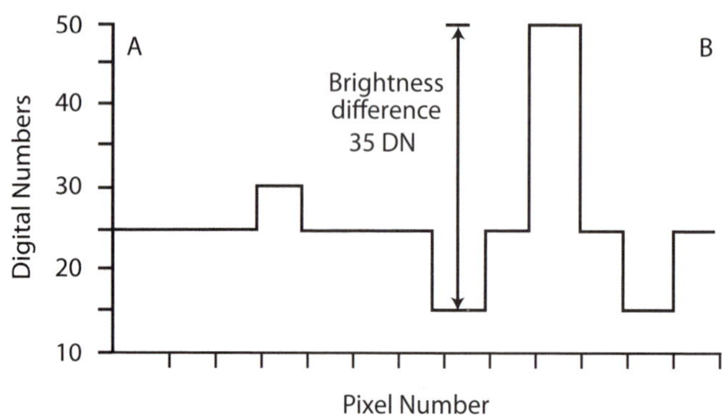

F. Northwest-trending edges enhanced.

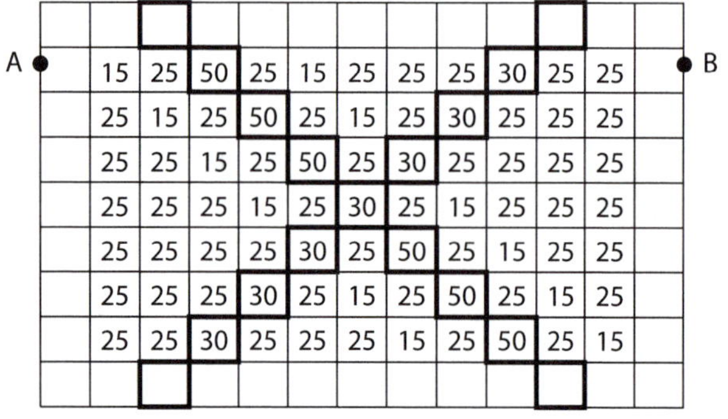

	15	25	50	25	15	25	25	25	30	25	25	
	25	15	25	50	25	15	25	30	25	25	25	
	25	25	15	25	50	25	30	25	25	25	25	
	25	25	25	15	25	30	25	15	25	25	25	
	25	25	25	25	30	25	50	25	15	25	25	
	25	25	25	30	25	15	25	50	25	15	25	
	25	25	30	25	25	25	15	25	50	25	15	

A ● — ● B

G. Profile A–B of enhanced northwest-trending edges.

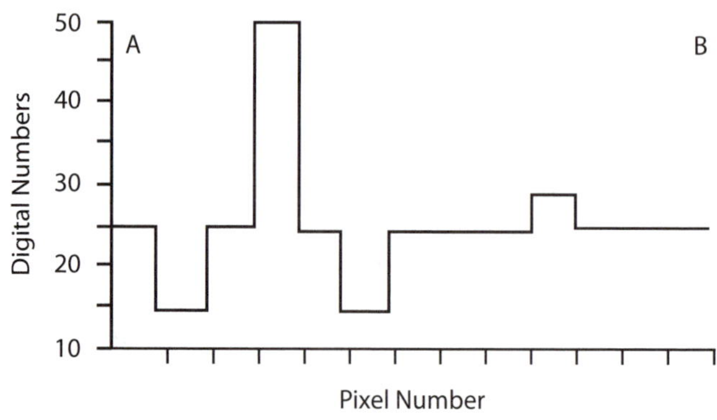

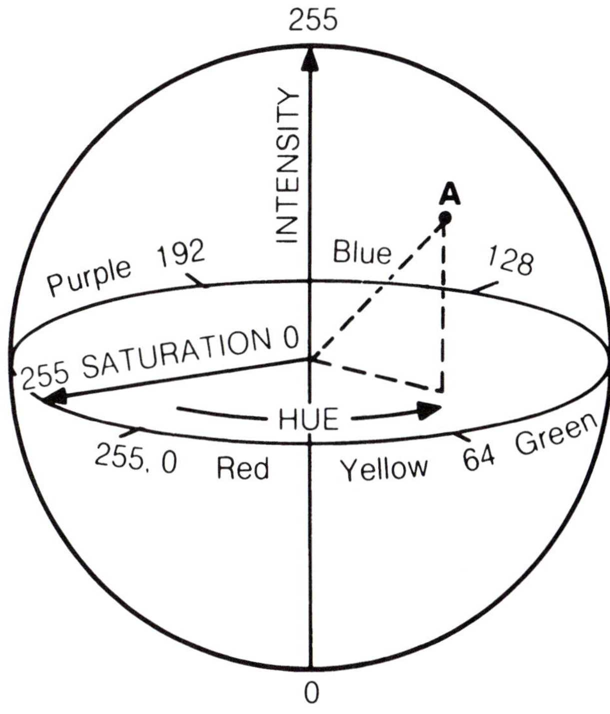

FIGURE 9-19 Coordinate system for the IHS transformation. The color at point A has the following values: I = 205, H = 75, S = 130 (undersaturated).

extracted from this diagram for expressing either system in similar terms. In Figure 9-20 the circle represents a horizontal section through the equatorial plane of the IHS sphere, with the intensity axis passing vertically through the plane of the diagram. The corners of the equilateral triangle are located at the positions of the red, green, and blue hues. Hue changes in a counterclockwise direction around the triangle, from red (0), to green (1), to blue (2), and again to red (3). Values of saturation start at 0 at the center of the triangle and increase to a maximum of 1 at the corners. Any perceived color can be described by a unique set of IHS values; in the BGR system, however, different combinations of additive primaries can produce the same color. The IHS values can be derived from BGR values for the interval $0 < H < 1$, extended to $1 < H < 3$ through the following transformation equations:

$$I = R + G + B \quad \textbf{(9-3)}$$

$$H = I - 3B \quad \textbf{(9-4)}$$

$$S = I \quad \textbf{(9-5)}$$

After enhancing the saturation image, IHS values can be converted back into BGR values by inverse equations.

The traditional method of enhancing edges on multispectral (TM) images is to enhance each of the three bands separately. It is more efficient to transform the three TM bands into IHS components, and then to apply edge enhancement to the intensity component. When the image is transformed back into BGR components, the edge enhancement is applied to each of the bands. This operation may be combined with the saturation enhancement using the following steps:

1. The 3-band BGR image is transformed into the IHS color space.
2. The intensity is replaced with the panchromatic band.
3. The hue and saturation bands are resampled to the panchromatic pixel size using the nearest neighbor, bilinear, or cubic convolution technique.
4. The pan-H-S image is transformed back into BGR color space.

When any three spectral bands of OLI or TM (or other multispectral data) are combined in the BGR system, the resulting color images typically lack saturation, even though the bands have been contrast stretched. **Digital Image 9-1A** is a normal color image prepared from OLI bands 2-3-4 of the Thermopolis, Wyoming, subscene. The individual bands were contrast enhanced, but the color image has the pastel appearance that is typical of many Landsat images. The undersaturation is due to the high degree of correlation between spectral bands (Figure 9-4A and Table 9-2). High reflectance values in the green band, for example, are accompanied by high values in the blue and red bands, so pure colors are not produced.

Thermopolis, Wyoming, Example

The IHS transformation was applied to OLI bands 2-3-4 in B-G-R of the Thermopolis subscene to produce the intensity, hue, and saturation images illustrated in Figure 9-21.

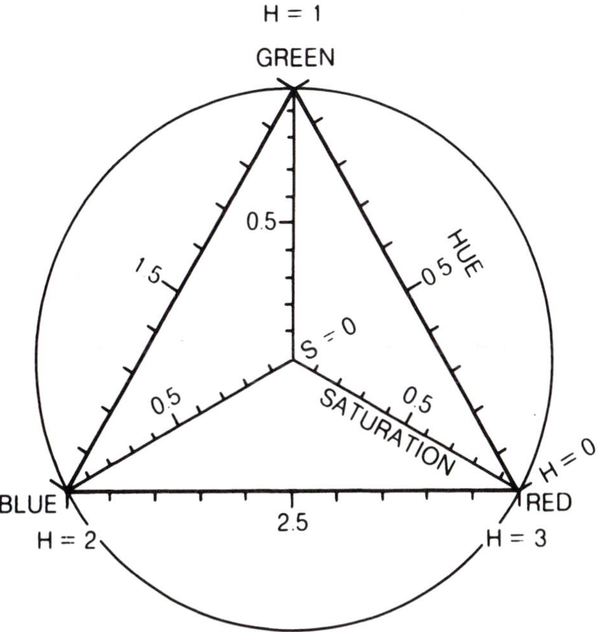

FIGURE 9-20 Diagram showing the relationship between the BGR and IHS systems.

The intensity image (Figure 9-21A) is dominated by albedo and topography. Sunlit slopes have high intensity values (bright tones), and shadowed areas have low values (dark tones). The water in the Wind River has low values (see Figure 3-11H for location). Vegetation has intermediate intensity values, as do most of the rocks. In the hue image (Figure 9-21B), red beds of the Chugwater Formation have conspicuous dark tones caused by low values assigned to red hues. Vegetation has intermediate to light gray values assigned to the green hue. The original saturation image (Figure 9-21C) is dark overall because of the lack of saturation in the original OLI data. A linear contrast stretch was applied to enhance the original saturation image (Figure 9-21D). Note the overall increased brightness of the enhanced image. Also note the improved discrimination between terrain types and agricultural fields.

A. Intensity image.

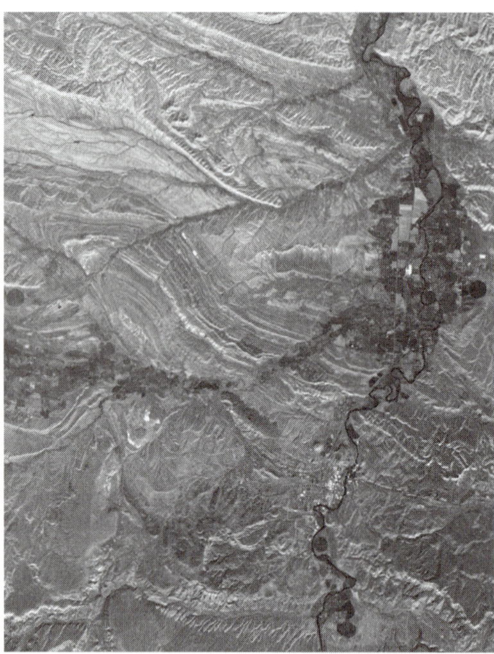

B. Hue image.

C. Saturation image, original.

D. Saturation image, stretched.

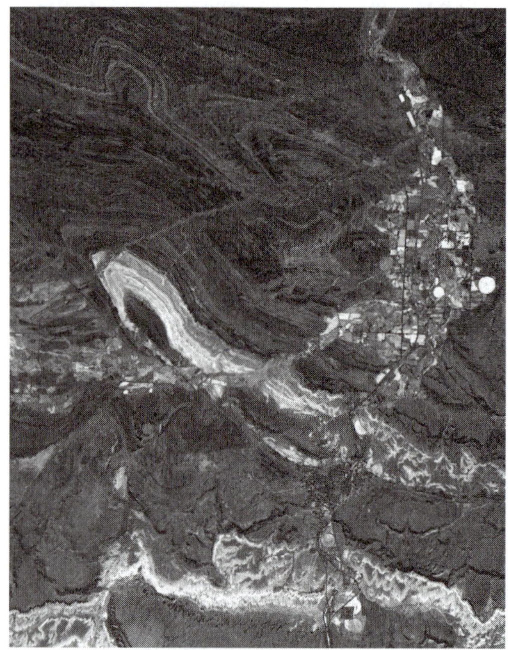

FIGURE 9-21 Images created by the IHS transformation of OLI bands 2-3-4 (B-G-R), Thermopolis, Wyoming, subscene.

Lastly, the intensity, hue, and enhanced saturation images from the IHS system were transformed back into the BGR system. These enhanced BGR images were used to prepare the color composite image of **Digital Image 9-1B**, which is a significant improvement over the original version in **Digital Image 9-1A**. In the IHS version, note the wide range of colors and improved discrimination between colors. The red beds of the Chugwater Formation are more distinguished and fallow agricultural fields and other rock units have a wider range of color tones.

Western Bolivia Example

Saturation enhancement is also effective in other regions and other combinations of spectral bands. **Digital Image 9-1C** is a TM bands 2-4-7 in B-G-R image of western Bolivia. The right-hand portion of the image consists of volcanic features of the Andes and the left-hand portion is alluvial deposits. The bands were contrast enhanced but are dominated by monotonous undersaturated hues of brown, tan, and gray. **Digital Image 9-1D** is the same image after saturation enhancement. Subtle color differences are emphasized, and much more geologic information can be interpreted from the enhanced image.

Pan-Sharpening and Combination Images

Pan-sharpening is a routine enhancement process for systems that acquire both panchromatic and multispectral images, such as Landsat TM and OLI (Figure 3-6), commercial high resolution satellites, and aerial panchromatic + multispectral cameras. The higher spatial resolution panchromatic band is acquired at the same time as the coarser multispectral bands, minimizing land cover and shadow changes associated with merging images acquired at different times.

Several methods have been developed for sharpening lower resolution multispectral images with a higher resolution grayscale image (ENVI, 2017; Hexagon, 2019). Some algorithms only operate on 3-band color images while others can sharpen multiple bands. As already discussed, the IHS transform is one method that sharpens 3-band color images. Other techniques include the Brovey (color normalization) transform, the principal component (PC) method, and the Gram–Schmidt method.

The Brovey transform uses a formula integrating the DNs of the three color bands prior to interacting with the panchromatic band to visually increase contrast in the high and low ends of an image's histogram. The Brovey method alters the original spectral information of the 3-band color image.

The PC and Gram–Schmidt methods can sharpen more than three bands while preserving the original spectral information of the multispectral data. Multispectral data sharpened with these two methods (or other methods that retain original spectral information) can generate reliable spectral classification maps (including land cover) because the sharpened multispectral bands retain their original spectral character. Lazaridou and Karagianni (2016) spectrally classified pan-sharpened 15 m Landsat bands and the original 30 m bands to generate land cover maps of an urban area in Greece. The land cover map built from the pan-sharpened bands was more accurate.

The PC transformation is explained in detail in the next section. For the pan-sharpening application, the PC transformation is performed on the lower resolution multispectral bands and the first PC image that is generated is replaced with the higher resolution panchromatic band. The panchromatic band is scaled to match the histogram of the PC image, so little distortion of the spectral information occurs. Gram–Schmidt is computationally intensive but recommended as an excellent method for preserving the original spectral information of the multispectral data (ENVI, 2017). Gram–Schmidt builds a simulated panchromatic image from the spectral characteristics of the original multispectral data, input from commercial satellites, and other sources prior to integrating the real panchromatic image. Both methods resample multispectral bands to a high resolution pixel size using a nearest neighbor, bilinear, or cubic convolution technique.

Different images that have been digitally merged are called *combination images*. For example, Landsat TM color images display three spectral bands but have a relatively coarse spatial resolution (30 m). SPOT pan images have good resolution (10 m) but display only a single spectral band. A combination image can display the spectral characteristics of the TM color image with the spatial resolution of a SPOT pan image. This technique is used to combine other images, such as radar.

Chavez and others (1991), Jensen (2016), and Welch and Ehlers (1987) describe methods of merging images, as do many user guides for image processing software.

Digital Mosaics

Georeferenced (or orthorectified) images with different spatial resolutions, brightness histograms, and sizes can be digitally mosaicked with advanced image processing software. The images can overlap or be floating in their own map space. Images that overlap can have their mosaic seamlines feathered to provide a gradation in color or grayscale between the images, minimizing the visual distraction of the seamline. If the scenes have the same spectral and temporal characteristics (e.g., 8-bit, BGR bands, same season of the year), the remote sensing analyst can choose a base image and match the histograms of the surrounding images to the base image to align color balance between the scenes and improve the visual appearance of the mosaic.

Images are loaded sequentially into a digital mosaic frame; typically the last image that is loaded covers previously loaded images where they overlap. Therefore, it is recommended that the best images are loaded last so they are on top of less visually appealing images.

The margins of multispectral bands can have one or more lines or rows with noise, missing data, and irregular edges. Mosaicking images with noisy edges creates distracting linear zones filled with pixels of different colors across the mosaic if the image with the noisy edges is placed on top of other images. Landsat images often display noise of varying widths (from 1 to over 10 pixels) along the east–west margins at the edge of the wedges with no data (Figure 9-1). Distracting multicolor strips along the west and east margins of the image are visible when three bands are combined into a color composite. Noise along the north–south margins of a Landsat MSS and TM scene is mostly limited to one or two lines. The noisy margins need to be eliminated by clipping or masking.

MASKING

Masking focuses the image processing effort on features of interest by removing unwanted pixels in the scene from the analysis. Masking can also be used to cover noisy pixels along the margins of multispectral bands so the problematic pixels are not included in the analysis and to improve mosaics. Masks can be generated from manually drawn polygons, spectrally classified features (e.g., water bodies, sand dunes, vegetation), or a DN range specified by the analyst. The mask is built and then applied to the image so that pixels within the mask are given a value that will not interfere or be included in subsequent processing. Masks are discussed and displayed in the following section.

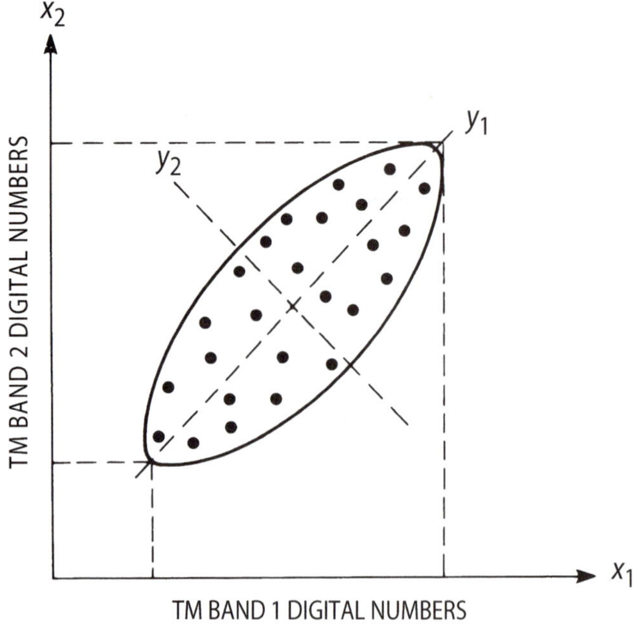

FIGURE 9-22 Plot of DNs for TM band 1 (x_1) and band 2 (x_2) showing correlation between these bands. The PC transformation was used to generate a new coordinate system (y_1, y_2). After Swain and Davis (1978, Figure 7.9).

INFORMATION EXTRACTION

Preprocessing and enhancement utilize computers to provide corrected and improved images with coordinates and a map projection that can be used for study by human interpreters and input into knowledge-based expert systems. The knowledge and experience of the remote sensing team, quality of the data, and capabilities of the image-processing software and hardware enable new geospatial information to be discovered about the environment and man-made features.

PRINCIPAL COMPONENT IMAGES

If we compare individual OLI bands, we note a strong similarity. Areas that are bright or dark in one band tend to be bright or dark in the other bands. This relationship is shown diagrammatically in Figures 9-4A and 9-22, in which DNs for the Landsat TM/OLI blue band are plotted versus the Landsat TM/OLI green band. Data points are distributed in an elongate band, which shows that as DNs increase for one band, they increase for the other band. If for any pixel we know the DN for band 1, we can predict the approximate value in band 2. The data are said to be strongly correlated. This correlation means that there is much redundancy of information in a multispectral data set. If this redundancy is reduced, the amount of data required to describe a multispectral image is compressed.

The PC transformation, originally known as the Karhunen–Loève transformation, is used to compress multispectral data sets by calculating a new coordinate system. For the two bands of data in Figure 9-22, the transformation defines a new axis (y_1) oriented in the long dimension of the distribution and a second axis (y_2) perpendicular to y_1. The mathematical operation makes a linear combination of pixel values in the original coordinate system that results in pixel values in the new coordinate system:

$$y_1 = a_i x_1 + a_i x_2 \qquad \textbf{(9-6)}$$

$$y_2 = a_i x_1 + a_i x_2 \qquad \textbf{(9-7)}$$

where:

(x_1, x_2) = pixel coordinates in the original system
(y_1, y_2) = coordinates in the new system
a_i = constants

In Figure 9-22 note that the range of pixel values for y_1 is greater than the ranges for either of the original coordinates, x_1 or x_2, and that the range of values for y_2 is relatively small. In this 2-band example, y_1 is the first principal component (PC image 1) and y_2 is the second principal component (PC image 2).

The PC transformation may be carried out for multispectral data sets consisting of any number of bands. Additional coordinate directions are defined sequentially. Each new coordinate is oriented perpendicular to all the previously defined directions and in the direction of the remaining maximum density of pixel data points. For each pixel, new DNs are determined relative to each of the new coordinate axes. A set of DN values is determined for each pixel relative to the first principal component. These DNs are then used to generate an image of the first principal component. The same procedure is used to produce images for the remaining principal components (see Swain and Davis, 1978; Vincent, 1997).

A PC transformation was performed on the three visible and three reflected IR bands of OLI data for the Thermopolis, Wyoming, subscene (Figure 3-11). Each pixel was assigned six new DNs corresponding to the first through the sixth PC coordinate axes. Figure 9-23 illustrates the six PC images from Landsat 8 OLI, which have been contrast enhanced. As noted earlier, each successive principal component accounts for a progressively smaller proportion of the variation of the original multispectral data set. These percentages of variation are indicated in Figure 9-23 and Table 9-3 and are plotted graphically in Figure 9-24. The first three PC images contain 98.8% of the variation of the original six OLI bands, which is a significant compression of data.

A. OLI PC image 1 (78.4%).

B. OLI PC image 2 (16.7%).

C. OLI PC image 3 (3.7%).

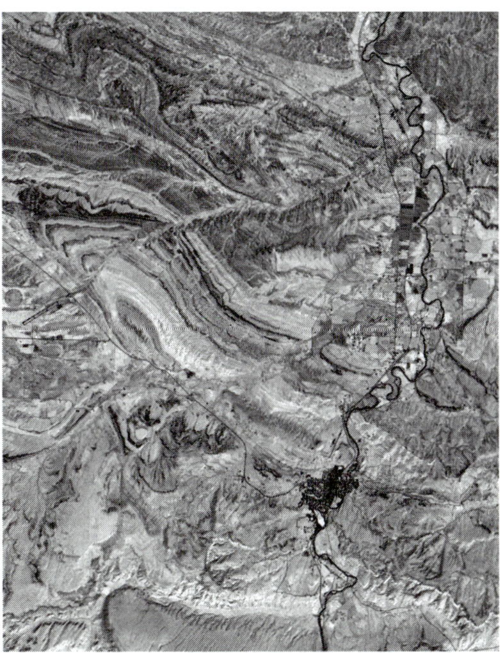

D. OLI PC image 4 (0.7%).

Figure 9-23 PC images of the Thermopolis, Wyoming, subscene. PC images were generated from the six visible, NIR, and SWIR bands of the OLI data. (G) and (H) are PC images from Landsat TM data over two decades old. The percentage of variance represented by each PC image is provided. Landsat courtesy USGS. *(cont.)*

Visual inspection of the six OLI PC images in Figure 9-23 reveals patterns that delineate and help interpret different features (rock types, water, agriculture, settlements). Evaluating the correlation matrix that compares Landsat bands with PC images reveals how each PC is affected by each Landsat band, improving the understanding of what each PC image represents. The gray cells in Table 9-4 indicate which bands have the most positive or negative correlation with the PC image (values range between –1 and 1). PC image 1 (Figure 9-23A) is dominated by topography, expressed as highlights and shadows, that is negatively correlated in all six of the original OLI bands. PC image 2 (Figure 9-23B) is similar in appearance to the NIR band (OLI band 5) (Figure 3-11D). The correlation table quantifies the visual observation—PC

E. OLI PC image 5 (0.4%).

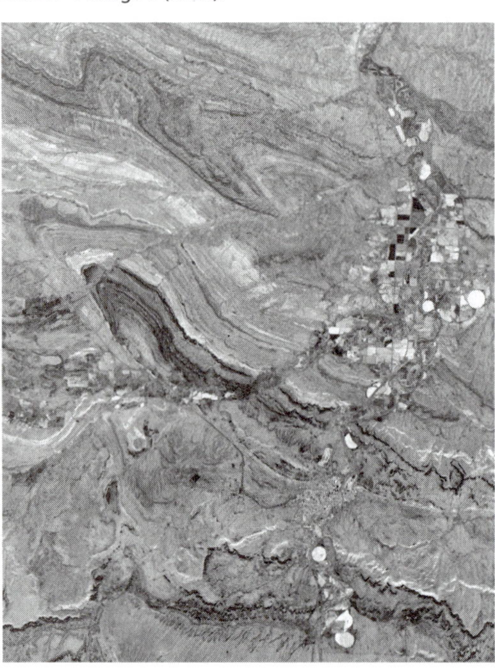

F. OLI PC image 6 (0.06%).

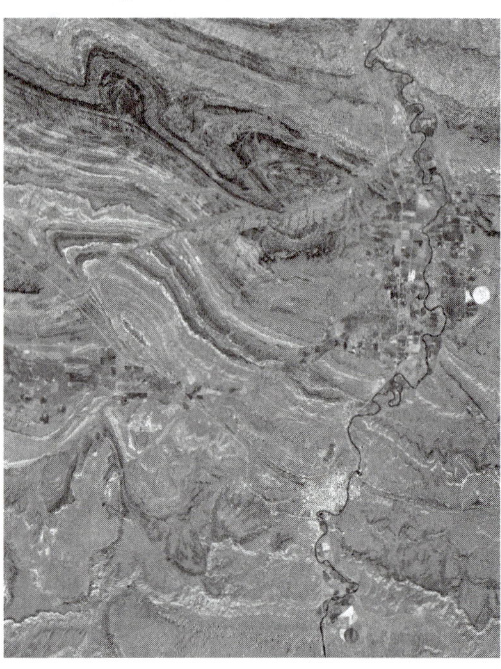

G. TM PC image 5 (0.5%).

H. TM PC image 6 (0.3%).

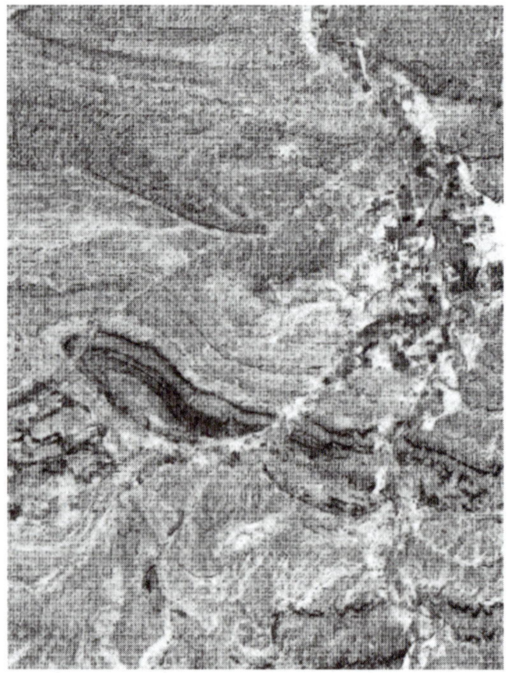

FIGURE 9-23 *Continued.*

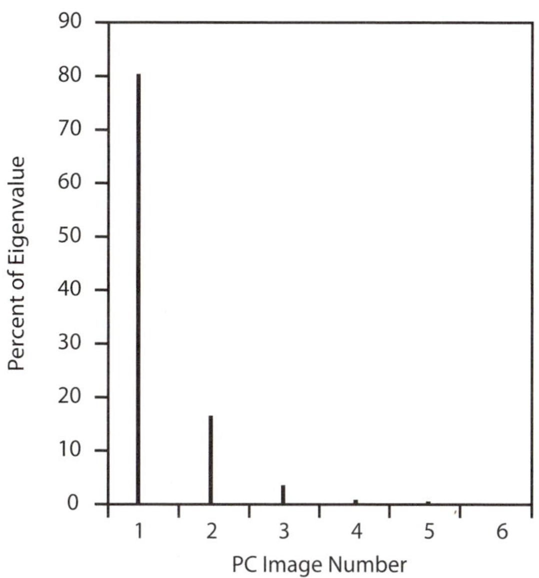

Figure 9-24 Plot showing the percentage of variance represented by PC images of the Thermopolis, Wyoming, subscene (Figure 9-23; Table 9-3).

image 2 has a very high positive correlation (0.943) with band 5. PC image 5 (Figure 9-23E) displays parallel arcuate dark bands in the outcrop belt of the Chugwater Formation, which may represent bedding that is more red in color.

Table 9-3 PC Eigenvalues and percent of variance.

PC	Eigenvalue	Variance (%)
1	13,426,995	78.37
2	2,866,950	16.73
3	635,430	3.71
4	125,885	0.73
5	68,405	0.40
6	9,921	0.06

This interpretation is based on a high negative correlation (–0.765) between PC image 5 and the red band 4. Chugwater red beds that are brighter in Landsat OLI band 4 are darker in PC image 5.

Figures 9-23G and 9-23H show PC images 5 and 6 that were generated over two decades ago using TM data. The older TM images cover much of the same geographic area as the newer Landsat OLI images. Noise dominates the TM PC images 5 and 6 in contrast to the almost noise-free OLI PC images 5 and 6.

Of interest is the visual similarity between noisy TM PC image 6 and the clear OLI PC image 5 (compare Figure 9-23E and Figure 9-23H). Both highlight the Chugwater Formation with dark arcuate bands. PCs are scene-dependent, so PCs derived from data collected by a multispectral sensor of the same area on different dates can be very different. Visual inspection of the PC images and examination of the correlation matrix and percent of variance for the PC eigenvalues must be done prior to using PCs acquired on different dates for change detection and other applications.

Any three PC images can be combined in red, green, and blue to create a color image. Plate 27 was produced by combining PC images from Figure 9-23A–E in the following manner:

- **Plate 27A**: PC 2 = red, PC 3 = green, PC 4 = blue. PC image 1 was not used in order to minimize topographic effects. As a result, the color PC 234 image displays a great deal of spectral variation in the vegetation and rocks. The three images constitute 21.2% of the variation of the original data set.
- **Plate 27B**: PC 4 = red, PC 5 = green, PC 6 = blue. Again, PC image 1 was not used in order to minimize topographic effects. PC image 456 displays new spectral patterns in agricultural and rocky terrain compared with PC image 234, although the three images constitute only 5.3% of the variation or the original data set.
- **Plate 27C**: PC 2 = red, PC 3 = green, PC 4 = blue. This PC was generated over two decades ago using TM data. The increase in noise and the reduced range of colors with the older data is striking when comparing the TM and OLI PC 234 images.

Table 9-4 Correlation matrix for Landsat bands and PC images.

	Landsat 8 OLI Bands					
	Band 2 (Blue)	Band 3 (Green)	Band 4 (Red)	Band 5 (NIR)	Band 6 (SWIR1)	Band 7 (SWIR2)
PC 1	–0.203	–0.300	–0.403	–0.287	–0.574	–0.543
PC 2	–0.064	–0.030	–0.111	0.943	–0.073	–0.297
PC 3	–0.435	–0.541	–0.434	0.011	0.531	0.217
PC 4	–0.279	–0.201	0.128	0.157	–0.612	0.683
PC 5	0.482	0.262	–0.765	0.045	–0.096	0.320
PC 6	0.675	–0.712	0.187	0.044	–0.021	0.002

RATIO IMAGES

Ratio images are prepared by dividing the DN value in one band by the corresponding DN value in another band for each pixel. The resulting values are plotted as a ratio image. Figure 9-25 illustrates two ratio images prepared from the OLI and TM bands of the Thermopolis, Wyoming, subscene. In a ratio image the black and white extremes of the grayscale represent pixels having the greatest difference in reflectivity between the two spectral bands. The darkest signatures are areas where the denominator of the ratio is greater than the numerator. Conversely, the numerator is greater than the denominator for the brightest signatures. Where the denominator and the numerator are the same, there is no difference between the two bands.

A. OLI ratio image 4/2 (red/blue).

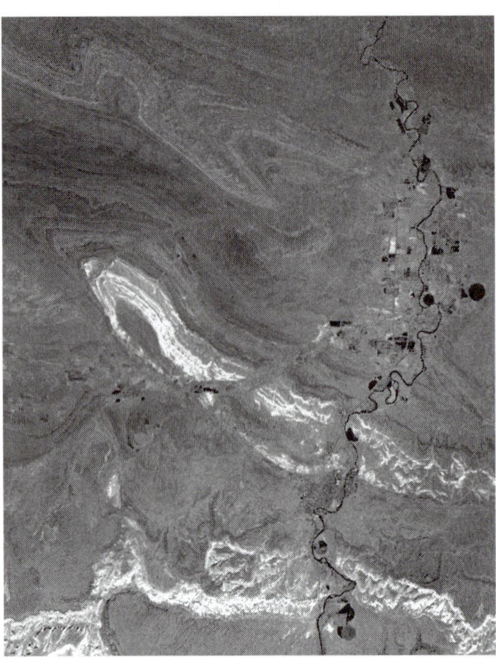

B. TM ratio image 3/1 (red/blue).

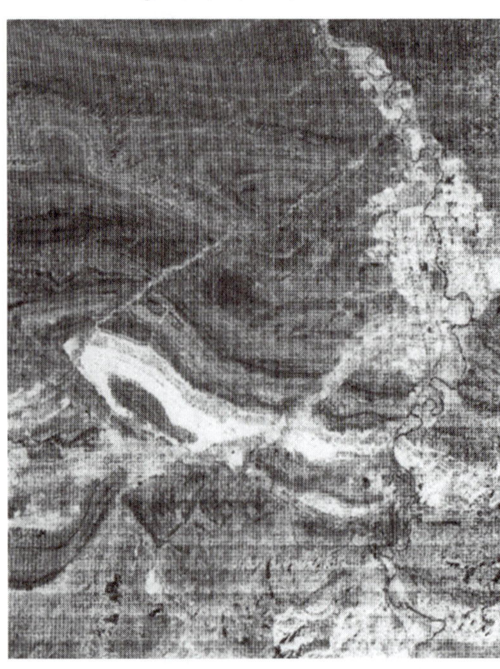

C. OLI ratio image 6/7 (SWIR1/SWIR2).

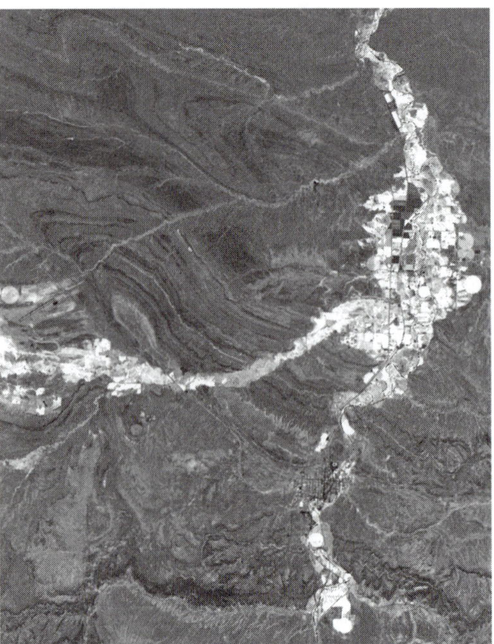

D. TM band ratio 5/7 (SWIR1/SWIR2).

FIGURE 9-25 Landsat OLI band ratios compared to Landsat TM band ratios.

For example, the red beds of the Chugwater outcrops have a high reflectance in the red band (OLI band 4 and TM band 3) and a relatively low reflectance in the blue band (OLI band 2 and TM band 1). A ratio image of OLI 4/2 or TM 3/1 (red/blue) displays the red beds with very bright signatures (Figures 9-25A and 9-25B). The ratio image of OLI 6/7 and TM 5/7 of the Thermopolis area highlights the clay-rich soils associated with irrigated agriculture (Figures 9-25C and 9-25D). Clay minerals preferentially absorb SWIR2 wavelengths compared to SWIR1 light, therefore, the SWIR2 band is darker compared to the SWIR1 band when the pixels contain clay-rich soils. The noise difference between band ratios generated from Landsat OLI and the older Landsat TM data is very significant.

An advantage of ratio images is that they extract and emphasize differences in spectral reflectance of materials. A disadvantage of ratio images is that they suppress differences in albedo; materials that have different albedos but similar spectral properties may be indistinguishable in ratio images. Another disadvantage is that any noise is emphasized in ratio images. Image DNs must be converted to reflectance values prior to the ratioing.

Ratio images also minimize differences in illumination conditions, thus suppressing the expression of topography. In Figure 9-26 a red siltstone bed crops out on both the sunlit and shadowed sides of a ridge. In the individual Landsat TM bands 1 and 3, the DNs of the siltstone are lower in the shadowed area than in the sunlit outcrop, which makes it difficult to follow the siltstone bed around the hill. Values of the ratio image 3/1, however, are identical in the shadowed and sunlit areas; thus, the siltstone has similar signatures throughout the ratio image. Highlights and shadows are notably lacking in the ratio images of Figure 9-25.

TIR emissivity bands are also ratioed to highlight features of interest. ASTER TIR emissivity bands 12, 13, and 14 are ratioed for geologic mapping as follows (Mars and Rowan, 2011):

- 14/12 for quartz-rich rocks,
- 13/14 for carbonate-rich rocks, and
- 12+14/13 for mafic-rich rocks.

ASTER TIR bands are processed as band ratios to improve geologic maps (see example in Chapter 13 on Afghanistan).

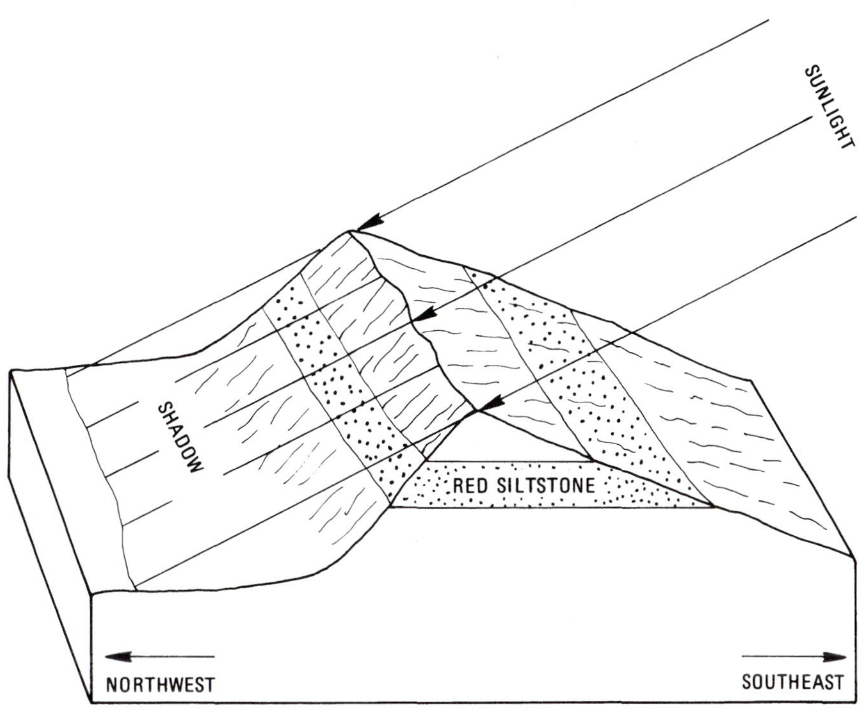

SILTSTONE REFLECTANCE

ILLUMINATION	TM BAND 3	TM BAND 1	RATIO 3/1
Sunlight	94	42	2.24
Shadow	76	34	2.23

Figure 9-26 Suppression of illumination differences on a ratio image.

In addition to ratios of individual bands, a number of other ratios are computed for specific applications. An individual band may be divided by the average for all the bands, resulting in a normalized ratio image. Another ratio combination is produced by dividing the difference between two bands by their sum; for example, (band 4 – band 5) / (band 4 + band 5).

Normalized Difference Vegetation Index

The most widely used band ratio is the *Normalized Difference Vegetation Index* (NDVI):

$$\text{NDVI} = \text{NIR} - \text{red} / \text{NIR} + \text{red} \qquad \textbf{(9-8)}$$

Vegetation indices are dimensionless, radiometric measures that indicate the relative abundance and activity of green vegetation, including percentage green cover, chlorophyll content, and green biomass (Jensen, 2016). The value of the NDVI index ranges from –1 to +1. The common range for green vegetation is 0.2 to 0.8. NDVI is often interpreted in terms of vegetation vigor or stress.

Figure 9-27A is the Landsat 8 NDVI image for September 21, 2015 of the semiarid Thermopolis, Wyoming, area. The brightest pixels (highest NDVI values) are in the irrigated agricultural areas along the Wind River and its tributaries. The NDVI image is similar to the clay band ratio image (Figure 9-25C) confirming the clay-rich nature of the irrigated soils.

MODIS generates vegetation indices at 250 and 500 m spatial resolution. There are many remote sensing vegetation indices. Several are sensor specific and others are designed for narrow bandwidth hyperspectral data cubes (Jensen, 2016; Vincent, 1997).

Masking

Vegetation can detract from image processing and interpretation when the focus is water, rocks, man-made features, etc. An NDVI image is an excellent tool for highlighting pixels covered with some percentage of vegetation. The remote sensing analyst can adjust a threshold on the NDVI grayscale image using contrast enhancement or density slice tools to reduce or expand the number of pixels designated as vegetation covered. A binary image (black and white with only 0 and 1 DN values) is generated where black pixels are interpreted to be vegetated to some degree and white pixels are without vegetation (based on the analyst's threshold). This binary image is converted to a mask (Figure 9-27B). This mask can eliminate pixels with vegetation from the processing (Figures 9-27C and 9-27D), or be reversed and enable processing to involve only pixels that are vegetated. Contrast enhancement of the OLI ratio image 6/7 with the vegetation mask (Figure 9-27D) provides more visual information about the nonvegetated terrain compared with the same ratio image without a vegetation mask (Figure 9-25C). The PC method and spectral classification benefit significantly when irrelevant pixels are masked and eliminated from the analysis.

Density Slice

Density slices of band ratio images target areas of interest for fieldwork and more advanced imaging. The OLI band ratio 4/2 shows pixels with red beds as brighter shades of gray (Figure 9-25A). Red beds can be indicative of iron-rich rocks that are of interest for mineral exploration and field sampling. A three-color density slice of the brightest pixels in the ratio image visually highlights rocks with the reddest color and significantly reduces the geographic area that needs to be covered for finding potential iron-rich rock samples in the field (Plate 28).

MULTISPECTRAL CLASSIFICATION

Multispectral classification is an information extraction process that analyzes spectral signatures and then assigns pixels to classes based on similar signatures. The process involves statistics and decisions by the remote sensing analyst for the following:

- Number and type of classes to be mapped,
- Use of supervised or unsupervised techniques,
- Hard or soft (fuzzy) classes,
- Segmentation of the scene,
- Inclusion of object-based classification, and
- Accuracy assessment.

Supervised and unsupervised techniques are major approaches to multispectral classification:

1. **Supervised classification**: The analyst defines on the image areas, called *training sites*, which are representative of each class. Spectral values for each pixel in the training sites are used to define the decision space for that class. The classification algorithm uses the statistics within each decision space to assign the remaining pixels in the scene as in or out of the defined classes.
2. **Unsupervised classification**: The computer program clusters the pixels into natural groupings based on the spectral characteristics of the pixels with no direction from the analyst, except for setting basic parameters such as number of classes, number of iterations performed by the program, and minimum number of pixels per class.

A raster thematic map is generated by classification algorithms where each pixel is assigned a value correlating to one of the classes in the map. Fuzzy classes, segmentation, and object-based classification techniques are not covered in this textbook. Jensen (2016) and Lillesand and others (2015) provide more information on various multispectral classification approaches. Accuracy assessment is reviewed with an example in Chapter 14.

Classification Basics

Landsat OLI data acquired on September 23, 2014 of Martinez, California, is used as an example for multispectral classification (Figure 9-28). NAIP aerial color and color IR orthoimages of the east-central portion of the classified area are shown in Plate 5. The area is characterized by water, wetlands, industry, suburbs, and widespread dry grass in nonirrigated terrain. The Sacramento River flows across the upper margin of the area. The size of the Landsat scene is 350 by 600 for 210,000 pixels and covers 10.5 by 18 km with 30 m spatial resolution.

The six VNIR-SWIR bands were chosen for both supervised and unsupervised classifications because of their relatively low correlation (except for the blue and green bands) and visual inspection of two-band feature space plots, which indicate that the bands have excellent potential for informative classification mapping. The PC transformation was performed on the six bands and PC images 3 and 4 were

A. NDVI ratio image.

B. Vegetation mask created from NDVI ratio image.

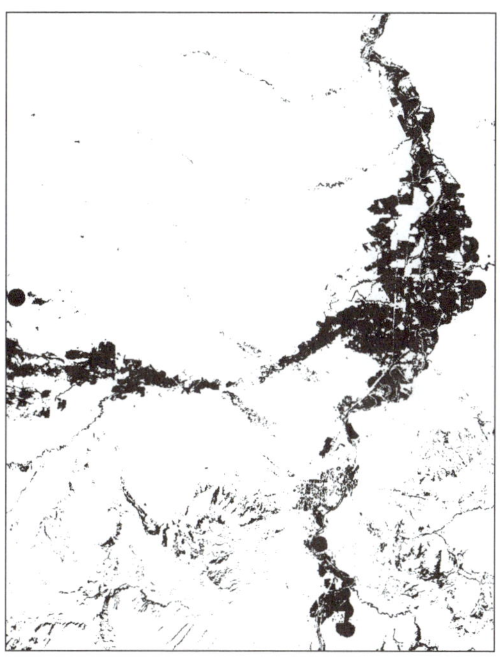

C. OLI ratio image 4/2 (iron) with NDVI mask.

D. OLI ratio image 6/7 (clay) with NDVI mask.

Figure 9-27 Landsat ratio images with a vegetation mask generated from NDVI. Landsat courtesy USGS.

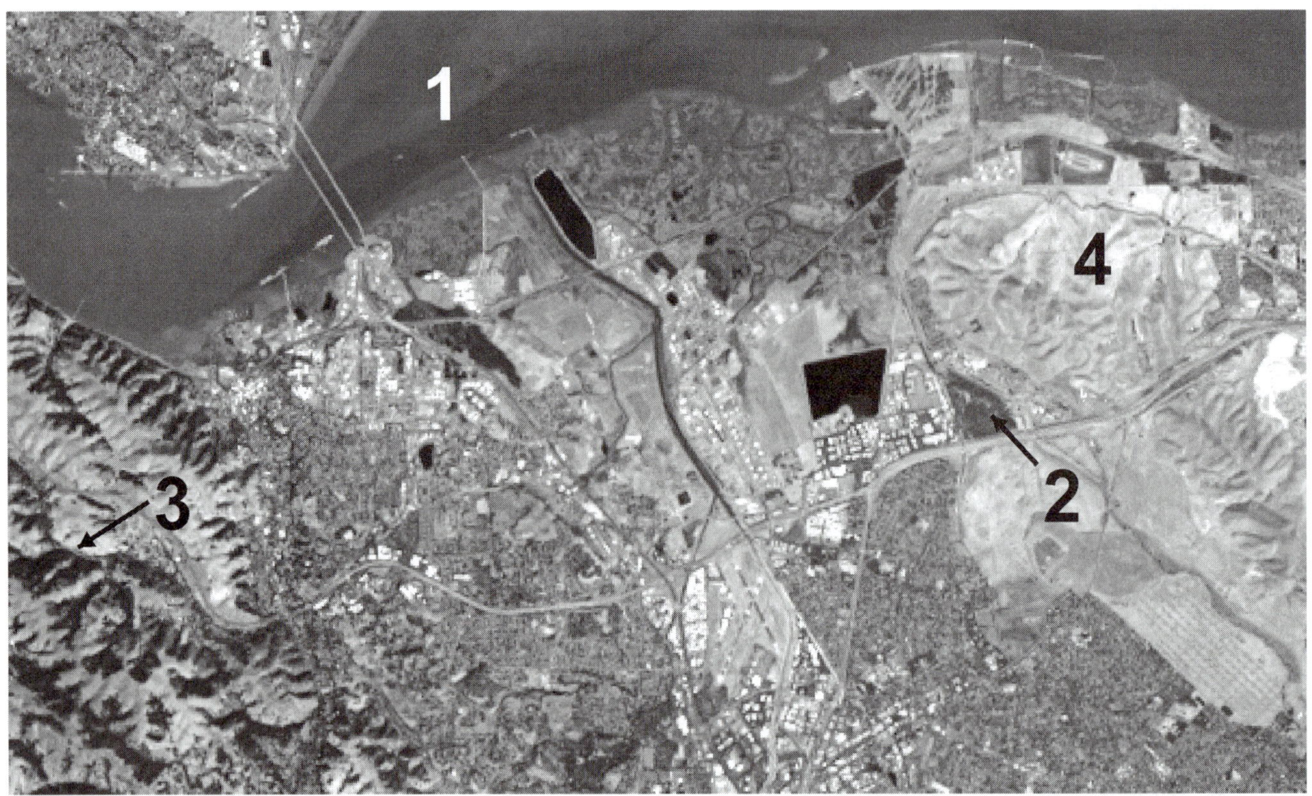

FIGURE 9-28 Landsat OLI band 5 (NIR), Martinez, California. 1 = water, 2 = irrigated vegetation, 3 = trees, and 4 = dry grass. Landsat courtesy USGS.

selected to include in both supervised and unsupervised classifications because these two PC images highlight the industrial hardscape, suppress topographic shadowing in the hills, and have very low correlations with the Landsat bands (Table 9-5). The six bands and two PC images used in the classifications are shown in Figure 9-29.

For each pixel in the Landsat OLI image, the spectral brightness is recorded for six spectral bands in the visible and reflected IR regions. A pixel may be characterized by its *spectral signature*, which is determined by the relative DN brightness in the different wavelength bands (Figure 1-20). Spectral signatures of a single pixel that is representative of water, irrigated vegetation, trees, and dry grass are shown in Figure 9-30. The vertical axis of the spectral signature plot represents the DN brightness values of Landsat OLI's 16-bit bands (ranging between approximately 5,000 to 20,000) while the horizontal axis displays the wavelength of the six OLI bands. The DN brightness values of the spectral signatures shown in Figure 9-30 are listed in Table 9-6. The DN brightness values from this table for any three bands can be plotted in a three-dimensional coordinate system (Figure 9-31). The position of the four dots on Figure 9-31 is based on the DN values of single pixels representing the land cover classes. However, conceptually these dots can also represent

TABLE 9-5 Correlation matrix of Landsat bands and PC images 3 and 4.

	Band	Band 2 (Blue)	Band 3 (Green)	Band 4 (Red)	Band 5 (NIR)	Band 6 (SWIR1)	Band 7 (SWIR2)	PC 3	PC 4
Blue	2	1.0000							
Green	3	0.9677	1.0000						
Red	4	0.8736	0.9415	1.0000					
NIR	5	0.3515	0.4338	0.5844	1.0000				
SWIR1	6	0.4542	0.5572	0.7757	0.8086	1.0000			
SWIR2	7	0.5398	0.6229	0.8161	0.7339	0.9734	1.0000		
PC 3		0.0909	−0.0166	−0.0764	0.0145	−0.0532	0.1240	1.0000	
PC 4		−0.6615	−0.6139	−0.3824	−0.1622	0.1490	0.1209	0.0001	1.0000

A. OLI band 2 (blue).

B. OLI band 3 (green).

C. OLI band 4 (red).

D. OLI band 5 (NIR).

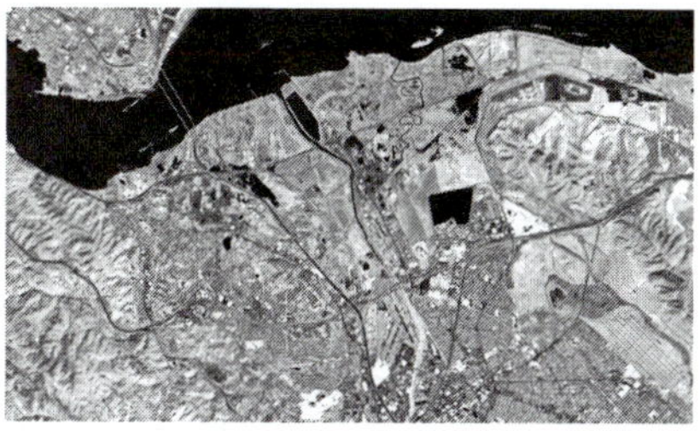

E. OLI band 6 (SWIR1).

F. OLI band 7 (SWIR2).

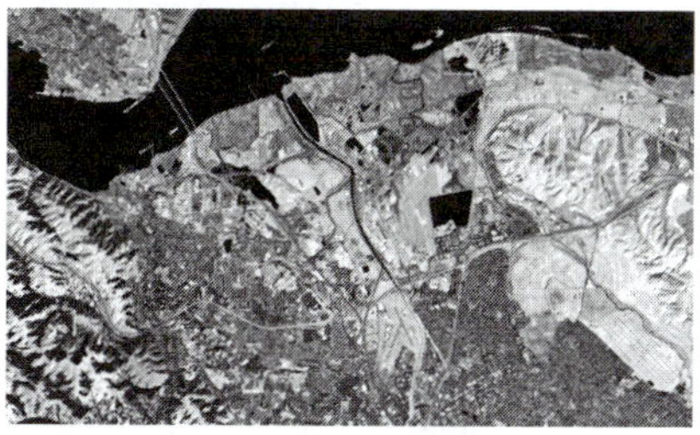

G. PC image 3.

H. PC image 4.

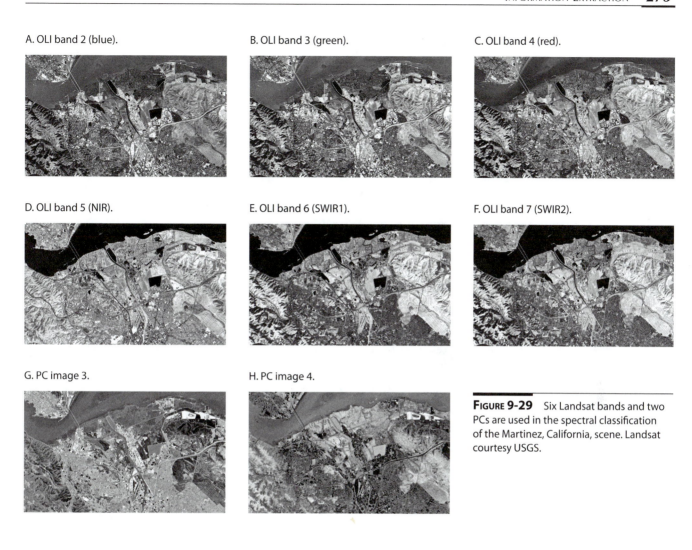

Figure 9-29 Six Landsat bands and two PCs are used in the spectral classification of the Martinez, California, scene. Landsat courtesy USGS.

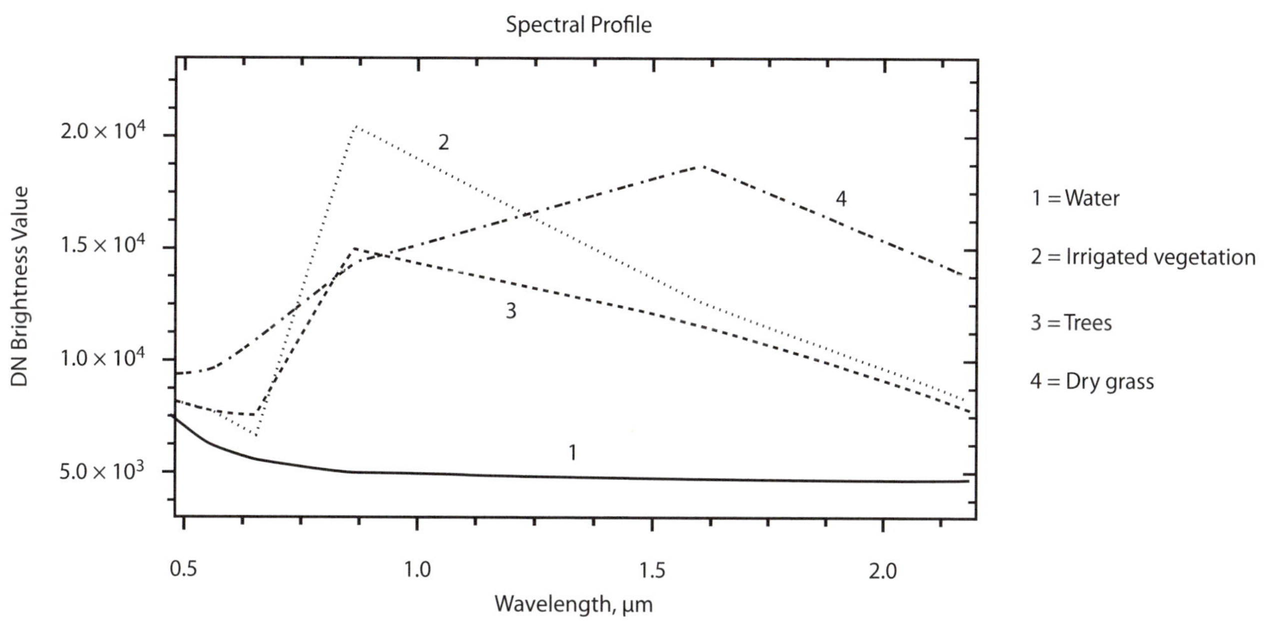

Figure 9-30 Spectral signatures of four land cover classes.

TABLE 9-6 DN brightness values of spectral signatures.

	OLI Band	(1) Water	(2) Irrigated Vegetation	(3) Trees	(4) Dry Grass
Blue	2	8,989	8,534	7,961	9,864
Green	3	8,501	8,158	7,080	9,758
Red	4	7,477	7,218	6,373	10,862
NIR	5	5,550	19,042	11,281	14,706
SWIR1	6	5,108	12,081	7,586	19,507
SWIR2	7	5,074	8,283	6,076	14,507

the mean DN value of pixels assigned to each land cover class while developing training sites for supervised classification or the mean DN value of pixels within natural clusters of pixels that are generated during unsupervised classification.

In Figure 9-31 the brightness ranges of the red, NIR, and SWIR1 bands (OLI bands 4-5-6) form the axes of the three-dimensional coordinate system. The solid dots are the loci of the four landscape types represented by the spectral signature of one pixel (Table 9-6). Plotting additional pixels from the four landscape types would produce irregularly shaped, three-dimensional clusters around these dots. These clusters can overlap, especially if the multispectral bands are highly correlated, the landscape types are mixed within pixels, or the landscape types are not spectrally well differentiated with multispectral data. Overlap between clusters creates spectral confusion and reduces the accuracy of the classification map.

Spectral classification programs assign pixels from the clusters into a class based on statistical analyses of each band's mean, standard deviation, variance, and range of DN values. For the sake of simplicity, only three bands are considered to be involved in the discussion below, as shown by the three axes in Figure 9-32. In actual practice, the programs employ a separate axis for each band involved in the classification.

Figure 9-32 is a simplistic visualization of spectral classification. The classification of pixels in three bands forms a *decision boundary* in the shape of an ellipsoid, which encloses the statistical criteria that characterizes each of the four classes. The mean DN value of pixels within the ellipsoid is shown as a dot in Figure 9-32. The volume inside the decision boundary is called the *decision space*. Classification programs differ in their criteria for defining decision boundaries. In many programs the analyst is able to modify the boundaries to achieve optimum results.

Once the boundaries for each *spectral class* are defined, the computer retrieves the spectral value for each pixel in the scene and determines its position in the classification space. Should the pixel fall within one of the ellipsoids, it is classified accordingly. Pixels that do not fall within an ellipsoid are considered unclassified. In practice, the computer calculates the distance or mathematical probability that a pixel belongs to a class; if the probability exceeds a designated threshold (represented spatially by the decision boundary) or is within the specified distance, the pixel is assigned to that class. Hard classes are defined as those where pixels are assigned either in or out of a class. The examples of supervised and unsupervised land cover classifications that follow use hard classes.

In the most basic terms, with the supervised classification method the remote sensing analyst manually determines the land cover feature(s) included in classes and the number of classes. In contrast, the unsupervised classification method

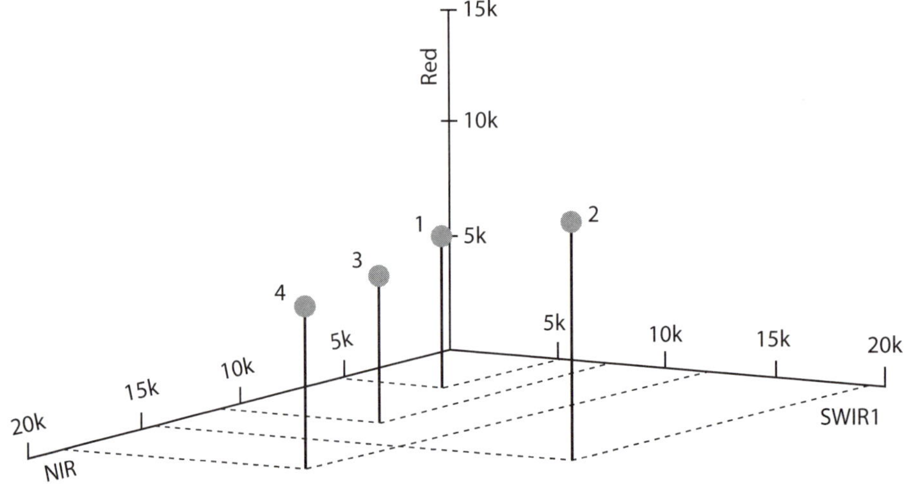

FIGURE 9-31 Three-dimensional coordinate system for red, NIR, and SWIR1 Landsat OLI bands with each solid dot representing the DN brightness values for a pixel in land cover classes 1, 2, 3, and 4 (Table 9-6).

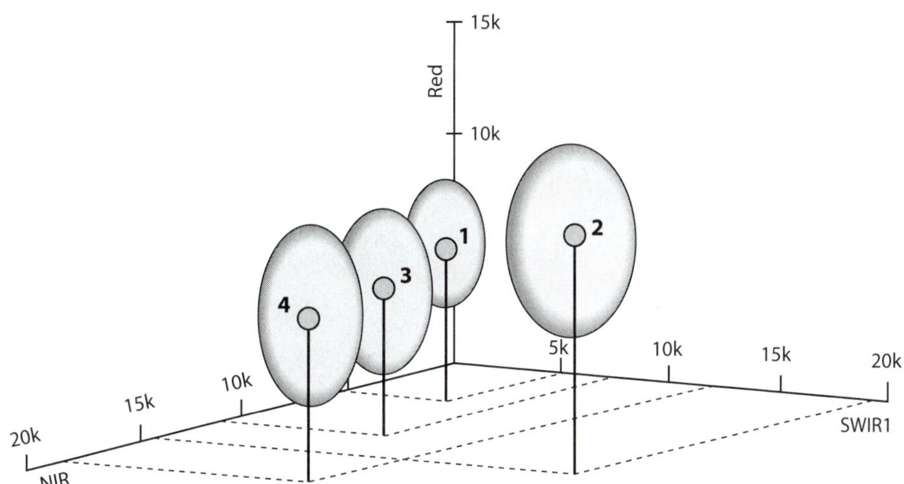

Figure 9-32 Schematic of pixels in the Martinez, California, Landsat image that are spectrally classified into one of the four land cover classes based on the statistical characteristics of red, NIR, and SWIR1 bands.

has the computer program determine what feature(s) are included in classes after the analyst establishes the number of classes. In order to compare supervised and unsupervised classification maps that used the same data, the remote sensing analyst must identify the features in the classes and assign similar names to classes that contain similar features. Classes with similar names in the unsupervised and supervised maps will not contain the exact same features or cover the same geographic area because the methods used to develop them are different (as demonstrated below). Consolidating classes into more generalized land cover groups (such as water, vegetation, developed, and bare soil) usually minimizes differences between land cover maps developed by the two methods.

Supervised Classification

The first steps in supervised classification are to define the land cover classes and then to select training sites for each class. Polygons are digitized by the remote sensing analyst around pixels representative of each land cover class. Training sites for each class can be represented by more than one polygon so that the variability within each class is available for the statistical analysis. With advanced software, training sites can also be created with two-band feature space plots (Figure 9-4) where pixels identified on the plot are highlighted on the image and saved as training sites.

For our analysis, we will use the Martinez, California, Landsat image. The supervised classification uses the maximum likelihood technique, where each pixel is assigned to a class that has the highest probability. In this example, the probability threshold was set so all 210,000 pixels in the image are classified into one of 17 classes (Table 9-7). Figure 9-33 illustrates the locations of the training sites. For class 17 (dry grass: dark shadows), 12 training site polygons were defined to capture the varying degrees of shadowing and dry grass orientation. Four polygons in two industrial ponds and two reservoirs characterize class 1 (water: ponded) while seven polygons are digitized on the Sacramento River to include water with varying amounts of sediment load (class 2). Acreage determination for the different classes is an important contribution of land cover maps (Table 9-7). Names of classes can be changed and classes combined after examination of the map.

The accuracy of the map can be visually estimated by comparing known features on the imagery with the land cover designation on the map (Plate 29). The supervised classification raster thematic map is converted to a vector polygon map (process discussed in Chapter 10) that distributes the 17 classes into 19,126 polygons (Plate 29B). River water, ponded water, and irrigated vegetation are accurately mapped. Wetlands are well mapped in the lowlands near the river, however, wetland classes are also found in the hills to the southwest where the spectral character of shadowed dry grass is confused with the wetland spectra. Segmenting the image based on elevation so the wetland class is confined to low elevation terrain would eliminate this error. A formal accuracy assessment (discussed in Chapter 14) should be completed based on field visits, reliable maps, verification sites, and/or examining high resolution satellite imagery (Congalton, 1991).

Unsupervised Classification

The isodata algorithm was set up to delineate a maximum of 25 clusters in 20 or less iterations. The 25 clusters were visually evaluated and combined into 12 land cover classes (Plate 29C). The names of the classes were aligned with the names used in the supervised classification so that the two maps can be compared. Table 9-8 lists the acreage of classes in the unsupervised map. The unsupervised land cover map has 19,228 polygons. Dry grass, irrigated vegetation, and water are mapped reasonably well based on visual inspection (compare Plates 29A and 29C). The acreages for dry grasses, wetlands, irrigated vegetation, and white roofs/concrete are relatively similar (within ~10%) between the supervised and unsupervised maps, however, the supervised map has more industry (22% vs 16%) and the unsupervised map has more suburbs, irrigated vegetation, and trees (43% vs 54%) (Tables 9-7 and 9-8).

Increasing the number of classes in the unsupervised classification may enable separation of water into two classes, as was done manually with the training sites in the supervised classification map. The suburb class in the unsupervised map has many errors as it is located in the wetlands and in the hills to the southwest. Spectral confusion between the suburb cluster and other classes containing nonirrigated vegetation (wetlands and trees) is apparent. Both unsupervised and supervised maps place wetlands in shadowed dry grass terrain with more extensive error in the unsupervised map. The white roofs and concrete class is visually more coherent in the unsupervised map compared to the supervised map, where shadows bring in the suburb class.

Generalizing a Classification

Both the supervised and unsupervised land cover maps have a "salt and pepper" or "speckled" appearance as there are single pixels and narrow strings of pixels assigned to one class that are surrounded by many pixels that are assigned into one or two other land cover classes. Land cover maps are *generalized* using a majority filter. The filter is applied to the thematic raster map and replaces the class designa-

TABLE 9-7 Land cover classes and acreage for supervised classification analysis.

Class	Class Name	Acreage (km^2)
1	Water: ponded	1.95
2	Water: river	26.85
3	Wetlands	11.44
4	Dry grass: light	16.50
5	Dry grass 1: dark	10.52
6	Suburbs	31.74
7	Trees	9.31
8	Grass: not irrigated	11.30
9	Irrigated vegetation	1.85
10	Dark soils: wet?	1.11
11	Industrial 2	7.29
12	White roofs: concrete	6.79
13	Industry 1	14.77
14	Shrubs-grass: not irrigated	23.57
15	Construction site: exposed white soil	0.43
16	Dry grass 2: dark	3.40
17	Dry grass: dark shadows	10.19

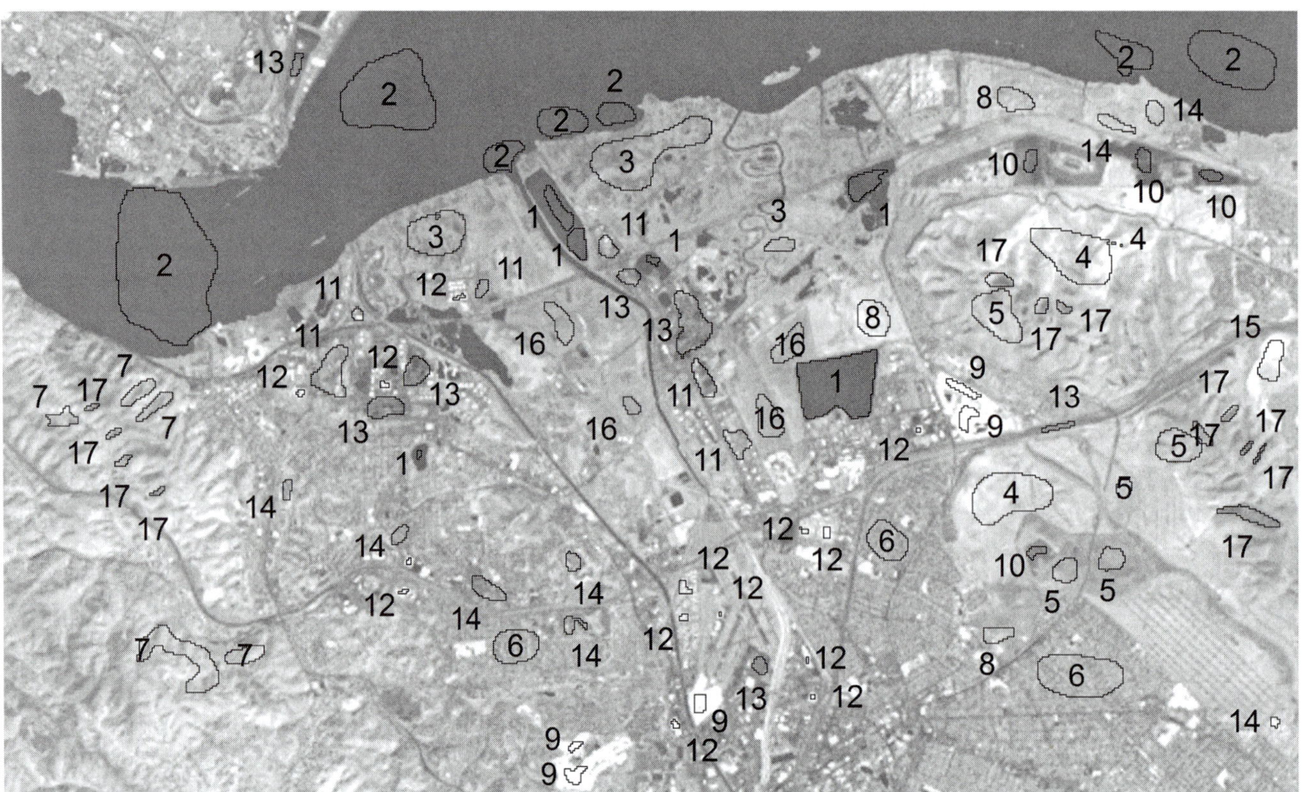

FIGURE 9-33 Training sites for 17 land cover types (Table 9-7).

tion of single pixels and narrow strings of pixels with the class identity of the majority of pixels that surround these isolated pixels.

For example, Figure 9-34 shows an example of a 3 by 3 majority filter being applied to 9 pixels in a land cover classification image. Three land cover classes are shown: 3, 5, and 7. Class 3 is water, class 5 is grass, and class 7 is bare soil. In the original classification image, water (3) is isolated in the center pixel, grass (5) is assigned to 5 pixels, and bare soil (7) is assigned to 3 pixels (Figure 9-34A). The majority filter is applied. Because grass is the majority land cover class in the 3 by 3 filter, grass replaces water in the center pixel (Figure 9-34B). The filter moves one column to the right and the process is repeated. The filter is applied to the rows and columns of the thematic raster classification image. The 3 by 3 majority filter was applied to the supervised thematic raster map (Plate 29B), resulting in a less detailed, smoother-appearing land cover map (Plate 30), especially within the wetlands, suburbs, and industry classes.

The original classification map (Plate 29B) has 19,126 polygons while the majority-filtered classification map (Plate 30) has only 6,225 polygons. Computer performance during the raster to vector conversion is significantly improved when the raster classification map has first undergone generalization with a majority filter.

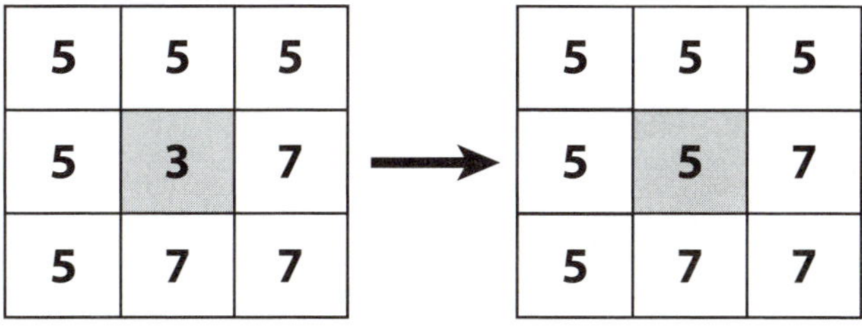
A. Isolated class 3 pixel surrounded by five pixels of class 5 and three pixels of class 7.
B. Majority filter applied and class 3 is changed to majority class 5.
5 5 5
5 3 7
5 7 7
5 5 5
5 5 7
5 7 7

Figure 9-34 Three land cover classes (3, 5, and 7) with majority filter (3 by 3) used to reduce speckled appearance of classification images.

Change-Detection Images

Change-detection images provide information about seasonal, land cover, or other changes. The information is extracted by comparing two or more images of an area that were acquired at different times. The images must be coregistered and have the same pixel size. The DNs of one image are subtracted from those of an image acquired during a different time period. The resulting values for each pixel are positive, negative, or zero; the latter indicates no change. The next step is to plot these values as an image in which a neutral gray tone represents no change. Black and white tones represent the maximum negative and positive differences, respectively. Contrast stretching is employed to emphasize the differences.

The change-detection process is illustrated here with Landsat TM images from the late summer of 1986 and 2010 of Brentwood, California, a fast growing suburban community east of Oakland (Figures 9-35A and 9-35B). TM band 3 (red) was selected for change detection. The DN in each pixel of the 1986 image is subtracted from the DN of the corresponding registered pixel in the 2010 image. The resulting values are linearly stretched and displayed as the change-detection, or difference, image (Figure 9-35C). The location map (Figure 9-35D) aids in understanding signatures in the difference image. Neutral gray tones representing areas of little change are concentrated in the dry, hilly grasslands in the southwest portion of the image. The darkest pixels are located in the center of the image where suburban sprawl moves from the west toward the east–southeast between 1986 and 2010. In addition, tidal changes in the northeast corner, where the Sacramento River is shallow, cause darker pixels in the 2010 image compared to the 1986 image, resulting in a negative change in the difference image. A positive increase in brightness occurs in the southeast portion due to loss of irrigated farmland (dark pixels in the 1986 image) and clearing of grass-covered land for developments between 1986 and 2010. About twice as many pixels decrease in brightness compared with those that increase in brightness between 1986 and 2010 (Figure 9-35E). Change-detection processing is also used to produce difference images for other remote sensing data, such as between nighttime and daytime TIR images (Chapter 5).

Table 9-8 **Acreage of classes in unsupervised classification map.**

Class	Class Name	Acreage (km^2)
1	Dry grass	30.38
2	Dry grass & exposed soil?	5.82
3	Grass: not irrigated	24.60
4	Industry 1	6.41
5	Industry 2	9.46
6	Irrigated vegetation	1.95
7	Shrubs-grass: wet?	7.45
8	Suburbs	38.32
9	Trees	13.74
10	Water	32.97
11	Wetlands	10.79
12	White roofs and concrete	7.11

Hyperspectral Processing

Hyperspectal data requires advanced image processing software that can preprocess, reduce the number of bands, and manage spectral libraries (Figure 9-36). Between approximately 10 to over 400 overlapping, contiguous bands comprise a hyperspectral data cube (Plate 2). For applications using spectral libraries the data must be in reflectance.

A. 1986 Landsat TM band 3.

B. 2010 Landsat TM band 3.

C. Brightness changes for band 3 (red) from 1986 to 2010.

D. USGS topographic map of the Brentwood, California, area.

E. Brightness change chart.

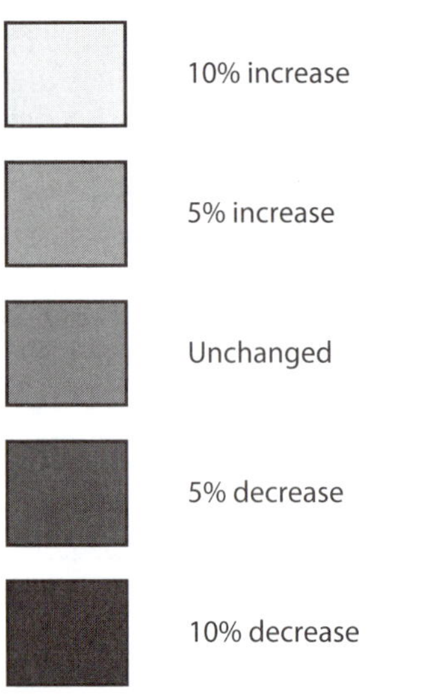

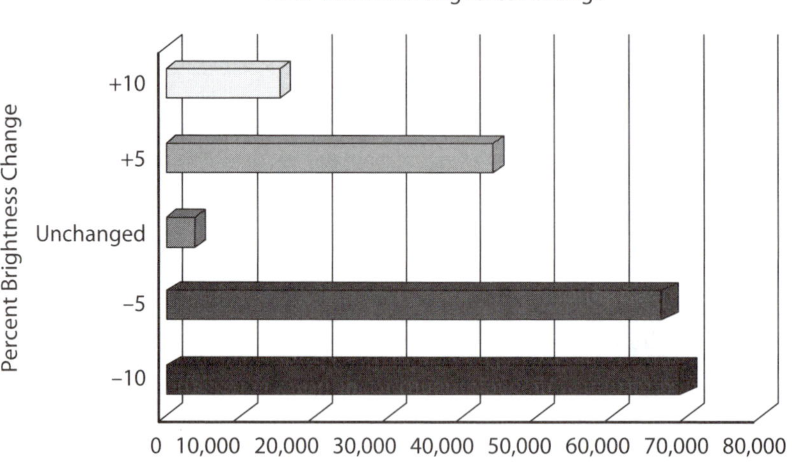

FIGURE 9-35 Change detection using Landsat images from 1986 and 2010, Brentwood, California. Landsat courtesy USGS.

Masking pixels with features of no interest will improve the image enhancement and information extraction. Kruse and others (1993) developed a knowledge base to support an expert system that automatically identified Earth surface materials based on their hyperspectral characteristics.

Advanced image processing software enables the user to display spectral signatures where absolute increases and decreases in reflectance (or brightness) as measured at the sensor are shown on the vertical *y*-axis and wavelength along the horizontal *x*-axis (Figures 1-20, 3-2, and 4-3). This format is excellent for showing increased chlorophyll absorption of blue and red light compared with green light and the red edge between red and NIR wavelengths (Figure 9-37A). The *continuum removal* procedure normalizes reflectance spectra (the value varies between 0 and 1) so you can compare individual absorption features from a common baseline (ENVI, 2017). Figure 9-37B shows the chlorophyll absorption feature (seen as the red edge in Figure 9-37A) generated by the continuum removal procedure. The vertical axis in this chart varies from ~0.2 to 1.0 reflectance units. The depth of the absorption feature is almost 0.7 reflectance units, as measured by the vertical double-ended arrow on Figure 9-37B. This quantitative measure of a feature's spectral absorption depth enables more detailed interpretation and documentation of the feature of interest.

The wavelength interval that contains the absorption features of interest is subset from the data's overall spectral signature during processing to achieve optimum results. For example, a VNIR-SWIR spectral signature would be subset over the SWIR 2.1 to 2.4 µm interval for examining the shape and measuring the depth of clay absorption features with the continuum removal process.

Different types of maps are generated with hyperspectral data, including anomaly detection, target detection, and material mapping (distribution and abundance) (ENVI, 2017; Hexagon, 2019) (Figure 9-36). Mixed pixels are an issue that can be addressed with subpixel algorithms that attempt to calculate the abundance of a material in a pixel. For example, a subpixel algorithm provides abundance maps that can be examined over any given pixel or groups of pixels to indicate the relative abundance of different materials within that pixel (Boardman and Kruse, 2011). Preparing the derived maps for GIS and a general audience is key to expanding awareness of the power of hyperspectral remote sensing for environmental, renewable resources, and infrastructure applications.

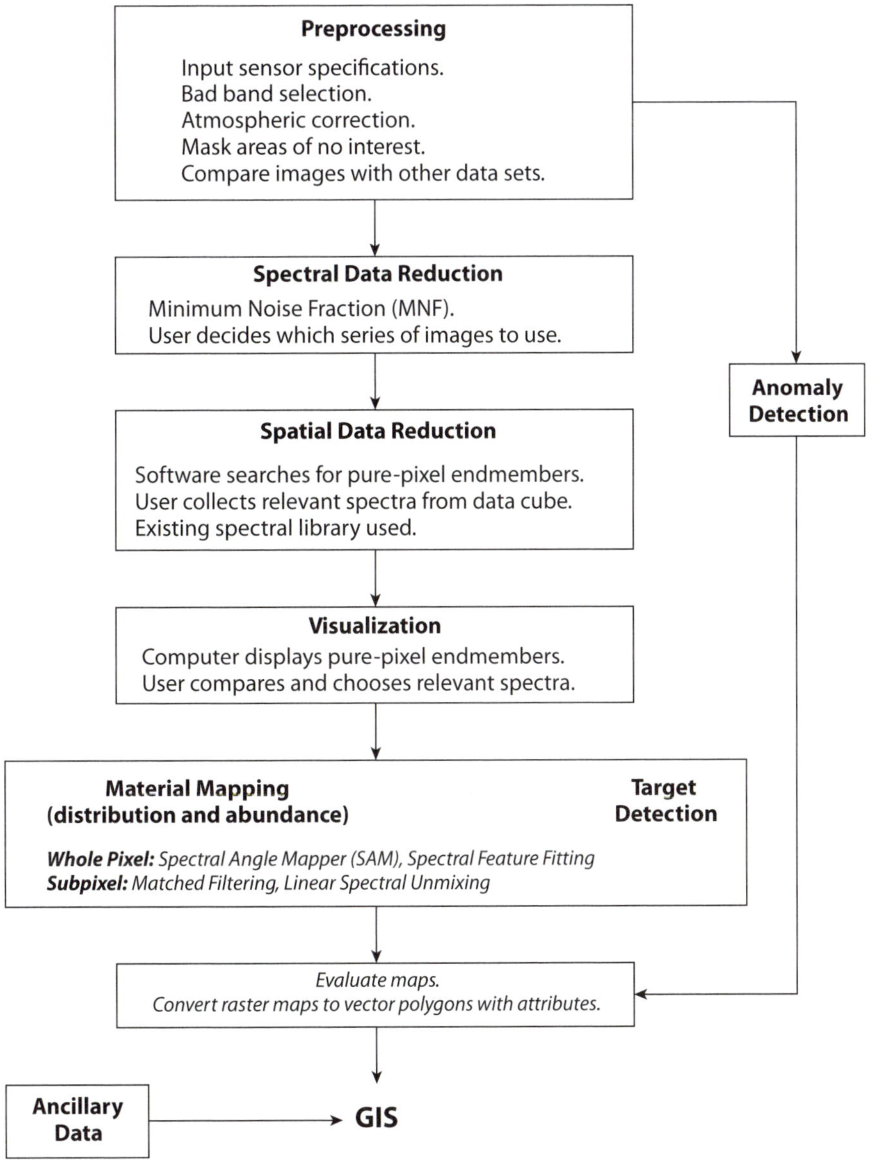

IMAGE PROCESSING

Image processing continues to rapidly evolve as new hardware and software, along with increased Internet connectivity and capacity, becomes available. Users have options for managing and processing remote sensing data that were not available a few years ago. Desktop image processing is still the standard mode of operation, with software loaded onto a workstation, data downloaded from multiple sources via the Internet, and tools for programming to speed up repetitive processing or to create new processing algorithms (Figure 9-38).

Figure 9-36 Hyperspectral workflow.

However, cloud-based services offer online subscriptions and turn-key solutions that reduce the need for remote sensing users to have their own desktop image processing software. Drone operators who are not able or inclined to learn remote sensing and manage their own software can upload acquired imagery with metadata to cloud-based services that will process and send back via the Internet image mosaics and maps. For example, the agricultural sector has several cloud-based solutions for drone operators. The main question for users is how to ensure the quality of the returned results given that all the processing decisions are no longer in their control.

Google Earth Engine is an open source, cloud computing platform designed to store and process huge data sets (at petabyte scale) for interactive analysis and decision making. Google archived all the Landsat data, some other satellite data, and select GIS layers (including social, demographic, weather, and DEMs). Scientists, students, and hobbyists can use an easily accessible and user-friendly front end to mine this massive warehouse of data for detecting change, mapping trends, and inventorying resources on the Earth's surface. Kumar and Mutanga (2019) provide an overview and 22 case histories that include vegetation monitoring, cropland mapping, and ecosystem assessment.

DESKTOP HARDWARE

Remote sensing desktops do not have to be high-end gaming computers, but they do need fast components to enable timely processing of very large remote sensing files. A brief list for an appropriate desktop includes:

- 64-bit operating system,
- > 24 GB random access memory (RAM) to support fast processing,
- > 3 GHz multi-core central processing unit (CPU),
- L2 or L3 cache to speed up transfer of data between the processor, memory, and components in the computer,
- Fast front side bus on the motherboard (400 MHz),
- Solid state drives (SSD),
- A video card with its own CPU and memory along with a fast interface,
- USB ports,
- Dual monitors (one for the imagery and the other for the menus),
- Large internal drives (1 TB), and
- Fast Internet connection.

A. Normal spectral signature with the red edge (indicated by the arrow) showing marked increase in reflectance values between the visible red and NIR wavelengths.

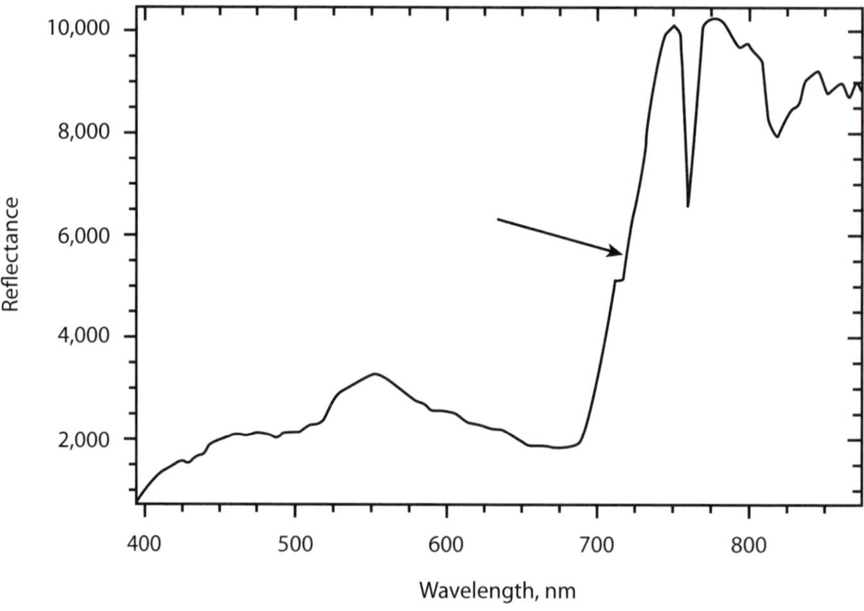

B. Continuum removal applied to the spectral signature showing the depth of the chlorophyll absorption feature (the depth is measured along the length of the arrow).

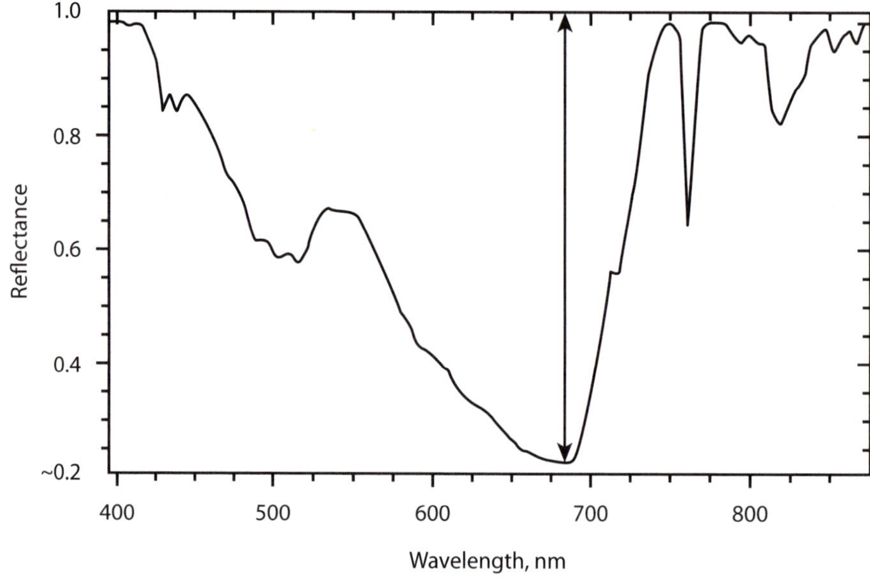

FIGURE 9-37 Comparison of two display methods for the same spectral signature of healthy vegetation. Data courtesy Galileo Goup, Inc. and AISA.

Desktop Software

Software is available from commercial and open sources. Trial versions can usually be downloaded. Many of the software packages are very large, exceeding a gigabyte in size.

Commercial

Advanced image processing software with many modules for specific applications and data types are available from companies. The advantage of commercial software in the rapidly evolving field of remote sensing is the support provided by the companies to input new satellite data formats, fix bugs, embrace new technology (e.g., drones, lidar, high resolution satellite imagery), and offer extensive help manuals, tutorials, and sample data sets. Examples of some of the available commercial software packages include:

- ArcGIS Spatial Analyst extension
- eCognition
- ENVI
- ERDAS Imagine
- ER Mapper
- PCI Geomatica
- TerrSet
- TNTmips

Searching the Internet for software descriptions and downloading trial versions of the software is recommended before purchasing the software.

Open Source

Open source software is built in an open, collaborative manner and is available to the public. The recommended path to finding appropriate, open source software is to use the Internet and read the descriptions for the different packages, choose a few, and download them to see if they will work for you. Support for open source is dependent on voluntary contributions or upgrades to more commercial versions, so users may have to have some patience and programming skills to make the software work on their computer.

GISGeography (gisgeography.com) annually publishes a list of open source remote sensing software packages. Originally developed by the European Space Agency, LEOWorks is a solid, image processing package supported by Terrasigna (leoworks.terrasigna.com). MultiSpec is an advanced, research-oriented software package from the Purdue Research Foundation (engineering.purdue.edu/~biehl/MultiSpec). It will process multispectral and hyperspectral data. Google Earth Engine (earthengine.google.com/platform) is a computing platform that allows users to run geospatial analysis on Google's infrastructure.

Project Workflow

A successful remote sensing project workflow starts with taking time to understand the project objectives and then finding geospatial data that will achieve the project objectives (Figure 9-39). Ancillary data includes integrating DEMs, published maps, field observations, and GIS maps and images to provide a framework for the new products that will

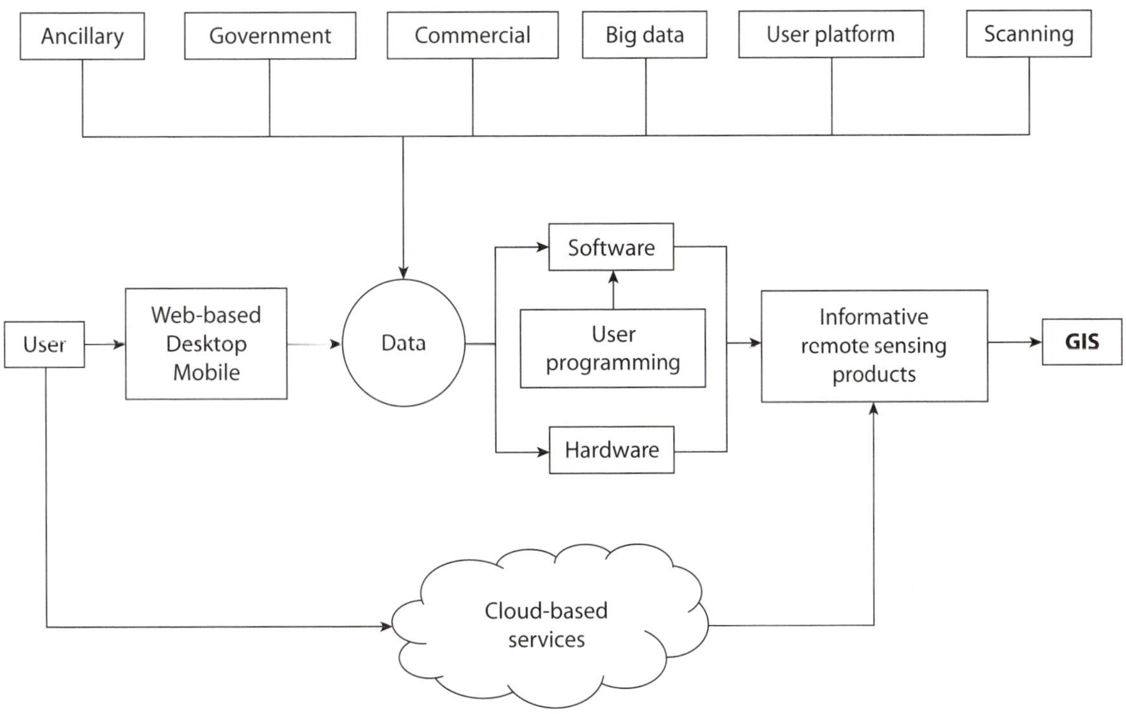

Figure 9-38 Image processing workflow.

be created during the project. The processing steps reviewed in this chapter will generate new images and maps that will need to be evaluated. If these images and maps are found to be not acceptable, the analyst should return to the image enhancement step and reassess the resulting products until they meet project objectives. After the images and maps are approved, metadata needs to be attached that documents at least the author, date of processing, processes involved, and limitations of the new maps and images. A report should also be written to document what worked and what didn't work to improve the performance of future projects. Remote sensing images and maps are most useful when loaded into a GIS and integrated with other geospatial layers. Within the GIS environment, spatial analysis tools are applied to the remote sensing data to extract more information (Chapter 10).

QUESTIONS

1. Which two Landsat bands have the highest correlation with the SWIR2 band in Table 9-2?
2. Using Figure 9-4B, describe how the brightness of pixels in the OLI NIR band 5 varies as the brightness of the OLI red band 4 changes from dark pixels to bright pixels.
3. Why would you correct Landsat radiance DN values to reflectance values?
4. What atmospheric conditions cause more atmospheric scattering? What Landsat bands are most affected by atmospheric scattering?
5. Describe the UTM coordinate system.
6. Why do cities and counties use the State Plane Coordinate System?
7. What are the three methods of resampling? How many pixels does each method use from the input image to calculate the brightness value for each pixel in the output image?

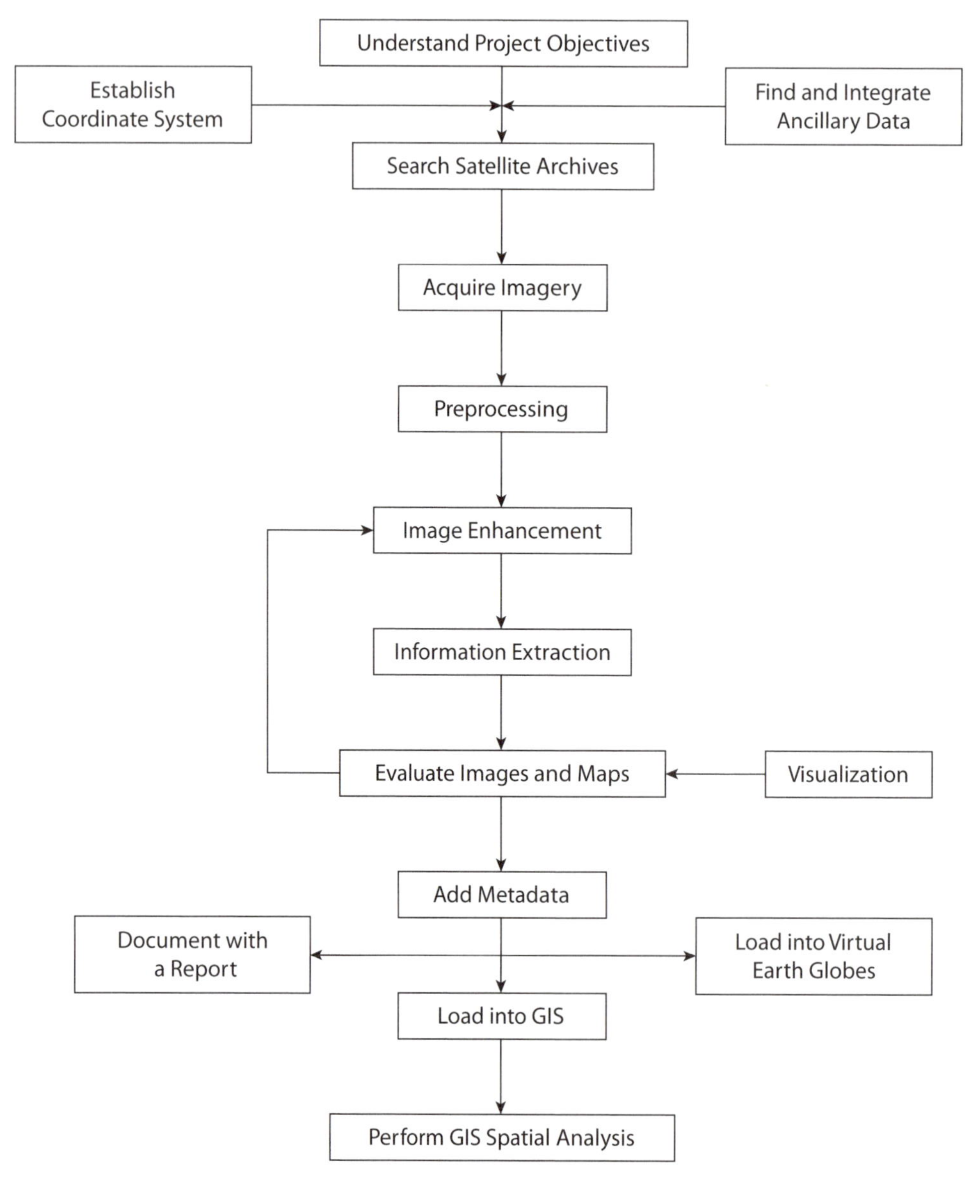

FIGURE 9-39 Project workflow.

8. Why is the saturated contrast stretch method used to enhance many images?
9. Using Figure 9-14A, replace all the 35s with 47s and all the 45s with 38s. Produce the following enhanced results:
   a. Filtered data set.
   b. Profile of the revised original data.
   c. Profile of your filtered data.
10. Identify each component of the IHS system. What are the measurement systems and range of values used for IHS images?
11. What are two advantages and two disadvantages of band ratios?
12. What is the band ratio for NDVI?
13. Why mask an image prior to image enhancement and information extraction?
14. Describe the two major techniques for multispectral classification.
15. Why do you generalize a raster classification land cover map?
16. What types of maps are generated from hyperspectral data cubes?

REFERENCES

ASPRS. 2015. ASPRS positional accuracy standards for digital geospatial data. *Photogrammetric Engineering & Remote Sensing*, 81(3).

Boardman, J. W., and F. A. Kruse. 2011. Analysis of imaging spectrometer data using *n*-dimensional geometry and a mixture-tuned matched filtering approach. *Transactions on Geoscience and Remote Sensing*, 49(11), 4,138–4,152. http://ieeexplore.ieee.org/stamp/stamp.jsp?tp=&arnumber=5985518 (accessed January 2018).

Bolstad, P. 2016. *GIS Fundamentals: A First Text on Geographic Information Systems* (5th ed.). Ann Arbor, MI: XanEdu Publishing.

Chavez, P. S. 1975. Atmospheric, Solar, and MTF Corrections for ERTS Digital Imagery. Proceedings of Annual Meeting of the American Society for Photogrammetry and Remote Sensing, Phoenix, AZ.

Chavez, P. S., S. C. Sides, and J. A. Anderson. 1991. Comparison of three different methods to merge multiresolution and multispectral data: Landsat TM and SPOT panchromatic. *Photogrammetric Engineering and Remote Sensing*, 57, 295–303.

Congalton, R. G. 1991. A review of assessing the accuracy of classifications of remotely sensed data. *Remote Sensing of the Environment*, 37, 35–46.

Dempsey, C. 2013, March 13. Statistical Surfaces in GIS. GIS Lounge. http://www.gislounge.com/statistical-surfaces-in-gis (accessed January 2018).

ENVI. 2017. ENVI Help. Harris Geospatial Solutions. http://www.harrisgeospatial.com/docs/AboutENVI.html (accessed January 2018).

FEATool Multiphysics. 2016, September 9. Implementing Custom and User Defined Finite Element Shape and Basis Functions. Precise Simulation Ltd. http://www.featool.com/tutorial/2016/09/09/User-Defined-Finite-Element-FEM-Shape-Functions-with-FEATool.html (accessed January 2018).

Federal Geographic Data Committee. 1998. *Geospatial Positioning Accuracy Standards. Part 3: National Standards for Spatial Data Accuracy* (FGDC-STD-007.3-1998). http://www.fgdc.gov/standards/projects/FGDC-standards-projects/accuracy/part3/chapter3 (accessed January 2018).

Haralick, R. M. 1984. Digital step edges from zero crossing of second directional filters. *IEEE Transactions on Pattern Analysis and Machine Intelligence*, v. PAMI-6, 58–68.

Hexagon. 2019. Producer Field Guide. http://hexagongeospatial.fluidtopics.net/reader/uOKHREQkd_XR9iPo9Y_Ijw/T4Jgg0x~vh4KPhNZBjstmg (accessed February 2019).

Jensen, J. R. 2016. *Introductory Digital Image Processing: A Remote Sensing Perspective* (4th ed.). Upper Saddle River, NJ: Pearson.

Kruse, F. A., A. B. Lefkoff, and J. B. Dietz. 1993, May–June. Expert system-based mineral mapping in northern Death Valley, California/Nevada using the Airborne Visible/Infrared Imaging Spectrometer (AVIRIS). *Remote Sensing of Environment*, 44, 309–336.

Kumar, L., and O. Mutanga (Eds.). 2019. Google Earth Engine Applications. *Remote Sensing*, 11(5), 591. Also published by MDPI Books. https://doi.org/10.3390/books978-3-03897-885-5

Lazaridou, M. A., and A. Ch. Karagianni. 2016. Landsat 8 multispectral and pansharpened imagery processing on the study of civil engineering issues. *The International Archives of the Photogrammetry, Remote Sensing, and Spatial Information Sciences*, XLI-B8, 941–945. doi:10.5194/isprsarchives-XLI-B8-941-2016

Lillesand, T. M., K. W. Kiefer, and J. W. Chipman. 2015. *Remote Sensing and Image Interpretation* (7th ed.). Hoboken, NJ: John Wiley.

Mars, J. C., and L. C. Rowan. 2011. ASTER spectral analysis and lithologic mapping of the Khanneshin carbonatite volcano, Afghanistan. *Geosphere*, 7(1), 276–289.

Richards, J. A. (Ed.). 2013. *Remote Sensing Digital Image Analysis* (5th ed.). Berlin, Germany: Springer.

Russ, J. C. 2011. *The Image Processing Handbook* (6th ed.). Boca Raton, FL: CRC Press.

Sabins, F. F. 1997. *Remote Sensing: Principles and Interpretation* (3rd ed.). Long Grove, IL: Waveland Press.

Schowengerdt, R. A. 1983. *Techniques for Image Processing and Classification in Remote Sensing*. New York: Academic Press.

Schowengerdt, R. A. 2007. *Remote Sensing Models and Methods for Image Processing* (3rd ed.). Burlington, MA: Elsevier.

Snyder, J. P. 1987. *Map Projections—A Working Manual* (Professional Paper 1395). Washington, DC: US Geological Survey.

Swain, P. H., and S. M. Davis. 1978. *Remote Sensing—the Quantitative Approach*. New York: McGraw-Hill.

US Geological Survey. 2013. Teaching about and Using Coordinate Systems. http://education.usgs.gov/lessons/coordinatesystems.pdf (accessed January 2018).

Vincent, R. K. 1997. *Fundamentals of Geologic and Environmental Remote Sensing*. Upper Saddle River, NJ: Prentice Hall.

Welch, R., and W. Ehlers. 1987. Merging multiresolution SPOT HRV and Landsat TM data. *Photogrammetric Engineering and Remote Sensing*, 53(3), 301–303.

Geographic Information Systems

A few decades ago satellite and aerial images were manually interpreted by laying a photographic print or film on a light table and taping tracing paper, Mylar, or some other transparent sheet on top of the print. The interpretation map was then laid on top of other maps (topography, roads, parcels, parks, etc.) at the same scale and map projection to build a hardcopy geospatial database. The same principles used with hardcopy stacks of maps and images apply to layers of digital files in a geographic information system (GIS) (Figure 10-1). GIS is a more complete solution, as it integrates hardware, software, and data for capturing, managing, analyzing, and displaying all forms of geographically referenced information (ESRI, 2017; Ryerson and Aranoff, 2010).

Remote sensing images and DEMs benefit from being integrated with other geospatial layers and processed with sophisticated spatial analysis tools in a GIS. Integration with GIS also increases the visibility of remote sensing as GIS is used by a wide range of disciplines, professions, and organizations. This chapter focuses on how remote sensing images and DEMs are processed and symbolized within a GIS. Green and others (2017) provide case histories and discussion integrating imagery into GIS analysis and maps. Bolstad (2016), Bonham-Carter (1994), and Wade and Sommer (2006) provide more comprehensive information on GIS. Productive online sources of information include: GIS Lounge (gislounge.com), GIS Geography (gisgeography.com), and ESRI (esri.com/en-us/home).

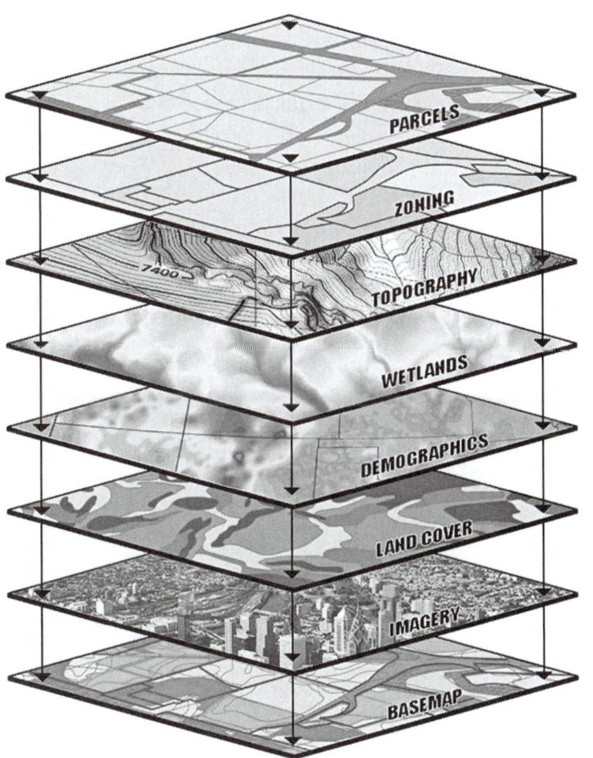

 **Figure 10-1** GIS data layers visualization. Courtesy Ontario County, New York, GIS Program.

GIS DATA MODELS

GIS presents real-world features (trees, buildings, lakes, etc.) as vector or raster models (Figure 10-2). Vector (also named map or feature) data contain the location and shape of geographic features with points, lines, and areas (polygons). Vector graphics are composed of vertices and paths with vector points representing an *x*, *y* coordinate. Lines are connected dots (vertices with a path). Polygons are closed lines and represent a two-dimensional area (building footprint, agricultural field, parcel, etc.). Polygons have an inside and outside defined by the closed line. Shapefiles are a common format for storing geometric location and attribute information of GIS vector data (ESRI, 2017).

Feature Type
Vector Model
Raster Model
Point feature
Building
Line feature
Road
Area feature
1
2
3
4
Land-use

FIGURE 10-2 GIS feature types (point, line, and area or polygon) and their representations in vector and raster data models. From Zhu (2014, Figure 2), https://creativecommons.org/licenses/by-nc-sa/3.0/.

Raster data is an array or grid of cells that represents an image, DEM, or classification map. Each cell has a value. Continuous raster layers are images and DEMs where each cell has a brightness DN or elevation value. Discrete raster layers are thematic raster maps where each cell has a value that relates to a class or type of feature. The spectral classification map (Plate 29), the density slice (Plate 28), and the mask (Figure 9-27B) generated from Landsat imagery are examples of thematic raster maps. Spectral classification maps in the thematic raster format are capable of displaying 256 classes with different symbology and names. Spatial analysis computations are very fast and efficient with thematic raster maps because of their grid structure, which has a value assigned to each cell.

Vector features can be converted to raster data and raster data can be converted to vector features. For example, in Figure 10-3A, the background is a Landsat OLI band 5 (NIR) image and six classes from the thematic raster map were outlined as polygons. Note that the class outlines match exactly the edges of the 30-m pixels. In Figure 10-3B, these six raster classes were converted to vector polygons and each class was labeled with a class ID (dry grass, dry grass and exposed soil, grass—not irrigated, industry 1, industry 2, and white roofs and concrete). Several classes have more than one polygon in the selected area while dry grass has only one polygon that surrounds the other 14 polygons (Figure 10-3B).

A. Six classes outlined as polygons on Landsat image with 30-m pixels.

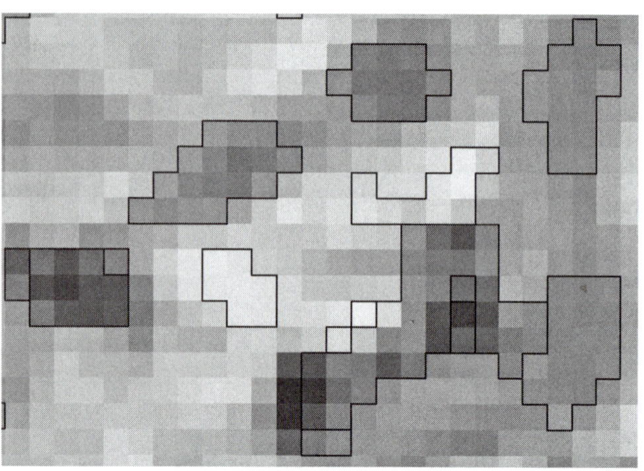

B. Six classes as filled polygons with class ID. Dry grass (class 1) surrounds the other 14 polygons.

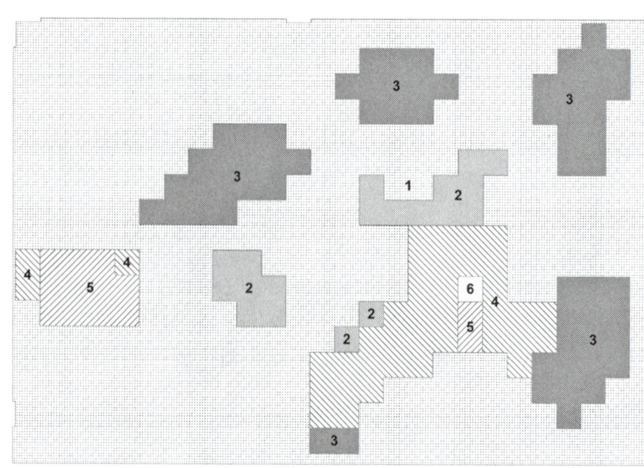

FIGURE 10-3 Conversion of six classes on a thematic raster map to vector polygons.

Attribute (tabular) data is the descriptive data that GIS links to map features (points, lines, and polygons). Attribute tables are powerful tools for symbolizing, analyzing, and improving understanding of maps. An attribute table was created for Figure 10-3 and is included in Table 10-1. A simple GIS summation tool applied to the 15 polygons calculates that dry grass (including areas with exposed soil) covers 75 acres while industry 1 and 2 cover 11.2 acres. Thematic raster maps are often restricted to displaying a single attribute field (gisgeography.com), therefore, they are converted to vector maps to enable the addition of many new fields to the attribute table. The ability to expand the attribute table with a vector map supports more complex analysis of pixel-based classes, integration with other attribute tables, and improved symbology.

RASTER IMAGES

Raster images (typically in 24-bit RGB and 8-bit panchromatic formats) are static layers in a GIS that provide an easily understood background for other geospatial layers. Basic contrast and brightness enhancement can be done on images, with more advanced GIS software capable of performing many image processing functions. Today satellite and aerial images are streamed over the Internet into online mapping applications such as Google Maps, Google Earth, ArcGIS Online, and navigation apps for display in smartphones, mobile devices, laptops, and desktop computers. These enhanced, natural-looking color images typically cannot be altered by users. Panchromatic, color infrared, and other color combinations are normally not provided.

Table 10-1 Attribute table of the GIS map in Figure 10-3.

Class Name	Class ID	m^2	Acres
Dry grass	1	287,304	71.0
Dry grass and exposed soil?	2	9,000	2.2
Dry grass and exposed soil?	2	6,300	1.6
Dry grass and exposed soil?	2	1,800	0.4
Grass—not irrigated	3	13,500	3.3
Grass—not irrigated	3	9,900	2.4
Grass—not irrigated	3	15,300	3.8
Grass—not irrigated	3	15,300	3.8
Grass—not irrigated	3	1,800	0.4
Industry 1	4	31,500	7.8
Industry 1	4	1,800	0.4
Industry 1	4	900	0.2
Industry 2	5	9,900	2.4
Industry 2	5	1,800	0.4
White roofs and concrete	6	900	0.2

DIGITAL ELEVATION MODELS

Digital elevation models (DEMs) have x, y coordinates and surface z elevations. DEMs can be continuous raster grids with x, y coordinates and z elevations attached to each cell or triangular irregular network (TIN). TINs model terrain with a vector-based network of triangles where the node of each triangle has an x, y coordinate and a surface z elevation (Chapter 7). TINs are often converted to raster-based DEMs for projects covering large areas that require less accuracy. TINs are also converted to raster-based DEMs to enable GIS spatial analyses with other raster grid data. DEMs and TINs are typically visualized as clouds with elevation coded by a grayscale or color ramp and hillshade models (Figures 7-4F and 7-4G and Plates 21 to 23 and 26). The GIS terrain analysis tools discussed below extract more information from DEMs that is useful for many applications, including engineering, natural hazard detection, water management, and geologic structure mapping.

Hillshade

Artificial hillshade of a DEM provides visual clues of geologic fracture patterns, scarps, and deposits related to landslides and slumps, parcels remaining in shadows during different times of the day, etc. The hillshade tool calculates the hypothetical illumination of a surface by determining illumination values for each cell in relation to neighboring cells in a raster DEM (ESRI, 2017). The user sets the azimuth and the altitude above the horizon of the illumination source.

In Figure 10-4A, a National Elevation Dataset (NED) DEM with 8-m cells of Concord, California, is shown. In Figure 10-4B, the image was processed for hillshade. The illumination source for the hillshade is from the northeast and 30° above the horizon. This DEM is in the northeast corner of the Landsat 8 scene in Plate 29A and along the eastern portion of the NAIP aerial imagery in Plate 5. Rugged topographic ridges and stream channels are clearly displayed on the elevated terrain in the center of the hillshade DEM (Figure 10-4B). Across the lowlands to the north, west, and south of the elevated terrain, subtle linear features represent roads, streams, and the shoreline while small rectangular features are buildings in the hillshade DEM. The hillshade algorithm requires the DEM coordinate system to be projected (units in meters or feet) and not in a geographic coordinate system (latitude and longitude degrees). For more information, see ESRI (2017) and Burrough and others (2015).

Slope and Aspect

Slope and aspect are important terrain properties that impact hydrology, vegetation, site planning, and infrastructure. For example, steeper slopes are often associated with unstable hillsides while the aspect of slope determines the flow direction of surface water and the amount of sunshine a parcel will receive. Slope is the rise of terrain relative to horizontal distance and aspect is the compass direction of the steepest slope in a grid cell. Slope is either in percent or degrees (no slope = 0% or 0°, vertical slope = 100% or 90°). Calculating slope is complex as each cell in a raster DEM can be considered a center cell with slope for that cell calculated by integrating the elevations of the eight neigh-

A. DEM.

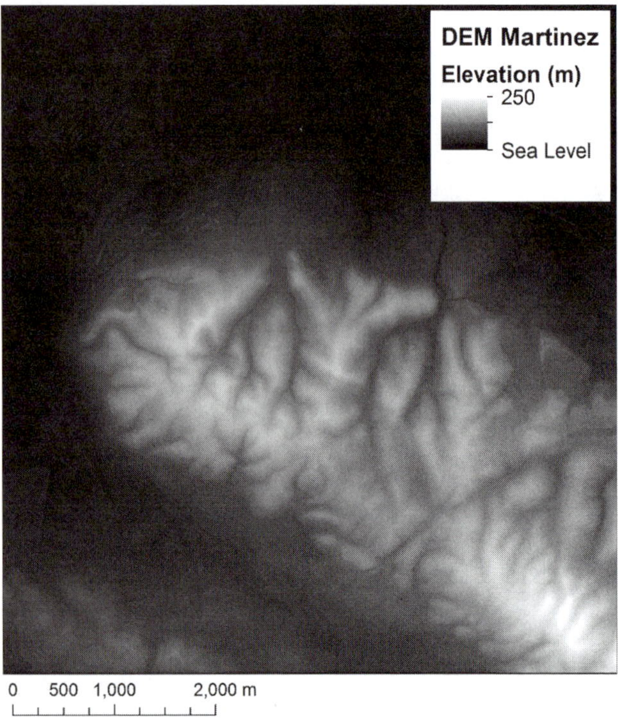

B. Hillshade DEM illuminated from the northeast.

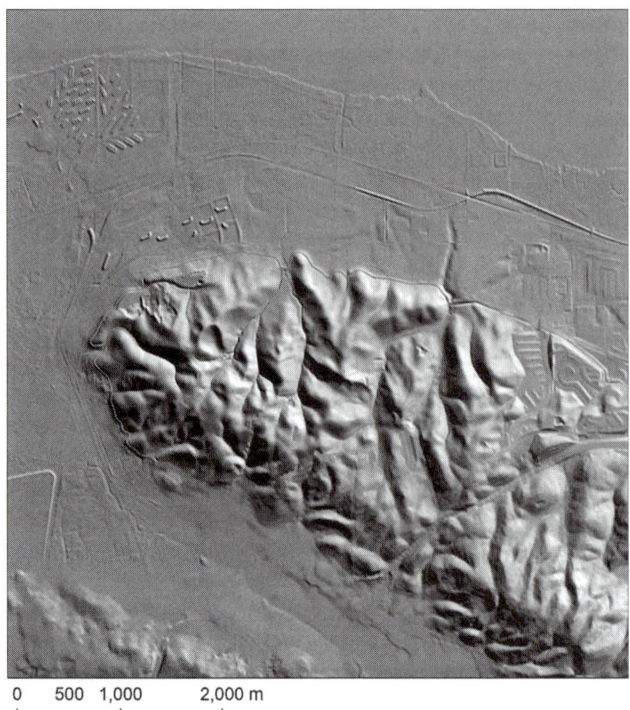

C. Slope in five classes (degrees).

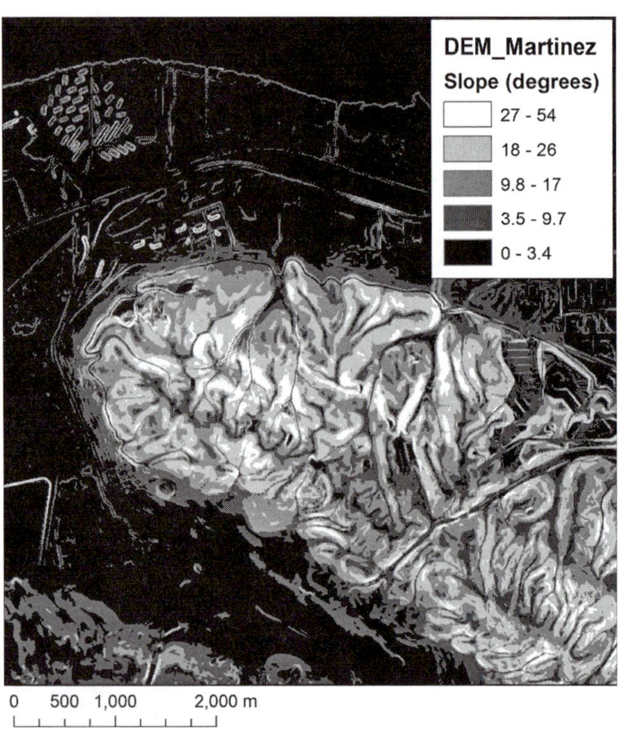

D. Aspect in four quadrants.

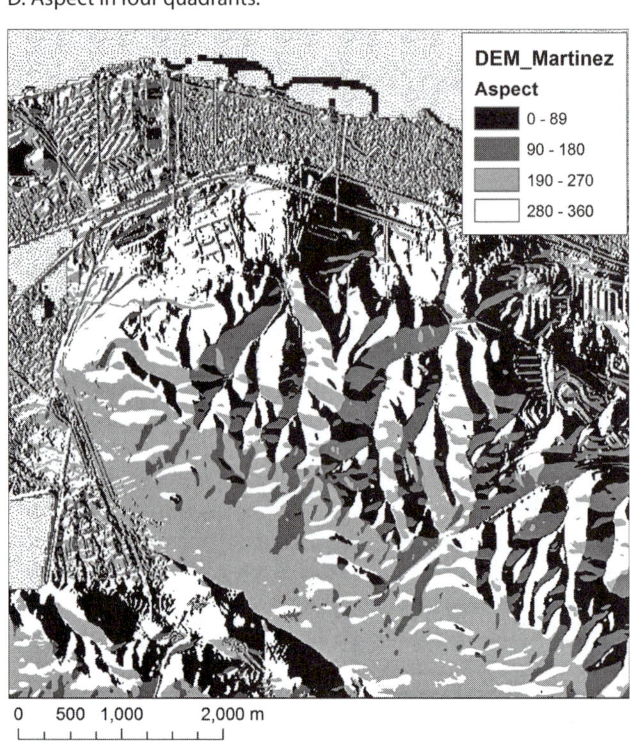

FIGURE 10-4 DEM enhancements include hillshade, slope, and aspect. Naval Weapons Station area, Concord, California. DEM courtesy USGS.

boring cells with the center cell (Bolstad, 2016). Aspect is the direction of the steepest downhill direction. There are many methods to determine aspect. TIN models can also be used to calculate slope and aspect in projects requiring high precision as the three nodes of each TIN triangle have elevation attributes and the three edges have slope and direction attributes (Peucker and others, 1978).

Slope is calculated for the NED DEM in degrees (Figure 10-4C). The slope map is classified into five classes, with the steepest slopes (27 to 54°) displayed as white pixels. Slope maps relate directly to the cost of new construction as the steeper the slope, the more expensive the work. Engineers reduce the impact of slope on construction costs by cut and fill, as shown on the northeast side of the hilly terrain where man-made flat topography with man-made linear and circular terraces and roads can be seen.

Aspect is calculated with the direction of the face of the slopes simplified into four quadrants: northeast, southeast, northwest, and southwest (Figure 10-4D). Slopes facing the northeast quadrant (0 to 89°) are black while those facing the southwest quadrant (190 to 270°) are light gray. Water in the Sacramento River along the northern margin and in a reservoir and pond along the western margin have a stippled pattern.

The environment for proposed developments in this central California climate (hot and dry summers, cool and wet winters) can be evaluated with the aspect map. Buildings constructed on slopes facing the northeast quadrant would be the coolest in the summer, while buildings constructed on slopes facing the southwest quadrant would be the hottest in the summer. In winter, construction on slopes facing the southwest quadrant would be drier compared with those in the northeast-facing quadrant. To save air-conditioning and heating energy, contractors could modify building footprint orientation, roof overhang, insulation design, and window type and placement based on the aspect of slopes where buildings are constructed.

Drainage Network

Hydrology uses DEMs for delineating watersheds, building drainage networks, and documenting drainage parameters such as area, elevation drop, and length of each stream segment. Published topographic maps can display inaccurate elevations due to new developments that change topography with earth movers, building slabs, and road networks; subsidence due to groundwater withdrawal; and land movement. Geologists interpret subsurface structure from surface drainage anomalies. Using the NED DEM in Figure 10-4A, Rivix RiverTools software was applied to generate a drainage network of the northeastern portion of the hilly terrain. The results are seen in Figure 10-5. Circular and linear drainage can be seen on the east side of the image, which reflects suburban developments with drainage ditches around a cul-de-sac (circular feature), cut and fill terraces, and roads in subdivisions. The valleys in the hilly terrain and the drainage divides separating the watersheds are clearly delineated by the drainage network.

Figure 10-5 Drainage network derived from the NED DEM, Naval Weapons Station area, Concord, California. DEM courtesy USGS.

Perspective Views

Visualizing the terrain from a perspective view can improve understanding of features on the Earth's surface. Any raster image or vector map can be draped on a DEM. If the terrain is flat, vertical exaggeration can be increased to accentuate subtle topographic relief. For example, the vector drainage network that was superimposed on the raster hillshade DEM (Figure 10-5) was exported as a combined grayscale image in GeoTIFF format with coordinates. This combination image was draped on the NED DEM in ESRI's ArcScene and viewed looking southwest (Figure 10-6). The vertical exaggeration was set at 2X. On the left side of the perspective view, the flat-topped terraces and the circular and linear drainage patterns associated with suburban developments are more easily seen and understood compared with the 2-D nadir view in Figure 10-5. The more natural stream channels in the undeveloped hilly terrain on the right side of the model stand in stark contrast to engineered surface drainage developed in the suburban areas.

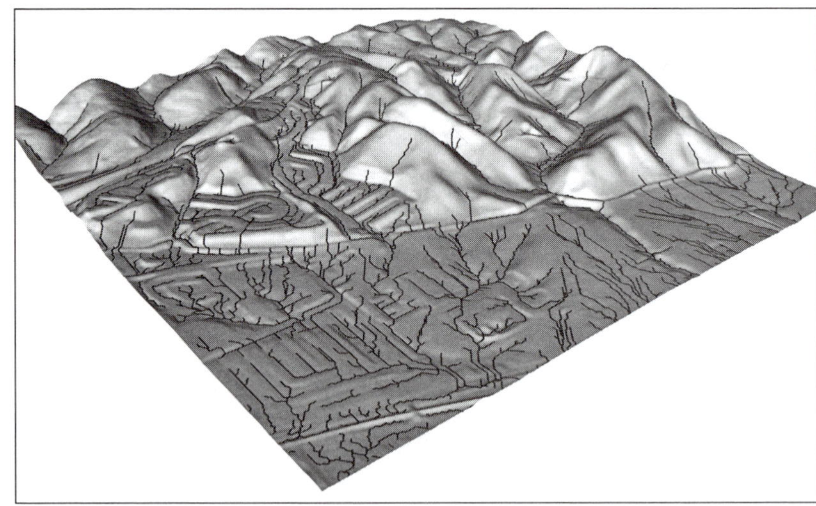

FIGURE 10-6 Drainage network draped on hillshade DEM with a perspective view and vertical exaggeration set at 2X.

STRIKE AND DIP

Measuring the slope and orientation (strike and dip) of geologic bedding from DEMs depends on the spatial resolution of the DEM, the vertical accuracy of the DEM, the length of an exposed bedding surface, and the dip of the bed (Figure 10-7). Steeply dipping beds are less accurately measured compared with gently dipping beds. A NED DEM of Split Mountain, Utah, with 8-m cells (Figure 10-7A) is compared with a SRTM DEM with 24-m pixels (Figure 10-7B). Steeply dipping beds at the bottom of the hillshade DEM cannot be clearly identified from the SRTM DEM, but can be interpreted as steeply dipping beds in the more detailed NED DEM. The more gently dipping beds in the center and top of the hillshade DEM would support strike and dip mea-

A. Hillshade NED DEM with an 8-m grid showing geologic beds dipping toward the south–southwest.

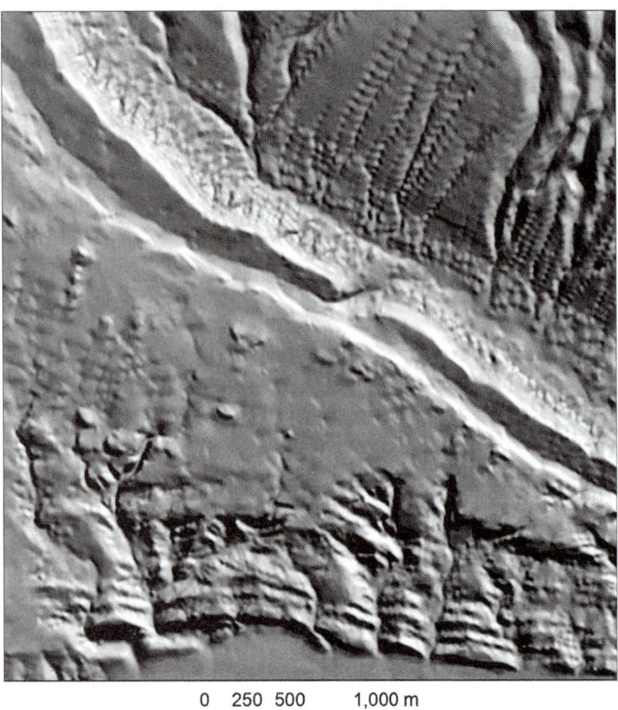

0 250 500 1,000 m

B. Hillshade SRTM DEM with a 24-m grid showing less visual information and detail than the NED DEM.

0 250 500 1,000 m

FIGURE 10-7 Geologic detail on an 8-m and 24-m hillshade DEM. Data courtesy USGS.

TABLE 10-2 Elevations and distances for three-point method.

Point	Relative Elevation	Elevation (m)	Line	Distance (m)	Difference in Elevation (m)
A	Low	1,735.0	A–B	163.5	20.0
B	Medium	1,755.0	B–C	193.2	17.8
C	High	1,772.8	A–C	219.6	37.8

surements in both DEMs, but the NED DEM measurements should be more accurate. Superimposing an enhanced satellite or aerial image with the same spatial resolution on the DEM will improve the interpretation of bedding planes and increase the accuracy of the strike and dip calculations.

Geologists typically use a graphical three-point method for determining strike and dip (Compton, 1961; Suppe, 1985). Using a geologic map, basic formulas, and graph paper, the geologist can generate a strike line, dip direction, and angle of dip. However, GIS software can not only collect the necessary data for the three-point method, it can also more accurately calculate the angle of dip compared to the graphical method. A GIS automatically records the x, y coordinates of each point. These coordinates are used to calculate the azimuth direction from the high to medium and high to low elevation points. As an example of the three-point method, in Figure 10-8A three points on a planar bedding surface of a DEM have been plotted. The distances between the points, elevations at each point, and differences in elevation are then measured (Table 10-2). Using this information, a geologist can generate a strike line (B–D in Figure 10-8B), dip direction (A–E in Figure 10-8B), and angle of dip (Allison, 2015). Dip measurements are more accurate at outcrops with longer bedding planes compared to those with less exposure on the slope of the bedding plane (Berger, 1994).

SPATIAL ANALYSIS

The value of thematic raster maps (land cover, density slice, and masks) and DEMs are significantly increased when they are integrated with other geospatial layers in a GIS and analyzed with spatial analysis tools. The density of features in a landscape, the cost or risk associated with transportation routes that cross varying topographic slopes and land cover types, the effects of solar radiation over a geographic area

A. Three points (A, B, C) plotted on a dipping geologic bed of a hillshade DEM. These points are used in the three-point method of calculating geologic strike and dip.

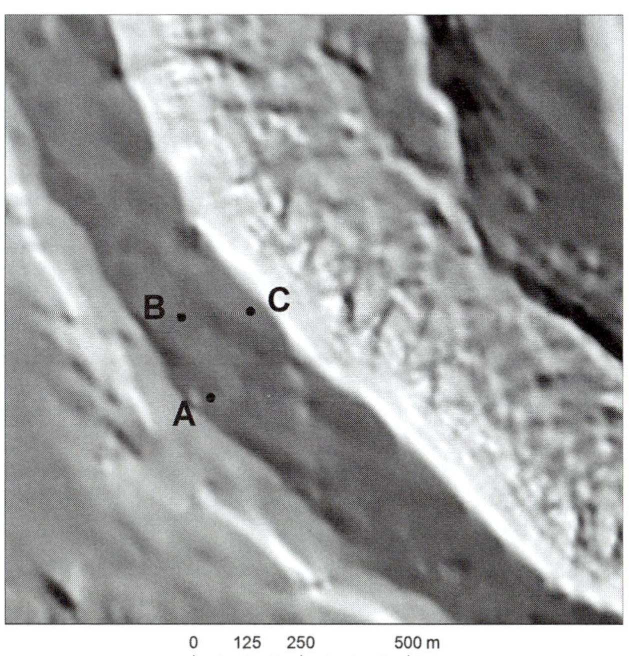

B. Three-point solution for strike and dip using the hillshade DEM.

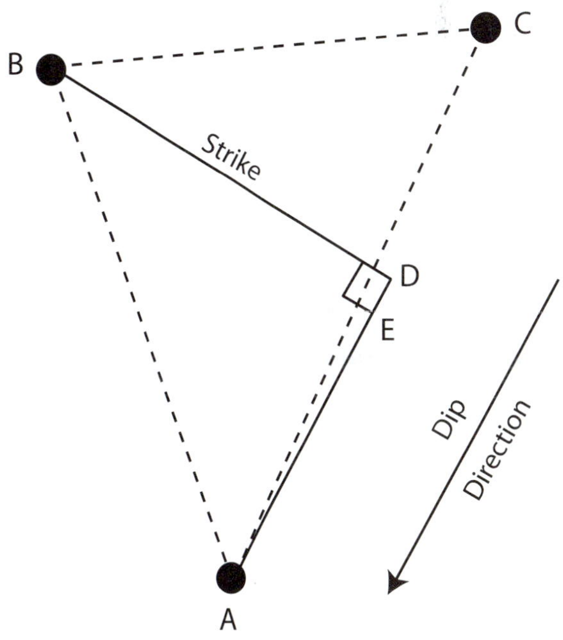

FIGURE 10-8 Graphical three-point method for determining strike and dip of a geologic outcrop, Split Mountain, Utah. DEM courtesy USGS.

for specific time periods, and models that construct lines of sight and viewsheds from DEMs are a few of the questions that can be answered with spatial analysis tools.

In order to illustrate how these tools can be used, we started with two maps of Martinez, California. The generalized Landsat land cover supervised classification map of Plate 30 was combined with the slope map of Figure 10-4C to calculate a least cost path for a hypothetical pipeline from the northeast shore, across a populated area, to a site on the hilly terrain in the southwest (Figure 10-9). The ArcGIS

A. Landsat NIR image.

B. Slope reclassified for cost.

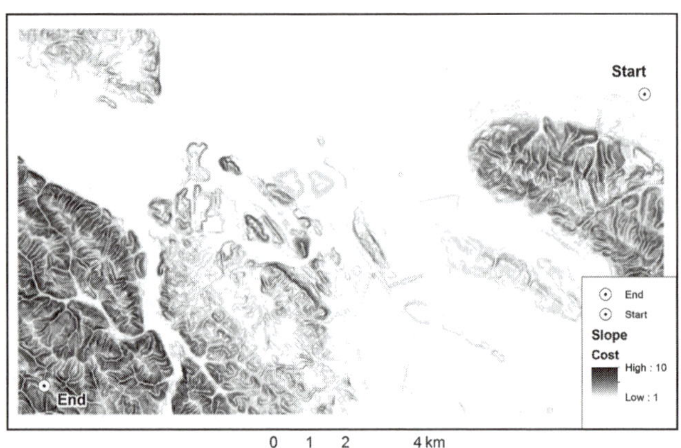

C. Classification map reclassified for cost with legend.

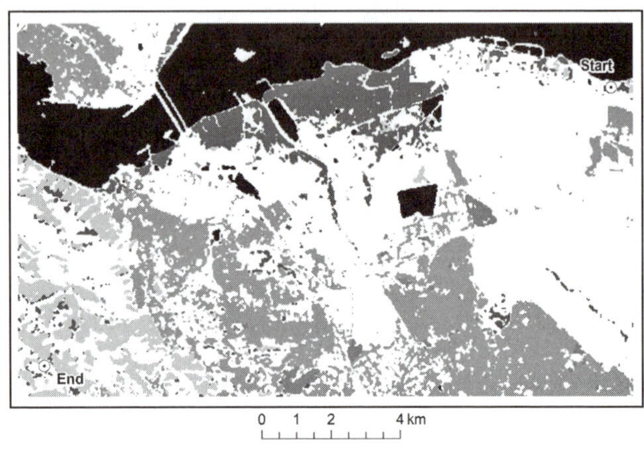

Start
End
Supervised Classification
Water
Wetlands
Irrigated vegetation
Suburbs
Trees
All Other - grasses, industry

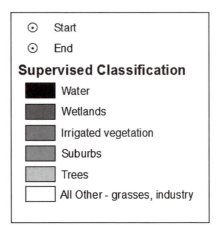

D. Land cover and slope combined with weighting.

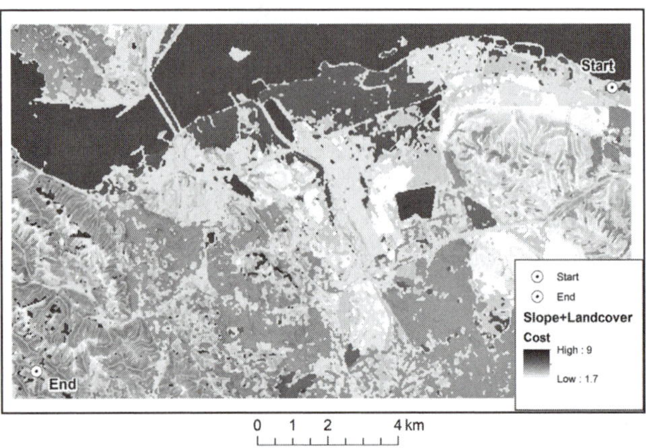

E. Least cost path for pipeline.

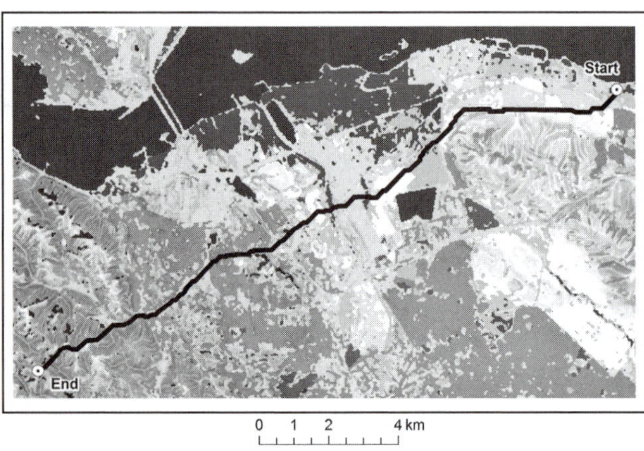

FIGURE 10-9 GIS spatial analysis to determine least cost path for a hypothetical pipeline, Martinez, California, from DEM slope map (Figure 10-4C) and Landsat land cover map (Plate 30).

Spatial Analyst tools were used with ModelBuilder to perform the many format and calculation steps (Figure 10-10). ModelBuilder visualizes the steps involved, and enables the user to run multiple iterations with different parameters and tools to achieve the desired results without reloading the data. For example, the user can reclassify or change values attached to the DEM slope and land cover classes to enable a calculation of cost.

In this example, the cost (or risk) associated with the pipeline route is determined by the steepness of topographic slope and the type of land cover. Each cell in the slope image is reclassified from the original degrees value to a cost value determined by the user. Cells with the highest cost are assigned a value of 10 while those with the lowest a value of 1. Steepest slopes are assigned a value of 10 and flat terrain a value of 1 (Figure 10-9B). The type of land cover in each pixel of the supervised classification map is simplified and reclassified on a cost scale of 1 to 10 as follows (Figure 10-9C):

Water (most sensitive)	10
Wetlands	8
Irrigated vegetation	8
Suburbs	6
Trees	4
All other (industry, dry grass, and nonirrigated vegetation)	2 and 3

Darker areas on the reclassified land cover map are associated with higher cost (Figure 10-9C). Land cover is determined to have twice as much risk/cost compared with topographic slope for a pipeline, so when the land cover and slope maps are combined, the land cover values are weighted twice as much as the slope values for each cell (the "Raster Calc" step in Figure 10-10). In the combined land cover and slope cost map, higher costs are associated with darker pixels (Figure 10-9D), which includes water and wetlands. As intended, the tree-covered hills in the southwest have more cost compared with the dry grass-covered hills in the northeast.

The final calculation determines the path for the pipeline with the least distance and cost from a start point to an end point. This path was superimposed on the combined and weighted land cover and slope map (Figure 10-9E). Visual inspection shows the least cost pipeline route avoids water and wetlands, and only has a short segment crossing the suburbs in the west-central portion of the map. The model can be rerun with other costs and weights associated with slope and land cover type. Other features can be included in the GIS model, including roads, schools, existing pipeline corridors, and soil type. GIS spatial analysis of thematic raster maps and DEMs adds significant value to remote sensing products.

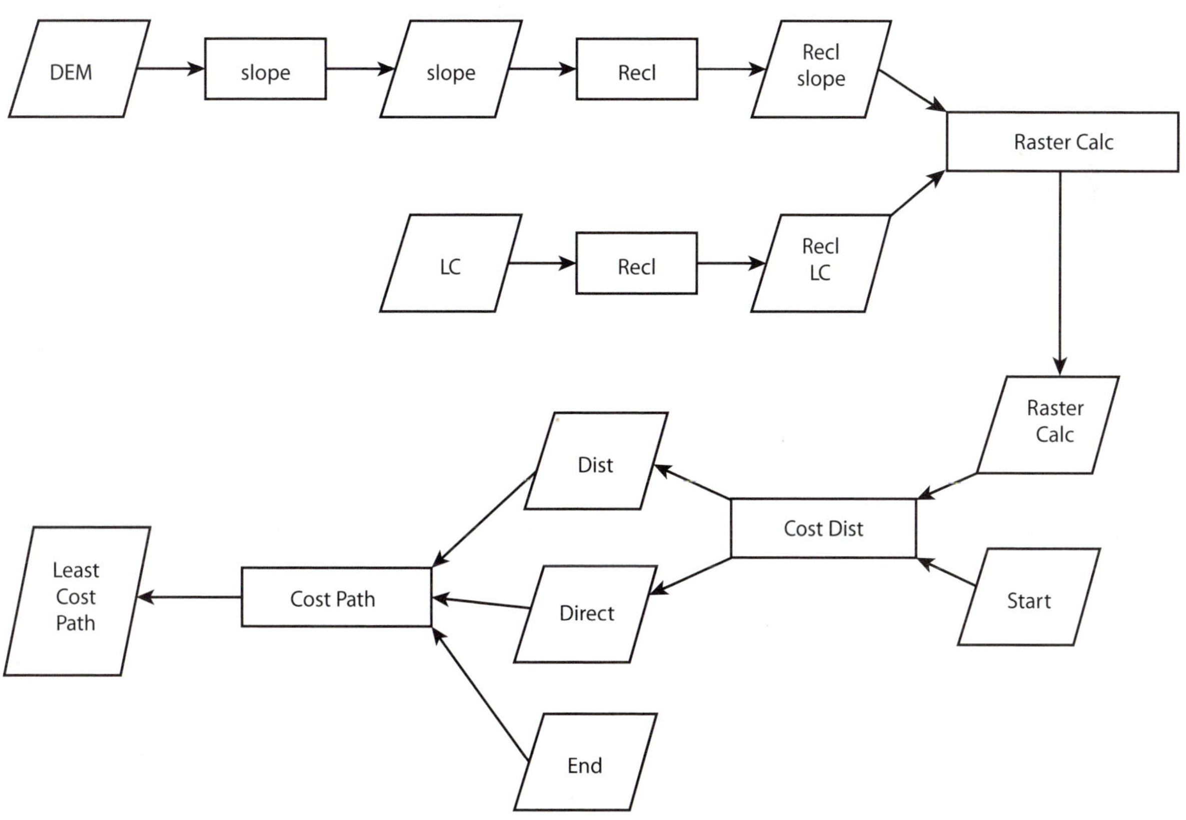

Figure 10-10 Sequential spatial analysis steps structured with ArcGIS ModelBuilder to determine least cost path for proposed pipeline (Figure 10-9). Recl = reclassify; slope = DEM slope; LC = land cover map; Raster Calc = raster calculation; and Cost Dist = cost distance.

HEADS-UP DIGITIZING

Heads-up interpretation of imagery and hillshade DEMs displayed on a GIS computer monitor enables integration with other geospatial layers while mapping. This integration increases the accuracy of the new interpretation maps. The size of the area to be mapped is a primary decision. The scale of mapping (sometimes referred to as the minimum mapping unit, or mmu) needs to be established so that the level of detail required for the project is achieved. A list of features to be mapped, along with a format and naming convention for each feature type, takes advantage of the many symbology options in GIS that require standardization.

Features to be mapped are modeled as points, lines, or polygons. Planning a field trip to collect field samples can be done in the office with imagery, a DEM, and GIS. The imagery is analyzed for suitable sampling locations near features of interest. The DEM is queried to record the elevation of each sampling site. An attribute table is built for each sample site with fields that are filled in while digitizing the sample points. The following is a field sampling example. Site ID, elevation, type of test, test number, comments, and UTM 10N WGS84 easting and northing (x, y) coordinates are entered into a GIS attribute table while digitizing the 12 points (Figure 10-11; Table 10-3). Other attributes such as location name, sampler name, confidence level, and parcel number can also be added to the attribute table. Area and perimeter for polygons, length for lines, and coordinates (x, y or latitude, longitude) for points are calculated by the GIS for each feature mapped, and entered into the attribute table.

Draping the imagery on the DEM and viewing a synthetic stereo DEM in the anaglyph format with red/cyan glasses improves the ability of the interpreter to perceive and draw features related to subtle topographic relief (see Plate 25). Buildings, trees, and other objects protruding above the terrain will be seen in 3-D if a digital surface model is used for the synthetic stereo model. Some features such as thrust faults and normal faults have standard symbology with hachures, triangles, or other short symbols protruding along one side of a line or polygon. The direction used for digitizing automatically places these unique symbols on the correct side of the line or polygon.

FIGURE 10-11 Twelve ground sampling points determined from imagery and a DEM. Attributes (listed in Table 10-3) were recorded while digitizing these points.

TABLE 10-3 Attribute table built while digitizing sampling points.

ID	Elevation (m)	x Coordinate (m)	y Coordinate (m)	Type of Test	Test Number	Comments
1	93	579602.1	4207596.2	Groundwater (O)	4	groundwater table 2 ft depth
2	81	579468.8	4207780.3	Groundwater (CO_2)	4	
3	60	579359.2	4207953.1	Particulate	6	moved point based on image
4	40	579262.3	4208097.4	Particulate	6	moved point based on topo
5	30	579075.1	4208269.3	Particulate sand	6	
6	16	578922.7	4208383.6	Particulate	6	near berm
7	64	579227.5	4207526.3	Acidity	2	
8	98	579083.9	4207772.9	Acidity	2	low priority
9	44	578929.0	4207850.2	Acidity	2	moved point based on DEM
10	38	578865.5	4208053.4	Groundwater (HCl)	4	
11	18	578808.4	4208174.0	Groundwater	4	
12	15	578713.1	4208320.1	Groundwater	4	near berm

FIELD TOOLS

Digital cameras and mobile devices that geotag photographs with latitude/longitude coordinates have significantly simplified adding ground photographs to a GIS database. These geotagged photographs pop up on the GIS monitor when activated, adding field evidence to features seen on images and interpreted on maps. The photographs can also be used for training and verification sites to support classification mapping of land cover, geology, and vegetation.

FIGURE 10-12 Mobile system with a surface map displayed on a GIS field tablet and subsurface reflections recorded by a GPR sensor. Courtesy Ioannis Kavouras (2016, Figure 3), Thessaloniki Water & Sewage Co., S.A.

GIS software is available for smartphones and mobile devices that display imagery and GNSS/GPS paths and collection points. Mobile GIS can be integrated with ground penetrating radar (GPR) equipment in the field to more accurately identify underground network assets (Figure 10-12). Most mobile GIS devices require an Internet connection to see streaming imagery and maps from a cloud service while in the field. The GIS Pro software works on iPads and iPhones and can download and store satellite imagery, topographic maps, and street maps, eliminating the need for an Internet connection while in the field (Figure 10-13). GIS Pro users can add points, lines, and polygons and add/edit attributes. Vector maps drawn on the device can be exported as ArcGIS shapefiles, kmz, and gpx files for integration into a GIS database. Mappt is mobile mapping software for Android devices that allows users to create, edit, store, and share geospatial data. Mappt is fully functional offline—it supports shapefiles, kmz, and gpx files—and allows users to document ground photos as points on a map with attributes.

VIRTUAL 3-D GLOBES

Virtual 3-D globes are used by millions of people for recreation, work, navigation, science, and communication. The most famous, and perhaps the most flexible for enabling integration of external remote sensing images, is Google Earth (google.com/earth). New georeferenced images created with desktop, mobile, or cloud-based remote sensing software can be imported into Google Earth and displayed on the 3-D virtual globe. The hillshade DEM with drainage generated from the DEM of the Naval Weapons Station area, Concord, California, was imported into Google Earth Pro.

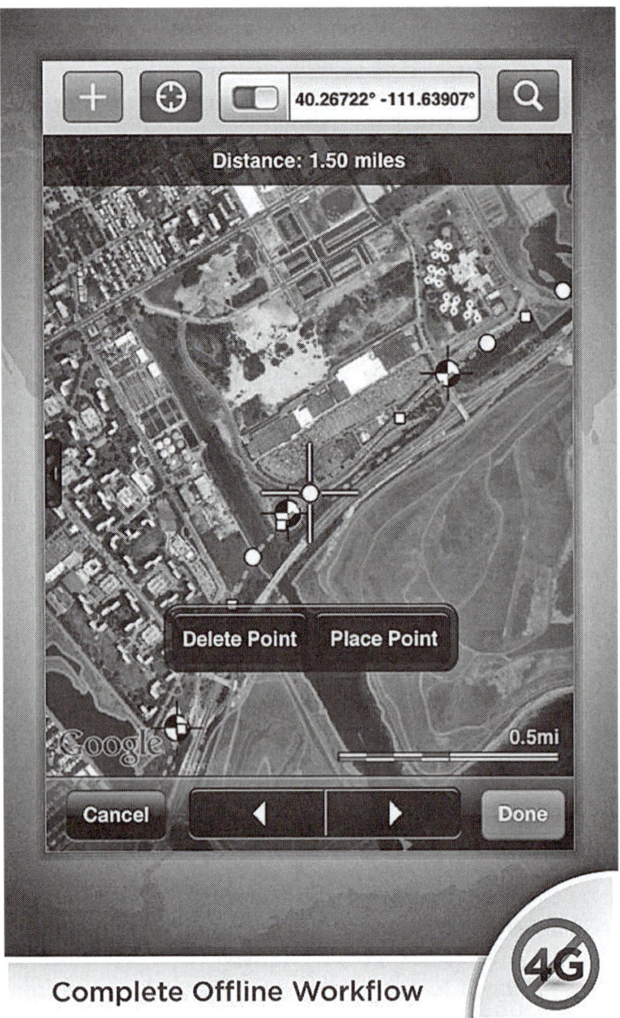

FIGURE 10-13 GIS Pro software running on mobile device for field data collection with imagery as a backdrop in the GIS display. Courtesy GIS Pro (garafa.com/wordpress).

Once imported this new layer can be shared globally by right-clicking on the file name displayed in the Google Earth Table of Contents (Places category) and choosing the option "Save Place As" (Figure 10-14). The raster image is then converted to a kml (keyhole mark-up language) or kmz (keyhole mark-up zipped) file. A kml/kmz file can be sent anywhere as an email attachment, placed into a website or ftp site for download, or imported and displayed in another 3-D virtual globe, a GIS, or in software programs such as Adobe Photoshop and AutoCAD. Kml/kmz is a powerful communication tool for the remote sensing community.

The kml/kmz format specifies a set of features (images, polylines, place marks, polygons, 3-D models) for display in geospatial software. Keyhole, Inc. created the kml format in 2004. In 2008, the Open Geospatial Consortium, Inc. made kml the international encoding standard for displaying geospatial data in Internet-based 2-D maps and 3-D virtual globes. Kml/kmz's status as the international standard has made it widely accepted as a digital format for exchanging geospatial information.

ESRI's ArcGIS Earth enables users to add basemaps (ESRI's streaming, raster-based layers of topography, bathymetry, images, and street maps) and maps made public by users of ArcGIS Online. The software also opens georeferenced raster images that have been converted to kml/kmz files, shapefiles, and web services (Figure 10-15). Points, lines, and polygons can be drawn and saved as kml/kmz files to be used with other applications or shared.

Marble is an open source geographical atlas and virtual globe that streams different overlays, including a classic topographic map, historic (1492 and 1689) maps, satellite imagery and orbital paths, average temperature and precipitation maps, and Earth's city lights at night. The free, street-level map of the world provided by the OpenStreetMap project is accessible on Marble's virtual globe (marble.kde.org/index.php).

GIS CONFIGURATIONS

GIS operations range from a single desktop connected to the Internet to complex enterprise solutions with multiple users, servers, and Internet-based services supporting mobile devices and web GIS. GIS users can access and display remote sensing images from local drives, servers on a network, and cloud-based streaming services.

WEB GIS

Web GIS is a type of distributed information system, comprising of at least a server and a client, where the server

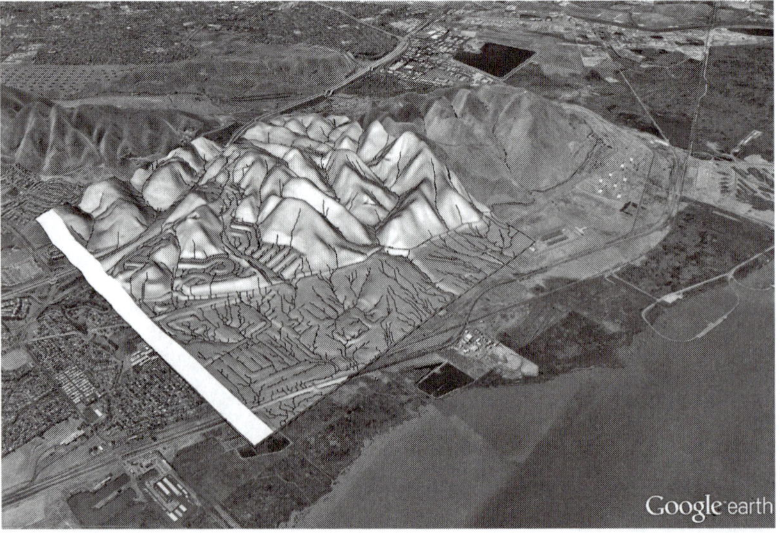

FIGURE 10-14 Drainage network draped on hillshade DEM (Figure 10-6) of Naval Weapons Station area, Concord, California, and imported into Google Earth and displayed on a 3-D virtual globe. Courtesy Google Earth.

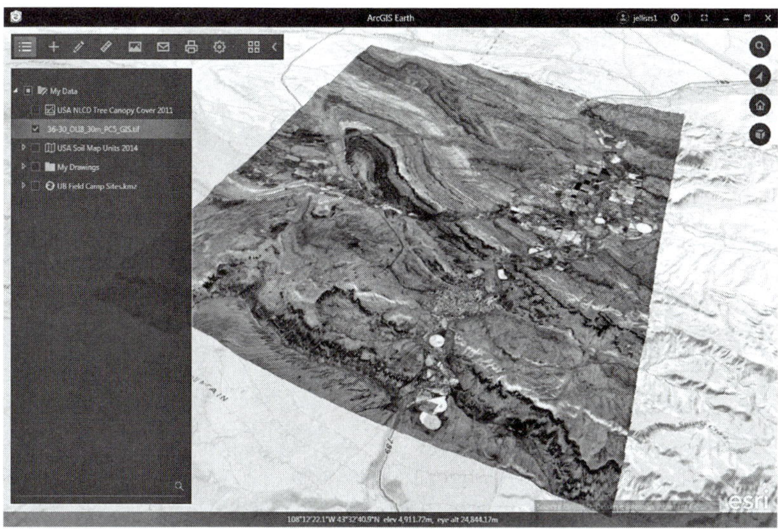

Figure 10-15 Landsat OLI PC image 5 of Thermopolis, Wyoming, subscene (Figure 9-23E) imported into ArcGIS Earth and displayed on a 3-D virtual globe. Content is the intellectual property of Esri and is used herein with permission.

is a GIS server and the client is a web browser, desktop application, or mobile application. In its simplest form, web GIS can be defined as any GIS that uses web technology to communicate between a server and a client (ESRI, 2017). Unlike desktop GIS, which is limited to a certain number of GIS professionals, web GIS can be used by everyone in an enterprise as well as the public at large. Web GIS has to be scalable to support a large number of users and ever increasing size of data sets.

Cities, counties, and states maintain web GIS systems that include parcels, roads, schools, zoning maps, and high resolution satellite or aerial imagery. The public systems often support downloading of orthorectified imagery with a simple registration by the requestor. Agencies have embraced web GIS as a cost effective way to communicate with their constituents. The Federal Emergency Management Agency uses the ArcGIS Online web GIS to distribute the official National Flood Hazard Layer.

ArcGIS Online is free with basic tools and services. A subscription adds extra capabilities and services. ArcGIS Online's "Living Atlas" streams over 3,800 layers that include global datasets and detailed geospatial data of smaller areas. For example, the USDA's NAIP imagery of the United States is streamed as natural color and color IR imagery and as the vegetation greenness index NDVI. Global coverage of Landsat 8 OLI (30 m) multispectral and multitemporal 8-band data is available along with pan-sharpened (15 m) natural color imagery. Environmental maps at global, continental, and regional scales are provided, including land cover, carbon biomass, population, human footprint, and soil maps. This online service removes the need for users to store, manage, or process these very large datasets.

Satellite and aerial imagery can be streamed into Internet-connected devices to support hundreds of applications. ESRI's ArcGIS and Blue Marble's Global Mapper stream "basemaps" from the cloud into their user's GIS displays (bluemarblegeo.com/products/global-mapper.php). These basemaps include static raster versions of hillshade topographic and bathymetric maps, imagery, and street maps. The USGS 3D Elevation Program (3DEP) streams a hillshade DEM with 1-m spatial resolution over select areas of the conterminous United States (usgs.gov/core-science-systems/ngp/3dep). The streaming basemaps are very useful as backdrops that provide a geographic framework for user-generated images and maps. Global Mapper streams DEM grids with elevations at each cell to support user-defined topographic profiles and perspective viewing of topography.

NASA developed the Giovanni web portal to access and visualize satellite-based Earth science data in the classroom (Lloyd, 2008). Users can easily try various combinations of parameters measured by different NASA satellites, download enhanced imagery, and have statistical analyses performed. Disciplines include aerosols, cryosphere, hydrology, ocean biology, and oceanography. Giovanni was designed to be an interactive online bridge between data and science (giovanni.gsfc.nasa.gov/giovanni).

Cloud Computing

Cloud computing furnishes technological capabilities—commonly maintained off premises—that are delivered on demand as a service via standard Internet protocols. Software as a service, platform as a service, and infrastructure as a service are the three core service models for cloud computing (Kouyoumjian, 2011). Precision Hawk's PrecisionMapper (precisionhawk.com/precisionmapper) and DroneDeploy (dronedeploy.com) are examples of cloud software and infrastructure as a service for the drone community. Both companies provide cloud-based tools for flight management, deliver final 2-D and 3-D products (orthorectified mosaics, DEMs, volume measurement, and agricultural maps), and archive the data in the cloud for drone operators. Figure 9-38 shows how users go directly to the cloud for their remote sensing products.

Big Data

A very significant challenge for remote sensing and GIS is how to manage the ever-increasing amount of data available for analysis. Businesses and public agencies are collecting data points at all times, from a growing number of connected and remote sensors located all over our planet. The Landsat 7 and 8 satellites alone collect 1,200 images per day at approximately 1 GB per image, and the European Space Agency's Copernicus Program Sentinel fleet soon will produce 10 TB per day of free and open data (Burton, 2016).

Big data is an accumulation of data that is too large and complex for processing by traditional database management tools. Big data is an evolving term that describes any voluminous amount of structured, semistructured, and unstructured data that has the potential to be mined for information (TechTarget, 2017). The tools described in this textbook have limited application for managing, processing, and extracting information from remote sensing images archived in big data (Chi and others, 2016).

SOFTWARE

COMMERCIAL

Sources of commercial GIS software include:

- ESRI ArcGIS (esri.com/en-us/arcgis/products/index)
- Geomedia (hexagongeospatial.com/products/power-portfolio/geomedia)
- GIS Geography (gisgeography.com)
- GIS Lounge (gislounge.com)
- IDRISI (clarklabs.org/terrset/idrisi-gis)
- Manifold (manifold.net)
- MapInfo (pitneybowes.com/us/location-intelligence/geographic-information-systems/mapinfo-pro.html)

OPEN SOURCE

An open source application by definition is software that you can freely access and modify the source code. Open source projects typically are worked on by a community of volunteer programmers. Open source GIS programs are based on different base programming languages. Three main groups of open source GIS (outside of web GIS) in terms of programming languages are: "C" languages, Java, and .NET. Open source GIS software is widely available and includes:

- GRASS (grass.osgeo.org)
- gvSIG (gvsig.com/en)
- QGIS (qgis.org/en/site)
- Whitebox GAT (uoguelph.ca/~hydrogeo/Whitebox)

QUESTIONS

1. What are three types of vector data?
2. What is a thematic raster image (map)?
3. Why are attribute tables important?
4. What does TIN mean? Describe the TIN model.
5. What is the difference between a DEM slope map and a DEM aspect map?
6. Why are GIS model builders important for productivity?
7. Define mmu and why is this parameter important?
8. What does kmz and kml mean? Why are these formats important?
9. What is web GIS?

REFERENCES

Allison, D. T. 2015. Structural Geology Lab Manual (4th ed.). University of South Alabama. http://www.usouthal.edu/geography/allison/GY403/StructuralGeologyLabManual.pdf (accessed January 2018).

Berger, Z. 1994. *Satellite Hydrocarbon Exploration: Interpretation and Integration Techniques* (pp. 165–167). Berlin: Springer-Verlag.

Bolstad, P. 2016. *GIS Fundamentals: A First Text on Geographic Information Systems* (5th ed.). Ann Arbor, MI: XanEdu Publishing.

Bonham-Carter, G. F. 1994. *Geographic Information Systems for Geoscientists: Modelling with GIS*. Kidlington, UK: Elsevier Science.

Burrough, P. A., R. A. McDonnell, and C. D. Lloyd. 2015. *Principles of Geographical Information Systems*. New York: Oxford University Press.

Burton, C. 2016, July 28. Earth observation and big data: Creatively collecting, processing and applying global information. *Earth Imaging Journal*. http://eijournal.com/print/articles/earth-observation-and-big-data-creatively-collecting-processing-and-applying-global-information (accessed January 2018).

Chi, M., A. Plaza, J. A. Benediktsson, S. Zhongyi, S. Jinsheng, and Z. Yangyong. 2016, November. Big data for remote sensing: Challenges and opportunities. *Proceedings of the IEEE*, 104(11), 2,207–2,219.

Compton, R. R. 1961. *Manual of Field Geology*. New York: John Wiley.

ESRI. 2017. Learn ArcGIS. http://learn.arcgis.com/en/ (accessed August 2019).

Green, K., R. G. Congalton, and M. Turkman. 2017. *Imagery and GIS: Best Practices for Extracting Information from Imagery*. Redlands, CA: ESRI Press.

Kavouras, I. 2016, Winter. Greek Utility Company Operates 24/7 with Mobile GIS. ESRI ArcGIS News. http://www.esri.com/esri-news (accessed January 2018).

Kouyoumjian, V. 2011. GIS in the Cloud: The New Age of Cloud Computing and Geographic Information Systems. ESRI. http://www.esri.com/library/ebooks/gis-in-the-cloud.pdf (accessed January 2018).

Lloyd, S. 2008. Using NASA's Giovanni Web Portal to Access and Visualize Satellite-Based Earth Science Data in the Classroom. NASA Goddard Earth Sciences DISC, 44 slides. http://irina.eas.gatech.edu/EAS6145_Spring2011/Giovanni-tutorial.pdf (accessed January 2018).

Open Geospatial Consortium, Inc. 2008, April 14. OGC Approves kml as Open Standard. Open Geospatial Consortium, Inc. http://www.opengeospatial.org/pressroom/pressreleases/857 (accessed January 2018).

Peucker, T. K., R. J. Fowler, J. J. Little, and D. M. Mark. 1978, May 9–11. The Triangulated Irregular Network (pp. 516–540). Proceedings, American Society of Photogrammetry, Digital Terrain Models (DTM) Symposium, St. Louis, MO.

Ryerson, R. A., and S. Aranoff. 2010. *Why "Where" Matters: Understanding and Profiting from GPS, GIS, and Remote Sensing*. Manotick, Ontario, Canada: Kim Geomatics Corporation.

Suppe, J. 1985. *Principles of Structural Geology*. Englewood Cliffs, NJ: Prentice-Hall.

TechTarget. 2017. Guide to Big Data Analytics Tools, Trends, and Best Practices. TechTarget. http://searchbusinessanalytics.techtarget.com/essentialguide/Guide-to-big-data-analytics-tools-trends-and-best-practices (accessed January 2018).

Wade, T., and S. Sommer. 2006. *A to Z GIS*. Redlands, CA: ESRI Press.

Zhu, X. 2014. GIS and urban mining. *Resources*, 3(1), 235–247. doi:10.3390/resources3010235

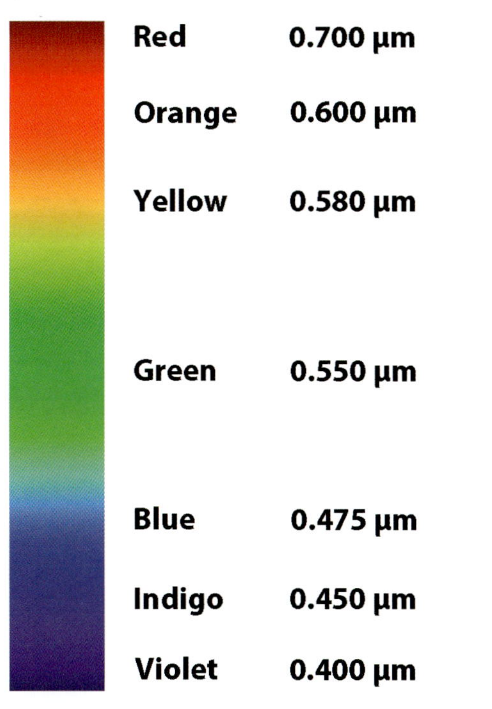

PLATE 1 (CHAPTER 1) Visible light spectrum with wavelengths.

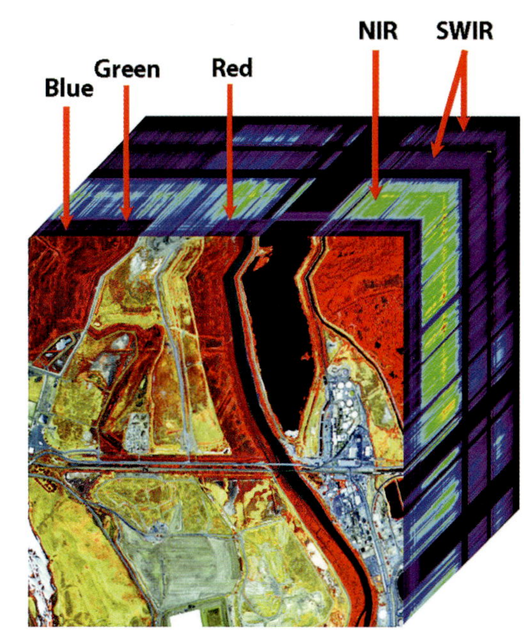

PLATE 2 (CHAPTER 1) Hyperspectral data cube. Courtesy GeoSat Committee.

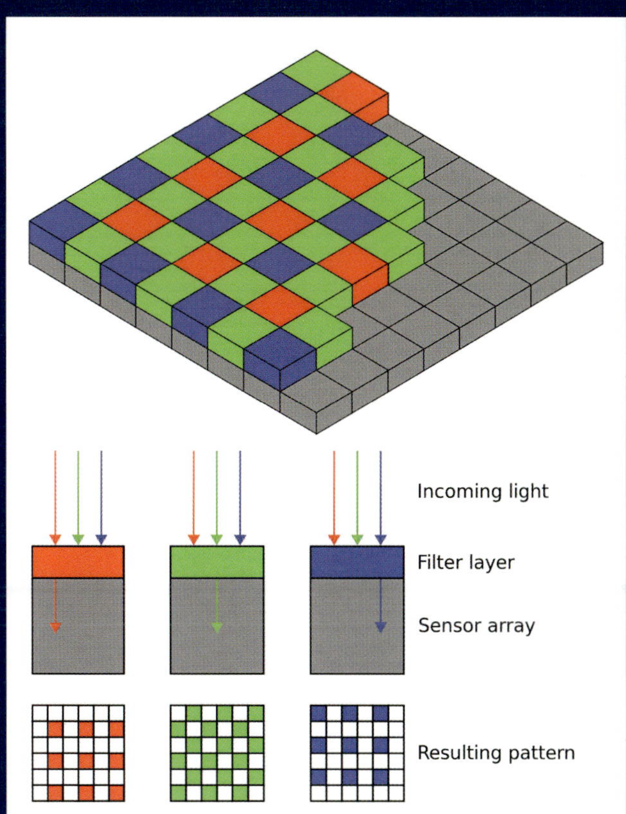

PLATE 3 (CHAPTER 2) Bayer pattern. Created by C. Burnett (December 2006), Wikipedia (http://creativecommons.org/licenses/by-sa/3.0/deed.en).

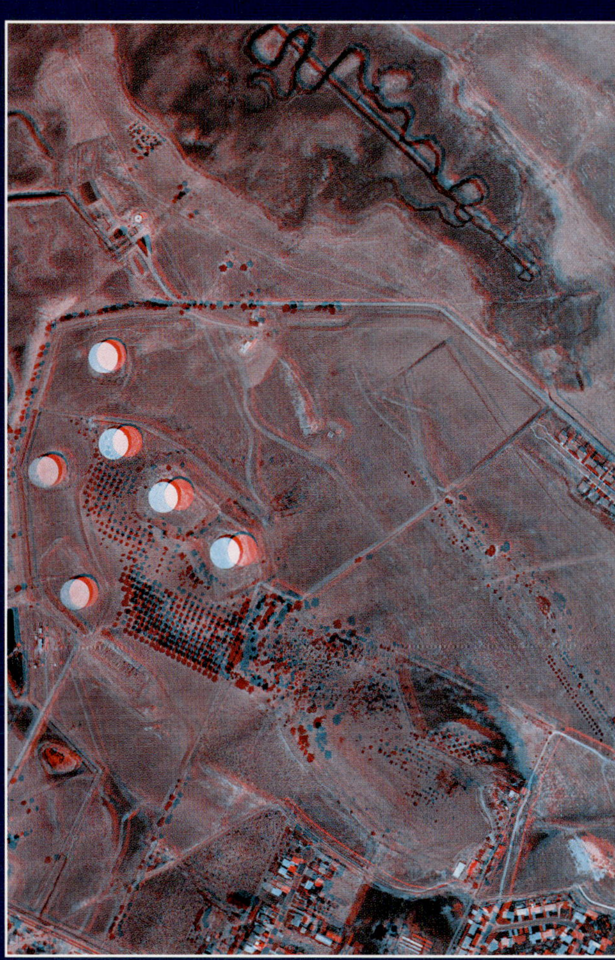

PLATE 4 (CHAPTER 2) Anaglyph of overlapping aerial photographs, Martinez, California (view with red/cyan glasses). Data courtesy of D. Ruiz, Quantum Spatial, Inc., Novato, California.

A. Color image.

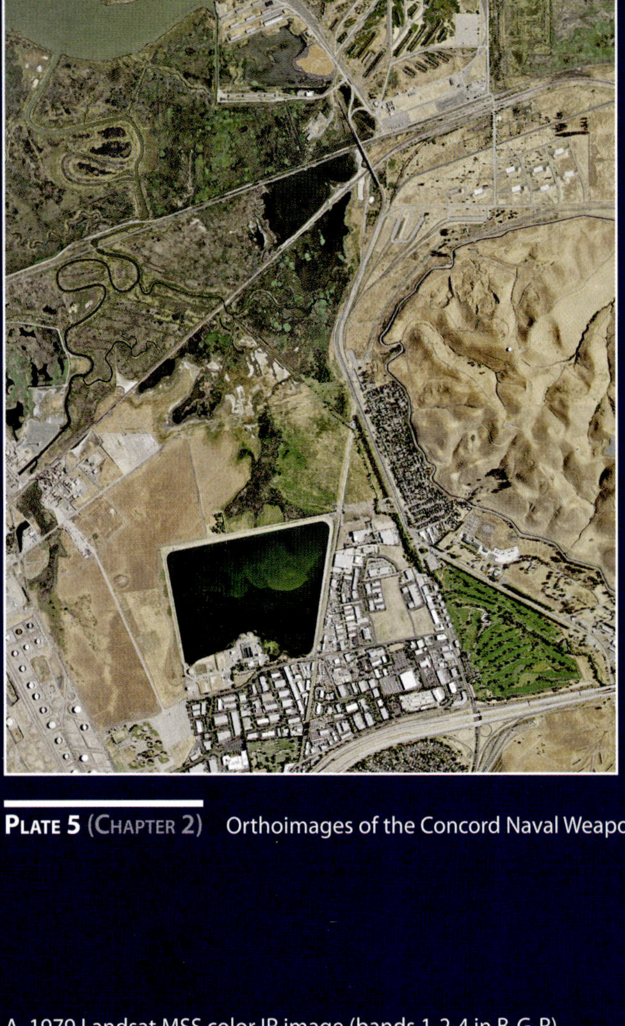

B. Color IR image.

Plate 5 (Chapter 2) Orthoimages of the Concord Naval Weapons Station, California. Courtesy USDA NAIP.

A. 1979 Landsat MSS color IR image (bands 1-2-4 in B-G-R).

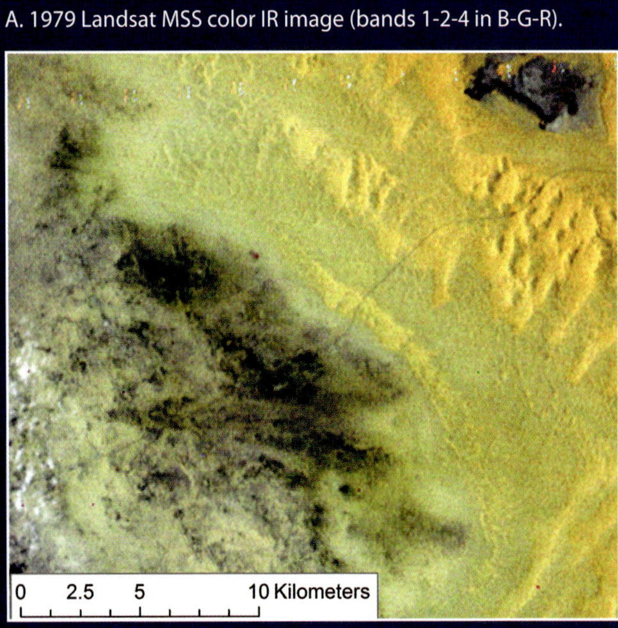

B. 2016 Landsat 8 OLI color IR image (bands 3-4-5 in B-G-R).

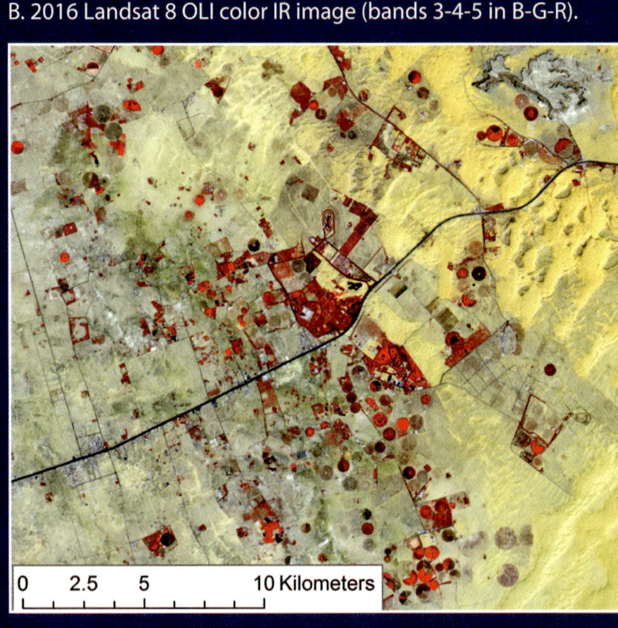

Plate 6 (Chapter 3) Landsat images showing temporal changes in Saudi Arabia. Irrigated vegetation shown as shades of red. Landsat courtesy USGS.

A. Natural color (bands 2-3-4 in B-G-R).

B. Color IR (bands 3-4-5 in B-G-R).

C. Total IR (bands 5-6-7 in B-G-R).

D. Enhanced color (bands 3-5-7 in B-G-R).

Plate 7 (Chapter 3) Color combinations of Landsat 8 OLI bands, Thermopolis, Wyoming. Landsat courtesy USGS.

A. Natural color (bands 1-2-3 in B-G-R).

B. Color IR (bands 2-3-4 in B-G-R).

C. Total IR (bands 4-5-7 in B-G-R).

D. Enhanced color (bands 2-4-7 in B-G-R).

Plate 8 (Chapter 3) Color combinations of Landsat TM bands, Mapia scene in West Papua, Indonesia. Landsat courtesy USGS.

A. Landsat TM color image (bands 1-2-3 in B-G-R).

B. Bathymetric DEM generated from TM band 1, color-coded for water depth, and hillshaded to highlight sand bars.

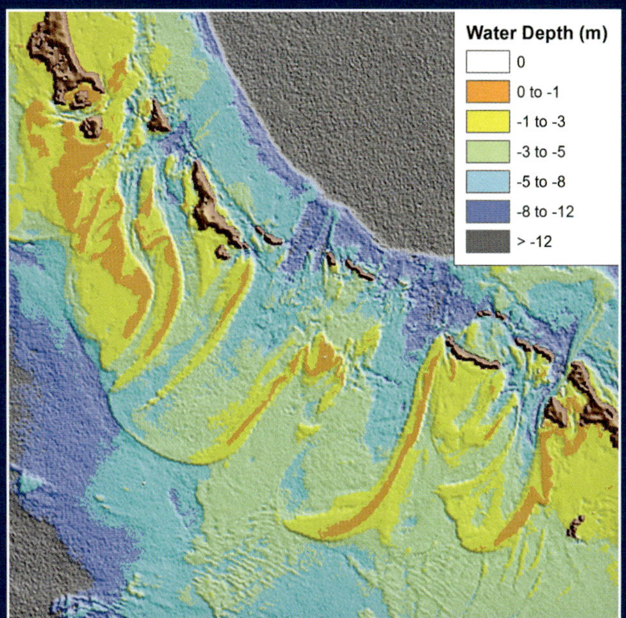

Plate 9 (Chapter 3) Submerged, shallow water carbonate sand bars in the Exumas Islands, The Bahamas. Landsat courtesy USGS.

A. Landsat image acquired December 23, 2016.

B. Landsat image acquired January 24, 2017.

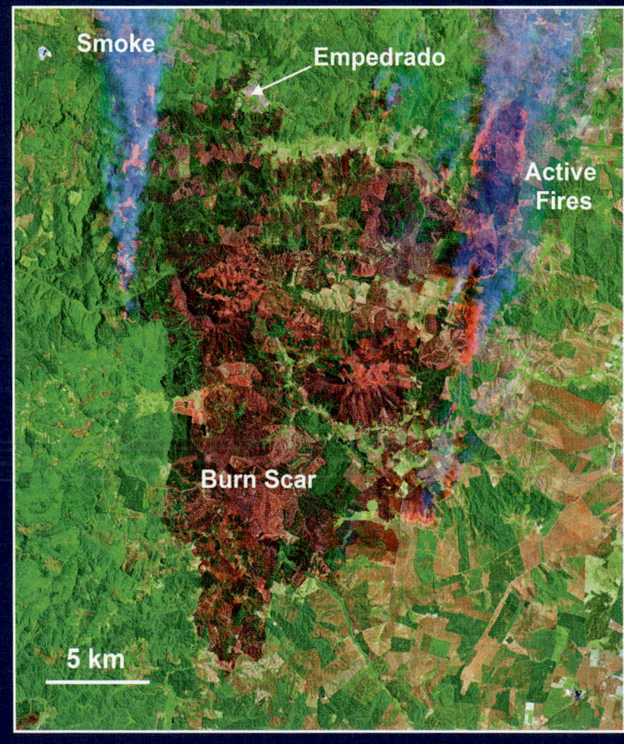

Plate 10 (Chapter 3) Fire detection and mapping, central Chile. Landsat OLI enhanced color images (bands 3-5-7 in B-G-R) before fire and during fire event. Landsat courtesy USGS.

Plate 11 (Chapter 3) Landsat TM enhanced color image (bands 2-4-7 in B-G-R), Saharan Atlas Mountains, Algeria. Landsat courtesy USGS.

Plate 12 (Chapter 4) SPOT XS color IR image of the Djebel Amour area, Saharan Atlas Mountains, Algeria. Courtesy SPOT.

A. Coral reef atoll, Cocos (Keeling) Islands. ALI bands 2-4-5 in B-G-R. Courtesy USGS.

B. Fringing reef off of Mo'orea Island. ASTER bands 1-3-2 (G-NIR-R) in B-G-R. Courtesy USGS.

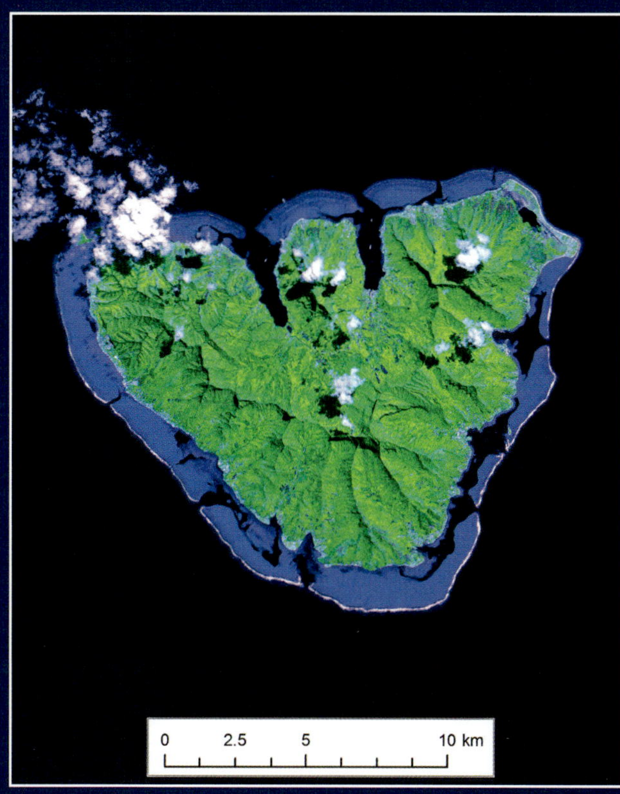

C. Offshore reef, Bora Bora. SPOT 6 bands 1-2-3 in B-G-R. Satellite Imaging Corporation, © Airbus DS (2019).

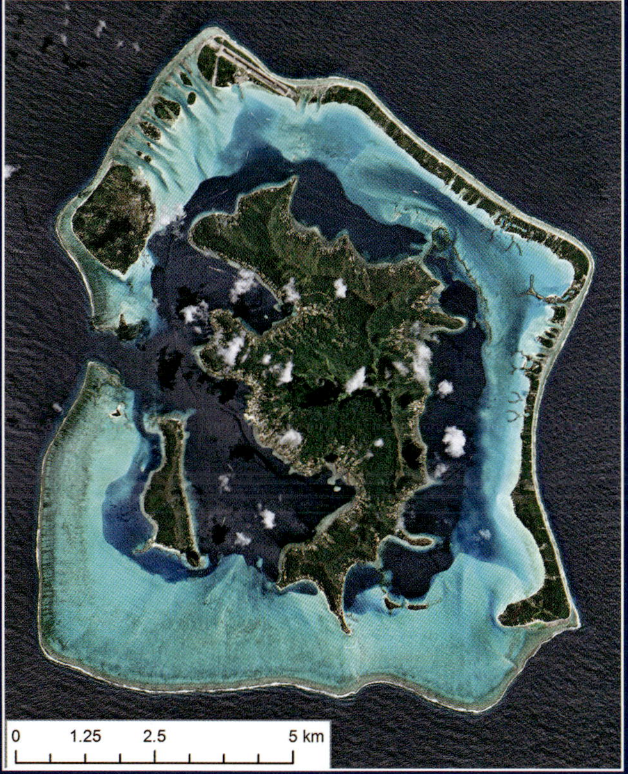

D. Atoll, Maldives. Sentinel-2 bands 2-3-4 in B-G-R. Courtesy ESA.

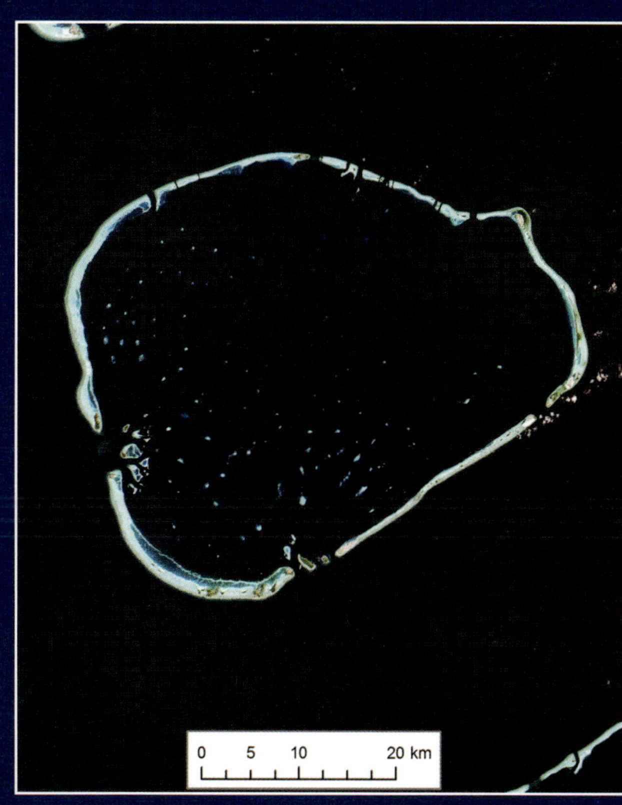

PLATE 13 (CHAPTER 4) Images from four satellite systems with medium to high spatial resolution. The images show stages in development of atolls that Darwin defined during the voyage of the HMS *Beagle* (1831 to 1836).

A. Smartphone with TIR attachment.

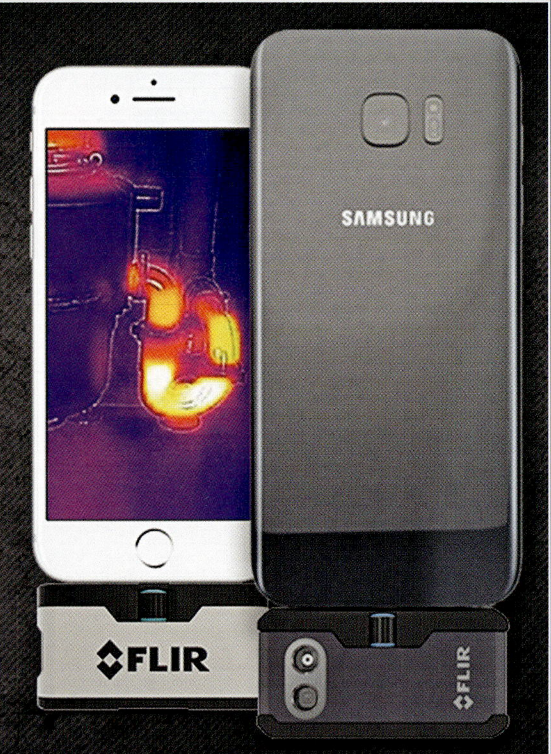

B. Thermal image.

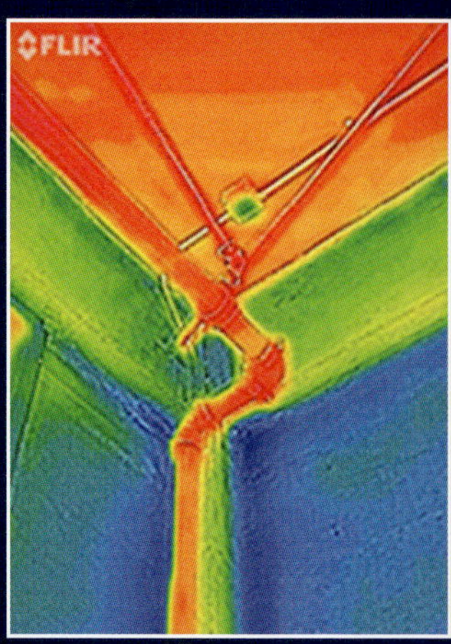

C. Visible image.

D. Extracted detail.

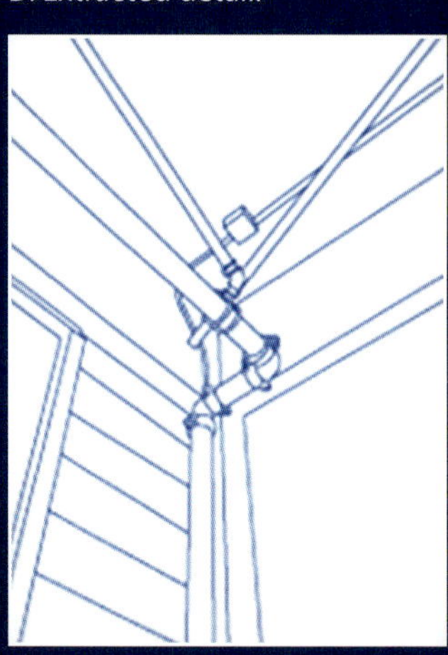

E. Combined image.

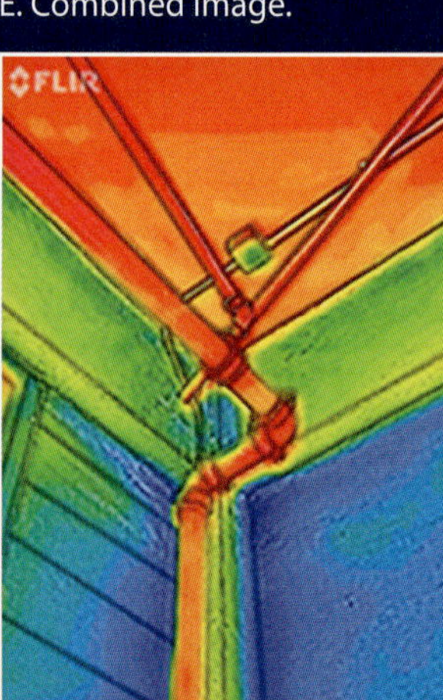

Plate 14 (Chapter 5) FLIR ONE TIR camera attachment for a smartphone. Courtesy FLIR Systems.

A. Nighttime surface temperature map.

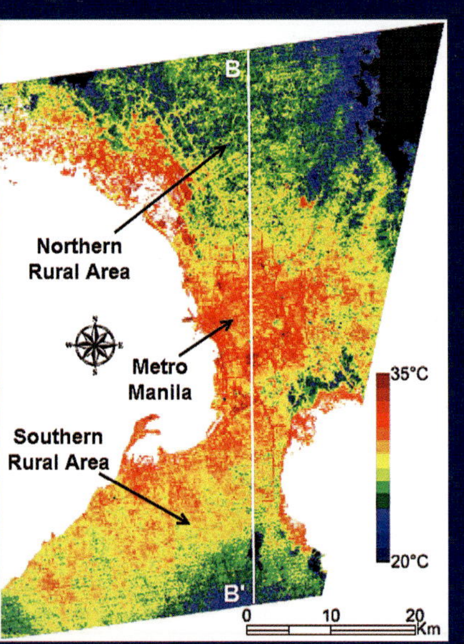

B. Nighttime surface temperature profile along B–B' transect.

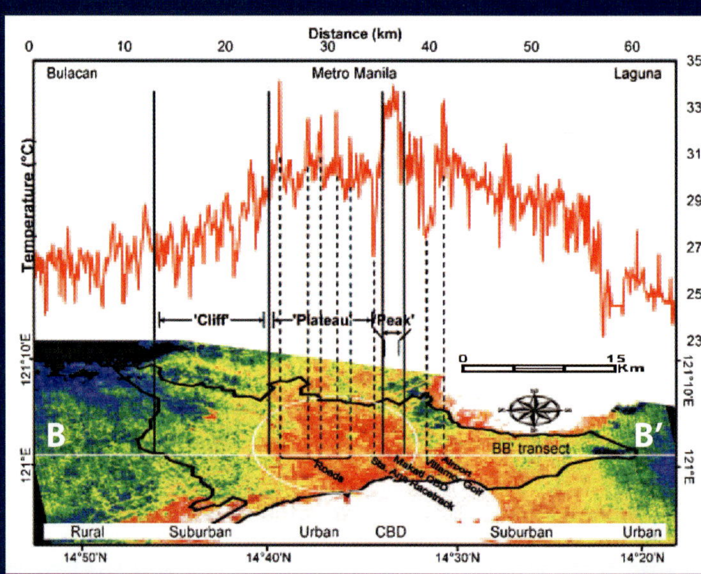

Plate 15 (Chapter 5) ASTER TIR urban heat island map and temperature profile, May 4, 2002, Manila, Philippines. Water bodies have been masked (white features). From M. Tiangco, A. M. F. Lagmay, and J. Argete. 2008. ASTER-based study of night-time urban head island effect in Metro Manila. *International Journal of Remote Sensing*, pp. 1–20 (Figures 3 and 5). Reprinted with permission of the publisher, Taylor & Francis Ltd.

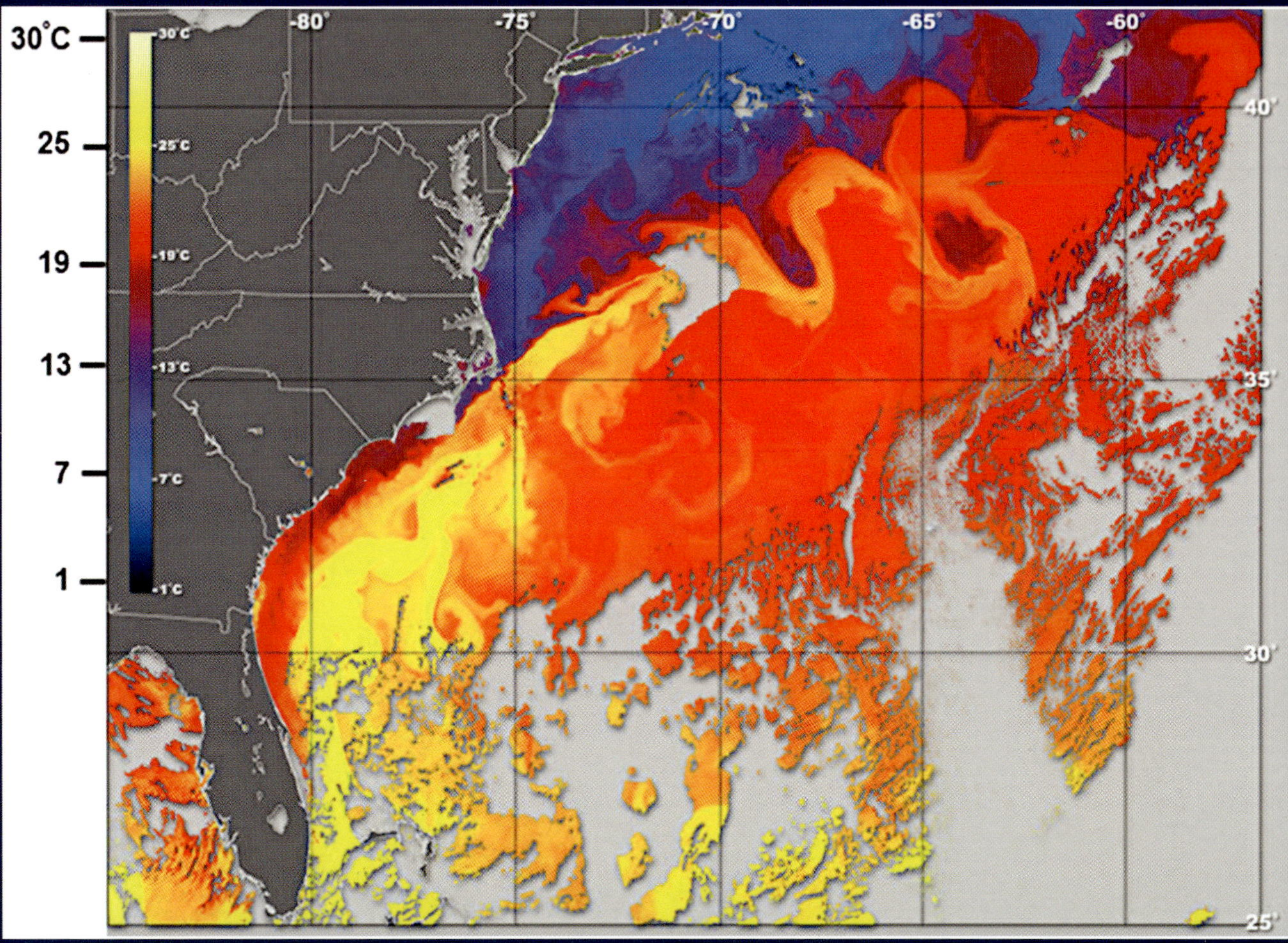

Plate 16 (Chapter 5) MODIS TIR image of the Gulf Stream and sea surface temperatures, northwestern Atlantic Ocean. Courtesy NASA GSFC.

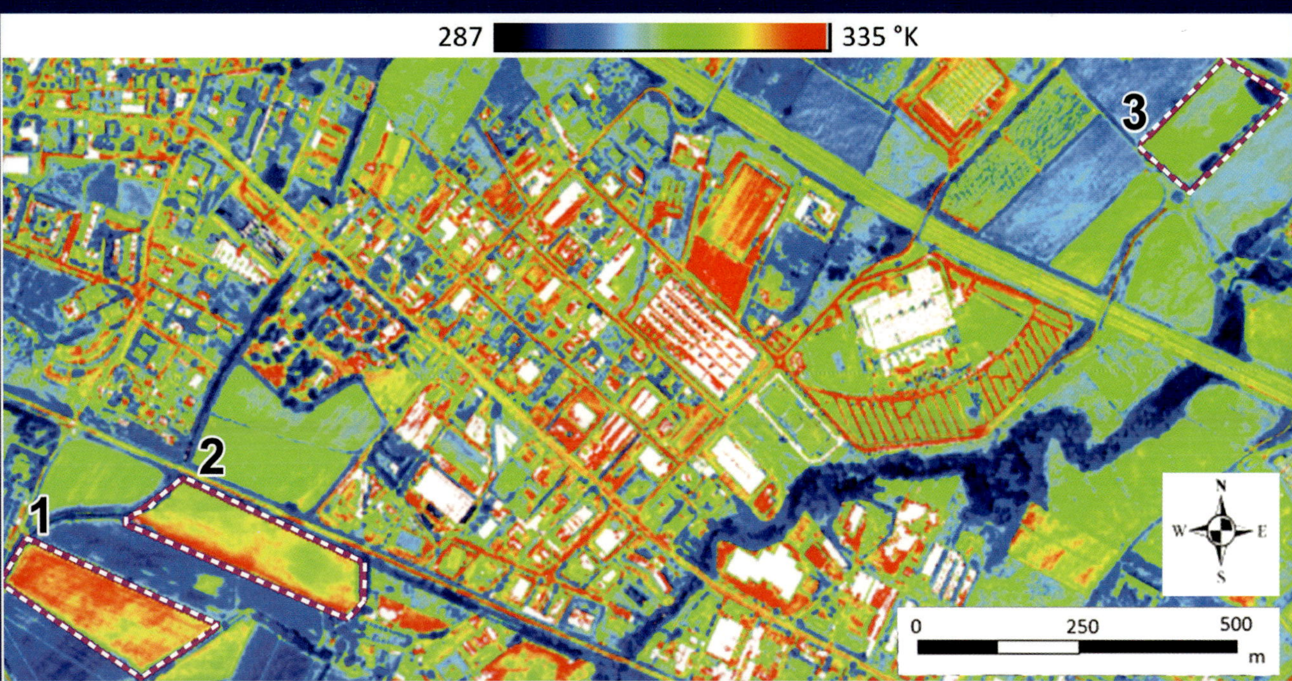

Plate 17 (Chapter 5) Airborne TIR hyperspectral temperature image of agricultural fields, Italy. From S. Pascucci, R. Casa, C. Belviso, A. Palombo, S. Pignatti, and F. Castaldi. 2014. Estimation of soil organic carbon from airborne hyperspectral thermal infrared data: A case study. *European Journal of Soil Science*, 65, 865–875 (Figure 1D). Reprinted with permission of the publisher, John Wiley and Sons.

PLATE 18 (CHAPTER 6) Radar image compared with VNIR-SWIR color composite, Death Valley, California. Courtesy JAXA and USGS.

A. 2016 PALSAR-2, L-band HH radar image.

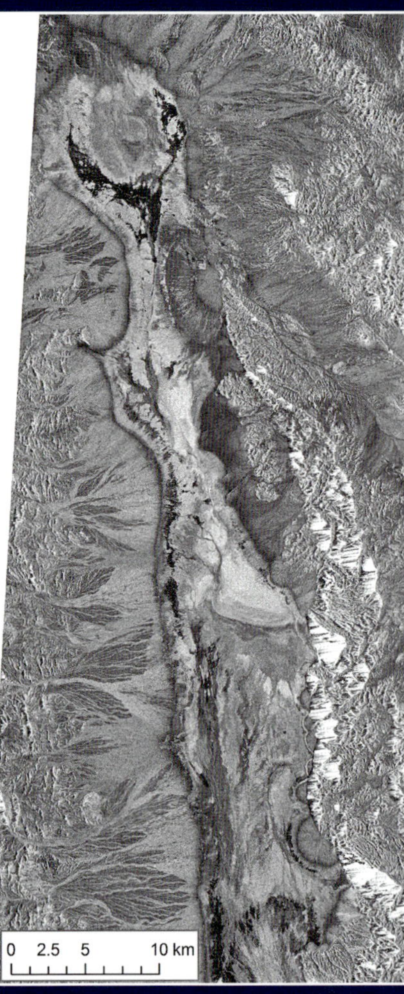

B. 2017 Landsat 8 OLI image (bands 3-5-7 in B-G-R).

A. Satellite color image of Hawaii. The white outline on (A) is the area shown in (B).

B. Interference fringes of Kilauea volcano, Hawaii.

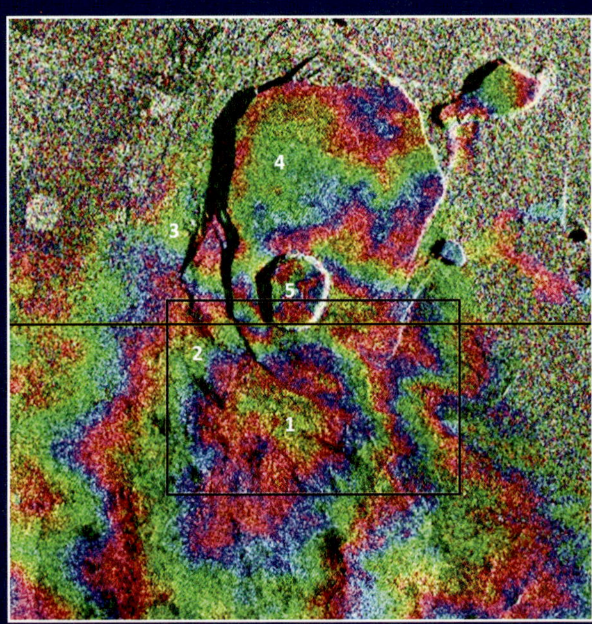

PLATE 19 (CHAPTER 6) Satellite and interference images of the Big Island of Hawaii. Courtesy USGS and NASA/JPL.

PLATE 20 (CHAPTER 6) SRTM DEM integrated with geology, Big Island of Hawaii. The black outline highlights Kilauea caldera (Plate 19 and Figure 6-43). DEM illuminated from the northwest.

A. SRTM DEM color-coded for elevation with volcanic deposits less than ~200 years old (red) and north–northeast to south–southwest topographic profile line. Courtesy USGS.

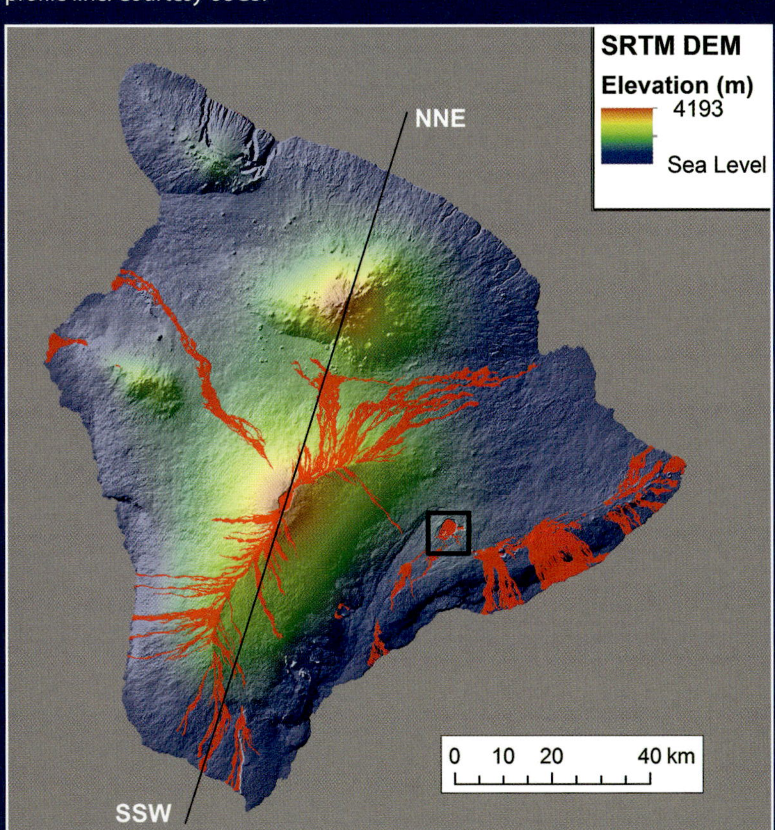

B. Geologic map draped on SRTM DEM. From D. R. Sherrod, J. M. Sinton, S. E. Watkins, and K. M. Brunt. 2007. Geologic Map of the State of Hawaii (Open-File Report 2007-1089). Washington, DC: US Geological Survey.

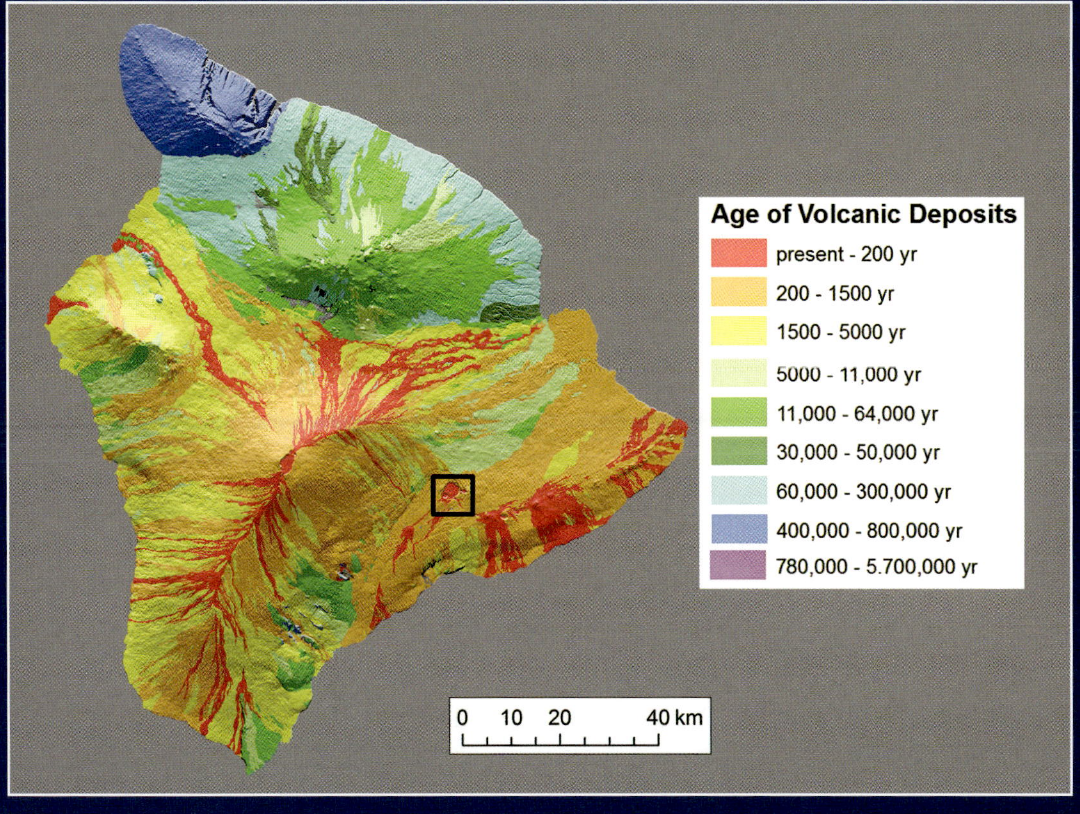

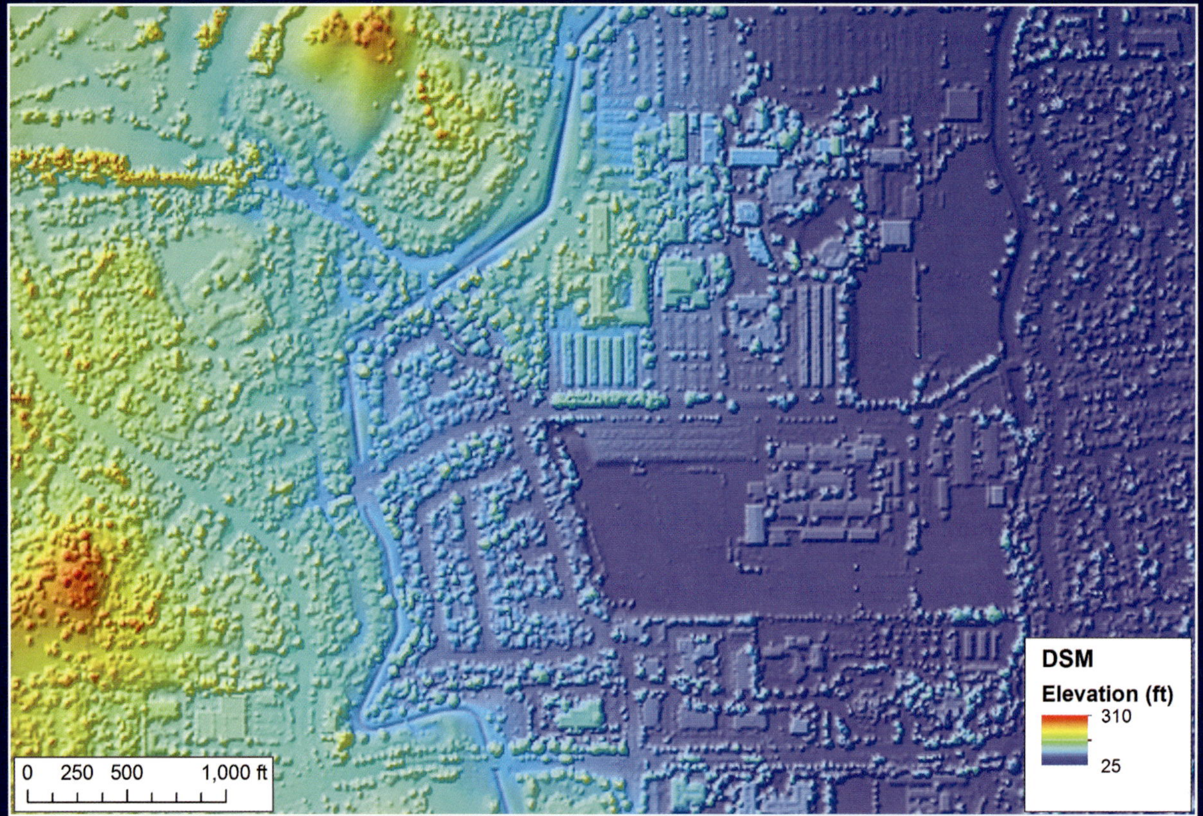

Plate 21 (Chapter 7) Hillshade image of Diablo Valley College and surrounding suburbs, California, faded and draped over a lidar DSM color-coded for elevation. Data courtesy Contra Costa County Orthoimagery Project.

Plate 22 (Chapter 7) Hillshade image of Diablo Valley College and surrounding suburbs, California, faded and draped over a lidar DTM color-coded for elevation. Data courtesy of Contra Costa County Orthoimagery Project.

A. GeoSAR X-band radar image.

B. DSM generated from X-band showing elevation of the top of the tree-shrub canopy and ground.

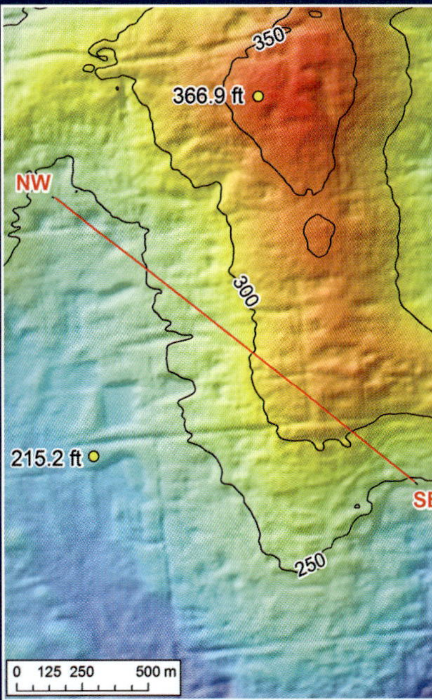

C. DTM derived from P- and X-bands showing elevation of bare earth.

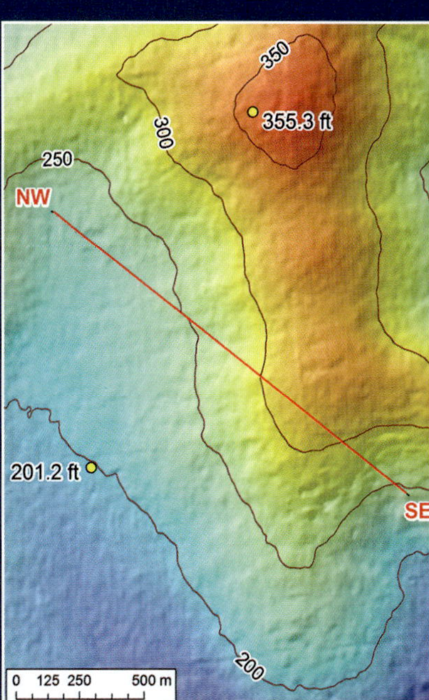

D. Northwest–southeast profile with elevation of X-band DSM and P-band DTM.

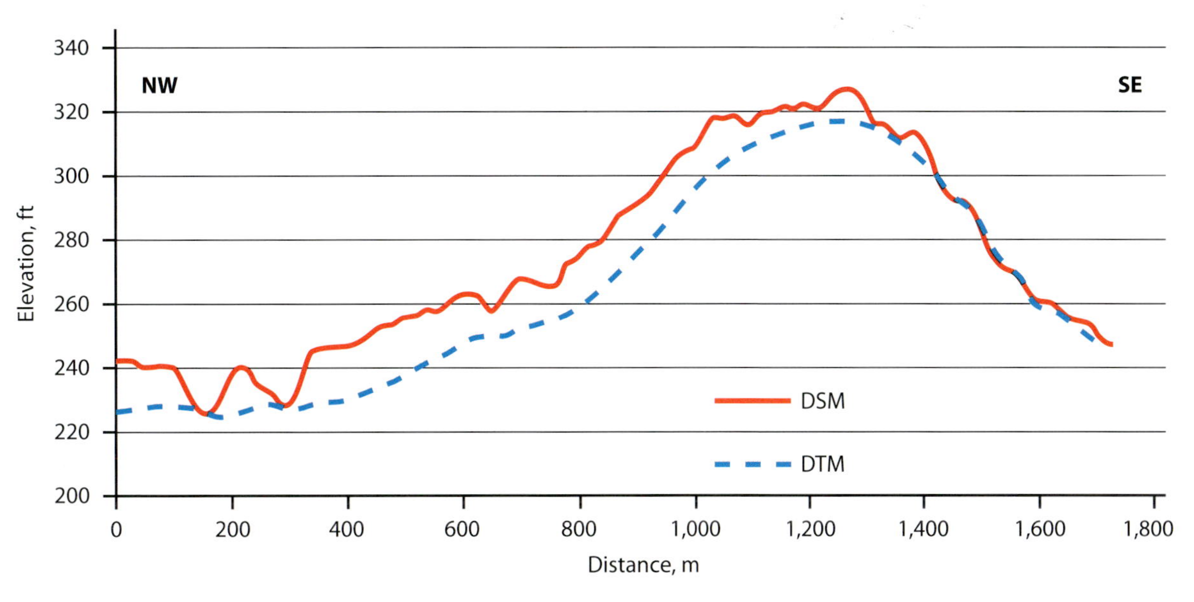

Plate 23 (Chapter 7) Airborne GeoSAR image and DEMs, Fairbanks, Alaska. Elevation points on DSM and DTM are shown by yellow dots. Northwest–southeast elevation profile is shown by red lines. Data courtesy State of Alaska, Division of Geological and Geophysical Surveys.

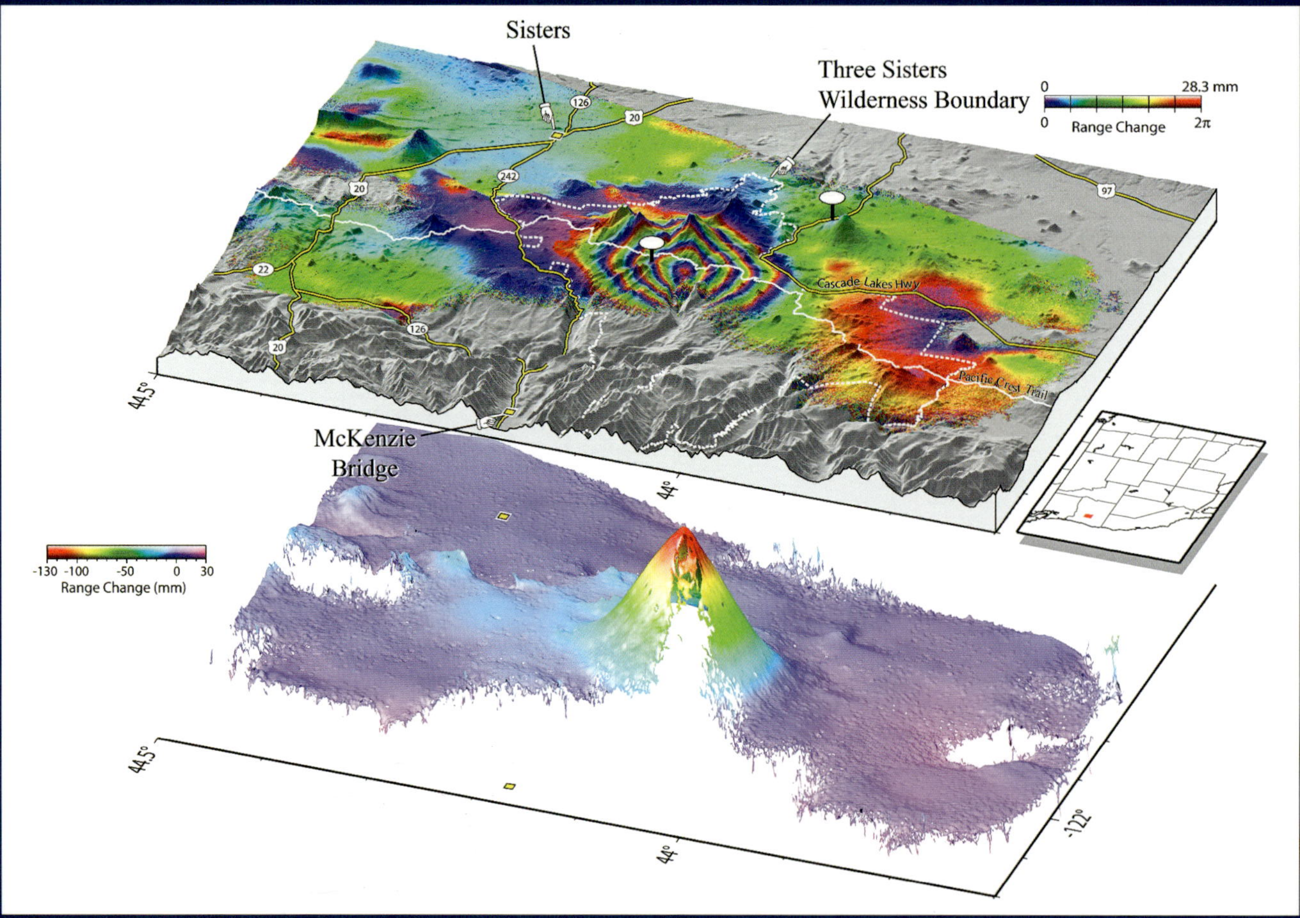

Plate 24 (Chapter 7) 1996–2000 interferogram draped over a 30-m DEM (top layer) with uplift (decrease in range to satellites) approaching 15 cm (bottom layer). From USGS. 2016. InSAR—Satellite-Based Technique Captures Overall Deformation "Picture." US Geological Survey. Volcanic Hazards Program. http://volcanoes.usgs.gov/vhp/insar.html (accessed May 2019).

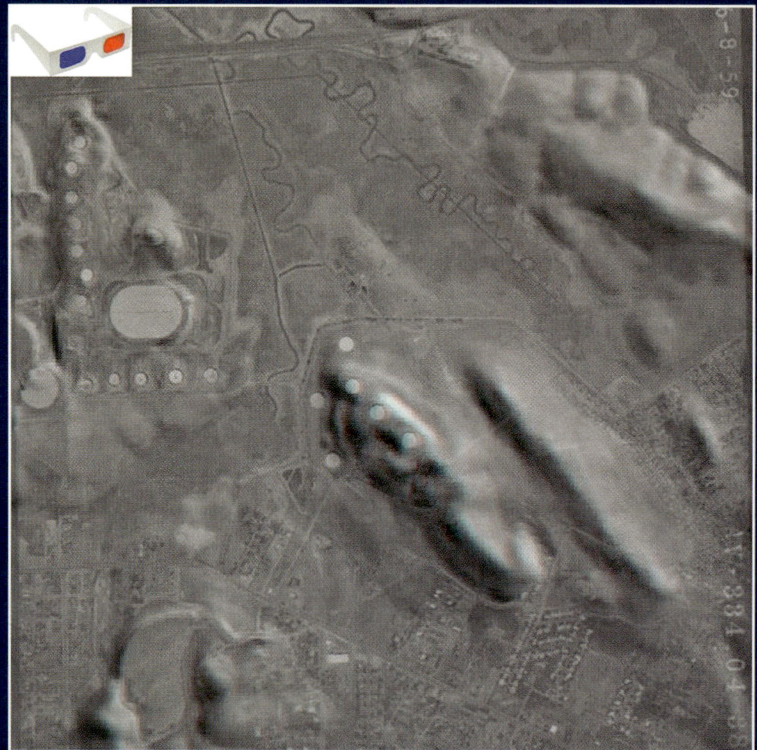

Plate 25 (Chapter 7) Red/cyan anaglyph of synthetic stereo DEM. Compare with stereo model from aerial photography in Plate 4. Data courtesy USGS and D. Ruiz, Quantum Spatial, Inc., Novato, California.

A. Ten flight lines and 236 photos were needed for 80% forward lap and 40% sidelap.

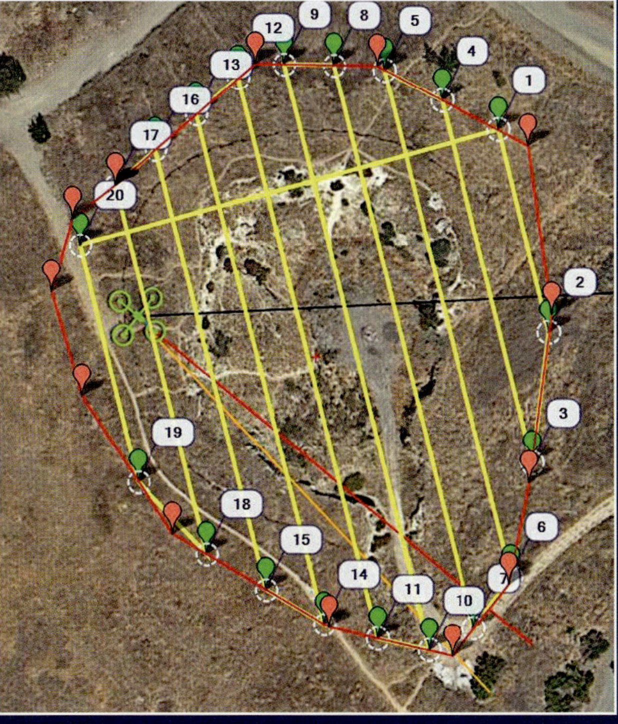

B. Color orthoimage mosaic with 1 cm pixels.

C. DTM generated from SfM photogrammetric technique.

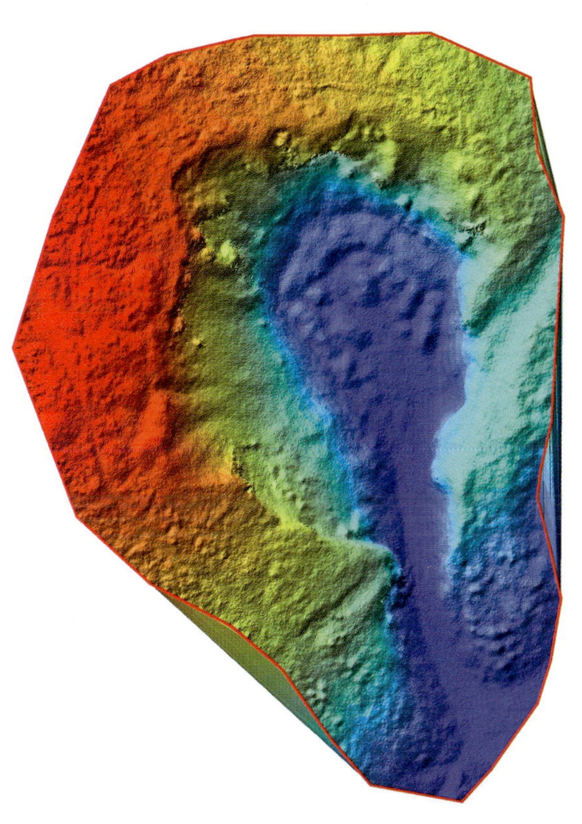

D. Topographic contours derived from DTM.

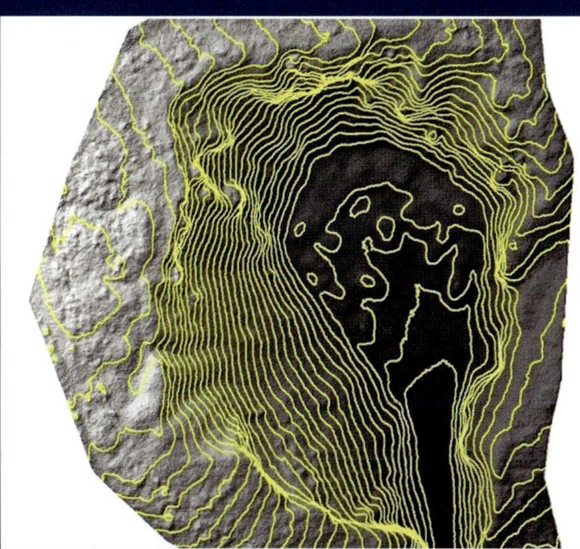

Plate 26 (Chapter 8) sUAS flight plan and three photogrammetrically derived products using an iPhone 5S and a quadcopter, Concord, California. From J. C. Miller. 2016. The acquisition of image data by means of an iPhone mounted to a drone and an assessment of the photogrammetric products. MA Thesis, California State University at East Bay. http://iphonedroneimagery.com/project-methodology (accessed January 2018).

A. Landsat 8 OLI PC 234 image.

B. Landsat 8 OLI PC 456 image.

C. Landsat 5 TM PC 234 image.

Plate 27 (Chapter 9) Landsat 8 OLI color PC images compared with Landsat 5 TM color PC image, Thermopolis, Wyoming, subscene. Landsat courtesy USGS.

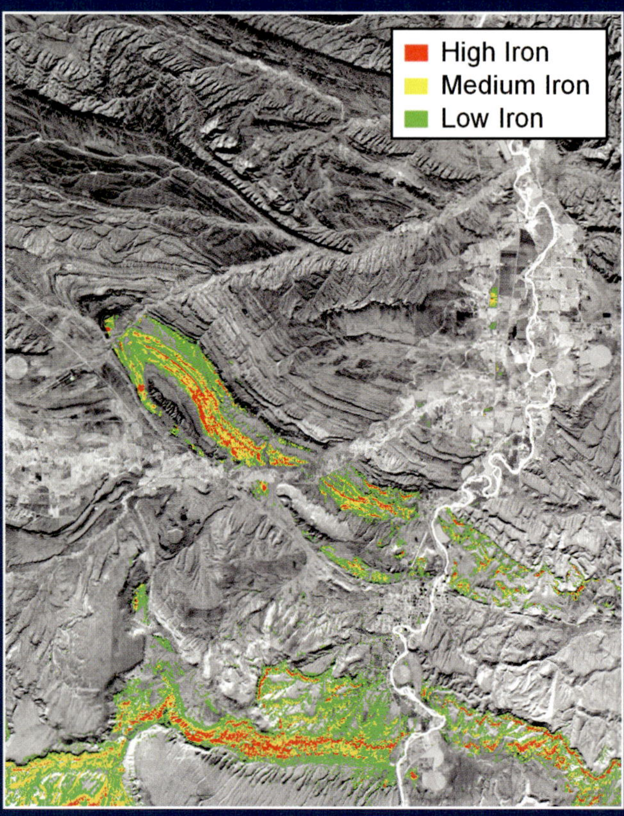

Plate 28 (Chapter 9) Density slice of Landsat 8 iron band ratio 4/2 (red/blue), Thermopolis, Wyoming. Landsat courtesy USGS.

Plate 29 (Chapter 9) Landsat 8 OLI imagery spectrally classified using the supervised (maximum likelihood) technique and unsupervised (isodata) technique, Martinez, California. Landsat courtesy USGS.

A. Landsat 8 CIR image.

B. Supervised classification with legend.

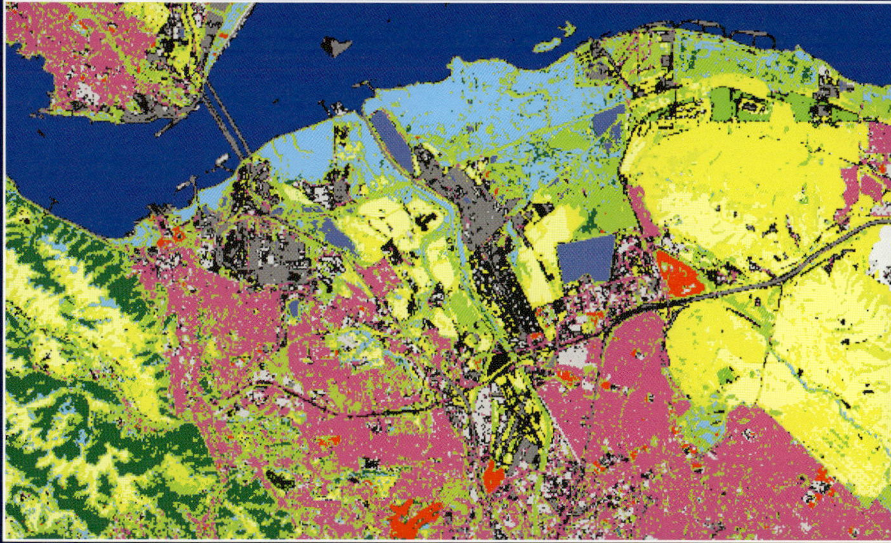

Supervised Classification

Class Name

- White roofs and concrete
- Construction site—white soils?
- Industry 1
- Industry 2
- Suburbs
- Irrigated vegetation
- Trees
- Dry grass light
- Dry grass 1—dark
- Dry grass 2—dark
- Dry grass—dark shadows
- Shrubs–grass—not irrigated
- Grass—not irrigated
- Dark soils—wet?
- Wetlands
- Water-ponded
- Water–river

C. Unsupervised classification with legend.

Unsupervised Classification

Class Name

- White roofs and concrete
- Industry 1
- Industry 2
- Suburbs
- Irrigated vegetation
- Trees
- Dry grass
- Dry grass and exposed soil?
- Grass—not irrigated
- Shrubs–grass—wet?
- Wetlands
- Water

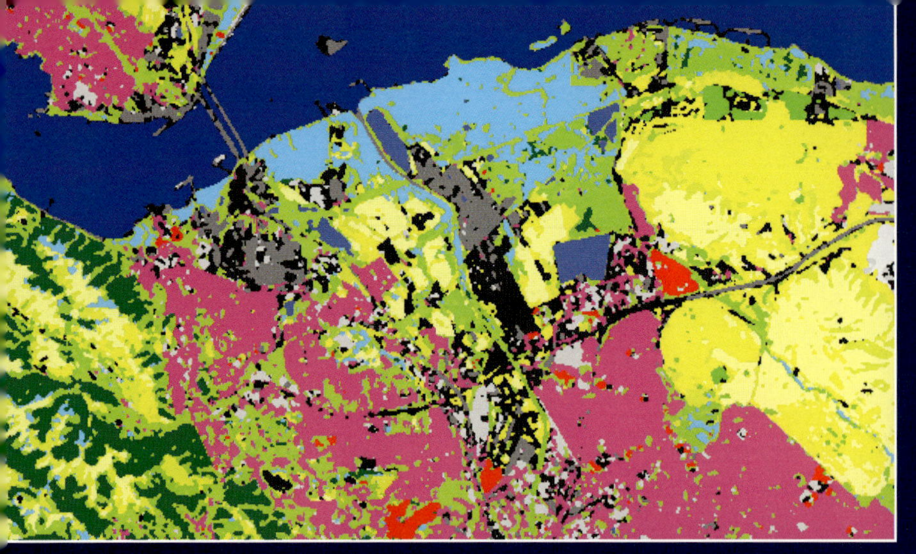

Plate 30 (Chapter 9) Generalization of supervised classification map using majority filter, Martinez, California. Landsat courtesy USGS.

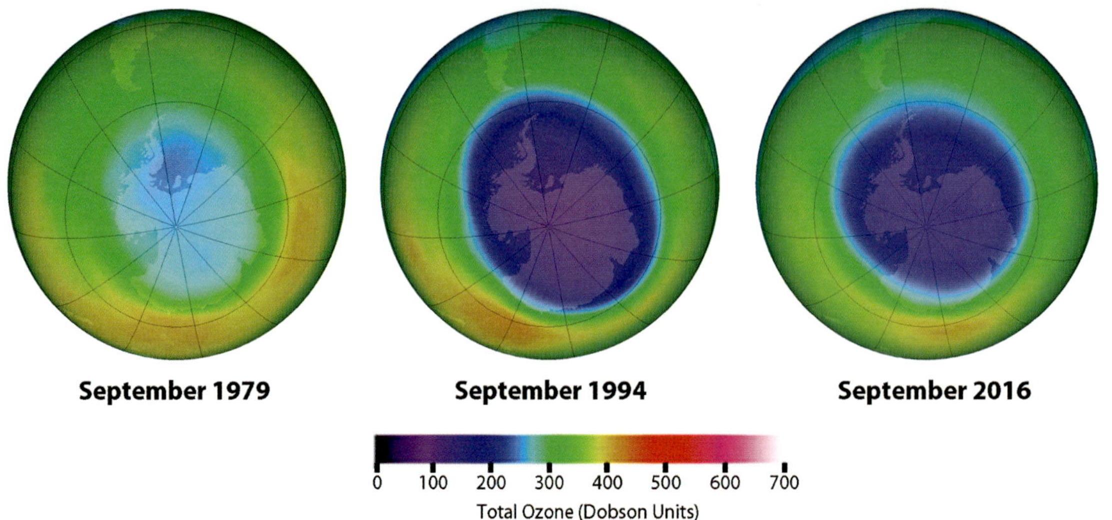

Plate 31 (Chapter 11) Ozone hole over the South Pole as mapped by TOMS and OMI data. Courtesy NASA GSFC Ozone Watch (ozonewatch.gsfc.nasa.gov/SH.html).

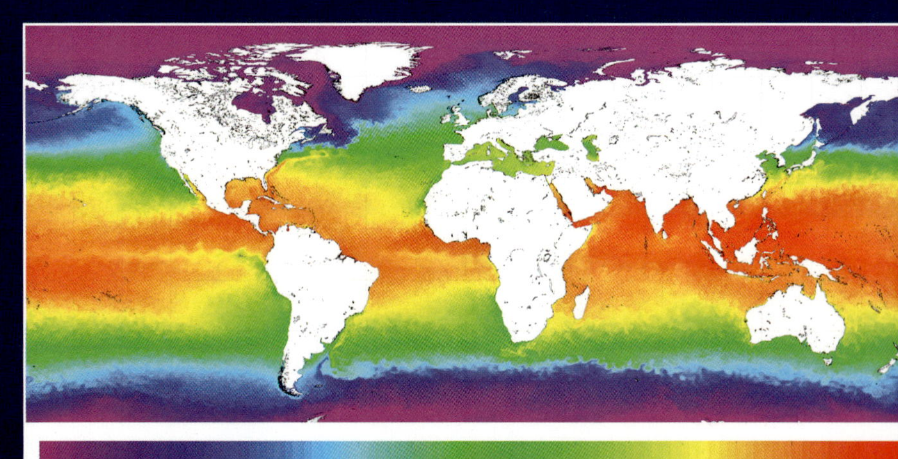

Plate 32 (Chapter 11) Sea surface temperatures for June 2, 2017. Courtesy NOAA

A. AVHRR sea surface temperature. B. SeaWiFS chlorophyll-a map.

AVHRR SST (°C)
18
16
14
12
10
Monterey Bay
Pacific
SeaWiFS Chlorophyll (mg m^{-3})
8
4
2
1
.5
.2
39°N
37°N
35°N
127°W
125°W
123°W
121°W
127°W
125°W
123°W
121°W

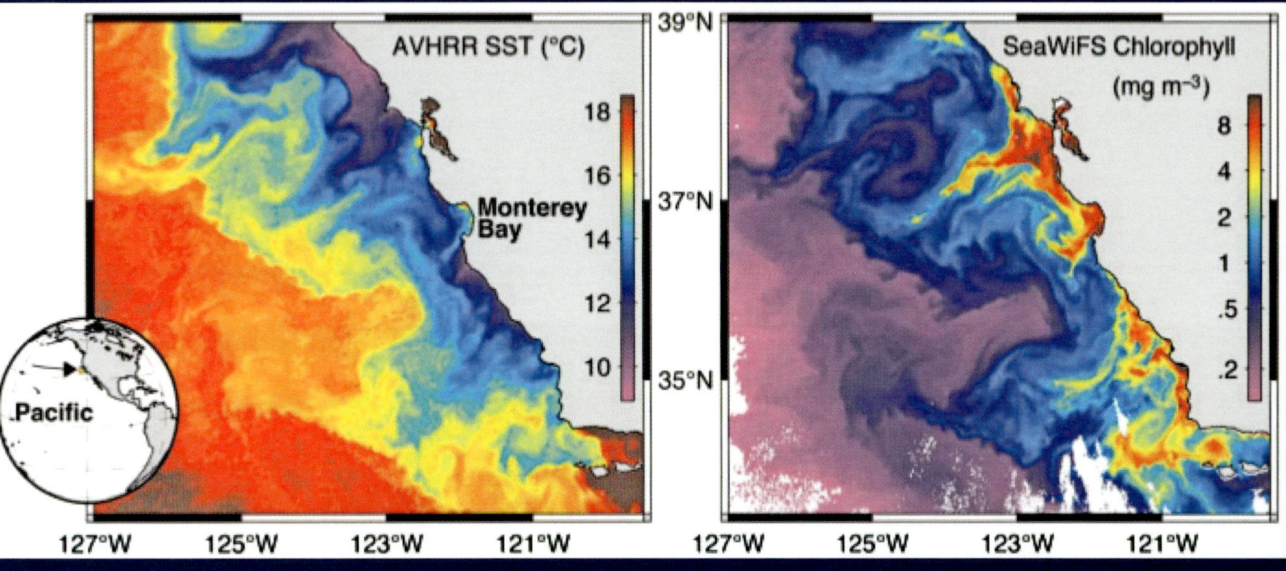

Plate 33 (Chapter 11) Comparison of sea surface temperature and ocean color. From J. P. Ryan, F. P. Chavez, and J. G. Bellingham. 2005. Physical-biological coupling in Monterey Bay, California: Topographic influences on phytoplankton ecology. *Marine Ecology Progress Series*, 287, 23–32. © Inter-Research 2005.

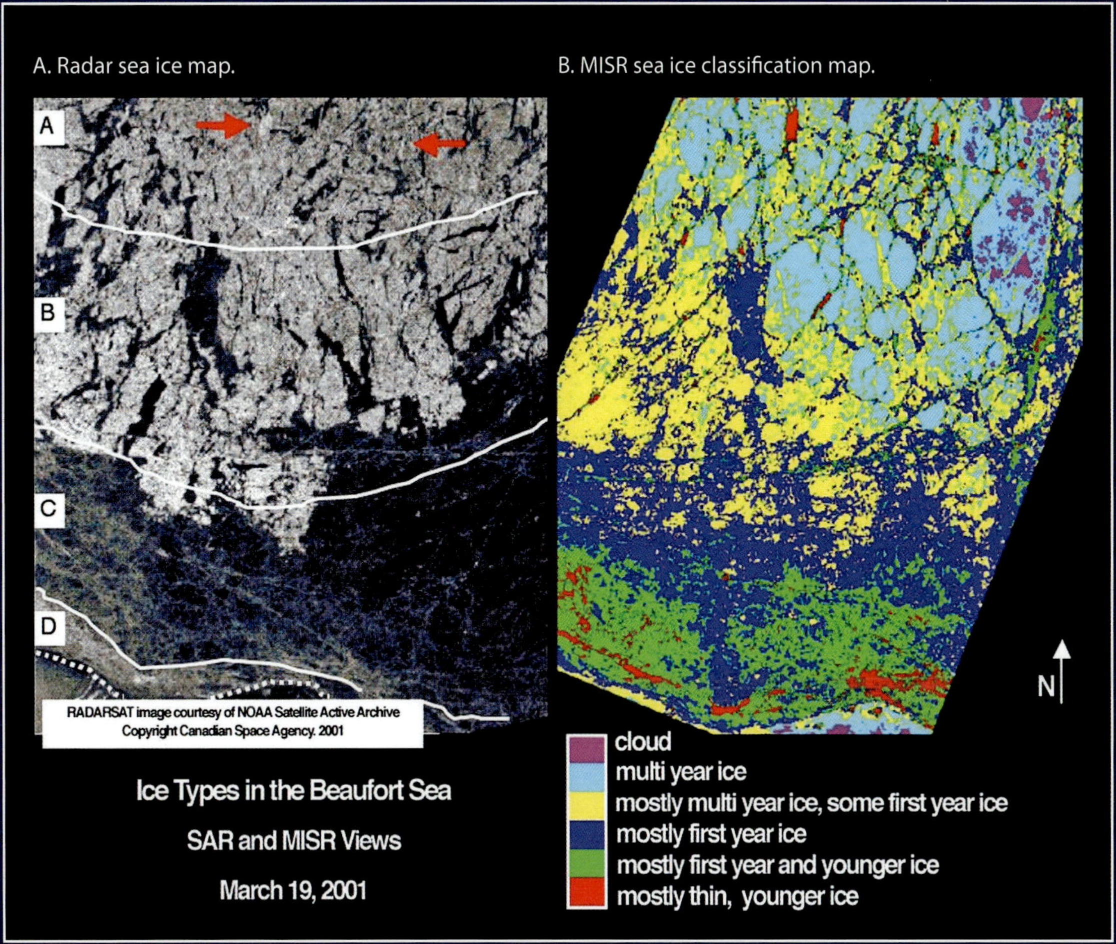

A. Radar sea ice map.
B. MISR sea ice classification map.
A
B
C
D
RADARSAT image courtesy of NOAA Satellite Active Archive
Copyright Canadian Space Agency. 2001
N
cloud
multi year ice
mostly multi year ice, some first year ice
mostly first year ice
mostly first year and younger ice
mostly thin, younger ice
Ice Types in the Beaufort Sea
SAR and MISR Views
March 19, 2001

Plate 34 (Chapter 11) Sea ice types mapped from SAR and MISR. © IEEE. From A. W. Nolin, F. M. Fetterer, and T. A. Scambos. 2002. Surface roughness characterizations of sea ice and ice sheets: Case studies with MISR data. *IEEE Transactions on Geoscience and Remote Sensing*, 40(7), 1,605–1,615.

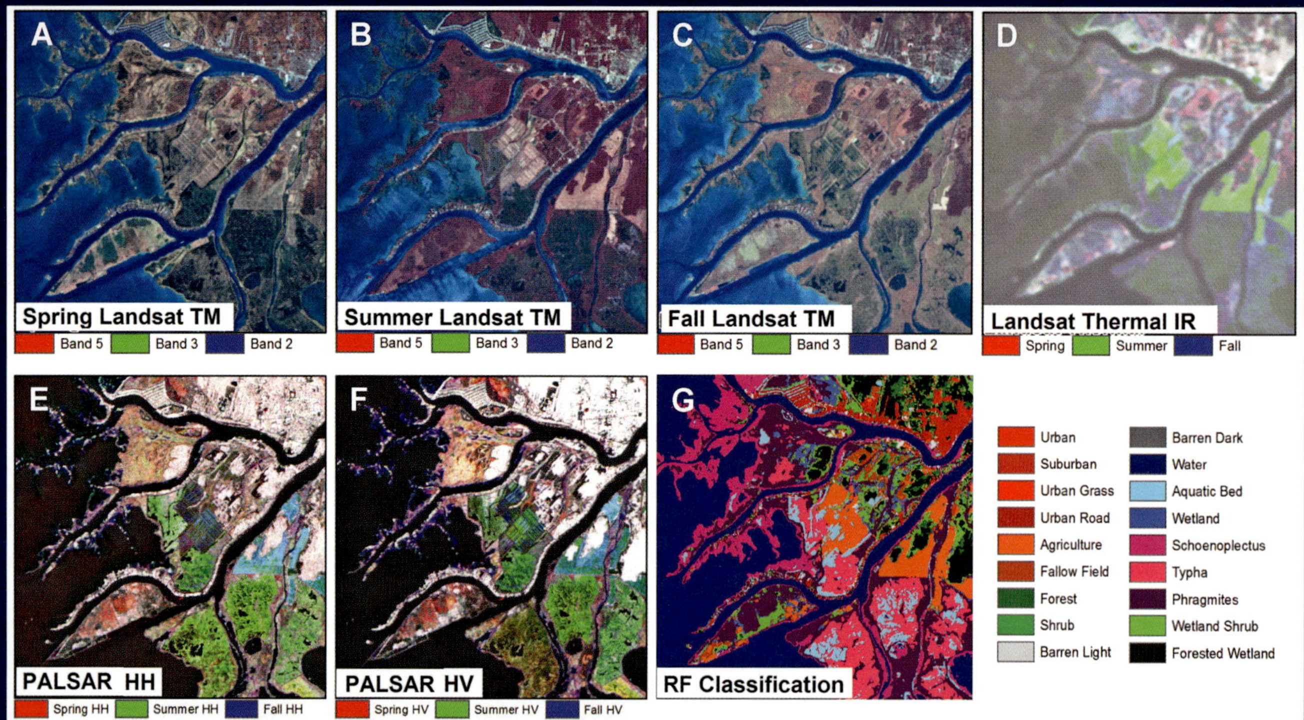

PLATE 35 (CHAPTER 11) Landsat and PALSAR imagery used to map land cover and wetlands types. HH and HV indicate polarization of PALSAR imagery. Land cover map from fused Landsat and PALSAR images. From L. Bourgeau-Chavez, S. Endres, M. Battaglia, M. E. Miller, E. Banda, Z. Laubach, P. Higman, P. Chow-Fraser, and J. Marcaccio. 2015. Development of a bi-national Great Lakes coastal wetland and land use map using three season PALSAR and Landsat imagery. *Remote Sensing*, 7, 8,655–8,682 (http://creativecommons.org/licenses/by/4.0).

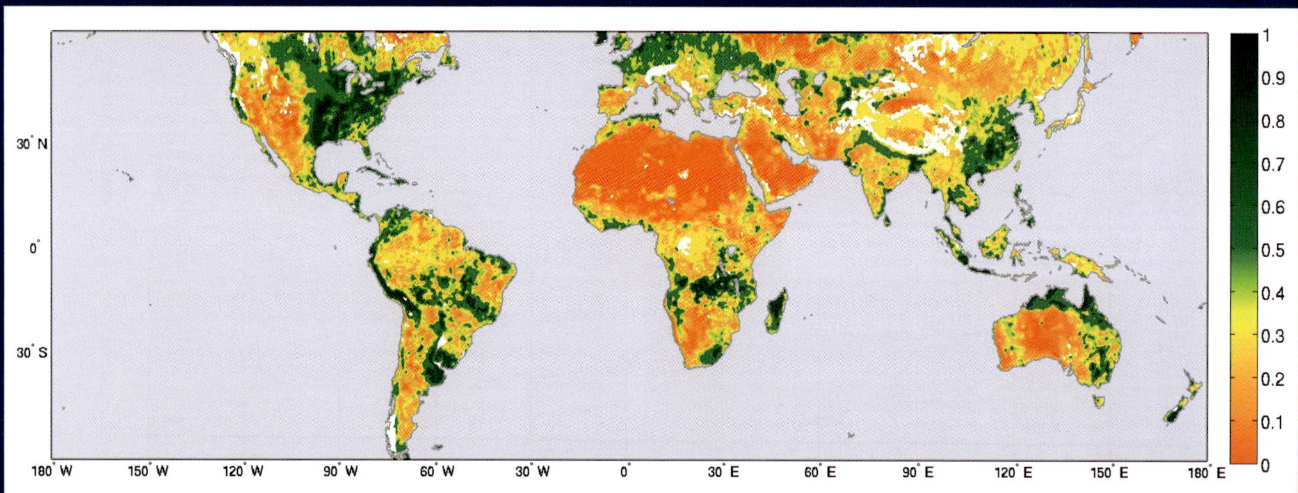

PLATE 36 (CHAPTER 11) SMOS global soil moisture map for April 2, 2012. From the Center for the Study of the Biosphere from Space. 2017. SMOS Global Drought Monitor. SMOS Blog. http://www.cesbio.ups-tlse.fr/SMOS_blog/?page_id=2589 (accessed January 2018).

A. Bands 1-2-3 in B-R-G.

B. Bands 4-5-7 in B-G-R.

C. Bands 2-4-7 in B-G-R.

D. TIR band 6 density-sliced. Red = warm; dark blue = cool.

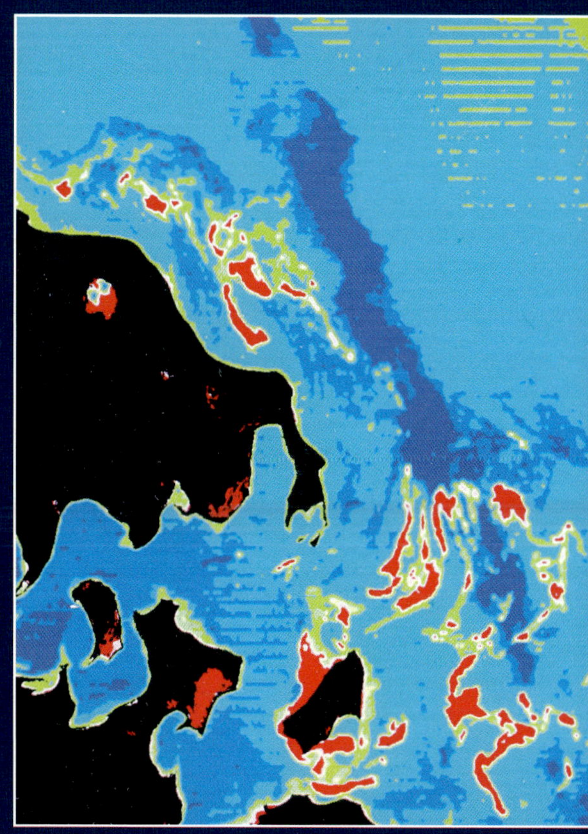

PLATE 37 (CHAPTER 11) Landsat TM subscenes of Arabian Gulf oil spill, acquired February 16, 1991.

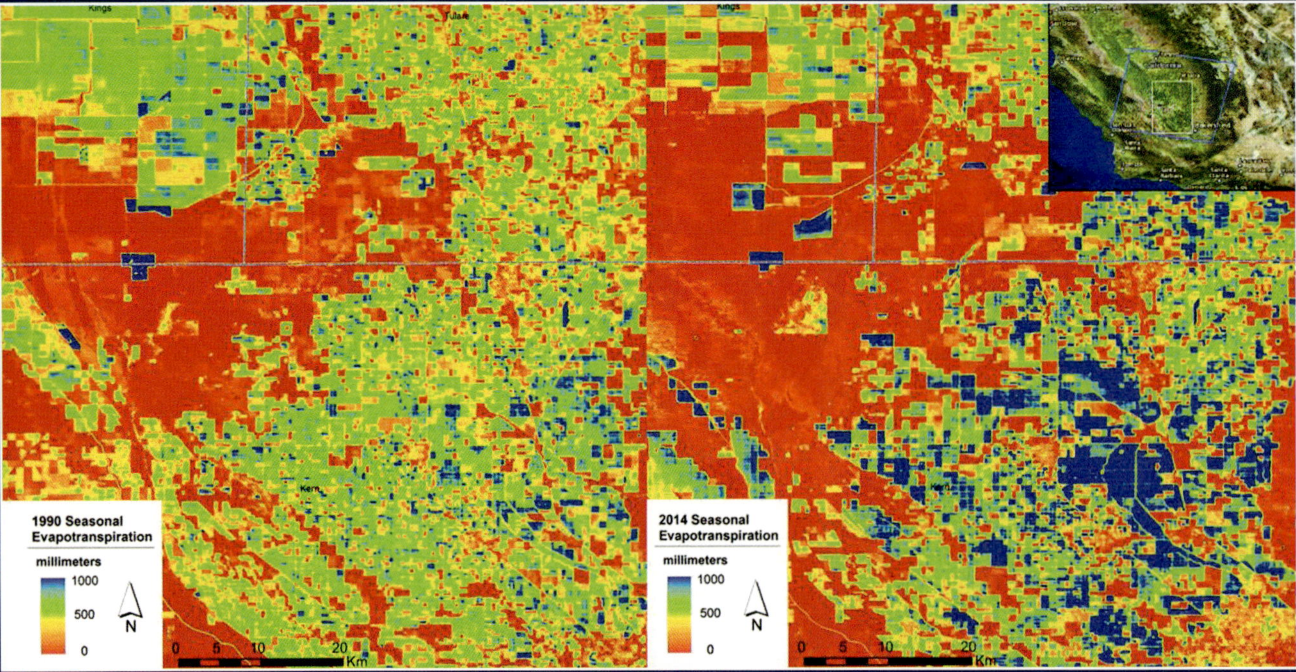

Plate 38 (Chapter 12) This pair of ET water-use maps shows crop water use in California's San Joaquin Valley in 1990 (left) and 2014 (right). From USGS. 2016. Mapping Water Use: Landsat and America's Water Resources. http://www.usgs.gov/news/mapping-water-use-landsat-and-americas-water-resources (accessed January 2018).

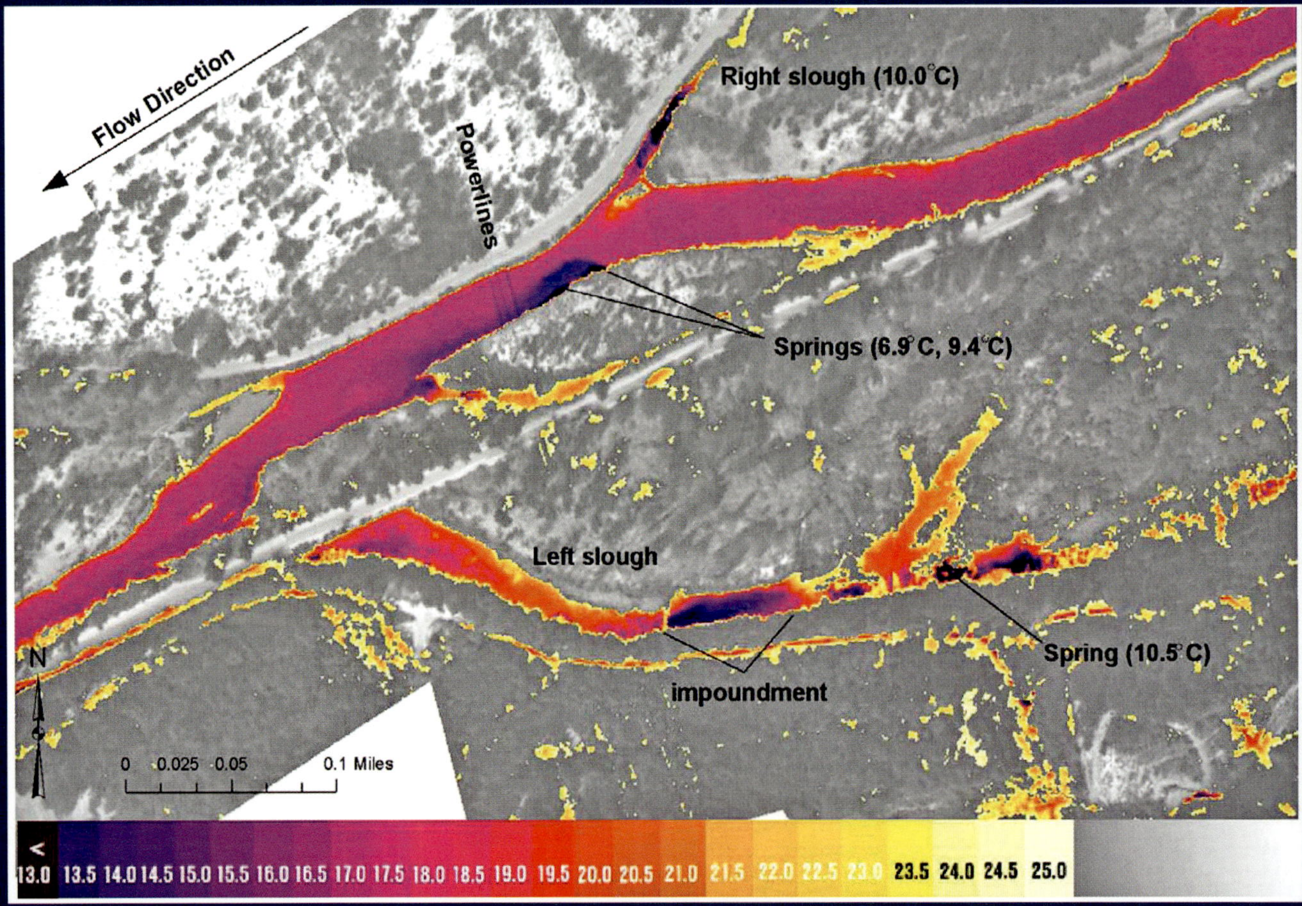

Plate 39 (Chapter 12) Airborne TIR image of Coeur d'Alene River, Idaho, with surface water temperatures. From US Forest Service. 2007. *Airborne Thermal Infrared Remote Sensing, Coeur d'Alene River, Idaho.* Report submitted by Watershed Sciences, Inc.

A. Shaib Thamamah feature training site.

B. Raghib discovery.

PLATE 40 (CHAPTER 13) Oil discovery in Saudi Arabia using Landsat TM images. Landsat courtesy USGS.

A. Landsat image. OLI bands 3-5-7 in B-G-R.

B. HRV alteration image.

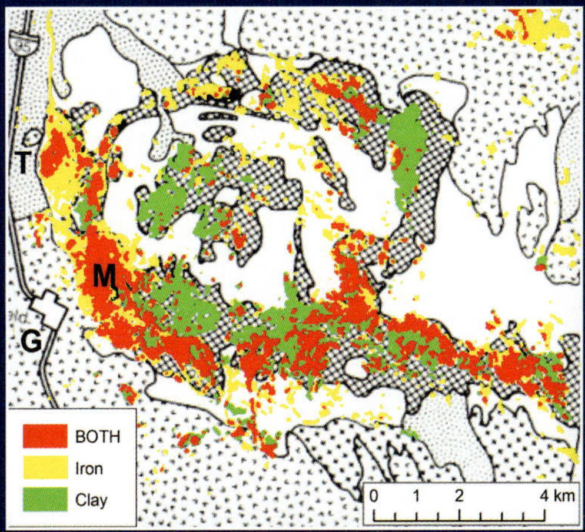

C. USGS Landsat mineral map.

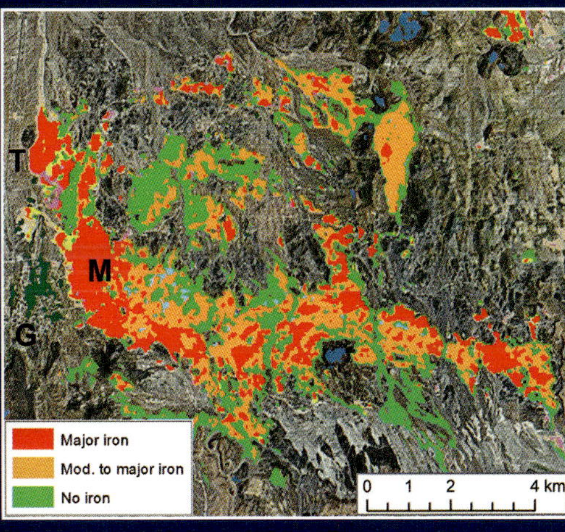

D. AVIRIS endmember abundance image.

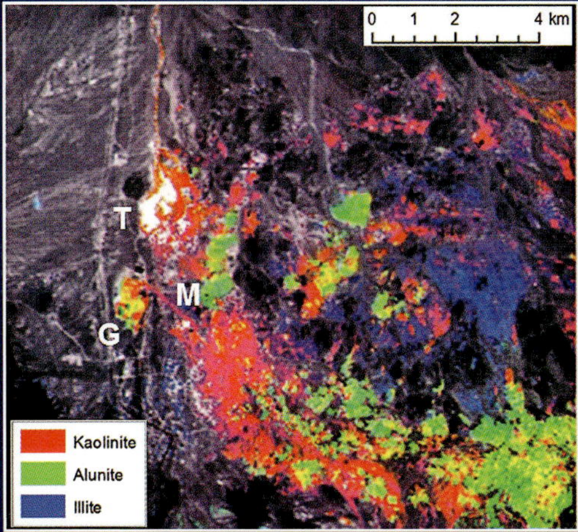

PLATE 41 (CHAPTER 13) Goldfield Mining District, Nevada. The district is the training site where we developed the HRV method for mineral exploration. Landsat courtesy of USGS. (C) From B. W. Rockwell, L. C. Bonham, and S. A. Giles. 2015. USGS National Map of Surficial Mineralogy. US Geological Survey Online Map Resource. http://cmerwebmap.cr.usgs.gov/usminmap.html (accessed January 2018). (D) Courtesy F. A. Kruse.

A. 1986 Landsat image before copper discoveries. No evidence of the deposits.

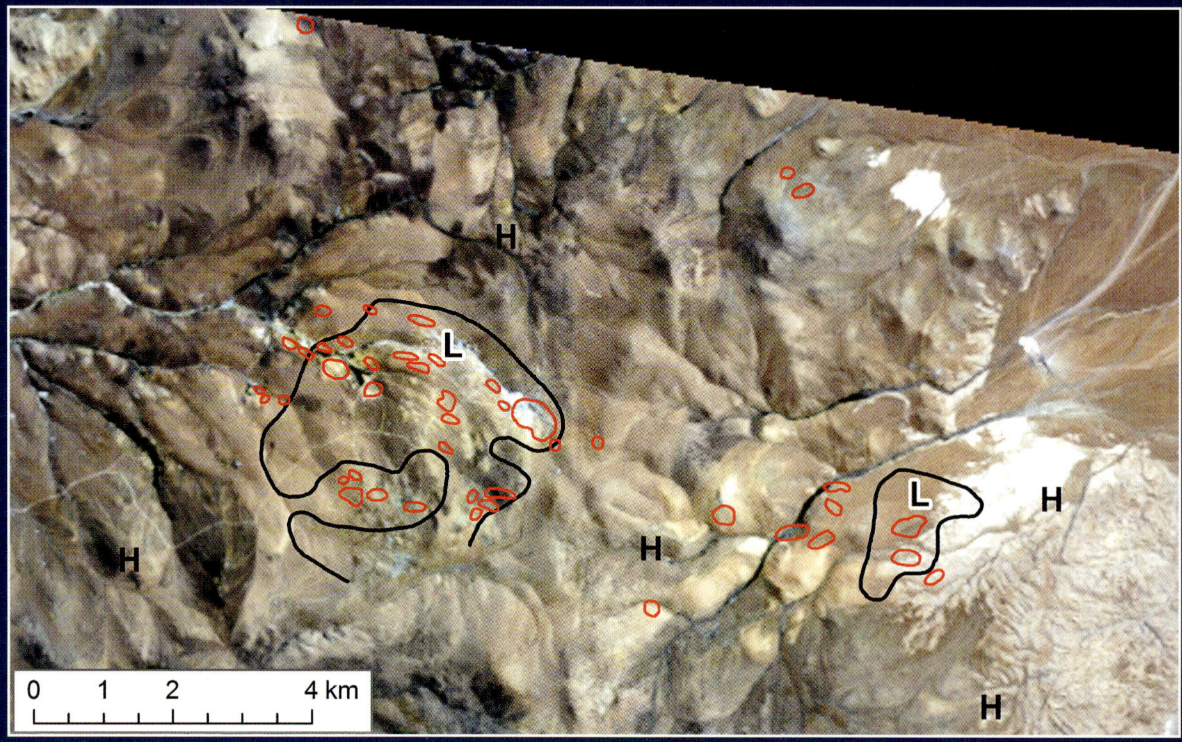

B. 2017 Landsat image after eight years of copper development.

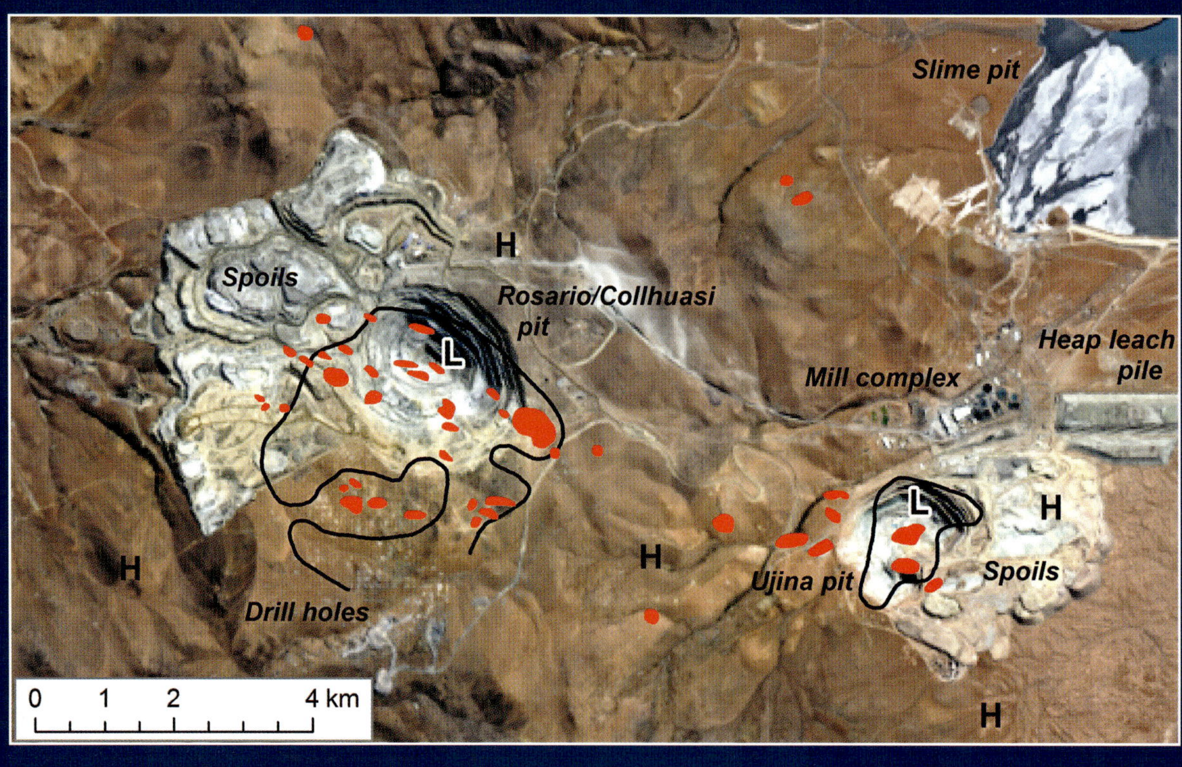

Low resistivity contours	L Low resistivity values (with mineral potential)	H High resistivity values (indicates barren ground)	Hydrothermal alteration (from HRV features)

Plate 42 (Chapter 13) Rosario/Collahuasi and Ujina, Chile world-class porphyry copper discoveries. Red outlines and red-filled polygons are HRV alteration features from early 1990s Landsat images. Landsat courtesy USGS.

Plate 43 (Chapter 13) Landsat images of Balkhab Valley, Afghanistan. Arrows indicate a known copper deposit. Landsat courtesy USGS.

A. Enhanced color image. White outlines are unconformable contacts between the VMS sequence and overlying Mesozoic strata.

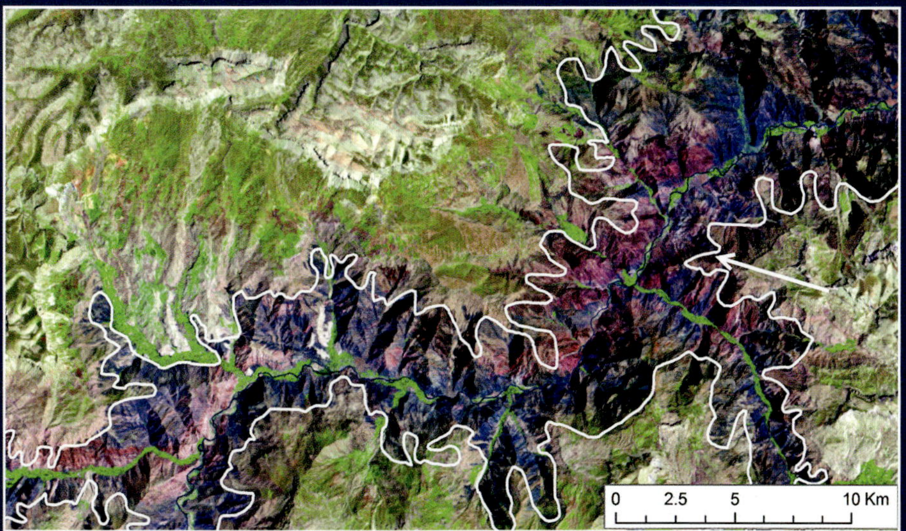

B. HRV alteration image. White outlines show the locations of seven exploration sites. Site 1 is shown in Plate 44.

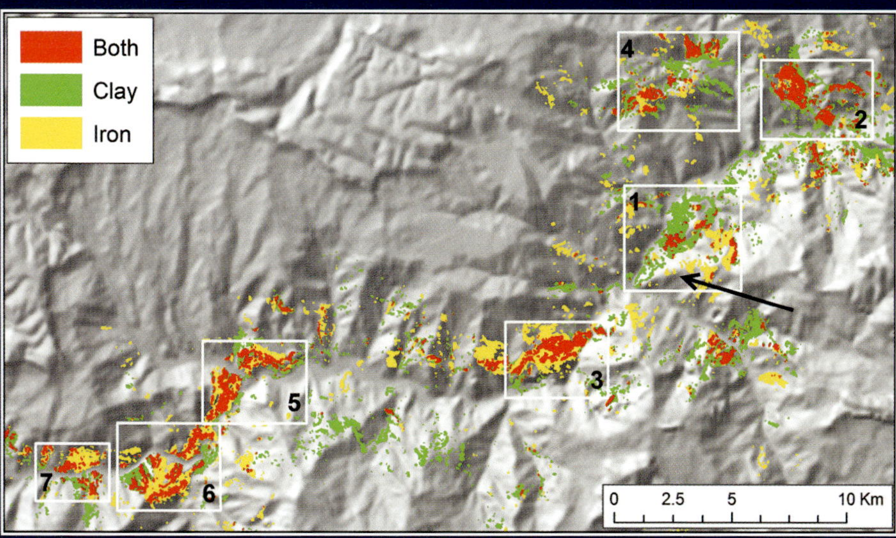

C. Unsupervised classification image.

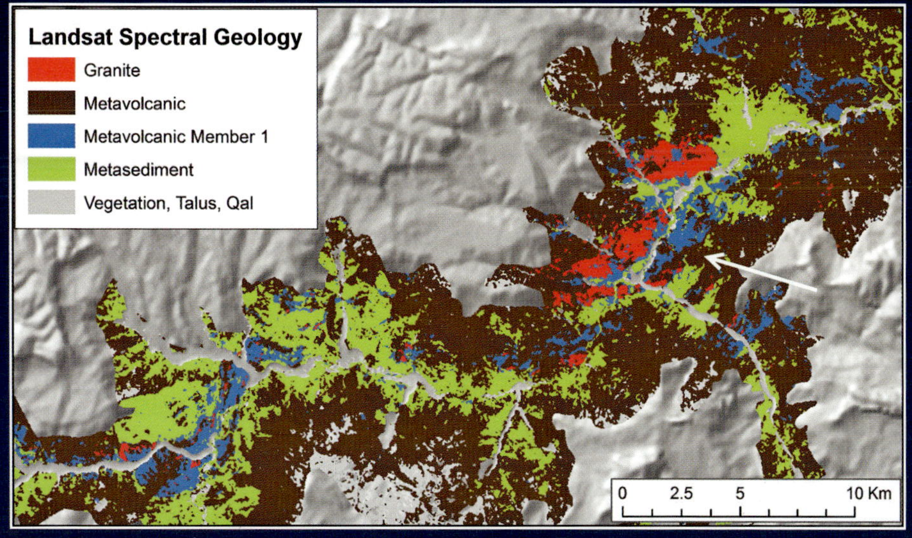

Plate 44 (Chapter 13) Balkhab Valley Site 1 from Plate 43B, comparing HRV and HyMap images. Mineral prospects X, Y, and Z are described in Chapter 13. HyMap data and Landsat courtesy USGS.

A. Landsat HRV alteration image.

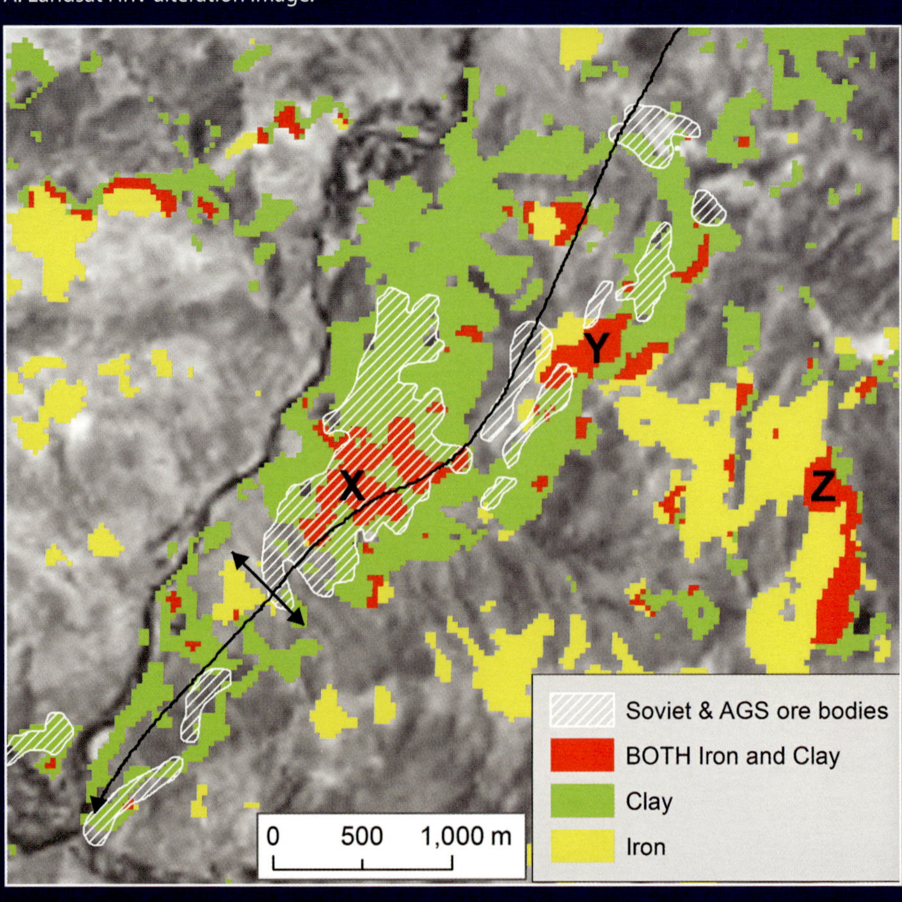

B. HyMap hyperspectral mineral map.

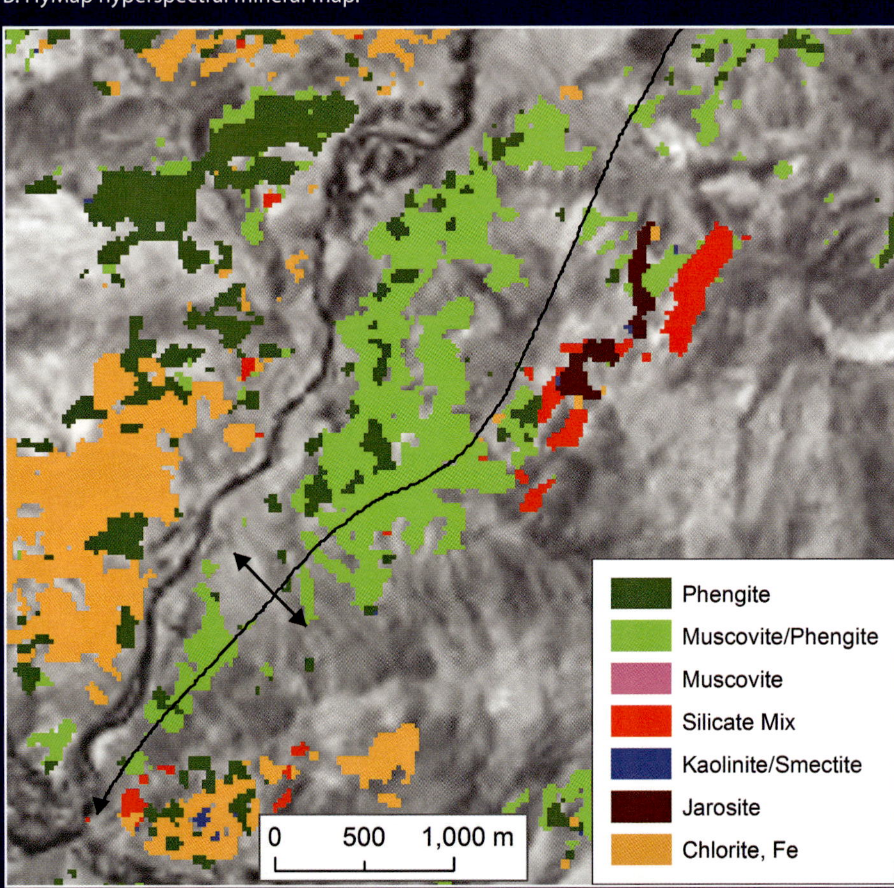

A. USGS geologic map.

Q3a Quaternary conglomerate, sandstone, sand
Q3loe Late Pleistocene loess (silt)
P2csh Paleocene clay and shale
P1bl Paleocene basalt lava

B. SRTM DEM color-coded for elevation.

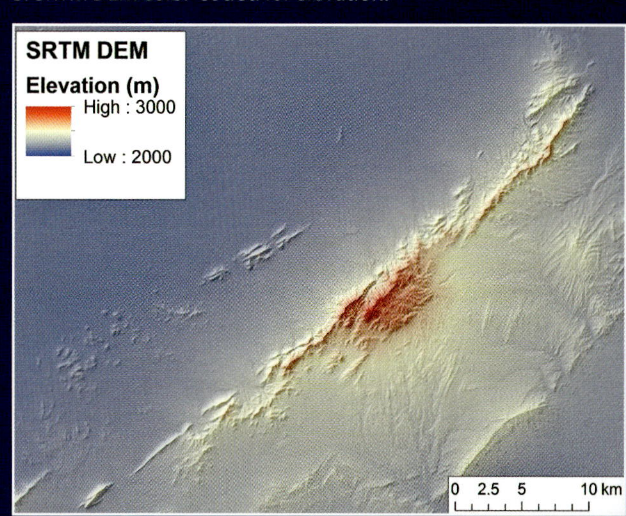

C. ASTER TIR quartz band ratio 14/12.

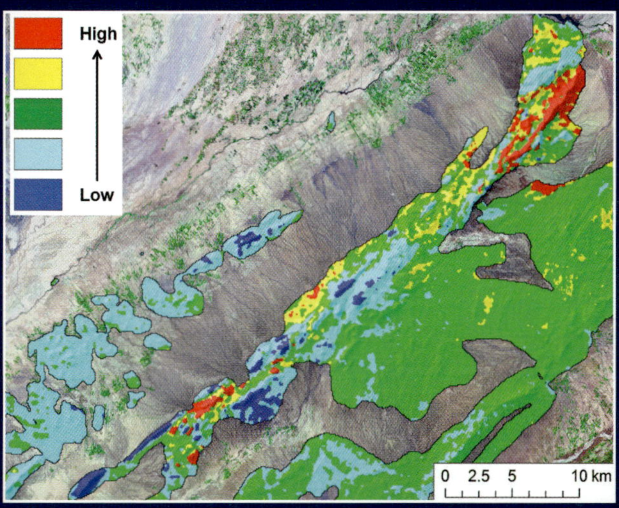

D. ASTER TIR carbonate band ratio 13/14.

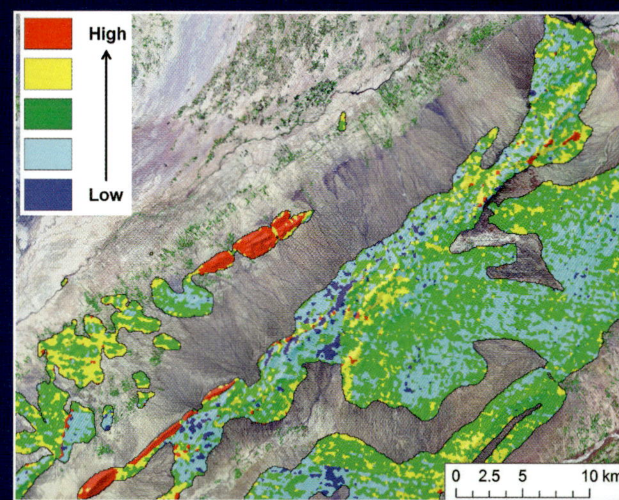

E. ASTER TIR mafic band ratio 12+14/13.

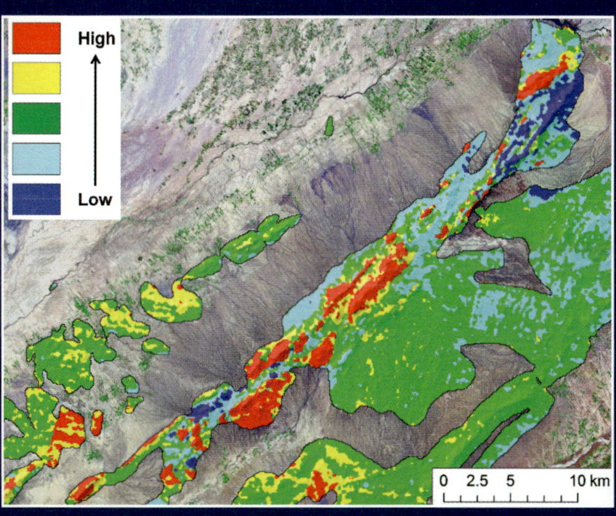

F. Geologic map from ASTER TIR band ratios.

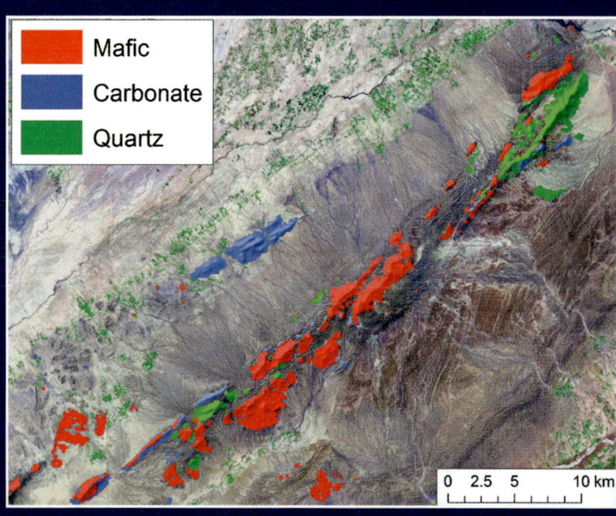

Plate 45 (Chapter 13) ASTER TIR emissivity band ratio images shown as color-coded density slices, Katawas area, Afghanistan. See Figure 13-39 for grayscale ASTER TIR band ratio images. DEM and ASTER data courtesy USGS. (A) From J. L. Doebrich and R. R. Wahl. 2006. *Geologic and Mineral Resource Map of Afghanistan* (Open File Report 2006-1038). Washington, DC: US Geological Survey.

A. Interferogram of JERS data collected 86 days apart. One rainbow color cycle corresponds with 0 to 11.75 cm of deformation.

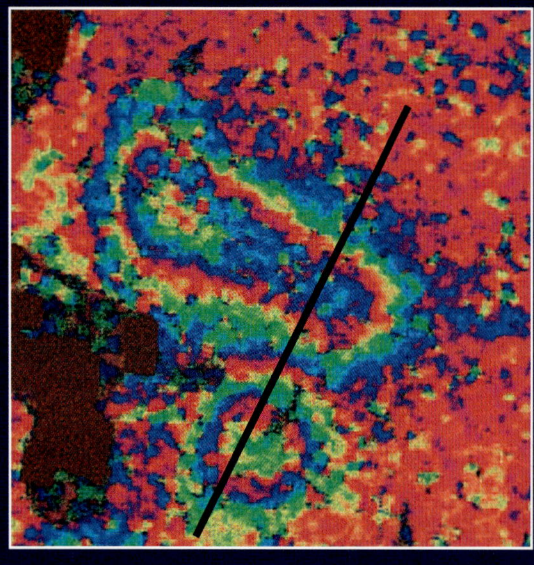

B. Deformation measurements in subarea A over a 43-day period.

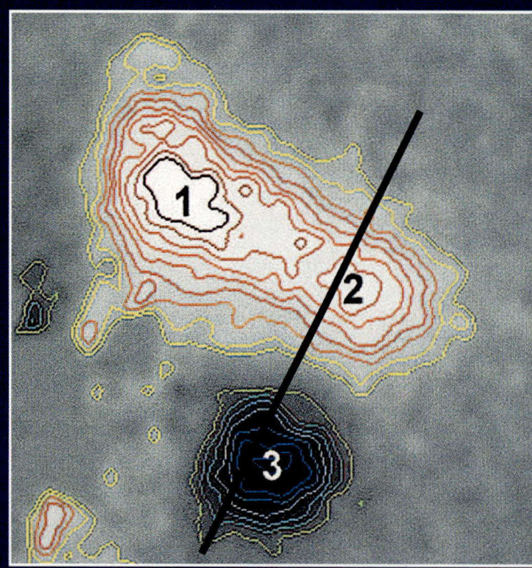

Plate 46 (Chapter 13) IfSAR monitoring of ground deformation, Cold Lake oil field, Alberta, Canada. Black line is location of profile shown in Figure 13-40. Modified from R. P. W. Stancliffe and M. W. A. van der Kooij. 2001. The use of satellite-based radar interferometry to monitor the production activity at the Cold Lake heavy oil field, Alberta, Canada. *American Association of Petroleum Geologists Bulletin*, 85(5), 781–793.

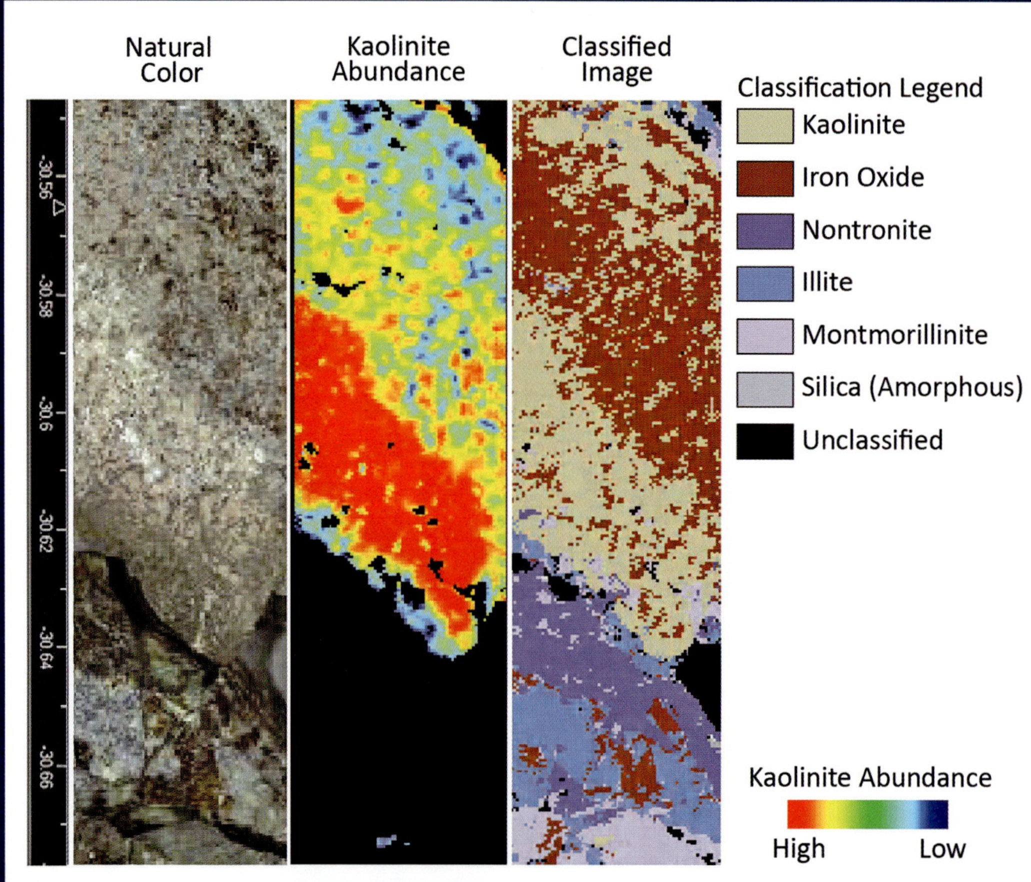

Plate 47 (Chapter 13) Core digitally scanned with natural color and hyperspectral sensors. The hyperspectral data is processed for minerals. Courtesy Corescan.

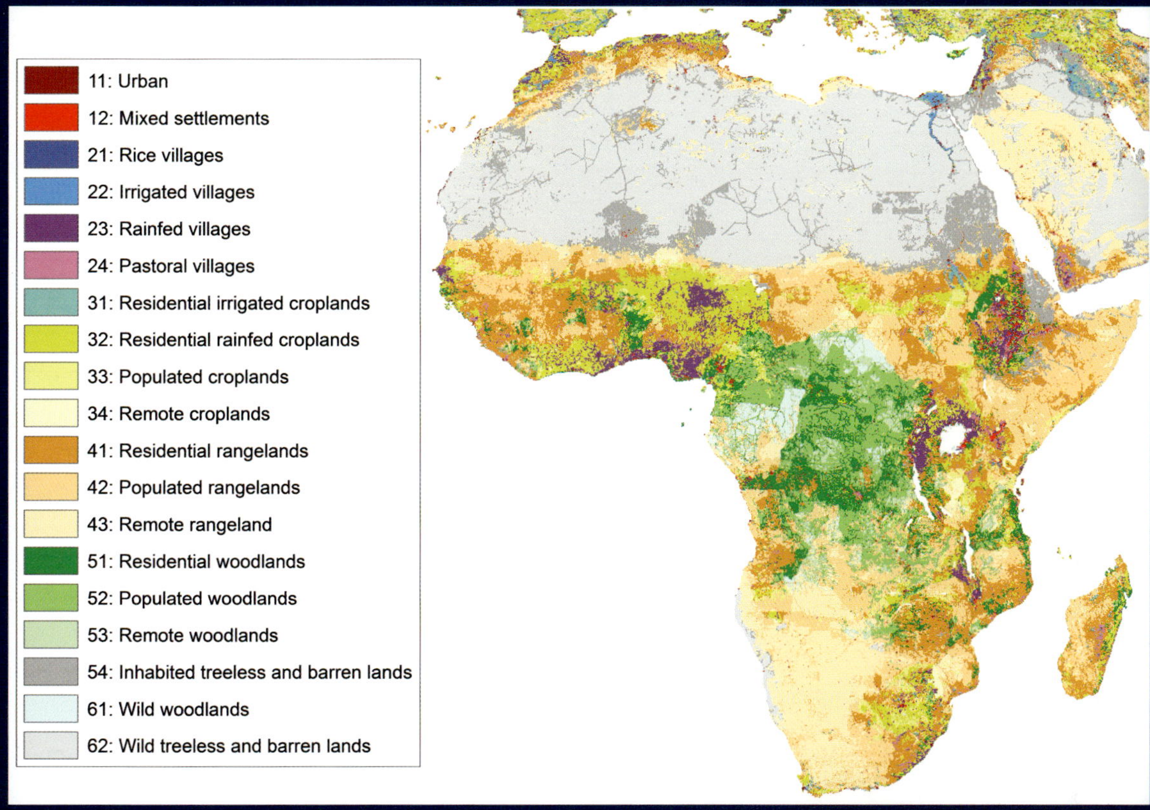

Plate 48 (Chapter 14) Anthropogenic biome map of Africa. From E. C. Ellis, K. K. Goldewijk, S. Siebert, D. Lightman, and N. Ramankutty. 2013. Anthropogenic Biomes of the World, Version 2: 2000. Palisades, NY: NASA Socioeconomic Data and Applications Center. http://sedac.ciesin.columbia.edu/data/set/anthromes-anthropogenic-biomes-world-v2-2000.

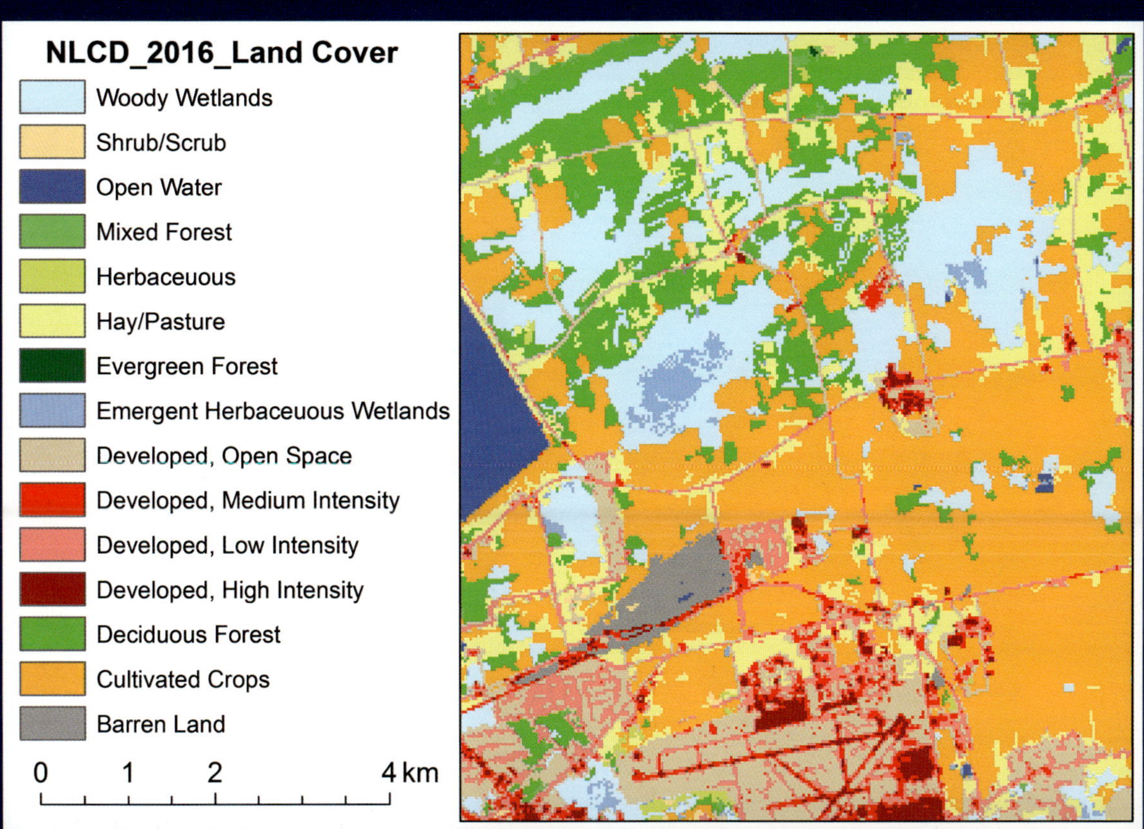

Plate 49 (Chapter 14) 2016 NLCD map of area east of Niagara Falls, New York (Figure 14-1). Courtesy USGS.

A. 2006 Washington D.C. – Baltimore developed (red).

B. 2050 Washington D.C. – Baltimore developed (red).

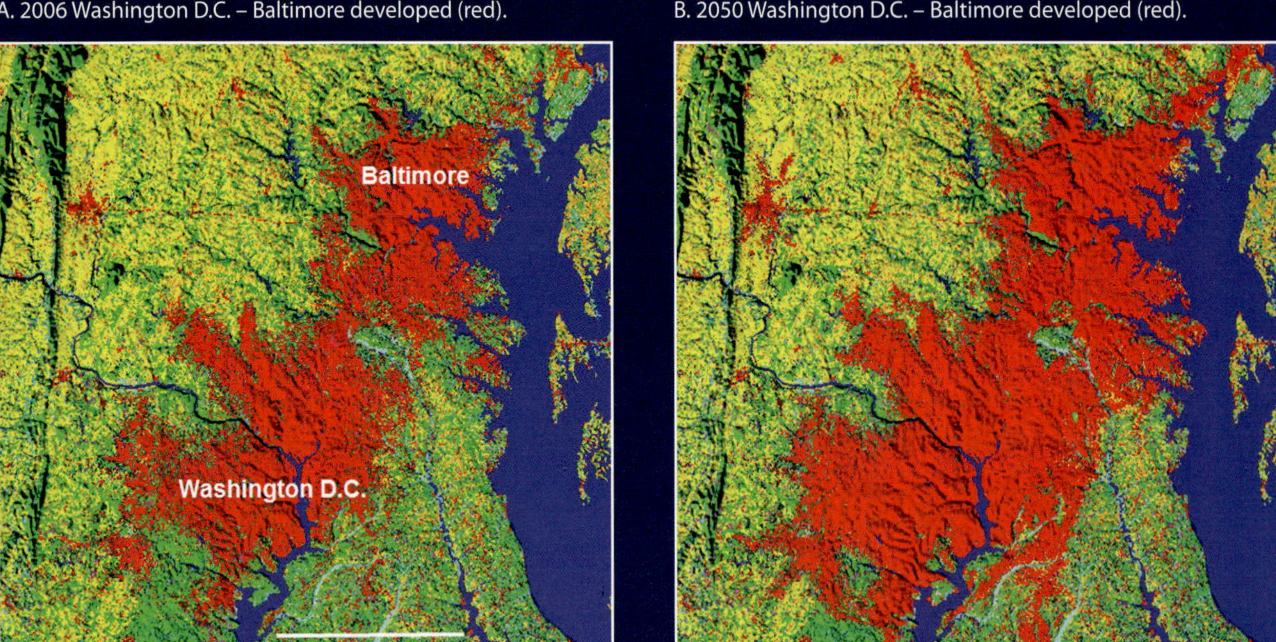

Plate 50 (Chapter 14) Projected development in the Washington, DC and Baltimore area, 2006 and 2050. From USGS. 2016. Land Use Land Cover Modeling. https://landcover-modeling.cr.usgs.gov/index.php (accessed January 2018).

A. Interferogram of the ground deformation from the Napa Valley earthquake.

B. Interferogram was processed to show total motion toward and away from the satellite.

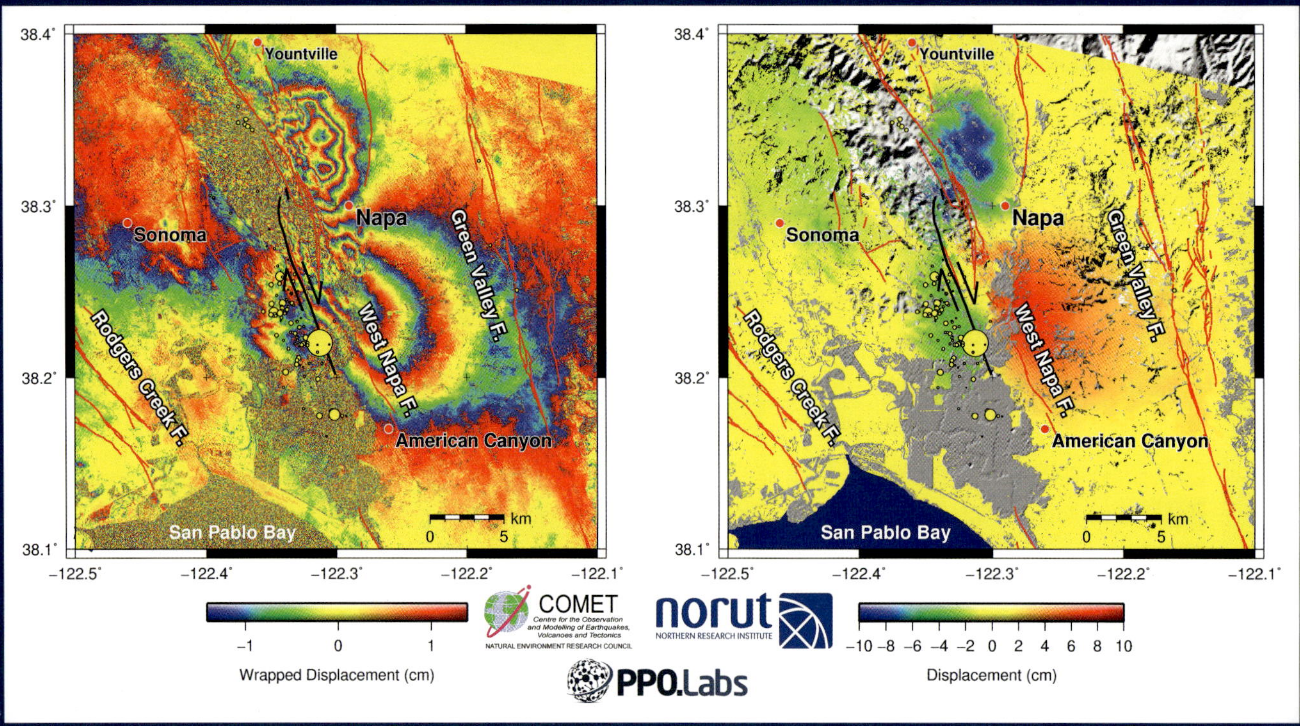

Plate 51 (Chapter 15) Interferograms from Sentinel-1A radar satellite images of the 2014 Napa Valley, California, earthquake. Copyright Copernicus data (2014)/ESA/PPO.labs-Norut-COMET-SEOM Insarap study.

A. Before (May 2012) Hurricane Sandy.

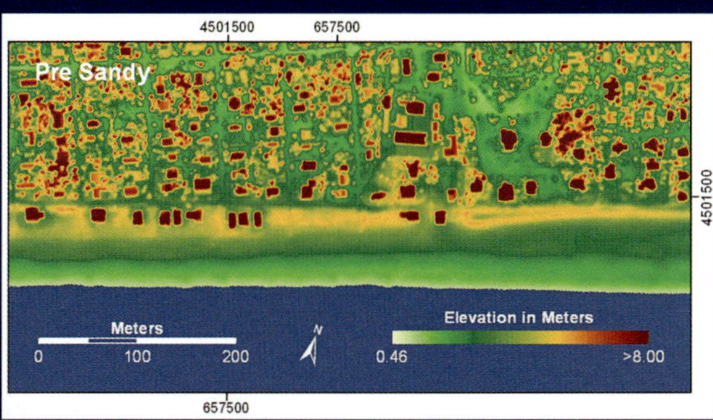

B. After (November 2012) Hurricane Sandy.

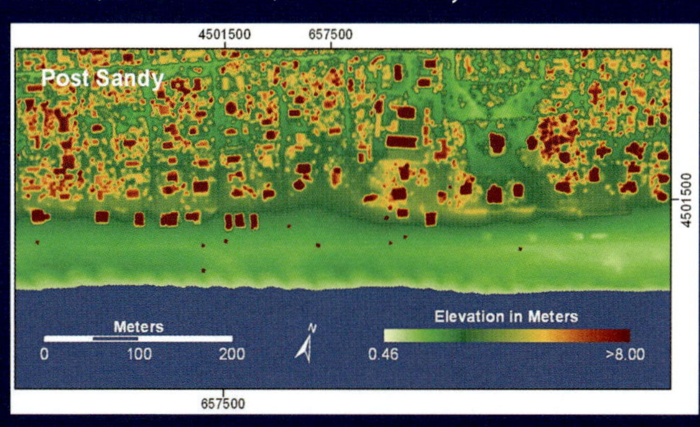

C. Elevation changes.

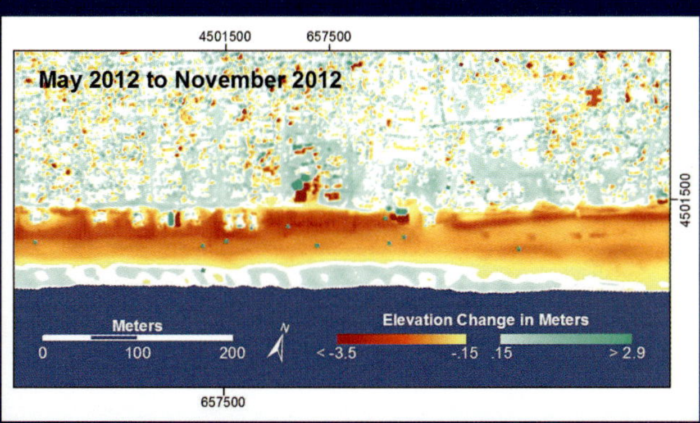

PLATE 52 (CHAPTER 15) Three-dimensional lidar topography of a populated area at Ocean Bay Park, Fire Island, New York (Figure 15-39). From K. L. Sopkin, H. F. Stockdon, K. S. Doran, N. G. Plant, K. L. M. Morgan, K. K. Guy, and K. E. L. Smith. 2014. *Hurricane Sandy: Observations and Analysis of Coastal Change* (Open-File Report 2014-1088). Washington, DC: US Geological Survey.

A. Before (May 2012) Hurricane Sandy.

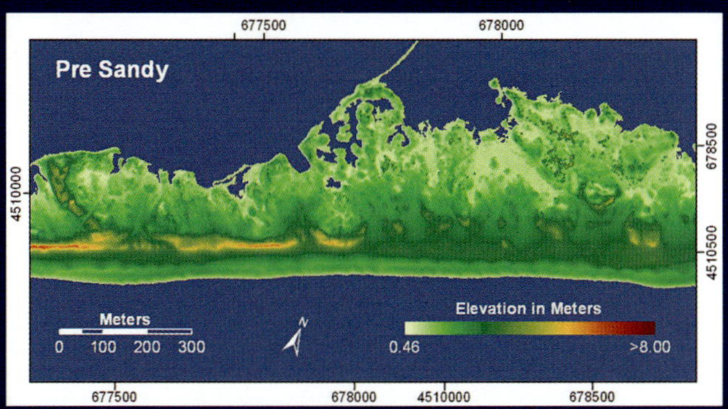

B. After (November 2012) Hurricane Sandy.

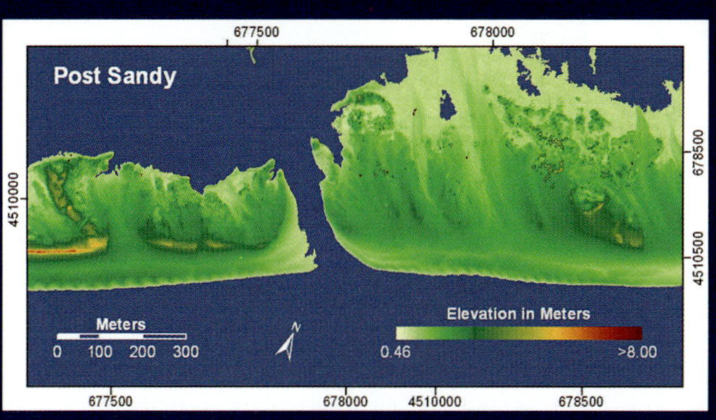

C. Elevation changes.

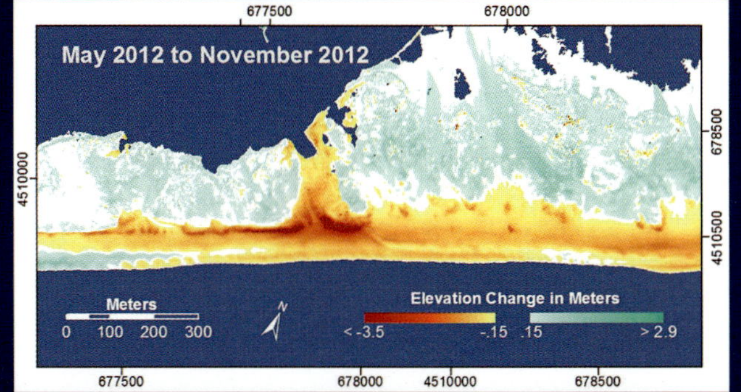

PLATE 53 (CHAPTER 15) Three-dimensional lidar topography of an unpopulated island at Old Inlet, Fire Island, New York (Figure 15-40). From K. L. Sopkin, H. F. Stockdon, K. S. Doran, N. G. Plant, K. L. M. Morgan, K. K. Guy, and K. E. L. Smith. 2014. *Hurricane Sandy: Observations and Analysis of Coastal Change* (Open-File Report 2014-1088). Washington, DC: US Geological Survey.

A. Difference between dust in the atmosphere in 2008 to 2012 and 2001 to 2005.

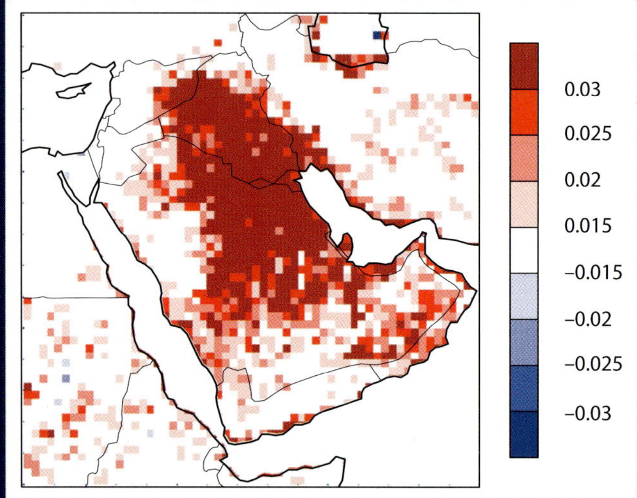

B. Ratio of dust during 2008 to 2012 to dust during 2001 to 2005.

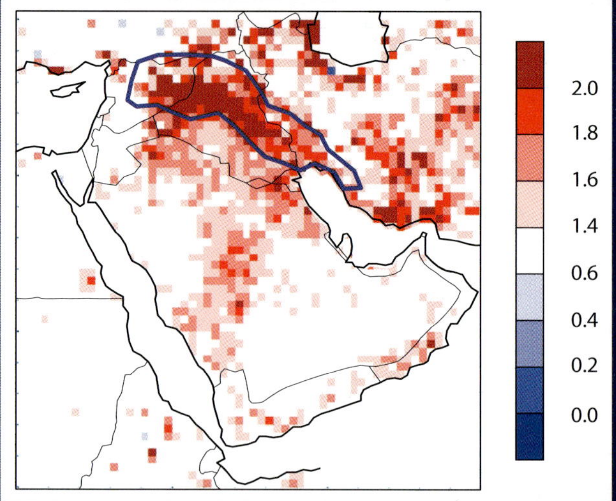

Plate 54 (Chapter 16) Dust in the atmosphere as measured with Terra's MISR instrument. The blue outline is an area of high vegetation loss during the 2007 to 2010 drought. Dust in units of DOAD. From M. Notaro, Y. Yu, and O. V. Kalashnikova. 2015. Regime shift in Arabian dust activity, triggered by persistent Fertile Crescent drought. *Journal of Geophysical Research: Atmospheres*, 120(19), 10,229–10,249.

A. Lidar profile of jungle canopy and understory.

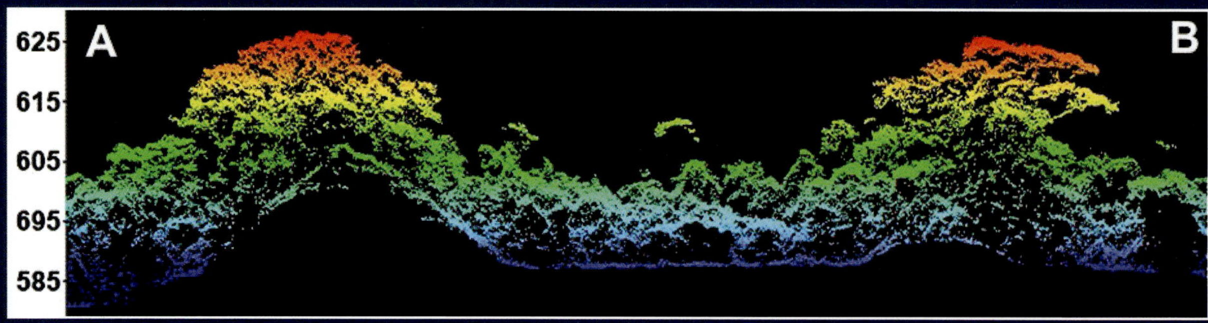

B. Lidar hillshade DTM.

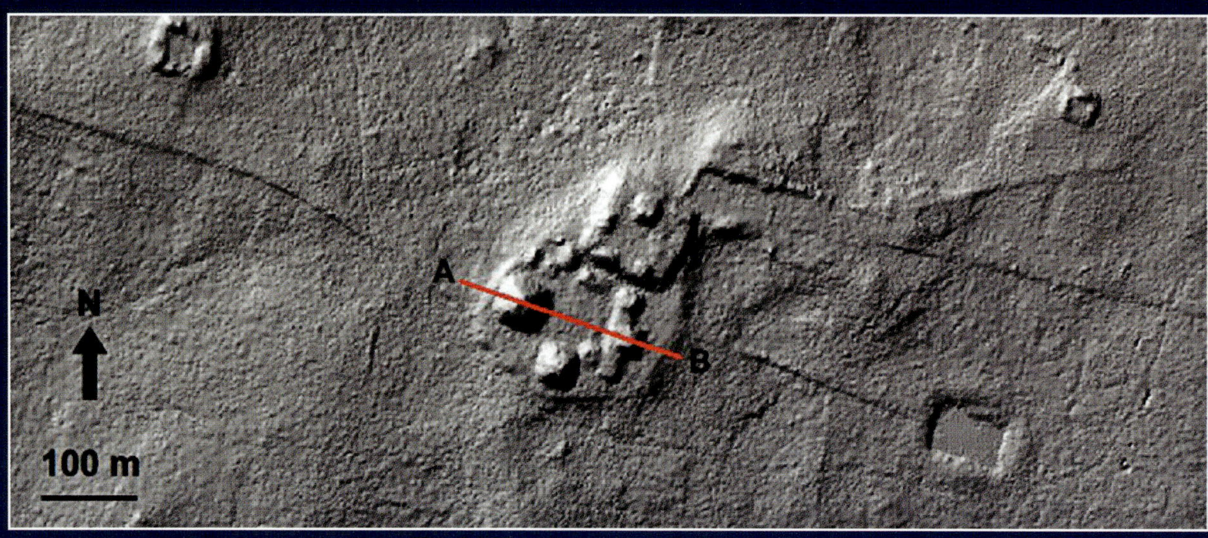

Plate 55 (Chapter 17) Small footprint lidar survey of Caracol, Belize, Central America. From A. F. Chase, D. Z. Chase, J. F. Weishampel, J. B. Drake, R. L. Shrestha, K. C. Slatton, J. J. Awe, and W. E. Carter. 2011. Airborne LiDAR, archaeology, and the ancient Maya landscape at Caracol, Belize. *Journal of Archaeological Science*, 38(2), 387–398.

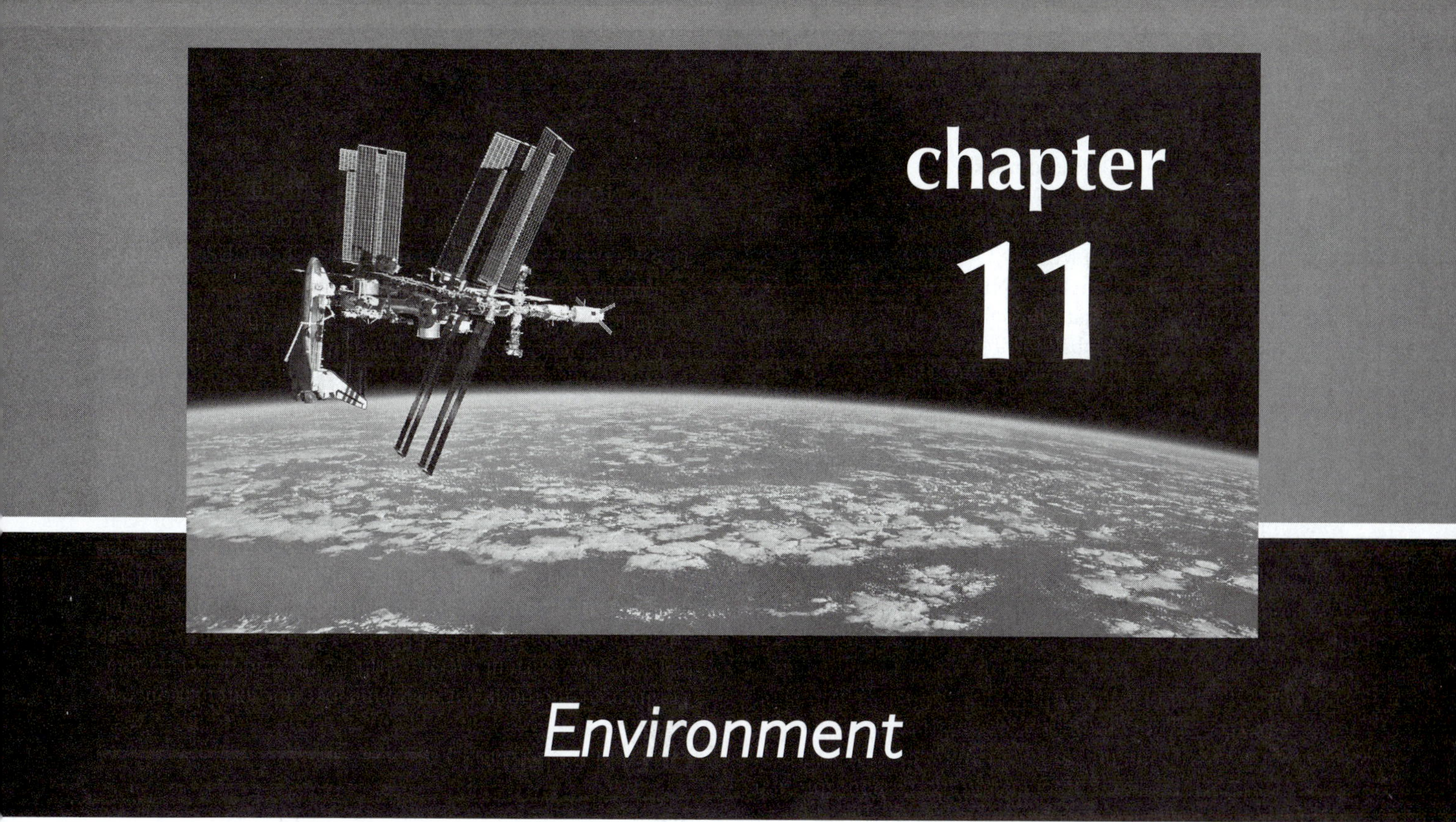

Chapter 11 Environment

Remote sensing is a valuable source of environmental information about the Earth's atmosphere, continents, oceans, and biosphere. In 1960, the United States launched the first unmanned satellites for environmental and meteorological monitoring. These early satellites carried miniature television cameras that produced low resolution, visible band images of cloud patterns for meteorological use. All wavelength regions of the electromagnetic spectrum, from ultraviolet (UV) through microwave, are useful. Chapter 4 included descriptions of low spatial resolution satellites (GOES, MODIS, AIRS, DMSP, and AVHRR) that were designed for global environmental applications, including meteorology and oceanography. Several more low spatial resolution satellites have been deployed by different countries and agencies for specific environmental applications. Table 11-1 lists select environmental satellites with their instruments. Some of the characteristics and applications for the instruments are provided in Table 11-2. Acronyms are used in the tables, however, formal names are also included in the text. These two tables can serve as references for more information regarding the satellites and instruments discussed throughout the chapter.

ENVIRONMENTAL SATELLITES

NOAA's operational weather satellite system is composed of two types of satellites: Geostationary Operational Environmental Satellites (GOES) for short-range warning (1 day) and polar orbiting for longer-term forecasting (3 to 7 days). The polar-orbiting system includes the Suomi National Polar-orbiting Partnership (Suomi NPP) and the Joint Polar Satellite System (JPSS). Suomi NPP and JPSS satellites carry five instruments to provide many measurements, including ocean productivity, sea temperature, ice motion, cloud properties, ozone, and atmosphere profiles. Soil moisture and ocean salinity are measured with NASA's Soil Moisture Active Passive (SMAP) and ESA's Soil Moisture Ocean Salinity (SMOS) satellites. Although not remote sensing satellites, NASA's Gravity Recovery and Climate Experiment (GRACE) and ESA's Gravity Field and Steady-State Ocean Circulation Explorer (GOCE) missions made remarkable discoveries about deep ocean currents and groundwater depletion until their decommission in 2017 and 2013, respectively. Follow-on twin GRACE-FO satellites were operational in 2019.

The Tropical Rainfall Measuring Mission (TRMM) collected precipitation data between 35°N and 35°S latitude for 17 years (1997 to 2015) and was replaced by the Global Precipitation Measurement (GPM) Mission. TRMM and GPM are joint missions between NASA and the Japanese space agency JAXA. Ocean wind measurements use radar instruments onboard the International Space Station's RapidSCAT; Defense Meteorological Satellite Program (DMSP) polar-orbiting satellites; and the European Organisation for the Exploitation of Meteorological Satellites (EUMETSAT) MetOP-A, MetOp-B, and MetOp-C satellites. The Multi-angle Imaging SpectroRadiometer (MISR) instrument maps sea ice from the Terra platform, which also carries ASTER and MODIS.

ICESat-2 carries the Advanced Topographic Laser Altimeter System (ATLAS) that measures the travel times of laser pulses to calculate the distance between the spacecraft and Earth's surface (described in Chapter 16). Sentinel-3 is focused on an ocean mission that is accomplished with two identical satellites launched in 2016 and 2018. The Sentinel-3 satellites measure temperature, color, and height of the sea surface as well as the thickness of sea ice. The Ocean and Land Colour Instrument (OLCI) collects 21 VNIR bands with a swath width of 1,270 km and a spatial resolution of 300 m.

In addition to the global environmental satellites discussed here, satellites, manned aircraft, and sUAS systems with much higher spatial resolutions monitor features in the Earth's environment using the VNIR, SWIR, and TIR bands as well as radar sensors. Differentiating land cover types, detecting invasive species, tracking thermal plumes in air and water, monitoring vegetation vigor, and documenting fauna and flora diversity in parks and wilderness are some of the many environmental applications routinely accomplished by remote sensing systems. This chapter summarizes a limited selection of remote sensing applications that improve our understanding of global, regional, and local environments. Dodge and Congalton (2013) provide several case studies where remote sensing is used to monitor environmental change and support environmental restoration.

UV RADIATION AND OZONE CONCENTRATION

Ozone (O_3), a molecule made up of three atoms of oxygen, forms a layer in the atmosphere from 25 to 60 km above

TABLE 11-1 Select global environmental satellites.

Satellite/ Mission	Operator	Orbit	Number of Instruments	Instruments
Suomi NPP and JPSS	NOAA	Polar	5	ATMS—scanner VIIRS—2-day revisit Cross-track Infrared Sounder (CrIS) Ozone Mapping Profiler Suite (OMPS) CERES—radiometer
SMAP	NASA	Polar	1	SMAP—radar and passive radiometer; 2- to 3-day revisit
SMOS	ESA	Polar	1	Microwave Imaging Radiometer using Aperture Synthesis (MIRAS)—3-day revisit
SAC-D	NASA Argentina	Polar	6	Aquarius
ICESat-2	NASA	Polar	1	ATLAS—laser altimetry
Sentinel-3	ESA	Polar	5	OLCI—temperature, radiometer, radar altimeter
Aura	United States Netherlands Finland	Polar	4	OMI—hyperspectral imaging, spectrometer
TRMM	United States Japan	Polar	5	PR VIIRS TMI CERES Lightning Imaging Sensor (LIS) (1997–2015)
GPM	United States Japan	Circular; inclined 65° to the equator	3	GMI DPR
QuikSCAT	NASA	Polar	2	SeaWinds—active scatterometer (1999–2002) NSCAT—passive scatterometer
ISS RapidSCAT	NASA	Equatorial	1	QuikSCAT SeaWinds backup—active
DMSP	DMSP	Polar	1	SSM/I, SSMIS—passive
MetOP-A and -B	EUMETSAT	Polar	1	ASCAT
Terra	NASA	Polar	5	MISR—9-day revisit

sea level that absorbs incoming UV radiation from the sun (Figure 11-1). The warm layer in the atmospheric temperature profile is caused by solar energy absorbed by the ozone layer. Without the ozone layer, incoming UV radiation from the sun would effectively sterilize the Earth of all forms of life. The Halley Research Station on the Antarctic coast makes ground-based total ozone measurements during the daylight months. Farman and others (1985) first reported

Table 11-2 Characteristics and applications of select global environmental satellites.

Instrument	Spectral Bands	Wavelength Range[a]	Spatial Resolution (km)	Swath Width (km)	Environmental Applications
ATMS	22	0.16 to 1.30 cm			Atmosphere profiles
VIIRS	22	0.415 to 12.5 μm	0.37, 0.74	3,040	Clouds, aerosols, ocean color, land and sea temperature
CrIS	1,035	SWIR, MWIR, LWIR	14	2,200	3-D atmosphere profiles
OMPS		0.25 to 0.38 μm	50 250	2,800	Ozone near troposphere; 15 to 60 km elevation
CERES	3	0.3 to 5 μm 0.3 to 200 μm 8 to 12 μm	35–50		Cloud properties
SMAP	1	L-band	9 (combine SAR + radiometer)	1,000	Soil moisture, salinity
MIRAS	1	L-band	35–50	360	Soil moisture, salinity
Aquarius	1	L-band	150	390	Ocean salinity
OMI	780	.27 to .50 μm	13 by 25	2,600	Ozone, nitrous dioxide, sulfur dioxide
PR	1	?	5	247	Rainfall, vertical profiles rain and snow
VIRS	5	0.63 to 12 μm	2.4	878	Rainfall, vertical profiles rain and snow
TMI	5	0.35 to 2.8 cm	?	833	Rainfall, vertical profiles rain and snow
GMI	13	0.16 to 3 cm	?	904	Precipitation intensity
DPR	2	Ka-band Ku-band	5 5	125 254	3-D structure of precipitation
QuikSCAT SeaWinds	1	Ku-band	25	1,800	Wind direction and speed
NSCAT	1	Ku-band	50?	?	Wind speed
ISS RapidSCAT	1	Ku-band	25?	900?	Wind direction and speed
SSM/I	4	0.35 cm, 0.81 cm, 1.35 cm, 1.55 cm	15 by 13 to 43 by 69	1,707	Ocean wind speed, land surface temperature
ASCAT	1	C-band	25 50	500 (2)	Wind direction and speed, sea ice, permafrost
MISR	4	0.44 μm, 0.55 μm, 0.67 μm, 0.86 μm	0.275	380	Sea ice and dust concentration

[a] Ka band = 2.3 cm, Ku band = 0.86 cm.

the unexpected depletion of ozone over the Antarctic, commonly called the "ozone hole." These measurements were subsequently confirmed and monitored over time by satellite remote sensing. Probably no other global environmental change has caused as much worldwide concern as ozone depletion, which results from complex chemical reactions. It is known that man-made chlorofluorocarbon gases (CFCs), used in refrigeration and other industries, are capable of depleting atmospheric ozone. In 1987, the Montreal Protocol was adopted to facilitate international cooperation for the control of CFCs. Global consumption of ozone-depleting substances was reduced by 98% between 1986 and 2008 and global production of CFCs ended by 2010 (Ozone Hole Inc., 2017). CFCs have been replaced by noninteractive substitutes in refrigeration and air-conditioning equipment.

OZONE MONITORING INSTRUMENT

Global measurements of ozone were first made by the Total Ozone Mapping Spectrometer (TOMS), which was first launched in 1978 on the polar-orbiting Nimbus-7. In 1991, the former Soviet Union launched a Meteor-3 satellite carrying a TOMS instrument provided by NASA. The last TOMS instrument ceased operations in 2006. The TOMS series was replaced by the Ozone Monitoring Instrument (OMI) that was built by agencies from the Netherlands and Finland. The OMI was successfully carried into orbit on NASA's Aura satellite in 2004. OMI employs hyperspectral imaging in a nadir-viewing, along-track mode to observe solar backscatter radiation in the visible and ultraviolet range (NASA OMI, 2017). OMI provides daily global coverage of ozone with a spatial resolution of 13 by 25 km.

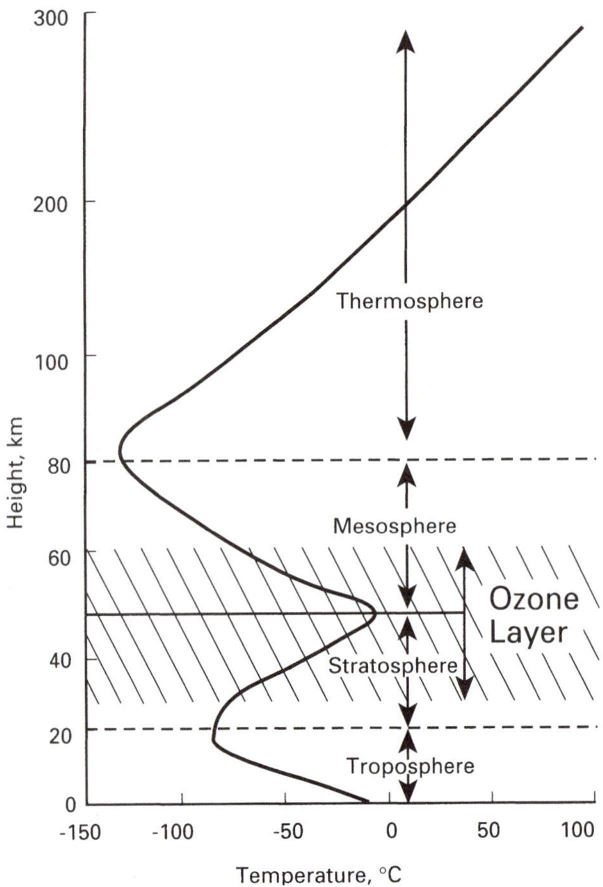

FIGURE 11-1 Vertical structure of the atmosphere.

OZONE DEPLETION TRENDS AND THE ANTARCTIC OZONE HOLE

Plate 31 shows a sequence of TOMS and OMI images (1979, 1994, 2016) for the September monthly mean ozone concentration levels. The projections are centered near the South Pole. Ozone concentration is shown using a color scale that is calibrated in *Dobson units* (DU), which represent the physical thickness the ozone layer would have if it were brought to the Earth's surface (100 DU = 1 mm thickness). The lowest ozone concentrations, shown in the colors black through blue, form the circular pattern centered at the South Pole that is the ozone hole. The images show the progressive development of the ozone hole, as the thickness of the ozone layer decreased from a 1979 high of 225 DU to a minimum in 1994 of 92 DU (Figure 11-2A). Since 1994, the ozone minimum has stabilized with a value of 124 DU in 2016. The size of the ozone hole increased from 1979 to a high of 27,000,000 km^2 in 2006 then generally decreased in size to 21,000,000 km^2 by 2016 (Figure 11-2B).

The depth and area of the ozone hole are primarily governed by the amounts of chlorine and bromine in the Antarctic stratosphere. Very low temperatures are needed to form polar stratospheric clouds. Chlorine gases react on the surface of these clouds to release chlorine into a form that can easily destroy ozone. The chlorine and bromine chemical catalytic reactions that destroy ozone need sunlight. Hence, the ozone hole begins to grow as the sun is rising over Antarctica at the end of the winter.

The ozone hole begins to grow in August and reaches its largest area in depth from the middle of September to early October. Prior to 1984 the hole was small because chlorine and bromine levels over Antarctica were low. Year-to-year variations in area and depth are caused by year-to-year variations in temperature. Colder conditions result in a larger area and lower ozone values in the center of the hole (NASA Ozone Watch, 2017).

AIR POLLUTION

The OMI UV-visible hyperspectral instrument onboard NASA's Aura satellite measures air pollutants, dust, smoke, and other aerosols (tiny solid and liquid particles suspended in the atmosphere).

NITROGEN DIOXIDE

Nitrogen dioxide (NO_2) is a yellow-brown gas that is a common emission from cars, power plants, and industrial activity. NO_2 can quickly transform into ground-level ozone, a major respiratory pollutant in urban smog. NO_2 hotspots, used as an indicator of general air quality, occur over most major cities in developed and developing nations (NASA, 2015). NO_2 has spectral absorption features in the 0.42 to 0.47 μm wavelength range that are measured by OMI (Popp and others, 2012). Global OMI observations made from 2005 to 2014 were analyzed by scientists at NASA's Goddard Space Flight Center for NO_2 trends in China, Japan, and the United States (**Digital Image 11-1**). Red colors indicate increased NO_2 emissions and blue colors indicate decreased NO_2 from 2005 to 2014.

China saw an increase of 20 to 50% in nitrogen oxide, much of it from the North China Plain. Three major Chinese metropolitan areas—Beijing, Shanghai, and the Pearl River Delta—saw nitrogen dioxide reductions of as much as 40%. Hong Kong and Japanese urban areas show a decrease in nitrogen dioxide pollution. The United States saw nitrogen dioxide decrease from 20 to 50%. Researchers concluded that the reductions are largely due to the effects of environmental regulations that require technological improvements to reduce pollution emissions from cars and power plants (NASA, 2015).

A. Average minimum ozone.

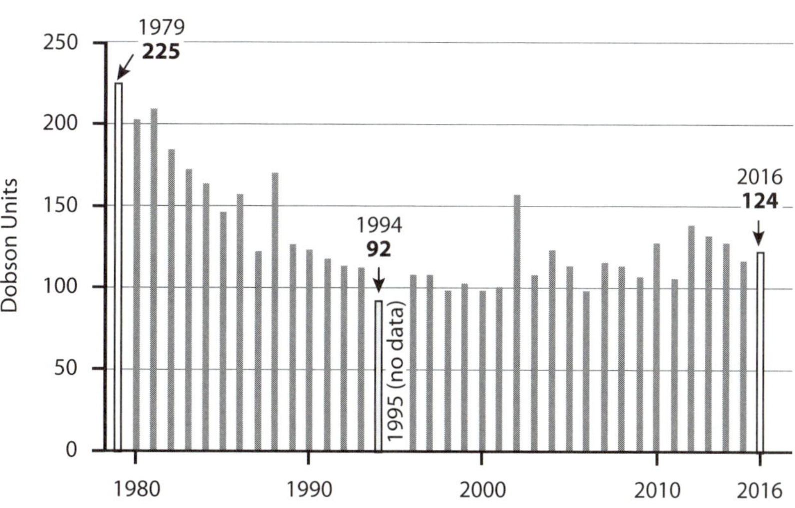

B. Average ozone hole area.

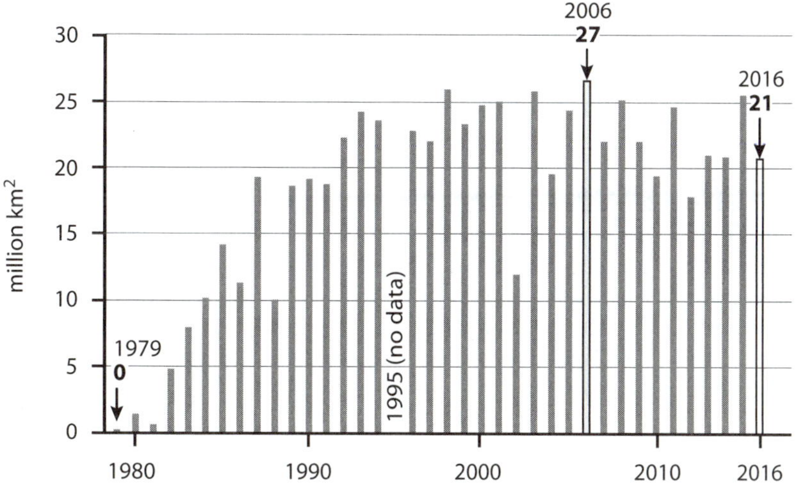

FIGURE 11-2 Development of the ozone hole over the South Pole from 1979 to 2016. Courtesy NASA Ozone Watch (2017).

SULFUR DIOXIDE

Sulfur dioxide (SO_2) is a known health hazard and contributor to acid rain and is one of six air pollutants regulated by the US Environmental Protection Agency (EPA). NASA Earth Observatory (2016) notes that one of the tools that the EPA uses to monitor sulfur dioxide is an emissions inventory; that is, researchers collect and analyze ground-based measurements of the gas, while estimating emissions from other activities that produce or emit it (such as motor vehicles and power generation). These inventories help guide regulatory policies for air quality and help anticipate future emissions that may occur with economic and population growth. In order to develop comprehensive and accurate inventories, industries, government agencies, and scientists first must know the location of pollution sources. The OMI satellite locates and measures SO_2 with data collected in the UV spectral range of 310 to 380 μm, which has strong and weak absorption features associated with SO_2 (Yang and others, 2009). OMI observations have a 150-fold improvement in sensitivity over TOMS SO_2 measurements (Krotkov and others, 2006).

A new emission-source detection algorithm was applied to OMI measurements of SO_2 on data collected from 2005 to 2014 (McLinden and others, 2016). The research compiled the first global, satellite-based emissions inventory of SO_2, and is completely independent of conventional information sources (NASA Earth Observatory, 2016). The satellite-based inventory found 39 previously unreported emission sources. Among them were clusters of coal-burning power plants, smelters, and oil and gas operations, most notably in the Middle East. Figure 11-3 is centered on the countries surrounding the Persian Gulf. The map is based on OMI data collected between 2007 and 2009. The small dark features are areas with increased SO_2 emissions. Dark features enclosed by a dashed polygon were missing in previous emission inventories.

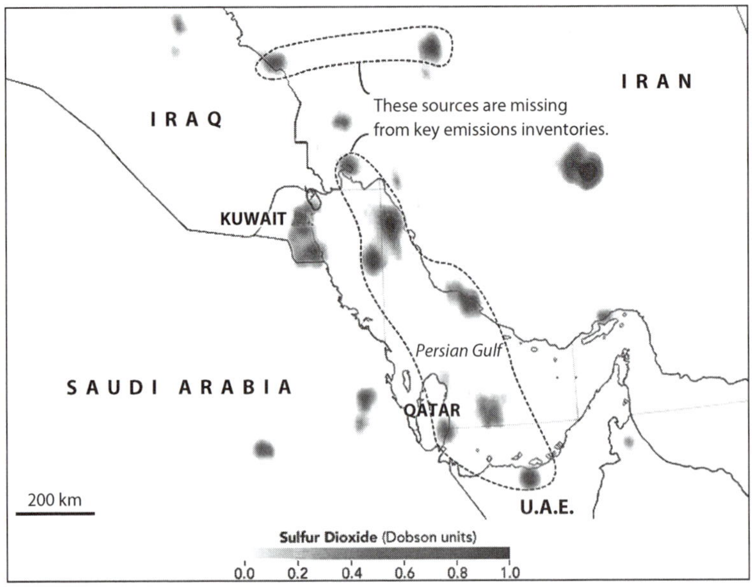

FIGURE 11-3 A regional map of the Middle East showing sulfur dioxide emissions as detected by OMI on NASA's Aura spacecraft. Courtesy NASA Earth Observatory (2016).

The OMI satellite measurements of emissions from some known sources were two to three times higher than what was being reported in ground-based estimates. Altogether, the unreported and underreported sources account for about 12% of all man-made emissions of sulfur dioxide. McLinden and others (2016) also located 75 natural sources of sulfur dioxide, many of them nonerupting volcanoes that slowly leak the gas throughout the year. While not necessarily unknown, many of these volcanic sources are in remote locations and not routinely monitored, so this satellite-based data set is the first to provide regular annual information on passive volcanic emissions.

CLIMATE AND WEATHER

Daily and seasonal changes in the heating of the Earth by the sun cause atmospheric motions that are called *weather*. The sum of the seasonal variations is called *climate*. Clouds, precipitation, wind, and temperature are elements of weather and climate that are described in this section.

CLOUD MAPPING

Clouds are aerosols of tiny water droplets that have condensed from water vapor that has been evaporated from the surface (mostly oceans); under favorable conditions, clouds precipitate as rain or snow. Changes in cloud patterns signal weather changes, especially the advent of precipitation. Properties of clouds largely determine the amount of sunlight that reaches the Earth and the amount of radiation that escapes; therefore, in addition to supplying water, clouds affect the surface temperature. Clouds are a key link in both the water and energy cycles that determine climate. Over very long time scales, climate is altered by natural causes, such as changes in the amount of sunlight, biologic changes of atmospheric composition, and geologic changes of continents and oceans. Human activities are affecting the natural balance of atmospheric constituents (such as carbon dioxide and ozone), which results in relatively fast climate changes.

Clouds may have a warming or cooling influence depending on their altitude, type, and when they form. Clouds reflect sunlight back into space, which causes cooling. But they can also absorb heat that radiates from the Earth's surface, preventing it from freely escaping to space. One of the biggest sources of uncertainty in computer models that predict future climate is how clouds influence the climate system and how their role might change as the climate warms.

Satellite remote sensing of clouds began with TIROS-1 in April 1960. Polar-orbiting NOAA satellites have routinely imaged the entire globe twice daily since the mid-1970s. Today the MODIS and Clouds and the Earth's Radiant Energy System (CERES) instruments on the Terra and Aqua satellites and the CERES and Visible Infrared Imaging Radiometer Suite (VIIRS) instruments on Suomi NPP and the JPSS series of satellites monitor cloud properties. Each CERES instrument has three channels: a shortwave channel for measuring reflected sunlight (0.3 to 5 μm), a longwave channel for measuring Earth-emitted thermal IR (TIR) radiation in the 8 to 12 μm region, and a channel for total radiation (0.3 to 200 μm). The CERES measurements are combined with images recorded by MODIS and VIIRS and other data to provide the properties of clouds, including optical thickness, effective particles size, cloud top parameters, cloud base height, and cloud layers (CERES, 2017).

MODIS derives cloud cover maps on a daily, global basis with a spatial resolution of 1 by 1 km. The daily maps are combined to generate monthly cloud cover maps that use cloud fraction on a scale of 0 to 1 to illustrate the amount of coverage. Cloud fraction is the portion of each pixel that is covered by clouds. The monthly cloud cover for April 2017 is shown in Figure 11-4. The darkest areas of northern Africa, Saudi Arabia, India, and the central west coast of South America indicate no clouds for the month of April. The brightest white areas of the Amazon Basin, western

Canada, equatorial Africa, north Atlantic, and Indonesia indicate total cloudiness for the month.

From month to month, a band of clouds girdles the equator. This band of persistent clouds is called the Intertropical Convergence Zone (ITCZ), the place where the easterly trade winds in the Northern and Southern Hemispheres meet. The meeting of the winds pushes warm, moist air high into the atmosphere. The air expands and cools, and the water vapor condenses into clouds and rain. The cloud band shifts slightly north and south of the equator with the seasons. In tropical countries, this shifting of the ITCZ is what causes rainy and dry seasons.

Another frequently cloudy place is the Southern Ocean. Although there is not as much evaporation in the high latitudes as in the tropics, the air is cold. The colder the air, the more readily any water vapor in the air will condense into clouds (NASA Earth Observations, 2017b).

Rainfall Mapping

Precipitation is the nearly universal ultimate source of the water that supports life on the Earth. A better understanding of precipitation patterns and how they are affected by global changes is essential for the health and survival of humankind. The process of precipitation can change atmospheric circulation, especially in the tropics, because condensation of water vapor releases the latent heat of condensation. Thus, the processes of atmospheric circulation, cloud formation, and precipitation are mutually interactive components of the global climate system.

Measuring precipitation is one of the most challenging tasks for meteorologists. Rain gauges are the earliest, and still most common, measuring device. The most serious drawback to rain gauges is that they record point measurements, whereas precipitation is highly variable in space. Even in developed regions, such as Europe and the United States, the number and distribution of gauges provides only an approximation of moisture patterns. Essentially no rain gauge measurements are made over the oceans. For these reasons various remote sensing techniques have been developed. The first was ground-based radar systems operating at wavelengths of millimeters, which are backscattered by rain and ice particles but not by cloud aerosols. As a rotating antenna scans the atmosphere, any backscattered return signal is interpreted in terms of precipitation. Estimating rainfall rates is difficult, however, because the relationship between backscatter and rain rate is highly nonlinear. Also, a radar beam does not curve with the surface of the Earth, which limits the maximum range to a few hundred kilometers and nearer to 100 km for quantitative data.

Meteorologists began to attempt estimations of precipitation from the very earliest satellite images. From 1997 to 2015, daily precipitation measurements were made between 35° north and south latitude by the TRMM satellite, which was jointly operated by NASA and the Japanese space agency JAXA (NASA TRMM, 2017). On February 27, 2014, the GPM Core Observatory was launched to serve as a calibration standard for other satellites collecting precipitation data (NASA GPM, 2017). GPM is another joint NASA and JAXA project, orbiting in a non-sun-synchronous orbit

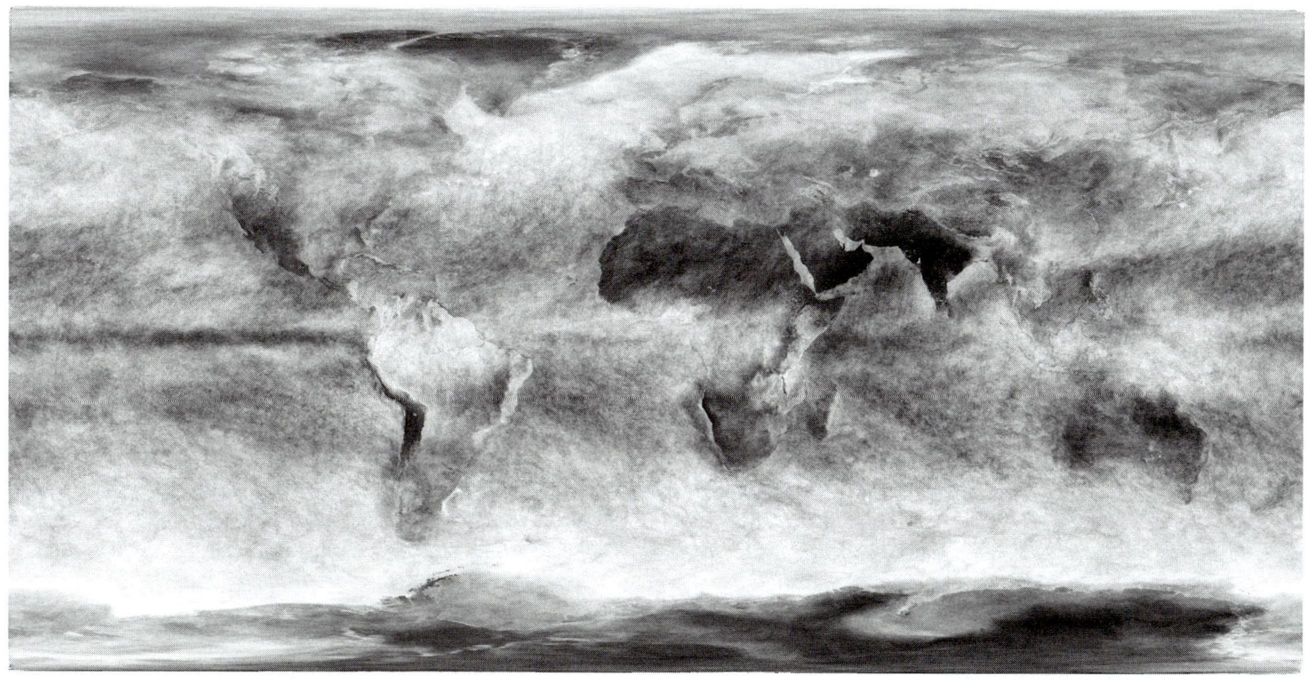

Figure 11-4 Cloud coverage for the month of April 2017. 0 = black = no clouds to 1.0 = white = total cloud cover. Courtesy NASA Earth Observations (2017b).

that covers the Earth from 65° north and south latitude (from about the Antarctic Circle to the Arctic Circle). Both TRMM and GPM use passive microwave radiometer and active radar instruments to measure precipitation.

Global Precipitation Images

Figure 11-5 is a TRMM global precipitation image with a nonlinear precipitation scale ranging from 1 to 2,000 mm (2 m). The white areas over southern Africa, the eastern Pacific Ocean west of North and South America, and the Middle East are regions of low precipitation during August 2016. High precipitation along the equator and ITCZ, northwest South America, southeast Asia, and the western Pacific Ocean is shown as darker shades of gray.

Precipitation Estimates from Passive Microwave Images

Passive microwave systems are cross-track scanners that use radiometers to detect energy radiated from the surface at microwave wavelengths. At these longer wavelengths the intensity of radiant energy is very low; therefore, ground resolution cells are large, on the order of many square kilometers.

There are presently two distinct methods for estimating precipitation from passive microwave data (J. E. Janowiak, NOAA/NWS, personal communication):

1. **Emission methods** use observations of the thermal emissions from precipitation to estimate rainfall rates over ocean surfaces, where surface emissivities are uniformly low. This method cannot estimate rainfall over land because of large variations in emissivity of different land surfaces. The method is capable of detecting both convective and "warm process" (stratiform, orographically induced) rainfall.

2. **Scattering methods** infer precipitation from the scattering of upwelling radiation due to precipitation-size ice particles above the rain layer in convective systems. The short wavelength microwaves (0.35 cm) are strongly scattered by ice present at the tops of many raining clouds (NASA TRMM, 2017). This method is useful for quantifying precipitation over both land and water, but it can only detect rainfall associated primarily with convective systems that penetrate the freezing level.

The Special Sensor Microwave Imager (SSM/I) is carried on DMSP polar-orbiting satellites since 1987. In 2003, the Special Sensor Microwave Imager/Sounder (SSMIS) was added to improve estimates of atmospheric temperature and moisture parameters from data collected by the DMSP satellites. SSM/I records four wavelengths of microwave energy (0.35, 0.81, 1.35, and 1.55 cm) with an image swath width of 1,400 km and a nominal ground resolution cell ranging between 13 by 15 km to 43 by 69 km. Both vertically (V) and horizontally (H) polarized energy is recorded, except for the 1.35 cm band, which records vertically polarized energy only. No cross-polarized energy is recorded because no energy is transmitted in this passive system. Grody (1991) developed a decision tree that identified precipitation from all other atmospheric and surface features by using the following four SSM/I bands: 1.55 cm (V), 1.55 cm (H), 1.35 cm (V), 0.35 cm (V). Weng and others (1994) developed an SSM/I scattering index that was calibrated against known radar rates of rainfall (mm · hr^{-1}). The scattering index is calculated for each pixel identified as precipitation to determine its rainfall rate.

The TRMM Microwave Imager (TMI) and the GPM Microwave Imager (GMI) are passive microwave radiometers based on the design of the SSM/I, but they collect additional microwave wavelengths compared to SSM/I. The TMI scans a swath 833 km across while the GMI scans a swath 904 km across. Both instruments collect precipitation intensities and horizontal patterns (NASA GPM, 2017).

Precipitation Estimates from Active Radar

The Precipitation Radar (PR) instrument onboard the TRMM satellite was the first spaceborne instrument designed to provide three-dimensional maps of storm structure. These measurements yield invaluable information on the intensity and distribution of the rain, on the rain type, on the storm depth, and on the height at which snow melts into rain. The estimates of the heat released into the atmosphere at different heights based on these measurements can be used to improve models of the global atmospheric circulation (NASA TRMM, 2017).

The PR has a horizontal resolution at the ground of about 3.1 mi (5 km) and a swath width of 154 mi (247 km). One of its most important features is its ability to provide vertical profiles of the rain and snow from the surface up to a height of about 12 mi (20 km). The Precipitation Radar is able to detect fairly light rain rates down to about .027 in · hr^{-1} (0.7 mm · hr^{-1}). It is able to separate out rain echoes for vertical sample sizes of about 820 ft (250 m) when looking straight down (NASA TRMM, 2017).

The GPM Mission carries the Dual-frequency Precipitation Radar (DPR), which provides a three-dimensional structure of precipitating particles (NASA GPM, 2017). The two frequencies of the DPR also allow the radar to infer the sizes of precipitation particles and offer insights into a storm's physical characteristics. The Ka-band frequency (0.86 cm wavelength) scans across a region of 125 km and is nested within the wider scan of the Ku-band frequency (2.3 cm wavelength) of 254 km. JAXA and Japan's National Institute of Information and Communications Technology (NICT) built the DPR. The DPR is more sensitive than its TRMM predecessor, especially in the measurement of light rainfall and snowfall in high latitude regions. Rain/snow determination is accomplished by using the differential attenuation between the Ku-band and the Ka-band frequencies (NASA GPM, 2017).

WIND PATTERNS OVER THE OCEANS

Accurate information on surface wind patterns over the oceans is essential because:

1. Winds drive the ocean currents.

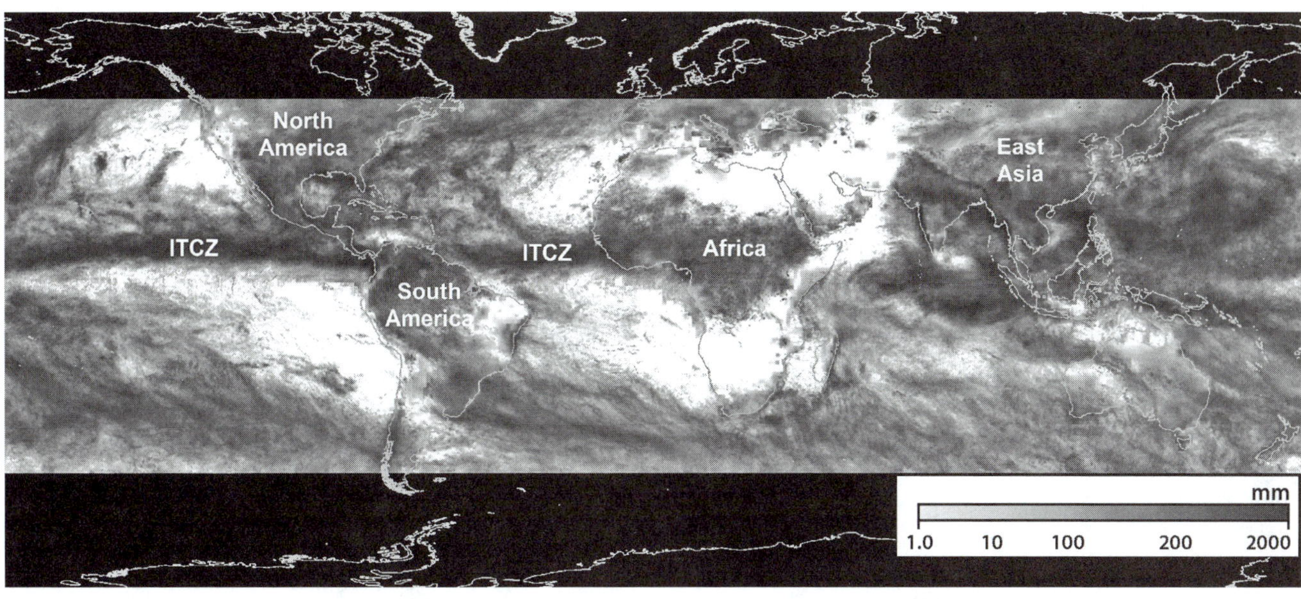

FIGURE 11-5 TRMM global rainfall for August 2016. During its operation, the TRMM satellite flew in an equatorial orbit, therefore, the data it collected only covered the equator and mid-latitudes. Courtesy NASA Earth Observations (neo.sci.gsfc.nasa.gov).

2. Fluxes of heat, moisture, and momentum across the air-sea boundary are important factors in forming, moving, and modifying water masses. These interactions also intensify storms near coasts and over the open ocean.

Prior to satellite remote sensing observations, ocean winds were measured by anemometers on ships and buoys, which lack coverage, accuracy, and timeliness. Unlike anemometers, remote sensing systems do not record winds directly. However, wind blowing across the ocean produces waves ranging in wavelength and height from capillary waves to ocean swells. The size and direction of waves is determined by the velocity and direction of the surface wind. Active and passive satellite systems operating in the microwave region record surface characteristics of the ocean (Table 11-2). These data are processed and interpreted to produce timely maps of wind patterns.

Passive Microwave Radiometer

Passive microwave systems record energy radiated from the ocean at microwave wavelengths. The intensity of microwave energy is the product of emissivity times kinetic temperature of the water. Emissivity varies with the roughness of the water and is used to estimate wind velocity. SSM/I records passive microwave data for measuring surface wind speeds, but not directions, over the oceans.

Active Microwave Scatterometer

The first scatterometer flew as part of the Skylab missions in 1973 and 1974. The 1978 Seasat and the 1991 ERS-1 satellites carried scatterometers to measure wind velocity. NASA's QuikScat was launched in 1999 with the SeaWinds scatterometer to measure surface wind speed and direction over the ice-free global oceans. Each day SeaWinds recorded over 400,000 measurements of wind speed and direction and was able to collect at least one vector wind measurement over 93% of the world's oceans each day (NASA JPL, 2017). The SeaWinds instrument operated until 2009 and was replaced by NASA in 2014 by attaching a back-up instrument to the ISS. This replacement was named the ISS-RapidScat; it operated for two years, and failed in August 2016. As of 2017, NASA does not have plans to launch another scatterometer mission, as ocean wind data is provided by other systems, including the Indian Space Research Organization's SCATSAT scatterometer, which was launched on September 26, 2016, and EUMETSAT's Advanced Scatterometer (ASCAT) instrument.

Wind Farms

Data from the QuikScat satellite is used for wind power density maps across the global oceans to reveal potential locations for offshore wind farms to convert wind energy to electric energy (NASA JPL, 2008). **Digital Image 11-2A** is the wind power density map for winter while **Digital Image 11-2B** is the map for summer. Red and white colors indicate high energy is available while blue colors indicate lower energy. Ocean areas with high winds could potentially harvest up to 500 to 800 W · m^{-2}, which is the equivalent of 10 to 15% of the future world energy requirements (NASA JPL, 2008).

An example of one such high-wind area is located off the coast of northern California near Cape Mendocino. The protruding land mass of the cape deflects northerly winds along the California coast, creating a local wind jet that blows year-round. Similar jets are formed from westerly winds blowing around Tasmania, New Zealand, and Tierra del Fuego in South America, among other locations. Areas with large-scale, high wind power potential also can be found in regions of the mid-latitudes of the Atlantic and Pacific Oceans, where winter storms normally track (NASA JPL, 2008).

WEATHER PREDICTION

NOAA's operational weather satellite system is composed of two types of satellites: Geostationary Operational Environmental Satellites (GOES) for short-range warning and polar-orbiting for longer-term forecasting.

Geostationary Satellites

Chapter 4 discussed NOAA's GOES-12 to -15, the significantly upgraded GOES-16, and GOES-17, which was launched in March 2018. Geostationary satellites provide continuous imagery and atmospheric measurements of the Earth. Water vapor in the middle and upper troposphere (Figure 11-1) is detected in the 6.7 to 7.3 μm wavelength range. Water vapor imagery is a very valuable tool for weather analysis and prediction because water vapor imagery shows moisture in the atmosphere, not just cloud patterns. This allows meteorologists to observe large-scale circulation patterns even when clouds are not present (NOAA SIS, 2017b).

Polar-Orbiting Satellites

The polar-orbiting system includes the Suomi NPP and the JPSS systems. Suomi NPP and JPSS satellites carry five instruments to provide many measurements, including ocean productivity, sea temperature, ice motion, cloud properties, ozone, and atmosphere profiles. These satellites circle the Earth from pole to pole and cross the equator about 14 times a day to provide full global coverage twice a day. They download their data to ground receivers every 50 minutes. These systems provide the majority of data used for weather forecasting in the United States and deliver critical observations during severe weather events such as hurricanes, tornadoes, and blizzards (NOAA SIS, 2017a).

OCEANOGRAPHY

Polar-orbiting and geostationary satellites are used to monitor the oceans for many parameters, including sea temperature, salinity, chlorophyll, and ocean currents. The NOAA Satellite Information System website is a central location for finding information about NOAA's environmental satellites (noaasis.noaa.gov/NOAASIS).

SEA SURFACE TEMPERATURE

Sea surface temperature (SST) is a measure of the energy due to the motion of molecules at the top layer of the ocean. Depending on the sensor, spaceborne measurements give us an unprecedented global measurement of SSTs every few days to a week. Temperatures are measured from approximately 10 μm below the surface (infrared region) to 1 mm (microwave region) depths using radiometers (NASA PODAAC, 2017).

Prior to the 1980s measurements of SST were derived from instruments on shorelines, ships, and buoys. Since the 1980s most of the information about global SSTs has come from polar-orbiting satellites such as NOAA's AVHRR, Europe's three MetOP systems, and Terra and Aqua with their MODIS instrument.

SST datasets provide global and regional coverage with a daily, monthly mean, daytime, and nighttime temporal resolution. NOAA also provides animations and SST datasets for monitoring coral bleaching (NOAA OSPO, 2017). Plate 32 is a global SST image for June 2, 2017 with 5 km spatial resolution. Temperatures range from –2 to 32.9°C on the color-coded temperature scale. In general, SST decreases from the equator to the poles, which is expected as the date of this image is close to the summer solstice when the sun's radiation is centered over the equator during the day. Deviations from this cooling poleward pattern are due to oceanic and atmospheric circulation patterns that move relatively colder or warmer water in the oceans.

In Plate 32, the warmest ocean surfaces are in the northern Indian Ocean and southwest Pacific Ocean. A thin, sinuous, and relatively cool zone trends east–west along the equator west of Africa and South America. These cool zones are caused by interaction between the Equatorial and Equatorial Counter Currents. A narrow wedge of cold water that flows north is seen along the southeast coast of South America. The Gulf Stream is displayed as a narrow band of warm water along the east coast of the United States (see Plate 16 for a more detailed illustration). The south-flowing Agulhas Current along the east coast of Africa drives warm water from the equator past Mozambique in contrast to the colder, north-flowing Benguela Current on the southwest coast of Africa. Some of the SST patterns seen on the June 2, 2017 map (Plate 32) may be, or may become, anomalies.

NOAA generates anomaly SST maps by subtracting the long-term mean SST (for that location during that time of the year) from the current value (NASA Earth Observatory, 2002). SST anomalies that persist over many years can be signals of regional or global climate change. In addition, anomalous temperatures (either cold or warm) can favor one organism in an ecosystem over another, causing populations of one kind of bacteria, algae, or fish to thrive or decline. The SSTs in the Pacific Ocean along the equator become warmer or cooler than normal at irregular intervals (roughly every 3 to 6 years). These anomalies are the hallmark of El Niño and La Niña climate cycles, which can influence weather patterns across the globe. SST anomaly maps have a spatial resolution of 50 km and are updated twice weekly (NOAA OSPO, 2017).

SALINITY

Small variations in ocean surface salinity (i.e., concentration of dissolved salts) can have dramatic effects on the water cycle and ocean circulation. Evaporation of ocean water and formation of sea ice both increase the salinity of the ocean. These "salinity raising" factors are continually counterbalanced by processes that decrease salinity, such as the continuous input of fresh water from rivers, precipitation of rain and snow, and melting of ice (NASA Science, 2017). Salinity is measured in parts per thousand (ppt), as a percentage, in $gm \cdot kg^{-1}$, and with the Practical Salinity Scale (PSS). For example, the ocean average of sea surface salinity is about 35 ppt, 3.5% salt, $0.2\ gm \cdot kg^{-1}$ seawater, and

35 PSS. Modern oceanography uses the PSS to derive salinity from precise instrument measurements of seawater electric conductivity, temperature, and pressure (depth) (Miklus and deCharon, 2011). Across the globe, sea surface salinity ranges from 30 to 40 ppt or 32 to 37 PSS.

Surface winds drive currents in the upper ocean, but deep below the surface, ocean circulation is primary driven by changes in seawater density, which is determined by salinity and temperature. Density-controlled circulation is key to transporting heat in the ocean and maintaining Earth's climate. Excess heat associated with the increase in global temperature during the last century is being absorbed and moved by the ocean. Studies suggest that seawater is becoming fresher in high latitudes and tropical areas dominated by rain, while in subtropical, high evaporation regions waters are getting saltier. Such changes in the water cycle could significantly impact not only ocean circulation but also the climate in which we live (NASA Science, 2017).

NASA's SMAP and ESA's SMOS detect faint microwave emissions from the Earth's surface using a passive radiometer to map levels of ocean salinity. These satellites also measure soil moisture, which is discussed later in this chapter. Both satellites use the radar L-band (21 cm) wavelength to map surface emissions as brightness temperature. SMAP also carries an active radar system that collects surface wind conditions and differentiates sea ice from open water. In June 2011, the Argentine spacecraft Satélite de Aplicaciones Cientificas (SAC-D) was launched with the Aquarius sensor, NASA's first instrument specifically built for studying the salt content of the ocean surface waters. Until June 2015, three Aquarius passive microwave sensors collected 300,000 measurements to generate global maps that can show salinity changes of 0.2 PSS, which is equivalent to about 2 parts in 10,000 (NASA Aquarius, 2017).

An Aquarius global map of average ocean salinity from May 27 to June 2, 2012 is shown in **Digital Image 11-3**. Lower values are represented in purples and blues; higher values are shown in shades of orange and red. Black areas occur where no data was available, either due to the orbit of the satellite or because the ocean was covered by ice, which Aquarius cannot see through. The Atlantic Ocean is saltier than the Pacific and Indian Oceans. Rivers such as the Amazon carry tremendous amounts of fresh runoff from land and spread plumes far into the sea. And in the tropics, particularly near the Pacific's ITCZ, cxtra rainfall makcs cquatorial waters somewhat fresher.

Near most coastlines and inland seas in the map, waters appear much fresher or saltier than in open-ocean locations. The Red Sea and the Mediterranean are saltier than average due to evaporation and other processes. The Black Sea, icy high latitudes, and around many islands and peninsulas of Southeast Asia the waters are fresher than average due to runoff from rivers and melting ice.

Chlorophyll and Ocean Color

The food chain in the world's oceans begins with simple one-celled microscopic plants called phytoplankton that are concentrated in the upper few tens of meters of sunlit water. The distribution patterns of phytoplankton are determined by concentrations of nutrients in seawater that in turn are determined by upwelling currents that bring cold, nutrient-rich water to the surface. Phytoplankton has the characteristic spectral signature of chlorophyll in the visible region: blue and red wavelengths are absorbed; green wavelengths are reflected. Pigments have evolved in chlorophyll that absorb and reflect specific wavelengths. *Chlorophyll-a* is a pigment that absorbs blue light at 0.43 μm and red light at 0.66 μm and is used by remote sensing to map phytoplankton concentration. The spectral signature of chlorophyll-a is shown in Figure 11-6. Note that the vertical axis in Figure 11-6 is percent absorption and not percent reflectance.

Near coastlines and the mouths of rivers, *ocean color* may be determined by the amount of suspended clay and silt. For most of the ocean, however, suspended organic constituents influence ocean color. Figure 11-7 shows laboratory radiance spectra in the visible region for four samples of seawater, in which chlorophyll content ranges from 0.09 to 60.40 mg · m^{-3}. Chlorophyll content increases by approximately an order of magnitude for each of the four water samples. In the blue band, water with the lowest chlorophyll content has a radiance 1,000 times that of water with the highest chlorophyll content. In the green band, however, the radiance difference from minimum to maximum chlorophyll is only 10 times. In the red band the radiance difference from minimum to maximum chlorophyll reverses with the 60.40 mg · m^{-3} water having 15 times more radiance compared with the clear water. These spectral differences are seen in the color of seawater, which is deep blue in the open ocean, where concentrations of phytoplankton are low. Water in coastal zones and zones of upwelling, however, is rich in nutrients that support phytoplankton. The resulting absorption of blue light by chlorophyll causes green hues in the water.

NASA has adopted a standardized algorithm that returns the near-surface concentration of chlorophyll-a in mg · m^{-3}; it is calculated using an empirical relationship derived from *in situ* measurements of chlorophyll-a and remote sensing reflectances in the blue to red region of the visible spectrum.

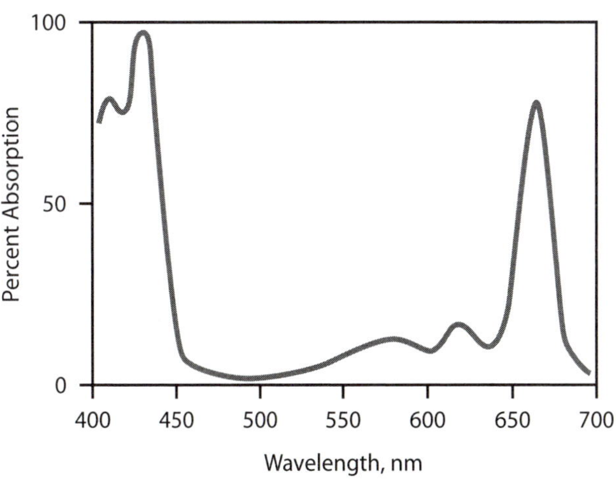

Figure 11-6 Absorption spectrum of the plant pigment chlorophyll-a. From Kerner (n.d.).

FIGURE 11-7 Spectra of four samples of seawater with various concentrations of chlorophyll. From Hovis and others. 1980. Nimbus-7 coastal zone color scanner: System description and initial imagery. *Science*, 210(4465), 60–63, Figure 1. Reprinted with permission from AAAS.

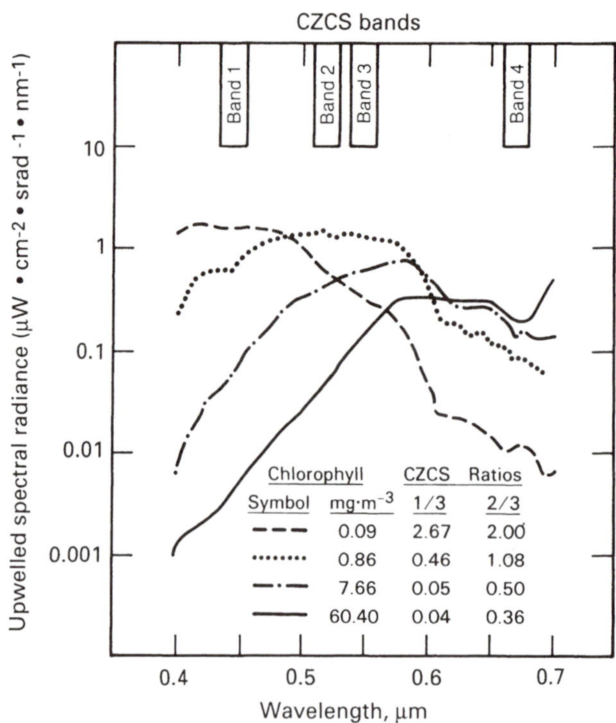

The implementation is contingent on the availability of three or more sensor bands spanning the 0.44 to 0.67 μm spectral range. The algorithm is applicable to all current ocean color sensors (NASA EarthData, 2017). NASA's MODIS and SeaWiFS (Chapter 4) and NOAA's Suomi NPP and JPSS satellites with the VIIRS instrument are satellites with bands in the 0.44 to 0.67 μm range that map chlorophyll-a concentration on the surface of water. Ocean color data have been deemed critical by the oceanographic community for the study of primary production in the ocean and global biogeochemistry (NASA SeaWiFS Project, 2017).

GLOBAL PRODUCTIVITY

Digital Image 11-4 is a MODIS chlorophyll-a seasonal composite for Spring 2014. The highest chlorophyll-a concentrations are shown in red while the lowest are colored blue and purple. Black areas have no data. The highest chlorophyll concentrations are in cold polar waters and in places where ocean currents bring cold water to the surface, such as along the shores of continents. In many coastal areas, the rising slope of the sea floor pushes cold water from the lowest layers of the ocean to the surface. The Spring 2014 image shows that the east and west coasts in the Northern Hemisphere have more extensive phytoplankton-rich zones compared with the equatorial region and the Southern Hemisphere, except for the coasts of southern South America and southwest and western Africa.

Ocean phytoplankton chemically fix carbon through photosynthesis, taking in dissolved carbon dioxide and producing oxygen. Through this process, marine plants capture about an equal amount of carbon as does photosynthesis by land vegetation. Changes in the amount of phytoplankton indicate the change in productivity of the oceans and provide a key ocean link for global climate change monitoring. Scientists use chlorophyll in modeling Earth's biogeochemical cycles such as the carbon cycle or the nitrogen cycle (NASA Earth Observations, 2017a).

PRODUCTIVITY AND UPWELLING ALONG THE CALIFORNIA COAST

Plate 33 is an AVHRR SST image and a SeaWiFS chlorophyll-a map of the central California coast (Ryan and others, 2005). The AVHRR color scale shows warm water in red, orange, and yellow hues and cold water in blue and magenta. The SeaWiFS color scale shows high chlorophyll concentrations in red, orange, and yellow hues and low concentrations are shown in blue and magenta. White patches are clouds.

The narrow belt of cool temperatures along the coast is caused by upwelling of cool, nutrient-rich water. Nutrient-rich water is concentrated in a narrow zone along the coast that contrasts with nutrient-poor water farther offshore. There is a very close correlation between the cool water and the high concentrations of chlorophyll in Plate 33. Even the narrow offshore filaments of cool water are accompanied by high chlorophyll concentrations. Along the southern California coast (not shown in Plate 33) periods of strong upwelling in the summer are accompanied by "red tides" caused by blooms of dinoflagellates in the nutrient-rich colder water.

OCEAN CURRENTS

Traditional methods for mapping oceanic circulation patterns employ current meters, drift floats, and direct temperature measurements. In addition to being expensive, these methods are hampered by the difficulty of obtaining simultaneous data over broad expanses of water. These problems are largely overcome by satellite remote sensing systems that provide essentially instantaneous images of circulation patterns over very large areas. Current systems are mapped by recognizing some property of a water current that differs from that of the surrounding water. Remote sensing systems record the following properties of water:

1. **Color** due to suspended material such as sediment and plankton. Satellite images acquired in the visible wavelength region are employed.
2. **Radiant temperature**: TIR bands of GOES, AVHRR, Landsat, MODIS, and others are used. Plate 16 shows the MODIS TIR map of the Gulf Stream.

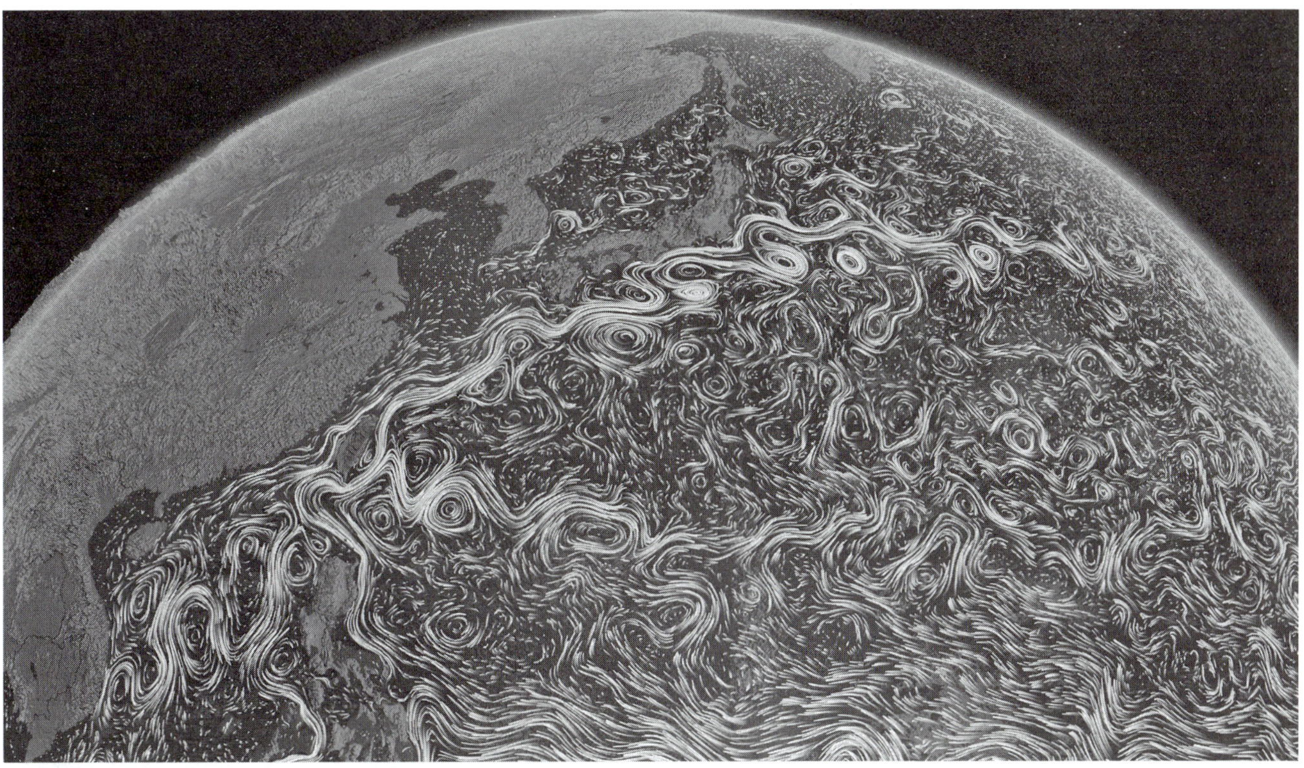

FIGURE 11-8 Ocean surface currents. From NASA Scientific Visualization Studio (2012).

3. **Surface roughness**: The faster moving water of a current system commonly has a pattern of small waves that differs from the pattern of adjacent water. These roughness differences are readily detected by their brightness patterns in radar images from various instruments, such as RADARSAT, Sentinel-1, and TerraSAR-X.
4. **Sea surface height** from satellite altimeters (Chapter 7).

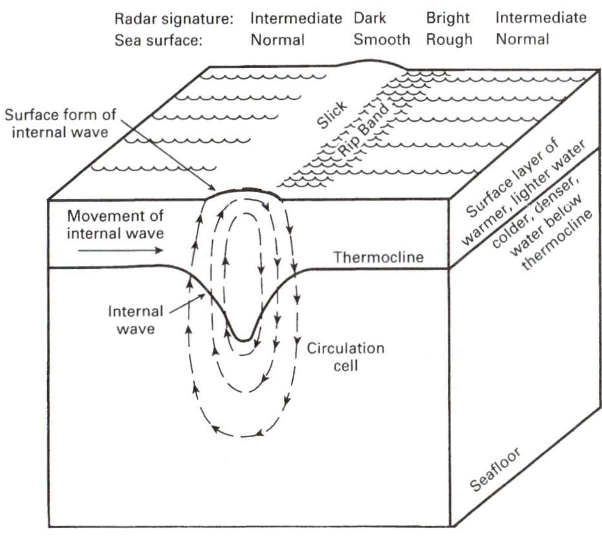

FIGURE 11-9 Model of an internal wave showing relationships between circulation pattern, surface roughness, and radar signatures. Modified from Osborne and Burch (1980, Figure 3).

MODELING OCEAN CURRENTS

NASA's Scientific Visualization Studio (2012) uses remote sensing data along with satellite gravity data and other parameters and integrates them into a complex numerical model that visualizes how ocean circulation evolves. In Figure 11-8, data from the month of May 2007 was extracted from this model to show ocean surface currents in the southwest Pacific Ocean. Thousands of ocean currents are revealed in the model, which is used to quantify the ocean's role in the global carbon cycle and to monitor time-evolving heat, water, and chemical exchanges across the ocean.

RADAR IMAGES OF INTERNAL WAVES

Waves form at the interface between fluids of different density; the well-known example is surface waves, or wind waves, which form at the interface between water and air. Within water bodies, the *thermocline* is the interface between the surface layer of warmer, less-dense water and the underlying layer of colder, denser water. In shallow seas, tidal currents that encounter seafloor irregularities, such as submarine canyons and breaks in slope, may cause waves at the thermocline (Figure 11-9). These waves are commonly called *internal waves* but are also known as *solitons* or *internal gravity waves*. The circulation pattern of an internal wave causes a low linear bulge at the surface that is accompanied by distinctive patterns of small-scale surface waves. The circulation cell rises toward the sea surface at the trailing edge of an internal wave, which decreases the roughness of the small-scale surface waves and results in a linear band of smooth water, called a *slick*, parallel with the crest of the internal wave (Figure 11-9). The circulation cell descends at the leading edge, causing a band of rougher water, called a *rip band*.

Figure 11-10A is an oblique aerial photograph of alternating slicks and rip bands accompanying internal waves in the Gulf of Georgia on the Pacific Coast of Canada. The rip bands have bright signatures because of increased sun glint from the rougher water, and the slicks are dark (Hughes and Gower, 1983). Figure 11-10B is an aircraft radar image that was acquired 8 min after the photograph. The rip bands have bright radar signatures, and the slicks are dark. The Canadian research vessel *Endeavor* is present in both images; the metal hull and superstructure produce the very bright radar signature.

The Strait of Gibraltar connects the Atlantic Ocean with the Mediterranean Sea. Within the strait, there are two layers of water with different densities. A difference in salinity, not temperature, is the cause of the density difference. The deep layer is dense, salty Mediterranean water and the upper layer is less dense and less salty Atlantic water. The boundary between these two layers is termed a *halocline*. The semidiurnal tidal flow across the rugged bathymetry of the strait gives rise to periodic deformations of the halocline, which causes internal waves. Radar detects internal waves that propagate eastward into the Mediterranean Sea (Figure 11-11).

SEA ICE

Increased shipping activity and petroleum exploration in Arctic waters have increased the need for information on sea ice conditions. Global weather predictions require information about the thermal conditions of the polar seas, which in turn are related to ice abundance. Thus, information on ice cover can aid meteorologists. Remote sensing, especially from satellites, is the only practical way to map sea ice on a regional, repetitive basis. Sea ice is measured from space using both passive and active sensors operating at a variety of wavelengths from visible to microwave.

The most important sea ice features are defined in Table 11-3, which is summarized from the more extensive nomenclature of the World Meteorological Organization. Most of these features are illustrated in the images of the following sections.

A. Oblique aerial photograph showing the research vessel *Endeavor*.

B. Aircraft L-band image (23.5 cm). The very bright spot near the center of the image is the *Endeavor*. Sea conditions are similar but not identical to those in (A).

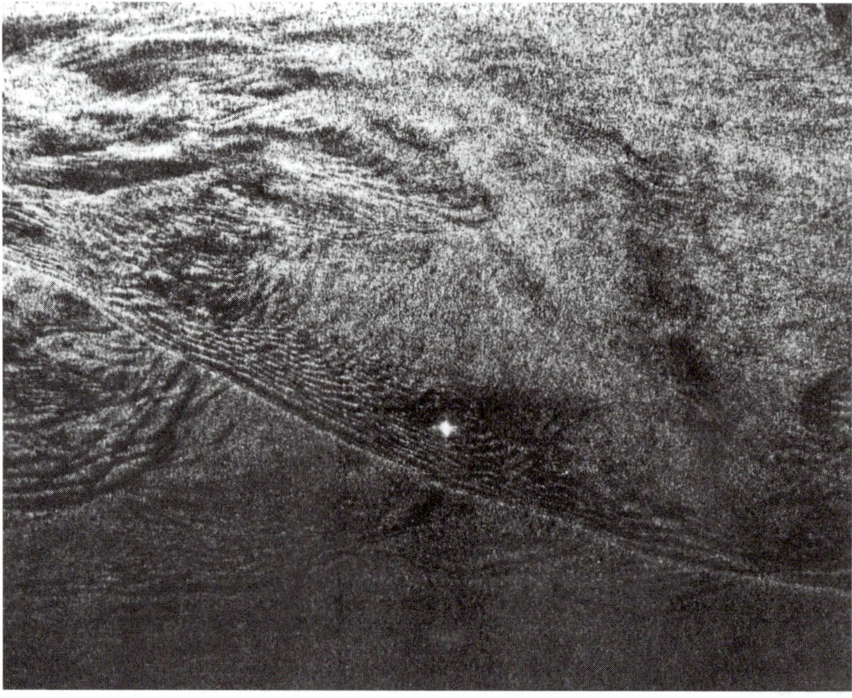

FIGURE 11-10 Internal waves in the Gulf of Georgia, British Columbia, Canada, July 1978. Courtesy J. F. R. Gower, Canada Institute of Ocean Sciences.

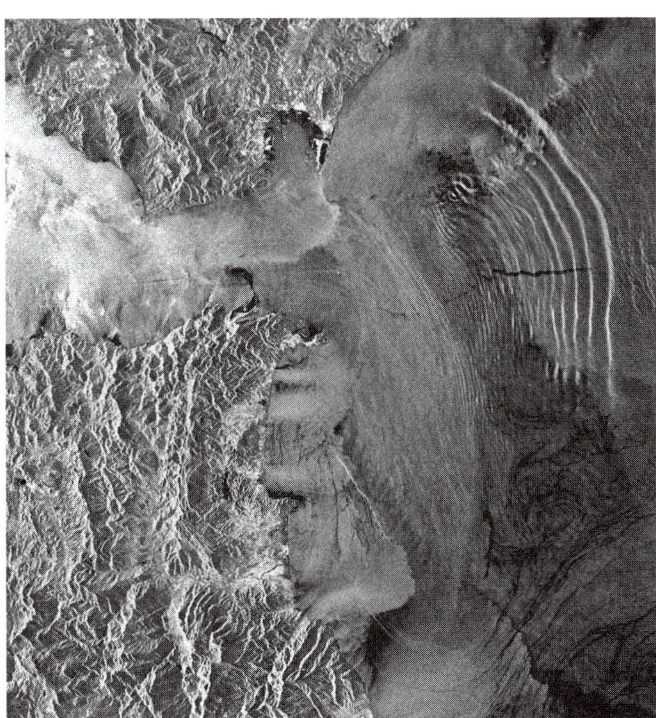

FIGURE 11-11 ERS-1 radar satellite image of the Strait of Gibraltar. An internal wave packet propagates eastward into the Mediterranean Sea. The dark line intersecting the wave packet originated from a ship passing through the Strait that discharged oil. Courtesy ESA (2015).

LANDSAT IMAGES

Figure 11-12 is a Landsat MSS band 4 (NIR2) image of Dove Bay, on the east coast of Greenland, which illustrates many of the features in Table 11-3. The prominent flaw lead separates the pack ice to the east from the fast ice to the west that is attached to the shore. The flaw lead and many other leads in the pack ice are refrozen, as indicated by their dark gray tone; open leads have dark signatures. Fragments of broken floes occur along the southern part of the flaw lead and between large floes of the pack ice. Individual floes have a wide range of sizes and shapes. This pack has a concentration of over 90% ice floes. The pack is classified as *consolidated* because the leads are refrozen and there is no open water. *Open pack ice* has approximately equal proportions of ice and water, and the floes are not in contact. The 1973 winter fast ice in the fjord extends southeast from Dove Bay (bottom of Figure 11-12) and engulfed icebergs that calved from coastal glaciers in the summer of 1972.

TABLE 11-3 Sea ice terminology.

Feature	Description
Fast ice	Ice that forms adjacent to and remains attached to the shore. May extend seaward for a few meters to several hundred kilometers from the coast.
Floe	Any relatively flat piece of sea ice 20 m or more across. Floes are classified according to size.
Ice concentration	Percentage of total sea-surface area that is covered by ice.
Pack ice	General term for any area of sea ice, other than fast ice, regardless of form or forms present. Pack ice is classified by concentration of the floes.
Lead	Any fracture or passageway through sea ice that is navigable by surface vessels. Leads may be open or refrozen. A flaw lead separates fast ice from pack ice.
First-year ice	Sea ice of not more than one winter's growth. Thickness ranges from 0.3 to 2.0 m.
Multiyear ice	Old ice that has survived more than one summer's melt. Compared to first-year ice, floes of multiyear ice are thicker and rougher and have rounder outlines.
Pressure ridge	Wall of broken ice forced up by pressure.
Brash ice	Accumulations of floating ice made up of fragments not more than 2 m across. The wreckage of other forms of ice.

Landsat orbits provide images as far north and south as 81° latitude, although above 70° latitude illumination is insufficient to acquire images from late October to late March. Convergence of orbits at these high latitudes provides up to three or four consecutive days of coverage of the same area during each 16- or 18-day cycle. These repeated images may be used to measure movement of sea ice. Figure 11-13 shows images acquired on March 20 and 21, 1973, in the Davis Strait between Greenland and Baffin Island. The images were registered to each other by the latitude and longitude grid. The eastern one-quarter of both images is covered with brash ice and very small floes. Most of the area is covered by large floes up to 40 km long. On the March 20 image (Figure 11-13A) there are few open leads, indicated by black

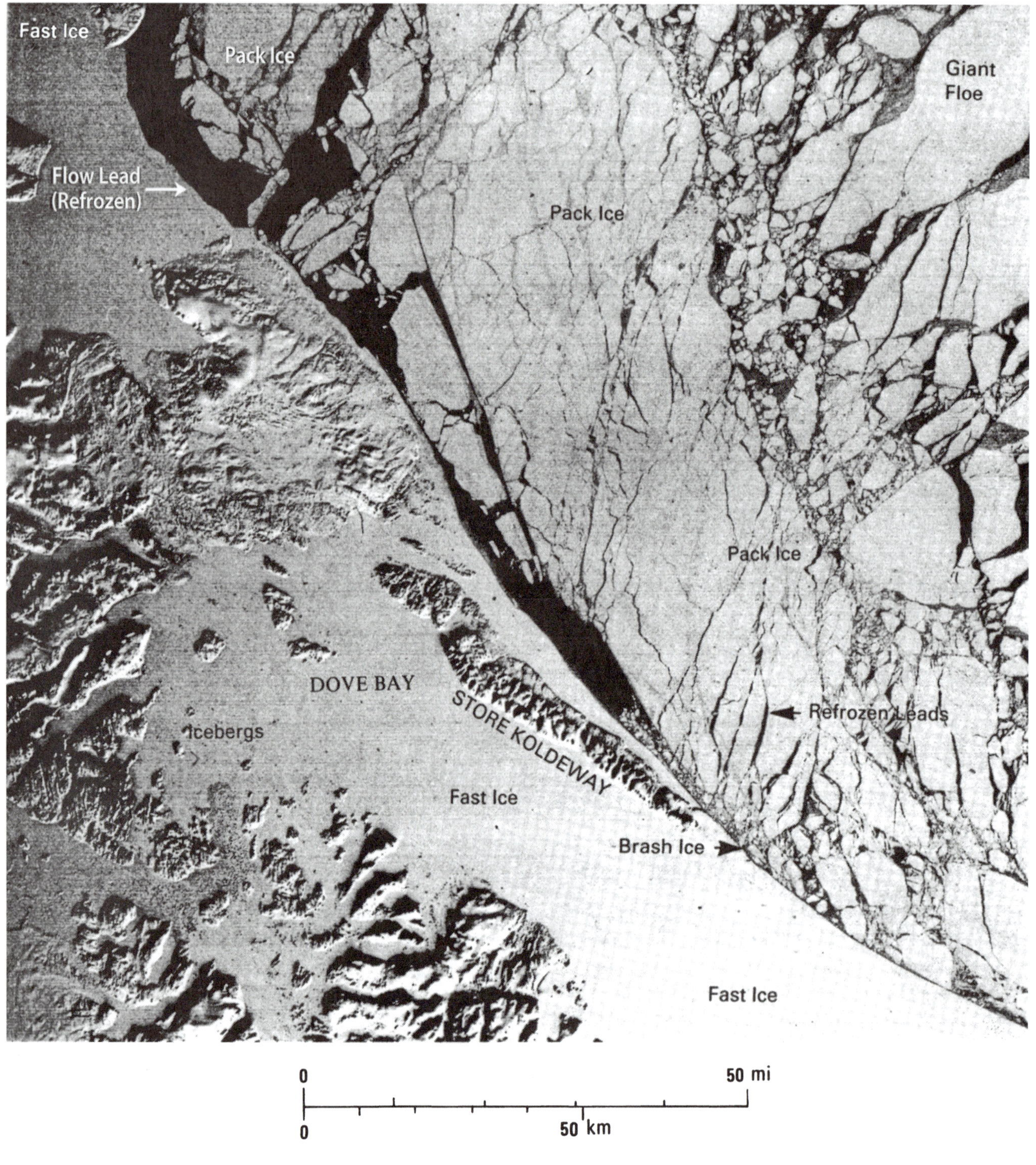

FIGURE 11-12 Landsat MSS band 4 (NIR2) image of sea ice in Dove Bay, east coast of Greenland, March 25, 1973. Landsat courtesy USGS.

signatures, and many of the leads are refrozen, shown by the gray signatures. The open leads are wider and more abundant on the March 21 image (Figure 11-13B), and some of the larger floes have broken up. The arrows in Figure 11-13B are vectors that show the direction and amount of ice movement during the 24-hour period. Ice movement was consistently southeastward at an average rate of 0.4 km · hr^{-1}. This rate assumes that the ice traveled a straight path; if it had followed an irregular course, the actual rate of movement would be higher.

TIR Images

TIR systems can acquire images during periods of polar darkness, but not when heavy clouds and fog are present. One can estimate relative ice thickness from TIR signatures. Figure 11-14 shows TIR images acquired in November 1983 with an aircraft cross-track scanner of the Arctic Ocean north of Alaska. The bright horizontal line records the aircraft flight path. The irregular bright line is a topographic profile of the surface recorded with a laser altimeter along the flight path; peaks directed toward the top of the page are topographic highs.

Open water has the highest radiant temperature and brightest signature in the images. Sea ice insulates the relatively warm water beneath it. Larger amounts of radiant energy are transmitted to the surface of thin ice and smaller amounts to the surface of thicker ice. As a result, thin ice appears warmer than thick ice. Figure 11-14A illustrates these relationships between ice thickness and radiant temperatures. The center of the image has an extensive area of open water with a bright signature (warm temperature). The right portion of the image is covered with a thin sheet of first-year ice with a gray signature (intermediate temperature). The left portion is a large rounded floe of thicker, multiyear ice with a dark signature (cool temperature). The open water contains both multiyear floes and first-year floes. The multiyear floes are larger and cooler and have rounded outlines. The first-year floes are smaller and warmer with angular outlines. The general relationship between radiant temperature and ice thickness can be altered by other factors, such as variations in emissivity and the presence of water on the ice surface during the summer melting period.

The laser profile shows a relationship between ice type and surface roughness. Patches of open water in the center of Figure 11-14A have smooth profiles; the first-year ice in the right portion has minor irregularities; the multiyear floes have rough surfaces caused by their history of fracturing, thawing, refreezing, and ablation by Arctic storms.

Figure 11-14B shows multiyear floes separated by leads that are largely refrozen and have intermediate temperatures. A few narrow open leads are recognizable by their relatively

A. March 20, 1973.

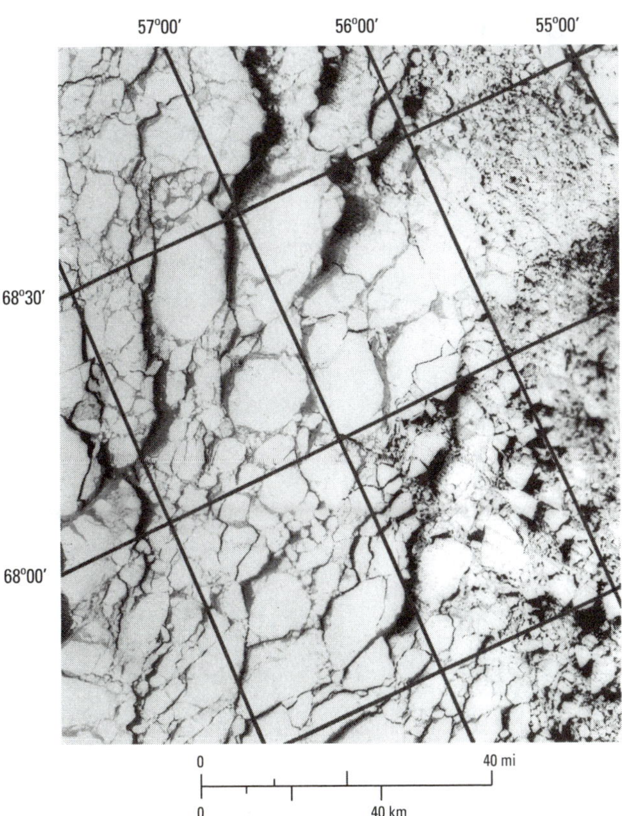

B. March 21, 1973.

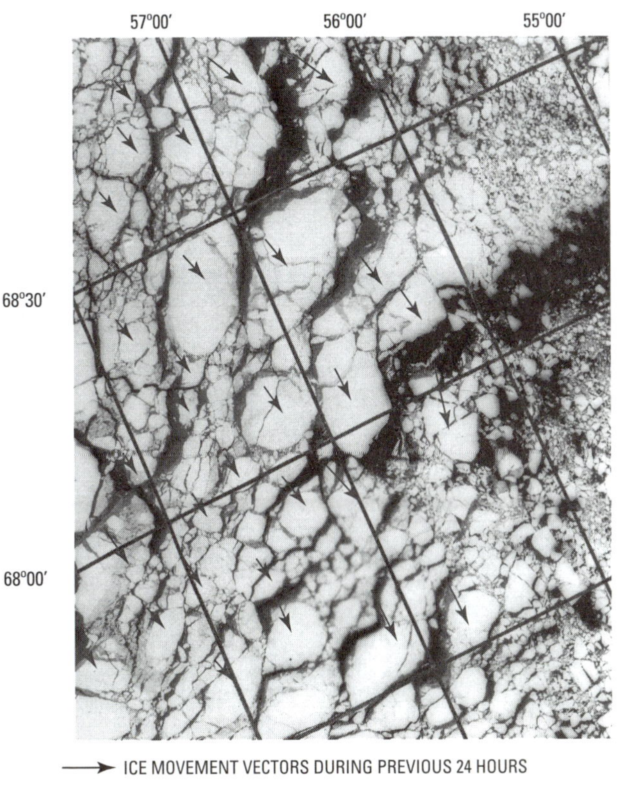

Figure 11-13 Movement of sea ice measured on Landsat MSS images of Davis Strait, west of Greenland. Landsat courtesy USGS.

A. First-year and multiyear ice with open water.

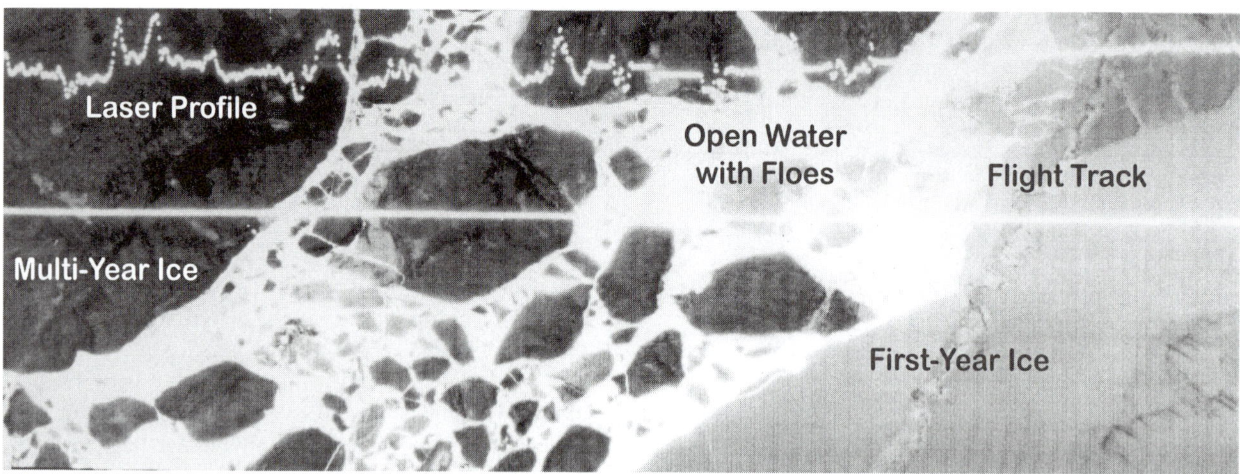

B. Floes of multiyear ice separated by open water and refrozen leads.

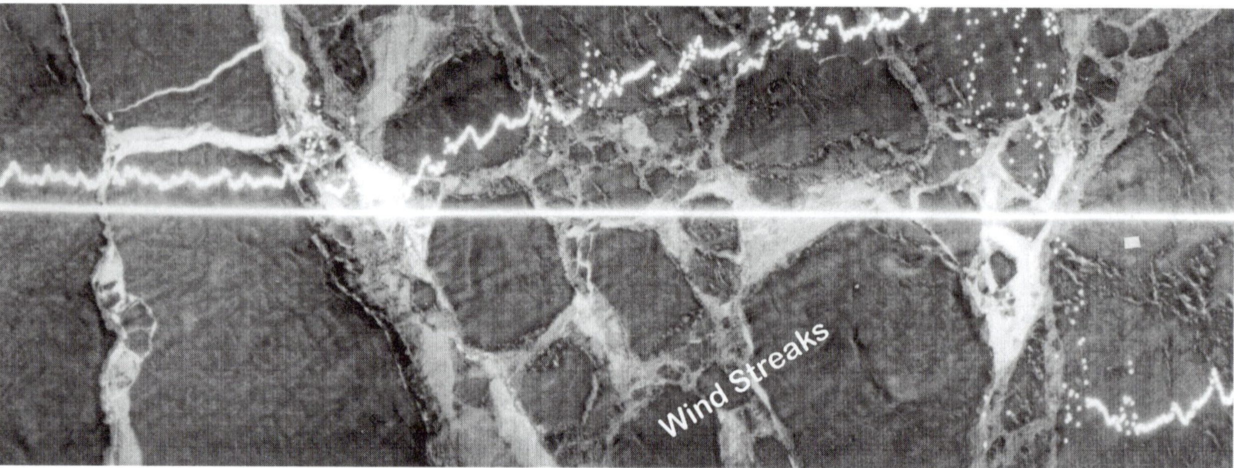

C. Multiyear ice with leads and pressure ridges.

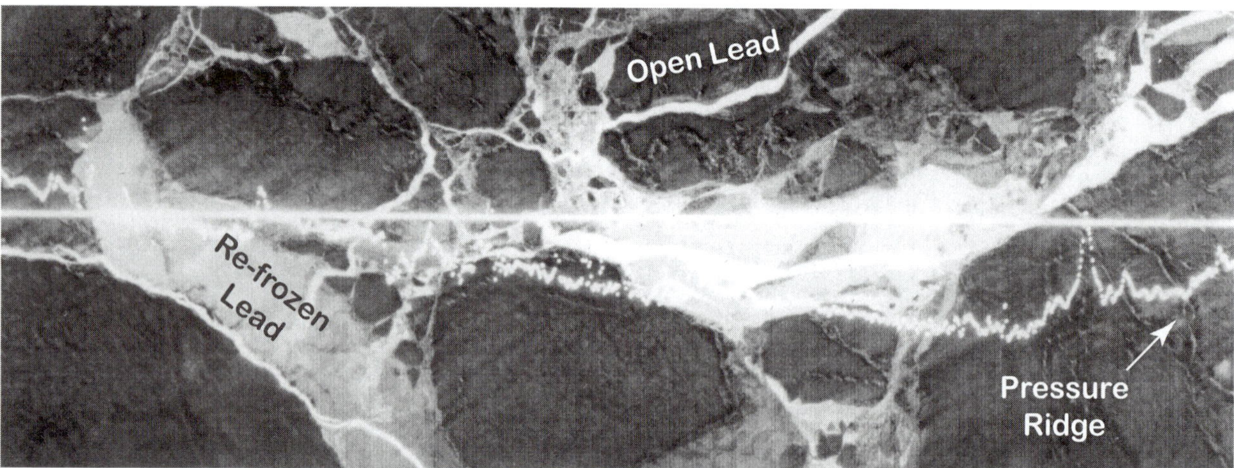

FIGURE 11-14 Aircraft LWIR TIR images of sea ice in the Arctic Ocean north of Alaska, November 1983. Bright signatures are warm; dark signatures are cool. Cross-track width is 480 m. Courtesy Chevron Corporation, Marathon Oil Company, Shell Development Company, and Sohio Petroleum Company.

warm temperature. The faint diagonal pattern from lower left to upper right results from wind streaks (Chapter 5). The laser profile drifts toward the top of the image because of changing aircraft altitude. Figure 11-14C shows extensive open and refrozen leads. The multiyear floe in the lower right corner of the image is cut by several pressure ridges, one of which is crossed by the profile that records several meters of elevation. The thicker ice of the ridge has an irregular narrow TIR signature that is cooler than the surrounding floe. Some ridges are accompanied by parallel fractures with warmer signatures.

TIR images acquired by satellite systems such as AVHRR are used to compile regional maps of ice concentration in polar regions and are especially useful during periods of winter darkness. The TIR bands of Landsat and ASTER are also useful for mapping ice.

RADAR IMAGES

Radar is capable of acquiring images during both darkness and bad weather, which is essential for investigating sea ice. Radar images record surface roughness as a function of radar wavelength and depression angle as defined by the roughness criteria discussed in Chapter 6.

Aircraft Radar Images

Radar signatures of sea ice features are shown in aircraft X-band images of the Beaufort Sea off northern Canada. Figure 11-15A shows an extensive sheet of first-year ice that encloses some multiyear floes and is cut by open and refrozen leads and by pressure ridges. The rough surface of multiyear ice causes strong backscatter and bright radar signatures. Calm water in open leads causes little or no backscatter and has very dark signatures. First-year ice has minor roughness and a dark gray signature. The very narrow, very bright lines crossing the first-year ice are highlights from pressure ridges. Broader bands with light gray signatures are leads filled with brash ice.

Figure 11-15B is located near the terminus of glaciers in the Arctic islands that produce icebergs similar to those in the Landsat image in Figure 11-12. The left two-thirds of Figure 11-15B show a sheet of first-year fast ice attached to a small island of bedrock in the northwest. The fast ice encloses a number of icebergs calved from nearby glaciers. Tabular icebergs in the upper left area have flat but rough surfaces; the other icebergs are irregular. The right portion of the image has floes of first-year ice separated by calm open water (dark signature) and by brash ice (bright signature). A few icebergs are included with the floes; three large tabular icebergs occur along the lower right margin of the image, and several irregular icebergs occur in the upper right area. This example demonstrates the value of radar for detecting and tracking bergs as they enter shipping lanes. Kirby and Lowry (1981) published radar images of bergs and discussed their interpretation.

As stated earlier, younger ice is generally smoother and has darker radar signatures; older ice is rougher and has brighter signatures. During the summer melting season, however, thin sheets of water may cover the ice and partially mask the roughness characteristics. Even under these circumstances, experienced interpreters can recognize ice types based on morphology and distribution patterns.

Satellite Radar Images

In 1978, Seasat acquired repetitive images of Arctic sea ice. The L-band Seasat system had a longer wavelength (23.5 cm) and steeper depression angle (70°) than do typical aircraft X-band systems. As a result, the Seasat smooth and rough criteria (1 cm and 6 cm, respectively) were higher than for the aircraft images in Figure 11-15. Figure 11-16 illustrates two Seasat images acquired at a 3-day interval in October 1978. Fu and Holt (1982) describe these and other images. The northwest corner of Banks Island in the Beaufort Sea occurs in the southeast corner of each image. A sheet of smooth fast ice is attached to the west coast of Banks Island. Smooth ice is distinguished from calm water by the presence of bright pressure ridges. The pack ice consists of floes of multiyear ice, many of which are aggregates of smaller floes that are separated by brash ice. The very rough brash ice has a brighter signature than the floes. A conspicuous floe is Fletcher's Ice Island, a tabular iceberg 7 by 12 km in size. The bright signature of much of the iceberg is attributed to low corrugated ice ridges and scattered rock debris inherited from its glacial origin on Ellesmere Island. This iceberg, which is also called T-3, was discovered in 1946 and has been tracked since then, remaining in the clockwise circulation pattern of the Beaufort Sea.

Considerable ice movement occurred during the 3-day interval in the acquisition of the two Seasat images. The vectors in Figure 11-16B were plotted by connecting positions of individual floes, using the technique applied to repetitive Landsat images. The pack directly north of Banks Island was stable, but on the west and northwest, floes moved southward approximately 20 km at an average rate of 0.3 km · hr^{-1}. In the earlier Seasat image (Figure 11-16A), the leads were narrow, but three days later they were extensive in the moving portion of the pack. Many of the leads in Figure 11-16B have dark signatures indicating smooth, calm water. The gray patches within the leads represent the formation of new ice as the leads began to freeze.

In other Seasat images of the Arctic seas, areas of open water commonly have bright signatures due to small-scale waves generated by wind; these areas should not be mistaken for patches of rough ice. Ketchum (1984) illustrated and described Seasat images of sea ice.

Radar Scatterometer Data

A *radar scatterometer* is a nonimaging active system for quantitatively measuring the radar backscatter of terrain as a function of the incidence angle. Scatterometer data are useful for characterizing the surface roughness of materials and are particularly useful for identifying types of sea ice. A scatterometer transmits a continuous radar signal directly along the flight path. The 3°-wide beam illuminates terrain both ahead and behind the aircraft, but for simplicity Figure 11-17 shows only the forward portion. The 0° incidence angle is directly beneath the aircraft, and the maximum incidence

A. First-year ice (dark) and multiyear ice (bright) with open and refrozen leads. Beaufort Sea north of Liverpool Bay, Canada, January 1984.

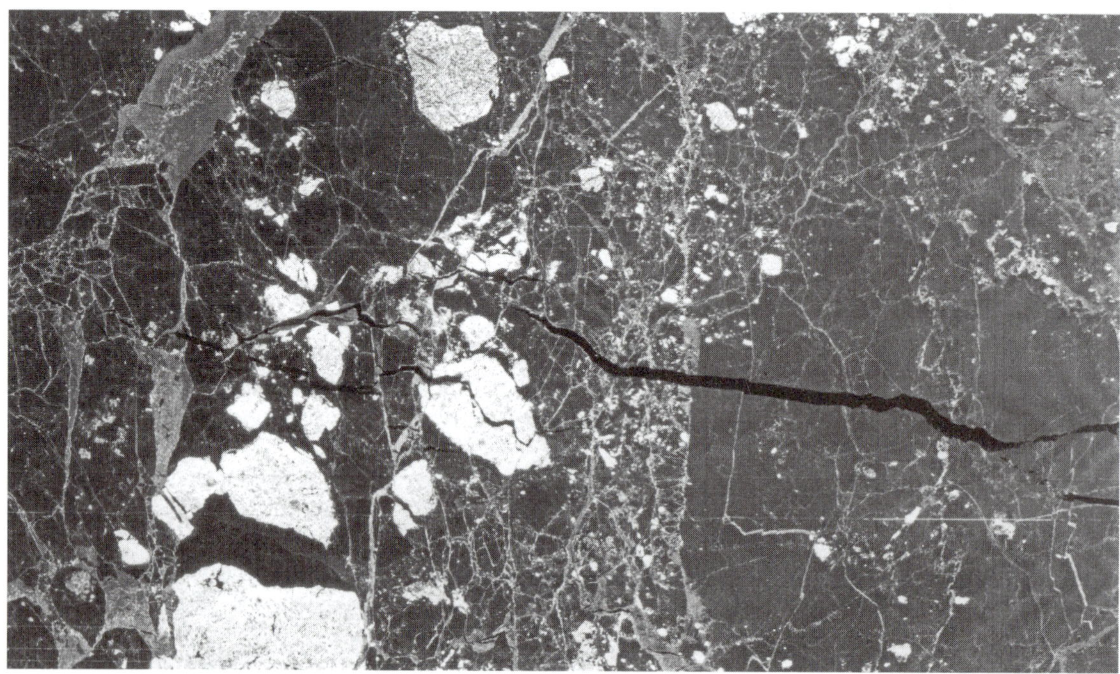

B. Fast ice and floes of first-year ice with icebergs (small bright signatures). Baffin Bay off the west coast of Greenland, June 9, 1984.

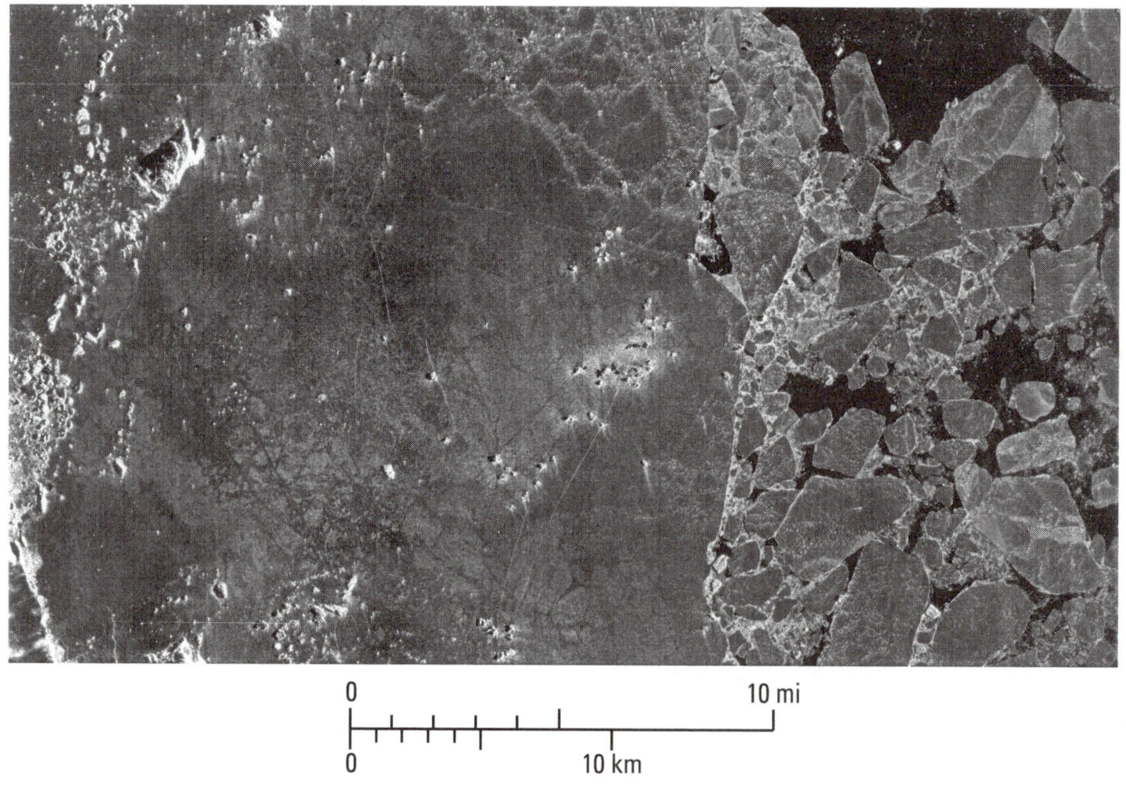

Figure 11-15 Aircraft X-band radar images (3 cm wavelength) of sea ice. Courtesy M. A. Wride, Intera Technologies, Ltd.

A. October 3, 1978.

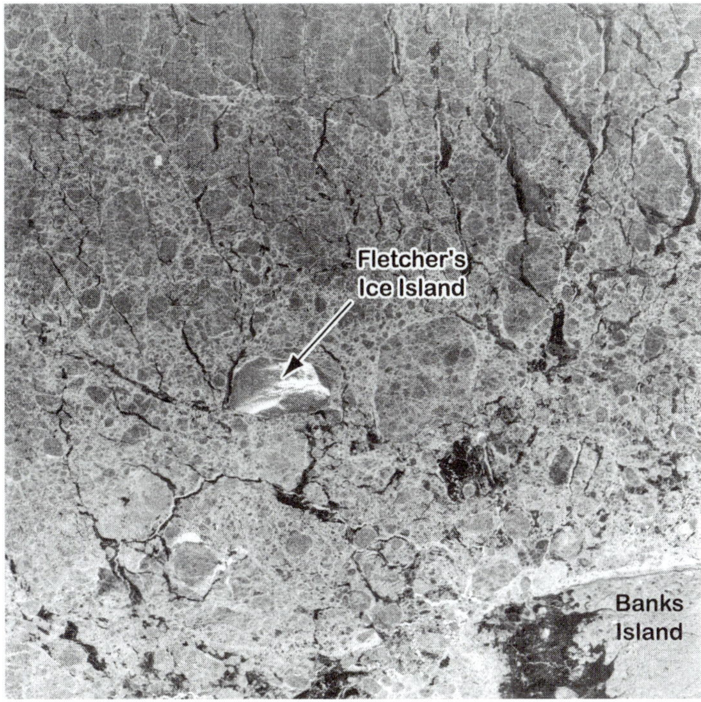

B. October 6, 1978. Arrows show movement during previous three days.

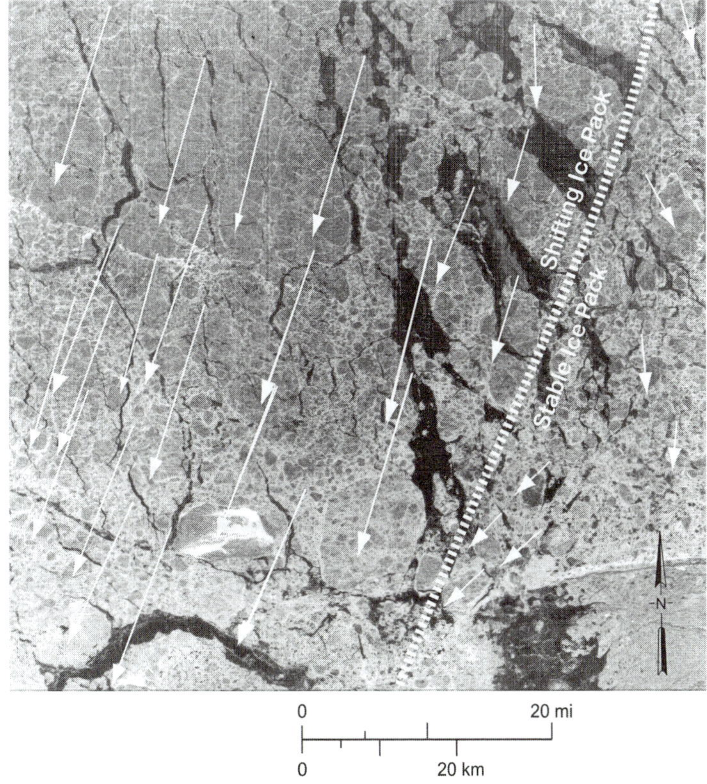

FIGURE 11-16 Movement of sea ice interpreted from Seasat images (23.5 cm wavelength). Beaufort Sea northwest of Baffin Island, Canada. Courtesy J. P. Ford, JPL (Retired).

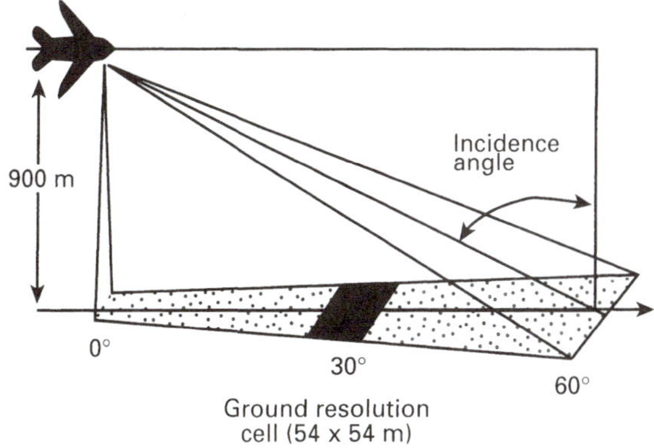

FIGURE 11-17 Geometry of the radar scatterometer system.

angle is 60°. At an altitude of 900 m, the ground resolution cell at a 30° incidence angle is a 54 by 54 m square. The return signal is a composite of the backscattering properties of all of the terrain features within the cell. The wavelength of this scatterometer system is 2.25 cm (X-band), and both the transmitted and received energy are vertically polarized. As the aircraft advances along the flight path, a ground resolution cell is illuminated initially at a 60° incidence angle and then at successively decreasing angles. The recorded amplitude and frequency of the successive returns are processed with Doppler techniques to obtain the scattering coefficient at each of several incidence angles. Details of scatterometer theory and operation are given in Moore (1983).

Scatterometer data may be displayed either as profiles or as scattering coefficient curves. Profiles display the scattering coefficient of the terrain along the flight line for a particular incidence angle. Scatterometer profiles and aerial photographs of sea ice off Point Barrow, Alaska, were acquired in 1967 and interpreted by Rouse (1968). In Figure 11-18, the incidence angles range from 2.5° (almost directly beneath the aircraft) to 52.0°. Scattering coefficient curves (Figure 11-19) display the returns at different incidence angles for an area on the ground.

The photomosaic (Figure 11-18A) shows the following features:

A and F	Open water.
A to C	Smooth first-year ice.
B	Narrow open leads.
C to F	Slightly ridged first-year ice.
C	Major pressure ridge separating the smooth ice and ridged first-year ice.
D	Floe of smooth ice within the ridged ice.
B, E, and F	Areas of smooth ice, ridged ice, and open water, respectively; scattering coefficient curves shown in Figure 11-19.

A. Photomosaic showing scatterometer flight line.

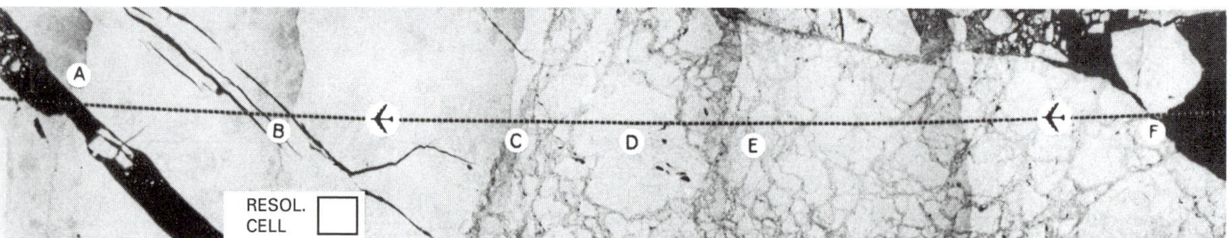

B. Scatterometer profiles.

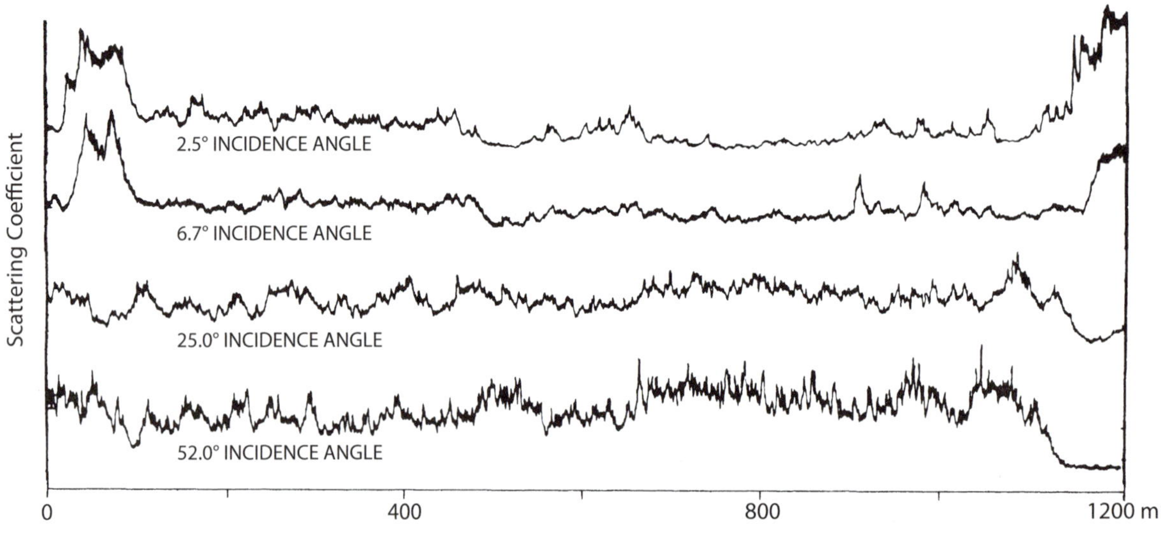

FIGURE 11-18 Scatterometer flight line and profiles of sea ice off Point Barrow, Alaska. From Rouse (1968, Figure 10).

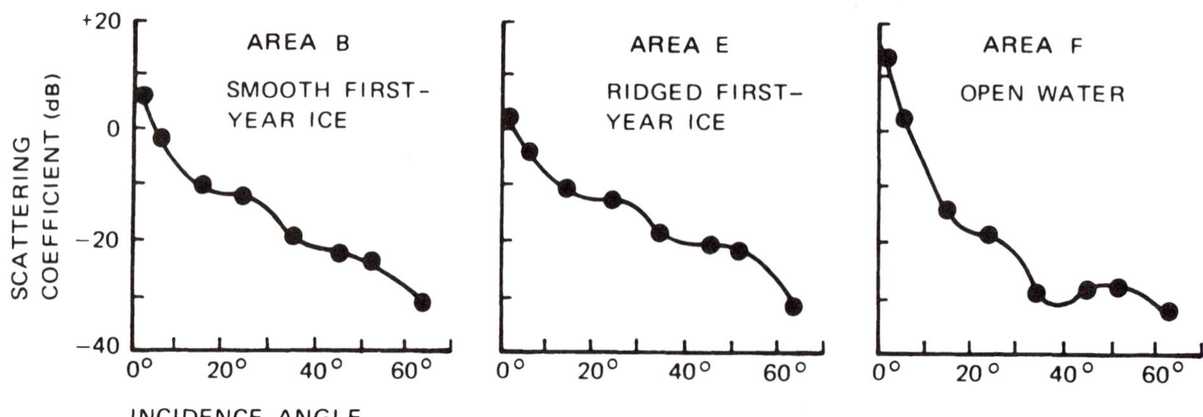

FIGURE 11-19 Scattering coefficient curves of various types of sea ice off Point Barrow, Alaska. Areas B, E, and F are localities shown in Figure 11-18A. From Rouse (1968, Figure 10).

The scatterometer profiles in Figure 11-18B were plotted at the same horizontal scale as the photomosaic. At low incidence angles (2.5° and 6.7°), open water has a strong return. Similarly the smooth ice has stronger returns than ridged ice at these low incidence angles. At the higher incidence angles (25.0° and 52.0°), most of the microwave energy encountering the water and smooth ice is specularly reflected away from the antenna and produces little return. Energy backscattered from the rough ridged ice, however, produces a relatively strong and characteristically spiked profile at these incidence angles. Note that the floe of smooth ice at Area D can be recognized on the 52.0° profile by its reduced return within the ridged ice.

Passive Microwave Images

Passive microwave images acquired by SSM/I and other satellite systems are used to prepare global maps of sea ice coverage (Chapter 16). In the microwave region, radiant temperature, or brightness temperature, is the product of kinetic temperature times emissivity at the wavelength band of the detector. At microwave wavelengths, brightness temperature is determined by the following equation, which is a version of Equation 5-8:

$$T_{rad} = \varepsilon \times T_{kin} \tag{11-1}$$

where:

T_{rad} = radiant temperature

ε = emissivity (at microwave wavelengths)

T_{kin} = kinetic temperature

Emissivity has a stronger influence on radiant temperature in the microwave region than in the TIR region. In the microwave region, ice has a higher emissivity than water; therefore, ice has a higher microwave brightness temperature than water. Different types of ice have different emissivity values. In the microwave region of 1 to 3 cm, using horizontal polarization, typical brightness temperatures are as follows: open ocean, 100 to 140°K; multiyear ice, 180 to 210°K; and first-year ice, 220°K.

Parkinson and Gloersen (1993) show seasonal images of ice cover in the Arctic and Antarctic regions that were acquired by several passive microwave systems.

MULTI-ANGLE IMAGING SPECTRORADIOMETER

The MISR instrument onboard the Terra satellite is a multispectral, multi-angle imaging camera (NASA Terra, 2012). It views the Earth at nine different angles. One camera points toward nadir, and the others provide forward and aft view angles of 26.1°, 45.6°, 60.0°, and 70.5° at the Earth's surface. As the instrument flies overhead, each region of the Earth's surface is successively imaged by all nine cameras in each of four wavelengths (blue, green, red, and NIR). MISR provides unique spectral information on the amount of sunlight that is scattered in different directions under natural conditions.

Nolin and others (2002) compared the classifications of types of sea ice identified from RADARSAT and MISR images acquired on March 19, 2001 in the Beaufort Sea off the north coast of Alaska. In the SAR image, white lines delineate different sea ice zones as identified by the National Ice Center (Plate 34A). Regions of mostly multiyear ice (A) are separated from regions with large amounts of first-year and younger ice (B through D), and the dashed white line at the bottom marks the coastline. In general, sea ice types that exhibit increased radar backscatter appear bright in SAR and are identified as rougher, older ice types. Younger, smoother ice types appear dark to SAR. Near the top of the SAR image, however, red arrows point to bright areas in which large, crystalline "frost flowers" have formed on young, thin ice, causing this young ice type to exhibit an increased radar backscatter. Frost flowers are strongly backscattering at radar wavelengths (cm) due to both surface roughness and the high salinity of frost flowers, which causes them to be highly reflective to radar energy.

Surface roughness is also registered by MISR, although the roughness observed is at a different spatial scale. Older, rougher ice areas are predominantly backward scattering to the MISR cameras, whereas younger, smoother ice types are predominantly forward scattering. In Plate 34B, the MISR map was generated using a statistical, unsupervised classification routine and analyzed using ice charts from the National Ice Center. Five classes of sea ice were found based upon the classification of MISR angular data. These are described, based on interpretation of the SAR image, by the image key. Very smooth ice areas that are predominantly forward scattering are colored red. Frost flowers are largely smooth to the MISR visible band sensor and are mapped as forward scattering. Areas mapped as blue are predominantly backward scattering, and the other three ice classes have statistically distinct angular signatures and fall within the middle of the forward/backward scattering continuum. Some areas that may be first-year or younger ice between the multiyear ice floes are not discernible to SAR, illustrating how MISR potentially can make a unique contribution to sea ice mapping (Nolin and others, 2002).

GLOBAL VEGETATION

The visible (0.63 μm) and reflected IR (0.85 μm) bands of AVHRR are highly effective for mapping vegetation and vegetation vigor on a global basis. The global Normalized Difference Vegetation Index (NDVI) is calculated from these two bands on a daily basis. Gutman and others (1995) developed a global map of the 5-year NDVI mean and standard deviation (σ) for the month of July between 1985 and 1991 (**Digital Image 11-5** 🌐). The July 5-year means on Digital Image 11-5 show, as expected, low values in desert areas and high values in the tropics and across much of Europe, northern Asia, Canada, Alaska, and the eastern United States. Of most interest to climatologists are the many areas with high interannual vegetation variability on the standard deviation map. The areas in central Siberia, southeast Australia, and northeast Brazil show high interannual NDVI variability.

Droughts caused by ENSO (El Niño–Southern Oscillation) during the five years of measurement are the accepted cause for the high variability in northeast Brazil (Gutman and others, 1995).

GLOBAL FIRES

Before AVHRR and MODIS, scientists had no way of mapping the global distribution of fire. The United States, Canada, and some countries in Europe had fairly robust fire monitoring systems that utilized ground-based networks of fire towers and aircraft surveillance, but large portions of the world had little or no monitoring capability. Even in the United States, there were huge gaps in remote parts of the West and throughout much of Alaska (NASA, 2011). The launch of the first MODIS instrument in 1999 on Terra and a second in 2002 on Aqua represent a major technological leap forward. MODIS is specifically designed to detect fire's thermal signature with four bands that are sensitive to fires. Two are located in the infrared portion of the electromagnetic spectrum at 4 and 11 μm with thresholds at 500 and 400°K, respectively, and can be used to monitor fires day and night. The other two are located at 1.6 and 2.1 μm for nighttime fire detection (NASA Earth Observatory, 1998).

Striking patterns have emerged from the MODIS data. "It's not an exaggeration to call Earth the fire planet," says Chris Justice of the University of Maryland, College Park, a scientist who leads NASA's effort to use MODIS data to study the world's fires. "On an average day in August, MODIS typically detects some 10,000 actively burning fires around the world." A full 30% of the land surface is affected by fire. And during any given year, MODIS has shown about 3% of the world's land surface has clear burn scars visible to the satellites (NASA Scientific Visualization Studio, 2011).

One of the most noticeable patterns to emerge is the sheer abundance of burning that occurs in Africa. **Digital Image 11-6** shows fires across Africa in July 2011 as observed by MODIS and displayed on a virtual 3-D globe. MODIS has demonstrated that some 70% of the world's fires occur in Africa, and more than 50% of the total area burned in the last two decades has occurred on that continent, due largely to the extensive burning of savanna grasslands during the dry season. During a fairly average burning season from July through September 2006, the visualizations show a huge outbreak of savanna fires in Central Africa driven mainly by agricultural activities, but also driven by lightning strikes (NASA Scientific Visualization Studio, 2011). NASA supports a wide range of fire-related research to advance understanding about Earth's climate system, air quality, ecosystem health, and the global carbon cycle.

WETLANDS

Wetlands are places where land is permanently or seasonally saturated with water, forming a distinct ecosystem that is both aquatic and land based. Wetland plants trap sediment, which stabilizes shorelines. They provide a buffer against waves and storm surges. Wetlands also absorb pollutants, preventing toxic elements from flowing downstream or percolating underground. Along parts of Lake Erie, for instance, there are no longer enough wetlands to filter agricultural runoff. Nitrogen and phosphorous now flow into the lake and produce toxic algal blooms that can cover up to 300 square miles (NASA Sensing Our Planet, 2016) (**Digital Image 3-2**).

The Michigan Tech Research Institute studies wetlands in the Great Lakes region of the northern United States (Bourgeau-Chavez and others, 2015). The prime focus of the Michigan Tech remote sensing study is to map wetland ecosystem types (such as emergent wetland and forested wetland), and differentiate peatlands (fens and bogs) from other wetland vegetation types. The study delineates wetland monocultures such as the invasive *Phragmites*, a type of reed that blocks sunlight, forces out native plants, and prevents birds from navigating due to its ability to produce thick and tall stands (Figure 11-20).

Bourgeau-Chavez and her team used Landsat and ALOS PALSAR radar imagery to identify differences between land and water, as well as different types of vegetation. Surface temperature data from Landsat helped distinguish wetlands from topographically higher and drier ground. They employed the PALSAR L-band to detect flooding beneath a vegetation canopy, monitor water levels and soil moisture, and record biomass and vegetation structure characteristics. HH polarization images can detect large stands of *Phragmites*, as this dense and tall vegetation type (up to 5 m high) (Figure

FIGURE 11-20 *Phragmites*, a species of common reed, can dominate wetlands. Photo by E. Banda (NASA Sensing Our Planet, 2016).

11-20) has enough biomass to strongly reflect the L-band wavelength (~24 cm) in comparison to other vegetation types. Field data and interpretation of aerial photographs supported the satellite remote sensing study.

Landsat and PALSAR imagery were acquired during spring, summer, and fall from 2007 through 2011 to capture wetlands in different stages of flooding and to improve mapping of vegetation, as different stages of plant growth (*phenology*) can be associated with unique and diagnostic spectral and SAR backscatter characteristics (Plate 35). The optical and SAR data were fused into one digital data set and spectrally classified using a machine learning algorithm, Random Forests. The fused optical and SAR imagery improve the accuracy of wetlands mapping. Across the top of Plate 35, Landsat TM color images (bands 5-3-2 as R-G-B) from spring, summer, and fall as well as a color Landsat thermal image generated from a spring, summer, and fall thermal band (as RGB) are displayed. The PALSAR images in the lower left of Plate 35 are color composites of the HH and HV polarized images acquired in spring, summer, and fall. The supervised classification of the fused Landsat and PALSAR images generated 18 land cover and wetlands classes. The map that was generated from the classification (Plate 35G) clearly identifies the invasive *Phragmites*.

The PALSAR Spring HH band (Plate 35E), which is sensitive to moisture/inundation, along with the Spring Landsat TM NDVI (Plate 35A) and TIR band (Plate 35D), are particularly important to the accuracy of the land cover classification map (Plate 35G). The importance of these three images is consistent across areas of interest with large regions of wetland cover (Bourgeau-Chavez and others, 2015). The overall accuracy of the coastal Great Lakes maps is 94%.

SOIL MOISTURE

NASA's SMAP and ESA's SMOS satellites measure soil moisture (and salinity, as discussed earlier) by detecting faint microwave emissions from the Earth's surface. SMAP was launched in 2015 and SMOS in 2009. The satellites measure the amount of water in the top 5 cm of soil everywhere on Earth. The variability in soil moisture is mainly governed by different rates of evaporation and precipitation. The importance of estimating soil moisture in the root zone is paramount for improving short- and medium-term meteorological modeling, hydrological modeling, the monitoring of plant growth, as well as contributing to the forecasting of hazardous events such as floods. The SMOS instrument can measure as little as 4% moisture in soil.

The Center for the Study of the Biosphere from Space (2017) notes that with a growing world population, the issue of food security is of global concern. Agricultural management and global food security can be more effective if crop yield is estimated accurately. Root-zone soil moisture data are now being used by the US Department of Agriculture in their crop-yield forecasting system. The impact of SMOS is especially significant over data-poor areas of the world that are prone to food insecurity and famine, such as southern Africa. It is also expected to have a positive effect in other areas such as the Horn of Africa, South America, and potentially India and Central Asia.

Monitoring soil moisture in the root zone allows the onset of drought to be detected and provides information on the water available to plants, which is particularly relevant for semiarid regions. Water shortages can then be predicted several weeks before vegetation is likely to suffer. Root-zone soil water products have already been used in drought monitoring systems developed by the International Hydrological Programme and Princeton University.

Plate 36 is a SMOS root-zone soil moisture map of the globe for April 2, 2012. The color legend correlates to soil moisture in m^3 of water per m^3 of soil. Saturated soils have a value of 1 (dark green) and dry soils have a value of 0 (orange). North Africa, western North America, southern Europe, and southeast England have very low soil moisture. The satellite-based measurement is confirmed by reports of drought conditions in these areas.

ANIMAL COUNTING

Anaho Island National Wildlife Refuge (NWR) in Nevada was established in 1913 to provide a secure nesting area for colonial waterbirds. It supports one of the largest American white pelican nesting colonies in the western United States (Figure 11-21). On average, 10,000 pelicans return to Anaho Island each spring for the nesting season. Management objectives for this refuge include monitoring the breeding populations of the various colonial nesting species, as they are vulnerable to multiple factors (e.g., predation, weather events, disease, lack of sufficient forage, human disturbance) that can influence their productivity and survival (USGS NUPO, 2017).

Monitoring of the waterbird nesting population is essential to detect and evaluate changes in waterbird distribution and abundance at Anaho Island NWR. The USGS, in coordination with the US Fish and Wildlife Service, Humboldt State University, and the Pyramid Lake Paiute Tribe, are evaluating various techniques of obtaining population data to select the best method for estimating population sizes at the refuge. These colonial nesting birds are sensitive to human disturbance and will abandon their nests if appro-

FIGURE 11-21 American white pelican at the Anaho Island NWR. From USGS NUPO (2017).

priate methods are not utilized during population surveys. Unmanned aircraft systems (UAS) may prove to be a useful alternative to land-based and/or manned aircraft-based visual surveys and have the potential to be more economical, less obtrusive, safer, and a more efficient and versatile method to conduct surveys (USGS NUPO, 2017).

The sUAS platform was programmed to fly a photogrammetric survey of the nesting area. The imagery was recorded on low-cost digital cameras that record color and color IR images. Individual pelicans sitting on their nest can easily be seen on the sUAS imagery (Figure 11-22), enabling an accurate census count. Image frames were acquired with different look directions and overlap to enable an orthorectified image mosaic of the area and a 3-D model of the terrain to be created with SfM photogrammetric software (**Digital Image 11-7**) (see Chapters 7 and 8). The red pins on Digital Image 11-7 mark individual American pelicans on the island.

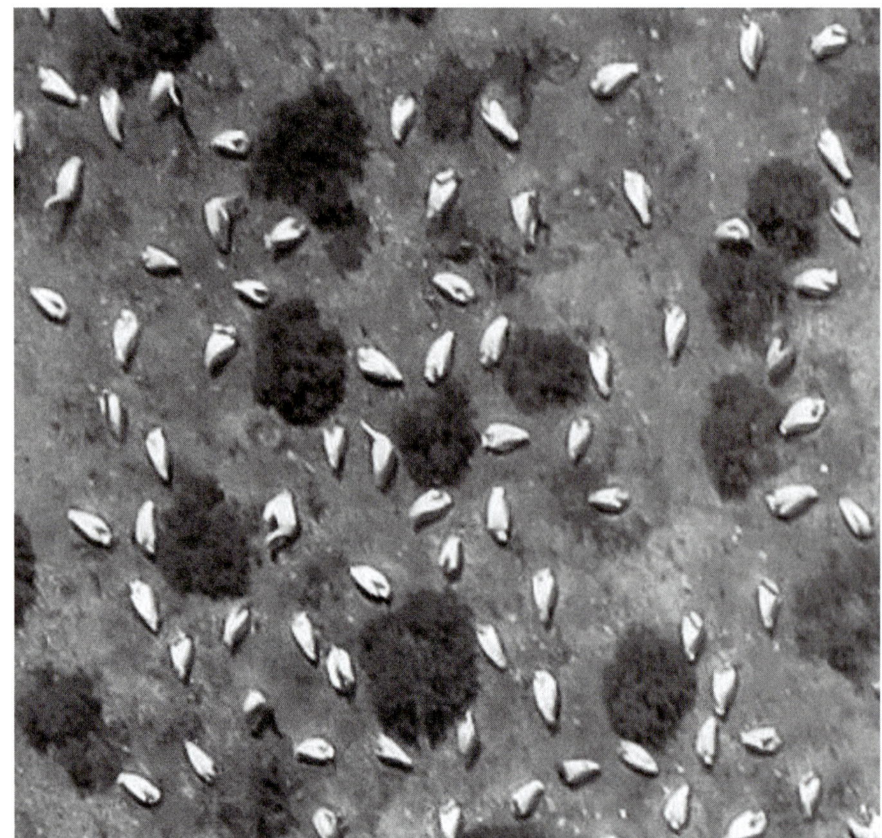

FIGURE 11-22 sUAS aerial image of nesting American white pelicans. From USGS NUPO (2017).

HAZARDOUS WASTE SITES

The term *hazardous waste* refers to by-products of human activity that range from mine dumps to nuclear waste repositories. At most sites the major problem is chemicals leaching from the dump due to rainfall and entering the surface and subsurface water supplies. Major applications of remote sensing are to

1. Locate unreported waste sites;
2. Map the extent and severity of damage to adjacent areas; and
3. Monitor cleanup operations and recovery of the environment.

Many sites are covered by soil and vegetation, as are the adjacent areas. The actual chemicals are rarely detectable by remote sensing, but they can cause stress to vegetation that is initially detectable by decreased reflectance in the near IR (NIR) band. Vegetation stress can be recognized on color IR photographs and multispectral/hyperspectral images.

Color IR photographs were used to investigate a Superfund site in Michigan. *Superfund sites* have been designated as especially hazardous by the US EPA, which provides federal funds for cleanup. Industrial wastewater containing toxic organic chemicals was dumped into unlined ponds and contaminated surface water and groundwater. Herman and others (1994) analyzed a series of color IR photographs acquired from 1969 to 1986. The photographs were scanned and converted into green, red, and NIR bands of digital data. Ratio images were produced of the NIR band divided by the red band (NIR/R). On the ratio images, stressed vegetation has a darker signature than healthy vegetation. The stressed vegetation correlated with contaminated surface water and groundwater as shown by chemical tests. Dumping was stopped and the site was remediated by pumping the contaminated groundwater. Subsequent photographs showed a marked decrease in stressed vegetation.

Marsh and others (1991) investigated a waste site near Phoenix, Arizona. Color IR photographs and airborne video images were acquired 2.5 years apart and documented both cleanup efforts and additional dumping.

Another EPA Superfund site is the old Virginia City gold-mining district in north-central Nevada. The ore was ground and treated with mercury to extract the gold. The waste, called mill tailings, is contaminated with hazardous levels of mercury. During the past 130 years, runoff has spread the tailings up to 80 km from the mill sites. Fenstermaker and Miller (1994) used aircraft hyperspectral scanner images to map the redistributed tailings, which have a distinctive spectral signature.

THERMAL PLUMES

Many industrial plants withdraw water from lakes, rivers, and the ocean to cool their processes and then return it to the same water bodies at higher temperatures. The heated water discharges, called *thermal plumes*, may be monitored by airborne TIR scanners in the same manner as natural water currents of different temperatures. Nuclear and fossil-fuel electrical power plants, refineries, and chemical and steel plants use large volumes of water. Aside from any chemicals or suspended matter, the heated water affects the environment in two ways:

1. Excessively high temperatures may kill organisms or inhibit their growth and reproduction. In some areas, however, the heated discharge water is used for commercial cultivation of lobsters and oysters.
2. The heated water has a lower content of the dissolved oxygen that is essential for aquatic animals and for oxidation of organic wastes.

Environmental legislation has been enacted to regulate thermal discharges. In California coastal waters, for example, the maximum temperature of thermal discharges must not exceed the natural temperature of receiving waters by more than 11°C. At a distance of 300 m from the point of discharge, the surface temperature of the ocean must not increase by more than 2.2°C. The temperature of the discharge water may be lowered by passing it through cooling towers or by mixing it with cooler water before it is discharged. TIR surveys are an ideal way to monitor the temperature and pattern of the discharge outfalls.

Figure 11-23A is a calibrated aircraft TIR image of the heated plume discharged into Montsweag Bay, Maine, from the Maine Yankee Nuclear Power Plant. Figure 11-23B is a contour map of surface water temperatures that was

A. TIR image (8 to 14 μm).

B. Temperature map.

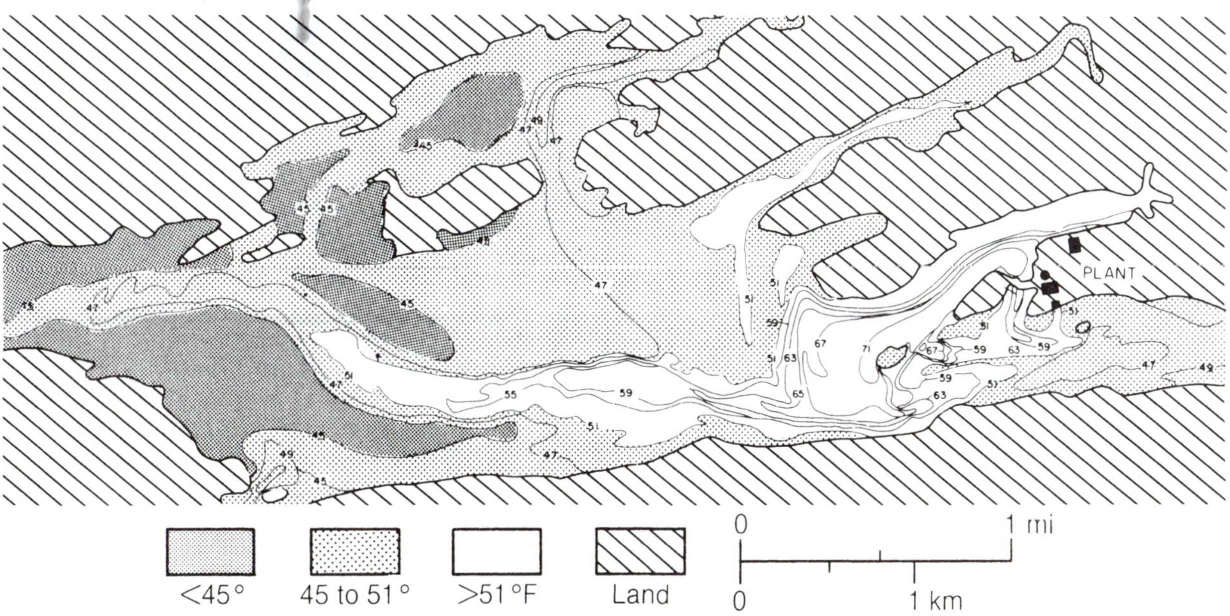

Figure 11-23 Thermal plume from the Maine Yankee Nuclear Power Plant, Montsweag Bay, Maine. Image was acquired at low tide. Courtesy Maine Yankee Power Company and Daedalus Enterprises, Inc.

derived from the image data. The branching or merging contour lines occur where the horizontal thermal gradient is so steep that contour lines coalesce. Both image and map show the location of the plume and the temperature distribution within it. Note, for example, that some parts of the bay are not affected by the plume during low tide. The upstream and downstream extent and temperature level of the plume are precisely shown. The images and map provide no information about the thickness of the plume because the TIR energy is radiated from the surface of the water.

To appreciate the practical value of monitoring thermal plumes using TIR surveys, imagine undertaking the following exercise. Design a monitoring system using conventional surface thermometers that will produce a thermal map, with the precision and detail of Figure 11-23B, throughout a tidal cycle. You must use several hundred surface thermometers, precisely positioned and located; you must calibrate them and record from them to an accuracy of 0.5°C; and they must all be read at the same times to provide simultaneous data for contouring. Finally, you must deploy and retrieve the thermometers during strong tidal currents. This survey would be impractical to conduct but is readily accomplished with several TIR scanning flights. Davies and Mofor (1993) used airborne density-sliced TIR images to monitor thermal plumes from power plants on the coast of Scotland.

OIL SPILLS

Major industrialized nations and regions of the world (such as the European Union, the United States, China, India, and Japan) obtain much of their energy from petroleum that is imported via tanker ships from overseas sources. The National Research Council (2003) used average annual releases of petroleum between 1990 and 1999 to estimate the amount of petroleum entering the oceans. They estimated that 1.3 million metric tons of petroleum enters the ocean annually from all known sources (Table 11-4); natural seeps account for 46% of oil entering the ocean. The following conclusions were made by the National Research Council (2003).

- Worldwide substantial steps have been taken to reduce transportation spills and operational discharges.
- Worldwide reduction in offshore production and pipeline spills is significant, and is believed to reflect changes in industry practice, especially in areas where stricter regulations have been implemented.
- Increase in produced water is significant, mostly reflecting maturing oil fields and increased offshore oil production.
- Natural seeps are a significant source in the report.
- The large annual amounts due to land-based sources (including municipal, industrial, urban, and river runoff) are a significant environmental concern at a variety of scales.

TABLE 11-4 Annual sources of oil in the oceans (1990–1999 estimates in metric tons).

Source	Best Estimate (millions)	Minimum (millions)	Maximum (millions)	Best Estimate (%)
Transportation	**150**	**120**	**260**	**12**
Pipeline spills	12	6.1	37	
Tank vessel spills	100	93	130	
Operational discharges (cargo washing)	36	18	72	
Other	5.3	2.6	16	
Consumption	**480**	**130**	**6,000**	**37**
Land-based (river and runoff)	140	6.8	5,000	
Spills (nontank vessels)	7.1	6.5	8.8	
Operational discharges (vessels $\geq$ 100 GT)	270	90	810	
Atmospheric deposition	52	23	200	
Jettisoned aircraft fuel	7.5	5	22	
Other	?	?	?	
Natural Seeps	**600**	**200**	**2,000**	**46**
Extraction	**38**	**20**	**62**	**3**
Platforms	0.086	0.029	1.4	
Atmospheric deposition	1.3	0.38	2.6	
Produced waters	36	19	58	
Estimated Total	1,300	470	8,300	98

Source: National Research Council (2003, Table 2-2).

Remote sensing is an effective tool for detecting and monitoring the two largest contributors of oil entering the ocean: land-based sources and natural seeps. In general, natural seep oil emissions are highly distinct from most oil spills and involve persistent and widely dispersed oil emissions that generally result in very thin sheens (MacDonald and others, 2002).

Extensive information on using remote sensing technology for oil spills is available from the Arctic Oil Spill Response Technology Joint Industry Programme (JIP). Their 2013 report describes state-of-the-art, surface remote sensing technologies used to monitor oil under varying conditions of ice and visibility. The JIP is managed under the auspices of the International Association of Oil and Gas Producers and is supported by nine international oil and gas companies—BP, Chevron, ConocoPhillips, Eni, ExxonMobil, North Caspian Operating Company, Shell, Statoil, and Total—making it the largest pan-industry program dedicated to this area of research and development. Leifer and others (2012) also review state-of-the-art satellite and airborne sensors for marine oil spills based on the experience gained with the 2010 BP *Deepwater Horizon* oil spill in the Gulf of Mexico. Airborne bathymetric lidar and spaceborne lidar both demonstrated an exciting new capability to remote sense near-surface, submerged oil (Leifer and others, 2012).

Characteristics and Interaction Mechanisms of Oil Spills

The EPA uses the following nonquantitative classification of oil spills, given in the order of decreasing thickness:

- **Mousse**: Brown emulsion of oil, water, and air that forms thick streaks and resembles the dessert called chocolate mousse.
- **Slick**: Relatively thick layer with a definite brown or black color.
- **Sheen**: Thin, silvery layer on the water surface with no black or brown color.
- **Rainbow**: Very thin iridescent multicolored bands visible on the water surface.
- **Rainbow/sheen**: Two categories commonly lumped together because they are volumetrically insignificant and are difficult to distinguish.

Oil spill volumes are provided in different units. A barrel of oil is equivalent to 42 US gallons, which is equal to 159 liters. Figure 11-24 shows the distribution and relative extent of these components in a typical oil spill, where 90% of the volume is concentrated in 10% of the area, largely in the form of slicks and mousse. Table 11-5 lists the interactions between oil and electromagnetic energy at wavelength regions ranging from UV through radar. Figure 11-25 shows these interactions diagrammatically for oil and water to explain their different signatures. The following sections describe the detection of oil slicks on images acquired in the various spectral regions.

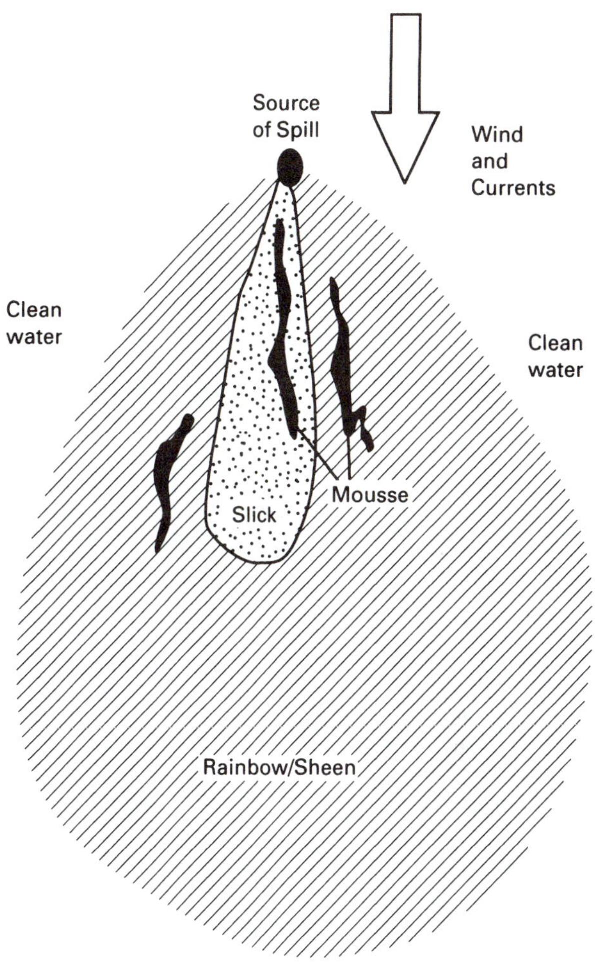

Figure 11-24 Diagram showing the characteristics of a typical oil spill.

UV Images

Figure 11-25A shows that incoming UV radiation from the sun stimulates fluorescence in oil. Figure 11-26 shows spectral radiance curves for water (solid) and a thin layer of oil (dashed). The higher radiance of oil at wavelengths of 0.30 to 0.45 μm (long wavelength UV to visible blue) is due to fluorescence. Individual crude oils and refined products have peak fluorescence at different wavelengths that depend upon composition and weathering of the hydrocarbons. UV images are the most sensitive remote sensing method for monitoring oil on water and can detect films as thin as 0.15 μm (Maurer and Edgerton, 1976). Daylight and a very clear atmosphere are necessary to acquire UV images. UV energy is strongly scattered by the atmosphere but usable images can be acquired at altitudes below 1,000 m. Floating patches of foam and seaweed have bright UV signatures that may be confused with oil (Table 11-5). Foam and seaweed can be recognized on images in the visible band acquired simultaneously with the UV images.

Laser fluorosensors cannot measure oil thickness greater than 10 to 20 μm, as UV laser light is completely absorbed

TABLE 11-5 Remote sensing of oil spills.

Spectral Region	Oil Signature	Oil Property Detected	Imaging Requirements	False Signatures
UV, passive (0.3 to 0.4 μm)	Bright	Fluorescence stimulated by sun	Day; good weather	Foam
UV, active (0.3 to 0.4 μm)	Bright	Fluorescence stimulated by laser	Day and night; good weather	Foam
Visible and reflected IR (0.4 to 3.0 μm)	Bright mousse Dark—slick Bright—sheen	Reflection and absorption of sunlight	Day; good weather	Wind slicks, discolored water
TIR (8 to 14 μm)	Bright mousse Dark—slick	Radiant temperature controlled by emissivity	Day and night; good weather	Warm and cool currents
Radar (3 to 30 cm)	Dark	Dampening of capillary waves	Day and night; all weather	Wind slicks, current patterns

A. UV region.

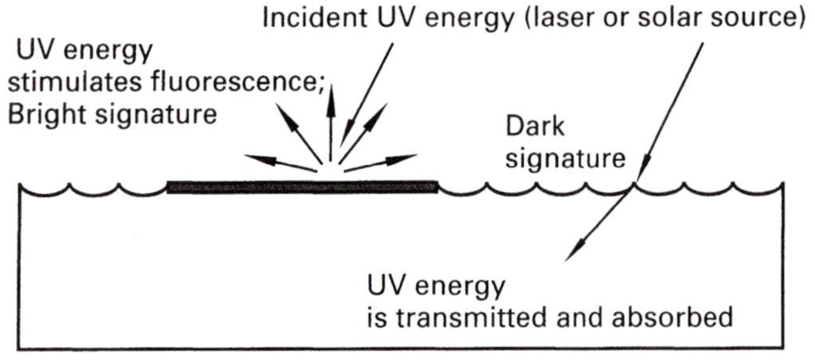

B. Visible and reflected IR regions.

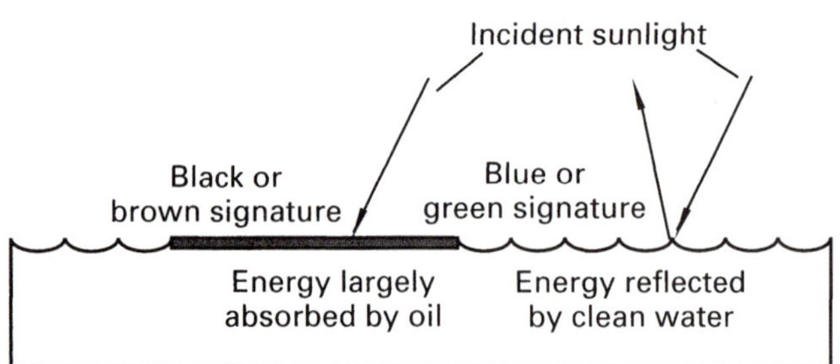

C. TIR region.

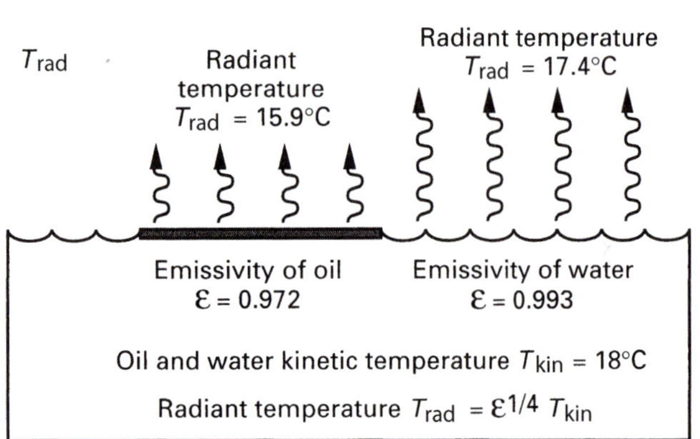

D. Radar region.

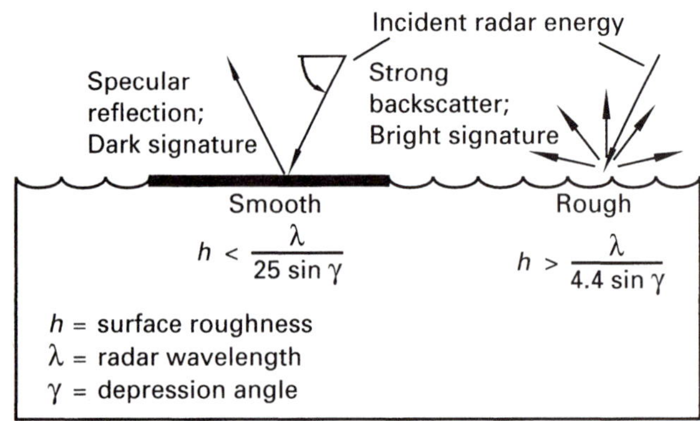

FIGURE 11-25 Interaction mechanisms among oil, water, and electromagnetic energy at different wavelength regions.

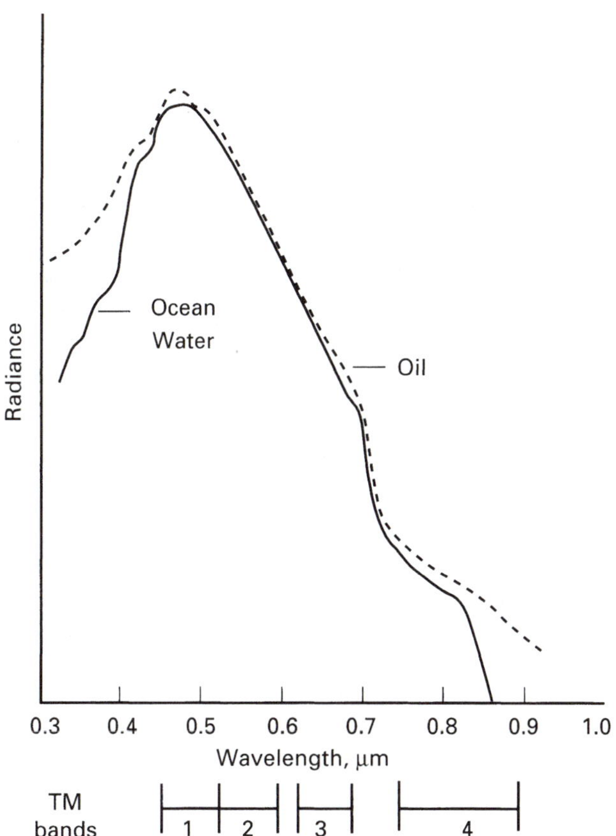

FIGURE 11-26 Spectral radiance for ocean water and a thin film of oil. From Vizy (1974, Figure 5). Data from R. Horvath, University of Michigan.

by oil and cannot reach the underlying water (Brown and Fingas, 2003a). The laser fluorosensor is the most useful and reliable instrument to detect oil on various backgrounds including water, soil, weeds, ice, and snow (Jha and others, 2008). They are the only reliable sensors to detect oil in the presence of ice or snow (Brown and Fingas, 2003b). The laser fluorosensor signals also contain information about some ecologically relevant properties, including seawater attenuation coefficients, phytoplankton, and gelbstoff concentrations (i.e., dissolved organic matter). These parameters are useful to describe the ecological state of coastal waters (Brown and Fingas, 2003b).

The British Petroleum Company used an active UV system, called *airborne laser fluorosensor* (ALF), for oil exploration in offshore basins around the world. Many oil accumulations leak hydrocarbons that form slicks in offshore basins, such as the Santa Barbara Basin. The ability to detect slicks by remote sensing in unexplored basins can contribute to an exploration project. British Petroleum and Pertamina (the Indonesian national oil company) used ALF to survey seven offshore basins in Indonesia (Thompson and others, 1991). Six of the basins had slicks; three of those basins were judged to have the highest exploration potential.

VISIBLE AND REFLECTED IR IMAGES

The interaction between oil and electromagnetic energy in the visible and reflected IR regions is determined by the absorption and reflection of sunlight (Figure 11-25B). A complicating factor is sunglint from calm water caused by the oil. During their invasion of Kuwait in January 1991, Iraqi forces deliberately released 4 to 6 million barrels of crude oil into the Arabian Gulf, where it covered approximately 1,200 km^2 of water and 500 km along the Saudi Arabian coastline. This spill may be the largest in history. For comparison, the massive 2010 *Deepwater Horizon* oil spill in the Gulf of Mexico released an estimated 4,900,000 barrels of oil (On-Scene Coordinator, 2011) while the 1989 *Exxon Valdez* spill in Alaska released over 262,000 barrels. The Iraqi forces also set fire to onshore oil wells, which created giant smoke plumes and spilled additional oil onto the desert. The smoke plumes were tracked on images from AVHRR, Meteosat, and Landsat (Cahalan, 1992; Limaye and others, 1992).

Figure 11-27 shows Landsat TM band images for a subscene of the Arabian Gulf and Saudi Arabian coast that were acquired February 16, 1991. These images show signatures of the spill in the visible, reflected IR, and TIR bands. Figure 11-28 is an interpretation map that Sabins prepared from the images and from personal experience with other spills; no field information was available. Figure 11-29 shows reflectance spectra of water, mousse, and oil sheen that were measured from the TM digital data. Mousse has a higher reflectance in band 5 than in band 4 or 7. Stringer and others (1992, Figure 14) showed similar spectra for the *Exxon Valdez* spill in Alaska.

The individual TM bands of the Arabian Gulf spill were digitally processed to extract additional information (Figure 11-30). Three different sets of principal component (PC) images were created: (1) six PC images from the six visible

A. Band 1, blue (0.45 to 0.52 μm).

B. Band 2, green (0.52 to 0.60 μm).

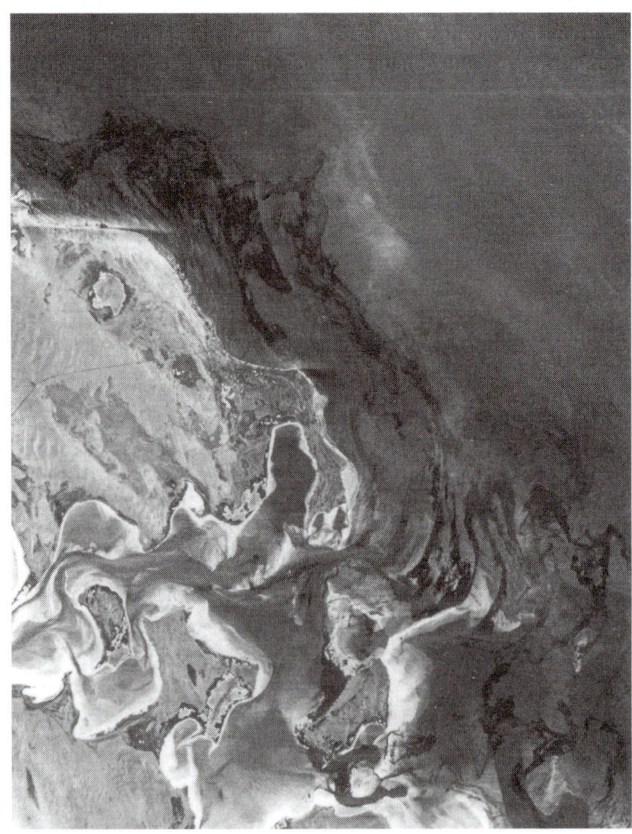

C. Band 3, red (0.63 to 0.69 μm).

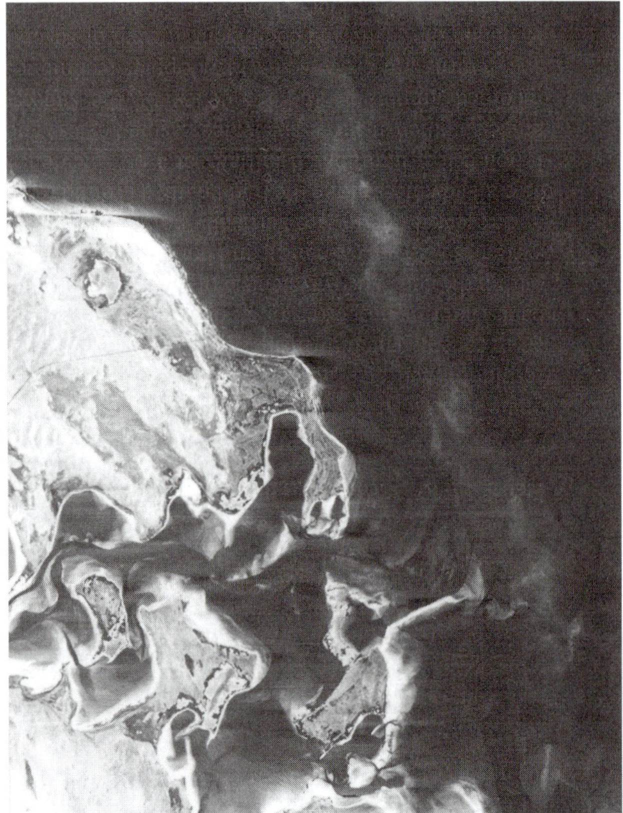

D. Band 4, NIR (0.76 to 0.90 μm).

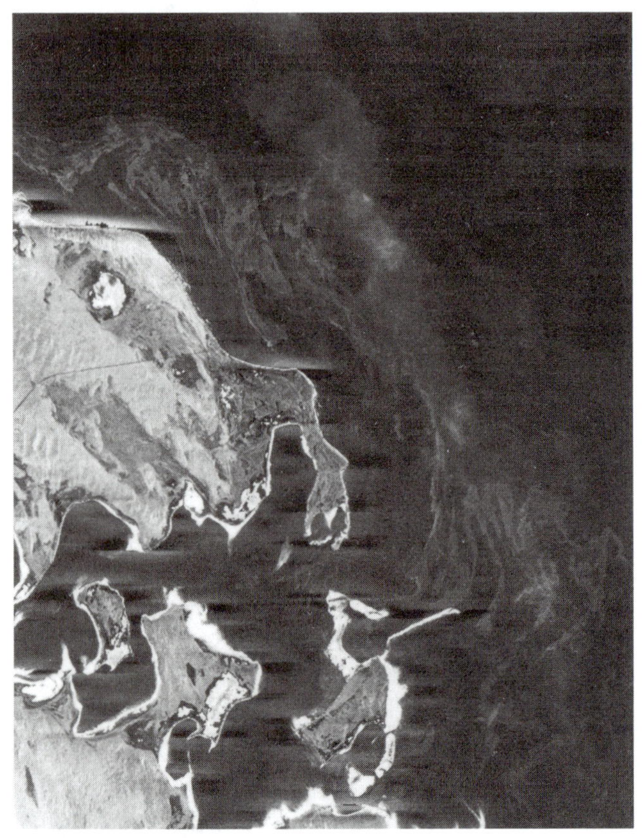

FIGURE 11-27 Landsat TM band images of an oil spill in the Arabian Gulf, February 16, 1991.

E. Band 5, SWIR1 (1.55 to 1.75 µm).

F. Band 6, TIR (10.40 to 12.50 µm).

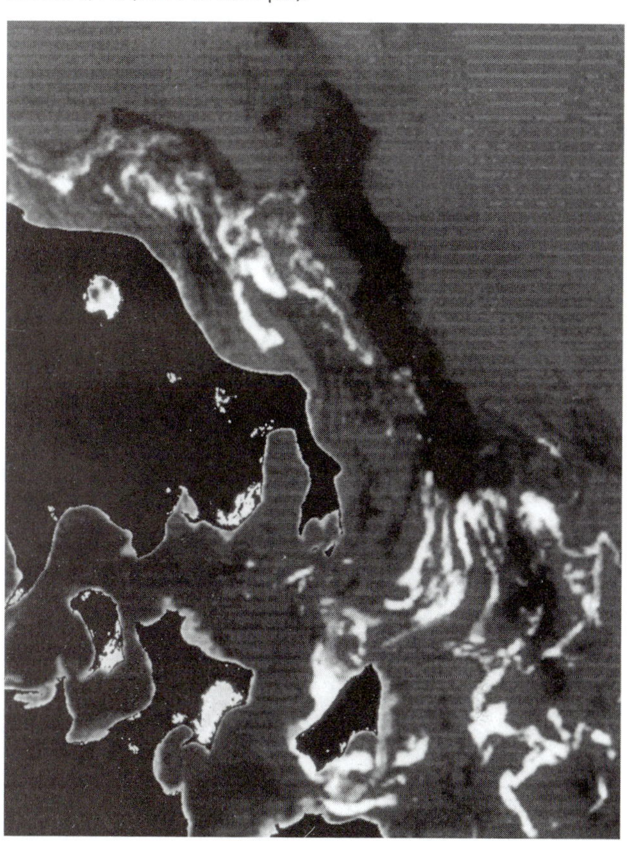

G. Band 7, SWIR2 (2.08 to 2.35 µm).

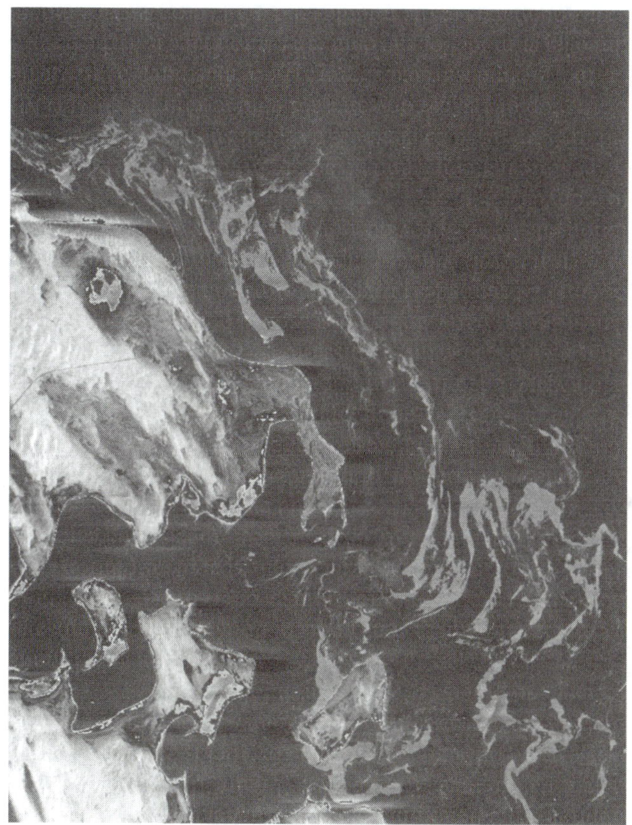

H. Light gray pattern represents shoals. Full legend in Figure 11-28.

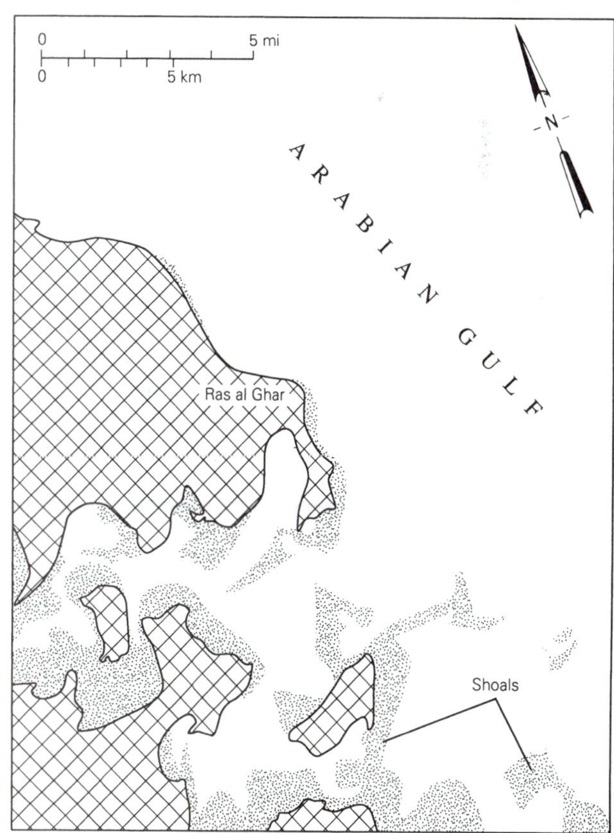

FIGURE 11-28 Map of the Arabian Gulf oil spill interpreted from the TM images in Figure 11-27.

and reflected IR bands, (2) three PC images from the three visible bands, and (3) three PC images from the three reflected IR bands. The PC images of the reflected IR bands 4, 5, and 7 extract the most information. This relationship is explained by the spectra in Figure 11-29, which have maximum differences in the reflected IR region. PC image 1 (Figure 11-30A) emphasizes the heaviest concentrations of mousse with bright signatures. PC image 2 emphasizes fine details of the mousse plus slick, as shown by the narrow black tendrils adjacent to the coast in the top portion of Figure 11-30B. Although PC image 3 (Figure 11-30C) is dominated by noise, the northwest-trending brighter patch is atmospheric haze or smoke. A number of band-ratio images were created, but in general these were not as effective as the PC images. In TM band ratio 4/3 (Figure 11-30D) the mousse is extracted with a bright signature that is comparable to PC image 1.

Plate 37 shows the combinations of individual TM bands that are most effective in discriminating the oil. In the TM band 1-2-3 normal color image (Plate 37A), the dark signature of mousse is difficult to discriminate from the dark blue of the water. Water penetration of these visible bands causes bright signatures from the shallow shoals (light gray pattern in Figure 11-27H), which complicates interpretation. In the TM band 4-5-7 image of reflected IR bands (Plate 37B), mousse has a distinctive bluish green signature. Mousse has a strong reflectance in band 5 (Figure 11-29), which is shown in green.

The National Geographic Society (Williams and others, 1991) published an extensive collection of images and photographs of the environmental destruction caused in Kuwait by the Iraqi invasion.

TIR IMAGES

Figure 11-25C shows the interaction of TIR energy with oil and water. Both liquids have the same kinetic temperature because they are in direct contact. Chapter 5 points out that radiant temperature is a function of both radiant temperature and emissivity. Table 5-3 shows that the emissivity of water is 0.993, but a thin film of petroleum reduces water's emissivity to 0.972. Radiant temperature is calculated from Equation 5-8 as

$$T_{rad} = \varepsilon^{1/4}\, T_{kin}$$

For water at a kinetic temperature of 291°K (18°C), the radiant temperature is

$$T_{rad} = 0.993^{1/4} \times 291°\text{K}$$
$$= 290.5°\text{K or } 17.5°\text{C}$$

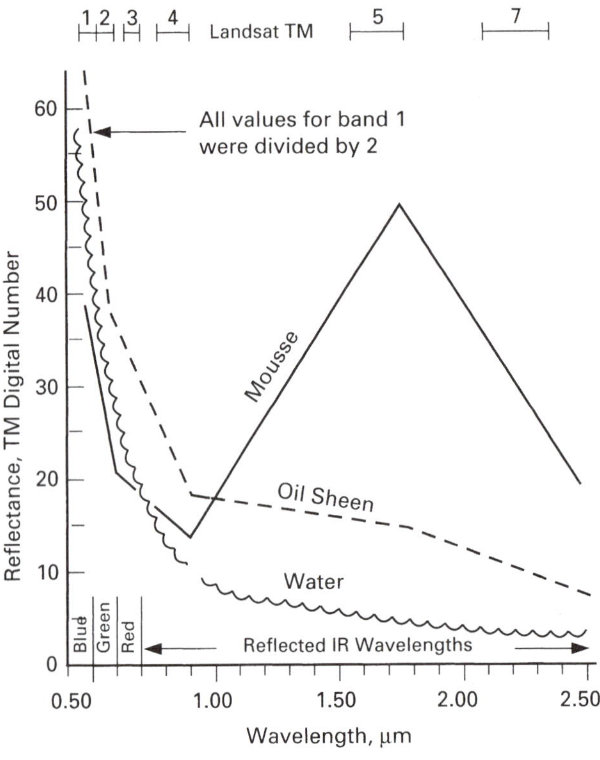

FIGURE 11-29 Reflectance spectra derived from TM bands of the oil spill in the Arabian Gulf (Figure 11-27).

A. PC image 1 of bands 4-5-7.

B. PC image 2 of bands 4-5-7.

C. PC image 3 of bands 4-5-7.

D. Ratio 4/3 image.

Figure 11-30 TM images that were digitally processed to extract information from the oil spill in the Arabian Gulf.

For an oil slick at the same kinetic temperature of 291°K, the radiant temperature is

$$T_{\text{rad}} = 0.972^{1/4} \times 291°\text{K}$$
$$= 288.9°\text{K or } 15.9°\text{C}$$

This difference of 1.6°C in radiant temperature between the oil (15.9°C) and water (17.5°C) is readily measured by TIR detectors, which are typically sensitive to temperature differences of 0.1°C. Figure 11-31 shows two images acquired by an aircraft multispectral scanner of an oil slick in the Gulf of Mexico off Galveston, Texas. The oil tanker *Burmah Agate* collided with another vessel, spilling crude oil. Figure 11-31B is a TIR image in which the oil slick has a cooler signature (darker) than the surrounding water (brighter signature). The warm streaks are caused by mousse, which reradiates absorbed sunlight at TIR wavelengths. The oil slick and mousse are identified much better in this TIR image than in the matching visible image (Figure 11-31A).

Figure 11-27F is a TM TIR band 6 image of the Arabian Gulf spill. The bright (warm) sinuous streaks are mousse. Plate 37D is a color density-sliced version of the TIR image. The mousse has warm signatures (red and yellow). A broad, northwest-trending cool band (dark blue signature) correlates with bright tones in the normal color image (Plate 37A) and is attributed to thin, high clouds. On the color density-sliced image, sinuous cool streaks associated with the warm mousse have dark signatures on the normal color image and are attributed to oil slicks.

Aircraft and TM images record a single radiant temperature value for oil over a broad spectral range. Salisbury and others (1993) measured TIR reflectance spectra for five oils, plus seawater, foam, and mousse. At wavelengths of 8 to 14 µm, spectra of the oils and mousse are essentially flat and featureless. Water and foam, however, have a broad reflectance minimum from 10 to 12 µm; this feature could be useful in discriminating oil from water on multispectral TIR images, such as images from ASTER.

The daytime and nighttime capability of TIR systems is valuable for surveillance around the clock. Rain and fog, however, prevent image acquisition. Also, the interpreter must be careful to avoid confusing cool water currents with oil slicks. This problem can be minimized by interpreting simultaneously acquired UV and TIR images.

RADAR IMAGES

Figure 11-25D shows that an oil slick eliminates the roughness caused by small-scale waves, which results in an area of low backscatter (dark signature) surrounded by the stronger backscatter (bright signature) from rough, clean water. Figure 11-32 shows the characteristic dark signatures of oil slicks on two ERS-1 images of European waters. ERS-1 radar images are particularly sensitive to differences in water roughness because of their steep depression angle and VV polarization.

Radar images may be acquired day or night under any weather conditions, which is an advantage over other remote sensing systems for monitoring oil spills. As with other images, however, radar images must be interpreted carefully, because dark streaks may be signatures of smooth water that is not caused by oil. Internal waves and shallow bathymetric features are two other possible causes of dark signatures. This problem can be reduced by comparing radar signatures with signatures on simultaneously acquired images in other wavelength regions. Leifer and others (2012) document that a low noise, fully polarimetric radar instrument can differentiate between different oil slick properties better than other radar sensors by separating the effects of thickness, surface coverage, and oil properties on the radar backscatter.

A. Green (0.52 to 0.60 µm).

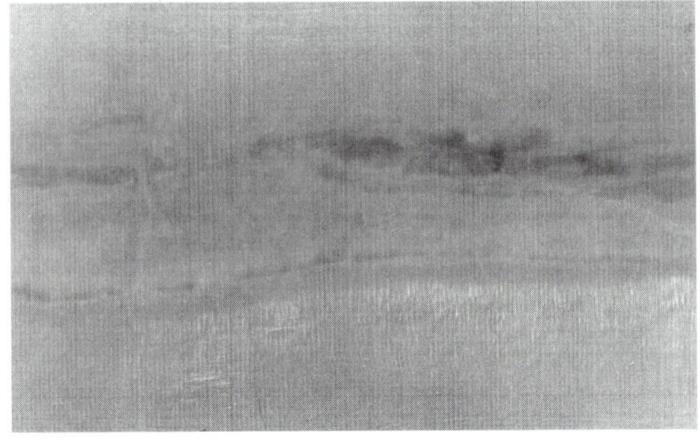

B. TIR (10.40 to 12.50 µm).

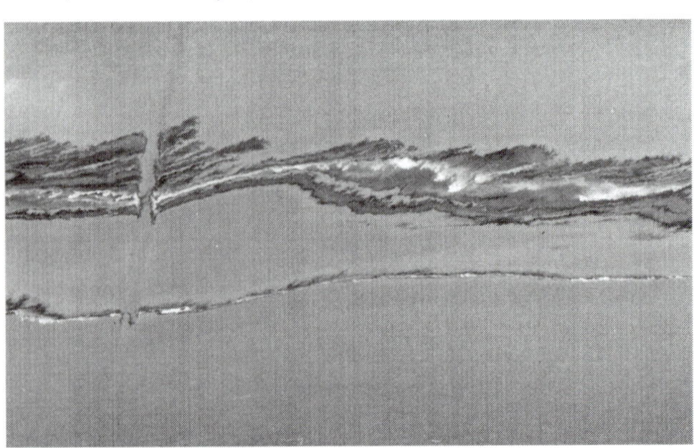

FIGURE 11-31 Aircraft multispectral scanner images of an oil spill in the Gulf of Mexico near Galveston, Texas. Courtesy NASA Johnson Space Center.

A. Offshore from Gothenberg, Sweden.

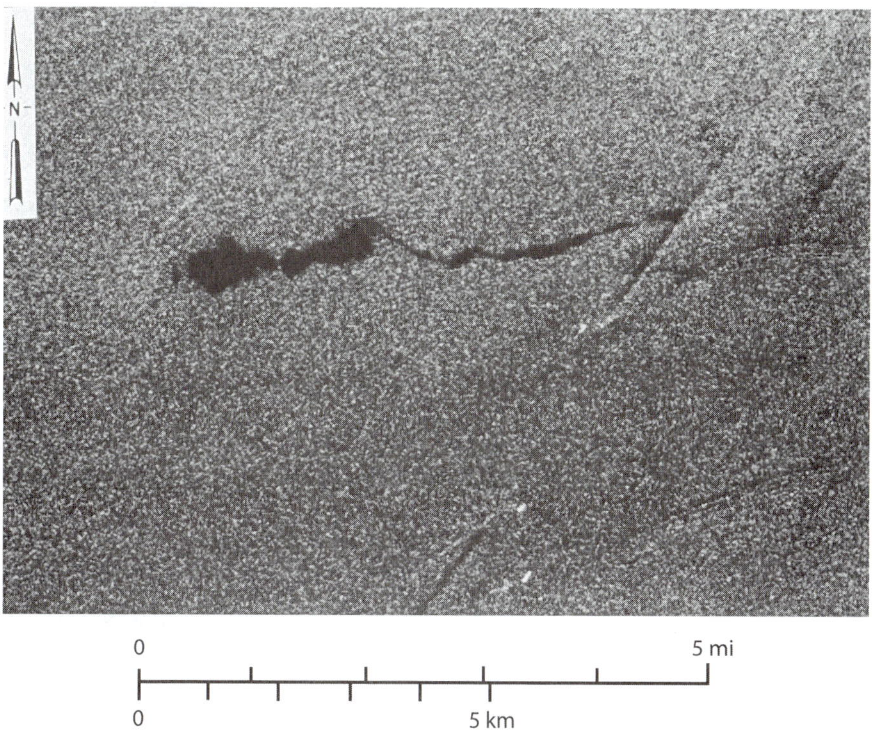

B. Oporto, Portugal, October 4, 1994.

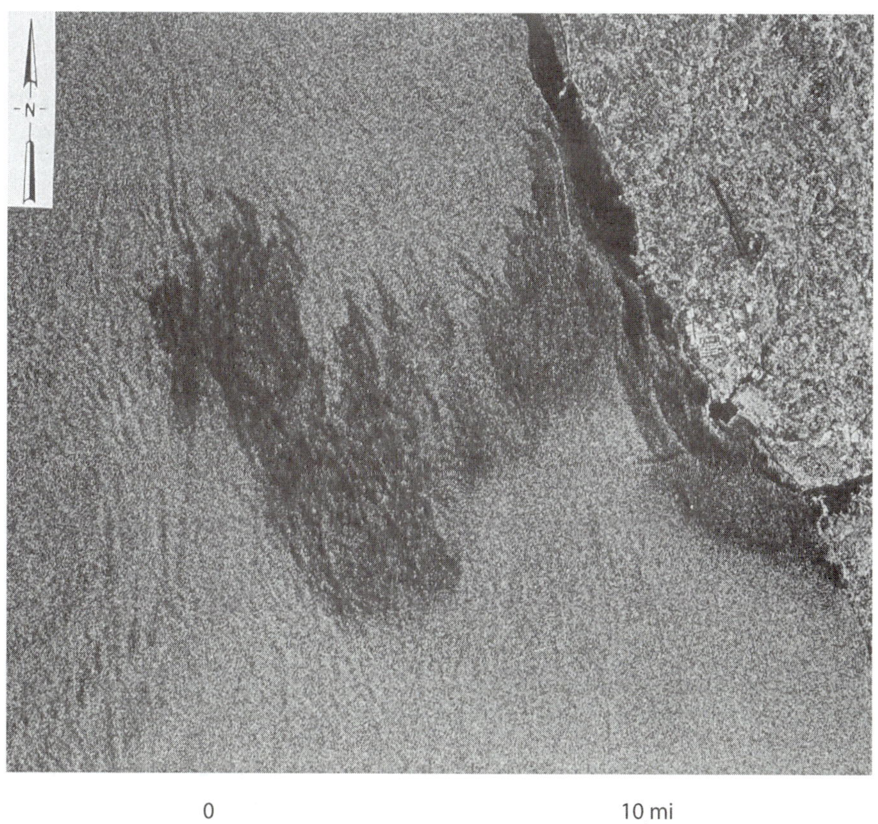

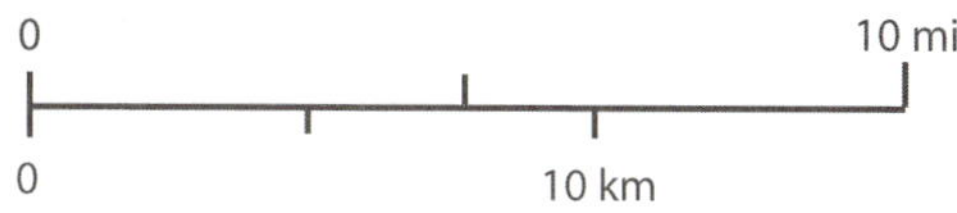

Figure 11-32 ERS-I radar images of oil spills. For these images, which were acquired at a wavelength of 5.7 cm and a depression angle of 67°, the smooth criterion is 0.25 cm and the rough criterion is 1.41 cm. Courtesy European Space Agency.

QUESTIONS

1. What are the five instruments on NOAA's Suomi NPP and JPSS satellites and what environmental application is each used for?
2. How does the OMI instrument onboard the Aura satellite measure ozone?
3. How does the OMI instrument measure sulfur dioxide in the atmosphere?
4. What cloud properties are collected by CERES, MODIS, and VIIRS instruments onboard the Terra, Aqua, Suomi NPP, and JPSS satellites?
5. Precipitation is measured by the DPR and GMI instruments onboard the GPM satellite. Describe the wavelengths and mode of operation for these two instruments.
6. What wavelength does GOES satellites use to detect water vapor? Why is water vapor an important component of the atmosphere to map?
7. How is ocean salinity measured by satellites?
8. What wavelength region is required for mapping "ocean color" and why is this range required?
9. What is a thermocline?
10. Why are TIR systems used to map sea ice in the polar regions?
11. What radar signature is acquired over sea ice areas with:
    a. Calm water in open leads.
    b. First-year ice.
    c. Pressure ridges.
    d. Multiyear ice.
    e. Brash ice.
12. What type of instrument is the MISR? How many angles and wavelengths are acquired as the instrument orbits over sea ice? Why does MISR collect this unique spectral information?
13. What four bands on the MODIS instrument are specifically designed to detect the thermal signature of fire?
14. Why did Bourgeau-Chavez and her team use both Landsat and ALOS PALSAR imagery to map wetlands around the Great Lakes?
15. What wavelengths are used by passive and active UV instruments for the detection of oil spills?
16. In a TIR image of oil floating on water, is the oil warmer or cooler than the adjacent water that is unaffected by the oil spill? Why?
17. Radar displays oil slicks floating on water as dark pixels. Why does this happen? What other conditions may cause radar to display dark signatures?
18. Describe how you would employ MODIS chlorophyll-a images and AVHRR TIR images to increase the catches of commercial fisheries in the oceans. (*Hint*: Upwelling occurs where deeper water comes to the surface carrying dissolved nutrients. Under the right conditions, upwelling will support plankton growth, which is the first link in the oceanic food chain.)

REFERENCES

Arctic Oil Spill Response Technology Joint Industry Programme. 2013. *Oil Spill Detection and Mapping in Low Visibility and Ice: Surface Remote Sensing* (Final Report 5.1). http://www.arcticresponsetechnology.org/wp-content/uploads/2013/10/Report%205.1%20-%20SURFACE%20REMOTE%20SENSING.pdf (accessed January 2018).

Bourgeau-Chavez, L., S. Endres, M. Battaglia, M. E. Miller, E. Banda, Z. Laubach, P. Higman, P. Chow-Fraser, and J. Marcaccio. 2015. Development of a bi-national Great Lakes coastal wetland and land use map using three season PALSAR and Landsat imagery. *Remote Sensing*, 7, 8,655–8,682. doi:10.3390/rs70708655

Brown, C., and M. Fingas. 2003a. Development of airborne oil thickness measurements. *Marine Pollution Bulletin*, 47, 485–492.

Brown, C., and M. Fingas. 2003b. Review of the development of laser fluorosensors for oil spill application. *Marine Pollution Bulletin*, 47, 477–484.

Cahalan, B. 1992. The Kuwait oil fires as seen by Landsat. *Journal of Geophysical Research*, 97(D13), 14,565–14,571.

Center for the Study of the Biosphere from Space. 2017. SMOS Global Drought Monitor. SMOS Blog. http://www.cesbio.ups-tlse.fr/SMOS_blog/?page_id=2589 (accessed January 2018).

CERES. 2017. CERES FM5 on S-NPP. http://ceres.larc.nasa.gov/npp_ceres.php (accessed January 2018).

Davies, P. A., and L. A. Mofor. 1993. Remote sensing observations and analyses of cooling water discharges from a coastal power station. *International Journal of Remote Sensing*, 14, 253–273.

Dodge, R. L., and R. G. Congalton. 2013. *Meeting Environmental Challenges with Remote Sensing Imagery*. Alexandria, VA: American Geosciences Institute.

ESA. 2015. Oceanic Internal Waves: Strait of Gibraltar. ESA Earth Online. http://earth.esa.int/web/guest/missions/esa-operational-eo-missions/ers/instruments/sar/applications/tropical/-/asset_publisher/tZ7pAG6SCnM8/content/oceanic-internal-waves-strait-of-gibraltar (accessed March 2019).

Farman, J. C., B. C. Gardiner, and J. D. Shanklin. 1985. Large losses of total ozone in Antarctica reveal seasonal ClO_x/NO_x interaction. *Nature*, 315, 207–210.

Fenstermaker, L. K., and J. R. Miller. 1994. Identification of fluvially redistributed mill tailings using high resolution aircraft data. *Photogrammetric Engineering and Remote Sensing*, 60, 989–995.

Fu, L. L., and B. Holt. 1982. *Seasat Views Oceans and Sea Ice with Synthetic Aperture Radar* (Publication 81-120). Pasadena, CA: Jet Propulsion Laboratory.

Grody, N. C. 1991. Classification of snow cover and precipitation using a special microwave imager. *Journal of Geophysical Research*, 96, 7,423–7,435.

Gutman, G., D. Tarpley, A. Ignatov, and S. Olson. 1995. The enhanced NOAA global land dataset from the Advanced Very High Resolution Radiometer. *Bulletin of the American Meteorological Society*, 76, 1,141–1,156.

Herman, J. D., J. E. Waites, R. M. Ponitz, and P. Etzler. 1994. A temporal and spatial resolution study of a Michigan Superfund site. *Photogrammetric Engineering and Remote Sensing*, 60, 1,007–1,017.

Hovis, W. A., D. K. Clark, F. Anderson, R. W. Austin, W. H. Wilson, E. T. Baker, D. Ball, H. R. Gordon, J. L. Mueller, S. Z. El-Sayed, B. Sturm, R. C. Wrigley, and C. S. Yentsch. 1980. Nimbus-7 coastal zone color scanner: System description and initial imagery. *Science*, 210(4465), 60–63.

Hughes, B. A., and J. F. R. Gower. 1983. SAR imagery and surface truth comparisons of internal waves in Georgia Strait, British Columbia, Canada. *Journal of Geophysical Research*, 88, 1,809–1,824.

Jha, M. N., J. Levy, and Y. Gao. 2008. Advances in remote sensing for oil spill disaster management: State-of-the-art sensors technology for oil spill surveillance. *Sensors*, 8(1), 236–255.

Kerner, N. n.d. Chem 125: Experiment II—Solution Color, Absorbance, and Beer's Law. Ann Arbor: University of Michigan. http://umich.edu/~chem125/softchalk/Exp2_Final_2/Exp2_Final_2_print.html (accessed January 2018).

Ketchum, R. D. 1984. Seasat SAR sea ice imagery—summer melt to autumn freeze-up. *International Journal of Remote Sensing*, 5, 533–544.

Kirby, M. E., and R. T. Lowry. 1981. Iceberg detectability problems using SAR and SLAR systems. In M. Deutsch, D. R. Weisnet, and A. Rango (Eds.), *Satellite Hydrology* (pp. 200–212). Minneapolis, MN: American Water Resources Association.

Krotkov, N. A., S. A. Carn, A. J. Krueger, P. K. Bhartia, and K. Yang. 2006. Band residual difference algorithm for retrieval of SO_2 from the Aura Ozone Monitoring Instrument (OMI). *IEEE Transactions on Geoscience and Remote Sensing*, 44(5), 1,259–1,266. doi:10.1109/TGRS.2005.861932

Leifer, I., W. J. Lehr, D. Simecek-Beatty, E. Bradley, R. N. Clark, P. E. Dennison, Y. Hu, S. Matheson, C. E. Jones, B. Holt, M. Reif, D. A. Roberts, J. Svejkovsky, G. A. Swayze, and J. M. Wozencraft. 2012. State of the art satellite and airborne marine oil spill remote sensing: Application to the BP *Deepwater Horizon* oil spill. *Remote Sensing of Environment*, 124, 185–209.

Limaye, S. S., S. A. Ackerman, P. M. Fry, M. Isa, H. Ali, G. Ali, A. Wright, and A. Rangno. 1992. Satellite monitoring of smoke from the Kuwait oil fires. *Journal of Geophysical Research*, 97(D13), 14,551–14,563.

MacDonald, I. R., I. Leifer, R. Sassen, P. Stine, R. Mitchell, and N. Guinasso Jr. (2002). Transfer of hydrocarbons from natural seeps to the water column and atmosphere. *Geofluids*, 2(2), 95–107.

Marsh, S. E., J. L. Walsh, C. T. Lee, and L. A. Graham. 1991. Multitemporal analysis of hazardous waste sites through the use of a new bi-spectral video remote sensing system and standard color-IR photography. *Photogrammetric Engineering and Remote Sensing*, 57, 1,221–1,226.

Maurer, A., and A. T. Edgerton. 1976. Flight evaluation of U.S. Coast Guard airborne oil surveillance system. *Marine Technology Society Journal*, 10, 38–52.

McLinden, C. A., V. Fioletov, M. W. Shephard, N. Krotkov, C. Li, R. V. Martin, M. D. Moran, and J. Joiner. 2016. Space-based detection of missing sulfur dioxide sources of global air pollution. *Nature Geoscience*, 9, 496–500. doi: 10.1038/ngeo2724

Miklus, N., and A. deCharon. 2011. *Aquarius/SAC-D: Sea Surface Salinity from Space* (NASA NP-2010-8-166-GSFC). NASA.

Moore, R. K. 1983. Radar fundamentals and scatterometers. In R. N. Colwell (Ed.), *Manual of Remote Sensing* (2nd ed., pp. 369–427). Falls Church, VA: American Society for Photogrammetry and Remote Sensing.

NASA. 2011. NASA Releases Visual Tour of Earth's Fires. Fire and Smoke. http://www.nasa.gov/mission_pages/fires/main/modis-10-overview.html (accessed January 2018).

NASA. 2015, December 14. New NASA Satellite Maps Show Human Fingerprint on Global Air Quality (Release 15-233). http://www.nasa.gov/press-release/new-nasa-satellite-maps-show-human-fingerprint-on-global-air-quality (accessed January 2018).

NASA Aquarius. 2017. Sea Surface Salinity from Space. http://aquarius.nasa.gov (accessed January 2018).

NASA EarthData. 2017. Chlorophyll a (chlor_a). OceanColor Web. http://oceancolor.gsfc.nasa.gov/atbd/chlor_a (accessed January 2018).

NASA Earth Observations. 2017a. Chlorophyll Concentration (1 Month—Aqua/MODIS). http://neo.sci.gsfc.nasa.gov/view.php?datasetId=MY1DMM_CHLORA (accessed January 2018).

NASA Earth Observations. 2017b. Cloud Fraction (1 Month—Terra/MODIS). http://neo.sci.gsfc.nasa.gov/view.php?datasetId=MODAL2_M_CLD_FR (accessed January 2018).

NASA Earth Observatory. 1998, May 29. NASA Demonstrates New Technology for Monitoring Fires from Space. http://earthobservatory.nasa.gov/Features/Fire (accessed January 2018).

NASA Earth Observatory. 2002. Sea Surface Temperature Anomaly. http://earthobservatory.nasa.gov/GlobalMaps/view.php?d1=AMSRE_SSTAn_M (accessed January 2018).

NASA Earth Observatory. 2016, June 7. Satellite Finds Unreported Sources of Sulfur Dioxide. http://earthobservatory.nasa.gov/images/88153/satellite-finds-unreported-sources-of-sulfur-dioxide (accessed January 2018).

NASA GPM. 2017. Precipitation Measurement Missions. Core Observatory. http://pmm.nasa.gov/GPM/flight-project/core-observatory (accessed January 2018).

NASA JPL. 2008, July 9. Ocean Wind Power Maps Reveal Possible Wind Energy Sources. Press Release. http://www.jpl.nasa.gov/news/news.php?release=2008-128 (accessed January 2018).

NASA JPL. 2017. Winds: Measuring Ocean Winds from Space. http://winds.jpl.nasa.gov/ (accessed January 2018).

NASA OMI. 2017. Ozone Monitoring Instrument. http://aura.gsfc.nasa.gov/omi.html (accessed January 2018).

NASA Ozone Watch. 2017. NASA Goddard Space Flight Center. http://ozonewatch.gsfc.nasa.gov/SH.html (accessed January 2018).

NASA PODAAC. 2017. Measurements—Sea Surface Temperature. NASA Physical Oceanography Distributed Active Archive Center. http://podaac.jpl.nasa.gov/SeaSurfaceTemperature (accessed January 2018).

NASA Science. 2017. Salinity. NASA Science Beta Programs—Research and Analysis. http://science.nasa.gov/earth-science/oceanography/physical-ocean/salinity (accessed January 2018).

NASA Scientific Visualization Studio. 2011, October 18. Global Fire Observations and MODIS NDVI. http://svs.gsfc.nasa.gov/vis/a000000/a003800/a003868/index.html (accessed January 2018).

NASA Scientific Visualization Studio. 2012, April 9. NASA Views Our Perpetual Ocean. https://www.nasa.gov/topics/earth/features/perpetual-ocean.html (accessed January 2018).

NASA SeaWiFS Project. 2017. Background of the SeaWiFS Project. http://oceancolor.gsfc.nasa.gov/SeaWiFS/BACKGROUND/SEAWIFS_BACKGROUND.html (accessed January 2018).

NASA Sensing Our Planet. 2016. Where the Wetlands Are: A New Map Breaks Down Conservation Borders. http://earthdata.nasa.gov/user-resources/sensing-our-planet/where-the-wetlands-are (accessed January 2018).

NASA Terra. 2012. Multi-angle Imaging SpectroRadiometer. Terra Instruments. http://terra.nasa.gov/about/terra-instruments/misr (accessed January 2018).

NASA TRMM. 2017. Tropical Rainfall Measuring Mission. http://trmm.gsfc.nasa.gov (accessed January 2018).

National Research Council. 2003. *Oil in the Sea III: Inputs, Fates, and Effects*. Washington, DC: The National Academies Press.

NOAA OSPO. 2017. Sea Surface Temperature (SST) Products. NOAA Office of Satellite and Product Operations. http://www.ospo.noaa.gov/Products/ocean/index.html (accessed January 2018).

NOAA SIS. 2017a. JPSS-1 Mission. NOAA Satellite Information System. http://www.nesdis.noaa.gov/content/jpss-1-mission (accessed January 2018).

NOAA SIS. 2017b. NOAA's Geostationary and Polar-Orbiting Weather Satellites. NOAA Satellite Information System. http://noaasis.noaa.gov/NOAASIS/ml/genlsatl.html (accessed January 2018).

Nolin, A. W., F. M. Fetterer, and T. A. Scambos. 2002. Surface roughness characterizations of sea ice and ice sheets: Case studies with MISR data. *IEEE Transactions on Geoscience and Remote Sensing*, 40(7), 1,605–1,615.

On-Scene Coordinator. 2011, September. On-Scene Coordinator Report *Deepwater Horizon* Oil Spill: Submitted to the National Response Team. http://www.uscg.mil/foia/docs/dwh/fosc_dwh_report.pdf (accessed January 2018).

Osborne, A. R., and T. L. Burch. 1980. Internal solitons in the Andaman Sea. *Science*, 208, 451–460.

Ozone Hole Inc. 2017. The Ozone Hole History. http://www.theozonehole.com/cfc.htm (accessed January 2018).

Parkinson, C. L., and P. Gloersen. 1993. Global sea ice coverage. In R. J. Gurney, J. L. Foster, and C. L. Parkinson (Eds.), *Atlas of Satellite Observations Related to Global Change* (pp. 141–163). Cambridge, UK: Cambridge University Press.

Popp, C., D. Brunner, A. Damm, M. Van Roozendal, C. Fayt, and B. Buchmann. 2012. High-resolution NO_2 remote sensing from the Airborne Prism EXperiment (APEX) imaging spectrometer. *Atmospheric Measurement Techniques*, 5, 2,211–2,225.

Rouse, J. W. 1968. *Arctic Ice Type Identification by Radar* (Technical Report 121-1). Lawrence: University of Kansas Center for Research.

Ryan, J. P., F. P. Chavez, and J. G. Bellingham. 2005. Physical-biological coupling in Monterey Bay, California: Topographic influences on phytoplankton ecology. *Marine Ecology Progress Series*, 287, 23–32.

Salisbury, J. W., D. M. D'Aria, and F. F. Sabins. 1993. Thermal remote sensing of crude oil slicks. *Remote Sensing of Environment*, 45, 225–231.

Stringer, W. J., K. G. Dean, R. M. Guritz, H. M. Garbeil, J. E. Groves, and K. Ahlnaes. 1992. Detection of petroleum spilled from the MV *Exxon Valdez. International Journal of Remote Sensing*, 13(5), 799–824.

Thompson, M. C., C. Remington, J. Pumomo, and D. Macgregor. 1991. Detection of Liquid Hydrocarbon Seepage in Indonesian Offshore Frontier Basins Using Airborne Laser Fluorosensor (ALF)—the Results of a Pertamina/BP Joint Study (pp. 663–689). Proceedings, 20th Annual Convention, Indonesian Petroleum Association, Jakarta, Indonesia.

USGS NUPO. 2017. Census of Ground-Nesting Colonial Waterbirds: Anaho Island National Wildlife Refuge in Nevada. USGS National Unmanned Aircraft Systems Project Office. http://uas.usgs.gov/mission/NV_AnahoIslandNWRPelicans.shtml (accessed January 2018).

Vizy, K. N. 1974. Detecting and monitoring oil slicks with aerial photos. *Photogrammetric Engineering*, 40, 697–708.

Weng, F., R. R. Ferraro, and N. C. Grody. 1994. Global precipitation estimations using Defense Meteorologic Satellite program F10 and F11 special sensor microwave imager data. *Journal of Geophysical Research*, 99, 14,493–14,502.

Williams, R. S., J. Heckman, and J. Schneeberger. 1991. *Environmental Consequences of the Persian Gulf War.* Washington, DC: National Geographic Society.

Yang, K., N. A. Krotkov, A. J. Krueger, S. A. Carn, P. K. Bhartia, and P. F. Levelt. 2009. Improving retrieval of volcanic sulfur dioxide from backscattered UV satellite observations. *Geophysical Research Letters*, 36(3), L03102. doi:10.1029/2008GL036036

Renewable Resources

A renewable resource is a natural resource that will replenish itself through reoccurring processes in a predictable amount of time. Using various methods, some of these natural, reoccurring processes can be detected by remote sensing, allowing for the useful application of data for multiple purposes. In this chapter, we will discuss examples of the applications of remote sensing to the following renewable resources: agriculture, forestry, wetlands, fisheries, fresh water, and alternative energy.

In agriculture, remote sensing can be used to monitor crop growth and water use and needs. Forestry applications include measuring tree dimensions and small gaps in a forest canopy, building 3-D models of individual trees, and understanding factors driving deforestation.

Remote sensing also supports resource management, conservation, and harvesting estimations. It can help measure parameters that affect distribution and abundance, such as surface temperature, vertical and horizontal circulation, and pollution.

Fresh water applications include discovering and monitoring changes such as shrinking or shifting water bodies and the seasonal bloom of nuisance algae. Remote sensing can also prove very useful for groundwater exploration in arid regions. With the expansion and growing need for renewable energy solutions, remote sensing is used to plan, help measure potential, and monitor renewable energy projects that utilize geothermal, solar, and wind energy sources.

AGRICULTURE

Remote sensing is most successful for mapping and monitoring renewable plant resources when image acquisitions are synchronized with the growth (phenological) cycle of the vegetation of interest. Interpreting plant communities, species, and health from remote sensing images requires an understanding of which phase of the annual growth cycle each plant of interest is experiencing. Pathogens that cause plant stress, unseasonably high or low temperatures and precipitation, and planting time for agricultural crops alter the phenological growth cycle and, therefore, should alter the timing of the remote sensing data acquisition (Jensen, 2007).

Imagery of agricultural fields in the bare soil phase of the growth cycle is best for photogrammetrists building DTMs and geologists interpreting soil texture and type. Deciduous forests (where the trees drop their leaves seasonally) provide a "leaf-off" and "leaf-on" phase during their growth cycle. Photogrammetrists building DTMs and geologists mapping outcrops want imagery over deciduous forests acquired during the leaf-off phase. In contrast, forest managers prefer imagery acquired during the leaf-on phase for most of their applications. Jensen (2007) reviews the annual phenological growth cycles of both natural vegetation systems and managed agricultural systems, demonstrating the wide range in timing of canopy closure, maximum biomass, budding, flowering, leaf development, and dormancy during the growth cycle.

Timing of image acquisition can also be understood by comparing the response of vegetation to a plot of NIR and red reflectance during the growth cycle of an annual agricultural crop. The gray shaded area in Figure 12-1 is the distribution of all of the pixels in a plot of NIR versus red reflectance for one growth cycle in an agricultural scene. The soil line extends between wet bare soil (dark pixels) and dry bare soil (bright pixels) with no presence of vegetation. Both NIR and red reflectance increase in value as the soil dries.

A. Pixels move along the soil line between wet and dry conditions with no vegetation and in a perpendicular direction as biomass and canopy closure changes.

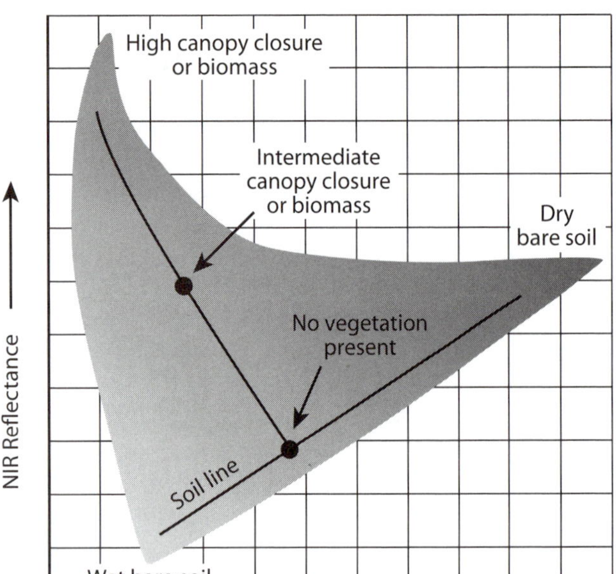

B. Movement of a single pixel in NIR and red spectral space during the growing season.

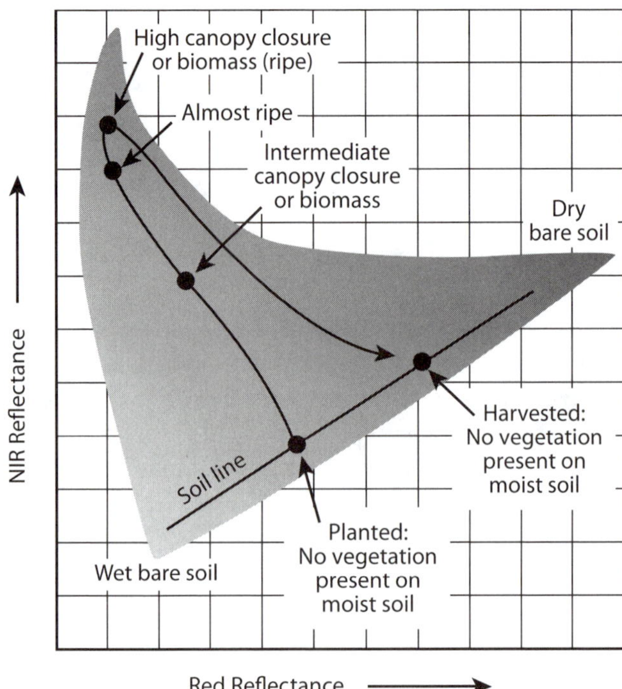

FIGURE 12-1 The gray shaded area is the distribution of all of the pixels in a plot of NIR versus red reflectance for one growth cycle in an agricultural scene. Modified from Jensen (2007, Figure 11-6).

The *direct* relationship between NIR reflectance and plant biomass and the *indirect* relationship between red light reflectance and plant biomass are illustrated in Figure 12-1. The greater the biomass (or canopy closure), the higher the NIR reflectance and the lower the red reflectance (due to absorption of red light by chlorophyll in the leaves). As biomass and canopy closure increases, the pixel's spectral value moves in a perpendicular direction from the soil line to higher NIR values and lower red values (Figure 12-1A). Figure 12-1B shows the movement of a single pixel in the agricultural scene from planting in moist bare soil to high canopy closure (a ripe crop with a high biomass) to harvest with no vegetation present in the pixel and a drier soil condition. Figure 12-1B can be used to plan the timing of image acquisitions so that spectral characteristics of the vegetation that are optimum for interpretation, identification, and measurement of growth are recorded by the sensor. Jensen (2007) notes that the inverse relationship between NIR and red reflectance of vegetation shown in Figure 12-1 has resulted in the development of numerous vegetation indices to improve measurement and modeling of various vegetation biophysical variables.

In addition to the ability to formulate indices, there are a growing number of remote sensing applications to solve problems in agriculture. In this chapter, we discuss the following examples: ground hyperspectral measurements of grapevines, flying an unmanned aerial system (UAS) with an RGB camera to generate 3-D models of corn plants to improve corn yield predictions, and using Landsat to monitor water use.

GRAPEVINE WATER STATUS

In the Duoro wine region of northern Portugal (Figure 12-2), irrigation has recently been introduced to regulate vineyard crop yield and quality. Grapevine predawn leaf water potential (Ψ_{pd}) is measured in the field and is considered a reliable indicator of crop water status in vineyards, which is used to implement irrigation schedules. Pôças and others (2015) evaluated the use of hyperspectral remote sensing to estimate Ψ_{pd} and minimize labor-intensive field measurements. This section is a brief summary of the research by Pôças and others (2015).

Spectral reflectance data in the form of vegetation indices (combinations of wavelength bands and ratios) are used to describe the biophysical and biochemical characteristics of vegetation, such as vegetation vigor and stress, leaf pigment content, canopy water content, etc. Hyperspectral data with narrow bandwidths (1 to 10 nm) enable more exact and sensitive vegetation indices to be developed. Pôças and others (2015) used handheld spectroradiometer measurements of visible and red edge reflectance to determine optimum wavelengths and vegetation indices for predicting grapevine Ψ_{pd}. Red edge is the sharp transition zone between red and NIR

FIGURE 12-2 Vineyard in Duoro Valley, Portugal (Agnieszka Skalska, Shutterstock).

spectral ranges (Figure 9-37A). Narrow band hyperspectral data is more easily obtained in the visible and red edge wavelengths because of its high energy content (Equation 1-2). Although there is no water absorption in the visible spectral region, hyperspectral data of visible and red edge spectra can be used to determine crop water status because there is a high correlation between crop water status and leaf pigments that are dependent on water (Pôças and others, 2015). Narrow band hyperspectral measurements detect subtle leaf pigment characteristics and changes that are associated with grapevine leaf water potential and crop water status.

A portable spectroradiometer recorded spectral data between 0.325 to 1.075 μm with a wavelength interval of 1 nm. The measurements were made during cloud free conditions within +/− one hour of local noon (high sun). The sensor was held 30 cm above the leaf canopy. Previous research documented the high correlation between midday and predawn leaf water potential (Ψ_{pd}). Figure 12-3 is an example of the reflectance spectral signatures for several classes of water deficit conditions in the vineyard, identifying the wavelengths (in nm) that show the best results for vegetation indices with vertical arrows. Canopies of grapevines with none to moderate stress have less reflectance compared with those with moderate to high stress.

Many vegetation indices were evaluated in the study by Pôças and others (2015). For low and moderate stress conditions, two indices (with specific wavelengths for reflected blue, green, and red light) predicted Ψ_{pd} most reliably. These

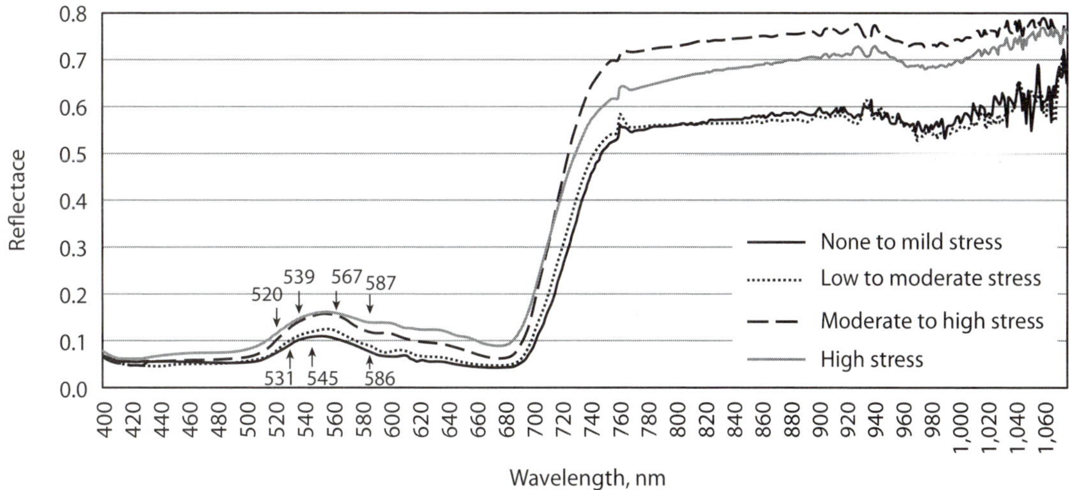

FIGURE 12-3 Reflectance spectra signatures obtained for different water deficit conditions in a vineyard. The vertical arrows represent the wavelengths selected for the optimal vegetation indices. From Pôças and others (2015, Figure 4) (http://creativecommons.org/licenses/by/4.0/).

two indices were the Visible Atmospherically Resistant Index (VARI) and the Normalized Difference Greenness Vegetation Index (NDGI). They are calculated as follows:

$$\text{VARI} = (R_{\text{green}} - R_{\text{red}})/(R_{\text{green}} + R_{\text{red}} - R_{\text{blue}})$$

where:

R = reflectance value
blue = 0.520 μm
green = 0.539 μm
red = 0.586 μm

$$\text{NDGI} = (R_{\text{green}} - R_{\text{red}})/(R_{\text{green}} + R_{\text{red}})$$

where:

R = reflectance value
green = 0.531 μm
red = 0.587 μm

The results indicate that VARI and NDGI (derived from narrow band, portable hyperspectral sensors that record spectra in the visible region) can be used to estimate grapevine crop water status and stress conditions. The use of VARI and NDGI with the optimum wavelength bands documented in this study is more practical and cost effective than conventional field measurements of grapevine leaf water potential (Pôças and others, 2015). Collecting these hyperspectral indices on aerial platforms across an entire vineyard provides additional support for the management of irrigation water.

CORN YIELD MODELS

Precision farming management strategies are commonly based on estimations of within-field yield potential based on vegetation indices calculated from remote sensing data. Yield estimates prior to harvest help determine input factors such as nutrients, pesticides, and water as well as planning for harvesting, drying, and storage. Geipel and others (2014) improved prediction models of corn yield by including crop height determined from a UAS (Figure 12-4). The research was carried out in experimental corn fields at the University of Hohenheim in southern Germany. A brief summary of Geipel and others (2014) research follows.

The UAS used in the study carried a consumer-grade RGB digital camera that collects limited spectral information when compared with multispectral, hyperspectral, and TIR sensors, but records very high spatial resolution data to generate detailed 3-D crop surface models. The camera's spatial resolution was set at 4,000 by 3,000 pixels to achieve a ground resolution of approximately 2 cm per pixel at an altitude of 50 m. The aerial images were acquired with an 80% along-track overlap and 60% cross-track overlap in nadir view. Each flight mission collected about 400 overlapping images. The flight missions were performed on three dates during early and mid-season crop development. The original 2 cm images provide information on fine corn structure (leaf level) while images resampled to a spatial resolution of 10 cm provide information on coarse structures (canopy level).

Feature matching and Structure from Motion (SfM) algorithms were used to generate a digital surface model (DSM) of the top of the canopy and a digital terrain model (DTM) of the ground surface. Figure 12-5 illustrates a grayscale three-layer model of the data: an orthoimage is at the base, a crop-soil map is in the middle, and a 3-D crop surface model (DSM) is at the top of the stack. Mean crop height is calculated as 0.96 m and displayed as a flat horizontal plane through the DTM (note the label "Height" on the left vertical edge). The vegetation index (ExG) is displayed as a horizontal plane in the middle portion of the stack with rows of corn depicted as irregular linear polygons symbolized with a dark fill parallel with the soil exposed between the rows of corn displayed with a light gray fill. Height of the corn canopy is calculated by subtracting the DTM elevation from

FIGURE 12-4 Ground photograph of a cornfield (Jeff Smith—Perspectives, Shutterstock).

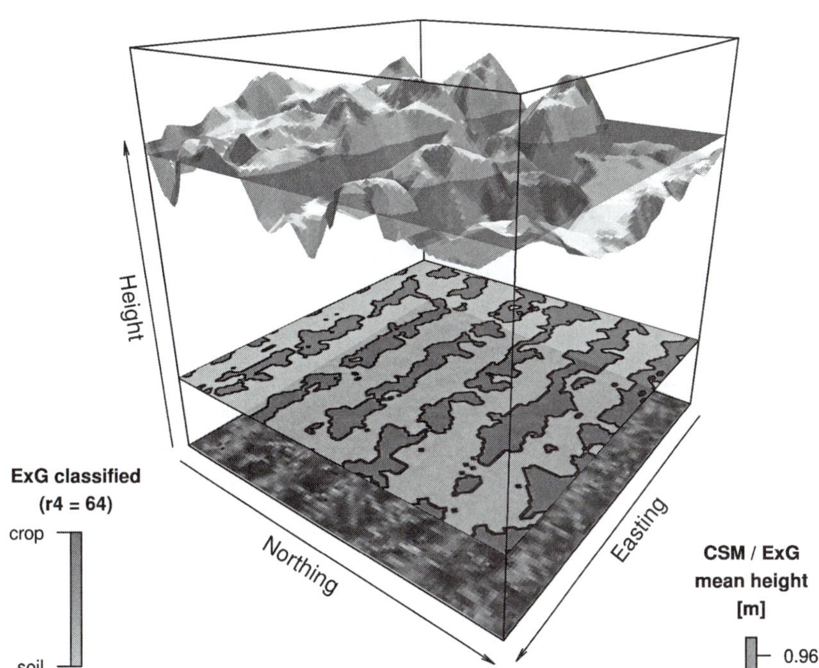

FIGURE 12-5 3-D model with the DSM as the top layer, the orthoimage as the bottom layer, and the crop-soil map in the middle. From Geipel and others (2014, Figure 4) (http://creativecommons.org/licenses/by/4.0/).

the DSM elevation. In addition to corn height, simple vegetation indices were estimated from the RGB image to differentiate crop from soil (middle layer in Figure 12-5).

Yield prediction proved difficult for early growth stages when corn stalks are small and soil exposure is extensive. Combined 3-D models of corn height and simple vegetation indices based on intermediate spatial resolution images (4 cm pixels) are most effective for mid-season yield prediction. The study proves the methodology to be suitable for mid-season corn yield prediction (Geipel and others, 2014).

LANDSAT: MAPPING AGRICULTURAL WATER USE

Changes in land use, climate, and population demographics place significant demands on US water supplies. Scientists with the USGS (2016) have helped refine an evapotranspiration (ET) water-use mapping technique to measure how much water crops use across landscapes and through time. These ET water-use maps are created using a computer model that integrates Landsat and weather data. Crucial to the process is Landsat's TIR band. Water-use maps are created at a scale detailed enough to show how much water crops are using at the level of individual fields anywhere in the country.

ET water-use maps can show how much water crops are using in a single day or during an entire growing season. Drawing on the vast Landsat satellite image archive, it is also possible to create maps that span decades to reveal long-term trends in water use. USGS scientists can map water use at different scales to address different water resource questions and concerns. Field-scale maps, for example, are powerful tools for estimating and managing water consumption on irrigated croplands. Basin-scale water-use maps assist in understanding water balance and availability in river basins and watersheds. These large area maps are useful for:

- Estimating water use by different sectors within a watershed.
- Resolving disputes regarding water rights and allocations.
- Evaluating aquifer depletions and quantifying net groundwater pumping.

Plate 38 is an example of ET water-use maps in California's agricultural San Joaquin Valley derived from Landsat TM and weather data by the USGS. The left image shows an ET water-use map from 1990 and the right image shows ET water use in 2014. Both ET water-use maps have seasonal values ranging from 0 (red) to 1,000 ml (blue). An increase in agricultural water use between 1990 and 2014 is shown by blue areas in the 2014 map. Land that was taken out of agricultural production is shown by irrigated areas that were green in 1990 and reddish brown in 2014. The changes in water use documented with Landsat ET maps contain valuable information to better manage water, one of the United States' most important natural resources.

FORESTRY

Forestry applications of remote sensing include airborne lidar for measuring tree dimensions, UAVs and photogrammetry for building 3-D models of individual trees, UAVs for imaging small gaps in a forest canopy, and satellite imagery for understanding factors driving deforestation. Comprehensive methodologies to map deforestation and forest degradation, estimate forest height, and monitor forest biomass using satellite radar are provided in *The SAR Handbook* (Flores-Anderson and others, 2019).

LIDAR FOR FORESTRY

Lidar technology provides horizontal and vertical information at high spatial resolutions and vertical accuracies (Lim and others, 2003). Canopy height can be directly retrieved from lidar data to model aboveground biomass and canopy volume.

Interactive Measurement of a Tree

McGaughey (2018), of the USDA Forest Service, Pacific Northwest Research Station, developed a free lidar analysis and visualization software package called FUSION. Utilizing acquired lidar data, this software can function as a digital training database. Figure 12-6 is a demo image that is available in the FUSION software. It is a vertical aerial image of individual trees, access roads, and an open area devoid of trees. The shadows of tree trunks and canopies are cast toward the bottom of the figure. One tree is selected for analysis, and is highlighted with a white circle.

As a next step, the selected tree is contained within a white vertical cylinder so that measurements of the tree can be determined. The lidar returns below and around the tree are extracted from the database and displayed with FUSION's LIDAR Data Viewer (Figure 12-7). The first returns that encountered tree canopy (outside of the cylinder of the selected tree) are displayed as gray balls and the last returns from the ground as black balls.

The tree is extracted in Figure 12-8 to reveal its shape and enable interactive height, canopy diameter, and location measurements to be recorded. The elevation of the tree (158.8 m) is shown on the scale along the left side of the LIDAR Data Viewer, where black is near ground level and white is the top of the tree. The tree top is measured as 1,267 m above sea level and the tree canopy is 28 m in diameter. Lidar data collected with adequate density with some

FIGURE 12-6 Aerial image of individual trees with white circle over tree analyzed with FUSION lidar software. From McGaughey (2018).

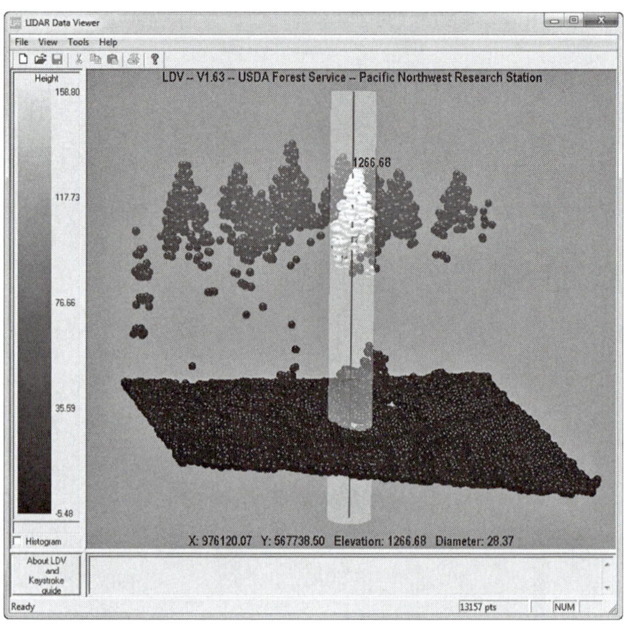

FIGURE 12-7 Perspective view of lidar returns with the selected tree highlighted in a vertical cylinder.

ground hits can result in accurate models of individual trees that support identification of the tree type (Figure 12-9).

Measuring the Tallest Trees in a Forest

The Great Smoky Mountains National Park in Tennessee and North Carolina is one of the most biologically diverse areas on the planet, with almost 40% virgin forest cover, rich soil, and ample precipitation (Strother and others, 2015). Documenting the unique biological resources in the park, includ-

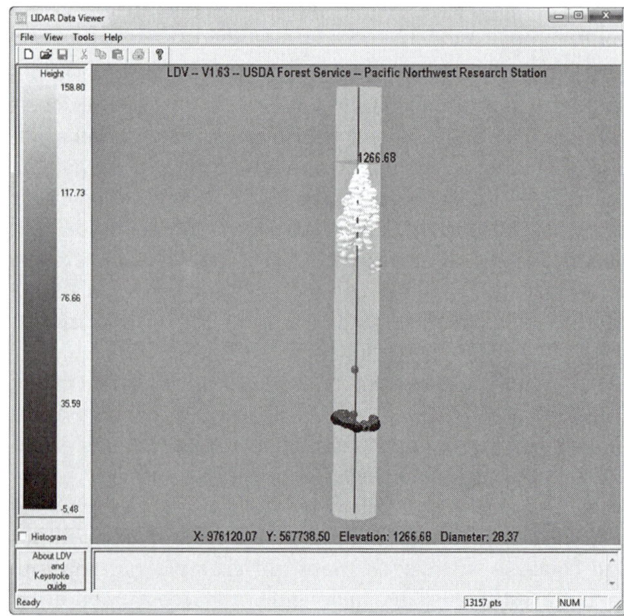

FIGURE 12-8 Perspective view of lidar returns for tree in white circle (Figure 12-6).

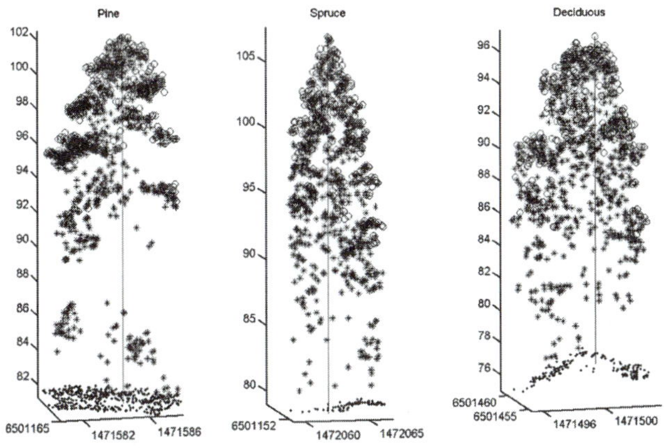

FIGURE 12-9 Lidar point returns from pine (left), spruce (middle), and deciduous (right) trees. From Tamás (2010, Figure 4.2.7), with E. Naesset.

ing the tallest trees, is important to understand and monitor environment change in this complex forest community.

Strother and others (2015) used lidar to detect and measure the ten tallest trees in the Tennessee portion of the park. Heights of trees in the park are difficult to determine with field measurements due to steep topographic slopes, persistent rainfall, and crown overlap. The thick tree canopy also makes seeing the ground on aerial photographs difficult, limiting the accuracy of bare earth DEMs generated with photogrammetry. The lidar survey discovered the two tallest trees (59 m high) ever measured in the eastern United States. The ten tallest tree sites varied in their tree species, general overstory community type, and terrain slope, but had similarities in regards to elevation and terrain aspect. The lidar-based geospatial framework supports arborists and park managers as they gather more environmental and ecological information about these unique trees (Strother and others, 2015).

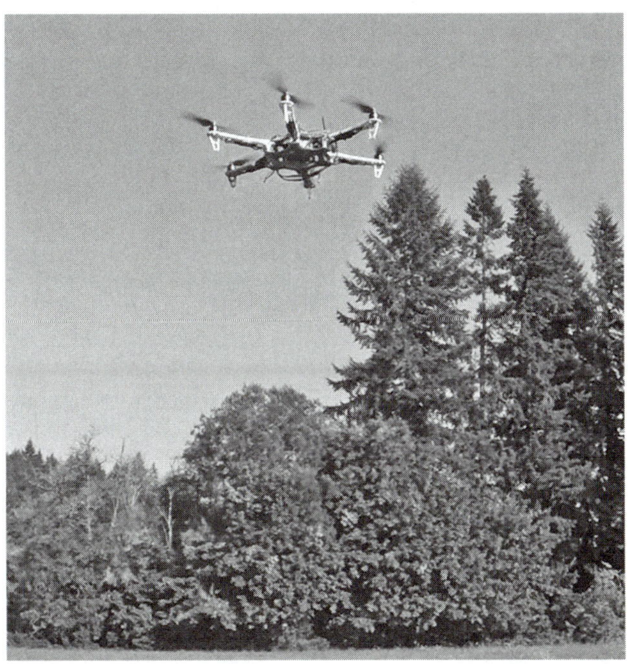

FIGURE 12-10 Hexacopter UAV used for imaging individual trees. From Gatziolis and others (2015, Figure 4).

PHOTOGRAMMETRY AND sUAVs FOR 3-D TREES

Gatziolis and others (2015) used UAVs and photogrammetry to measure the dimensionality, spatial arrangement, shape, and shape of components for individual trees. Conventional methods of acquiring tree canopy measurements include airborne, nadir-viewing cameras, and lidar. However, nadir-viewing sensors do not acquire much data on lower canopy components and the ground surface due to visual blockage or substantial attenuation of the lidar pulse energy in dense, multistory stands of trees. The accuracy of tree dimensionality measurements is limited with nadir-viewing systems, especially for the sides of tree crowns. Terrestrial lidar is expensive as multiple stations viewing the tree from different directions are required to minimize gaps in the final point cloud of the tree and its canopy.

Gatziolis and others (2015) employed slow-moving UAVs (Figure 12-10) that acquired images along predefined trajectories near and around targeted trees, and computer vision-based approaches that processed the images to obtain detailed tree reconstructions. They developed an affordable method and document a workflow that employs a sequence of computer programs for obtaining precise and comprehensive 3-D models of trees and small groups of trees. A brief summary of their research follows.

Terrestial lidar generates dense 3-D models of individual trees as a baseline for checking the accuracy of 3-D models built with the sUAS. The terrestrial lidar system acquires panchromatic imagery while scanning the trees. The imagery and lidar point cloud are merged and visualized as 3-D models (Figure 12-11A).

The flight path of the UAS needs to acquire adequate coverage of crown sides to support research on lateral crown competition for space and resources. Oblique and horizontal camera views and UAS trajectories around trees and tree groups at variable aboveground heights ensure gap-free representation of trees. Prior to flying the UAS, simulations were completed using highly accurate, 3-D terrestrial lidar models of trees. A simulated spiral flight path is plotted on the lidar 3-D model in Figure 12-11A. A 3-D reconstruction of the tree from synthetic images acquired along the simulated flight path is shown in Figure 12-11B. The simulation of UAS flight and imaging around 3-D lidar models demonstrated that sequential UAS images require approximately 70% overlap for optimal reconstruction of the tree. Less overlap leads to gaps in the coverage and a less complete UAS-based 3-D model.

Many flight path patterns were evaluated, including circular at a constant height, a stack of circles each at a different aboveground height (for tall trees more than 20 m high), spiral for trees with complex geometry, and vertical meandering. The spiral UAS trajectory was among the most reliable flight path for complete tree reconstruction as it provides varying vertical viewing angles of the same area after one rotation. For species with predominantly horizontal or angular branch arrangement and lower crown compaction rates, vertical viewing variability allows internal crown components to be represented adequately in the derived point cloud. For tree species with dense, uniform distribution of foliage and deeply shaded crown centers, the variability in vertical view angles offered by the spiral trajectory pattern may be unimportant.

A meticulously planned, image-acquisition mission that includes appropriate flight trajectory, UAS speed, and image acquisition frequency will deliver a comprehensive dense reconstruction of targeted vegetation, except in unfavorable sun illumination and wind conditions.

The accurate 3-D representations of individual trees using UAS and photogrammetry enable an accurate assessment of size, shape, and spatial distribution of interacting trees that improves understanding of tree growth, competition, and varying morphology within species and ecosystems (Gatziolis and others, 2015).

A. Perspective view of a point cloud acquired with terrestrial lidar and simulated UAS camera locations (white spheres) used to obtain simulated (virtual) images of the scene.

B. 3-D reconstruction model of the tree generated with simulated UAS images.

FIGURE 12-11 Perspective 3-D model of a tree obtained with terrestrial lidar and simulated UAS imaging. From Gatziolis and others (2015, Figure 3).

UAVs FOR MAPPING FOREST GAPS

Getzin and others (2014) used UAVs to quantify spatial gap patterns in forests. The following is a brief summary of their research. Gap distributions in forests reflect the spatial impact of man-made tree harvesting or naturally induced patterns of tree death caused by wind, intertree competition, disease, or senescence. Gap sizes can vary from large (> 100 m^2) to small (< 10 m^2) and they may have contrasting spatial patterns, such as being aggregated or regularly distributed. Gaps in forest canopies play a key role in the regeneration of trees and generally for the diversity of understory biota. The spatial distribution of gaps has implications for seed establishment and, therefore, the formation of future forest cover patterns. Forest management via thinning intensity may greatly influence canopy cover and, subsequently, species diversity and the cover of ground vegetation.

Remote sensing is the most efficient tool for mapping canopy gaps. However, the shape complexity of very small gaps (< 5 m^2) cannot be discerned from satellite or airborne sensors with spatial resolutions larger than 20 cm per pixel. Airborne lidar records very detailed representations of small gaps but it is expensive to deploy. UAVs offer a low cost solution to record the detailed shape of very small canopy openings.

In the Getzin and others (2014) study, ten (10), 1-ha (100 by 100 m) study sites were located within beech-dominated deciduous forests of Germany. The 10 plots included managed forests (age-class management type), less intensely managed selection cutting forests, and unmanaged forests. Very high spatial resolution RGB images (~7 cm pixels) were taken at the end of the summer in 2008 and 2009 with a fixed wing UAV flying above the centers of the forest plots at an altitude of approximately 250 m. All images were orthorectified and converted into binary (black and white) images. Gap polygons were manually interpreted, including gaps as small as 1 m^2 in size.

Figure 12-12 shows grayscale UAV images of two plots with their UTM coordinates and a scale bar. These two plots are within managed forests. Interpreted gaps are symbolized as black polygons with white stripes and outlines. Figure 12-12A is a UAV image of plot AEW20 acquired on September 2, 2009 with delineated gaps that range in size from 1.1 m^2 to 81.7 m^2. Plot AEW20 is an age-class managed forest with 56 gaps having a mean gap size of 17.5 m^2. Figure 12-12B is a UAV image of plot HEW31 acquired on October 8, 2008. Plot HEW31 is a selection cutting managed forest with 65 gaps having a mean gap size of 14.1 m^2 and a maximum gap size of 163.5 m^2.

The plots shown in Figure 12-12 are displayed in Figure 12-13 as binary images with the gaps shown as black and the canopy as white. The gap images were evaluated with spatial analysis techniques to better understand gap patterns (for example, random, aggregated, or regularly spaced). Gaps in managed forests ranged from 4.9 to 13.9% of the 1-ha plots, which reflects past intensities of logging. Managed forests have a larger number of gaps compared with the less intensely managed selection cutting and unmanaged forests.

The UAV 7 cm images enable gap sizes as small as 1 m^2 to be mapped. These smallest gaps are important determinants of regeneration, since the variability of diffuse radiation does influence understory biodiversity. The very small gaps (of a size < 5 m^2) made up the largest proportion of all gaps found in the study plots. This was particularly the case

A. UAV image of tree canopy in plot AEW20 with delineated gaps.

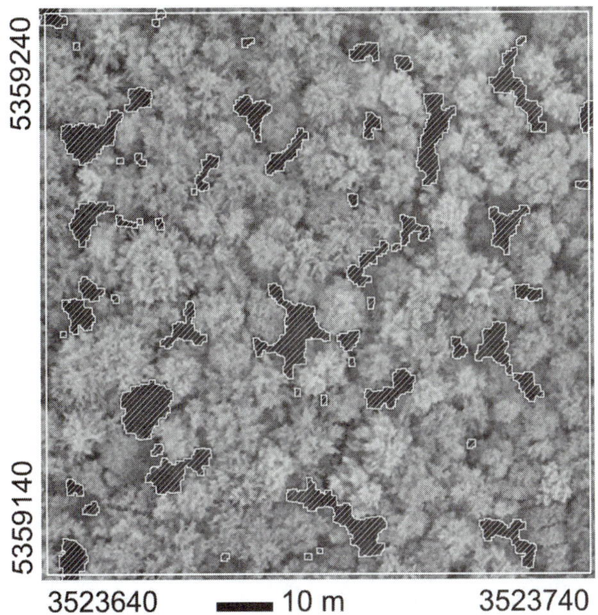

B. UAV image of tree canopy in plot HEW31 with delineated gaps.

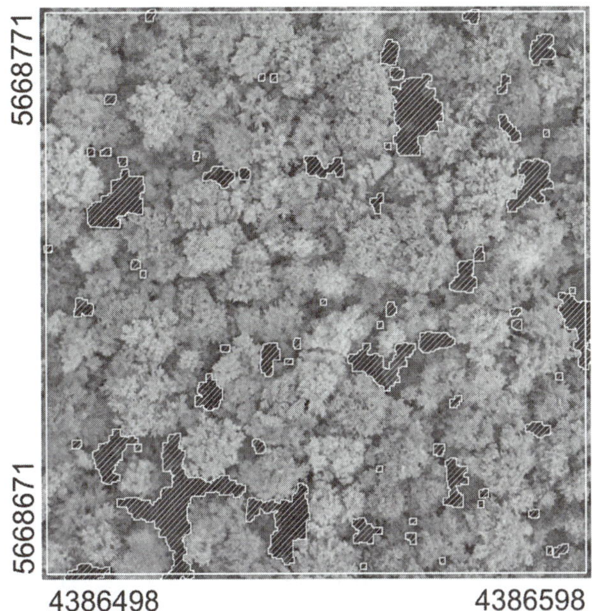

FIGURE 12-12 Images with 7 cm spatial resolution acquired with UAV of managed forest with gaps in the canopy delineated and symbolized as black polygons with white stripes and outline. Modified from Getzin and others (2014, Figure 1).

for the three unmanaged forests where gaps are naturally induced, mainly by disturbance.

Forest gap mapping is most reliable with UAV images acquired under relatively cloudy conditions without direct sunlight. Diffuse sky radiation helps avoid misclassifications of gaps caused by the hard shadows of neighboring trees, which appear as dark patches. Unmanned aerial vehicles are highly suitable tools for mapping small gaps, repeated gap formation, and spatial canopy structures in general (Getzin and others, 2014).

A. Delineated gaps in plot AEW20.

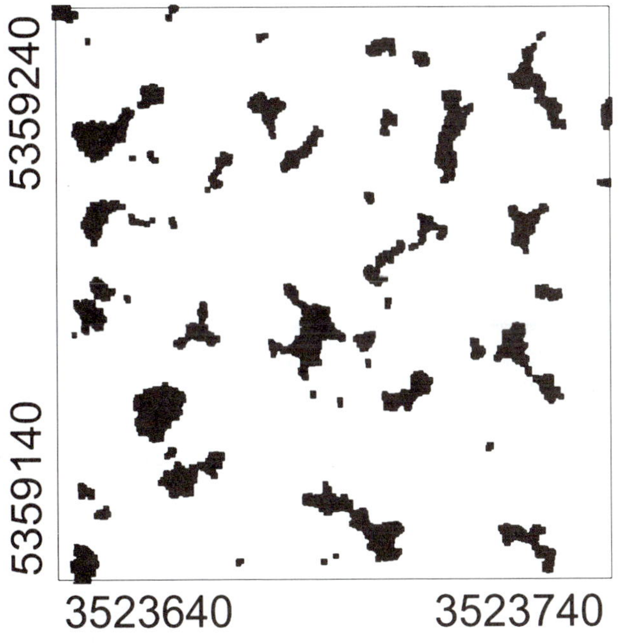

B. Delineated gaps in plot HEW31.

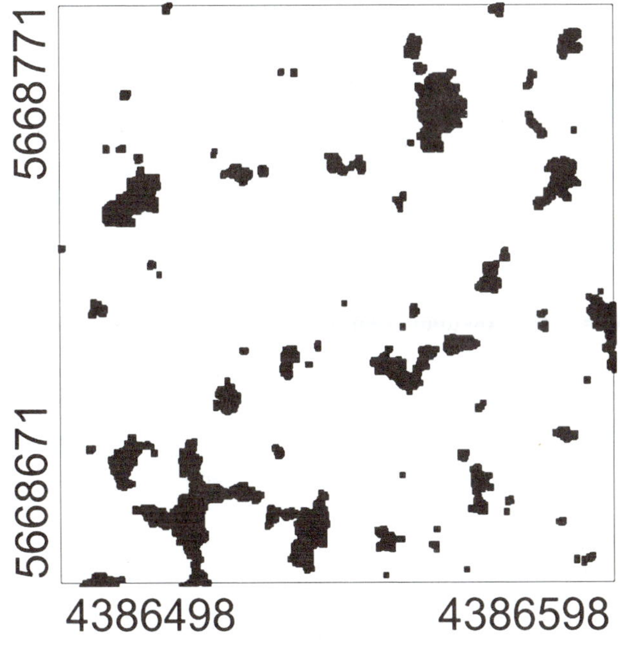

FIGURE 12-13 Binary (black and white) maps of gaps interpreted from UAV images in Figure 12-12. From Getzin and others (2014, Figure 2).

UNDERSTANDING DRIVERS OF DEFORESTATION

Baynard (2013) advocates and documents the application of remote sensing beyond the traditional mapping of land use and land cover change. One innovative application is the analysis of society's infrastructure and development footprint on the landscape. Baynard and others (2012a, 2012b) discuss how remote sensing is used to monitor forest integrity and measure deforestation as development proceeds under different policies in eastern Ecuador's tropical forest. The following is an overview of their work.

Oil exploration and production (E&P) activities in remote regions are often considered the catalyst for landscape change through the direct alterations created by infrastructure features, such as well pads, pits, pipelines, and seismic corridors, as well as through the accessibility created by roads built to reach oil fields. The construction, expansion, and improvement in transportation routes in isolated areas can attract newcomers and parallel economic activities such as logging, hunting, rubber tapping, gold mining, and agricultural colonization. These parallel activities can create their own set of roads and open up previously inaccessible areas. The development may affect indigenous territories and natural preserves.

Baynard and others (2012a, 2012b) ask whether these parallel economic activities and outcomes always accompany E&P development in remote regions, or can controlled access minimize landscape change and deforestation? Declassified Corona satellite imagery, Landsat imagery, land use/land cover maps, soils data, protected areas, and colonization zones were integrated into a GIS database to determine the spatial relationship between infrastructure patterns, disturbance regimes, agricultural conversion, and the type of road access for the year 2000 in four neighboring oil license blocks in eastern Ecuador's tropical forest. The Corona imagery was acquired in the 1960s and shows the tropical forest to be a continuous, close-canopy rainforest that covered hundreds of kilometers across the upper Amazon drainage basin.

The four license blocks (Figure 12-14) experienced three types of E&P development: public access (northern license), controlled access (licenses 16 and 14), and roadless (license 10). Roads were designated as oil (used by E&P to access well pads and facilities) or non-oil (not used by the E&P).

The oil field development roads in the northern E&P block had public access. Oil development began in the late 1960s and continues to the present day. The original oil field development field runs north–south down the approximate center of the northern license. A fishbone pattern of roads parallel to and located east–west of the primary oil development road can be seen in the public access block. These non-oil roads are tied to national policies promoting colonization and are constructed with an east–west spacing of approximately 2 km. Two designated protected areas in the public access block were preserved from colonization and road construction by the government; they remain mostly forested.

The oil field development road running through blocks 16 and 14 has a controlled access point at the Rio Napo. Additional roads have not developed along this main road, demonstrating that controlled access minimizes landscape disturbance and deforestation. Block 10 was the first roadless E&P development project in eastern Ecuador. Helicopters are used to reach this block; 98% of the forest has remained intact in this license during oil field development.

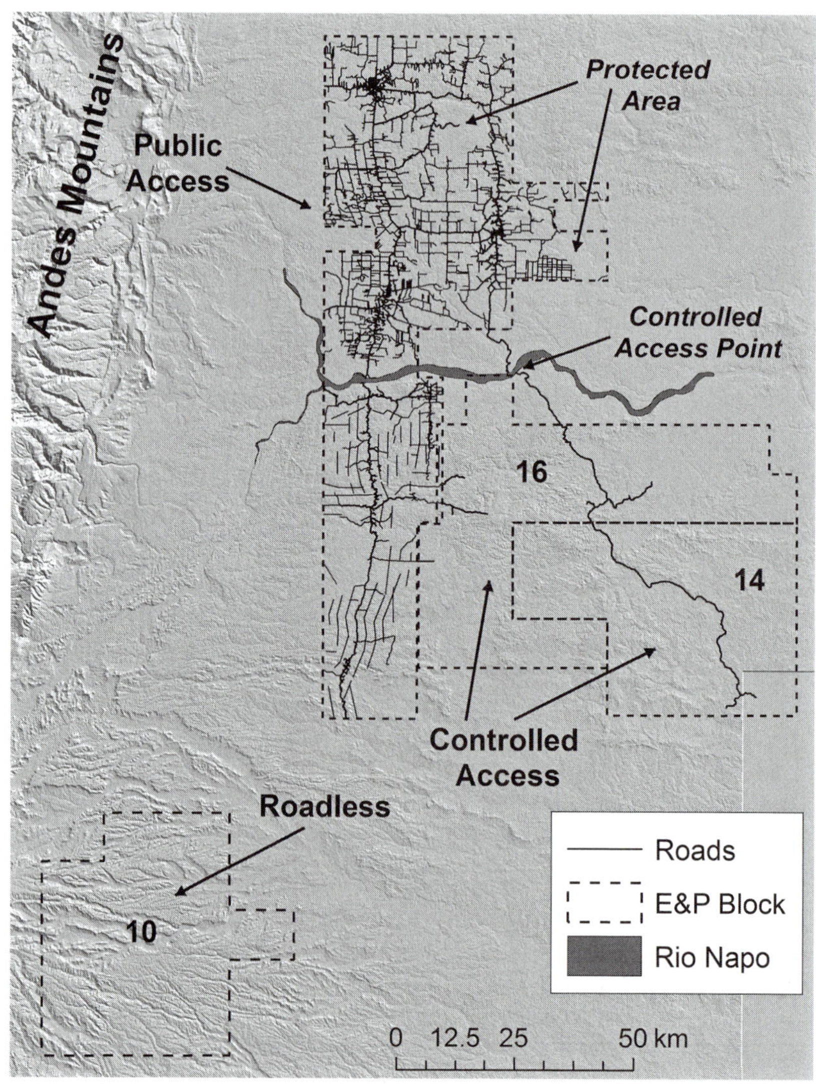

FIGURE 12-14 Map showing four E&P blocks with roads and the Rio Napo, eastern Ecuador. The roads are interpreted from satellite imagery and topographic and development maps. Modified from Baynard and others (2012b, Figure 4).

Spatial analysis findings indicate that areas of overlap where colonization zones, public access roads, and fertile soils meet are most prone to deforestation in terms of agricultural conversion, primarily due to a dense network of public access roads. A linear regression model of the data indicates that the presence of public access non-oil roads best explains deforestation, whereby a 1% increase in this type of road results in a 22% increase in agricultural conversion inside each 1 km^2 analysis block. A 1.0% addition of public access oil roads leads to a 6.0% increase in deforestation. For controlled access oil roads (blocks 16 and 14), the increased deforestation figure was 3.0%. The remote sensing study demonstrates that controlled access and roadless E&P are viable options to preserve forest where E&P development and land conservation are priorities.

FISHERIES

Remote sensing supports resource management, conservation, and harvesting of fisheries. Sonar is used for direct detection of fish and biomass estimation. Indirect detection of fishery resources measures parameters that affect fishery distribution and abundance, such as surface optical properties (including ocean color, chlorophyll-a, and total suspended matter), surface temperature, vertical and horizontal circulation, salinity, pollution, and sea state (Butler and others, 1988). Two examples are provided.

AIRBORNE TIR MAPPING OF RIVERS

The US Forest Service (2007) contracted with Watershed Sciences, Inc. to provide TIR imagery of a portion of the Coeur d'Alene River—from the South Fork of the Coeur d'Alene River upstream to the confluence of Shoshone Creek—located in Idaho. The objective of the image acquisition was to detect and map areas of cold water refuge for fish within the river floodplain resulting from subsurface upwelling and tributary inflows. Water temperature is an important driver of fish persistence and survival, and spatial patterns in water temperature can define species distributions and affect fish life histories (Fullerton and others, 2015). The data was successfully acquired on August 9, 2007 during the mid-afternoon hours. The following is summarized from US Forest Service (2007).

Images were collected with a FLIR Systems SC6000 LWIR sensor (8 to 9.2 μm) mounted on the underside of a Bell Jet Ranger helicopter. The sensor was co-located with a 12.4 megapixel digital camera and positioned to look vertically down (nadir) from the aircraft to maintain a consistent path length. TIR images were recorded directly from the sensor to an on-board computer as raw counts, which were then converted to radiant temperatures. The individual images were referenced with time, position, and heading information provided by a global positioning system.

The aircraft was flown longitudinally along the stream corridor in order to have the river in the center of the display (Figure 12-15). The objective was for the stream to occupy 30 to 60% of the image. The TIR sensor was set to acquire images at a rate of 1 image every 2 seconds, resulting in considerable overlap between images. The flight altitude of 670 m (2,200 ft) was selected to provide a pixel ground sample distance of just over 0.5 m. The airborne survey attempted to cover all surface water within the floodplain, including side channels and tributary junctions. If a side channel or other surface water was not captured in the image field of view, the side channel was flown separately so that all surface water was captured. Since water is essentially opaque to TIR wavelengths, the sensor was only measuring water surface temperature.

The Coeur d'Alene River survey began at the

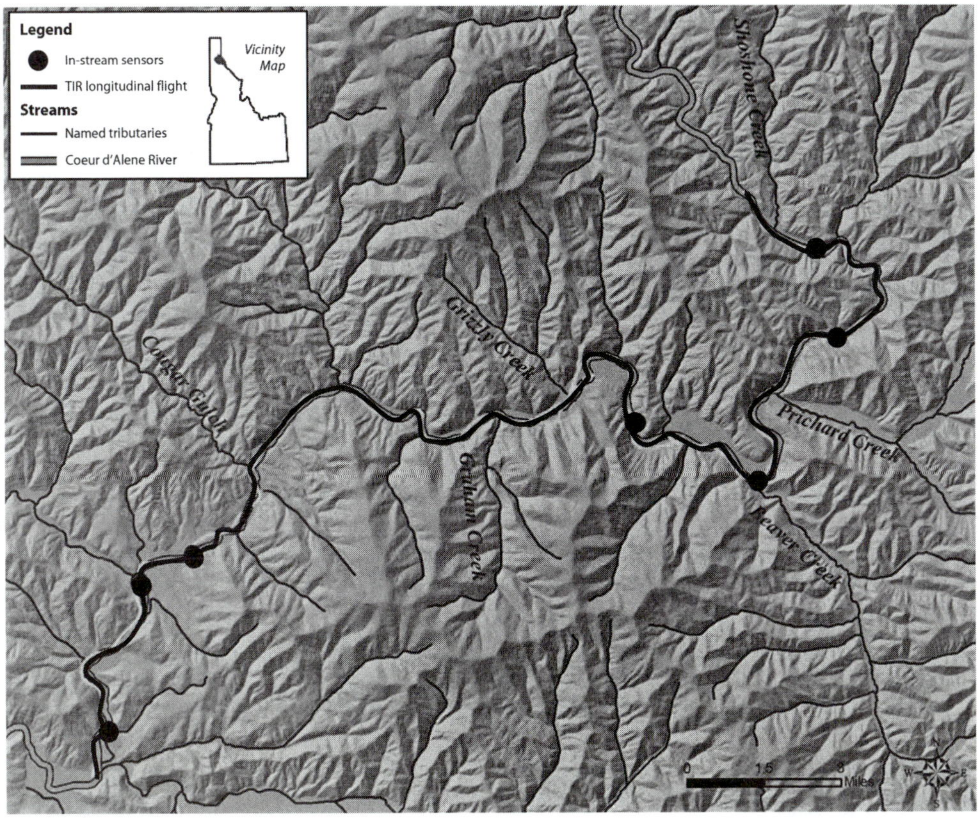

FIGURE 12-15 Approximate flight path of airborne TIR survey (black line) over the Coeur d'Alene River with in-stream sensors (black dots). Modified from US Forest Service (2007, Figure 1).

confluence with the South Fork of the Coeur d'Alene and was flown upstream for approximately 31 miles (Figure 12-15). Bulk water temperatures ranged between 17 to 21°C and generally decreased in the downstream direction. A large number of springs and cool water tributaries were detected along the stream gradient that contributed to the thermal complexity and structure of the river. Of the 28 sampled surface inflows, 14 contributed water that was significantly cooler than the Coeur d'Alene River. In addition, 35 spring inputs were sampled during the analysis. A spring was classified as any distinct discharge that was not associated with a tributary or other obvious surface inflow.

Plate 39 shows a color-coded TIR image of a section of the river designated as River Mile 15.7. The color symbology displays cold water as shades of dark blue to black (< 14°C), medium temperature water as magenta, and warm water as orange and yellow (> 20°C). The river is flowing from upper right to lower left. The white linear feature trending northeast–southwest is a 2-lane paved road. Two springs (6.9°C and 9.4°C) are visible along the left bank of the Coeur d'Alene River that flow directly into the main channel. Power lines are visible across the river just downstream of these springs. Two large side channels are also visible in the image. The slough along the right bank contains cool water (10.0°C), indicating a subsurface influence and the potential for off-channel habitat. The slough on the left bank shows a wide range of temperatures, including potential springs and an impoundment, but seems to have little effect on the bulk water temperature of the main channel.

The Bear River of Idaho, Montana, and Wyoming was mapped with airborne TIR imagery (Pacificorp Energy and Trout Unlimited, 2007). This study documents the impact of a reservoir, dam, and power plant discharge on water temperature in the river. Fullerton and others (2015) constructed longitudinal patterns of temperature from upstream to downstream in 53 large rivers (for a total length of 16,866 km) of the Pacific Northwest. Complex water patterns were revealed due to local climatic conditions such as coastal fog and cold water input from springs and tributaries. The temperature variability within and among the rivers suggests fish that depend on cold water may be able to use thermal diversity in rivers to survive in a warming climate (Fullerton and others, 2015).

MONITORING FISHING VESSELS

The Global Fishing Watch maintains an online, publically accessible, global fishing activity map that reveals the location and movement of commercial fishing fleets (globalfishingwatch.org). Objectives include accelerating scientific research and development of innovations that support improved fishery sustainability and partnering with countries to improve policy and compliance, seafood sourcing, and ocean conservation.

In January 2018, Global Fishing Watch entered into a new data-sharing partnership with NOAA to improve understanding of the nighttime activity of fishing vessels in Indonesian waters. Through the partnership, Global Fishing Watch and NOAA are matching Vessel Monitoring System (VMS) data from the Indonesian government with NOAA's satellite-based Visible Infrared Imaging Radiometer Suite (VIIRS), which can reveal the locations of brightly lit vessels at night. Two goals of the partnership are to identify fishing vessels that are not picked up by other monitoring systems and to test and refine the use of VIIRS for identifying and distinguishing different types of fishing vessels (Emmert, 2018). The VIIRS Day/Night Band used for this partnership is described in Chapter 15.

By cross matching VMS data from Indonesia with VIIRS, it was determined that roughly 80% of VIIRS detections could not be correlated to a vessel broadcasting VMS. There are several possible reasons for this large discrepancy. Some VIIRS detections could be from fishing vessels that are not required to carry VMS because they are under the 30 gross ton threshold established by the government of Indonesia. It is also possible that some vessels detected only by VIIRS meet the size requirement but have switched off their VMS or have a faulty device. In this case, vessels may be switching off their VMS and using bright lights to attract fish. Another possibility is that VIIRS is detecting foreign boats that are not carrying VMS because they are poaching from Indonesian waters. Ultimately, the results of the partnership indicate that the addition of VIIRS data can greatly enhance transparency in commercial fishing in Indonesia (Emmert, 2018).

FRESH WATER

Fresh water applications of remote sensing include the use of declassified satellite photographs and MODIS imagery to document the shrinking of bodies of water, the use of UAVs to monitor nuisance algae, and the use of Landsat to detect shifting rivers.

THE SHRINKING ARAL SEA

During the Cold War, the United States operated a series of classified, polar-orbiting satellites that acquired reconnaissance photographs of "areas of interest." Many of these photographs are now declassified, allowing for historical comparisons of areas or resources that are undergoing change.

For example, Figure 12-16A is a declassified photograph of the Aral Sea that was acquired August 22, 1964 by the US reconnaissance ARGON satellite, which employed film cameras with 124 m spatial resolution (NASA Earth Observatory, 2012). The quality of this ARGON photograph of the Aral Sea is impressive. When it was acquired in 1964, the Aral Sea was the world's fourth largest lake. It was fed by the Amu Darya in the south and the Syr Darya in the east. These rivers were fed by snowmelt and precipitation in their mountainous headwaters in the south. The Aral Sea was brimful in 1964 and supported numerous fishing communities.

In the 1950s and 1960s, the former Soviet Union began diverting water from the two rivers to support irrigated farming in the arid regions east and south of the sea. Dams, canals, and other projects were built to transform the surrounding desert and these actions led to the slow disappearance of the Aral Sea.

A. ARGON photograph acquired August 22, 1964.

B. MODIS image acquired on August 26, 2000.

C. MODIS image acquired on August 16, 2007.

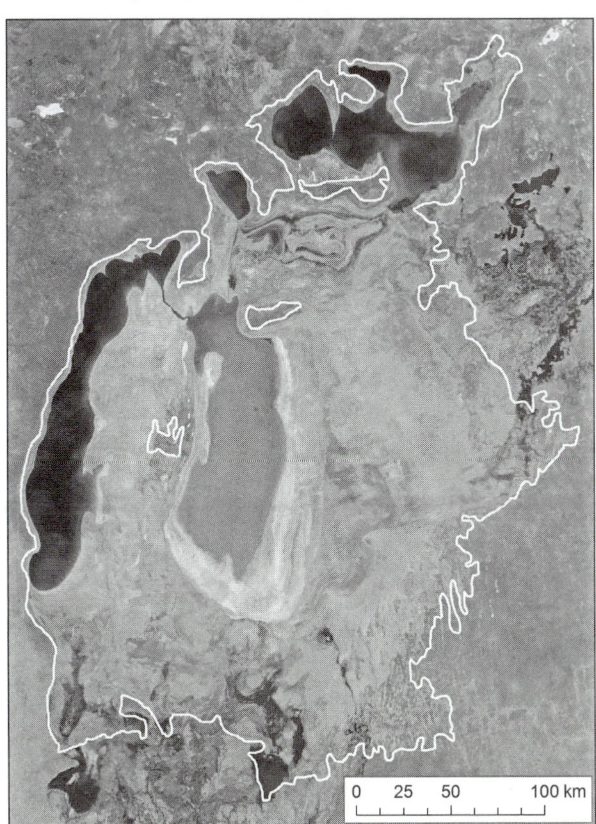

D. MODIS image acquired on August 21, 2016.

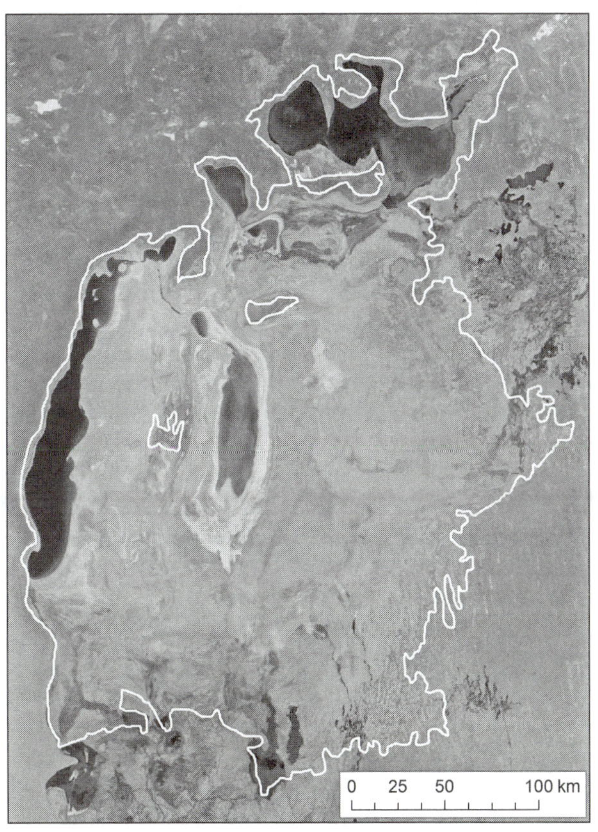

Figure 12-16 Aral Sea, 1964–2016. The white outline is the approximate 1964 Aral Sea shoreline. Courtesy NASA Earth Observatory (2012, 2019).

Figure 12-16 visually shows the impacts of water diversion on the Aral Sea using a sequence of MODIS images of the Aral Sea acquired from 2000 to 2016 (NASA Earth Observatory, 2019). Each image was acquired in August to match the late summer conditions of the 1964 ARGON photograph. The white outline in the MODIS images illustrates the approximate 1964 Aral Sea shoreline, before water from the two supporting rivers was diverted. The MODIS images are seen in color on **Digital Image 12-1** 🌐. By 2000, the shrunken Aral Sea was separated into the following three "lakes":

1. A northern lake that has maintained a shrunken, but constant, outline since 2000. It is maintained by intermittent stream flow from a distributary of the Syr Darya. A dam at the southeast corner of the lake prevents any outflow.
2. A narrow, linear lake that has maintained a constant western shoreline, as shown by the 1964 outline. The eastern shoreline has retreated between 2000 and 2016.
3. An eastern lake that has shown the maximum variation since 2000. MODIS images (not illustrated) show that the lake was completely dry in 2009 and 2014; both dry years were followed by an influx of water in 2010 and 2015. Seasonal snowmelt has increased the size of the eastern lake in 2017 and 2018 (NASA Earth Observatory, 2019).

The ability to track the dynamic changes of the Aral Sea illustrates the value of the temporal resolution provided by MODIS and other satellite images.

Mapping Nuisance Algae

Cladophora glomerata is considered a nuisance algae that grows attached to substrate in dense mats that disrupt benthic fauna and flora, fouls fishing lines and lures, and negatively affects both dissolved oxygen and pH in lakes and rivers. A number of remote sensing tools have been used to map *Cladophora* in lakes and coastal environments, including aerial photography from manned aircraft and multispectral satellite imagery. Field observations are frequently employed to locate and monitor *Cladophora* in rivers, but this approach is site specific and labor intensive. To evaluate the economic and mapping advantages of UAV remote sensing in this situation, Flynn and Chapra (2014) deployed an off-the-shelf, multirotor UAS with a consumer camera (RGB with an array of 3,000 by 4,000 pixels) equipped with a wide angle lens. The wide angle lens enables a large area to be imaged while flying at a low altitude. This UAS was equipped with a Bluetooth wireless connection so the field of view could be monitored during flight and the shutter activated remotely in real time via an iPhone app.

Flynn and Chapra (2014) studied the Clark Fork River in western Montana. Ground control points were established at six sites along the river corridor. Eighteen flights were completed during the summer of 2013 along a 1-km segment of the river. Each mission acquired 100 to 250 images. The images were georeferenced and spectral classification algorithms were applied to map *Cladophora* in the river. **Digital Image 12-2** 🌐 displays the pixels classified as *Cladophora* as bright green. The narrow gray rectangles with black hash-marked lines across the river channel upstream and downstream of the bright green pixels denote the geographic limits of the study area. The study documents algal cover ranges from less than 5% to greater than 50% during the summer of 2013. Results indicate that optical remote sensing with UAV holds promise for completing spatially precise and multitemporal measurements of algae or submerged aquatic vegetation in shallow rivers with low turbidity and good optical transmission. Flynn and Chapra (2014) note that the advantages of UAVs are numerous, but flexibility in flight planning and the ability to conduct repeat missions provide the greatest benefit.

Shifting River in Bolivia

Indigenous communities living along small rivers in remote tropical locations depend upon reliable access for water, food, and transportation. A summary of NASA's use of Landsat to monitor the course of a river in Bolivia follows.

In the Moxos Plains of northern Bolivia, a tropical area east of the Andes, several rivers meander through a swampy landscape of savanna, forests, and ponds. While studying three decades of satellite imagery of this area, geographer Umberto Lombardo of University Pompeu Fabra noticed a particularly striking example of rapid change on a stretch of the Maniqui River. Figure 12-17A shows the course of the river in September 2013, when it flowed in a northeasterly direction. By July 2016, sediment deposited by the river had filled in most of the pond and the river was charting a more easterly path (Figure 12-17B). Over those three years, vegetation growth started to cover up the channel that was full of water in 2013.

Lombardo's (2016) findings with Landsat imagery have on-the-ground implications. Authorities should be prepared for the possibility that indigenous communities living along small rivers may have to be resettled if rivers shift over time. Lombardo (2016) observes that a planned road linking Villa Tunari to San Ignacio de Moxos may be destroyed or regularly flooded without careful planning. He also states that shifting rivers have implications for biodiversity.

RENEWABLE ENERGY

Remote sensing is used to plan, help measure potential, and monitor renewable energy projects such as geothermal, solar, and wind (ESRI, 2010). Short summaries of examples from the alternative energy fields follow.

Geothermal

ASTER TIR imagery is used to detect elevated surface temperatures that may be associated with geothermal resources. The California Energy Commission (2010) commissioned a study of a region in the central part of eastern California with an emphasis on the Coso geothermal field. Nighttime ASTER scenes were most useful because of the significantly diminished effect of solar irradiation compared with daytime imagery. Several factors have to be corrected

A. Maniqui River in 2013.

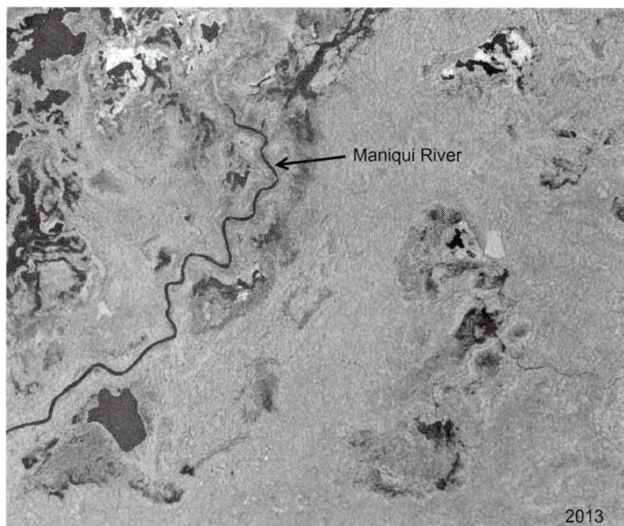

B. Maniqui River in 2016.

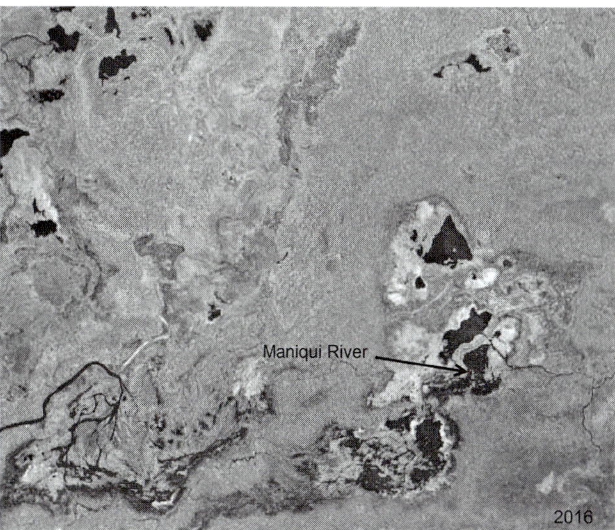

FIGURE 12-17 Landsat 8 images of the shifting Maniqui River in Bolivia. Courtesy NASA Earth Observatory (2016).

prior to using ASTER TIR imagery for mapping subtle differences in ground temperature, including:

- Aspect of topographic slopes (southern slopes are warmer than northern ones),
- Albedo (dark-colored surfaces are warmer than light-colored ones),
- Thermal inertia (surfaces with low thermal inertia cool off more at night),
- Elevation (areas at lower altitudes are warmer than those at higher altitudes), and
- Temperature inversions (local meteorological conditions sometimes cause cooler temperatures at the bottom of valleys and warmer temperatures higher up).

The integration of remote sensing with field observations is promising for future geothermal exploration studies (California Energy Commission, 2010).

SOLAR

Hammer and others (2003) used remote sensing data from geostationary satellites such as Meteosat to derive information on solar irradiance for a very large area in Germany. The goal was to reduce energy consumption in urban areas by designing more energy efficient buildings and cities. Several computational methods have been developed for estimating the downward solar irradiance from satellite observations for solar energy and daylight use. A good understanding of daylight climate is essential for optimal use of natural lighting in buildings and a reduction in overall energy consumption during the day. Meteosat's spatial resolution (of up to 2.5 km) and temporal resolution (30 minutes) are better and more available where solar sites are planned compared with ground station data. Solar irradiances have been calculated on a 30-minute basis and a spatial grid of 10 km for western and central Europe from Meteosat imagery. Hammer and others (2003) research introduces remote sensing data as a primary source for solar energy applications.

Germany's SUN-AREA research project uses aerial laser scanners to collect rooftop data and GIS spatial analysis tools to calculate rooftop outer form, roof slope inclination and orientation, and shadows cast by chimneys and other rooftops to determine the solar potential of all roof areas (ESRI, 2010). About 20% of the country's rooftops are suitable for solar power production. Preliminary findings estimate that these roofs could supply solar power that would meet the entire energy needs of homes throughout Germany.

WIND

Remote sensing is one source of input data for a geospatial approach to prioritizing wind farm development (Miller and Li, 2014). DEMs provide fundamental geospatial information about the terrain for planning the location of wind turbines (ESRI, 2010).

QUESTIONS

1. What is the phenological growth cycle?
2. Describe the movement of a pixel in the NIR and red spectral space during the growing season in a cultivated agricultural field. Compare the NIR and red reflectance values during the growing season (refer to Figure 12-1).
3. Calculate a Visible Atmospherically Resistant Index (VARI) when the reflectance values are as follows:
   $R_{blue} = 0.13$
   $R_{green} = 0.17$
   $R_{red} = 0.15$
4. What is ET? In general, how is the USGS generating ET water-use maps?
5. Why are the USGS ET water-use maps of river basins and watersheds useful?

6. Why do Gatziolis and others (2015) recommend a spiral UAS trajectory around a tree for imaging and supporting a virtual reconstruction of that tree?
7. Why are spatial gaps in forest canopies important?
8. During oil exploration and production (E&P) activities, what is the likely outcome to a closed stand of forest, rangeland, or wetland when the following public policies concerning roads are implemented:
   a. Public access allowed.
   b. Controlled access only.
   c. No roads allowed.
9. Describe the discoveries made by the USGS TIR mapping of the Coeur d'Alene River in Idaho regarding the distribution of temperature in the river. Why is this important for some types of fish?
10. Why did the Aral Sea almost disappear between 1964 and 2016?
11. What are important advantages to using UAVs for mapping nuisance aquatic vegetation in shallow rivers and lakes?

REFERENCES

Baynard, C. W. 2013. Remote sensing applications: Beyond land-use and land-cover change. *Advances in Remote Sensing*, 2(3), 228–241. doi: 10.4236/ars.2013.23025

Baynard, C. W., J. M. Ellis, and H. Davis. 2012a. Evaluating disturbance of E&P access roads. ESRI News for Petroleum, Spring, pp. 6–7. http://www.esri.com/library/newsletters/petroleum-perspectives/petrol-spring-2012.pdf.

Baynard, C. W., J. M. Ellis, and H. Davis. 2012b. Roads, petroleum, and accessibility: The case of eastern Ecuador. *GeoJournal*, 78(4), 675–695. doi 10.1007/s10708-012-9459-5

Butler, M. J. A., M. C. Mouchot, V. Barale, and C. LeBlanc. 1988. *The Application of Remote Sensing Technology to Marine Fisheries: An Introductory Manual* (FAO Fisheries Technical Paper 295). http://www.fao.org/docrep/003/t0355e/T0355E00.HTM#toc (accessed January 2018).

California Energy Commission. 2010. *Geothermal Exploration in Eastern California Using ASTER Thermal Infrared Data*. Public Interest Energy Research (PIER) Program, Final Project Report (CEC-500-2012-005). Imageair, Inc.

Emmert, S. 2018, January 9. Bright Lights Reveal the "Dark" Fleet. Global Fishing Watch. Press Release. http://globalfishingwatch.org/press-release/media-kit/bright-lights-reveal-dark-fleet (accessed January 2018).

ESRI. 2010, January. GIS for Renewable Energy. ERSI GIS Best Practices Series (G44363).

Flores-Anderson, A. I., K. E. Herndon, R. B. Thapa, and E. Cherrington (Eds.). 2019. *The Synthetic Aperture Radar (SAR) Handbook: Comprehensive Methodologies for Forest Monitoring and Biomass Estimation*. Huntsville, AL: SERVIR Global Science Coordination Office, National Space Science and Technology Center. http://gis1.servirglobal.net/TrainingMaterials/SAR/SARHB_FullRes.pdf.

Flynn, K. F., and S. C. Chapra. 2014. Remote sensing of submerged aquatic vegetation in a shallow non-turbid river using an unmanned aerial vehicle. *Remote Sensing*, 6(12), 12,815–12,836. http://www.mdpi.com/2072-4292/6/12/12815 (accessed January 2018).

Fullerton, A. H., C. E. Torgersen, J. J. Lawler, R. N. Faux, E. A. Steel, T. J. Beechie, J. L. Ebersole, and S. G. Leibowitz. 2015. Rethinking the longitudinal stream temperature paradigm: Region-wide comparison of thermal infrared imagery reveals unexpected complexity of river temperatures. *Hydrological Processes*, 29(22), 4,719–4,737. doi: 10.1002/hyp.10506

Gatziolis, D., J. F. Lienard, A. Vogs, and N. S. Strigul. 2015. 3D tree dimensionality assessment using photogrammetry and small unmanned aerial vehicles. *PLoS One*, 10(9). doi:10:1371/journal.pone. 0137765

Geipel, J., J. Link, and W. Claupein. 2014. Combined spectral and spatial modeling of corn yield based on aerial images and crop surface models acquired with an unmanned aircraft system. *Remote Sensing*, 6(11), 10,335–10,355.

Getzin, S., R. S. Nuske, and K. Wiegand. 2014. Using unmanned aerial vehicles (UAV) to quantify spatial gap patterns in forests. *Remote Sensing*, 6(8), 6,988–7,004.

Hammer, A., D. Heinemann, C. Hoyer, R. Kuhlemann, E. Lorenz, R. Muller, and H. G. Beyer. 2003. Solar energy assessment using remote sensing technologies. *Remote Sensing of Environment*, 86, 423–432.

Jensen, J. R. 2007. *Remote Sensing of the Environment: An Earth Resource Perspective* (2nd ed.). Upper Saddle River, NJ: Pearson.

Lim, K., P. Treitz, M. Wulder, B. St-Onge, and M. Flood. 2003. LiDAR remote sensing of forest structure. *Progress in Physical Geography*, 27(1), 88–106.

Lombardo, U. 2016. Alluvial plain dynamics in the southern Amazonian foreland basin. *Earth System Dynamics*, 7, 453–467.

McGaughey, R. J. 2018, August. FUSION/LDV: Software for LIDAR Data Analysis and Visualization Manual (FUSION Version 3.80). USDA Forest Service, Pacific Northwest Research Station.

Miller, A., and R. Li. 2014. A geospatial approach for prioritizing wind farm development in northeast Nebraska, USA. *ISPRS International Journal of Geo-Information*, 3, 968–979. doi:10.3390/ijgi3030968

NASA Earth Observatory. 2012, February 24. The Aral Sea, Before the Streams Ran Dry. http://earthobservatory.nasa.gov/IOTD/view.php?id=77193 (accessed January 2018).

NASA Earth Observatory. 2016, December 21. A Shape-Shifting River in Bolivia. http://earthobservatory.nasa.gov/IOTD/view.php?id=89266&src=eoa-iotd (accessed January 2018).

NASA Earth Observatory. 2019. World of Change: Shrinking Aral Sea. http://earthobservatory.nasa.gov/world-of-change/aral_sea.php (accessed March 2019).

Pacificorp Energy and Trout Unlimited. 2007. *Airborne Thermal Infrared Remote Sensing, Bear River Basin, ID/WY/UT*. Report submitted by Watershed Sciences, Inc.

Pôças, I., A. Rodrigues, S. Goncalves, P. M. Costa, I. Goncalves, L. S. Pereira, and M. Cunha. 2015. Predicting grapevine water status based on hyperspectral reflectance vegetation indices. *Remote Sensing*, 7, 16,460–16,479.

Strother, C. W., M. Madden, T. R. Jordan, and A. Presotto. 2015. Lidar detection of the ten tallest trees in the Tennessee portion of the Great Smoky Mountains National Park. *Photogrammetric Engineering & Remote Sensing*, 81(5), 407–413.

Tamás, L. 2010. Data Acquisition and Integration. Section 4: Laser Scanning. University of West Hungary. Digitális Tankönyvtár, Hungary. http://www.tankonyvtar.hu/en/tartalom/tamop425/0027_DAI4/ch01s02.html (accessed January 2018).

US Forest Service. 2007. *Airborne Thermal Infrared Remote Sensing, Coeur d'Alene River, Idaho*. Report submitted by Watershed Sciences, Inc.

USGS. 2016. Mapping Water Use: Landsat and America's Water Resources. http://www.usgs.gov/news/mapping-water-use-landsat-and-americas-water-resources (accessed January 2018).

chapter 13

Nonrenewable Resources

Since the early 1960s, geologists have applied remote sensing to energy and mineral exploration. The vast differences in geology, terrain, land cover, and weather found around the Earth require us to employ various technologies that are relevant to the goals of a project. Each project also requires a multidisciplinary team approach. Field and subsurface geology, geochemistry, and geophysics are essential disciplines. Geologists exploring for oil and minerals must be prepared to perform in this evolving technical environment.

Remote sensing is an integral part of exploration. Satellite and airborne images, along with DEMs, are used to discover and map new oil fields and mineral prospects. Hyperspectral and TIR data is processed to further refine areas for future fieldwork. Modern analogs and 3-D morphometry measurements have been developed with remote sensing images and DEMs to improve interpretation of ancient landforms intersected by subsurface seismic data and wells. In addition, oil field and mine development, environment, and infrastructure are monitored with remote sensing throughout the life cycle. Satellite radar interferometry is used to monitor and help plan oil production activity. Air quality and landscape changes are monitored with remote sensing in terrain that holds natural gas resources developed with fracking and horizontal drilling technologies. Ground-based remote sensing maps the geology and alteration on vertical mine walls to improve efficiency of production and mine output while core scanners improve mine development and ore processing.

In this chapter, case histories are described to show how the development and implementation of remote sensing technologies can be applied to oil and mineral exploration. We trust that our experience (and good luck!) may be useful to our colleagues and students. More examples of remote sensing for oil exploration are provided by Prost (2014), Berry and Prost (1999), and Berger (1994). Examples of mineral exploration include Bedell and others (2009), Kruse (1999), and *Economic Geology* (1983). The Environmental Research Institute of Michigan coordinated 14 remote sensing exploration and applied geology conferences between 1981 and 2000. The proceedings from these informative conferences are available from the Michigan Tech Research Institute, a research center of Michigan Technological University in Ann Arbor, Michigan (mtri.org). The Geological Remote Sensing Group (grsg.org.uk) routinely publishes articles on nonrenewable applications and hosted an Oil & Gas Remote Sensing Workshop in July 2018 in Boulder, Colorado.

OIL EXPLORATION AND DISCOVERY

The search for oil in unexplored onshore areas normally begins with regional geologic reconnaissance followed by progressively more detailed (and expensive) steps that culminate by drilling a wildcat well. *Wildcat wells* are exploratory tests in previously undrilled areas, whereas *development wells* are drilled to produce oil from a previously discovered field. A typical onshore exploration program proceeds as follows:

1. **Regional remote sensing reconnaissance**: Small-scale image mosaics of VNIR and SWIR images, radar images, and hillshade DEMs that may cover thousands of square kilometers are especially useful in this phase. The objectives are:

a. Define *sedimentary basins*, which are areas underlain by thick sequences of sedimentary rocks.

b. Map major geologic structures and trends within the basins.

This image reconnaissance is essential in oil exploration.

2. **Regional geophysical surveys**: *Aerial magnetic surveys* record the intensity of the Earth's magnetic field. Sedimentary basins have lower magnetic intensities than do areas underlain by basement rocks such as granite and metamorphic rocks. The aerial magnetic maps can confirm the presence of sedimentary basins. Surface *gravity surveys* record the intensity of the Earth's gravity field. Sedimentary rocks have a lower specific gravity than basement rocks, hence, sedimentary basins are shown by lower values on gravity maps. Gravity and magnetic maps may also show regional structural features.
3. **Detailed remote sensing interpretation**: Digitally processed images are interpreted to identify and map geologic structures, such as anticlines, domes, and faults, that may form oil traps. Promising structures are mapped in detail using stereo-pairs of images in the VNIR spectral region and DEMs. Radar images are used in regions of poor weather where it is difficult to acquire good VNIR images. Lidar images can show terrain that is obscured by vegetation. Once the images are interpreted, geologists go into the field to check the interpretation and collect samples of the exposed rocks.
4. **Seismic surveys**: Explosives or mechanical devices are used to transmit waves of sonic energy into the subsurface, where they are reflected by geologic structures. The reflected waves are recorded at the surface and processed to produce *seismic maps* and *cross sections* that show details of subsurface geologic structure. Remote sensing images and maps also facilitate logistics and navigation of seismic crews.
5. **Drilling**: Wildcat wells are drilled to test the *oil prospects*, or drilling targets, that are defined by the preceding steps. Because of the inevitable uncertainties of predicting geologic conditions thousands of kilometers below the surface, less than 20% of wildcat wells are successful.

The following case histories employed airborne and satellite remote sensing technologies for oil exploration. The Papua New Guinea Project used airborne radar images to support the discovery of the country's first oil fields. The Central Arabian Arch Project used Landsat TM images to discover a new oil trend in a region with very subtle surface expression of oil traps.

PAPUA NEW GUINEA PROJECT

When the Chevron Corporation acquired Gulf Oil Company in 1984, it also acquired Gulf's partnership share in two petroleum prospecting licenses (PPLs) located in the nation of Papua New Guinea (PNG), which occupies the eastern por-

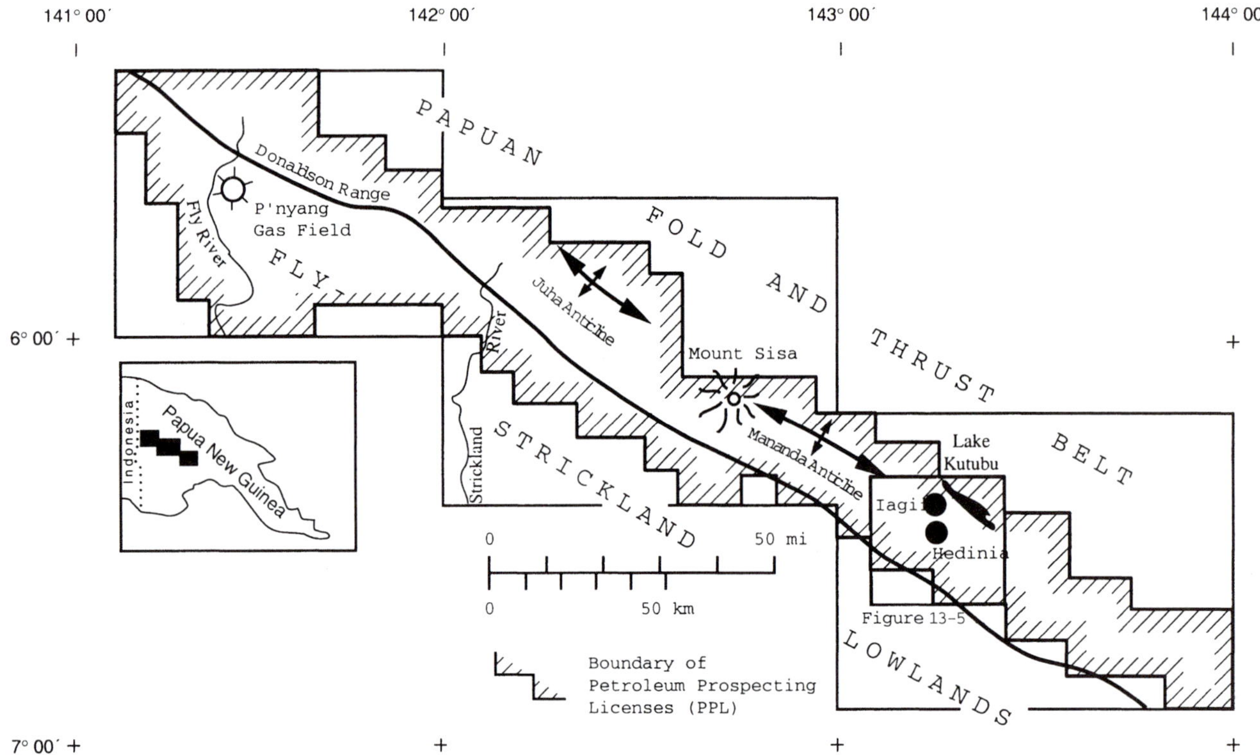

FIGURE 13-1 Location map of the boundaries of the Chevron PPLs in the PNG, which are indicated by hachured outlines. Topical and geologic features are identified. The area shown here corresponds to the area covered in Figure 13-2.

tion of the island of New Guinea. Chevron operated the project on behalf of the other partners who owned major interests.

Background

Figure 13-1 includes an index map of New Guinea and a location map of the PPLs, which extend for more than 350 km southeast from the border with Indonesia. The PPLs cover an area of approximately 14,400 km^2 that spans the boundary between the Fly-Strickland Lowlands in the southwest and the Papuan Fold and Thrust Belt in the northeast. Oil geologists have been interested in the region for many years because (1) numerous oil and gas seeps occur in the fold belt and (2) some wildcat wells had indications of hydrocarbons, which are called *shows*. Although a number of wells had been drilled, no commercial oil fields had been discovered in PNG when this project began.

Oil exploration in the PPLs was difficult for several reasons. The area was remote and inaccessible with few roads; transportation was largely by foot, by helicopter, or by fixed wing aircraft that could use the scattered airstrips. The area was covered by tropical rain forest. The climate was hot and humid with persistent cloud cover and heavy rainfall. The terrain was dominated by rugged linear ridges and steep stream channels. Much of the region is underlain by limestone that weathers to karst topography with pinnacles and steep-walled pits that cannot be crossed by vehicles and are dangerous to traverse on foot. Seismic surveys, which are the routine method for defining drilling prospects, could not be recorded in most of the PPLs because (1) the inaccessible and rugged topography resulted in seismic survey costs that exceeded $100,000 (in 1998 dollars) per linear kilometer and (2) subsurface caverns in the karst terrain strongly scattered the seismic energy, which resulted in very poor data. For these reasons, it was decided at the outset of the project to rely on radar images and limited fieldwork.

Radar Images

The persistent cloud cover precluded timely acquisition of Landsat or SPOT images. Earlier Chevron experience with SIR-A images of Indonesia had demonstrated that radar images were optimum for geologic interpretation in rain forest terrain. In 1985, Intera Technologies conducted an airborne radar survey of the PPLs. The X-band (3-cm wavelength), synthetic-aperture images had a spatial resolution of 12 m. Data were recorded digitally and processed into image strips that were combined to produce three mosaics. Figure 13-2 is a greatly reduced version of the mosaics that were produced at an original scale of 1:250,000. The side-lapping image strips could also be viewed and interpreted stereoscopically.

In Chapter 6, we demonstrated the importance of selecting the optimum radar look direction to enhance the expression of geologic structures. The Papuan Fold and Thrust Belt, which underlies most of the project area, consists of

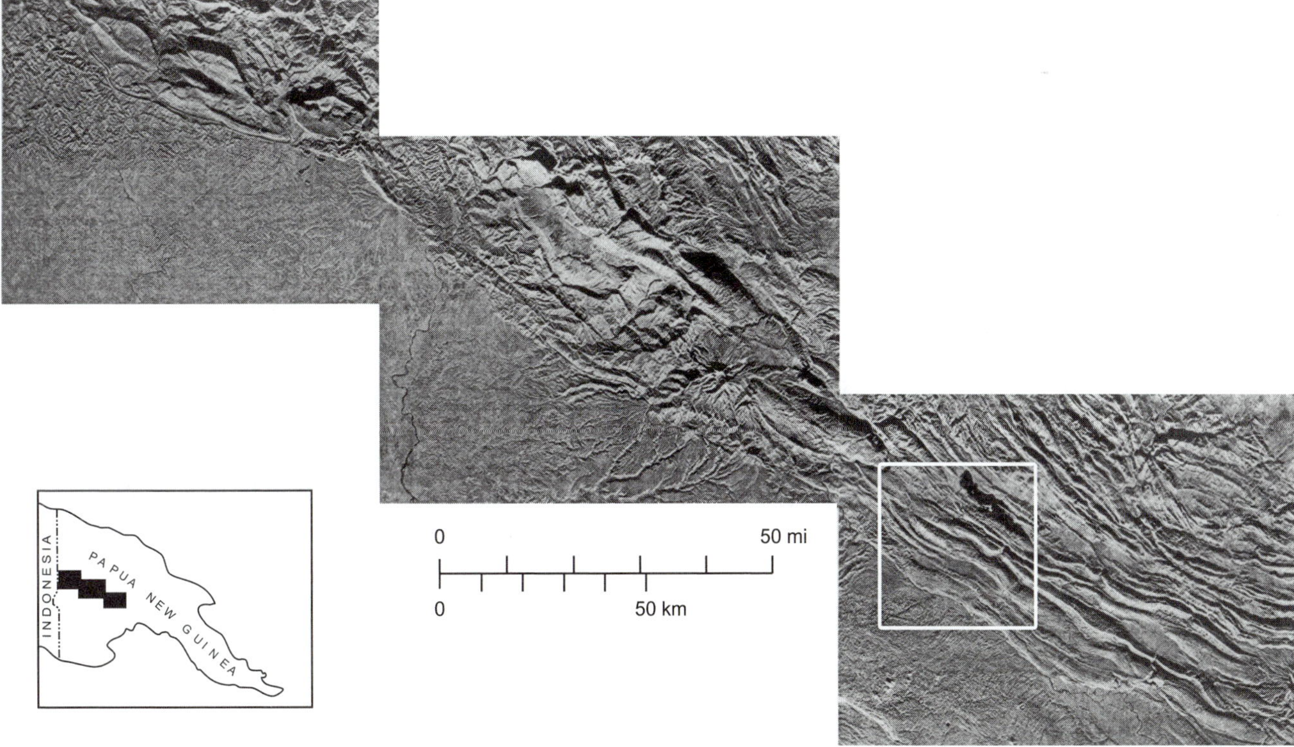

FIGURE 13-2 Mosaic of X-band (3-cm wavelength) aircraft radar images of the Papuan Fold and Thrust Belt. Radar look direction is toward the north, which is oblique to the northwest strike of the belt; average depression angle is 17°. This geometry is optimum for imaging the belt. The white outline shows the location of the Lake Kutubu region (Figure 13-3).

FIGURE 13-3 Radar image of the Lake Kutubu region, which includes oil fields discovered by the PNG Project (Figure 13-4 for interpretation map).

northwest-trending ridges that are the expression of anticlines and thrust faults. Most of the thrust faults are exposed along the southwest flanks of the anticlines, whereas the northeast flanks are predominantly dipslopes. We considered using a northeast look direction, which is normal to the regional trend. This look direction, however, would have resulted in extremely bright highlights from the southwest-facing fault scarps, which would have obscured important information. For this reason, the radar look direction was oriented toward the north, which resulted in high-quality images. Radar shadows are oriented toward the upper margin of the mosaic (Figure 13-2), which may cause topographic inversion for some viewers. Rotate the page 180° to correct this problem. Adjacent radar swaths were acquired with 60% sidelap, which produced stereo images. The radar mosaic and stereo image strips were interpreted using the criteria developed earlier in Indonesia (Chapter 6).

Exploration Program

The exploration program focused on the Papuan Fold and Thrust Belt because of the numerous surface anticlines and abundant oil and gas seeps (Ellis and Pruett, 1986). The initial wildcat well was located on the crest of the Mananda anticline, shown in Figure 13-1, which is a prominent feature on the mosaic. The well was dry, but there were encouraging shows of oil in the Toro Sandstone (Cretaceous).

Exploration then shifted southeast from Mananda to the area south of Lake Kutubu where several promising anticlines are expressed on the radar mosaic. Figures 13-3 and 13-4 are the image and geologic interpretation map for the Lake Kutubu area, shown at the 1:250,000 scale of the original mosaic. Most of the terrain in the image is underlain by the Darai Limestone (Middle Tertiary), which weathers to karst topography with a distinctive pitted signature on the image. Smooth terrain in the valleys is formed by alluvial deposits. A few outcrops of clastic strata of the Orubadi Formation (Late Tertiary) occur along southwest-facing slopes. Topography is generally a reliable expression of the underlying geologic structure because the region was deformed fairly recently and erosion has not obliterated structural landforms. Northwest-trending ridges are anticlines, and the intervening valleys are synclines.

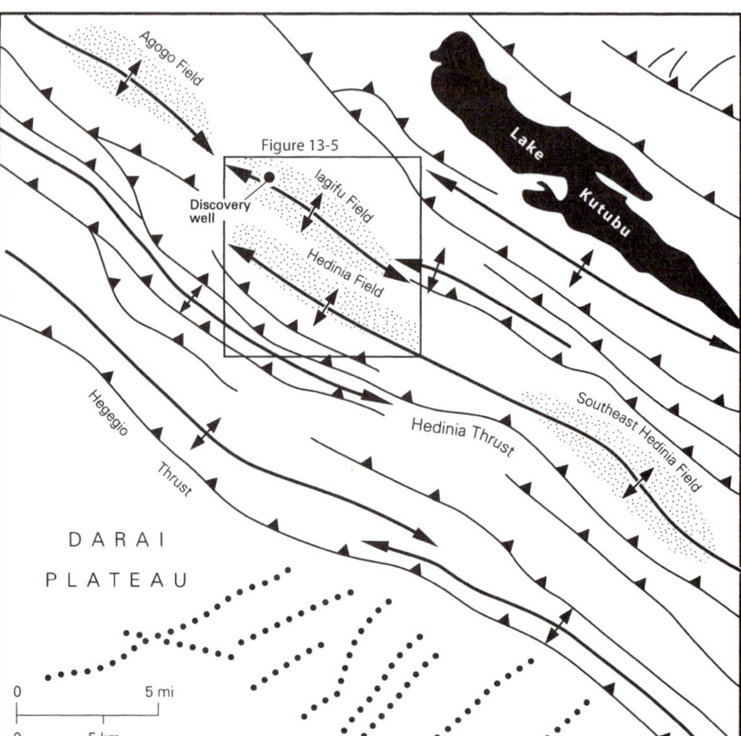

FIGURE 13-4 Structural map interpreted from the radar image of the Lake Kutubu region (Figure 13-3). Stippled patterns show locations of four oil fields discovered by the PNG Project.

The Darai Plateau in the southwest portion of the Lake Kutubu area is within the relatively undeformed Fly-Strickland Lowlands, which forms a generally uniform surface with karst topography. Dotted lines in the map are northeast-trending linears in the plateau that are fractures or faults that have been enlarged by solution of the limestone bedrock. The Hegegio thrust fault (Figure 13-4) is the structural boundary between the lowlands and the Papuan Thrust and Fold Belt to the northeast. The numerous anticlines of the belt are expressed on the image as slightly curving linear topographic swells that are outlined by highlights and shadows. The anticlines are capped by the Darai Limestone with its distinctive karst topography. The synclines are floored by alluvial deposits with a characteristic smooth texture on the image. Near the Hegegio thrust fault, the anticlines and synclines are of comparable width, but to the northeast, toward Lake Kutubu, the synclines become narrower because they are overridden by thrust plates that have moved relatively southwestward.

Oil Discoveries

The first commercial oil well in PNG was completed in 1986 on the crest of the Iagifu anticline, which is now the Iagifu field (Figure 13-4). Additional wells were drilled to delineate the Iagifu field. Next, the Hedinia anticline was successfully drilled and is now the Hedinia field. Subsequently, the Agogo field and Southeast Hedinia field were discovered. Figure 13-5 is a subsurface structure map of the Iagifu and Hedinia fields with contours on top of the Toro Sandstone, which is the oil-producing formation. Figure 13-6 is a geologic cross section across the Iagifu and Hedinia fields, which are trapped in the crests of anticlines located above the Hedinia thrust fault. On the cross section note that the subsurface anticlines are expressed as ridges on the topographic profile, which helps explain the accuracy of the radar interpretation. In 1990, Chevron used radar images to discover the P'nyang gas field in the northwest portion of the PPLs (Figure 13-1). Valenti and others (1996) show the image and interpretation map of the P'nyang anticline. A pipeline carries the oil southeast to the Arafura Sea, where an offshore terminal loads it onto tankers. After Chevron exited the project, a number of fields have been discovered throughout the Papuan Thrust and Fold Belt.

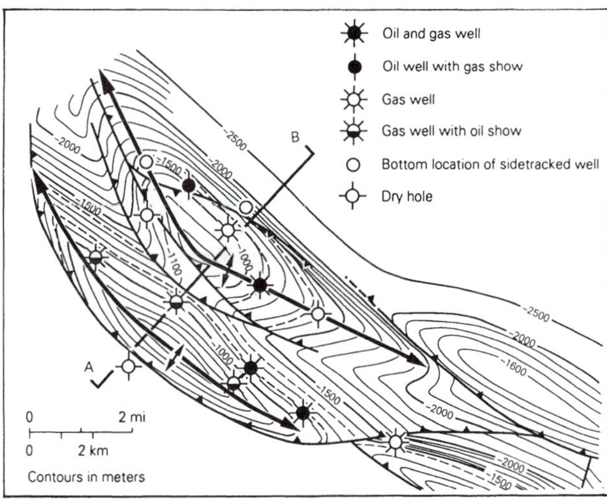

Figure 13-5 Subsurface structure map of the Iagifu (north anticline) and Hedinia (south anticline) oil fields. Contours are drawn on top of the productive Toro Sandstone. From Lamerson (1990, Figure 12).

Central Arabian Arch Project

The Kingdom of Saudi Arabia has the world's largest oil reserves and is a major exporter of oil. Therefore, it is surprising to realize that most of the Kingdom was relatively unexplored for oil until the late 1980s, for reasons given in the following section. The Central Arabian Arch Project discovered significant new oil fields in a new exploration trend.

Background

In 1933, the Chevron Corporation, then called Standard Oil Company of California, was a small West Coast company in need of new oil reserves. At that time Chevron obtained from King Ibn Saud the exploration rights to a huge area in eastern and central Saudi Arabia. In 1936, Texaco joined the venture to create the partnership that later became the Arabian American Oil Company (Aramco). Chevron geologists mapped the Dammam Dome at Dhahran (Figure 13-7). In 1939, after several dry holes, the first oil was discovered in the Kingdom at Dammam. Oil exploration and development halted during World War II but resumed shortly after the war. In 1948, Aramco used seismic surveys to discover the giant Ghawar field located in sand-covered terrain. Chevron and Texaco realized that large capital investments were required to develop this resource. Therefore, the Aramco partnership was expanded to include Exxon and Mobil. Aramco proceeded to discover a number of major fields both onshore and in the Persian Gulf. In 1972, the government of Saudi Arabia began to acquire the assets of Aramco from the four partners. By 1980 the oil reserves and equipment

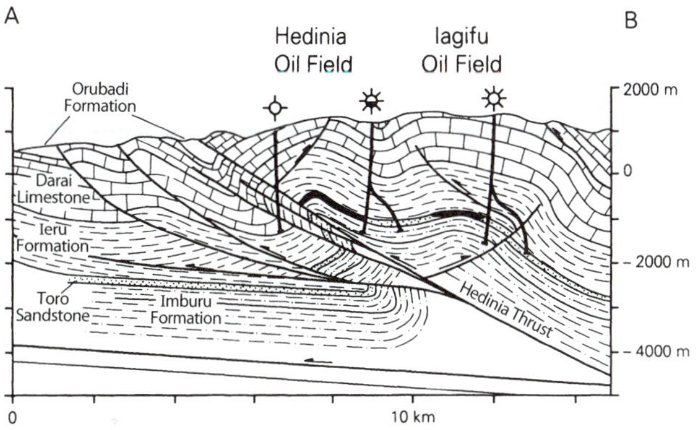

Figure 13-6 Cross section A–B across the Iagifu and Hedinia oil fields in Figure 13-5. The Toro Sandstone is the reservoir formation. From Lamerson (1990, Figure 10-13).

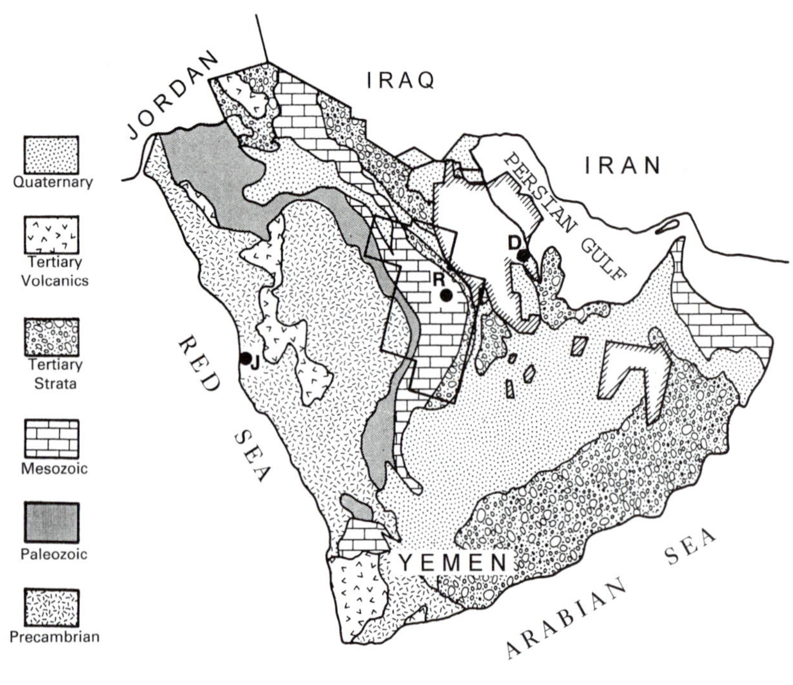

FIGURE 13-7 Geologic interpretation map of the Arabian Peninsula. Retained areas are indicated by hachured borders and the area of the Central Arabian Arch Project is outlined. D = Dhahran, J = Jiddah, R = Riyadh.

were wholly owned by the Kingdom. The company name was changed to Saudi Aramco in 1989.

For many years the Saudi government had restricted exploration to "retained areas" in the vicinity of the existing fields (Figure 13-7). The objective was to prevent an oversupply of crude oil on the world market. This restriction accounted for the lack of oil exploration in much of the Kingdom; no new fields were discovered within the retained areas. It wasn't until 1986 when the government instructed Saudi Aramco to explore the region for oil potential outside the retained areas. In 1987, the chief geologist of Saudi Aramco visited the original Aramco partners (Exxon, Mobil, Texaco, and Chevron) to solicit their recommendations for exploring this vast terrain. During his week at Chevron, Sabins gave the chief geologist a condensed version of his UCLA Geology 150 Remote Sensing Course complete with image interpretation projects. At the end of the week Sabins recommended an exploration project using computer-enhanced Landsat TM images. Sabins also strongly recommended that Saudi Aramco geologists visit Chevron's research laboratory and participate in all phases of the project: image selection, digital processing, interpretation, map making, and report writing. In early 1988 Saudi Aramco requested the Remote Sensing Research Group of Chevron to conduct a Landsat exploration project to aid the new exploration program. The project was conducted by L. E. Wender (formerly with Saudi Aramco and Exxon Mobil) and F. F. Sabins (Chevron). The initial study site was an area in the Central Arabian Arch (Figure 13-8).

Major objectives of the Central Arabian Arch project were to (1) map the regional geology at a scale of 1:250,000

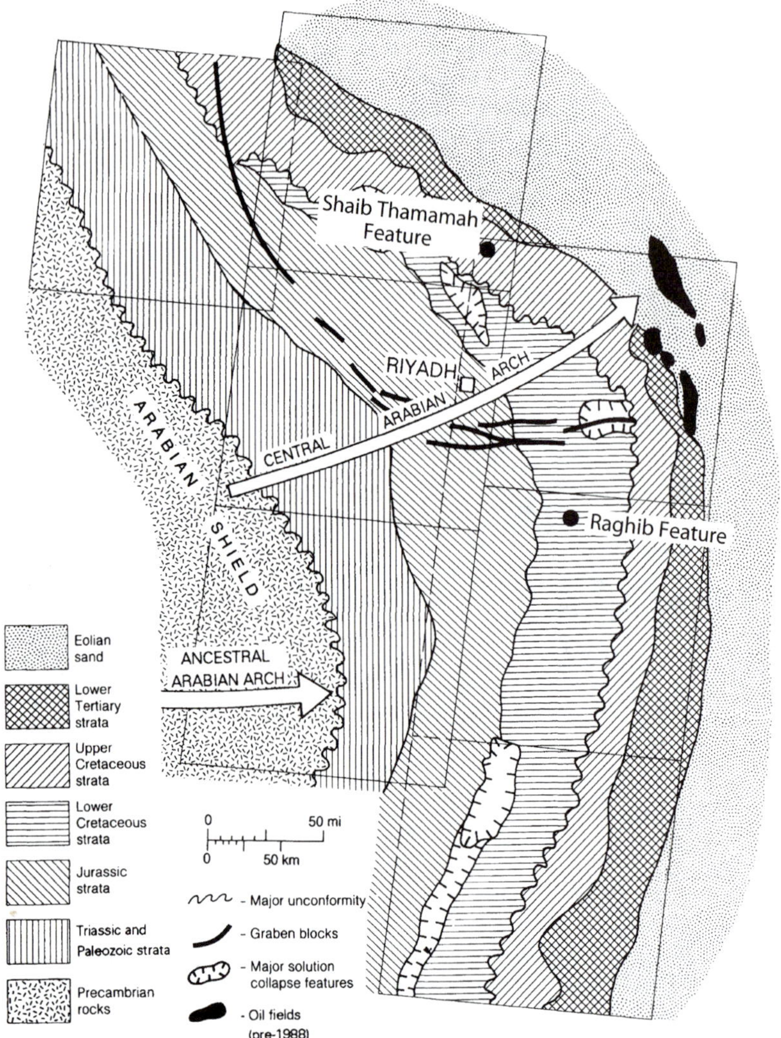

FIGURE 13-8 Geologic map of the Central Arabian Arch interpreted from TM images. Outlines of Landsat image are shown. From Wender and Sabins (1991, Figure 2).

and (2) interpret local features that may be expressions of subsurface structures that are potential oil fields. The TM images were digitally processed using techniques described in Chapter 9.

Regional Geologic Mapping

Seven Landsat TM images were digitally processed by the Chevron Remote Sensing Research Group (Sabins, 1991). The images were interpreted at a scale of 1:250,000 using the procedures described by Wender and Sabins (1991). The individual geologic maps were compiled into the generalized small-scale map in Figure 13-8, which shows locations of the individual images. The project area is located on the east flank of the Arabian shield, which is a regional uplift of Precambrian basement rocks and overlying strata. The basement rocks extend far to the east in the subsurface and provide a subsiding platform for additional strata; the strata were deposited as aggregate for a total thickness of nearly 5,500 m (18,000 ft) during the Paleozoic through Tertiary ages.

Structure in the Paleozoic and Mesozoic strata is dominated by a gentle regional dip of approximately 1° away from the Arabian shield. In the southern part of the project area, dips are toward the east; in the north, dips are toward the northeast. The area where the dip direction changes is called the Central Arabian Arch (Figure 13-8).

Landsat Features

The second objective of the project was to recognize features on the images that may be oil prospects. Prior to this project, Sabins' image interpretation experience was largely in regions of moderate to high structural relief, such as the Saharan Atlas Mountains, Papuan Thrust and Fold Belt, and the western United States. In these areas, surface dips of 5° or more are common and subsurface structures are expressed at the surface by opposing dip directions, arcuate outcrop patterns, and offset beds, as seen in the Thermopolis, Wyoming, images in Chapter 3.

In the Central Arabian Arch, however, surface dips are 1° or less and the regional structural pattern is dominated by broad, uniform dipslopes that extend for several hundred kilometers along the strike and up to several tens of kilometers in the dip direction. In order to recognize structural features in this region of low structural relief, we needed a three-dimensional model to show the relationship between subsurface structures and their expression on Landsat images. Within the retained areas, oil fields are simple drape anticlines in Mesozoic strata formed over high-angle faults that offset basement rocks and Paleozoic strata. The anticlines grade upward into flattening of the regional dip to form subtle structural terraces at the surface. Using this information, we developed the model shown in Figure 13-9. Erosion of the structural terrace produces the surface features of the model that are recognizable on Landsat images.

Shaib Thamamah Exploration Model

On the images we recognized the Shaib Thamamah feature, located 70 km north of Riyadh (Figure 13-8), which is an example of our structure model. Plate 40A is the Landsat image and Figure 13-10 is the geologic interpretation map of the feature, which occurs within the outcrop belt of the Aruma Formation (upper Cretaceous). Prior to this project, the Aruma Formation was mapped as a single unit. On the Landsat enhanced color image (TM bands 2-4-7 in B-G-R) (Plate 40A), however, it is readily divided into two members with the following characteristics:

1. **Upper member**: Resistant limestone that forms extensive dipslopes and steep antidip scarps with a dark, greenish brown image signature. In the field the color is medium to light tan.
2. **Lower member**: Poorly resistant shaley limestone that weathers to slopes with a very light blue image signature. In the field the color is medium to light tan.

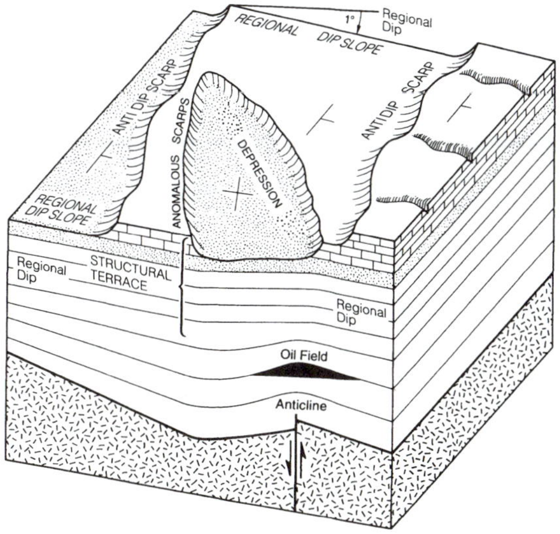

FIGURE 13-9 Geologic model for oil exploration in terrain with subhorizontal strata. The topographic and geologic features at the surface are caused by the same subsurface structure that forms the oil field.

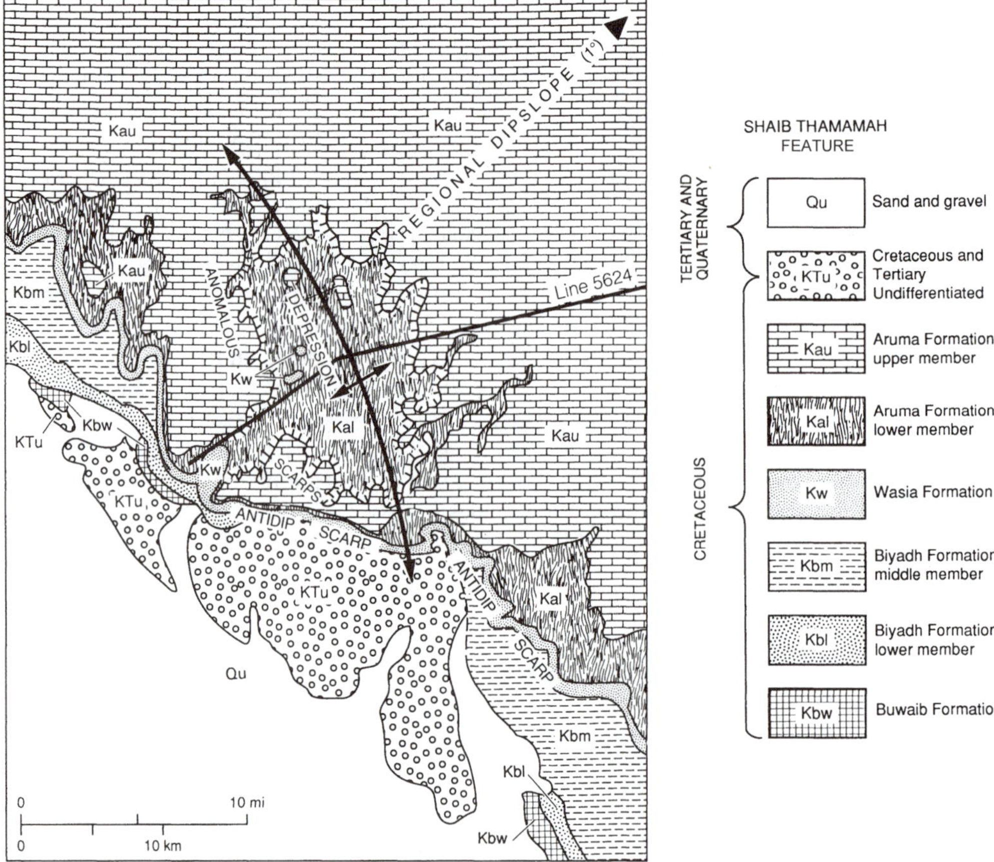

FIGURE 13-10 Geologic interpretation map of the Shaib Thamamah feature with legend (Plate 40A). From Wender and Sabins (1991, Figure 4).

The ability to separate the Aruma Formation into two members on the image is a major factor in recognizing the Shaib Thamamah feature. The feature is defined on the color image (Plate 40A) by geomorphology and outcrop patterns.

- **Geomorphology**: The regional dipslope of the upper member terminates as a steep, southwest-facing antidip scarp that is shown in the field photograph of Figure 13-11A. The feature is a flat-floored topographic depression, bounded by erosional scarps of the upper member. These scarps are anomalous because they face east, which opposes the westward orientation of the regional antidip scarps. Figure 13-11B is a field photograph showing the depression in the foreground bounded by anomalous scarps in the background. The geologic map in Figure 13-10 shows the scarps that enclose the depression, which forms the feature.
- **Outcrop pattern**: The normal outcrop pattern of the Aruma Formation is a broad uniform dipslope formed by the upper member. At Shaib Thamamah, however, this pattern is interrupted by an inlier of the lower member that is 20 km long and 10 km wide. In Plate 40A the light blue signature of the lower member distinguishes it from the upper member. In the field we observed two small exposures of the Wasia Sandstone, shown by small stippled patterns and the symbol Kw in Figure 13-10, that underlies the lower member. The presence of these older rocks within the inlier emphasizes the structural significance of the Shaib Thamamah feature. The feature is so large that it is difficult to recognize in the field, but it is easily identified in the Landsat image.

In the geologic model (Figure 13-9) the surface feature is underlain by an anticline that formed over a fault. In order to evaluate our interpretation of subsurface structure at Shaib Thamamah, Saudi Aramco recorded seismic line 5624 at the location shown in Figure 13-10. The seismic section (Figure 13-12) shows that the feature is underlain by closely spaced vertical faults that offset the Khuff Formation (Permian) and adjacent beds. The offset dies out upward into a gentle arching of the Jilh Formation (Triassic). Higher in the section the arch becomes a structural terrace, or flattening of the northeast regional dip, at the Arab Formation (Jurassic) and overlying units. We were naturally gratified that the seismic section confirmed our model. The surface profile of the seismic line shows the depression that coincides with the geomorphic feature at Shaib Thamamah. Elsewhere in the project area, new seismic lines and found relationships were similar to those at Shaib Thamamah.

Oil Discoveries

We used the model concept to interpret a number of features on the images, which we then checked in the field. A few features were eliminated as potential structures, but most were confirmed as targets for additional evaluation. Saudi Aramco followed up with seismic surveys and drilling of the most promising features.

Plate 40B is the Landsat enhanced color image (TM bands 2-4-7 as B-G-R) and Figure 13-13 is the geologic interpretation map of the Raghib feature, which is located 110 km southeast of Riyadh (Figure 13-8). Limestones of the Arab and Sulaiy Formations crop out in the west portion of the image and are eroded to an irregular surface. The blue signature of the limestones is locally obscured by patches of yellow windblown sand. The green circles in the northwest corner are wheat fields with centerpoint irrigation systems. Because of solution and collapse of the limestones, it is locally difficult to separate the upper Sulaiy and lower Yamama Formations. The Yamama Formation (Cretaceous) consists of limestone with a medium brown signature. The Buwaib Formation (Cretaceous) is a thin sequence of lime-

A. View east along antidip scarp formed by the upper member of the Aruma Formation. Note the gentle dipslope.

B. View west across the center of Shaib Thamamah. The depression in the foreground is underlain by the lower member of the Aruma Formation. The anomalous scarps in the background are formed by the upper member of the Aruma Formation.

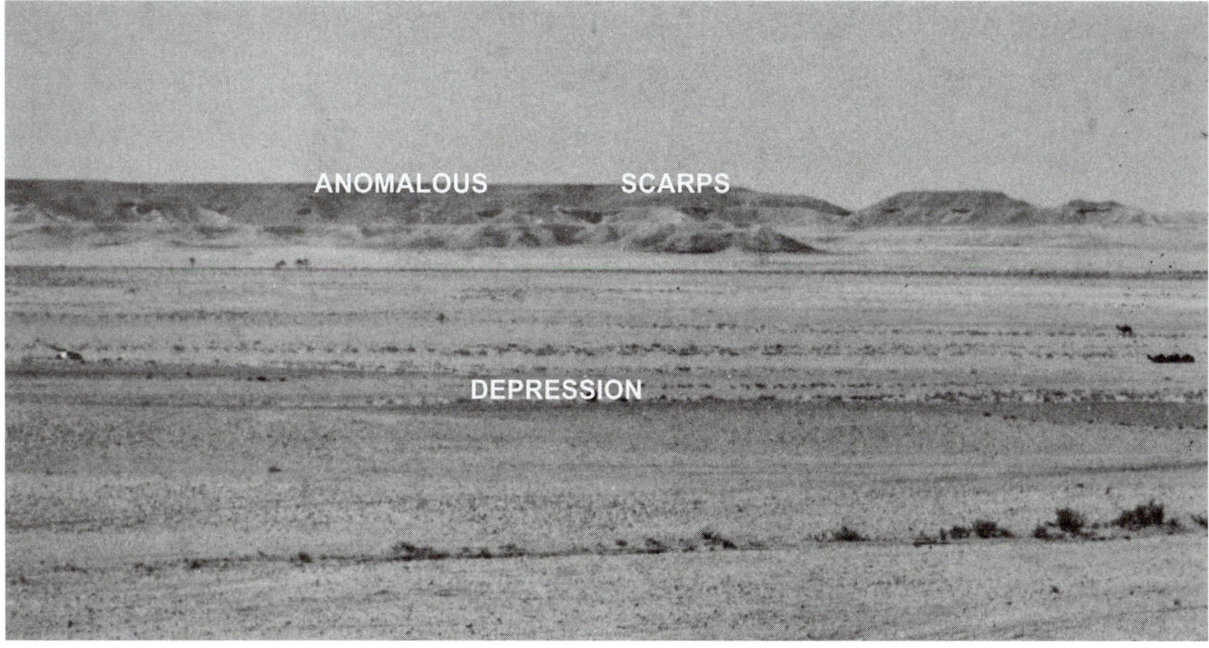

Figure 13-11 Field photographs of the Shaib Thamamah feature.

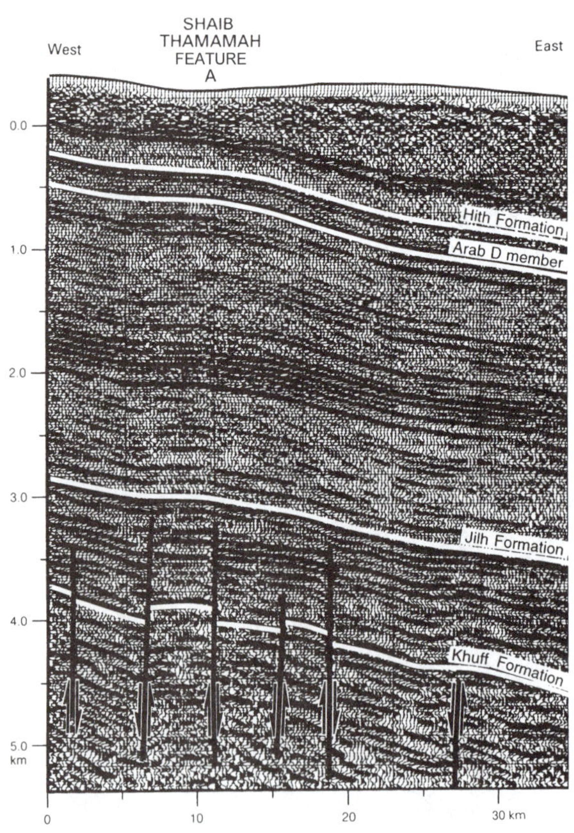

FIGURE 13-12 Seismic section 5624 across Shaib Thamamah. Formation offsets along the deep faults pass upward into an arch that becomes a flattening of regional dip at the surface. Location of the section is shown in Figure 13-10.

stone with alternating image signatures of light blue and medium brown. The lower member of the Biyadh Formation is a resistant sandstone with a heavy coating of desert varnish and a dark brown signature. The middle member is nonresistant, light to medium gray sandstone with irregular dark brown patches. Windblown sand covers much of the middle member.

In this region south of the Central Arabian Arch, strata dip east at 1°; erosion produces linear north-trending antidip scarps. At Raghib this regional pattern is interrupted in the Sulaiy outcrops by arcuate scarps that are concave to the west (Plate 40B). The depression on the west side of the arcuate scarps is mantled by windblown sand. On the geologic interpretation map the scarps are shown by bold hachured lines with the hachures pointing toward the depression. These features are the erosional remnants of a feature that originally resembled Shaib Thamamah. Erosion has removed the updip western margin of the feature, leaving the central depression and the arcuate scarps on the downdip eastern margin.

In late 1988 we checked the Landsat interpretation in the field. Early in 1989 Saudi Aramco completed a seismic

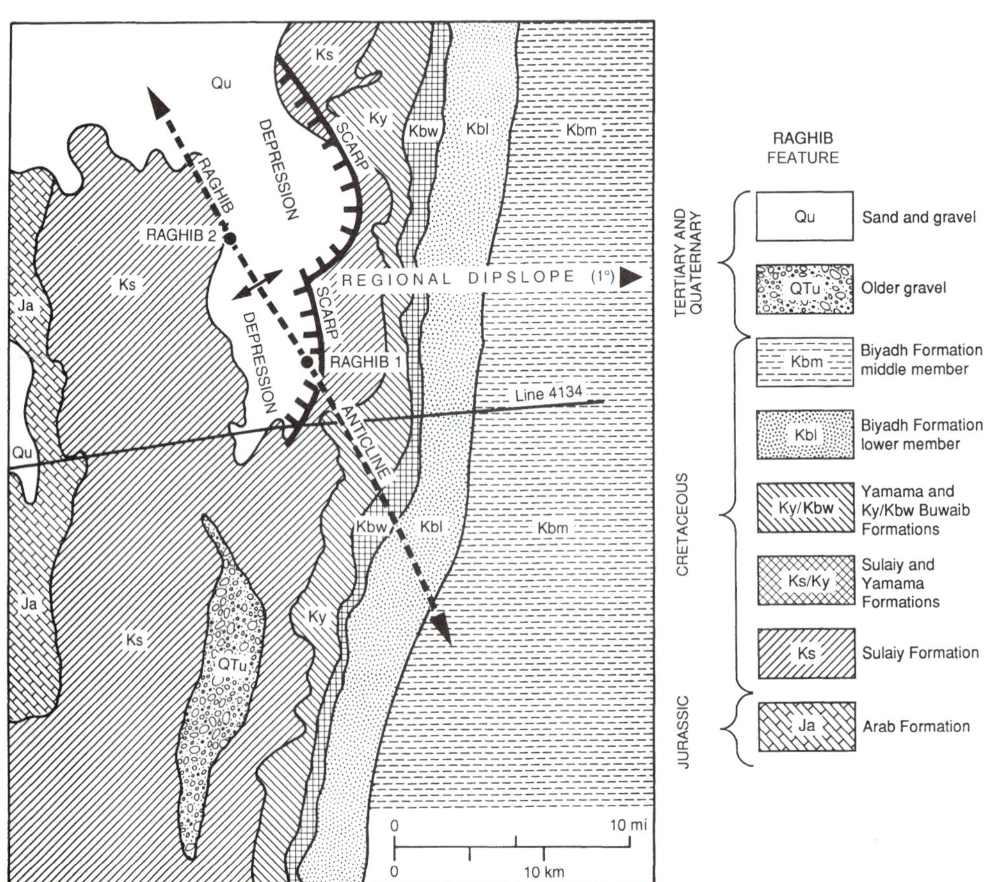

FIGURE 13-13 Geologic interpretation map of the Raghib feature with legend. Raghib 1 and 2 are the discovery well and confirmation well for the Raghib oil field. From Wender and Sabins (1991, Figure 7).

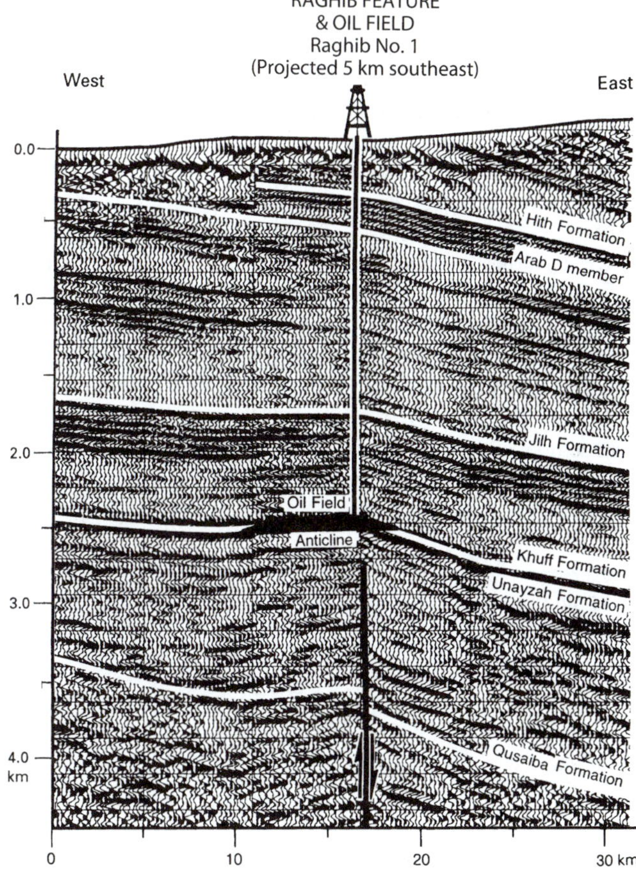

FIGURE 13-15 Seismic section 4134 across the Raghib feature. Location of the line is shown in Figures 13-13 and 13-14. Sandstone of the Unayzah Formation (Pennsylvanian and Permian) is the reservoir for the field. Organic-rich shale of the Qusaiba Formation (Silurian) is the source of the oil.

survey of the Raghib area. Figure 13-14 is a seismic structure map that covers the same area as the Landsat image (Plate 40B) and the geologic interpretation (Figure 13-13). Seismic structure contours are analogous to topographic contours but show the elevations of a subsurface rock formation rather than topography. Contours in Figure 13-14 are drawn on top of the Khuff Formation. Contour values are shown as two-way travel times in seconds (relative to sea level datum). Shorter travel times indicate higher elevations of the Khuff Formation. Contours with travel times less than 1.020 sec define a northwest-trending, doubly plunging anticline that is 35 km long and 10 km wide. The axis of the subsurface anticline is annotated on the geologic interpretation map, which shows that the east flank of the anticline underlies the anomalous arcuate scarps on the image. The depression west of the scarps coincides with the structurally highest part of the anticline. Figure 13-15 is seismic section 4134, which crosses the Raghib feature at the location shown in Figure 13-14. The seismic section shows that the anticline at the Khuff Formation overlies a vertical fault that offsets older strata. The anticline dies out upward into a structural terrace at the surface. Erosion of the terrace caused the Raghib feature shown in the Landsat image. These relationships match those of the structure model (Figure 13-9).

In late 1989 Saudi Aramco drilled a Raghib 1 wildcat well located at the base of the surface scarp that coincides with the crest of the subsurface anticline. Location of the well is shown in the geologic interpretation map (Figure 13-13) and the seismic map (Figure 13-14). Commercial hydrocarbons were discovered in sandstones of the Unayzah Formation (Pennsylvanian and Permian) that directly underlies the Khuff Formation (Figure 13-15). Sandstone of the Unayzah reservoir has an average porosity greater than 20%, and permeabilities of several darcies are common (McGillivray and Husseini, 1992). Production tests flowed 2,984 barrels of oil, 1,180 barrels of condensate, and 26.2 million ft^3 of gas per day. Similar results were obtained for the Raghib 2 confirmation well drilled 10 km northwest of Raghib 1. Organic-rich shale of the Qusaiba Formation of Silurian age is the source of the oil in the Raghib field (Abu-Ali and others, 1991). The oil probably migrated upward along the fault and was trapped in the Raghib anticline. A number of wells

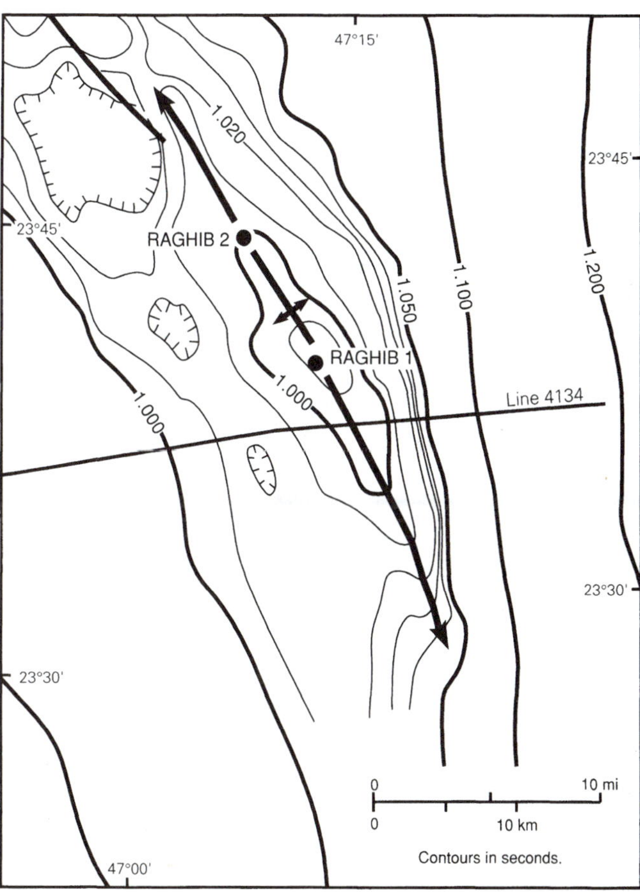

FIGURE 13-14 Seismic structure map of the Raghib feature. The area coincides with Plate 40B and Figure 13-13. Contours are drawn on top of the Khuff Limestone (Permian) and show two-way travel time in seconds. From Wender and Sabins (1991, Figure 8).

have been drilled to develop the full extent of the field, which is approximately defined by the deepest closed contour at 1.010 sec below sea level (Figure 13-14). After the Raghib discovery, the Mulayh and Nuyyam fields were discovered 50 km and 100 km, respectively, to the south along the new Unayzah Trend (Wender and Sabins, 1991).

MODERN ANALOGS

Oil and gas exploration includes interpreting seismic records and well logs of subsurface geologic formations, which can include ancient landforms such as sandbars, reefs, river channel deposits, and sand dunes. Remote sensing images and DEMs of landforms may serve as modern analogs to improve interpretation of ancient landforms intersected by seismic data and wells. This application is based upon a fundamental geological concept: "The present is the key to the past."

Interest in 3-D mapping of modern analogs for carbonate sand reservoirs is based on the substantial number of reservoirs that produce from carbonate grainstones and packstones. It is estimated that more than 60% of the world's oil and 40% of the world's gas reserves are held in subsurface carbonate reservoirs (Schlumberger, Ltd., 2017). The main reservoir of the world's largest oil field, the Ghawar in Saudi Arabia, is composed of carbonate sediment deposited as sandbars and shoals.

More than 150 oil and gas fields in west Texas and southeast New Mexico produce from dolomites of Late Permian age. A majority of these fields are situated on platforms or shelves (Figure 13-16). Many of the productive anticlinal structures appear to have formed by differential compaction over and around carbonate sandbars (Craig, 1990). The best

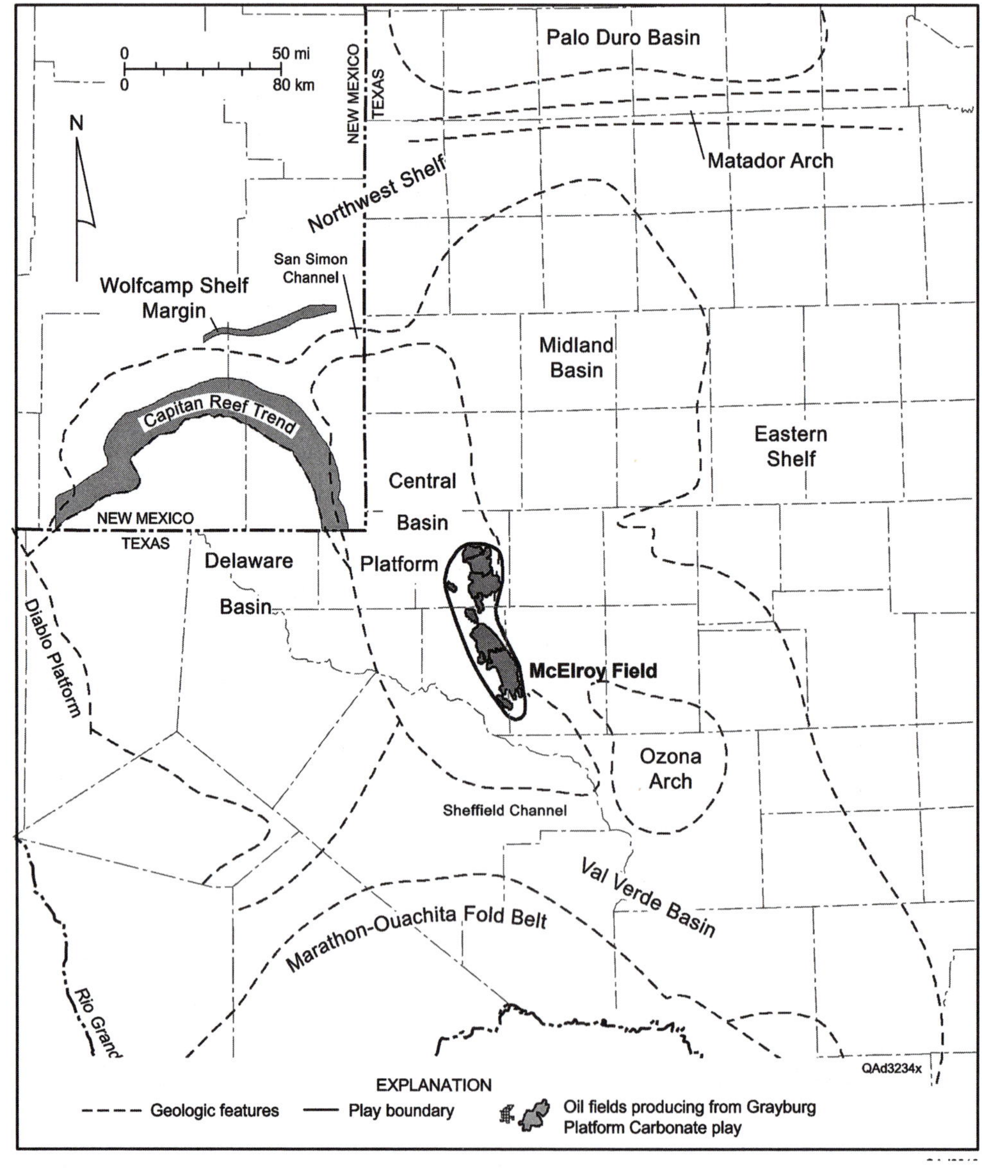

FIGURE 13-16 Major subdivisions and boundaries of the Permian basin in west Texas and southeast New Mexico showing the McElroy field on the eastern edge of the Central Basin Platform. Modified from Dutton and others (2004).

porosity and permeability in the McElroy field in west Texas occur in carbonate sandbars along the crest of the present-day structure (Harris and Walker, 1990). Figure 13-16 shows the oil fields producing from the Grayburg Platform Carbonate play (Dutton and others, 2004). The McElroy field is located along the southern portion of the play on the eastern edge of the Central Basin Platform (Figure 13-16). The ~250 million year old Central Basin Platform is similar to the modern-day Great Bahama Bank southeast of Florida. The location of the McElroy field is similar to that of the Schooners sand body; both are situated at the edge of the platform (Figure 13-17).

Harris and others (2010) interpreted satellite imagery and bathymetric DEMs derived from Landsat data of modern carbonate sand deposits in the Great Bahama Bank to measure landform morphometry (size, shape, topographic relief, density, orientation, etc.). Morphometry measurements of modern carbonate analogs demonstrate that certain aspects of these depositional systems behave in a systematic, and hence predictable, manner (Harris and others, 2011). Modern carbonate sand bodies in the Bahamas match the morphometry of sandbars, channels, and barrier bars mapped in outcrops of the Miami oolite (fossilized carbonate sands) in south Florida. The Miami oolite has been exposed for 115,000 years, validating the application of morphometric studies of modern landforms to improve the interpretation of ancient rocks (Purkis and Harris, 2017).

Large carbonate sand bodies have developed on the Holocene Great Bahama Bank where a change in bathymetry of the sea bottom coincides with strong tidal currents or wave action. The Great Bahama Bank is "the" modern day example of a flat-topped, isolated carbonate platform. It rises thousands of meters above the ocean floor southeast of Florida and north of Cuba (Figure 13-17) and is employed extensively as an aid to interpret deposits of ancient, subsurface carbonate platforms (Purkis and Harris, 2016).

The black polygons in Figure 13-17 are three large carbonate sand bodies: Schooners, Exumas, and Toto. This analysis focuses on Schooners. Schooners sand body is a zone of high carbonate production that extends over 60 km in length and covers 716 km^2. Schooners was built just north of the platform's edge (black line in Figure 13-18). The carbonate sands at Schooners are white, providing a uniform surface to calculate water depth using the brightness value of a Landsat TM band 1 (blue). The calculated water depth at Schooners varies from sea level to –15 m.

Schooners is a high energy environment where significant winds and currents interact with the carbonate sands to form sandbars with varying shapes and sizes. Figure 13-19 shows a subarea along the south-central portion of Schooners with sandbars mapped as dark polygons superimposed on the bathymetric DEM. Sandbars represent areas in a sand body with significant wave action that results in better sediment sorting, and therefore, better reservoir porosity and permeability potential.

Figure 13-20 illustrates the sandbar found in Figure 13-19 and highlighted with a black arrow. Area, total length, distance made good (DMG), maximum width, estimated average width, and perimeter are dimensions measured at each sandbar from imagery and the bathymetric DEM. These size measurements develop shape indices for each sandbar. The following shape indices, form factor, sinuosity, and aspect, are calculated as follows:

$$\text{form factor} = \frac{(4 \times 3.140845 \times \text{area})}{(\text{perimeter} \times \text{perimeter})}$$

$$\text{sinuosity} = \frac{(\text{total length} - \text{DMG})}{\text{total length}}$$

$$\text{aspect} = \frac{(\text{total length} - \text{average width})}{\text{total length}}$$

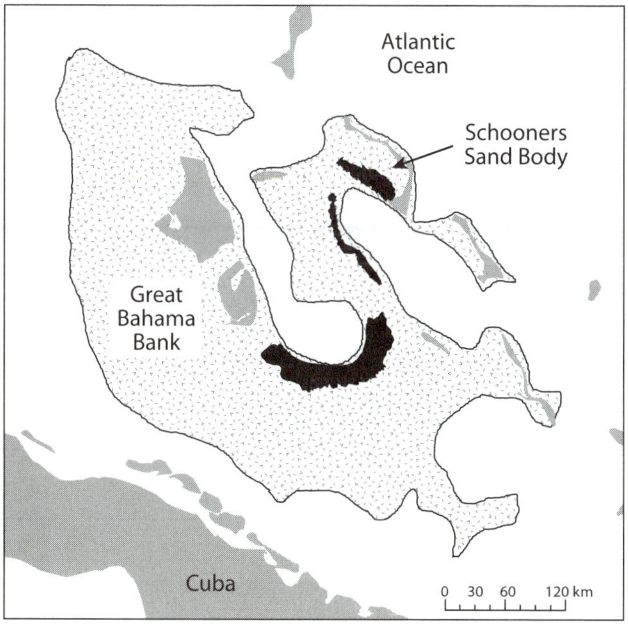

FIGURE 13-17 Location map of Great Bahama Bank and Schooners sand body. Light gray features are islands.

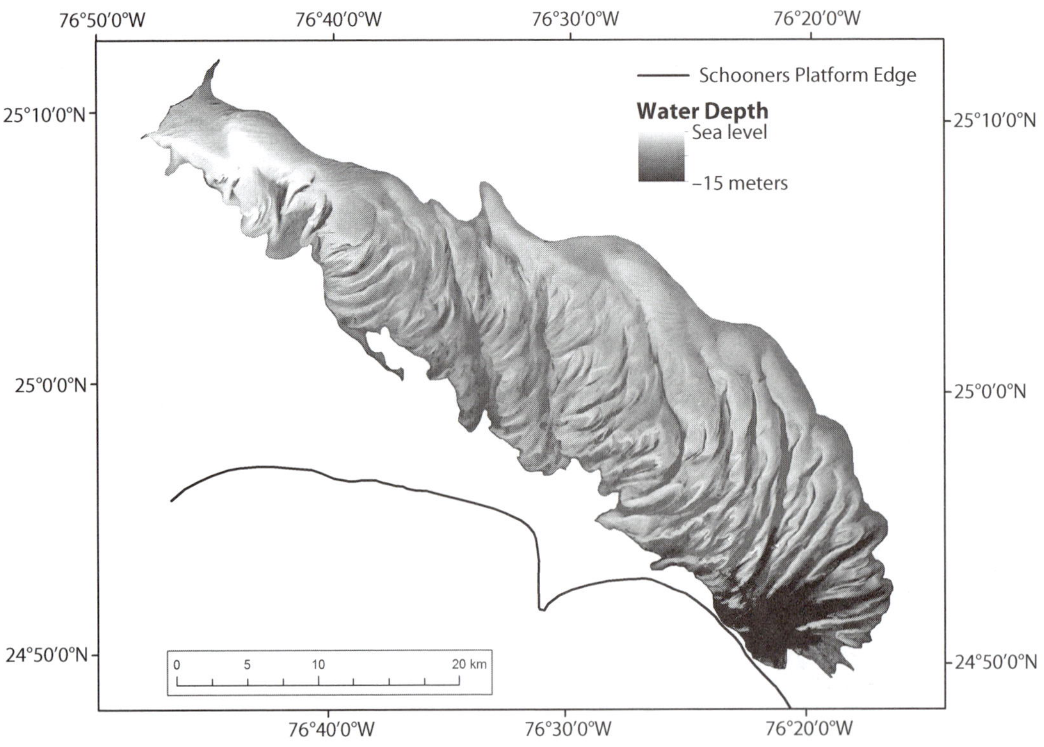

FIGURE 13-18 Bathymetric DEM derived from a Landsat TM band 1 (blue) image for the Schooners sand body. Water depth varies between sea level and −15 m. The platform edge is depicted as a black line in the lower portion of the figure.

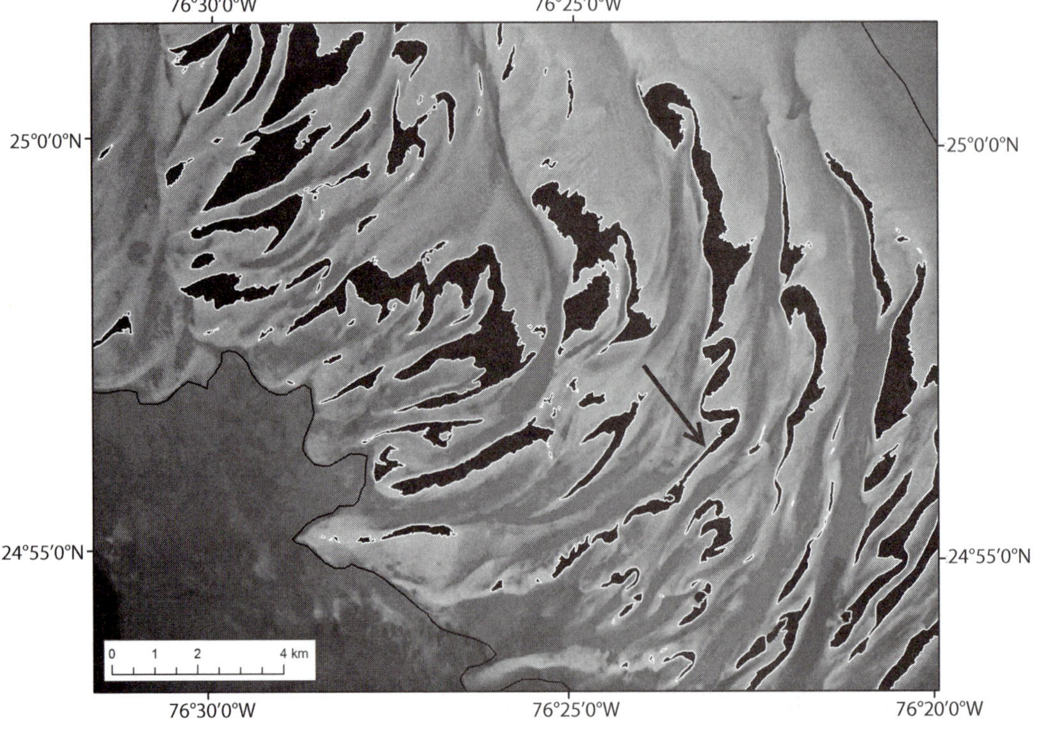

FIGURE 13-19 Individual sandbars are highlighted as black polygons. The sandbars were extracted from the south-central portion of the DEM (Figure 13-18).

A circular shaped sandbar will have a form factor of 1 while an elongated sandbar will have a form factor approaching 0. The form factor for each sand bar at Schooners was mapped with GIS in Figure 13-21. The GIS map symbology uses a 5-level grayscale legend that ranges from black to white. A sandbar with a form factor between 0.04 and 0.17 is black while a sandbar with a form factor greater than 0.56 is white. Figure 13-22 shows a plot of form factor versus sandbar area for Schooners, Exumas, and Toto. Small sandbars tend to be rounded, whereas large sandbars (> 1 km^2) are exclusively elongate. When interpreting seismic and well logs of buried sand bodies with sandbars, an explorationist can have more confidence knowing that if the subsurface data indicates that the sandbar is large, it is also probably an elongated landform.

Using the Landsat bathymetric DEM (Figure 13-18), the depth to the crest of sandbars was measured. Sandbar depths at Schooners were plotted and symbolized with GIS (Figure 13-23). The black line in Figure 13-23 is the edge of the platform. The distance from each sandbar to the platform edge was measured. A plot of sandbar depth versus distance to platform edge at Schooners is shown in Figure 13-24. The tops of sandbars in close proximity to the platform edge tend to be, on average, deeper than those found farther from the platform edge. When interpreting seismic and well logs of sandbars with a platform edge nearby, an explorationist can have more confidence knowing that if the edge is relatively far away, the sandbar depth may be shallower compared to sandbars closer to the edge. Shallower sandbars are exposed to more wave action, better sediment sorting, and have greater reservoir porosity and permeability potential compared with deeper sandbars.

The Bahamas study demonstrates how morphometric analyses of modern landforms using remote sensing images and DEMs can yield quantitative information about complex depositional environments such as sand bodies and sandbars. The approach allows for a detailed description to be condensed to a small number of shape indices. The application of modern day shape indices to improve subsurface interpretations and models may be validated by analyzing patterns in seismic and well data of oil and gas fields with carbonate sandbars along the margin of platforms, such as the McElroy field of west Texas.

Direct Detection of Oil and Gas

The PNG and Central Arabian Arch discoveries represent a broad category of oil fields that are expressed on images by their geologic structures and surface features. Another category of fields is expressed by seeps of oil and gas that have leaked to the surface. Exploration based on recognizing oil and gas seeps is called *direct detection*. Much research has been done with broadband multispectral sensors such as Landsat

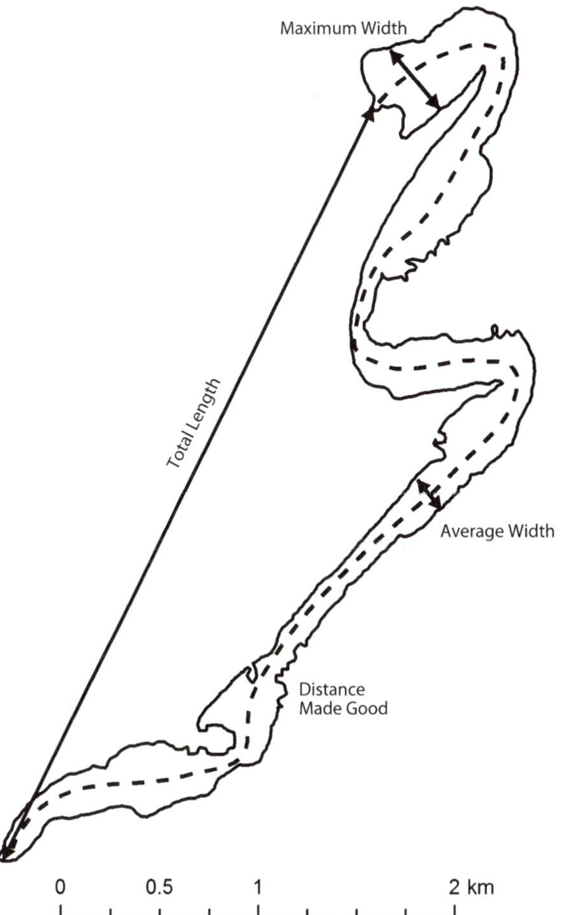

Figure 13-20 An irregular-shaped sandbar with examples of measuring total length, maximum width, distance made good, and average width to develop shape indices for each sandbar.

TM to detect altered soil and geobotanical changes that can be associated with seeps (Donovan, 1974; Yang and others, 2000). The results have been difficult to extrapolate globally.

Hyperspectral VNIR-SWIR sensors map onshore oil seeps and spills by detecting the spectral absorption feature of bitumen at a wavelength of 2.3 μm, together with the more subtle 1.703 μm absorption feature (Ellis and others, 2001). The spectral signature of oil, however, resembles the spectral signature of other organic materials, plastics, and asphalt. The imagery must be processed to minimize man-made features that generate false signatures.

Noomen (2007) used hyperspectral VNIR-SWIR sensors with spatial resolution less than 1 m to document that gas leaks cause changes in vegetation reflectance as early as two weeks after gas leakage starts. The source of anomalous concentrations of natural gas could be natural microseepages or underground gas pipelines. The study was designed for detecting leaks in grassland and crop vegetation. False anomalies were minimized by removing built-up areas and forests in the imagery. Seeps in Noomen's research area had a ring of high biomass around a pit of no biomass. To minimize false anomalies, Noomen used a high-pass filter to search for seeps

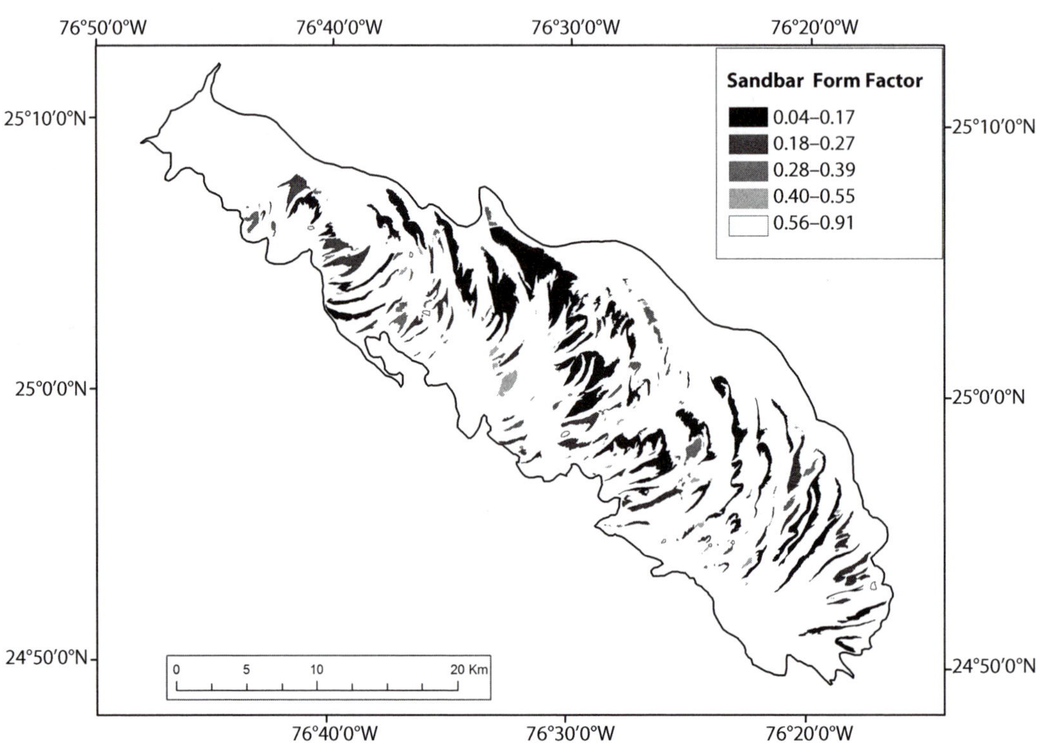

FIGURE 13-21 GIS map showing the form factor of sandbars at Schooners. Larger sandbars are more elongated.

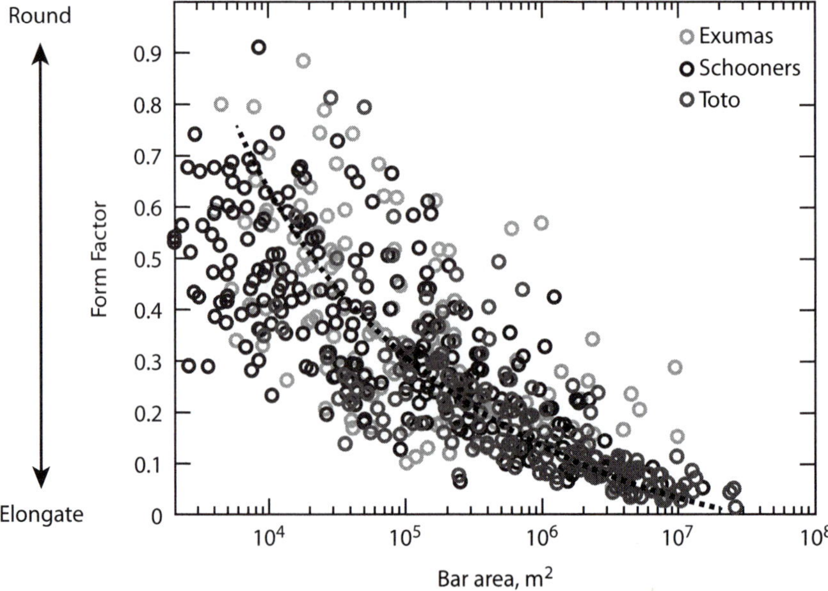

FIGURE 13-22 Plot showing the area of sandbars (logarithmic scale) versus their form factor with a best fit line (dashed). A circular-shaped sandbar will have a form factor of 1 while an elongated sandbar will have a form factor approaching 0.

with a high biomass ring. The high-pass filter searches for the specific spectral and spatial characteristics of these seeps.

MINERAL EXPLORATION AND DISCOVERY

In the 1980s the major oil companies diversified by adding mineral exploration divisions to their organizations. Some companies, such as Chevron, enjoyed profitable world-class mineral discoveries, but by the end of the century the companies had exited the mineral business. Until his retirement in 1992, Sabins was employed at Chevron's research subsidiary, where using remote sensing for mineral exploration was an obvious opportunity to apply the experience and technology we had gained from oil exploration. This process has been called technology transfer. There are at least two methods for applying existing technology to new exploration projects.

1. **Models**: Models illustrate patterns of alteration minerals associated with veins and with porphyries. Later in this chapter we illustrate each model and its use in exploration.

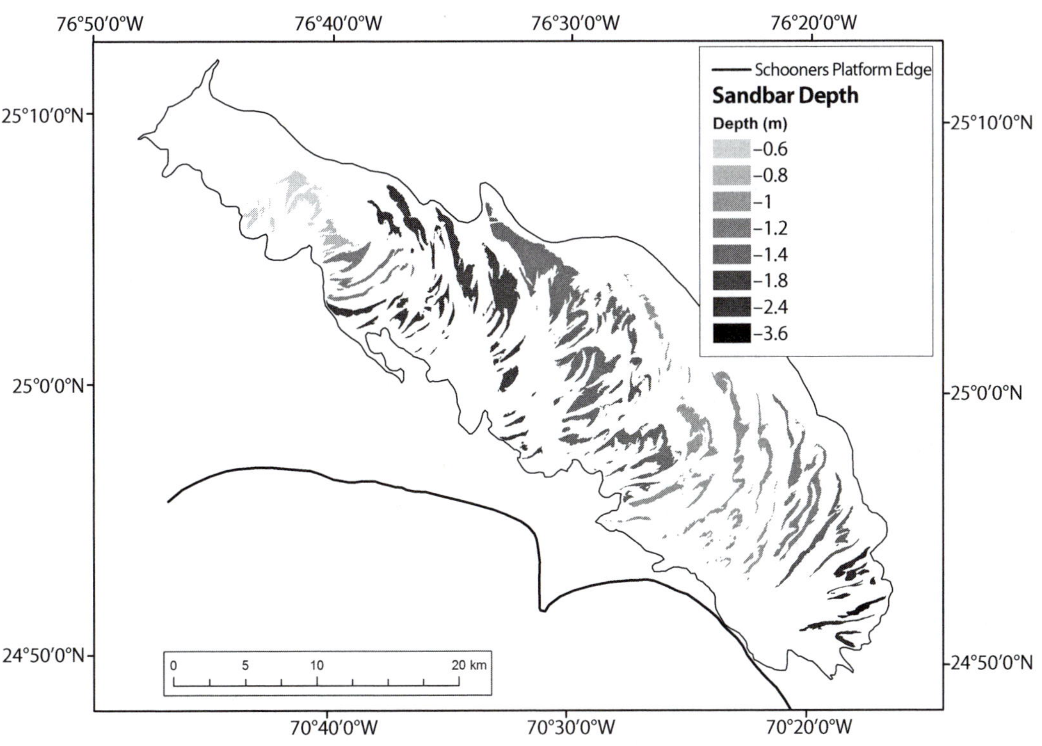

FIGURE 13-23 Spatial distribution of mean depth of sandbars at Schooners sand body. The minimum distance from each sandbar polygon to the platform edge is measured with GIS.

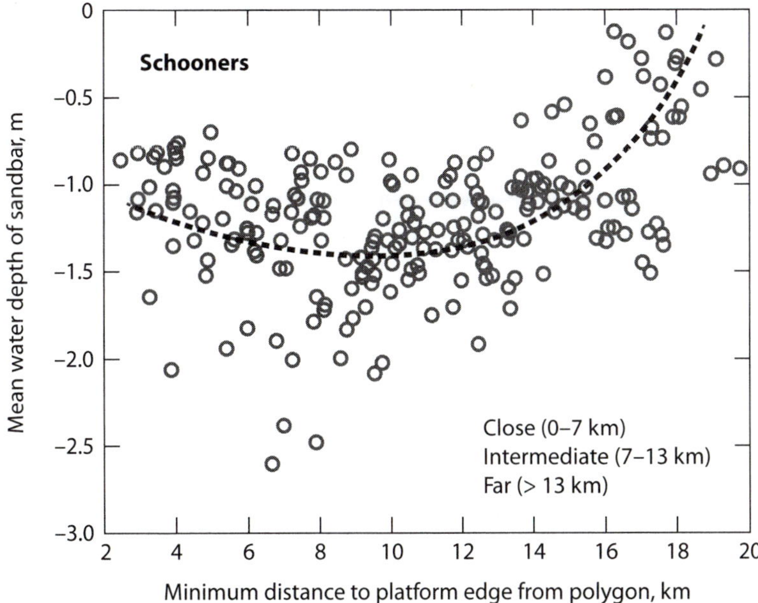

FIGURE 13-24 Plot showing minimum distance to platform edge from sandbar polygons (Figure 13-23) versus mean water depth of sandbar. Deeper sandbars tend to be closer to the platform edge.

2. **Case histories**: Case histories describe the exploration steps/procedures that lead to a mine discovery. In this chapter, Goldfield, Nevada, is a case history for copper vein deposits. The world-class Collahuasi Mine, Chile is a case history for porphyry copper discoveries. We describe these case histories as guides for future mineral prospectors.

MAPPING HYDROTHERMALLY ALTERED ROCKS

Many ore bodies are formed by a process called *hydrothermal alteration*. The ore bodies are deposited by hot aqueous fluids, called *hydrothermal solutions*, that invade, circulate through, and interact with the host rock, or *country rock*. During formation of the ore deposits, these solutions also interact chemically with the country rock to alter the mineral composition for considerable distances beyond the site of ore deposition. The hydrothermally altered country rocks contain distinctive assemblages of secondary minerals, called *alteration minerals*, that replace the original minerals. Alteration minerals commonly occur in distinct sequences, or *zones of hydrothermal alteration*, relative to the ore body. These zones are caused by changes in temperature, pressure, and chemistry of hydrothermal solutions at progressively greater distances from the ore body. At the time of ore deposition, the zones of altered country rock may not extend to the surface of the ground. Later uplift and erosion expose successively deeper alteration zones and eventually the ore body itself. Not all alteration is associated with ore bodies, and not all ore bodies are marked by alteration zones, but these zones are valuable indicators of possible deposits.

Hydrothermally altered country rocks that contain ore deposits are the target of mineral exploration. The two major categories of ore deposits are vein deposits and porphyry deposits; each with its own geology and pattern of hydrothermal alteration.

GOLDFIELD MINING DISTRICT, NEVADA

The Goldfield Mining District is located in southwest Nevada on Highway 95 midway between the cities of Las Vegas and Reno. Goldfield was noted for the richness of its ore. Over 4 million troy ounces (130,000 kg) of gold plus silver and copper were produced, largely in the boom period between 1903 and 1910. During peak production the town had a population of 15,000, but today the town of Goldfield is considerably smaller.

Geology and Ore Deposits

Figure 13-25 shows the geology and hydrothermal alteration of the district that was thoroughly mapped and analyzed by the USGS (Ashley, 1974, 1979). This information, plus the lack of extensive soil and vegetation cover, make Goldfield an excellent locality to develop and test methods for mineral exploration. The district is readily accessible and we have made a number of trips to check our image interpretations and collect samples.

Volcanism at Goldfield began in the Oligocene epoch with the eruption of rhyolite and quartz latite flows and the formation of a small caldera and ring-fracture system. Hydrothermal alteration and ore deposition occurred during a second period of volcanism in the Early Miocene epoch when the dacite and andesite flows that host the ore deposits were extruded. Heating associated with volcanic activity

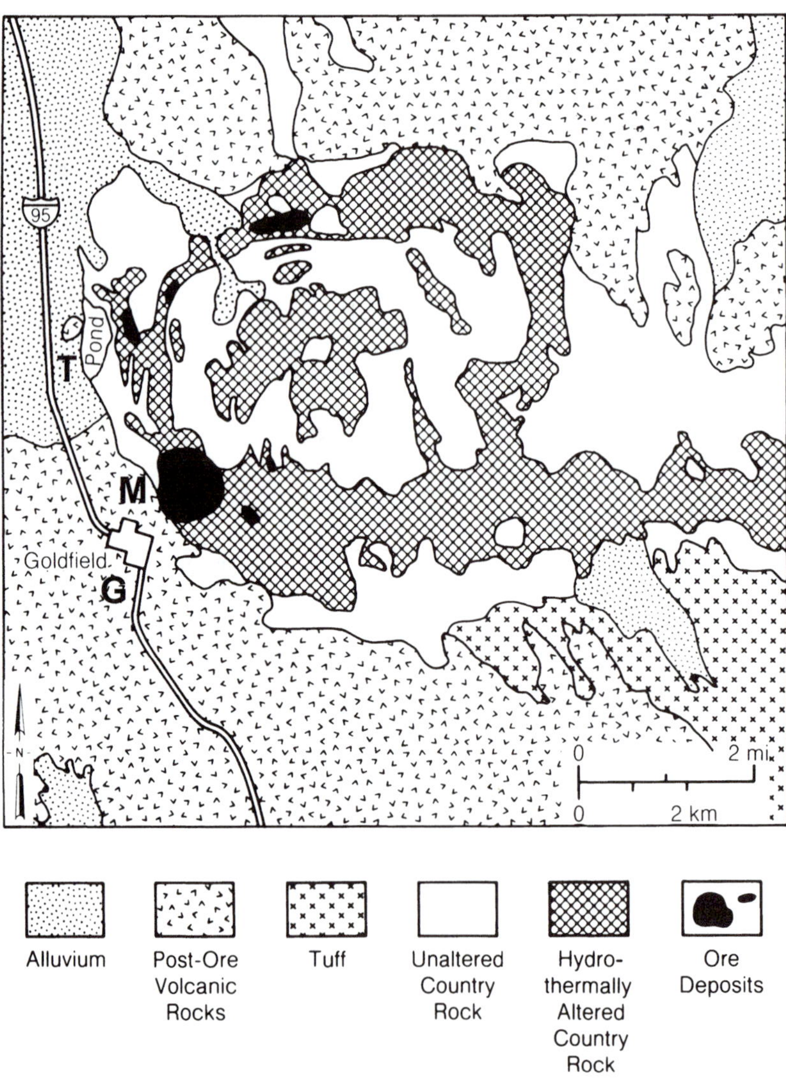

FIGURE 13-25 Map showing geology and hydrothermal alteration of the Goldfield Mining District, Nevada. T = tailings pond (dry) of the old Goldfield Consolidation mill where gold and silver were separated from the ore. M = main mining district. G = Town of Goldfield. Map is generalized from Ashley (1979, Figures 1 and 8).

at depth caused convective circulation of hot, acidic, hydrothermal solutions through the rocks. Fluid movement was concentrated in the fractures and faults of the ring-fracture system, where ore was deposited in silica veins and the adjacent volcanic host rocks were altered to clays and iron minerals. Following ore deposition, the area was covered by younger volcanic flows. Later doming and erosion exposed the older volcanic rim and center with altered rocks and ore deposits.

In Figure 13-25, altered rocks are cross-hatched and the unaltered country rocks are white. Alluvial deposits and post-ore volcanic rocks (those that formed after ore deposition) are shown. Altered rocks underlie approximately 40 km^2 of the area, but less than 2 km^2 of the altered rocks contain mineral deposits, which are shown in black. The major deposit is less than 1 km north of Goldfield. The large oval belt of altered rocks was controlled by a circular ring-fracture system, with a linear extension toward the east. The central patch of alteration shown in Figure 13-25 was controlled by closely spaced faults and fractures but is not mineralized.

Hydrothermal Alteration Model

Figure 13-26 is a cross section through a typical ore-bearing vein and the associated altered rocks at Goldfield. Alteration is concentrated at faults and fractures where hydrothermal solutions penetrated the country rocks and deposited gold-bearing silica veins. Intensity of alteration decreases laterally away from the vein and results in a sequence of alteration zones and subzones that are characterized by different secondary clay minerals. **Digital Image 13-1** 🌐 shows the alteration diagram plus images of the rock sequence from a highly altered vein to unaltered country rock. Characteristics of the alteration sequence are summarized as follows:

- **Silicic zone**: Predominantly quartz, which replaces the ground mass of host rock; subordinate amounts of alunite and kaolinite replace feldspar phenocrysts. Fresh rock from this zone, which is gray and resembles chert, is resistant to erosion and weathers to ridges with conspicuous dark coatings of desert varnish. Contact with adjacent argillic zones is sharp. All ore deposits occur in veins of the silicic zone, but not all veins contain ore.
- **Argillic zone**: Alteration minerals are predominantly clays. The argillic zone is divided into three subzones (Figure 13-26) based on the predominant clay species. Disseminated grains of pyrite (iron sulfide) are present that weather to secondary iron oxide minerals. The argillic rocks generally have a bleached appearance, but the secondary iron oxide minerals (limonite, jarosite, and goethite) may impart local patches of red, yellow, and brown to the outcrops. At Goldfield, no ore deposits occur in the argillic zone, but the presence of these altered rocks may be a clue to the occurrence of veins.
    - **Alunite-kaolinite subzone**: Relatively narrow and locally absent. In addition to alunite and kaolinite, some quartz is present.
    - **Illite-kaolinite subzone**: Marked by the occurrence of illite.
    - **Montmorillonite subzone**: Montmorillonite is the dominant clay mineral in this subzone, which has a pale yellow color due to jarosite, an iron sulfate mineral.
- **Propylitic zone**: These rocks represent a regional alteration of lower intensity than the argillic and silicic zones. Chlorite, calcite, and antigorite are typical minerals in this zone and impart a green or purple color to the rocks. Propylitic alteration is absent at numerous localities in Goldfield where the argillic zone grades directly into unaltered rocks.

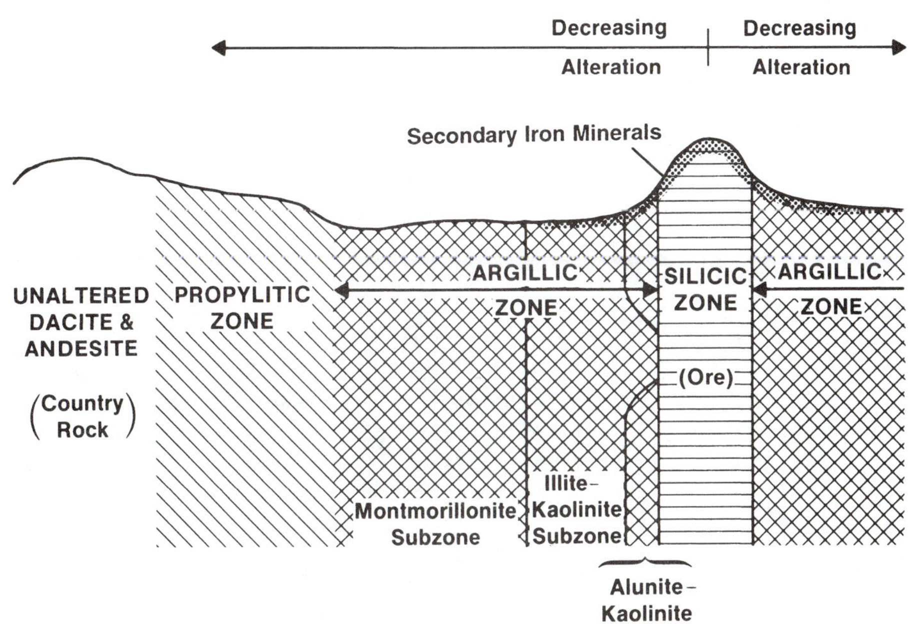

Figure 13-26 Cross-section model of typical hydrothermal alteration zones at the Goldfield Mining District. Dikes (or veins) of the silicic zone have a maximum width of a few meters. The argillic zone may extend for several tens of meters on each side of the silicic zone. Not to scale. After Ashley (1974) and Harvey and Vitaliano (1964).

- **Country rock**: Dacite and andesite. These unaltered gray volcanic rocks are hard and resistant to erosion. Country rocks, shown by the white areas in the geologic map (Figure 13-25), surround the outcrops of altered rocks.
- **Secondary iron minerals**: Pyrite (iron sulfide) is deposited along with ore minerals in and adjacent to the silicic zone. Weathering oxidizes pyrite to form limonite and hematite (iron oxide), which have distinctive red and brown signatures. Jarosite (iron sulfate) is also common in hydrothermally altered rocks and has a yellow signature.

The orderly sequence of subzones shown in Figure 13-26 may be complex in the field, because many veins are so closely spaced that subzones coalesce and overlap to form the altered outcrops shown in the map. Prospectors have long been aware of the association between hydrothermally altered rocks and ore deposits. Many mines were discovered by recognizing outcrops of altered rocks, followed by assays of rock samples. Prior to remote sensing, altered rocks were recognized by their appearance in the visible spectral bands. Today remote sensing and digital image processing provide valuable images of nonvisible bands for mineral exploration.

Recognizing Hydrothermal Alteration on Landsat Images

Plate 41A is our "arid land" color composite image (Chapter 3) of Landsat OLI bands (green-NIR-SWIR2 or 3-5-7 in B-G-R), which is superior to natural color or color IR images of the district. The town of Goldfield (G) is located along the left-central edge of the image and is characterized by a road network grid. A light blue patch directly northeast of the town of Goldfield is caused by the mine dumps and disturbed ground of the main mineralized area (M). A light blue patch 3 km north of Goldfield is the dry tailings pond (T) of the abandoned Columbia Mill, where gold was concentrated from the altered host rock. The tailings pond is a useful reference standard because it contains a concentration of altered rock material. Both the pond and the intensely mined mineralized area are man-made features that are absent in frontier exploration regions that have not been mined. The dark signatures in the margins of the image are volcanic rocks that are younger than the ore deposits and altered rocks.

In the 1970s Landsat MSS was the initial Landsat satellite. It recorded the following four spectral bands: 4 (green), 5 (red), 6 (NIR1), and 7 (NIR2). (A blue band was not recorded. Bands 1, 2, and 3 were assigned to an alternate imaging system that failed.) In a seminal and historic paper, Rowan and others (1974) used Landsat MSS images of Goldfield to introduce ratio images for detection of hydrothermally altered rocks. They recognized that ratios of certain MSS bands identified particular groups of alteration minerals and rock types. Color variations seen in the color-ratio composites represent spectral-reflectance differences.

Today Landsat TM and OLI cover a broader spectral range with more bands and with higher spectral resolution than the original MSS of some 55 years ago. Kneeper (2010) used three Landsat TM band ratios to generate a color composite image that revealed potential hydrothermally altered rocks. Sophisticated image processing of multispectral data for mineral exploration employs multiband ratios, measurement of shape and depth of spectral absorption features (continuum removal), comparison of actual with model band response, and library unmixing (D. Taranik, 2018, personal communication).

High Ratio Value Method

While attempting to use the traditional three-color ratio combination, Sabins recognized two limitations:

1. OLI band ratio 4/2 covers iron minerals and ratio 6/7 covers clays, which are the two relevant groups of alteration minerals. Adding a third ratio, such as 4/6, to complete a color image introduces 33% irrelevant data.
2. The full spectral range of each ratio (0 to 255) is conventionally used in a color ratio composite image. Ore deposits are typically hosted in the most strongly altered rocks, such as the top 10 to 15% of the spectral range of the iron and clay ratios. Using the full spectral ranges of the ratio images introduces up to 85% irrelevant data.

These limitations led to reservations about the three-color ratio image method. Therefore, we developed the high ratio value (HRV) method to address these reservations and to facilitate exploration for minerals. The HRV method resolves the reservations as follows:

1. Only the iron ratio and the clay ratio images are employed and
2. Only the highest range of values of the iron and clay ratio images are employed. The value range differs from image to image and is determined by the interpreter who interacts with the image to identify the optimum range.

Figure 13-27 is a flowchart illustrating two paths for the HRV method. The left path shows the sequence of images for recognizing iron. The right path shows the sequence for clay. The following discussion illustrates the complete sequence of creating a HRV alteration image using OLI images of the Goldfield Mining District. It begins with the implementation of the iron path on the left side of the HRV flowchart.

Iron Minerals on HRV 4/2 Ratio Images Iron minerals are a major indicator of hydrothermally altered rocks. Pyrite (iron sulfide) is a significant accessory mineral in many ore deposits, including the deposits at Goldfield. Where pyrite-bearing rocks are exposed at the surface, weathering converts the pyrite to secondary iron oxide minerals, such as goethite and hematite, which have hues of red or brown. Continued weathering disseminates the iron minerals over the surface, as shown in the alteration model (Figure 13-26), and the clay subzones typically acquire a subtle tan or yellow hue. Recognition and mapping of iron-stained outcrops is a vital step in mineral exploration and Goldfield is an excellent training site.

Figure 13-28A shows reflectance spectra for three common iron minerals that have similar spectra. The gray bars show band 2 (blue) (0.4 to 0.5 μm), which is absorbed, and band 4 (red) (0.6 to 0.7 μm), which is reflected by the iron minerals. This difference in reflectance explains why iron minerals have high ratio values and bright signatures on ratio images that are used to identify these minerals. Figures 13-29A and 13-29B show images for OLI band 2 (blue) and band 4 (red). Except for the bright signatures at the mining area and tailings pond, neither band shows a distinct iron signature.

The HRV path for iron begins with creating an OLI band ratio 4/2 image (Figure 13-29C), which shows bright signatures for iron minerals that agree with the alteration pattern shown in the geologic map (Figure 13-25). The image shows the full range of ratio values from 0 to 255. For the next

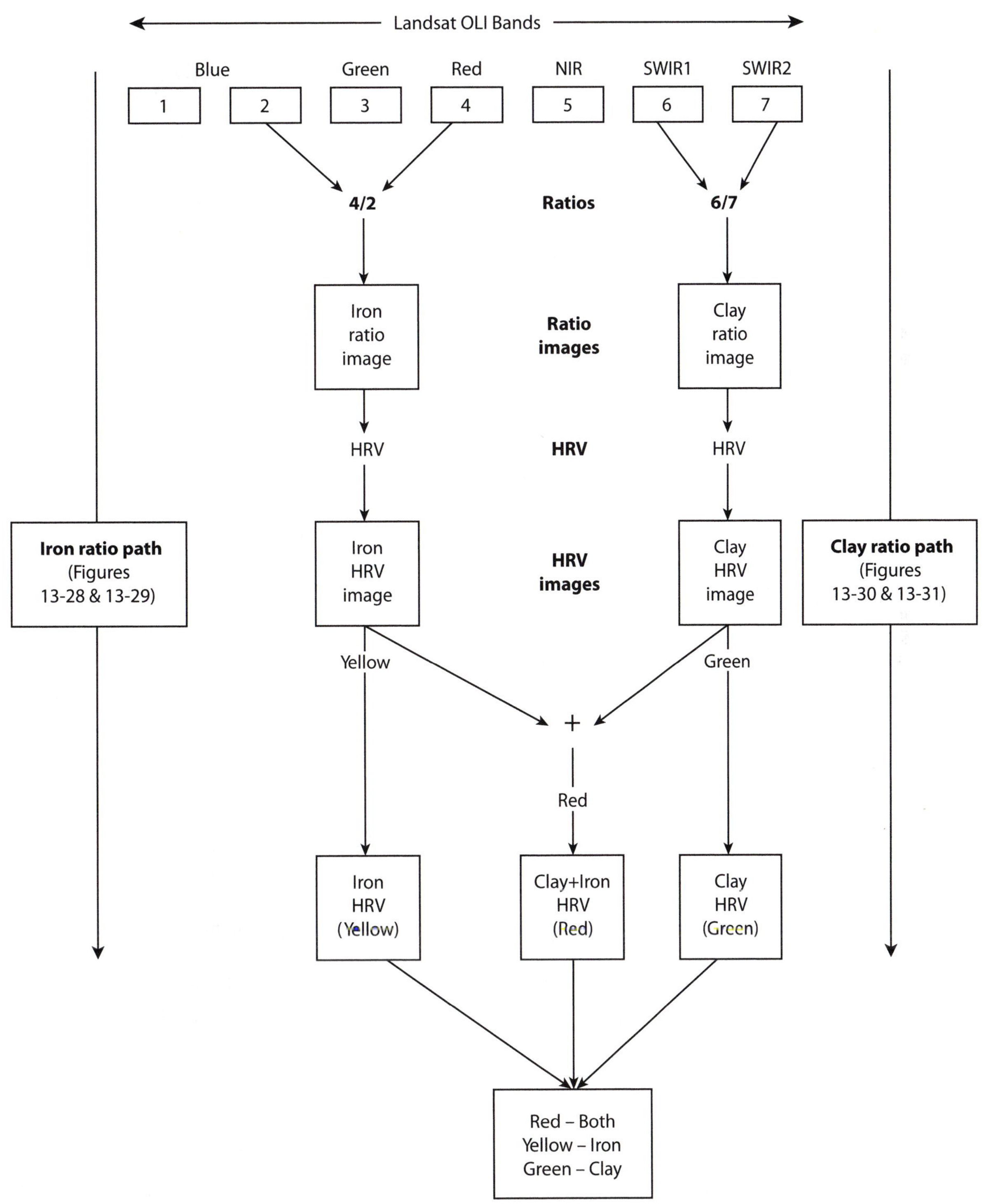

Figure 13-27 HRV flowchart for computing the sequence of iron (left path) and clay (right path) alteration images.

step in the HRV path we density sliced the 4/2 ratio image to show the highest iron values in white. This is an interactive process. We consulted with the available information to ensure that any training localities, such as known mines, are included within the density slice. At Goldfield, for example, we ensured that the tailings pond and mining area were included within the density slice. The histogram (Figure 13-28B) shows the full range of 4/2 ratio values. In order to use only the highest values, we selected values > 150, which are shown with a bright signature in the HRV density slice image (Figure 13-29D).

Clay Minerals on HRV 6/7 Ratio Images Figure 13-30A shows reflectance spectra for three common clay minerals and alunite, which have similar spectra. The gray bars show SWIR1 band 6 (1.57 to 1.65 µm), which is reflected by clay, and SWIR2 band 7 (2.11 to 2.29 µm), which is absorbed by clay. This difference in reflectance explains why clays have high ratio values and bright signatures on SWIR1/SWIR2 (OLI band 6/7) ratio images. Figures 13-31A and 13-31B show images for SWIR1 and SWIR2. Except for the bright signatures at the mining area and tailings pond, neither band shows a clay signature.

The HRV path for clay begins with creating a ratio OLI 6/7 image (Figure 13-31C), which shows bright signatures for clay minerals that agree with the alteration pattern shown in the geologic map (Figure 13-25). The image shows the full range of ratio values from 0 to 255. For the next step in the HRV clay path, we density sliced the 6/7 ratio image to show the highest clay values in white. This is an interactive process. We consulted with the available information to ensure that any training areas were included within the density slice. At Goldfield we ensured that the tailings pond and the mining area were included within the density slice. The histogram (Figure 13-30B) shows the full range of 6/7 ratio values. We selected values > 145, which constitute the HRV density slice that is shown in Figure 13-31D.

HRV Alteration Color Image The final step is to create a HRV alteration color image by combining in color the HRV density slices for iron (Figure 3-29D) and for clay (Figure 13-31D). We use the mask method to produce a color HRV alteration image that combines only the high values of the iron and clay ratios. The HRV density slice for iron is plotted as a mask that shows a single hue of yellow. The HRV density slice for clay is plotted as a mask that shows a single hue of green. The yellow and green masks are superimposed to generate a new image where pixels with values for both iron and clay (yellow and green pixels overlap) are plotted in red. Plate 41B shows the resulting HRV alteration color image overlain on the Goldfield geologic map. Both the main mining area (M) and the tailings pond (T) are shown in red. Red signatures are common in the southern half of the alteration band and the eastern extension. Old diggings in these areas suggest that ore may have been produced there, but not in the bonanza quantities of the main mining area.

The yellow, green, and red pixels are essentially confined to the oval rim of altered rocks and their eastern extension. The unmineralized central patch of alteration shows only a few scattered red and green patches. Although Goldfield has been extensively explored, it may be worthwhile to take a new look at the areas with red signatures, unless they are currently claimed. We recommend this with the prospector's eternal and essential optimism, but note the following precaution. Ensure that the areas are not claimed. Claim holders are known to be inhospitable to strangers messing with their claim.

A. Laboratory spectra.

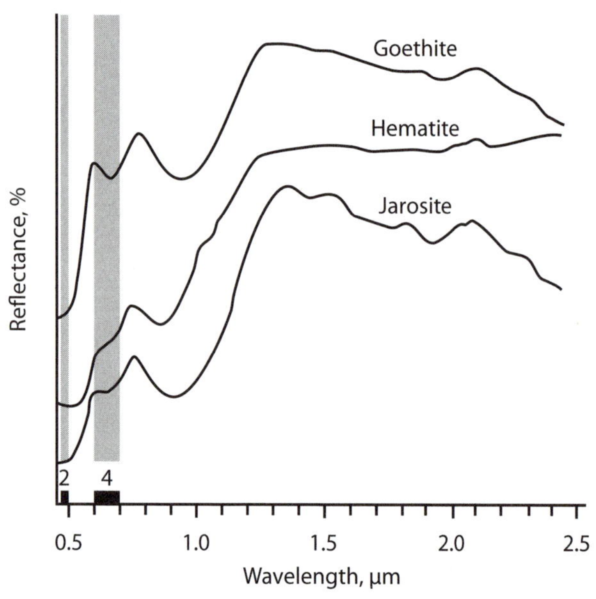

B. Histogram for the OLI 4/2 ratio image. Values > 150 are used in the iron HRV alteration image.

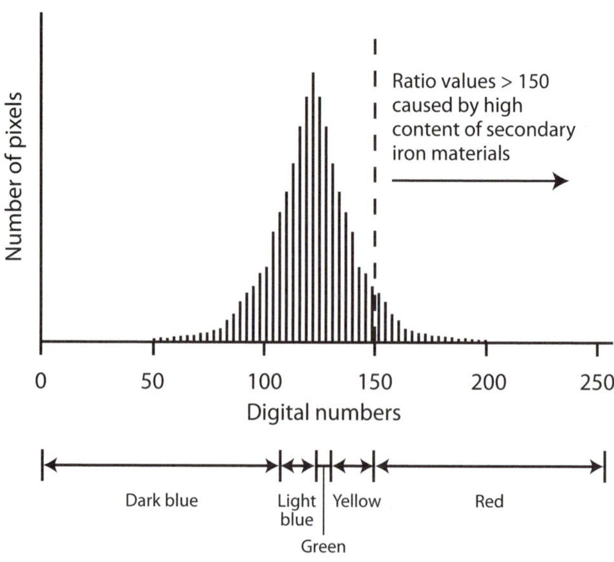

Figure 13-28 Spectra and histogram for iron minerals in the HRV method.

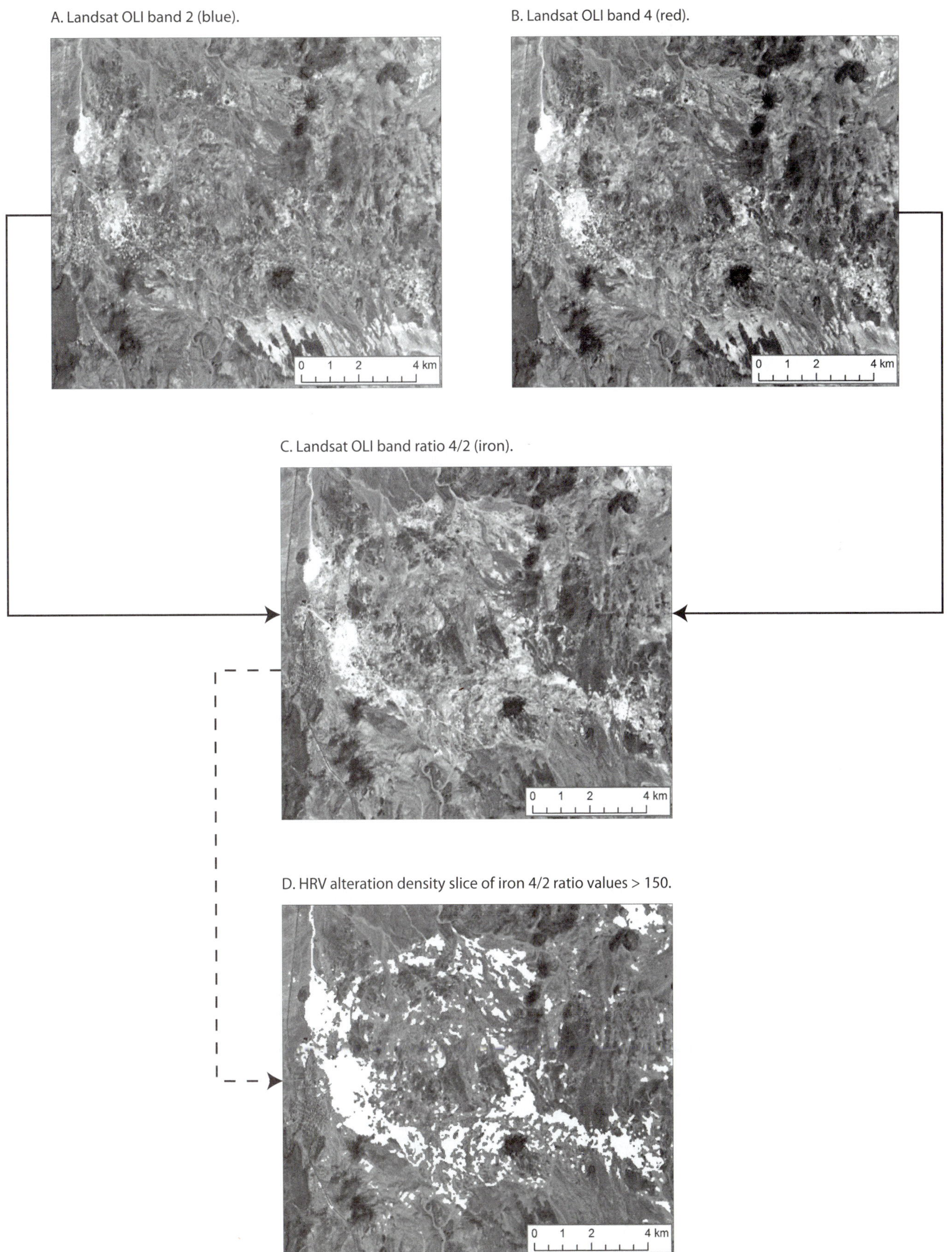

A. Landsat OLI band 2 (blue).
B. Landsat OLI band 4 (red).
C. Landsat OLI band ratio 4/2 (iron).
D. HRV alteration density slice of iron 4/2 ratio values > 150.
0 1 2 4 km

FIGURE 13-29 Sequence of images used to create an iron HRV alteration image of the Goldfield Mining District. The final image (13-29D) is shown in yellow in Plate 41B. Figure 13-28 shows iron spectra and a histogram for an OLI ratio 4/2 image.

USGS National Map of Surficial Mineralogy

Rockwell and others (2015) of the USGS developed maps of exposed surface mineral groups derived from automated spectral analysis of Landsat and ASTER data for the United States and its territories. These maps may have potential for (1) discovering mineral deposits and (2) recognizing environmental effects of mining and/or unmined, hydrothermally altered rocks. The maps over the conterminous United States are available online (minerals.cr.usgs.gov/cmerwebmap/usminmap.html). The underlying map services can be accessed using ArcMap for integration with other geospatial data.

Plate 41C is the USGS surficial mineralogy map of Goldfield with a simplified legend showing iron content. Note that the iron categories also include clay. The major and moderate iron categories (red and orange signatures) correlate with the iron (yellow) and iron plus clay (red) categories of the HRV alteration map (Plate 41B). The tailings pond and main mining area are shown in red.

Goldfield Regional Spectra

To bridge the gap between laboratory spectra and outcrop observations, Rowan and others (1977) used a portable spectrometer in the field to record spectra of several hundred representative outcrops of altered and unaltered rocks at Goldfield. Figure 13-32 summarizes their results as average reflectance curves for altered and for unaltered outcrops. Because of the averaging effect, these curves lack the fine spectral detail of laboratory spectra (Figures 13-28A and 13-30A), but the two curves clearly show the differences between altered and unaltered rocks. The altered rocks have distinctly lower reflectance in TM band 7 than in TM band 5, which results in a low value for TM clay band ratio 5/7. The TM band ratio 5/7 is essentially unity for unaltered rocks. Iron minerals in altered rocks cause high red reflectance (TM band 3). The TM iron band ratio 3/1 is higher for altered rocks than for unaltered rocks.

AVIRIS HYPERSPECTRAL SYSTEM AND GOLDFIELD ALTERATION

AVIRIS is operated by JPL and is used for experimental projects by a wide range of investigators. Typical ground resolution cells measure 20 by 20 m. The system records 224 spectral bands in the visible and SWIR spectral regions. Figure 13-33 shows the spectral range and the position of the 50 individual bands in the SWIR spectral region. In areas of complex geology, such as Goldfield, Nevada, the 400 m^2 of a cell includes a range of different minerals. The resulting pixel is called a *mixed pixel* because its spectrum is a mixture of the spectra for the different minerals that crop out in the ground resolution cell. For processing purposes these individual mineral species are called *spectral endmembers*. Digital *unmixing* programs are used to derive the spectra of the endmembers for each mixed pixel. For each endmember, an *endmember abundance* image is derived that shows the relative abundance of the endmember.

Figure 13-34 shows unmixed AVIRIS endmember images for kaolinite, illite, and alunite in the western two-thirds of Goldfield. These images were digitally processed at Analytical Imaging and Geophysics, LLC using a spectral unmixing program of Boardman (1993) and Kruse (1999). Note the white signature of the tailings pond, which means that all three endmembers were mixed in the host rock that

A. Laboratory spectra.

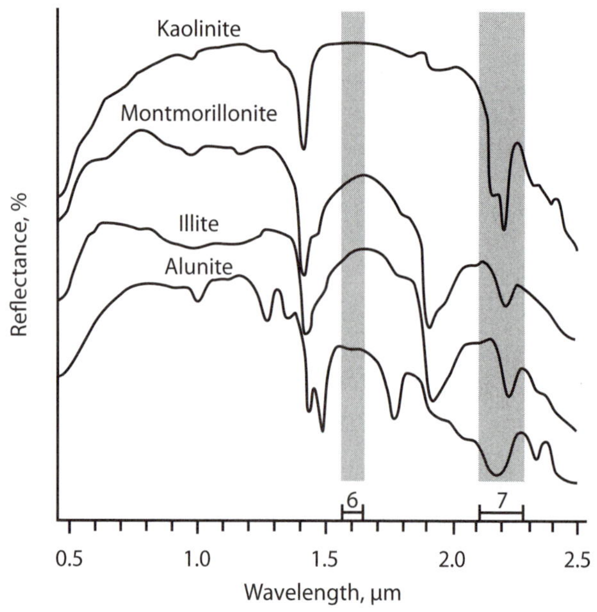

B. Histogram for an OLI 6/7 ratio image.

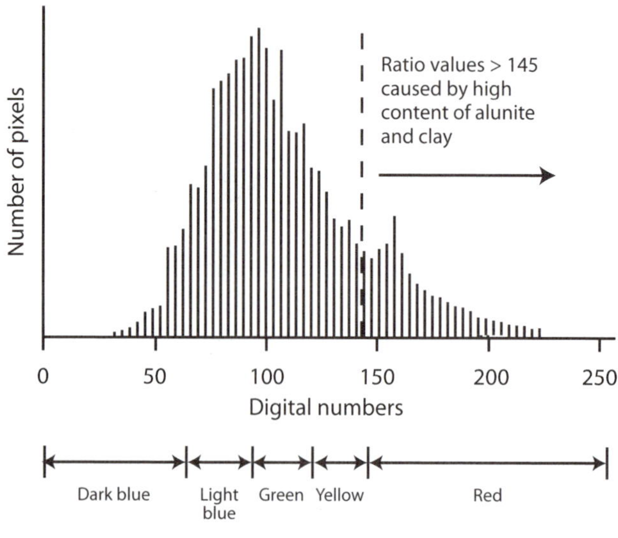

FIGURE 13-30 Spectra and histogram for clay minerals in the HRV method. Values >145 are used in clay HRV alteration image.

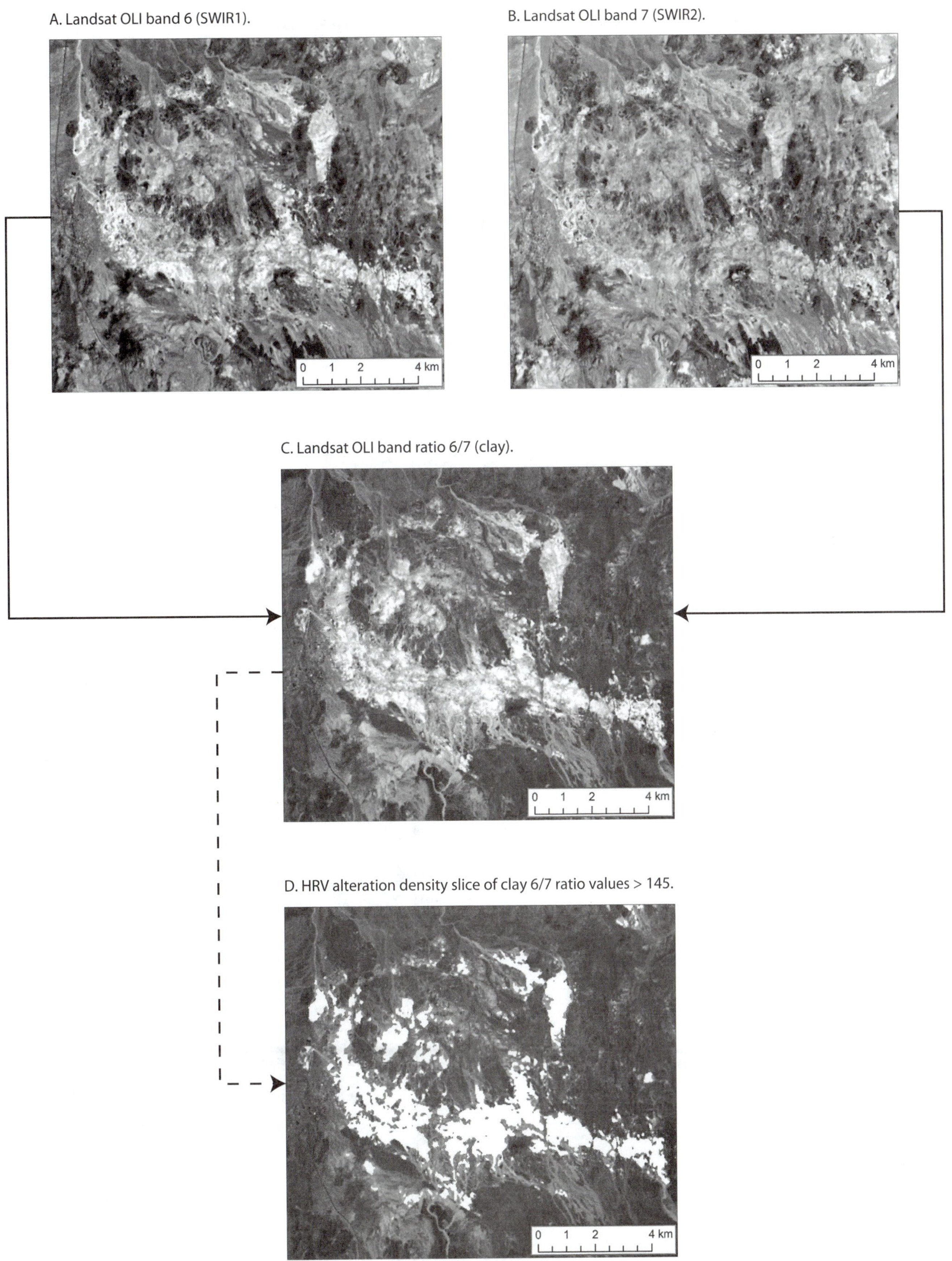

Figure 13-31 Sequence of images used to create a clay HRV alteration image of the Goldfield Mining District. The final image is shown in green in Plate 41B. Figure 13-30 shows clay spectra and a histogram for a 6/7 ratio image.

was processed at the mill and dumped in the pond. For each spectral endmember, the brightness signature for each pixel is proportional to the abundance of that endmember. Figure 13-34A is the endmember abundance image for kaolinite, which is the most abundant clay mineral at Goldfield. The kaolinite image accurately shows the pattern of hydrothermal alteration in the geologic map for this area (Figure 13-34D). Figure 13-34B is the endmember abundance image for illite, which is less abundant than kaolinite. The highest concentration of illite occurs in the central area of altered rocks that are not mineralized. Figure 13-34C is the endmember abundance image for alunite, which is the least abundant mineral. Patches of alunite occur in the south and west portions of the alteration belt that surrounds the district. A few patches of alunite also occur within the central area of alteration.

Plate 41D is a color composite image that shows the endmember abundance of illite in blue, alunite in green, and kaolinite in red. The black-and-white base is AVIRIS band 30 (visible red). The primary colors show areas with high concentrations of the assigned clay mineral. Other colors indicate mixtures of endmember minerals. The white signature of the tailings pond accurately shows that it contains a mixture of the three minerals in the legend.

The AVIRIS color image (Plate 41D) covers the western two-thirds of the Landsat images shown in Plate 41A–C. It is instructive to compare these images. The green signatures of the HRV alteration image (Plate 41B) show the aggregate distribution of clays and alunite. The colors of the AVIRIS image show the distribution of individual alteration minerals. In summary, Landsat images show the broad pattern of hydrothermal alteration and AVIRIS images show the distribution of the individual alteration minerals.

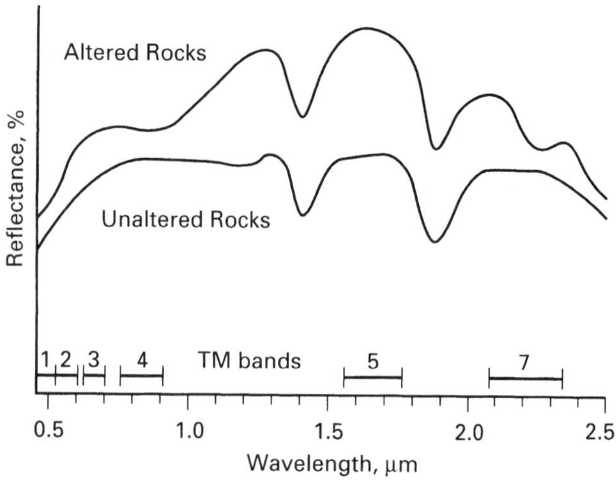

FIGURE 13-32 Field spectra (averaged) of altered and unaltered rocks at the Goldfield Mining District. The TM clay ratio 5/7 is essentially 1.0 for unaltered rocks, but is a high value for altered rocks with their high clay content. From Rowan and others (1977, Figure 2A).

PORPHYRY COPPER DEPOSITS

Most of the world's copper is mined from *porphyry deposits*, which occur in a different geologic environment from vein deposits such as Goldfield. Porphyry deposits are named for the *porphyritic* texture of the granitic host rock, in

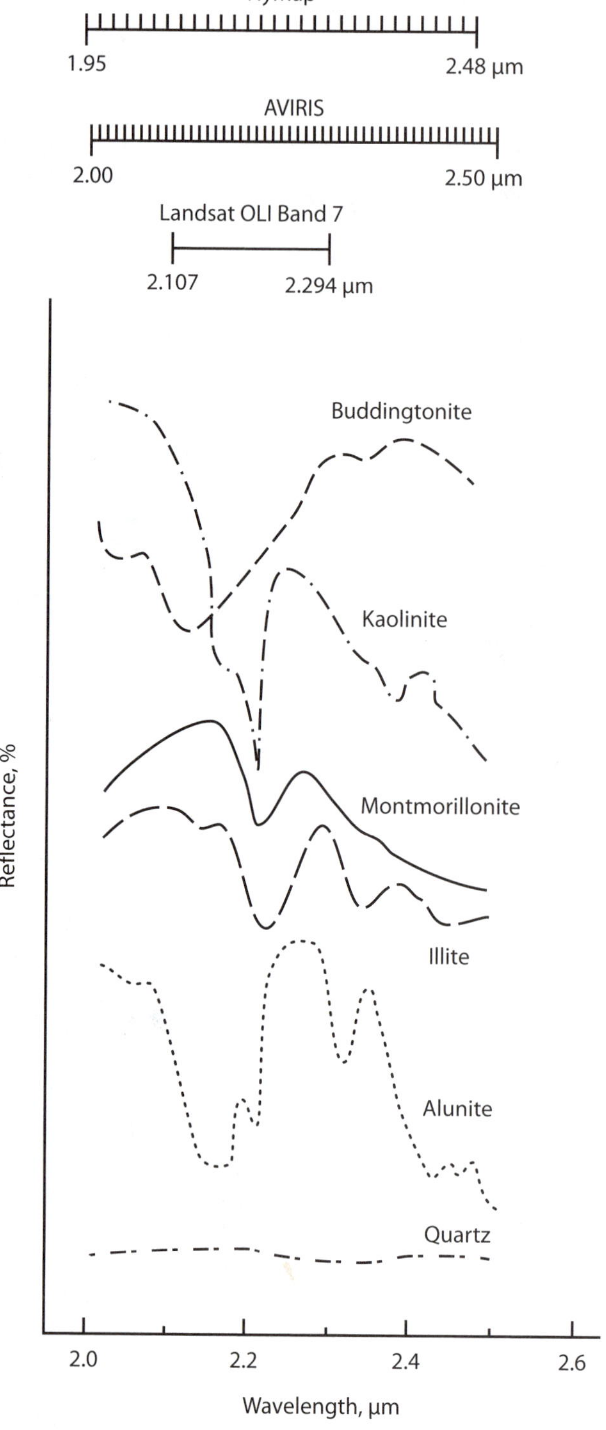

FIGURE 13-33 Laboratory spectra of alteration minerals in the SWIR spectral region. Band passes of AVIRIS and HyMap hyperspectral systems and Landsat OLI band 7 are shown.

A. Kaolinite image.

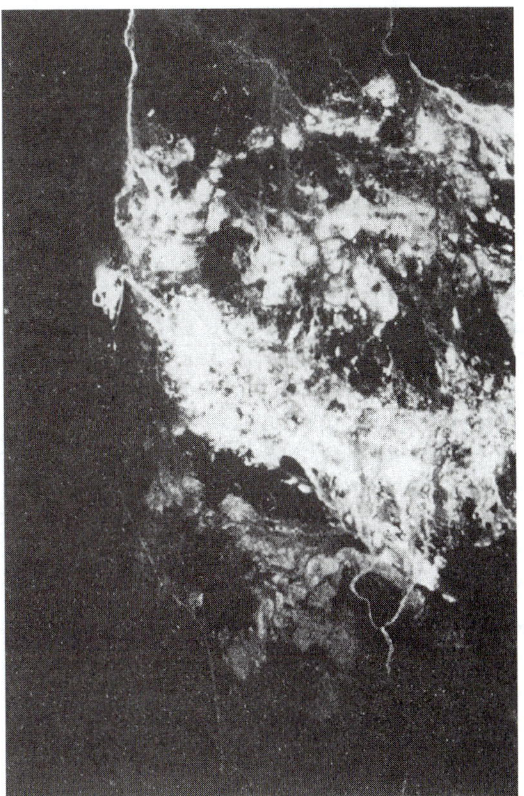

B. Illite image.

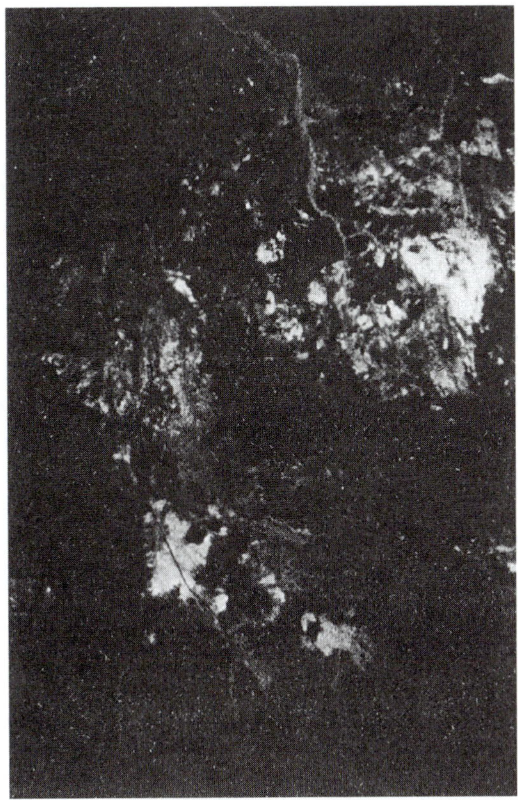

C. Alunite image.

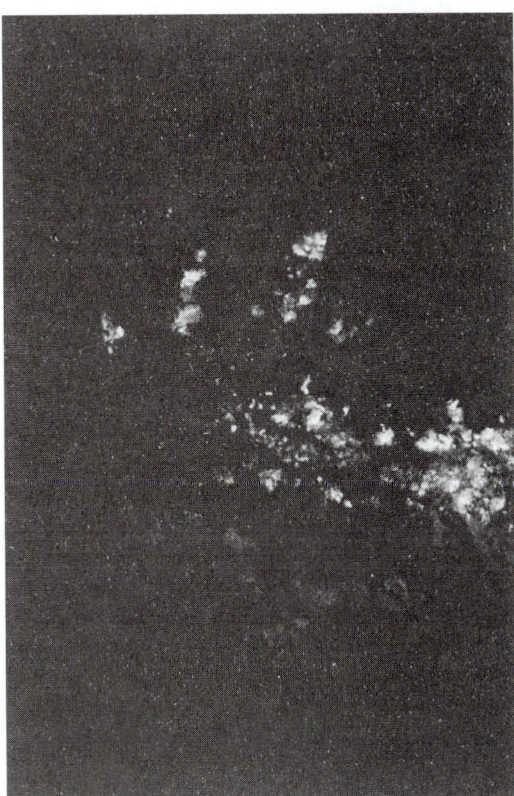

D. Alteration map. Symbols are explained in Figure 13-25.

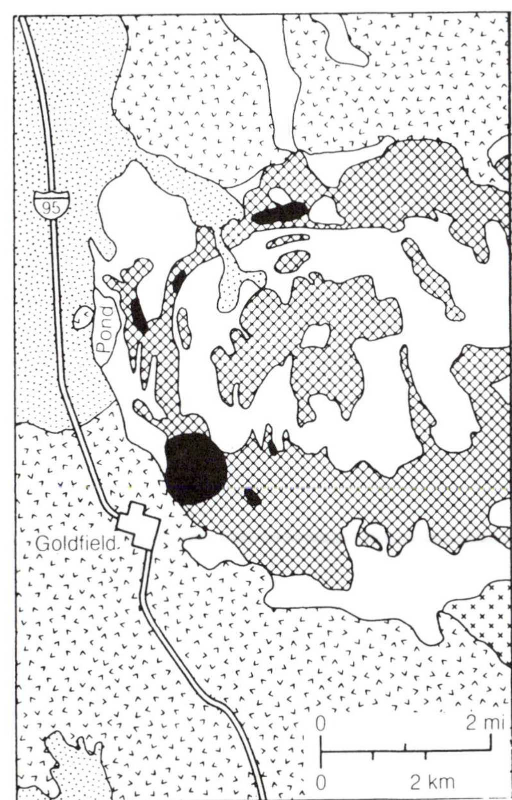

Figure 13-34 Endmember abundance images derived from AVIRIS data of the Goldfield Mining District. Brighter signatures represent higher concentrations of the mineral. See Plate 41D for a color composite image. Courtesy F. A. Kruse, Analytical Imaging and Geophysics, LLC, Boulder Colorado.

which large feldspar crystals (phenocrysts) are surrounded by a fine-grained matrix of quartz and other minerals. Granitic porphyries occur as plugs (or *stocks*) up to several kilometers in diameter that intrude the older country rock and reach within several kilometers of the surface. Intensive fracturing of the porphyry and country rock occurs during the emplacement and cooling of the stock. Heat from the magma body causes convective circulation of hydrothermal fluids through the fracture system to form ore deposits and alter the porphyry stocks and adjacent country rocks.

Porphyry Copper Alteration Model

Figure 13-35 is a model of hydrothermal alteration of porphyry copper deposits that was developed by Lowell and Guilbert (1970) and is widely used in exploration projects. The most intense alteration occurs in the core of the porphyry body and diminishes radially outward in a series of zones, as described below.

- **Potassic zone**: This innermost zone contains the most intensely altered rocks in the core of the stock. Characteristic minerals are quartz, sericite, biotite, and potassium feldspar.
- **Phyllic zone**: Quartz, sericite, and pyrite are common.
- **Ore zone**: This zone consists of disseminated grains of chalcopyrite, molybdenite, pyrite, and other metal sulfides. Much of the ore occurs in a cylindrical *ore shell* near the boundary between the potassic and phyllic zones. Copper typically constitutes 1% or less of the rock, but the large volume of ore is suitable for open pit mining. Where the ore zone is exposed by erosion, pyrite oxidizes to form a red to brown limonitic crust called a *gossan*. Gossans can be useful indicators of underlying mineral deposits, although not all gossans are associated with ore deposits.
- **Argillic zone**: Quartz, kaolinite, and montmorillonite are characteristic minerals of the argillic zone in porphyry deposits, just as they are associated with the argillic zone at Goldfield and elsewhere.
- **Propylitic zone**: Epidote, calcite, and chlorite occur in these weakly altered rocks. Propylitic alteration may be of broad extent and have little significance for ore exploration.
- **Country rock**: Unaltered rock.

Few porphyry deposits have the symmetry and completeness of the model in Figure 13-35. Structural deformation, erosion, and deposition commonly conceal large portions of the system. Nevertheless, recognition of small patches

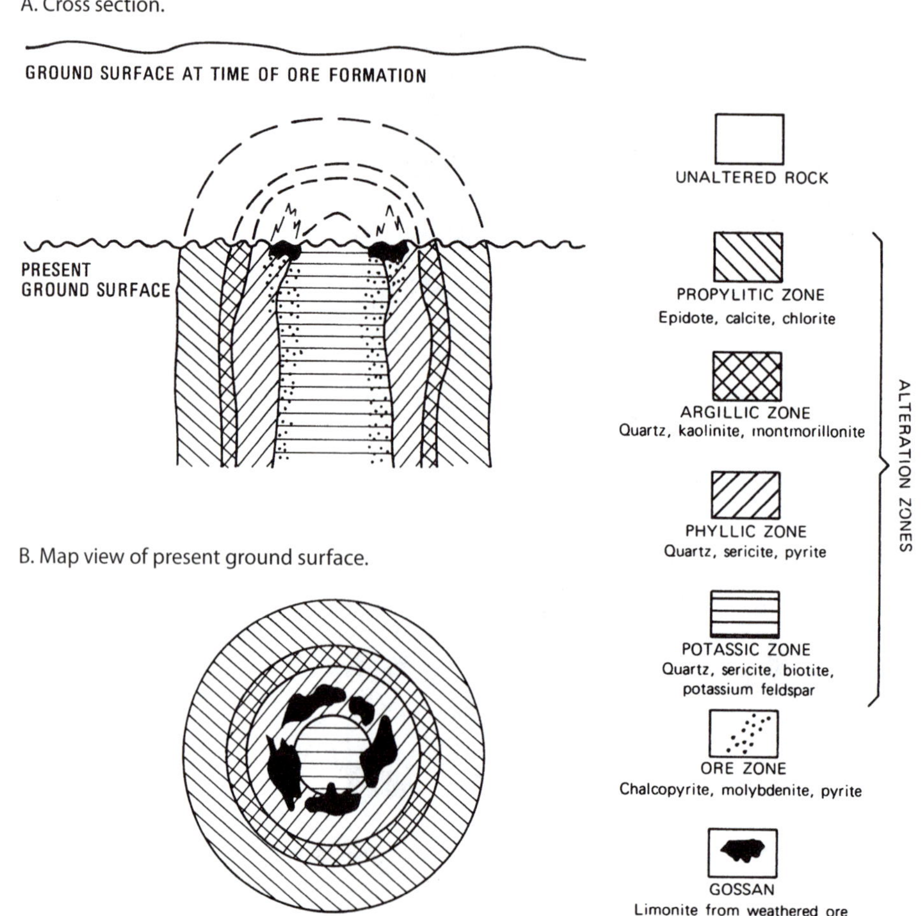

FIGURE 13-35 Model of hydrothermal alteration zones associated with porphyry copper deposits. From Lowell and Guilbert (1970, Figure 3).

of altered rock on remote sensing images can be a valuable exploration clue. In the early 1980s, NASA and the Geosat Committee evaluated satellite and airborne multispectral images for porphyry copper deposits in southern Arizona. At the Silver Bell Mining District, Abrams and Brown (1985) used color ratio images to separate the phyllic and potassic alteration zones from the argillic and propylitic zones. A supervised classification map defined the outcrops of altered rocks.

Collahuasi Copper Mine, Chile

The Collahuasi Mine is located in northern Chile, 180 km southeast of the coastal city of Iquique. The mine lies within a north-trending belt of porphyry copper deposits and includes the Rosario/Collahuasi and Ujina hydrothermal systems and deposits. Mineral production in the district began in the late 1800s when copper was mined from veins located at Rosario (Figure 13-36). During the 1930s, these veins were Chile's third largest producer of copper. Modern exploration began in 1976 when a joint venture between Superior Oil Company and Falconbridge Limited acquired the Rosario/Collahuasi properties. In 1985, ownership of the district changed to a three-way partnership of Falconbridge Limited, Shell Oil Company, and Chevron Corporation. From 1985 to 1990 the partners explored for additional veins in the Rosario system with limited success. However, because of their research results at the Goldfield Mining District, Chevron recommended a Landsat HRV analysis of Rosario and the surrounding Collahuasi concession that was held by the partnership (Dick and others, 1993). Figure 13-36 is a geologic map that shows the distribution of the Macata, Capella, and Collahuasi Formations, which are of Jurassic and Cretaceous age. These country rocks are intruded by granitic stocks of Late Cretaceous to Early Tertiary age that are hosts for the porphyry copper deposits.

HRV Alteration Images

The Remote Sensing Research Group of Chevron used the HRV alteration method to process a TM image of the Rosario/Collahuasi area and the Ujina area to the east, which was also part of the partners' concession. Northern Chile is well suited for this project because vegetation, soils, and clouds are virtually absent in this arid environment

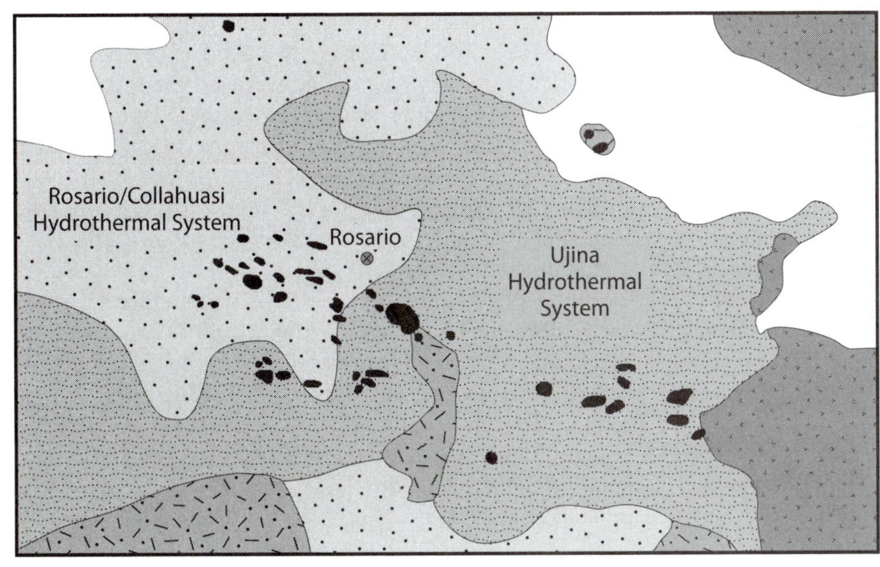

Figure 13-36 Geologic map of the Collahuasi Mine, Chile. Black signatures are HRV alteration features that led to the discovery of two world-class copper porphyry deposits: Rosario/Collahuasi in the northwest and Ujina in the southeast. Geology modified from Vergara (1978a, 1978b).

of the high Andes Mountains. Plate 42A is a Landsat TM natural color image acquired in 1986. We used the HRV method on early 1990s Landsat images to produce the alteration features that are outlined in red on the 1986 image. There is no evidence of the porphyry deposits inside or around the red outlines. The red alteration features form two distinct clusters:

1. A western cluster almost 8 km in diameter that is called the Rosario/Collahuasi porphyry. The old Rosario vein deposits, with a diameter of 1.5 km, occupy only a small portion of the north margin of the HRV cluster. The remainder of the cluster was largely unexplored.
2. A second smaller cluster of HRV alteration features is located 8 km to the east in the Ujina area, which was essentially unexplored for minerals.

Based on the strong alteration shown by the HRV images, the partners decided to conduct geophysical surveys.

Geophysical Surveys

The geophysical surveys recorded subsurface properties of the rocks at the Rosario/Collahuasi and Ujina areas (Dick and others, 1993). The entire district was covered by a helicopter-borne aeromagnetic survey that mapped subsurface geologic structures and the distribution of magnetic minerals. The Ujina system has a circular rim of high magnetic values that is interpreted as an ore shell within the granitic pluton similar to that shown in the porphyry model (Figure 13-35). A ground-based survey measured resistivity of the rocks. Unmineralized rocks typically have high resistivity values. Metallic minerals, such as copper, have low resistivity. The porphyry model shows that porphyry copper deposits have circular rims with copper minerals, which lower the resistivity. Figure 13-37 is a contour map of the resistivity survey that covers the same area shown in the satellite image (Plate 42A) and geologic map (Figure 13-36). High resistivity values are shown by H; the very important low values are shown by L.

Results of the resistivity survey are outstanding. Circular patterns of low resistivity contours match the HRV alteration features that define both the Rosario/Collahuasi and Ujina hydrothermal systems. These patterns are analogous to those of classic porphyry copper deposits. At Rosario/Collahuasi, the resistivity pattern is 5 km in diameter. The lowest values form a marginal rim that probably represents the ore shell of the porphyry model. The very low overall resistivity of the Collahuasi system is interpreted as an extensive development of veinlet mineralization within the porphyry igneous stock.

The Ujina hydrothermal system has a circular pattern of low resistivity contours 3 km in diameter. The geologic map (Figure 13-36) shows outcrops of the Ujina Tuff, which postdates the hydrothermal activity and covers the eastern portion of the Ujina resistivity feature. The Landsat features coincide with the exposed western portion of the feature.

Core Drilling and Ore Discoveries

Core holes were drilled to test the two hydrothermal systems outlined by the remote sensing and geophysical studies. The first drill holes tested the low resistivity values on the north rim of the Rosario/Collahuasi system, and found zones of structurally controlled copper mineralization. Drilling at Ujina confirmed a second major new porphyry copper deposit. By early 1993, drilling had outlined over 150 million tons of enriched ore with a grade of 1.8% copper. Production started in late 1998 and should continue for 45 years. Total current mineable reserves are 14 million tons of copper with a value of $36.4 billion at recent copper prices. Remote sensing played a key role in discovering these two valuable properties; the remote sensing work that contributed so much to the increased value of the property cost less than $50,000 of intra-Chevron charges.

Plate 42B is a 2016 Landsat image that shows the two major open-pit mines that were developed following the HRV alteration images and the resistivity survey. The two clusters of red HRV alteration features that we interpreted in the early 1990s encompass the open pits at Rosario/Collahuasi and Ujina, which demonstrates the effectiveness of this mineral exploration method. Advanced satellite-based remote sensing continues around the Rosario/Collahuasi and Ujina mines. ASTER (five SWIR bands with 30-m pixels) and WorldView-3 (eight SWIR bands with 7.5-m pixels) multispectral data have been processed to spectrally measure

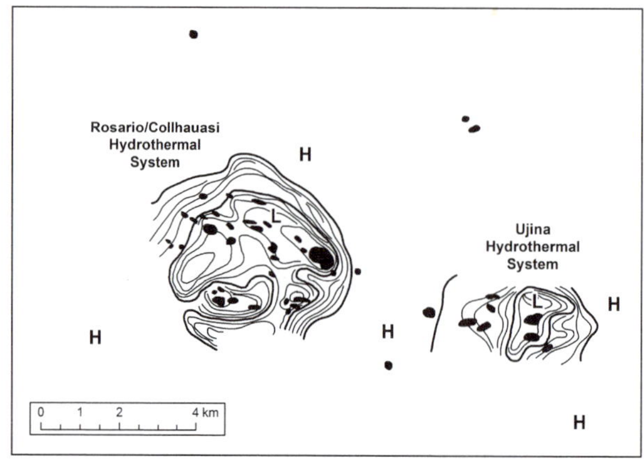

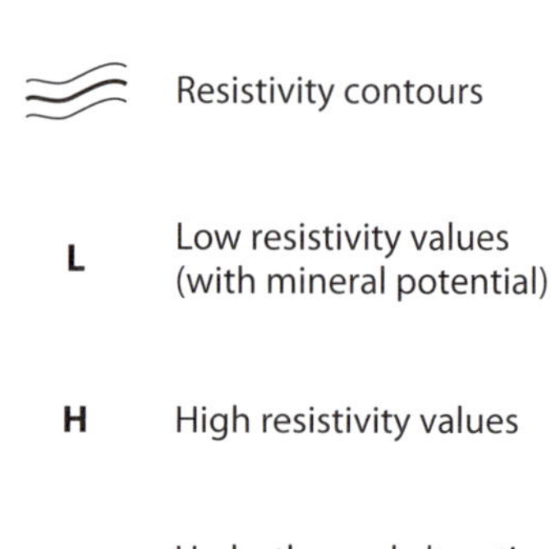

FIGURE 13-37 Contour map of resistivity values of the Rosario/Collahuasi and Ujina hydrothermal systems. L = low resistivity values due to copper minerals. H = high resistivity values due to unmineralized terrain. Black signatures are HRV alteration features. This map confirmed the HRV alteration map.

the diagnostic absorption bands associated with clays, carbonates, silica, and iron oxides. Maps of minerals associated with hydrothermal alteration have been generated that identify several areas recommended for future mining activity (Lattus, 2017).

Balkhab Valley, Afghanistan

Between 2010 and 2013, Sabins and Ellis were tasked by the US Department of Defense Task Force for Business and Stability Operations (TFBSO) to explore for mineral prospects in Afghanistan using remote sensing. One objective of the TFBSO was for Afghans to use the remote sensing prospects to locate and develop mines that would provide jobs and capital for developing the country. The mines could combat unemployment and poverty, which are major recruiting environments for terrorist groups. In cooperation with the USGS, we selected 16 sites that represented a range of mineral commodities and deposit types. We processed data acquired by Landsat TM and OLI, ASTER TIR, SRTM DEMs, HyMap hyperspectral, IKONOS, Quickbird, and WorldView-2 systems. From the 16 sites we interpreted a total of 416 "hard rock" exploration prospects that included volcanogenic massive sulfide (VMS), gold veins, chromite, copper porphyry, and pegmatites. An unpublished report (Sabins and Ellis, 2013b) summarizes the 17 individual reports. The USGS (Peters and others, 2011) published several of our prospects. Some sites were covered by more than one report as additional types of images became available. The VMS terrain in the Balkhab Valley is an example of one of our projects.

Volcanogenic Massive Sulfide Prospects

The Balkhab Valley in north-central Afghanistan is a VMS terrain where we located mineral prospects. VMS deposits form at or near the seafloor where circulating hydrothermal fluids are quenched by mixing with seawater and deposit copper sulfide and other metallic minerals (Shanks and Koski, 2010). Stratiform (layered) deposits are common as are alteration minerals, similar to those of porphyry deposits (Figure 13-35). In Phase I of the Balkhab Valley Project we processed and interpreted Landsat TM images (TBFSO, 2011). Phase II employed airborne HyMap hyperspectral and ASTER TIR images (Sabins and others, 2012).

Plate 43A is an enhanced color TM image (bands 2-4-7 as B-G-R) of the valley, which is cut by the northeast-flowing Balkhab River and its tributaries. A narrow thread of green vegetation marks the course of the river in the center of the valley. The predominantly purple rocks that border the river are the VMS sequence of Paleozoic age. The white lines mark the unconformable contact between the contorted VMS sequence and the overlying, essentially horizontal strata of Mesozoic age with a sparse cover of vegetation.

Plate 43B is a Landsat TM HRV alteration image of the valley with the standard signatures: yellow = iron, green = clay, red = both iron and clay. Altered rocks are not uniformly distributed in the valley but are concentrated in distinct clusters and trends. In the HRV alteration image, we outlined seven sites that include almost all of the altered rocks. Our Phase I report (Sabins and Ellis, 2010) included an intensive analysis of each site and an interpretation of mineral prospects. The following sections summarize our work on Site 1 in the east-central portion of Plate 43B (shown by the black arrow).

Plate 43C is an unsupervised Landsat classification image of the valley from a combination of principal component and band ratio images. Twenty-five original classes were combined into the five classes shown. For discussion purposes we arbitrarily call the green class metasediments and the brown class metavolcanics. The blue class Metavolcanic Member 1 is recognized as the host rock because its distribution matches that of much of the altered rocks in Plate 43B.

Site 1 Landsat Geology and HRV Prospects

We selected Site 1 from Plate 43B as a training site because it includes the only known copper deposit in the Balkhab Valley. Plate 44A is a Landsat HRV alteration image of Site 1 with the copper deposits shown by the white diagonal line pattern. Copper was mined in the past, as shown by old surface and underground workings (Peters and others, 2011, Chapter 4). The ore deposits are stockworks that are a complex system of structurally controlled and disseminated veins. In Plate 44A we interpret the green signature as a stratiform (layered) unit that is folded into a southwest-plunging anticline along the southeast bank of the Balkhab River. The anticline is asymmetric with a broad northwest flank and a narrow southeast flank. We use the term anticline as a convenient descriptive term, but not a genetic term. In these massive rocks we lack attitude (dip and strike) data or relative ages that would prove or disprove the anticline title.

The nose and both limbs of the anticline are altered to clay, as shown by the green signatures in Plate 44A. Scattered yellow signatures are iron minerals. Areas with red signatures are altered to both iron and clay and our exploration targets are labeled X, Y, and Z. Target X on the crest of the anticline is located within a red alteration zone that is within the ore body (white diagonal lines) mapped by the Soviets and the Afghan Geological Survey (AGS). Target Y is adjacent to known ore deposits on the southeast limb of the anticline. Target Z is a weak feature near the fold axis in the northeast portion of the image.

Site 1 HyMap Hyperspectral Image and Prospects

HyMap is an airborne hyperspectral imaging system built by the Integrated Spectronics Pty. Ltd. of Australia and is used commercially by the HyVista Corporation (hyvista.com/technology/sensors). HyMap records 124 spectral bands in the VNIR through SWIR spectral regions. Figure 13-33 shows the position of the 30 HyMap bands that span the SWIR spectral region. Bandwidth of HyMap in the SWIR region is comparable to that of AVIRIS. Bandwidth of individual HyMap bands ranges from 15 to 16 nm in the visible, NIR, and SWIR1 regions. Bandwidth is 18 to 20 nm in the SWIR2 region.

In 2007, the USGS leased a HyMap system that was installed in a NASA RB57 aircraft. Under the direction of the USGS, the RB57, flying at 12.5 km above the terrain, acquired a mosaic of side-lapping, north–south image strips of Afghanistan (Kokaly and others, 2008). The entire country was covered except for a 60-km wide buffer zone just inside the international boundary where the aircraft maneuvered at the beginning or end of each line. King and others (2011) processed and interpreted the mineralogy of a large number of Afghan sites.

Because of our work in Afghanistan we were keenly interested in the HyMap images. The USGS Spectroscopy Laboratory provided digital data (with 26.1 m pixels) for the HyMap imagery that covered the Balkhab Valley in north-central Afghanistan. Plate 44B is a HyMap image of Site 1 that we processed with ENVI's minimum noise fraction (MNF) algorithm, which is a version of the principal component transform (Chapter 9). MNF segregates system noise and reduces the HyMap 124 bands into fewer bands that contain most of the significant spectral variance. We selected 12 MNF bands for further processing. Applied Spectral Imaging input the MNF data into ENVI's mixture tuned matched filter (MTMF) routine, which unmixes the complex spectra of each pixel into some individual mineral spectra. The MTMF process identified a range of minerals that we reduced to the seven minerals shown in Plate 44B.

The color signatures in Plate 44B show the HyMap pixels that have the highest concentration of one of the seven minerals selected. Phengite is a high silica and magnesium version of muscovite that is concentrated in the northwest. It is also scattered along the northwest limb of the anticline. Muscovite/phengite forms the plunge and the northwest limb of the anticline. The southeast limb is altered to jarosite and silicate mix.

Figure 13-38 is the reflectance spectra of the seven minerals determined by spectral unmixing of the HyMap MNF images. Phengite, muscovite/phengite, and muscovite each have a unique SWIR absorption feature. The muscovite absorption feature (2,202 nm) shifts to longer wavelengths for phengite (2,235 nm). For the muscovite/phengite mixture the absorption feature is at an intermediate wavelength (2,221 nm).

A number of Australian investigators report that VMS ore deposits are accompanied by transitional white micas, such as phengite (Cudahy and others, 2000). Muscovite/phengite (light green signatures) occurs in the known copper deposit on Plate 44B, which indicates that the Australian correlation of transitional white micas and VMS ore deposits is applicable for Balkhab.

GEOLOGIC MAPPING WITH ASTER TIR

A major advance for mapping geology is the ASTER TIR module that records five multispectral TIR bands (numbered 10 to 14) (Table 5-8). We process these bands as emissivity images to recognize quartz, carbonate, and mafic minerals using ratios of the TIR bands (Mars and Rowan, 2011). Quartz has a diagnostic signature with an absorption feature at 9.1 μm in the TIR region recorded by ASTER. Carbonates exhibit a weak 11.2 μm absorption feature and many

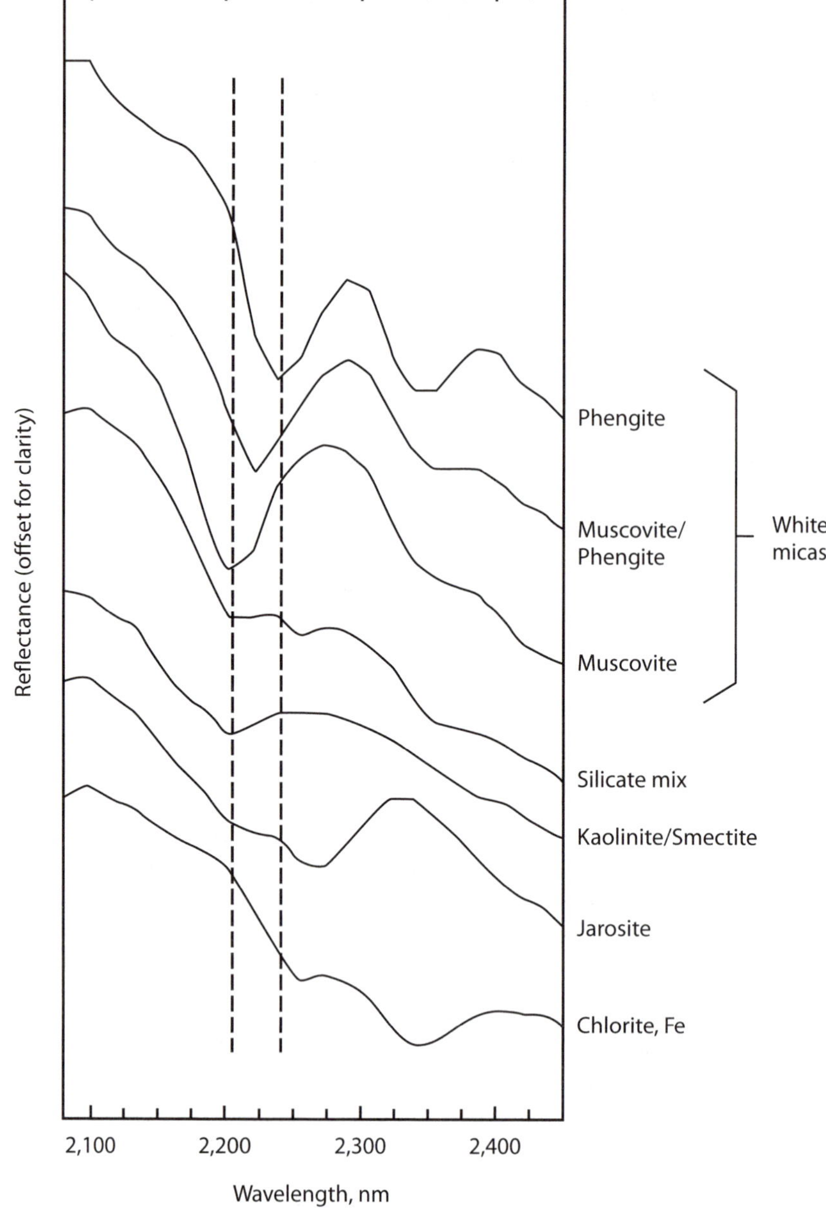

FIGURE 13-38 HyMap SWIR spectral signatures for seven minerals identified at Site 1 in the Balkhab Valley, Afghanistan (Plate 44B). Dashed vertical lines highlight narrow wavelength range of absorption (35 nm) between phengite and muscovite.

mafic-rich rocks have a 10.6 μm absorption feature (Mars and Rowan, 2011). ASTER TIR emissivity bands 12, 13, and 14 are ratioed for geologic mapping as follows (Mars and Rowan, 2011):

- 14/12 for quartz-rich rocks,
- 13/14 for carbonate-rich rocks, and
- 12+14/13 for mafic-rich rocks.

We utilized ASTER TIR band ratios to improve the geologic mapping of the Katawas outcrops in central Afghanistan (Sabins and Ellis, 2013a). Katawas is approximately 310 km southeast of the Balkhab Valley. Plate 45A is an enlargement of a 1:850,000 scale geologic map (Doebrich and Wahl, 2006) that shows a narrow, northeast-trending central ridge of Paleocene basalt lava (P_1bl) that is bounded by faults and surrounded by Quaternary alluvial (Q_3a) and eolian (Q_3loe) plains and Paleocene (P_2csh) clays and shales. Plate 45B is a DEM color coded for elevation revealing the narrow, northeast-trending ridges that are surrounded by low-lying terrain (see also Figure 13-39A).

Figure 13-39B–D displays the ASTER TIR grayscale band ratios for quartz, carbonate, and mafic rocks. Brighter pixels in the band ratio images indicate higher concentrations of quartz, carbonates, or mafic rocks at the Earth's surface. The images are mosaics of three ASTER TIR scenes acquired on different dates. Seam lines between the mosaicked scenes can be seen trending north–northeast and east–southeast. East–southeast striping caused by excessive noise in the data can also be seen, especially in Figure 13-39C. A mask was developed to focus image processing on the outcrops. The mask minimizes terrain covered with alluvial fans, agriculture, and unconsolidated sediments in low-lying areas.

The masked grayscale band ratios are density sliced into five classes to highlight those outcrops with quartz-rich, carbonate-rich, and mafic-rich rocks. In Plate 45C–E, red

A. SRTM 30 m DEM illuminated from the northwest.

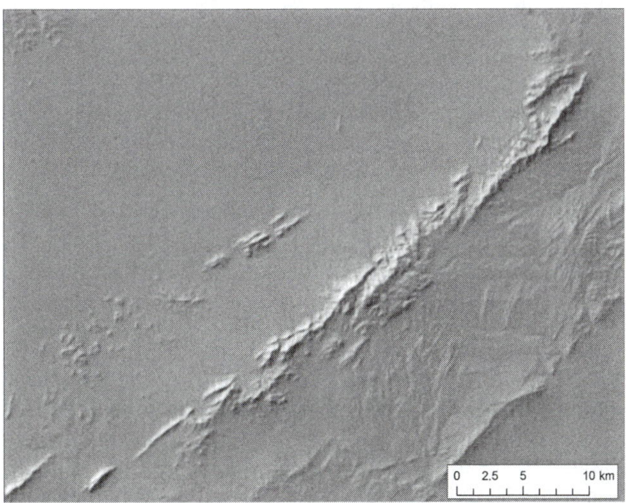

B. ASTER TIR quartz band ratio 14/12.

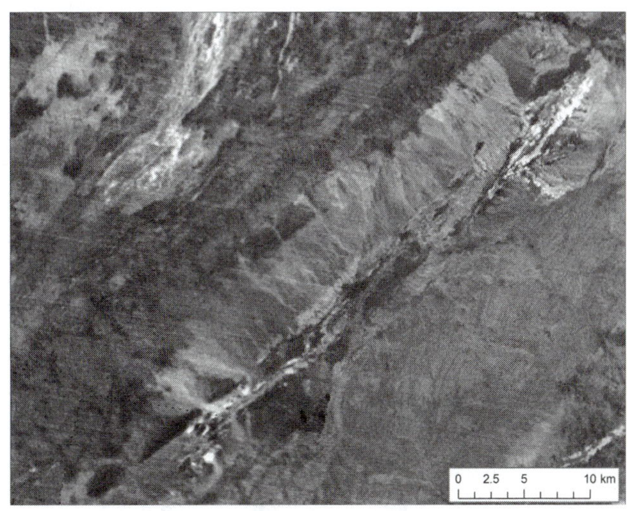

C. ASTER TIR carbonate band ratio 13/14.

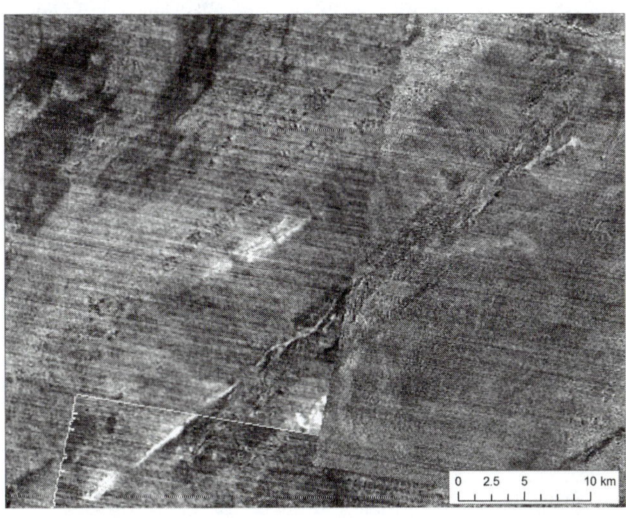

D. ASTER TIR mafic band ratio 12+14/13.

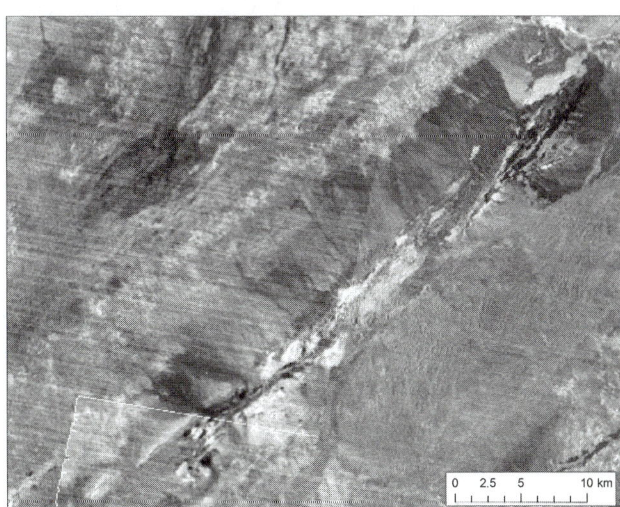

Figure 13-39 SRTM DEM and ASTER TIR emissivity band ratios. The grayscale band ratios are density sliced and color coded in Plate 45C–E.

signatures indicate high ratio values. Quartz-rich rocks are concentrated in the northeast portion of the image (Plate 45C). The small scattered red signatures may indicate wind-blown quartz sand. Carbonate-rich rocks are mapped in two areas along narrow ridges (Plate 45D). Mafic-rich rocks are concentrated in the basalt in the center of the image with a small occurrence of volcanic rocks in the far northeast portion of the image (Plate 45E). The robustness of this technique is demonstrated by the separation of the red signatures for the three classes and the relative richness between the three classes at specific sites. For instance, where the mafic ratio is high, the quartz ratio is low. Where the carbonate ratio is high, the quartz and mafic ratios are low.

The highest density slice for each band ratio image (red in Plate 45C–E) are combined in Plate 45F to generate a geologic map displaying mafic-rich rocks in red, carbonate-rich rocks in blue, and quartz-rich rocks in green. The ASTER TIR band ratios provide another unique source of information for mapping geology in remote and difficult to access areas. Based on the ASTER TIR band ratios, the narrow ridges protruding above the surrounding alluvial fans and low-lying floodplains and agricultural plots at Katawas have a more complex geology compared with that captured on the country-wide geologic map (Plate 45A). Hubbard (2011, Figure 4.8.1) used ASTER VNIR-SWIR-TIR bands to classify a mineral alteration anomaly at Katawas.

MONITORING DEVELOPMENT WITH REMOTE SENSING

Remote sensing is used to monitor production activity and environmental impacts as oil and gas fields are developed, infrastructure is built, and facilities are maintained. Leaking gas is detected by TIR cameras (Chapter 5). Drones fly around offshore platforms to assess the integrity of infrastructure that is difficult and dangerous to access. Land use/land cover maps are developed from imagery to plan development and monitor landscape changes through time (Chapters 12 and 14).

Mining operations use remote sensing technologies to monitor the volume of rock excavated and deposited as spoils and piles, locate areas in open pits that have prime commercial potential (Kruse and others, 2000), scan drill cores and cuttings to improve mine development and ore processing, and document infrastructure conditions and environmental impacts.

The following discussion provides examples of measuring ground deformation during oil production and core scanning.

RADAR INTERFEROMETRY TO MONITOR STEAM INJECTION

Steam is injected into fields with heavy oil to increase oil production. Stancliffe and van der Kooij (2001) investigated the use of satellite radar interferometry (IfSAR) as a less expensive technique to monitor ground deformation, interpret the movement of steam, and avoid areas of faults and fractures as the production area changes. They used radar data collected by ERS (Europe), JERS (Japan), and RADARSAT (Canada) satellites of the Cold Lake oil field, northeast Alberta, Canada. The field produces a heavy viscous oil from oil sand deposits located approximately 415 to 470 m below the surface. Steam is injected into the reservoir to reduce the viscosity of the oil and enable it to flow faster, enhancing the production rate. The steam is injected for several months into the reservoir, left to soak for several months, and then the less viscous oil is pumped to the surface.

The surveyed area is forested. Pairs of IfSAR radar data were collected with a 43-day difference between JERS images, 35-day difference between ERS images, and 24-day difference between RADARSAT images. Temporal decorrelation between the IfSAR data pairs caused by changes in the structure of the surface between the times of data acquisition is the largest source of error for measuring vertical ground movement. Surface structure is changed by snow, rain, spring runoff swelling of the muskeg, grass and tree growth, and alteration of the land surface by humans, including forestry, farming, and fire. The long wavelength, L-band radar (23.5 cm) emitted by the JERS satellite is advantageous for this project, as the longer wave is commonly preserved in forested terrain due to deeper penetration through the canopy and reflection from solid, stable targets such as tree trunks.

The best interferograms show the vertical movement over the Cold Lake oil field can total over 36 cm during a month. Plate 46A is a subarea of the JERS interferogram from data acquired 86 days apart. Each interferometric cycle is represented by a rainbow color scheme (blue to green to yellow to red) and represents 0 to 11.75 cm of deformation. The elliptical to circular patterns of deformation highlight areas with active steam injection and oil production.

Interferometric cycles have been contoured on Plate 46B to show the magnitude of swelling (positive values) and subsidence (negative values) over a 43-day period. Location 2 shows an area where steam was injected over the 43-day period, causing an elevation increase of + 0.20 cm. Location 3 is an area brought onto production, causing subsidence of –21 cm. Stancliffe and van der Kooij (2001) explain the production processes causing the rise and fall of the surface. A profile of the ground deformation over the 43-day period is shown in Figure 13-40.

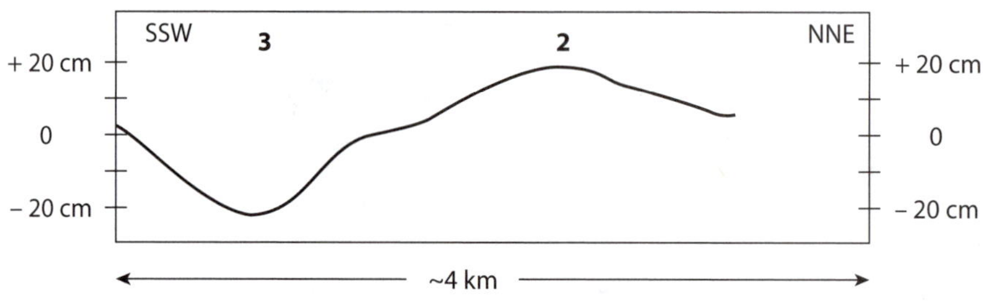

FIGURE 13-40 South-southwest–north-northeast profile of ground deformation over a 43-day period (Plate 46), Cold Creek oil field, Alberta, Canada. Interpreted from Stancliffe and van der Kooij (2001).

Ground deformation studies are inherently more precise and accurate in arid areas without vegetation and on large man-made structures compared with forested terrain. Integrating land use, vegetation level/type, climate, and soil type with L-band InSAR data of vegetated terrain improves ground deformation measurements.

Drill Cores

The acquisition (drilling) and analysis of drill cores is widely used in both the oil and mining industries. In the Collahuasi Mine located in Chile, drill cores were used to confirm our remote sensing mineral discoveries.

Coring Equipment

Figure 13-41A is a portable core drill rig that is powered by a diesel fuel engine. Figure 13-41B are drill bits for coring with hollow centers that cut a cylinder of rock (core). The drill bit is attached to a core barrel that temporarily stores the core. Between the bit and the barrel a core spring or lifter retains the core in the barrel. During coring, air, water, or a mixture of water and clay (called drilling mud) is pumped down the drill stem, through the core barrel, and out the opening in the core bit. The fluids lubricate and cool the bit. The fluids return to the surface via the space between the hole wall and the exterior of the core barrel (the annulus). The returning drilling mud carries rock chips (cuttings) to the surface where they are removed and the drilling mud is recirculated.

When the barrel is filled with core it is retrieved to the surface, where the core is removed and labeled by depth and locality. A "quick and dirty" lithologic description (core log) is commonly done at the well site before the cores are packed and transported to a central facility, called the core shed. There a geologist does a more detailed core log and samples are taken at regular intervals, typically between 1 and 3 m depth, for laboratory analyses. In the oil industry the samples are tested for porosity, permeability, and oil staining. X-ray fluorescence is used to identify mineralogy. In the mining industry the tests are for commercial minerals (primarily metals) and for secondary alteration minerals (primarily clays and iron minerals).

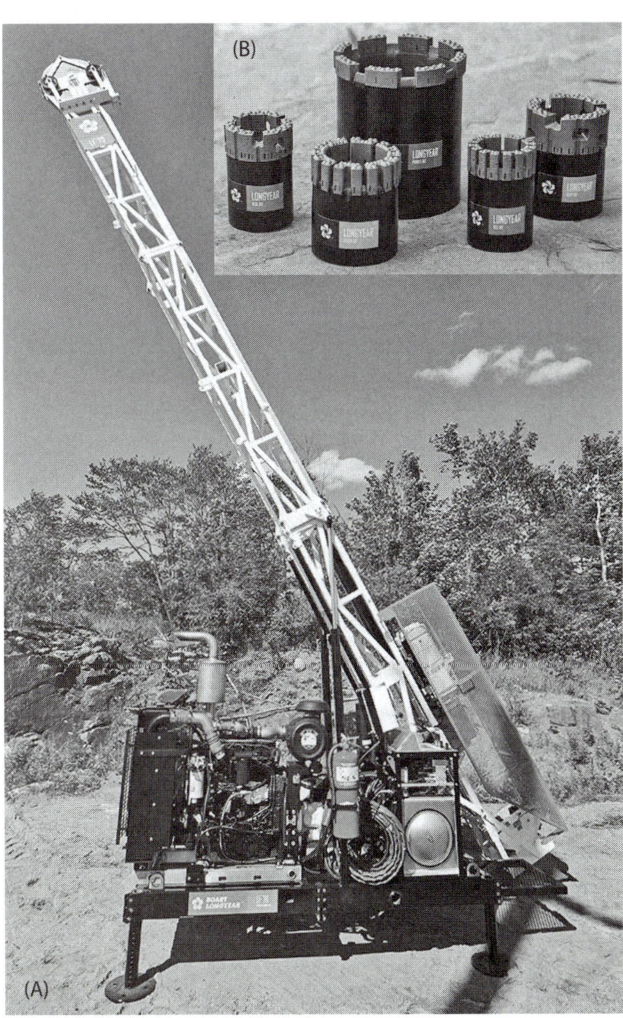

Figure 13-41 (A) Portable core drill rig and (B) core drill bits. Photos courtesy of Boart Longyear.

Commercial Core Scanning

In 1986, Frank Pruett and Floyd Sabins were granted a US patent for a "High Resolution Geologic Sample Scanning Apparatus and Process of Scanning Geologic Samples." Their insight was to remove an airborne multispectral scanner from a plane and mount it on a stand 1 m above a core that was moved beneath the scanner by a simple screw-drive mechanism. By the early twenty-first century, several service companies designed and built core scanners for sale or lease. Some companies will transport their system and an operator to the drill site, where the cores are scanned, processed, and interpreted promptly. Figure 13-42 shows the Corescan HCI-3 system. The scanner is the large box mounted above the roller system that transports cores beneath the scanner. The system can scan 300 to 500 m of core per day (corescan.com.au). Hyperspectral data are collected across a wavelength range of 0.45 to 2.5 μm with a 0.004 μm spectral resolution and 500 μm (0.5 mm) spatial resolution. Terracore provides hyperspectral imaging from the visible to the TIR wavelengths (terracoregeo.com). In a typical remote mineral exploration project where a number of drill rigs are operating, it is more efficient and timely to analyze cores on-site rather than crate the cores, ship them to a distant facility, and await results while drillers are waiting to know where their next holes should be located.

Analysis of Core Scans

A major advantage of a core scanner system is that it provides a continuous pixel by pixel record of mineralogy, whereas the conventional sample method only provides mineralogy for samples separated by 1 m or more along the length of the core. Plate 47 shows images acquired by a commercial core scan system. The left column is a natural color image of the core. The depth scale on the left margin shows that the core extends from a depth of 30.55 m down to a depth of 30.67 m for a total length of 0.12 m, or 12 cm. Down to a

FIGURE 13-42 Portable core scanner laboratory operating at a remote locality. The hyperspectral scanner and illumination sources are mounted above a track that transports the cores beneath the scanner. The long flat tray holds three cores that are being scanned sequentially. Courtesy Corescan.

depth of 30.64 m the core is a very fine-grained, unfractured granitic rock with a faint layering (gneissic texture) that is inclined (dips) from the left down to the right. The lower portion of the core is strongly fractured and altered, as shown by the iron staining. The kaolinite abundance image in the center of Plate 47 shows relative abundance of that mineral, which is keyed to the color scale. Kaolinite is absent from the fractured lower portion of the core. Similar abundance images can be generated for any minerals specified for classification. The classification image in the right column shows two distinct mineral assemblages. The unfractured upper portion of the core is altered to kaolinite and iron oxides in a layered pattern. The fractured lower portion is altered to illite, montmorillonite, and nontronite (an iron-rich clay mineral) with iron oxide staining along the fractures.

QUESTIONS

1. You are employed by an international oil exploration company that plans to evaluate the petroleum potential of the western portion of Kazakhstan in preparation for negotiating an exploration concession. Because of limited accessibility, deadlines, and competitor pressure, concession areas must be selected solely on the basis of remote sensing evaluations. Prepare such a plan for your management, giving reasons for each step.
2. Assume that western Kazakhstan is completely covered by Landsat 8 OLI/TIR images, Sentinel-1 radar images, and high spatial resolution satellite panchromatic images. List the advantages and disadvantages of each kind of image and the data it could provide for your project.
3. Why are modern analogs studied with remote sensing for energy exploration?
4. What are three shape indices used to describe modern landforms? Explain each with a diagram of the landform (not a sandbar) being studied.
5. What are the spectral absorption features of hydrocarbons at the Earth's surface that can be detected by a hyperspectral VNIR-SWIR?
6. How is a HRV image generated from Landsat 8 OLI bands?
7. What are the ASTER TIR emissivity band ratios used for detecting three different types of rock? What is the TIR spectral absorption feature associated with each rock type?
8. The wavelength of the deepest spectral absorption feature associated with the six minerals on Figure 13-33 is as follows:

   Buddingtonite: 2.120 µm
   Kaolinite: 2.210 µm
   Montmorillonite: 2.220 µm
   Illite: 2.230 µm
   Alunite: 2.170 µm
   Quartz: none

   You have airborne hyperspectral SWIR data (2.0 to 2.4 µm wavelength range) that has 0.03 µm bandwidth and band centers starting at 2.00 µm. Which minerals can you differentiate with this data? Why or why not? (Look at the shape of the six absorption features on Figure 13-33.)
9. The six minerals in question 8 above are in a core, and the core scanner in Figure 13-42 is used (spectral resolution described in the text). Which minerals can you differentiate with this data? Why or why not?

References

Abrams, M. J., and D. Brown. 1985. Silver Bell, Arizona, porphyry copper test site. Joint NASA/Geosat Test Case Study, Section 4. Tulsa, OK: American Association of Petroleum Geologists.

Abu-Ali, M. A., V. A. Franz, J. Shen, F. Monnier, M. D. Mahmoud, and T. M. Chambers. 1991. Hydrocarbon generation and migration in the Paleozoic sequence of Saudi Arabia. In Proceedings, Middle East Oil Show, Bahrain (SPE 21376-MS, pp. 345–356). Society of Petroleum Engineers, Dallas, TX.

Ashley, R. P. 1974. Goldfield mining district. In *Guidebook to the Geology of Four Tertiary Volcanic Centers in Central Nevada* (Report 19, pp. 49–66). Nevada Bureau of Mines and Geology.

Ashley, R. P. 1979. Relation between volcanism and ore deposition at Goldfield, Nevada. In *Papers on Mineral Deposits of Western North America* (Report 33, pp. 77–86). Nevada Bureau of Mines and Geology.

Bedell, R., A. P. Crósta, and E. Grunsky (Eds.). 2009. Remote sensing and spectral geology. *Reviews in Economic Geology* (Vol. 16). Littleton, CO: Society of Economic Geologists, Inc.

Berger, Z. 1994. *Satellite Hydrocarbon Exploration: Interpretation and Integration Techniques*. Berlin, Germany: Springer-Verlag.

Berry, J. L., and G. L. Prost. 1999. Hydrocarbon exploration. In A. N. Rencz (Ed.), *Remote Sensing for the Earth Sciences: Manual of Remote Sensing* (Vol. 3) (3rd ed., pp. 449–508). New York: John Wiley.

Boardman, J. W. 1993. Automated spectral unmixing of AVIRIS data using convex geometry concepts. In Summaries, Fourth Annual JPL Airborne Geoscience Workshop (Vol. 1, pp. 11–14), Pasadena, CA.

Craig, D. H. 1990. Yates and other Guadalupian (Kazanian) oil fields, U.S. Permian Basin. *Classic Petroleum Provinces* (Special Publications, Vol. 50) (pp. 249–263). London, England: Geological Society of London.

Cudahy, T. J., K. Okada, and C. Brauhart. 2000, November. Targeting VMS-style Zn mineralization at Panorama, Australia using airborne hyperspectral VNIR-SWIR HyMap data. In Proceedings, ERIM Fourteenth International Conference on Applied Geologic Remote Sensing (pp. 395–402), Las Vegas, NV.

Dick, L. A., G. Ossandon, R. G. Fitch, C. M. Swift, and A. Watts. 1993. Discovery of blind copper mineralization at Collahuasi, Chile. In S. B. Romberger and D. I. Fletcher (Eds.), *Integrated Methods in Exploration and Discovery* (pp. AB21–23). Denver, CO: Society of Economic Geologists.

Doebrich, J. L., and R. R. Wahl. 2006. *Geologic and Mineral Resource Map of Afghanistan* (Open File Report 2006-1038). Washington, DC: US Geological Survey.

Donovan, T. J. 1974. Petroleum microseepage at Cement field, Oklahoma evidence and mechanism. *Bulletin of the American Association of Petroleum Geologists*, 58, 429–446.

Dutton, S. P., E. M. Kim, R. F. Broadhead, C. L. Breton, W. D. Raatz, S. C. Ruppel, and C. Kerans. 2004. *Play Analysis and Digital Portfolio of Major Oil Reservoirs in the Permian Basin: Application and Transfer of Advanced Geological and Engineering Technologies for Incremental Production Opportunities*. Bureau of Economic Geology, University of Texas and New Mexico Bureau of Geology and Mineral Resources, New Mexico Institute of Mining and Technology.

Economic Geology. 1983, July. Techniques and results of remote sensing for mineral exploration. 78(4), 573–784.

Ellis, J. M., H. H. Davis, and J. Zamudio. 2001, September. Exploring for onshore oil seeps with hyperspectral imaging. *Oil and Gas Journal*, 99(37), 49–58.

Ellis, J. M., and F. D. Pruett. 1986. Application of synthetic aperture radar (SAR) to southern Papuan Fold Belt exploration. In Proceedings, Fifth Thematic Conference on Geologic Remote Sensing (pp. 15–34), Environmental Research Institute of Michigan, Ann Arbor, MI.

Harris, P. M., J. M. Ellis, and S. J. Purkis. 2010. Delineating and Quantifying Depositional Facies Patterns of Modern Carbonate Sand Deposits on Great Bahama Bank (DVD Set). Society for Sedimentary Geology Short Course Notes 54 (ISMB 978-1-56576-301-2).

Harris, P. M., S. J. Purkis, and J. M. Ellis. 2011. Analyzing spatial patterns in modern carbonate sand bodies from Great Bahama Bank. *Journal of Sedimentary Research*, 81(3), 185–206.

Harris, P. M., and S. D. Walker. 1990. McElroy Field: Development geology of a dolostone reservoir, Permian Basin, West Texas. In D. G. Bebout and P. M. Harris (Eds.), *Geologic and Engineering Approaches in Evaluation of San Andres/Grayburg Hydrocarbon Reservoirs—Permian Basin* (pp. 275–296). University of Texas Bureau of Economic Geology.

Harvey, R. D., and C. J. Vitaliano. 1964. Wall-rock alteration in the Goldfield District, Nevada. *Journal of Geology*, 72, 564–579.

Hubbard, B. E. 2011. *Identification of Mineral Anomalies in Afghanistan Using Advanced Spaceborne Thermal Emission and Reflection Radiometer* (Open-File Report OF 2011-1229, Chapter 4). Washington, DC: US Geological Survey.

King, T. V. V., T. M. Hoefen, M. R. Johnson, R. F. Kokaly, and K. E. Livo. 2011. *Mapping Anomalous Mineral Zones Using HyMap Imaging Spectrometer: Data for Selected Areas of Interest in Afghanistan* (Open-File Report OF 2011-1229, Chapter 5). Washington, DC: US Geological Survey.

Kneeper, D. H. 2010. *Distribution of Potential Hydrothermally Altered Rocks in Central Colorado Derived from Landsat Thematic Mapper Data: A Geographic Information System Data Set* (Open-File Report 2010-1076). Washington, DC: US Geological Survey.

Kokaly, R. F., T. V. V. King, and K. E. Livo. 2008. *Airborne Hyperspectral Survey of Afghanistan 2007: Flight Line Planning and HyMap Data Collection* (Open-File Report OF 2008-1235). Washington, DC: US Geological Survey.

Kruse, F. A. 1999. Visible and infrared: Sensors and case studies. In A. N. Rencz (Ed.), *Remote Sensing for the Earth Sciences: Manual of Remote Sensing* (Vol. 3) (3rd ed., pp. 567–612). New York: John Wiley.

Kruse, F. A., J. W. Boardman, A. B. Lefkoff, J. M. Young, K. S. Kierein-Young, T. D. Cocks, R. Jenssen, and P. A. Cocks. 2000. *HyMap: An Australian Hyperspectral Sensor Solving World Problems—Results from USA HyMap Data Acquisitions*. Boulder, CO: Analytical Imaging and Geophysics. Sydney, Australia: Integrated Spectronics Pty. Ltd.

Lamerson, P. R. 1990. Evolution of Structural Interpretations in Iagifu/Hedinia Field, Papua New Guinea. In Proceedings, First PNG Petroleum Convention, G. J. Carman and Z. Carman (Eds.), Petroleum Exploration in Papua New Guinea (pp. 283–300). PNG Chamber of Mines and Petroleum, Port Moresby, Papua New Guinea.

Lattus, J. M. 2017, June 28. SRGIS Uses Remotely Sensed Data and ENVI to Target Locations for Mineral Excavation. Harris Geospatial Solutions Case Studies. Available http://www.harrisgeospatial.com/Learn/CaseStudies.aspx (accessed November 2017).

Lowell, J. D., and J. M. Guilbert. 1970. Lateral and vertical alteration-mineralization zoning in porphyry ore deposits. *Economic Geology*, 65, 373–408.

Mars, J. C., and L. C. Rowan. 2011. ASTER spectral analysis and lithologic mapping of the Khanneshin carbonatite volcano, Afghanistan. *Geosphere*, 7(1), 276–289.

McGillivray, J. G., and M. I. Husseini. 1992. The Paleozoic petroleum geology of central Arabia. *American Association of Petroleum Geologists Bulletin*, 76, 1,473–1,490.

Noomen, M. F. 2007. Hyperspectral Reflectance of Vegetation Affected by Underground Hydrocarbon Gas Seepage (ITC Dissertation No. 145). International Institute for Geo-Information Science & Earth Observation, Enschede, the Netherlands.

Peters, S. G., T. V. V. King, T. J. Mack, M. P. Chornack (Eds.), and the USGS Afghanistan Mineral Assessment Team. 2011. *Summaries of Important Areas for Mineral Investment and Production Opportunities of Nonfuel Minerals in Afghanistan* (Open File Report 2011-1204, Vol. 1). Washington, DC: US Geological Survey.

Prost, G. 2014. *Remote Sensing for Geoscientists: Image Analysis and Integration* (3rd ed.). Boca Raton, FL: CRC Press.

Purkis, S. J., and P. M. Harris. 2016. The extent and patterns of sediment filling of accommodation space on Great Bahama Bank. *Journal of Sedimentary Research*, 86, 294–310.

Purkis, S. J., and P. M. Harris. 2017. Quantitative interrogation of a fossilized carbonate sand body—the Pleistocene Miami oolite of south Florida. *Sedimentology*, 64, 1,439–1,464.

Rockwell, B. W., L. C. Bonham, and S. A. Giles. 2015. USGS National Map of Surficial Mineralogy. US Geological Survey Online Map Resource. http://cmerwebmap.cr.usgs.gov/usminmap.html (accessed January 2018).

Rowan, L. C., A. F. H. Goetz, and R. P. Ashley. 1977. Discrimination of hydrothermally altered and unaltered rocks in the visible and near infrared. *Geophysics*, 42(3), 533–535.

Rowan, L. C., P. H. Wetlaufer, A. F. H. Goetz, F. C. Billingsley, and J. H. Stewart. 1974. *Discrimination of Rock Types and Detection of Hydrothermally Altered Areas in South Central Nevada by the Use of Computer-Enhanced ERTS Images* (Professional Paper 883). Washington, DC: US Geological Survey.

Sabins, F. F. 1991. Digital processing of satellite images of Saudi Arabia. In Proceedings, Middle East Oil Show, Bahrain (SPE 21376-MS, pp. 207–212). Society of Petroleum Engineers, Dallas, TX.

Sabins, F. F., and J. M. Ellis. 2010. *Landsat Analysis of Mineral Anomalies, Balkhab Region, Afghanistan*. Report to US Department of Defense Task Force for Business and Stability Operations.

Sabins, F. F., and J. M. Ellis. 2013a. *Remote Sensing of Mineral Exploration Targets in Katawas AOI, Afghanistan*. Report to US Department of Defense Task Force for Business and Stability Operations.

Sabins, F. F., and J. M. Ellis. 2013b. *Summary of 17 Afghan Remote Sensing Reports by Remote Sensing Enterprises, Inc*. Report to US Department of Defense Task Force for Business and Stability Operations.

Sabins, F. F., J. M. Ellis, and J. Zamudio. 2012. *Identify Mineral Exploration Targets at Balkhab AOI with HyMap and ASTER Images*. Report to US Department of Defense Task Force for Business and Stability Operations.

Schlumberger, Ltd. 2017. Carbonate Reservoirs. http://www.slb.com/services/technical_challenges/carbonates.aspx (accessed January 2018).

Shanks III, W. C. P., and R. A. Koski (Eds.). 2010. Introduction. In W. C. P. Shanks III and R. Thurston (Eds.), *Volcanogenic Massive Sulfide Occurrence Model* (Scientific Investigations Report 2010-5070-C). Washington, DC: US Geological Survey.

Stancliffe, R. P. W., and M. W. A. van der Kooij. 2001. The use of satellite-based radar interferometry to monitor the production activity at the Cold Lake heavy oil field, Alberta, Canada. *American Association of Petroleum Geologists Bulletin*, 85(5), 781–793.

TFBSO. 2011. Mineral Resource Team 2010 Activities Summary (pp. 25–28). Task Force for Business Stability Operations. Washington, DC: US Department of Defense. http://www.dtic.mil/dtic/tr/fulltext/u2/a545347.pdf.

Valenti, G. L., I. C. Phelps, and L. I. Eisenberg. 1996. Geological remote sensing for hydrocarbon exploration in Papua New Guinea. In Proceedings, Eleventh Thematic Conference on Geologic Remote Sensing (pp. 1–97), Environmental Institute of Michigan, Ann Arbor, MI.

Vergara, H. 1978a. Cuadrángulo Ujina: Carta Geológica de Chile, no. 33, Escala 1:50,000. Santiago, Chile: Instituto de Investigaciones Geológicas.

Vergara, H. 1978b. Cuadrángulo Quehuita y sector occidental del cuadrángulo Volcán Mino: Carta Geológica de Chile, no. 32, Escala 1:50,000. Santiago, Chile: Instituto de Investigaciones Geológicas.

Wender, L. E., and F. F. Sabins. 1991. Geologic interpretation of satellite images, Saudi Arabia. In Proceedings, Middle East Oil Show, Bahrain (SPE 21358-MS, pp. 213–218). Society of Petroleum Engineers, Dallas, TX.

Yang, H., F. V. D. Meer, J. Ahang, and S. B. Kroonenberg. 2000. Direct detection of onshore hydrocarbon microseepages by remote sensing techniques. *Remote Sensing Reviews*, 18(1), 1–18. doi:10.1080/02757250009532381

chapter 14

Land Use/Land Cover

Land use and land cover maps range from global to local, and cover the past, present, and future. *Land use* describes how an area of land is used (such as for agriculture, residences, or industry), whereas *land cover* describes the materials (such as vegetation, rocks, or buildings) that are present on the surface. For example, the land cover of an area may be grass, but the land use may be residential, city park, grasslands preserve, golf course, sod farm, or combinations of activities. Land use maps require higher spatial resolution imagery and ancillary information such as parcel, zoning, and census data. Green and others (2017) discuss best practices for extracting information from imagery. Assessment techniques measure the accuracy of land use and land cover maps to ensure that the maps are reliable and may be confidently used for different applications. Congalton and Green (2008) provide extensive information on concepts, design. and implementation of accuracy assessment techniques.

Remote sensing methods are important for mapping land use and land cover (LULC) for the following reasons:

1. Large areas can be imaged quickly and repetitively,
2. Images can be acquired with a spatial resolution that matches the degree of detail required for the survey,
3. LULC maps cover global, national, and local scales,
4. Remote sensing images eliminate the problems of surface access that often hamper ground surveys,
5. Images provide a perspective that is lacking for ground surveys,
6. Image interpretation is faster and less expensive than conducting ground surveys, and
7. Images provide an objective, permanent dataset that may be interpreted for a wide range of specific land uses and land covers, such as forestry, agriculture, and urban growth.

There are some disadvantages to remote sensing surveys:

1. Some types of land use may not be distinguishable on images and
2. Most images lack the horizontal perspective that is valuable for identifying many categories of land use.

GLOBAL MAPS

Global land cover maps enable international and governmental groups to monitor and measure changes to land and water resources, agriculture, forests, and development that can be used to address issues regarding food security and environmental sustainability. Global land cover maps and databases are generated from remote sensing images by several organizations, typically using images with spatial resolutions greater than 300 m. The entire GIS database for the Earth is available from the NASA Socioeconomic Data and Applications Center (sedac.ciesin.columbia.edu). The grid cell size varies from 85 km^2 per cell at the equator to 11 km^2 per cell at the poles. The spatial resolution generates a 4,320 by 2,160 global grid.

FOOD AND AGRICULTURE ORGANIZATION OF THE UNITED NATIONS

The Food and Agriculture Organization (FAO) of the United Nations, along with many other international organizations and scientific institutions, needs timely and reliable information on land cover and its changes at global, regional, and country levels. Land cover information is fundamental in fulfilling the mandates of many institutional programs; it supports the formulation of evidence-based, sustainable land development and land use policy at various scales. The dynamics of land cover assessment are essential requirements in understanding and recognizing variations in natural phenomena such as climate change (Latham and others, 2014).

One significant hurdle to using existing national and subnational maps for global analysis is a lack of standardization of land cover terminology, definitions, and categories (land cover types). FAO supports the Land Cover Classification System (LCCS) to enable integration of various available land cover databases. LCCS provides a valuable, universal land cover language for building cover legends and comparing existing legends (Herold and others, 2006). In 2014, the Land and Water Division of the FAO created Global Land Cover SHARE (GLC-SHARE), a new land cover database formed in partnership with contributions from various partners and institutions (Latham and others, 2014). GLC-SHARE is aligned with LCCS and provides a set of 11 major thematic land cover types (Table 14-1) for global mapping.

The FAO has collected a large selection of national, regional, and subnational land cover databases and translated their legends into these 11 classes, developing a global land cover database generated with ~1 km spatial resolution. This database provides an estimate of the percentage of different land cover types across the Earth's continents (Table 14-2). It also produces GIS datasets, along with interactive maps and satellite imagery. Complete, free, and open access to the data and metadata products is available at fao.org/geonetwork.

EUROPEAN SPACE AGENCY

From 1992 to 2015, the European Space Agency's (ESA) Climate Change Initiative Land Cover (CCI-LC) Project offered an online, global land cover map at 300 m spatial resolution on an annual basis (see **Digital Image 14-1** 🌐 for an example). The maps are available online at maps.elie.ucl.ac.be/CCI/viewer/index.php. Many interactive tools are available at this site to change the legend, download data, and display annual meteorological and environmental charts for a user-selected location (e.g., NDVI, burned areas, snow seasonality charts). The ESA map legend uses an expanded version of FAO's standardized LCCS with 22 categories (Table 14-3).

The CCI-LC Project used different satellite systems to build the 1992 to 2015 sequence of land cover maps. More recent maps are available as downloads. ESA's polar-orbiting ENVISAT satellite collected 300 m and 1 km VNIR imagery with the MERIS instrument from 2003 to 2012. The AVHRR system collected 1 km imagery between 1992 and 1999, the SPOT VGT recorded VNIR imagery with 1.15 km spatial resolution between 1999 and 2013, and ESA's Proba-V mission collected 300 and 600 m VNIR-SWIR imagery in a 2,250 km wide swath starting in 2013. The imagery was integrated to produce the annual land cover maps at 300 m spatial resolution. These maps are useful for many applications, including modeling climate change extent and impacts, conserving biodiversity, and managing natural resources.

GLOBE

Global Learning and Observations to Benefit the Environment (GLOBE) is an international science and education program that provides students and the public with the opportunity to contribute meaningfully to our understanding of the global environment by participating in data collection and the scientific process. GLOBE coordinates the work of students (aged 5 to 18), teachers, and scientists from over 100 countries on five continents.

TABLE 14-1 GLC-SHARE land cover legend.

Label	Land Cover Type
01	Artificial surfaces
02	Cropland
03	Grassland
04	Tree covered areas
05	Shrubs covered areas
06	Herbaceous vegetation, aquatic, or regularly flooded
07	Mangroves
08	Sparse vegetation
09	Bare soil
10	Snow and glaciers
11	Water bodies

TABLE 14-2 Distribution of land cover classes globally.

Generalized Land Cover Type	Percent
Artificial surfaces	0.6
Cropland	12.6
Grassland/shrubs/herbaceous/sparse vegetation	31.5
Tree covered area	27.7
Bare soil	15.2
Snow and glaciers + Antarctica	9.7
Water bodies/mangroves	2.7
Total	100

Accurate ground reference data is fundamental to the use of remotely sensed data for land cover classification and mapping. However, very little ground reference data has been collected, so GLOBE scientist-student collaboration can help fill this void. The GLOBE Program uses the Modified UNESCO Classification (MUC) System, a classification system that follows international standards and uses ecological terminology for the identification of specific land cover classes (GLOBE Program, 2010).

In the MUC System, there are 10 classification levels (Table 14-4). Each level contains classes. There is a class for each type of land cover and every system is *mutually exclusive*, meaning there is only one appropriate class with its own unique identifier or numerical code for each land cover. Therefore, a detailed classification of a land cover type will be identified by a string of numbers. For example, the high level *closed forest* category has a designation of MUC 0. The more specific *closed forest, mainly evergreen, subtropical wet, lowland* category has a unique MUC 0141 identification code. The more detailed *cultivated land, nonagriculture, parks and athletic fields* has a unique MUC 821 code.

If scientists are to use student data, it is important that those data be as accurate as possible to ensure reliability of research results. The GLOBE Project provides workshops, protocol training, and web tutorials for its participants (globe.gov/en). The MUC Field Guide includes definitions of words used to describe the MUC classes, drawings,

Table 14-4 GLOBE Modified UNESCO Classification System.

Level	Land Cover Type
MUC 0	Closed forest
MUC 1	Woodland
MUC 2	Shrubland or thicket
MUC 3	Dwarf-shrubland or dwarf-thicket
MUC 4	Herbaceous vegetation
MUC 5	Barren land
MUC 6	Wetland
MUC 7	Open water
MUC 8	Cultivated land
MUC 9	Urban

Table 14-3 ESA CCI-LC Project land cover legend.

Label	Land Cover Type
0	No data
10	Cropland, rainfed
20	Cropland, irrigated or post-flooding
30	Mosaic cropland (> 50%) / natural vegetation (tree, shrub, herbaceous cover) (< 50%)
40	Mosaic natural vegetation (tree, shrub, herbaceous cover) (> 50%) / cropland (< 50%)
50	Tree cover, broadleaved, evergreen, closed to open (> 15%)
60	Tree cover, broadleaved, deciduous, closed to open (> 15%)
70	Tree cover, needleleaved, evergreen, closed to open (> 15%)
80	Tree cover, needleleaved, deciduous, closed to open (> 15%)
90	Tree cover, mixed leaf type (broadleaved and needleleaved)
100	Mosaic tree and shrub (> 50%) / herbaceous cover (< 50%)
110	Mosaic herbaceous cover (> 50%) / tree and shrub (< 50%)
120	Shrubland
130	Grassland
140	Lichens and mosses
150	Sparse vegetation (tree, shrub, herbaceous cover) (< 15%)
160	Tree cover, flooded, fresh or brackish water
170	Tree cover, flooded, saline water
180	Shrub or herbaceous cover, flooded, fresh/saline/brackish water
190	Urban areas
200	Bare areas
210	Water bodies
220	Permanent snow and ice

and rules for accurately assigning a class to a unique land cover type. Becker and others (1998) found that reference land cover data collected by students who followed the GLOBE protocols are accurate enough to support rigorous scientific investigations.

HUMAN IMPACT

As described by Beitler (2011), ecologists use *biomes* to classify the global patterns of ecology on land based on vegetation types that correspond to global patterns in climate. The different biomes—lush tropical forests, hot arid deserts, grasslands, or cold, arid tundra—support unique kinds and amounts of life. In contrast to the natural biome classification system, Ellis and Ramankutty (2008) developed a new biome system, with human-dominated ecosystems that they name *anthropogenic biomes*, to describe the ecology that people create and sustain over a long period of time. They used global data to reveal and visualize the impact of humans on the ecology of the Earth. Ellis and Ramankutty (2008) identified 21 anthropogenic biomes that represent heterogeneous landscape mosaics defined by population density and vegetation cover. The biomes are grouped into six major categories (Table 14-5):

1. **Dense settlements**: Dense settlements with substantial urban area.
2. **Villages**: Dense agricultural settlements.
3. **Cropland**: Annual crops mixed with other land uses and land covers.
4. **Rangeland**: Livestock grazing; minimal crops and forests.
5. **Forested**: Forests with human populations and agriculture.
6. **Wildland**: Land without human populations or agriculture.

Ellis and Ramankutty (2008) integrated global data sets on population, land use, and land cover to quantify the significance of anthropologic ecosystems. Their analysis produced two major insights: (1) up to 77% of the world's landscape is an anthropogenic biome and (2) human-influenced ecosystems are not just classic cropland or urban area biomes, but are environments rich in trees that promote carbon storage. Only 4 of the 22 biomes on the anthropogenic ecology map are wild (class 53, 54, 61, and 62). Ellis and Ramankutty (2008) state:

> Anthropogenic biomes point to a necessary turnaround in ecological science and education. . . . The biosphere has long been depicted as being composed of natural biomes, perpetuating an outdated view of the world as "natural ecosystems with humans disturbing them. . . ." Anthropogenic biomes tell a completely different story, one of "human systems, with natural ecosystems embedded within them. . . ."

Digital Image 14-1 🌐 is a "natural" biome land cover map of the world. Plate 48 is the anthropogenic biome map of Africa for the year 2000. When comparing the two maps for Africa, the standard land cover map (**Digital Image 14-1** 🌐) is dominated by tree, shrubland, grassland, and cropland land cover in contrast to the anthropogenic map (Plate 48) that identifies residential, populated, and remote woodlands, rangelands, and croplands along with extensive terrain mapped as settlements and villages. The anthropogenic biome map clearly reveals the geographic human impact on global ecosystems.

USGS NATIONAL MAPS

In the United States, land use and land cover maps are created by national agencies for many applications. The USGS leads a consortium of federal agencies to build the US national land cover map.

NATIONAL LAND COVER DATABASE

The National Land Cover Database (NLCD) supports a wide variety of federal, state, local, and nongovernmental applications that seek to assess ecosystem status and health, understand the spatial patterns of biodiversity, predict effects of climate change, and develop land management policy. NLCD products are created by the Multi-Resolution Land Characteristics (MRLC) Consortium, which is a partnership of federal agencies led by the USGS. All NLCD data products are available for download (mrlc.gov). The NLCD provides spatial reference and descriptive data for characteristics of the land surface such as thematic class (e.g., urban, agriculture, and forest), percent impervious surface, and percent tree canopy cover. NLCD data are thematic raster images with 30 m pixels and 256 categories in available land cover, percent impervious surface, and percent tree canopy cover.

In May 2019, the USGS released NLCD 2016 for the conterminous United States. This new edition of products includes the following:

- Twenty-eight different land cover products characterizing land cover and land cover change across seven editions of the database from 2001 to 2016.
- Urban imperviousness and urban imperviousness change across four editions of the database from 2001 to 2016.
- Tree canopy and tree canopy change across two editions of the database from 2011 to 2016.
- Western US shrub and grassland areas for 2016.

Data are available on the MRLC website either as prepackaged products or custom product areas that can be interactively chosen using the MRLC Viewer (mrlc.gov/viewer). NLCD 2016 represents the most comprehensive land cover database ever produced by the USGS and was specifically developed to meet the rapidly growing demand for land cover change data. As with the previous six NLCD land cover maps, which covered 2001 to 2013, NLCD 2016 keeps the same 16-class land cover classification scheme that has been applied consistently across the lower 48 states (Table 14-6). Several land cover categories unique to Alaska are included in the NLCD legend.

A new product for NLCD 2016 is the NLCD Land Cover Change Index. This index provides a simple and comprehensive way to visualize change from all seven editions of land cover data in a single layer. The change index was designed to assist NLCD users to understand complex land cover change with a single product.

Multispectral Landsat is the primary data source for the NLCD land cover maps. A Landsat 8 image of an area with development and agriculture east of Niagara Falls, New York, is shown in Figure 14-1. The grayscale image was taken in the NIR band (OLI band 5) and acquired on September 16, 2015 with 30 m spatial resolution. The black polygon on the west margin is the Niagara Power Project water reservoir. Along the southern margin the US Niagara Falls Air Force Base runways are visible as gray lines oriented in different compass directions. Narrow linear and curvilinear

Table 14-5 Anthropogenic biome descriptions.

	Group	Biome	Description
Dense Settlements	11	Urban	Dense built environments with very high populations
	12	Dense settlements	Dense mix of rural and urban populations, including both suburbs and villages
Villages	21	Rice villages	Villages dominated by paddy rice
	22	Irrigated villages	Villages dominated by irrigated crops
	23	Cropped and pastoral villages	Villages with a mix of crops and pasture
	24	Pastoral villages	Villages dominated by rangeland
	25	Rainfed villages	Villages dominated by rainfed agriculture
	26	Rainfed mosaic villages	Villages with a mix of trees and crops
Cropland	31	Residential irrigated cropland	Irrigated cropland with substantial human populations
	32	Residential rainfed mosaic	Mix of trees and rainfed cropland with substantial human populations
	33	Populated irrigated cropland	Irrigated cropland with minor human populations
	34	Populated rainfed cropland	Rainfed cropland with minor human populations
	35	Remote croplands	Cropland with inconsequential human populations
Rangeland	41	Residential rangelands	Rangelands with substantial human populations
	42	Populated rangelands	Rangelands with minor human populations
	43	Remote rangelands	Rangelands with inconsequential human populations
Forested	51	Populated forests	Forests with minor human populations
	52	Remote forests	Forests with inconsequential human populations
Wildland	61	Wild forests	High tree cover, mostly boreal and tropical forests
	62	Sparse trees	Low tree cover, mostly cold and arid lands
	63	Barren	No tree cover, mostly deserts and frozen land

Source: Ellis and Ramankutty (2008).

TABLE 14-6 NLCD land cover classification legend.

Class	Value	Classification Description
Water	11	Open Water—areas of open water, generally with less than 25% cover of vegetation or soil.
	12	Perennial Ice/Snow—areas characterized by a perennial cover of ice and/or snow, generally greater than 25% of total cover.
Developed	21	Developed, Open Space—areas with a mixture of some constructed materials, but mostly vegetation in the form of lawn grasses. Impervious surfaces account for less than 20% of total cover. These areas most commonly include large-lot, single-family housing units, parks, golf courses, and vegetation planted in developed settings for recreation, erosion control, or aesthetic purposes.
	22	Developed, Low Intensity— areas with a mixture of constructed materials and vegetation. Impervious surfaces account for 20 to 49% of total cover. These areas most commonly include single-family housing units.
	23	Developed, Medium Intensity—areas with a mixture of constructed materials and vegetation. Impervious surfaces account for 50 to 79% of the total cover. These areas most commonly include single-family housing units.
	24	Developed, High Intensity—highly developed areas where people reside or work in high numbers. Examples include apartment complexes, row houses, and commercial/industrial. Impervious surfaces account for 80 to 100% of the total cover.
Barren	31	Barren Land (Rock/Sand/Clay)—barren areas of bedrock, desert pavement, scarps, talus, slides, volcanic material, glacial debris, sand dunes, strip mines, gravel pits, and other accumulations of earthen material. Generally, vegetation accounts for less than 15% of total cover.
Forest	41	Deciduous Forest—areas dominated by trees generally greater than 5 m tall, and greater than 20% of total vegetation cover. More than 75% of the tree species shed foliage simultaneously in response to seasonal change.
	42	Evergreen Forest—areas dominated by trees generally greater than 5 m tall, and greater than 20% of total vegetation cover. More than 75% of the tree species maintain their leaves all year. Canopy is never without green foliage.
	43	Mixed Forest—areas dominated by trees generally greater than 5 m tall, and greater than 20% of total vegetation cover. Neither deciduous nor evergreen species are greater than 75% of total tree cover.
Shrubland	51*	Dwarf Scrub—Alaska only areas dominated by shrubs less than 20 cm tall with shrub canopy typically greater than 20% of total vegetation. This type is often co-associated with grasses, sedges, herbs, and nonvascular vegetation.
	52	Shrub/Scrub—areas dominated by shrubs; less than 5 m tall with shrub canopy typically greater than 20% of total vegetation. This class includes true shrubs, young trees in an early successional stage, or trees stunted from environmental conditions.
Herbaceous	71	Grassland/Herbaceous—areas dominated by gramanoid or herbaceous vegetation, generally greater than 80% of total vegetation. These areas are not subject to intensive management such as tilling, but can be utilized for grazing.
	72*	Sedge/Herbaceous—Alaska only areas dominated by sedges and forbs, generally greater than 80% of total vegetation. This type can occur with significant other grasses or other grass like plants, and includes sedge tundra, and sedge tussock tundra.
	73*	Lichens—Alaska only areas dominated by fruticose or foliose lichens generally greater than 80% of total vegetation.
	74*	Moss—Alaska only areas dominated by mosses, generally greater than 80% of total vegetation.
Planted/ Cultivated	81	Pasture/Hay—areas of grasses, legumes, or grass-legume mixtures planted for livestock grazing or the production of seed or hay crops, typically on a perennial cycle. Pasture/hay vegetation accounts for greater than 20% of total vegetation.
	82	Cultivated Crops—areas used for the production of annual crops, such as corn, soybeans, vegetables, tobacco, and cotton, and also perennial woody crops such as orchards and vineyards. Crop vegetation accounts for greater than 20% of total vegetation. This class also includes all land being actively tilled.
Wetlands	90	Woody Wetlands—areas where forest or shrubland vegetation accounts for greater than 20% of vegetative cover and the soil or substrate is periodically saturated with or covered with water.
	95	Emergent Herbaceous Wetlands—areas where perennial herbaceous vegetation accounts for greater than 80% of vegetative cover and the soil or substrate is periodically saturated with or covered with water.

Note: Values with an * are categories unique to Alaska.

gray lines are roads. The rectangular features with various shades of gray fill are agricultural fields in different stages of growth.

The NLCD 2016 land cover map shows that the southern portion of the area is dominated by different levels of development ranging from developed, open space to developed, high intensity (Plate 49). The runways and buildings at the Air Force base are mapped as high-intensity development. The power project reservoir is mapped as open water with woody wetlands located to the east across the northern portion of the area. Pasture/Hay, Cultivated Crops, and Deciduous Forest land cover categories dominate the central and nonwetland northern portions of the area.

USGS Anderson Classification System

The NLCD legend was modified from the USGS Anderson Land Cover Classification System (Anderson and others, 1976). The Anderson system is hierarchal with four levels. The features that can be mapped in each level within the system are defined by the spatial resolution of the remotely sensed data that is acquired.

- Level I uses spatial resolution imagery of 20 to 100 m.
- Level II uses spatial resolution imagery of 5 to 20 m.
- Level III uses spatial resolution imagery of 1 to 5 m.
- Level IV uses spatial resolution imagery of 0.25 to 1 m. Level IV features cannot be mapped with imagery that has 5 m pixels.

A detailed chart and table of minimum remote sensing resolutions for mapping urban/suburban features is provided by Jensen (2007, Figure 13-2 and Table 13-2).

The Anderson system is well suited for detailed land cover mapping of urban/suburban features. It is a hierarchy system with established numerical designations for each category in each level. The name of each category describes the classification. The final attribute for a location is a string of numbers, where each number was selected from a category from each level (Anderson and others, 1976). An example follows for inventorying a house in an urban residential landscape (Table 14-7). We begin the classification with Level I, which has nine categories. We select the category Urban or Built-up Land (1). We now proceed to Level II, which has seven categories. For our example, we select the Residen-

Figure 14-1 2015 Landsat 8 OLI band 5 (NIR) image of a developed and rural area east of Niagara Falls, New York (see Plate 49 for NLCD 2016 land cover map).

tial category (11). In Level III, Single-Family Residential is selected (111), followed by the more detailed Level IV, which includes House, houseboat, hut, tent (1111).

NLCD Percent Developed Impervious Map

A wide range of urban ecosystem studies, including urban hydrology, urban climate, land use planning, and resource management, require current and accurate geospatial data of urban impervious surfaces. Monitoring the extent of imperviousness assists in efforts to mitigate the negative effects of impervious cover, such as the risks related to storm water runoff, urban heat islands, and natural hazards, such as flooding or drought. Areas with a high percentage of impervious surfaces have a high rate of surface runoff during precipitation events and amplify the urban heat island effect.

An example of the NLCD 2016 percent developed impervious surface map for San Francisco is shown in Figure 14-2B. The grayscale ranges from 0% impervious surface area (white) to 100% impervious surface area (black) in Figure 14-2B. A recent Landsat image is displayed in Figure 14-2A for comparison with the impervious map. Downtown San Francisco is in the upper right portion of Figure 14-2. The dense network of paved roads and buildings results in a dark tone on the impervious map. In contrast, Presidio Park at the north end of the San Francisco peninsula, Golden Gate Park on the west side, and other open terrain are displayed as large white areas with a sparse network of paved roads (gray lines). New for NLCD 2016 is an impervious surface descriptor layer. This descriptor layer identifies types of roads, core urban areas, and energy production sites for each impervious pixel to allow deeper analysis of developed features.

Other National Maps

Other national maps include the USGS National Map of Surficial Minerology (Rockwell and others, 2015) (see Plate 41C). This map identifies exposed minerals in the lower 48 states. The map was generated from 1,630 ASTER scenes, 447 Landsat 7 ETM+ scenes, and data acquired from airborne hyperspectral surveys. The US Fish and Wildlife Service (2017) provides the National Wetlands Inventory as a web map service, a kml file for viewing with Google Earth, and as a GIS database.

TABLE 14-7 Example of USGS Anderson land cover levels for urban or built-up land category.

Level I	Land Cover/Land Use Category
1	Urban or built-up land
2	Agricultural land
3	Rangeland
4	Forest land
5	Water
6	Wetland
7	Barren land
8	Tundra
9	Perennial snow or ice
Level II	**Land Cover/Land Use Category**
11	Residential
12	Commercial and services
13	Industrial
14	Transportation, communications, and utilities
15	Industrial and commercial complexes
16	Mixed urban or built-up
17	Urban or built-up land
Level III	**Land Cover/Land Use Category**
111	Single-family residential
112	Multiple-family residential
Level IV	**Land Cover/Land Use Category**
1111	House, houseboat, hut, tent
1112	Mobile home

REGIONAL MAPPING

Slonecker and others (2012) used USDA NAIP aerial imagery collected between 2004 and 2010 and the USGS NLCD of 2001 to map and measure changes in land cover and land use in Bradford and Washington Counties, Pennsylvania. The area has extensive drilling and production associated with hydraulic fracturing of the Marcellus Shale and with extraction of coal bed methane.

Washington County has 56% forest cover that supports a diverse population of birds and other animals. Agriculture covers 27% and developed areas cover 14% of the county. This study's landscape assessment utilizes spatially explicit imagery and GIS data on land cover, elevation, roads, hydrology, vegetation, and in situ sampling results to compute a suite of numerical indicators known as landscape metrics to assess ecosystem condition. Metrics such as average patch size, fragmentation, and interior forest dimension capture spatial characteristics of habitat quality and potential change effects on critical animal and vegetation populations.

The Slonecker and others (2012) report shows that gas extraction infrastructure changed the landscape with consequences for the ecosystems, wildlife, and human populations co-located with extraction activities. Figure 14-3 displays a 2010 aerial image of forested landscape in McKean County, Pennsylvania, revealing the spatial effects of roads, well pads, and pipelines related to natural gas development in the area. Forest is shown with a rough medium dark gray texture, agricultural fields with a smooth medium gray tone, and clearing for infrastructure and development is shown as white lines.

Natural gas exploration and development result in spatially explicit patterns of landscape disturbance involving the construction of well pads and impoundments, roads,

A. 2013 Landsat 8 band 6 (SWIR1) image.

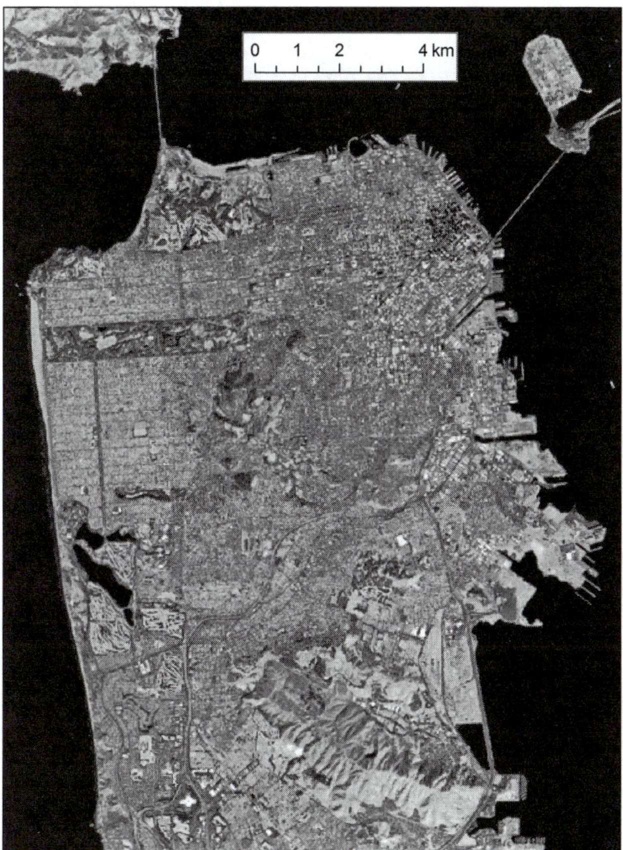

B. NLCD 2016 percent developed impervious surface area map.

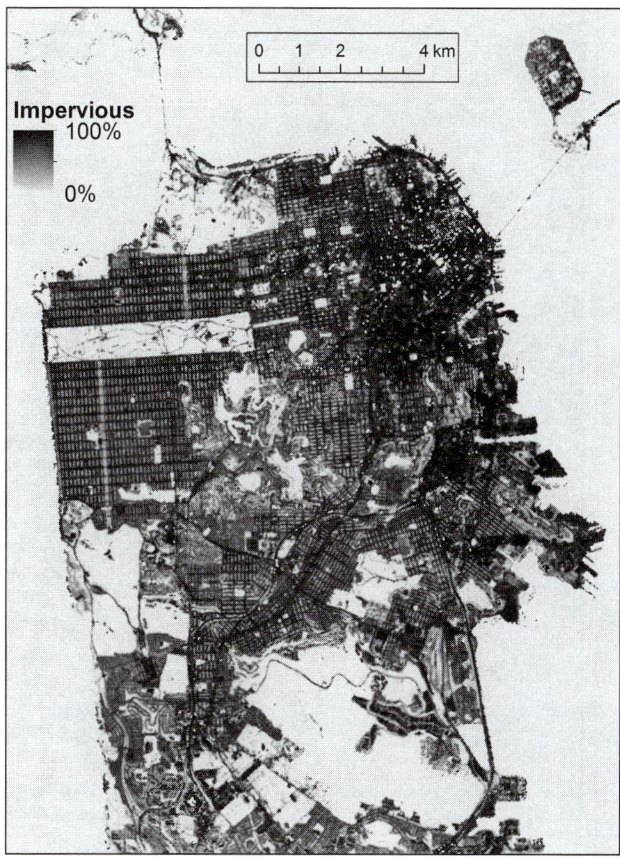

Figure 14-2 Landsat image compared with NLCD 2016 impervious surface area map, San Francisco, California. Landsat and NLCD data courtesy USGS.

pipelines, and disposal activities that impact the landscape. Forest fragmentation, forest edge, integrity of interior forest (areas that are at least 100 m from the forest edge), and other metrics were developed with imagery and GIS. The imagery was interpreted for distinct signs of disturbance related to oil and gas development. The disturbed areas were manually digitized as line and polygon features in a GIS. The disturbance footprints were superimposed on the 2001 NLCD land cover map to develop a new class: gas extraction disturbance.

Figure 14-4 shows gas extraction disturbance identified between 2004 and 2010 in Washington County. The disturbance occurs in two general clusters: the northwest, which is mostly Marcellus Shale development, and the southeast, which is mostly coal bed methane development. The detailed inset shows the disturbance footprints (black features) in the context of a simplified, two-class (forest and agriculture) NLCD land cover map. The white lines are non-gas field roads.

Infrastructure metrics for development in Washington County (223,000 hectares) between 2004 and 2010 include:

- A total of 671 disturbance sites with 170 Marcellus and 510 coal bed methane sites.
- Footprint disturbance from all infrastructure totaled 1,847 hectares (4,623 acres).
- 277 km of road and 216 km of pipeline were constructed.
- The mean size of disturbance per site is 1.3 hectares (13,000 m^2 or 3.2 acres). Marcellus sites are larger (3.0 hectares) compared with coal bed methane sites.
- Forest declined by 0.42%, which contributed to a 0.96% loss of interior forest and a gain of 0.38% in edge forest.
- Forest patches increased from 3,660 to 4,644 with 505 of these patches attributed to pipeline construction.

Land cover metrics were chosen for their overall indication of human impacts on the landscape and environment, including along edges between dissimilar land covers, which increased by 1,160.9 km for Washington County, with the largest amount attributable to pipeline construction. Pipeline construction was the source of most of the increase in forest fragmentation as measured by the number and size of new forest patches. Pipeline construction was the major contributor to forest loss, with a reduction in interior forest and an increase in edge forest in both counties.

Slonecker and others (2012) used imagery and GIS to map how agricultural and forested areas are converted to a new land cover class: natural gas extraction disturbance. Energy companies can apply lessons learned in this study to help plan natural gas development that minimizes land-

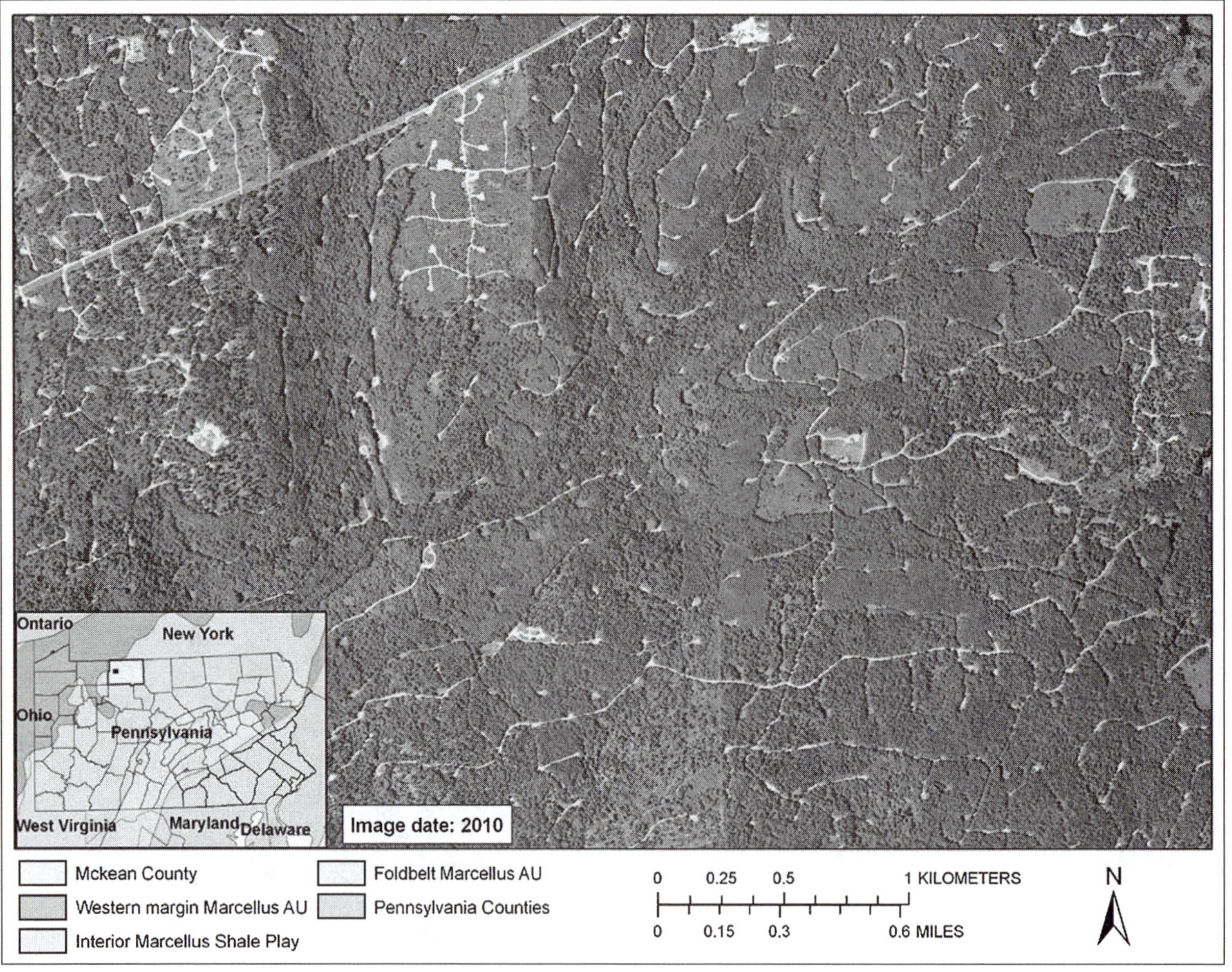

FIGURE 14-3 Forested landscape in McKean County, Pennsylvania, showing the spatial effects of roads, well pads, and pipelines related to natural gas development. Modified (subscene) from Slonecker and others (2012, Figure 2).

scape disturbance. Imagery, an accurate and up-to-date GIS database on existing infrastructure, and GIS spatial analysis tools can be proactively employed to model future landscape consequences associated with different construction and development scenarios.

LOCAL MAPPING

High spatial and spectral resolution remote sensing supports the creation of more detailed and site-specific land cover and land use maps. Airborne hyperspectral imagery of wetlands enables unique spectral libraries of vegetation to be built and new land cover classification maps of plant species and communities to be generated. These local land cover maps and the spectral libraries are designed to answer specific questions about the environment. Zomer and others (2009) collected 1,336 spectral measurements in the field of plant species, mixed vegetation, and soil in coastal wetland communities of California, Texas, and Mississippi. The hyperspectral library is supported by field observations and is used to guide image processing and land cover classification. In addition, the field-based spectral library is used to validate airborne hyperspectral signatures of plant communities and other environmental features.

The processes involved in building a local land cover map of the Pacheco Creek wetlands in Martinez, California (shown in the upper central portion of Figure 9-28 and Plate 29A) is documented by Zomer and others (2009). A handheld spectrometer that recorded VNIR-SWIR bands was used to build the spectral library. The airborne hyperspectral system collects 128 VNIR-SWIR bands at 5 m spatial resolution, 12-bit radiometric resolution, and 12 to 13 nm bandwidth. The spectral library is applied to airborne hyperspectral data to produce a vegetation map that displays the geographical distribution of plant species within the marsh, including saltgrass, cattails, phragmites, bulrush, and pickleweed.

This study demonstrates that very detailed land cover maps of specific areas are developed with hyperspectral remote sensing that includes field-based spectral libraries and observations. Zomer and others (2009) recommend a standardized classification system (similar to the FAO's LCCS described earlier) be adopted for wetland communities. The standardized system would build on spectral libraries and

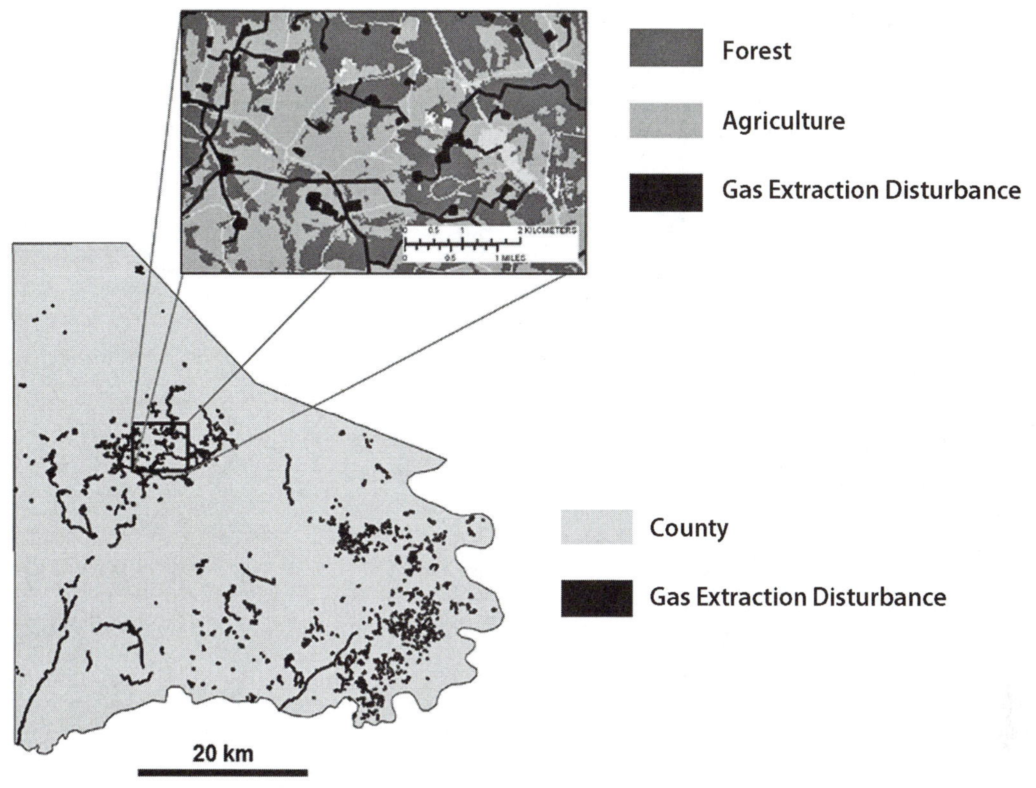

Figure 14-4 Gas extraction-related disturbance identified between 2004 and 2010 in Washington County, Pennsylvania. Disturbance footprints are depicted in black. Simplified from Slonecker and others (2012, Figure 10).

facilitate a global approach to implementing new advanced remote sensing technologies for mapping and conservation of wetlands (Zomer and others, 2009).

ACCURACY ASSESSMENT

Land cover maps derived from remote sensing imagery are subjected to an accuracy assessment to enable the map to be confidently used for different applications. In general, an accuracy assessment of a land cover classification map compares the map to a more detailed, independently collected sample set named *verification* or *reference data*. The verification data can be based on field observations and visual interpretation of higher spatial resolution imagery. Congalton and Green (2008) provide extensive information on concepts, design, and implementation of accuracy assessment techniques.

Fifty samples for each land cover category is a good rule of thumb to obtain a statistically sound assessment (Congalton and Green, 2008). However, 50 samples per category are often impractical or economically unfeasible if collected solely in the field. More samples per land cover category can be achieved by including higher spatial resolution imagery that enables accurate identification of the land cover types. The location and number of verification sites for each land cover category should be included with the map.

Verification Sites (Reference Data)

In order to demonstrate the accuracy assessment methodology of Congalton and Green (2008), six land cover classes were created and mapped. The resulting polygons on Figure 14-5A were classified and labeled with a land cover category number (1 through 6). A limited number of verification sites were obtained for each of these land cover classes and plotted as black dots on the land cover map (Figure 14-5B). Visual inspection of the spatial distribution of the verification sites in Figure 14-5B shows that only the northern half of the map can be assessed for accuracy. The southern portion of the map lacks verification sites, perhaps because it is inaccessible to field crews or there is a lack of imagery with a higher spatial resolution than that used for the classification map. This accuracy assessment shortcoming should be documented in the map's metadata.

Thirty-one verification sites were used for the accuracy assessment. The number of verification sites per land cover category ranges from 4 to 6. The land cover category documented at each verification site was compared to the category mapped at the corresponding pixel(s) in the classification map. The site was correctly mapped if the verification site and map category agree. The site was incorrectly mapped (an error) if the verification site and map category do not agree. Table 14-8 lists the number of sites that agree with the map and the number of sites that do not agree for each category. A total of

FIGURE 14-5 Demonstration of the accuracy of the assessment methodology of Congalton and Green (2008).

Class	Land cover category
1	Water
2	Wetlands
3	Industry
4	Trees
5	Grass
6	Suburbs

A. Land cover map with six classes.

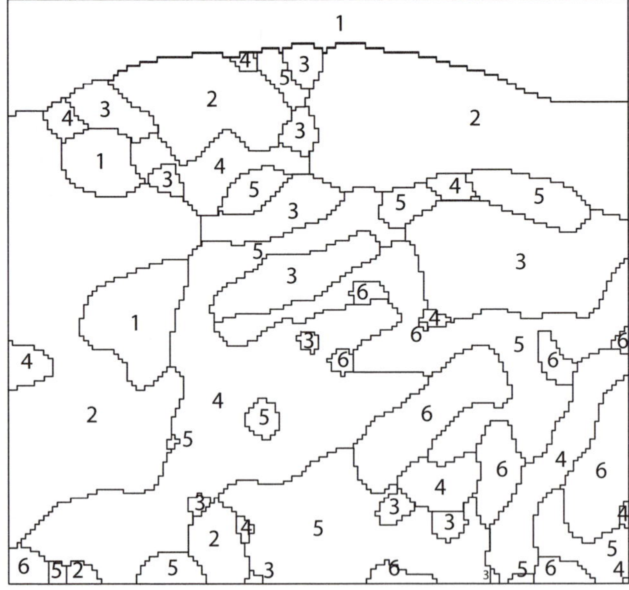

B. Land cover map with verification sites added (black dots) (Table 14-8).

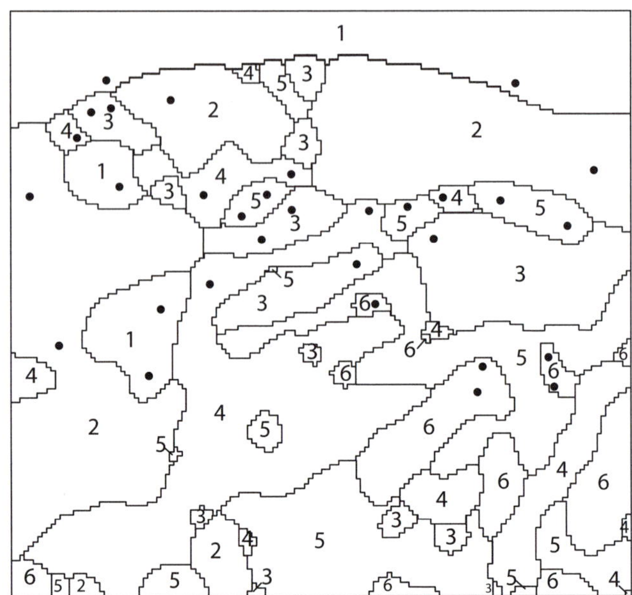

21 sites were correctly mapped while 10 sites were incorrectly mapped. The verification site IDs are listed Table 14-8.

Table 14-9 provides information on which verification sites agree (C) and disagree (B) with the land cover map. It shows the correct category (A) for those verification sites that were incorrectly mapped. For example, verification site ID 6 was categorized as wetlands, but it was mapped as water. Suburbs, grass, and trees have many incorrect (B) and misclassified (A) entries in Table 14-9, indicating spectral similarities that results in classification errors on the map.

ERROR MATRIX

An accuracy assessment *error matrix* quantifies the correct and incorrect categories on a land cover map in a standardized manner (Congalton and Green, 2008). Table 14-10 is the error matrix for the land cover map and verification sites shown in Figure 14-5. The error matrix compares information from verification sites to information on the map for all of the sample sites. The matrix is a square array of numbers set out in rows and columns that express the labels of samples assigned to a particular category in the verification data relative to the labels of samples assigned to a particular category on the map.

Within the error matrix, Column Total is the total number of samples in each column or category, is assumed to be correct, and is termed the *reference data.* Row Total displays the total number of samples in each category on the map and is termed the *classified data.* If the classified map was 100% accurate, the integers found in the Column Total and

TABLE 14-8 Characteristics of verification sites.

Class	Land Cover Category	Number of Verification Sites	Map and Verification Sites Agree (Correct)	Map and Verification Sites Do Not Agree (Error)	Verification Site ID
1	Water	5	5	0	1–5
2	Wetlands	4	3	1	6–9
3	Industry	6	4	2	10–15
4	Trees	6	3	3	16–21
5	Grass	5	2	3	22–27
6	Suburbs	5	4	1	28–31
Total		31	21	10	

the Row Total cells would be equal for each category. The numbers in the Column Total and Row Total cells always sum to the same value—in this example to 31—that is, the total number of verification sites in our accuracy assessment. Error matrices are very effective representations of map accuracy, because the individual accuracies of each map category are plainly described. The overall accuracy is simply the sum of the major diagonal (i.e., the correctly classified pixels or samples) divided by the total number of pixels or samples in the error matrix.

Table 14-9 Summation of correct and incorrect areas (pixels) on land cover classification map.

Verification Site ID	Water	Wetlands	Industry	Trees	Grass	Suburbs
1	C					
2	C					
3	C					
4	C					
5	C					
6	B	A				
7		C				
8		C				
9		C				
10			A		B	
11			C			
12			C			
13			A			B
14			C			
15			C			
16				C		
17				A		B
18				A		B
19				C		
20				A		B
21				C		
22					C	
23					A	B
24					A	B
25		B			A	
26					C	
27						C
28						C
29						C
30				B		A
31						C
Total	6	4	4	4	3	10
Grand Total = 31						

A = Right answer for incorrectly (B) classified pixel(s) based on verification data.

B = Incorrect: Pixel(s) mapped as this category *does not agree* with verification data.

C = Correct: Both classified map and verification data *agree* on category.

Total = number of correct (C) and incorrect (B) in each category. (A) is not included.

Grand Total = number of correct (C) and incorrect (B) areas (pixels) on classified map for all categories. (A) is not included.

TABLE 14-10 Accuracy assessment error matrix.

	What is the Category According to the Classification Map?	Reference Data: What is the Category According to the Verification Sites?						
		Water	**Wetlands**	**Industry**	**Trees**	**Grass**	**Suburbs**	**Row Total**
Classified Data	Water	5						5
	Wetlands	1	3					4
	Industry			4		1	1	6
	Trees				3		3	6
	Grass		1			2	2	5
	Suburbs				1		4	5
	Column Total	6	4	4	4	3	10	31

Black box (correct) = classification map and verification site have the same category (C in Table 14-9).

Gray box (error) = classification map and verification site are different (B in Table 14-9).

Total correct = 5 + 3 + 4 + 3 + 2 + 4 = 21.

Overall accuracy = 21 / 31 = 68%.

The overall accuracy for the land cover map (Figure 14-5A) is 68%. An overall accuracy of 85% is acceptable in many applications. The error matrix indicates that trees and grass are the least accurately mapped categories. In addition, many categories are misclassified as suburbs. Standardized accuracy measures of *individual* land cover categories, such as trees, grass, and suburbs, can be extracted from the error matrix (Congalton and Green, 2008). Improving the training sites for supervised classification, reducing the number of land cover categories (generalization), changing the parameters for unsupervised classification, manual editing of the map's attribute table, and employing a sensor with higher spectral and spatial resolution are techniques used to increase the map accuracy of individual classes.

DASYMETRIC MAPS

Human population distributions are commonly displayed using census data. However, these data are aggregates of geographic units (census tracts or block groups) whose boundaries do not always reflect the natural distribution of human populations. A *dasymetric* mapping technique is one potential solution for mapping population density relative to residential land use. Dasymetric mapping depicts quantitative areal data using boundaries that divide the area into zones of relative homogeneity with the purpose of better portraying the population distribution (USGS, 2017a).

Census tracts with population density are shown in Figure 14-6A. For this example, the population density is provided

A. Population density data on census tracts.

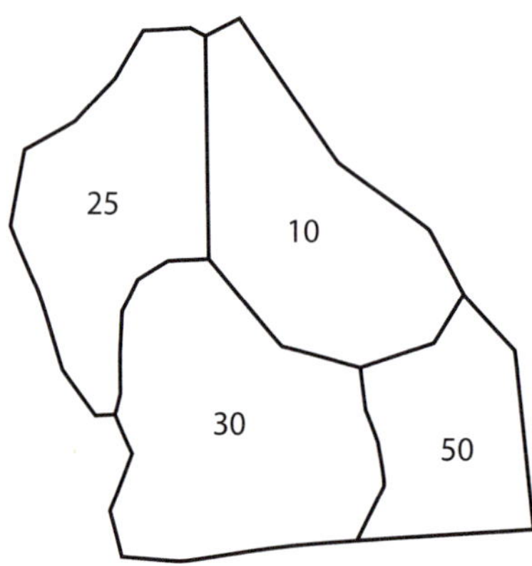

B. Land cover map with residential areas.

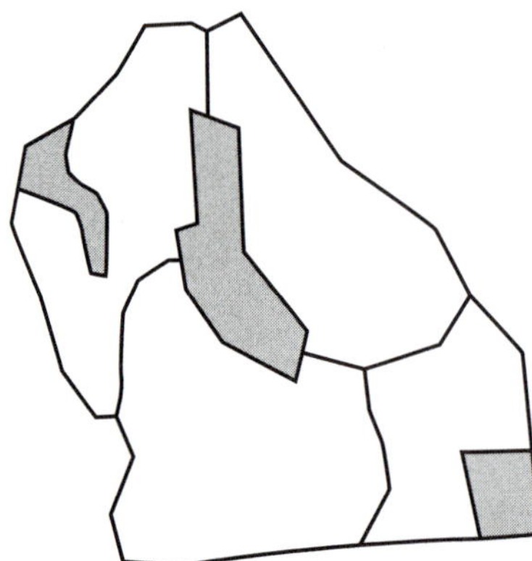

C. Dasymetric map with census data transformed to only residential areas.

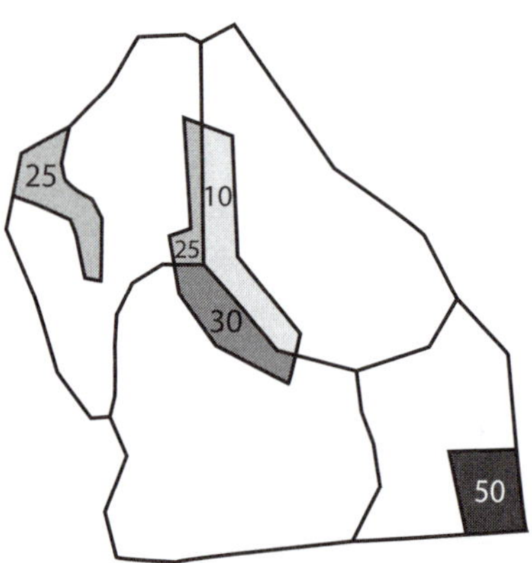

FIGURE 14-6 Transformation of population density data on census tracts to a dasymetric map incorporating land cover maps of residential areas.

as the number of individuals · km^{-2}. The population density ranges between 10 and 50 individuals · km^{-2}. A land cover map of residential areas (gray polygons) is shown in Figure 14-6B. The residential areas are derived from remote sensing imagery and superimposed on the census tracts. People do not live outside the residential polygons. The population density of each census block is assigned to the overlapping residential polygon in the dasymetric process (Figure 14-6C). The northwestern census tract (Figure 14-6A) has residents located in only two areas that together average a population density of 25 individuals · km^{-2}. The northeastern census tract has residents located along the western edge with a population density of 10 individuals · km^{-2}. The dasymetric map (Figure 14-6C) is a more realistic interpretation of populated areas and population density compared with the original census tract map (Figure 14-6A).

The USGS (2017a) provides examples and step-by-step methodology of the dasymetric technique for integrating the NLCD with US Census Bureau data. The NLCD classification system has four levels for the developed category (open space, low intensity, medium intensity, high intensity) (Table 14-6). Therefore, the USGS dasymetric technique would subdivide the "residential" polygons in Figure 14-6 into four levels of population density with a spatial resolution of 30 m. Mennis (2003) and Eicher and Brewer (2001) provide more examples and methods of dasymetric mapping.

MODELING THE PAST AND FUTURE

Land cover maps based on remote sensing images can be integrated through time to document change. Land cover maps can be extended into the future to predict change using existing maps, socioeconomic and population data, and computer modeling techniques. USGS research with past and future land cover maps are summarized below.

PAST LAND COVER CHANGE

The USGS Land Cover Trends Project focused on understanding the rates, trends, causes, and consequences of contemporary US land use and land cover change from 1999 to 2011 (USGS, 2017b). Ongoing research is being conducted as part of the Land Change Research Project. The USGS report *Status and Trends of Land Change in the Western United States* (Sleeter and others, 2012) relied on Landsat satellite imagery—the longest continuous and consistent dataset of synoptic Earth observations—to characterize changes across 11 primary LULC classes between 1973 and 2000 (Table 14-11). These classes are based on the LULC classes from the 2001 NLCD and the Anderson Land Cover Classification System (Anderson and others, 1976), but include two transitional disturbance categories: mechanically disturbed and nonmechanically disturbed.

The report divided the western United States into six geographic regions (Figure 14-7). These six regions contain 30 ecoregions. Dominant patterns of change have been associated with urbanization, wildfire, forest cutting for timber production, and shifts in agricultural production. The *overall* spatial change across the western United States, that is, the amount of land area changed at least one time over the 27-year study period, was only 5.8%. However, considerable variability exists with spatial change. For example, the Puget lowland ecoregion in the Marine West Coast Forests region underwent a 28% spatial change while the Chihuahuan Desert ecoregion in the Warm Deserts region underwent an estimated 0.5% change.

Ecoregions where timber harvesting is common experienced the highest rates of land cover change. Urbanization and expansion of new developed land was most common in coastal ecoregions. Coastal central and southern California experienced 2,234 km^2 conversion to new developed land over the 27-year study period. Agricultural land was converted to urban uses and farmers relocated to peripheral areas, changing the spatial footprint and location of the agricultural land cover class. The results of the 1973 to 2000 study provide useful, if not essential, information for understanding climate change, biodiversity, resource management and planning, resource security, and disaster planning (Sleeter and others, 2012).

FUTURE LAND COVER CHANGE

Projecting future land cover change allows for the optimization and mitigation of potential consequences on numerous ecosystem processes, such as biodiversity, water quality, and climate. Projecting these changes for an area requires modelers to account for the driving forces behind land cover change, and how these driving forces interact over space and time. Modelers must also operate at scales from local ("bottom-up") to global ("top-down"). As a result, there is a high level of uncertainty associated with predicting future developments in complex socio-environmental systems (USGS, 2016).

Scientists at USGS Earth Resources Observation and Science (EROS) use scenario-based modeling approaches to represent a wide range of plausible future conditions. EROS developed the Forecasting Scenarios of Land-use Change (FORE-SCE) modeling framework to provide spatially explicit projections of future land use and land cover change. It also uses a modular approach to handle large-scale (national to global) and small-scale (local) drivers of change. Based on IPCC (2007) global economic and population scenarios (A1B, A2, B1, and B2), future scenario models have been developed to assess carbon and greenhouse gas fluxes and emissions (2006 to 2050), bird species distribution (2001 to 2075), and High Plains groundwater availability (2008 to 2050). Modeled annual forest and land cover maps of the conterminous 48 states have also been produced.

Plate 50 is a FORE-SCE model of projected urban growth between 2006 and 2050 in the Washington, DC and Baltimore, Maryland, area in the eastern United States. The IPCC (2007) economic and population scenario A2 was used for this simulation. In the IPCC A2 scenario, high population growth, an emphasis on economic growth over environmental conservation, and high demand for food, fiber, and energy resources leads to an expansion of the human footprint on the landscape. "Anthropogenic" land cover classes, those representing intense human use of the landscape (i.e., urban development, agricultural land, mining, and forestry), expand significantly in the A2 scenario, while "natural" landscapes (i.e., forest, grassland, shrubland, and wetland) decline significantly.

In the Washington, DC/Baltimore area, high population growth results in a significant expansion in urban development between 2006 and 2050. Both metropolitan areas expand substantially into the surrounding landscape, resulting in a loss of forested and agricultural lands. With a loss of prime agricultural land, but continued high global demand for agricultural products, agricultural activities expand in the region to increasingly marginal lands. High demand for wood and fiber products results in an increasingly managed forested landscape, with intensive management practices designed to maximize forest productivity. The relative prosperity of the region allows for a moderate regional emphasis on environmental conservation, but protected lands and other natural refugia become increasingly fragmented (USGS, 2016).

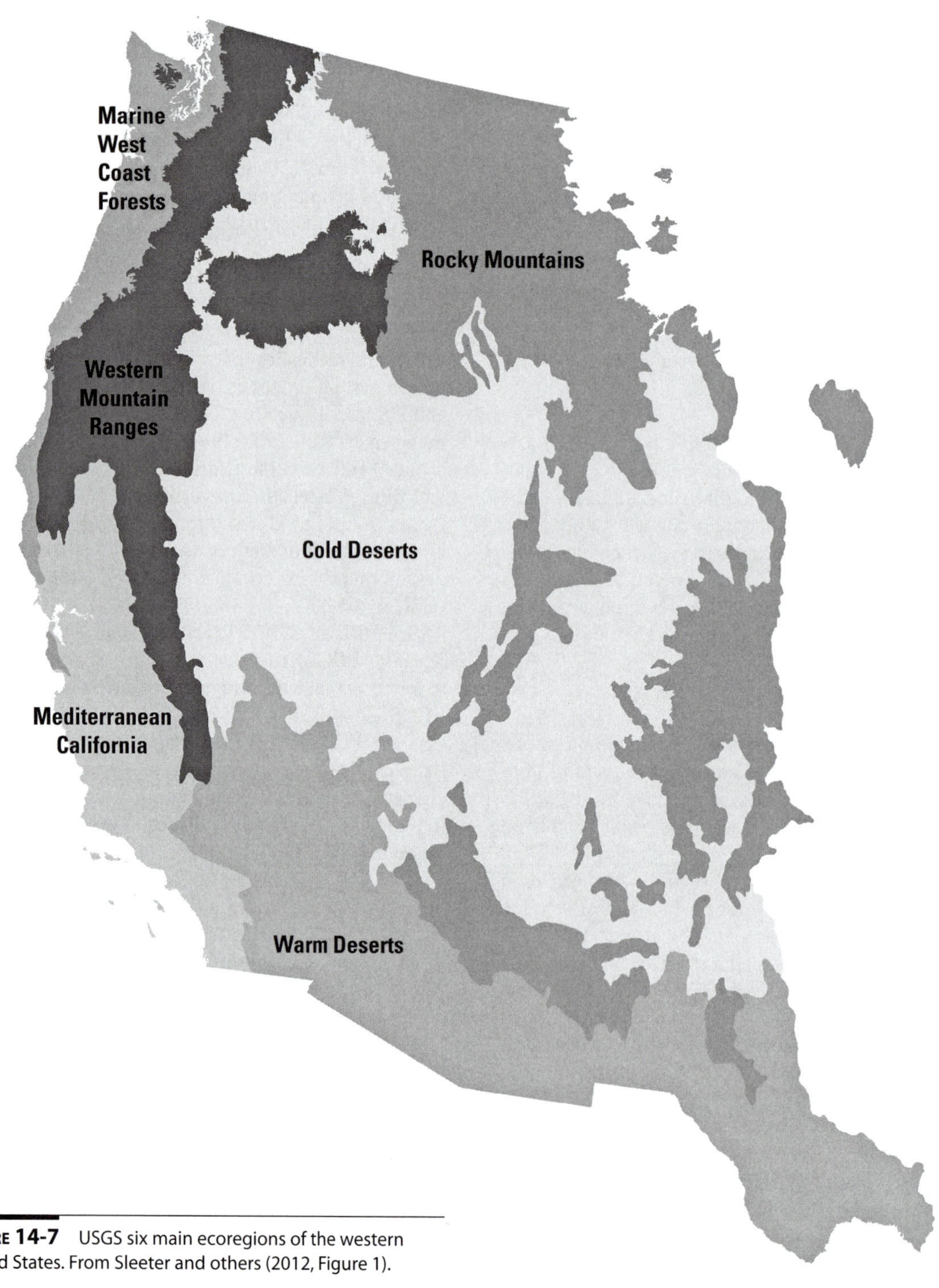

FIGURE 14-7 USGS six main ecoregions of the western United States. From Sleeter and others (2012, Figure 1).

TABLE 14-11 Eleven LULC classes used for Land Cover Trends Project classification system.

Class	Description
Water	Areas persistently covered with water, such as streams, canals, lakes, reservoirs, bays, or oceans.
Developed/urban	Areas of intensive use with much of the land covered with structures (e.g., high-density residential, commercial, industrial, transportation, mining, confined livestock operations) or less intensive uses where the land cover matrix includes both vegetation and structures (e.g., low-density residential, recreational facilities, cemeteries, etc.), including any land functionally attached to the urban or built-up activity.
Mechanically disturbed*	Land in an altered and often nonvegetated state that, due to disturbances by mechanical means, is in transition from one cover type to another. Mechanical disturbances include forest clear-cutting, earthmoving, scraping, chaining, reservoir drawdown, and other similar human-induced changes.
Barren	Land comprised of natural occurrences of soils, sand, or rocks where less than 10% of the area is vegetated.
Mining	Areas with extractive mining activities that have a significant surface expression. This includes (to the extent that these features can be detected) mining buildings, quarry pits, overburden, leach, evaporative, tailing, or other related components.
Forests/woodlands	Tree-covered land where the tree-cover density is greater than 10%. Note that cleared forest land (i.e., clear-cut logging) will be mapped according to current cover (e.g., disturbed or transitional, shrubland/grassland).
Grassland/shrubland	Land predominately covered with grasses, forbs, or shrubs. The vegetated cover must comprise at least 10% of the area.
Agriculture	Cropland or pastureland in either a vegetated or nonvegetated state used for the production of food and fiber. Note that forest plantations are considered as forests or woodlands regardless of the use of the wood products.
Wetland	Lands where water saturation is the determining factor in soil characteristics, vegetation types, and animal communities. Wetlands are comprised of water and vegetated cover.
Nonmechanically disturbed*	Land in an altered and often nonvegetated state that, due to disturbances by nonmechanical means, is in transition from one cover type to another. Nonmechanical disturbances are caused by wind, floods, fire, animals, and other similar phenomena.
Ice/snow	Land where the accumulation of snow and ice does not completely melt during the summer period.

Sources: USGS (2017b); Sleeter and others (2012).

QUESTIONS

1. What is the difference between land cover and land use? Which is more difficult to interpret from remote sensing images and why?
2. What are three reasons why global land cover classification systems need to be standardized?
3. Go to the ESA's CCI-LC online viewer (maps.elie.ucl.ac.be/CCI/viewer/index.php). The viewer automatically provides land cover data along with greenness, snow, and burned areas seasonality charts for the globe. In the upper left corner of the map display, under the zoom controls is a small gray icon with a white arrow pointing toward the southeast. Click on the icon. Empty cells for entering longitude/latitude appear. Enter the following: longitude 140.50 and latitude 36.41. Click on Go to Coordinates. You should zoom into Tokyo, Japan.

   The online viewer does not place the cursor at the 140.50E/36.41N coordinate—you have to manually do that. Note that the longitude, latitude for the cursor is shown at the bottom of the legend on the left side (scroll down to see the coordinates). An upside-down teardrop is placed on the point where you click your mouse. Click around until you are very close to the 140.50E/36.41N point on the map. Then answer the following questions:

   a. Land Cover Map 2015 is the default map (shown on the top margin of the map). What is the land cover designation for Tokyo in 2015? In the Land Cover Map drop-down menu, choose 1992. What is the land cover designation for Tokyo in 1992?

   b. Along the upper margin is a drop-down menu for Land Surface Seasonality. Choose Snow and Vegetation in the drop-down menu. Right click your mouse over Tokyo to have charts pop up on the map. What is the maximum greenness seasonality value and what month does it occur in 2015? What is the approximate probability of snow in January and February 2015?

   c. What is the spatial resolution of the CCI-LC online map?
4. How does GLOBE support students collecting reference land cover data that is used in land cover research investigations?

5. Go to the NASA SEDAC Map Viewer, Version 2 (sedac.ciesin.columbia.edu/mapping/viewer). Four maps are displayed—we will use the one in the lower left corner. The Anthropogenic Biomes v2: 2000 map enables you to see the impact of humans across the globe. Open the Legend, which will help you to answer the following questions:
   a. What anthropogenic biome characterizes northeastern India?
   b. Eastern Portugal is dominated by a biome colored light green. What is this biome?
   c. The most northern portion of Russia and Siberia is dominated by what biome?
   d. What are five anthropogenic biomes around urban Mexico City?
   e. What is the spatial resolution (number of km^2 per pixel) of the online Anthropogenic Biome map (*Hint*: it varies from the equator to the poles).
6. What does the USGS NLCD class value 24 mean?
7. What is the land cover/land use category for the USGS Anderson classification designation 1112?
8. Using remote sensing and GIS, what ecological impacts (beyond the physical footprint) can be measured on a forest due to pipeline construction?
9. Why is accuracy assessment done on land cover maps?
10. What comparison does an error matrix enable?
11. Why are dasymetric maps that integrate residential land use maps more appropriate for emergency response compared with population density by census tract maps?

REFERENCES

Anderson, J. R., E. E. Hardy, J. T. Roach, and R. E. Witmer. 1976. *A Land Use and Land Cover Classification System for Use with Remote Sensor Data* (Professional Paper 964). Washington, DC: US Government Printing Office.

Becker, M. L., R. G. Congalton, R. Budd, and A. Fried. 1998. A GLOBE collaboration to develop land cover data collection and analysis protocols. *Journal of Science Education and Technology*, 7(1), 85–96.

Beitler, J. 2011, October 19. Repatterning the World. NASA Earthdata. Sensing Our Planet. http://earthdata.nasa.gov/user-resources/sensing-our-planet/repatterning-the-world (accessed January 2018).

Congalton, R. G., & K. Green. 2008. *Assessing the Accuracy of Remotely Sensed Data: Principles and Practices* (2nd ed.). Boca Raton, FL: CRC Press.

Eicher, C. L., and C. A. Brewer. 2001. Dasymetric mapping and areal interpolation: Implementation and evaluation. *Cartography and Geographic Information Science*, 28(2), 125–138.

Ellis, E. C., and N. Ramankutty. 2008. Putting people in the map: Anthropogenic biomes of the world. *Frontiers in Ecology and the Environment*, 6(8), 439–447.

GLOBE Program. 2010. MUC Field Guide: A Key to Land Cover Classification. https://www.globe.gov/documents/355050/355097/MUC+Field+Guide/5a2ab7cc-2fdc-41dc-b7a3-59e3b110e25f (accessed January 2018).

Green, K., R. G. Congalton, and M. Turkman. 2017. *Imagery and GIS: Best Practices for Extracting Information from Imagery*. Redlands, CA: ESRI Press.

Herold, M., J. S. Latham, A. Di Gregorio, and C. C. Schmullius. 2006. Activities for evolving standards in land cover characterization. *Journal of Land Use Science*, 1(2–4), 157–168.

IPCC. 2007. Summary for Policymakers. In S. Solomon, D. Qin, M. Manning, Z. Chen, M. Marquis, K. Averyt, M. Tignor, H. L. Miller, and Z. Chen (Eds.), *Climate Change 2007: The Physical Science Basis*. Contribution of Working Group I to the Fourth Assessment Report of the Intergovernmental Panel on Climate Change. Cambridge, UK: Cambridge University Press.

Jensen, J. R. 2007. *Remote Sensing of the Environment: An Earth Resource Perspective* (2nd ed.). Upper Saddle River, NJ: Pearson.

Latham, J., R. Cumani, I. Rosati, and M. Bloise. 2014. Global Land Cover SHARE (GLC-SHARE) Database. Food and Agriculture Organization of the United Nations.

Mennis, J. 2003. Generating surface models of population using dasymetric mapping. *The Professional Geographer*, 55(1), 31–42.

Rockwell, B. W., L. C. Bonham, and S. A. Giles. 2015. USGS National Map of Surficial Mineralogy. US Geological Survey Online Map Resource. https://cmerwebmap.cr.usgs.gov/usminmap.html (accessed January 2018).

Sleeter, B. M., T. S. Wilson, and W. Acevedo (Eds.). 2012. *Status and Trends of Land Change in the Western United States—1973 to 2000* (Professional Paper 1794-A). Washington, DC: US Geological Survey.

Slonecker, E. T., L. E. Milheim, C. M. Roig-Silva, A. R. Malizia, and G. B. Fisher. 2012. *Landscape Consequences of Natural Gas Extraction in Bradford and Washington Counties, Pennsylvania, 2004–2010* (Open File Report 2012-1154). Washington, DC: US Geological Survey.

US Fish and Wildlife Service. 2017. Geospatial Services—USFWS National GIS Data. https://www.fws.gov/gis/data/national/index.html (accessed January 2018).

USGS. 2016. Land Use Land Cover Modeling. https://landcover-modeling.cr.usgs.gov/index.php (accessed January 2018).

USGS. 2017a. Dasymetric Mapping: An Alternative Approach to Visually and Statistically Enhancing Population Density. USGS Western Geographic Center. https://geography.wr.usgs.gov/science/dasymetric (accessed January 2018).

USGS. 2017b. Land Cover Trends. https://www.usgs.gov/centers/wgsc/science/land-cover-trends?qt-science_center_objects=0#qt-science_center_objects (accessed April 2019).

Zomer, R. J., A. Trabucco, and S. L. Ustin. 2009. Building spectral libraries for wetlands land cover classification and hyperspectral remote sensing. *Journal of Environmental Management*, 90(7), 2,170–2,177.

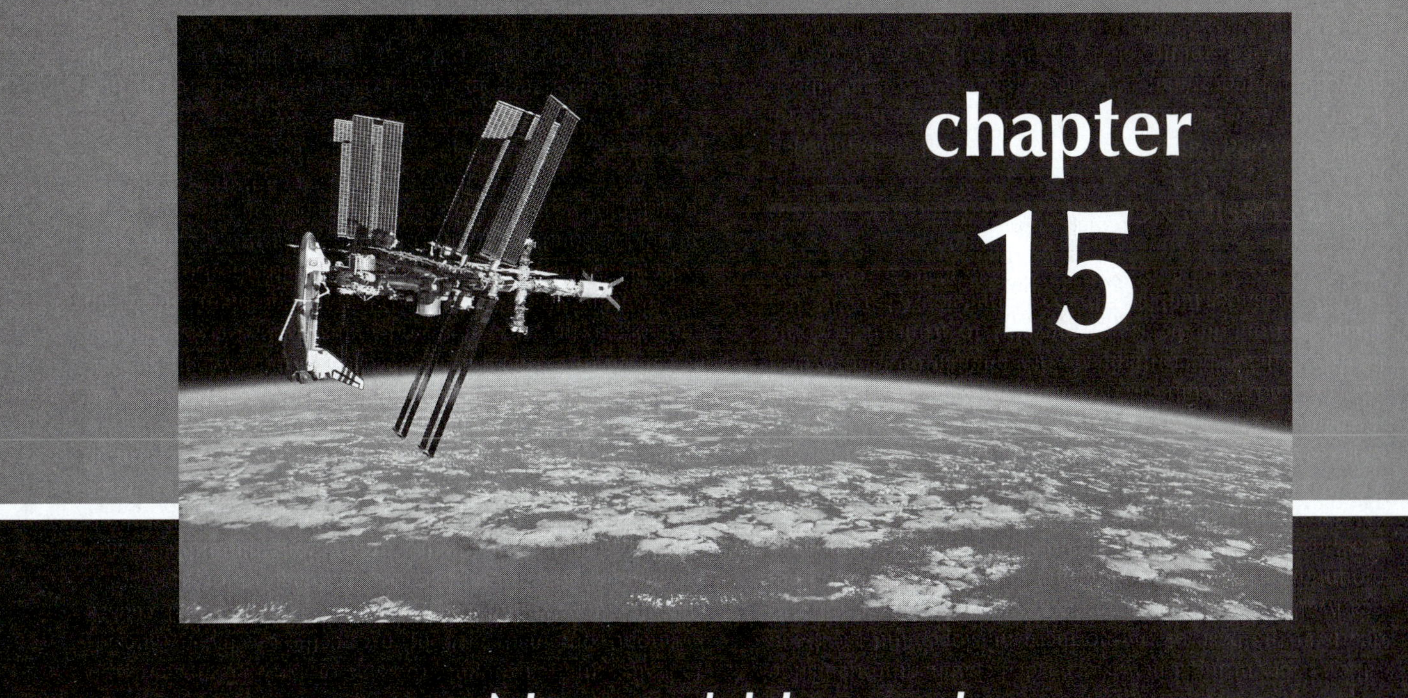

chapter 15

Natural Hazards

Earthquakes, landslides, volcanic eruptions, fires, droughts, and floods are natural hazards that typically kill tens of thousands of people and destroy tens of billions of dollars of habitat and property each year. Floods are the number one natural disaster in the United States and they account for 40% of all natural disasters worldwide. Flooding is the leading cause of weather-related deaths in the United States (NASA JPL, 2019). The year 2017 was the most expensive year in the decade due to a series of powerful hurricanes across the United States and Caribbean. Hurricane Harvey, Hurricane Irma, and Hurricane Maria cost $95 billion, $80.7 billion, and $69.7 billion, respectively. When looking at types of events, 2017 was characterized by a higher number of reported storms (127) compared to the annual average (98). Similar patterns were seen with wildfires, with 15 compared to the annual average of 9, and 25 landslides compared to the annual average of 17 (CRED, 2018). These losses will rise as population increases and more people reside in areas that are subject to these hazards. Dams can control some flood hazards, and proper engineering design can reduce landslide risks. Aside from steps such as these, there is little that people can do to prevent the occurrence of natural hazards. However, the following actions will minimize their effects:

1. Analyze the risk that natural hazards will occur in a given area. Examples are to identify faults or volcanoes that have the potential for earthquakes or eruptions. In addition to recognizing hazards, risk analysis should delineate areas on the basis of their relative susceptibility to damage.
2. Provide advance warning for specific hazardous events, which is not practical for many hazards. However, satellite images provide timely warnings of floods and severe weather. Volcanic eruptions in Hawaii and elsewhere have been predicted on the basis of ground movements.
3. Assess the damage caused by a hazardous event. An early evaluation of damage caused by floods and earthquakes is essential for carrying out rescue, relief, and rehabilitation efforts.

Remote sensing is a valuable tool for analyzing, warning about, and assessing damage related to natural hazards (Lewis, 2009).

EARTHQUAKES

Earthquakes are caused by the abrupt release of strain that has built up in the Earth's crust. Most zones of maximum earthquake intensity and frequency occur at the boundaries between the moving plates that form the crust of the Earth. Major earthquakes also occur within the interior of crustal plates, such as those in China, Russia, and the southeast United States. Much research has been done to predict earthquakes using non-remote sensing technologies, but results to date are inconclusive. *Seismic risk analysis,* however, is an established discipline that estimates the geographic distribution, frequency, and intensity of seismic activity without attempting to predict specific earthquakes. This analysis is essential for locating and designing dams, power plants, and other projects in seismically active areas.

One method of seismic risk analysis is based on the study of *historic earthquakes,* which are those recorded by humans. These records cover some 2,000 years in Japan and 3,000 years in China but are less extensive in other regions. In southern California, for example, the earliest historic earthquake was recorded in 1769. Beginning in the 1930s, earthquakes have been recorded by instruments called *seismographs.* Both the historic and instrumental records are too brief to make valid predictions of earthquakes.

The second method of seismic risk analysis is based on the recognition of *active faults,* which are defined as breaks along which movement has occurred in late Quaternary, or Holocene, time (the past 11,700 years). Remote sensing analyses and field studies of active faults provide a geologic record that extends our instrumental and historic records. Surface faulting during large shallow earthquakes is universal; analysis of this geomorphic evidence and radiometric age dating of earlier events are two techniques that are utilized (Allen, 1975). Remote sensing images are now facilitating the recognition and measurement of active faults, as shown by examples from California, China, and the seafloor.

SOUTHERN CALIFORNIA

Southern California is an ideal region to demonstrate remote sensing for seismic risk analysis because:

1. The region is seismically active.
2. The Southern California Seismic Network (SCSN) records details of earthquakes, with an online computer catalog extending back to 1932.
3. Many active faults are well exposed in the mountains and desert.

Figure 15-1 is a map of the region showing the major faults, many of which are active. The heavy lines identify fault segments that have ruptured in historic time (1872 to 1994). The dates and magnitudes of major historic earthquakes

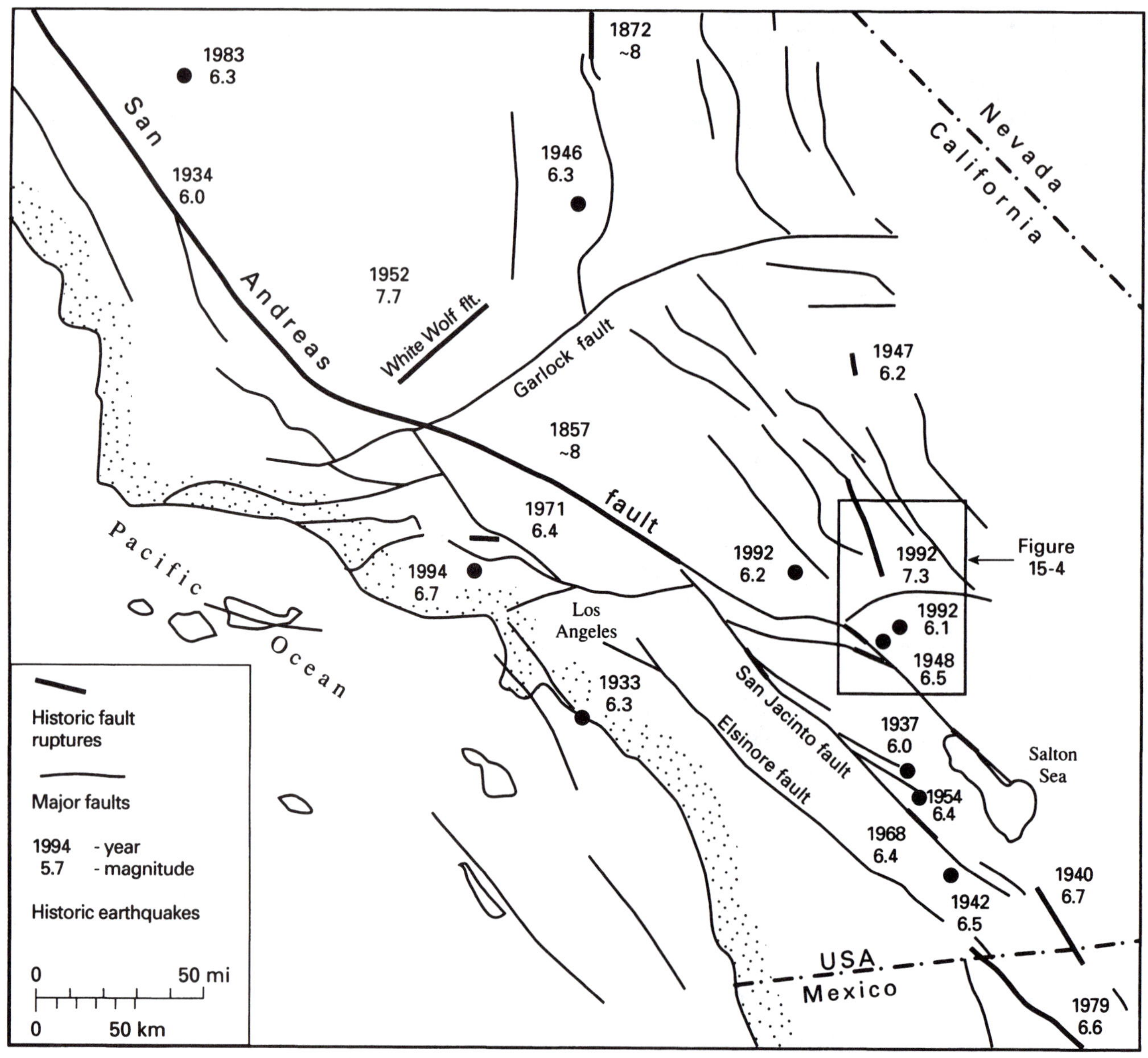

FIGURE 15-1 1872 to 1994 fault map of southern California showing dates and magnitudes of historic earthquakes (magnitude > 6.0). Darker lines indicate historic fault ruptures. Dots indicate earthquakes with no surface ruptures. Modified from Hutton and others (1991, Figure 2).

(having magnitudes of 6.0 and greater on the Richter scale) are shown. Many of these earthquakes are associated with the active fault breaks. Dots indicate earthquakes that did not rupture the surface, although several occurred along the traces of active faults.

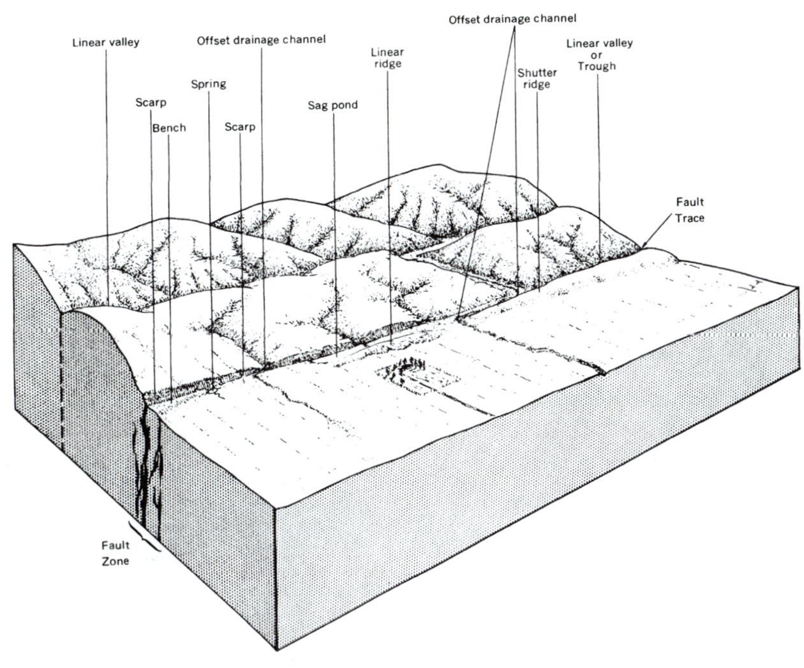

FIGURE 15-2 Typical topographic features along an active strike-slip fault. From Vedder and Wallace (1970).

Characteristics of Active Faults

Evidence for Holocene movement includes (1) historic earthquakes, (2) rock units younger than 11,700 years that are faulted, and (3) certain topographic features caused by faulting. Figure 15-2 is a diagram of an active fault zone that may be hundreds of kilometers in length and several kilometers in width. A *fault trace* is the surface expression of an individual fault. Figure 15-2 also shows typical topographic features formed by active strike-slip faults, while Figure 15-3 is a satellite photograph of typical features along the active Garlock strike-slip fault in the Mojave Desert of California. These topographic features are formed by horizontal and vertical displacements along faults. *Sag ponds* are lakes that occupy structural depressions within the fault zone. In arid environments these are dry lakes. *Shutter ridges* are topographic ridges that have been offset laterally to shut off drainage channels. Other narrow fault blocks are called *benches* and *linear ridges. Scarps* are the surface expression of fault planes; they may cut a topographic ridge to form a *faceted ridge. Springs* form where faults block the movement of groundwater, causing it to emerge at the surface. In arid terrain, *vegetation anomalies* are strips of vegetation that are concentrated along faults because of shallower groundwater. *Linear valleys* result from increased erosion of fractured rocks along a fault. *Offset drainage channels* are especially significant because they also indicate the sense and amount of lateral displacement along a strike-slip fault. The left-lateral, strike-slip displacement of the Garlock fault is indicated by the offset drainage channels in Figure 15-3. The presence of these topographic features shows that a fault is active; had the features formed before the Holocene epoch, most of them would have been obliterated by erosion and deposition.

Satellite images are well suited for recognizing the continuity and regional relationships of faults as well as many local details. Stereo viewing of overlapping images provide detailed information on topographic features formed by faulting. Highlights and shadows on low sun angle aerial photographs can emphasize topographic scarps associated with active faults, as shown in Chapter 2. TIR images of arid and semiarid areas may record the presence of active faults with little or no surface expression, such as the Superstition Hills fault, which was interpreted in Chapter 5. Radar images have highlights and shadows that enhance the expression of faults, even in forested terrain, as shown in Chapter 6. DTMs reveal subtle topographic relief associated with faults and scarps with hillshade models that are illuminated from different compass directions and elevation above the horizon (Chapter 7). IfSAR enables measurement of ground deformation associated with earthquakes.

Relationship between Faults and Earthquakes

Allen (1975) noted the following relationships between faulting and earthquakes, using California as an example:

1. Virtually all large earthquakes (with magnitudes greater than 6.0) have resulted from ruptures along faults that had been recognized by field geologists prior to the events. A notable exception is the 1994 Northridge earthquake, which occurred along a fault that is concealed beneath a cover of sedimentary rocks.
2. All of these faults have a history of earlier displacements in Quaternary and possibly Holocene times.
3. All the earthquakes have been relatively shallow, not exceeding about 20 km in depth. Most earthquakes larger than magnitude 6.0 have been accompanied by surface faulting, as have many of the smaller events.
4. The larger earthquakes have generally occurred on the longer faults, although there has been sufficiently wide variation to indicate caution in blindly applying any single formula for this relationship.
5. Generally only a small segment of the entire length of a fault zone will break during any single earthquake, although there are some conspicuous and significant exceptions.

These relationships are clearly seen on images that have been merged with earthquake data and maps of active faults.

FIGURE 15-3 Satellite photograph of the topographic features along the left-lateral, strike-slip Garlock fault, in the Mojave Desert of southern California. The following features indicate that the fault is active: D = depression, SR = shutter ridge, OC = offset channel, LR = linear ridge, LV = linear valley, FR = faceted ridge. From Merifield and Lamar (1975, Figure 2). Courtesy P. M. Merifield, UCLA.

Landsat Regional Mosaic

Digital Image 15-1 is a mosaic of Landsat TM images of southern California. The base image is a color composite of TM band 7 shown in red, band 4 in green, and the average of bands 1 and 2 shown in blue. GIS methods were used to merge two attributes with the mosaic:

1. Traces of active faults were digitized from the map by Jennings (1975) and registered to the mosaic, where they are shown as yellow lines.
2. A digital database of earthquakes recorded from 1970 to 1995 by the SCSN was registered to the mosaic. In the earthquake scale, magnitudes of 5 or greater are shown by white circles of increasing diameter, magnitudes of 2 to 4 in red circles, and a magnitude of 1 in red dots.

Digital Image 15-1 shows the relationship between active faults and earthquakes that was summarized earlier. Most of the major earthquakes (large white circles) are associated with active faults (shown in yellow), which form topographic linears on the image. A few large earthquakes occurred in the greater Los Angeles Basin, where the faults are concealed beneath relatively young sediments. Many of the smaller earthquakes (shown in red) are aftershocks of major events and are clustered around the major epicenters. Many other small events are concentrated in linear belts along faults and represent small releases of energy.

LANDERS, CALIFORNIA, EARTHQUAKE

Early on the morning of June 28, 1992, millions of southern Californians were awakened by the largest earthquake since 1952 in the western United States. The magnitude 7.3 quake began at the town of Landers in the Mojave Desert (Figure 15-4; **Digital Image 15-1**) and caused ruptures to the north and northwest along active faults. Fortunately, the strongest shaking occurred in sparsely inhabited regions, but one person was killed and 400 were injured. Property damage exceeded $100 million. Three hours after the Landers earthquake, a second damaging quake (magnitude 6.5) occurred on a separate fault near Big Bear Lake, 35 km west of Landers. These events were preceded on April 23, 1992, by the Joshua Tree earthquake and succeeded by many aftershocks. The SCSN provided a detailed record of the seismic activity (Hauksson and others, 1993).

Surface Faulting

Figure 15-4 shows the active faults and the earthquakes (magnitude greater than 1.8) recorded from January 1 to August 18, 1992 in the Landers region. The correspondence between earthquakes and faults agrees with Allen's assessments, cited earlier. The Landers earthquake resulted from right-lateral slip on faults within a broad zone that is 70 km long. The total length of the overlapping fault strands is 85 km. All but the Landers fault had been mapped before the quake. Landforms that are characteristic of active faults (Figure 15-2) were present along the faults prior to the earthquake. These landforms are less common and more eroded than seismic landforms along the San Andreas fault, where ruptures occur every one to two centuries. This comparison suggests that the last major ruptures in the Landers region occurred at least several thousand years ago. This long interval, during which stress accumulated, may account for the high stress drop of the Landers earthquake. All of the Landers fault and most of the Homestead Valley fault ruptured, but only portions of the other faults were offset in the Landers earthquake. Up to several meters of right-lateral offset were typical; the maximum slip of 6 m equals that of the largest surficial strike-slip dislocation of the twentieth century in the Western Hemisphere. Vertical displacements were also common and in several places exceeded 1 m (Sieh and others, 1993).

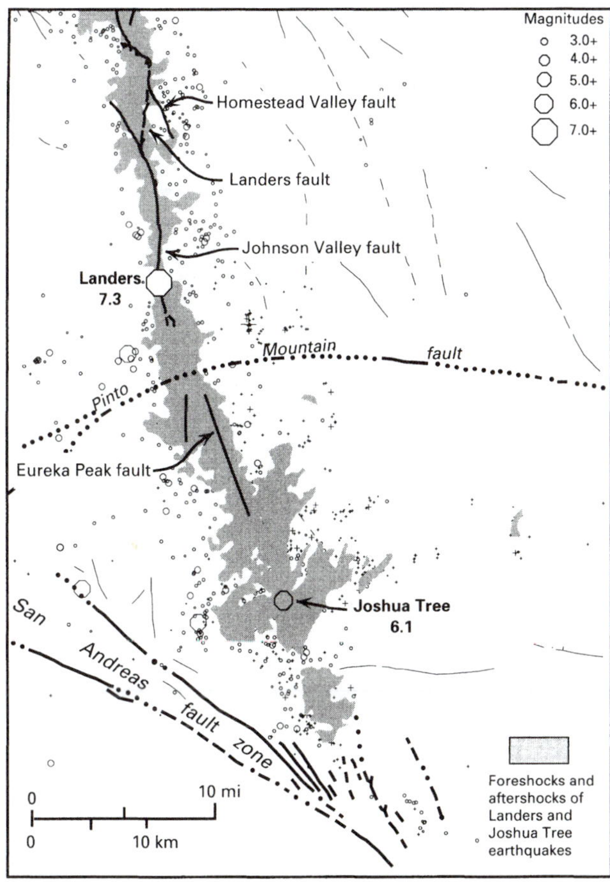

FIGURE 15-4 Seismicity (January 1 to August 18, 1992) for the Landers earthquake of June 28, 1992. Solid lines are exposed faults; dotted lines are faults concealed by young deposits; dashed lines are inferred faults. Modified from Yeats, Sieh, and Allen (1996, Figure 8-51). Courtesy K. Sieh, California Institute of Technology.

Aside from these large breaks, most of the surface offsets were smaller and occurred as multiple cracks in zones up to several hundred meters wide that crossed the desert floor (Mori and others, 1992). SPOT and radar images were digitally processed to help analyze surface effects of the earthquake.

SPOT Images

Figure 15-5 shows subscenes of SPOT panchromatic images that are located north of the area shown in Figure 15-4. The images cover a portion of the Emerson fault, which ruptured during the Landers earthquake. The image in Figure 15-5A was acquired 11 months before the earthquake and shows no evidence of the Emerson fault. A geologic sketch of the area (Figure 15-6) shows the alluvial deposits that concealed the fault trace. The image in Figure 15-5B was acquired one month after the Landers earthquake and was digitally processed by R. E. Crippen to enhance the trace of the rupture as a dark line that trends northwest across the image. Figure 15-5C is an enlargement of the central portion of the post-earthquake image. The fault ruptures were very subtle in the field, where Crippen found that individual cracks ranged from millimeters to only a few centimeters in width with little, if any, vertical displacement. The cracks are visible on the image, however, despite the 10-m resolution of SPOT. Their visibility is attributed to the digital image enhancement and to the abundance of cracks and the near vertical view of SPOT, which recorded the shadowed interiors of the open cracks.

Crippen and Blom (1992) also processed the SPOT images to analyze the dynamics of the Landers earthquake. The images were digitally matched on one side of the fault at subpixel levels by shifting the post-quake image relative to the pre-quake image using correlation analysis methods. By rapidly alternating the enlarged images on a computer display screen, subpixel displacements along the fault appeared as motion, which confirmed the fault location and right-lateral sense of displacement. By using statistical measurements at several points in the images, Crippen and Blom (1992) were also able to quantify the displacements, thereby measuring offsets along the fault.

The ready availability of satellite images makes this a practical technique for analyzing earthquakes with surface ruptures. Crippen (1992) gives details of the processing technique.

A. Image acquired July 27, 1991 (before earthquake).

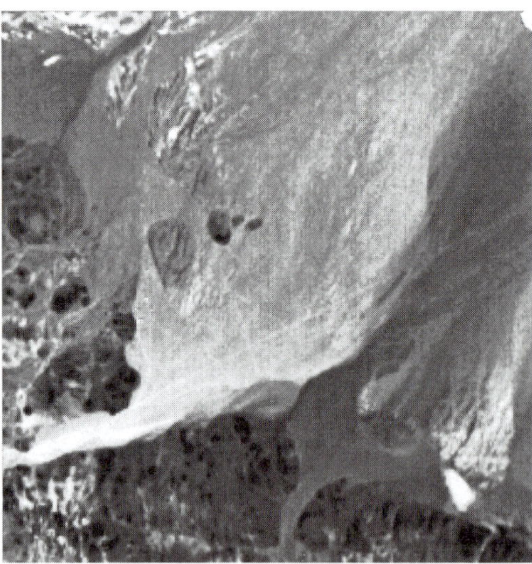

B. Image acquired July 25, 1992 (after earthquake).

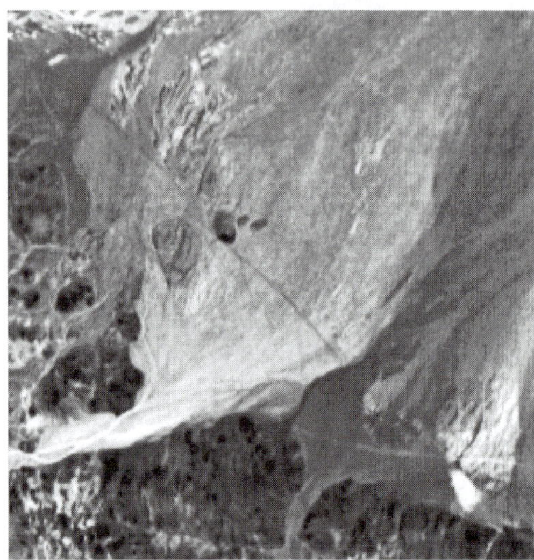

C. Enlarged central portion of (B).

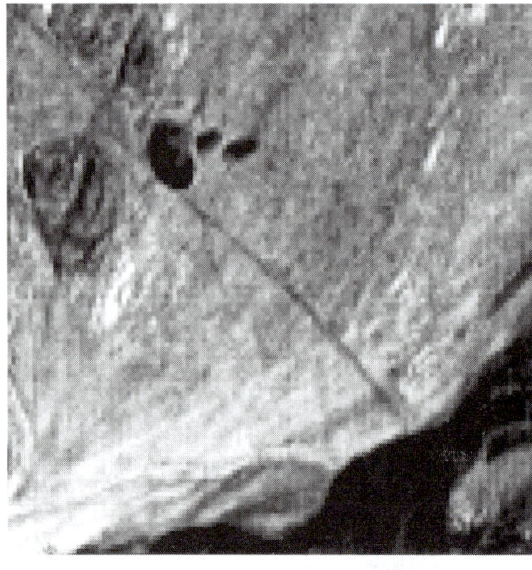

FIGURE 15-5 SPOT pan images of the Emerson fault, which ruptured during the Landers earthquake of June 28, 1992. Courtesy R. E. Crippen, JPL (Retired).

Radar Interferograms

Chapter 6 describes how interferograms showing topography are generated from radar images that are simultaneously acquired by two spatially separated antennas. Massonnet and others (1993) processed a pair of ERS-1 images of the Landers region, acquired on April 24 and August 7, 1992, to produce an interferogram of the displacement field caused by the earthquake. *Displacement field* refers to all of the vertical and horizontal shifts in the region of an earthquake, not just the fault ruptures.

The ERS-1 images provide adequate orbital separation for preparing an interferogram. Such an interferogram shows both the pre-earthquake topography plus the displacement field caused by the earthquake. Digital terrain data obtained before the earthquake were used to remove the topographic information from the interference pattern. The result is a *residual interferogram,* which shows only the displacement field caused by the earthquake. Figure 15-7 is the residual interferogram, which is a contour map of the change in range distance, or the component of displacement that points toward the satellite. Each grayscale fringe corresponds to 28 mm of displacement. For comparison, Massonnet and others (1993) used field data and a dislocation model to calculate a synthetic interferogram (Figure 15-8) that models the predicted displacement field. The synthetic stereogram closely matches the actual interferogram; the two versions agree to within two fringes (56 mm).

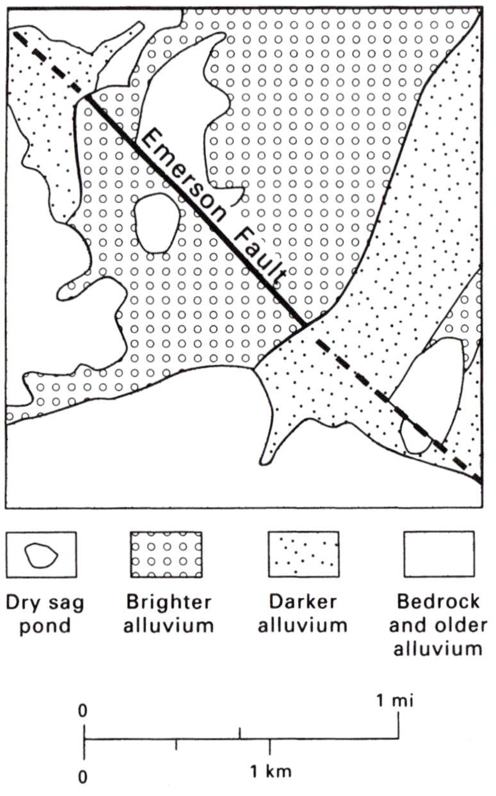

FIGURE 15-6 Geologic sketch of the SPOT image of the Emerson fault acquired after the Landers earthquake (Figure 15-5B).

FIGURE 15-7 Radar interferogram of the Landers, California, region prepared from ERS-I images acquired April 24, 1992 (before quake) and August 7, 1992 (after quake). White lines that show fault breaks were added from field maps. Each cycle of gray shading represents a radar range difference of 28 mm between the two image dates. There are at least 20 cycles (560-mm total range difference) in the interferogram patterns on either side of the fault. From Massonnet and others (1993, Figure 3A). Courtesy G. Peltzer, JPL.

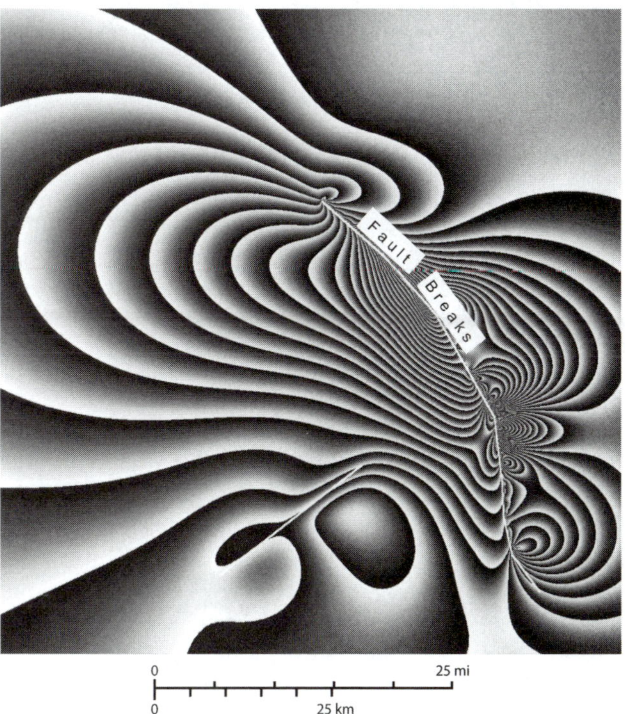

NAPA VALLEY, CALIFORNIA, EARTHQUAKE

Northern California was shaken by a magnitude 6.0 earthquake on August 24, 2014. The epicenter was located at a depth of around 11 km in the middle of the Napa Valley, a wine growing area extending about 40 km north of San Francisco. Buildings and roads were damaged (Figures 15-9 and 15-10), but there was no loss of life.

ESA's Sentinel-1A radar satellite acquired an image of Napa Valley before (August 7) and after (August 31) the earthquake. An interferogram (Plate 51) was created from the pre- and post-earthquake images by collaborators at the Centre for Observation & Modelling of Earthquakes, Volcanoes, and Tectonics (COMET) (comet.nerc.ac.uk). In Plate 51, the black line trending north–northwest to south–southeast is the surface rupture mapped in the field by scientists from University of California at Davis. The arrows on the left and right of the black line indicate this fault as a right-lateral strike slip. The red lines are faults as mapped by the USGS. The interferogram shows this earthquake's fault slip continues further north than the extent of the mapped rupture at the surface (black line in Plate 51).

The interferogram was processed to show total motion toward and away from the satellite. Each fringe cycle represents 28 mm of vertical movement. Two areas on the east side of the surface rupture line experienced the largest ground deformation as revealed by several nested cycles. The total-motion interferogram (Plate 51B) reveals that the southeast side of the rupture moved towards the satellite by about 10 cm, whilst the northern portion moved away by 10 cm. While most of the motion in a strike-slip earthquake is horizontal in the direction of the fault, the ground motion at the end of these fault ruptures is a combination of perpendicular and vertical faulting. Plate 51B shows up-down and to a lesser extent east–west motions. On the east side of the fault, these motions are in the same direction, either both toward or both away from the satellite, resulting in a large signal. However, on the western side of the fault, the east–west and vertical motions are in the opposite sense, cancelling each other out and explaining the asymmetry seen in the deformation pattern across the fault. The small surface displacements measured in the interferogram (Plate 51A) agree with the small offsets measured in the field by geologists surveying the fault rupture, who found displacements in roads and sidewalk curbs of about 10 to 20 cm (Earthquakes without Frontiers, 2014).

FIGURE 15-8 Synthetic interferogram modeled from fault displacements caused by the Landers earthquake. The displacement data were measured in the field. Each cycle of gray shading represents the same interval as the radar interferogram (28 mm) shown in Figure 15-7. The two interferograms agree to within two gray cycles (56 mm). From Massonnet and others (1993, Figure 3B). Courtesy G. Peltzer, JPL.

FIGURE 15-9 Damage suffered by many buildings in Napa, including the 1910 Alexandria Hotel, was documented with drones. Photo by Jim Heaphy (CCO 1.0 License).

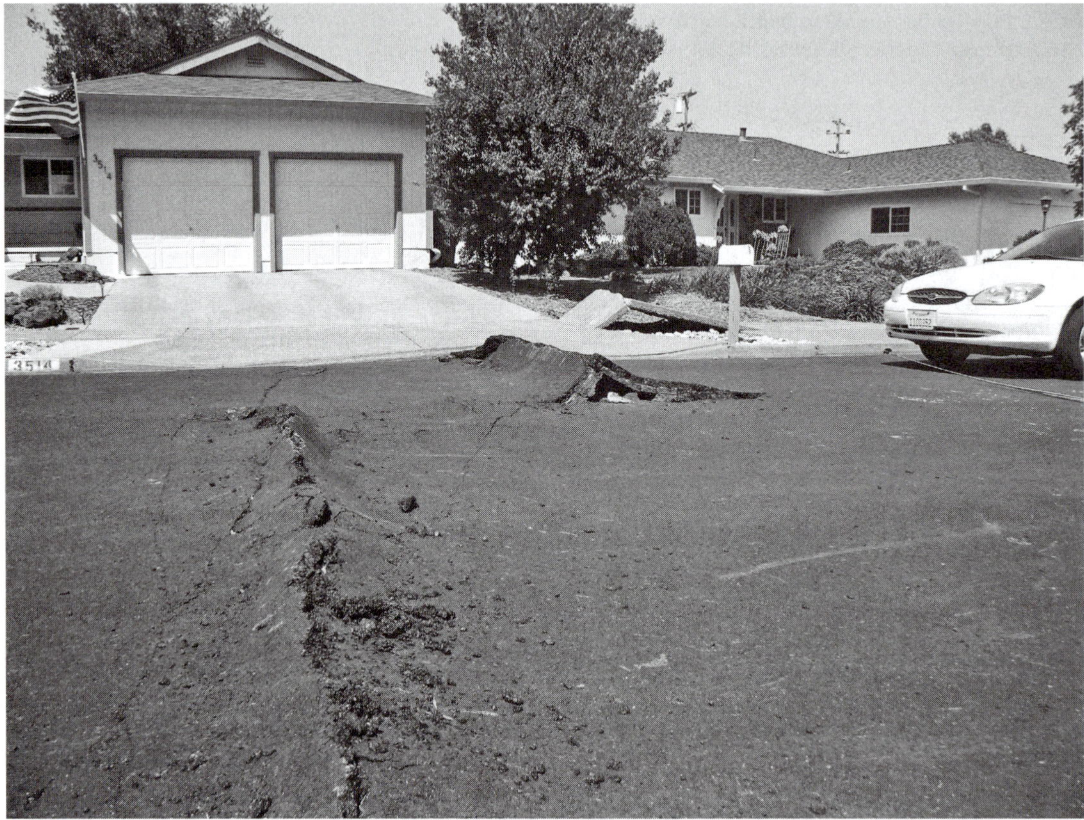

FIGURE 15-10 The fault rupture ran through homes and across roads, buckling the tarmac surface and pavements. Copyright Austin Elliott, University of California, Davis.

RADAR IMAGE OF CALICO ACTIVE FAULT, CALIFORNIA

The Calico fault is one of the active faults that strike northwest across the western Mojave Desert and has right-lateral, strike-slip displacement. Figure 15-11A is a Landsat image of the western portion of the Troy Valley, which is crossed by the Calico fault. The trace of the fault is obscure on the Landsat image and is also obscure on an enlarged TM color image (not shown). Figure 15-11B is a Seasat radar image of the same area. The Calico fault is a distinct northwest-trending radar linear formed by the boundary between brighter signatures on the west and darker signatures on the east. Radar roughness criteria (Chapter 6) for Seasat predict a surface roughness of greater than 6 cm for the bright terrain and less than 1 cm for the dark terrain. In the field the bright terrain consists of hummocky sand dunes that have been stabilized by desert shrubs. The dunes terminate abruptly eastward against relatively smooth desert terrain. The map in Figure 15-12 shows these relationships.

Figure 15-13 is a west to east section across the fault that explains the linear in Figure 15-11B. The linear is not caused by surface displacement along the fault; rather, it is due to the effect of the fault on the water table, which is the boundary between dry soil near the surface and deeper soil that is permanently saturated by groundwater. In the Troy Valley, unobstructed groundwater would flow eastward in the direction of the surface slope. The Calico fault, however, forms a barrier to groundwater movement, and the water table is shallow along the west side of the fault. The shallow water table supports an unusually dense growth of desert shrubs and trees (creosote bush and mesquite) along the west side

A. Landsat image acquired 1978.

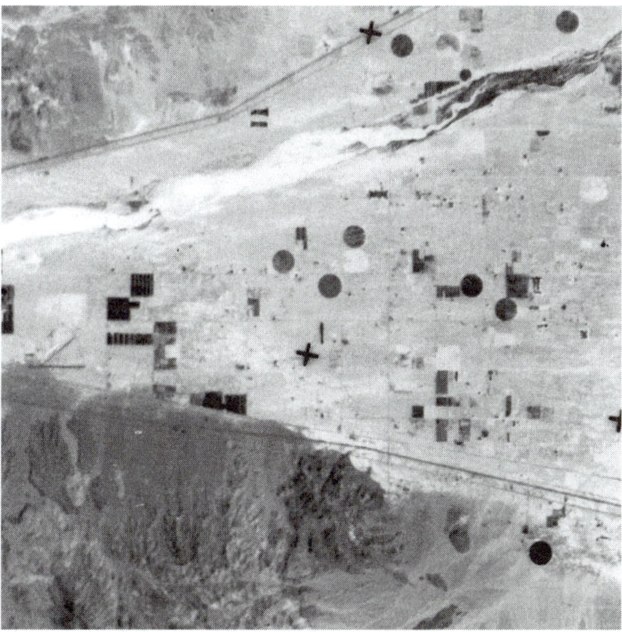

B. Seasat radar image acquired 1978.

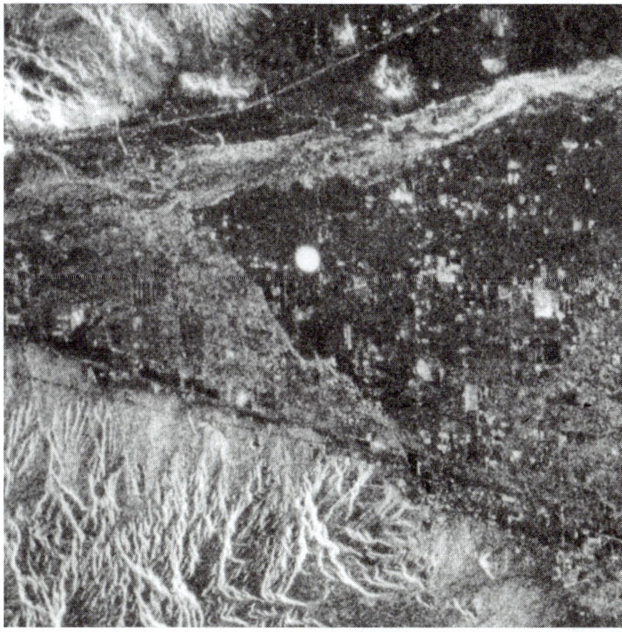

FIGURE 15-11 Images of the Calico fault in the Troy Valley, Mojave Desert, California.

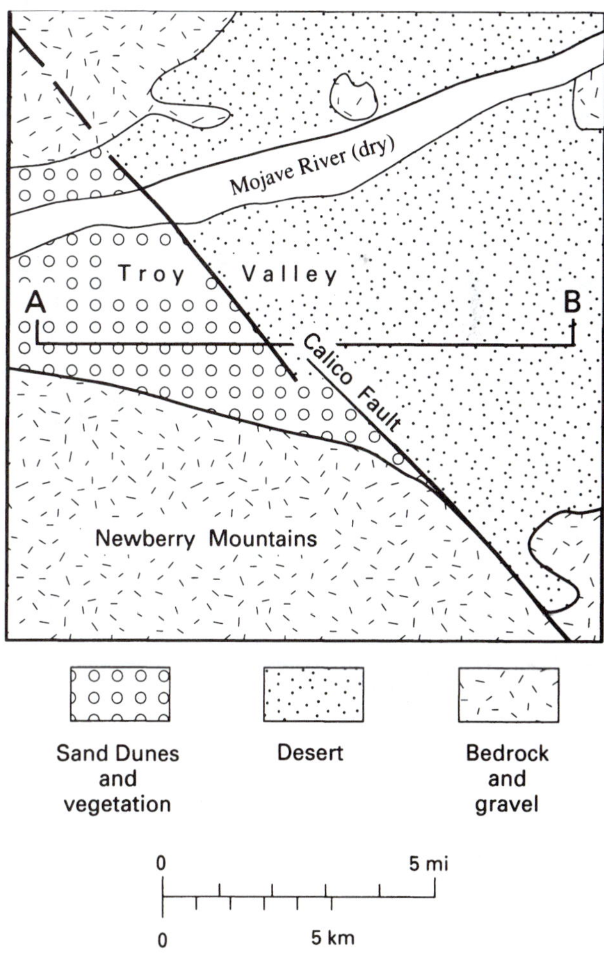

FIGURE 15-12 Interpretation map for the Seasat image of the Calico fault, Troy Valley, California (Figure 15-11B).

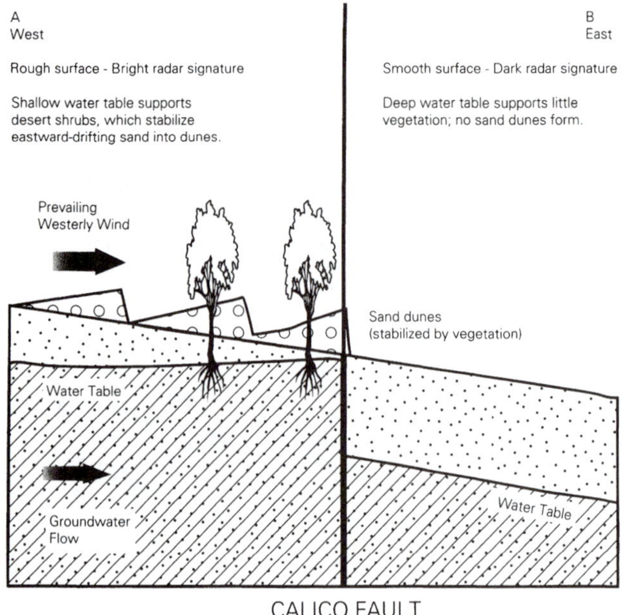

FIGURE 15-13 Diagrammatic section across the Calico fault to explain the tonal linear feature on the Seasat image of the area (Figure 15-11B).

of the fault. The prevailing westerly wind moves sand along the desert surface. The vegetation interrupts the wind flow, causing the sand to accumulate in dunes along the vegetated west side of the fault trace. Continued growth of vegetation stabilizes the dunes.

The complex relationship that formed the radar linear is not unique to the Calico fault. An identical relationship is shown in radar images and in the field at the Mesquite Lake fault near the town of Twenty Nine Palms, California.

CHINA

Much of China is seismically active. The Sichuan earthquake in 2008 killed over 87,000 people (Rodriguez and others, 2009). Tapponnier and Molnar (1977, 1979) used mosaics of Landsat MSS images to interpret a number of major faults that appear to be active. For example, the active Altyn Tagh fault trends northeast for almost 2,000 km along the south margin of the Tarim Basin. Major left-lateral displacement accommodates much of the northeastward movement of the India plate relative to the Asia plate. Little else was known about this important fault until SPOT images became available. Peltzer and others (1989) selected seven SPOT panchromatic images based on the earlier Landsat MSS interpretations. The SPOT images of the Altyn Tagh fault show spectacular long-term, left-lateral offsets of drainage channels, alluvial fans, and glacial deposits that imply slip rates of 2 to 3 cm per year.

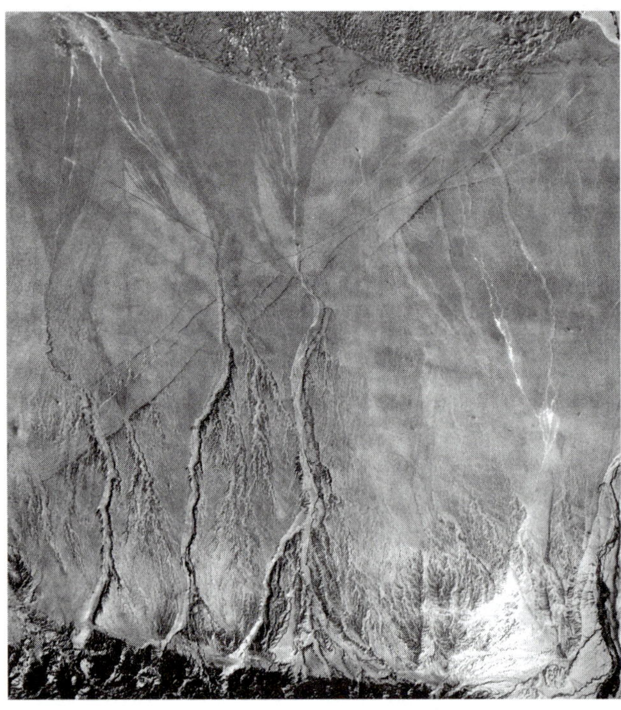

FIGURE 15-14 SPOT pan image of the Hotan-Qira zone of active faults in southern Xinjiang, China. The sun was shining toward the northwest (upper left comer). From Avouac and Peltzer (1993, Figure 3A). Courtesy G. Peltzer, JPL.

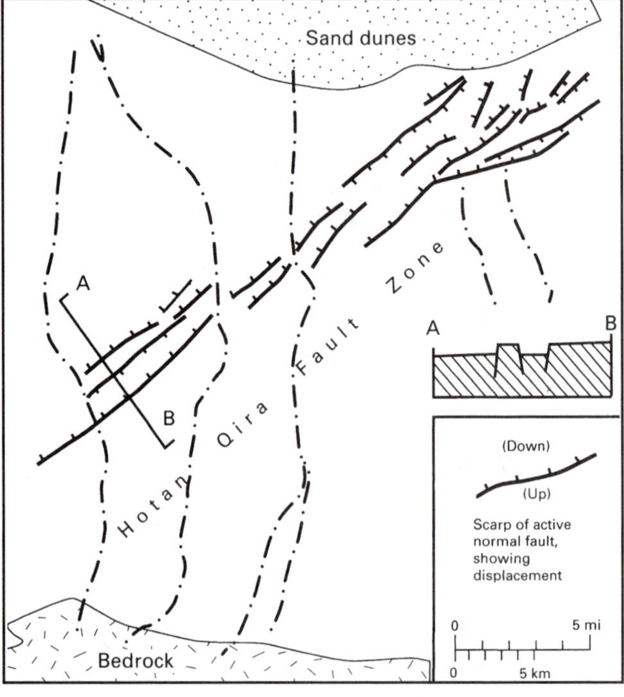

FIGURE 15-15 Interpretation map and cross section of the SPOT image of the Hotan-Qira fault system (Figure 15-14). Modified from Avouac and Peltzer (1993, Figure 3A).

Avouac and Peltzer (1993) interpreted SPOT panchromatic images of the south margin of the Tarim Basin, which is north of the Altyn Tagh fault. Figure 15-14 is an image of the Hotan-Qira fault system, which cuts young alluvial fan surfaces that slope northward into the Tarim Basin. The sun is shining from the southeast at a relatively low elevation. The northeast-trending faults are indicated by linear highlights and shadows. The shadows are caused by northwest-facing fault scarps, and the highlights are caused by southeast-facing scarps. Enlargements of the image (not shown) depict many smaller faults that are not visible at the scale of Figure 15-14. Figure 15-15 is an interpretation map and cross section for the image.

Sequences of river terraces and alluvial fan surfaces are seen in the image. Offsets of these features are used to determine the amount and rate of movement along the faults. Field investigations (Avouac and Peltzer, 1993) show that the highest scarps reach 20 m with recent offsets about 2 m high. Total vertical offset of the Hotan-Qira fault system is 70 m. The minimum rate of subsidence is 3.5 ± 2 mm · yr^{-1}.

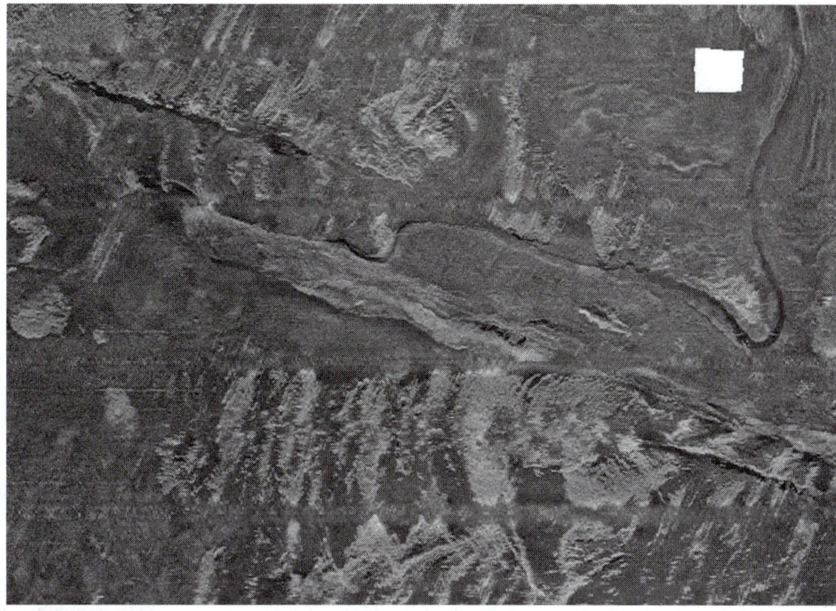

FIGURE 15-16 Mosaic of side-scanning sonar images of the Blanco fault zone in the Pacific Ocean off the coast of Oregon. From USGS (2010).

ACTIVE FAULTS ON THE SEAFLOOR

Active faults on the seafloor are imaged by side-scan sonar systems that are described in Chapter 7. Figure 15-16 is a mosaic of Gloria II side-scan sonar images of a portion of the Blanco fault zone in the Pacific Ocean 200 km off the coast of Oregon (USGS, 1991). The image was acquired by the USGS (2010) as part of the survey of the offshore Exclusive Economic Zone (EEZ). Water depths are approximately 3,000 m. Bright signatures are strong sonar returns, and dark signatures are weak returns.

The Blanco fracture zone is a linear trench trending west–northwest that is bounded by steep fault scarps shown in the map and cross section of Figure 15-17. The south-facing scarp along the north margin is marked by a narrow linear shadow. Bathymetric contours (not shown) indicate that the scarp reaches a height of 300 m. The south margin of the trench is formed by a series of north-facing scarps with strong highlights. The seafloor adjacent to the fracture zone is crossed by narrow ridges trending north–northeast that represent small fractures. Seamounts, of possible volcanic origin, have circular to irregular outlines. The sinuous Cascadia Channel flows southward into the Blanco Trench, follows the north margin for 70 km, then follows the south margin for 50 km. The dark signature of the channel is caused by the fine-grained sediments, which do not scatter energy back to the sonar detector.

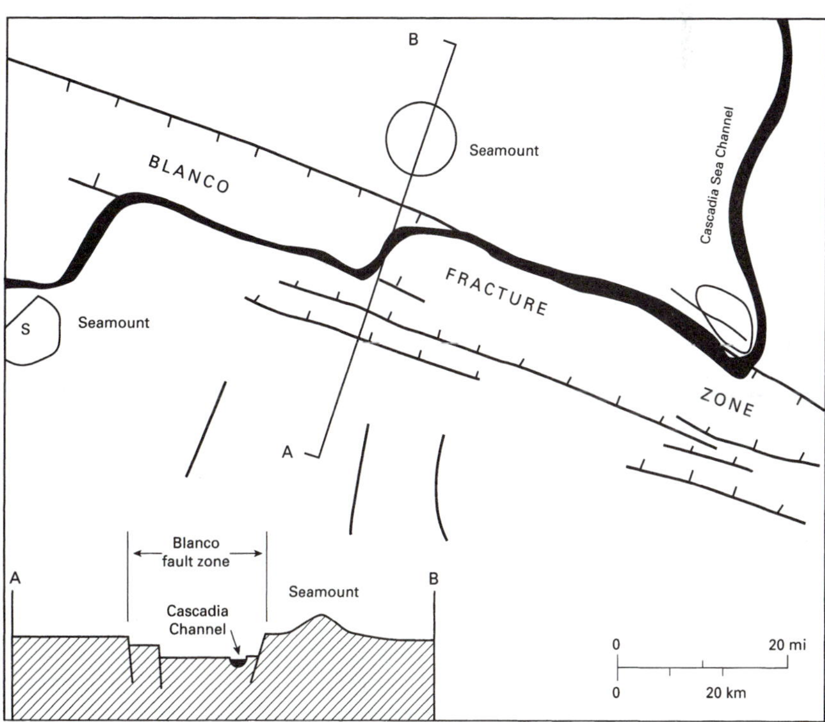

FIGURE 15-17 Interpretation map and cross section of the Blanco fault zone (Figure 15-16).

LANDSLIDES

Landslides occur on the land and on the seafloor in areas underlain by unstable materials. Each year landslides cause extensive property damage and death. Landslides killed over 32,000 people between 2003 to 2010 worldwide, mostly in densely populated mountainous areas with intense rainfall, such as in Asia and Latin America (Wartman, 2016). Individual landslide events are difficult or impossible to predict by remote sensing methods. Research on landslide prediction in hazard zones mapped with remote sensing technology includes real-time monitoring of hillslope movement with GNSS/GPS units embedded in the ground, along with continuous measurements of rainfall, soil water content, and soil water pressure integrated into real-time mathematical models (Reid and others, 2012). The most effective way to prevent landslide damage is to identify unstable areas and avoid building in their vicinity. Where construction must be done in unstable areas, knowledge of the potential for landslides can be used to stabilize the foundations.

Stereo-pairs of aerial photographs (black and white, normal color, and color IR) have long been used to recognize slides and slide-prone terrain. Rib and Liang (1978) published an extensive collection of stereo-pairs together with interpretation criteria. High soil moisture lubricates unstable material and is a major factor in landslides. TIR images have been used to recognize damp ground associated with landslides in California (Blanchard and others, 1974, Figure 1). Evaporative cooling of the damp ground produces a cool signature on aircraft TIR images. Lidar, radar, and side-scanning sonar have expanded our capability to recognize unstable terrain where slides occur, as shown by the following examples.

BLACKHAWK LANDSLIDE, SOUTHERN CALIFORNIA

The prehistoric Blackhawk landslide originated on the north flank of Blackhawk Ridge in the San Bernardino Mountains and moved northward for 9 km into the Mojave Desert. The slide has a maximum width of 3.2 km and includes a volume of 2.7×10^9 m^3 of crushed rock (Shreve, 1968). In form and structure the Blackhawk slide is similar to the smaller and well-known historic slides at Elm in Switzerland, at Frank in Alberta, Canada, on the Sherman Glacier in Alaska, and to the great prehistoric slide at Saidmarreh in Iran. These similarities make the Blackhawk slide a good example for remote sensing analysis.

Figure 15-18 shows two remote sensing images and an interpretation map of the Blackhawk slide. The slide resulted from erosion that over-steepened the north slope of Blackhawk Ridge and caused the highly fractured bedrock to collapse along an arcuate scarp (Figure 15-18C). The falling rock debris trapped a cushion of air during its descent that lubricated the mass, enabling it to flow several kilometers onto the desert floor (Shreve, 1968). In the satellite image (Figure 15-18A), the headward scarp of the original debris fall is shadowed, which enhances the characteristic crescent shape. The distal lobe of the slide has a hummocky appearance that is typical for landslide deposits. The toe and lateral margins of the slide form a pressure ridge 15 to 30 m high that stands slightly above the surface of the slide. A northwest-flowing ephemeral stream has deposited younger alluvium over the central portion of the slide. In the satellite photograph, the slide debris is darker than the stream and desert alluvium. The radar image (Figure 15-18B) has a coarser resolution than the photograph but portrays major features of the slide. The coarse debris and hummocky surface are rough and produce a bright signature relative to the darker signature of the finer-grained alluvium. The pressure ridges have narrow, bright signatures.

SUBMARINE LANDSLIDES

The delta of the Mississippi River consists of unconsolidated, water-saturated, fine-grained sediment. The submerged, gentle depositional slopes around the margin of the delta are unstable and may collapse to form submarine landslides and related features. Because the delta is an oil-producing region, it is important to map slide-prone areas before installing production platforms and seafloor pipelines. The turbid water prevents remote sensing at visible wavelengths. The Coastal Studies Institute of Louisiana State University and the USGS acquired parallel strips of side-scanning sonar images that were digitally processed and compiled into the mosaic of Figure 15-19A. In this mosaic bright signatures record weak sonar returns from shadow zones or smooth surfaces. Dark signatures record strong returns from scarps facing the sonar pulse and from rough surfaces.

The interpretation map (Figure 15-19B) shows that the slides originate at lobate slump areas where the sediment collapses to form irregular blocks. The slump material moves downslope through narrow, steep-sided channels (chutes) that merge at junctions. At the junctions, the slump material may leave the chute and form spillover deposits. Extensive systems of slumps and chutes have been mapped on sonar images of the submerged portions of the Mississippi Delta.

Piper and others (1985) acquired and interpreted sonar images near the epicenter of the 1929 Grand Banks earthquake on the seafloor south of Newfoundland. The earthquake triggered a major submarine landslide and turbidity currents. The sonar images clearly show the following features: landslide scarps, slumps, debris flows, a lineated seafloor, channels, and gullies. Sonar images can identify unstable areas that should be avoided for offshore engineering projects.

OSO MUDSLIDE, WASHINGTON

A hillside above Oso, Washington, collapsed on March 22, 2014, unleashing a torrent of mud and debris that buried the community of Steelhead Haven. Forty-three people lost their lives, making it one of the deadliest landslide disasters in US history (Wartman, 2016).

Figure 15-20 includes *before* and *after* oblique aerial photographs of the Oso landslide. The black arrows point to the same features on both photographs. The landslide is within a 200-m high hillslope comprised of unconsolidated glacial

A. Satellite image.

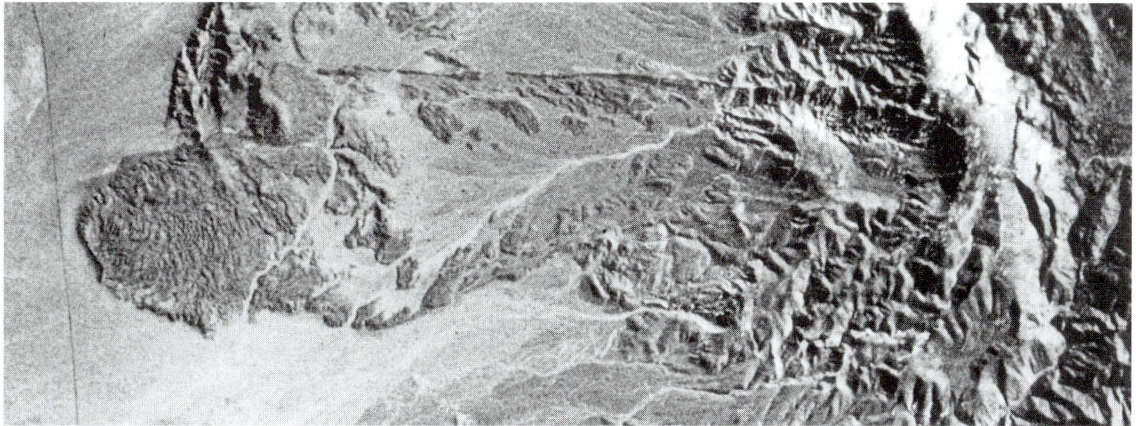

B. Aircraft X-band radar image.

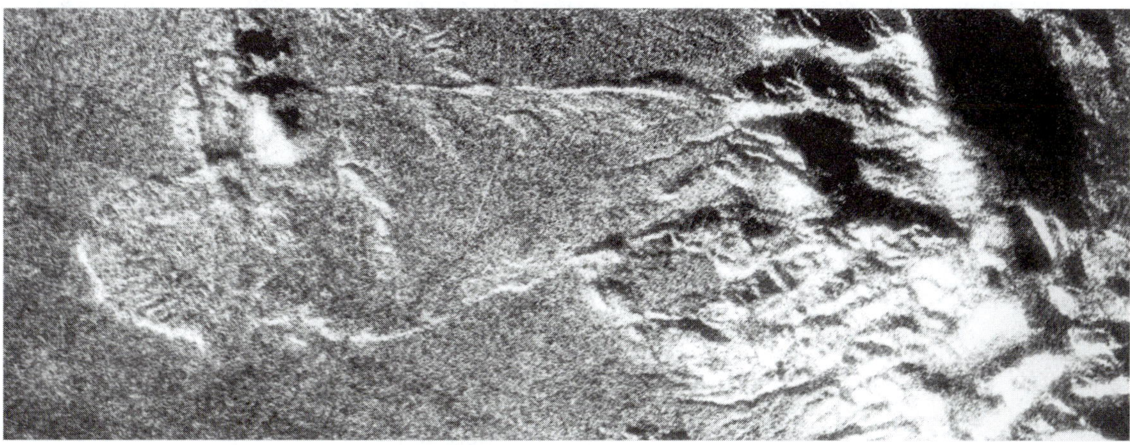

C. Interpretation map.

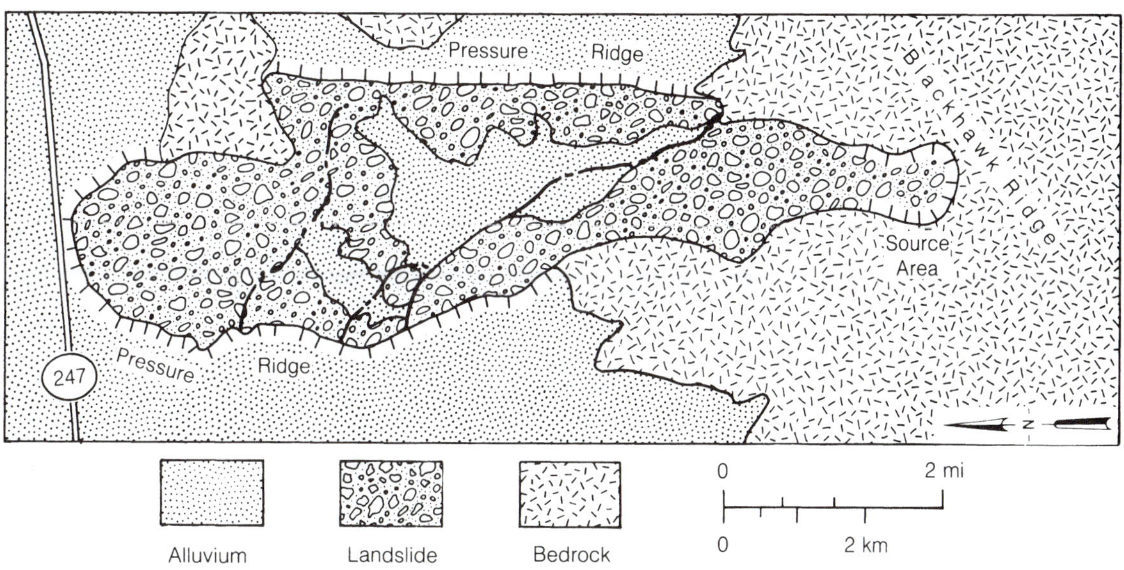

Figure 15-18 Blackhawk landslide, San Bernardino Mountains, California.

A. Mosaic of side-scanning sonar images. From Prior and others (1979, Figure 2). Courtesy D. B. Prior, Louisiana State University.

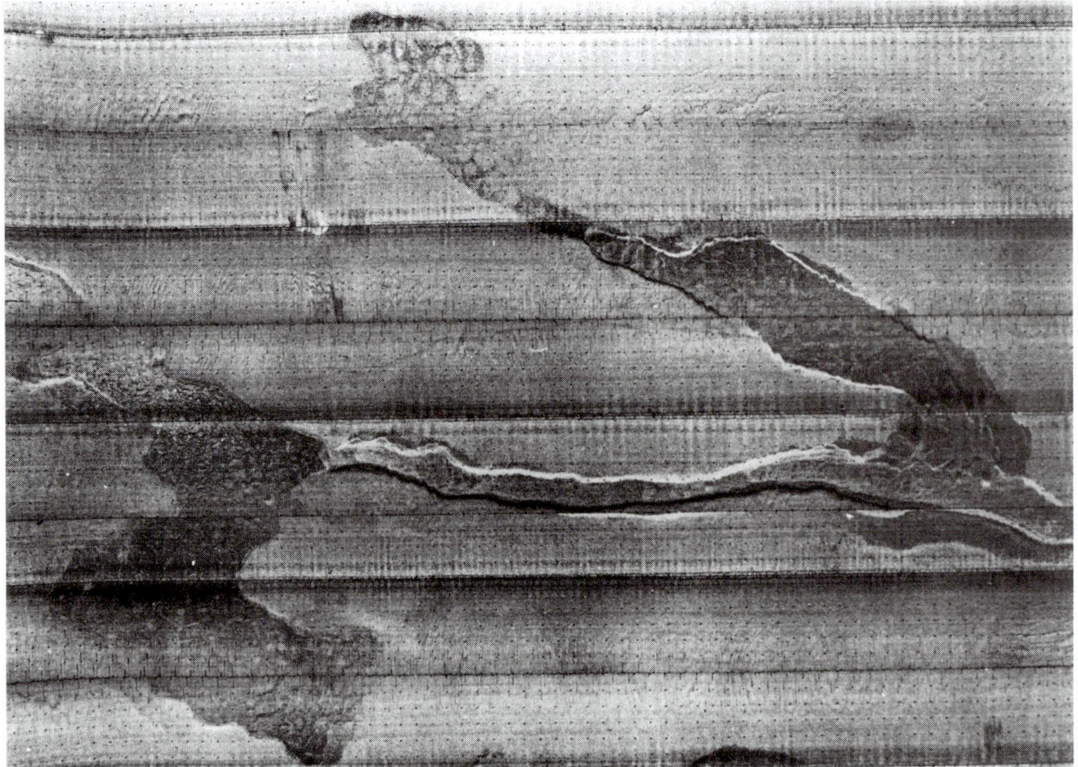

B. Interpretation map.

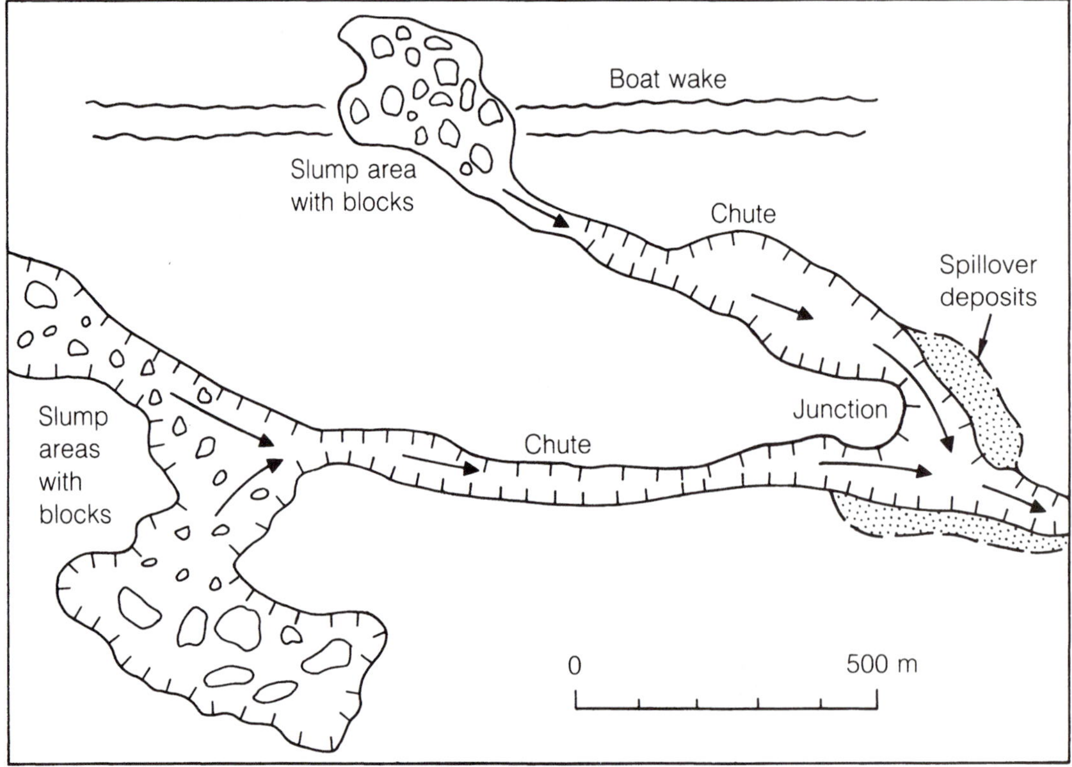

Figure 15-19 Submarine landslides, offshore Mississippi Delta, Gulf of Mexico.

Figure 15-21 View of Oso landslide scarp with white arrow pointing to airborne helicopter for scale. From Burns (2014, slide 19).

clays, sands, and gravels (LaHusen and others, 2016). The headward scarp is approximately 100 m high and seen as a light-tone arcuate feature in Figure 15-20B. In Figure 15-21, a photograph looking northwest at the top of the headward scarp reveals stratification in the unconsolidated glacial sediment that affects the distribution of groundwater in the subsurface. In the field, active groundwater seeps were seen along the face of the scarp (Keaton and others, 2014). The distribution of groundwater in the hillside combined with three weeks of extreme rainfall that preceded the landslide, as well as saturated soils, are key factors in the initiation of the Oso landslide. Burns (2014) includes a steep slope, weak soils, a river undercutting the base of the slope, and clear cutting of trees at the top of the slope as factors also affecting the landslide.

The landslide traveled more than a kilometer across the valley floor, burying approximately 600 m of highway under 6 m of debris (Figure 15-20). An airborne lidar survey with 0.9 m spatial resolution was flown over the Oso landslide and nearby terrain. The headwall scarp and hummocky surface of the Oso landslide are clearly depicted in the bare earth (DTM) lidar image (Figure 15-22). Many other landslide bodies and headward scarps can be interpreted from this image that flowed into the valley from the northern and southern plateaus. For example, an older, large slide similar to the Oso event is seen immediately to the left of the 2014 landslide. As landslide debris bodies age they become more topographically smooth (LaHusen and others, 2016). The detailed topographic relief captured on the lidar DTM of different landslides (Figure 15-22) enabled LaHusen and others (2016) to correlate surface roughness from lidar with the age of the landslide. They determined a landslide occurs every 140 years in this region, demonstrating the inherent instability of the landscape and the dangers faced by people downslope of potential landslide terrain.

A. Before the landslide.

B. After the landslide.

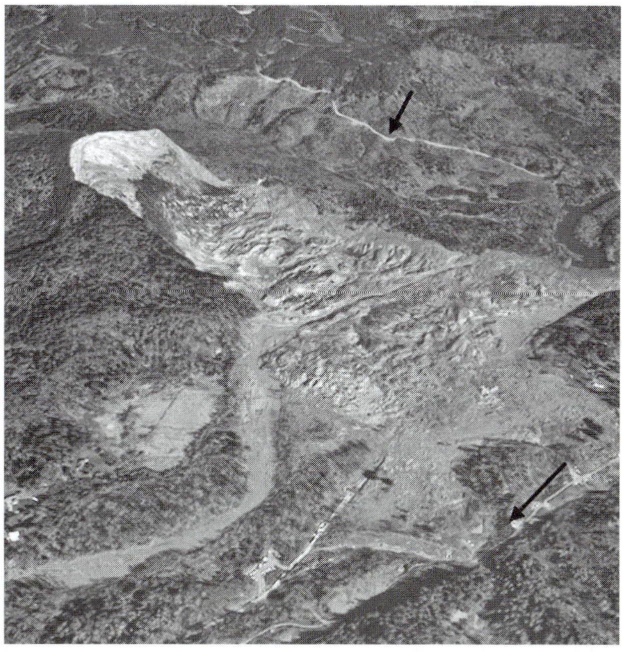

Figure 15-20 Oblique aerial images illustrating the effects of the Oso landslide in 2014. The black arrows point to the same feature in both images. From Wartman (2016) (https://creativecommons.org/licenses/by-nd/4.0/).

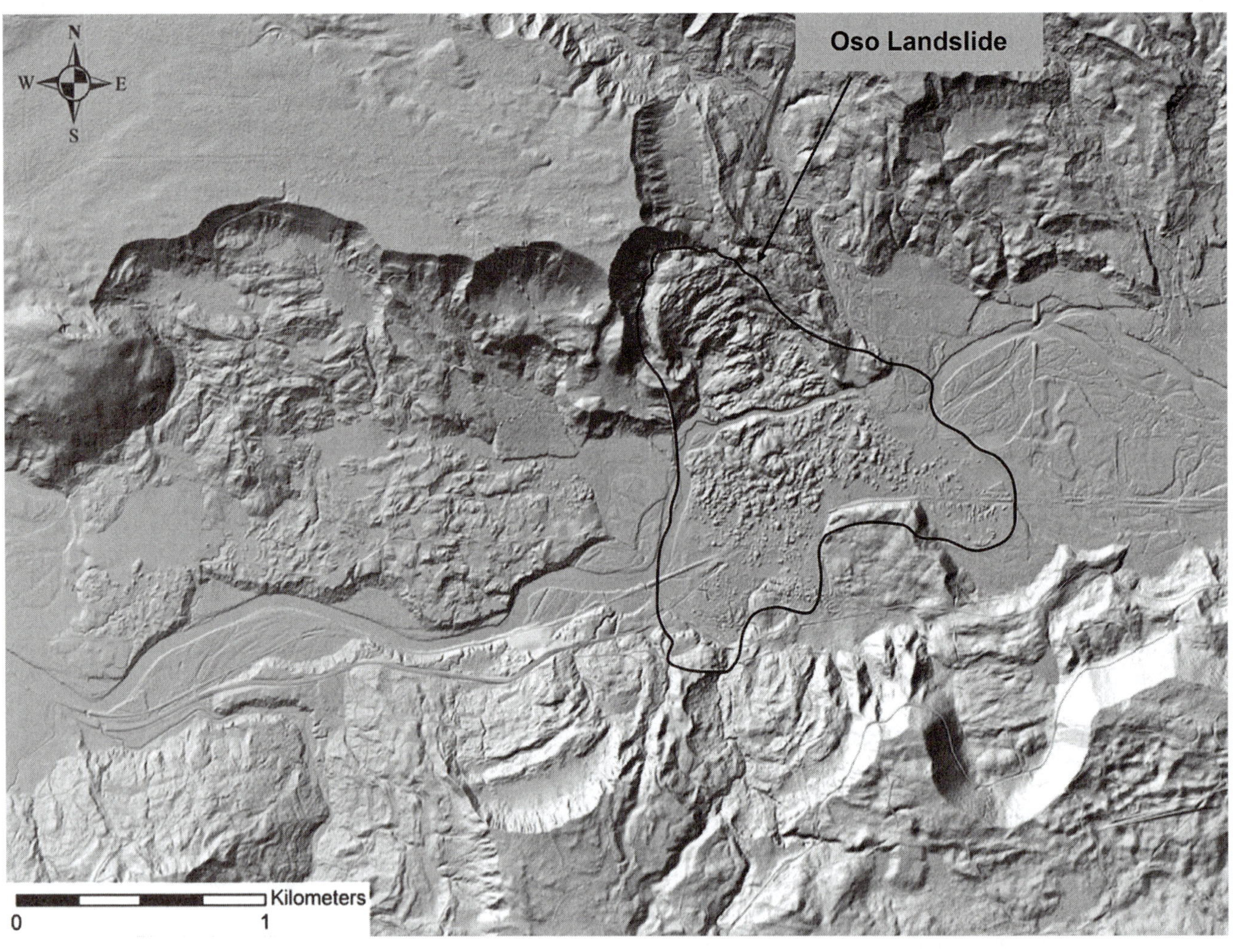

FIGURE 15-22 Lidar DTM (bare earth) image of the Oso landslide (outlined in black) and surrounding landslide-prone terrain. From Wartman (2016), with data provided by the Puget Sound LIDAR Consortium (https://creativecommons.org/licenses/by-nd/4.0).

A close-up of the Oso landslide lidar DTM (Figure 15-23) displays the buried road toward the toe of the landslide, the valley river cutting a new, steep-sided channel through slide debris at the base of the northern hillslope, and the steep headwall scarp. Lidar penetrates vegetation in the hilly to mountainous and forested terrain of the State of Washington, providing critical and previously unknown morphological information about landslides and landslide-prone terrain. The Washington Geological Survey has acquired lidar over 25 to 35% of the state to detect, map, and better understand landslides. The State of Oregon has similar topography, climate, and vegetation cover as Washington. Over 41,000 landslides have been mapped with lidar in Oregon over a small portion of the state. Previous to using lidar, only 12,000 landslides had been mapped (Luccio, 2015).

LAND SUBSIDENCE

Land subsidence, which is downward vertical movement, is a different type of hazard from landslides, in which lateral movement predominates. Land subsidence is a gradual settling or sudden sinking of the Earth's surface owing to subsurface movement of earth materials. Subsidence is a global problem. In the United States more than 44,000 km^2 in 45 states, an area roughly the size of New Hampshire and Vermont combined, have been directly affected by subsidence (Galloway and others, 1999). The principal causes are aquifer system compaction, drainage of organic soils, underground mining, hydrocompaction, natural compaction, sinkholes, and thawing permafrost (National Research Council, 1991). More than 80% of the identified subsidence in the United States is a consequence of the exploitation of underground water (Galloway and others, 1999).

Surface subsidence of up to several meters in depth can be caused by extraction of subsurface fluids. Where subsidence occurs in populated and industrialized coastal areas, the resulting flooding can be a major problem. A classic example is the subsidence associated with the Long Beach oil field in southern California, which was stopped through injection of water to replace the extracted oil.

HOUSTON, TEXAS

Land subsidence in metropolitan Houston, Texas, first occurred in the early 1900s. Extensive subsidence, caused mainly by groundwater pumping but also by oil and gas extraction, has increased the frequency of flooding, caused extensive damage to industrial and transportation infrastructure, motivated major investments in levees, reservoirs, and surface water distribution facilities, and caused substantial loss of wetland habitat (Coplin and Galloway, 1999). By 1979, up to 3 m of subsidence had occurred east of downtown Houston. Subsidence has ceased on the coastal plain east of Houston as the urbanized area converted from groundwater to imported surface water supplies. In contrast, the fast-growing inland areas north and west of Houston, which still rely on groundwater, are experiencing continued land subsidence and movement along faults caused by groundwater level decline. Lidar DTMs, interpretation of aerial images, and fieldwork document the growth of active faults and subsidence.

Clanton and Verbeek (1981) acquired oblique aerial photographs with normal color and color IR film and interpreted them using the following criteria to identify traces of active faults:

1. Fault scarps;
2. Sag ponds along fault traces;
3. Differences in drainage patterns on opposite sides of faults;
4. Linear tonal anomalies caused by higher soil moisture on the downthrown sides of faults; and
5. Vegetation anomalies.

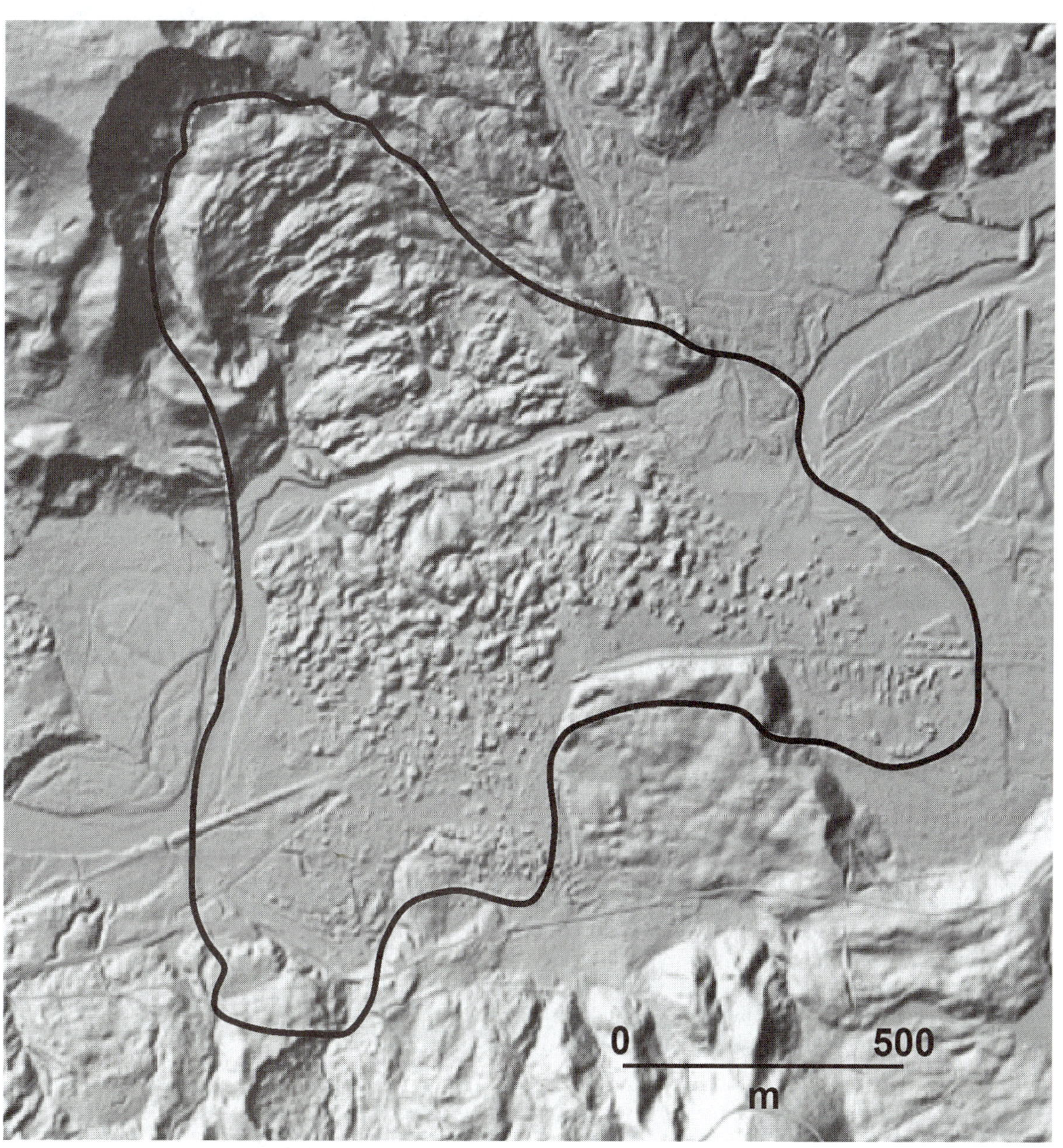

Figure 15-23 Close-up of Oso landslide lidar DTM (Figure 15-22). From Wartman (2016), with data provided by the Puget Sound LIDAR Consortium (https://creativecommons.org/licenses/by-nd/4.0).

FIGURE 15-24 Photograph of a sinkhole in Winter Park, Florida, that collapsed in the spring of 1981, swallowing a house, part of a municipal swimming pool, and part of a car dealership. Photo by Anthony S. Navoy, USGS.

FLORIDA

In areas of limestone terrain, groundwater can dissolve the bedrock to produce underground caverns that may collapse and cause surface subsidence. In Florida these collapse areas, called *sinkholes,* are up to several hundred meters wide and tens of meters deep and may form in a few hours. Areas of incipient sinkholes have anomalously high surface moisture. A few aircraft TIR images show cool areas that may be indicators of future sinkholes.

On May 8, 1981, a sinkhole developed in suburban Winter Park, Florida, that by the next day swallowed a three-bedroom house, part of a municipal swimming pool, and five automobiles (Figure 15-24). By the time the sinkhole was stabilized, it was 107 m wide, 23 m deep, and caused $4 million in damages (Florida History Network, 2017). Sinkholes such as the one at Winter Park are often triggered by groundwater level declines caused by pumping and by enhanced percolation of water through the susceptible limestone rock (Galloway and others, 1999).

VOLCANOES

Based on their activity, volcanoes may be classified as follows:

- **Active**: Erupted at least once in historic time.
- **Dormant**: No historic eruptions, but probably capable of erupting.
- **Extinct**: Incapable of further eruptions (dormant for over 10,000 years) (B. Hausback, personal communication, 2017).

The following describes the categories of volcanic hazards:

1. **Pyroclastic eruption**: Explosive eruptions that produce clouds of ash and coarser ejecta. Of particular danger are glowing cloud avalanches, such as the eruption that destroyed Martinique (Table 15-1). Ash plumes (even very dilute ash plumes) are a hazard to jet aircraft.
2. **Slope failure**: The over-steepened flanks of volcanoes can result in massive, rapid landslides. The slope failures may trigger eruptions by removing overburden that confined the magma.
3. **Debris flows** (also called lahars): Thick accumulations of ash on steep flanks of volcanoes mix with rain or snowmelt to produce high-velocity mudflows that destroy everything in their path.
4. **Toxic gas**: Carbon dioxide, sulfur gases, carbon monoxide, and other lethal gases are by-products of eruptions.

Table 15-1 lists representative occurrences of these hazards. Despite the danger, many regions of active volcanoes are densely populated, including in countries such as Japan, Java, and Italy. Forecasting eruptions months or years in advance has not been particularly successful. Predictions a few days or hours in advance are notably more exact. Figure 15-25 shows the four major warning criteria, which are summarized below:

1. Earthquakes caused by expanding magma chamber are detected by seismometers.
2. Changes in the shape of a volcano caused by an expanding or contracting magma chamber are recorded by tiltmeters and by repeated GNSS/GPS surveys. Radar interferometry and lidar are also used.
3. Increased emission of heat and gas at fractures can be detected on TIR images from aircraft or satellites.
4. Increased eruption of ash is monitored by visual observation and images.

Table 15-1 Volcanic hazards.

Pyroclastic Eruption
El Chichón, Mexico (1982), 1,000 dead
Mt. Pelee, Martinique (1902). 29,025 dead
Vesuvius, Italy (AD 79), 2,000 dead
Mt. Nyiragongo, Democratic Republic of the Congo (2002), over 100 dead
Mt. Merapi, Indonesia (2010), over 300 dead
Mt. Ontake, Japan (2014), 63 dead
Mt. Calbuco, Chile (2015), 0 casualties[a]
Fuego, Guatemala (2018), over 62 dead
Slope Failure
Mount St. Helens, Washington (1980), 57 dead[a]
Ili Werung, Indonesia (1979), 500 dead
Unzen, Japan (1792), 14,524 dead
Debris Flows (lahars)
Ruiz, Colombia (1985), 22,000 dead[b]
Irazu, Costa Rica (1964), 1 dead[a]
Kelut, Indonesia (1919), 5,000 dead
Mt. Pinatubo, Philippines (1991), over 800 dead[a]
Toxic Gas
Lake Nyos, Cameroon (1986), 1,700 dead
Lake Monoun, Cameroon (1984), 37 dead
Sinila crater, Indonesia (1979), 142 dead

[a] Hazard warning acted upon locally.
[b] Hazard warning issued but not acted upon.
Sources: Rothery (1989); Science X Network (2018).

The last three criteria may be detected by remote sensing systems. A fifth criteria that is often used is the ratio of sulfur dioxide to carbon dioxide gas in volcanic emissions. As the sulfur dioxide percentage increases it is considered that the magma is becoming more active or closer to the surface, and hence more likely to erupt (B. Hausback, personal communication, 2017). Satellite, aerial, and terrestrial UV and TIR sensors monitor emissions of sulfur dioxide and other gases from volcanoes (Textor and others, 2003).

Mount St. Helens, Washington

Mount St. Helens, in the southwestern part of the State of Washington, erupted several times in the 1800s and has long been recognized as an active volcano. The major explosive eruption of May 18, 1980 was thoroughly monitored and analyzed. The first warning was a strong earthquake at a shallow depth beneath the volcano on March 20, 1980. On March 27, explosive hydrothermal activity began at the summit, accompanied by the formation of a small crater, ground fracturing, and the beginning of a topographic bulge high on the north flank. Strong seismic activity and relatively mild steam-blast eruptions continued intermittently into mid-May. During that time the new crater gradually enlarged and the bulge became larger. On the morning of May 18, an earthquake caused great avalanches of rock debris at the over-steepened bulge on the north flank. This unloading led to a northward-directed lateral blast, partly driven by steam explosions, which devastated an area of nearly 600 km^2. A vertical column of ash extended more than 25 km high, causing ash falls more than 1,500 km to the east. Pyroclastic flows occurred on the north flank. Melting snow and ice contributed to catastrophic mudflows. Smaller eruptions have occurred intermittently since the main blast. USGS reports by Lipman and Mullineaux (1981) and by Foxworthy and Hill (1982) document the events.

Judged by its volume of ejecta (0.6 km^3), Mount St. Helens was a minor event compared with the 1883 eruption of Krakatoa, Indonesia (26 km^3) or the 1815 eruption of Tambora, Indonesia (46 km^3). Because of the advance warnings, authorities had restricted access to Mount St. Helens, and only about 60 people were killed in the eruption. Property damage is estimated at $2 to $3 billion. The fact that the relatively moderate eruption of Mount St. Helens caused so much damage is reason to continue improving methods of volcano monitoring and prediction.

A. Earthquakes caused by expanding magma chamber.

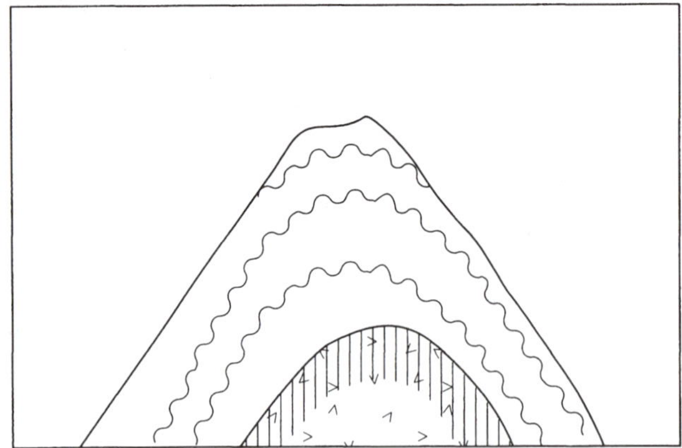

B. Changes in shape from expansion or contraction of chamber.

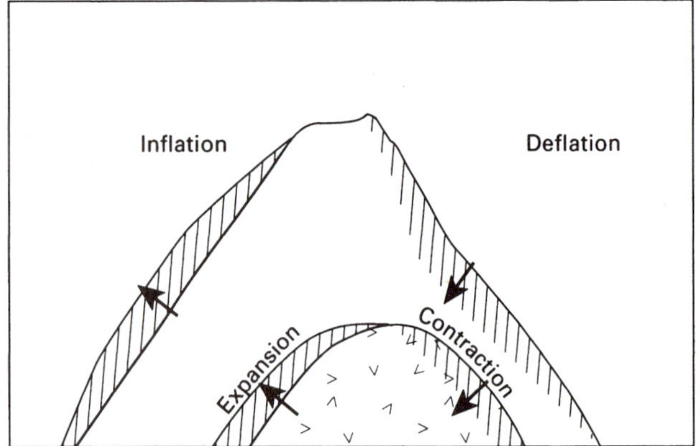

C. Increased heating of fractures.

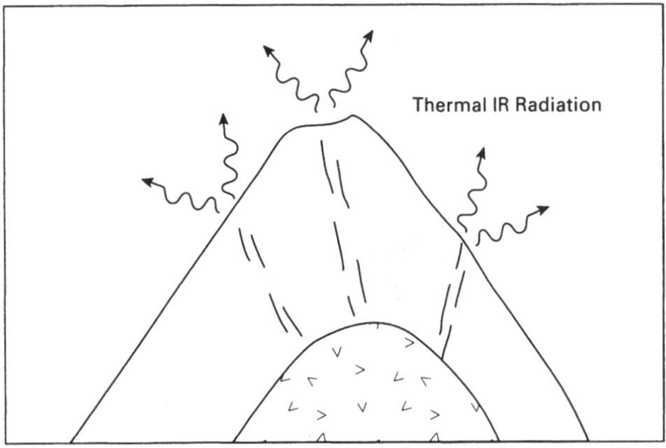

D. Increased eruption of ash and gas.

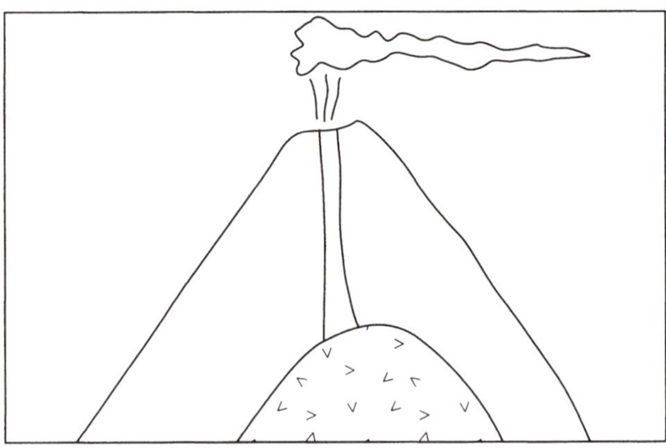

FIGURE 15-25 Warning criteria for monitoring active volcanoes. From Foxworthy and Hill (1982, Figure 3).

Aerial Photographs

Numerous remote sensing images were acquired before and after the main eruption. A pre-eruption aerial photograph (Figure 15-26A) shows the symmetric form of the mountain, which lacked a summit crater. The May 18 blast created a large crater and blew out the north rim, as shown in the post-eruption photograph (Figure 15-26B). The interpretation map (Figure 15-27) shows the major destructional features and volcanic deposits. Timber north of the volcano was flattened by the blast. The deposits blocked the outlet of Spirit Lake; as a result, the lake is larger in the post-eruption photograph and has a gray signature, which is caused by floating logs. The east, south, and west flanks of the volcano were modified by mudflows. The upper slopes were scoured by the flowing mud, and the lower slopes were covered by mudflow deposits. In the post-eruption photograph (Figure 15-26B), steam clouds conceal a lava dome that formed on the floor of the crater.

TIR Images

Aircraft TIR images (8 to 14 μm) were acquired before and after the main eruption. The initial explosive event occurred on March 27, 1980. Figure 15-28A is a TIR image of the resulting crater acquired on May 16, two days before the main blast. The image has not been geometrically corrected for scanner distortion. The image was processed to emphasize the hottest ground temperatures (> 12°C), which have very bright signatures. The background temperature away from the bright thermal features ranged between –13 to –8°C. Figure 15-28B is an interpretation map. Dark patches are the hot spots that coincide with deep pits, many of which were the source of steam plumes. Kieffer and others (1981) inferred from the image data that a temperature of 400°C occurred at a depth of only 40 m and was caused by intense hydrothermal circulation. The cluster of hot spots on the north flank of the volcano mark the bulge that was subsequently obliterated by the May 18 blast.

Figure 15-28C is a post-eruption image that has not been corrected for scanner distortion; hence the semicircular crater has an oval outline in the image. Thermal patterns

FIGURE 15-26 Aerial photographs of Mount St. Helens, Washington, acquired before and after the May 18, 1980 eruption.

in the image correlate with features shown in the geologic map (Figure 15-27). Flanks of the volcano are cool (dark signature). Pyroclastic flow deposits with intermediate temperatures (gray signatures) fill the crater and the depression on the north flank that was created by the blast. The highest radiant temperatures (brightest signatures) were identified by density slicing the data and are shown in black in the interpretation map (Figure 15-28D). The lava dome formed in August is the hot oval on the crater floor. A semicircular pattern of hot spots occurs along the inner wall of the crater. The irregular hot spots in the southeast floor of the crater are fumaroles of steam and gas. A linear hot feature extending south from the rim is an igneous dike 15 m wide. Other hot linear features occur in the new pyroclastic deposits and represent fractures probably caused by compaction. A northwest-trending alignment of hot spots is probably related to a major structural feature cutting across the core of the volcano (Friedman and others, 1981).

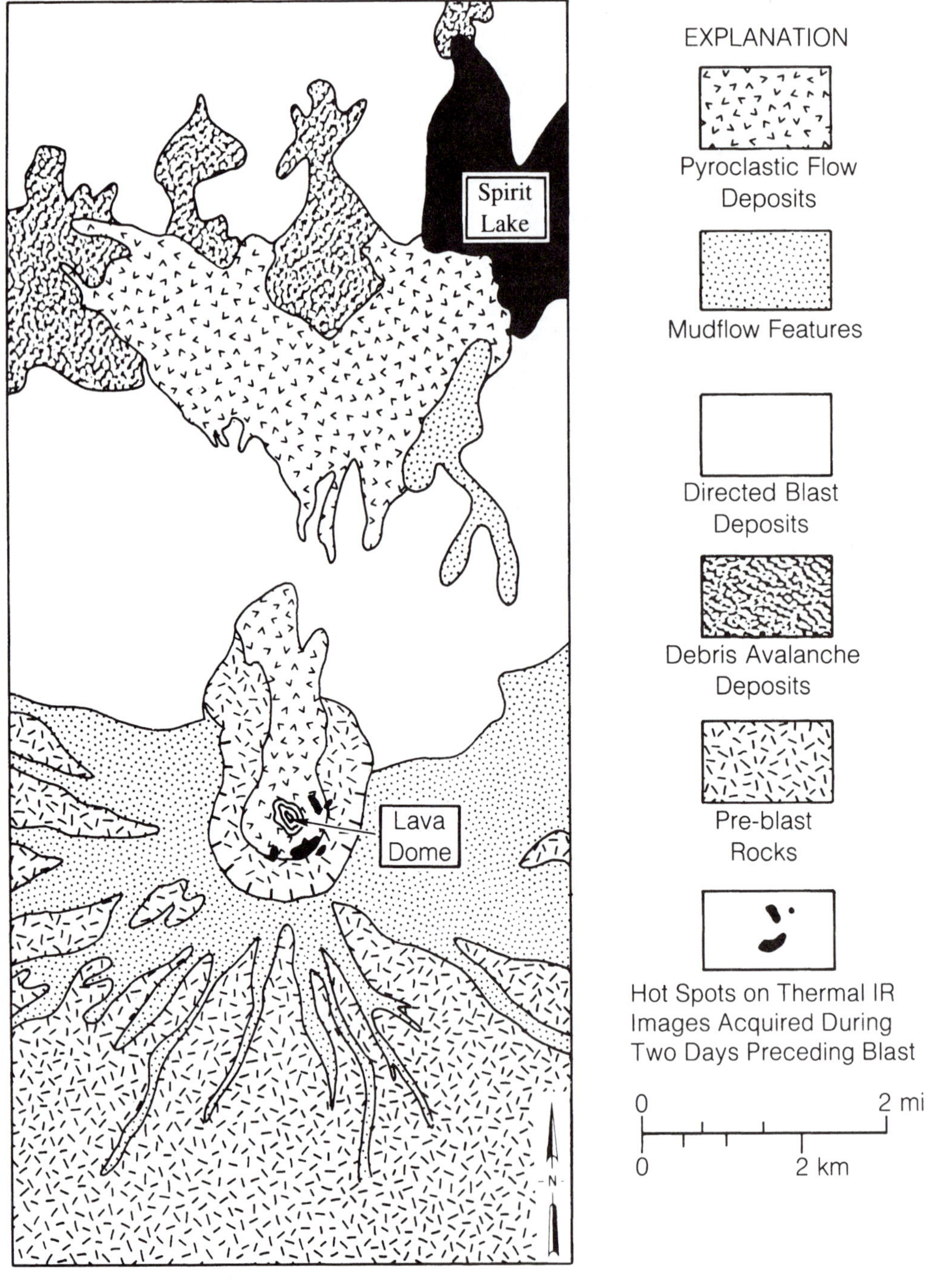

FIGURE 15-27 Volcanic features and deposits of the 1980 eruption of Mount St. Helens (Figure 15-26B). From Lipman and Mullineaux (1981, Plate 1).

A. May 16, 1980 image.

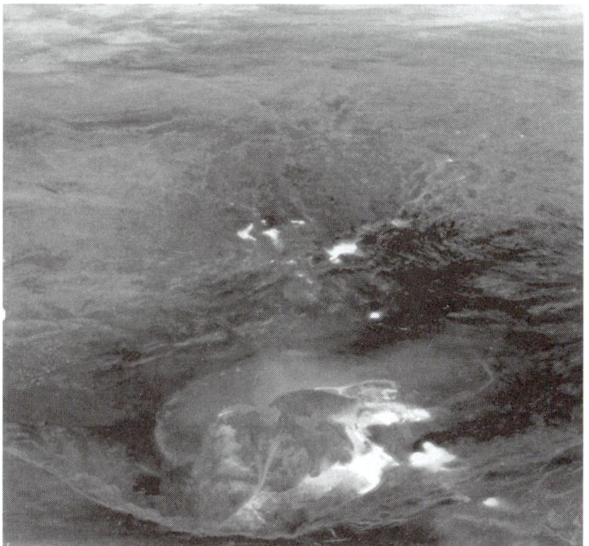

B. May 16, 1980 map.

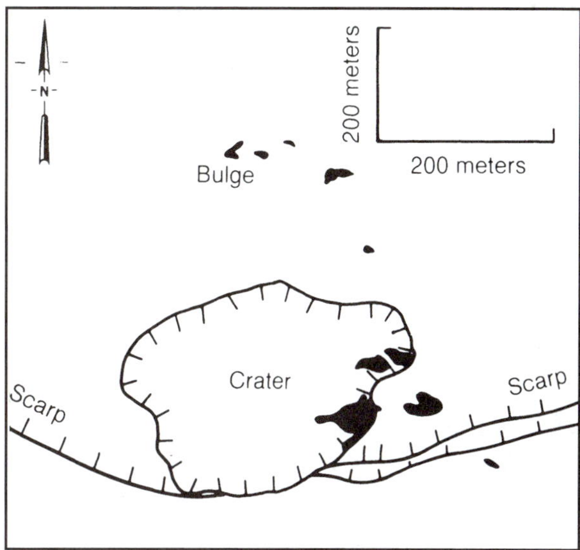

C. August 20, 1980 image.

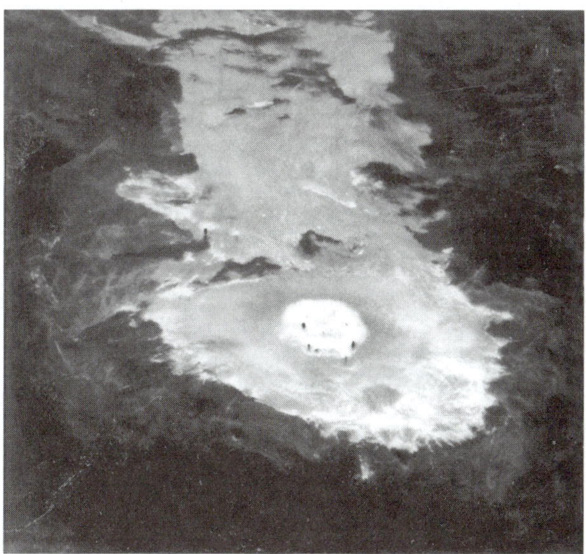

D. August 20, 1980 map.

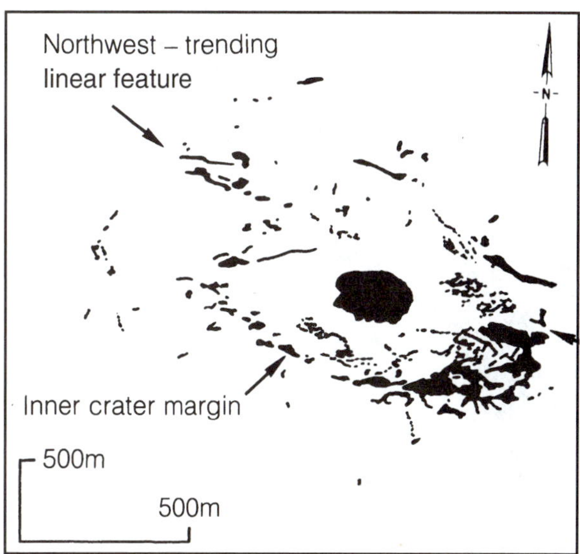

FIGURE 15-28 Nighttime TIR images (8 to 14 μm) of the summit of Mount St. Helens acquired before and after the eruption. Maps show highest temperatures in black. From Kieffer and others (1981, Figures 158 and 174). Courtesy H. H. Kieffer and J. D. Friedman, USGS.

MONITORING VOLCANOES THROUGH TIME

Volcano eruptions around the globe are monitored by multispectral and radar satellites orbiting the Earth. The Smithsonian Institution's Global Volcanism Program maintains an online global volcanism database that provides a catalog of volcanoes and eruptions from the past 10,000 years. The program's Eruptions, Earthquakes, & Emissions web application includes time-lapse animation of volcanic gas emissions since 1978—the year NASA launched the TOMS instrument (Chapter 11) (volcano.si.edu). New volcanic activity is documented with satellite, aerial, and ground images. Satellite and aerial remote sensing enable volcanologists and emergency response teams to better understand the location and extent of deposits and plumes to improve research, forecasting, and evacuations.

Villarrica is one of Chile's most active volcanoes. The University of Hawaii (Institute of Geophysics and Planetology) and NASA have processed TIR imagery of the Villarrica volcano in Chile since 2000 to provide temperature measurements of the crater several times a day. The data are collected by the MODIS system onboard the Terra and Aqua satellites and are available for viewing at the Proyecto Observación Volcán Villarrica Internet (POVI) (povi.cl/modvolc.html). Villarrica is a stratovolcano with the rim at an elevation of 2,840 m. The resort city of Pucón lies 16 km north at an elevation of 225 m. Pucón has a population of approximately 30,000 that doubles or triples during the tourist season.

During the week of April 3–9, 2019, there was an increase of explosions at Villarrica with incandescent material ejected to 50 m above the crater rim. During the week of September 4–10, 2019, there was increased seismicity indicating fluctuating lava lake activity at Villarrica. The Chilean National Geology and Mining Service raised the Alert Level to orange, the second highest on a four-color scale. The month before the increase in activity the east edge of the summit crater collapsed.

Thousands of people were evacuated around the volcano when it erupted two times during March 2015. NASA Earth Observatory (2015a, 2015b) published Landsat 8 OLI and Advanced Land Imager (ALI) images that were acquired before and during the eruptions (Figure 15-29). (See Chapter 4 for ALI characteristics.)

On February 22, 2015 the volcano summit was covered with a clean coat of snow (Figure 15-29A). The ALI system documented a large volume of volcanic material ejected during an eruption that was mostly deposited on the mountain's eastern slope (Figure 15-29B). The deposit is interpreted to have covered over 7 km^2. On March 18, 2015 the ALI system recorded an ash plume drifting toward the southeast (Figure 15-29C). An interpretation of the deposit and plume can be found in the hillshade SRTM DEM (Figure 15-29D).

AIRCRAFT TIR IMAGES

Figure 15-30 is a map of the town of Rabaul and vicinity, at the northeast end of the island of New Britain in Papua New Guinea. This region has a long history of volcanic eruptions. Simpson Harbor occupies a large collapse caldera formed by a series of ignimbrite eruptions over the past 18,000 years. *Ignimbrite eruptions* are incandescent clouds of ash and ejecta that move rapidly down the flanks of a volcano with destructive effects. Dating of the ignimbrite deposits shows the interval between eruptions ranges from 2,000

A. Snow-covered volcano, February 22, 2015. From NASA Earth Observatory (2015a).

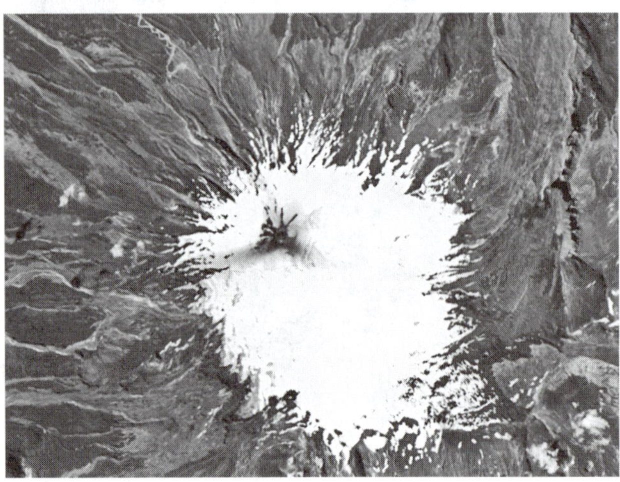

B. Erupted debris on snow, March 5, 2015. From NASA Earth Observatory (2015a).

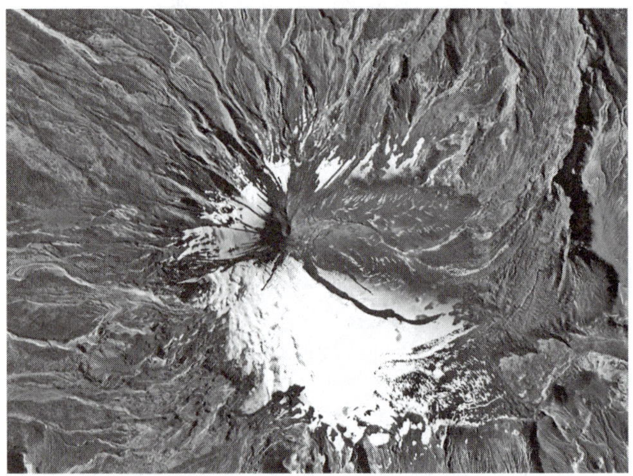

C. Ash plume, March 18, 2015. From NASA Earth Observatory (2015b).

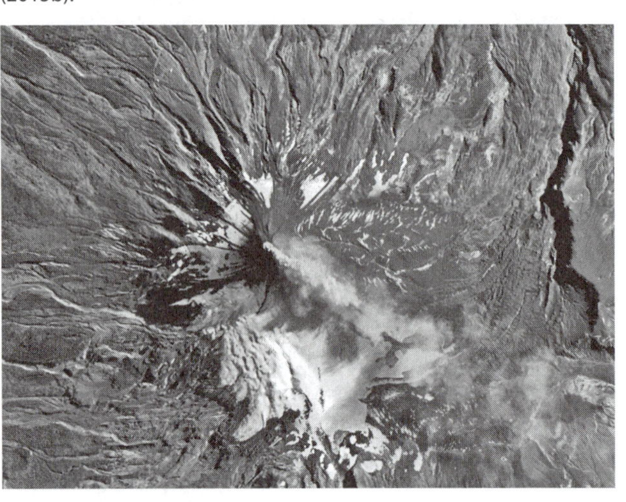

D. SRTM hillshade DEM with interpretation of erupted debris and ash plume. Courtesy USGS.

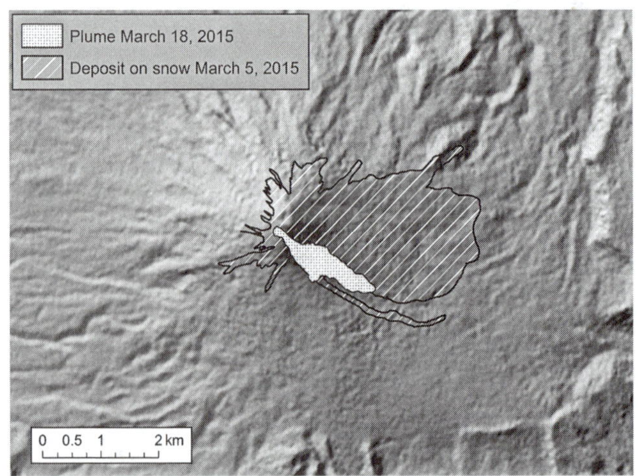

FIGURE 15-29 Villarrica volcano, Chile.

to 3,600 years. The last major eruption occurred about 1,400 years ago (Johnson and others, 1995). A number of volcanic centers are associated with the caldera (Figure 15-30). Both Tavurvur and Vulcan erupted in 1878 and 1937. Poorly documented eruptions occurred in 1767 and 1791 (Johnson and others, 1995). Because of this history the Australian Bureau of Mineral Resources conducted an aircraft TIR survey of the area in 1973 (Perry and Crick, 1976). Figure 15-31 shows images of the two volcanoes that have significant warm signatures. The horizontal lines are the flight paths. The wavy lines are profiles of the radiant temperatures along the flight paths that were measured with a radiometer aimed downward from the survey aircraft. Rabalankaia volcano (Figure 15-31A) has concentric warm rings that coincide with bare ground on the floor and walls of the summit crater. Tavurvur volcano (Figure 15-31B) has an irregular pattern of warm spots that partially correspond to bare ground. The radiometer profile crossed the warm areas and recorded radiant temperatures up to 27°C.

On September 19, 1994, explosive eruptions occurred at Tavurvur and Vulcan. Ash falls up to 1-m thick caused massive damage to Rabaul, which was evacuated. The 1973 image (Figure 15-31B) clearly shows thermal features at Tavurvur, but none were present at Vulcan. Rabalankaia has conspicuous thermal signatures, but no record of historic eruptions, which should not downgrade its potential for future eruptions. The survey conducted by the Australian Bureau of Mineral Resources points out both the promise and uncertainty of forecasting eruptions. The 1973 TIR survey identified one of the two future eruptive sites, but two decades elapsed before the eruption occurred.

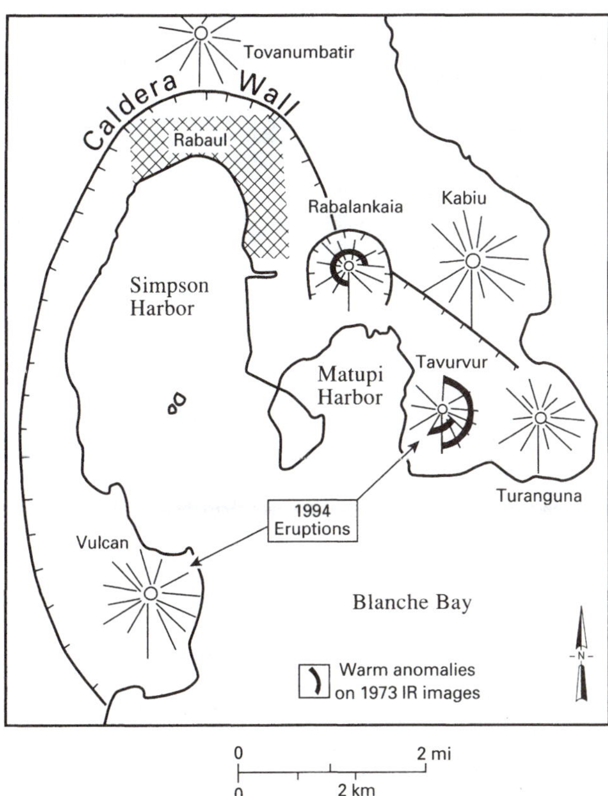

FIGURE 15-30 Map of Rabaul and vicinity, New Britain, Papua New Guinea.

SATELLITE TIR IMAGES

Landsat TM images are widely used to monitor volcanoes. Representative studies include Erta Ale, Ethiopia (Rothery, 1989); Barren, Barren Island, Andaman Sea (Reddy and others, 1993); Stromboli and Vulcano, Italy (Gaonac'h and others, 1994); and Mount Erebus, Antarctica (Rothery and Francis, 1990). The repeated coverage by Landsat was used to analyze cycles of eruption at Lascar volcano.

Lascar is the most active volcano in the central Andes of northern Chile. The region is sparsely populated and the record of volcanic activity is meager. Local residents recall

A. Rabalankaia volcano.

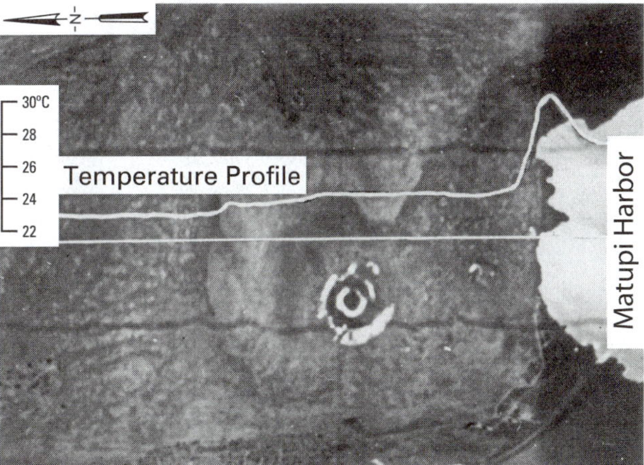

B. Tavurvur volcano.

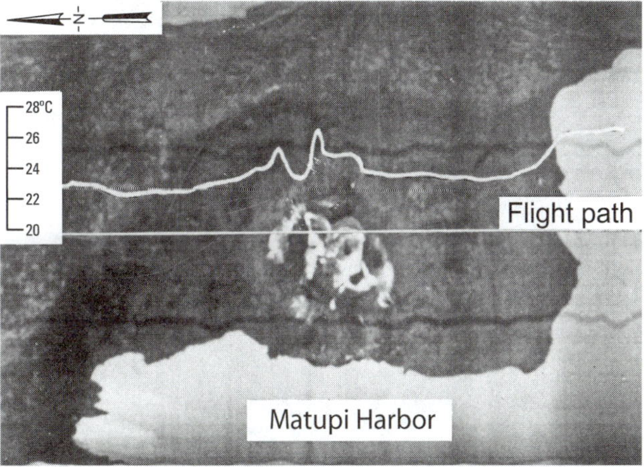

FIGURE 15-31 Nighttime TIR images (8 to 14 µm) of volcanoes near Rabaul, New Britain, Papua New Guinea. Bright signatures are warm radiant temperatures. The images are 5 km wide from left to right. From Perry and Crick (1976, Figure 5). Courtesy W. J. Perry, Bureau of Mineral Resources, Australia.

periodic events dating back to 1955. Glaze and others (1989) relied largely on satellite images to document an eruption in 1986. Oppenheimer and others (1993) analyzed volcanism at Lascar using Landsat TM bands 5 and 7 in the SWIR region. Chapter 5 pointed out that these bands record the high radiant temperatures typical of fresh lava. Fifteen images acquired between December 1984 and April 1992 recorded the evolution of thermal features and spanned two eruptions. Figure 15-32 shows four TM band 7 images (three daytime, one nighttime) of the summit that were recorded in 1989. In all the images, a large central cluster of very bright pixels is a lava dome. Figure 15-33 is a map based on the images that shows the dome and other features of the summit. Field observations in early 1990 showed the dome to be a circular body of blocky lava approximately 200 m in diameter. At night the dome was peppered with many glowing sites in arcuate chains and clusters located mostly near the margins. TIR radiometers recorded radiant temperatures ranging from 500° to almost 800°C for the incandescent sites. The arrows on the images in Figure 15-32 indicate three persistent small hot spots beyond the margins of the dome that are probably high-temperature fumaroles. The map (Figure 15-33) shows the relationship of the hot spots to the dome and the crater walls.

Figure 15-34 plots the radiance of band 7 for the central dome for the images acquired from 1984 to 1992. Variations in radiance correspond to periods of dome growth that were punctuated by explosive eruptions (1986, 1990) that produced major ash columns. High radiance values in late 1989 include measurements from the images of Figure 15-32. Similar high values in late 1987 indicate the dome was present then, although it was not confirmed by field observations until early 1989. High values in early 1985 indicate an earlier appearance of the dome. Aerial photographs taken in January 1987, however, show no evidence of a dome, which suggests that it had been destroyed by the 1986 eruption. These results at Lascar demonstrate the capability and value of satellites with SWIR bands for repetitive monitoring of active volcanoes in remote areas.

A. October 27, 1989, daytime.

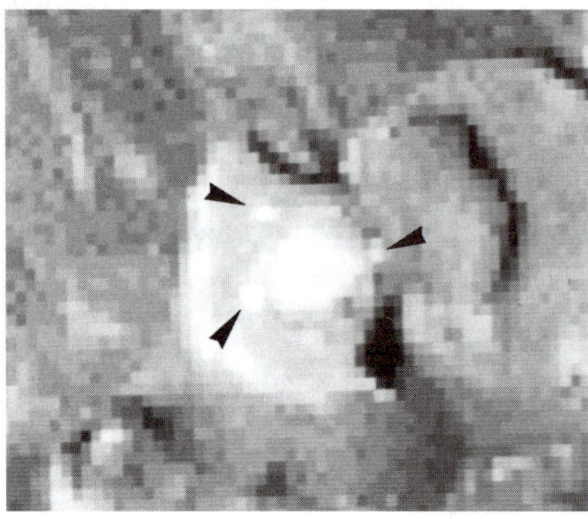

B. November 17, 1989, nighttime.

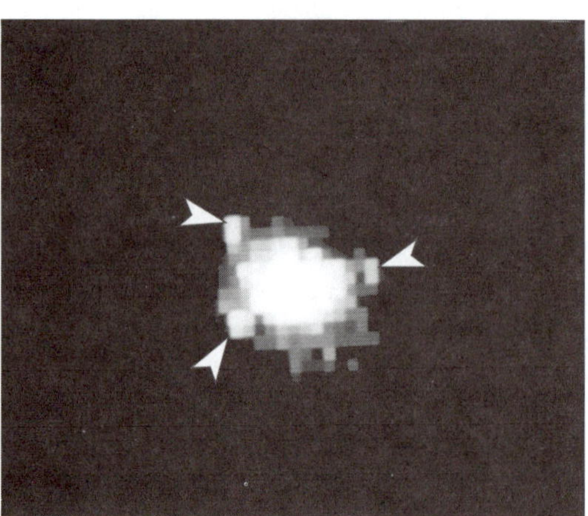

C. November 28, 1989, daytime.

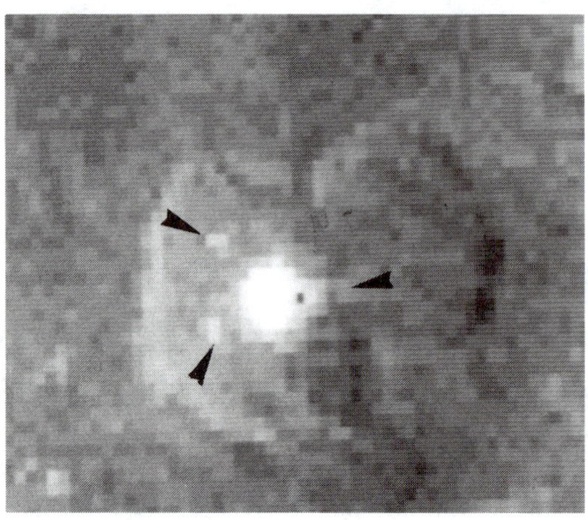

D. December 14,1989, daytime.

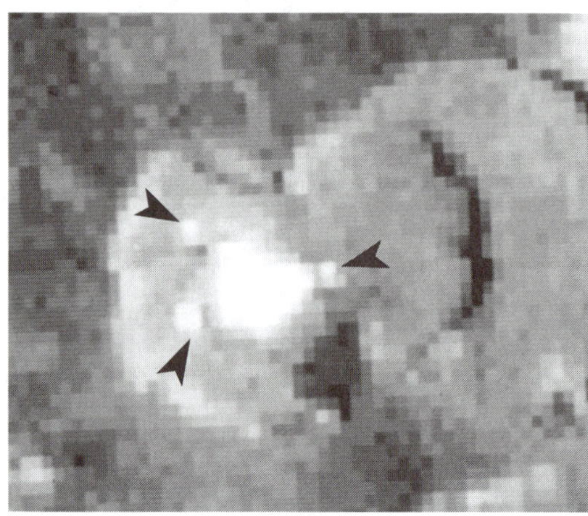

FIGURE 15-32 Landsat TM band 7 images of the summit of the Lascar volcano, Chile. From Oppenheimer and others (1993, Figure 13). Courtesy C. Oppenheimer, The Open University and L. Glaze, NASA Goddard Space Center.

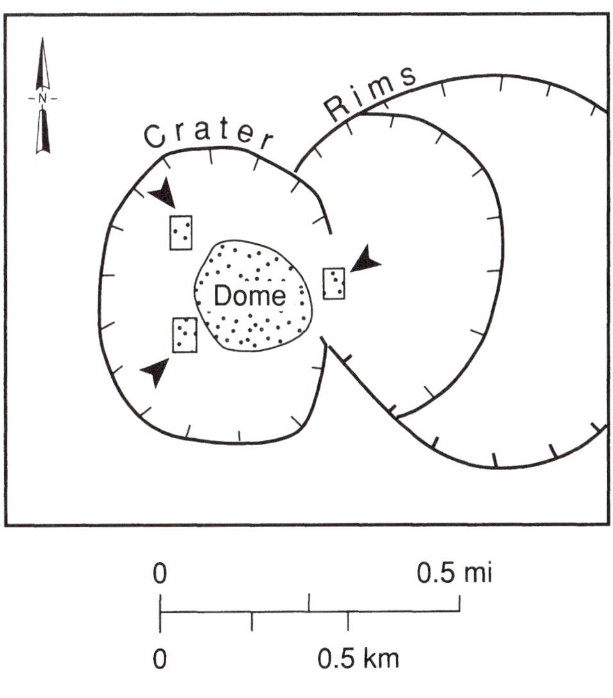

FIGURE 15-33 Map of the summit of the Lascar volcano, Chile.

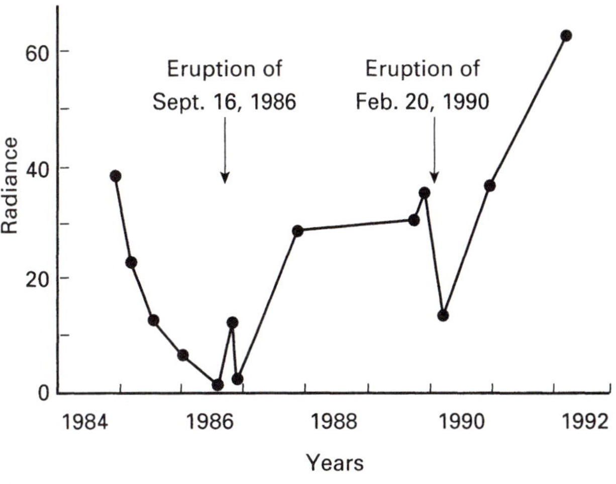

FIGURE 15-34 Spectral radiance of TM band 7 for the summit of the Lascar volcano from 1984 to 1992. Modified from Oppenheimer and others (1993, Figure 18).

RADAR INTERFEROGRAMS OF MOUNT ETNA

Figure 15-25B shows that expansion and contraction of the magma chamber cause inflation and deflation of the volcano. Monitoring these changes in shape can be used to predict eruptions. Making timely and precise measurements of these minor topographic differences is routine with interferograms from satellite radar data. The compilation of radar interferograms was described in Chapter 6. In Figure 15-7, an interferogram was combined with digital topographic data to create a residual interferogram that shows the terrain displacement caused by the Landers earthquake. Massonnet and others (1995) used this method to monitor deformation of Mount Etna, Italy, during a recent eruption.

Mount Etna is one of the world's most active and best studied volcanoes. A major eruption began December 14, 1991 and ended October 23, 1993, for a total of 473 days. Lava flowed at a constant rate during this period and erupted a volume of approximately 3×10^8 m^3, which caused deflation of the volcano. A stereo-pair of SPOT panchromatic images were used to compile a digital elevation model of the volcano prior to the eruption. A series of interferograms of the west flank of Mount Etna were compiled from 13 ERS-1 satellite radar images acquired on northbound orbits; 16 images of the east flank were acquired on southbound orbits. The interferograms were combined with the digital elevation data to produce a series of 12 reliable measurements of deformation that cover the second half of the eruption.

Massonnet and others (1995) illustrate their interferograms and describe the limitations and applications of this technique. The west flank deformed approximately 14 cm over 225 days. The error of individual measurements was 1 cm or less. Interferograms acquired after the eruption stopped showed no deformation.

Radar interferograms can help predict eruptions by monitoring deformation prior to an eruption. A NASA animation of Mount Etna that spans 1992 to 2001 shows how changes in the volcano's magna chamber deform the ground around the mountain (Pappas, 2016). The visualization is based on interferometric measurements by ERS-1 and ERS-2 and documents rising and subsiding ground movement that spans 25 cm over the decade.

VOLCANIC ASH PLUMES

Volcanic eruptions can inject huge volumes of ash into the atmosphere. After some eruptions, such as Krakatoa and more recently Mount Pinatubo, the ash remains in the atmosphere for years and causes worldwide cooling. The ash plumes are also hazards for aviation. The tiny particles of volcanic glass are ingested by jet engines, where they melt and are deposited on turbine vanes, which stalls the engine. Acidic aerosol particles also etch windshields and other surfaces. At altitudes flown by commercial aircraft the plumes are difficult to see and are not detected by aircraft radar. One incident occurred at midday on December 15, 1989 when a KLM Boeing 747 en route from Amsterdam to Anchorage at an altitude of 7.5 km flew through an ash plume from the Redoubt volcano, Alaska. All four engines shut down for 12 minutes, and the jet descended steeply for 4 km before the crew managed to restart the engines after seven or eight tries. The aircraft came within about 1.5 km of the mountaintops before landing safely at Anchorage. Damage to the jet reportedly exceeded $50 million (Kienle and others, 1990). Several dozen such incidents have occurred over the past two decades in regions of active volcanism. The ash plume from the 1982 eruption of the Galunggung volcano on the Indonesian island of West Java caused engine failures and emergency landings of two Boeing 747 commercial flights. In 2010, the ice-covered stratovolcano Eyjafjallajökull in

Iceland underwent explosive eruptions when water from the melting glacier ice cap encountered the molten lava. The steam explosions shattered the lava into tiny knife-edged shards that are highly abrasive to aircraft windows, flight surfaces, and jet engines. The volcanic ash rose several kilometers into the atmosphere and disrupted air travel in northwest Europe for weeks.

Ash plumes are detectable on satellite images by their distinctive shape and orientation relative to volcanic sources. Satellite images, including AVHRR and MODIS images, are particularly useful because of their daily worldwide coverage with daytime and nighttime images in the visible and TIR bands. Wen and Rose (1994), using AVHRR bands 4 and 5, developed a method for estimating the total mass and size range of particles in plumes. Figure 15-35A is an AVHRR TIR image of the ash plume from the Redoubt volcano that was acquired one day after the KLM aircraft encountered the same plume. Air traffic controllers use images such as this to direct aircraft away from these hazards.

Figure 15-35B is an AVHRR TIR image of the ash plume from the September 1994 eruption at Rabaul. The Rabaul plume exceeds 500 km in length in this image, whereas the Redoubt plume is only 50 km in length. In both images the thickest portions of the plumes have the coldest temperatures (brightest signatures), and the thinner margins have apparently warmer temperatures. In reality, the thinner portions of the plumes have the same cold temperature as the thicker portions; thermal energy radiated from the Earth's surface is transmitted through the thinner plumes to produce an apparent warmer temperature.

Many other plumes have been monitored on satellite images. The Tolbachik volcano on the Kamchatka Peninsula of eastern Russia erupted on July 6, 1975. Jayaweera and others (1976) measured the length and orientation of the plume on 14 NOAA images acquired from July 9 through August 17, 1975. The most spectacular image was acquired July 18, when the length of the plume was at least 960 km. From the distance between the plume and its shadow, which was measured from the visible image, and from the known sun elevation, the height of the plume on July 28 was estimated at 6.5 km.

Sulfur Dioxide

Significant amounts of sulfur dioxide (SO_2) can be emitted by volcanoes. Sulfur dioxide is a colorless gas that irritates skin and the tissues and mucous membranes of the eyes, nose, and throat. High concentrations of SO_2 produce volcanic smog (VOG), which causes persistent health problems for populations downwind of an erupting volcano. Depending on the concentration, SO_2 is potentially hazardous to people, animals, agriculture and natural vegetation, and infrastructure. SO_2 ejected greater than 10 km into the stratosphere is converted to sulfate aerosols, which reflect sunlight and have a cooling effect on the Earth's climate. Sulfate aerosols also deplete ozone (USGS, 2017).

A. Redoubt volcano, Alaska. December 16, 1989, at 1223 GMT.

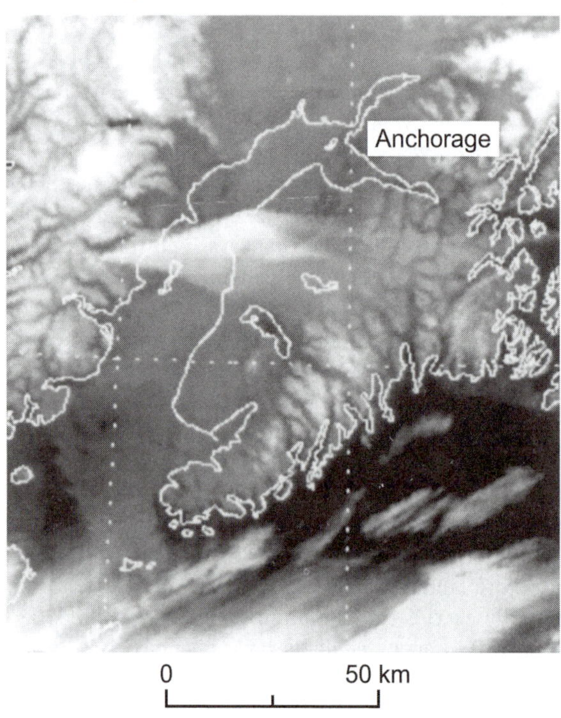

B. Rabaul volcanoes, Papua New Guinea. September 19, 1994, at 0859 GMT. From Rose and others (1995, Figure 2).

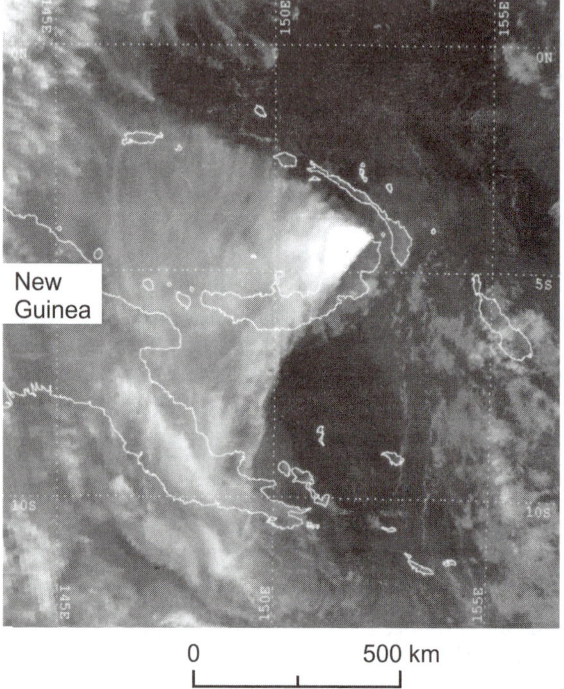

FIGURE 15-35 TIR images of volcanic ash plumes recorded by AVHRR band 4 (10.3 to 11.3 μm). Bright signatures represent cool temperatures. Times given as Greenwich mean time (GMT). Courtesy W. I. Rose and D. J. Schneider, Michigan Technological University.

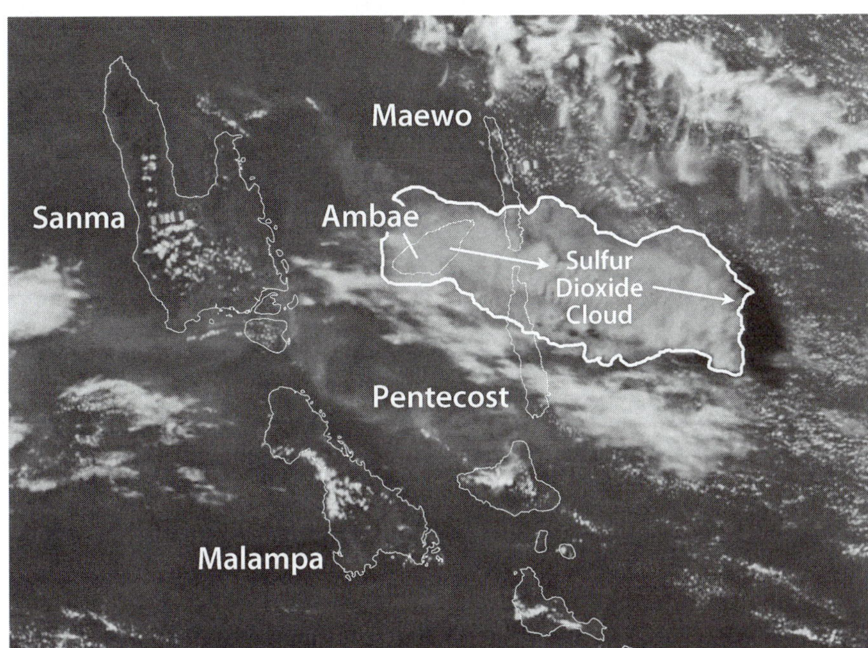

FIGURE 15-36 This satellite image, acquired on July 27, 2018 by VIIRS on Suomi NPP, shows the large sulfur dioxide eruption in 2018 from the Manaro Voui volcano on the island of Ambae, Vanuatu, South Pacific Ocean. Photo by Lauren Dauphin, NASA Earth Observatory.

The images of volcanic plumes (Figure 15-35) include meteorologic clouds that have the same range of temperatures as the plumes, which can make identification difficult. Prata (1989) noted that volcanic plumes contain droplets of sulfuric acid (H_2SO_4) formed by the hydration of sulfur dioxide gas (SO_2). Droplets of sulfuric acid absorb energy at the shorter wavelengths of AVHRR band 4 (10.3 to 11.3 μm) more strongly than at the longer wavelengths of band 5 (11.5 to 12.5 μm); therefore, band 4 minus band 5 results in a negative value for volcanic plumes. For clouds, droplets of water do not absorb the shorter wavelengths; therefore, band 4 minus band 5 is zero or a positive value. These relationships are used to differentiate volcanic plumes from clouds on digitally processed AVHRR images.

The Ozone Mapping Profiler Suite (OMPS) is carried on the Suomi NPP and NOAA-20 satellites (Chapter 11). OMPS contains hyperspectral ultraviolet sensors, which map volcanic clouds and measure sulfur dioxide emissions by observing reflected sunlight. Sulfur dioxide and other gases like ozone each have their own spectral absorption signature, that is, their unique fingerprint. OMPS measures these signatures, which are then converted, using complicated algorithms, into the number of SO_2 gas molecules in an atmospheric column.

During 2018, OMPS instruments revealed that the Manaro Voui volcano on the island of Ambae, Vanuatu in the South Pacific Ocean injected 600,000 tons of sulfur dioxide into the upper troposphere and stratosphere. This release was three times the amount released from all combined worldwide eruptions in 2017. During the most active eruption period in July 2018, the volcano expelled 400,000 tons of sulfur dioxide (NASA Earth Observatory, 2019a). A visible light image was acquired by VIIRS onboard Suomi NPP on July 27, 2018 during an active eruption (Figure 15-36) (see Chapter 11 for VIIRS system characteristics).

In Figure 15-36, clouds and the sulfur dioxide cloud obscure the islands of Ambae, Maewo, and Pentecost. Westerly winds are moving the sulfur dioxide cloud toward the east. During the series of eruptions at Ambae in 2018, volcanic ash blackened the sky, buried crops, and destroyed homes. Acid rain turned the rainwater, the island's main source of drinking water, cloudy and metallic. Over the course of the year, the island's entire population of 11,000 was forced to evacuate.

NASA produces maps of stratospheric SO_2 concentration within three hours of a satellite's overpass. These are used at volcanic ash advisory centers to predict the movement of volcanic clouds and reroute aircraft as needed. Scientists are trying to understand the collective impact of volcanoes like Ambae and others on the climate.

FLOODS

NASA's Soil Moisture Active Passive (SMAP) satellite measures soil moisture by detecting faint microwave emissions from the Earth's surface (see Chapter 11 for discussion). The system is capable of measuring the amount of water in the upper 5 cm of soil. SMAP data is used to generate maps of soil moisture anomalies in the United States (**Digital Image 15-2**). The National Weather Service (NWS) uses SMAP measurements, along with other data, to improve flash flood and flood warnings. SMAP's soil moisture anomaly maps enable the NWS to determine how wet the soil is, and which areas are saturated, before a rainstorm. Saturated soil significantly increases the probability of flooding. SMAP updates soil moisture readings of the United States every three days. These satellite measurements provide an important flood prediction tool for the NWS.

On March 21, 2019, NOAA noted that some parts of the central United States had already received 200% of the average amount of rain and snow that usually accumulates through late March. The NWS Climate Prediction Center's calculated soil moisture anomaly map for March 2019 reflects the substantial increase in precipitation announced by the NWS (**Digital Image 15-2B**). Much of the Mississippi River watershed had soil moisture 100% above the long-term average in March 2019.

MISSISSIPPI RIVER

Periodic floods in the Mississippi River Valley have been monitored by successive generations of remote sensing technology. The 1973 floods were analyzed with Landsat MSS images by the US Geological Survey (Deutsch and Ruggles, 1974). A decade later the floods of 1983 were analyzed by comparing Landsat TM images acquired before and during the high water (Sabins, 1987). Unusually severe flooding in 1993 (47 flood-related deaths and damages of $12 billion) was mapped by Gumley and King (1995) with an airborne simulator of the MODIS satellite system and the operational satellite-based Landsat TM system. After MODIS was launched in 1999, they developed an automatic identification algorithm for water using a NIR band on the airborne MODIS system and TM band 5 (SWIR1) on the Landsat system to determine 396 km^2 of typically dry land was flooded near St. Louis, Missouri.

In late February 2019, intense storms over three days caused major flooding along the Middle Mississippi River (NASA Earth Observatory, 2019b). The Landsat 8 OLI system acquired images of the region. Figure 15-37A shows typical winter conditions for the area south of Memphis, Tennessee, along the Mississippi River. An image of the same area was acquired on February 25, 2019 during the flooding (Figure 15-37B). The dark pattern along the Mississippi River shows the extent of the flooding when compared with typical conditions. Normally the Mississippi River adjacent to Memphis is contained within a channel approximately 1 km wide. During the February 2019 flood the river jumped its banks and expanded approximately 4 km across the floodplain to the west.

A. Typical winter conditions on February 27, 2014.

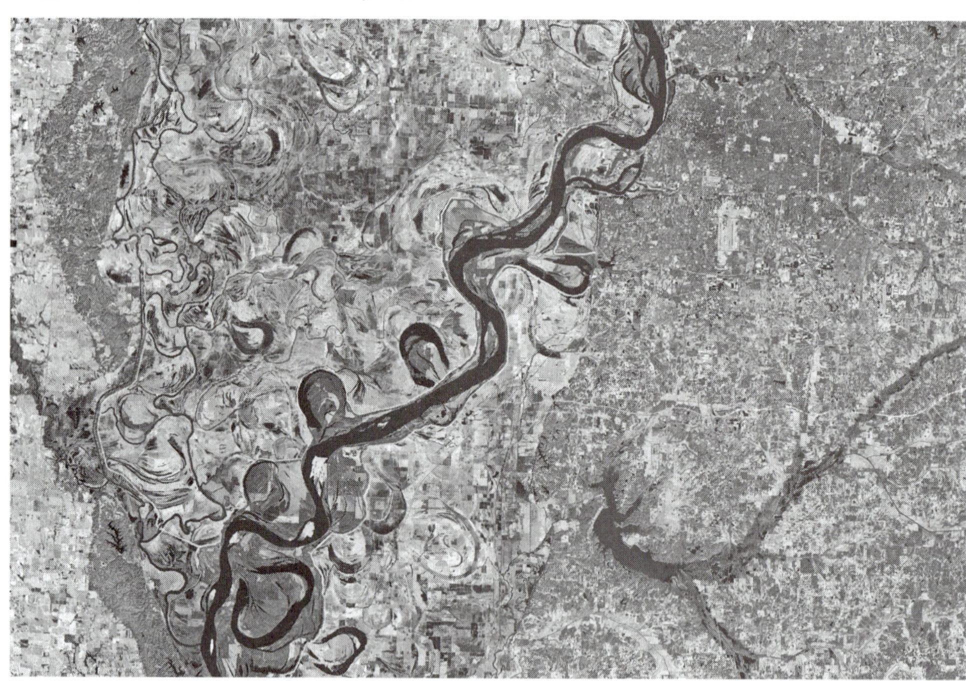

B. Major flooding on February 25, 2019.

FIGURE 15-37 Middle Mississippi River, Memphis, Tennessee. From NASA Earth Observatory (2019b).

STORM SURGE

The US Climate Resilience Toolkit (2017) defines storm surge as abnormally high water levels generated by severe storms that can produce sea levels much higher than normal high tide, resulting in extreme coastal and inland flooding. If storm surges coincide with high tide, they can raise water levels by 20 ft (6 m) or more above mean sea level. As a result of global sea level rise, storm surges that occur today are 8 in (0.2 m) higher than they would have been in 1900. By 2100, storm surges will happen on top of an additional 8 in to 6.6 ft (0.2 to 2 m) of global sea level rise.

Hurricane Sandy, the largest Atlantic hurricane on record, made landfall on October 29, 2012, and generated a storm surge along a long swath of the US Atlantic coastline. The barrier islands were breached in a number of places and beach and dune erosion occurred along most of the Mid-Atlantic Coast. As a part of the National Assessment of Coastal Change Hazards Project, the USGS collected post-Hurricane Sandy oblique aerial photography and lidar topographic surveys to document the changes that occurred as a result of the storm. Comparisons of post-storm photographs to those collected prior to Sandy's landfall were used to characterize the nature, magnitude, and spatial variability of hurricane-induced coastal changes. Analysis of pre- and post-storm lidar elevations was used to quantify magnitudes of change in shoreline position, dune elevation, and beach width (Sopkin and others, 2014).

The USGS assessed the morphological impacts of Hurricane Sandy on the beach and dune system at Fire Island National Seashore, Long Island, New York. Fire Island National Seashore is a east–northeast to west–southwest trending barrier island on the southeast side of central Long Island. Ocean Bay Park and Old Inlet are located in the western and eastern portions of the Fire Island National Seashore (Figure 15-38).

At the height of the storm, a record significant wave height of 9.6 m was recorded at the wave buoy offshore of Fire Island (Hapke and others, 2013). During the storm, beaches were severely eroded and dunes extensively overwashed. Fire Island was breached in three locations, and the coastal infrastructure, including many private residences, was heavily damaged.

The Hurricane Sandy storm surge washed over the dunes at Ocean Bay Park, an urbanized barrier island, resulting in the transport of large volumes of sand inland from the beach system, a severe decrease of dune height, and significant property damage. The May 21, 2009 pre-storm oblique photograph looking northwest from the ocean toward the interior bay shows the intact housing, infrastructure, and trees at Ocean Bay Park (Figure 15-39A). A wide beach with sand dunes is established along the seaward side of Ocean Bay Park. The post-storm aerial photograph reveals that overwash from the beach and narrow dunes along the ocean shoreline carried sand inland and towards the bayside of the island (Figure 15-39B). The morphological changes due to the storm surge were quantified with lidar (Plate 52A). Sand was removed from a 50-m wide swath of the beach and dunes. The sand beach and dunes were severely eroded with topographic elevation reduced by up to 3.5 m. The surge deposited sand toward the interior of the island. Some sites accumulated over 2 m of sand.

Breaching of the barrier island occurred at several locations along Fire Island, including at Old Inlet, an unpopu-

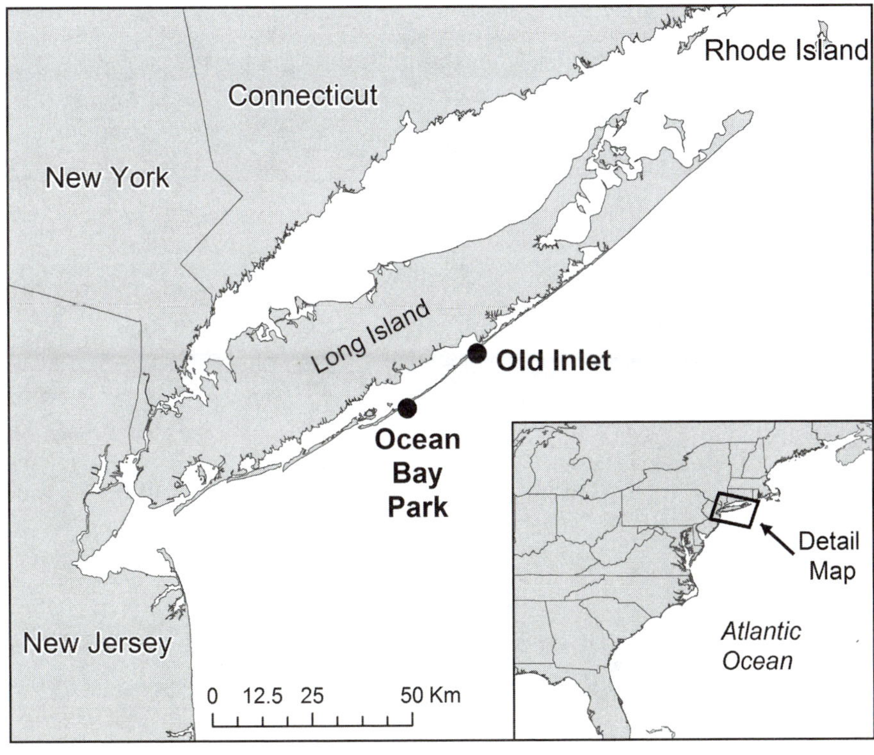

FIGURE 15-38 Location map for pre- and post-storm lidar survey maps at Fire Island, New York.

lated barrier island, where volume loss and shoreline retreat are apparent (Figure 15-40). A road across the island was destroyed during the storm surge and a new inlet was created. A fishing shack remained standing despite the breach. The lidar pre- and post-storm DEMs clearly show that the breach cut through 4-m high sand dunes (Plate 53). Beach and sand dunes outside of the breach and along the seaward side of the island were eroded with a decrease in elevation while the interior of the island accumulated up to approximately 2.5 m of sand washed over the island during the surge.

Hapke and others (2013) summarize the morphological changes at Fire Island. The beaches and dunes on Fire Island were severely eroded during Hurricane Sandy, and the island breached in three locations on the eastern segment of the island. The lidar measurements and fieldwork documented a landward shift of the upper portion of the beach that averaged 19.7 m, but varied substantially along the coast. The elevation of the shoreline was lowered by as much as 3 m. Shoreline change was highly variable, but the shoreline prograded during the storm by an average of 11.4 m due to the deposition of material eroded from the upper beach and dunes onto the lower portion of the beach. The beaches and dunes lost 54.4% of their pre-storm volume, and the dunes experienced overwash along 46.6% of the island. The inland overwash deposits account for 14% of the volume lost from the beaches and dunes, indicating that the majority of material was moved offshore. Hurricane Sandy reduced the topographic elevation of the barrier island, leaving the island vulnerable to future storms.

FIRE

Images acquired at night with the Day/Night Band of the VIIRS instrument onboard the NASA/NOAA Suomi NPP satellite capture low-light emissions under varying illumination conditions with 370 m spatial resolution (Blumenfeld, 2019). The Day/Night Band detects light in a range of wavelengths from green to NIR and uses filtering techniques to observe dim signals such as city lights, gas flares, auroras, wildfires, and reflected moonlight (Cole and others, 2012). Each Suomi NPP Day/Night Band has a swath width of

A. Oblique pre-storm aerial photograph.

B. Oblique post-storm aerial photograph.

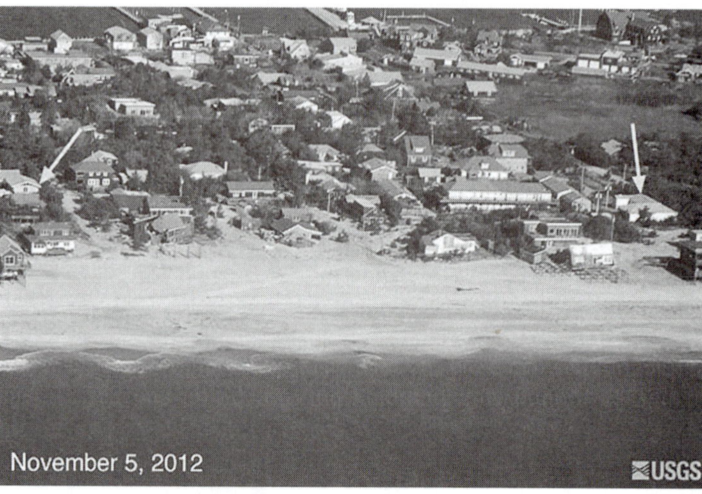

FIGURE 15-39 Ocean Bay Park, Fire Island, New York. The view is looking northwest across Fire Island towards Great South Bay. The white arrows in each image point to the same features. See Plate 52. From Sopkin and others (2014, Figure 32).

3,040 km (Chapter 11), enabling regional scenes that can provide nighttime images of many fires. These images are available generally within three hours of acquisition.

The Day/Night Band images effectively track nighttime fire fronts across large areas on a daily basis, informing emergency response teams and residents in harm's way of the fire front's size, shape, and direction(s) of movement. Day/Night Band images were acquired around 2 AM on the nights of December 4 through 8, 2017 of the huge Thomas Fire in Ventura County (Figure 15-41). Fierce Santa Ana winds, very low humidity, and dried out vegetation caused the front of the Thomas Fire to rapidly spread outward and toward the west during the five days, scorching terrain across an area greater than 35 km across.

The image in Figure 15-41B is a portion of the Day/Night Band scene acquired on December 4, approximately 16 hours before the Thomas Fire started. The white arrow points to the location of a brushfire that was reported at 6:28 PM on December 4, approximately 30 km northeast of Ventura (O'Neal, 2017). Figure 15-41C was acquired two hours before the fire had consumed 31,000 acres with over 150 structures destroyed and 27,000 people evacuated. The bright glows in the center of Figure 15-41C show the hot spots along the fire front. On December 6 (Figure 15-41D), the fire had three hot spots and had reached the coast. On December 7 (Figure 15-41E), the fire evolved into a two-pronged front: the southern front moving along the coast and the northern front tracking west in the mountains northeast of Santa Barbara. A few hours later the fire had burned 96,000 acres with over 400 homes lost. The December 8 image shows two long fire fronts in the mountains and some hotspots near the coast (Figure 15-41F). By December 8 the Thomas Fire had scorched 150,000 acres and is estimated to have cost $17 million (O'Neal, 2017).

The Suomi NPP Day/Night Band images supplement higher spatial resolution and daytime visible and reflected IR images by filling the nighttime gap with regional daily coverage of evolving fire fronts. The Suomi NPP MWIR bands (3.75 μm and 4.05 μm) also provide fire temperatures (Bachmeier, 2017). The large, wind-driven Thomas Fire was very hot: the maximum recorded temperature on December 5 was 434.6°K (161.6°C or 322.6°F).

A. Oblique pre-storm aerial photograph.

B. Oblique post-storm aerial photograph.

FIGURE 15-40 Old Inlet, New York. The view is looking northwest across Fire Island toward Great South Bay. The island breached during Hurricane Sandy, creating a new inlet. Despite the breach, a fishing shack (white arrow) remained standing. See Plate 53. From Sopkin and others (2014, Figure 34).

A. Location map. The general area of the Thomas Fire burn scar is highlighted by two arrows.

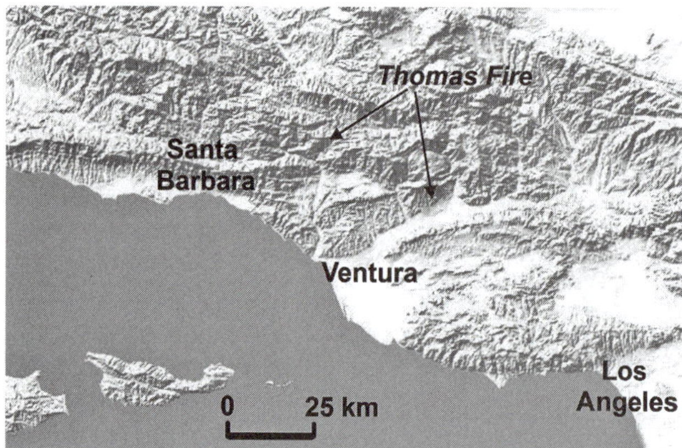

B. December 4, 2017.

C. December 5, 2017.

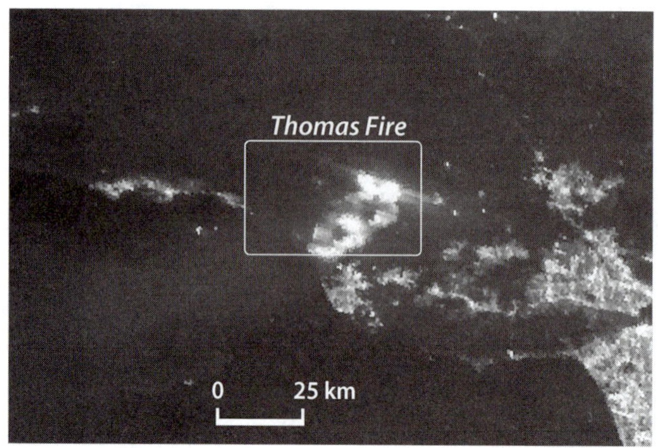

D. December 6, 2017.

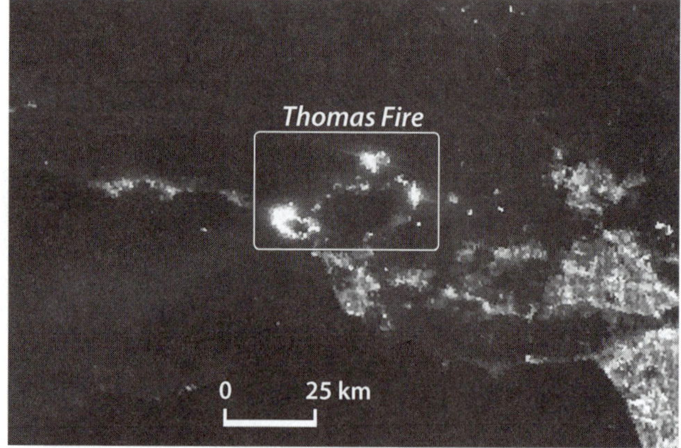

E. December 7, 2017.

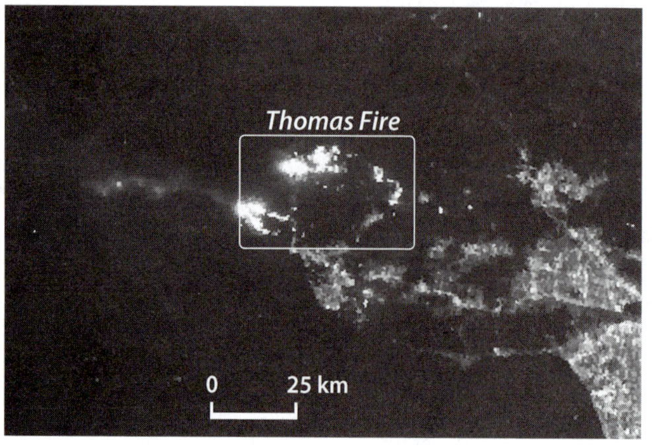

F. December 8, 2017.

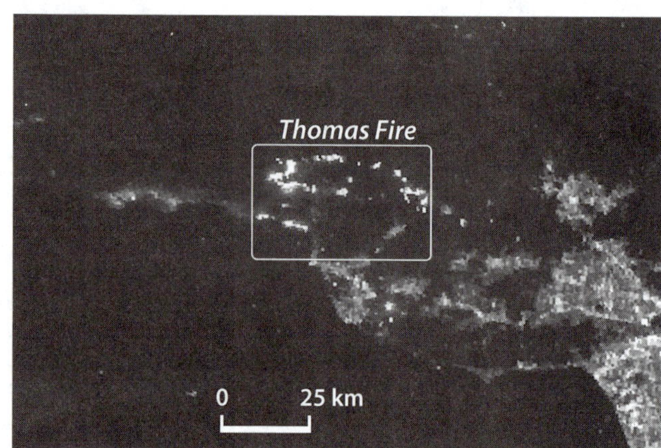

FIGURE 15-41 Suomi NPP Day/Night Band images of the Thomas Fire in southern California. From NASA Earth Observatory (2017).

Questions

1. Why is southern California an excellent laboratory for evaluating remote sensing technology used for the detection and monitoring of earthquakes?
2. Name and describe four topographic/geomorphic indicators of an active fault trace that can be interpreted from remote sensing imagery.
3. How do interferograms reveal ground displacements caused by earthquakes?
4. What are two methods used to generate real-time predictions of landslides in hazard zones?
5. What are (a) two infrastructure and (b) three geomorphic/topographic impacts that can be seen with remote sensing imagery of surface subsidence caused by groundwater and oil and gas extraction?
6. Name three remote sensing technologies used to predict volcanic eruptions. Describe how each technology is used to support the prediction.
7. During a volcanic eruption, how are two AVHRR bands processed to detect droplets of sulfuric acid (H_2SO_4) in a volcanic plume? How can volcanic plumes be differentiated from meteorological clouds? What wavelengths do these bands span?
8. What wavelengths are involved with the Day/Night Band onboard the NASA/NOAA Suomi NPP satellite? What are three applications for the Day/Night Band?

References

Allen, C. R. 1975. Geological criteria for evaluating seismicity. *Geological Society of America Bulletin*, 86, 1,041–1,057.

Avouac, J. P., and G. Peltzer. 1993. Active tectonics in southern Xinjiang, southern China: Analysis of terrace riser and normal fault scarp degradation along the Hotan-Qira fault system. *Journal of Geophysical Research*, B12, 21,773–21,807.

Bachmeier, S. 2017, December 5. Wildfires in Southern California. University of Wisconsin–Madison CIMSS Satellite Blog. http://cimss.ssec.wisc.edu/goes/blog/archives/26255 (accessed January 2018).

Blanchard, M. B., R. Greeley, and R. Goettleman. 1974. Use of Visible, Near-Infrared, and Thermal Infrared Remote Sensing to Study Soil Moisture. Proceedings of the Ninth International Symposium of Remote Sensing of Environment (pp. 693–700). Environmental Research Institute of Michigan, Ann Arbor.

Blumenfeld, J. 2019, April 5. Bringing Light to the Night: New VIIRS Nighttime Imagery Available through GIBS. EarthData. http://earthdata.nasa.gov/learn/articles/viirs-dnb (accessed April 2019).

Burns, S. 2014, March 22. Oso Landslide, Washington. Keynote Address (54 slides), Geotechnical Society of Edmonton. http://www.geotechnical.ca/Docs/Burns-Oso_landslide_keynote.pdf (accessed January 2018).

Centre for Research on the Epidemiology of Disasters (CRED). 2018. *Natural Disasters 2017*. Brussels, Belgium: CRED. http://cred.be/sites/default/files/adsr_2017.pdf.

Clanton, U. S., and E. R. Verbeek. 1981. Photographic portrait of active faults in the Houston metropolitan area, Texas. In E. M. Etter (Ed.), *Houston Area Environmental Geology: Surface Faulting, Ground Subsidence, and Hazard Liability* (pp. 70–113). Houston, TX: Houston Geological Society.

Cole, S., J. Leslie, R. Gran, and A. Keck. 2012, December 5. NASA–NOAA Satellite Reveals New Views of Earth at Night. NASA Headquarters Press Release No. 12-422. http://www.nasa.gov/mission_pages/NPP/news/earth-at-night.html.

Coplin, L. S., and D. Galloway. 1999. Houston–Galveston, Texas: Managing coastal subsidence. In D. Galloway, D. R. Jones, and S. E. Ingebritsen (Eds.), *Land Subsidence in the United States* (Circular 1182, pp. 35–48). Washington, DC: US Geological Survey. http://pubs.usgs.gov/circ/circ1182/pdf/07Houston.pdf (accessed January 2018).

Crippen, R. E. 1992. Measurements of subresolution terrain displacements using SPOT panchromatic imagery. *Episodes*, 15, 56–61.

Crippen, R. E., and R. G. Blom. 1992. The first visual observations of fault movements from space—the 1992 Landers earthquake. *Eos*, Transactions, American Geophysical Union, 73, 364.

Deutsch, M., and F. H. Ruggles. 1974. Optical data processing and projected applications of the ERTS-1 imagery covering the 1973 Mississippi River floods. *Water Resources Bulletin*, 10, 1,023–1,039.

Earthquakes without Frontiers. 2014, September 2. New Satellite Maps Out Napa Valley Earthquake. Blog Post. http://ewf.nerc.ac.uk/2014/09/02/new-satellite-maps-out-napa-valley-earthquake (accessed January 2018).

Florida History Network. 2017. May 9, 1981—Sinkhole Swallows House, Five Porsches in Winter Park. On This Day. http://www.floridahistorynetwork.com/may-9-1981---sinkhole-swallows-house-five-porsches-in-winter-park.html (accessed January 2018).

Foxworthy, B. L., and M. Hill. 1982. *Volcanic Eruptions of 1980 at Mount St. Helens: The First 100 Days* (Professional Paper 1249). Washington, DC: US Geological Survey.

Friedman, J. D., D. Frank, H. H. Kieffer, and D. L. Sawatzky. 1981. Thermal infrared surveys of the May 18 crater, subsequent lava domes, and associated deposits. In P. W. Lipman and D. L. Mullineaux (Eds.), *The 1980 Eruption of Mount St. Helens, Washington* (Professional Paper 1250, pp. 279–293). Washington, DC: US Geological Survey.

Galloway, D., D. R. Jones, and S. E. Ingebritsen (Eds.). 1999. *Land Subsidence in the United States* (Circular 1182). Washington, DC: US Geological Survey. http://pubs.usgs.gov/circ/circ1182 (accessed January 2018).

Gaonac'h, H., J. Vandemeulebrouck, J. Stix, and M. Halbwachs. 1994. Thermal infrared satellite measurements of volcanic activity at Stromboli and Vulcano. *Journal of Geophysical Research*, 99(B5), 9,477–9,485.

Glaze, L. S., P. W. Francis, S. Self, and D. W. Rothery. 1989. The 16 September 1986 eruption of Lascar volcano, north Chile: Satellite investigations. *Bulletin of Volcanology*, 51, 149–160.

Gumley, L. E., and M. D. King. 1995, June. Remote sensing of flooding in the U.S. upper Midwest during the summer of 1993. *Bulletin of the American Meteorological Society*, 76(6), 933–943.

Hapke, C. J., O. Brenner, R. Hehre, and B. J. Reynolds. 2013. *Coastal Change from Hurricane Sandy and the 2012–2013 Winter Storm Season: Fire Island, New York* (Open-File Report 2013-1231). Washington, DC: US Geological Survey.

Hauksson, E., L. M. Jones, K. Hutton, and D. Eberhart-Phillips. 1993. The 1992 Landers earthquake sequence seismological observations. *Journal of Geophysical Research*, 98, 19,835–19,858.

Hutton, L. K., L. M. Jones, E. Hauksson, and D. D. Given. 1991. Seismotectonics of southern California. In D. B. Slemmons, E. R. Engdahl, M. D. Zoback, and D. B. Blackwell (Eds.), *Neotectonics of North America* (pp. 133–152). Boulder, CO: Geological Society of America.

Jayaweera, K. O. L. F., R. Seifert, and G. Wendler. 1976. Satellite observations of the eruption of Tolbachik volcano. *Eos*, Transactions, American Geophysical Union, 57, 196–200.

Jennings, C. W. (Comp.). 1975. Fault Map of California with Locations of Volcanoes, Thermal Springs, and Thermal Wells (California Geologic Data Map Series Map 1). Sacramento: California Division of Mines and Geology.

Johnson, R. W., C. McKee, S. M. Eggins, J. D. Woodhead, J. Arculus, and B. W Chappell. 1995. Taking petrologic pathways toward understanding Rabaul's restless caldera. *Eos*, Transactions, American Geophysical Union, 76(17), 171–188.

Keaton, J. R., J. Wartman, S. Anderson, J. Benoit, J. deLaChapelle, R. Gilbert, D. R. Montgomery. 2014. The 22 March 2014 Oso Landslide, Snohomish County, Washington. Geotechnical Extreme Events Reconnaissance Association and National Science Foundation. http://snohomishcountywa.gov/DocumentCenter/View/18180 (accessed January 2018).

Kieffer, H. H., D. Frank, and J. D. Friedman. 1981. Thermal infrared surveys at Mount St. Helens prior to the eruption of May 18. In P. W. Lipman and D. L. Mullineaux (Eds.), *The 1980 Eruption of Mount St. Helens, Washington* (Professional Paper 1250, pp. 257–277). Washington, DC: US Geological Survey.

Kienle, J., K. G. Dean, H. Garbiel, and W. I. Rose. 1990. Satellite surveillance of volcanic ash plumes, application to aircraft safety. *Eos*, Transactions, American Geophysical Union, 71, 266.

LaHusen, S. R., A. R. Duvall, A. M. Booth, and D. R. Montgomery. 2016. Surface roughness dating of the long-runout landslides near Oso, Washington (USA), reveals persistent postglacial hillslope instability. *Geology*, 44(2), 111–114.

Lewis, S. 2009, November 11. Remote Sensing for Natural Disasters: Facts and Figures. SciDev.Net. http://www.scidev.net/global/earth-science/feature/remote-sensing-for-natural-disasters-facts-and-figures.html (accessed January 2018).

Lipman, P. W., and D. L. Mullineaux (Eds.). 1981. *The 1980 Eruption of Mount St. Helens, Washington* (Professional Paper 1250). Washington, DC: US Geological Survey.

Luccio, M. 2015. See the light: Lidar data help expose landslide hazards. *Earth Imaging Journal*, September 27. http://eijournal.com/print/articles/see-the-light-lidar-data-help-expose-landslide-hazards (accessed October 2019).

Massonnet, D., P. Briole, and A. Arnaud. 1995. Deflation of Mount Etna monitored by spaceborne radar interferometry. *Nature*, 375, 567–570.

Massonnet, D., M. Rossi, C. Carmona, F. Adragna, G. Peltzer, K. Feigl, and T. Rabaute. 1993. The displacement field of the Landers earthquake mapped by radar interferometry. *Nature*, 364, 138–142.

Merifield, P. M., and D. L. Lamar. 1975. Active and Inactive Faults in Southern California Viewed from Skylab. NASA Earth Resources Survey Symposium (TM X-58168, vol. 1, pp. 779–797).

Mori, J., K. Hudnut, L. Jones, E. Hauksson, and K. Hutton. 1992. Rapid response to Landers quake. *Eos*, Transactions, American Geophysical Union, 73, 417–418.

Mühlbauer, S., 2014, First InSAR deformation maps of Sentinel-1A from the recent Napa Valley earthquake: Geoawesomeness – Remote Sensing, September 9, 2014, http://geoawesomeness.com/first-insar-deformation-maps-of-sentinel-1a-from-the-recent-napa-valley-earthquake (Accessed January 2018).

NASA Earth Observatory. 2015a. Eruption of Villarrica Volcano. http://earthobservatory.nasa.gov/images/85465/eruption-of-villarrica-volcano (accessed April 2019).

NASA Earth Observatory. 2015b. Villarrica Volcano Awakens. http://earthobservatory.nasa.gov/images/85550/villarrica-volcano-awakens (accessed April 2019).

NASA Earth Observatory. 2017, December 9. Fast-Moving Fires Sweeps through Ventura County. http://earthobservatory.nasa.gov/images/91400/fast-moving-fire-sweeps-through-ventura-county (accessed April 2019).

NASA Earth Observatory. 2019a. The Biggest Eruption of 2018 Was Not Where You Think. http://earthobservatory.nasa.gov/images/144593/the-biggest-eruption-of-2018-was-not-where-you-think (accessed April 2019).

NASA Earth Observatory. 2019b. Early Flooding along the Mississippi. http://earthobservatory.nasa.gov/images/144598/early-flooding-along-the-mississippi (accessed April 2019).

NASA JPL. 2019. SMAP Soil Moisture Active Passive Mission—Why It Matters. http://smap.jpl.nasa.gov/mission/why-it-matters/#floods (accessed April 2019).

National Research Council. 1991. *Mitigating Losses from Land Subsidence in the United States*. Washington, DC: National Academy Press.

NOAA. 2019, March 21. 2019 National Hydrologic Assessment. http://www.nws.noaa.gov/oh/2019NHA.html (accessed April 2019).

O'Neal, C. 2017, December 13. Thomas Fire: Timeline Dec. 4 through Dec. 13. VCReporter. http://www.vcreporter.com/2017/12/thomas-fire-timeline-dec-4-through-dec-13 (accessed January 2018).

Oppenheimer, C., P. W. Francis, D. A. Rothery, and R. W. T. Carlton. 1993. Infrared image analysis of volcanic thermal features: Lascar volcano, Chile, 1984–1992. *Journal of Geophysical Research*, 98, 4,269–4,286.

Pappas, S. 2016. "Breathing" Volcano: How Scientists Captured This Awesome Animation. Live Science. http://www.livescience.com/54697-mount-etna-breathes-in-amazing-animation.html (accessed January 2018).

Peltzer, G., P. Tapponnier, and R. Armijo. 1989. Magnitude of late Quaternary left-lateral displacements along the north edge of Tibet. *Science*, 246, 1,285–1,289.

Perry, W. J., and I. H. Crick. 1976. Aerial thermal infrared survey, Rabaul area, New Britain, Papua New Guinea, 1973. In R. W. Johnson (Ed.), *Volcanism in Australasia*. New York: Elsevier.

Piper, D. J. W., A. N. Shor, J. A. Farre, S. O'Connell, and R. Jacobi. 1985. Sediment slides and turbidity currents on the Laurentian Fan: Sidescan sonar investigations near the epicenter of the 1929 Grand Banks earthquake. *Geology*, 13(8), 538–541.

Prata, A. J. 1989. Observations of volcanic ash clouds in the 10–12 pm window using AVHRR/2 data. *International Journal of Remote Sensing*, 10, 751–761.

Prior, D. B., J. M. Coleman, and L. E. Garrison. 1979. Digitally acquired undistorted side-scan sonar images of submarine landslides, Mississippi River Delta. *Geology*, 7, 423–425.

Reddy, C. S. S., A. Bhattacharya, and S. K. Srivastav. 1993. Nighttime TM short wavelength infrared data analysis of Barren Island volcano, South Andaman, India. *International Journal of Remote Sensing*, 14, 783–787.

Reid, M. E., R. G. LaHusen, R. L. Baum, J. W. Kean, W. H. Schulz, and L. M. Highland. 2012, February. *Real-Time Monitoring of Landslides* (Fact Sheet 2012-3008). http://pubs.usgs.gov/fs/2012/3008/contents/FS12-3008.pdf (accessed January 2018).

Rib, H. T., and T. Liang. 1978. Recognition and identification. In R. L. Schuster and R. J. Krizek (Eds.), *Landslides Analysis and Control* (Special Report 176, ch. 3, pp. 34–69). Washington, DC: National Academy of Sciences.

Rodriguez, J., F. Vos, R. Below, and D. Guha-Sapir. 2009. *Annual Disaster Statistical Review 2008: The Numbers and Trends*. Brussels, Belgium: Centre for Research on the Epidemiology of Disasters. http://www.cred.be/sites/default/files/ADSR_2008.pdf (accessed January 2018).

Rose, W. I., D. J. Delene, D. J. Schneider, G. J. S. Bluth, A. J. Krueger, I. Sprod, C. McKee, H. L. Davies, and G. G. J. Ernst. 1995. Ice in the 1994 Rabaul eruption cloud: Implications for volcano hazard and atmospheric effects. *Nature*, 375, 477–479.

Rothery, D. A. 1989. Volcano monitoring by satellite. *Geology Today*, 5, 128–132.

Rothery, D. A., and P. W. Francis. 1990. Short wavelength infrared images for volcano monitoring. *International Journal of Remote Sensing*, 11(10), 1,665–1,667.

Sabins, F. F. 1987. *Remote Sensing: Principles and Interpretation* (2nd ed.). New York: W. H. Freeman.

Science X Network. 2018, June 4. The Deadliest Volcanic Eruptions of the Past 25 Years. http://phys.org/news/2018-06-deadliest-volcanic-eruptions-years.html.

Shreve, R. L. 1968. *The Blackhawk Landslide* (Special Paper No. 108). Boulder, CO: Geological Society of America.

Sieh, K., L. Jones, E. Hauksson, K. Hudnut, D. Eberhart-Phillips, T. Heaton, S. Hough, K. Hutton, H. Kanamori, A. Lilje, S. Lindvall, S. F. McGill, J. Mori, C. Rubin, J. A. Spotila, J. Stock, H. K. Thio, J. Treiman, B. Wernicke, and J. Zachariasen. 1993. Near-field investigations of the Landers earthquake sequence, April to July, 1992. *Science*, 260, 171–176.

Sopkin, K. L., H. F. Stockdon, K. S. Doran, N. G. Plant, K. L. M. Morgan, K. K. Guy, and K. E. L. Smith. 2014. *Hurricane Sandy: Observations and Analysis of Coastal Change* (Open-File Report 2014-1088). Washington, DC: US Geological Survey.

Tapponnier, P., and P. Molnar. 1977. Active faulting and tectonics in China. *Journal of Geophysical Research*, 82, 2,905–2,930.

Tapponnier, P., and P. Molnar. 1979. Active faulting and Cenozoic tectonics of the Tien Shan, Mongolia, and Baykal regions. *Journal of Geophysical Research*, 84, 3,425–3,459.

Textor, C., H-F. Graf, C. Timmreck, and A. Robock. 2003. Emissions from volcanoes. In C. Granier, C. Reeves, and P. Artaxo (Eds.), *Emissions of Atmospheric Compounds* (pp. 269–303). Dordrecht, Netherlands: Kluwer.

US Climate Resilience Toolkit. 2017. Storm Surge. http://toolkit.climate.gov/topics/coastal/storm-surge (accessed January 2018).

USGS. 1991. Composite Mosaic of the California, Oregon, Washington (COW) GLORIA Sidescan Sonar Mosaics (white background) (cow_250m_geo_NAD27-white.tif). http://coastalmap.marine.usgs.gov (accessed January 2018).

USGS. 2010. United States Pacific West Coast Mosaic Index Map (Open File Report 2010-1332). http://pubs.usgs.gov/of/2010/1332/htmldocs/pc/pc_indexmap.html (accessed April 2019).

USGS. 2017. Volcanic Gases Can be Harmful to Health, Vegetation, and Infrastructure. USGS Volcano Hazards Program. http://volcanoes.usgs.gov/vhp/gas.html (accessed April 2019).

Vedder, J. G., and R. E. Wallace. 1970. Map Showing Recently Active Breaks Along the San Andreas and Related Faults between Cholame Valley and Tejon Pass, California. US Geological Survey, Miscellaneous Geologic Investigations, Map 1-574.

Wartman, J. 2016. What We've Learned from the Deadly Oso, Washington Landslide Two Years On. The Conversation. http://theconversation.com/what-weve-learned-from-the-deadly-oso-washington-landslide-two-years-on-56528 (accessed January 2018).

Wen, S., and W. I. Rose. 1994. Retrieval of sizes and total masses of particles in volcanic clouds using AVHRR bands 4 and 5. *Journal of Geophysical Research*, 99, 5,421–5,431.

Yeats, R. S., K. Sieh, and C. A. Allen. 1996. *Geology of Earthquakes*. New York: Oxford University Press.

chapter 16

Climate Change

*O*verall global warming is now a fact that is not only acknowledged by an overwhelming majority of the scientific community, but also by most national and international decision makers and the public (Trögler and Lingner, 2012). The *Climate Science Special Report: Fourth National Climate Assessment* states,

> The last few years have also seen record-breaking, climate-related weather extremes, and the last three years have been the warmest years on record for the globe. These trends are expected to continue over climate timescales. In addition to warming, many other aspects of global climate are changing, primarily in response to human activities. Thousands of studies conducted by researchers around the world have documented changes in surface, atmospheric, and oceanic temperatures; melting glaciers; diminishing snow cover; shrinking sea ice; rising sea levels; ocean acidification; and increasing atmospheric water vapor. . . . Since 1980, the cost of extreme events for the United States has exceeded $1.1 trillion; therefore, better understanding of the frequency and severity of these events in the context of a changing climate is warranted. (USGCRP, 2017)

Satellite remote sensing is an important component of climate system observations. It has gradually become the leading research method in climate change studies since the first space observation of solar irradiance and cloud reflection was made with radiometers onboard the Vanguard-2 satellite in 1959 (Yang and others, 2013). Tsang and Jackson (2010) discuss major initiatives by national and intergovernmental space agencies to improve the quality and broaden the range of satellite observations to monitor water, carbon, and global climate change. For over three decades NASA solar-climate satellites have monitored sunspot activity and changes in solar output energy.

Remote sensing delivers a wide-ranging collection of Earth observation data that is applied to climate change studies. ESA's (2017) CryoSat radar altimeter built a 3-D model of the Antarctica Ice Sheet surface based on 250,000,000 measurements between 2010 and 2016. Satellite altimetry data was used to monitor the level of 204 major lakes across the globe from 2002 to 2010 (Tan and others, 2017). It was also used to measure ocean surface topography: From 1992 to 2010 there was a global mean sea level rise of 3.2 ± 0.8 mm $\cdot$ yr^{-1}. In all ocean basins, NOAA's AVHRR satellites documented an increase in sea surface temperature (SST) of 0.28°C from 1984 to 2006 (Yang and others, 2013). Select examples of glaciers, sea ice, permafrost, and desertification are provided in this chapter to demonstrate how remote sensing contributes to our measurement and understanding of climate change.

In this chapter, the section on Glaciers, which summarizes remote sensing monitoring of two outlet glaciers of the Greenland Ice Sheet, is written by B. Csatho and T. Schenk (Department of Geology, University at Buffalo).

GLACIERS

The global cryosphere is an essential part of Earth's climate system. The fluctuations of ice sheets and glaciers dramatically affect the water cycle—locking up fresh water as they grow and advance, causing the sea level to rise as they thaw and retreat. Loss of ice sheet mass is a major contributor to current sea level rise, and is expected to continue as global warming proceeds (e.g., Church and others, 2013). Ongoing changes in the Arctic regions, including the decreasing sea ice cover and the continuing mass loss of the Greenland Ice Sheet (GrIS), are at the center of attention of scientists and the mainstream media alike. Despite its potential to contribute dramatically to sea level rise in response to ongoing and projected climate change, resulting in significant socioeconomic impacts in the United States and around the globe, the future behavior of the GrIS remains mostly uncertain. This uncertainty stems from our incomplete knowledge of the patterns and timescales of the ice sheet's response to climate change.

REMOTE SENSING AND GLACIERS

Glacier histories extending back in time prior to the instrumental record must be based on geomorphological and geological geomorphological information retrieved from formerly glaciated regions. The traditional approach was to map such features during detailed field investigations. The use of remote sensing allows for a more extensive spatial coverage to map glacial geological features with a reduced cost.

Many remote sensing tools are applied to the study of glaciers, including historic aerial photographs, satellite imagery, DEMs, and airborne and satellite altimetry measurements. NASA has employed the Airborne Topographic Mapper (ATM), an airborne scanning laser altimetry system, for monitoring the Greenland and Antarctic Ice Sheets since the early 1990s. The ATM system was designed, developed, and operated by NASA's Wallops Flight Facility. With flying heights between 500 and 1,500 m, the typical range of the swath width is between 400 and 1,200 m and the footprint varies from 1 to 3 m. The Land, Vegetation, and Ice Sensor (LVIS) is an airborne laser scanning altimeter, developed by NASA in the late 1990s for vegetation mapping, including the determination of canopy heights. The system was first used for cryospheric research in 2007 with flights over parts of the GrIS. The flying height is about 8,000 m, producing a swath width of 1.3 km. It has a footprint size of 25 m and pulse repetition rate of up to 500 Hz.

DEMs generated from stereo imagery collected by satellite systems, most notably from Declassified Intelligence Satellite Photography (DISP), ASTER, SPOT, and the different WorldView satellites, provide continuous coverage to fill the gaps between laser altimetry measurements. On the one hand, laser altimetry systems are much more accurate than DEMs derived by correlation from imaging systems. On the other hand, DEMs usually have a superior spatial resolution and extended spatial coverage. Thus, it makes eminent sense to fuse (combine) DEMs and laser altimetry to broaden the spatial coverage of surface elevation and change rates.

To accurately monitor ice sheet and sea ice changes, NASA launched the Ice, Cloud, and Land Elevation Satellite (ICESat) on January 13, 2003 (Zwally and others, 2002). ICESat orbited the Earth on a near circular orbit at 94° inclination and approximately 600-km altitude, providing coverage between ±86° latitude. After firing the last of nearly two billion shots on October 11, 2009, the satellite was decommissioned on August 14, 2010.

ICESat carried the Geoscience Laser Altimeter System (GLAS) with the primary goal of measuring elevation changes of the polar ice sheets to sufficient accuracy to assess their impact on global sea level. GLAS had three lasers that operated sequentially, with two to three campaigns per year. GLAS consisted of three neodymium-doped yttrium aluminum garnet crystal (Nd:YAG) lasers, operating at a wavelength of 1,064 nm (infrared) for determining surface elevation. With a 40 Hz pulse frequency and 600 km mean altitude, the laser footprints (laser spots) had approximately 170 m center-to-center spacing. The footprint size was about 70 m.

The ICESat follow-on mission, ICESat-2, was launched in September 2018 (icesat-2.gsfc.nasa.gov) (Markus and others, 2017). In contrast to ICESat, which carried a single laser, the Advanced Topographic Laser Altimeter System (ATLAS) on the ICESat-2 satellite operates with three pairs of beams, each pair separated by about 3 km cross-track with a pair spacing of 90 m. Each of the beams has a nominal 17 m diameter footprint with an along-track sampling interval of 0.7 m. The beam pair configuration of ICESat-2 allows for the determination of local cross-track slope, a significant factor in measuring elevation change for the outlet glaciers surrounding the Greenland and Antarctica coasts. The multiple beam pairs also provide improved spatial coverage. ICESat-2 employs a novel instrumentation concept, a photon-counting laser altimetry system that emits 10,000 pulses of 532 nm (green) laser light per second and uses single-photon sensitive detectors to measure range. About 20 trillion photons leave ATLAS with each pulse; only about a dozen hit the Earth's surface and return to the satellite's telescope (icesat-2.gsfc.nasa.gov/space-lasers).

During the last two decades surface elevation data have been gathered over the GrIS from a variety of different sensors, including aerial photogrammetry, spaceborne and airborne laser altimetry, as well as from stereo satellite imaging systems. The spatio-temporal resolution, the accuracy, and the spatial coverage of all these data differ widely. To cope with the data complexity and the computation of elevation change histories from these disparate data sets, Schenk and Csatho (2012) designed the Surface Elevation Reconstruction and Change (SERAC) System. SERAC simultaneously determines the ice sheet surface shape and the time series of elevation changes for surface patches whose size depends on the ruggedness of the surface and the point distribution of the sensors involved. By incorporating different sensors, SERAC is a real fusion system that generates the best plausible result (time series of elevation changes)—a result that is better than the sum of its parts.

MULTISENSOR AND MULTITEMPORAL STUDY OF GREENLAND OUTLET GLACIERS

As a response to increasing temperature since the Little Ice Age (LIA), outlet glaciers fringing the GrIS started to retreat in the mid- to late 1800s. This retreat has been observed and documented using the historical record, including painting, photographs, and maps. However, satellite and airborne measurements, offering the first opportunity to quantify glacier extent, elevation, mass loss, and changes in ice dynamics, only started a few decades ago (Csatho and others, 2017) (Figure 16-1).

Jakobshavn Isbræ and Kangerlussuaq Glacier are the two fastest flowing outlet glaciers of the GrIS (Figure 16-2). Jakobshavn Isbræ is the largest outlet glacier on the west coast of Greenland, draining approximately 5% of the ice sheet. Kangerlussuaq Glacier in east Greenland drains about 6% of the ice sheet. Taken together, the two glaciers contributed more than 30% to the 2000 to 2012 GrIS mass loss (Enderlin and others, 2014).

Rapid thinning and velocity increase on most Greenland outlet glaciers during the last two decades may indicate that these glaciers became unstable with terminus retreat leading to increased discharge from the interior and consequent further thinning and retreat. To place recent changes in the broader context of retreat and thinning since the LIA the longer-term records of Jakobshavn Isbræ and Kangerlussuaq Glacier were reconstructed with the focus on determining surface elevation and ice thickness changes (Csatho and others, 2008; Schenk and others, 2014). While terminus positions may offer a general picture of glacier behavior, they cannot be used to accurately reconstruct the history of mass changes, since during much of their retreat the glaciers' termini were likely to have been floating and thus susceptible to small and short-lived climate perturbations. In order to obtain accurate mass change histories, different types of elevation observations were fused to correct DEM errors and determine how outlet glacier mass loss evolved in time.

Jakobshavn Isbræ (1880s to 2016)

For the Jakobshavn region, field mapping, aerial photographs, airborne and satellite laser altimetry, and optical satellite images have been used to measure changes in surface elevation and ice thickness since the LIA.

Vegetation Trimline and Lateral Moraines (1880s) The highest ice-elevation stand reached during the LIA is readily observed in the field as a sharp boundary, called the trimline, separating vegetated terrain and rocks that were stripped bare of any vegetation during glacier advance (Figure 16-3). The ice sheet reached its maximum LIA extent in the

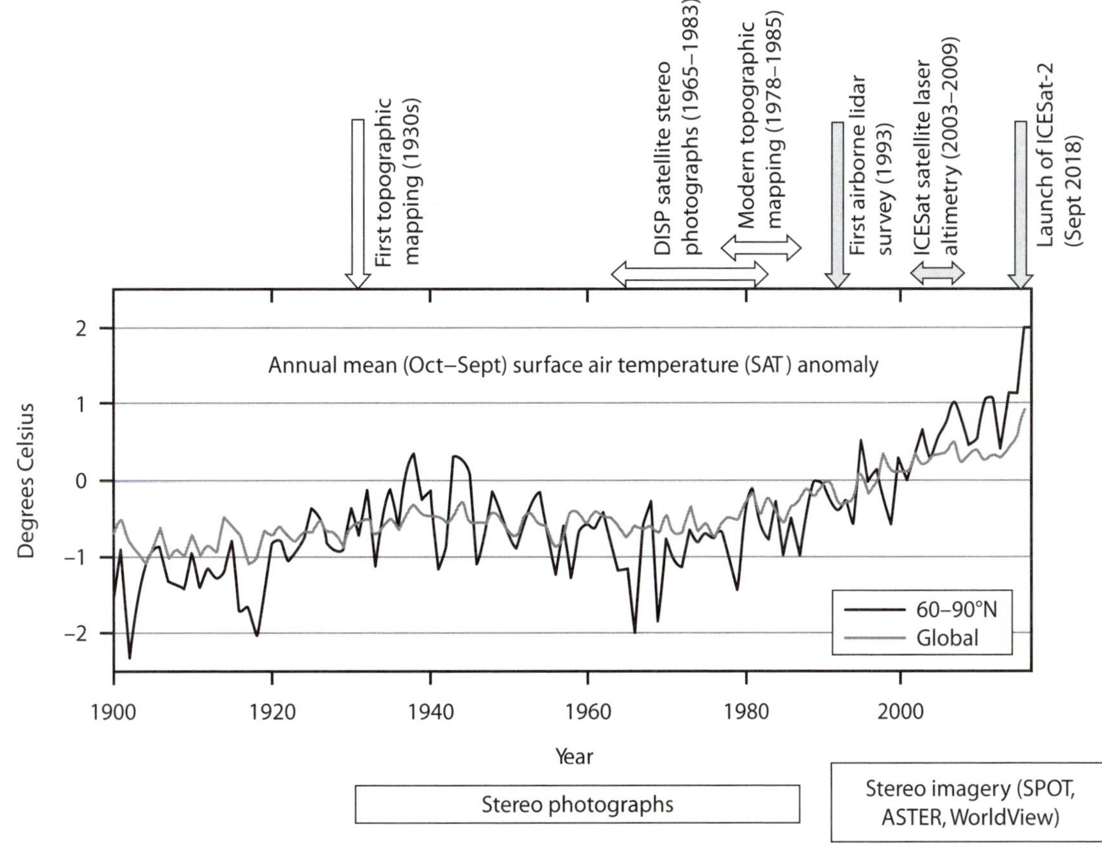

FIGURE 16-1 Remote sensing tools used through time to map and monitor the GrIS. Global (gray line) and Arctic (black line) annual mean surface air temperature changes since the Little Ice Age are shown. Courtesy NOAA (2016).

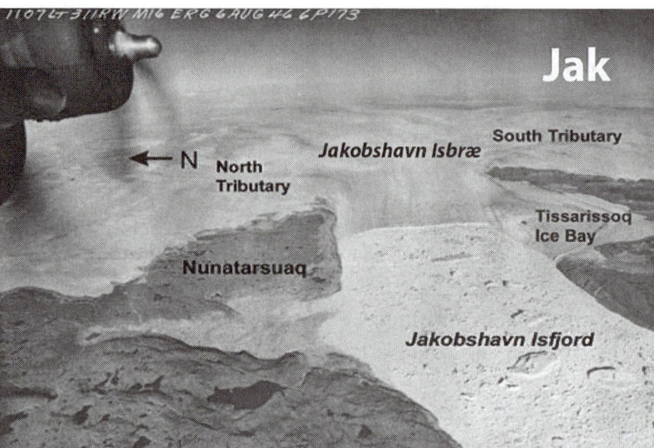

FIGURE 16-2 Overview of study sites on the GrIS. Left: Shaded relief topography of Greenland showing the locations of Jakobshavn Isbræ (Jak) and Kangerlussuaq Glacier (Kang). Right: Annotated perspective views of the glaciers on trimetrogon photographs collected by the US Air Force on August 6, 1946 (Jak) and June 6, 1947 (Kang). DEM courtesy University at Buffalo Remote Sensing Lab.

Jakobshavn region in the 1880s. Trimlines can be mapped using multispectral satellite images because exposed bedrock and freshly deposited sediments have distinctly different spectral-reflectance characteristics to surfaces covered with vegetation or lichens (van der Veen and Csatho, 2005). Csatho and others (2005) applied a maximum likelihood supervised classification procedure to a Landsat Enhanced Thematic Mapper Plus (ETM+) image to distinguish between the trimline zone and other surface classes (such as different snow facies, lichen and vegetation-covered surfaces, and open water). The authors identified 14 different surface classes based on their spectral signatures. Figure 16-3B shows a simplified version of the classification map presented in Csatho and others (2005), indicating the four major surface types, i.e., fresh rock surfaces and debris in the trimzone, tundra vegetation, frozen water (snow and ice), and liquid water (ocean, fjords, and lakes). The classification map distinguishes the recently exposed surfaces devoid of lichens and vegetation from surfaces that deglaciated earlier. The inland boundary of the region classified as trimzone corresponds to the position of the LIA trimline. To determine the change of glacier elevation through time the elevation of the trimline was measured manually from 1985 stereo aerial photographs using a softcopy photogrammetric workstation. A succession of lateral moraines were mapped near the July 7, 2001 calving front (Figure 16-3A) that provide evidence of glacial standstills over the past approximately 140 years as the glacier retreated from its LIA maximum.

Historical Aerial Photographs (1940s to 1985) For the Jakobshavn region, the aerial photography record dates back to the 1930s. Between the 1930s and 1985 various missions were flown by the Danish Geodata Agency (Geodatastyrelsen, or GST), as well as the US Air Force and French glaciological expeditions (Csatho and others, 2008). Photographs collected by these missions were used to reconstruct a time series of ice sheet retreat and surface-elevation changes between the 1940s and 1985. Figure 16-4 illustrates the flight lines of aerial photogrammetry flights over Jakobshavn Isfjord during this period. It also highlights the area of the profile A–A′ found in Figure 16-5. In Figure 16-5A, the small glaciers flowing from north to southeast of Camp-2 disappeared between 1944 and 1959 and a new proglacial lake formed north of Camp-2 (black circle). The aerial photographs also depict seasonal lake ice in 1944 and 1964 over a pair of lakes that are not connected to the ice sheet (white circle), and in 1964 over the proglacial lake that formed as the ice sheet retreated (black circle). Figure 16-5B shows the lowering of the main trunk of Jakobshavn Isbræ from 1944 to 1964 along the transect of A–A′. The convex glacier surface shape indicates that the glacier was grounded near its calving front in 1944. The glacier rapidly thinned between 1944 and 1959 and developed a floating tongue by 1959. The lifting of the calving front and the developing depression upstream indicate rotation of the front section that ultimately led to ice failure and calving in 1944 and 1964.

Airborne Laser Altimetry (1993 to Present) Repeat ATM airborne laser altimeter surveys have been conducted along a 120-km long profile over the northern part of Jakobshavn Isbræ since 1993 (Figure 16-6, transect A–A″). Comparison of elevations from a 1985 photogrammetry survey and repeat laser altimetry in 1993, 1997, 2001, and 2003 in Figure 16-7 reveals that the glacier thickened between 1985 and 1997, followed by rapid thinning that led to the disintegration of the floating tongue of Jakobshavn Isbræ. The calving front retreated to the fjord head by 2004 (Figure 16-8). These observations indicate major ongoing changes in this principal drainage route for inland ice, possibly triggered by relatively warm ocean water originating from the Irminger Sea arriving to the Jakobshavn Isfjord in 1997 (Holland and others, 2008).

A. Landsat ETM+ image acquired on July 7, 2001. Landsat courtesy USGS.

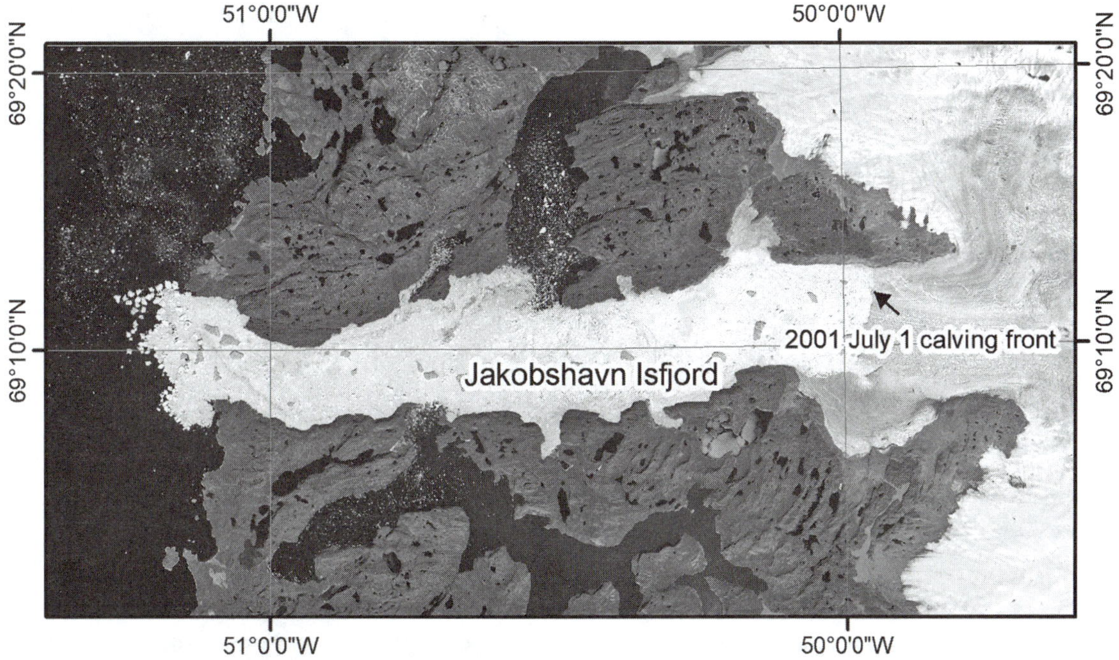

B. Supervised classification of Landsat imagery to distinguish vegetated and nonvegetated rocks, ice sheet, and water (ocean, lakes, rivers). The trimzone is highlighted with a black arrow. Modified from Csatho and others (2005).

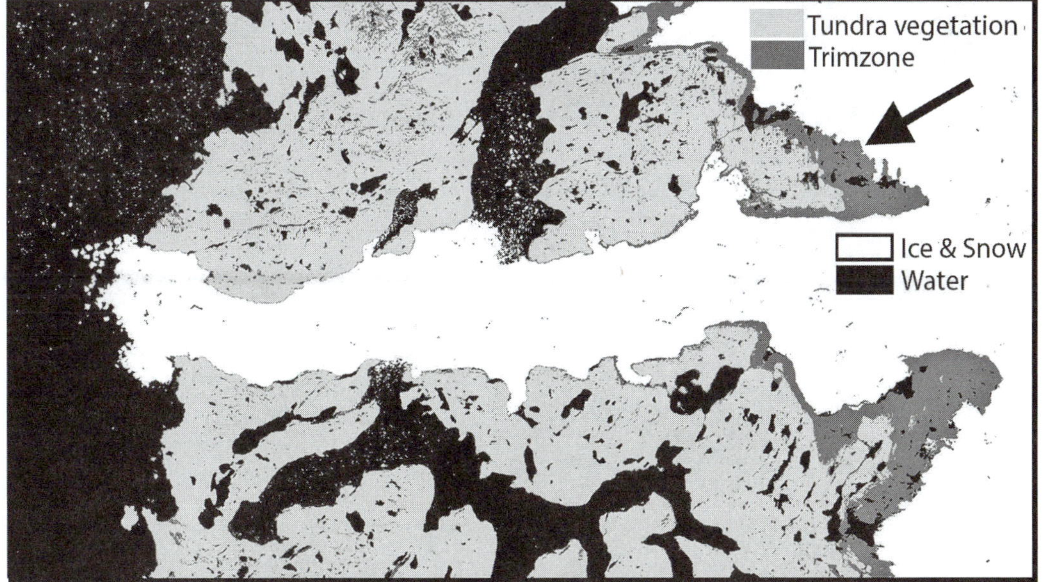

FIGURE 16-3 Trimzone (i.e., the region devoid of vegetation due to LIA ice coverage) around Jakobshavn Isbræ. Jakobshavn Isbræ is flowing from right to left. Jakobshavn Isfjord is packed with small fragments of inland ice (called *sikussaq* in Greenlandic) surrounding a few large icebergs.

FIGURE 16-4 Index map of aerial photogrammetry flights (1944?, 1953, 1959, 1964, and 1985) over Jakobshavn Isfjord, shown on an ETM+ satellite image acquired on July 7, 2001. Circles on the flight lines mark the exposure centers of the aerial photographs. The black boxed area is shown in Figure 16-5. Jakobshavn Isbræ is flowing from right to left with calved icebergs floating in the fjord to the left. Landsat courtesy USGS. From Csatho and others (2008).

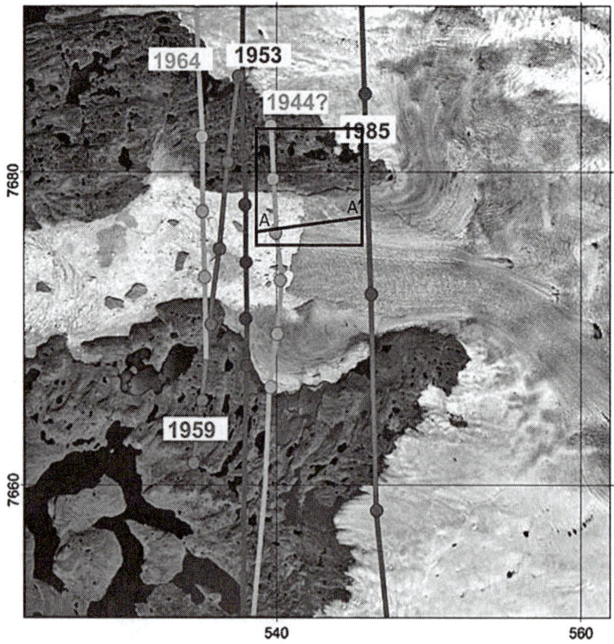

A. The glacier along profile A–A′ flows from right to left (Figure 16-4). The white circle marks a permanent lake outside of the LIA trimzone, and the black circle marks a lake formed between 1959 and 1964 in a depression exposed by the retreating ice sheet. Aerial photographs courtesy GST.

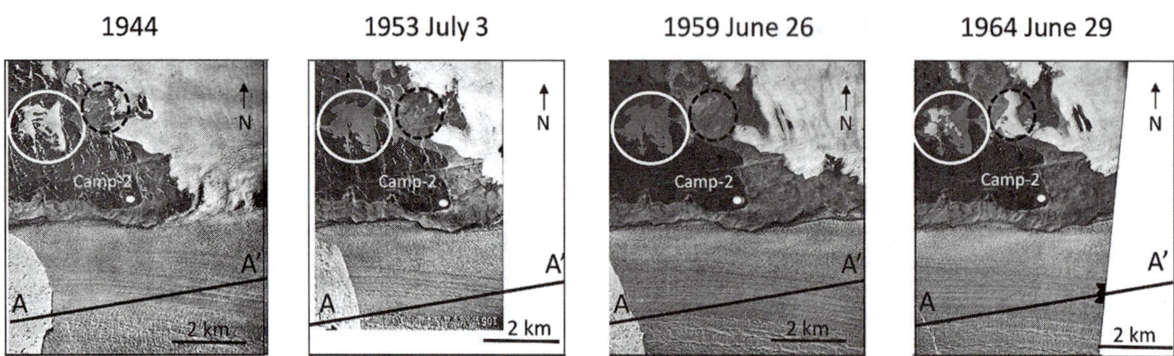

B. Surface elevation profiles.

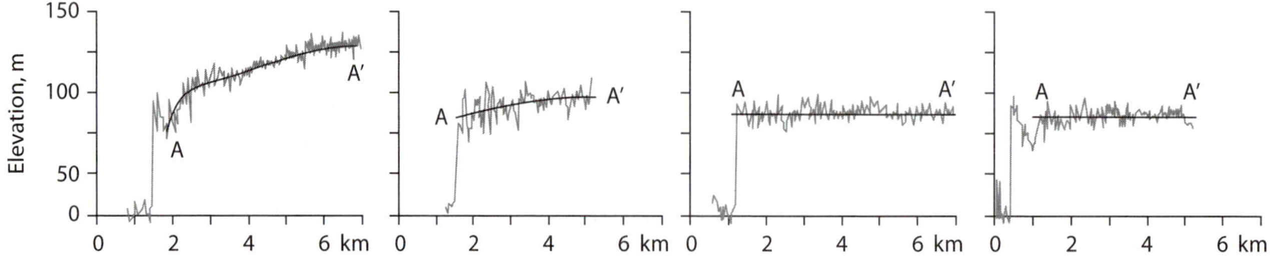

FIGURE 16-5 Ice sheet retreat and glacier thinning around Camp-2 between 1944 and 1964. Adapted from Csatho and others (2008).

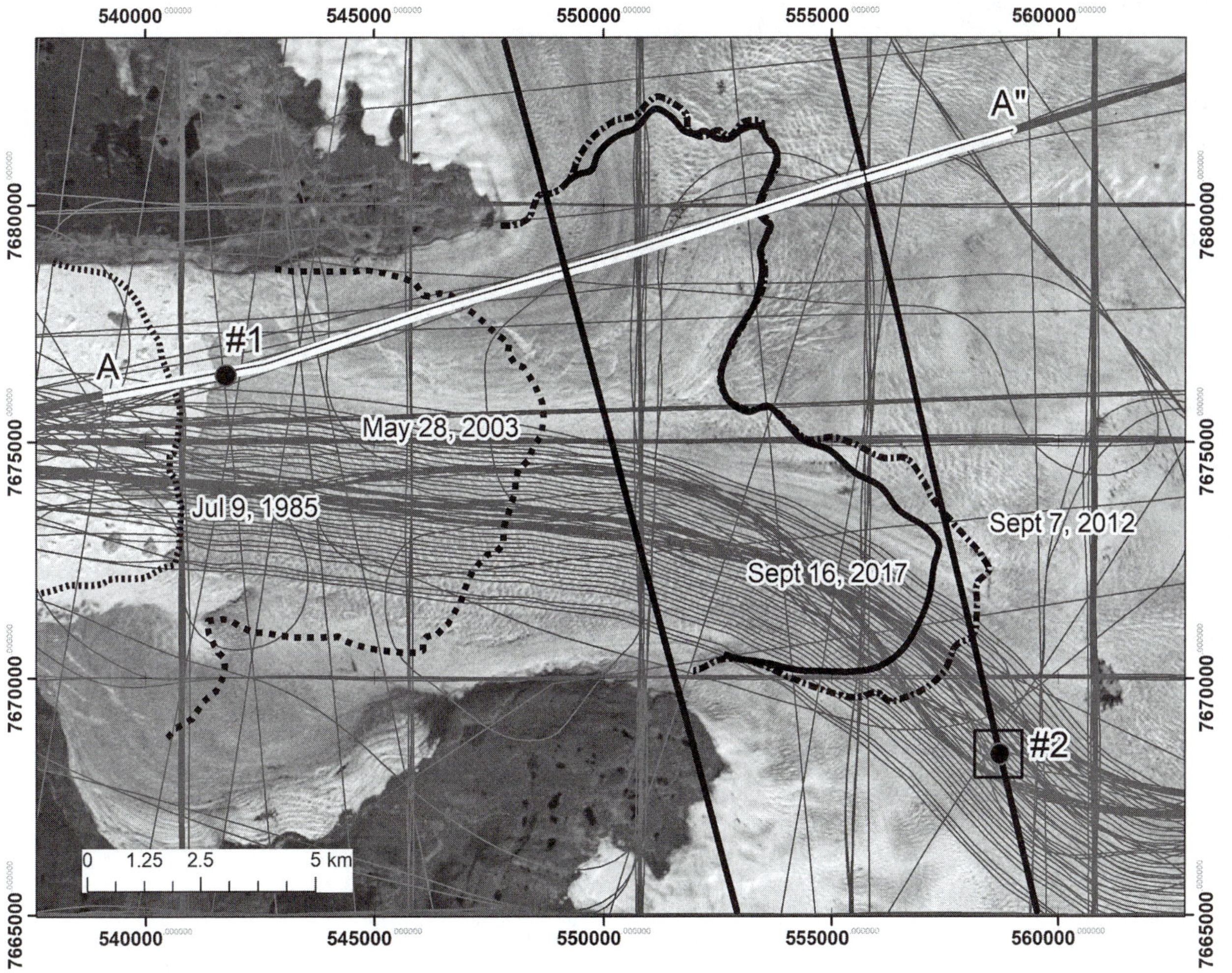

Figure 16-6 Airborne laser altimetry flight lines (ATM and LVIS) (gray lines) and ICESat satellite laser altimetry ground tracks (thick, straight, black lines) over the lower part of the Jakobshavn drainage basin. Calving fronts (edge of the glacier) in 1985, 2003, 2012, and 2017 are also shown. Background image is Landsat imagery acquired on July 7, 2001. Elevation change along the A–A″ transect is shown in Figure 16-7, and a time series of elevation changes at Site #1 and Site #2 are shown in Figure 16-9. This figure will also be used for highlighting the need of a new fusion approach to obtain time series. Landsat courtesy USGS.

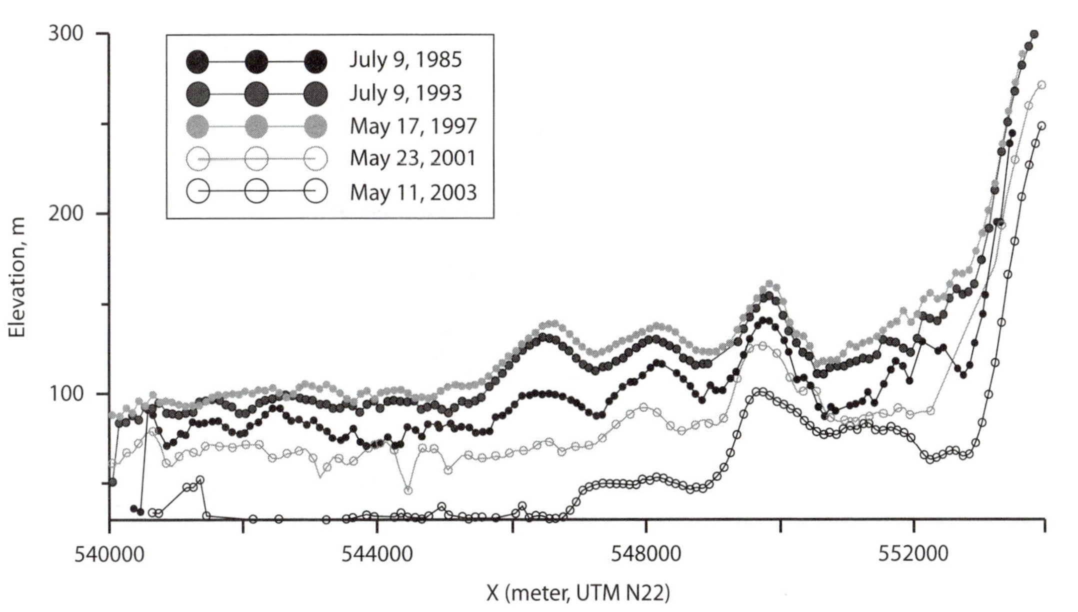

Figure 16-7 ATM airborne laser altimetry detected glacier thickening in 1985 to 1997, followed by rapid thinning in 1997 to 2003 along A–A″ (see Figure 16-6 for location). ATM data courtesy of NSIDC.

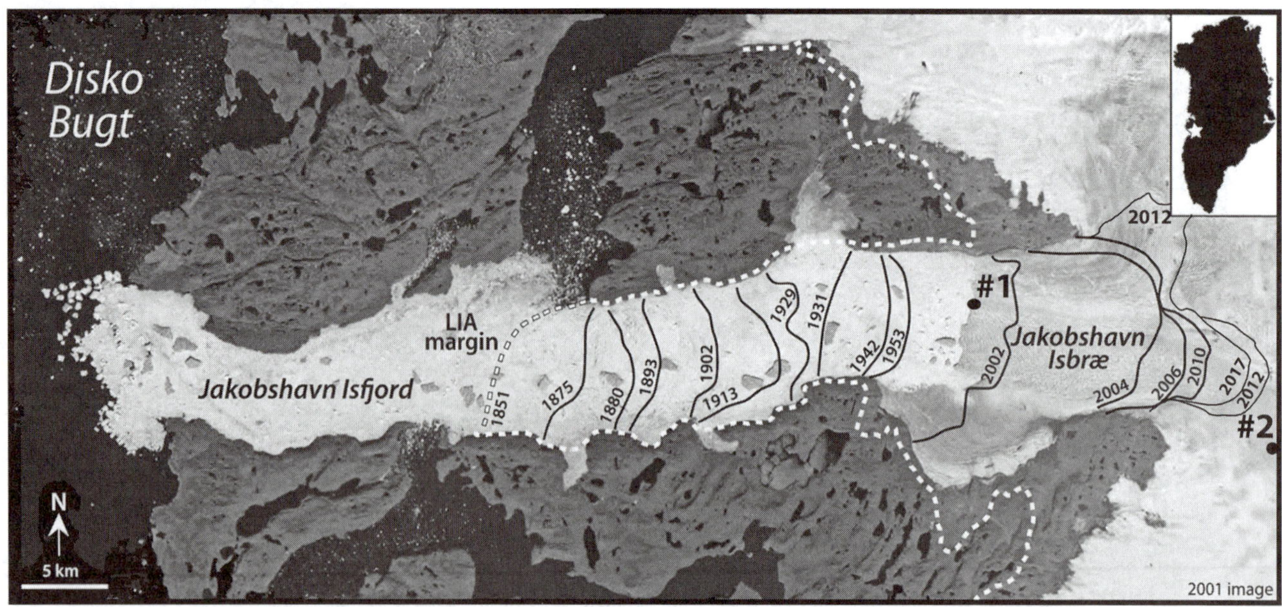

FIGURE 16-8 Retreat of Jakobshavn Isbræ from 1851 to 2017 (with 2012 to 2017 calving fronts added). Site #1 and Site #2 show the locations for the time series in Figure 16-9. Landsat image courtesy USGS; modified by Csatho and others (2008, Figure 5A) and Briner and others (2011).

Time Series of Elevation Changes (1880s to 2016) Figure 16-9 shows an elevation change record of Jakobshavn Isbræ since the 1880s. Elevation changes from laser altimetry observations (from 1993 to 2016 using ATM, LVIS, and ICESat) were determined by SERAC. To extend the time series to the past, Csatho and others (2008, Table 5) added elevations determined from historical aerial photographs, field notes, and trimline elevation measurements at Site #1. As the glacier retreated upstream to Site #1 in 2001, they combined two time series (Sites #1 and #2). The second site is located 20 km upstream from Site #1. The thinning of Jakobshavn Isbræ, shown in Figure 16-9, shows striking similarities to the annual mean temperature changes measured at the nearby town of Ilulissat, situated on the coast north of the fjord mouth. Increasing temperatures from the LIA to the 1940s resulted in glacier thinning and retreat, followed by a relatively stable glacier surface elevation and calving front location between 1960 to 1980, followed by a slight thickening between 1985 and 1997 (Figures 16-7, 16-8, and 16-9). Rapid thinning and retreat started in 1997 and has continued to the present.

Kangerlussuaq Glacier (1930s to 2013)

The Kangerlussuaq Glacier in east Greenland was also monitored by NASA's airborne and satellite laser altimetry flights (Figure 16-10) (see also Figure 16-2). However, the nearest ICESat ground tracks (thick black lines) were about 30 km upstream from the calving front, and most repeat ATM flight lines followed the central flowline (white lines), leaving large areas uncovered. Additional observations were collected by LVIS (straight, southwest–northeast lines, medium gray).

Removal of Systematic Errors An important prerequisite of fusing different data sets is the removal of any systematic errors. Traditional approaches rely on GCPs on stable terrain to remove these errors. However, the distribution of GCPs is poor along the ice sheet margin, where stable terrain is limited to the rocky terrain surrounding the ice sheet. Therefore, Schenk and others (2014) used elevation change time series derived from laser altimetry by SERAC to correct

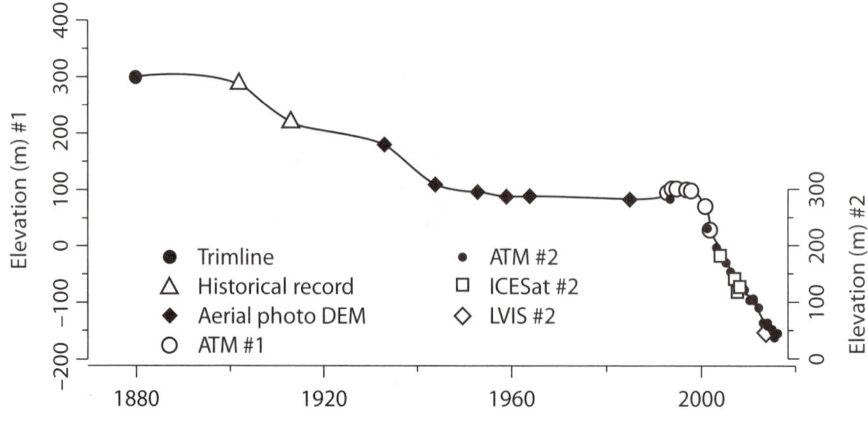

FIGURE 16-9 A time series of elevation change from 1880 to 2017 reconstructed by using the SERAC approach from historical, cartographic, and remote sensing data at Site #1 and Site #2 (Figure 16-8).

Figure 16-10 Kangerlussuaq Glacier study area on a SPOT image from July 28, 2007. Shown as black lines are ICESat ground tracks with only one crossover in the upper drainage basin. ATM swaths are in white, LVIS in medium gray. Sites 1 through 9 are locations where extended time series were determined and utilized for correcting stereoscopic DEMs. Sites 6 and 7 are about 1 km up-glacier from the calving front. Schenk, T., B. Csatho, C. van der Veen, and D. McCormick. 2014. Fusion of multi-sensor surface elevation data for improved characterization of rapidly changing outlet glaciers of Greenland. *Remote Sensing of Environment*, 149, 239–251, with permission of Elsevier.

the DEMs. The processing started by determining discrete elevation time series at the control points (Figure 16-10) and approximating them with continuous functions (Figure 16-11A). In the next step the differences of the DEM points to the fitted curve were determined in order to compute the DEM correction values. Finally, Figure 16-11B presents the time series of elevation changes after correcting the stereoscopic DEMs. The curve in Figure 16-11B is the same analytic curve as in Figure 16-11A, fitted through the laser altimetry points. Schenk and others (2014) observed that the DEM points are now randomly distributed about the curve.

Time Series of Elevation Changes (1933 to 2013) The combined data set, comprised of the laser altimetry observation, aerial and satellite photogrammetry DEMs, and a digitized topographic map, was used to determine the lowering of the Kangerlussuaq Glacier since 1933 and the spatial pattern of glacier elevation change (**Digital Image 16-1**). At Site 5 (shown in Figure 16-10), the glacier was grounded for the entire duration of the record. The reconstruction by Schenk and others (2014) shows a gradual, slow lowering of the glacier's surface between 1933 and 2001 with an average rate of -1.5 ± 0.5 m · year^{-1} (Figure 16-12). Thinning rates started to increase in 2001, reaching a maximum rate of 64 m · yr^{-1} in 2005. Thinning ceased abruptly in 2006. After a brief period of stability, the ice surface started to lower again in 2010 with an average thinning rate of 15 m · yr^{-1}.

Volume and Mass Change (2001 to 2010) After performing the planimetric and the height adjustments of the DEMs, six ASTER DEMs and one SPOT DEM were selected that are more or less regularly distributed in time between the period 2001 to 2010. **Digital Image 16-1** shows the evolution of elevation change rates from 2001 to 2010 for the area enclosed by the dashed line in Figure 16-10. The uncorrected elevation change rates show large temporal fluctuations that are not consistent with the behavior of the Kangerlussuaq Glacier as reconstructed from altimetry data. For example, the uncorrected maps depict tens of meters of thinning at higher elevations between 2003 and 2004 and 2005 to 2006 and thickening with similar magnitude in 2004 to 2005 and after 2006 (**Digital Image 16-1A**). After applying the corrections, Schenk and others (2014) obtained a smooth sequence of elevation changes, depicting the diffusive response of the glacier to a rapid thinning near the grounding line. However, contrary to the expectations, this thinning of the lower region started in 2003 to 2004, i.e., prior to the retreat of the terminus. Lower elevation thinning continued and intensified in 2004 to 2005, followed by a diffusive propagation of thinning toward the upper part of the drainage basin in 2005 to 2007. Thinning rates decreased by 2007 when the fast-flowing part of the glacier quickly adjusted to a new equilibrium state. The evolution of annual mass loss rates (Figure 16-13A) and the cumulative mass loss (Figure 16-13B) of the region between 2001 and 2010 is obtained by integrating annual changes over the lower part of the drainage basin (region outlined by dashed line in Figure 16-10). All mass loss is assumed to be caused by ice dynamics; an ice density of 917 kg · m^{-3} is used in the calculations. In Figure 16-13, mass loss is represented by the gray squares and dashed polyline; the data were obtained from the uncorrected DEMs. The black dots and solid polyline are results from the corrected DEMs, including error bars of the mass changes (Figure 16-13). The mass loss reached its maximum in 2006, contributing 0.022 mm · yr^{-1} to sea level rise, followed by a rapidly diminishing mass loss that became negligible by 2010.

GLACIER RETREAT IN THE BROOKS RANGE, ALASKA

Glaciers have occupied north-facing cirques and valleys in the Brooks Range of northern Alaska for the past ~4,500

A. Uncorrected elevations.

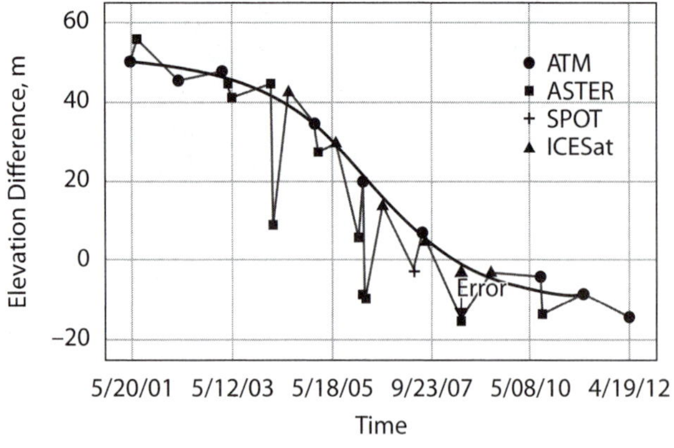

B. Time series after the correction of the SPOT and ASTER DEMs.

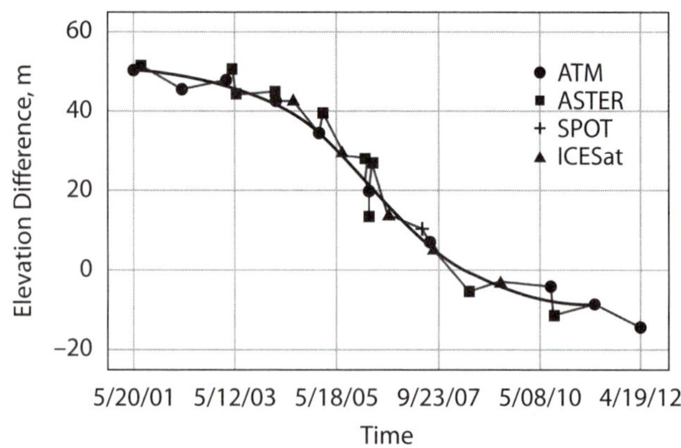

FIGURE 16-11 Time series of elevation changes at Site 8 (see Figure 16-10 for site location). A sigmoid curve is fitted through the laser altimetry points (ICESat, ATM) and the differences between the fitted curve and the DEMs (labeled error) are introduced as the control information to correct the DEMs. Elevations are given relative to August 31, 2006, when the ice sheet elevation was 1,116.88 m. Schenk, T., B. Csatho, C. van der Veen, and D. McCormick. 2014. Fusion of multi-sensor surface elevation data for improved characterization of rapidly changing outlet glaciers of Greenland. *Remote Sensing of Environment*, 149, 239–251, with permission of Elsevier.

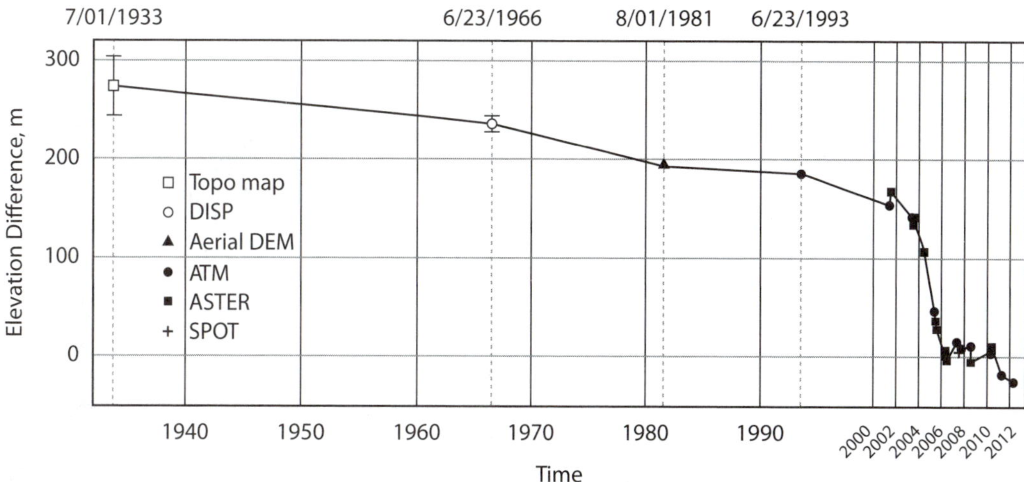

FIGURE 16-12 Extended time series ice sheet elevation change between 1933 and 2012 at Site 5 (see Figure 16-10 for location). Included in the glacier surface elevation change calculation is a digitized topographic map from 1933 with an estimated error of ±30 m, as well as a DEM derived from DISP with an estimated error of ±8 m. Other data sources (DEMs from stereo aerial photographs, ASTER and SPOT imagery, ATM laser altimetry data) have smaller error estimates—too small to be plotted in the graph. Schenk, T., B. Csatho, C. van der Veen, and D. McCormick. 2014. Fusion of multi-sensor surface elevation data for improved characterization of rapidly changing outlet glaciers of Greenland. *Remote Sensing of Environment*, 149, 239–251, with permission of Elsevier.

years (Ellis and Calkin, 1984). Buffalo Glacier, a large cirque glacier in the central Brooks Range, and Okpilak Glacier, a valley glacier in northeast Brooks Range, have well-preserved downslope moraines that were deposited by the glaciers during cold, cloudy, and/or snowy periods during the past 4,500 years (Figure 16-14). The last major advance was coincident with Europe's "Little Ice Age" and spanned 1410 to 1600. Glaciers across the central Brooks Range expanded and built large moraines over the ~200-year time interval. The glaciers stayed close to their Little Ice Age maximum extent until approximately 1640 to 1750, and then began to slowly retreat upslope as the climate became warmer, less

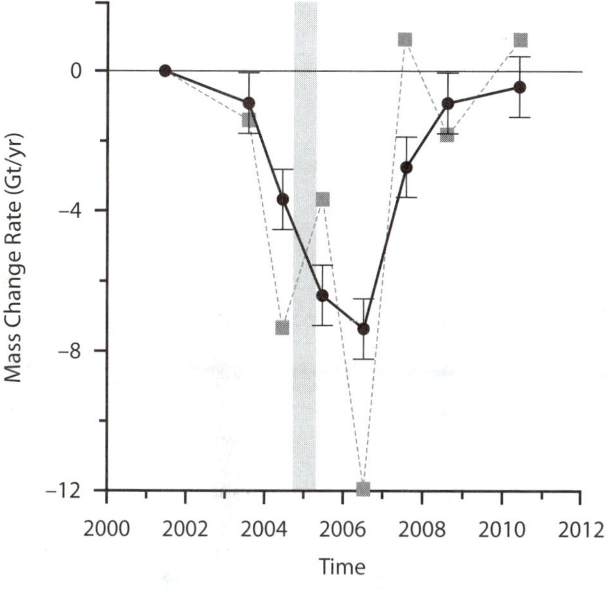

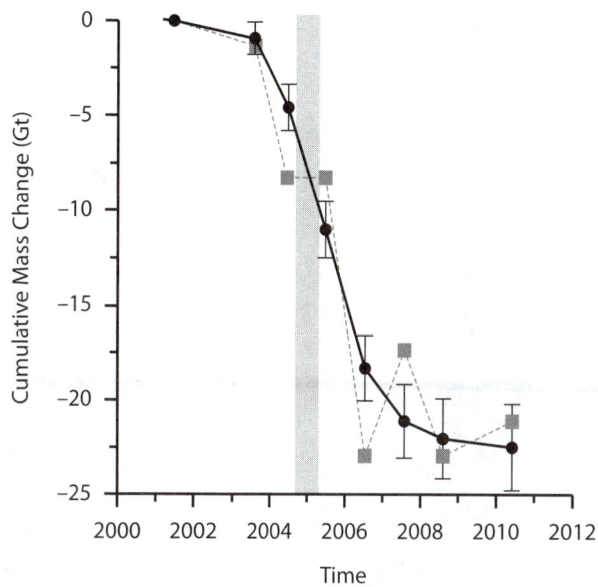

FIGURE 16-13 Mass change rates of the lower part of the Kangerlussuaq Glacier drainage basin (area outlined by dashed line in Figure 16-10). Dashed polyline is computed from uncorrected ASTER DEMs, while solid polyline shows the final estimate of mass loss computed from the corrected ASTER DEMs with corresponding error bars. Gray shaded area marks the period of abrupt retreat between July 1, 2004 and March 8, 2005. Schenk, T., B. Csatho, C. van der Veen, and D. McCormick. 2014. Fusion of multi-sensor surface elevation data for improved characterization of rapidly changing outlet glaciers of Greenland. *Remote Sensing of Environment*, 149, 239–251, with permission of Elsevier.

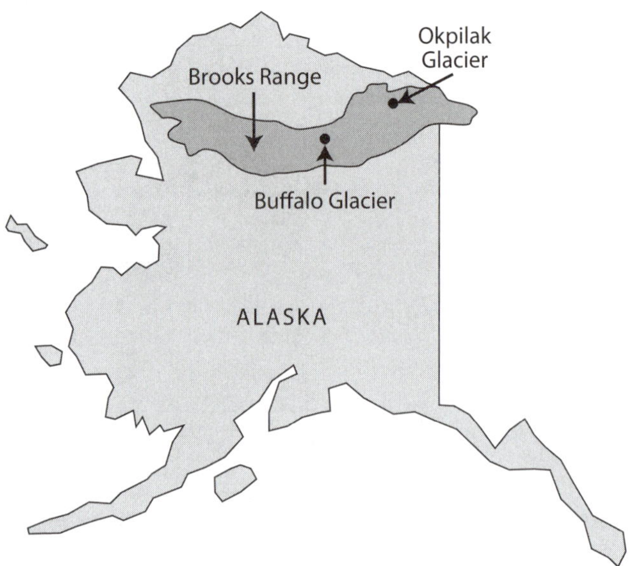

FIGURE 16-14 Location map of Buffalo and Okpilak Glaciers, Brooks Range, Alaska.

cloudy, and/or less snowy (Ellis and Calkin, 1984). The age of glacial deposits is determined by measuring the size of slow-growing lichen species on stable boulders and dating buried organic material with the radiocarbon technique (Calkin and Ellis, 1980).

Buffalo Glacier

Buffalo Glacier is one of the largest cirque glaciers in the central Brooks Range. The ice formed in two coalescing cirques and flowed almost 3 km downhill during the Little Ice Age to cover an area of 2.4 km^2. The maximum Little Ice Age extent of Buffalo Glacier was mapped in the field by Ellis in 1979 (black outline in Figure 16-15). The glacier remained within 100 m of its maximum Little Ice Age extent until about 150 years ago (Calkin and Ellis, 1980, Figure 8).

In 1979, the USGS acquired overlapping, color IR aerial photographs of Buffalo Glacier. The 1979 aerial photographs were scanned, orthorectified, and draped on the USGS DEM of the area (Figure 16-15A). A 2007 high-resolution satellite image was integrated into a GIS to create the perspective views in Figure 16-15B.

Area and length were measured to determine the size of the glacier in 1750, 1979, and 2007 (Table 16-1). Length was determined along the centerline of the glacier, from the

A. 1979 aerial photograph of Buffalo Glacier looking west. Photograph courtesy USGS.

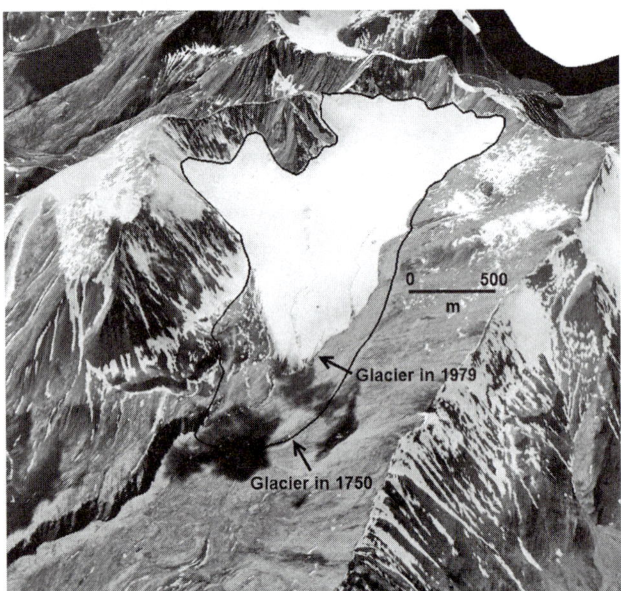

B. 2007 satellite image of Buffalo Glacier looking west. Google Earth and DigitalGlobe (2007).

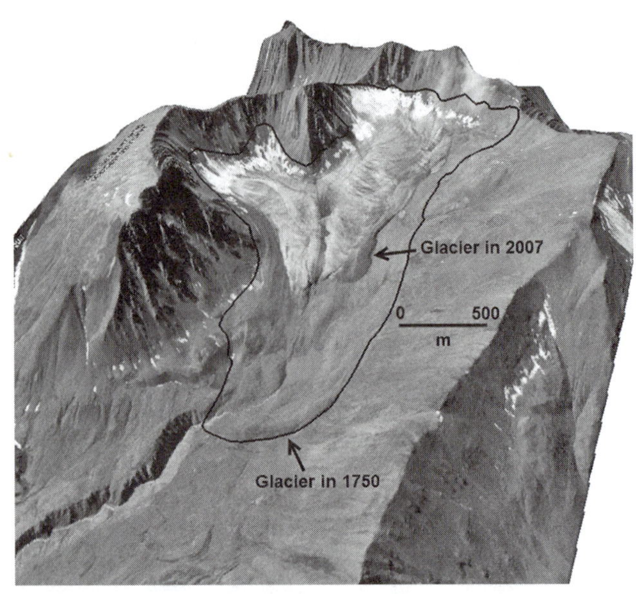

FIGURE 16-15 Perspective view of Buffalo Glacier with 1750, 1979, and 2007 margins.

TABLE 16-1 Buffalo Glacier dimensions 1750 to 2007.

Date	Perimeter (m)	Area (m^2)	Area (%)	Length (m)	Length (%)
1750	8,184	2,422,902	100.0	2,800	100.0
1979	7,063	1,890,531	78.0	2,250	80.4
2007	5,843	1,202,403	49.6	1,700	60.7

base of the exposed headwall to the toe of the downslope 1750 moraine (black outline on Figure 16-15) and the nose of the ice in the 1979 and 2007 images. The percent area and length were based on the Little Ice Age as 100% and more recent dimensions as a percentage of the Little Ice Age size (Table 16-1).

Buffalo Glacier's rate of retreat has significantly accelerated between 1979 and 2007. Buffalo Glacier lost 22% of its area and 20% of its length in the two centuries between the Little Ice Age and 1979. During the 28-year period of 1979 to 2007, Buffalo Glacier lost another 28% of the area and 20% of length attained during the Little Ice Age. As of 2007, Buffalo Glacier is 50% smaller and 40% shorter than in the Little Ice Age. Figure 16-16 plots the three dates and areas of Buffalo Glacier in Table 16-1 to graphically display the glacier's retreat from its maximum expansion in the Little Ice Age. Rapid acceleration of glacier melting after 1979 is evident.

Area and length dimensions are only two dimensional and as glaciers shrink in area and length they also decrease in the third dimension—topographic elevation. Geck and others (2013) used DEMs reconstructed from 1970 and 1973 USGS topographic maps and a 2001 interferometric SAR DEM to calculate volume and mass changes for 107 glaciers in the central Brooks Range. Over the period 1970 to 2001 approximately 0.5 m of ice thickness per year was lost from the surface of the glaciers in the central Brooks Range.

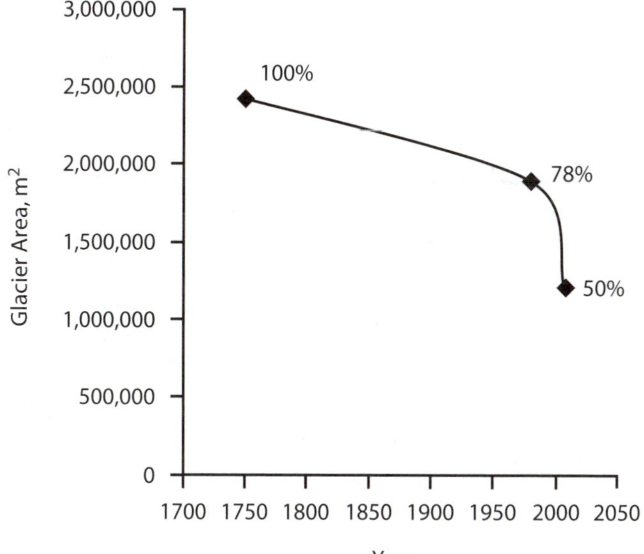

FIGURE 16-16 Buffalo Glacier area reduction from the Little Ice Age maximum to 2007.

Okpilak Glacier

Okpilak Glacier is fed by ice accumulation in several cirques that coalesce downstream into a valley glacier that was approximately 10.5 km long during the Little Ice Age. Okpilik is one of the few glaciers in the Brooks Range that was photographed over a century ago. Leffingwell (1919) photographed the glacier in 1907. In 1981, Ellis reoccupied Leffingwell's camera position and photographed the retreating glacier (Figure 16-17). In 1958, Sable photographed and surveyed the glacier (Sable, 1961), and Nolan reoccupied Leffingwell's camera position in 1994, 2004, and 2007 (for Sable and Nolan's images, see Pelto, 2010). Landsat images of the glacier were acquired in 2001 and 2017. These satellite images were draped on a USGS DEM and displayed with a perspective view; the outlines of past margins of the glacier from 1750 to 2017 are shown (Figure 16-18).

Table 16-2 quantifies the dimensions seen in Figures 16-17 and 16-18. The area for each date includes the upslope coalescing cirque glaciers and the downslope valley glacier while length was measured from the base of the headwall at the largest tributary to the toe of the downslope glacier. Okpilak retreated 3,400 m, or 32% of its total length, since the Little Ice Age. Between 1958 and 2017 (a span of 59 years), the glacier retreated 2,400 m, or an average of 40 m · yr^{-1}. Glacier recession from the Little Ice Age to 2017 has reduced the glacier's area by approximately 36%.

A. 1907 photograph of Okpilak Glacier looking west–southwest, northeast Brooks Range, Alaska. From Leffingwell (1919).

B. 1981 photograph of Okpilak Glacier looking west–southwest, northeast Brooks Range, Alaska. From Ellis (ellis-geospatial.com/brooksrange.html).

FIGURE 16-17 1907 and 1981 ground photographs of Okpilak Glacier, Alaska. The white arrows point to the same feature in the 1907 and 1981 photographs.

TABLE 16-2 Okpilak Glacier dimensions 1750 to 2017.

Date	Area (m^2)	Area (%)	Length (m)	Length (%)
1750	14,869,116	100.0	10,500	100.0
1907	13,282,364	89.3	10,000	95.2
1958	12,517,372	84.2	9,500	90.5
1981	10,965,436	73.7	8,800	83.8
2001	9,776,233	65.7	8,200	78.1
2017	9,503,065	63.9	7,100	67.6

A. 2001 Landsat 7 TM satellite image of Okpilak Glacier looking west.

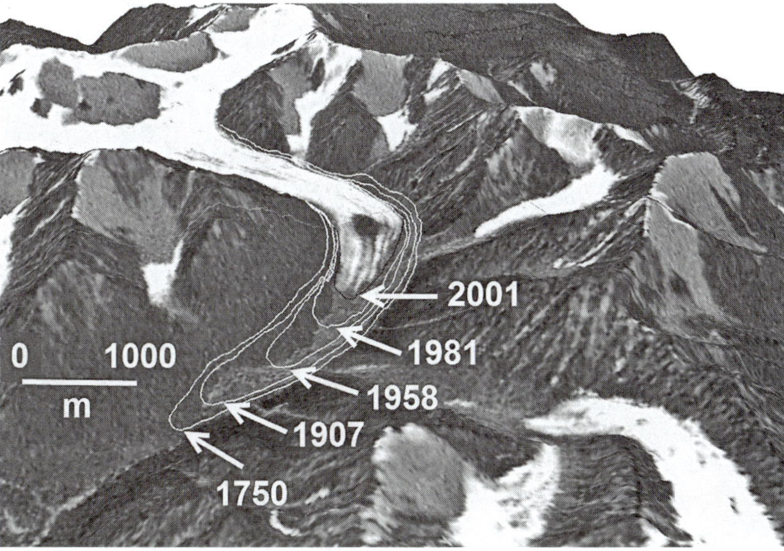

B. 2017 Landsat 8 OLI satellite image of Okpilak Glacier looking west.

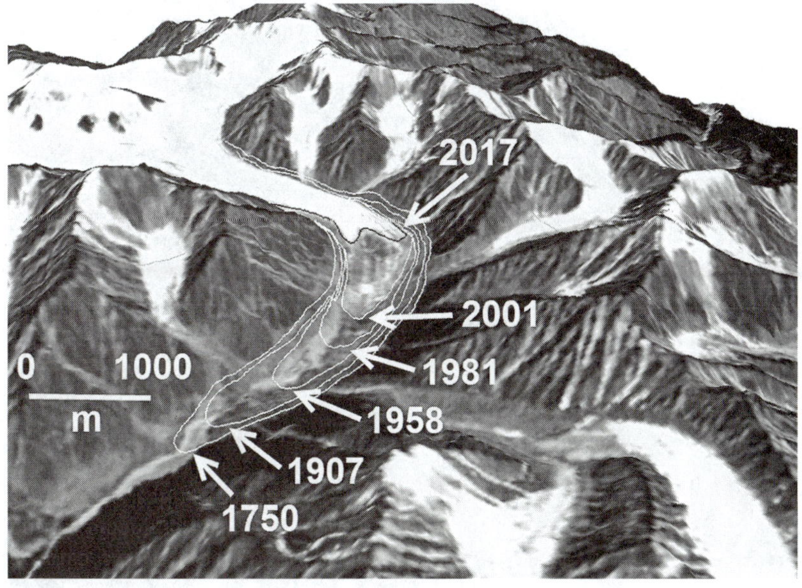

FIGURE 16-18 2001 and 2017 Landsat images of Okpilak Glacier with outlines of past ice margins labeled by year. Landsat courtesy USGS.

Figure 16-19 plots the Okpilak Glacier's length versus time. This plot displays the same dramatic acceleration of glacial shrinkage since the mid-1950s as was mapped at Buffalo Glacier 250 km to the southwest (Figure 16-16). Rabus and Echelmeyer (1998) calculated the topographic elevation loss at Okpilak Glacier as 51 cm · yr^{-1} between 1973 and 1993. They also determined a 420 m retreat during this 20 year interval for a rate of 21 m · yr^{-1}. The impact of shrinking glaciers on downstream ecosystems (fish, birds, floodplains, and estuaries) in the Arctic National Wildlife Refuge, which spans across the east-central to northeastern Brooks Range and includes the eastern Alaska North Slope, is discussed by Nolan and others (2011).

SEA ICE

Sea ice is frozen ocean water. It forms, grows, and melts in the ocean. In contrast, icebergs, glaciers, and ice shelves float in the ocean but originate on land. For most of the year, sea ice is typically covered with snow (NSIDC, 2017b).

In order to detect sea ice, remote sensors use passive microwaves. Objects at the Earth's surface emit not only thermal infrared radiation, they also emit microwaves at relatively low energy levels. When a sensor detects microwave radiation naturally emitted by the Earth, that radiation is called passive microwave. Clouds do not emit much microwave radiation when compared to sea ice. Thus, microwaves can penetrate clouds and be used to detect sea ice during the day and night, regardless of cloud cover (NSIDC, 2017c).

Microwave emission is not as strongly tied to the temperature of an object compared to thermal infrared. Instead, the object's physical properties, such as atomic composition and crystalline structure, determine the amount of microwave radiation it emits. The crystalline structure of ice typically emits more microwave energy than the liquid water in the ocean. Thus, sensors that detect passive microwave radiation can easily distinguish sea ice from the ocean (Equation 11-1). A major drawback to measuring passive microwave radiation is that the energy level is quite low. As a result, the radiation must be collected over a larger region with the spatial resolution measured in thousands of meters. Active radar technology has a higher spatial resolution and is

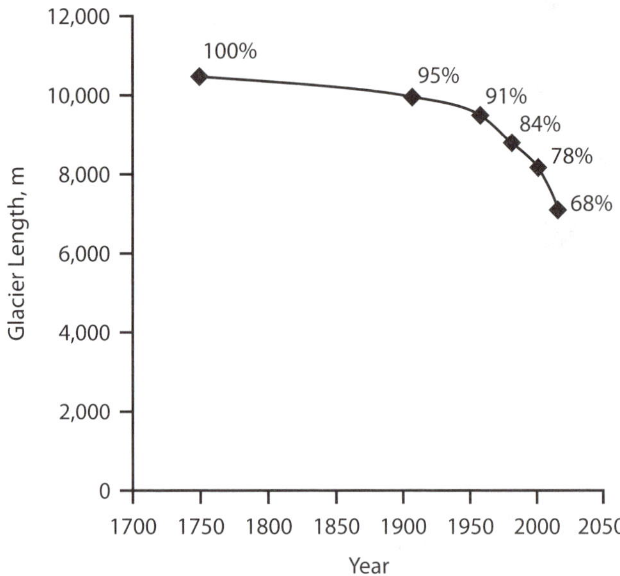

FIGURE 16-19 Okpilak Glacier length reduction from the Little Ice Age maximum to 2017.

used to map sea ice details such as leads, sea ice type, and ice bergs (Chapters 6 and 11).

Passive microwave sensors provide nearly complete images of all sea ice covered regions every day. Satellite passive microwave sensors include the Special Sensor Microwave/Imager (SSM/I) on the DMSP satellites and the Advanced Microwave Scanning Radiometer on EOS (AMSR-E) onboard NASA's Aqua and Japan Aerospace Exploration Agency's (JAXA) Global Change Observation Mission (GCOM-W1) satellites. Sea ice has been detected from satellites with passive microwave sensors since 1972. These sensors provide the most complete, long-term observations of sea ice.

Passive microwave satellite images of Arctic sea ice for each September since 1979 are available from the National Snow and Ice Data Center (nsidc.org). In September the Arctic sea ice is at its minimum extent. Figure 16-20 shows the extent of the Arctic sea ice in September 1979 and September 2018. The view in Figure 16-20 is looking down on the North Pole; the Arctic sea ice and the Greenland Ice Sheet are white features. The satellite images show open water has increased dramatically along the northern coast of

A. Minimum sea ice extent for September 1979.

B. Minimum sea ice extent for September 2018.

FIGURE 16-20 Minimum extent of the Arctic sea ice. From NSIDC (2017a).

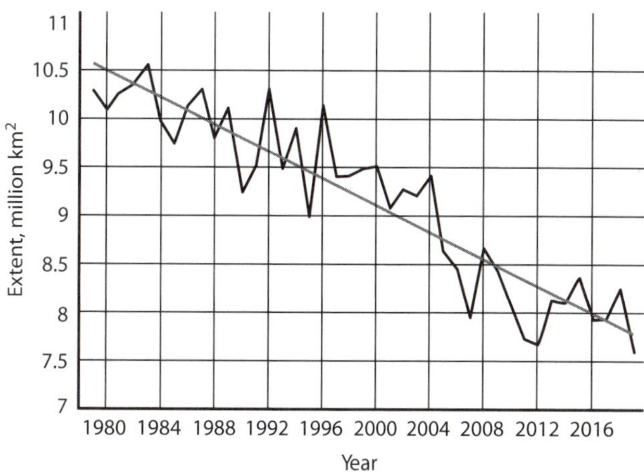

FIGURE 16-21 Annual minimum Arctic sea ice extent from July 1979 to July 2019. From NSIDC (2019).

Russia, Alaska, and northwest Canada during the past three decades. Figure 16-21 plots the July sea ice extent from 1979 to 2019 (NSIDC, 2019). Monthly July sea ice extent for 1979 to 2010 shows a decline of 7.32% per decade.

NSIDC (2017b) states a small temperature increase at the poles leads to still greater warming over time, making the poles the most sensitive regions to climate change on Earth. According to scientific measurements, both the thickness and summer sea ice extent in the Arctic have shown a dramatic decline over the past 30 years. This is consistent with observations of a warming Arctic. The loss of sea ice also has the potential to accelerate global warming trends and to change climate patterns.

PERMAFROST

Global warming is rapidly thawing permafrost along the Arctic Coastal Plain of northern Alaska, with a cascading effect that results in coastal subsidence, inundation by saltwater, and subsequent changes to habitat distribution and quality (Pearce, 2016).

HABITAT MAPPING

Pearce (2016) reports as part of the USGS Changing Arctic Ecosystem Initiative, scientists at the Alaska Science Center are using WorldView-2 and -3 satellite imagery to map changes taking place to goose habitat. Habitat maps are created using all eight WorldView-3 spectral bands, along with NDVI and Red Edge vegetation indices. The location and areal extent of salt-tolerant graminoid habitats across the full extent of the Arctic Coastal Plain are generated. These maps complement ongoing research to (1) estimate the biomass and nutrients available to waterfowl in salt-tolerant graminoid habitat, (2) plan for the location of coastal subsidence, inundation, and habitat change monitoring stations, and (3) retroactively analyze and forecast future coastal habitat change for US Department of Interior management agencies and local governments and residents.

GEESE POPULATION

Flint and others (2014) document how sea ice decline and permafrost thaw benefit the Pacific black brant geese on the Arctic Coastal Plain as follows. Loss of sea ice has increased ocean wave action, leading to erosion and saltwater inundation of coastal habitats. Figure 16-22 shows coastal erosion

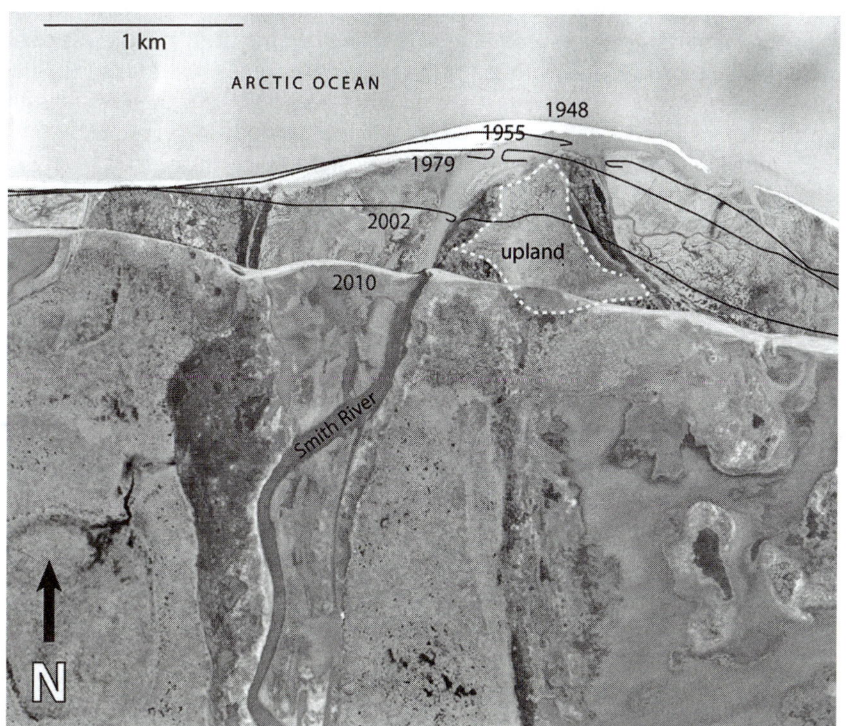

FIGURE 16-22 Time series orthoimagery showing coastal erosion from 1948 to 2010 at the Smith River on the Arctic Coastal Plain of Alaska. From Tape and others (2013, Figure 6).

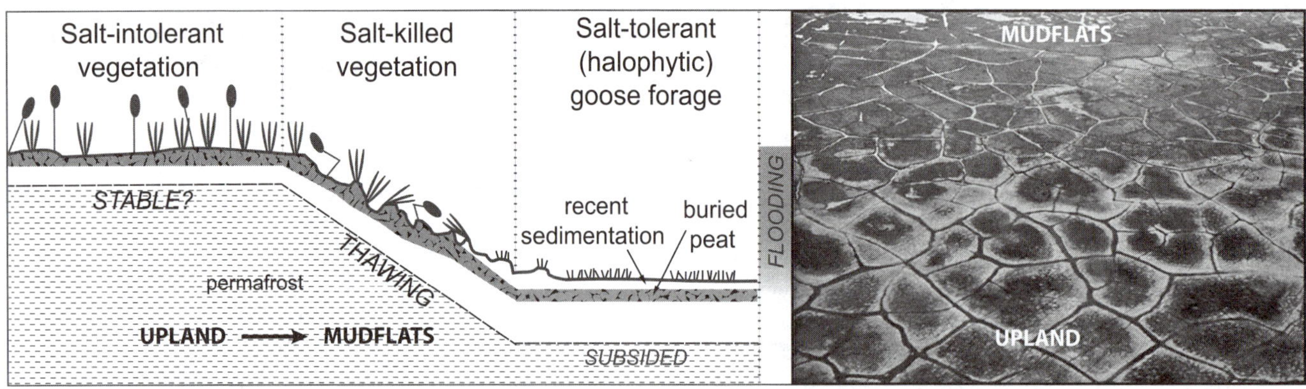

FIGURE 16-23 Diagram showing the process by which thawing permafrost, subsidence, saltwater flooding, and sedimentation facilitate increased habitat conversion from upland to mudflats. The July 2012 photograph shows an aerial view of this habitat conversion. Photograph by Ken Tape, University of Alaska, Fairbanks. From Tape and others (2013, Figures 4 and 7).

at the Smith River on the Arctic Coastal Plain as interpreted from aerial and satellite images acquired between 1948 and 2010. Orthorectified images from five years were superimposed over the 1948 aerial photograph. The 1955, 1979, 2002, and 2010 images record the shoreline moving farther inland (migrating south) over the 62-year time interval. Field work and remote sensing confirm salt-tolerant graminoid plants thrive in the areas inundated by saltwater.

The Pacific black brant geese migrate to the Arctic Coastal Plain each summer to breed and undergo wing molt, an annual event during which the birds are flightless for three to four weeks. The molt period requires high-quality food as well as open-water areas where the birds can escape predators. Permafrost thaw and subsidence of upland tundra along the coast, along with more saltwater storm surges that bury non-salt-tolerant vegetation, are creating a new coastal habitat of mudflats with salt-tolerant, goose-favored forage plants. Figure 16-23 is a diagram showing the process by which thawing permafrost, subsidence, saltwater flooding, and sedimentation facilitate increased habitat conversion from upland to mudflats. The July 2012 photograph in Figure 16-23 shows an aerial view of this habitat conversion.

The USGS has documented an increase of 50% in the population size of the Pacific black brant population along the Arctic Coastal Plain since the 1970s (Figure 16-24). This shift in distribution is due to an increase in high-quality forage along the coast brought about by reduced sea ice that transforms inland upland areas into salt marshes (Tape and others, 2013). In contrast, declining sea ice has negative effects on habitats of ice-dependent animals, such as polar bear and walrus. Flint and others (2014) observe that a key finding of the USGS research is that as habitat conditions are altered, wildlife populations are responding with changes in both abundance and distribution.

DESERTIFICATION

The Tigris and Euphrates Rivers irrigate the Fertile Crescent, an arc of rich land situated in the midst of arid and barren deserts in the Middle East. These rivers fostered Mesopotamian and Middle Eastern cultures for thousands of years (Figure 16-25). By the twentieth century, however, this fertile region began to dry out. Rising temperatures and persistent drying are transforming this region, which is also struggling with political unrest and population pressures that have stressed water supplies (Naranjo, 2016). Understanding the relationship between environmental changes mapped with remote

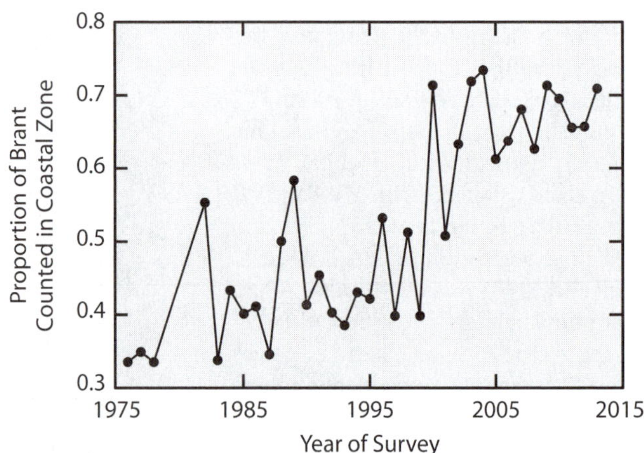

FIGURE 16-24 Increase in molting black brant geese along the Arctic Coastal Plain where saltwater flooding has created a new foraging habitat. Data provided by US Fish and Wildlife Service Migratory Bird Management. From Flint and others (2014, Figure 3).

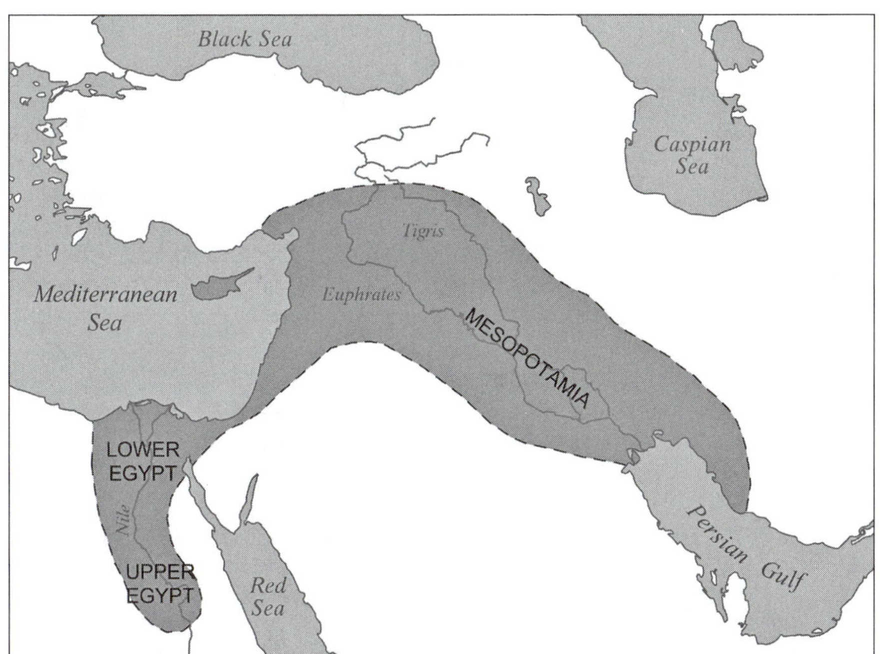

FIGURE 16-25 Location of the Fertile Crescent, Middle East.

sensing technology, political and social unrest and conflict, and climate circulation and change models is an ultimate goal.

VEGETATION AND GROUNDWATER CHANGE

In Syria, a severe, three-year drought from 2007 to 2010, along with several less severe, multiyear droughts in the 1980s and 1990s, dried up rivers and wells that were used to irrigate crops and water herds of sheep and cattle (Figure 16-26). The severity of the drought caused a mass migration of farmers and shepherds to urban centers. In addition, demands for water increased by 2010 as over 1 million refugees from Iraq entered Syria. The agricultural economy of Syria was crippled by 2011. In March 2011, the Syrian uprising started. Kelley and others (2015) perceive that the drought had a catalytic effect that contributed to political unrest.

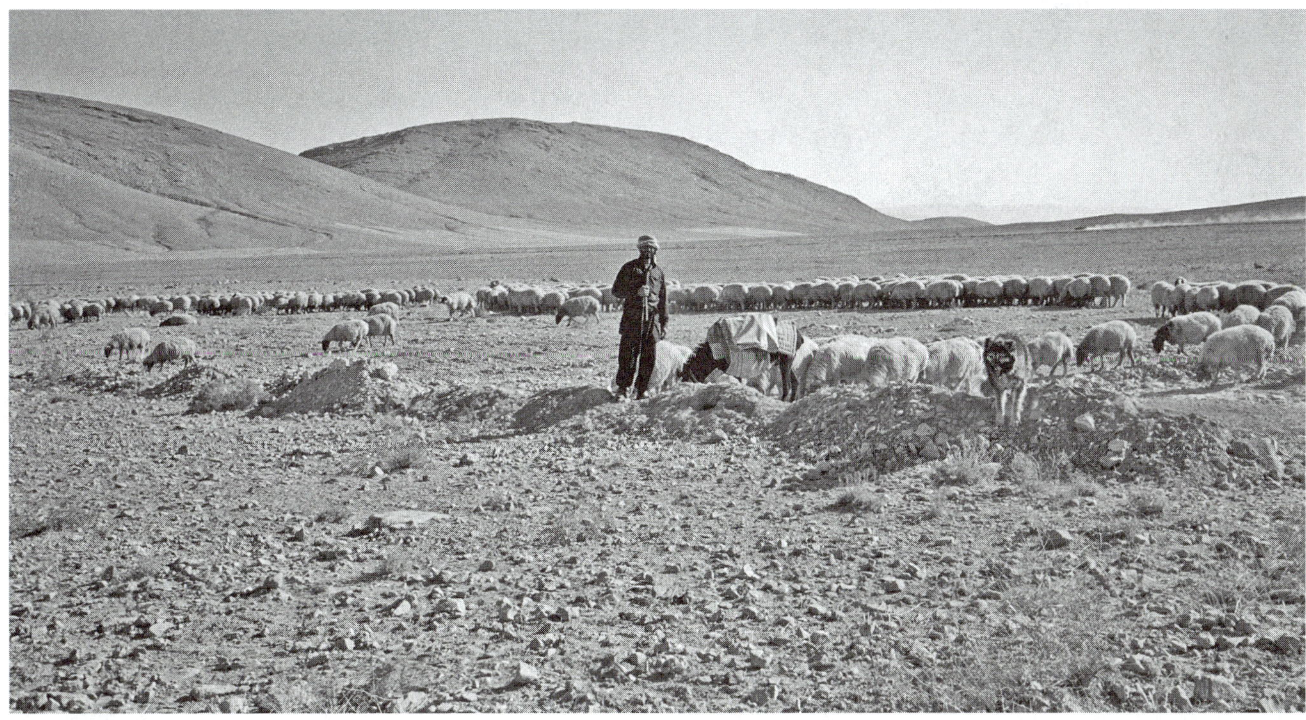

FIGURE 16-26 A Bedouin shepherd tends his sheep amid a parched landscape in Syria. Photograph by Jeff Werner (http://creativecommons.org/licenses/by-nc-sa/2.0).

Kelley and others (2015) searched for remote sensing evidence of the drought because ground measurements of agricultural production were sparse. To address the question of whether the drought was made more severe due to a contribution from long-term trends, they determined the long-term change in winter rainfall. Winter rainfall in mm per month from 1931 to 2008 is shown in **Digital Image 16-2A** 🌐. During this period, the Fertile Crescent experienced a decrease in winter rainfall of up to 30 mm per month (**Digital Image 16-2B** 🌐), confirming a longer term drying trend that was punctuated by multiyear droughts.

In addition to surface water for irrigation, Syrian farmers rely heavily on groundwater. To determine if groundwater supplies were depleted during the 2007 to 2010 drought, Kelley and others (2015) obtained data from the Gravity Recovery and Climate Experiment (GRACE) satellites, which can detect groundwater by measuring small gravity changes. **Digital Image 16-2C** 🌐 is the GRACE map where the liquid water equivalent (LWE) thickness relates to the water stored at and near the Earth's surface. **Digital Image 16-2C** 🌐 was generated by subtracting the mean GRACE LWE thickness values measured between 2002 and 2007 from the 2008 LWE thickness values (acquired during the second year of the drought) and reveals a significant decline in groundwater across the Fertile Crescent.

Kelley and others (2015) used MODIS vegetation greenness (NDVI) images as a proxy for crops and vegetation. The MODIS NDVI map (**Digital Image 16-2D** 🌐) was generated by subtracting the mean NDVI values measured between 2001 and 2007 from the 2008 NDVI values (acquired during the second year of the drought) and reveals a significant decline in vegetation (including crops and pastures) in the Fertile Crescent. Remote sensing confirms the severity and location of vegetation/agriculture decline and groundwater depletion during the severe 2007 to 2010 drought in Syria.

DUST CHANGE

The Middle East is second to the Sahara Desert as a source of sand and dust. Notaro and others (2015) used remote sensing technology to confirm that an abrupt increase in dust activity across the region was triggered by the severe Fertile Crescent drought of 2007 to 2010. The following is a summary of their findings. Dried soils and diminished vegetation cover in the Fertile Crescent, as evident through remotely sensed enhanced vegetation indices (**Digital Image 16-2D** 🌐), supported greater dust generation and transport to the Arabian Peninsula in 2007 to 2013 (Figure 16-27). Weather stations observed an increase in dust days during the six years. Notaro and others (2015) used the Multi-angle

🌐 **FIGURE 16-27** A massive sandstorm rolls over Al Asad, Iraq and closes in a nearby military camp on April 27, 2005. Photograph by Alicia M. Garcia, US Marine Corps (Wikimedia Commons).

Imaging SpectroRadiometer (MISR) instrument on NASA's Terra satellite to detect and measure the concentration of atmospheric dust between 2001 and 2012. MISR can distinguish dust, which typically consists of nonspherical particles, from spherical particles, such as water droplets and other chemical particles in the atmosphere.

The concentration of dust in the atmosphere is measured as *dust aerosol optical depth* (DAOD). The greater the DAOD value, the greater the concentration of dust. The MISR instrument measures DAOD with the 0.558 µm (green) band. NASA provides monthly total DOAD as a grid with a spatial resolution of 0.5° by 0.5° (latitude by longitude). Plate 54 shows plots of MISR DOAD across the Arabian Peninsula that include data from 2001 to 2005 (a period of reduced dust activity) and 2008 to 2012 (a period of enhanced dust activity). Plate 54A is the difference in mean DAOD between these two periods. The northern Arabian Peninsula experienced much higher levels of dust in the atmosphere during the Fertile Crescent drought.

Plate 54B is the ratio in mean DOAD between the two periods. The greatest change in dust between the two time periods was found in the northern Fertile Crescent (Figure 16-25), where agriculture around the Tigris and Euphrates Rivers collapsed during the drought. The area outlined in blue experienced the greatest decline in vegetation during the 2008 to 2010 drought period, as measured with a MODIS enhanced vegetation index (EVI) that is similar to NDVI but is more sensitive to vegetation canopy structural variations. The loss of crops and pasture during the drought within this area increased the amount of dust coming from this once rich, agricultural area and the Tigris and Euphrates River floodplains.

Implications for Climate Change

In their study of drought in the Fertile Crescent, Kelley and others (2015) stated:

> Century-long observed trends in precipitation, temperature, and sea level pressure, supported by climate model results, strongly suggest that anthropogenic forcing has increased the probability of severe and persistent droughts in this region, and made the occurrence of a 3-year drought as severe as that of 2007–2010 two to three times more likely than by natural variability alone. We conclude that human influences on the climate system are implicated in the current Syrian conflict.

Notaro and others (2015) linked the onset of drought in the Fertile Crescent to large-scale climate factors, in particular the El Niño Southern Oscillation and the Pacific Decadal Oscillation. Notaro and others (2015) observed clear implications for even greater dust activity for the Arabian Peninsula because of a "super high-resolution" global climate model by Kitoh and others (2008), which projects the possible collapse of Fertile Crescent agriculture during the twenty-first century in response to anthropogenic climate change and associated drought and water shortages.

Questions

1. What are five environmental changes associated with climate change that can be detected and measured with remote sensing technology?
2. How are each of the following remote sensing tools used to measure changes in the size of the Greenland Ice Sheet?
   a. Field mapping.
   b. Historic aerial photographs.
   c. Airborne laser altimetry.
   d. Satellite laser altimetry.
   e. Overlapping, optical satellite images.
3. What does SERAC do?
4. How is the length of the Buffalo coalescing cirque glacier in the Brooks Range determined?
5. What property of sea ice enables passive microwave sensors to differentiate sea ice from ocean water?
6. Why do passive microwave sensors require large footprints to generate an image?
7. Why is the Pacific black brant geese population increasing in the Arctic Coastal Plain?
8. Why are MODIS vegetation greenness (NDVI) images used as a proxy for agricultural production in the Middle East?
9. How does MISR detect atmospheric dust and differentiate it from other particles in the atmosphere?
10. Describe DAOD and how MISR measures atmospheric dust.

References

Briner, J. P., N. E. Young, E. K. Thomas, H. A. M. Stewart, S. Losee, and S. Truex. 2011. Varve and radiocarbon dating support the rapid advance of Jakobshavn Isbræ during the Little Ice Age. *Quaternary Science Reviews*, 30(19–20), 1–11. http://doi.org/10.1016/j.quascirev.2011.05.017

Calkin, P. E., and J. M. Ellis. 1980. A lichenometric dating curve and its application to Holocene glacier studies in the central Brooks Range, Alaska. *Arctic and Alpine Research*, 12(3), 245–264.

Church, J. A., P. U. Clark, A. Cazenave, J. M. Gregory, S. Jevrejeva, A. Levermann, M. A. Merrifield, G. A. Milne, R. S. Nerem, P. D. Nunn, A. J. Payne, W. T. Pfeffer, D. Stammer, and A. S. Unnikrishnan. 2013. Sea level change. In T. F. Stocker, D. Qin, G. K. Plattner, M. Tignor, S. K. Allen, J. Boschung, A. Nauels, Y. Xia, V. Bex, and P. M. Midgley (Eds.), *Climate Change 2013: The Physical Science Basis. Contribution of Working Group I to the Fifth Assessment Report of the Intergovernmental Panel on Climate Change* (pp. 1,137–1,216). New York: Cambridge University Press.

Csatho, B., T. Schenk, and C. van der Veen. 2017. Local Processes and Regional Patterns—Interpreting Greenland Ice Sheet Changes Since Little Ice Age. IGS Symposium Polar Ice, Polar Climate, Polar Change, Boulder, CO.

Csatho, B., T. Schenk, C. J. van der Veen, and W. B. Krabill. 2008. Intermittent thinning of Jakobshavn Isbræ, West Greenland, since the Little Ice Age. *Journal of Glaciology*, 54(184), 131–144.

Csatho, B., C. J. van der Veen, and C. Tremper. 2005. Trimline mapping from multispectral Landsat ETM+ imagery. *Géographie physique et Quaternaire*, 59(1), 49–52.

Ellis, J. M., and P. E. Calkin. 1984. Chronology of Holocene glaciation, central Brooks Range, Alaska. *Geological Society of America Bulletin*, 95, 897–912.

Enderlin, E. M., I. M. Howat, S. Jeong, M. J. Noh, J. H. Angelen, and M. R. Broeke. 2014. An improved mass budget for the Greenland ice sheet. *Geophysical Research Letters*, 41(3), 866–872. http://doi.org/10.1002/(ISSN)1944-8007

ESA. 2017. CryoSat Reveals Antarctica in 3D. European Space Agency. http://www.esa.int/Our_Activities/Observing_the_Earth/CryoSat/CryoSat_reveals_Antarctica_in_3D (accessed January 2018).

Flint, P., M. Whalen, and J. Pearce. 2014, August. *Changing Arctic Ecosystems—Sea Ice Decline, Permafrost Thaw, and Benefits for Geese* (Fact Sheet 2014-3088). Washington, DC: US Geological Survey.

Geck, J., R. Hock, and M. Nolan. 2013. Geodetic mass balance of glaciers in the central Brooks Range, Alaska, U.S.A., from 1970 to 2001. *Arctic, Antarctic, and Alpine Research*, 45(1), 29–38.

Holland, D. M., R. H. Thomas, B. De Young, M. H. Ribergaard, and B. Lyberth. 2008. Acceleration of Jakobshavn Isbræ triggered by warm subsurface ocean waters. *Nature Geoscience*, 1(10), 659–664. http://doi.org/10.1038/ngeo316

Kelley, C. P., S. Mohtadi, M. A. Cane, R. Seager, and Y. Kushnir. 2015. Climate change in the Fertile Crescent and implications of the recent Syrian drought. *Proceedings of the National Academy of Sciences of the United States of America*, 112(11), 3,241–3,246. doi:10.1073/pnas.1421533112

Kitoh, A., A. Yatagai, and P. Albert. 2008. First super-high-resolution model projection that ancient "Fertile Crescent" will disappear in this century. *Hydrological Research Letters*, 2, 1–4. doi:10.3178/HRL.2.1

Leffingwell, E. de K. 1919. *The Canning River Region, Northern Alaska* (Professional Paper 109). Washington, DC: US Geological Survey.

Markus, T., T. Neumann, A. Martino, W. Abdalati, K. Brunt, B. Csatho, S. Farrell, H. Fricker, A. Gardner, D. Harding, M. Jasinski, R. Kwok, L. Magruder, D. Lubin, S. Luthcke, J. Morison, R. Nelson, A. Neuenschwander, S. Palm, S. Popescu, C. K. Shum, B. E. Schutz, B.Smith, Y. Yang, and J. Zwally. 2017. The Ice, Cloud, and Land Elevation Satellite-2 (ICESat-2): Science requirements, concept, and implementation. *Remote Sensing of Environment*, 190, 260–273. http://doi.org/10.1016/j.rse.2016.12.029

Naranjo, L. 2016. Crisis in the Crescent: Drought Turns the Fertile Crescent Into a Dust Bowl. NASA EarthData. http://earthdata.nasa.gov/user-resources/sensing-our-planet/crisis-in-the-crescent (accessed January 2018).

NOAA. 2016. Arctic Report Card 2016. http://www.arctic.noaa.gov/Report-Card/Report-Card-2016 (accessed January 2018).

Nolan, M., R. Churchwell, J. Adams, J. McClelland, K. D. Tape, S. Kendall, A. Powell, K. Dunton, D. Payer, and P. Martin. 2011, September 26–30. Predicting the impact of glacier loss on fish, birds, floodplains and estuaries in the Arctic National Wildlife Refuge. In C. N. Medley, G. Patterson, and M. J. Parker (Eds.), *Observing, Studying, and Managing for Change* (USGS Scientific Investigations Report 2011-5169, pp. 49–54). The Fourth Interagency Conference on Research in the Watersheds, Fairbanks, AL.

Notaro, M., Y. Yu, and O. V. Kalashnikova. 2015. Regime shift in Arabian dust activity, triggered by persistent Fertile Crescent drought. *Journal of Geophysical Research: Atmospheres*, 120(19), 10,229–10,249. doi:10.1002/2015JD023855

NSIDC. 2017a. NSIDC Data on Google Earth. National Snow and Ice Data Center. http://nsidc.org/data/google_earth (accessed August 2019).

NSIDC. 2017b. Quick Facts on Arctic Sea Ice. National Snow and Ice Data Center. Quick Facts. http://nsidc.org/cryosphere/quickfacts/seaice.html (accessed August 2019).

NSIDC. 2017c. Remote Sensing: Passive Microwave. National Snow and Ice Data Center. All about Sea Ice. http://nsidc.org/cryosphere/seaice/study/passive_remote_sensing.html (accessed August 2019).

NSIDC. 2019, August 6. Europe's Heat Wave Moves North. Arctic Sea Ice News & Analysis. https://nsidc.org/arcticseaicenews/2019/08/europe-heat-wave-moves-north (accessed August 2019).

Pearce, J. 2016. Sea Ice Decline, Permafrost Thaw, and Benefits for Geese in the Alaskan Arctic. US Department of Interior. Remote Sensing Activities 2016. http://eros.usgs.gov/doi-remote-sensing-activities/2016/sea-ice-decline-permafrost-thaw-and-benefits-geese-alaskan-arctic (accessed January 2018).

Pelto, M. S. 2010, June 25. Okpilak Glacier Retreat, Brooks Range, Alaska. AGU Blogosphere. From a Glacier's Perspective. http://blogs.agu.org/fromaglaciersperspective/2010/06/25/okpilak-glacier-retreat-brooks-range-alaska (accessed January 2018).

Rabus, B. T., and K. A. Echelmeyer. 1998. The mass balance of McCall Glacier, Brooks Range, Alaska, U.S.A.: Its regional relevance and implications for climate change in the Arctic. *Journal of Glaciology*, 44(147), 333–351.

Sable, E. G. 1961. Recent recession and thinning of Okpilak Glacier, northeastern Alaska. *Arctic*, 14(3), 176–187.

Schenk, T., and B. Csatho. 2012. A new methodology for detecting ice sheet surface elevation changes from laser altimetry data. *IEE Transactions on Geoscience and Remote Sensing*, 50(9), 3,302–3,316.

Schenk, T., B. Csatho, C. van der Veen, and D. McCormick. 2014. Fusion of multi-sensor surface elevation data for improved characterization of rapidly changing outlet glaciers of Greenland. *Remote Sensing of Environment*, 149, 239–251.

Tan, C., M. Minggua, and K. Honghai. 2017. Spatial-temporal characteristics and climatic responses of water level fluctuations of global major lakes from 2002 to 2010. *Remote Sensing*, 9(2), 150. doi:10.3390/rs9020150

Tape, K., P. L. Flint, B. W. Meixell, and B. Gaglioti. 2013. Inundation, sedimentation, and subsidence creates goose habitat along the Arctic coast of Alaska. *Environmental Research Letters*, 8. doi:10.1088/1748-9326/8/4/045031

Trögler, M., and S. Lingner. 2012. *Remote Sensing and Regional Climate Change* (Report 41). Vienna, Austria: European Space Policy Institute.

Tsang, L., and T. Jackson. 2010. Satellite remote sensing missions for monitoring water, carbon, and global climate change. *Proceedings of the IEEE*, 98(5), 645–648.

USGCRP. 2017. *Climate Science Special Report: Fourth National Climate Assessment*, Volume I. D. J. Wuebbles, D. W. Fahey, K. A. Hibbard, D. J. Dokken, B. C. Stewart, and T. K. Maycock (Eds.). Washington, DC: US Global Change Research Program. doi: 10.7930/J0J964J6. http://science2017.globalchange.gov (accessed January 2018).

van der Veen, C. J., and B. M. Csatho. 2005. Spectral characteristics of Greenland lichens. *Géographie physique et Quaternaire*, 59(1), 63–73.

Yang, J., P. Gong, R. Fu, M. Zhang, J. Chen, S. Liang, B. Xu, J. Shi, and R. Dickinson. 2013, September 15. The role of satellite remote sensing in climate change studies. *Nature Climate Change*, 3, 875–883. http://www.nature.com/nclimate/index.html (accessed January 2018). doi:10.1038/nclimate1908

Zwally, H. J., B. Schutz, W. Abdalati, J. Abshire, C. Bentley, A. Brenner, J. Bufton, J. Dezio, D. Hancock, D. Harding, T. Herring, B. Minster, K. Quinn, S. Palm, J. Spinhirne, and R. Thomas. 2002. ICESat's laser measurements of polar ice, atmosphere, ocean, and land. *Journal of Geodynamics*, 34(3), 405–445.

chapter 17

Other Applications

Remote sensing is used successfully by many applications that are not covered in previous chapters (Table 17-1). Remote sensing is able to deliver high spatial resolution images that fulfill the stringent requirements of these applications because the temporal resolution of the technology has rapidly increased as more satellites, aircraft, and drones are deployed. Faster computer processing and delivery of finished products via online web mapping services have also contributed to the expanded use of remote sensing by a more diverse group of commercial, government, and academic users. In this chapter, brief discussions of humanitarian and infrastructure examples are followed by several archeological and public health applications.

HUMANITARIAN

More than 3 million people have been uprooted from their homes by wars and persecutions in South Sudan, Nigeria, Niger, and Mali. The violence has caused widespread displacement both within and across national borders. Drones are being used by humanitarian aid organizations like UNHCR, the UN Refugee Agency, to plan relief responses and save lives. Aerial technology can map populations of displaced people, enabling assistance organizations to assess their needs and determine how best to get assistance to them. Aerial images also are used to evaluate environmental damage caused by the displacement. "There are numerous peaceful applications of this technology, whether in human rights, aid delivery, or settlement mapping," says Andrew Harper, head of UNHCR's Innovation Unit (UNHCR, 2016).

For the Sayan Forage refugee camp, UNHCR turned to a self-taught Nigerian drone maker, Aziz Kountche, to help understand the dynamics of the population movements. He created a simple but effective drone that looks like a model airplane. The T-800 M, which has government authorization to operate in a frontline area, captured video and still images that were converted to accurate maps of the new settlements. These efforts will be crucial in supporting the humanitarian response across an area the size of Belgium (Figure 17-1).

"With the use of the drone images, we want to provide a new level of mapping to strengthen our analysis of the context," said UNHCR External Relations Officer Benoit Moreno. The images enable the UN Refugee Agency and its partners to visualize the situation in the sites and identify and meet the needs of multiple services, including water systems, latrines, education facilities, and health care. They also aid registration of the displaced.

UNHCR recently piloted a four-propeller drone over the sprawling Goudoubo camp, home to some 9,640 refugees near the town of Dori, Burkina Faso. A video camera on the drone filmed a 12-km long and 5-km wide area to map shelters, a primary school, market, health center, and the road to Dori. According to Alpha Oumar, head of the UNHCR field office in Dori:

> Aerial views and camp mapping can help reshape our ability to respond to short-term and long-term needs. For instance, we could track the evolution of the locations of the shelters and the movements within the camps, but also document the evolution of the environmental context and the available natural resources in and around the camps. This would also help better prevent and mitigate the risks of natural disasters. (UNHCR, 2016)

TABLE 17-1 Examples of remote sensing applications.

Application	Examples
Humanitarian	Deliver medicine in remote areas with drones.
Infrastructure	Measure road surface conditions.
Archeology	Reveal hidden structures under jungle canopy.
Business Intelligence	Count cars and trucks in parking lots. Construction status.
Law Enforcement	Use drones in dangerous situations.
Journalism	Use 3-D models of terrain and buildings.
Emergency Response	Post-earthquake flyovers of disaster areas.
Search and Rescue	Sonar for detecting downed airplanes in the oceans. Use TIR to find lost hikers in parks.
Tax Revenues	Identify new construction/building alterations. Check permit status.
Telecommunications	Plan antenna placement.
Real Estate	Drones provide 3-D models and fly-through videos.
Health	Bill and Melina Gates proposal for malaria mapping in real time. Identify abandoned swimming pools as mosquito breeding grounds.
Regional Economic Activity	Night lighting correlates with gross domestic product.
Animal Migration	Thermal imaging of walruses in the Bering Sea, Alaska. Fire load and modeling.
Insurance	Flood risk. Fraudulent building damage and crop claims.
Planetary	Missions to the planets and sun.

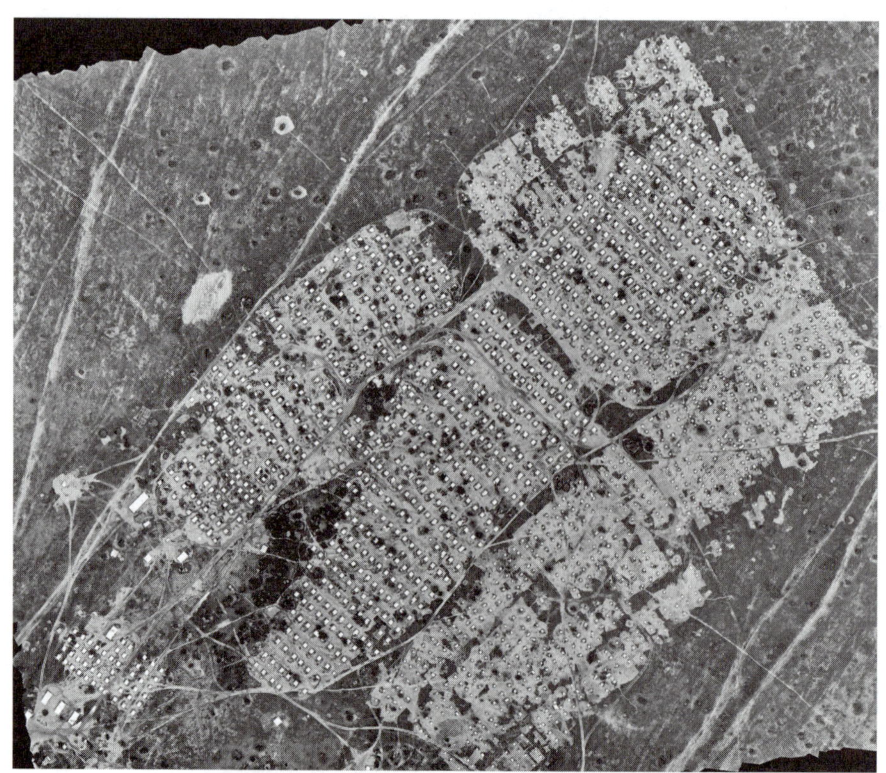

FIGURE 17-1 Mosaic of aerial images collected by the T-800 M drone. These images were used to map shelters housing displaced people at the Sayan Forage refugee camp in Niger. © UNHCR.

For UNHCR, the projects in hand are likely just the beginning. "We must recognize technological opportunities for the now, and more importantly for the future," Harper says. "This is one example of technology coming online that we must utilize for the organization. If we can harness the potential of these interventions, we will not only do our job more efficiently but have a greater impact on persons of concern."

INFRASTRUCTURE

Engineering, public safety, architecture, and urban planning use remote sensing to extract dimensions of buildings and condition of utilities, roads, and bridges. These factors are important to the viability and success of infrastructure planning and projects.

BUILDING OUTLINES

Efficiently and accurately classifying and extracting building outlines, roads, and other man-made features from high resolution imagery is an active, remote sensing research area (Theng, 2006). Spectral classification on a pixel-by-pixel basis (demonstrated in Chapter 9) of man-made objects does not achieve sufficient map accuracy for many applications. Algorithms have been developed that *segment* pixels in an image based on object points, edges, and regions. More advanced object-based image analysis programs analyze the physical characteristics (i.e., shape, size, orientation) and contextual information of features with the spectral characteristics of each pixel in the feature. Pixels that have been segmented inherently have more spectral information about an object compared to single pixels (Blaschke, 2010).

Higher spatial resolution imagery increases the probability that there will be many pixels within objects of interest (buildings, roads, sidewalks, etc.). Object-oriented classification programs will increase in accuracy as higher spatial resolution imagery is acquired over the features of interest (i.e., more pixels are within objects of interest). Vector polygons enclosing the extracted buildings enable measurements and spatial analysis to provide average size, minimum and maximum size, proximity, density, etc. Attributes can be assigned to each polygon with coordinates, area, use, and other characteristics for GIS applications. Extracting building polygons from airborne lidar point clouds is discussed by Widyaningrum and others (2019) and Albers and others (2016).

The Urban Remote Sensing Section of the journal *Remote Sensing* is a good source of articles on using remote sensing to extract buildings (see Awrangjeb and others, 2019). Applications of building extraction include real estate, city planning, homeland security, automatic solar potential estimation, change detection, and disaster management.

BRIDGE INSPECTIONS

In the United States, 76,000 railroad bridges are owned and maintained by railroad companies (GAO, 2007). Twenty percent of the bridges are too high to inspect from the ground or are over bodies of water, requiring lifts, barges, or personnel on hoists. In addition, traditional inspection methods generally require rail traffic interruptions (Unmanned Experts, 2017). UAS technology is a new way to complete bridge inspections that comply with industry standards and federal reporting requirements. UAS technology acquires detailed images from any viewpoint, including from underneath bridge structures, reduces costs and time, and generates 3-D models. Figure 17-2A shows a quadcopter hovering beneath a railroad bridge and Figure 17-2B is a close-up image of a railroad bridge joint with an enlargement of the upper hip pin.

The Tusten railroad bridge was built in 1904 and annually carries 1 million gross tons of freight across the Delaware River in southeastern New York State. Two teams of trained and certified sUAS operators came together after weeks of careful planning, site surveys, shot list preparations, and rail safety briefs to fly a complete, detailed inspection of the

A. Quadcopter sUAS adjacent to railroad bridge. Courtesy AirShark.

B. Upper hip pin on a railroad bridge imaged by sUAS. From Unmanned Experts (2017).

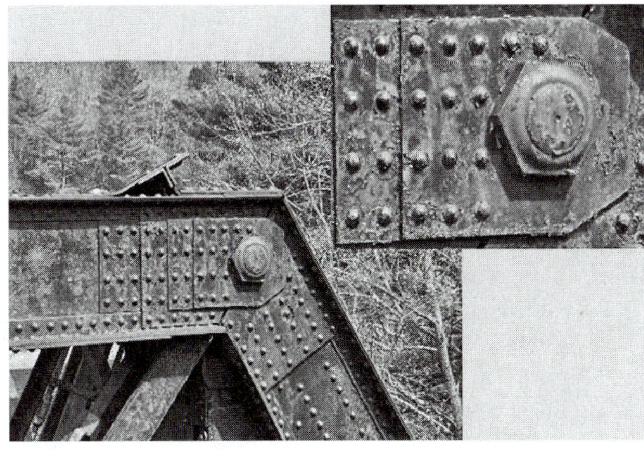

FIGURE 17-2 sUAS enables bridge inspections and detailed images of rivets and bolts.

bridge's three through-truss spans and one through-plate girder span. Unmanned Experts, a UAS training and services company based in Colorado, took several thousand HD photos plus videos of every side of the bridge deck, each pier and abutment, and the rail approaches. AirShark, a New Hampshire based team of specialist UAV pilots, brought a uniquely modified drone with a "look up" capability to photograph the state of the girders and ties underneath the decking (sUAS News, 2007).

The project acquired over 2,500 images and videos totaling 60 gigabytes of digital data. The imagery was reviewed and locations identified on the bridge structure for inclusion in a federally compliant report. UAS projects can be designed to complete other railroad tasks such as vegetation management, track inspection, documentation on encroachment, and imaging places that are precarious to visit, reducing the risk of injury for railroad workers (Senese, 2017).

ARCHEOLOGY

In the context of this book, archaeology may be defined as the study of ancient land use. Remote sensing is an accepted method for locating and mapping archaeological sites. Wernke and others (2014) advocate the integration of low altitude photogrammetry from drone and balloon platforms with mobile GIS systems to build a high spatial resolution data registry of archeological features, sites, and landscapes that approaches archeology's proven and very detailed cartographic techniques for documenting microscale phenomena (e.g., excavation pits, posthole molds, textural changes in house floors, and pottery). For this chapter, examples of satellite and aerial archeological remote sensing are provided from the Arabian Peninsula, northern Arizona, England, Central America, and Peru.

THE LOST CITY OF UBAR

Ubar is the name of a legendary "lost" city in the Arabian Peninsula that was regarded as a myth by some Western historians. In the Islamic world, however, accounts of the city go back for thousands of years. According to those accounts Ubar was the major trading center for frankincense, a fragrant balm obtained from the sap of a desert tree. Frankincense was a highly prized commodity and was a gift from the Magi to the Christ child. It was used in perfume, medicine, incense, and especially in preparing bodies for cremation. Ubar virtually monopolized the frankincense trade, and the traders became fabulously wealthy, with a luxurious and decadent lifestyle. The Koran describes it as a "many-towered city . . . whose like has not been built in the entire land." In the fourth century AD, however, the Emperor Constantine and the Roman Empire converted to Christianity; cremation went out of fashion and the lucrative frankincense market collapsed. According to the Koran, a windstorm buried the city under sand sometime between the first and fourth centuries AD.

Early in the twentieth century, desert adventurers made futile attempts to locate Ubar. An Englishman named Bertram Thomas searched for the city in the Empty Quarter of the Arabian Peninsula on the basis of Bedouin folktales. Thomas drew a partial map of the caravan route he believed the original frankincense traders must have followed south from Mesopotamia. Thomas found traces of the caravan tracks, which he recorded on maps, but was unable to find the city. In the early 1950s Wendell Phillips of Phillips Petroleum Company attempted to search for Ubar in the dune fields but was defeated by the terrain. In the 1980s a documentary filmmaker named Nicholas Clapp studied Thomas's maps plus medieval Muslim histories and ancient maps, especially those of the Greek geographer Claudius Ptolemy, who lived in Egypt around AD 150. According to these sources, Ubar lay in a region along the border between Oman and Saudi Arabia.

Clapp learned of the sand-penetration capability of SIR-A radar images and approached Jet Propulsion Laboratory (JPL) scientists for help. During the SIR-B mission, images were acquired of the area, but the data were of marginal quality. The images of the gravel plains showed traces of caravan tracks that occurred in the area identified by Thomas. It was determined that these tracks must be ancient because portions were overlapped by dunes that required centuries to reach heights up to 200 m. However, because of the narrow swath width and low quality of the SIR-B images, the tracks could not be traced for any distance. An initial field expedition found some ancient tracks and concluded that, contrary to legend and Bertram Thomas, Ubar must be located south of the dune fields. At this point, it was decided to employ Landsat images.

Figure 17-3A is a Landsat TM subscene of the small village of Ash Shisar in west-central Oman that was digitally processed by R. Blom and R. Crippen of JPL. Figure 17-3B is an interpretation map. The terrain consists of alluvial deposits and rock outcrops. Three modern roads, shown by pronounced bright lines, intersect at Ash Shisar. Of much greater significance are the very fine, bright lines that converge on Ash Shisar. These fine lines represent both modern tracks and ancient trails. The number and extent of converging trails indicate a volume of ancient traffic that greatly exceeds the present activity. The initial field expedition had passed through Ash Shisar and noted that the town's ancient well is a large sinkhole with a vast amount of rubble collapsed in its center. Groves of frankincense trees grow in the Qara Mountains 160 km to the south.

These clues led to a second expedition to Ash Shisar that excavated the rubble pile. Archaeologists excavated the foundations of a fortress complex surrounded by a massive eight-sided wall with 30-ft towers at each corner. A small city, dating earlier than 2000 BC, was located within the walls. Figure 17-4 is an archaeological reconstruction of the fortress. Part of the complex was built over a limestone cavern that collapsed and destroyed Ubar, which perhaps gave rise to the legends. The ages of the youngest artifacts recovered indicate the city collapsed between AD 300 and 400. Most of this account was provided courtesy of R. J. Blom of JPL, who participated in the two expeditions. Aspaturian (1992) and Williams (1992) describe the history and discovery of Ubar.

A. Landsat TM subscene. Very fine bright lines are modern and ancient caravan trails. Courtesy R. J. Blom and R. E. Crippen, JPL.

B. Interpretation map of Landsat TM image.

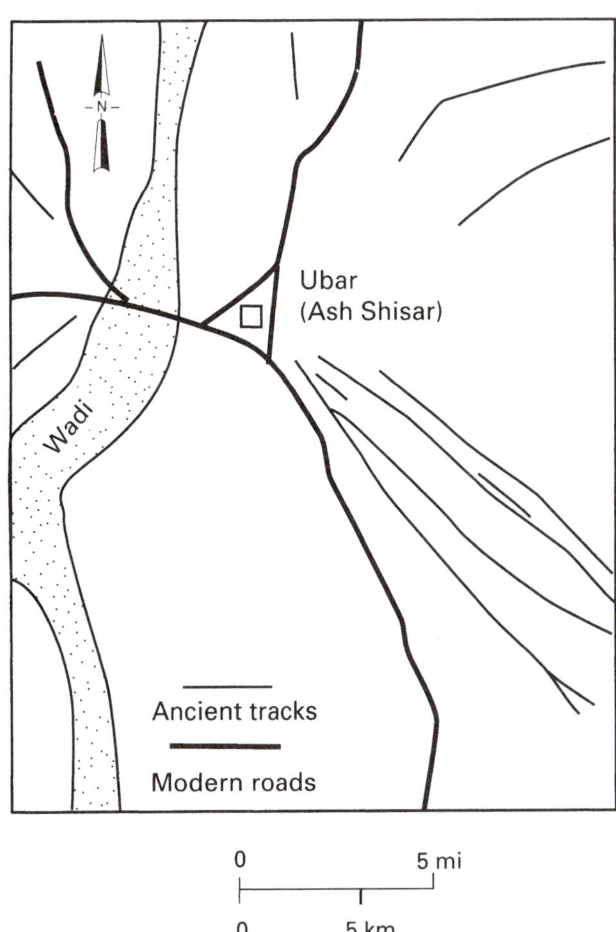

FIGURE 17-3 Satellite image and interpretation map, west-central Oman, that cover the site of Ubar and the present village of Ash Shisar. Most of the area is a rocky gravel plain.

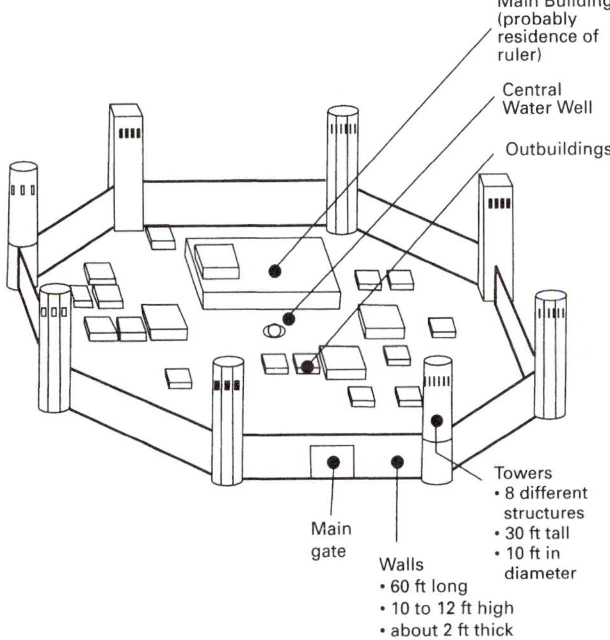

FIGURE 17-4 Archaeologic reconstruction of Ubar. Courtesy *Los Angeles Times*.

The story of Ubar has several interesting facets. The actual remains of the fortress are not expressed on the TM image. Instead, the location was indicated by the pattern of ancient trails converging on the site that were revealed on the digitally enhanced TM image (see Figure 17-3A). The early maps and accounts of Ubar, which had largely been discounted as myths, proved to be surprisingly accurate. Remote sensing provided the key evidence for locating the ancient site.

NORTHERN ARIZONA

Berlin and others (1977) documented their discovery of prehistoric Native American cornfields in northern Arizona using airborne thermal imagery and fieldwork. In 1966, NASA obtained daytime, single band thermal (LWIR, 8 to 14 μm) aerial imagery for the USGS in the eastern section of the San Francisco volcanic field (approximately 40 km northeast of Flagstaff, Arizona). The thermal sensor was a cross-track instrument with 120° angular field of view perpendicular to the flight path (see Figure 1-16). The temperature resolution of the instrument was 0.25°C.

Sunset Crater in the San Francisco volcanic field last erupted on AD 1065 or 1085 (Elson and others, 2011), covering the landscape with volcanic ash. The Sinagua people lived in the area between 1065 and 1250. Cornfields cultivated by Sinagua people have been recognized in the pattern of alternating warm and cool strips of land in TIR images. Fieldwork by Berlin and others (1977) showed that the warm strips correlate with bands of volcanic ash used as mulch on ridges and the cool strips correlate with shallow swales or depressions. The low-density ash on the ridges has a lower thermal inertia than adjacent soil in the swales and therefore warmer daytime radiant temperatures.

Within a few generations, most of the good farmland was utilized and much of it may have been depleted of nutrients essential for successful corn growing. By the early 1200s, a cooler and drier climatic cycle was underway and by 1250 the locality had been abandoned as an agricultural field. The alternating warm and cool, thermal striped pattern produced by the Sinagua agricultural practices over 750 years ago has persisted in the desert landscape to the present day.

HISTORIC ENGLAND

In Europe, many towns and roads of Roman and earlier times are now covered by agricultural fields. These sites commonly cause differences in character, moisture content, vegetation cover, and topographic relief of the overlying soils that are recognizable in remote sensing images and DEMs.

In non-wooded areas, vegetation growing above earthworks can have different vigor, type, density, and rate of growth compared with vegetation growing away from the buried archeological features. Patterns in vegetation caused by subsurface archeological features and seen on remote sensing imagery are termed crop marks. Figure 17-5 shows a grayscale NIR image of an agricultural field in Gloucestershire County, southwestern England, acquired with a camera on a kite. Roman ruins are adjacent to the field. The field supports a cereal crop (probably wheat) that has a uniform height and relatively closed canopy. A modern and ancient pattern can be seen in the image of the field. Pairs of narrow parallel lines are agricultural machinery tracks that cut into the growing crop. Many of the deep track grooves are characterized with a thin light line adjacent to a thin dark line. Because the sun's illumination is from the bottom of the figure, the side of a track depression that faces the bottom of the figure is illuminated by the sun and bright while the side that faces the top of the figure is cast in shadow and dark.

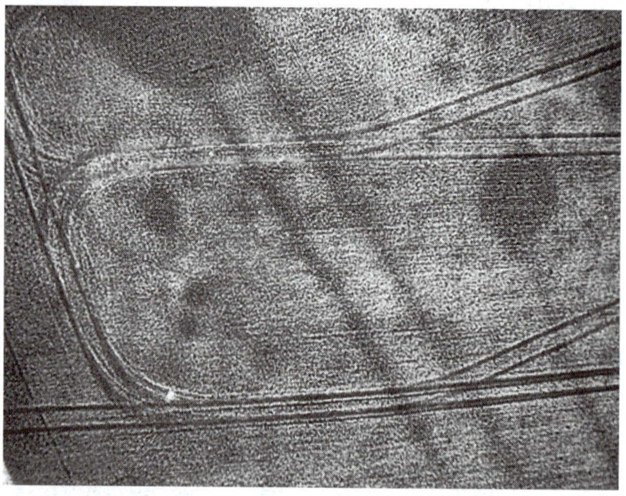

FIGURE 17-5 In an agricultural field in Gloucestershire County, England, modern agricultural machinery tracks can be seen along with crop marks. The crop marks appear as dark gray, circular to elliptical and linear features. Photo by Dr. John Wells, West Lothian Archeology Group (westlothianarchaeology.org.uk) (creativecommons.org/licenses/by/3.0/us).

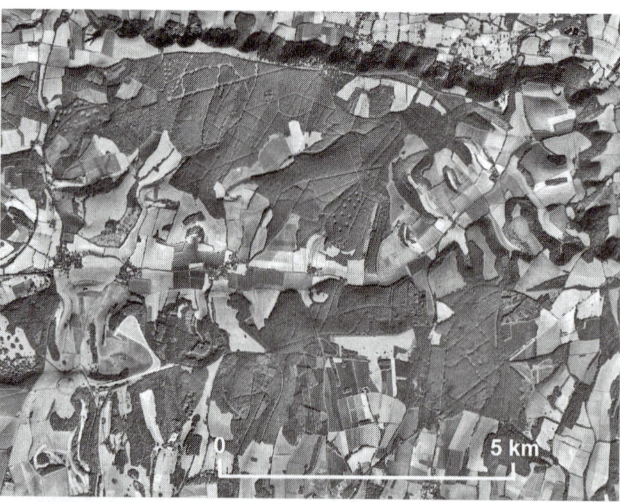

FIGURE 17-6 Sentinel-2 NIR (band 8) satellite image of the South Downs National Park, England. Woodlands have a textured, dark gray tone with roads and trails seen as narrow linear features. Agricultural fields have a smooth, light to medium gray tone. Image courtesy the European Space Agency.

In contrast to vehicle tracks, crop marks here are not associated with visible indentation of the vegetation's surface. Instead, subsurface archeological features have changed the spectral response of the homogeneous vegetation cover. The crop marks are displayed as dark gray, circular to elliptical features and diffuse lines that cross from the upper left to the lower right. The crop marks may be associated with buried foundations, middens, ditches, and trails.

England's National Mapping Programme (NMP), a component of the Heritage Lottery Fund, supported South Downs National Park Authority's Secrets of the High Woods Project. The High Woods area of West Sussex and eastern Hampshire is remarkable in terms of the range, extent and time depth of the archaeological earthworks preserved in the woodland.

The High Woods Project area is a landscape of woodland interspersed with fields and small villages (Figure 17-6). The large areas of woodland led to the preservation of extensive archeological earthworks. Subtle variations in topographic relief under the woodland canopy reveal the pattern and scale of archeological earthworks (Carpenter and others, 2016).

A key part of the project was an airborne lidar survey, which provided a highly accurate 3-D model of archaeological features surviving as earthworks or structures in open land and woodland. Figure 17-7 displays an aerial photograph, lidar DSM, lidar DTM, and lidar local relief model

(LRM) visualization of the same geographic area. The DSM reveals the texture of the woodland canopy and displays the tops of buildings. The DTM was processed to have the first and intermediate returns removed so that only the last lidar return ("bare earth") is displayed. In the DTM and LRM, lidar beams penetrated the woodland canopy and revealed subtle topographic relief, earthworks, roads, pits, and other terrain features on the Earth's surface.

The LRM visualization is mainly used for mapping archeological features. Hesse (2010) specifies how to generate an LRM. The LRM is derived by applying a low-pass filter to the original DTM, which can be considered a high frequency model. This low-pass filter approximates the large-scale landforms. A default neighborhood size of 11 by 11 is used as the low-pass filter. The low-pass filtered DTM is subtracted from the original high frequency DTM to obtain

A. Aerial photograph.

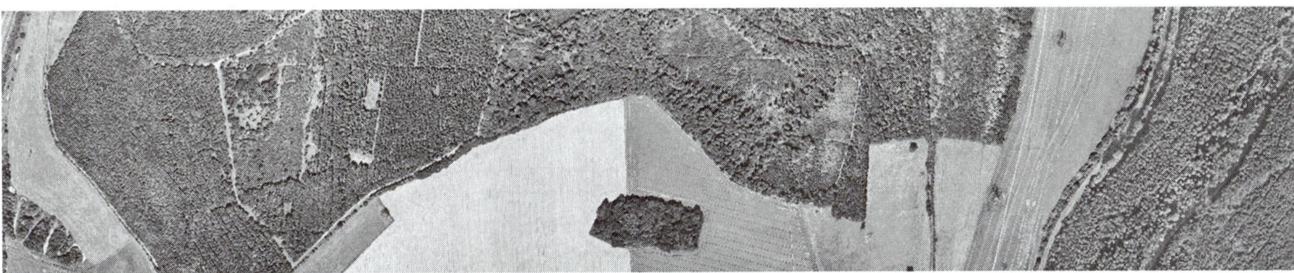

B. Lidar DSM.

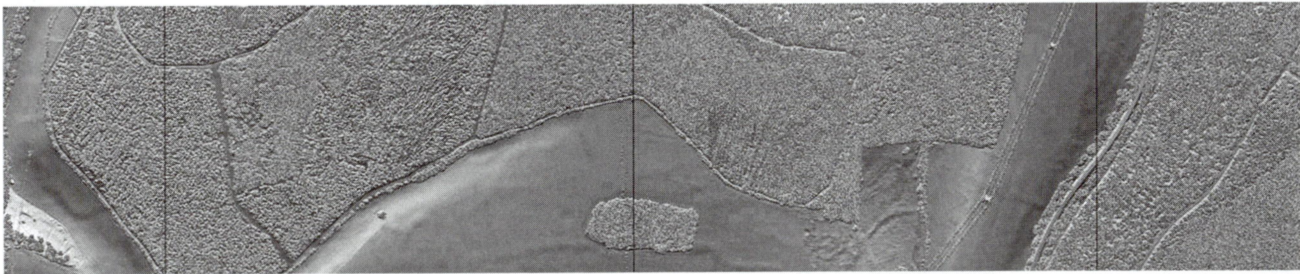

C. Lidar DTM.

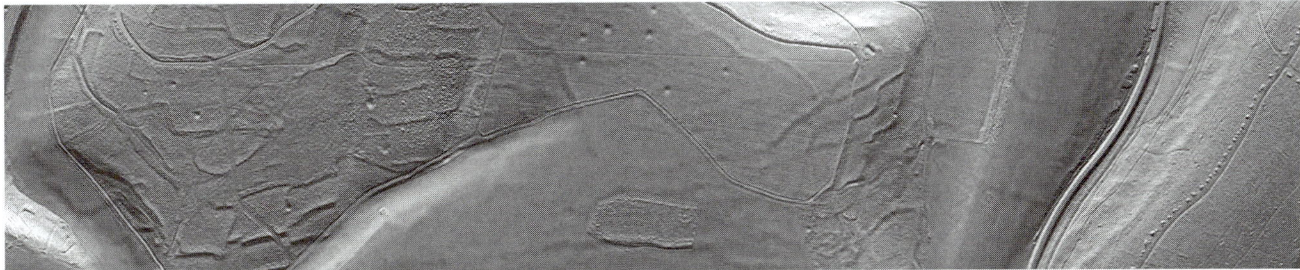

D. Lidar LRM.

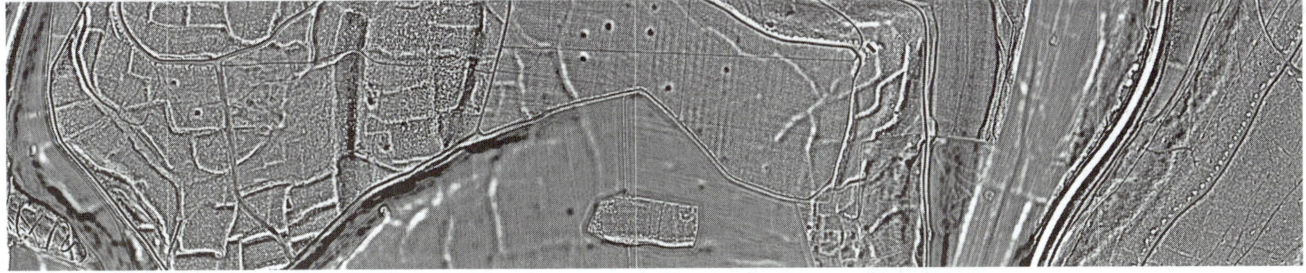

FIGURE 17-7 Airborne lidar survey of Queen Elizabeth Country Park, at the western edge of the High Woods Project area. Each image included here is of the same geographic area. Modified from Carpenter and others (2016, Figure 10). © Fugro Geospatial and South Downs National Park Authority.

local relief. Further processing removes the large-scale landforms to expose only local relief in the final LRM model.

Hesse (2010) notes that the LRM greatly enhances the visibility of small-scale, shallow topographic features irrespective of the illumination angle and allows their relative elevations, as well as their volumes, to be measured directly. The LRM raster map of local positive and negative relief variations can be used for the mapping and prospection of archaeological features such as burial mounds, linear and circular earthworks, sunken roads, agricultural terraces, ridge and furrow fields, kiln podia, and mining/quarrying sites. The data processing involved in LRM allows the regular patterning of the very low earthwork banks that are very difficult to see on the ground to be visualized, recognized, and mapped (Carpenter and others, 2016).

The Secrets of the High Woods Project documents the extensive nature of the archeological remains in the area. The lidar DTM demonstrates that prehistoric or Roman fields visualized in open terrain (Figure 17-8) are part of a much larger field system that extends into the surrounding woodland. Political, environmental, and social issues arise as the extent of archeological preservation and protection is debated. Analysis and mapping of the NMP's lidar and aerial photography survey provides a significantly enhanced level of information on the extent, form, and interpretation of archeological features in the High Woods area, especially in the woodland. Combined with other data, it will inform future planning and decisions by local communities, researchers, and managers of the historic environment (Carpenter and others, 2016).

CENTRAL AMERICA

Aerial photography, passive multispectral imaging, and ground investigations of tropical, jungle-covered terrain are hampered by swampy terrain with dense vegetation cover and persistent cloud cover. Adams and others (1981) used airborne radar, with its active beam of microwave energy, to reveal an extensive network of canals in the Yucatan Peninsula of Central America by which the Maya drained swamps for cultivation. The region is now overgrown with jungle, but on the radar images the canals are expressed as narrow gray lines surrounded by bright signatures of tropical vegetation. The canals are also covered by vegetation, but the canopy height is lower there than in the surrounding area, which accounts for the darker signature. Radar is not able to penetrate a closed jungle canopy but does collect returns from the ground and buildings if there are large openings in the canopy (see Chapters 6 and 7).

In contrast to radar, small footprint lidar is able to penetrate small gaps and spaces in the forest canopy to provide measurements of the ground and built structures. Chase and others (2011) describe a 2009 lidar survey flown over Caracol, Belize (south of the Yucatan Peninsula) and the archeological discoveries that followed. The project also utilized high resolution IKONOS satellite imagery, but the multispectral imagery did not provide any definitive detection of ancient construction under the closed jungle canopy.

The lidar instrument emitted 100,000 pulses per second. The survey was flown in April at the end of the dry season to maximize the number of leaves that would be off.

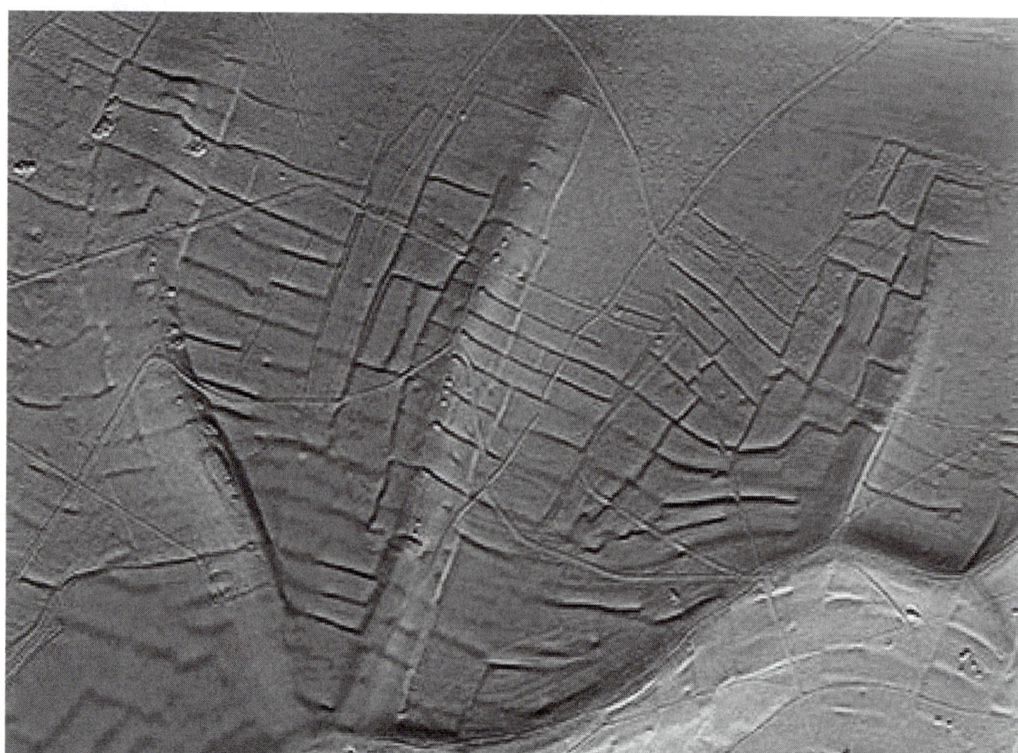

FIGURE 17-8 A lidar visualization of prehistoric or Roman fields at Lamb Lea. From Carpenter and others (2016, Figure 85). © Fugro Geospatial and South Downs National Park Authority.

A. Lidar canopy DSM.

B. Hillshade bare earth DTM.

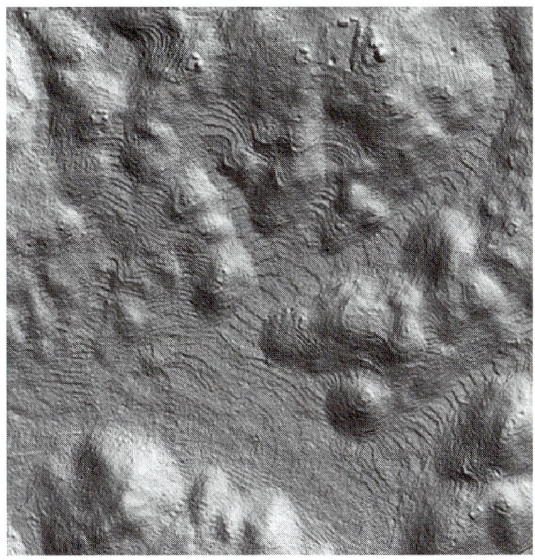

C. Rectified on-the-ground map.

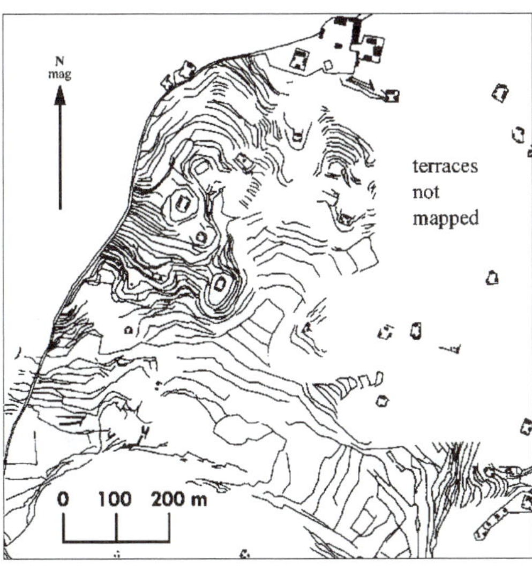

FIGURE 17-9 Comparison of DSM and DTM of Caracol, Belize. Modified from Chase and others (2011, Figure 3).

Seventy-two north–south and 60 east–west flight lines were flown over 200 km^2. The survey had 200% overlap between flight strips to enable collection of 20 laser shots per square meter. On average, 1.35 laser shots per square meter were able to reach the ground. A DSM was generated from the survey (Figure 17-9A), and as with the IKONOS imagery, ancient construction was not revealed on the DSM. A DTM was also generated with 1 m spatial resolution and a vertical accuracy of 5 to 30 cm (Figure 17-9B). The hillshade DTM clearly shows built structures, including terracing, agricultural fields, and buildings. An interpretation of built features seen on the DTM is shown in Figure 17-9C.

The height of the vegetation is measured between the first and last return of lidar energy. Plate 55A shows a lidar profile along a west to east transect (A–B). The profile is approximately 300 m long and displays an elevation range between 585 and 625 m above sea level. Plate 55B is a hillshade DTM of the area that contains the vegetation profile. On the western side of the profile a pyramid is seen on the hillshade DTM. The sides of the pyramid are seen on the profile as a triangular shape supporting vegetation that is up to 20 m high. A building about 4 to 5 m high is seen on the eastern part of the profile that supports vegetation almost 30 m in height.

The entire landscape surrounding the ancient Mayan city of Caracol, abandoned since AD 900, is imaged in the 200 km^2 survey. The lidar uncovers Caracol's urban sprawl, documenting unmapped buildings, agricultural fields, and causeways that supported a peak population of 115,000. Chase and others (2011) demonstrate that small footprint lidar provides the horizontal and vertical resolution needed to detect and accurately measure below canopy archeological features.

ADOBE STRUCTURES, PERU

Adobe is a sun-dried earth material dating back to 8000 BC that has been widely used for thousands of years, mainly in arid and semiarid lands, to build homes, communities, and temples (Masini and others, 2009). Earthen constructions have a long history in the Andean coast of South America and southern Peru, where the hyperarid climate promotes preservation of archaeological features. Masini and others (2009) used high resolution satellite imagery to detect and identify shallow and outcropping adobe walls, platforms, and terraces in the largest adobe ceremonial center in the world that was built between 200 BC and AD 400 by the Nasca people of southern Peru.

Pan-sharpened Quickbird multispectral (VNIR, 4 bands) satellite imagery acquired with a spatial resolution of 0.6 m was processed with the principal component (PC) transformation to enhance the low contrast between the archeological features and the surrounding area. Rectangular shapes that were mainly related to soil marks caused by outcropping walls, platforms, and terraces had a distinctive black tone on PC image 2 that was not seen in the rest of the subscene. The pan-sharpened red band image highlighted linear features caused by permeability differences and moisture content variations between the more compact buried adobe structures and the surrounding soil. The pan-sharpened NIR band image showed bright features related to vigorous vegetation in a paleo river channel. The vegetation index NDVI was calculated using the pan-sharpened red and NIR bands in terrain that was vegetated and revealed subtle crop marks.

A color composite of the pan-sharpened red band, panchromatic band, and NDVI images in RGB provided the most detailed archeological and paleoenvironmental information. A paleo river channel, rectilinear and intersecting linear features, crop marks, and irregular shapes were more clearly seen in the color composite compared with the grayscale bands and images. Additional processing and interpretation of satellite imagery was followed by geomagnetic surveys and ground penetrating radar to delineate the subsurface characteristics of the adobe structures, including tombs (Lasaponara and others, 2011). Noninvasive remote sensing and geophysical techniques can build prediction models useful to support decision making and to plan archeological excavations (Masini and others, 2016).

CROWDSOURCING

In the twenty-first century, the looting of archeological sites has become an economic issue. A combination of terrorist organizations looting for profit along with collectors driving demand has regulators scrambling to update antiquity laws to protect a country's archeological relics (Explorers Journal, 2016). In response, Sara Parcak, a National Geographic Fellow and founding director of the Laboratory for Global Observation at the University of Alabama, created the crowdsourcing platform GlobalXplorer (globalxplorer.org). This interactive tool enlists an army of amateur archaeologists to study high spatial resolution satellite images for signs of looting and destruction and to spur the discovery of new archeological sites (Clynes, 2017).

The imagery is subdivided into tens of millions of small tiles and displayed to users in random order without the ability to navigate or pan out (Figure 17-10). The tiles do not contain any location references or coordinate information (Clynes, 2017). This design prevents looters from using the platform to find new sites. The platform provides a tutorial to enable participants to effectively look for indications of looting, construction, or other encroachment; a training video; and numerous images of looting to help users improve their

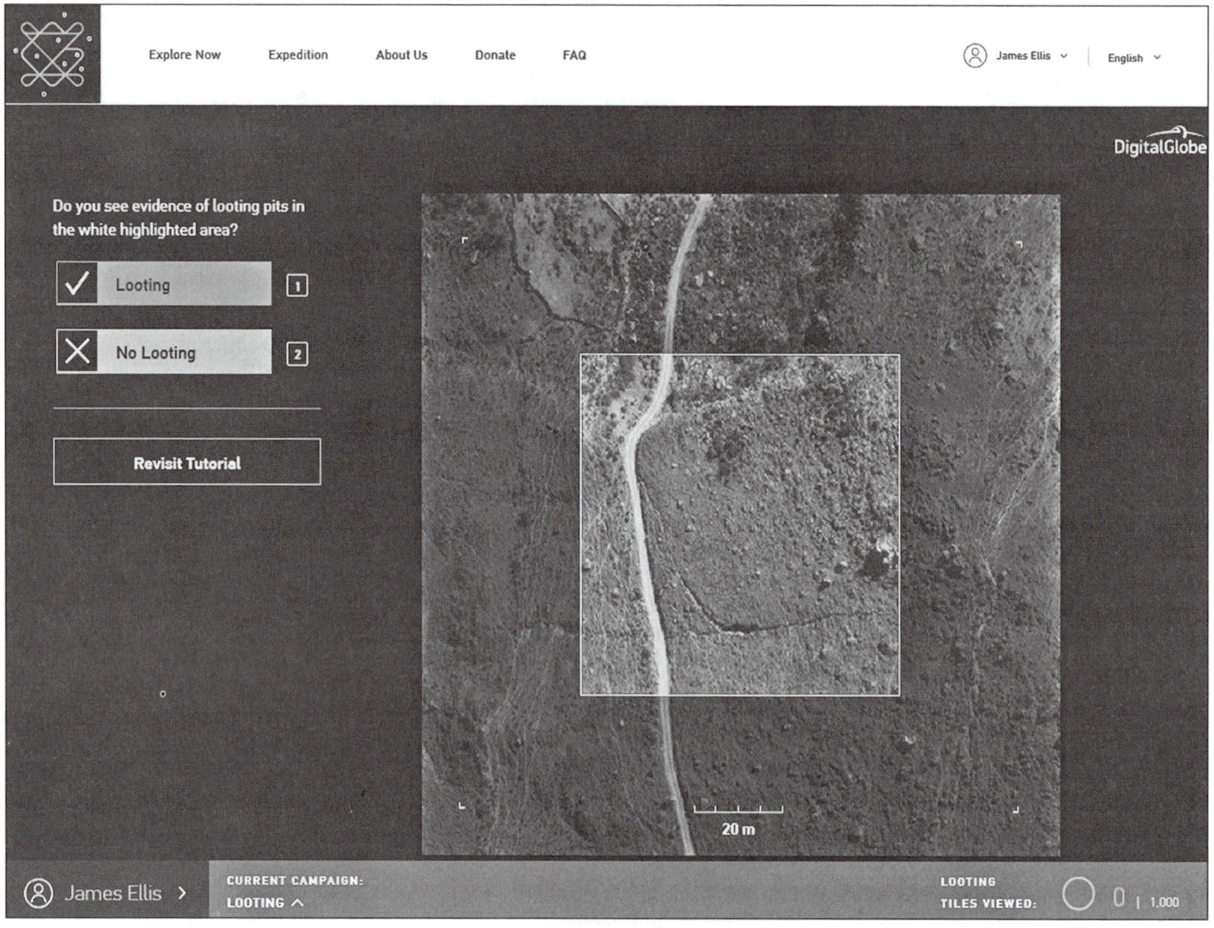

FIGURE 17-10 GlobalXplorer website showing a tile of a location in Peru for crowdsourced evaluation of looting. Courtesy GlobalXplorer (globalxplorer.org).

understanding of different patterns of looting seen on satellite imagery and to improve the accuracy of user mapping.

The GlobalXplorer crowdsourcing project started in January 2017. DigitalGlobe provided imagery that covers approximately 100,000 square miles of farms, towns, and countryside in Peru. By April 2017, 10,000,000 image tiles had been evaluated by 45,000 amateur archeologists. The users found almost 3,000 archaeological features in one small area north of Lima, Peru (Williams, 2017). Parcak hopes GlobalXplorer will help catalyze a modern age of discovery and preservation—one that will help preserve remnants of the ancient world and our human heritage.

In Figure 17-11, a high spatial resolution satellite image captures the extent of looting in the ancient site of South Dashur in Egypt. The circular features on the satellite image are looting pits characterized by a dark center (the pit hole) surrounded by a mound of soil excavated from the pit. A ground photograph of a looting pit from South Dashur is shown in Figure 17-12.

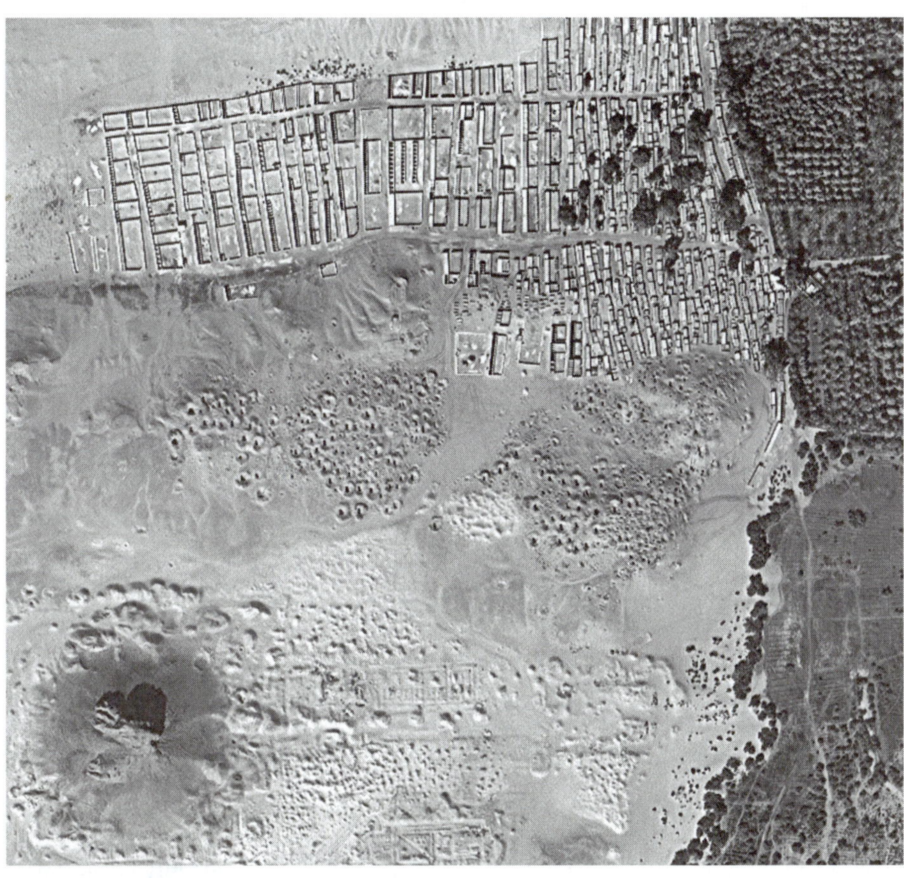

Figure 17-11 Satellite image of looting pits, South Dashur, Egypt. Satellite image © 2019 Maxar Technologies.

Figure 17-12 A looting pit, South Dashur, Egypt. Courtesy Sarah Parcak.

PUBLIC HEALTH

Remote sensing has been used for more than 50 years to examine the environmental factors associated with airborne, vectorborne, soilborne, and waterborne diseases. A special issue of the *ISPRS International Journal of Geo-Information* (Faruque, 2018) contains 15 papers covering applications of remote sensing and geospatial technologies to public health, including air pollution, vectorborne diseases, water quality, demographic factors, and emerging analytical techniques. Remote sensing enables mapping the environment for public health applications across a wide range of spatial, spectral, radiometric, and temporal resolutions. Map scales that range from global to neighborhood and change-detection studies that span decades are made possible with remote sensing.

The application of remote sensing to public health has commonly involved the NDVI and moisture indices and image spectral classification to map discrete landscape patches that provide habitat to disease vectors (Kelly and others, 2011). Kelly and others (2011) advocate object-based image analysis that first segments an image into objects based on spatially connected pixels with similar spectral properties, and then classifies these objects based on their spectral, spatial, and contextual attributes. The object-based approach leads to an increase in classification accuracy, especially with higher spatial resolution imagery. Hartfield and others (2011) integrated 1-m multispectral imagery and lidar DSMs and DTMs (with a density of 1 point · m^{-2}) in urban Tucson, Arizona, to improve land use/land cover (LULC) classification and understanding of the relationship between neighborhood composition and adult mosquito abundance data.

Research using remote sensing to improve understanding of malaria, air quality, and pesticide use are summarized below.

MALARIA

Malaria is transmitted by mosquitoes. The geographic and temporal distribution of malaria is determined by climate and other factors such as land cover, local land use practices, and the extent of deforestation.

Land Use/Land Cover Maps

Stefani and others (2013) note that the challenge when studying malaria is to identify all the natural factors (such as seasonality, rainfall, temperature, humidity, surface water, and vegetation) and anthropogenic elements (such as agriculture, irrigation, deforestation, urbanization, and movements of populations) of the study area. Linking these factors and elements with either the incidence of disease or the presence of vectors, while also integrating temporal and spatial variations, would enable the identification of risk factors from the set of possible environmental parameters. They reviewed 17 studies covering nine countries in the Amazon forest that link malaria and LULC characteristics to help identify malaria risk factors. The nine countries accounted for 89% of all malaria cases reported in the Americas in 2008.

The 17 studies used different satellite data, including Landsat TM, AVHRR, SPOT 5, Quickbird, JERS-1 SAR, and MERIS ENVISAT. Stefani and others (2013) summarized eight LULC classes associated with malaria risk, including water and wetlands, savannah and steppes, agricultural areas, dense forest, and deforestation. Deforestation was associated with high malaria risk. A generic model of the relationship between deforestation and malaria transmission risk emerged from the literature (Figure 17-13). The importance of time and the type of LULC landscape fragmentation that follows deforestation for malaria transmission risk is shown in the model. After deforestation, the conversion of a cleared area to urban or pasture provides less landscape fragmentation and less risk of malaria compared with a cleared area that evolves into secondary forest growth and small agricultural plots.

The authors recommend refinement and standardization of LULC classes associated with malaria to enable different studies to be directly compared and integrated. When landscape indicators of higher malaria transmission risk have been identified and mapped, prevention measures and treatment can be more efficiently deployed into remote communities.

DEMs and Drainage Maps

Moss and others (2011) used remote sensing to target hotspots of malaria transmission to support more efficient, cost-effective, and accelerated malaria control efforts in southern Zambia. Overall the prevalence of malaria has decreased since 2004 due to the introduction of medicine and long-lasting insecticidal nets. The decline in malaria cases has led to an interest in malaria elimination and eradication. Such a project would require identification of individual, household, and environmental factors associated with the transmission of malaria. Moss and others (2011) determined that remote sensing could identify environmental risk factors with sufficient detail and validity to support elimination of malaria.

Satellite images were used to construct a sampling framework for the selection of households to be interviewed and sampled for malaria. The SRTM DEM with 90 m cells was used to generate topography and to model water flow in order to calculate an index of topographic wetness. The

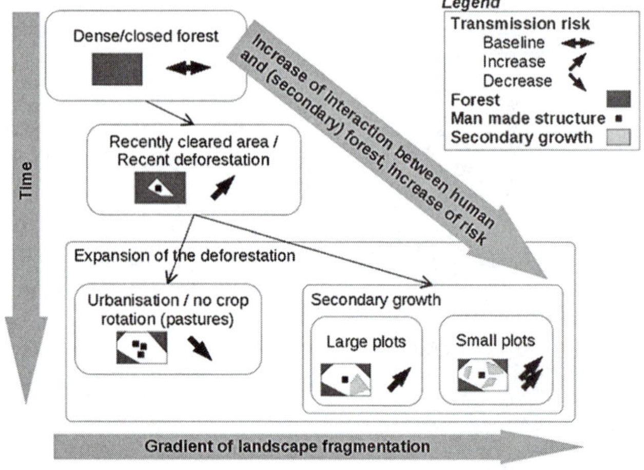

FIGURE 17-13 Landscape indicators that may increase or decrease malaria transmission risk as a function of time and landscape fragmentation. From Stefani and others (2013, Figure 3).

modeled drainage network was categorized according to the Strahler stream order classification. The classification is hierarchical. For example, a second order stream is formed where two upstream, first order streams join. A third order stream is formed where two second order streams join. The aspect, distance from streams, slope position, and landform type for surveyed households were derived from the DEM and drainage network.

The study utilized a 0.6 m pan-sharpened Quickbird satellite image to identify 8,751 households in the study area. For over 21 months, 768 individuals from 128 randomly selected households in the study area were surveyed for malaria. Of the surveyed households, 117 individuals (15.2%) tested positive for malaria. Lower elevation, lesser slope, and proximity to third order streams were associated with increased risk of malaria (**Digital Image 17-1** 🌐). Yellow and orange areas on **Digital Image 17-1** 🌐 are modeled as high risk while blue shading indicates regions of low risk based on prevalent cases. Based on the malaria risk map, targeting households in the top 80th percentile of malaria risk would require malaria control interventions directed to only 24% of the households, enabling a more efficient use of resources for malaria elimination.

Air Quality

Satellite data of atmospheric pollutants are employed for estimating emissions, tracking pollutant plumes, supporting air quality forecasting activities, providing evidence for "exceptional event" declarations, monitoring regional long-term trends, and evaluating air quality model output (Duncan and others, 2014). Respiratory and cardiovascular health has increased risk with increased particulate matter (PM) concentration in the air. $PM_{2.5}$ refers to atmospheric particulate matter that has a diameter of less than 2.5 μm (about 3% the diameter of a human hair). $PM_{2.5}$ particles are associated with combustion (from power plants, motor vehicles, and wood burning), volcanic eruptions, dust storms, and interaction between gases and particles in the atmosphere. $PM_{2.5}$ particles tend to stay longer in the air than heavier particles. $PM_{2.5}$ particles are able to penetrate deep into the lungs and some may even enter the circulatory system. Studies have linked exposure to PM with premature death in people with heart and lung disease, aggravated asthma, decreased lung function, and difficulty breathing (US EPA, 2018).

The aerosol optical depth (AOD) sensor onboard a variety of satellites, including MODIS, Suomi NPP, OMI, MISR, and GOES, is used to assess $PM_{2.5}$ concentrations. The AOD instrument measures the degree to which aerosols prevent the transmission of light by absorption or scattering of light through the entire vertical column of the atmosphere from the ground to the satellite's sensor. Duncan and others (2014) note that:

1. Satellite observations complement data collected by surface air quality (AQ) monitors,
2. Satellite data provide an overview of the regional buildup and the long-range transport of pollution, which can degrade AQ far downwind,
3. Satellite data has limitations (e.g., the lack of information on the vertical distribution and chemical composition of the aerosols, and gaps in spatial coverage due to clouds), and
4. There is a complicated relationship between AOD, relative humidity, and $PM_{2.5}$, meaning that high AOD values do not necessarily translate to high surface $PM_{2.5}$ levels.

Mirzaei and others (2018) observe that there is not a perfect correlation between AOD and $PM_{2.5}$, because $PM_{2.5}$ is a measure of particle mass that concentrates near the surface, while AOD represents the aerosol content distributed within a column of air from the Earth to the top of the atmosphere.

Dust Storms

A 2004 to 2008 collaboration between NASA's Applied Sciences Program, Center for Disease Control, University of New Mexico, and University of Arizona focused on using remote sensing data to study airborne dust in the Southwest United States (Morain and Sprigg, 2008). The project had three components:

1. Improving forecasts of fine particulates by assimilating monthly masks of PM into NOAA's National Weather Service (NWS) Dust Regional Atmospheric Model (DREAM) to simulate and predict the onset of dust storms and the 3-D-size concentration of the resulting airborne dust clouds,
2. Incorporating MODIS AOD data with modeled fine particulates from DREAM into the EPA's Community Multiscale Air Quality Model (CMAQ) program to estimate aerosol concentrations over the four corners region of Utah, Arizona, Colorado, and New Mexico, and
3. Integrating the products from these data and models into the New Mexico public health tracking system for analysis and dissemination.

The PM sources were derived from a MODIS 16-day NDVI land cover product combined with an improved land cover classification algorithm to inventory land patterns that alternate between cropped and barren ground. The project demonstrated that the timing of dust storms in the desert Southwest could be accurately forecast with 48 hours notice, 67% of the time.

Figure 17-14 is a MODIS image of a dust storm that struck north-central Texas on April 6, 2006 (NASA Visible Earth, 2006). In this image, white clouds are located in the northwest and along the eastern margin. Dust plumes appear as light gray tones and were carried toward the east–northeast by winds. The dust plumes originated from burn scars left by fires in March 2006. The approximate locations of the burn scars are shown by the black dots in Figure 17-14. Fires and dust storms in this region had been exacerbated by drought and low humidity, which created a large stockpile of dry vegetation (brush, grasses, and trees). The drought conditions caused more than 10,000 wildfires that burned at least 1.6 million hectares (4 million acres) between December 2005 and April 2006. The smoke and the dust that was transported downwind from the burn scars caused major health and navigation concerns.

Wildfire Smoke

Wildfires burned more than 18 million acres (7.3 million hectares) of land in 2017 and 2018 across the western United States, resulting in record levels of damages (NEON, 2019). In the Alberta province of western Canada, almost 900,000 hectares of area burned in 2019, nearly four times the five-year average (CTV News Edmonton, 2019). Wildfires need fuel to grow and spread. This fuel is primarily provided by biomass, or organic matter in the form of vegetation growing in the ecosystem. The amount, type, and moisture content of the vegetation influence how large a wildfire can get and how fast it can spread. The type of fuel the fire burns is also directly related to the chemical composition of the particulates and aerosol emissions it produces.

A 2018 study integrated airborne remote sensing data acquired *during* and *after* four wildfires in northwestern United States to characterize biomass availability and smoke plumes (NEON, 2019). Observations were carried out during the peak activity of the wildfires with an airborne platform flying through well-formed smoke plumes to measure total carbon flux as well as the chemical constituents of the plume, including carbon monoxide (CO), ammonia (NH_3), and ozone (O_3). The NEON platform (see Chapter 8 for a system description) flew after the fire was extinguished to measure the extent of the burned areas and to use data col-

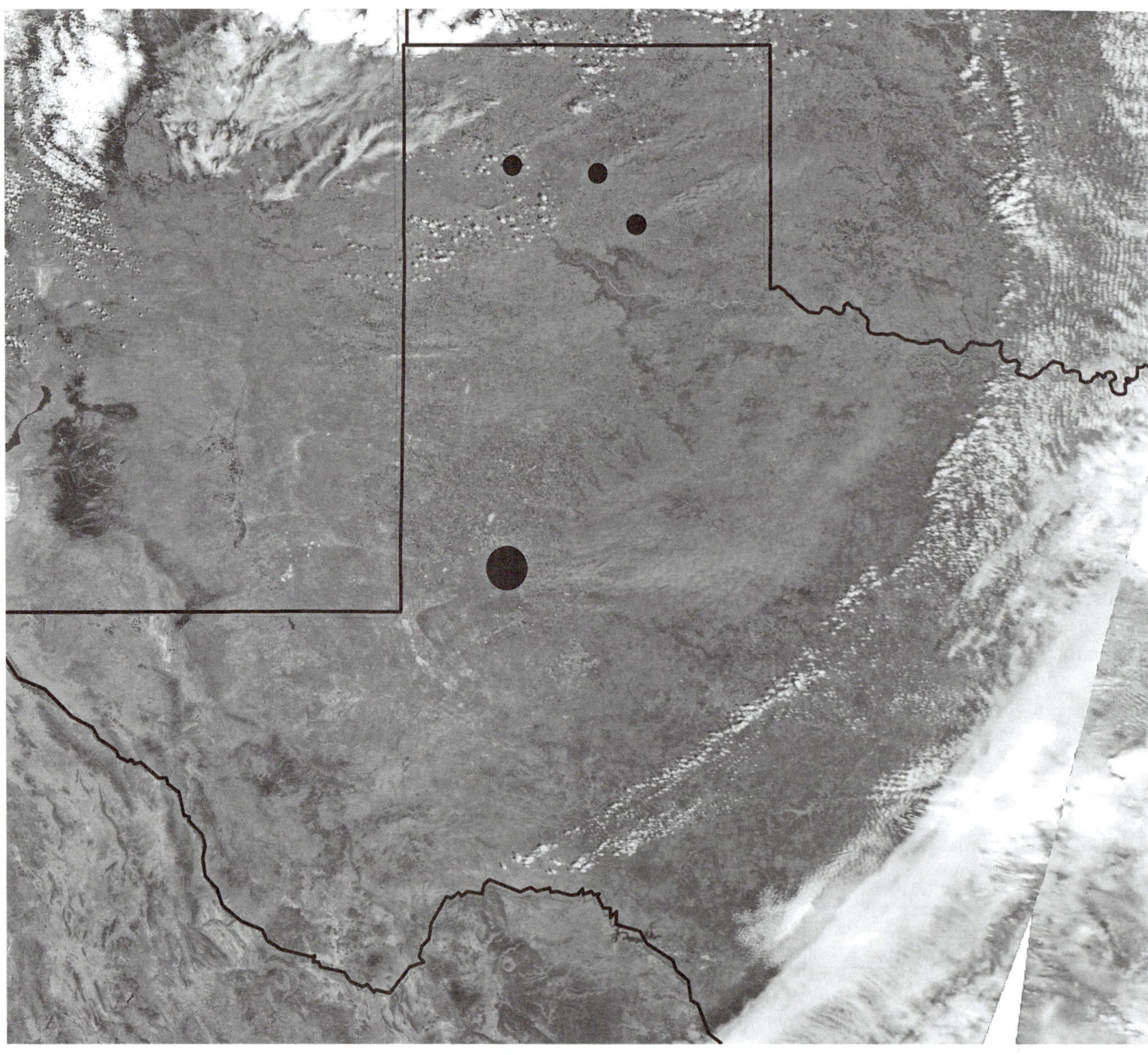

FIGURE 17-14 MODIS image of dust plumes, Texas. The approximate locations of burn scars that were the source of the dust plumes are shown with black dots. From NASA Visible Earth (2006).

lected over unburned vegetation on the land surrounding the fire-impacted areas to extrapolate the total biomass contained in each area prior to the wildfires. The research questions being asked in this ongoing study are:

- How do ecosystem characteristics and total available biomass impact the growth and spread of wildfires?
- How is the chemical composition of smoke plumes produced by wildfires related to the volume and type of biomass found in the ecosystem?

NASA and NOAA have a joint campaign named Fire Influence on Regional to Global Environments and Air Quality (FIREX-AQ) that is targeting broad questions about the chemical and physical properties of fire smoke, how it is measured, and how it changes from the moment of combustion to its final fate hundreds or thousands of miles downwind (NASA, 2019a). Flights over wildfires in the western United States sample smoke plumes and their changing chemistry along with weather dynamics. Small-scale plume dynamics are measured with flights over small agricultural fires in the southeastern United States that are often undetected by satellites but are closely situated to and impact downwind population centers. The aircraft will observe plume injection heights and integrate these observations with other data to improve smoke chemistry models.

NASA's Terra satellite carries the MISR instrument (system characteristics are described in Chapter 11). MISR measurements during the 2018 Camp Fire showed the plume was made of large, nonspherical particles over Paradise, California, an indication that buildings were burning. Smoke particles from the burning of the surrounding forest, on the other hand, were smaller and mostly spherical. Research has established that smoke from a burning building leads to larger and more irregularly shaped particles than particles found in smoke from wildfires. MISR's measurements also showed the fire had lofted smoke nearly 2 miles into the atmosphere and carried it about 180 miles downwind, toward the Pacific Ocean (NASA, 2019b).

Satellites have limitations for fire detection due to the following: heat signatures are averaged over pixels, which make accurate fire location and size difficult to determine, smoke above the fire can diminish the signal, and smoldering fires might not radiate as much energy as flaming fires (NASA, 2019b). To overcome these limitations, the US Forest Service deploys aircraft with TIR sensors that image a 6-mi swath of land and can map 300,000 acres $\cdot$ hr^{-1}. From an altitude of 10,000 ft, the sensor can detect a hotspot just 6 in across, and place it within 12.5 ft on a map. The data from each pass are recorded, compressed, and immediately downlinked to an FTP site, where analysts create maps that firefighters can access directly on a phone or tablet in the field. The TIR instruments are flown at night when there is no sun glint to compromise the measurements, the background is cooler, and the fires are less aggressive.

The NWS integrates remote sensing data collected by satellite, aircraft, and ground instruments to support atmospheric chemical transport models and to develop air quality forecasts (airquality.weather.gov). Several maps are generated by NWS on an hourly basis, including surface smoke, vertical smoke integration, surface dust, and vertical dust integration. **Digital Image 17-2** 🌐 shows a surface smoke map for 3:00 PM (EST) on October 24, 2019. The US EPA assigns an air quality index category of "Unhealthy" for $PM_{2.5}$ concentrations equal to or greater than 55.5 μm $\cdot$ m^{-3}. On October 24, 2019 the northwest United States and southwest Canada have several areas with unhealthy levels of surface smoke.

Lentile and others (2006) provide a comprehensive review of current and potential remote sensing methods used to assess fire behavior and effects and ecological responses to fire. They clarify the terminology to facilitate development and interpretation of comprehensible and defensible remote sensing products, present the potential and limitations of a variety of approaches for remotely measuring active fires and their post-fire ecological effects, and discuss challenges and future directions of fire-related remote sensing research. Koman and others (2019) declare that today and into the future the real value of satellite sensing for air quality lies in its integration with in-situ measurements to adjust and calibrate atmospheric chemical transport models such as the EPA's CMAQ.

Pesticide Use

Maxwell (2011) used Landsat images acquired and mapped during the growing season in the Central Valley of California to improve the spatial resolution of pesticide use maps. The following is a brief summary of her findings.

Exposure to pesticides has been associated with increased risk of adverse health effects such as cancer, birth defects, and Parkinson's disease. In California paraquat is a highly toxic herbicide widely used to control weeds in crop fields such as orchards and vineyards and to defoliate green vegetation on crops (primarily cotton) prior to harvest. During the years 1991 through 2009, paraquat was applied to between 423,000 and 736,000 hectares annually in California.

California maintains a comprehensive pesticide use database but the data are only recorded on a coarse geographic scale of an approximately 2.6 km^2 area (US Public Land Survey sections of approximately 1 mi^2). This spatial resolution is not adequate for residential-level exposure assessment. In addition, the pesticide use database is produced once every 7 to 10 years based on field observations at one time during the growing season, so crop rotation and multi-crop fields are frequently misclassified. Landsat imagery for 10 dates during the 1994 growing season were processed to provide NDVI maps that displayed the stage of seasonal crop growth (phenology) in agricultural fields. A phenological, spectral signature library was used to map crop growth on the 10 Landsat images during the growing season. Principal components and object-based image classification, along with manual editing, were used to define crop field boundaries.

Landsat successfully downscaled the pesticide use maps from section-level to crop field-level, enabling significant improvement in residential-scale, pesticide exposure estimation. Landsat identified specific fields or group of fields where paraquat was applied and also identified potential errors in the pesticide use database. Grapes and cotton fields were spectrally distinguished. The study identified the necessity for cloud-free, multispectral imagery during the growing season, a phenological spectral library, and higher spatial resolution imagery for mapping pesticide use on crops growing within small fields.

QUESTIONS

1. Choose five applications from Table 17-1 and provide a new example of how each application uses remote sensing.
2. How is a lidar LRM created?
3. What surface features are enhanced with a lidar LRM?
4. Why is lidar more effective compared with radar for revealing ancient construction under a tropical forest canopy?
5. Why would an interactive, archeological crowdsourcing website only provide ungeoreferenced tiles randomly presented to their users?
6. How does the AOD instrument assess $PM_{2.5}$ concentrations?
7. How are smoke particles from a burning building different than those released from a wildfire?
8. What would be some characteristics of a spectral signature library that track seasonal crop phenology?

REFERENCES

Adams, R. E.W., W. E. Brown, and T. Culbert. 1981. Radar mapping, archaeology, and ancient Maya land use. *Science*, 213, 1,457–1,463.

Albers, B., K. M. Kada, and A. Wichmann, 2016. Automatic Extraction and Regularization of Building Outlines from Airborne Lidar Point Clouds. The International Archives of the Photogrammetry, Remote Sensing, and Spatial Information Services, Vol. XLI-B3 (pp. 555–560). XXIII ISPRS Congress, July 12–19, Prague, Czech Republic.

Aspaturian, H. 1992. The road to Ubar. *Caltech News*, 26, 1–8.

Awrangjeb, M., X. Hu, B. Yang, and J. Tian (Eds.). 2019. Special issue "Remote sensing based building extraction": *Remote Sensing*, https://www.mdpi.com/journal/remotesensing/special_issues/Building_Detection.

Berlin, G. L., J. R. Ambler, R. H. Hevly, and G. G. Schaber. 1977. Identification of a Sinagua agricultural field by aerial thermography, soil chemistry, pollen/plant analysis, and archaeology. *American Antiquity*, 42, 588–600.

Blaschke, T. 2010. Object base image analysis for remote sensing. *Journal of Photogrammetry and Remote Sensing*, 65(1), 2–16.

Carpenter, E., F. Small, K. Truscoe, and C. Royall. 2016. *South Downs National Park: The High Woods from above NMP* (Research Report Series No. 14-2016). London, England: Historic England.

Chase, A. F., D. Z. Chase, J. F. Weishampel, J. B. Drake, R. L. Shrestha, K. C. Slatton, J. J. Awe, and W. E. Carter. 2011. Airborne LiDAR, archaeology, and the ancient Maya landscape at Caracol, Belize. *Journal of Archaeological Science*, 38(2), 387–398. http://doi.org/10.1016/j.jas.2010.09.018

Clynes, T. 2017, January 30. How to Become a Space Archeologist. National Geographic. Watch. http://news.nationalgeographic.com/2017/01/archaeologists-parcak-globalxplorer-looting-ted-prize (accessed January 2018).

CTV News Edmonton. 2019, October 22. Area Burned in 2019 Wildfires Was Nearly 4 Times the 5-Year Average. https://edmonton.ctvnews.ca/area-burned-in-2019-wildfires-was-nearly-4-times-the-5-year-average-1.4649803 (accessed October 2019).

Duncan, B. N., A. I. Prados, L. N. Lamsal, Y. Liu, D. G. Streets, P. Gupta, E. Hilsenrath, R. A. Kahn, J. E. Nielsen, A. J. Beyersdorf, S. P. Burton, A. M. Fiore, J. Fishman, D. K. Henze, C. A. Hostetler, N. A. Krotkov, P. Lee, M. Lin, S. Pawson, G. Pfister, K. E. Pickering, R. B. Pierce, Y. Yoshida, and L. D. Ziemba. 2014. Satellite data of atmospheric pollution for U.S. air quality applications: Examples of applications, summary of data end-user resources, answers to FAQs, and common mistakes to avoid. *Atmospheric Environment*, 94, 647–662. http://dx.doi.org/10.1016/j.atmosenv.2014.05.061.

Elson, M. D., M. H. Ort, P. R. Sheppard, T. L. Samples, K. C. Anderson, and E. M. May. 2011. A.D. 1064 No More? Re-Dating the Eruption of Sunset Crater Volcano, Northern Arizona. 76th Annual Meeting of the Society for American Archaeology. Sacramento, CA. https://www.academia.edu/11795079/A.D._1064_No_More_Re-Dating_the_Eruption_of_Sunset_Crater_Volcano_Northern_Arizona?auto=download (accessed January 2018).

Explorers Journal. 2016, May 19. Lust for Loot: Collecting is Driving the Demand for Plunder. National Geographic. Online Explorers Journal. http://voices.nationalgeographic.org/2016/05/19/lust-for-loot-collecting-antiquities-looting-egypt (accessed January 2018).

Faruque, F. S. (Ed.). 2018. Remote sensing and geospatial technologies in public health. *ISPRS International Journal of Geo-Information*, 7(8), 303. http://www.mdpi.com/2220-9964/7/8/303 (accessed October 2019).

GAO. 2007, August. *Railroad Bridges and Tunnels: Report to Congressional Requesters* (GAO-07-770). Washington, DC: US Government Accountability Office.

Hartfield, K. A., K. I. Landau, and W. J. D. van Leeuwen. 2011. Fusion of high resolution aerial multispectral and lidar data: Land cover in the context of urban mosquito habitat. *Remote Sensing*, 3, 2,364–2,383. doi:10.3390/rs3112364

Hesse, R. 2010. Lidar-derived local relief models—a new tool for archaeological prospection. *Archaeological Prospection*, 17, 67–72.

Kelly, M., S. D. Blanchard, E. Kersten, and K. Koy. 2011. Terrestrial remotely sensed imagery in support of public health: New avenues of research using object-based image analysis. *Remote Sensing*, 3, 2,321–2,345. doi:10.3390/rs3112321

Koman, P. D., M. Billmire, K. R. Baker, R. de Majo, F. J. Anderson, S. Hoshiko, B. J. Thelen, and N. H. F. French. 2019. Mapping modeled exposure of wildland fire smoke for human health studies in California. *Atmosphere*, Special Issue: Air Quality and Smoke Management, 10(6), 308. http://doi.org/10.3390/atmos10060308.

Lasaponara, R., N. Masini, E. Rizzo, and G. Orefici. 2011. New discoveries in the Piramide Naranjada in Cahuachi (Peru) using satellite, ground probing radar and magnetic investigations. *Journal of Archeological Science*, 38, 2,031–2,039. doi:10.1016/j.jas.2010.12.010

Lentile, L. B., S. A. Holden, A. M. S. Smith, M. J. Falkowski, A. T. Hudak, P. Morgan, S. A. Lewis, P. E. Gessler, and N. C. Benson. 2006. Remote sensing techniques to assess fire characteristics and post-fire effects. *International Journal of Wildland Fire*, 15(3), 319–345.

Masini, N., R. Lasaponara, and G. Orefici. 2009. Addressing the challenge of detecting archeological adobe structures in southern Peru using QuickBird imagery. *Journal of Cultural Heritage*, 10S, e3–e9. doi:10.1016/j.culher.2009.10.005

Masini, N., E. Rizzo, L. Capozzoli, G. Leucci, A. Pecci, G. Romano, M. Sileo, and R. Lasaponara. 2016. Remote sensing and geophysics for the study of the human past in the Nasca drainage. In R. Lasaponara, N. Masini, and G. Orefici (Eds.), *The Ancient Nasca World: New Insights from Science and Anthropology* (chapter 20, pp. 469–527). Springer International.

Maxwell, S. 2011. Downscaling pesticide use data to the crop field level in California using Landsat satellite imagery: Paraquat case study. *Remote Sensing*, 3, 1,805–1,816. doi:10.3390/rs3091805

Mirzaei, M., S. Bertazzon, and I. Couloigner. 2018. Modeling wildfire smoke pollution by integrating land use regression and remote sensing data: Regional multi-temporal estimates for public health and exposure models. *Atmosphere*, Special Issue: Impacts of Air Pollution on Human Health, 9(9), 335. http://doi.org/10.3390/atmos9090335.

Morain, S. A., and W. A. Sprigg. 2008. Public Heath Applications in Remote Sensing: Final Benchmark Report (February 2004–September 2008). Agreement NNSO4AA19A. http://phairs.unm.edu/publ/Final%20Benchmark%20v14b%209-30-08.pdf (accessed October 2019).

Moss, W. J., H. Hamapumbu, T. Kobayashi, T. Shields, A. Kamanga, J. Clennon, S. Mharakurwa, P. E. Thuma, and G. Glass. 2011. Use of remote sensing to identify spatial risk factors for malaria in a region of declining transmission: A cross-sectional and longitudinal community survey. *Malaria Journal*, 10:163. http://malariajournal.biomedcentral.com/articles/10.1186/1475-2875-10-163 (accessed October 2019.

NASA. 2019a, July 22. Tracking Smoke from Fires to Improve Air Quality Forecasting. http://www.nasa.gov/feature/goddard/2019/nasa-tracks-wildfires-from-above-to-aid-fire-fighters-below (accessed October 2019).

NASA. 2019b, July 29. NASA Tracks Wildfires from Above to Aid Firefighters Below. http://www.nasa.gov/feature/goddard/2019/tracking-smoke-from-fires-to-improve-air-quality-forecasting (accessed October 2019).

NASA Visible Earth. 2006, April 17. Dust Storm in Texas. MODIS Image of the Day. http://visibleearth.nasa.gov/images/75539/dust-storm-in-texas (accessed October 2019).

NEON. 2019, May 15. Answering Burning Questions about Wildfire Fuel and Emissions. Observatory Blog. http://www.neonscience.org/observatory/observatory-blog/answering-burning-questions-about-wildfire-fuel-emissions (accessed October 2019).

Senese, K. 2017, February. Giving the railroad industry a bird's eye view. *Railway Track and Structures*, 113(2), 24–27. www.rtands.com (accessed January 2018).

Stefani, A., I. Dusfour, A. Corrêa, M. Cruz, N. Dessay, A. Galardo, C. Galardo, R. Girod, M. Gomes, H. Gurgel, A. Lima, E. Moreno, L. Musset, M. Nacher, A. Soares, B. Carme, and E. Roux. 2013. Land cover, land use and malaria in the Amazon: A systematic literature review of studies using remotely sensed data. *Malaria Journal*, 12:192. http://malariajournal.biomedcentral.com/articles/10.1186/1475-2875-12-192 (accessed October 2019).

sUAS News. 2017, May 18. Unmanned Experts Teams with AirShark in Ground-Breaking Aerial Bridge Inspection. https://www.suasnews.com/2017/05/unmanned-experts-teams-airshark-ground-breaking-aerial-bridge-inspection (accessed January 2018).

Theng, L. B. 2006, November 4. Automatic building extraction from satellite imagery. *Engineering Letters*, 13(3).

UNHCR. 2016, November 21. UNHCR Uses Drones to Help Displaced Populations in Africa. http://www.unhcr.org/news/latest/2016/11/582dc6d24/unhcr-uses-drones-help-displaced-populations-africa.html (accessed September 2019).

Unmanned Experts. 2017. Unmanned Experts Teams with AirShark in Ground-Breaking Aerial Bridge Inspection. Unmanned Experts News. https://unmannedexperts.com/unmanned-experts-teams-airshark-ground-breaking-aerial-bridge-inspection (accessed January 2018).

US EPA. 2018, November 14. Particulate Matter (PM) Basics. http://www.epa.gov/pm-pollution/particulate-matter-pm-basics#PM (accessed October 2019).

Wernke, S. A., J. A. Adams, and E. R. Hooten. 2014. Capturing complexity toward an integrated low-altitude photogrammetry and mobile geographic information system archaeological registry system. *Advances in Archeological Practices*, August, 147–163.

Widyaningrum, E., B. Gorte, and R. Lindenbergh. 2019. Automatic building outline extraction from ALS point clouds by ordered points aided Hough Transform. *Remote Sensing*, 11(14), 1727.

Williams, A. R. 2017, August. Ancient sites as seen from space. *National Geographic*, 232(2), 140.

Williams, R. J. 1992. In search of a legend-the lost city of Ubar. *Point of Beginning*, 17, 10–18.

Basic Geology for Remote Sensing

On images of land areas, the terrain is a direct expression of the geology of the area. Many interpreters are concerned with nongeologic subjects (such as land use, environment, or forestry), but an understanding of the geology will contribute to the overall understanding of the image. This brief review emphasizes the major rock types and geologic structures that are expressed on images. Additional information is given in general geology texts, such as Grotzinger and Jordan (2014) and Tarbuck and others (2017).

ROCK TYPES

Rocks belong to three major categories—sedimentary, igneous, and metamorphic—which are described in the following sections.

Sedimentary Rocks

Material that has been transported and deposited by water or wind forms sedimentary rocks characterized by layers, called *beds* or *strata*, formed during deposition. The surfaces separating strata are called *bedding planes*. Outcrops of sedimentary strata typically have a banded appearance on images. *Sedimentary rocks* are divided into the broad categories of clastic and chemical rocks.

Clastic Rocks

Erosion of older rocks produces fragments and particles that are transported and deposited to form *clastic rocks*. After deposition, the fragments are compressed and cemented to form rocks. The consolidated rocks are classified on the basis of particle size before consolidation, as follows:

Consolidated Rocks	Unconsolidated Particles
Conglomerate	Boulders and gavel
Sandstone	Sand
Siltstone	Mud
Shale	Clay

Clastic rocks differ in their resistance to erosion; sandstone and conglomerate typically form ridges, but shale and siltstone form valleys.

Chemical Rocks

Minerals dissolved in water may be removed from solution to form *chemical rocks*, either by chemical precipitation or by uptake into organisms whose shells and skeletons form sediments after death. Algae and shellfish remove calcium carbonate from seawater as the mineral calcite, which accumulates to form the rock called *limestone*. Half of the calcium atoms in calcite may be replaced by magnesium to form the mineral and rock called *dolomite*. Evaporation of seawater produces deposits of gypsum, anhydrite, and salt. Because of its low density, salt may migrate upward into the overlying strata to form cylindrical plugs called *salt domes*. Open spaces between the grains of sedimentary rocks are called *pores* and contain fresh- or saltwater or, less commonly, oil and gas.

IGNEOUS ROCKS

Igneous rocks are rocks that have cooled from molten material, called *magma*. Igneous rocks are assigned to the classes of intrusive and extrusive rocks.

Intrusive Rocks

Magma that invades country rocks (areas of older rock) cools to form intrusive rocks. Based on the relationship to the country rock, intrusive rocks are classed as batholiths, dikes, or sills (Figure A-1). Batholiths are large, irregularly shaped masses of intrusive rock that cut across the structure of the country rock. Erosion of the overlying country rocks exposes outcrops of the batholith rock. The outcrops are commonly cut by intersecting fractures, called joints, that give a distinctive appearance to these rocks on images. Typical batholithic rocks crop out in the Peninsular Ranges of southern California, which are illustrated on Landsat images in Chapter 3.

As shown in Figure A-1, country rocks are commonly layered or stratified. Tabular bodies of igneous rock intruded parallel with the layers are sills, such as the Palisades sill that crops out along the west bank of the Hudson River. Tabular bodies of intrusive rock that cut across the layers of country rock are called dikes. Erosion of the overlying country rock may expose dikes in the form of ridges, such as the Chinese Wall at Yellowstone National Park. Some dikes are less resistant to erosion than the surrounding country rock and weather to form depressions. On images, dikes are recognized by their linear shape and cross-cutting relationships to the country rock.

As intruded magma slowly cools, silicate minerals form relatively coarse crystals that are visible to the unaided eye.

The two major types of intrusive rocks are classed according to their silica content.

Granite — High silica content and typically light gray to pink.

Gabbro — Relatively low silica content and high content of iron- and magnesium-bearing minerals. Gabbro is dark and has a higher density than granite.

Extrusive Rocks

Magma may reach the surface as a liquid called *lava*, which cools to form lava flows. The lava may also be explosively ejected into the air as cinders and ash that can accumulate as volcanoes, or *cinder cones* (Figure A-1). Because lava cools rapidly at the surface, the resulting *extrusive rocks* are very fine grained. Three major categories of extrusive rocks are as follows:

Rhyolite — The extrusive equivalent of granites. Rhyolite has a high silica content and is typically pink.

Andesite — A rock having intermediate silica content. Andesite volcanoes form the "Ring of Fire" around the Pacific Ocean; they also make up the Andes of South America and the Cascade Range of the northwestern United States. Mount St. Helens and Pinatubo (Chapter 15) are andesite volcanoes.

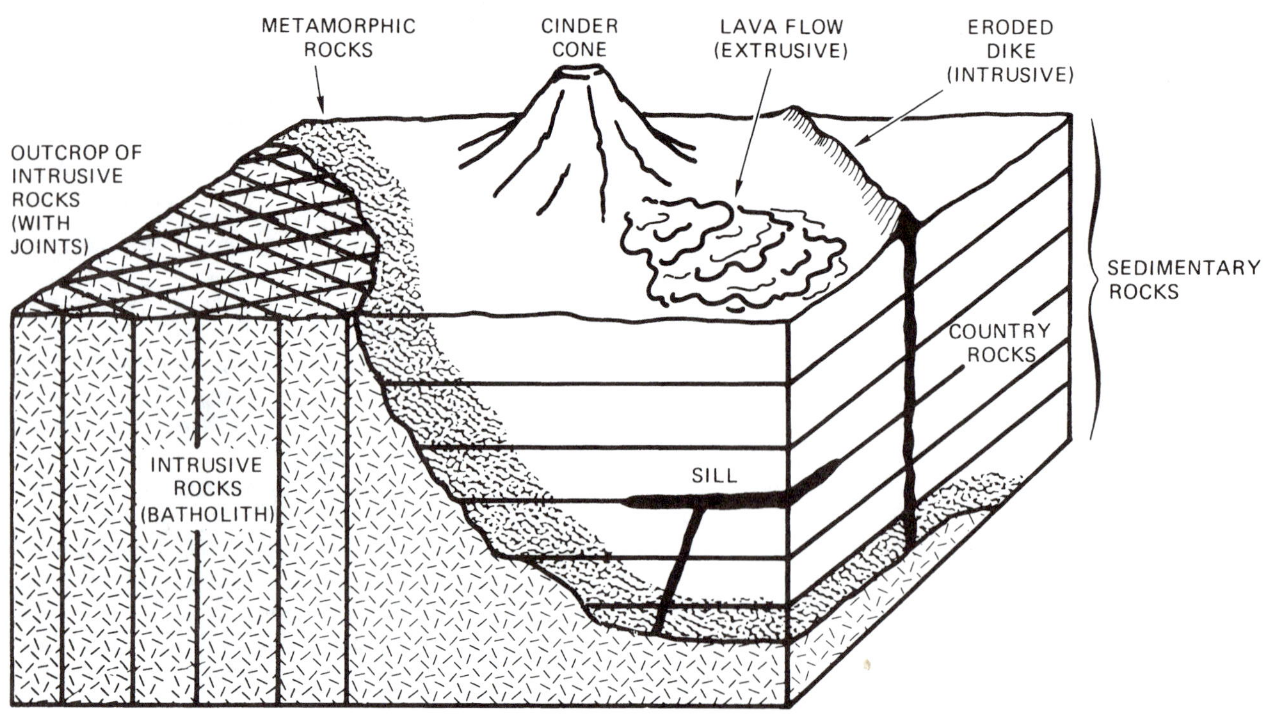

FIGURE A-1 Intrusive and extrusive rocks.

Basalt — The extrusive equivalent of gabbro. Basalt has a low silica content and a high content of iron and magnesium-bearing minerals. Basalt is dark and has a relatively high density. Flows of basalt with rough surfaces are called *aa*; those with smooth, ropy surfaces are called *pahoehoe*.

METAMORPHIC ROCKS

Heat and pressure may transform igneous and sedimentary rocks into *metamorphic rocks* (Figure A-1). The heat and pressure associated with intrusive rocks may convert original country rock into the following kinds of metamorphic rocks:

Original Rock	Metamorphic Equivalent
Sandstone	Quartzite
Shale and siltstone	Schist
Limestone	Marble

Intensive regional compression and deep burial can also produce metamorphic rocks.

STRUCTURAL GEOLOGY

Stresses within the Earth's crust may deform the rocks to produce geologic structures that may be mapped on various types of images. Most sedimentary rocks were deposited as horizontal strata, but later uplift and tilting caused the strata to become inclined.

STRIKE AND DIP

Geologists use the terms *strike* and *dip* to describe the orientation and degree of inclination. The line formed by the intersection between an inclined surface and a horizontal plane is the *strike* of the inclined surface (Figure A-2). The orientation of the strike is described by its geographic azimuth; in this case of Figure A-2, it is N45°E. *Dip* is measured in the vertical plane oriented normal to the strike and measures the inclination below horizontal of the inclined surface in degrees. In Figure A-2, the dip is 30° toward the southeast. The strike and dip symbol is used to record the *attitudes* of structures on geologic maps. Exposed bedding planes of dipping strata are called *dipslopes*. Erosion of dipping strata produces ledges called *antidip scarps* (Figure A-2) that face the opposite direction from dipslopes.

The ability to recognize strike and dip is fundamental for interpreting geologic structure from remote sensing images. Except for highly deformed areas, beds generally dip less than 45°, and the following criteria apply. Dipslopes are relatively broad and are traversed by relatively long streams that flow in the direction of dip. Antidip scarps are narrow and have a few short drainage channels that flow opposite to the dip direction. The orientation of shadows and highlights is an important key for interpreting images acquired with inclined illumination, such as low sun-angle aerial photographs (Chapter 2) and radar images (Chapter 6). If the scene in Figure A-2 were photographed in the morning with the sun shining from the southeast, the dipslope would form a bright expanse and the shadowed antidip scarp would form a narrow dark band; in the afternoon the highlights and shadows would be reversed.

FOLDS

Compressive stresses form folds called anticlines and synclines. As shown in Figure A-3, in *anticlines* the beds dip away from the axis; in *synclines* the beds dip toward the axis. Note the orientation of the strike and dip symbols and the attitude of the dipslopes. The *plunge* ("nose") of a fold is marked by arcuate outcrop patterns. Erosion exposes older beds in the center of anticlines and younger beds in the center of synclines. On geologic maps the axes of folds are shown by long lines with short crossing arrows that point away from the crest of anticlines and toward the center of synclines (Figure A-3). Chapter 2 shows aerial photographs of the Alkali anticline and syncline in Wyoming.

Organic matter contained in shales (*source rocks*) generates oil and gas, which migrate into the pore spaces of sandstones and limestones (*reservoir rocks*). Rocks in the subsurface are also saturated with water. Because hydrocarbons are less dense than water, oil and gas float on the water, becoming concentrated in the high points of structures, such as crests of anticlines. Such concentrations may form oil fields (Figure A-3). Chapter 3 shows a Landsat TM image of anticlinal oil fields in the Thermopolis area of Wyoming. Chapter 13 shows radar images of anticlinal oil fields in Papua New Guinea. Oil fields also occur along faults and in ancient reefs, sandbars, and river channels.

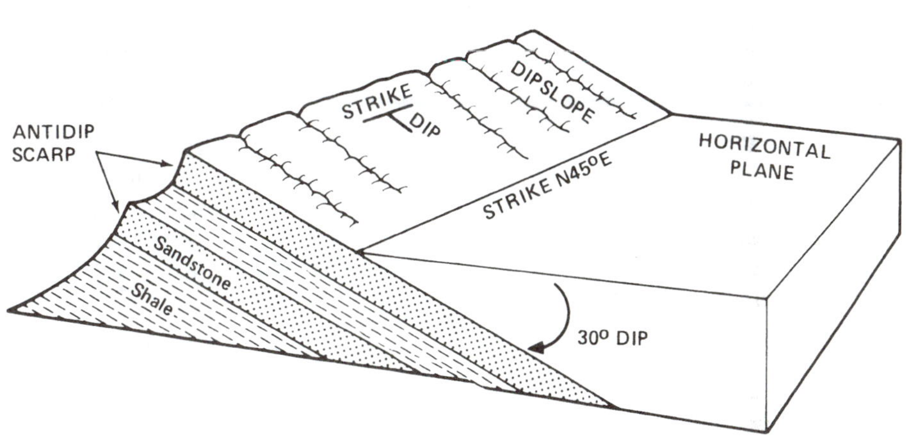

FIGURE A-2 Strike and dip of inclined beds.

Figure A-3 Anticline, syncline, and gas and oil fields.

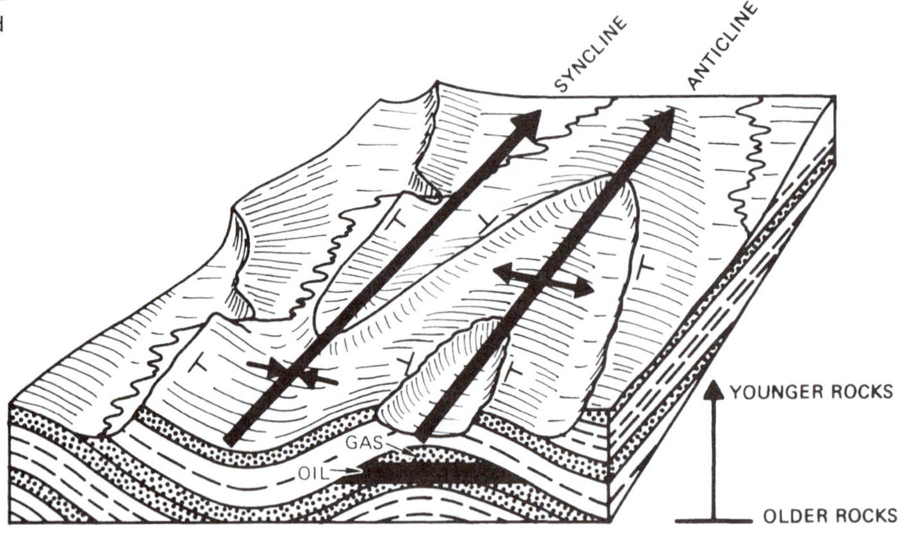

Joints and Faults

Joints, mentioned earlier in connection with batholiths, are intersecting sets of fractures along which no movement has occurred (Figure A-1). Erosion along joints produces linear depressions that are readily interpreted on images. *Faults* are fractures along which appreciable movement has occurred. Strike and dip terminology is also used to describe the attitude of fault surfaces. Faults are assigned to the following categories according to the kind of relative movement: normal faults, thrust faults, and strike-slip faults.

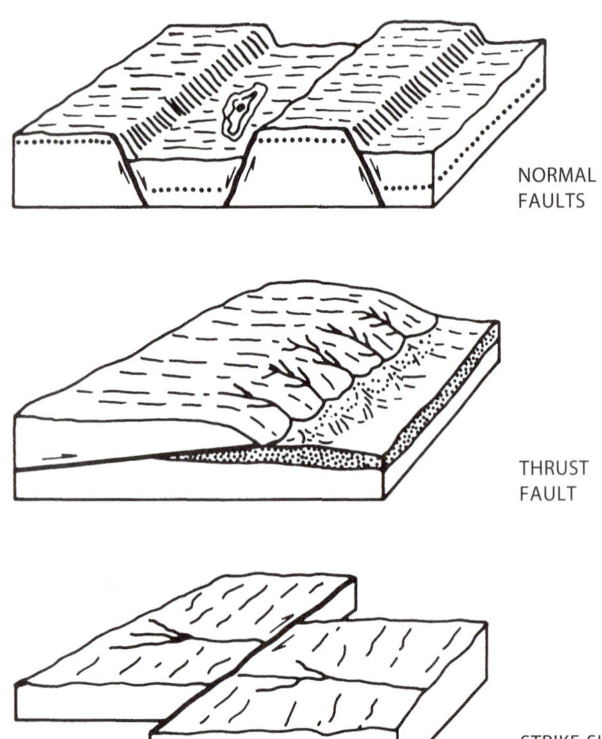

Figure A-4 Types of faults.

Normal Faults

The relative movement of rocks on either side of a fault is indicated by arrows, as shown in Figure A-4. *Normal faults* dip less than 90°, and the rocks on the upper side of the fault have moved downward relative to those on the lower side. Topographic escarpments caused by faults are called *fault scarps*. In the Basin and Range Province of the United States, normal faults separate the uplifted mountain ranges from the intervening down-dropped basins. The uplifted blocks are called *horsts*, and the down-dropped basins are called *graben*.

Thrust Faults

Thrust faults are horizontal or gently dipping faults in which the rocks overlying the fault have moved up and over the rocks below the fault (Figure A-4). Thrust faults are common in the Papuan Fold and Thrust Belt (Chapter 13) and in West Papua, Indonesia (Chapter 6).

Strike-Slip Faults

Faults along which the rocks have moved laterally are *strike-slip faults* (Figure A-4). Because of their steep dip, strike-slip faults have linear surface traces, as seen by the traces of the San Andreas, Garlock, and other faults in the Landsat mosaic of southern California (**Digital Image 15-1**). The offset of features on opposite sides of a strike-slip fault is used to determine the relative displacement. In Figure A-4 an observer standing on the stream and looking across the fault at the offset channel will note that the lateral displacement of the stream is toward the right, making this a right-lateral, strike-slip fault. In left-lateral faults the features are displaced to the left.

References

Grotzinger, J., and T. H. Jordan. 2014. *Understanding Earth* (7th ed.). New York: W. H. Freeman.

Tarbuck, E. J., F. K. Lutgens, and D. G. Tasa. 2017. *Earth: An Introduction to Physical Geology* (12th ed.). New York: Pearson.

appendix B

Location Maps

In this appendix a world map and a map of the United States have been provided for the reader's use in conjunction with the text as well as the additional materials and resources provided online (see waveland.com/Sabins-Ellis). These maps illustrate the location of images found in the figures, plates, and digital images referenced throughout the text. Beside each location noted on the map we have identified all of the relevant images associated with the site. Also, in combination with the many case studies discussed throughout the Fourth Edition, these maps can provide the opportunity for students to identify areas that they may want to study during the course of their careers.

11-12
11-16
11-15
11-17
11-14
Pl. 34 (Ch. 11)
16-22
16-17, 16-18
16-2 to 16-8
16-10, D.I. 16-1
16-15
3-21
11-13
Pl. 23 (Ch. 7)
15-35A
Pl. 46 (Ch. 13)
D.I. 6-2
6-42
17-5
17-6
17-7
17-8
11-10
D.I 6-1
15-16
11-12
12-2
11-32B
11-11
Pl. 16 (Ch. 5)
Pl. 12 (Ch. 3)
Pl. 11 (Ch. 3)
3-16, 13-18, 13-19, Pl. 9 (Ch. 3)
Pl. 19, 20 (Ch. 6)
7-14
17-9, Pl. 55 (Ch. 17)
6-16
6-20
12-14
4-23, 4-24, Pl. 13 B,C (Ch. 4)
4-8
12-17
15-29
D.I. 9-1 C,D
Pl. 42 (Ch. 13)
15-32
7-36
Pl. 10 (Ch. 3)
15-29

11-32A
2-12
12-16, D.I. 12-1
Pl. 17 (Ch. 5)
3-15
D.I. 16-2
Pl. 43, 44 (Ch. 13)
15-14
4-9
5-6
13-39, Pl. 45 (Ch. 13)
17-11
9-16
11-27, 11-30, Pl. 37 (Ch. 11)
Pl. 54 (Ch. 16)
Pl. 40 (Ch. 13)
6-34
Pl. 6 (Ch. 3)
17-3
17-1
Pl. 15 (Ch. 5)
D.I. 3-1
4-18
4-25, Pl. 13 D (Ch. 4)
3-13, 6-22, Pl. 8 (Ch. 3)
7-42
15-31, 15-35B
5-19A
13-2, 13-3
Pl. 13 A (Ch. 4)
D.I. 17-1
5-19B
15-37
D.I. 2-1
D.I. 3-3
5-27
4 7

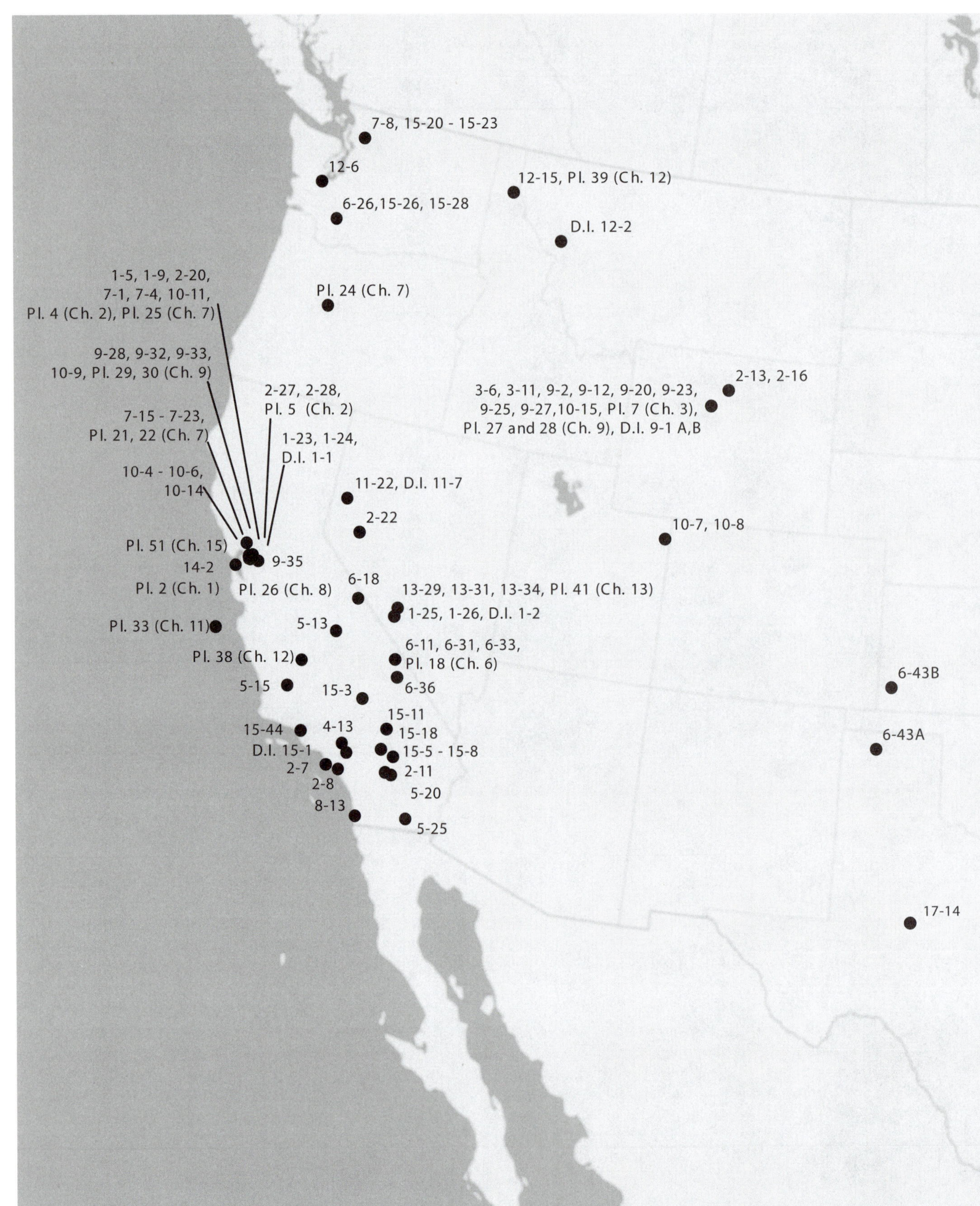
7-8, 15-20 - 15-23
12-6
12-15, Pl. 39 (Ch. 12)
6-26,15-26, 15-28
D.I. 12-2
1-5, 1-9, 2-20,
7-1, 7-4, 10-11,
Pl. 4 (Ch. 2), Pl. 25 (Ch. 7)
Pl. 24 (Ch. 7)
9-28, 9-32, 9-33,
10-9, Pl. 29, 30 (Ch. 9)
2-13, 2-16
3-6, 3-11, 9-2, 9-12, 9-20, 9-23,
9-25, 9-27,10-15, Pl. 7 (Ch. 3),
Pl. 27 and 28 (Ch. 9), D.I. 9-1 A,B
2-27, 2-28,
Pl. 5 (Ch. 2)
7-15 - 7-23,
Pl. 21, 22 (Ch. 7)
1-23, 1-24,
D.I. 1-1
10-4 - 10-6,
10-14
11-22, D.I. 11-7
2-22
10-7, 10-8
Pl. 51 (Ch. 15)
9-35
14-2
Pl. 2 (Ch. 1)
Pl. 26 (Ch. 8)
6-18
13-29, 13-31, 13-34, Pl. 41 (Ch. 13)
1-25, 1-26, D.I. 1-2
Pl. 33 (Ch. 11)
5-13
6-11, 6-31, 6-33,
Pl. 18 (Ch. 6)
Pl. 38 (Ch. 12)
5-15
15-3
6-36
6-43B
15-11
15-44
4-13
15-18
6-43A
D.I. 15-1
15-5 - 15-8
2-7
2-11
2-8
5-20
8-13
5-25
17-14

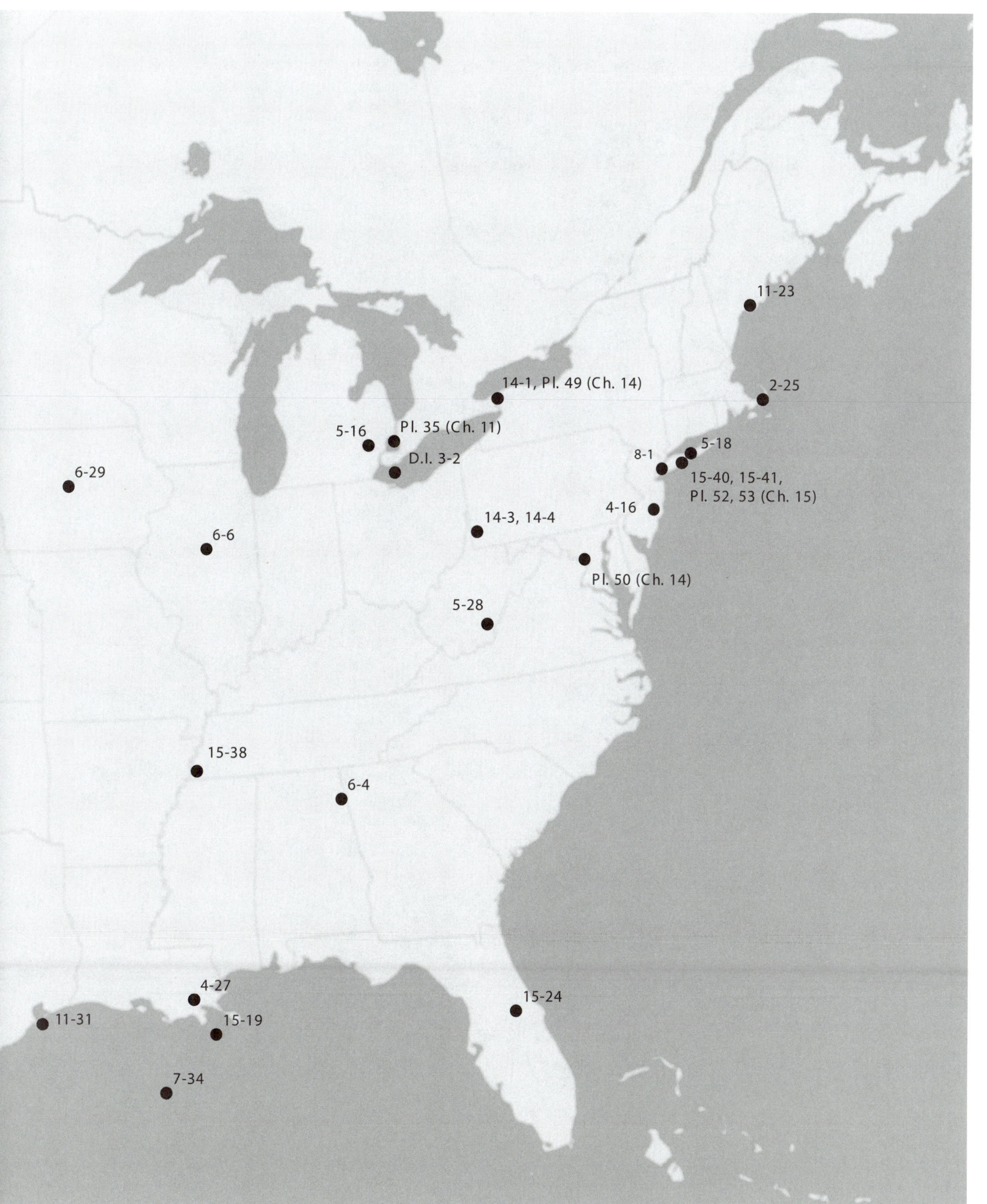
11-23
14-1, Pl. 49 (Ch. 14)
2-25
5-16
Pl. 35 (Ch. 11)
5-18
8-1
D.I. 3-2
6-29
15-40, 15-41,
Pl. 52, 53 (Ch. 15)
4-16
14-3, 14-4
6-6
Pl. 50 (Ch. 14)
5-28
15-38
6-4
4-27
15-24
11-31
15-19
7-34

Remote Sensing Digital Database

Easy access to ready-to-use remote sensing data can be a challenge for instructors and students. In conjunction with the Fourth Edition of *Remote Sensing: Principles, Interpretation, and Applications*, we provide a user-friendly Remote Sensing Digital Database so students can digitally explore 27 examples of satellite and airborne imagery. The database includes descriptions, georeferenced images, DEMs, maps, and metadata so users can display, process, and interpret images with open-source and commercial image processing and GIS software.

The remote sensing examples we provide highlight the *technology* discussions found in Chapters 1 through 10. The examples also demonstrate many of the *applications* discussed in Chapters 11 through 17. Additional images are provided in the database to jumpstart further exploration. Instructors can integrate their own images and maps to improve the learning experiences of students at various levels by creating step-by-step lab exercises using preferred image processing and GIS software. Instructors are encouraged to download the database and pick and choose topics that meet their course objectives (waveland.com/Sabins-Ellis). Also included in the database is a suggested course outline for lectures, labs, and special projects to help instructors organize a semester-long course in remote sensing.

An Introduction to the Remote Sensing Digital Database and an Instructional Video on how to use the Remote Sensing Digital Database are also provided. These ancillary materials explain the structure of the database and file naming conventions used. Tips and guidelines regarding the management of the database folders are included. We highly recommend you explore and read the contents of these introductory materials before proceeding to download individual folders containing data.

Each of the 27 examples provided in the Remote Sensing Digital Database include screen captures in ArcMap that are annotated with sensor type, geographic location, data source, and references to the relevant figures and plates found in the textbook, as well as the digital images provided online for download (waveland.com/Sabins-Ellis). The database also contains geospatial data, an ArcGIS .mxd project (versions 10.2 and 10.6) for rapid loading with appropriate GIS symbology and legend organization, ancillary information on the sensor, licensing guidelines (as needed), and proposed lab exercises. The geospatial data is courtesy of USGS, NASA, ESA, JAXA, NOAA, Airbus DS, Maxar Technologies, JPL, D. Ruiz of Quantum Spatial, Inc. (from their Pacific Aerial Surveys historical archive), Galileo, and Contra Costa County.

The images in the database are in standard GeoTIFF format. This format carries georeferencing information so the images can be displayed in their correct geographic location in a GIS. Most of the GeoTIFF images can be seen with basic image display software. GeoTIFF formats do not carry the wavelengths associated with multispectral and hyperspectral bands, so tables of band and wavelength are provided for images with more than five bands. Multispectral data with more than four bands and hyperspectral datacubes are in both GeoTIFF and ENVI formats (ENVI preserves the band and wavelength information).

The following is a list of the images and DEMs in the database along with relevant figures, plates, and digital images from the textbook noted.

CHAPTER 1 INTRODUCTION TO CONCEPTS AND SYSTEMS

- Examples of spatial resolution, Martinez, California.
  Figures 1-5 and 1-9

CHAPTER 2 AERIAL AND SATELLITE PHOTOGRAPHS

- Stereo-pair, Martinez, California.
  Figure 2-20
- Stereo red/cyan anaglyph, Martinez, California.
  Plate 4
- NAIP aerial multispectral imagery, Concord, California.
  Plate 5

CHAPTER 3 LANDSAT IMAGES

- Landsat 8 OLI, Thermopolis, Wyoming.
  Figures 3-11 and 9-23
  Plates 7 and 27
- Landsat 7 ETM+, glaciers with DEM, northeast Afghanistan.
  Figure 3-15
- Landsat 5 TM, Exumas Islands, The Bahamas.
  Figure 3-16
  Plate 9
- 1979 Landsat 3 MSS and 2016 Landsat 8 OLI, southwest of Riyadh, Saudi Arabia.
  Plate 6

CHAPTER 4 MULTISPECTRAL SATELLITES

- Pleiades 50 cm color image, Melbourne, Australia.
  Figure 4-7
- WorldView-3 30 cm pan + 4-band VNIR, Tokyo, Japan.
  Figure 4-9
- ESA Sentinel-2 10 m VNIR + 20 m SWIR, Benue, Nigeria.
  Figure 4-18
- WorldView-3 SWIR with map, Cuprite, Nevada.
  Location shown by Digital Image 1-2
- ASTER VNIR-SWIR 9-band reflectance, Thermopolis, Wyoming.
  Database

CHAPTER 5 THERMAL INFRARED IMAGES

- ASTER TIR radiance and emissivity, Ghazni, Afghanistan.
  Figures 5-6B and 5-6C
- ASTER TIR surface temperature, Ghazni, Afghanistan.
  Figure 5-6D

CHAPTER 6 RADAR IMAGES

- ESA Sentinel-1 SAR, Houston, Texas.
  Database
- JAXA PALSAR-2 and ESA Sentinel-1 SAR, Death Valley, California.
  Figure 6-11
  Plate 18
- JAXA PALSAR-2 SAR with forest maps, Bridger National Forest, Wyoming.
  Database

CHAPTER 7 DIGITAL ELEVATION MODELS AND LIDAR

- Photogrammetric DEM and TIN with USGS maps, Martinez, California.
  Figure 7-4
- Airborne lidar with orthoimage, Pleasant Hill, California.
  Plates 21 and 22
- GeoSAR X- and P-band interferometric DSM and DTM, Fairbanks, Alaska.
  Plate 23
- DEMs from ASTER, SRTM, and TerraSAR-X/TanDEM-X, Al Qassim, Saudi Arabia.
  Database
- Subscene of NOAA global DEM with Landsat, Red Sea.
  Database

CHAPTER 8 DRONES AND MANNED AIRCRAFT IMAGING

- USGS UAS DEM and orthoimage, West Fork Mine, Missouri.
  Database

CHAPTER 9 DIGITAL IMAGE PROCESSING

- AISA VNIR 128-band hyperspectral data cube, Douglas County, Colorado.
  Database
- AVIRIS 50- and 224-band hyperspectral data cubes with Landsat, Cuprite, Nevada.
  Figure 1-25
  Digital Image 1-2
- Landsat spectral classification maps with DEM, Martinez, California.
  Figures 9-28, 9-29, and 9-33
  Plates 29 and 30

CHAPTER 10 GEOGRAPHIC INFORMATION SYSTEMS

- kmzs of Landsat 8 OLI, Thermopolis, Wyoming.
  Figures 3-11 and 9-23
  Plates 7 and 27
- kmzs of Landsat 8 OLI images and classification maps, Martinez, California.
  Plates 29 and 30
- GIS spatial analysis with DEM and land cover map, Martinez, California.
  Figures 10-4, 10-5, and 10-9

appendix D

Digital Image Processing Lab Manual

Over the past two decades, GPS, GIS, and sensor advances have significantly expanded the user community and availability of remote sensing images. Constantly evolving tools, such as automation, cloud-based services, drones, and artificial intelligence, continue to expand and enhance the discipline. These new and far-reaching developments compel faculty, students, agencies, and industry to understand the basic principles, constraints, and potential of remote sensing so that they can successfully implement the new tools. For the Fourth Edition of *Remote Sensing: Principles, Interpretation, and Applications*, we created the Digital Image Processing Lab Manual so that instructors and students have ready-to-use exercises to support computer analysis, visualization, integration, and interpretation of data acquired by a broad range of sensors. The lab manual provides 12 step-by-step exercises and is available for download at waveland.com/Sabins-Ellis. It uses data found in the Remote Sensing Digital Database, along with some additional data provided with 10 of the 12 exercises. Each lab exercise includes references to discussions, figures, and plates in the textbook.

The labs use ENVI image processing software by L3Harris. Contingent upon US laws and regulations, students are offered the opportunity to access a 30-day, free ENVI license to complete the exercises found in the lab manual and to process other imagery found in the Remote Sensing Digital Database. Contact James Ellis via email (jellis@ellis-geospatial.com) to facilitate coordination with the L3Harris Academic & NGO Account Manager for the license. Prior to launching the license, we recommend self-taught students read Chapters 1 through 10 to become familiar with remote sensing principles and technology and to learn about the many examples that are discussed in the textbook and used in the lab manual.

Each lab handout averages 15 pages with screen captures of properly configured software menus, questions for the students to answer, and requests for student-processed files to be uploaded to the instructor. An answer key is provided to instructors for each exercise.

The instructor can add or delete sections in each exercise to accommodate their lab schedule. The instructor can retain the structure and concepts in each exercise while replacing the step-by-step directions designed for ENVI software with instructions specific to their lab's image processing software. The lab handouts and assignments are provided to the instructor as Microsoft Word documents to facilitate editing and revisions.

The lab manual contains the following exercises:

Lab 1 Introduction to ENVI

- Use data from Chapters 1 and 2.
- Interact with five ENVI Display tools:
  1. Review drop-down menus and tools
  2. Metadata
  3. Radiometric and spatial resolution
  4. Contrast stretching
  5. Spectral and arbitrary profile (transect)

Lab 2 Landsat Multispectral Processing

- Use data from Chapter 3.
- Interact with seven ENVI display tools:
  1. Evaluate individual bands
  2. Quick stats (overview of dataset statistics)
  3. Compare an enhanced image with original USGS data
  4. Spectral profile
  5. Evaluate different color composites in four display views
  6. Regions of interest (ROIs)
  7. Scatter plots (feature space)

Lab 3 Image Processing 1

- Use data from Chapters 3 and 4.
- Interact with five ENVI Toolbox tools:
  1. Statistics (complete, including correlation matrix)
  2. Resize data
  3. Rotate/flip data
  4. Layer stacking
  5. Edge enhancement

Lab 4 Image Processing 2

- Use data from Chapters 3, 5, and 8.
- Interact with four ENVI Toolbox tools:
  1. Band math
  2. Density slice
  3. Mosaicking
  4. Masking

Lab 5 Band Ratios and Principal Components (PCs)

- Use data from Chapters 3 and 9.
- Landsat 8 OLI iron and clay band ratios, NDVI vegetation index.
- Landsat 8 OLI PC grayscale images and color composites.

Lab 6 Georeferencing

- Use portions of air photos discussed in Chapters 1, 2, and 7.
- Use ASTER satellite image with San Francisco Estuary Institute 1880 and 1980 maps.

Lab 7 DEMs and Lidar

- Use data from Chapter 7.
- Airborne radar X-band DSM and P-band DTM.
- Airborne tile of lidar DSM and DTM along with orthorectified color image.

Lab 8 IHS and Image Sharpening

- Use data from Chapter 4.
- ASTER satellite 9-band (VNIR-SWIR) multispectral data.
- WorldView-3 120-cm multispectral and 30-cm panchromatic data.

Lab 9 Unsupervised Classification

- Use data from Chapter 3.
- Landsat TM mapping of islands with and without a water mask.
- Generalize raster classes.
- Convert thematic raster image to vector polygon GIS file.

Lab 10 Supervised Classification

- Use data from Chapter 3.
- Landsat 8 OLI bands (6) with 2 PC images.
- Develop training sites.
- Drape classification map on hillshade DEM.
- Import classification map to Google Earth.

Lab 11 Hyperspectral

- Use data from Chapter 9.
- Spectral profiles from a 128-band VNIR data cube.
- Noise reduction with the minimum noise transform (MNF).
- Spectral signatures of USGS lab data and airborne 50-band SWIR data cube.
- Process with ENVI's Spectral Hourglass Wizard.
- Spectral Analyst and Spectral Angle Mapper (SAM).

Lab 12 Change Detection and Radar

- Use data from Chapters 3 and 6.
- Landsat 8 OLI before and after fire images.
- Landsat TM urban sprawl images.
- Change Detection Difference Map.
- Image change workflow.
- Radar polarization band ratio and color composites.

For questions or comments contact James Ellis (jellis@ellis-geospatial.com).

glossary

The following glossary includes some terms, abbreviations, and acronyms commonly employed in remote sensing. The definitions refer to the applications for which the terms are used in this text and omit applications outside the field of remote sensing. Tasking online search engines with terms used in this textbook will generate huge lists of informative articles and links to improve your understanding of remote sensing definitions, concepts, and systems. Definitions of geologic and geographic terms may be found in any standard text on those subjects or in the *Glossary of Geology* (Fifth Edition, Revised), edited by K. K. E. Neuendorf, J. P. Mehl, Jr., and J. A. Jackson and published by the American Geosciences Institute.

absorption band Wavelength interval within which electromagnetic radiation is absorbed by the atmosphere or by other substances.

absorption features Downward excursions in spectral reflectance curves.

absorptivity Capacity of a material to absorb incident radiant energy.

accuracy assessment Compares a LULC, land use, or land cover classification map to a more detailed, independently collected sample set named verification or reference data. The verification data can be based on field observations and visual interpretation of higher spatial resolution imagery.

active remote sensing Remote sensing methods that provide their own source of electromagnetic radiation to illuminate the terrain. Radar is one example.

additive primary colors Blue, green, and red. Filters of these colors transmit the primary color of the filter and absorb the other two colors.

adiabatic cooling Refers to decrease in temperature with increasing altitude.

AGC Automatic gain control. Electronic device that reduces saturation of image data.

air base Ground distance between optical centers of successive overlapping aerial photographs.

albedo (*A*) Ratio of the amount of electromagnetic energy reflected by a surface to the amount of energy incident upon the surface.

along-track scanner Scanner with a linear array of detectors oriented normal to the flight path. The IFOV of each detector sweeps a path parallel with the flight direction.

alteration Changes in color and mineralogy of rocks surrounding a mineral deposit that are caused by the solutions that formed the deposit. Suites of alteration minerals commonly occur in zones.

amplitude For waves, the vertical distance from crest to trough.

anaglyph Red/cyan stereo image with the right image of a stereo-pair printed in red and the left image printed in cyan. Viewed in stereo with red/cyan anaglyph glasses.

angular beam width In radar, the angle subtended in the horizontal plane by the radar beam.

angular field of view Angle subtended by lines from a remote sensing system to the outer margins of the strip of terrain that is imaged by the system.

angular resolving power Minimum separation between two resolvable targets, expressed as angular separation.

anomaly An area on an image that differs from the surrounding normal area. For example, a concentration of vegetation within a desert scene constitutes an anomaly.

antenna Device that transmits and receives microwave energy in a radar system.

aperture Opening in a remote sensing system that admits electromagnetic radiation to the film or detector.

apparent thermal inertia (ATI) An approximation of thermal inertia calculated as 1 minus albedo divided by the difference between daytime and nighttime radiant temperatures.

array Two-dimensional array of photosensitive detectors that capture an image in digital format.

atmosphere Layer of gases that surrounds some planets.

atmospheric correction Image processing procedures that compensates for effects of selectively scattered light in multispectral and hyperspectral data.

atmospheric scattering Multiple interactions between light rays and the gases and particles in the atmosphere.

atmospheric window Wavelength interval within which the atmosphere readily transmits electromagnetic radiation.

attitude Angular orientation of a remote sensing system with respect to a geographic reference system.

automatic gain control (AGC) Electronic device that reduces saturation of image data.

azimuth Geographic orientation of a line given as an angle measured in degrees clockwise from north.

azimuth direction In radar images, the direction in which the aircraft or spacecraft is heading. Also called *flight direction.*

azimuth resolution In radar images, the spatial resolution in the azimuth direction.

background Area on an image or the terrain that surrounds an area of interest or target.

backscatter In radar, the portion of the microwave energy scattered by the terrain surface directly back toward the antenna.

backscatter coefficient A quantitative measure of the intensity of energy returned to a radar antenna from the terrain.

band A subdivision within an electromagnetic region. For example, the visible region is subdivided into the blue, green, and red bands.

banding A defect in scanner images in which alternating groups of scan lines are brighter or darker overall than adjacent groups.

bandwidth The wavelength interval recorded by a detector. Also called *spectral resolution.*

base-height ratio Air base divided by aircraft height. This ratio determines vertical exaggeration on stereo models.

bathymetry Configuration of the seafloor.

beam A focused pulse of energy.

beam divergence The narrow cone of energy that leaves the active sensor transmitter, expanding with distance from the transmitter.

BIL Bands interleaved in the multispectral or hyperspectral file in a band interleaved by line format.

bilinear A common resampling technique that uses the value of the four nearest cells from the input image to interpolate the value of the georeferenced/orthorectified cell. Results in a smoother appearing image compared with the nearest neighbor technique.

binary Numerical system using the base 2.

BIP Bands interleaved in the multispectral or hyperspectral file in a band interleaved by pixel format.

bit Contraction of *binary digit*, which in digital computing represents an exponent of the base 2.

blackbody An ideal substance that absorbs all the radiant energy incident on it and emits radiant energy at the maximum possible rate per unit area at each wavelength for any given temperature. No actual substance is a true blackbody, although some substances, such as lampblack, approach its properties.

BRDF Bidirectional reflectance distribution function. A measurement of the impact caused by different angles of incoming and reflected energy on the interpretation of features in imagery.

breaklines Photogrammetrically generated lines that represent sharp topographic edges such as ridge crests, valley channels, and edges of berms and levees.

brightness Magnitude of the response produced in the eye by light.

brute-force radar See **real-aperture radar**.

BSQ Bands interleaved in the multispectral or hyperspectral file in a band sequential format.

byte A group of eight bits of digital data.

calibration Process of comparing an instrument's measurements with a standard.

calorie Amount of heat required to raise the temperature of 1 g of water by 1°C.

camera Framing system that records images on photographic film.

cardinal point effect In radar, very bright signatures caused by optimally oriented corner reflectors, such as buildings.

change-detection images A difference image prepared by digitally comparing images acquired at different times. The gray tones or colors of each pixel record the amount of difference between the corresponding pixels of the original images.

charge-coupled detector (CCD) A device in which electrons are stored at the surface of a semiconductor.

chlorophyll absorption feature Visualized with the continuum removal process where the wavelength range between red and NIR (including the red edge) is normalized to a common baseline so the difference in reflectance value between the red and NIR wavelengths can be measured. Healthier vegetation has a deeper chlorophyll absorption feature.

chlorosis Yellowing of plant leaves resulting from an imbalance in the iron metabolism caused by excess concentrations of copper, zinc, manganese, or other elements in the plant.

circular scanner Scanner in which a faceted mirror rotates about a vertical axis to sweep the detector IFOV in a series of circular scan lines on the terrain.

classification Process of assigning individual pixels of an image to categories, generally on the basis of spectral reflectance characteristics.

cloud Returns are collected by the lidar system as a *cloud* of data points, referred to as *mass points*, each with an *x*, *y*, *z* location. The horizontal location of the mass points is aligned with the acquisition scan lines. Also refers to data stored and accessed over the Internet.

color composite image Color image prepared by combining three individual images in blue, green, and red.

color IR image NIR-red-green bands illuminated with RGB light.

color IR photograph Color photograph in which the red imaging layer is sensitive to photographic NIR wavelengths, the green imaging layer is sensitive to red light, and the blue imaging layer is sensitive to green light. Also known as *camouflage detection photographs* and *false-color photographs.*

color ratio image Color composite image prepared by combining three ratio images.

combination image A color image composited from two or more different images, such as from Landsat and SPOT.

compilation Photogrammetric mapping of features in a stereo model (roads, buildings, sidewalks, etc.) as points, lines, and polygons with *x*, *y* coordinates and *z* (elevation).

complementary colors Two primary colors (one additive and the other subtractive) that produce white light when added together. Red and cyan are complementary colors.

conduction Transfer of electromagnetic energy through a solid material by molecular interaction.

continuum removal An advanced image processing procedure for multispectral and hyperspectral data that normalizes reflectance spectra so individual absorption features can be compared and measured from a common baseline.

contrast enhancement Image processing procedure that improves the contrast ratio of images. The original narrow range of digital values is expanded to utilize the full range of available digital values. Also called *contrast stretch*.

contrast ratio On an image, the ratio of reflectance between the brightest and darkest parts of the image.

convection Transfer of heat through the physical movement of heated matter.

convergence Pointing the UAS camera at the terrain or object of interest so sequential images overlap and view the feature from several to many off-nadir directions so that a reliable 3-D model and orthomosaic can be constructed by the SfM process.

corner reflector Cavity formed by two or three smooth planar surfaces intersecting at right angles. Electromagnetic waves (especially radar) entering a corner reflector are reflected directly back toward the source.

cross-polarized A radar return in which the electric field vibrates in a direction normal to the direction of the transmitted pulse. Cross-polarized images may be HV (horizontal transmit, vertical return) or VH (vertical transmit, horizontal return).

cross-track scanner Scanner in which a faceted mirror rotates about a horizontal axis to sweep the detector IFOV in a series of parallel scan lines oriented normal to the flight direction.

cubic convolution A common resampling technique that uses the weighted average of the 16 nearest cells around a cell in the input image to interpolate the value of the georeferenced/orthorectified output cell. Results in a smooth-looking output image.

cycle One complete oscillation of a wave.

dasymetric maps Depicts quantitative areal data using boundaries that divide the area into zones of relative homogeneity with the purpose of better portraying the population distribution.

data cube A description of the many bands recorded for each pixel with hyperspectral imagery.

ΔT Difference between maximum and minimum radiant temperatures during a diurnal cycle.

DEM Digital elevation model. A general term that describes the Earth's topography with *x* and *y* coordinates (longitude and latitude or easting and northing) and *z* values specifying elevation at each *x*, *y* point.

density, of materials (ρ) Ratio of mass to volume of a material, typically expressed as grams per cubic centimeter.

density slicing Process of converting the continuous gray tones of an image into a series of density intervals, or slices, each corresponding to a specific digital range. The density slices are then displayed either as gray tones or as colors.

depolarized Refers to a change in polarization of a transmitted radar pulse as a result of various interactions with the terrain surface.

depression angle (γ) In radar, the angle between the imaginary horizontal plane passing through the antenna and the line connecting the antenna and the target.

detectability Measure of the smallest object that can be discerned on an image.

detector Component of a remote sensing system that converts electromagnetic radiation into a recorded signal.

dielectric constant Electrical property of matter that influences radar returns. Also called *complex dielectric constant*.

difference image Image prepared by subtracting the digital values of pixels in one image from those in a second image to produce a third set of pixels. This third set is used to form the difference image.

differential parallax (dp) distance between the bottom of an object and the top of the object in a stereo model, measured along the flight line direction, that is directly proportional to the height of the object.

diffuse reflector Surface that scatters incident radiation nearly equally in all directions.

digital display A form of data display in which values are shown as arrays of numbers.

digital image processing Computer manipulation of digital images.

digital mosaic Mosaic generated by a computer from digital records of images.

digital number (DN) Value assigned to a pixel in a digital image.

digital perspective images Three-dimensional views produced by merging digital topographic data with image data.

digitization Process of converting an analog display into a digital display.

digitizer Device for scanning an image and converting it into numerical format.

directional filter Mathematical filter for image processing that enhances linear features oriented in a designated direction.

discrete return Lidar system that records one to five returns for each pulse during a flight.

distortion On an image, changes in shape and position of objects with respect to their true shape and position.

diurnal Daily.

divergence Pointing the UAS camera at the terrain or object of interest in a planar (nadir) manner or with minimum overlap between sequential images so that a reliable 3-D model and orthomosaic cannot be constructed by the SfM process.

Doppler principle Describes the change in observed frequency of electromagnetic or other waves caused by movement of the source of waves relative to the observer.

drone An unmanned aircraft that is operated without the possibility of direct human intervention from within or on the aircraft. UAS or UAV are the preferred FAA terms for drones.

DSM Digital surface model. A DEM that displays top of vegetation canopy, buildings, cars, etc.

DTM Digital terrain model. A DEM of the bare earth.

dual polarization Radar systems that transmit the signal with one polarization and simultaneously receive both polarizations of the signal (for instance, send H and receive both H and V, generating an HH and HV image).

dwell time Time required for a detector IFOV to sweep across a ground resolution cell.

edge enhancement Image processing technique that emphasizes the appearance of edges and lines.

electromagnetic energy Energy that travels at the speed of light in a harmonic wave pattern.

electromagnetic radiation Energy propagated in the form of an advancing interaction between electric and magnetic fields.

electromagnetic spectrum Continuous sequence of electromagnetic energy arranged according to wavelength or frequency.

emission Process by which a body radiates electromagnetic energy. Emission is determined by kinetic temperature and emissivity.

emissivity (ε) Ratio of radiant flux from a body to that from a blackbody at the same kinetic temperature.

energy flux Radiant flux. The amount of energy reflected or radiated from a surface.

enhancement Process of altering the appearance of an image so that the interpreter can extract more information.

EO Earth observation.

error matrix Used in accuracy assessment to compare information from verification sites to information on the map for a number of sample areas. The matrix is a square array of numbers set out in rows and columns.

evaporative cooling Temperature drop caused by evaporation of water from a moist surface.

***f*-number** Representation of the speed of a lens determined by the focal length divided by diameter of the lens. Smaller numbers indicate faster lenses.

***f*-stop** Focal length of a lens divided by the diameter of the lens's adjustable diaphragm. Smaller numbers indicate larger openings, which admit more light to the film.

FAA US Federal Aviation Administration. Provides regulations for safe and efficient operation of the US air space. Defines UAS operator rules and guidelines.

far range The portion of a radar image farthest from the aircraft or spacecraft flight path.

feature space plot A 2-D plot of two bands where the axes represent the range of DNs for each band.

film Light-sensitive photographic emulsion and its base.

filter, digital Mathematical procedure for modifying values of numerical data.

filter, optical A material that, by absorption or reflection, selectively modifies the radiation transmitted through an optical system.

filter kernel A digital filter consisting of an odd number of lines and pixels.

first return Lidar beam that only encounters one surface, and is reflected back to the receiver with no other earlier or later returns. Most often associated with bare earth and tops of buildings and trees. Also called *last return*.

fixed wing UAS with a horizontal wing and propeller(s) that move the aircraft in a forward direction.

flight path Line on the ground directly beneath a remote sensing aircraft or spacecraft. Also called *flight line.*

fluorescence Emission of light from a substance stimulated by exposure to radiation from an external source.

focal length In cameras, the distance from the optical center of the lens to the plane at which the image of a distant object is brought into focus.

focal plane In a remote sensing system, the plane at which the image is sharply defined.

footprint spacing The nominal distance between the centers of consecutive lidar beam footprints along the scan lines and across the scan lines. Also called *nominal point spacing* (NPS).

foreshortening In radar images, the geometric displacement of the top of objects toward the near range relative to their base. Also called *layover.*

forward overlap The percent of duplication by successive photographs along a flight line.

framing system Remote sensing system that instantaneously acquires an image of an area, or frame, of terrain.

frequency (υ) Number of wave oscillations per unit time or the number of wavelengths that pass a point per unit time.

full waveform Lidar system that records the returned energy in a series of equal time intervals that yields a vertical summation of the returns from a pulse.

GCP Ground control point.

generalization A filter (often a majority filter) is applied to a classification map to smooth the appearance and minimize the "salt and pepper" or "speckled" appearance caused by isolated single class pixels and narrow strings of single class pixels.

geodesy The science of measuring and monitoring the size and shape of the Earth, which supports the development of spatial reference systems so satellite and aerial images can be accurately located on the surface of the Earth.

geographic coordinate system A spatial reference system that uses latitude and longitude to locate features on the Earth's surface.

Geographic Information System (GIS) Integrated computer hardware, software, and data for capturing, storing, analyzing, and displaying geographically referenced information.

geometric correction Image processing procedure that corrects spatial distortions in an image.

georeferencing Remote sensing images and scanned maps are georeferenced to a standard map projection to enable their use with other geospatial layers in a GIS. Georeferencing adds *x*, *y* or latitude, longitude coordinates to each pixel in the image or scanned map. Also called *rectification*. Distortions due to topography are not corrected.

geostationary Refers to satellites traveling at the angular velocity at which the Earth rotates; as a result, they remain above the same point on Earth at all times.

GMT Greenwich mean time. A universal 24-hour system for designating time.

GNSS Global Navigation Satellite System. Collects accurate horizontal and vertical (x, y, z) coordinates for moving or fixed platforms. The United States deployed the first GNSS named NAVSTAR Global Positioning System (GPS).

GOES Geostationary Operational Environmental Satellite.

gossan Outcrop of iron oxide formed by the weathering of metallic sulfide ore minerals.

GPS Global Positioning System deployed by the United States (the first GNSS).

gravity survey Geophysical technique that measures variations in density of rocks in the subsurface.

grayscale A sequence of gray tones ranging from black to white.

ground control point (GCP) A geographic feature of known location that is recognizable on images and can be used to determine geometric corrections.

ground deformation Measuring and monitoring the movement of the Earth's land surface in vertical and horizontal directions with SAR satellite interferometry techniques.

ground range On radar images, the distance from the ground track to an object.

ground-range image Radar image in which the scale in the range direction is constant.

ground receiving station Facility that records image data transmitted by a satellite, such as Landsat.

ground resolution The ability to resolve terrain features on images.

ground resolution cell Area on the terrain that is covered by the IFOV of a detector.

ground swath Width of the strip of terrain that is imaged by a scanner system.

harmonic Refers to waves in which the component frequencies are whole number multiples of the fundamental frequency.

heat capacity (c) Ratio of heat absorbed or released by a material to the corresponding temperature rise or fall. Expressed in calories per gram per degree centigrade. Also called *thermal capacity*.

highlights Areas of bright tone on an image caused by strong reflections from topographic features that face the energy direction.

histogram Statistical chart showing the distribution of data points as a function of some attribute, such as brightness.

hue In the IHS system, represents the dominant wavelength of a color.

hyperspectral sensor System that collects a continuous spectrum of reflectance at many narrow, contiguous, and closely spaced wavelength bands.

IFOV Instantaneous field of view.

IHS Intensity, hue, and saturation system of colors.

image A portrayal of a scene or subject that is acquired by a digital system.

image swath See **ground swath**.

imaging spectrometer Synonym for hyperspectral scanner.

incidence angle In radar, the angle formed between a line normal to the target and another connecting the antenna and the target.

incident energy Electromagnetic radiation impinging on a surface.

index of refraction (n) Ratio of the wavelength or velocity of electromagnetic radiation in a vacuum to that in a substance.

INS Inertial Navigation System. Determines where the sensor is pointed by recording the sensor's pitch, roll, and heading. Includes an inertial measurement unit (IMU).

instantaneous field of view (IFOV) Solid angle through which a detector is sensitive to radiation. In a scanning system, the solid angle subtended by the detector when the scanning motion is stopped.

intensity In the IHS system, brightness ranging from black to white.

intensity image An image created with the maximum returned echo from all the lidar pulse returns.

interactive processing Method of image processing in which the operator views preliminary results and can alter the instructions to the software to achieve desired results.

interferogram In radar, images that record interference patterns created by superposing images acquired by two antennas that are separated by a short distance.

interferometry The field of physics that deals with the interaction between superposed wave trains.

intermediate return Lidar beam that is reflected from surfaces below the first return (top of canopy) and last return (bare earth). Often associated with branches and understory.

interpolation The transformation of each cell in an input raster image to the coordinate space of the master or reference image during georeferencing and orthorectification. Typically the polynomial transformation is performed with a 1st order, 2nd order, or 3rd order equation.

interpretation The process in which a person extracts information from an image.

interpretation key Characteristic or combination of characteristics that enables an interpreter to identify an object on an image.

irradiance The amount of radiant flux incident upon (or received by) a surface per unit area.

isotherm Contour line connecting points of equal temperature. Isotherm maps are used to portray surface temperature patterns of water bodies.

kernel Two-dimensional array of digital numbers used in digital filtering.

kinetic energy The ability of a moving body to do work by virtue of its motion. The molecular motion of matter is a form of kinetic energy.

kinetic temperature Internal temperature of an object determined by random molecular motion. Kinetic temperature is measured with a contact thermometer.

L-band Radar wavelength region from 15 to 30 cm.

land cover Describes the materials (such as vegetation, rocks, or developed) that are present at the surface.

Landsat ETM+ Enhanced Thematic Mapper Plus. On Landsat 7.
Landsat MSS MultiSpectral Scanner. On Landsat 1, 2, and 3.
Landsat OLI Operational Land Imager. On Landsat 8.
Landsat TIRS Thermal Infrared Sensor. On Landsat 8.
Landsat TM Thematic Mapper. On Landsat 4 and 5.
land use Describes how an area of land is used (such as crops, golf course, industry, or mobile home park).
Laplacian filter A digital filter used in nondirectional edge enhancement.
large format camera (LFC) A camera with a film size of 23 by 46 cm that was carried on the Space Shuttle in October 1984.
last return Lidar beam that is reflected from the bare earth after passing through the top of a canopy and intermediate features (branches and understory).
layover In radar images, the geometric displacement of the top of objects toward the near range relative to their base. Also called *foreshortening*.
lens One or more pieces of glass or other transparent material that form an image by refraction of light.
LEO Low earth orbit.
LFC Large format camera.
light meter Device for measuring the intensity of visible radiation and determining the appropriate exposure of photographic film in a camera.
linear Alignment of a topographic or tonal feature on the terrain and on images, maps, and photographs that may represent a zone of structural weakness. Equivalent to "lineament."
line-pair Pair of light and dark bars of equal widths. The number of such line-pairs aligned side by side that can be distinguished per unit distance expresses the resolving power of an imaging system.
lineation The one-dimensional alignment of internal components of a rock that cannot be depicted as an individual feature on a map.
look direction Direction in which pulses of microwave energy are transmitted by a radar system. The look direction is normal to the azimuth direction. Also called *range direction*.
low sun-angle photograph Aerial photograph acquired in the morning, evening, or winter when the sun is at a low elevation above the horizon.
LULC Land use/land cover classification map.
luminance Quantitative measure of the intensity of light from a source.
LWIR Long wavelength infrared (8 to 14 μm).

magnetic survey Geophysical technique for measuring variations in magnetism of rocks in the subsurface.
map projection A systematic representation of the curved surface of the Earth on a plane.
mass points Photogrammetrically determined ground elevations at specific points with *x*, *y*, *z* values. Used to generate a DEM.
microbolometer A specific type of resistor, often vanadium oxide (VO_x), that is used as a detector in many 8 to 14 μm TIR cameras.
microwave Region of the electromagnetic spectrum in the wavelength range from 0.1 to 30 cm.
minimum ground separation Minimum distance on the ground between two targets at which they can be resolved on an image.
modulate To vary the frequency, phase, or amplitude of electromagnetic waves.
mosaic Composite image or photograph made by piecing together individual images or photographs covering adjacent areas.
multibeam sonar Uses a specially designed transducer that transmits a fan-shaped pulse of acoustic energy that is narrow along-track and wide across-track.
multipolarization image Color image composited from radar images of three different polarizations.
multirotor UAV that has more than one propeller. Quadcopters are the most common type. Also called *rotating wing*.
multispectral classification Identification of land cover categories by digital processing of data acquired by multispectral scanners.
multispectral system Framing or scanning system that simultaneously acquires bands of different wavelengths that can have different bandwidths. The bands can be contiguous, separate, or overlap.
multitemporal spectrum Plot showing variation of a characteristic as a function of time. For example, a multitemporal NDVI spectrum shows variations of NDVI during a growing season.
multivariate statistic Statistics developed from more than one band in a multispectral or hyperspectral dataset to determine which bands are highly correlated (similar spectral pattern) and which are most different.
MWIR Medium wavelength infrared (3 to 5 μm).

nadir Point on the ground directly in line with the remote sensing system and the center of the Earth.
NAIP USDA Farm Service Agency's National Agriculture Imagery Program that plans to acquire VNIR, 4-band multispectral images every three years across the conterminous United States.
navigation satellite A satellite that transmits data used by global positioning systems to determine latitude and longitude.
NDVI Normalized Difference Vegetation Index.
nearest neighbor A common and fast resampling technique appropriate for categorical or thematic raster data as it does not alter the values of the input cells. Results in a blocky appearance of the resampled image.
near infrared (NIR) Infrared region of the electromagnetic spectrum that includes wavelengths from 0.7 μm to 1 mm.
near range Refers to the portion of a radar image closest to the aircraft or satellite flight path.
NEON The National Ecological Observatory Network airborne observation platform includes a VNIR-SWIR hyperspectral instrument with 428 bands, a lidar system, and a high spatial resolution color camera for three-dimensional studies of ecosystems.

NLCD National Land Cover Database that has created land cover products in 2001, 2006, 2011, and 2016 for the United States.

noise Random or repetitive events that obscure or interfere with the desired information.

nondirectional filter Mathematical filter that enhances all orientations of linear features equally.

nonsystematic distortion Geometric irregularities on images that are not constant and cannot be predicted from the characteristics of the imaging system.

Normalized Difference Vegetation Index (NDVI) A measure of vegetation vigor computed from multispectral and hyperspectral data.

oblique photograph A photograph acquired with the camera intentionally directed at some angle between horizontal and vertical orientations.

orbit Path of a satellite around a body such as the Earth under the influence of gravity.

orthoimage Images that have been computer processed to remove radial and topographic distortion.

orthorectification A DEM is included in the rectification process to minimize input image distortions due to topographic relief.

overlap Extent to which adjacent images or photographs cover the same terrain along a flight line.

panchromatic (pan) image Grayscale image that typically spans a large portion of the visible and sometimes NIR wavelengths to maximize signal to noise ratio and generate a high spatial resolution band.

parallax (Ø) The angle between the same object in a stereo model and the camera position when each photograph was acquired.

parallel-polarized Describes a radar pulse in which the polarization of the return is the same as that of the transmission. Parallel-polarized images may be HH (horizontal transmit, horizontal return) or VV (vertical transmit, vertical return).

passive remote sensing Remote sensing of energy naturally reflected or radiated from the terrain.

path-and-row index System for referencing and locating Landsat and other images acquired in sun-synchronous orbits.

pattern Regular repetition of tonal variations on an image or photograph.

periodic line dropout Defect on scanner images in which no data are recorded at repetitive intervals, causing a pattern of black lines on the image.

photodetector Device for measuring energy in the visible region.

photogeology Interpretation of geologic features from images.

photogrammetery Art and science of obtaining reliable information about physical objects and the environment through the process of recording, measuring, and interpreting photographic images and other remote sensing data.

photograph A portrayal of a scene or subject acquired by an analog system that employs a lens and a photo-sensitive film medium.

photon Minimum discrete quantity of radiant energy.

picture element In a digitized image, the area on the ground represented by each digital number. Commonly contracted to *pixel*.

pitch Rotation of an aircraft about the horizontal axis normal to its longitudinal axis that causes a nose-up or nose-down attitude.

pixel Contraction of *picture element*.

polarimetric images A sequence of radar images that record a range of polarizations from parallel-polarized to cross-polarized.

polarization The direction in which the electrical field vector of electromagnetic radiation vibrates.

polar satellites These satellites circle the Earth in sun-synchronous orbits.

previsual symptom Vegetation stress that is recognizable on NIR film before it is detected by the eye or normal color photographs. Stressed vegetation initially loses its ability to reflect NIR energy, which reduces the red signature on color IR images and reduces vegetation indices, such as NDVI values.

primary colors A set of three colors that in various combinations will produce the full range of colors in the visible spectrum. There are two sets of primary colors, additive and subtractive.

principal-component (PC) image Digitally processed image produced by a transformation that recognizes maximum variance in multispectral images.

principal point Optical center of an aerial photograph.

projected coordinate system Systematic methods and mathematical transformations to transfer or "project" a map from the Earth's spherical surface onto a flat surface.

pulse Short burst of electromagnetic radiation transmitted by a radar antenna.

pulse length Duration of a burst of energy transmitted by a radar antenna (measured in microseconds) and from a lidar instrument (measured in nanoseconds).

pulse repetition frequency (PRF) Pulses of energy emitted per second by an active sensor. For lidar PRF can range from 10,000 to > 150,000 kHz (pulses per second).

push broom scanner An alternate term for an along-track scanner.

quad polarization Radar systems that transmit the signal as alternate pulses with H and V polarizations and receive the signal simultaneously for both polarizations. HH, VV, HV, and VH radar images are generated.

radar Acronym for *radio detection and ranging*. Radar is an active form of remote sensing that operates in the microwave and radio wavelength regions.

radar altimetry Nonimaging systems carried on aircraft and satellites to measure altitude with great precision. A pulse of microwave energy is transmitted vertically downward.

radar shadow Dark signature on a radar image representing no signal return. A shadow extends in the far-range direction from an object that intercepts the radar beam.

radian Angle subtended by an arc of a circle equal in length to the radius of the circle; 1 rad = 57.3°.

radiance Digital number measured by the sensor that includes atmospheric scattering and target reflectance.
radiant energy peak (λ max) Wavelength at which the maximum electromagnetic energy is radiated at a particular temperature.
radiant flux Rate of flow of electromagnetic radiation measured in watts per square centimeter.
radiant temperature Concentration of the radiant flux from a material.
radiation Propagation of energy in the form of electromagnetic waves.
radiometer Device for quantitatively measuring radiant energy, especially thermal radiation.
radiometric resolution The number of subdivisions, or bits, that an imaging system records for a given range of values. As the number of bits increases, the radiometric resolution increases.
random line dropout In scanner images, the loss of data from individual scan lines in a nonsystematic fashion.
range direction See **look direction**.
range resolution In radar images, the spatial resolution is in the range direction, which is determined by the pulse length of the transmitted microwave energy.
raster array Arrangement of digital data in lines and pixels.
ratio image An image prepared by processing digital multispectral data as follows: for each pixel, the value for one band is divided by that of another. The resulting digital values are displayed as an image.
Rayleigh criterion In radar, the relationship among surface roughness, depression angle, and wavelength that determines whether a surface will respond in a rough or smooth fashion to the incident radar pulse.
real-aperture radar Radar system in which azimuth resolution is determined by the transmitted beam width, which in turn is determined by the physical length of the antenna and by the wavelength.
real time Refers to images or data made available for inspection simultaneously with their acquisition.
recognizability Ability to identify an object on an image.
rectification Another term for georeferencing.
rectilinear Refers to images with no geometric distortion. The scales in the orthogonal directions are identical.
red edge The unique spectral line associated only with the spectra of vegetation that connects the absorption feature at the red wavelength with the reflectance feature at the NIR wavelength.
reflectance Value of brightness in each pixel of a band that is corrected for atmospheric scattering. Ratio of the radiant energy reflected by a body to the energy incident on it.
reflectance peaks Upward excursions in spectral reflectance curves.
reflectance spectrometer Instrument that records percent reflectance as a function of wavelength.
reflected energy peak Wavelength (0.5 μm) at which a maximum amount of energy is reflected from the Earth's surface. Also is the wavelength of the maximum amount of energy radiated from the sun.
reflected IR Electromagnetic region from 0.7 to 3.0 μm that consists primarily of reflected NIR and SWIR solar radiation.
reflectivity Ability of a surface to reflect incident energy.
refraction Bending of electromagnetic rays as they pass from one medium into a medium with a different index of refraction.
registration Process of geometrically adjusting two images so that equivalent geographic points coincide.
relief Vertical irregularities of a surface.
relief displacement Geometric distortion on vertical aerial photographs. The tops of objects appear in the photograph to be radially displaced from their bases outward from the photograph's center point.
remote sensing The science of acquiring, processing, and interpreting images that record the interaction between electromagnetic energy and matter.
repeat cycle For Earth satellites in sun-synchronous orbits, the number of days between repeated orbits.
resampling Values are assigned to each empty cell in the georeferenced or orthorectified image by assigning the cell values in the ungeoreferenced data set, typically based on the nearest neighbor, bilinear interpolation, or cubic convolution technique.
resolution target Series of regularly spaced alternating light and dark bars used to evaluate the resolution of images.
resolving power The ability to distinguish closely spaced targets.
reststrahlen band The absorption of TIR energy as a function of silica content.
return In radar, a pulse of microwave energy reflected by the terrain and received at the radar antenna. The strength of a return is referred to as *return intensity*.
roll Rotation of an aircraft that causes a wing-up or wing-down attitude.
roll compensation system Component of an airborne scanner system that measures and records the roll of the aircraft. This information is used to correct the imagery for distortion due to roll.
rough criterion In radar, the relationship among surface roughness, depression angle, and wavelength that determines whether a surface will scatter the incident radar pulse in a rough or intermediate fashion.
roughness In radar, the average vertical relief of small-scale irregularities of the terrain surface. Also called *surface roughness*.

SAR Synthetic aperture radar.
satellite An object in orbit around a celestial body.
saturation In the IHS system, saturation represents the purity of color. It is also the condition in which energy flux exceeds the sensitivity range of a detector.
scale Ratio of distance on an image to the equivalent distance on the ground.
scan line Narrow strip on the ground that is swept by the IFOV of a detector in a scanning system.
scan skew Distortion of scanner images caused by forward motion of the aircraft or satellite during the time required to complete a scan.

scanner See **scanning system**.
scanner distortion Geometric distortion that is characteristic of cross-track scanner images.
scanning system An imaging system in which the IFOV of one or more detectors is swept across the terrain.
scattering Multiple reflections of electromagnetic waves by particles or surfaces.
scattering coefficient curves Display of scatterometer data in which relative backscatter is shown as a function of incidence angle.
scatterometer Nonimaging radar device that quantitatively records backscatter of terrain as a function of incidence angle.
scene Area on the ground that is covered by an image or photograph.
seismic surveys Geophysical technique for mapping subsurface geology. Mechanical devices or explosions transmit sonic energy into the subsurface, where it is reflected from rock layers. Reflected energy is recorded and displayed as maps and cross sections.
sensitivity Degree to which a detector responds to incident electromagnetic energy.
sensor Device that detects electromagnetic radiation and converts it into a signal that can be recorded and displayed as either numerical data or an image.
SfM Structure from Motion. Multiray photogrammetry that uses multiple overlapping images of the same feature from different angles and distances to generate a 3-D model.
shadows Dark signatures caused by topographic features that block incident energy. Shadows are prominent on radar images and low sun-angle photographs.
shortwave IR (SWIR) Reflected IR region from 0.9 to 3 μm that is employed in remote sensing.
sidelap Extent to which images acquired on adjacent flight lines or orbits cover the same terrain.
side-looking airborne radar (SLAR) An airborne side-scanning system for acquiring radar images.
side-scanning sonar Active system for acquiring images of the seafloor using pulsed sound waves.
side-scanning system A system that acquires images of a strip of terrain parallel with the flight or orbit path but offset to one side.
signal Response of a detector to incident energy.
signature Set of characteristics by which a material or an object may be identified on an image or photograph.
skylight Component of light that is strongly scattered by the atmosphere and consists predominantly of shorter wavelengths.
slant range In radar, an imaginary line running between the antenna and the target.
slant-range distance Distance measured along the slant range.
slant-range distortion Geometric distortion of a slant-range image.
slant-range image In radar, an image in which objects are located at positions corresponding to their slant-range distances from the aircraft flight path. On slant-range images, the scale in the range direction is compressed in the near-range region.
small unmanned aerial system (sUAS) Also called small unmanned aerial vehicle (sUAV). "Aircraft" may be substituted for "aerial." The FAA states a sUAS weighs less than 25 kg (55 lb).
smooth criterion In radar, the relationship among surface roughness, depression angle, and wavelength that determines whether a surface will scatter the incident radar pulse in a smooth or intermediate fashion.
software Programs that control computer operations.
sonar Acronym for *sound navigation ranging*. Sonar is an active form of remote sensing that employs sonic energy to image the seafloor.
spatial resolution The ability to distinguish between closely spaced objects on an image. Commonly expressed as the most closely spaced line-pairs per unit distance that can be distinguished.
spectral reflectance Reflectance of electromagnetic energy at specified wavelength intervals.
spectral reflectance curves Plots that show percent reflectance as a function of wavelength.
spectral resolution Range of wavelengths recorded by a detector. Also called *bandwidth*.
spectral sensitivity Response, or sensitivity, of a film or detector to radiation in different spectral regions.
spectrometer A device designed to detect, measure, and analyze the intensity of electromagnetic radiation as a function of wavelength by using an optical grating or prism to disperse the radiation.
spectroradiometer A spectrometer with light intensity calibration that quantitatively measures spectral radiance in watts per (cm^2/nm/sr) and/or irradiance in watts per (cm^2/nm) of wavelength bands. Information can be reported for multiple (multispectral) or many contiguous (hyperspectral) bands, depending on the instrument.
spectrum Continuous sequence of electromagnetic energy arranged according to wavelength or frequency.
specular Refers to a surface that is smooth with respect to the wavelength of incident energy.
State Plane Coordinate System (SPCS) In the United States each state has its own SPCS that uses a projected coordinate system and varying number of SPCS zones to minimize distortions and increase the accuracy of measurements. Aerial and high spatial resolution satellite imagery are often projected into this system for county and city use.
stationary scanner Scanners that operate from a fixed position, rather than a moving aircraft or satellite.
Stefan–Boltzmann constant 5.68×10^{-12} W · cm^{-2} · °K^{-4}.
Stefan–Boltzmann law States that radiant flux of a blackbody is equal to the temperature to the fourth power times the Stefan–Boltzmann constant.
stereo base Distance between a pair of correlative points on a stereo-pair that are oriented for stereo viewing.
stereo model Three-dimensional visual impression produced by viewing a pair of overlapping images through a stereoscope.
stereo-pair Two overlapping images or photographs that may be viewed stereoscopically.

stereoscope Binocular optical device for viewing overlapping images or diagrams. The left eye sees only the left image, and the right eye sees only the right image.

subscene A portion of an image that is used for detailed analysis.

subtractive primary colors Yellow, magenta, and cyan. When used as filters for white light, these colors remove blue, green, and red light, respectively.

sunglint Bright reflectance of sunlight caused by ripples on water.

sun-synchronous orbit Satellite polar orbit pattern that covers the Earth during a cycle that lasts from days to weeks. On subsequent cycles the orbit paths are repeated at the same local sun time.

supervised classification Digital information extraction technique in which the operator provides training site information that the computer uses to assign pixels to categories.

surface phenomenon Interaction between electromagnetic radiation and the surface of a material.

surface roughness See **roughness**.

SWIR Reflected shortwave IR region (0.9 to 3 μm).

synthetic-aperture radar (SAR) Radar system in which fine azimuth resolution is achieved by storing and processing data on the Doppler shift of multiple return pulses in such a way as to give the effect of a much longer antenna.

synthetic stereo DEM Artificial parallax is introduced into a second, synthetic version of a DEM to create a stereo model with variable vertical exaggeration.

synthetic stereo image Stereo images constructed through digital processing of a single image. Topographic data are used to calculate parallax.

system Combination of components that constitute an imaging device.

systematic distortion Geometric irregularities on images that are caused by known and predictable characteristics.

target Object on the terrain of specific interest in a remote sensing investigation.

telemeter To transmit data by radio or microwave links.

temporal resolution The time interval between successive images.

terrain Surface of the Earth.

texture Frequency of change and arrangement of tones on an image.

thermal capacity (c) See **heat capacity**.

thermal conductivity (K) Measure of the rate at which heat will pass through a material, expressed in calories per centimeter per second per degree centigrade.

thermal crossover On a plot of radiant temperature versus time, the point at which temperature curves for two different materials intersect.

thermal diffusivity (k) Governs the rate at which temperature changes within a substance, expressed in centimeters squared per second.

thermal inertia (P) Measure of the response of a material to temperature changes, expressed in calories per square centimeter per square root of second.

thermal IR (TIR) IR region from 3 to 14 μm that is employed in remote sensing. This spectral region spans the radiant power peak of the Earth.

thermal model Mathematical expression that relates thermal and other physical properties of a material to its temperature. Models may be used to predict temperature for given properties and conditions.

thermography Medical applications of TIR images. Images of the body, called *thermograms*, have been used to detect tumors and monitor blood circulation.

TIN Triangular irregular network. A vector-based representation of the terrain, composed of a network of triangles.

TIR Thermal IR wavelength region (3 to 14 μm).

TIR image Image acquired by a scanner that records radiation within the TIR region.

tone Each distinguishable shade of gray from white to black on an image.

topographic contour Lines of equal elevation.

topographic inversion An optical illusion that may occur on images with extensive shadows. Ridges appear to be valleys, and valleys appear to be ridges. The illusion is corrected by orienting the image so that shadows trend from the top margin of the image to the bottom.

topographic reversal A geomorphic phenomenon in which topographic lows coincide with structural highs and vice versa. Valleys are eroded on crests of anticlines to cause topographic lows, and synclines form topographic highs.

trade-off As a result of changing one factor in a remote sensing system, there are compensating changes elsewhere in the system; such a compensating change is known as a trade-off.

training site Area of terrain with known properties or characteristics that is used in supervised classification.

transmissivity Property of a material that determines the amount of energy that can pass through the material.

transparency Image on a transparent photographic material.

transpiration Expulsion of water vapor and oxygen by vegetation.

travel time In radar, the time interval between the generation of a pulse of microwave energy and its return from the terrain.

triangular irregular network (TIN) A vector-based representation of the terrain, composed of a network of triangles.

UAS Unmanned aerial system. Equivalent to UAV. "Aircraft" may be substituted for "aerial." UAS weigh more than 25 kg (55 lb).

UHI Urban heat island.

univariate statistic Statistics developed from a single band for evaluating individual band quality and potential for generating informative maps.

unmanned aerial system (UAS) Also called unmanned aerial vehicle (UAV). "Aircraft" may be substituted for "aerial." UAS weigh more than 25 kg (55 lb).

unsupervised classification Digital information extraction technique in which the computer clusters pixels into natural groupings based on the spectral characteristics of the pixels with no instructions from the operator except for setting basic parameters.

UTM Universal Transverse Mercator projection. Common projected coordinate system for satellite imagery.

UV Ultraviolet region of the electromagnetic spectrum, ranging in wavelengths from 0.01 to 0.4 μm.

vegetation anomaly Deviation from the normal distribution or properties of vegetation. Vegetation anomalies may be caused by faults, trace elements in soil, or other factors.

vertical exaggeration In a stereo model, the extent to which the vertical scale appears larger than the horizontal scale.

vertical take-off and landing (VTOL) UAS that combines the best attributes of fixed-wing and multirotor aircraft.

visible energy Energy at wavelengths from 0.4 to 0.7 μm that is detectable by the human eye. Also called *light*.

VNIR Visible-near infrared (0.4 to 0.9 μm).

volume phenomenon Interaction between electromagnetic energy and the internal properties of matter.

volume scattering In radar, interaction between electromagnetic radiation and the interior of a material.

watt (W) Unit of electrical power equal to rate of work done by one ampere under a potential of one volt.

wavelength (λ) Distance between successive wave crests or other equivalent points in a harmonic wave.

Wien's displacement law Describes the shift of the radiant power peak to shorter wavelengths as temperature increases.

X-band Radar wavelength region from 2.4 to 3.8 cm.

yaw Rotation of an aircraft about its vertical axis so that the longitudinal axis deviates left or right from the flight line.

index